OPTICAL TECHNIQUES
FOR SOLID-STATE
MATERIALS CHARACTERIZATION

OPTICAL TECHNIQUES
FOR SOLID-STATE
MATERIALS CHARACTERIZATION

Edited by
Rohit P. Prasankumar
Antoinette J. Taylor

CRC Press
Taylor & Francis Group
Boca Raton London New York

CRC Press is an imprint of the
Taylor & Francis Group, an **informa** business

Cover Image: Volker Steger

CRC Press
Taylor & Francis Group
6000 Broken Sound Parkway NW, Suite 300
Boca Raton, FL 33487-2742

First issued in paperback 2020

© 2012 by Taylor & Francis Group, LLC
CRC Press is an imprint of Taylor & Francis Group, an Informa business

No claim to original U.S. Government works

Version Date: 20110705

ISBN-13: 978-0-367-57692-9 (pbk)
ISBN-13: 978-1-4398-1537-3 (hbk)

Library of Congress Cataloging-in-Publication Data

Techniques for solid-state materials characterization / edited by Rohit P. Prasankumar and Antoinette J. Taylor.
  p. cm.
Includes bibliographical references and index.
ISBN 978-1-4398-1537-3
  1. Solids--Optical properties. 2. Optical measurements. 3. Spectrum analysis. I. Prasankumar, Rohit P. II. Taylor, Antoinette J., 1956- III. Title.

QC176.8.O6T43 2011
530.4'12--dc22   2010050670

Visit the Taylor & Francis Web site at
http://www.taylorandfrancis.com

and the CRC Press Web site at
http://www.crcpress.com

To Anuradha (Rohit P. Prasankumar)

CONTENTS

PREFACE

You can't study the darkness by flooding it with light.

Edward Abbey

Or can you? From the earliest beginnings of humankind, darkness has evoked emotions of fear and uncertainty in nearly all of us. The control of fire and invention of the light bulb helped alleviate these feelings, but to this day, when entering a dark room, we usually respond by turning on the light. This reaction speaks to the primacy of vision among our senses, perhaps because we can immediately obtain a massive amount of information by simply opening our eyes and looking at the world around us. In much the same manner, one can rapidly learn a lot about a material by simply shining light on it. Accordingly, it is not surprising that optics has been used to study both natural and man-made phenomena from the beginning of recorded history.

In fact, light is virtually an ideal tool for studying nearly any material or phenomenon, as optical techniques have several notable advantages over other methods. Paramount is their noncontact and nondestructive nature, along with the ability to broadly tune from x-ray to terahertz (THz) frequencies, generate attosecond (10^{-18} s) duration pulses, and probe nanoscale dimensions (<10 nm). Therefore, by using one or more of the myriad optical techniques that have been developed over the last century for characterizing materials, substantial insight into phenomena ranging from high-temperature superconductivity to protein folding can be obtained.

The purpose of this book is thus to describe both basic and advanced experimental optical techniques that are commonly used to study materials. To the best of our knowledge, a single volume that describes the essential experimental techniques for optically characterizing materials does not exist, and relatively few references discuss any of these methods in detail. This then presents an opportunity to address these issues by describing established optical experiments in the same volume as more recently developed temporally and spatially resolved methods. More specifically, we aim to describe these techniques in enough detail for researchers with different levels of experience to build and/or use a working setup, acquire data on both simple and complex materials, and analyze and interpret these data to obtain insight into fundamental material properties with minimal reliance on other sources.

To accomplish this, we solicited contributions from experts who have pioneered these techniques and extensively applied them to characterize complex materials in their own laboratories. Although we have attempted to include the most widely known optical techniques for characterizing materials, this book is not meant to be exhaustive; several important methods have undoubtedly been left out, either inadvertently or due to space limitations. In addition, although our intention is to describe each of these techniques in extensive detail as mentioned above, in a book of this nature we are bound to have overlooked some potentially important issues, for which we apologize in advance. It is worth pointing out that all of the optical techniques described here can be used, with some (usually minor) variations, to characterize non-solid-state materials (e.g., soft matter, gases); we chose to concentrate on solid-state materials to avoid diluting the content and remain within our primary area of expertise.

Part I provides background information on light–matter interactions (Chapter 1), semiconductors (Chapter 2), and metals (Chapter 3). This knowledge will be critical in understanding the experimental techniques and results described later in the book. We assume that the reader has an undergraduate-level understanding of optics, electromagnetics, quantum mechanics, and solid-state physics, along with some knowledge of basic components used in experimental optics (e.g., lenses, wave plates, filters). However, many readers may be experts in one of these areas but only somewhat familiar with other subjects, particularly regarding the information that is most relevant for understanding data obtained with different optical probes. Our aim in this part is therefore to provide a broad knowledge base for understanding the experiments and results discussed later in the book, with additional, more specialized details contained in the appropriate chapters.

Part II focuses on linear, time-integrated optical experiments used to measure basic optical properties such as transmission, reflectivity, and luminescence. These experiments can provide essential information on characteristic physical quantities, such as the dielectric function and optical conductivity, that are intimately linked to a material's microscopic properties (e.g., the electronic and vibrational structure). They can also reveal the properties of elementary excitations including excitons, phonons, and magnons. To this end, Chapter 4 discusses standard methods for obtaining the optical constants of a material (e.g., the refractive index and absorption coefficient), focusing on the techniques of visible-ultraviolet (UV) spectroscopy and Fourier transform infrared spectroscopy (FT-IR). Space limitations prevented us from a detailed description of ellipsometry, another established technique for measuring optical constants, but several references are cited for the interested reader.

Chapter 5 discusses methods for obtaining the optical constants after continuous wave (cw) laser excitation, focusing on photoluminescence (PL), photoluminescence excitation (PLE), and photocurrent (PC) spectroscopies. These measurements reveal the transitions involved in light absorption and emission, as well as energy transfer between different electronic levels. Finally, Chapter 6 treats Raman scattering, describing how one can obtain symmetry, energy, and lifetime information on a wide range of elementary excitations, even under extreme conditions (e.g., high pressure and magnetic fields).

Time-resolved optical spectroscopy is discussed in Part III, with a focus on optical techniques that can characterize the dynamic properties of materials with femtosecond (or better) time resolution. To a large extent, each of these techniques is capable of temporally resolving photoinduced changes in one or more of the properties discussed in Part II (e.g., absorption, optical conductivity) while providing insight on properties that can only be accessed with dynamic measurements (e.g., dephasing and population relaxation times). All of these methods depend on ultrashort pulsed lasers to achieve femtosecond time resolution, and therefore the basics of ultrashort pulse generation and measurement are covered in Chapter 7, with a particular focus on understanding the principles of femtosecond laser operation and practical knowledge necessary for using these lasers in the laboratory. In addition, the disparate phenomena experienced by quasiparticles after femtosecond photoexcitation (carrier–carrier/carrier–phonon scattering, radiative/nonradiative recombination, defect/surface trapping, etc.) depend on the material being studied, but are independent of the technique used to probe them. Chapter 8 describes these processes in semiconductors and metals, along with some examples in more complex materials, to provide a basis for the experimental results discussed later in this part.

The remaining chapters in this part provide detailed descriptions of the most common time-resolved optical techniques, each capable of probing different phenomena. Chapter 9 describes pump–probe spectroscopy, arguably the simplest time-resolved technique, that can temporally resolve photoinduced transmission and reflection changes and relate these changes back to dynamical material properties (e.g., carrier population and transient electronic temperature). Four-wave mixing spectroscopy, capable of measuring dephasing processes and many-body interactions between different excitations, is discussed in Chapter 10. Chapter 11 then describes terahertz spectroscopy, which can probe both the static and dynamic properties of low-energy excitations. Time-resolved photoluminescence spectroscopy, discussed in Chapter 12, is used to measure radiative recombination and energy transfer processes. Spin dynamics in complex materials can be measured using time-resolved magneto-optical spectroscopy, described in Chapter 13. Finally, Chapter 14 discusses time-resolved Raman scattering, which can directly address the dynamical properties of low-energy excitations such as phonons and magnons, in a manner complementary to the other techniques described in this part.

The ability to perform optical measurements with high spatial resolution is important when characterizing materials, particularly due to the predominance of both naturally occurring and man-made materials with nanoscale features among materials of great current interest (e.g., quantum-confined nanostructures, metamaterials, and transition metal oxides). Part IV thus contains three chapters on conventional optical microscopy, micro-optical techniques, and near-field scanning optical microscopy. Chapter 15, on conventional optical microscopy, somewhat differs from the other chapters in this book, since a multitude of references already exist on the practical aspects of optical microscopy. Therefore, this chapter begins with a detailed description of resolution and image formation, followed by an overview of different microscopic techniques and ending with a detailed description of imaging interferometric microscopy. Micro-optical techniques, in which diffraction-limited spatial resolution can provide additional information on a given sample beyond the measurements discussed in Part II, are discussed in Chapter 16. Finally, Chapter 17 describes near-field scanning optical microscopy, a powerful tool for characterizing a variety of materials with nanometer spatial resolution.

Part V deals with more advanced optical techniques for characterizing materials that are, arguably, insufficiently developed to merit their own chapters, but worthy of inclusion in this volume, especially since they will likely develop into more established methods in the coming years. This part contains only one chapter, which describes advanced time-resolved optical techniques for resolving ultrafast photoinduced changes in a material's lattice and electronic structure, and then discusses wide-field and scanning optical techniques that combine high temporal and spatial resolution in a single experiment. It is worth noting that Chapter 15 gives an overview of advanced spatially resolved techniques, which therefore are not discussed in this part.

Certain chapters contain information useful for performing nearly any optical experiment. Chapters 4, 5, 6, and 16 discuss (in varying levels of detail) common light sources, optical components, detectors, spectrometers, and cryostats used in time-integrated optical experiments, which is useful background for almost all of the experiments discussed in this book. There is some redundancy between these chapters; however, our overall approach was to include all of the information essential for understanding and implementing a given technique in the relevant chapter, with references to other chapters or other sources that give more detailed information on a given aspect of each method. In general, we thought it was best to err on the side of including too much rather than too little information.

Chapter 4 provides a brief procedure for polishing samples, which is relevant for optical measurements on samples with rough surfaces. Table 4.1 also summarizes the spectral ranges covered by different IR sources, beam splitters, detectors, and optical window materials. Chapter 5 includes a discussion of laser safety practices, important reading for anyone working with lasers. The use and properties of metal contacts on a sample are also described in the context of photocurrent measurements in this chapter. Finally, it also discusses how to perform a simple knife edge scan for determining the width of a Gaussian beam, which is then discussed in more detail in Chapter 16. Chapter 6 includes a description of Raman scattering measurements under high pressure and magnetic field, which, to a large extent, can be adapted for many of the other optical experiments discussed here.

Chapter 9 is worth examining for any reader unfamiliar with time-resolved experiments, since it contains a great deal of information that is common to all of the techniques discussed in Part III; in some respects, the other experiments described in this part are all variations and extensions of the basic pump–probe technique described in Chapter 9. Finally, in Part IV, Chapter 16 includes a short procedure for aligning optical beams, which is critical for all of the experiments discussed in this book.

Optics has long been one of the first techniques applied to study new systems exhibiting novel properties, and we expect that the use of optical techniques to characterize materials will only grow in importance in the coming years. To this end, our goal was to put together a comprehensive reference for anyone using optics to experimentally characterize both simple and complex materials. We wish to empower a wide range of researchers, from experts in experimental optics who want to implement a new technique in their laboratories to those outside of optics who want to understand the basics of a particular optical technique that has been applied to a material of interest, to use and implement any of the techniques described here. Therefore, we hope that this book contributes to the vast body of literature demonstrating that, contrary to the quotation at the beginning of this preface, in fact you *can* study the darkness by simply turning on the light!

Rohit P. Prasankumar

Antoinette J. Taylor

ACKNOWLEDGMENTS

Any book of this nature, by definition, requires the collaboration of many people, all of whom have extremely busy schedules. Therefore, we deeply appreciate their efforts in contributing to this project amid their other deadlines and commitments, and apologize in advance to anyone that we unintentionally omit here. This book would not have been possible without the dedicated effort of the chapter authors, who have without exception been a pleasure to work with; their patience and perseverance during the long process of putting together this book has made it much more enjoyable. We would also like to thank Luna Han, our editor at Taylor & Francis, who helped formulate the initial idea for this book and has been extremely encouraging, responsive, and considerate throughout the whole process. We also appreciate the efforts of Kari Budyk and others at Taylor & Francis in the production stages of this project.

We have been very fortunate to work with several postdoctoral researchers who have been understanding and supportive of our efforts to put together this book, despite the fact that it limited the time we could spend advising their research. We would also like to thank our colleagues around the world and at Los Alamos National Laboratory, particularly within the Center for Integrated Nanotechnologies and Laboratory for Ultrafast Materials and Optical Science, for stimulating discussions and collaborations that helped us pursue exciting new scientific directions while also guiding the content of this book.

Finally, we would most of all like to thank our families. They have uncomplainingly put up with the amount of time that this project required, particularly in the countless nights and weekends spent in front of a computer screen instead of with them. This book would not have been possible without their love, support, and understanding.

Rohit P. Prasankumar

Antoinette J. Taylor

EDITORS

Rohit P. Prasankumar received his BS in electrical engineering from the University of Texas at Austin (1997) and his MS (1999) and PhD (2003) in electrical engineering from the Massachusetts Institute of Technology. He joined Los Alamos National Laboratory as a postdoctoral research associate in August 2003, focusing on the measurement of ultrafast mid-to-far-infrared dynamics in semiconductor nanostructures and strongly correlated compounds. He then became a technical staff member at the Center for Integrated Nanotechnologies at Los Alamos National Laboratory in February 2006, with research interests principally directed toward the measurement of dynamics in complex functional materials with high temporal and spatial resolution over a broad spectral range. He has coauthored over 30 publications, including 2 invited review papers.

Antoinette (Toni) J. Taylor is the leader of the Materials Physics and Applications Division at Los Alamos National Laboratory (LANL). Prior to this position, she was director of the Center for Integrated Nanotechnologies, a joint Sandia/LANL Nanoscience Research Center funded through the Department of Energy, Office of Basic Energy Sciences. Her research interests include the investigation of ultrafast dynamical nanoscale processes in materials and the development of novel optics-based measurement techniques for the understanding of new phenomena. She has published over 250 papers in these areas, written 2 book chapters, and edited 3 books. She has been a director-at-large of the Optical Society of America, a topical editor of the *Journal of the Optical Society B: Optical Physics*, and a member of the Solid State Science Committee, Board of Physics and Astronomy, and the National Academies; she has also chaired the National Academies' Committee on Nanophotonics Applicability and Accessibility. Currently, she is a member-at-large of the Division of Laser Science of the American Physical Society and the Optical Society of America's representative to the Joint Council of Quantum Electronics. She is an LANL fellow and a fellow of the American Physical Society, the Optical Society of America, and the American Association for the Advancement of Science. In 2003, Toni won the inaugural Los Alamos Fellow's Prize for Outstanding Leadership in Science and Engineering.

CONTRIBUTORS

Peter Abbamonte
Department of Physics
University of Illinois at Urbana–Champaign
Urbana, Illinois

Marc Achermann
Department of Physics
University of Massachusetts
Amherst, Massachusetts

Richard D. Averitt
Department of Physics
Boston University
Boston, Massachusetts

Harini Barath
Department of Physics
University of Illinois at Urbana–Champaign
Urbana, Illinois

Steven R.J. Brueck
Center for High Technology Materials
University of New Mexico
Albuquerque, New Mexico

Diego Casa
Department of Physics
University of Illinois at Urbana–Champaign
Urbana, Illinois

Xiaoqian Chen
Department of Physics
University of Illinois at Urbana–Champaign
Urbana, Illinois

Cesar E. Chialvo
Department of Physics
University of Illinois at Urbana–Champaign
Urbana, Illinois

S. Lance Cooper
Department of Physics
University of Illinois at Urbana–Champaign
Urbana, Illinois

Steven T. Cundiff
JILA
University of Colorado

and

National Institute for Standards and
    Technology
Boulder, Colorado

Marcelo Davanço
Center for Nanoscale Science and
    Technology
National Institute of Standards and
    Technology
Gaithersburg, Maryland

and

Maryland Nanocenter
University of Maryland
College Park, Maryland

Jeffrey Davis
Centre for Atom Optics and Ultrafast
    Spectroscopy
Swinburne University of Technology
Victoria, Australia

Thomas Dekorsy
Department of Physics
University of Konstanz
Konstanz, Germany

Jure Demsar
Department of Physics
University of Konstanz
Konstanz, Germany

Daniele Fausti
Department of Physics
University of Trieste
Trieste, Italy

Y. Gan
Department of Physics
University of Illinois at Urbana–Champaign
Urbana, Illinois

Jordan Gerton
Department of Physics and Astronomy
University of Utah
Salt Lake City, Utah

David J. Hilton
Department of Physics
University of Alabama–Birmingham
Birmingham, Alabama

Chennupati Jagadish
Department of Electronic Materials
    Engineering
Australian National University
Canberra, Australia

Young-Il Joe
Department of Physics
University of Illinois at Urbana–Champaign
Urbana, Illinois

Jessica Johnston
Department of Physics and Astronomy
University of Utah
Salt Lake City, Utah

Robert A. Kaindl
Materials Sciences Division
Lawrence Berkeley National Laboratory
Berkeley, California

John F. Karpus
Department of Physics
University of Illinois at Urbana–Champaign
Urbana, Illinois

Minjung Kim
Department of Physics
University of Illinois at Urbana–Champaign
Urbana, Illinois

Andrew Kowalevicz
Technology Solutions
BAE Systems
Columbia, Maryland

Yuliya Kuznetsova
Center for High Technology Materials
University of New Mexico
Albuquerque, New Mexico

Paul H.M. van Loosdrecht
Zernike Institute for Advanced Materials
University of Groningen
Groningen, the Netherlands

Ben Mangum
Department of Physics and Astronomy
University of Utah
Salt Lake City, Utah

Nadya Mason
Department of Physics
University of Illinois at Urbana-Champaign
Urbana, Illinois

Alexander Neumann
Center for High Technology Materials
University of New Mexico
Albuquerque, New Mexico

Hidekazu Okamura
Department of Physics
Kobe University
Kobe, Japan

Willie J. Padilla
Department of Physics
Boston College
Chestnut Hill, Massachusetts

Rohit P. Prasankumar
Center for Integrated Nanotechnologies
Los Alamos National Laboratory
Los Alamos, New Mexico

Matthew T. Rakher
Center for Nanoscale Science and
    Technology
National Institute of Standards and
    Technology
Gaithersburg, Maryland

James P. Reed
Department of Physics
University of Illinois at Urbana–Champaign
Urbana, Illinois

Sajan Saini
Department of Physics
Queens College of The City University of
    New York
Flushing, New York

Eyal Shafran
Department of Physics and Astronomy
University of Utah
Salt Lake City, Utah

C.S. Snow
Department of Physics
University of Illinois at Urbana–Champaign
Urbana, Illinois

Kartik Srinivasan
Center for Nanoscale Science and
    Technology
National Institute of Standards and
    Technology
Gaithersburg, Maryland

Antoinette J. Taylor
Materials Physics and Applications
Los Alamos National Laboratory
Los Alamos, New Mexico

Jigang Wang
Ames Laboratory and Department of
    Physics and Astronomy
Iowa State University
Ames, Iowa

LIST OF ABBREVIATIONS

Time

Femtosecond	fs
Picosecond	ps
Nanosecond	ns

Distance

Nanometer	nm
Micrometer (micron)	μm
Millimeter	mm
Centimeter	cm
Meter	m

Energy

Femtojoule	fJ
Picojoule	pJ
Nanojoule	nJ
Microjoule	μJ
Millijoule	mJ
Milli-electron volt	meV
Electron volt	eV

Frequency

Kilohertz	kHz
Megahertz	MHz
Gigahertz	GHz
Terahertz	THz
Ultraviolet	UV
Visible	VIS
Infrared	IR
Alternating current	AC
Direct current	DC

Physical constants

Speed of light in a vacuum $c = 2.99792458 \times 10^8$ m/s
Permittivity of free space $\varepsilon_0 = 8.8542 \times 10^{-12}$ F/m
Permeability of free space $\mu_0 = 1.2566 \times 10^{-6}$ H/m
Electron charge $e = 1.602 \times 10^{-19}$ C
Electron mass $m_e = 9.01 \times 10^{-31}$ kg
Planck's constant $h = 6.626 \times 10^{-34}$ J s
Boltzmann constant $k_B = 1.38066 \times 10^{-23}$ J/K

PART I

BACKGROUND

Chapter 1

LIGHT–MATTER INTERACTIONS

Willie J. Padilla

CONTENTS

This chapter overviews the interactions of electromagnetic waves with matter—the study of which has a long and illustrious history. Inquiries into the nature of light began thousands of years ago and seem, not surprisingly, to have been first during the Classical and Hellenistic periods in ancient Greece (Pomeroy 1999). For example, the Greek philosopher Pythagoras (570–495 BC) was the first to propose a fifth element (in addition to earth, water, air, and fire), which he titled "aether," to describe the properties of light. Democritus, Empedocles, Plato, Aristotle, and others also developed several theories of light. In Euclid's book *Catoptrics* (300 BC), he described both the rectilinear propagation and reflection of light. Heron of Alexandria (AD 10–70), a Greek mathematician, described both of these phenomena by explaining that light traverses the shortest allowed path between two points.

We do not intend to overview the entire history of light–matter interactions here, but would like to simply point out that, although ever-increasing insight regarding light and its interaction with matter has taken place over the last several thousand years, the theory of electromagnetic waves was placed on firm theoretical ground during the nineteenth century with the development of Maxwell's equations. In fact, all classical descriptions of the interactions between light and matter can be described by Maxwell's equations and the constitutive relations. Thus, we begin this chapter by describing the relationship between what is known—Maxwell's equations—and what can be measured.

The most experimentally accessible optical quantities of measure are the frequency-dependent reflectance $R(\omega)$ and transmittance $T(\omega)$. On the other hand, quantities most directly related to the electronic and magnetic structure of materials are the electrical permittivity $\varepsilon(\omega)$ and the magnetic permeability $\mu(\omega)$. Maxwell's equations facilitate an understanding of the relationship between the measured quantities and the electromagnetic structure of materials, that is, $[R,T] \leftrightarrow [\varepsilon,\mu]$. Maxwell's equations are really a collection of theories put forward by several scientists, namely, Gauss, Ampere, and Lorentz. They were unified by Maxwell in 1865 (Maxwell 1865) and put into their modern form by Gibbs and Wilson (1929) and Heaviside (1892).

Before we begin, we pause to note that throughout this chapter we use the *Système International* (SI) units, also called the rationalized meter–kilogram–second (mks) system. There are, however, four other common systems of electromagnetic units in use, all connected to the centimeter–gram–second (cgs). These four are the Gaussian system, Heaviside–Lorentz system, cgs electrostatic system, and cgs electromagnetic system. Historically the cgs electrostatic (esu) and cgs electromagnetic (emu) systems were developed first—during the middle of the nineteenth century—and were the systems of choice for the development of Maxwell's equations. Toward the end of the nineteenth century, the Heaviside–Lorentz system and the Gaussian system were developed, followed shortly by the rationalized mks system.

There can be ambiguity and confusion when converting back and forth between different systems of units. These difficulties seem to arise due to the number of fundamental dimensions the systems are based on. During the development of the electromagnetic theory in the nineteenth century, the esu, emu, and Gaussian systems were naturally based on the three fundamental dimensions of *length*, *mass*, and *time*. The rationalized mks system of units, however, recognizes a fourth fundamental "electromagnetic" dimension of *charge*, and a process of "rationalization" eliminates factors of 4π from Maxwell's equations (Cohen 2001). The rationalized mks system of units is the most common one employed today for electromagnetism and is used in nearly every introductory text.

1.1 MAXWELL'S EQUATIONS

1.1.1 Microscopic and Macroscopic Forms

Maxwell's equations describe the properties of electromagnetic waves and their interactions with matter. In free space, they take the form

$$\bar{\nabla} \cdot \bar{\mathbf{E}} = \frac{1}{\varepsilon_0}\rho, \tag{1.1}$$

$$\bar{\nabla} \cdot \bar{\mathbf{B}} = 0, \tag{1.2}$$

$$\bar{\nabla} \times \bar{\mathbf{E}} = -\frac{\partial \bar{\mathbf{B}}}{\partial t}, \tag{1.3}$$

$$\bar{\nabla} \times \bar{\mathbf{B}} = \mu_0 \bar{\mathbf{J}} + \mu_0 \varepsilon_0 \frac{\partial \bar{\mathbf{E}}}{\partial t}, \tag{1.4}$$

where
   $\bar{\mathbf{E}}$ is called the electric field strength
   $\bar{\mathbf{B}}$ is the magnetic flux density
   ρ is the charge density
   $\bar{\mathbf{J}}$ is the current density
   ε_0 and μ_0 are the permittivity and permeability of free space

The permeability is defined to be exactly $\mu_0 = 4\pi \times 10^{-7}$ H/m, and thus the permittivity has a value of $\varepsilon_0 = 10^7/4\pi c^2$ F/m, that is, the uncertainty in ε_0 is the uncertainty in the velocity of light. Equation 1.1 is known as Gauss's law for electric fields, (1.2) as Gauss's law for magnetic fields, (1.3) as Faraday's induction law, and (1.4) as Ampere's law. Maxwell modified Ampere's law by adding the $\mu_0 \varepsilon_0 \, \partial \vec{\mathbf{E}}/\partial t$ term, and thus the particular forms of (1.1) through (1.4) are known as Maxwell's microscopic equations.

Matter may be polarized in the presence of electric and/or magnetic fields. If we assume a linear response and isotropic media, the electric and magnetic polarizations are taken to be proportional to the external electric and magnetic fields. Additionally, an explicit factor of ε_0 and/or μ_0 appears. The polarization fields plus the incident fields together form the total fields, called the electric displacement $\bar{\mathbf{D}}$ and the magnetic field strength $\bar{\mathbf{H}}$. These relations are derived explicitly below. The electric displacement is defined as

$$\bar{\mathbf{D}} \equiv \varepsilon_0 \bar{\mathbf{E}} + \bar{\mathbf{P}}, \tag{1.5}$$

where

$$\bar{\mathbf{P}} \equiv \varepsilon_0 \chi_e \bar{\mathbf{E}}, \tag{1.6}$$

with $\bar{\mathbf{P}}$ the polarization and χ_e the electric susceptibility. The electric displacement can thus be written as

$$\bar{\mathbf{D}} = \varepsilon \bar{\mathbf{E}}, \tag{1.7}$$

where

$$\varepsilon \equiv \varepsilon_0 (1 + \chi_e) = \varepsilon_0 \varepsilon_r. \tag{1.8}$$

The magnetic field strength is defined as

$$\bar{\mathbf{H}} \equiv \frac{1}{\mu_0} \bar{\mathbf{B}} - \bar{\mathbf{M}}, \tag{1.9}$$

where

$$\bar{\mathbf{M}} \equiv \chi_m \bar{\mathbf{H}}, \tag{1.10}$$

$\bar{\mathbf{M}}$ is the magnetization
χ_m is the magnetic susceptibility

The magnetic flux density can thus be written as

$$\bar{\mathbf{B}} = \mu \bar{\mathbf{H}}, \tag{1.11}$$

where

$$\mu \equiv \mu_0(1 + \chi_m) = \mu_0 \mu_r. \tag{1.12}$$

Using the definitions of $\bar{\mathbf{D}}$ and $\bar{\mathbf{B}}$ above, Maxwell's equations are rewritten as shown below.

$$\bar{\nabla} \cdot \bar{\mathbf{D}} = \rho. \tag{1.13}$$

$$\bar{\nabla} \cdot \bar{\mathbf{B}} = 0. \tag{1.14}$$

$$\bar{\nabla} \times \bar{\mathbf{E}} = -\frac{\partial \bar{\mathbf{B}}}{\partial t}. \tag{1.15}$$

$$\bar{\nabla} \times \bar{\mathbf{H}} = \bar{\mathbf{J}} + \frac{\partial \bar{\mathbf{D}}}{\partial t}. \tag{1.16}$$

Equations 1.13 through 1.16 are termed Maxwell's macroscopic equations, in contrast to (1.1) through (1.4), the microscopic form.

The microscopic Maxwell's equations completely describe the properties of electromagnetic fields; that is, for specified sources ρ and $\bar{\mathbf{J}}$ Maxwell's microscopic equations give the fields $\bar{\mathbf{E}}$ and $\bar{\mathbf{B}}$ everywhere in space. If this is the case, then why do we need the form of Maxwell's equations presented in (1.13) through (1.16)? We can answer this by considering electromagnetic wave propagation in matter. In this case, the number of sources—electrons—may be on the order of 10^{23}. Solving for the electric and magnetic fields in matter is thus of such complexity that is clearly not feasible. However, by introduction of the optical constants, via Maxwell's macroscopic equations, we may describe the average of fields and sources over scales much larger than individual atomic sizes (Jackson 1998). Thus, the optical constants facilitate the description of interactions between electromagnetic waves and matter.

Before proceeding, we pause to specify one other form of Maxwell's equations that is particularly useful for studying harmonic phenomena. We consider regions of space in which there are no charges or currents ($\rho = \bar{\mathbf{J}} = 0$), and that all fields in (1.13) through (1.16) vary as $\exp[i(\mathbf{k} \cdot \mathbf{r} - \omega t)]$, where $\bar{\mathbf{k}}$ is called the wave vector or the propagation vector, ω is the angular frequency, $\bar{\mathbf{r}}$ is the position vector, and t is time. In this case we find that Maxwell's equations can be written as

$$\bar{\mathbf{k}} \cdot \varepsilon \bar{\mathbf{E}} = 0. \tag{1.17}$$

$$\bar{\mathbf{k}} \cdot \mu \bar{\mathbf{H}} = 0. \tag{1.18}$$

$$\bar{\mathbf{k}} \times \bar{\mathbf{E}} = \omega\mu\bar{\mathbf{H}}. \tag{1.19}$$

$$\bar{\mathbf{k}} \times \bar{\mathbf{H}} = -\omega\varepsilon\bar{\mathbf{E}}. \tag{1.20}$$

Thus $\bar{\mathbf{E}}$ and $\bar{\mathbf{H}}$ are mutually perpendicular and perpendicular to $\bar{\mathbf{k}}$.

1.1.2 Boundary Conditions

Maxwell's equations may be used to derive the boundary conditions—that is, the relationship between the fields at an interface between two different types of media described by their respective optical constants. We will not do this explicitly here, and instead simply list the results. The interested reader is referred to Jackson (1998), who has derived these:

$$\bar{\mathbf{n}} \cdot \left(\bar{\mathbf{D}}^{(2)} - \bar{\mathbf{D}}^{(1)} \right) = 0, \tag{1.21}$$

$$\bar{\mathbf{n}} \cdot \left(\bar{\mathbf{B}}^{(2)} - \bar{\mathbf{B}}^{(1)} \right) = 0, \tag{1.22}$$

$$\bar{\mathbf{n}} \times \left(\bar{\mathbf{E}}^{(2)} - \bar{\mathbf{E}}^{(1)} \right) = 0, \tag{1.23}$$

$$\bar{\mathbf{n}} \times \left(\bar{\mathbf{H}}^{(2)} - \bar{\mathbf{H}}^{(1)} \right) = 0, \tag{1.24}$$

where $\bar{\mathbf{n}}$ is a unit vector perpendicular to the interface and the superscripts (1) and (2) refer to the two different media which form the interface. The specific form of the boundary conditions shown above is valid in regions of space in which there are no charges or currents ($\rho = \mathbf{J} = 0$). Furthermore, (1.21) through (1.24) allow us to explicitly state the relationships between tangential and normal components of the field at the interface. For example, (1.21) and (1.22) allow us to specify the relationship between the normal components of $\bar{\mathbf{D}}$ and $\bar{\mathbf{B}}$ as

$$\bar{\mathbf{D}}_n^{(1)} = \bar{\mathbf{D}}_n^{(2)} \quad \text{and} \quad \bar{\mathbf{B}}_n^{(1)} = \bar{\mathbf{B}}_n^{(2)} \tag{1.25}$$

and (1.23) and (1.24) allow us to specify the relationship between the tangential components of $\bar{\mathbf{E}}$ and $\bar{\mathbf{H}}$ as

$$\bar{\mathbf{E}}_t^{(1)} = \bar{\mathbf{E}}_t^{(2)} \quad \text{and} \quad \bar{\mathbf{H}}_t^{(1)} = \bar{\mathbf{H}}_t^{(2)}. \tag{1.26}$$

where the subscripted n's in (1.25) stand for normal components and the subscripted t's in (1.26) stand for the tangential components.

Thus we may summarize the boundary conditions by stating that the normal components of the electric displacement and the magnetic flux density are continuous at the interface, and that the tangential components of the electric field strength and the magnetic field strength are continuous at the interface.

1.1.3 Optical Constants

The relation between the microscopic and macroscopic forms of Maxwell's equations was first investigated by Lorentz (Choy 1999, Sihvola 1999) and is a subject detailed in various condensed matter texts (Ashcroft and Mermin 1976, Kittel 1996). However, an earlier connection between the atomic polarizability and the electric permittivity was developed by Mossotti in 1850, and independently by Clausius in 1879. This relationship, known as the Clausius–Mossotti equation, established that for any given material, the dielectric constant ε should be proportional to the density of atoms

$$\frac{\varepsilon - 1}{\varepsilon + 2} = \frac{1}{3\varepsilon_0} \sum N_j \alpha_j, \tag{1.27}$$

where
    N_j is the concentration
    α_j is the atomic polarizability

Equation 1.27 thus provides an important connection between macroscopic and microscopic theories. The atomic polarizability would be calculated by a microscopic theory, which could then be connected to a dielectric function, which would describe the response of the ions to the local field acting on them.

The dielectric constant (also called the electric permittivity) ε and the magnetic permeability μ are called the optical constants. Although in some cases it may be suitable to treat them as constant, they are, however, functions of frequency and, in general, may also depend on the wave vector, that is, $\varepsilon = \varepsilon(\omega, \mathbf{k})$, and $\mu = \mu(\omega, \mathbf{k})$. Here we denote angular frequency (cycles per unit time) as ω, which has units of rad/s, and the wave vector (wavelengths in a length of 2π) as $\mathbf{k}$, which has units of 1/m. The relationship between the angular frequency and frequency is $\omega = 2\pi f$. It is typical to refer to ω as "frequency," as it is encountered more often than f.

As they have been derived here the optical constants are linear response functions that obey causality. Thus we may write them as complex functions $\tilde{\varepsilon} = \varepsilon_1 + i\varepsilon_2$ and $\tilde{\mu} = \mu_1 + i\mu_2$, where the subscripts 1 and 2 denote the real and imaginary parts, respectively. The real part of the complex optical constants describes the amplitude "response," and imaginary portions describe the damping or loss, due to external electric or magnetic fields.

Notice that ε and μ are the only two parameters in Maxwell's macroscopic equations which describe interactions of electromagnetic waves and matter. Other quantities that describe electromagnetic–matter interactions, called material parameters, have simple algebraic relations to the optical constants. If we assume a time-harmonic convention $e^{-i\omega t}$ (which we use throughout), the wave impedance (Z), conductivity (σ), and refractive index (n) are related to ε and μ by

$$\tilde{n} = n_1 + in_2 = \sqrt{\tilde{\varepsilon}_r \tilde{\mu}_r}. \tag{1.28}$$

$$\tilde{Z} = Z_1 + iZ_2 = \sqrt{\frac{\tilde{\mu}}{\tilde{\varepsilon}}}. \tag{1.29}$$

$$\tilde{\sigma} = \sigma_1 + i\sigma_2 = i\omega(\varepsilon_0 - \tilde{\varepsilon}). \tag{1.30}$$

The impedance (Z) of a medium to an electromagnetic wave is closely related to energy propagation. By analogy with a one-dimensional transmission line, the key parameter was shown to be the ratio of the total electric field to the total magnetic field (Schelkunoff 1938, Stratton 1941, Ramo et al. 1994). Thus, if we consider a wave propagating along the z-direction (with its electric field in the x-direction and the magnetic field in the y-direction), the load presented to the wave at any point z is given by

$$\tilde{Z}(z) = \frac{\tilde{E}_x(z)}{\tilde{H}_y(z)}, \tag{1.31}$$

which has units of resistance (Ohms). For isotropic media (1.29) and (1.31) are equal.

The ability of a material to carry an electric current is given by the real conductivity (σ_1), which has units of *siemens per meter* (S/m). Equivalently, materials are often characterized by their resistivity (ρ), which is the inverse of the conductivity $\rho = 1/\sigma_1$. The units of resistivity are ohm-meters, and thus $1\,\mathrm{ohm} = (1\,\mathrm{S})^{-1}$.

When light enters a medium it may appear to be bent. The material parameter which describes this effect is the real part of the index of refraction (n_1). In addition to (1.28), which describes the connection between the optical constants and n_1, the index of refraction is the ratio between the speed of light c in vacuum and the speed of light v in that medium $n_1 = c/v$. We develop this idea later in this chapter.

Although (1.28) through (1.30) are a compact way of presenting the relationships between the optical constants and the material parameters, they offer little physical insight. Thus we expand each of these terms and list their explicit dependence in both real and imaginary parts in Table 1.1.

From Table 1.1 we can observe that the real (imaginary) part of the dielectric function is connected to the imaginary (real) part of the conductivity. It was previously mentioned that

TABLE 1.1

Relationships between the Optical Constants and the Material Parameters

	Dielectric Constant, $\tilde{\varepsilon}$	Electrical Conductivity, $\tilde{\sigma}$	Index of Refraction and Wave Impedance, $\tilde{n}$, $\tilde{Z}$
$\tilde{\varepsilon}$	$\tilde{\varepsilon} = \varepsilon_1 + i\varepsilon_2$	$\varepsilon_1 = \varepsilon_0 - \dfrac{1}{\omega}\sigma_2$	$\varepsilon_1 = c\,\dfrac{n_1 Z_1 + n_2 Z_2}{Z_1^2 + Z_2^2}$
		$\varepsilon_2 = \dfrac{1}{\omega}\sigma_1$	$\varepsilon_2 = c\,\dfrac{n_2 Z_1 - n_1 Z_2}{Z_1^2 + Z_2^2}$
$\tilde{\sigma}$	$\sigma_1 = \omega\varepsilon_2$	$\tilde{\sigma} = \sigma_1 + i\sigma_2$	$\sigma_1 = \omega c\,\dfrac{n_2 Z_1 - n_1 Z_2}{Z_1^2 + Z_2^2}$
	$\sigma_2 = \omega(\varepsilon_0 - \varepsilon_1)$		$\sigma_2 = \omega\left[\varepsilon_0 - c\,\dfrac{n_1 Z_1 + n_2 Z_2}{Z_1^2 + Z_2^2}\right]$
$\tilde{n}$, $\tilde{Z}$	$n_1 = \varepsilon_1 Z_1 - \varepsilon_2 Z_2$	$n_1 = c\left[Z_1\left(\varepsilon_0 - \dfrac{1}{\omega}\sigma_2\right) - Z_2\left(\dfrac{1}{\omega}\sigma_1\right)\right]$	$\tilde{n} = n_1 + in_2$
	$n_2 = \varepsilon_1 Z_2 + \varepsilon_2 Z_1$	$n_2 = c\left[Z_1\left(\dfrac{1}{\omega}\sigma_1\right) + Z_2\left(\varepsilon_0 - \dfrac{1}{\omega}\sigma_2\right)\right]$	$\tilde{Z} = Z_1 + iZ_2$

the imaginary portions of the optical constants characterize the loss of electromagnetic fields in a material. Thus we see that the real part of the electrical conductivity σ_1, which governs how well a material conducts electricity, is directly proportional to the quantity ε_2, which characterizes the loss of the electric component of an electromagnetic wave in a material. This connection will be made clear when we discuss the Drude model later in this chapter and in Chapter 3.

Although Table 1.1 is useful for converting between the conductivity and the dielectric function, the equations describing the relations between the index of refraction and the impedance to other material parameters and optical constants are less clear. This complexity arises due to the inclusion of the magnetic permeability. If we are in a regime of the electromagnetic spectrum where the magnetic response is weak (i.e., $\mu = \mu_0$), we are considering a material that has only an electric response, or we are at sufficiently high frequencies that magnetic response may be ignored (Landau et al. 1984)—these relations may be further simplified. Equations used to convert from the index of refraction to the dielectric function and back, with $\mu = \mu_0$, are

$$\varepsilon_1 = \varepsilon_0(n_1^2 - n_2^2),$$
$$\varepsilon_2 = 2\varepsilon_0 n_1 n_2, \tag{1.32}$$

$$n_1 = \left\{ \frac{\varepsilon_1 + \sqrt{\varepsilon_1^2 + \varepsilon_2^2}}{2\varepsilon_0} \right\}^{1/2},$$
$$n_2 = \left\{ \frac{-\varepsilon_1 + \sqrt{\varepsilon_1^2 + \varepsilon_2^2}}{2\varepsilon_0} \right\}^{1/2}, \tag{1.33}$$

where, again, the subscripts 1 and 2 denote the real and imaginary portions, respectively, of the complex optical constants and material parameters.

1.1.4 Wave Equation

In the late nineteenth century, scientists believed that light needed a medium to be propagated in—the so-called luminiferous aether. In 1887, Michelson and Morley showed that the velocity of light was a constant c, which was the first strong evidence against the aether hypothesis. We now know that light is an electromagnetic wave—the speed of light emerging from Maxwell's equations is a crowning achievement of Maxwell's electromagnetic theory. We next derive the equations that govern the propagation of electromagnetic waves from Maxwell's equations.

If we assume there are no charges or currents ($\rho = \bar{\mathbf{J}} = 0$), then using the vector identity

$$\bar{\nabla} \times \left(\bar{\nabla} \times \bar{\mathbf{E}} \right) = \bar{\nabla} \left(\bar{\nabla} \cdot \bar{\mathbf{E}} \right) - \nabla^2 \bar{\mathbf{E}}, \tag{1.34}$$

we take the curl of (1.15) and, plugging in (1.13) and (1.16), we obtain

$$\nabla^2 \bar{\mathbf{E}} = \mu\varepsilon \frac{\partial^2 \bar{\mathbf{E}}}{\partial t}. \tag{1.35}$$

We may perform a similar operation for (1.16) and plug in (1.14) and (1.15) to obtain

$$\nabla^2\bar{\mathbf{B}} = \mu\varepsilon\frac{\partial^2\bar{\mathbf{B}}}{\partial t^2}. \tag{1.36}$$

Equations 1.35 and 1.36 are second-order partial differential equations, which describe the propagation of electromagnetic waves through a medium with a velocity $v = (\varepsilon\mu)^{-1/2}$, and are termed the wave equations. They have the general solution

$$\begin{Bmatrix} \bar{\mathbf{E}} \\ \bar{\mathbf{B}} \end{Bmatrix} = \begin{Bmatrix} \bar{\mathbf{E}}_0 \\ \bar{\mathbf{B}}_0 \end{Bmatrix} \exp i(\bar{\mathbf{k}}\cdot\bar{\mathbf{r}} - \omega t), \tag{1.37}$$

where
$\bar{\mathbf{k}}$ is called the wave vector or the propagation vector
ω is the frequency
$\bar{\mathbf{r}}$ is the position vector
$\bar{\mathbf{E}}_0, \bar{\mathbf{B}}_0$ are the amplitudes

In Figure 1.1 we plot an electromagnetic wave.

Upon substituting the general solution (1.37) into the wave equation (1.35) or (1.36) we find

$$k^2 = \omega^2\mu\varepsilon, \tag{1.38}$$

where k is the magnitude of the wave vector. Equation 1.38 is known as a dispersion relation and it describes the connection between the spatial dispersion and frequency dispersion of an electromagnetic wave, and its relation to the ratio of the speed of the electromagnetic wave to its speed in vacuum. We further develop the ideas of electromagnetic wave velocity next.

1.1.5 Group and Phase Velocities

The general solution (1.37) to Maxwell's equations describes a plane wave. For a particular wave vector $\bar{\mathbf{k}}$, a phase front is determined by $\bar{\mathbf{k}}\cdot\bar{\mathbf{r}}$ = constant. The phase front is a plane and is perpendicular to the wave vector. The phase velocity of an electromagnetic wave is defined to be

$$v_{\mathrm{p}} = \frac{\omega}{k}. \tag{1.39}$$

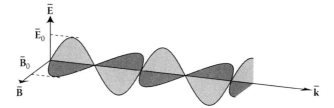

Figure 1.1 Transverse electromagnetic (TEM) wave. The wave is said to be "polarized" in the direction determined by the vector **E**.

This terminology originates from the sinusoidal dependence of waves, that is, the argument of a sinusoid is called its phase, so the velocity for which the phase is constant is called the phase velocity (Ramo et al. 1994). The group velocity is defined to be

$$v_g = \frac{d\omega}{dk}. \tag{1.40}$$

The definition of the group velocity in (1.40) is for a particular direction $\bar{\mathbf{k}}$, which we have not specified. However, (1.40) holds in Cartesian coordinates and thus we may express it in vector form as

$$\bar{\mathbf{v}}_g = \bar{\nabla}_k \omega(k). \tag{1.41}$$

We can put the group velocity (1.40) in a more useful form by writing our dispersion relation (1.38) as

$$k = \frac{n_1}{c}\omega = \frac{n_1}{\lambda}2\pi \tag{1.42}$$

where
  $c = 1/\sqrt{\varepsilon_0 \mu_0}$ is the speed of light in vacuum
  n_1 is the real part of the index of refraction, defined in (1.28)
  λ is the wavelength

If $n_1 = n_1(\omega)$, then we can write (1.42) as

$$dk = \frac{1}{c}d[n_1(\omega)\omega] = \frac{1}{c}[n_1(\omega)d\omega + \omega dn_1(\omega)]. \tag{1.43}$$

Dividing $d\omega$ by (1.43), we can write the group velocity as

$$v_g = \frac{c}{n_1(\omega) + \omega(dn_1(\omega)/d\omega)}. \tag{1.44}$$

The group velocity is the velocity with which a signal composed of a narrow band of frequency components propagates, that is, it is the transmission velocity of a wave packet (Collin 1991, Kittel 1996). For wave propagation in frequency-dispersive media, one may define a number of other velocities, for example, signal velocity and energy velocity, as well as the Sommerfeld precursor and the Brillouin first and second precursors (Sommerfeld 1950, Brillouin 1960). We will not elaborate on these other velocities and phenomena here, but refer to several texts where this is detailed (Brillouin 1960, Collin 1991). We do note, however, that the velocity of energy transfer is determined by the Poynting vector, which we discuss next.

1.1.6 Poynting Theorem

Conservation of energy is a law of physics which states that *for any closed (isolated) system, the energy is conserved over time.* Of course the implication of this is that energy cannot be created or destroyed, but instead can only be transferred back and forth between different forms.

In electricity and magnetism conservation of energy can be found from Maxwell's equations by dotting $\bar{\mathbf{H}}$ into (1.15) and subtracting $\bar{\mathbf{E}}$ dotted into (1.16). This yields

$$\bar{\mathbf{H}} \cdot \bar{\nabla} \times \bar{\mathbf{E}} - \bar{\mathbf{E}} \cdot \bar{\nabla} \times \bar{\mathbf{H}} = -\left(\bar{\mathbf{E}} \cdot \bar{\mathbf{J}} + \bar{\mathbf{H}} \cdot \frac{\partial \mathbf{B}}{\partial t} + \bar{\mathbf{E}} \cdot \frac{\partial \mathbf{D}}{\partial t} \right). \tag{1.45}$$

Using the identity $\bar{\nabla} \cdot \left(\bar{\mathbf{E}} \times \bar{\mathbf{H}} \right) = \bar{\mathbf{H}} \cdot \bar{\nabla} \times \bar{\mathbf{E}} - \bar{\mathbf{E}} \cdot \bar{\nabla} \times \bar{\mathbf{H}}$, (1.45) can be written as

$$\bar{\nabla} \cdot \left(\bar{\mathbf{E}} \times \bar{\mathbf{H}} \right) = -\left(\bar{\mathbf{E}} \cdot \bar{\mathbf{J}} + \bar{\mathbf{H}} \cdot \frac{\partial \bar{\mathbf{B}}}{\partial t} + \bar{\mathbf{E}} \cdot \frac{\partial \bar{\mathbf{D}}}{\partial t} \right). \tag{1.46}$$

Thus we observe that the conservation of energy in electricity and magnetism is determined by (1.46). One may also define the Poynting vector as

$$\bar{\mathbf{S}} = \bar{\mathbf{E}} \times \bar{\mathbf{H}}, \tag{1.47}$$

which has the units of (energy/area × time), and thus represents the power flux density. The quantity

$$\bar{\mathbf{H}} \cdot \frac{\partial \bar{\mathbf{B}}}{\partial t} + \bar{\mathbf{E}} \cdot \frac{\partial \bar{\mathbf{D}}}{\partial t} \equiv \frac{\partial u}{\partial t} \tag{1.48}$$

represents the time rate of change of the stored electric and magnetic energy ($u = u_{\mathrm{e}} + u_{\mathrm{m}}$) over a volume, and $\bar{\mathbf{E}} \cdot \bar{\mathbf{J}}$ is the power supplied by a current density $\bar{\mathbf{J}}$ over some volume. We may also define, over some volume, the electric energy $u_{\mathrm{e}} = \bar{\mathbf{E}} \cdot \bar{\mathbf{D}}$ and the magnetic energy $u_{\mathrm{m}} = \bar{\mathbf{B}} \cdot \bar{\mathbf{H}}$. Thus finally we may write (1.46) as

$$\bar{\nabla} \cdot \bar{\mathbf{S}} + \frac{\partial u}{\partial t} = -\bar{\mathbf{J}} \cdot \bar{\mathbf{E}}. \tag{1.49}$$

The meaning of (1.49) is that the energy transported by the fields per unit area per unit time, plus the time rate of change of the energy density, is equal to the negative of the total work done by the fields on the sources. It should be noted that (1.49) is a statement of the conservation of energy, but is valid only for linear responses (electric and magnetic) and is not valid for hysteretic materials.

Since the physical meaning of $\bar{\mathbf{S}}$ is that it is the energy flow, it must be real. In (1.47) we may specify that we will only consider the real parts of each vector $\bar{\mathbf{S}}, \bar{\mathbf{E}}, \bar{\mathbf{H}}$. Further, we are interested in waves that have a harmonic dependence $e^{-i\omega t}$. Thus we show, without proof, that if we define $\bar{\mathbf{S}}$ over one period, then we may write (Jackson 1998)

$$\bar{\mathbf{S}} = \frac{1}{2} \mathrm{Re} \left[\bar{\mathbf{E}} \times \bar{\mathbf{H}}^{*} \right], \tag{1.50}$$

where $\bar{\mathbf{H}}^{*}$ is the complex conjugate of the magnetic field strength. The usefulness of the Poynting vector shown in (1.50) will be made clear when we discuss the Fresnel equations later in this chapter.

The particular form of the time rate of change of the electric energy density in (1.48) permits us to consider the response of a Drude metal (discussed in more detail later and in Chapter 3), here an electron gas, to an additional electron moving within the solid. This moving electron produces a field $\bar{\mathbf{D}}$, which we assume is time varying as $\bar{\mathbf{D}} = \bar{\mathbf{D}}_0 \exp i(\bar{\mathbf{k}} \cdot \bar{\mathbf{r}} - \omega t)$, and thus the electronic energy density absorbed per unit volume is

$$\frac{\partial u_{\mathrm{e}}}{\partial t} = \frac{\bar{\mathbf{D}}}{\tilde{\varepsilon}} \cdot \frac{\partial \bar{\mathbf{D}}}{\partial t} = -\frac{i\omega}{\tilde{\varepsilon}} |D_0|^2 \,. \tag{1.51}$$

If we consider the real and imaginary parts of (1.51), then upon equating real parts we find

$$\mathrm{Re}\left(\frac{\partial u_{\mathrm{e}}}{\partial t}\right) = -\omega |D_0|^2 \; \mathrm{Re}\left(\frac{i}{\tilde{\varepsilon}}\right) = \omega |D_0|^2 \; \mathrm{Im}\left(\frac{1}{\tilde{\varepsilon}}\right). \tag{1.52}$$

Thus we find that the imaginary part of $1/\tilde{\varepsilon}$ describes the energy loss due to an electron moving through the medium. Specifically we associate the term

$$\mathrm{Im}\left(\frac{1}{\tilde{\varepsilon}}\right) = -\frac{\varepsilon_2}{\varepsilon_1^2 + \varepsilon_2^2} \tag{1.53}$$

with this loss, and thus (1.53) is called the loss function. Although we have left out its explicit frequency dependence, in general the loss function is a function of frequency. It can be observed that the loss function peaks where the real part of the dielectric function is near zero and where the imaginary part of the dielectric function (losses) is small. The loss function is important in optics, where it determines the center frequency of longitudinal optical phonons (and other longitudinal resonant phenomena) (Burns 1990), and in plasmonics where it determines the frequency of surface, geometrical, and bulk plasmons (Cottam and Tilley 1989, Kittel 1996).

1.2 CONSTITUTIVE RELATIONS

Maxwell's macroscopic equations, as shown in (1.13) through (1.16) are a set of four equations involving the components of the four fields $\bar{\mathbf{E}}, \bar{\mathbf{B}}, \bar{\mathbf{D}}, \bar{\mathbf{H}}$. As we have shown in a simply linear case, a medium may be "polarized" by incident $\bar{\mathbf{E}}, \bar{\mathbf{H}}$ fields, resulting in a total $\bar{\mathbf{D}}, \bar{\mathbf{B}}$ field. Equations that govern the relationships between the incident and total fields are called constitutive equations. It is useful to characterize materials by the specific form of the constitutive relations, and the symmetry they posses. In a very general way one may express the constitutive relations as

$$\bar{\mathbf{D}} = \bar{\mathbf{D}}\left[\bar{\mathbf{E}}, \bar{\mathbf{H}}\right], \tag{1.54}$$

$$\bar{\mathbf{B}} = \bar{\mathbf{B}}\left[\bar{\mathbf{E}}, \bar{\mathbf{H}}\right], \tag{1.55}$$

where the square brackets indicate that the relations between the given fields and the resultant fields are not necessarily simple, and may depend upon the entire past history of the sample (hysteresis), may be nonlinear, or may be bianisotropic, etc. (Jackson 1998).

1.2.1 Magneto-Optical Permittivities

For isotropic, anisotropic, or more complicated types of media, we may express the constitutive relations generally as (Cheng and Kong 1968a,b, Kong 1990)

$$\begin{pmatrix} \overline{\mathbf{D}} \\ \overline{\mathbf{B}} \end{pmatrix} = \begin{pmatrix} \overline{\overline{\varepsilon}} & \overline{\overline{\xi}} \\ \overline{\overline{\zeta}} & \overline{\overline{\mu}} \end{pmatrix} \cdot \begin{pmatrix} \overline{\mathbf{E}} \\ \overline{\mathbf{H}} \end{pmatrix}. \tag{1.56}$$

The quantities $\overline{\overline{\varepsilon}}$ and $\overline{\overline{\mu}}$, the dielectric permittivity and the magnetic permeability, have already been introduced. The double bar above each indicates that it is a tensor quantity, in this particular case of rank 2. Thus, for example, we may write the dielectric permittivity as

$$\overline{\overline{\varepsilon}} = \begin{pmatrix} \varepsilon_{xx} & \varepsilon_{xy} & \varepsilon_{xz} \\ \varepsilon_{yx} & \varepsilon_{yy} & \varepsilon_{yz} \\ \varepsilon_{zx} & \varepsilon_{zy} & \varepsilon_{zz} \end{pmatrix}. \tag{1.57}$$

A similar form may be written for the other quantities $\left(\overline{\overline{\mu}}, \overline{\overline{\xi}}, \overline{\overline{\zeta}}\right)$ in (1.56). It should be stated that all of the quantities in (1.56) may depend on frequency (ω) and wave vector (k) but, for simplification, we drop this explicit dependence. Typically in crystal optics, the magnetic permeability does not exhibit frequency dependence and its value may be assumed to be equal to that of free space. Under these conditions, one may further distinguish between different types of crystals by considering the symmetry of the dielectric tensor. For example, usually the off-diagonal terms in (1.57) are zero such that

$$\overline{\overline{\varepsilon}} = \begin{pmatrix} \varepsilon_{xx} & 0 & 0 \\ 0 & \varepsilon_{yy} & 0 \\ 0 & 0 & \varepsilon_{zz} \end{pmatrix}. \tag{1.58}$$

In the case of (1.58), if we have the specific relationship $\varepsilon_{xx} = \varepsilon_{yy} \neq \varepsilon_{zz}$ between diagonal components of the permittivity, then the crystal is termed "uniaxial." For the specific case where the relationship is $\varepsilon_{xx} \neq \varepsilon_{yy} \neq \varepsilon_{zz}$, the crystal is called "biaxial." By considering the simple relationship between the dielectric permittivity and the index of refraction (1.28), notice that for a uniaxial crystal there exist two different refractive indices. These two different refractive indices are termed the "ordinary index" (n_o) if light is polarized in the $\hat{x}$ or $\hat{y}$ direction, and the "extraordinary index" (n_e) for light polarized in the $\hat{z}$ direction.

For the more complicated case where we cannot ignore the magnetic response of materials we must consider the full constitutive equations (1.56). Beyond having just a magnetic and electric response we also allow a magneto-electric response. In order to describe such responses it is useful to define the terms $\overline{\overline{\xi}}$ and $\overline{\overline{\zeta}}$, which are called the magneto-optical permittivities. They describe coupling of the magnetic-to-electric response $\overline{\overline{\xi}}$ and electric-to-magnetic response $\overline{\overline{\zeta}}$. Multiferroics are one example of a class of materials in which (1.56) facilitates a description of their electromagnetic response (Eerenstein et al. 2006). Just as in the case of the symmetry of the dielectric permittivity tensor, we may describe various media based on the symmetry of the magneto-optical permittivity tensors, shown in (1.56). This is summarized in Table 1.2.

TABLE 1.2

Description of Materials Based on the Symmetry of the Magneto-Optical Permittivity Tensor (1.56)

Material Type	$\bar{\bar{\varepsilon}}, \bar{\bar{\mu}}$	$\bar{\bar{\xi}}, \bar{\bar{\zeta}}$
Isotropic	$\propto I$	$\bar{\bar{\xi}}, \bar{\bar{\zeta}} = 0$
Anisotropic	$\bar{\bar{\varepsilon}}$ and/or $\bar{\bar{\mu}}$ not $\propto I$	$\bar{\bar{\xi}}, \bar{\bar{\zeta}} = 0$
Bi-isotropic	$\propto I$	$\propto I$
Bi-anisotropic	All other cases	All other cases

1.3 FRESNEL EQUATIONS

If an isotropic opaque material may be described by the macroscopic form of Maxwell's equations (1.13 through 1.16), we may calculate the reflectance of the intensity of a plane wave from the material with the Fresnel equations. We begin by deriving the appropriate equations for two different types of polarization—the so-called transverse electric (TE) and the transverse magnetic (TM) cases.

1.3.1 Reflection and Transmission at an Interface

In Figure 1.2 we show a plane wave incident on the interface between a medium with optical constants $\tilde{\varepsilon}^{(1)}$, $\tilde{\mu}^{(1)}$ and a medium (shaded gray region) described by the optical constants $\tilde{\varepsilon}^{(2)}$, $\tilde{\mu}^{(2)}$, where the wave vector forms an angle θ_i with the surface normal and the superscripted (1) and (2) refer to the first and second media, respectively. $\bar{E}$ represents the electric field, $\bar{H}$ the magnetic field, and $\bar{k}$ the wave vector of the incident wave. The subscripted i, r, and t on each of the fields and wave vectors stand for the incident, reflected, and transmitted waves, respectively. The incoming and outgoing wave vectors form the "plane of incidence," and TE and TM denote the relation of the electric field vector and magnetic field vector with respect to the plane of incidence. TE means that the electric field vector is perpendicular to the plane

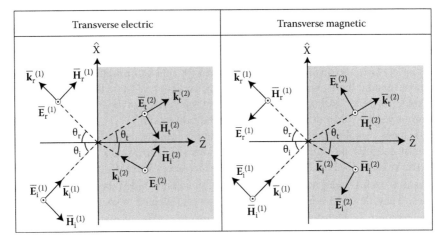

Figure 1.2 Schematic depicting the transmission and reflection of an electromagnetic wave at an interface. The left and right panels show the TE and TM polarization cases, respectively. In both cases, waves are incident from both sides of the interface.

of incidence (with the magnetic field vector parallel to the plane of incidence). TM is the opposite, with the magnetic field vector perpendicular and the electric field vector parallel to the plane of incidence.

We first discuss the TE case, as depicted on the left portion of Figure 1.2. The interface lies in the xy-plane and thus the electric field is polarized only in the y-direction. The magnetic field, however, has two components lying in the x- and z-directions. The boundary conditions (1.21) through (1.24) permit us to specify the relationship of the electric and magnetic fields across the interface. For example, (1.26) dictates that the parallel (tangential) components of the electric field must be the same on either side of the interface, that is,

$$E_i^{(1)} + E_r^{(1)} = E_t^{(2)} + E_i^{(2)}. \tag{1.59}$$

Likewise the same must hold for the magnetic field,

$$H_{ix}^{(1)} + H_{rx}^{(1)} = H_{tx}^{(2)} + H_{ix}^{(2)}, \tag{1.60}$$

where we have added a subscripted x to denote that we are only considering the components parallel to the interface. We can use Maxwell's equation (1.19) to write the magnetic field in terms of the electric field, that is,

$$-k_{iz}^{(1)} E_i^{(1)} = \omega\mu^{(1)} H_{ix}^{(1)}; \quad k_{rz}^{(1)} E_r^{(1)} = \omega\mu^{(1)} H_{rx}^{(1)}; \quad -k_{tz}^{(2)} E_t^{(2)} = \omega\mu^{(2)} H_{tx}^{(2)}. \tag{1.61}$$

The first equation in (1.61) is for the incident components (i subscripts), the second for the reflected components (r subscripts), the third for the transmitted components (t subscripts), and there is a similar equation for the wave incident on the interface from medium (2). Plugging in (1.61) into our boundary condition for the magnetic field (1.60) we see

$$-\frac{k_z^{(1)}}{\mu^{(1)}} E_i^{(1)} + \frac{k_z^{(1)}}{\mu^{(1)}} E_r^{(1)} = -\frac{k_z^{(2)}}{\mu^{(2)}} E_t^{(2)} + \frac{k_z^{(2)}}{\mu^{(2)}} E_i^{(2)}. \tag{1.62}$$

Our boundary conditions are now all expressed in terms of the electric field. The system of Equations 1.59 and 1.62 may thus be expressed in matrix form as

$$\begin{pmatrix} 1 & 1 \\ -\dfrac{k_z^{(1)}}{\mu^{(1)}} & \dfrac{k_z^{(1)}}{\mu^{(1)}} \end{pmatrix} \begin{pmatrix} E_i^{(1)} \\ E_r^{(1)} \end{pmatrix} = \begin{pmatrix} 1 & 1 \\ -\dfrac{k_z^{(2)}}{\mu^{(2)}} & \dfrac{k_z^{(2)}}{\mu^{(2)}} \end{pmatrix} \begin{pmatrix} E_t^{(2)} \\ E_i^{(2)} \end{pmatrix}. \tag{1.63}$$

We may express (1.63) in terms of the electric fields on one side of the interface in terms of the other by

$$\begin{pmatrix} E_t^{(2)} \\ E_i^{(2)} \end{pmatrix} = \mathbf{M}^{\text{TE}} \begin{pmatrix} E_i^{(1)} \\ E_r^{(1)} \end{pmatrix}, \tag{1.64}$$

where $\mathbf{M}^{\text{TE}}$ is called the transfer matrix and has the explicit form

$$\mathbf{M}^{\text{TE}} = \frac{1}{2} \begin{pmatrix} 1 + \dfrac{\mu^{(2)}}{\mu^{(1)}} \dfrac{k_z^{(1)}}{k_z^{(2)}} & 1 - \dfrac{\mu^{(2)}}{\mu^{(1)}} \dfrac{k_z^{(1)}}{k_z^{(2)}} \\ 1 - \dfrac{\mu^{(2)}}{\mu^{(1)}} \dfrac{k_z^{(1)}}{k_z^{(2)}} & 1 + \dfrac{\mu^{(2)}}{\mu^{(1)}} \dfrac{k_z^{(1)}}{k_z^{(2)}} \end{pmatrix}. \tag{1.65}$$

Equation 1.65 is the general form for the transmission matrix in the TE polarization case when waves are incident from both sides of the interface. In the case that waves are only incident from the left side of the interface we have $E_i^{(2)} = 0$, and we can write (1.64) as

$$E_t^{(2)} = M_{11}^{TE} E_i^{(1)} + M_{12}^{TE} E_r^{(1)}$$
$$E_i^{(2)} = M_{21}^{TE} E_i^{(1)} + M_{22}^{TE} E_r^{(1)} = 0. \tag{1.66}$$

The transmission and reflection coefficients for the electric field are defined as

$$t_{TE} = \frac{E_t^{(2)}}{E_i^{(1)}} \quad \text{and} \quad r_{TE} = \frac{E_r^{(1)}}{E_i^{(1)}}. \tag{1.67}$$

Using (1.64) we may solve the system of equations (1.66) to give

$$t_{TE} = \frac{\det \mathbf{M}^{TE}}{M_{22}^{TE}} = \frac{2\mu^{(2)} k_z^{(1)}}{\mu^{(1)} k_z^{(2)} + \mu^{(2)} k_z^{(1)}} \tag{1.68}$$

and

$$r_{TE} = -\frac{M_{21}^{TE}}{M_{22}^{TE}} = \frac{\mu^{(2)} k_z^{(1)} - \mu^{(1)} k_z^{(2)}}{\mu^{(2)} k_z^{(1)} - \mu^{(1)} k_z^{(2)}}. \tag{1.69}$$

The intensity transmittance is given by the ratio of the energy flow in the two media that make up the interface. Using our equation for the description of energy flow for harmonic phenomena (1.50) we have

$$S^{(1)} = \frac{1}{2} \operatorname{Re} \left[E_i^{(1)} H_{ix}^{(1)*} \right] = \frac{1}{2\omega} \operatorname{Re} \left[\frac{k_z^{(1)}}{\mu^{(1)}} \right] \left| E_i^{(1)} \right|^2 \tag{1.70}$$

$$S^{(2)} = \frac{1}{2} \operatorname{Re} \left[E_t^{(2)} H_{tx}^{(2)*} \right] = \frac{1}{2\omega} \operatorname{Re} \left[\frac{k_z^{(2)}}{\mu^{(2)}} \right] \left| E_t^{(2)} \right|^2 \tag{1.71}$$

where $S^{(1)}$, $S^{(2)}$ are the energy flows in the first and second media, respectively. Since $E_t^{(2)} = t_{TE} E_i^{(1)}$ we may finally write the transmittance (or transmissivity) and reflectance (or reflectivity) as

$$T_{TE} = \frac{S^{(2)}}{S^{(1)}} = \frac{\operatorname{Re}\left[k_z^{(2)} / \mu^{(2)} \right]}{\operatorname{Re}\left[k_z^{(1)} / \mu^{(1)} \right]} \left| t_{TE} \right|^2 \tag{1.72}$$

and

$$R_{TE} = \left| r_{TE} \right|^2. \tag{1.73}$$

Next we discuss the TM case, as depicted on the right portion of Figure 1.2. The interface lies in the xy-plane and thus the magnetic field is polarized only in the y-direction. The electric field, however, has two components lying in the x- and z-directions. Using the same boundary conditions as for the TE case, (1.21) through (1.24), we have

$$H_i^{(1)} + H_r^{(1)} = H_t^{(2)} + H_i^{(2)}. \tag{1.74}$$

Likewise the same must hold for the electric field

$$E_{ix}^{(1)} + E_{rx}^{(1)} = E_{tx}^{(2)} + E_{ix}^{(2)}, \tag{1.75}$$

where we have added a subscripted x to denote that we are only considering the components parallel to the interface. We can use Maxwell's equation (1.20) to write the electric field in terms of the magnetic field, that is,

$$k_z^{(1)} H_i^{(1)} = \omega \varepsilon^{(1)} E_{ix}^{(1)}; \quad -k_z^{(1)} H_r^{(1)} = \omega \varepsilon^{(1)} E_{rx}^{(1)}; \quad -k_z^{(2)} H_t^{(2)} = \omega \varepsilon^{(2)} E_{tx}^{(2)}. \tag{1.76}$$

We obtain, in matrix form, the relationship

$$\begin{pmatrix} 1 & 1 \\ \dfrac{k_z^{(1)}}{\varepsilon^{(1)}} & -\dfrac{k_z^{(1)}}{\varepsilon^{(1)}} \end{pmatrix} \begin{pmatrix} H_i^{(1)} \\ H_r^{(1)} \end{pmatrix} = \begin{pmatrix} 1 & 1 \\ -\dfrac{k_z^{(2)}}{\varepsilon^{(2)}} & \dfrac{k_z^{(2)}}{\varepsilon^{(2)}} \end{pmatrix} \begin{pmatrix} H_t^{(2)} \\ H_i^{(2)} \end{pmatrix}. \tag{1.77}$$

We may express (1.77) in terms of the magnetic fields on one side of the interface in terms of the other by

$$\begin{pmatrix} H_t^{(2)} \\ H_i^{(2)} \end{pmatrix} = \mathbf{M}^{\text{TM}} \begin{pmatrix} H_i^{(1)} \\ H_r^{(1)} \end{pmatrix} \tag{1.78}$$

where $\mathbf{M}^{\text{TM}}$ has the explicit form,

$$\mathbf{M}^{\text{TM}} = \frac{1}{2} \begin{pmatrix} 1 + \dfrac{\varepsilon^{(2)}}{\varepsilon^{(1)}} \dfrac{k_z^{(1)}}{k_z^{(2)}} & 1 - \dfrac{\varepsilon^{(2)}}{\varepsilon^{(1)}} \dfrac{k_z^{(1)}}{k_z^{(2)}} \\ 1 - \dfrac{\varepsilon^{(2)}}{\varepsilon^{(1)}} \dfrac{k_z^{(1)}}{k_z^{(2)}} & 1 + \dfrac{\varepsilon^{(2)}}{\varepsilon^{(1)}} \dfrac{k_z^{(1)}}{k_z^{(2)}} \end{pmatrix} \tag{1.79}$$

Equation 1.79 is the general form for the transmission matrix in the TM polarization case when waves are incident from both sides of the interface. In the case that waves are only incident from the left side of the interface we have $H_i^{(2)} = 0$, and we can write the transmission and reflection coefficients for the magnetic field as

$$t_{\text{TM}} = \frac{\det \mathbf{M}^{\text{TM}}}{M_{22}^{\text{TM}}} = \frac{2\varepsilon^{(2)} k_z^{(1)}}{\varepsilon^{(1)} k_z^{(2)} + \varepsilon^{(2)} k_z^{(1)}} \tag{1.80}$$

and

$$r_{\text{TM}} = -\frac{M_{21}^{\text{TM}}}{M_{22}^{\text{TM}}} = \frac{\varepsilon^{(2)} k_z^{(1)} - \varepsilon^{(1)} k_z^{(2)}}{\varepsilon^{(2)} k_z^{(1)} - \varepsilon^{(1)} k_z^{(2)}}. \tag{1.81}$$

As with the TE case, the intensity transmittance is given by the ratio of the energy flow in the two media that make up the interface, and thus we have

$$S^{(1)} = \frac{1}{2} \text{Re}\left[E_{ix}^{(1)} H_i^{(1)*} \right] = \frac{1}{2\omega} \text{Re}\left[\frac{k_z^{(1)}}{\varepsilon^{(1)}} \right] \left| H_i^{(1)} \right|^2 \tag{1.82}$$

$$S^{(2)} = \frac{1}{2} \text{Re}\left[E_{tx}^{(2)} H_t^{(2)*} \right] = \frac{1}{2\omega} \text{Re}\left[\frac{k_z^{(2)}}{\varepsilon^{(2)}} \right] \left| H_t^{(2)} \right|^2 \tag{1.83}$$

We may write the transmittance and reflectance for TM waves as

$$T_{\text{TM}} = \frac{S^{(2)}}{S^{(1)}} = \frac{\text{Re}\left[k_z^{(2)} / \varepsilon^{(2)} \right]}{\text{Re}\left[k_z^{(1)} / \varepsilon^{(1)} \right]} \left| t_{\text{TM}} \right|^2 \tag{1.84}$$

and

$$R_{\text{TM}} = \left| r_{\text{TM}} \right|^2 . \tag{1.85}$$

The reflectance for both TE, (1.72) through (1.73), and TM, (1.84) and (1.85), waves can easily be expressed in more common forms found in many textbooks. For example, the z-component of our incident and transmitted wave vectors may be expressed in terms of the incident and transmitted angles by

$$k_z^{(1)} = k^{(1)} \cos\theta_i \quad \text{and} \quad k_z^{(2)} = k^{(2)} \cos\theta_t. \tag{1.86}$$

If the interface on the left is free space, then $\varepsilon^{(1)} = \varepsilon_0$, $\mu^{(1)} = \mu_0$. Further we may use (1.42) to express the wave vector in terms of the index of refraction. With these substitutions and using Snell's law, $n^{(1)} \sin\theta_1 = n^{(2)} \sin\theta_2$, we find

$$R_{\text{TE}} = \left| r_{\text{TE}} \right|^2 = \left| \frac{\cos\theta - \mu_r^{-1}\sqrt{n^2 - \sin^2\theta}}{\cos\theta + \mu_r^{-1}\sqrt{n^2 - \sin^2\theta}} \right|^2 , \tag{1.87}$$

$$R_{\text{TM}} = \left| r_{\text{TM}} \right|^2 = \left| \frac{\varepsilon_r \cos\theta - \sqrt{n^2 - \sin^2\theta}}{\varepsilon_r \cos\theta + \sqrt{n^2 - \sin^2\theta}} \right|^2 \tag{1.88}$$

where
   θ is the angle of incidence
   ε_r, μ_r, and n are the relative permittivity, relative permeability, and index of refraction of the medium, respectively

Here we assume the combination of absorption and material thickness is sufficient such that the electromagnetic wave is attenuated before reaching the other side of the material, that is,

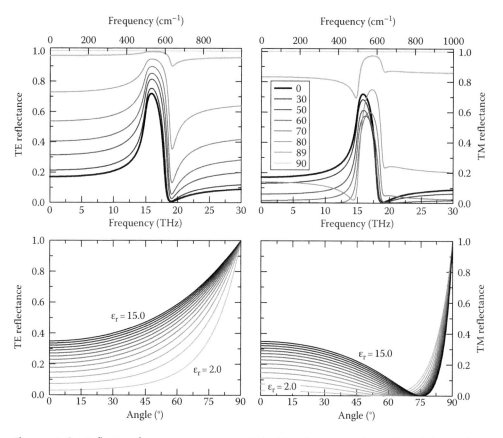

Figure 1.3 Reflection from an opaque material with a frequency-dependent dielectric function (top panels) and as a function of various constant dielectric values (bottom panels). The top panels are the reflectance (1.87) and (1.88) as a function of frequency, plotted in both THz and cm^{-1} units, and assume a single Lorentz oscillator (1.130), with parameters ε_∞ = 4.0, ω_p = $2\pi \times$ 20 THz, ω_0 = $2\pi \times$ 15 THz, and γ = $2\pi \times$ 0.75 THz. The bottom panels show the TE (1.87) and TM (1.88) reflectance for a material of varying dielectric constant, ranging ε_r = 2.0–15, as a function of angle.

there is no reflection or transmission from the outgoing interface. Figure 1.3 plots (1.87) and (1.88) for both frequency-dependent optical constants (top panels) and for constant optical constants (bottom panels).

For normal incidence these equations may be further simplified and expressed as

$$R = |r|^2 = \left| \frac{\mu_r - n}{\mu_r + n} \right|^2 = \left| \frac{Z - Z_0}{Z + Z_0} \right|^2, \tag{1.89}$$

where $Z_0 = \sqrt{\mu_0/\varepsilon_0}$ is the impedance of free space, and we have dropped the TE and TM subscripts, since these are undefined for $\theta = 0$. The phase of the reflected wave ϕ_r is given by

$$\tan(\phi_r) = \frac{\text{Im}(r)}{\text{Re}(r)} = \frac{-2n_2\mu_r}{\mu_r^2 - n_1^2 - n_2^2} = \frac{2Z_0 Z_2}{Z_1^2 + Z_2^2 - Z_0^2}, \tag{1.90}$$

where Im and Re are the imaginary and real portions of the complex reflection coefficient r. Likewise at normal incidence the transmittance of a layer can be simplified using (1.42) and (1.86) and expressed as

$$T = \mathrm{Re}\left[\frac{n}{\mu_r}\right]\left|\frac{2\mu_r}{n+\mu_r}\right|^2 = \mathrm{Re}\left[\frac{n}{\mu_r}\right]\left|\frac{2Z}{Z+Z_0}\right|^2 \tag{1.91}$$

1.3.2 Brewster's Angle

Notice that the reflectances given in (1.69) and (1.81) for TE and TM polarizations, respectively, both involve a numerator that is the difference between two terms. This leads us to ask whether there is a particular combination of optical constants and incident angle at which the reflectivity goes to zero. By setting each of the two terms, in the numerator of both (1.69) and (1.81), equal to each other, and using (1.42) and Snell's law we find

$$\cos^2\theta_{B,TE} = \frac{\varepsilon^{(2)}/\varepsilon^{(1)} - \mu^{(1)}/\mu^{(2)}}{\mu^{(2)}/\mu^{(1)} - \mu^{(1)}/\mu^{(2)}} \tag{1.92}$$

and

$$\cos^2\theta_{B,TE} = \frac{\mu^{(2)}/\mu^{(1)} - \varepsilon^{(1)}/\varepsilon^{(2)}}{\varepsilon^{(2)}/\varepsilon^{(1)} - \varepsilon^{(1)}/\varepsilon^{(2)}}, \tag{1.93}$$

where the subscripted B,TE and B,TM refer to Brewster's angle (also known as the polarizing angle) for TE and TM polarizations, respectively. At the particular angles specified by (1.92) and (1.93), the reflectivity of light is zero and the wave is entirely refracted. If the materials under consideration do not exhibit a magnetic response, such that $\mu^{(1)} = \mu^{(2)} = \mu_0$, then no Brewster angle exists for TE polarization; however, (1.93) for TM polarization may be expressed as

$$\tan\theta_{B,TM} = \frac{n^{(2)}}{n^{(1)}}. \tag{1.94}$$

The effect of Brewster's angle can be observed in Figure 1.3, where we plot the TM reflectance as a function of incident angle for media consisting of different dielectric constants.

1.3.3 Reflection and Transmission for a Slab

We next consider the TE transmission and reflection from a flat slab of material, as shown in Figure 1.4. The transmission matrices for the two interfaces of the slab have already been found in (1.65), and the transmission matrix that describes propagation across a material with optical constants $\varepsilon^{(2)}$, $\mu^{(2)}$ is

$$\mathbf{M}^{slab} = \begin{pmatrix} e^{ik_z^{(2)}d} & 0 \\ 0 & e^{-ik_z^{(2)}d} \end{pmatrix} \tag{1.95}$$

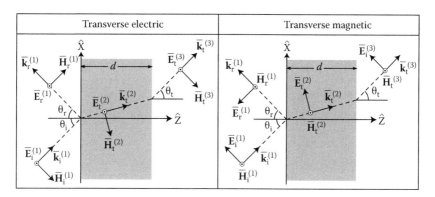

Figure 1.4 Schematic depicting the transmission and reflection of an electromagnetic wave from a slab of material. The left and right panels show the TE and TM polarization cases, respectively.

We would like to find a transmission matrix that describes the total transmission and reflection of the two interfaces plus the slab. This is given by the transmission matrix composition law (Markos and Soukoulis 2008) as

$$\mathbf{M}^{\text{total}} = \mathbf{M}_{12} \begin{pmatrix} e^{ik_z^{(2)}d} & 0 \\ 0 & e^{-ik_z^{(2)}d} \end{pmatrix} \mathbf{M}_{23}, \tag{1.96}$$

where

$\mathbf{M}_{12}$ is the transmission matrix for the interface between media (1) and (2)

$\mathbf{M}_{23}$ is the transmission matrix for the interface between media (2) and (3)

More generally if we have N total slabs and interfaces, we may write

$$\mathbf{M}^{\text{total}} = \mathbf{M}_N \mathbf{M}_{N-1} \cdots \mathbf{M}_2 \mathbf{M}_1. \tag{1.97}$$

For TE polarization we may write

$$\begin{pmatrix} E_t^{(3)} \\ E_i^{(3)} \end{pmatrix} = \mathbf{M}^{\text{total}} \begin{pmatrix} E_t^{(1)} \\ E_i^{(1)} \end{pmatrix}. \tag{1.98}$$

By performing the matrix multiplication specified in (1.96) we find the specific matrix elements

$$M_{11}^{\text{total}} = \frac{1}{2}\left[1 + \frac{\mu^{(3)}}{\mu^{(1)}}\frac{k_z^{(1)}}{k_z^{(3)}}\right]\cos\left(k_z^{(2)}d\right) + \frac{i}{2}\left[\frac{\mu^{(3)}}{\mu^{(2)}}\frac{k_z^{(2)}}{k_z^{(3)}} + \frac{\mu^{(2)}}{\mu^{(1)}}\frac{k_z^{(1)}}{k_z^{(2)}}\right]\sin\left(k_z^{(2)}d\right). \tag{1.99}$$

$$M_{12}^{\text{total}} = \frac{1}{2}\left[1 - \frac{\mu^{(3)}}{\mu^{(1)}}\frac{k_z^{(1)}}{k_z^{(3)}}\right]\cos\left(k_z^{(2)}d\right) + \frac{i}{2}\left[\frac{\mu^{(3)}}{\mu^{(2)}}\frac{k_z^{(2)}}{k_z^{(3)}} - \frac{\mu^{(2)}}{\mu^{(1)}}\frac{k_z^{(1)}}{k_z^{(2)}}\right]\sin\left(k_z^{(2)}d\right). \tag{1.100}$$

$$M_{22}^{\text{total}} = \frac{1}{2}\left[1+\frac{\mu^{(3)}}{\mu^{(1)}}\frac{k_z^{(1)}}{k_z^{(3)}}\right]\cos\left(k_z^{(2)}d\right) - \frac{i}{2}\left[\frac{\mu^{(3)}}{\mu^{(2)}}\frac{k_z^{(2)}}{k_z^{(3)}}+\frac{\mu^{(2)}}{\mu^{(1)}}\frac{k_z^{(1)}}{k_z^{(2)}}\right]\sin\left(k_z^{(2)}d\right). \quad (1.101)$$

$$M_{21}^{\text{total}} = \frac{1}{2}\left[1-\frac{\mu^{(3)}}{\mu^{(1)}}\frac{k_z^{(1)}}{k_z^{(3)}}\right]\cos\left(k_z^{(2)}d\right) - \frac{i}{2}\left[\frac{\mu^{(3)}}{\mu^{(2)}}\frac{k_z^{(2)}}{k_z^{(3)}}-\frac{\mu^{(2)}}{\mu^{(1)}}\frac{k_z^{(1)}}{k_z^{(2)}}\right]\sin\left(k_z^{(2)}d\right). \quad (1.102)$$

As in the case of the single interface, we can obtain the transmission and reflection coefficients as

$$t_{\text{TE}} = \frac{E_t^{(3)}}{E_i^{(1)}} = \frac{\det \mathbf{M}^{\text{total}}}{M_{22}^{\text{total}}}, \quad (1.103)$$

$$r_{\text{TE}} = \frac{E_r^{(1)}}{E_i^{(1)}} = -\frac{M_{21}^{\text{total}}}{M_{22}^{\text{total}}} \quad (1.104)$$

and thus the transmittance and reflectance are

$$T = \frac{S^{(3)}}{S^{(1)}} = \frac{\mu^{(1)}}{\mu^{(3)}}\frac{\operatorname{Re} k_z^{(3)}}{\operatorname{Re} k_z^{(1)}}\left|t_{\text{TE}}\right|^2. \quad (1.105)$$

$$R = \left|r_{\text{TE}}\right|^2. \quad (1.106)$$

The case for TM polarization follows in a straightforward manner. We simply list here the matrix elements for the total scattering matrix

$$M_{11}^{\text{total}} = \frac{1}{2}\left[1+\frac{\varepsilon^{(3)}}{\varepsilon^{(1)}}\frac{k_z^{(1)}}{k_z^{(3)}}\right]\cos\left(k_z^{(2)}d\right) + \frac{i}{2}\left[\frac{\varepsilon^{(3)}}{\varepsilon^{(2)}}\frac{k_z^{(2)}}{k_z^{(3)}}+\frac{\varepsilon^{(2)}}{\varepsilon^{(1)}}\frac{k_z^{(1)}}{k_z^{(2)}}\right]\sin\left(k_z^{(2)}d\right). \quad (1.107)$$

$$M_{12}^{\text{total}} = \frac{1}{2}\left[1-\frac{\varepsilon^{(3)}}{\varepsilon^{(1)}}\frac{k_z^{(1)}}{k_z^{(3)}}\right]\cos\left(k_z^{(2)}d\right) + \frac{i}{2}\left[\frac{\varepsilon^{(3)}}{\varepsilon^{(2)}}\frac{k_z^{(2)}}{k_z^{(3)}}-\frac{\varepsilon^{(2)}}{\varepsilon^{(1)}}\frac{k_z^{(1)}}{k_z^{(2)}}\right]\sin\left(k_z^{(2)}d\right). \quad (1.108)$$

$$M_{22}^{\text{total}} = \frac{1}{2}\left[1+\frac{\varepsilon^{(3)}}{\varepsilon^{(1)}}\frac{k_z^{(1)}}{k_z^{(3)}}\right]\cos\left(k_z^{(2)}d\right) - \frac{i}{2}\left[\frac{\varepsilon^{(3)}}{\varepsilon^{(2)}}\frac{k_z^{(2)}}{k_z^{(3)}}+\frac{\varepsilon^{(2)}}{\varepsilon^{(1)}}\frac{k_z^{(1)}}{k_z^{(2)}}\right]\sin\left(k_z^{(2)}d\right). \quad (1.109)$$

$$M_{21}^{\text{total}} = \frac{1}{2}\left[1-\frac{\varepsilon^{(3)}}{\varepsilon^{(1)}}\frac{k_z^{(1)}}{k_z^{(3)}}\right]\cos\left(k_z^{(2)}d\right) - \frac{i}{2}\left[\frac{\varepsilon^{(3)}}{\varepsilon^{(2)}}\frac{k_z^{(2)}}{k_z^{(3)}}-\frac{\varepsilon^{(2)}}{\varepsilon^{(1)}}\frac{k_z^{(1)}}{k_z^{(2)}}\right]\sin\left(k_z^{(2)}d\right). \quad (1.110)$$

Equations 1.99 through 1.102 and 1.107 through 1.110 completely describe the transmission and reflection of an electromagnetic wave from a material $\varepsilon^{(2)}$, $\mu^{(2)}$, where the wave originates in a material $\varepsilon^{(1)}$, $\mu^{(1)}$ and is transmitted into a different material $\varepsilon^{(3)}$, $\mu^{(3)}$.

We next consider the special case where the slab is a nonmagnetic dielectric and is embedded in free space such that $\mu^{(1)} = \mu^{(2)} = \mu^{(3)} = \mu_0$ and $\varepsilon^{(1)} = \varepsilon^{(3)} = \varepsilon_0$. Since our slab is embedded within the same type of material we may also explicitly write $k_z^{(1)} = k_z^{(3)}$. The matrix elements for the TE case become

$$M_{11}^{\text{total}} = \cos\left(k_z^{(2)}d\right) + \frac{i}{2}\left[\frac{k_z^{(2)}}{k_z^{(1)}} + \frac{k_z^{(1)}}{k_z^{(2)}}\right]\sin\left(k_z^{(2)}d\right) \tag{1.111}$$

$$M_{12}^{\text{total}} = \frac{i}{2}\left[\frac{k_z^{(2)}}{k_z^{(1)}} - \frac{k_z^{(1)}}{k_z^{(2)}}\right]\sin\left(k_z^{(2)}d\right). \tag{1.112}$$

$$M_{22}^{\text{total}} = \cos\left(k_z^{(2)}d\right) - \frac{i}{2}\left[\frac{k_z^{(2)}}{k_z^{(1)}} + \frac{k_z^{(1)}}{k_z^{(2)}}\right]\sin\left(k_z^{(2)}d\right). \tag{1.113}$$

$$M_{21}^{\text{total}} = -\frac{i}{2}\left[\frac{k_z^{(2)}}{k_z^{(1)}} - \frac{\mu^{(2)}}{\mu^{(1)}}\frac{k_z^{(1)}}{k_z^{(2)}}\right]\sin\left(k_z^{(2)}d\right). \tag{1.114}$$

We obtain the transmission and reflection coefficients as

$$t_{\text{TE}} = \frac{1}{M_{22}^{\text{total}}}, \tag{1.115}$$

$$r_{\text{TE}} = -\frac{M_{12}^{\text{total}}}{M_{22}^{\text{total}}}, \tag{1.116}$$

and thus the transmittance and reflectance are

$$T = \left|t_{\text{TE}}\right|^2, \tag{1.117}$$

$$R = \left|r_{\text{TE}}\right|^2. \tag{1.118}$$

Using (1.111) through (1.118), it is then possible to characterize the complex dielectric properties, $\tilde{\varepsilon}(\omega) = \varepsilon_1(\omega) + i\varepsilon_2(\omega)$, of single-layer materials. In this case, one must assume that the material exhibits only an electric response and that the magnetic permeability is equal to the free space value. The transmittance $T(\omega)$, reflectance $R(\omega)$, and thickness d can be measured and then used to calculate the dielectric function using least-squares nonlinear curve-fitting algorithms (Chapter 4).

1.3.4 Beer's Law

The form we have obtained for the transmittance of TE polarized light (1.117) completely describes, as we have mentioned, the transmittance. However, we would like to derive a simpler form valid for normal incidence that is easier to use for calculation.

When light passes through a slab of material, its absorption is described by Beer's law as

$$I(z) = I_0 e^{-\alpha z}, \tag{1.119}$$

where
  z is the direction the wave propagates
  $I(z)$ is the intensity of the wave
  I_0 is the intensity of the wave at $z = 0$
  α is the absorption coefficient

Beer's law describes the amount of power absorbed in a medium per unit length, that is, α has units of 1/m. We may also express the electric field of an electromagnetic wave traveling along the z-direction within a medium as

$$\overline{\mathbf{E}}(z) = \overline{\mathbf{E}}_0 e^{i(kz-\omega t)} = \overline{\mathbf{E}}_0 e^{i(k_1 z-\omega t)} e^{-k_2 z}, \tag{1.120}$$

where we have expressed the wave vector in terms of its real and imaginary parts, that is, $\tilde{k} = k_1 + i k_2$. The intensity of this electromagnetic wave is thus

$$I(z) = I_0 e^{-2k_2 z}, \tag{1.121}$$

where $I_0 = |E_0|^2$. Upon equating (1.121) through (1.119) we may write the relationship between the absorption coefficient and the imaginary part of the wave vector and imaginary index of refraction as

$$\alpha = 2k_2 = 2n_2 \frac{\omega}{c}. \tag{1.122}$$

Thus if light is incident upon a medium of thickness d, we may express the total transmittance as the transmittance of the first interface (1.91), multiplied by the power lost within the medium (1.119) or (1.121), then finally multiplied by the transmittance of the outgoing interface (1.91). (We assume the medium is reciprocal and thus the transmittance of both interfaces is identical.)

We may then express the total transmittance as

$$T^{total} = \left(\mathrm{Re}\left[\frac{n}{\mu_r} \right] \left| \frac{2Z}{Z+Z_0} \right|^2 \right)^2 e^{-2k_2 d} \tag{1.123}$$

Although this form is simpler than (1.117), it does not account for multiple reflected and transmitted paths that may occur due to the parallel interfaces of the material. Both (1.117) and (1.123) are plotted in Figure 1.5.

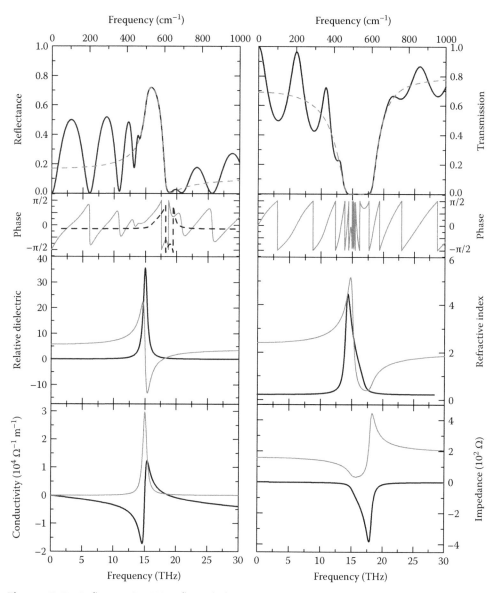

Figure 1.5 Reflection (1.118), reflected phase (ϕ_R = arctan(Im r_{TE}/Re r_{TE})), transmission (1.117), transmitted phase (ϕ_T = arctan(Im t_{TE}/Re t_{TE})), dielectric function (1.130), index of refraction (1.28), impedance (1.29), and the conductivity (1.30) for a transparent material of thickness d = 10 μm. Here the material is modeled to consist of a single Lorentz oscillator (1.130), with parameters ε_∞ = 4.0, ω_p = 2π × 20 THz, ω_0 = 2π × 15 THz, and γ = 2π × 0.75 THz. The real and imaginary portions of the optical constants are plotted as gray and dark gray, respectively, for the bottom four panels. The reflectance and reflected phase for an opaque material with the same parameters is also plotted as a dashed light gray curve and dashed dark gray curve, respectively. The transmission is also plotted (dashed light gray curve top right) via (1.123), which neglects multiple reflections within the material.

1.4 ELECTROMAGNETIC RESPONSE OF MATERIALS

1.4.1 Drude–Lorentz Model

Much resonant phenomena in materials can be expressed in terms of Drude–Lorentz oscillators. We consider the electrons within a material to be tightly bound to the atoms, and in an externally applied electric field their motion can be described by

$$m\frac{d^2\overline{\mathbf{r}}}{dt^2} + m\gamma\frac{d\overline{\mathbf{r}}}{dt} + m\omega_0^2\overline{\mathbf{r}} = -e\overline{\mathbf{E}}, \tag{1.124}$$

where

m and $-e$ are the mass and charge of the electron
γ is the damping
$\overline{\mathbf{r}}$ is the displacement
ω_0 is the resonant frequency
$\overline{\mathbf{E}}$ is the external electric field

The loss term is general, as required by causality. Assuming an electric field space and time dependence of $e^{i(\overline{\mathbf{k}}\cdot\overline{\mathbf{r}} - \omega t)}$, we find the solution

$$\overline{\mathbf{r}} = -\frac{e\overline{\mathbf{E}}}{m}\frac{1}{\omega_0^2 - \omega^2 - i\gamma\omega}. \tag{1.125}$$

Physically this means that under the action of a harmonic electric field $\overline{\mathbf{E}}$, charges are displaced from their equilibrium positions in accord with (1.125). Thus the polarization can then be expressed as

$$\overline{\mathbf{P}} = -ne\overline{\mathbf{r}} = \frac{ne^2\overline{\mathbf{E}}}{m}\frac{1}{\omega_0^2 - \omega^2 - i\gamma\omega}, \tag{1.126}$$

where n is the number density (number of charges per unit volume). Using (1.6) and (1.8) we may then write the relative dielectric function as

$$\tilde{\varepsilon}_r(\omega) = 1 + \frac{\omega_p^2}{\omega_0^2 - \omega^2 - i\gamma\omega}, \tag{1.127}$$

where we have defined

$$\omega_p^2 \equiv \frac{ne^2}{\varepsilon_0 m}. \tag{1.128}$$

ω_p^2 is known as the oscillator strength, or the square of the plasma frequency. Equation 1.127 is known as the Drude–Lorentz equation and describes a resonant response to a driving force, in this case, a time-varying electric field.

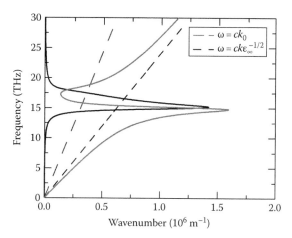

Figure 1.6 Dispersion relation (1.38) for a Lorentz oscillator (1.130), with parameters ε_∞ = 4.0, $\omega_p = 2\pi \times 20\,THz$, $\omega_0 = 2\pi \times 15\,THz$, and $\gamma = 2\pi \times 0.75\,THz$. The real and imaginary portions are plotted as gray and dark gray, respectively. The long dashed and short dashed lines are the free space light line and the light line in the material, respectively.

If we have the case that there are n oscillators, then Equation 1.127 may be generalized to

$$\tilde{\varepsilon}_r(\omega) = 1 + \sum_n \frac{\omega_{p,n}^2}{\omega_{0,n}^2 - \omega^2 - i\gamma_n\omega}. \tag{1.129}$$

However, it is more convenient to modify (1.128) even further when considering the spectroscopy of actual materials. For example, one may characterize phonons for a solid in the far-infrared frequency regime, but contributions to the dielectric function may occur due to higher-energy modes. Thus typically (1.128) is written as

$$\tilde{\varepsilon}_r(\omega) = \varepsilon_\infty + \sum_n \frac{\omega_{p,n}^2}{\omega_{0,n}^2 - \omega^2 - i\gamma_n\omega}. \tag{1.130}$$

where ε_∞ is called "epsilon infinity" and describes contributions to the dielectric function at frequencies higher than the region of consideration (Chapter 3).

In Figure 1.5 we plot the reflection, transmission, reflected phase, transmitted phase, dielectric function, refractive index, conductivity, and impedance for a Lorentz oscillator using (1.130). One may calculate the index of refraction from the relative dielectric function through (1.28) and then further relate this to the wave vector through the dispersion relation (1.38). Figure 1.6 plots the dispersion relation—real and imaginary portions of the wave vector versus frequency—for a Lorentz oscillator (1.130).

1.4.2 Drude Model

If our resonator is centered at zero frequency $\omega_0 = 0$, (1.127) becomes

$$\tilde{\varepsilon}_r(\omega) = 1 - \frac{\omega_p^2}{\omega^2 + i\gamma\omega}. \tag{1.131}$$

We can express this in terms of the conductivity by substituting (1.127) into (1.30) with the help of (1.8). We find

$$\tilde{\sigma}(\omega) = \varepsilon_0 \frac{\omega_p^2}{\gamma - i\omega}. \tag{1.132}$$

We can express (1.132) explicitly in real and imaginary parts to find

$$\tilde{\sigma}(\omega) = \varepsilon_0 \omega_p^2 \left[\frac{\tau}{1 + \omega^2 \tau^2} + i \frac{\omega \tau^2}{1 + \omega^2 \tau^2} \right], \tag{1.133}$$

where we have used the scattering time in place of the damping

$$\tau = \frac{1}{\gamma}. \tag{1.134}$$

The Drude–Lorentz model, although based on an ad hoc model, accurately describes the response of many dielectric and metallic materials; several examples are given in Chapter 3. The Drude model (1.132) is useful for characterizing the response of good metals to electromagnetic waves, whereas the Drude–Lorentz model (1.130) may properly describe "bound" resonant phenomena, such as optically active phonons.

1.4.3 Extended Drude Model

Although the Drude model is extremely successful in describing the electromagnetic response of metals to electromagnetic waves, there are some cases where a more complex model may elucidate the underlying physics (Timusk and Tanner 1989). For example, the response of carriers to external time-varying electric fields and the relationship to dissipative effects on carriers through scattering by phonons and defects can be described generally by a transport equation. The Drude model for metals (1.132) may be derived in a much more rigorous way from the Boltzmann transport equation (Huang 1987), with the approximation of a collision term that is described by a single collision frequency $\gamma = 1/\tau$ (Ashcroft and Mermin 1976, Kittel 1996). However, the Boltzmann transport formalism permits a much more valid description of the electromagnetic response of metals than just simply reproducing the Drude equation. The scattering rate (1.134) and carrier mass (1.128), under the Boltzmann formalism, become frequency dependent and can be described in terms of the real and imaginary parts, respectively, of a complex memory function

$$\tilde{\gamma}(\omega) = \gamma_1(\omega) + i\gamma_2(\omega). \tag{1.135}$$

Defining a dimensionless parameter

$$\lambda(\omega) = -\frac{\gamma_2(\omega)}{\omega}, \tag{1.136}$$

and plugging (1.135) through (1.136) into the Drude equation (1.132), we find

$$\tilde{\sigma}(\omega) = \varepsilon_0 \frac{\omega_p^2 / [1 + \lambda(\omega)]}{\gamma_1(\omega)/[1 + \lambda(\omega)] - i\omega}. \tag{1.137}$$

By defining a renormalized plasma frequency

$$\omega_p^{*2}(\omega) = \frac{\omega_p^2}{1+\lambda(\omega)}, \tag{1.138}$$

and damping rate

$$\gamma^*(\omega) = \frac{\gamma_1(\omega)}{1+\lambda(\omega)}, \tag{1.139}$$

Equation 1.137 can be rewritten in a more familiar "Drude" form (compared to (1.132)):

$$\tilde{\sigma}(\omega) = \varepsilon_0 \frac{\omega_p^{*2}}{\gamma^*(\omega)-i\omega}. \tag{1.140}$$

Renormalization of the plasma frequency (1.123) occurs via introduction of an effective mass

$$\frac{m^*(\omega)}{m} = 1+\lambda(\omega). \tag{1.141}$$

By rearranging (1.136) we can describe the scattering rate and effective mass in terms of the conductivity

$$\frac{1}{\tau(\omega)} = \gamma(\omega) = \varepsilon_0 \omega_p^2 \, \mathrm{Re}\left(\frac{1}{\tilde{\sigma}(\omega)}\right), \tag{1.142}$$

$$\frac{m^*(\omega)}{m} = -\varepsilon_0 \omega_p^2 \omega^{-1} \, \mathrm{Im}\left(\frac{1}{\tilde{\sigma}(\omega)}\right). \tag{1.143}$$

The conductivity is a causal response function and thus the frequency-dependent damping and effective mass are also causal and related to each other by the Kramers–Kronig relations.

1.4.4 Kramers–Kronig Relations

The Kramers–Kronig relations are equations that connect the real and imaginary parts of any analytic complex function (Kronig 1926, Kramers 1927). Typically, in spectroscopy, they are used to infer the real (or imaginary) part of a response function, given the imaginary (or real) part. The connection between the Kramers–Kronig relations and response functions exists because causality implies analyticity, and conversely analyticity implies causality (Toll 1956). A system that responds to an external stimulus may be described by a response function and, in general, for a linear system we may write

$$X(\mathbf{r},t) = \int_{-\infty}^{\infty} G(\mathbf{r},\mathbf{r}',t,t')f(\mathbf{r}',t')d\mathbf{r}'dt'. \tag{1.144}$$

This describes the response $X(\mathbf{r}, t)$ of the system at location $\mathbf{r}$ and time t to a stimulus $f(\mathbf{r}', t')$, which acts at a time t' and a location $\mathbf{r}'$. The quantity $G(\mathbf{r}, \mathbf{r}', t, t')$ is called the response function.

In our case the stimulus will be an electromagnetic wave, which we shall assume has a sufficiently long enough wavelength that we neglect spatial dispersion, that is, we assume that the response functions do not depend on the wave vector. Thus, we make the local approximation—what happens at a particular location depends only on the fields existing at that place. Thus, our response function can be written as

$$G(\mathbf{r}, \mathbf{r}', t, t') = \delta(\mathbf{r} - \mathbf{r}')G(t - t'), \tag{1.145}$$

and (1.144) becomes

$$X(t) = \int_{-\infty}^{\infty} G(t - t')f(t')dt'. \tag{1.146}$$

The requirement of causality states that there can be no response before the stimulus occurs, that is,

$$G(t - t') = 0, \quad t < t'. \tag{1.147}$$

It is now convenient to transform to the frequency domain, as we may take advantage of the convolution theorem. The Fourier transforms may be written as

$$f(\omega) = \int f(t)\exp(i\omega t)dt. \tag{1.148}$$

$$X(\omega) = \int X(t)\exp(i\omega t)dt. \tag{1.149}$$

$$G(\omega) = \int G(t - t')\exp[i\omega(t - t')]dt. \tag{1.150}$$

Plugging (1.146) into (1.149) we find

$$X(\omega) = \int dt \exp(i\omega t)\left[\int G(t - t')f(t')dt'\right]$$
$$= \int dt'f(t')\left[\int G(t - t')\exp(i\omega t)dt\right]$$
$$= \int dt'f(t')(\exp i\omega t)\left[\int G(t - t')\exp i\omega t(t - t')dt\right]. \tag{1.151}$$

Substituting in (1.142) and (1.144) we find

$$X(\omega) = G(\omega)f(\omega). \tag{1.152}$$

Thus in the frequency domain we see that the response $X(\omega)$ is simply equal to multiplication of a monochromatic stimulus $f(\omega)$ by a number $G(\omega)$, termed the generalized susceptibility.

Generally, $G(\omega)$ is complex, where the real part describes the amplitude of the response and the imaginary part describes the phase difference between the stimulus and the response.

If we consider a complex frequency $\tilde{\omega} = \omega_1 + i\omega_2$, then (1.150) becomes

$$G(\tilde{\omega}) = \int G(t - t') \exp[i\omega_1(t - t')] \exp[-\omega_2(t - t')] dt. \qquad (1.153)$$

This integral can be evaluated using Cauchy's integral theorem (Brown and Churchill 1996). However, first we must determine the half space in which the integral is to be determined. The factor $\exp[i\omega_1(t - t')]$ is bounded everywhere, and the factor $\exp[-\omega_2(t - t')]$ is bounded in the upper half plane for $(t - t') > 0$, or bounded in the lower half plane for $(t - t') < 0$. Our requirement of causality (1.147) thus requires that (1.153) be evaluated in the upper half plane.

If we let the contour be a semicircle, as shown in Figure 1.7, with a small bump around the frequency ω, then from Cauchy's theorem we may write

$$\tilde{G}(\omega) = \frac{1}{i\pi} \mathbf{P} \int_{-\infty}^{\infty} \frac{\tilde{G}(\omega')}{\omega' - \omega_0} d\omega', \qquad (1.154)$$

where $\mathbf{P}$ stands for the principal value. We integrate over the contour such that contributions of the semicircle to $G(\omega)$ go to zero as the radius of the semicircle approaches infinity. If we now express our response function as real and imaginary parts, that is, $\tilde{G}(\omega) = G_1(\omega) + iG_2(\omega)$, then we may, now dropping the ω_0 notation, write

$$G_1(\omega) = \frac{1}{\pi} \mathbf{P} \int_{-\infty}^{\infty} \frac{G_2(\omega')}{\omega' - \omega} d\omega' \qquad (1.155)$$

$$G_2(\omega) = -\frac{1}{\pi} \mathbf{P} \int_{-\infty}^{\infty} \frac{G_1(\omega')}{\omega' - \omega} d\omega'. \qquad (1.156)$$

Notice that (1.156) tells us that if we have determined $G_1(\omega')$ over all frequencies, then we can find $G_2(\omega)$. Thus we observe that the real and imaginary parts of our response function $\tilde{G}(\omega)$ are not independent, but are related by formulas (1.155) and (1.156), known as dispersion relations.

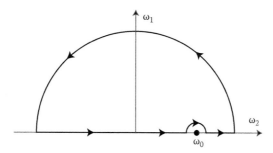

Figure 1.7 Contour for obtaining Equation 1.154.

We now consider a specific example of the response of a system to an external electric field. In (1.126) we found the polarization of an electron due to an external electric field. Upon comparison to (1.152) and (1.6) we see

$$\overline{P}(\omega) = G(\omega)\overline{E}(\omega),$$ (1.157)

where

$$G(\omega) = \varepsilon_0 \chi_e = \frac{ne^2}{m} \frac{1}{\omega_0^2 - \omega^2 - i\gamma\omega}.$$ (1.158)

From inspection of (1.155) and (1.156), and considering a complex electric susceptibility $\tilde{\chi}(\omega) = \chi_1(\omega) + i\chi_2(\omega)$ (dropping the subscripted "e") we may thus write

$$\chi_1(\omega) = \frac{1}{\pi} \mathbf{P} \int_{-\infty}^{\infty} \frac{\chi_2(\omega')}{\omega' - \omega} d\omega'.$$ (1.159)

$$\chi_2(\omega) = -\frac{1}{\pi} \mathbf{P} \int_{-\infty}^{\infty} \frac{\chi_1(\omega')}{\omega' - \omega} d\omega'.$$ (1.160)

We are interested in physical systems with real inputs and real outputs. This means we must have

$$\tilde{G}(\omega) = \tilde{G}^*(-\omega),$$ (1.161)

or in terms of the real and imaginary parts of the susceptibility

$$\chi_1(\omega) = \chi_1(-\omega).$$ (1.162)

$$\chi_2(\omega) = -\chi_2(-\omega).$$ (1.163)

From (1.8) we see that $\tilde{\chi}(\omega) = \tilde{\varepsilon}_r(\omega) - 1$. Thus we may write

$$\varepsilon_{1,r}(\omega) - 1 = \frac{2}{\pi} \mathbf{P} \int_{0}^{\infty} \frac{\omega' \varepsilon_{2,r}(\omega')}{\omega'^2 - \omega^2} d\omega'.$$ (1.164)

$$\varepsilon_{2,r}(\omega) = -\frac{2\omega}{\pi} \mathbf{P} \int_{0}^{\infty} \frac{\varepsilon_{1,r}(\omega') - 1}{\omega'^2 - \omega^2} d\omega'$$ (1.165)

Equations 1.164 and 1.165 are known as the Kramers–Kronig dispersion relations.

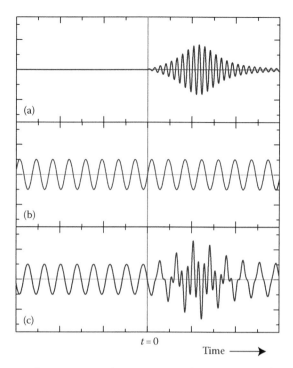

Figure 1.8   Diagram demonstrating the connection between causality and dispersion. (a) Time pulse incident upon a system at $t = 0$. (b) Single frequency component absorbed by the system. (c) Remaining electromagnetic pulse after subtraction of (b) from (a).

Although we have used a purely mathematical treatment to derive the relation between the real and imaginary portions of a causal response function, a simple intuitive argument presents the same picture (Toll 1956). Consider a pulse of electromagnetic radiation which impinges on a system at $t = 0$ as shown in Figure 1.8a. Let us assume this system can only absorb a single frequency, as shown in Figure 1.8b. Then upon absorption of this one component, what remains of our electromagnetic pulse is shown in Figure 1.8c, that is, the subtraction of B from A. No real physical (causal) system can have a response before the arrival of the signal, and thus what is shown in Figure 1.8c is nonsense. Therefore, absorption of the single frequency shown in Figure 1.8b must be accompanied by a phase shift of the other frequency components such that, for $t < 0$, the time signal shown in Figure 1.8c vanishes. In terms of physical systems, this means that a response function cannot describe the absorption alone but must also describe the phase shift and therefore must consist of real and imaginary parts.

1.4.5 Sum Rules

Equation 1.164 may be written as

$$\varepsilon_{1,r}(\omega) - 1 = \frac{2}{\pi} \mathbf{P} \int_0^{\omega_c} \frac{\omega' \varepsilon_{2,r}(\omega')}{\omega'^2 - \omega^2} d\omega' + \frac{2}{\pi} \mathbf{P} \int_{\omega_c}^{\infty} \frac{\omega' \varepsilon_{2,r}(\omega')}{\omega'^2 - \omega^2} d\omega', \tag{1.166}$$

where ω_c denotes a cutoff frequency above which there is no absorption. Thus according to (1.165), $\varepsilon_{2,r}(\omega) = 0$ for $\omega > \omega_c$. At frequencies much greater than the cutoff frequency, that is,

$\omega \gg \omega_c$, we can neglect ω' in the denominator of the first integral in (1.166), and the second integral is zero since $\varepsilon_{2,r}(\omega) = 0$ for $\omega \gg \omega_c$. Under these conditions, (1.166) becomes

$$\varepsilon_{1,r}(\omega) = 1 - \frac{2}{\pi\omega^2} \int_0^{\omega_c} \omega' \varepsilon_{2,r}(\omega')d\omega', \quad \omega \gg \omega_c. \tag{1.167}$$

Notice also that if we are at sufficiently high frequencies, then (1.132) can be written as

$$\tilde{\varepsilon}_r(\omega) = 1 - \frac{\omega_p^2}{\omega^2}. \tag{1.168}$$

Thus equating (1.167) and (1.168) we see

$$\int_0^\infty \omega \varepsilon_{2,r}(\omega)d\omega = \frac{\pi}{2}\omega_p^2, \tag{1.169}$$

where we have extended the upper limit of integration since $\varepsilon_{2,r}(\omega) = 0$ for $\omega > \omega_c$. It is more common to express (1.169) in terms of the conductivity. Using (1.30) and (1.128) we may write

$$\int_0^\infty \sigma_1(\omega)d\omega = \frac{\pi}{2}\omega_p^2 = \frac{ne^2\pi}{2\varepsilon_0 m}. \tag{1.170}$$

1.5 SUMMARY

We have provided an abbreviated overview of classical electromagnetic wave theory. Although we were only able to touch upon a limited number of topics, we hope this chapter will provide useful knowledge, not only for the rest of this book but also as a general reference on electricity and magnetism.

ACKNOWLEDGMENTS

I would like to acknowledge the Office of Naval Research for support under contract N00014-07-1-0819.

REFERENCES

Ashcroft, N.W. and Mermin, N.D. 1976, *Solid State Physics*, Brooks Cole, Pacific Grove, CA.
Brillouin, L. 1960, *Wave Propagation and Group Velocity*, Academic Press, New York.
Brown, J.W. and Churchill, R.V. 1996, *Complex Variables and Applications*, 6th edn., McGraw Hill, London, U.K.
Burns, G. 1990, *Solid State Physics*, Academic Press, Boston, MA.
Cheng, D.K. and Kong, J.A. 1968a, Covariant descriptions of bianisotropic media, *Proc. IEEE*, 56, 248.
Cheng, D.K. and Kong, J.A. 1968b, Time-harmonic fields in source-free bianisotropic media, *J. Appl. Phys.*, 39, 5792.
Choy, T.C. 1999, *Effective Medium Theory: Principles and Applications*, Oxford University Press, New York.

Cohen, D.L. 2001, *Demystifying Electromagnetic Equations*, SPIE–The International Society for Optical Engineering, Bellingham, Washington.

Collin, R.E. 1991, *Field Theory of Guided Waves*, 2nd edn., John Wiley & Sons, Inc., New York.

Cottam, M.G. and Tilley, D.R. 1989, *Introduction to Surface and Superlattice Excitations*, Cambridge University Press, Cambridge, U.K.

Eerenstein, W., Mathur, N.D., and Scott, J.F. 2006, Multiferroic and magnetoelectric materials, *Nature*, 442, 759.

Gibbs, W. and Wilson, E.B. 1929, *Vector Analysis. A Text-Book for the Use of Students of Mathematics and Physics*, Yale University Press, New Haven, CT.

Heaviside, O. 1892, On the forces, stresses, and fluxes of energy in the electromagnetic field, *Phil. Trans. R. Soc. A*, 183, 423–480.

Huang, K. 1987, *Statistical Mechanics*, 2nd edn., Wiley, Indianapolis, IN.

Jackson, J.D. 1998, *Classical Electrodynamics*, 3rd edn., John Wiley & Sons, Inc., Indianapolis, IN.

Kittel, C. 1996, *Introduction to Solid State Physics*, 7th edn., John Wiley & Sons, Inc., Indianapolis, IN.

Kong, J.A. 1990, *Electromagnetic Wave Theory*, 2nd edn., John Wiley & Sons, Inc., Indianapolis, IN.

Kramers, H.A. 1927, La diffusion de la lumiere par les atomes, *Atti Cong. Intern. Fisica*, 2, 545–557.

Kronig, R. de L. 1926, On the theory of the dispersion of x-rays, *J. Opt. Soc. Am.*, 12, 547–557.

Landau, L.D., Pitaevskii, L.P., and Lifshitz, E.M. 1984, *Electrodynamics of Continuous Media*, 2nd edn., Butterworth-Heinemann, London, U.K.

Markos, P. and Soukoulis, C.M. 2008, *Wave Propagation*, Princeton University Press, Princeton, NJ.

Maxwell, J.C. 1865, A dynamical theory of the electromagnetic field, *R. Soc. Trans.*, 155, 459–512.

Pomeroy, S.B. 1999, *Ancient Greece: A Political, Social, and Cultural History*, Oxford University Press, Oxford, U.K.

Ramo, S., Whinnery, J.R., and Van Duzer, T. 1994, *Fields and Waves in Communication Electronics*, 3rd edn., John Wiley & Sons, Inc., Indianapolis, IN.

Schelkunoff, S.A. 1938, The impedance concept and its application to problems of reflection, refraction, shielding and power absorption, *Bell System Tech. J.*, 17, p. 17.

Sihvola, A.H. 1999, *Electromagnetic Mixing Formulas and Applications*, The Institution of Engineering and Technology, Herts, U.K.

Sommerfeld, A. 1950, *Lectures on Theoretical Physics*, Academic Press, New York.

Stratton, J.A. 1989, *Electromagnetic Theory*, McGraw-Hill Book Co., New York.

Timusk, T. and Tanner, D.B. 1941, Optical properties of high-temperature superconductors, in *The Physical Properties of High Temperature Superconductors*, ed. D.M. Ginsberg, World Scientific, Singapore, pp. 363–459.

Toll, J.S. 1956, Causality and the dispersion relation: Logical foundations, *Phys. Rev.*, 104, 1760–1770.

Chapter 2

SEMICONDUCTORS AND THEIR NANOSTRUCTURES

Jeffrey Davis and Chennupati Jagadish

CONTENTS

2.1 BASIC SEMICONDUCTOR PHYSICS

A semiconducting material is defined as one that is electrically insulating at a temperature of absolute zero but that becomes conducting for temperatures below its melting point. The property that first attracted the attention of physicists in the early nineteenth century was the decreasing resistivity with increasing temperature, in stark contrast to metals where the opposite is true (Faraday 1839, Smith 1964, Ashcroft and Mermin 1976, p. 563). Both of these effects occur in semiconductors because it is possible to thermally excite charge carriers from a full energy band that prevents conduction to an unoccupied band that then allows electrons to move freely. The structure of these energy bands is used to explain most of the properties of semiconductors. In this chapter, we therefore begin with a brief introduction to semiconductor band structures and then continue with our discussion on some of the important and interesting properties observed in semiconductors and their nanostructures.

2.1.1 Band Structure

Consider a single silicon atom; the solution to the Schrödinger equation reveals discrete electronic orbitals, with four electrons in the valence shell. This silicon atom is able to covalently bond with four other silicon atoms, which in turn are able to bond with four silicon atoms, and so on, forming a crystal lattice. In a crystal lattice, the electronic energy structure changes. To a first approximation, the atomic orbitals are replaced by filled bonding and empty antibonding orbitals, with pairs of electrons localized between the two nuclei they are binding. In the absence of interactions between different pairs of electrons, every electron in the lattice will have an identical set of discrete states. In practice, however, this is a strongly interacting system and the Hamiltonian needs to include the electromagnetic interactions between all the valence electrons. The effect of this is that the Pauli exclusion principle prevents more than two electrons from existing in the same quantum state, lifting the degeneracy and generating an additional state for each additional Si–Si bond, with all states split by an amount that is determined by the strength of the interaction. In other words, a single Si–Si bond will have a single molecular orbital that is doubly degenerate (i.e., it can contain two electrons of opposite spin); a pair of Si–Si bonds will have two molecular orbitals (each doubly degenerate) that are split in energy due to the electromagnetic interaction; three Si–Si bonds will have three molecular orbitals that are split in energy; and so on, until, in a semi-infinite crystal lattice, there are a semi-infinite number of Si–Si bonds, and a semi-infinite number of states, split by a semi-infinitesimal amount, as depicted in Figure 2.1a. An alternative way of looking at the formation of bands is depicted in Figure 2.1b, where the energy structure is shown as a function of atom separation. As the nuclei become closer, the discrete atomic states shift and form bands as the interaction between electrons increases (Allison 1971). In a semi-infinite lattice, rather than a series of discrete states, there is, therefore, a quasi-continuum of states that is referred to as an energy band. In a tetravalent system, such as Si, there are actually four valence bands, corresponding to the four molecular bonding orbitals for each Si atom, and another four "conduction" bands for the corresponding antibonding orbitals. At low temperature, the valence bands are all fully occupied, and the conduction bands are empty. The properties of these bands and splitting between them are determined largely by the type of atoms, the crystal structure, and relevant momentum and angular momentum considerations.

The vast majority of semiconductors are tetravalent; however, trivalent, heptavalent, and hexavalent semiconductors also occur, with corresponding crystal structures. Regardless, the periodic nature of semiconductor crystals allows one to introduce the concepts of the unit cell and reciprocal lattice. The unit cell is defined as a volume of the lattice that is repeated within the crystal, completely filling the space with no overlap, and usually represents the full symmetry of the lattice. In this way, the unit cell, together with the associated lattice vectors

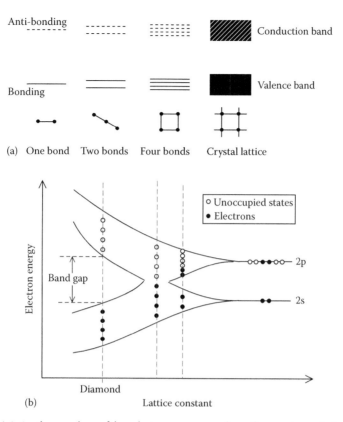

Figure 2.1 (a) As the number of bonds increases, so does the number of distinct bonding and antibonding states. In a crystal lattice, these distinct states form a quasi-continuum of states, leading to the formation of energy bands. Alternatively in (b), as the separation between carbon atoms decreases, and interactions between their electrons increase, the electron energies transition from discrete atomic states to bands separated by a band gap. (Adapted from Allison, J., *Electronic Engineering Materials and Devices*, McGraw Hill, London, 1971. With permission.)

(i.e., the vectors that define the unit cell) can be used to describe the entire crystal lattice. The choice of unit cell is not unique, and different unit cells may be defined for different purposes. In addition to the conventional unit cell described earlier, one particularly useful and uniquely defined unit cell is the Wigner–Seitz unit cell (Ashcroft and Mermin 1976). The advantage of this representation is that it defines a primitive unit cell (i.e., it contains exactly one repeating unit per cell) and represents the full symmetry of the crystal. The Wigner–Seitz unit cell is shown together with the conventional unit cell for a face-centered cubic lattice in Figure 2.2.

The reciprocal lattice is defined by the set of wave vectors, **k**, which satisfy $e^{-i\mathbf{k}\cdot\mathbf{R}} = 1$ (defining a set of plane waves that represent the symmetry of the direct lattice), where **R** are the lattice vectors of the real space Bravais lattice.* Just as we defined a primitive unit cell in the real lattice, an equivalent primitive cell of the reciprocal lattice can be defined. One particular formation of such a primitive cell is referred to as the first Brillouin zone and is determined in the same way as the Wigner–Seitz primitive cell in the real lattice. More details on these fundamental concepts can be found in any solid state textbook, for example, Ashcroft and Mermin (1976) and Kittel (2005).

---

* A Bravais lattice is defined as an infinite array of discrete points with an arrangement and orientation that appear exactly the same regardless of which point an array is viewed from. Where this is not the case, a basis that contains multiple atoms can usually be defined, which can then be represented as a Bravais lattice.

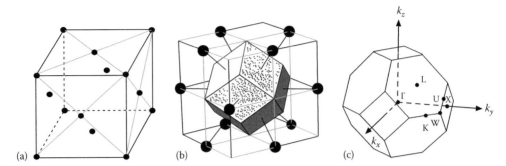

Figure 2.2 (a) The conventional unit cell for a face-centered cubic lattice (f.c.c.). (b) The corresponding Wigner–Seitz unit cell (note the cubic lattice depicted here is different from that in (a)). (c) The first Brillouin zone of an f.c.c. crystal, with the points of high symmetry labeled.

The unique definition of the first Brillouin zone allows several points of high symmetry to be defined, as depicted in Figure 2.2c for a face-centered cubic lattice. The most obvious is at the zone center and is given the label Γ; others are found at the edges of the Brillouin zone at different vertices and plane centers. To explain the purpose of introducing the reciprocal lattice and these high symmetry points, consider again an electron in the crystal lattice. The energy bands described earlier assume stationary electrons; however, one of the key properties of semiconductors (and metals) is that the electrons are not always stationary, and so the kinetic energy (KE) needs to be included. In a crystal, however, this is not simply KE = $\frac{1}{2}mv^2$, with m the mass and v the velocity, as the electrons interact with the nuclei (leading to a change in their effective mass), and the extent of this interaction is dependent on the electron's speed and direction of motion. In order to fully describe the energy of an electron, it then becomes necessary to define the kinetic energy for motion in different directions. This is where the reciprocal lattice and key symmetry points become useful.

The defined points of high symmetry in the reciprocal lattice correspond to wave vectors that define the momentum of electrons in the real crystal lattice. Plotting the energy of the electrons for these points in the reciprocal lattice, and along the lines joining these points, thereby provides a good description of the total energy landscape.

This type of representation is known as a dispersion relation, examples of which are shown in Figure 2.3. The horizontal axis in these figures is labeled by the high symmetry points in the reciprocal lattice. Between these points the position is varied linearly along the line connecting them in the reciprocal lattice. In real space, this can be thought of as changing either the direction or magnitude of the electron wave vector/momentum. The Γ-point is located at the center of the Brillouin zone and corresponds to zero momentum. The regions between the Γ-point and any other point, or between two points on the surface of the Brillouin zone, therefore correspond to increasing the magnitude of the wave vector/momentum in the direction defined by the other high symmetry point, or changing the direction of the wave vector/ momentum, respectively.

The complete band structure of semiconductors is usually defined in this way, and many of their properties are described with respect to this representation. For example, the effective mass of electrons in the crystal lattice is determined by the curvature of the relevant band, where a steeper dispersion curve corresponds to a smaller effective mass. Conversely, if the band structure is not known, the effective mass can be determined experimentally, as described in Section 2.2.6, and this helps to define the band structure.

Conduction, thermal, and optical properties are also described within the band structure formalism, and similarly can also be used to experimentally determine aspects of the band

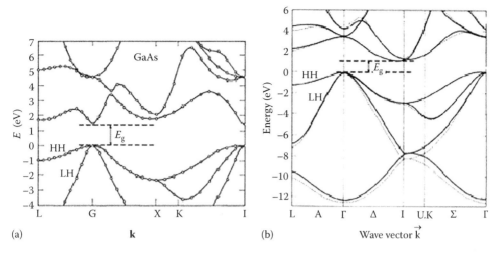

Figure 2.3 (a) The band structure of GaAs (Reprinted figure with permission from Cohen, M.L. and Bergstresser, T.K., *Phys. Rev.*, 141, 789, 1966. Copyright (1966) by the American Physical Society.) and (b) the band structure of Si (Reprinted figure with permission from Chelikowsky, J.R. and Cohen, M.L., *Phys. Rev. B*, 10, 5095, 1974. Copyright (1966) by the American Physical Society.) The horizontal axis represents the magnitude and direction of the wave vector by placing symmetry labels that correspond to specific points of high symmetry in the reciprocal lattice, and in this way identifies the orientation of the wave vector, while between these labels the magnitude is varied.

structure. The band structure can also be calculated by solving the Schrödinger equation for the crystal lattice, which is often achieved practically by making use of the $\mathbf{k} \cdot \mathbf{p}$ formalism. For details of this type of calculation, we refer the reader to Haug and Koch (2004), for example.

Perhaps, the single most important property of semiconductors is the band gap, so it is worth taking the time to detail exactly what is meant by this term. In the dispersion relations in Figure 2.3, it can be seen that the energy gap between the valence band and conduction band varies as a function of wave vector, and so it is impossible to give a single value. "The" band gap of a material is defined as the energy difference between the highest maximum of the highest energy valence band and the lowest minimum of the lowest energy conduction band. In the case of direct gap semiconductors, as shown in Figure 2.3a for GaAs, this is straightforward, as they both occur at the same point in $\mathbf{k}$-space (i.e., at the center of the Brillouin zone, $\mathbf{k} = 0$). In some materials, however, the valence band maximum and conduction band minimum can be displaced from each other in momentum space, as shown in Figure 2.3b for silicon. In this case, the band gap is still defined as the energy difference between the valence band maximum and conduction band minimum, and is referred to as an indirect gap.

2.1.2 Conduction Properties

At low temperature and in pure, intrinsic, semiconductors, the valence band is fully occupied and the conduction band is empty. In other words, all electrons are bound to their nuclei, and in order to move to another atomic site the energy of the band gap would need to be overcome. This is very unlikely at low temperatures, meaning there are no free electrons and the material is insulating. If, however, an electron can in some way be excited to the conduction band, there is a continuum of empty states at the same or similar energies, so no excess energy is required to move throughout the crystal lattice, allowing it to conduct electricity.

In addition, the vacancy created in the valence band can be filled by another valence band electron, thereby moving the vacancy to another site, where another electron can move into it, and so on, thereby allowing the free movement of charge. Rather than thinking of this in terms of many electrons moving one at a time in sequence, it is helpful to think of the vacancy moving. In this way, by exciting the electron from the valence band to the conduction band, a quasiparticle in the valence band, referred to as a hole, is generated. The hole then has the opposite properties to an electron, including a positive charge, and it can be treated as a charge carrier that conducts electricity.

In thermal equilibrium, the probability of an electron being excited from the valence band into the conduction band is given by Fermi–Dirac statistics. This introduces the concept of the chemical potential or Fermi level. In semiconductors, the definition of the Fermi level is somewhat different from metals, since it usually lies in the band gap. The Fermi level, or chemical potential, defines the energy at which the probability for occupation is 0.5. At absolute zero, where the valence (conduction) band is completely full (empty), the Fermi level is in the middle of the band gap. At finite temperatures, the Fermi level lies within the order of $k_B T$ (where k_B is the Boltzmann constant and T is the temperature) of the center of the band gap. For intrinsic semiconductors where the band gap is large compared to $k_B T$, the Fermi level remains well within the band gap, and the probability of finding conduction electrons remains low.

In order to promote electrons from the valence band to the conduction band, energy greater than or equal to the band gap needs to be absorbed. For semiconductors, the band gap can range from a few hundred meV (e.g., InSb) up to ~6 eV (e.g., AlN) and often corresponds to optical photon energies (1–4 eV), meaning the absorption of visible light can lead to a transient conductivity. This is the basis for many light-detection and light-harvesting devices. For many applications, however, it is also desirable to have permanent conductivity. This can be achieved by doping the crystal lattice with impurities that have more (or fewer) valence electrons than the host semiconductor. For example, if an atom with five valence electrons replaces an atom with four valence electrons (e.g., P in Si), then four of the electrons will simply take the role of those in the original atom, leaving one extra electron. The energy level occupied by this electron is typically located just below the conduction band, as illustrated in Figure 2.4a. It can then be easily promoted thermally into the conduction band, allowing free movement of the charge. This is known as n-type doping and the dopant atoms are known as donors. In other words, the Fermi level is raised into or close to the conduction band, allowing free conduction. On the other hand, if an atom with three valence electrons is included in the lattice (e.g., B in Si), then an empty state is generated just above the valence band, into which an electron from the valence band can be excited, leaving a hole in the valence band that is

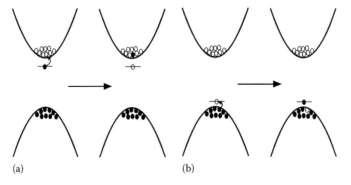

(a) (b)

Figure 2.4 Schematic showing the incorporation of (a) n-type and (b) p-type dopants, and the excitation that allows them to conduct electrons and holes, respectively.

free to move and conduct electrical current, as shown in Figure 2.4b. This is known as p-type doping, with the dopant atoms known as acceptors, and corresponds to lowering the Fermi level close to the valence band. Most consumer electronics are made up of combinations of intrinsic (pure) and extrinsic (doped) semiconductor regions, which make use of both p- and n-type doping and the interface between such regions. For more information on p–n junctions and semiconductor devices, we refer the reader to Ashcroft and Mermin (1976), Kittel (2005), Sze and Kwok (2007), and Seeger (2004).

2.1.3 Compound Semiconductors

To date, we have been assuming that semiconductors are pure homogeneous materials made up of one single type of atom (with the exception of impurities). This need not be the case, however, as it is possible to form ionic crystals that are also semiconducting. These types of semiconductors are typically made by combining group III atoms with group V atoms (e.g., GaAs), or group II with group VI atoms (e.g., ZnO). In this way, the number of valence electrons remains the same as if all atoms were from group IV, and the bonding remains tetrahedral. The fact that the nuclei are no longer identical does not affect most properties, as it is the crystal structure and average number of electrons per site that determine such properties, and these are unchanged.

One example of a situation in which the local environment and the heterogeneous nature of the unit cell can play an important role is at boundaries and interfaces. Since the unit cell in compound semiconductors contains at least two different types of atoms, it is likely that in at least one direction, taken in isolation, it will be electrically polarized, as the electron affinity of the two atoms will almost certainly not be the same. This spontaneous polarization is not seen in an infinite lattice, as the fields from neighboring unit cells counter each other. At a surface, however, there is no neighboring unit cell, and an excess of charge can exist, leading to an electric field. This type of charge imbalance can also be found at an interface between two different materials, where the spontaneous polarization of the two unit cells is different. This can be particularly relevant in semiconductor nanostructures.

The major reason compound semiconductors are of particular interest is the flexibility they allow. Take GaAs, for example; by replacing a fraction, x, of the Ga atoms with Al atoms (another group III element) to form $Al_xGa_{1-x}As$, the band gap can be increased by an amount that is dependent on the Al concentration. Alternatively, the band gap can be decreased by replacing Ga with In to form $In_xGa_{1-x}As$. This ability to tune the band gap introduces an extra degree of freedom that is particularly useful when designing semiconductor devices and nanostructures, especially those that emit or absorb light.

2.1.4 Phonons

Much like any other crystalline solid, vibrations within semiconductors are quantized; in classical mechanics, this is similar to the normal modes of vibration. When a bulk system is treated quantum mechanically, however, these "normal modes" have particle-like properties and can be treated as quasiparticles* known as phonons. It then becomes possible to talk about the number or density of phonons, which increases with increasing temperature. For

---

* A quasiparticle is an elementary excitation that exists only in a solid (e.g., phonons, excitons, holes) with particle-like properties like position, momentum, and mass that are modified by interactions with its environment. This contrasts with "real" particles like electrons, protons, and photons that can exist in vacuum (Prasankumar et al. 2009).

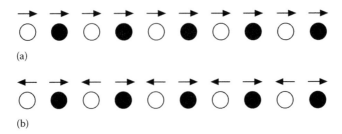

Figure 2.5 (a) Longitudinal acoustic and (b) longitudinal optical phonon modes for a 1D lattice. For acoustic modes the atoms in the unit cell move in phase, while for optical modes they move out of phase.

details regarding the derivation of phonons, see, for example, Ashcroft and Mermin (1976, pp. 437–443) and Ridley (1997).

As an example, consider the simplest case of a one-dimensional (1D) lattice. In this case, there are two general types of vibrations possible, as depicted in Figure 2.5. In the first, (a), the two different atoms that make up the unit cell move in phase, while in the second, (b), they move out of phase. The former case represents acoustic phonons while the latter corresponds to optical phonons, so called because in an ionic crystal this oscillating separation is in fact an oscillating dipole, which makes this type of phonon optically active. Extending this picture to three dimensions is not so straightforward conceptually, but the basic principles remain the same: Acoustic phonons correspond to intercellular interactions (i.e., nuclei within the unit cell move in phase), while optical phonons correspond to intracellular interactions (i.e., nuclei within the unit cell move out of phase) (Ashcroft and Mermin 1976).

One further consideration is the orientation of the vibrational motion relative to the phonon propagation direction. Consistent with the quasiparticle picture, phonons do not exist across the entire lattice simultaneously, but rather are localized, and move through the lattice with a given velocity. This defines the propagation direction and allows the definition of transverse and longitudinal modes. In general, one can define three orthogonal orientations for the vibrational motion: one that is parallel to the propagation direction (longitudinal) (Figure 2.5), and two that are orthogonal to the propagation direction (transverse) (Figure 2.6). This definition of transverse and longitudinal is not always going to be strictly true, as it depends on the direction of propagation relative to the principal axes of the crystal lattice; nonetheless, this terminology remains useful and continues to be used.

A corollary of phonon propagation is that they transfer momentum through the crystal lattice. The momentum of a phonon is manifested in the vibrational motion of the nuclei, and is transferred throughout the lattice as the phonon propagates. The phonon's momentum is one

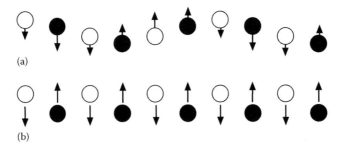

Figure 2.6 (a) Transverse acoustic and (b) transverse optical phonon modes for a 1D lattice. Again, for acoustic modes the atoms in the unit cell move in phase, while for optical modes they move out of phase.

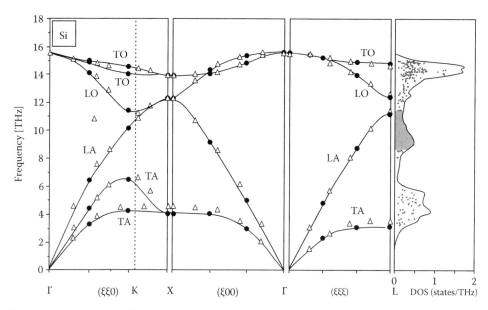

Figure 2.7 Example of phonon dispersion curve and density of states for Si. (Reprinted figure with permission from Savrasov, B., *Phys. Rev. B,* 54, 16470, 1996. Copyright (1966) by the American Physical Society.)

of its key properties, and similar to electrons moving throughout the crystal lattice, phonons propagating in different directions will have different energies depending on the magnitude and direction of their wave vector with respect to the principal axes of the lattice. For this reason, it is useful to introduce phonon dispersion curves that plot phonon energy as a function of momentum (or wave vector), an example of which is shown in Figure 2.7 (Savrasov 1996). It is evident in this figure that the dispersion of optical phonons is rather flat, whereas the energy of the acoustic phonons is strongly dependent on momentum and zero at the zone center (i.e., $\mathbf{k} = 0$). This is typical for most crystals, as is the observation that for a given wave vector, the energy of optical phonons is generally greater than for acoustic phonons, particularly close to the zone center. This means that at low temperatures it is possible to "freeze out" the optical phonons, allowing the effects of acoustic phonons to be isolated. Optical phonon energies are typically in the range of a few tens of meV, which is comparable to the thermal energy at room temperature (25 meV), thus as the temperature is reduced, the prevalence of optical phonons rapidly decreases. At even lower temperatures, it becomes possible to minimize the acoustic phonon density, but because the linear dispersion curve goes to zero at the Γ point, there can always be a finite acoustic phonon population at finite temperatures; under certain conditions, however, zero-phonon effects can still be observed (Borri et al. 2001).

A corollary of treating phonons as quasiparticles is that they are able to interact and scatter with other particles and quasiparticles. For example, phonons can scatter with electrons, holes, and photons in either elastic or inelastic collisions, and can transfer momentum. The effects of these interactions may include alteration of the electrical conductivity, non-radiative recombination of optically excited carriers, or dephasing of a coherently excited system. Changes to these properties as a function of temperature are therefore usually attributed to phonon interactions, as the phonon density (corresponding to the number of lattice vibrations) increases with temperature (Shah 1999).

Interactions of phonons with photons can be divided into two different cases: infrared (IR) absorption and Raman scattering. In order for an IR photon to be absorbed by a crystal as a phonon, both energy and momentum must be conserved. From the phonon dispersion curve, it is clear that the only place where a photon will have the same energy and momentum as an acoustic phonon is at $\mathbf{k} = 0$, corresponding to a static electric field. The photon dispersion will however overlap with the optical phonon branches at finite values of E and $\mathbf{k}$. In order for the electromagnetic field of the photon to interact with the phonon modes, the crystal must have a permanent dipole moment, that is, the bonds must be at least a little polar, as will be the case in compound semiconductors, but not monotonic semiconductors such as Si. Furthermore, since the electric field of the photons is transverse, it will interact only with the TO phonon modes. Phonons that satisfy these selection rules are said to be IR active (Klingshirn 2007).

Raman scattering and Brillouin scattering (which is equivalent to Raman scattering, but with acoustic phonons) are discussed in greater detail later in this chapter (and in Chapters 6 and 14). At this point, however, we point out that all phonon modes can potentially be Raman active depending on the crystal structure. The precise selection rules to determine whether a phonon mode is Raman active is somewhat more complicated, requiring group theory and a more detailed analysis of the symmetry of the crystal and corresponding phonon modes. In general, however, if the crystal structure has an inversion center, then the phonon modes will be either IR or Raman active; if the crystal lacks such inversion symmetry, then some phonon modes may be both IR and Raman active. Fox (2001) gives a more detailed summary of the optical properties of phonons.

2.1.5 Other Lattice Excitations

In addition to phonons, there are other forms of lattice interactions that can play an important role in semiconductor physics. We briefly mention two of these here: polarons and magnons.

Polarons are formed by the interaction of the electric field of an excited electron (or hole) with that of the nuclei. This interaction causes a deformation of the lattice in the vicinity of the charge carrier, which leads to a local polarization of the lattice. As the free charge carrier moves throughout the lattice, the deformation and polarization move with it. This combination of the charge carrier and its polarization field is treated as a quasiparticle known as a polaron. One important effect of the local induced field is that it hinders the motion of the electron through the lattice, thus reducing its mobility (Klingshirn 2007). This often leads to a "hopping"-type conductivity when polarons are the dominant charge carriers.

Magnons are quasiparticles that represent quantized spin waves and are completely analogous to phonons. Where phonons represent excitations of the nuclear lattice and carry momentum, magnons represent excitations of the spin lattice and carry angular momentum. Similarly, just as phonons can have an important role in energy loss mechanisms, so too can magnons in magnetic semiconductors (Klingshirn 2007).

2.2 OPTICAL PROPERTIES OF SEMICONDUCTORS

The optical response of semiconductors is one of their most useful characteristics and one of the most important tools for determining their band structure. In discussing the optical response, the most important property is the size of the band gap and whether it is direct or indirect. The presence of a band gap between the full valence band and empty conduction band means that a photon with energy greater than the band gap can be readily absorbed, whereas photons with energy less than the band gap will typically be transmitted. For many semiconductors of interest, the band gap is around the optical region of the electromagnetic spectrum, and so it is not surprising that semiconductors now form the basis for many light emitting and detecting devices.

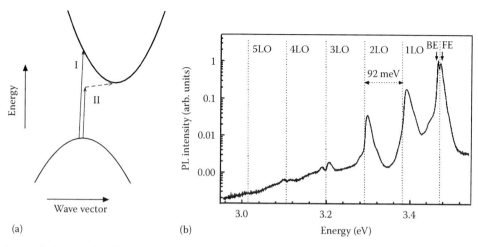

Figure 2.8 (a) The direct transition (I) in an indirect gap material requires far greater photon energy than the phonon-assisted transition (II) between the band edges. In the phonon-assisted transition, the phonon (dashed line) provides the additional momentum to allow the transition. (b) Similar processes can also occur in direct gap materials such as GaN and lead to the formation of phonon replica peaks labeled by the dashed lines and the labels indicating the number of phonons involved in the transition. (Reprinted figure with permission from Kovalev, D. et al., *Phys. Rev. B*, 54, 2518, 1996. Copyright (1966) by the American Physical Society.)

Knowledge only of the band gap, however, is not enough to fully describe the optical properties. In addition to energy conservation, which can be specified with respect to the band gap, momentum and angular momentum must also be conserved.

The momentum carried by a photon ($p = \hbar\mathbf{k} = h/\lambda$) is small compared to the momentum of electrons/holes, and so optical transitions where the change in momentum (or electron/hole wave vector) is close to zero tend to dominate. This is particularly the case for direct gap semiconductors. For indirect gap semiconductors, a valence to conduction band transition with minimum photon energy can occur only if a phonon is absorbed or emitted simultaneously to facilitate conservation of momentum. This type of phonon-assisted transition tends to dominate for indirect gap materials, such as silicon, causing the optical properties to become increasingly temperature dependent. This type of phonon-assisted transition is depicted in Figure 2.8a, where path I represents the direct transition, which requires a photon with energy greater than the band gap, and path II represents the phonon-assisted transition. In this case, a photon and phonon are absorbed simultaneously to allow the indirect transition between the band edges, conserving both energy and momentum.

Phonon-assisted transitions can also occur in direct gap semiconductors and appear as additional "phonon-replica" peaks in the absorption/emission spectrum, separated from the "zero-phonon" peak by multiples of the phonon energy. The presence of phonon replicas is significantly enhanced for strongly polar crystals due to the strong interaction of the TO phonons with incident photons (Fox 2001). An emission spectrum for GaN, a very polar material, is shown in Figure 2.8b, with five phonon replica peaks present.

Another important consideration in determining the optical properties is that photons always carry exactly one unit of angular momentum, and so only transitions that result in such a change of angular momentum are allowed. In order to assess the allowed transitions, we again refer to the band structure. Figure 2.9 shows the three highest valence bands, corresponding to the atomic p-orbitals, with orbital angular momentum $l = 1$. The sixfold

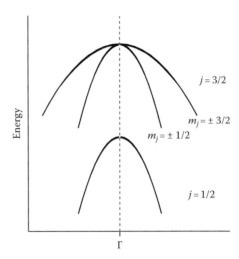

Figure 2.9 The three highest valence bands are shown. The SO band, with $j = 1/2$, is split from the $j = 3/2$ bands due to the spin–orbit interaction. The two $j = 3/2$ bands are also split except at the point of highest symmetry.

degenerate p-like valence band is split by the spin–orbit interaction* into a fourfold degenerate band with total angular momentum $j = 3/2$ and doubly degenerate band with $j = 1/2$. The lowest band, with $j = 1/2$, is referred to as the spin–orbit split-off (SO) band and in most cases does not significantly contribute to the optical properties. At points of high symmetry, the $j = 3/2$ band is fourfold degenerate; however, away from the zone center (i.e., the Γ-point) a splitting occurs, as can be seen in Figure 2.9. The bands are split according to the magnitude of the z-projection of the angular momentum, or magnetic quantum number, m_j, so that the states with $m_j = \pm 1/2$ form one band and those with $m_j = \pm 3/2$ form the other. The dispersion relation of a free particle is parabolic, since $E = p^2/2m$, and the curvature is related to the particle mass. The different curvature of the two $j = 3/2$ bands is interpreted as the effective mass of the holes that are generated when electrons are excited. The band with $m_j = \pm 1/2$, which has greater curvature, is therefore termed the light-hole (LH) band, and that with $m_j = \pm 3/2$, the heavy-hole (HH) band. The physical reason for the different curvatures is that HHs, with a greater magnetic quantum number, and hence greater magnetic moment, interact more strongly with the lattice, giving the appearance of being "heavier" and therefore leading to a shallower dispersion relation.

The angular momentum of the conduction band is determined entirely by the electron spin, as the orbital angular momentum, l, equals zero for the lowest conduction band. The total angular momentum is, therefore, $j = 1/2$ and the band is doubly degenerate with $m_j = \pm 1/2$. If we represent each of these energy bands as discrete states, as shown in Figure 2.10, the allowed transitions can be determined by the requirement for a change of one unit of angular momentum in absorbing a photon. The states involved in these transitions are all pure spin states and require a change in m_j of ± 1, which means that these pure transitions are circularly polarized, with σ^+ polarized photons adding 1 to m_j and σ^- subtracting 1 from m_j. Excitation with linearly polarized light would, in contrast, excite a linear superposition of the

---

* The spin and orbital angular momentum of charged particles generate finite magnetic moments that interact with each other, leading to a splitting of otherwise degenerate states. This interaction is referred to as the spin–orbit interaction.

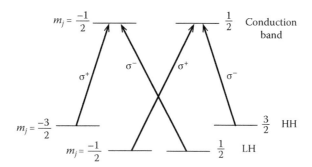

Figure 2.10 The requirement for conservation of angular momentum limits the allowed transitions from the $j = 3/2$ bands to those indicated, each of which are circularly polarized.

two circularly polarized transitions. This can be a useful tool in determining the properties of semiconductor states and identifying or selecting transitions under various conditions.

In addition to the optically allowed transitions described earlier, there are transitions that involve the addition of zero (e.g., $m_j = 1/2$ to $m_j = 1/2$) or two (e.g., $m_j = -3/2$ to $m_j = 1/2$) units of angular momentum. These transitions are one-photon forbidden, but two-photon allowed. In two-photon absorption, the angular momentum contribution from the photons can be added or subtracted, leading to a change of two or zero units of angular momentum in the sample. The other requirement for two-photon absorption is that the sum of the photon energies equals the energy of the transition. These principles can be extended to higher order multiphoton absorptions, and allow access to other "dark" states.

2.2.1 Excitons

To date, we have discussed electrons and holes as free carriers that do not interact; however, this is not always the case. The treatment of holes as particles with positive charge implies that electrons and holes should interact via the Coulomb potential. This is indeed the case, and the Coulombic attraction can lead to bound electron–hole pairs, for which two limiting cases are typically considered. The first is that of strong electron–hole attraction, in which the electron and hole are tightly bound to each other and are located within the same unit cell. This is referred to as a Frenkel exciton, and these are commonly observed in organic and molecular systems; however, in most semiconductors they are not observed, as the Coulomb interaction between the electron and hole is strongly screened by the dielectric crystal lattice. In such cases, the electron and hole are only weakly bound and extend over many unit cells. These excitons are known as Wannier–Mott excitons, and we will focus our attention on them since this description usually applies to excitons observed in semiconductors (Shah 1999, Fox 2001, Klingshirn 2007).

Wannier excitons are analogous to an electron and proton "orbiting" about each other in a hydrogen atom, and many of their properties can be derived from the well-known properties of the hydrogen atom, simply by accounting for changes in effective masses and dielectric constant. This bound pair state is lower in energy than the unbound electron and hole states, and the amount of energy required to separate the exciton into its constituent parts is referred to as the exciton binding energy. In common semiconductors, the binding energy is typically in the range of tens of meV, which means, particularly at low temperatures, that excitons are often the dominant excitation.

Continuing the analogy with the hydrogen atom, in addition to the ground state of the exciton, there are a series of quantized excited states, with energy given by

$$E_{bn} = \frac{E_{b0}}{n^2},$$ (2.1)

where
E_{b0} is the binding energy of the lowest exciton state
n is the principal quantum number

The exciton transition energy is then given by the difference between the band gap and the exciton binding energy:

$$E_{xn} = E_g - E_{bn}.$$ (2.2)

The Bohr radius, which represents the average separation between the bound electron and hole, varies from ~3 to 35 nm in common bulk semiconductors and is defined as

$$a_0 = \frac{\hbar^2 \varepsilon_0}{m_{eff} e^2}.$$ (2.3)

To ascertain the role of excitons in the optical properties of semiconductors, consider the oscillator strength, which depends linearly on the overlap of the electron and hole wave functions, and is directly proportional to the absorption magnitude. It is clear that the wave function overlap between bound electrons and holes (i.e., excitons) will be far greater than between free electrons and holes, and therefore that the oscillator strength for excitons will exceed that for free electron–hole pairs. As a result, excitons appear in absorption spectra as a series of sharp, intense peaks decreasing in intensity as energy is increased from the ground state exciton to the band-edge transition energy (which can also be renormalized by the electron–hole Coulomb interaction, discussed in more detail in Chapter 8). It then becomes clear that at temperatures where the thermal energy is less than the exciton binding energy and excitons are abundant, they dominate the optical response of semiconductors (Klingshirn 2007).

Exciton binding energies vary over more than an order of magnitude in common semiconductors, from 4.3 meV in GaAs (corresponding to a temperature of 50 K, below which excitons begin to dominate) to 60 meV for ZnO (corresponding to 696 K, indicating that excitons dominate at and above room temperature) (Jagadish and Pearton 2006). A list of the exciton binding energy, Bohr radius, and band gap for many common bulk semiconductors can be found in Table 2.1. The absorption spectrum of high-quality GaAs at 1.2 K is shown in Figure 2.11, where peaks corresponding to the first three exciton states can be identified.

The optical selection rules that govern excitation of electrons from the valence band to the conduction band also apply for excitons, and to date we have been considering just the bright, optically allowed excitons. The electron and hole states involved in the one-photon forbidden transitions discussed in the previous section can also form excitons. These "dark excitons" have angular momentum zero or two, meaning they cannot be generated in one-photon processes. In addition to generating these dark excitons with multiphoton absorption, they can also be formed from bright excitons following a spin flip of either the electron or hole. Such spin-flip processes can be quite common under certain conditions, and thus present an important relaxation pathway for bright excitons. Furthermore, because dark excitons are not optically active, radiative recombination is closed as a relaxation pathway, and so they tend to have significantly longer lifetimes.

TABLE 2.1

The Band Gap (E_g), Exciton Binding Energy, and Exciton Bohr Radius
for Several Common Semiconductors

Material	Band Gap (eV)	Binding Energy (meV)	Bohr Radius (nm)
GaN (Fox 2001)	3.5	23	3.1
ZnO (Madelung et al. 2002, Jagadish 2006)	3.4	60	2.3
ZnSe (Fox 2001)	2.8	20	4.5
CdS (Fox 2001)	2.6	28	2.7
CdSe (Fox 2001)	1.8	15	5.4
GaAs (Fox 2001)	1.5	4.2	13
InP (Fox 2001)	1.4	4.8	12
Si (Smith et al. 1975)	1.1	14.7	5
GaSb (Fox 2001)	0.8	2.0	23
Ge (Madelung et al. 2002, Smith et al. 1975)	0.66	3.6	18
InAs (Madelung et al. 2002)	0.36	1.0	35

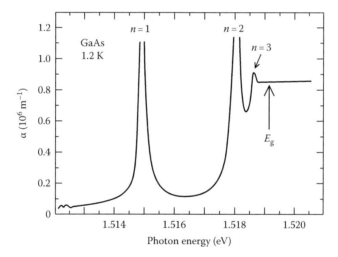

Figure 2.11 The absorption spectrum of high-quality GaAs at 1.2 K. Peaks corresponding to the first three exciton states can be seen. (Reprinted from *J. Luminescence*, 30, Fehrenbach, G.W. et al., 154. Copyright 1985, with permission from Elsevier.)

Just as optical transitions between unbound electrons and holes predominantly occur near the center of the Brillouin zone (i.e., $\mathbf{k} = 0$), so too excitons with $\mathbf{k} \sim 0$ dominate optical spectra because of the minimal momentum carried by photons. Excitons with $\mathbf{k} \neq 0$ can, however, be involved in optical transitions where absorption or emission of a phonon is involved, leading to phonon replicas, as discussed previously. The coupling of excitons and optical phonons is stronger than that between unbound carriers and phonons because the stronger dipole moment of excitons enhances the interaction. This means that phonon-assisted excitonic transitions are more prevalent than the equivalent transition between unbound electron/hole states. Similarly, phonon-assisted transitions allow the formation of indirect excitons in indirect gap materials such as Si and Ge. This is in contrast to direct excitons in such materials, which can be formed by excitation high up into the conduction band, but which are short lived due to the rapid relaxation of the electron. Indirect excitons have a finite momentum that must be conserved, which leads to reduced relaxation pathways, and an increase in their lifetime from nanoseconds to microseconds, compared to excitons in direct gap materials (Timusk 1976).

Under certain conditions, other bound particles can be detected. These include charged excitons, which consist of either two electrons and one hole, or two holes and one electron, and which can dominate for n- or p-doped semiconductors, respectively. Another excitation that is commonly observed in intrinsic semiconductors is the biexciton, which consists of two electrons and two holes. Biexcitons become increasingly prevalent as the number of carriers injected into the material increases (biexciton luminescence increases quadratically with excitation intensity), and typically have a binding energy of a few meV (Klingshirn 2007).

2.2.2 Defect States

There are many different types of defects that can be incorporated in semiconductor materials. Some may be deliberately included to change the band gap, or introduce free carriers, as discussed previously. Others may arise unintentionally and possess properties that prohibit proper device operation.

In general, defects can be classified as point defects, which usually involve isolated atoms or sites in the crystal lattice, or line defects, which include dislocations in the crystal lattice. We concentrate on the former, as they tend to have a greater effect on the optical properties.

Point defects can be further classified according to the nature of the fault. The most common types of point defects are vacancies, created by a missing atom; interstitials, created by an atom occupying an interstitial site; and substitutional defects, where an atom is replaced by an atom of a different type. Here we briefly discuss the optical properties of such defects, but for a complete review of the electrical, thermal, and optical properties of defects and their prevalence, we refer the reader to references Lannoo and Bourgoin (1981), Watts (1977), Pantelides (1978), and Klingshirn (2007).

Each of the point defects mentioned earlier can act as either a donor or acceptor (depending on the bulk crystal structure, and the atom/s involved), and it is frequently this property, rather than the specific type of defect, that determines the effect on the optical properties. One of the greatest effects occurs when these defects are able to bind an exciton (i.e., the bound electron and hole are both localized at the defect). The binding energy for such a system is typically greatest for a neutral acceptor,[*] followed by a neutral donor and then by an ionized donor, while ionized acceptors are not usually able to bind excitons (Klingshirn 2007). These bound excitons have zero degrees of translational freedom, and so interact less with charge and lattice excitations, leaving few available broadening mechanisms. Such states appear as very sharp peaks in luminescence and absorption spectra, redshifted from the free exciton peak by the defect binding energy. The magnitude of this binding energy is typically less than 10 meV, which means that defect-bound excitons can be observed only at low temperatures, where the thermal energy is insufficient to promote escape from the defect.

Another corollary of the lack of translational degrees of freedom for excitons bound to defect states is that the requirements for momentum conservation are reduced.[†] This means that these excitons can easily couple to acoustic phonons in direct gap semiconductors, leading to acoustic wings in absorption and emission spectra. Similarly, for indirect semiconductors, defect states can be observed without the participation of phonons.

---

[*] A neutral acceptor (donor) is a p-type (n-type) defect that has not given up its extra hole (electron), while an ionized acceptor (donor) is one that has.

[†] The Heisenberg uncertainty principle states that $\Delta x \Delta p > \hbar/2$, and so the accurately known position (i.e., Δx is small) leads to a large uncertainty in momentum, Δp. This then means that the strict requirement for conservation of momentum can be relaxed.

The effects of defect states have so far been discussed with respect to their spectral signatures, and indeed, the spectra of defects can be rather complicated, and can include additional processes involving excited bound states, multiple defects, free-to-bound recombination, or ionization of defects. There exists, however, a range of defects that are not optically active, and instead assist non-radiative recombination processes, thereby quenching optical luminescence. Dislocations are a prime example of this type of defect, where the electronic energy is transferred to the lattice. Another example is so-called deep centers (as opposed to the "shallow" defect states discussed to date) where the defect state lies in the middle of the band gap. In this case, they can interact with both the conduction and valence bands and frequently lead to rapid non-radiative recombination (Pantelides 1986, Klingshirn 2007). In some cases, however, they can act as radiative centers, and lead to absorption and emission peaks well below the band-edge transition (Pantelides 1986). The prevalence of these "deep centers" increases as the band gap increases, and so they are more common in wide gap semiconductors, such as ZnO and GaN (Reshchikov and Morkoc 2005), and in insulators.

2.2.3 Free Carrier Absorption

When additional carriers introduced by defects can be thermally ionized, they behave as free carriers, and their interaction with light fields becomes very similar to free carriers in metals. This is particularly relevant for n- or p-doped materials and small-gap semiconductors, where the defect states tend to be very shallow and ionize easily. To understand these interactions, we can treat the material as a simple metal and analyze the properties on the basis of the Drude model, as discussed in Chapters 1 and 3. The main difference between metals and free carriers in semiconductors is that the carrier concentration can be changed in semiconductors by controlling the dopant density, and typically is significantly less than the free carrier concentration in metals. As a result, the plasma frequency (Chapters 1 and 3), which determines many of the optical properties, is in the infrared range for doped semiconductors, compared to the visible/UV for metals.

In semiconductors, since either electrons or holes can be the majority carrier, both can contribute to the optical properties in different ways. When electrons are the majority carrier, free carrier absorption is typically due to intraband transitions, and typically involves a phonon in order to conserve momentum, as shown in Figure 2.12a. In this case, free carrier absorption can be used as an important tool for measuring phonon scattering rates and interactions (Fan 1967, Yu and Cardona 2005). When holes are the majority carrier, interband transitions are more common, involving transitions between SO, LH, and HH bands, as shown in Figure 2.12b.

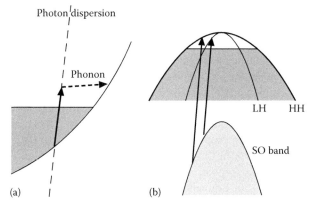

Figure 2.12   (a) Free carrier transitions in the conduction band are limited to intra-band transitions and usually require the involvement of a phonon. (b) Free carrier transitions in the valence bands can be inter-band transitions, between SO, LH, and HH bands.

Using this process, the ratios of the effective mass at the top of the valence bands can be determined by comparing the absorption spectra for transitions between different bands (Fan 1967, Yu and Cardona 2005).

One final difference between the optical properties of free carriers in semiconductors and metals is that the plasma resonances can be much sharper in semiconductors. The reason for this is that for frequencies above the plasma frequency the photon energy is often less than the band gap in semiconductors, whereas in metals, it may still be sufficient to allow inter-band transitions, thereby broadening the plasma edge.

2.2.4 Optical Emission (Radiative Recombination)

To this point, we have concentrated predominantly on the interaction of semiconductors with an incoming light field. What happens after this initial interaction is, however, at least as important (Shah 1999). Many applications of semiconductors utilize the recombination of electrons and holes and subsequent emission of light, for example, as luminescence in LEDs, or stimulated emission in lasers. The dynamic relaxation processes that can occur between excitation and recombination, together with the processes that can affect both radiative and non-radiative recombination dynamics are discussed in detail in later chapters (e.g., Chapter 8). In this section, we limit the discussion to a brief description of the interactions between semiconductors and outgoing light fields.

Emission of light from a semiconductor is referred to generally as luminescence, occurring when an electron and hole recombine, most often through an electronic transition from the conduction band to the valence band. This transition usually occurs from a state close to the conduction band minimum to a state close to the valence band maximum, regardless of where in the band structure the carriers are initially generated. The reason for this is that carriers generated at higher energies undergo intraband relaxation, or cooling, at a much faster rate than they radiatively recombine (Chapter 8, Shah 1999). As a result, emission spectroscopies are a sensitive probe of low energy states. This type of band-edge emission is particularly dominant in direct gap semiconductors; in indirect gap semiconductors, however, it requires the participation of phonons and is therefore much less efficient than in direct gap semiconductors, as is the case for absorption.

Depending on how the electrons and holes are initially created in the sample, this type of radiative recombination is referred to as photoluminescence, for carriers excited optically (Chapter 5); thermoluminescence, for carriers excited thermally; electroluminescence, for carriers injected into the sample by an external current; or cathodoluminescence, for carriers generated by electron bombardment.

Regardless of the manner of excitation, the emission of light from semiconductors can generally be treated as the inverse process to absorption. In the case of absorption, a semiclassical approach is normally sufficient. If we follow this same procedure, however, we find that in order for emission to occur, an electromagnetic wave needs to be present (Yu and Cardona 2005). This process is known as stimulated emission; however, our experience tells us that it is also possible for emission to occur in the absence of an electromagnetic wave. To explain this, a full quantum mechanical treatment of the light field is required, which identifies the probability of emission as being proportional to $1 + N_{\rm p}$, which is clearly nonzero even when $N_{\rm p}$, the photon number, is zero. In this case, the "spontaneous emission" can be thought of as being stimulated by the "zero-point amplitude" of photons,* which is independent of $N_{\rm p}$, while "stimulated emission" corresponds to the part proportional to $N_{\rm p}$.

---

* "Zero-point" energy is a fundamental outcome of quantum mechanics that identifies the lowest energy state as still having a finite energy. This is evident even in the simplest case of a quantum harmonic oscillator, where the ground state is higher in energy than the potential minimum.

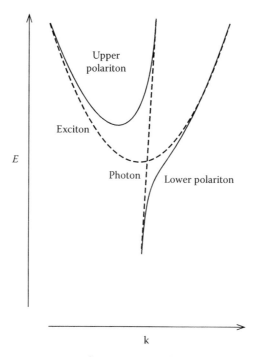

Figure 2.13 The strong interaction of an exciton and photon leads to the creation of polaritons, which have a band structure that is generated by combining the exciton and photon bands with an avoided crossing at their intersection.

As with any rate equation, the rate of spontaneous emission is dependent on the concentration of initial states (i.e., the density of electrons in the conduction band) and the concentration of unoccupied final states (i.e., the density of holes in the valence band) (Chapter 5). An additional consideration is the overlap of the carriers in both reciprocal and real space. In reciprocal space, recombination typically occurs between carriers at the center of the Brillouin zone, where the change in momentum is small. In real space, the recombination rate is proportional to the overlap of the electron and hole wave functions. For free electron–hole pairs, the spatial overlap is averaged over the entire crystal and is dependent on the concentration of electrons and holes. For excitons, the electron–hole overlap is far greater, and so the recombination rate is faster. With regard to luminescence from excitons, however, one further consideration is required, as identified by Toyozawa (1959). To accurately reproduce experimental excitonic spectra, it is essential to include the strong coupling between exciton and photon, which combine to create a polariton, as indicated in the band structure in Figure 2.13. The dispersion curve of the lower polariton does not have a minimum, and so one might not expect to see an emission peak. However, by considering the lifetimes of polaritons at different points on the dispersion curve, it can be shown that a bottleneck occurs near the transverse exciton energy. This leads to an asymmetric peak that adequately reproduces experiments. The full derivation of the exciton polariton spectrum is beyond the scope of this chapter, and so we refer the reader to Toyozawa (1959) and Yu and Cardona (2005).

2.2.5 Inelastic Scattering

The final type of interaction between semiconductors and optical light fields that we will discuss involves the inelastic scattering of an incident light field, resulting in emission that is shifted in energy. Normally, we think of light scattering off a static defect or edge; however, it is also possible for light to scatter from a transient change in the susceptibility of the material, such as those generated by phonons. This type of inelastic scattering between photons and phonons is referred to as Raman scattering, or in the specific case of acoustic phonons, Brillouin scattering.

The interaction of a light field with a phonon, or more specifically with the modulated susceptibility induced by the phonon, can generate a photon with energy that has been either increased or decreased by the phonon energy. The two Raman signals are then observed at $\omega - v$ corresponding to Stokes scattering, and at $\omega + v$ corresponding to anti-Stokes scattering, where ω is the frequency of the light and v is the frequency of the phonon.

In addition to conserving energy, the Raman scattering process is also required to conserve momentum. Where photons are incident on the material, the magnitude of the change in momentum is limited to less than twice the photon momentum (Chapter 6). This means that for optical photons, only phonons close to the zone center, that is, those with little momentum, can scatter in this way with the light field. For acoustic phonons, this also limits the energy of the phonons as, unlike optical phonons, the dispersion curve increases almost linearly with increasing momentum. A useful implication of this is that the velocity of acoustic phonons can be determined by careful analysis of Brillouin scattering (Benedek and Fritsch 1966).

The requirement that only phonons with little momentum can participate in Raman scattering is relaxed when two-phonon Raman scattering processes are considered. In this case, the two phonons can provide roughly equal and opposite momentum, thereby allowing momentum to be conserved, even though the photon momentum still only changes a small amount. This means that two-phonon Raman scattering can involve all phonons, independent of momentum, and provides a means for measuring the phonon density of states (Uchinokura et al. 1974).

2.2.6 Optical Properties in a Magnetic Field

A free electron in a magnetic field will rotate about the magnetic field with a frequency given by $\omega_c = e\mathbf{B} \, m_{eff}$, where ω_c is the cyclotron frequency and m_{eff} is the effective mass (Ashcroft and Mermin 1976). In semiconductors, where there are free electrons in the conduction band, or free holes in the valence band, this is also the case. Indeed experiments making use of this phenomenon are used to determine the effective mass of electrons and holes by measuring the cyclotron frequency using resonant microwaves in a manner similar to nuclear magnetic resonance (NMR) (Yu and Cardona 2005).

In addition to this common effect, there are three main effects of an external magnetic field that can be seen in the optical response of semiconductors: Zeeman splitting, diamagnetic shift, and the formation of Landau levels. A detailed understanding of these effects allows the experimentalist to measure various properties of semiconductor materials.

2.2.6.1 Zeeman Splitting

Excitons, electrons, and holes all possess spin and in some cases orbital angular momentum. This means that they, as charge carriers, also have a finite magnetic dipole moment, proportional to the z-projection of the angular momentum, m_j. When an external magnetic field is applied to such a dipole moment, its energy is shifted by an amount proportional to $g\mu_B\mathbf{B} \cdot \mathbf{j}$ ($=g\mu_B\mathbf{B}m_j$), where g is the Lande g-factor, μ_B is the Bohr magneton, $\mathbf{B}$ is the magnetic field, and $\mathbf{j}$ is the angular momentum vector, with the scalar product, $\mathbf{B} \cdot \mathbf{j}$, replaced by the magnitude of the magnetic field and m_j, the projection of the angular momentum in the direction of $\mathbf{B}$. This shift is known as the Zeeman shift and causes otherwise degenerate levels to become split by an amount linearly dependent on the applied field strength. Figure 2.14

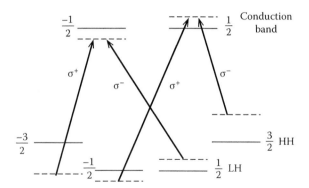

Figure 2.14 The Zeeman splitting proportional to m_j is shown, with the solid lines the unshifted states and the dashed lines the Zeeman shifted states. The splitting between the HH states is larger than that of the LH and conduction band states; however, the change in each of the transition energies is given by $\pm g\mu_B B$.

shows the splitting of the HH, LH, and conduction band states that were previously shown in Figure 2.10 and the allowed transitions between them. From this figure, it can be seen that the two allowed HH transitions are split by the same amount as the two allowed LH transitions. The reason for this becomes more evident by considering the total angular momentum of the excitons, as shown in Figure 2.15. In this case, the ground state is the zero exciton state and the optically allowed excited states are the one exciton states with total angular momentum projections of ±1. In this representation, the optically allowed LH-excitons and HH-excitons both have angular momentum vectors ±1, and both are therefore split by the same amount in an applied magnetic field. The exciton states with $m_j = 0, \pm 2$ are dark states and correspond to the forbidden transitions that require a change of zero or two units of angular momentum, respectively.

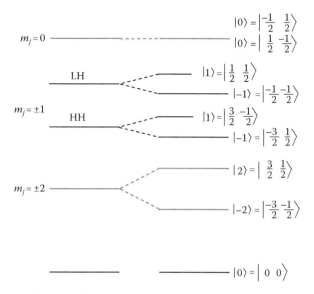

Figure 2.15 By switching to the exciton representation, it becomes clear that the bright HH and LH excitons are split by the same amount by the Zeeman effect. The lowermost |0> state is the zero exciton ground state, and the dark excitons are also shown for completeness.

2.2.6.2 Diamagnetic Shift

The application of a magnetic field to an electron or hole causes it to rotate in an orbit, thereby giving it additional angular momentum. Similarly, the application of a magnetic field to an exciton causes a change in the relative motion of the electron and hole, and a change in the overall angular momentum. If we treat the exciton states as atomic orbital-like states, then this deformation can be thought of as adding a component of a different state, the amount of which is proportional to the field strength. For example, an exciton in a purely s-like state ($l = 0$) will develop some p-like character ($l = 1$) in a magnetic field. This causes a change in the total angular momentum of the exciton, with magnitude linearly dependent on the field strength. The orientation of the angular momentum is such that the exciton develops a magnetic dipole moment in the opposite direction to the applied field, as predicted by Lenz's rule. The magnitude of this magnetic dipole is then linearly dependent on field strength. The energy of such a magnetic dipole, $\mathbf{\mu}$, in an external magnetic field, $\mathbf{B}$, is given by $\mathbf{\mu} \cdot \mathbf{B}$, that is, it is linearly dependent on the field strength and on the magnitude of the dipole moment, as observed in the Zeeman effect. Since the magnitude of the field-induced dipole moment is also linearly dependent on the magnetic field strength, the energy of an exciton in an applied magnetic field increases quadratically with the field strength. This is known as a diamagnetic shift.

2.2.6.3 Landau Levels

The presence of the cyclotron orbits discussed at the beginning of this section can be predicted from classical mechanics. However, a quantum mechanical treatment of this interaction between free carriers and an external magnetic field reveals additional behavior.

Solving the Hamiltonian for an electron in a magnetic field reveals that parallel to the field the wave function is unchanged and the electron is free. Perpendicular to the field, however, the Hamiltonian resembles that for a simple harmonic oscillator and gives eigenvalues:

$$E = \left(n + \frac{1}{2}\right)\hbar\omega_c \quad \text{with } n = 0, 1, \dots, \tag{2.4}$$

where ω_c is the cyclotron frequency ($\omega_c = e\mathbf{B}/m_{\text{eff}}$). As a result, the carrier is free to move in the direction parallel to the field, but confined perpendicular to it into concentric circles, which correspond to the different states of the harmonic oscillator, the so-called Landau levels. This confinement in two dimensions generates a quasi-1D system, similar to a quantum wire, with the energies of the Landau levels given by

$$E(n, \mathbf{k}_{\parallel}) = E_0 + \left(n + \frac{1}{2}\right)\omega_c + \frac{\hbar^2 \mathbf{k}_{\parallel}^2}{2m_{\text{eff}}}, \tag{2.5}$$

where

    n in the Landau quantum number
    E_0 is the state energy in the absence of the magnetic field
    $\mathbf{k}_{\parallel}$ is the wave vector parallel to the applied magnetic field

Equation 2.5 shows that Landau levels are equally spaced in energy, and the spacing varies linearly with field strength, since $\omega_c \propto \mathbf{B}$. The radius of the lowest Landau level is given by $l_b = [\hbar c/(e\mathbf{B})]^{1/2}$ and is closely related to the density of states for each level, which resembles that of the 1D systems discussed in the next section. Transitions between Landau levels are allowed for a change of $\Delta n_l = \pm 1$, which allows the levels to be identified as long as their spacing is greater than their broadening. The splitting of the Landau levels arises as a result of the quantization of the frequency of the cyclotron orbits induced by the magnetic field. This is in contrast to Zeeman splitting, which occurs as a result of the quantization of the total angular momentum intrinsic to the carrier/s, and its interaction with the applied magnetic field.

2.2.6.4 Other Magneto-Optical Effects

In addition to the three main effects discussed earlier, under certain conditions, other magneto-optical effects can be observed, and we mention two of these here.

The Faraday effect arises when an external magnetic field causes circularly polarized light with different orientations (i.e., σ^+ and σ^-) to propagate at different speeds, thereby causing a change in the phase of the rotating electric field. For linearly polarized light, this effect causes a rotation of the polarization axis.

A closely related effect is the magneto-optical Kerr effect; the only real difference between the Faraday and Kerr effects is that the Kerr effect occurs in reflection, while the Faraday effect occurs in transmission. A corollary of this difference is that in the Kerr effect an additional directionality is imposed, that of the plane of reflection. This means that there are additional geometries that can be established between the incoming light field, the plane of reflection, and the magnetic field direction, each leading to slightly different changes to the reflected polarization (Kato et al. 2004, Klingshirn 2007). These effects are often used to study the polarization of spins in semiconductors, which is particularly significant for proposed spintronic devices, as detailed in Chapter 13 and in references Kikkawa and Awschalom (1998), Atatüre et al. (2007), and Gupta et al. (1999).

2.2.7 External Electric Fields

The application of an electric field to an isolated atom causes a reduction of its symmetry and leads to a field-dependent splitting of otherwise degenerate levels. This is known as the DC Stark effect and is the analog of the Zeeman effect for magnetic fields. In semiconductors, a similar effect is seen for excitons. The origin of this can be seen by examining the exciton potential in the z-direction, with and without an applied electric field, as shown in Figure 2.16. Without an applied electric field, the potential well responsible for exciton formation is shown (Figure 2.16a), with the $n = 1$ wave function also represented. The application of an electric field in the z-direction places a constant gradient on the potential, which deforms the exciton wave function and shifts the exciton energy by an amount that is usually quadratically dependent on field strength. In practice, however, it is difficult to observe this shift in bulk semiconductors because the magnitude needs to be larger than the spectral width of the

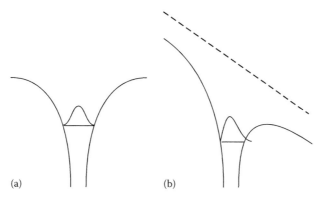

(a)   (b)

Figure 2.16   The potential profile of an atom or exciton (a) is altered by the application of an electric field (b). As a result of the electric field, the energy of the localized state is lowered, and tunnelling through the potential barrier gains a finite probability.

exciton transition. In most semiconductors, this means fields greater than 10^6 Vm^{-1} would be required (Klingshirn 2007). An additional effect of such strong fields, however, is to broaden, or destroy, the excitonic resonance. From Figure 2.16b, it can also be seen that the application of an electric field makes it possible for the electron to tunnel out of the Coulomb barrier, ionizing the exciton. Increasing the field increases the probability of this occurring, and the presence of a significant tunneling probability leads to further broadening of the transition. The width of the exciton transition is also enhanced due to increased collisions with high energy free carriers that can be accelerated by a strong electric field. The Stark effect has however been observed for excitons in bulk semiconductors, using modulation spectroscopies with weak electric fields (Cardona 1969). The full derivation of the Stark shift in semiconductors can be found in Haug and Koch (2004).

Another effect of an applied electric field that is observed in bulk semiconductors, in the absence of excitons, is the Franz–Keldysh effect. Once again, the derivation of this result can be found in Haug and Koch (2004) where it is revealed that the perturbation introduced by the applied field leads to the solutions of the Schrödinger equation being Airy functions rather than plane waves. The effect of this is that rather than a sharp cutoff at the band edge, an exponential low energy tail is present in the absorption spectrum. The extent of this tail increases with increasing field strength, and can be thought of as a photon-assisted, field-induced tunneling of an electron across the band gap. At energies above the band edge, the Airy functions are oscillatory, and lead to oscillations in the absorption spectrum, known as Franz–Keldysh oscillations. Where excitons are dominant, however, the Franz–Keldysh effects are typically modified by large excitonic effects (Haug and Koch 2004).

Two further effects of external electric fields are analogous to the magneto-optical Faraday and Kerr effects. These are known as the electro-optical Kerr effect and the Pockels effect. In these processes, a birefringence* is introduced, which can then rotate the polarization of an electromagnetic field. In the Pockels effect, a crystal without inversion symmetry is required, and the amount of birefringence, and hence amount of rotation, is linearly proportional to the field strength. In contrast, the birefringence introduced in the DC-Kerr effect (which is present in all materials to some extent) varies quadratically with field strength (Klingshirn 2007). Both the Kerr effect and Pockels effect are used extensively as tools in optical spectroscopy to control and modulate pulses of light due to their fast response times.

2.2.8 Strain Fields

In addition to magnetic and electric fields, external mechanical strain fields can be applied that may alter the optical properties of crystalline semiconductors. Physically, a strain field causes a shifting of the nuclei from their equilibrium positions in the crystal lattice. Depending on the direction of the strain, this can lead to a shift and/or splitting of peaks in the absorption spectrum. In order to quantify the effect of a strain field, one typically refers to the deformation potential, which is the amount of energy the band extrema move for a given deformation. For the conduction band and valence band, it is given by $a \times dE_{c,v}/dx$, where a is the lattice constant and $dE_{c,v}$ is the change in the band-edge energy caused by a length variation of the unit cell by an amount dx.

The deformation energy is typically defined for strain applied in different directions relative to the key directions in the unit cell. Where the strain is applied in a direction such that it alters the symmetry of the unit cell, a splitting of degenerate states occurs in addition to the shift caused by altering the lattice constant. For example, the HH and LH bands are

---

* Birefringence is a property of some materials whereby the velocity of an electromagnetic wave is different for orthogonal polarization orientations.

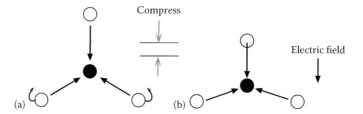

Figure 2.17   In the unperturbed system (a), the symmetry of the polar bonds ensures the net polarization is zero. When the system is compressed (b), the system loses this symmetry and a net polarization across the unit cell results.

degenerate at $\mathbf{k} = 0$ in unstrained bulk materials (Figure 2.9); this degeneracy is lifted by mechanical strain on the lattice. As a result of this splitting, it becomes possible to distinguish the two in optical experiments where transitions near $\mathbf{k} = 0$ dominate, allowing selective excitation of specific spin states in the conduction band using circularly polarized light, as identified in Figure 2.10.

In compound semiconductors, a strain field also has the potential to induce an electric field. In an effect similar to the spontaneous polarization present in ionic crystals, a strain field that alters the symmetry of the crystal structure can also alter the electric field across the unit cell. This is represented schematically in Figure 2.17 where the strain field reduces the symmetry and fields that previously canceled each other cease to do so. This is the basis for piezoelectric materials.

The role played by strain fields has become increasingly important as semiconductor nanostructures have been developed. These nanostructures typically rely on interfaces between different semiconductor materials, which often have different lattice constants, therefore introducing strain at the interfaces. The following section is devoted to such nanostructures and their optical properties.

2.3 LOW-DIMENSIONAL SEMICONDUCTOR NANOSTRUCTURES

2.3.1 How They Are Realized

Nanostructures are defined as materials that have at least one dimension smaller than 100 nm. In semiconductors, they are usually not isolated structures, as is the case, for example, in metal nanoparticles (Chapter 3), but rather nanoscale regions of one type of semiconductor surrounded by a different semiconductor. The two different regions will usually have a different band gap, which will lead to a potential barrier for electrons and/or holes at the interfaces. An example of this is shown schematically in one dimension in Figure 2.18. In this case, the band gap of the barrier material is larger than that of the nanostructure, creating a potential well for electrons in the conduction band and holes in the valence band. This means that electrons and holes will be confined to this region, and will, when this region is smaller than the de Broglie wavelength, exhibit properties that are well described by the well-known quantum mechanical problem of a "particle in a box." In practice, however, we are dealing with carriers that are confined in one (quantum wells [QWs]), two (quantum wires [QWrs]), or three (quantum dots [QDs]) dimensions, but which behave as free particles in the other two, one, or zero dimensions, respectively, as represented in Figure 2.19. We will come to the effects of confinement in each dimensionality in due course, but first there are some general considerations to take into account.

The type of confinement depicted in Figure 2.18 is characteristic of a type-I nanostructure, where both electrons and holes are confined in the same region. That is, using Figure 2.18

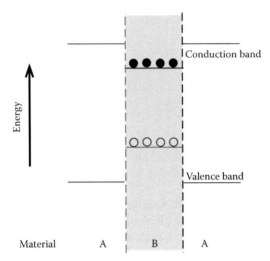

Figure 2.18 The different band gaps of materials A and B lead to confinement of electrons in the conduction band and holes in the valence band in this example of a type-I nanostructure.

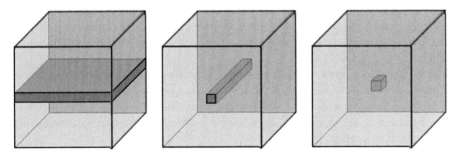

Figure 2.19 The confinement in one, two, and three dimensions shown leads to the formation of quantum wells, quantum wires, and quantum dots, respectively.

as an example, the conduction band of material B is lower in energy, and the valence band higher, than those of material A. This is not necessarily the case whenever two different materials form a nanostructure, as the alignment of the bands is also important and is not simply dependent on the respective band gaps. The factor that controls the relative band alignment of the two different regions is the electron affinity (i.e., how strongly the nuclei attract excess electrons). Depending on the choice of materials, it then becomes possible to create type-II and type-III nanostructures, as depicted in Figure 2.20, where both valence band and conduction band are greater in B than A, confining electrons and holes to opposite sides of the barrier. Nanostructures that will confine just electrons or just holes can then be designed. The difference in type-III nanostructures is that the valence band of material B is higher in energy than the conduction band of material A. This leads to electrons from the valence band of material B relaxing into the conduction band of A without excitation, thereby confining intrinsic electrons to material A and holes to material B. For the remainder of this chapter, however, we focus on type-I nanostructures and confinement of both electrons and holes, as they are much more common.

Making such nanostructures is not as straightforward as simply placing the different materials together, and has only become possible over the past two decades due to great advances in epitaxial growth technologies, particularly molecular beam epitaxy (MBE) (Tsao 1993)

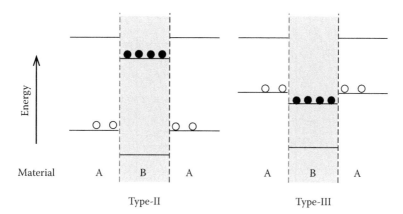

Figure 2.20 Type-II and type-III confinements lead to the localization of electrons and holes in different regions of the nanostructure. Type-III is a special case of type-II, where the conduction band of material B is lower than the valence band of material A.

and metal-organic chemical vapor deposition (MOCVD) (Herman et al. 2004). The ability to tune the band gap of compound semiconductors by varying the concentration of its constituents is essential to control the potential profiles and ensuing properties of semiconductor nanostructures.

The biggest challenge in growing semiconductor nanostructures is to ensure the interfaces between different materials are of high quality. If this is not the case, there will be a large number of defect and interface states that will dominate the electronic and optical properties, and the desired confinement effects will be obscured. One of the most important considerations is to select materials that have similar lattice constants (i.e., the spacing between atoms in the two materials is similar), as large lattice mismatch at the interfaces will lead to dislocations and a high concentration of defects. For thin films, it is, however, possible to grow high-quality interfaces even when the lattice constants are not perfectly matched. This is achievable because a thin layer of one material can be grown with a lattice constant that is different from its natural value and instead matches that of the material onto which it is being deposited. This thin layer is then under strain, but depending on the extent of lattice mismatch and thickness of the layer, it can form an interface free of dislocations, allowing high-quality QWs to be grown.

The strain present at hetero-interfaces in epitaxial layers is in some instances essential to growing nanostructures of a given dimensionality. The Stranski–Krastanow growth method, for example, utilizes the strain (and other material properties) to form quantum dots when the growth changes from being 2D to 3D. This process generates islands (quantum dots) on top of a thin (1–2 monolayers) 2D "wetting layer" (Stranski and Krastanow 1939, Bauer 1958, Seeger 2004).

The selective growth of quantum wells, quantum wires, and quantum dots is a highly evolved science, and depends on a wide variety of growth and material parameters. Nonetheless, it is now possible to grow many of these structures controllably and repeatably, and for a review of these growth techniques, we refer the reader to Zhang et al. (2007).

2.3.2 Effects of Reduced Dimensionality

The particle in a box problem is one of the first concepts usually encountered in a course on quantum mechanics, and the solutions of the Schrödinger equation are well known and easily

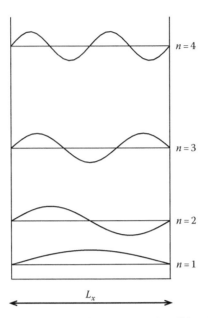

Figure 2.21 Confining a particle in an infinite potential well leads to the formation of quantized energy levels (as given by Equation 2.7) and wave functions that go to zero at the barrier (as given by Equation 2.6).

derived. The effects of electron and/or hole confinement are quantitatively similar to this problem, and so we begin with a brief review of the particle in a box problem.

Consider a 1D potential well, as depicted in Figure 2.21, with infinitely high barriers. It is straightforward to show that the solution to the Schrödinger equation $\mathbf{H}\Psi = E\Psi$ gives

$$\Psi = \sin\left(\frac{n\pi x}{L_x}\right), \tag{2.6}$$

$$E = \frac{(\hbar\pi n)^2}{2mL_x^2}, \quad n = 1,2,3\ldots, \tag{2.7}$$

where
  L_x is the width of the well
  m is the mass of the particle
  n is an integer that represents the principal quantum number

It is evident from Equation 2.7 that the energy becomes quantized and proportional to the square of the principal quantum number, as indicated in Figure 2.21 where the first four states are shown for the particle in a 1D box.

This corresponds to 1D confinement in QWs, and this problem can be extended relatively easily to 2D and 3D confinements, corresponding to quantum wires and dots, respectively. One of the biggest differences between this case and that of realistic confinement in semiconductor nanostructures is that the barriers are not infinitely high. This means that the wave functions do not go exactly to zero at the barriers, and the probability of the carriers tunneling out of the potential well becomes finite, as does the probability of escaping from the well due to thermal or other excitations.

TABLE 2.2
Number of Carriers and Density of States for Infinite 1D, 2D, and 3D Systems

Dimensionality of Infinite Lattice	Density of States	Integrated Density of States
1 (quantum wire)	$D(E) = (L/\pi)(2m/\hbar^2)^{1/2}E^{-1/2}$	$N(E) = (2L/\pi)(2m/\hbar^2)^{1/2}E^{1/2}$
2 (quantum well)	$D(E) = (A/2\pi)(2m/\hbar^2)$	$N(E) = (A/2\pi)(2m/\hbar^2)E$
3 (bulk)	$D(E) = (V/2\pi^2)(2m/\hbar^2)^{3/2}E^{1/2}$	$N(E) = (V/3\pi^2)(2m/\hbar^2)^{3/2}E^{3/2}$

L, A, and V are the length, area, and volume of the unit cell, respectively.

Additionally, in real semiconductor nanostructures, along with the confinement in one, two, or three dimensions, there is a semi-infinite system in the other two, one, or zero dimensions, respectively. Hence, to fully define the states of quantum wells and wires, it is necessary to also consider the properties of carriers in 2D and 1D systems.

The different dependence on wave vector of the energy of carriers in one, two, and three dimensions can be conceptualized by considering the density of carriers in inverse space. For a given number of carriers, the density in a 1D system will be proportional to $2\mathbf{k}$, in two dimensions proportional to $\pi\mathbf{k}^2$, and in three dimensions to $4\pi\mathbf{k}^3/3$. By combining these relationships with the fact that the energy $E = \hbar^2k^2/(2m)$, the density of states, $D(E)$ (which effectively gives the number of states at a given energy), and the integrated density of states up to a given energy E, $N(E)$, for each dimensionality can be obtained, as shown in Table 2.2 (Poole et al. 2003).

In considering carriers in semiconductor nanostructures, it then becomes necessary to combine the solutions for the particle in a box problem with the solutions for the corresponding system of reduced dimensionality. Figure 2.22 shows the number of carriers and density of states as a function of energy for each dimension of confinement. It can be seen from the density of states that for quantum dots, the confinement in three dimensions leads to the creation of truly discrete energy levels, similar to atomic energy levels. In contrast, quantum wells, with confinement in one dimension show a step-like density of states, with each step defining subbands corresponding to the discretization caused by confinement in one dimension and the flat density of states between steps due to the kinetic energy the carriers can have in the other two dimensions (as revealed by $D(E)$ for the 2D system shown in Figure 2.22). Similarly for QWrs, discrete steps defining the subbands are seen with a density of states between steps determined by the kinetic energy in a 1D system. QDs, on the other hand, show subbands consisting of truly discrete peaks, as in a 0D system there is no kinetic energy.

Also shown in Figure 2.22 is a representation of the band structure for each level of confinement. In bulk systems, the energy band changes as a function of wave vector in all three dimensions and is represented on a dispersion curve, as discussed previously. In a quantum well, the confinement in one dimension means that the corresponding wave vector is ill-defined, and the symmetry of the system is now dependent more on the confinement than the crystal lattice. As a result, the energy for each QW state varies approximately as $k_x^2 + k_y^2$, which corresponds to the kinetic energy of the carriers in the plane of the well. Quantum wires are similar; the wave vector is ill-defined in two dimensions, and the energy varies as k_x^2. Quantum dots then present the other extreme, where the wave vector is ill-defined in all three dimensions, and there is no kinetic energy, leading to atomic-like discrete energy states.

The band structures depicted in Figure 2.22 assume perfect nanostructures; however, in practice this is rarely achieved. Not only are the shapes rarely uniform, but there is often strain induced at the interfaces, caused by lattice mismatch. The effect of this strain is to

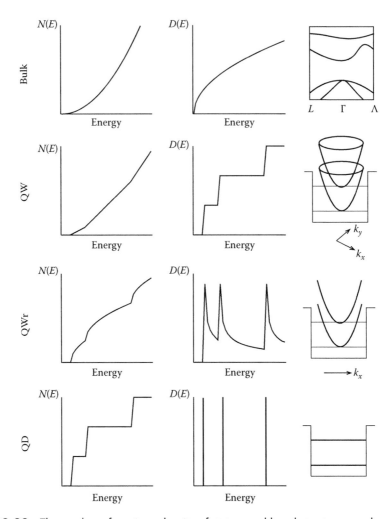

Figure 2.22 The number of carriers, density of states, and band structure are shown for bulk semiconductors, quantum wells, quantum wires, and quantum dots.

alter the band structure, meaning that the confinement potential profile no longer appears as a square well, but can be rather more complicated. The specific effect of the strain can be as varied as the size and shape of the QW, QWr, or QD. This is particularly the case for QDs, where many different shapes and orientations are possible. For a more detailed discussion of the effects of strain in QDs, we refer the reader to Yoffe (2001).

The properties of semiconductor nanostructures can then be well described by reference to a combination of the density of states and the dispersion relations or band structures depicted in Figure 2.22. In the following section, we pay particular attention to the optical properties of such nanostructures, and the implications of confinement in one, two, or three dimensions, but first we discuss briefly some of the other effects of reduced dimensionality.

One of the most important considerations for electrons and holes in bulk semiconductors is the extent and nature of their interactions with other quasiparticles. These include carrier–carrier and carrier–phonon interactions and can be responsible for altering many of the properties that make semiconductors so useful. In bulk systems, electrons and holes are free to move in all directions and interact with many other carriers and phonons. When carriers are confined, however, their ability to interact with phonons and other carriers is greatly reduced.

As a result, for example, in QDs at low temperatures and low carrier densities, very long coherence times are observed (Meier et al. 2007), a useful and necessary requirement for quantum information processing.

On the other hand, some types of interactions may actually be increased as a result of the confinement. Because the wave vector is ill-defined in one, two, or three dimensions for carriers in nanostructures, the requirement to conserve momentum can be relaxed and is only required in the unconfined directions. This can be useful, for example, in four-wave mixing experiments (Chapter 10) or detrimental due to increased scattering and reduced lifetimes.

2.3.3 Implications for Optical Properties

Most of the optical properties in bulk semiconductors are defined by the crystal structure, its symmetry, and the ensuing band structure. For semiconductor nanostructures, however, the significance of the crystal symmetry is reduced, and the shape and symmetry of the nanostructure become important. This is evidenced by the density of states and band structures depicted in Figure 2.22. As a result, the optical properties of nanostructures of different dimensionality can vary greatly. In this section, we discuss some of the general implications of confinement on the optical properties and some considerations that only become important at low dimensions.

One of the most fundamental effects of reduced dimensionality is that the exciton size is reduced. For QDs, it is clear that the size will be reduced since the wave functions are confined in three dimensions. For QWs, however, it is not so clear; the confinement perpendicular to the well compresses the wave function in this direction, which also leads to a 2D Bohr radius in the plane of the well which is smaller than the 3D value (this may seem counterintuitive, but stems from the fact that the wave function tries to conserve its spherical symmetry and is clearly revealed in the complete treatment of the problem (Haug and Koch 2004)). The size of the exciton is thereby reduced both perpendicular and parallel to the plane. An important corollary of the reduced size of electron and hole wave functions confined in type-I nanostructures is that there is greater overlap between them. This in turn means that the oscillator strength is enhanced, and brighter transitions are observed. This makes semiconductor nanostructures especially attractive for light-emitting and absorbing applications. The corollary of increased overlap and oscillator strength is that the radiative lifetime is reduced further for each dimension of confinement.

Where electrons and holes are confined in the same region (i.e., in type-I nanostructures), the exciton binding energy is, in most cases,* also increased over that in the bulk material. The increased binding energy results from the increased confinement, which implies that the carriers have fewer dimensions in which to move, and so require greater energy to overcome the Coulomb attraction between electron and hole. It follows therefore that the greater the degree of confinement, the greater the exciton binding energy. As a result, excitons tend to dominate the optical properties, even more so than for bulk semiconductors, and much of our discussion will revolve around excitons.

Another general effect of confinement is that the reduced dimensionality of nanostructures causes the wave vector in the direction of confinement to become ill-defined. This has several implications for the optical properties, particularly for processes where conservation of momentum prevents them from occurring. In particular, absorption and emission from indirect gap semiconductor materials become for more probable, and indeed there is significant

---

* This is not always the case, as, for example, when an internal electric field is present in the nanostructure, which may induce electron–hole separation.

effort around the world to develop nanostructures in silicon-based semiconductors, with the aim of being able to integrate optically active devices with current, highly evolved silicon processing technologies. Additionally, the collapse of requirements for momentum conservation also alters the interactions with phonons, which in some cases can lead to enhanced exciton–LO phonon interactions that can result in sidebands.

In addition to these very general effects of confinement, the optical spectrum of nanostructures also changes, and in the coming pages we discuss these effects for each dimension in turn.

For quantum wells, the density of states appears as a step-like function, with each step corresponding to different quantum states in the direction of confinement. One might initially expect that the absorption and emission spectrum would then closely replicate this density of states. In practice, however, much like in bulk semiconductors, emission is predominantly from the band edge due to rapid intraband relaxation. Furthermore, the excitonic transition is seen below the band edge and tends to dominate due to the greater oscillator strength, especially at low temperatures. Figure 2.23 shows the absorption spectrum for GaAs as it changes from a bulk to a 2D system. In bulk GaAs, the spectrum shows the exciton peak, followed by the featureless absorption band at higher energies. As the thickness of the GaAs layer is reduced, discrete peaks are seen corresponding to excitons at the edge of each step in the density of states. The observation of these confinement effects first occurs for layer thicknesses less than twice the Bohr radius ($a_B = 13.6\,\text{nm}$ in GaAs). As the width is reduced further, the $n = 1$ transition shifts to higher energies, as expected from Equation 2.7 (which shows that the

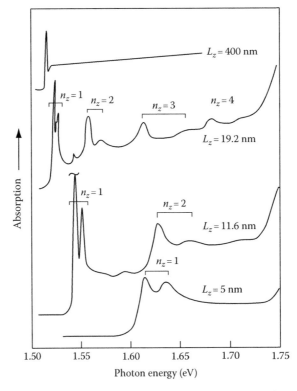

Figure 2.23 The transition from bulk GaAs to a thin GaAs QW shows the effects of the transition from bulk to quantum-confined systems (Gossard 1983, Gobel and Ploog 1990). The Bohr radius in bulk GaAs is 13.6 nm; however, this value changes as the extent of confinement changes. (Reprinted with kind permission from Springer Science+Business Media: Klingshirn, C.F., *Semiconductor Optics*, 2nd edn., Springer-Verlag, Berlin, Germany, 2007.)

energy is inversely proportional to well width), and the spacing between levels also increases. This blue shift of the $n = 1$ peak as compared to the band edge of the bulk material is often referred to as the confinement potential. The ability to change the confinement energy, and hence the transition energy, by changing the size of the nanostructures is similarly applicable to QDs and QWrs. This provides a means of tuning the transition energy without changing the material content and is one property that makes nanostructures so important and useful for optoelectronic applications.

Another noticeable feature that appears as the well width decreases is the splitting of the exciton peak into LH and HH excitons. This splitting is due to the removal of the degeneracy of the HH and LH valence bands at $\mathbf{k} = 0$, where most optical transitions occur in direct gap materials. This occurs as the symmetry of the system becomes more dependent on the nanostructure shape and less on the crystal lattice; as a result, the symmetry of the system decreases with decreasing well width, thereby increasing the splitting of the HH and LH bands.

It can also be seen in Figure 2.23 that as the well width decreases, the width of the excitonic peaks increases. This is due largely to inhomogeneous broadening, or in other words, due to variations in the thickness of the well. In a bulk semiconductor, the inhomogeneous linewidth is determined predominantly by the quantity and nature of defects, and in high-quality samples, the linewidth can be very small. In quantum wells, where the shape of the well is an important feature, fluctuations in the well width lead to fluctuations in the exciton energy, and an increase in inhomogeneous broadening. The strong dependence on well width can then be explained by considering, for example, the change in energy caused by fluctuations of 0.5 nm in a 5 nm well compared to the same size fluctuations in a 20 nm well. Clearly, the relative change to the width is much greater in the narrower well, and given that the energy is inversely proportional to well width, the absolute variation in energy (or inhomogeneous broadening) caused by the 0.5 nm fluctuations in well width is four times greater in the narrower well. The spectral width of the exciton is therefore greater for narrower wells. Broadening due to defects in QWs is also increased for narrower wells for similar reasons. The inhomogeneous linewidth seen for QWs is, therefore, greatly dependent on the quality of the sample, particularly the interfaces.

Quantum wires are confined in two dimensions, and while their size in these directions is ideally uniform, this is rarely the case in practice. Not only are there fluctuations in width, leading to inhomogeneous broadening, but the cross section of these wires is not usually circular, or even square. The extent of these fluctuations is highly dependent on the method of their growth, and can drastically alter the optical properties (Zhu et al. 2006). For example, some growth techniques will give wires that are rectangular, with confinement in one direction greater than in the other, while other techniques can give triangular cross sections, almost circular cross sections, or anything in between (Zhang et al. 2007). The specific properties of each of these types of quantum wires subsequently vary greatly, and we will not detail each specific case. Instead, we refer the reader to Zhang et al. (2007) and the references therein, and suggest that the general properties of absorption and emission spectra can be extrapolated from the properties of quantum wells and quantum dots discussed here.

Quantum dots can be made by many different methods, take many different shapes, and possess vastly different properties. Three of the most common types of semiconductor quantum dots are colloidal QDs, which tend to be close to spherical in shape and have high potential barriers; interfacial QDs, which are formed by monolayer fluctuations in QWs and tend to be flat with relatively small barriers; and Stranski–Krastanow self-assembled quantum dots, which are typically pyramidal in shape, with intermediate barrier heights. These different types of QDs possess the same general properties to a large extent, and with one exception we shall treat them as such here.

The density of states for QDs, as shown in Figure 2.22, suggests that they should exhibit narrow, discrete, atomic-like peaks in the absorption/emission spectrum. In practice, however, one typically looks at an ensemble of dots, and the observed linewidths are significantly increased due to large inhomogeneous broadening. The extent of the broadening can be explained by extension of the discussion on the broadening of QW spectral peaks. Confinement in three dimensions means that the confinement energy is high, and even the slightest change in the size or shape of the dots will lead to a large shift in energy. Much work has been done to refine growth techniques to reduce the distribution of size and shape of QDs, but inhomogeneous broadening remains a significant issue in QD ensembles, with spectral widths anywhere from 10 nm to >100 nm depending on the band gap and growth technique. If, however, one is able to study an isolated QD through, for example, the spatially resolved techniques discussed in Part IV of this book, sharp atomic-like peaks are indeed observed. In Stranski–Krastanow self-assembled QDs, the wetting layer may also contribute to the optical spectrum, acting effectively as a narrow quantum well and appearing slightly blueshifted from the QD peaks.

In quantum dots, the behavior of the system is highly dependent on the symmetry of the dot, and very little on the crystal structure, as evidenced primarily by the large inhomogeneous broadening. The degeneracy of the HH and LH bands at $\mathbf{k} = 0$ is usually lifted as a result of the asymmetry of QDs by an amount greater than that discussed for QWs. In interfacial and self-assembled QDs, which tend to be highly asymmetric, the HH–LH splitting is usually very large, to the extent that the HH band becomes the highest occupied state and dominates the optical properties.

It was discussed earlier that the allowed optical transitions from the HH band can be divided into two circularly polarized transitions, σ^+ and σ^-. It has, however, been shown that in quantum dots that are elongated in one dimension, the two degenerate HH exciton states become mixed, and a splitting between linearly polarized transitions is induced by the anisotropy, as indicated in Figure 2.24 (Tartakovskii et al. 2004). This presents a clear example of how the properties of nanostructures, and quantum dots in particular, are strongly dependent on the size and shape of the nanostructure, more so than the crystal structure.

In addition to transitions between valence band and conduction band states, the discretization and formation of subbands presents another possible type of transition: intraband or inter-subband transitions. Intraband absorption is an important process, particularly for infrared photodetectors, and typically requires free electrons in the conduction band in the initial state, which are generally provided by doping. Excitation to higher confined electron states or

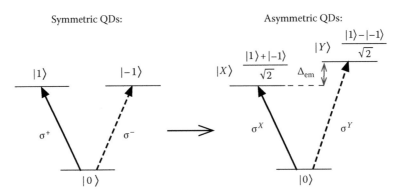

Figure 2.24 In asymmetric quantum dots, angular momentum ceases to be a good quantum number, and mixed HH exciton states are formed, split by the anisotropy energy Δ_{em}, typically up to ~20 meV. The allowed transitions then become linearly polarized.

unconfined states of the barrier is then typically in the range of 10–200 meV and can be varied by altering the size of the nanostructure. These processes can occur in QWs, QWrs, and QDs; however, the selection rules state that the electric field must be parallel to the confinement direction. For quantum dots, this is not a problem; for quantum wells, however, this precludes excitation incident normal to the QW plane, as would usually be convenient, instead requiring light incident on the edge of the structure, or bent into the QW by a diffraction grating grown on the surface. For this reason, QDs are expected to have an enhanced infrared detector performance over their QW counterparts. However, in reality, the volume of the material plays an important role in terms of infrared absorption (Kelsall et al. 2005).

To date, much of the discussion has revolved around the effects of confinement on optical absorption, and while many of these properties can be applied equally to emission, there are several processes that only become important for emission. Many of these processes depend on where the excited carriers are initially generated, both in real space and in terms of the energy states, regardless of how they got there. In general, there are three possibilities: they can be generated in the barrier; in high energy, but still confined states within the nanostructure; or in the lowest excited states of the nanostructure.

Where excited carriers are initially in the barrier region of the nanostructure, emission from the QW, QWr, or QD is dependent on the carriers relaxing into or being "captured" by the nanostructure. The dynamics of these processes depends on many different factors, including the spatial proximity to the quantum structure and carrier mobility, the energy difference between the barrier and confined states, and the availability of phonons to assist the process, among others, as discussed in Chapter 8. Under these conditions, it is also possible to capture electrons before holes or vice versa, which can further complicate the dynamics, but also reveal specific electron or hole processes.

When carriers are in higher excited levels in the nanostructure, either because they were generated there or because they were captured into these subbands from the barrier, they will typically relax to the lowest subband before recombining radiatively. This inter-subband relaxation is typically mediated by phonons, and the dynamics are often revealed through Raman spectroscopies, as discussed in Chapter 6.

In QWs and QWrs, inter-subband relaxation typically occurs rapidly, because the energy and momentum conservation requirements of the phonon-induced transitions are easily met. This can be explained by again referring to the band structures depicted in Figure 2.22. For quantum wells and wires, the transitions are truly between bands, and so there is a large number of possible energy and momentum combinations that will facilitate the inter-subband transition. As a result, there will be many phonon modes available to participate in the transition.

In QDs, however, despite the removal of any momentum conservation requirements, the subbands are actually discrete sub-states, and require phonons of exactly the right energy to facilitate the inter-subband relaxation. If there are no phonon modes available at this energy, the relaxation into the lowest subband is drastically slowed, and the subsequent emission efficiency reduced (Yoffe 2001). This phenomenon, which can be beneficial or detrimental depending on the desired application, is known as the "phonon bottleneck," and while the logic may be infallible, in practice this is not always the case. For example, other processes resulting from the non-ideality of real semiconductor quantum dots may provide alternative relaxation pathways. This is often the case in self-assembled QDs, where the necessary presence of a wetting layer provides access to a high density of nearby electronic and phonon states, and the expected lengthening of the lifetime is not observed. Experiments reporting reduced emission intensity in QDs have often been reported as evidence of a phonon bottleneck, which has in turn been refuted and attributed to other non-radiative recombination pathways, due to surface effects and/or Auger recombination (Yoffe 2001). There have also been experiments identifying the extended lifetime of excited carries in QDs predicted for the phonon bottleneck (Urayama et al. 2001). In these cases, it seems clear that a phonon bottleneck is present, suggesting that the presence or absence of this effect is highly sample dependent.

2.3.3.1 Electro- and Magneto-Optical Effects

In addition to altering optical effects, carrier confinement also alters some of the electro- and magneto-optical effects discussed previously. The most common of these effects caused by confinement is the quantum-confined Stark effect (QCSE) (Miller et al. 1984), which can be present in nanostructures of all dimensionality, but which for simplicity we will discuss in terms of a 1D quantum well.

As discussed previously (Figure 2.16), the application of an electric field induces a "tilt" on the band structure, and for a quantum well, the effects can be quite significant. There are two main effects associated with the QCSE, both of which can be identified in Figure 2.25. Firstly, the tilt imparted on the quantum well causes one side to be lower, and the other higher. As a result, the lowest level in the conduction band is lowered, and the highest state in the valence band is raised. The transition between these states is thus redshifted and in some cases beyond the transition energy of the corresponding bulk material. The other major effect is that the electron wave function is pushed to one side of the well, while the hole wave function is pushed to the other. The effect of this is to reduce the exciton binding energy, reduce the oscillator strength, and increase the radiative lifetime.

In some semiconductor nanostructures, there can be an internal electric field present without application of an external field. In these cases, the electric field is caused by the difference in spontaneous polarization and/or piezoelectric polarization at the interface of the nanostructure and the barrier. The shape of the potential in these materials is then always sloped, and QCSEs are always present, as is the case, for example, in ZnO-based quantum wells (Davis and Jagadish 2009) and GaN-based quantum wells (Leroux et al. 1998).

One effect of an applied magnetic field, discussed previously, is that by creating Landau levels, the field is able to reduce the dimensionality of a bulk semiconductor to a pseudo 1D system. This effect is enhanced in quantum wells, as it is able to reduce the dimensionality from a 2D system to a 0D system. The effect is much the same as going from quantum wells to quantum dots, with reductions in the interactions with the environment, and increases in coherence times; by varying the field strength the effect of this change can then be studied systematically (Davis et al. 2007).

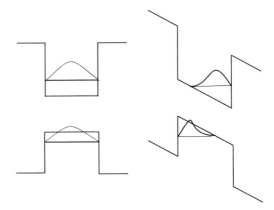

Figure 2.25   The application of an external electric field leads to the quantum-confined Stark effect, which redshifts the transition energy and pushes the electron and hole to opposite sides of the potential well.

One final effect of an applied magnetic field on nanostructures is in reference to the case of asymmetric QDs discussed earlier (i.e., where the states become mixed and the allowed transitions linearly polarized). In this case, the interaction of the carriers with the applied field is dependent on angular momentum, and the previously mixed states are forced to become pure spin states once again. As a result, the optically allowed transitions once again become circularly polarized and are split by the Zeeman effect discussed earlier.

2.4 SUMMARY

The material covered in this chapter could easily be expanded to fill an entire book, or even several volumes. Indeed, there are several excellent references devoted entirely to semiconductor physics, nanostructures, and their optical properties. As a result, this chapter can be neither fully inclusive nor include all the details of this subject. In this chapter on semiconductors and their nanostructures, we do however hope to have provided a brief introduction to the many and diverse properties of semiconductors as a basis for understanding and interpreting the data from the experimental techniques described in the subsequent chapters of this book. For those seeking further information, we refer the reader to the many excellent references mentioned throughout this chapter.

REFERENCES

Allison, J. 1971. *Electronic Engineering Materials and Devices*. London U.K.: McGraw Hill.

Ashcroft, N.W. and Mermin, N.D. 1976. *Solid State Physics*. New York: Thomson Learning.

Atatüre, M., Dreiser, J., Badolato, A., and Imamoglu, A. 2007. Observation of Faraday rotation from a single confined spin. *Nature Physics*. 3: 101.

Bauer, E. 1958. Phaenomenologische theorie der kristallabscheidung an oberflaechen I. *Zeitschrift für Kristallographie*. 110: 372–394.

Benedek, G.B. and Fritsch, K. 1966. Brillouin scattering in cubic crystals. *Physical Review*. 149: 647.

Borri, P., Langbein, W., Schneider, S. et al. Ultralong dephasing time in InGaAs quantum dots. *Physical Review Letters*. 87: 157401.

Cardona, M. 1969. *Modulation Spectroscopy*. New York: Academic Press.

Chelikowsky, J.R. and Cohen, M.L. 1974. Electronic structure of silicon. *Physical Review B*. 10: 5095.

Cohen, M.L. and Bergstresser, T.K. 1966. Band structures and pseudopotential form factors for fourteen semiconductors of the diamond and zinc-blende structures. *Physical Review*. 141: 789.

Davis, J.A., Wathen, J.J., Blanchet, V., and Phillips, R.T. 2007. Time-resolved four-wave mixing spectroscopy of excitons in a single quantum well. *Physical Review B*. 75: 035317.

Davis, J.A. and Jagadish, C. 2009. Ultrafast spectroscopy of ZnO/ZnMgO quantum wells. *Laser and Photonics Review*. 3: 85.

Fan, H.Y. 1967. Effects of free carriers on the optical properties. *Optical Properties on III-V Compounds: Semiconductors and Semimetals*. New York: Academic Press.

Faraday, M. 1839. *Experimental Researches in Electricity*. London, U.K.: Taylor & Francis.

Fehrenbach, G.W., Schafer, W., and Ulbrich, R.G. 1985. Excitonic versus plasma screening in highly excited gallium arsenide. *Journal of Luminescence*. 30: 154.

Fox, A.M. 2001. *Optical Properties of Solids*. New York: Oxford University Press.

Gobel, E.O. and Ploog, K. 1990. Fabrication and optical properties of semiconductor quantum wells and superlattices. *Progress in Quantum Electronics*. 14: 289.

Gossard, A.C. 1983. *Thin Films, Preparation and Properties*. New York: Academic Press.

Gupta, J.A., Awschalom, D.D., Peng, X., and Alivisatos, A.P. 1999. Spin coherence in semiconductor quantum dots. *Physical Review B*. 59: R10421.

Haug, H. and Koch, S.W. 2004. *Quantum Theory of the Optical and Electronic Properties of Semiconductors*, 4th edn. Singapore: World Scientific Publishing.

Herman, M.A., Richter, W., and Sitter, H. 2004. *Epitaxy: Physical Foundation and Technical Implementation*. Berlin, Germany: Springer.

Jagadish, C. and Pearton, S.J. 2006. *Zinc Oxide Bulk, Thin Films and Nanostructures*. Oxford, U.K.: Elsevier.

Kato, Y.K., Myers, R.C., Gossard, A.C., and Awschalom, D.D. 2004. Observation of the spin hall effect in semiconductors. *Science*. 306: 1910.

Kelsall, R.W., Hamley, I.W., and Geoghegan, M. 2005. *Nanoscale Science and Technology*. Chichester, U.K.: John Wiley & Sons.

Kikkawa, J.M. and Awschalom, D.D. 1998. Resonant spin amplification in n-type GaAs. *Physical Review Letters*. 80: 4313.

Kittel, C. 2005. *Solid State Physics*, 8th edn. Hoboken, NJ: John Wiley & Sons.

Klingshirn, C.F. 2007. *Semiconductor Optics*, 2nd edn. Berlin, Germany: Springer-Verlag.

Kovalev, D., Averboukh, B., Volm, D., Meyer, B.K., Amano, H., and Akasaki, I. 1996. Free exciton emission in GaN. *Physical Review B*. 54: 2518.

Lannoo, M. and Bourgoin, J. 1981. *Point Defects in Semiconductors I, Theoretical Aspects*, Springer Series on Solid-State Science, Vol. 22. Berlin, Germany: Springer.

Leroux, M., Grandjean, N., Laugt, M. et al. 1998. Quantum confined stark effect due to built-in internal polarization fields in (Al,Ga)N/GaN quantum wells. *Physical Review B*. 58: 13371.

Madelung, O., Rössler, U., and M. Schulz. 2002. *Group IV Elements, IV-IV and III-V Compounds. Part B—Electronic, Transport, Optical and Other Properties*. Berlin, Germany: Springer-Verlag.

Meier, T., Thomas, P., and Koch, S.W. 2007. *Coherent Semiconductor Optics*. Berlin, Germany: Springer-Verlag.

Miller, D.A.B., Chemla, D.S., Damen,T.C. et al. 1984. Band-edge electroabsorption in quantum well structures: The quantum-confined stark effect. *Physical Review Letters*. 53: 2173.

Pantelides, S. 1978. The electronic structure of impurity and defect states in semiconductors. *Reviews of Modern Physics*. 50: 797.

Pantelides, S. 1986. *Deep Centre in Semiconductors, A State of the Art Approach*. New York: Gordon and Breach.

Poole, C.P. Jr. and Owens, F.J. 2003. *Introduction to Nanotechnology*. Hoboken, NJ: John Wiley & Sons.

Prasankumar, R.P., Upadhya, P.C., and Taylor, A.J. 2009. Ultrafast carrier dynamics in semiconductor nanowires. *Physics Status Solidi (B)*. 246: 1973.

Reshchikov, M.A. and H. Morkoc. 2005. Luminescence properties of defects in nanostructures. *Journal of Applied Physics*. 97: 061301.

Ridley, B.K. 1997. *Electrons and Phonons in Semiconductor Multilayers*. Cambridge, U.K.: Cambridge University Press.

Savrasov, B. 1996. Linear-response theory and lattice dynamics: A muffin-tin-orbital approach. *Physical Review B*. 54: 16470.

Seeger, K. 2004. *Semiconductor Physics: An Introduction*, 9th edn. Vienna, Austria: Springer-Verlag.

Shah, J. 1999. *Ultrafast Spectroscopy of Semiconductors and Semiconductor Nanostructures*, 2nd edn. Berlin, Germany: Springer-Verlag.

Smith, D.L., Pan, D.S., and McGill, T.C. 1975. Impact ionization of excitons in Ge and Si. *Physical Review B*. 12: 4360.

Smith, R.A. 1964. *Semiconductors*. Cambridge, U.K.: Cambridge University Press.

Stranski, I.N. and Krastanow, L. 1939. Abhandlungen der mathematisch-naturwissenschaftlichen klasse. *Akademie der Wissenschaften und der Literatur in Mainz*. 146: 797.

Sze, S.M. and Kwok, K. Ng. 2007. *Physics of Semiconductor Devices*. Hoboken, NJ: John Wiley & Sons.

Tartakovskii, A.I., Cahill, J., Makhonin, M.N. et al. 2004. Dynamics of coherent and incoherent spin polarizations in ensembles of quantum dots. *Physical Review Letters*. 93: 057401.

Timusk, T. 1976. Far-infrared absorption study of exciton ionization in germanium. *Physical Review B*. 13: 3511.

Tsao, J.Y. 1993. *Materials Fundamentals of Molecular Beam Epitaxy*. San Diego, CA: Academic Press.

Toyozawa, Y. 1959. On the dynamical behavior of an exciton. *Progress of Theoretical Physics, Supplement* 12: 111.

Urayama, J., Norris, T.B., Singh, J., and Bhattacharya, P. 2001. Observation of phonon bottleneck in quantum dot electronic relaxation. *Physical Review Letters*. 86: 4930.

Uchinokura, K., Sekine, T., and Matsuura, E. 1974. Critical-point analysis of the two-phonon Raman spectrum of silicon. *Journal of Physics and Chemistry of Solids*. 35:171.

Watts, R.K. 1977. *Point Defects in Crystals*. New York: Wiley-Interscience.

Yoffe, A.D. 2001. Semiconductor quantum dots and related systems: Electronic, optical, luminescence and related properties of low dimensional systems. *Advances in Physics*. 50: 1–208.

Yu, P.Y. and Cardona, M. 2005. *Fundamentals of Semiconductors*, 3rd edn. Berlin, Germany: Springer-Verlag.

Zhang, L., Fang, X., and Ye, C. 2007. *Controlled Growth of Nanomaterials*. Singapore: World Scientific.

Zhu, Q., Pelucchi, E., Dalessi, S., Leifer, K., Dupertuis M.A., and Kapon, E. 2006. Alloy segregation, quantum confinement and carrier capture in self-ordered pyramidal quantum wires. *Nano Letters* 6: 1036.

Chapter 3

THE OPTICAL PROPERTIES OF METALS: FROM WIDEBAND TO NARROWBAND MATERIALS

Richard D. Averitt

CONTENTS

3.1 INTRODUCTION

Metals are ubiquitous and play an important role in society including structure, beauty, and function. While semiconductors have played a leading role in the transformation of society toward ever increasing technological sophistication, without metals and their generally large conductivities, the information age would hardly have been possible.

In the realm of optics, metals have played a transformative role during the past century in terms of enabling the generation, detection, and manipulation of electromagnetic radiation. Their high reflectivity over a broad swath of the electromagnetic spectrum—spanning from radio frequencies through the visible—is, in combination with techniques to form high-quality surfaces and structures, the primary reason for the importance of metals in optics.

Generally, when students think of high-conductivity metals, Al, Cu, Ag, and Au come to mind. These are indeed excellent metals as judged by the metric of the magnitude of the DC conductivity, with room temperature values on the order of 5×10^6 $(\Omega\,cm)^{-1}$ (Johnson and Christy 1972, Ashcroft and Mermin 1976, Burns 1985, Marder 2000). The same physics that leads to the high conductivity of these metals leads to their excellent optical properties and, in simple terms, can be summed up in terms of a large number of electrons with a comparatively small scattering rate. The phenomenological Drude model, as described in the following, embodies these simple yet important characteristics. Nonetheless, even amongst these "good" metals, there are differences in their optical properties. This is, in large part, because of the details of the band structure, which in turn derives from the orbital character of the bands that cross the Fermi energy E_f and those that are within a photon energy $h\nu$ (where ν is the photon frequency and h is Planck's constant) of E_f, with $0.001 < h\nu < 10\,\text{eV}$ as an approximate range of interest for "optical" spectroscopy.

The term "good" metal necessarily implies that there are "poor" or even "bad" metals with regard to the magnitude of the DC conductivity. Further, to the extent that good metals can be understood (or at least characterized) in terms of their Drude response and band structure and corresponding optical response, we might expect that the properties of poor or bad metals to manifest in the optical response in a manner that can be semiquantitatively compared with good metals. This is indeed the case and is, at an elementary level, the topic of this chapter. Namely, good metals such as Au, Al, Ag, etc., serve as an excellent starting point toward developing insight into other metals; and optical spectroscopy, broadly defined, is a powerful experimental tool to carry out fundamental investigations of all types of metals.

The terms "poor" or "bad" in reference to the characteristics of a metal are, of course, colloquial and should not be equated with noninteresting or useless. In fact, from a contemporary view, such metals are of great intellectual and technological interest. A better term is perhaps complex metal, though the term correlated metal, or more generally, correlated electron material, is in common usage for certain classes of materials (Imada et al. 1998, Degiorgi 1999, Tsuda et al. 2000, Dagotto 2003). Let us consider a couple of examples to make the discussion more concrete. In comparison to silver, lead would be considered a poor metal in that its room-temperature DC conductivity is approximately 12 times smaller than silver. This is because the electrons in lead interact more strongly with the lattice, leading to enhanced scattering. However, this strong electron–lattice coupling, while limiting the room temperature conductivity, is what allows Pb to superconduct at ~7 K. In contrast, Ag and Au are not superconducting, though Al does become a superconductor at 1.2 K. Another example is iron, which has a DC conductivity that is merely six times smaller than Ag. While this value of the conductivity makes Fe a poor choice as a wire, Fe is very interesting in that it is ferromagnetic below 1043 K. These facts are not unrelated, and evidently there is a correlation where, as a function of decreasing conductivity, there is an increase in complexity of the phenomena observed in metals.

Continuing along the path from good to bad metals, we mention transition metal oxides (TMOs) where the DC conductivity (in the metallic state) can be more than a factor of 100 smaller than a metal such as Ag. Many TMOs including the manganites, cuprates, and vanadates exhibit complex phenomena such as colossal magnetoresistance, superconductivity, and metal–insulator transitions in response to external stimuli such as temperature, magnetic field, or pressure (Tokura 2000, Dagotto 2003). The hallmark of these correlated electron materials is a competition between the charge, lattice, orbital, and spin degrees of freedom leading to extreme susceptibility to small perturbations. The microscopic origin of many of these phenomena remains unexplained, with high-temperature superconductivity as perhaps the most famous unsolved problem in condensed matter physics. The optical spectroscopy of "bad" metals is amazingly rich, where very dramatic changes (e.g., transfer of spectral weight) can manifest in comparison to other metals resulting, primarily, from on-site Coulomb repulsion (Imada et al. 1998, Degiorgi 1999, Cooper 2001, Dressel and Grüner 2002, Basov and Timusk 2005).

This highly simplified discussion might then suggest that we have a decent understanding of good metals (we do), and, therefore, it is actually the bad and poor metals that provide the greatest intellectual challenge and thus deserve our primary attention. There is some truth to this, but we again note that in developing an understanding of complex or correlated metals, simple metals are a starting point. Of equal importance is that highly conductive metals such as Ag and Au continue to be of extreme interest in optical spectroscopy. This is because these are the metals of choice for plasmonics and metamaterials (Krebig and Vollmer 1995, Barnes et al. 2003, Maier 2007, Solymar and Shamonina 2009). Plasmonics and metamaterials research aims to control and manipulate light on a subwavelength scale using metallic structures that have been patterned using either top-down or bottom-up fabrication techniques. The functionality of these metallic structures is based upon the resonant response of the conduction electrons to an incident electromagnetic field. Thus, low conductivity metals (and their associated high loss) are not suitable as the main constituents of metamaterial or plasmonic devices (though the judicious combination of "good" and "bad" metals offers exciting opportunities—see Driscoll et al. 2009 for a recent example).

Again, the goal of this chapter is to provide some simple insights into the optical spectroscopy of metals using the Drude model and interband transitions to guide the discussion. Hopefully, this will provide some background for some of the other chapters and motivate the reader to investigate topics of greater interest in detail using the references cited. It will be useful to keep the following concept in mind as we proceed through our discussion. The itinerancy of an electron, or its kinetic energy, is directly related to the width of the band in which it resides. For example, it can be shown in a tight-binding model that $W = zt$, where W is the bandwidth, z is the number of nearest neighbors, and t is the overlap integral (Ashcroft and Mermin 1976, Harrison 1980, Marder 2000). Thus, the best metals have the widest bands, which derive from s-orbitals and can strongly overlap in the solid state. This is because s-orbitals, in comparison to d-orbitals, extend further from the nucleus leading to greater orbital overlap. The larger the overlap integral, the easier it is for an electron to hop from site to site in a lattice. In other words, wideband metals are the "good" metals, where the carriers have a large kinetic energy, making this the dominant energy scale. In contrast, electrons in a narrow band that is typically derived from d-orbitals have a smaller kinetic energy. If the kinetic energy is small enough (~1 eV) then exchange and on-site Coulomb repulsion become essential or even dominant ingredients in understanding the properties of a material. A simple schematic in Figure 3.1 highlights this simple point-of-view, which will guide the presentation.

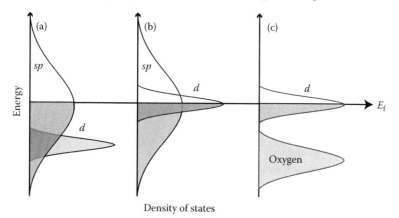

Figure 3.1 Schematic of DOS as a function of energy for various metals with increasing complexity, going from left to right. (a) In typical "good" metals, the conductivity is due to itinerant carriers in wide sp-derived bands. There may also be filled bands of d-orbital character within a few electron volts of E_f. (b) In "poor" transition metals, electrons still occupy the s bands, but now the d-bands may also cross E_f. The narrower bandwidth coupled with the higher DOS of d-band electrons increases the complexity of the metal due to enhanced exchange and can result, for example, in ferromagnetism. (c) In TMOs or "bad" metals, the electronegativity of oxygen can lead to the "disappearance" of s-like electrons from E_f (e.g., they move to lower energies and gain O $2p$ character). The physics is dominated by narrow d-bands and Coulomb repulsion results in strong interaction and competition between the spin, lattice, orbital, and charge degrees of freedom. This can lead to enormously rich physics, including metal-to-insulator transitions due to the competition between itinerancy and localization.

This chapter will proceed as has the introductory discussion, proceeding from good metals to bad metals. The following section will introduce the Drude model in terms of the conductivity, reflectivity, and dielectric response and indicate how interband transitions lead to deviations from the canonical Drude response. Simple examples showing how high-conductivity metals yield a useful plasmonic response and what ultrafast optical spectroscopy probes in good metals will be presented. Then, we will briefly discuss transition metals with an emphasis on magnetism. Namely, why and how does magnetism manifest in the optical properties of transition metals? This will be followed with a brief introduction to the very active field of femtomagnetism (Chapter 13 contains a more detailed discussion). In the final section, we will consider some simple examples of TMOs and how these d-band metals can exhibit dramatic changes in their optical response as a function of the local environment. The vanadates and the manganites will be presented as examples followed by a simple discussion of the electromagnetic response of superconductors.

3.2 WIDEBAND METALS

The simplest description of free carriers in metals (or doped semiconductors, etc.) in response to an applied electric field is given by the Drude model (derived in Chapter 1). In this model, carriers elastically scatter with a rate τ^{-1}. The complex Drude conductivity is given as follows:

$$\sigma(\omega) = \frac{\sigma_{DC}}{1 - i\omega\tau} \tag{3.1}$$

where the DC conductivity σ_{DC} is given by

$$\sigma_{DC} \equiv \frac{ne^2\tau}{m^*} = ne\mu = \varepsilon_0\omega_p^2\tau \tag{3.2}$$

In this equation
  n is the carrier density
  τ is the collision time
  m^* is the effective mass
  μ is the mobility
  ε_0 is the free space permittivity
  $\omega_p = ne^2/\varepsilon_0 m$ is the bare plasma frequency

While the Drude response can be derived from very simple considerations, it is important to emphasize that its validity is much broader. In fact, the Drude response can be derived from the semiclassical Boltzmann transport equation (Chapter 1) and even more rigorously from Kubo–Greenwood formalism. For example, using a Boltzmann approach, it is easy to show that

$$\sigma_{DC} = \frac{e^2}{12\pi^3\hbar}\langle\lambda_F\rangle S_F \tag{3.3}$$

where
  $\langle\lambda_F\rangle$ is the mean free path averaged over the Fermi surface
  S_F is the area of the Fermi surface (Wooten 1972)

For a spherical Fermi surface this is equivalent to the Drude result, highlighting that the relevant electrons are those at the Fermi surface with velocity v_F.

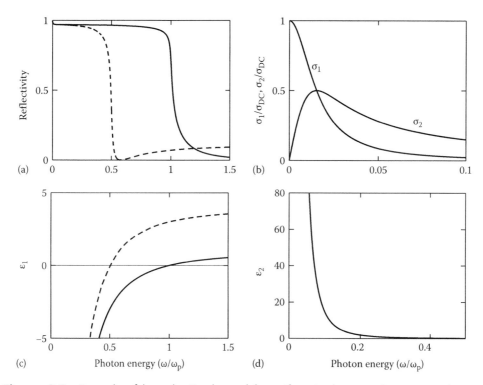

Figure 3.2 Example of how the Drude model manifests in the optical response, showing (a) the reflectivity, (b) the conductivity, (c) the real part of the dielectric function, and (d) the imaginary part. For this model calculation, the values typical of metals were used: $\hbar\omega_p = 9.0\,\text{eV}$, $t = 5\,\text{fs}$. The x-axis is normalized by the unscreened plasma frequency (i.e., 9.0 eV). The dashed lines in (a) and (c) show what happens with ε_∞ greater than one ($\varepsilon_\infty = 4$).

It is important to understand how the Drude response appears—as a function of frequency—in the optical response of metals. We will start with a model calculation taking $\hbar\omega_p = 9.0\,\text{eV}$ and $\tau = 5\,\text{fs}$, which are typical values for a good metal. Figure 3.2a shows the reflectivity, which is often the experimental quantity that is measured. Below the plasma frequency, the reflectivity approaches unity. In the vicinity of $\hbar\omega_p$, the reflectivity rapidly drops to zero as the electrons no longer respond to the incident electromagnetic field (we will return to the dashed line in Figure 3.2a and c momentarily). Figure 3.2b shows the corresponding complex conductivity $\sigma(\omega) = \sigma_1(\omega) + i\sigma_2(\omega)$ (normalized by σ_{DC}). At low frequencies, $\sigma_1(\omega)$ tends to σ_{DC} while $\sigma_2(\omega)$ approaches zero. Importantly, $\sigma_2(\omega)$ increases and crosses $\sigma_1(\omega)$ at the frequency $1/\tau$ (at $1/\tau$, $\sigma_1(\omega) = \sigma_2(\omega) = \sigma_{DC}/2$), providing an intuitive measure of the scattering rate. We also note, as is evident in Figure 3.2b, all of the frequency variation in the conductivity occurs well below the plasma frequency.

Quite often, it is more useful to consider the complex dielectric response, $\varepsilon(\omega) = \varepsilon_1(\omega) + i\varepsilon_2(\omega)$, which can be written in terms of the conductivity as $\varepsilon(\omega) = \varepsilon_\infty + i\sigma(\omega)/\varepsilon_0\omega$ (Chapter 1). Looking at Figure 3.2c we see that $\varepsilon_1(\omega)$ (solid line, taking $\varepsilon_\infty = 1$; note: ε_∞ will be defined in the following paragraph) is negative below $\hbar\omega_p$ and positive above, crossing zero at the plasma frequency. In Figure 3.2d, $\varepsilon_2(\omega)$ shows a dramatic increase with decreasing frequency.

Figure 3.2 thus summarizes how the Drude response manifests in the optical response. Of course, this completely neglects interband transitions. Interband transitions arise from the photoexcitation of electrons from one band to another. Thus, energy is absorbed from the incident light field. This absorptive response appears in $\varepsilon_2(\omega)$. An interband response would thus show up as a Lorentzian-like response yielding nonzero values of $\varepsilon_2(\omega)$ at higher frequencies. There is a corresponding dispersive component as necessitated by causality that gives a contribution to $\varepsilon_1(\omega)$. Away from the interband resonance, this yields a constant offset. Thus, the totality of the interband transitions is the origin of ε_∞. Typically, values of ε_∞ range from ~1 to 6 in various metals. This might seem to be of little consequence given that, well below $\hbar\omega_p$, the magnitudes of $\varepsilon_1(\omega)$ and $\varepsilon_2(\omega)$ are much larger than ε_∞. However, ε_∞ does have pronounced effects. As shown by the dashed line in Figure 3.2c, an increase in ε_∞ (from 1 to 4) shifts where $\varepsilon_1(\omega)$ crosses zero to lower frequencies. Correspondingly, this causes the roll-off in the reflectivity (see the dashed line in Figure 3.2a) to occur at lower frequencies and may be thought of as renormalizing the plasma frequency ($\omega_p = \rightarrow \omega_p/(\varepsilon_\infty)^{1/2}$). Note also that as the frequency increases, the reflectivity has a pronounced dip and then increases up to the value $(1 - (\varepsilon_\infty)^{1/2})^2/(1 + (\varepsilon_\infty)^{1/2})^2$. If $\varepsilon_\infty = 1$, then the reflectivity simply monotonically decreases to zero. Finally, we mention that ε_∞ plays an important role in determining the plasmonic properties of a given metal precisely because of the renormalization of ω_p.

Of course, these are just model calculations. A very reasonable question is "how well does the Drude model describe the optical response of good metals such as Au, Ag, and Al?" The answer is that it does a good job and the Drude response is clearly observable even in the presence of interband transitions. Figure 3.3 shows the experimentally measured reflectivity as a function of photon energy for Al, Ag, and Au (Johnson and Christy 1972, Rakic 1995). Aluminum is an interesting case. It has a fairly flat reflectivity out to ~16 eV, corresponding to the roll-off in the Drude response (i.e., $\omega_p = 16$ eV in Al). The monotonic decrease suggests that $\varepsilon_\infty \approx 1$. However, there is a slight dip at approximately 1.5 eV. This is actually due to a weak interband transition. Nonetheless, Drude-like behavior dominates over a broad spectral range. The data for Ag also displays a clear Drude-like response with a high reflectivity, which drops off dramatically at ~4.0 eV. For Ag, $\hbar\omega_p = 9.0$ eV, but the roll-off occurs at 4.0 eV

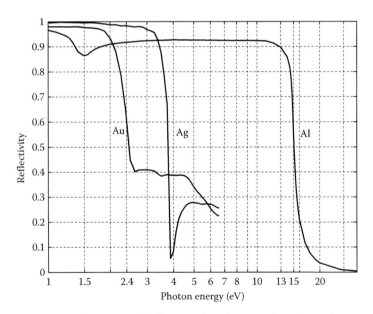

Figure 3.3 Experimentally measured reflectivity for Al, Au, and Ag. (Data for Au and Ag from Johnson, P.B. and Christy, R.W., *Phys. Rev. B*, 6, 4370, 1972, and data for Al from Rakic, A. D., *Appl. Opt.*, 34, 4755, 1995.)

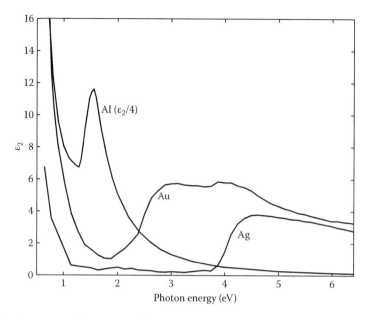

Figure 3.4 Experimentally measured $\varepsilon_2(\omega)$. (Data for Au and Ag from Johnson, P.B. and Christy, R.W., *Phys. Rev. B, 6,* 4370, 1972 and data for Al from Rakic, A. D., *Appl. Opt., 34,* 4755, 1995.)

because of the renormalization effects of the interband transitions. This is even more dramatic for Au where the Drude-like reflectivity is interrupted by the onset of interband transitions. Similar to Ag, in Au $\hbar\omega_p = 8.7\,\text{eV}$, but the roll-off in the reflectivity occurs at a considerably lower photon energy (i.e., ~2.4 eV). Below $\hbar\omega_p$, Ag exhibits a higher reflectivity than Au because of a smaller scattering rate. The lower value in Al is primarily because of absorption due to the aforementioned interband transitions.

The onset of interband transitions is more clearly observed in the spectral dependence of $\varepsilon_2(\omega)$, which is presented in Figure 3.4. For Al, $\varepsilon_2(\omega)$ displays a clear Lorentzian-like response (Chapter 1) at 1.5 eV on top of the Drude response while Au and Ag show the onset of interband transitions as more clearly separated from the Drude roll-off in $\varepsilon_2(\omega)$. Clearly, these results show the Drude model, in combination with a simple picture of interband transitions, is a useful starting point to understand the optical response of good metals. Of course, the details of the optical properties of these good metals have been investigated in much greater detail than we can provide in this introduction, and we refer the reader to some of the literature for more details (Johnson and Christy 1972, Wooten 1972, Rakic 1995, Dressel and Grüner 2002).

The examples of good metals we have presented show that optical spectroscopy can provide insight into the fundamental optical and electronic properties of metals. That is, response is embodied in $\varepsilon(\omega)$, and, therefore, measurements of the dielectric function allow, in turn, the calculation of the transmission, reflectivity, and other quantities of interest. But, is that it? The answer is no because having $\varepsilon(\omega)$ in hand, along with insights from theory, is the starting point for further quantitative investigations. We will briefly provide two examples to highlight this point of view. First we will consider plasmonics, where $\varepsilon(\omega)$ is crucial for accurate calculations of the response of surface plasmon polaritons (we shall just use the term surface plasmons [SPs]), including propagating two-dimensional plasmons at a metal dielectric interface and particle plasmons. The second example we will briefly introduce is femtosecond optical spectroscopy, which is a form of modulation spectroscopy (Cardona 1969) (Part III), with the specific advantage that it provides sub-100 fs temporal resolution, which is sufficient to probe the dynamics of electrons and temporally resolve how they shed excess energy through the emission of phonons.

Plasmonics is an active field of research with the goal of manipulating and controlling electromagnetic radiation on a subwavelength scale using surface plasmon polaritons. For visible light, this corresponds to the nanoscale, and thus developments in nanotechnology and in plasmonics go hand-in-hand. There are several recent books on plasmonics to which we refer the reader (Krebig and Vollmer 1995, Maier 2007, Solymar and Shamonina 2009). The success of plasmonics derives from the properties of good metals; Au and Ag are thus the canonical metals for plasmonics research because of their low losses in comparison to other metals. Accurate calculations require detailed knowledge of the bulk $\varepsilon(\omega)$ used to fabricate the plasmonic structure, though we note that in nanoscale composites surface scattering provides additional loss that must be taken into account with a phenomenological approach. We will not go into these details and instead focus on the basic ideas, using the bulk value of $\varepsilon(\omega)$ in our examples.

The basic idea underlying plasmonics is the concept of dielectric confinement where, at an interface, $\varepsilon_1(\omega)$ changes sign. Recall from our previous discussion that, for metals, $\varepsilon_1(\omega)$ is negative below the renormalized plasma frequency, meaning that metal/dielectric interfaces exhibit dielectric confinement (because ε_1 is positive in a dielectric). The physics of this is that in the presence of an electric field (with a component normal to the interface), the electrons in the metal will be driven toward the dielectric. A restoring force from the positive ions then pulls the electrons back, and this induced surface polarization leads to a collective oscillation of the electrons, which is the surface plasmon polariton. This basic idea holds whether we are considering SPs associated with planar two-dimensional metal/dielectric interfaces or metal nanoparticles in a dielectric environment.

Let us briefly consider SPs at a planar metal/dielectric interface. Determining the electromagnetic response is essentially a boundary value problem using Maxwell's equations as demonstrated in many texts (Krebig and Vollmer 1995, Maier 2007, Solymar and Shamonina 2009). A common setup to generate SPs is depicted in Figure 3.5a. To obtain a component of the electric field normal to the surface, p-polarized light is utilized. In addition, to generate SPs with light, momentum matching must be achieved. The problem is that, for a given photon energy, light has a smaller momentum than a surface plasmon. To achieve momentum matching between the incident light and the surface plasmon, a prism can be utilized (dielectric constant ε_p). A thin metal film (e.g., ~50 nm) with dielectric function $\varepsilon_m(\omega)$ is deposited on the surface of this prism, with a dielectric layer ε_d deposited on top of the metal. The reflectivity of good metals is quite high (Figure 3.3), but at an angle Θ_{sp}, efficient plasmon generation will yield a dramatic reduction of the reflectivity. The SPs at the interface are of evanescent character, with the electric field strength decaying exponentially away from the surface and propagating along the interface where the electric field has transverse and longitudinal components, as depicted in Figure 3.5a. From the solution of Maxwell's equations and using the appropriate constitutive relations, numerous quantities can be calculated, including Θ_{sp}, the evanescent decay length, the in-plane propagation length, and the local field enhancement (Maier 2007, Solymar and Shamonina 2009). All of these depend on $\varepsilon_m(\omega)$ and as such vary as a function of the incident photon energy. For example, we have

$$\Theta_{sp} = \sin^{-1}\sqrt{\frac{\varepsilon_m \varepsilon_d}{(\varepsilon_m + \varepsilon_d)\varepsilon_p}} \tag{3.4}$$

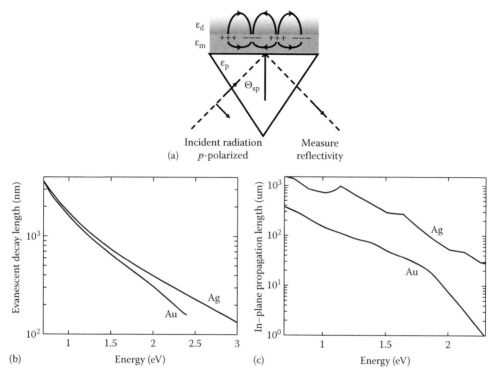

Figure 3.5 (a) Surface plasmon schematic. (b) Evanescent decay length and (c) in-plane plasmon propagation distance for Ag and Au using experimentally measured values for the metal dielectric function. This glass prism is assumed to be SF11 glass, and the dielectric above the metal film is $\varepsilon_d = 1$.

for the plasmon resonance angle and

$$k_{sp} = \frac{2\pi}{\lambda} \sqrt{\frac{\varepsilon_m \varepsilon_d}{(\varepsilon_m + \varepsilon_d)}} \tag{3.5}$$

for the surface plasmon wave vector k_{sp} (λ is the free space wavelength of the incident radiation). As mentioned, these vary as a function of frequency because of $\varepsilon_m(\omega)$. Of course, any frequency dependence of ε_d and ε_p will also show up in the surface plasmon response. In particular, SPs are very sensitive to the local dielectric environment (ε_d) and have been developed for biosensing applications. Figure 3.5b and c shows calculations using the bulk values of the dielectric functions of Ag and Au discussed earlier. Figure 3.5b plots the evanescent decay length (normal to the interface). The field is confined within a few hundred nanometers of the surface. The confinement is subwavelength, on the order of ~100 nm in the visible and increasing for lower photon energies. Figure 3.5c shows the in-plane decay length ($1/(2\mathrm{Im}(k_{sp}))$), which is the characteristic length scale over which plasmons can propagate. The lower the losses (as determined by $\varepsilon_2(\omega)$), the longer the propagation length, which is why Ag and Au are the best surface plasmon metals (Figure 3.4). As Figure 3.5c indicates, Ag is generally better than Au in accordance with its lower losses.

Metal nanoparticles also display a pronounced surface plasmon resonance that leads to a dramatic resonant absorption. As with planar SPs, this arises because of dielectric confinement and the electromagnetic response can be determined from Maxwell's equations.

A complete solution can be found for a sphere of arbitrary size using Mie scattering theory (Krebig and Vollmer 1995). However, in the limit of sufficiently small particles (diameter less than ~50 nm in the visible) the dominant response is dipolar absorption, which can be determined in the quasistatic limit by solving Poisson's equation (Krebig and Vollmer 1995). Since we are in the dipole absorption limit, one can consider each nanoparticle as having a dipole moment p, which can be polarized in response to an electric field E, giving $p = \alpha E$, where α is the polarizability (Chapter 1). This yields for the polarizability

$$\alpha = 4\pi r^3 \varepsilon_0 \frac{\varepsilon_m - \varepsilon_d}{\varepsilon_m + 2\varepsilon_d} \tag{3.6}$$

where r is the radius (ε_0 is the free space permittivity). Thus, the polarizability is proportional to the particle volume and a resonance occurs when $\varepsilon_m(\omega) = -2\varepsilon_d$, which is a manifestation of the dielectric confinement alluded to earlier. This resonance condition is obtained by looking at the denominator of Equation 3.6. It is easy to show that the surface plasmon resonance frequency occurs at $\omega_{sp} = \omega_p/(\varepsilon_\infty + 2)^{1/2}$ (Krebig and Vollmer 1995). Figure 3.6 shows calculations of the absorption cross section ($\sigma_{abs} = (k/\varepsilon_0)\mathrm{Im}(\alpha)$) for Au and Ag using the experimentally measured bulk values of $\varepsilon(\omega)$. A very strong resonance is observed for Ag at approximately 370 nm. Gold exhibits a pronounced resonance at ~520 nm, which is a factor of ~20 smaller in magnitude and substantially broader than that for Ag, which is because Au is a more lossy metal (Figure 3.4). A dilute solution of Ag nanoparticles appears yellow, while a solution of Au nanoparticles is ruby red colored due to their resonant absorption of green light. A more accurate calculation would take into account damping by surface scattering and inhomogeneous broadening resulting from the particle size distribution. These effects would reduce σ_{abs} and lead to spectral broadening, respectively. However, these simple calculations provide reasonable agreement with experiments and highlight the importance of understanding $\varepsilon(\omega)$ and having available high-quality experimental measurements of this quantity.

As a second example, we consider ultrafast optical spectroscopy. In short, initial experiments on good metals—in particular, Au—have played an extremely important role in the

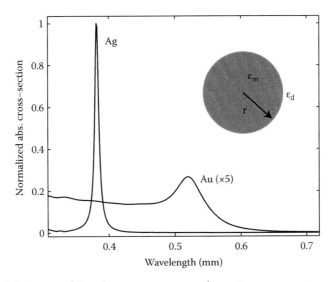

Figure 3.6 Calculation of the plasmon resonance absorption cross section σ_{abs} for Ag and Au nanoparticles using the bulk values of $\varepsilon(\omega)$, taking $\varepsilon_d = 1.78$ (corresponding to water) and $r = 15$ nm. The magnitudes have been normalized to the peak value of the Ag nanoparticle σ_{abs}.

development of femtosecond spectroscopy as an experimental tool to investigate quasiparticle dynamics in condensed matter (Chapter 8). These initial experiments focused on electron–lattice coupling, where the general idea is that ultrafast optical spectroscopy provides the possibility to temporally resolve phenomena at the fundamental timescales of atomic and electronic motion (Eesley 1983, Fujimoto et al. 1984, Allen 1987, Schoenlein et al. 1987, Brorson et al. 1990, Sun et al. 1993, Groeneveld et al. 1995).

In ultrafast optical experiments (Part III), an incident pump pulse perturbs (or prepares) a sample on a sub-100 fs time scale. This perturbation is probed with a second ultrashort pulse that, depending on the experimental setup, measures pump-induced changes in a desired optical property of the sample, such as the reflectivity, transmission, conductivity, or magnetization. In the majority of experimental studies on condensed matter to date, the pump pulse creates a nonthermal electron distribution fast enough that, to first order, there is no coupling to other degrees of freedom. During the first 100 fs, the nonthermal (and potentially coherent) distribution relaxes primarily by electron–electron scattering. Subsequently, the excited Fermi–Dirac distribution thermalizes through coupling to the other degrees of freedom (Chapter 8 gives a detailed discussion of these processes).

Formally, ultrafast optical spectroscopy is a nonlinear optical technique, and pump-probe spectroscopy (Chapter 9), the most widely used technique for studying carrier dynamics in metals, can be described in terms of the third-order nonlinear susceptibility. However, more insight is often obtained by considering ultrafast optical spectroscopy as a kind of modulation spectroscopy, where the self-referencing probe beam measures the induced change in reflectivity $\Delta R/R$ or transmission $\Delta T/T$ (Cardona 1969, Averitt and Taylor 2002). This provides an important connection with time-integrated optical spectroscopy, where the experimentally measured reflectivity and the extracted dielectric response $\varepsilon(\omega)$ are the starting point for interpreting and analyzing the results of measurements. For example, in femtosecond experiments, the dynamics can be interpreted using the following equation:

$$\frac{\Delta R}{R} = \frac{\partial \ln(R)}{\partial \varepsilon_1} \Delta \varepsilon_1 + \frac{\partial \ln(R)}{\partial \varepsilon_2} \Delta \varepsilon_2 \qquad (3.7)$$

where

R is the reflectivity

$\Delta \varepsilon_1(\omega)$ and $\Delta \varepsilon_2(\omega)$ are the induced changes in the real and imaginary parts of the dielectric function, respectively

Insights into the electronic properties obtained from $\varepsilon(\omega) = \varepsilon_1(\omega) + i\varepsilon_2(\omega)$ thus serve as a useful starting point to understand the quasiparticle dynamics measured using time-resolved techniques.

If it is possible to ascribe a specific process that contributes to the induced changes in $\varepsilon(\omega)$, this can aid in understanding the origin of the electron dynamics being measured. This can be quite difficult. For example, in probing the photo-induced reflectivity change $\Delta R/R$ in the vicinity of an interband transition, it is important to recall that the interband contributions to $\varepsilon(\omega)$ are determined by the joint density of states (JDOS) between occupied and unoccupied levels summed over all bands. Thus, the induced changes in $\varepsilon(\omega)$ likely arise from induced changes in the JDOS (though in principle changes in the matrix elements between states could be modified too). This is a nontrivial task. However, the relatively simple band structure of Au has enabled detailed calculations of the spectral dependence $\Delta R/R$ in the vicinity of interband transitions. Historically, this helped in understanding what ultrafast optical spectroscopy probes and provides insight into various contributions that can lead to changes in $\Delta R/R$ (Sun et al. 1993).

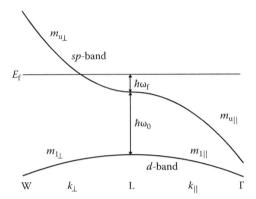

Figure 3.7 Band structure of Au in the vicinity of the L-point of the Brillouin zone. The onset of strong interband absorption at ~2.4 eV is due to a transition from occupied d-band states to unoccupied states in the vicinity of E_f. Heating of the electrons (with an ultrashort optical pulse or otherwise) changes the occupancy near E_f, thereby causing changes in the interband transitions.

Figure 3.7 shows a portion of the band structure of Au in the vicinity of the L-point of the Brillouin zone. The JDOS from this portion of the band structure makes the dominant contribution to $\varepsilon(\omega)$. Interband transitions arise when electrons are excited from occupied d-levels to above E_f. A pump pulse can heat the electrons, changing the occupancy in the vicinity of E_f, and a probe pulse that interrogates the interband transition will be quite sensitive to the changes in the occupancy. This can be calculated fairly accurately by treating the bands as parabolic in a constant matrix element approximation. For example, in Figure 3.7, each band is shown in the parabolic approximation, where $E = \hbar^2 k^2/2m$, where m is different for each band and can be negative for hole-like bands. We refer the reader to the literature for additional details (Rosei, 1972, Sun et al., 1993, Averitt et al., 1999).

However, in Figure 3.8 we show how heating of the electrons manifests in $\Delta R/R$ and $\Delta T/T$ as determined from calculation of changes in $\varepsilon(\omega)$. If the changes in occupancy are from a

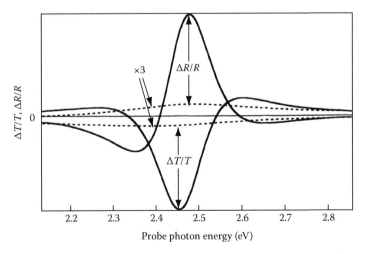

Figure 3.8 Calculated changes in the reflectivity $\Delta R/R$ and transmission $\Delta T/T$ at photon energies corresponding to the onset of interband transitions in gold. The solid lines are in the case of a Fermi-Dirac distribution (i.e., well-defined temperature) while the dashed lines are for a nonthermal electron distribution. These calculations are in reasonable agreement with experiments. (Reprinted with permission from Sun, C.K. et al., *Phys. Rev. B*, 48, 12365, 1993. Copyright 1993 by the American Physical Society.)

nonthermal distribution, it is expected that the change would be quite small and spectrally featureless (dashed lines in Figure 3.8). If the changes are thermal (i.e., can be described by a Fermi-Dirac temperature), then the changes in transmission or reflection show strong spectral variations in the vicinity of interband transitions. Pump-probe experiments on Au are in accord with the calculations in Figure 3.8 (Sun et al. 1993). At early times (~100 fs) following excitation, the distribution is nonthermal and looks like the dashed lines in Figure 3.8, while at longer times (100 fs–1 ps), the carriers reach a thermalized distribution and look more like the solid lines in Figure 3.8. While not depicted here, these spectral changes will gradually decrease as the electrons thermalize with the lattice on a picosecond timescale. These experiments and others highlighted the importance of $\varepsilon(\omega)$ and helped to show that both thermal and nonthermal effects can be observed in probing quasiparticle dynamics in metals.

Suffice to say, good metals play an important role in advancing our understanding of integrated and time-resolved optical spectroscopy and play an important role in plasmons and related areas such as metamaterials. Of course, as the complexity of the metals increases, so does the complexity of the band structure and thus $\varepsilon(\omega)$. Nonetheless, as the following sections show, optical spectroscopy still has much—and perhaps even more—to offer in the study of such bad and poor metals.

3.3 TRANSITION METALS

Many of the good metals we discussed in the previous section (Au, Ag, and Cu) are transition metals. As previously mentioned, in these cases the d-bands don't cross the Fermi level. Of course, in many other transition metals, the d-bands do cross E_f, and the reflectivity and $\varepsilon(\omega)$ (or $\sigma(\omega)$) become increasingly complex. This would be a boring sort of complexity if it was simply the case that numerous interband transitions obscured any spectroscopic hint of a Drude response, making it nearly impossible to elucidate optical properties and electronic structure from optical measurements. While true to a limited extent, what we have in mind are transition metals where the additional complexity leads to the emergence of ferromagnetism—a wonderful example of a broken-symmetry state that is of significant technological interest. This, in particular, includes Ni, Fe, and Co. Many textbooks describe the origin and properties of ferromagnetism (Hurd 1975, Ashcroft and Mermin 1976, Harrison 1980, Burns 1985, Marder 2000, Duan and Guojun 2005). Our focus in the present section will be on why and how magnetism is observed in the optical response. As with the other sections, our goal is to provide a bit of intuition from a highly simplified band perspective (i.e., the middle panel of Figure 3.1). We suggest the following books for further introductory discussion: Blundell (2001), Burns (1985), and Harrison (1980). There are also numerous texts on magneto-optics, but in terms of the fundamental aspects as to how light couples to spins in metals, the original research papers are quite insightful (Bennett and Stern 1965, Erskine and Stern 1973, Wang and Callaway 1974, Erskine and Stern 1975, Singh et al. 1975). This section has benefited from a recent comprehensive review article that, while quite advanced, offers a detailed overview along with numerous insights into the experimental observables of magneto-optics, and an up-to-date survey of advanced theory for calculating the magneto-optical response of transition metals and related compounds (Ebert 1996). Our presentation will be brief and highlight important aspects that are often not covered in standard texts dedicated to the optical spectroscopy of condensed matter (Wooten 1972, Dressel and Grüner 2002).

In learning about magnetism, the starting point is often a picture describing electrons localized at lattice sites that, through quantum-mechanical exchange, "talk" to one another leading to long-range magnetic order. This is an amazingly important perspective, but it is reasonable to question its validity in terms of transition metal ferromagnets, where we know there are itinerant electrons that lead to a poor, but significant, DC conductivity. If site-localized

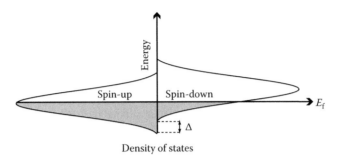

Figure 3.9 Schematic DOS versus energy, representative of ferromagnetic materials. In this simple picture, an energy shift Δ between spin-up and spin-down bands results in a net magnetic moment due to an imbalance in the carrier density (i.e., $n_\uparrow - n_\downarrow > 0$). The narrow d-bands and large DOS at E_f are favorable for the emergence of ferromagnetism.

electrons were solely responsible for the ferromagnetism, we would expect that measurements of the magnetization, M, would yield an integer number of magnetic moments per ion. This is not the case. For example, in Fe there is the equivalent of 2.2 spins per ion and in Ni 0.6. This suggests that itinerancy cannot be neglected.

A simple band approach can provide considerable insight into ferromagnetism in metals (Harrison 1980, Blundell 2001). Up to now, we have been considering metals with an electron density n. We can break this total carrier density into spin-up ($n_\uparrow$) and spin-down ($n_\downarrow$) fractions where $n = n_\uparrow + n_\downarrow$ and $M \propto n_\uparrow - n_\downarrow$. Figure 3.9 depicts the band structure with the spin-up states on the left and the spin-down states on the right. As shown, there is a shift of the bands, which would lead to a nonzero value of $n_\uparrow - n_\downarrow$ and hence a net magnetization. As a function of Δ, it is easy to imagine a case where a non-integer number of spins per ion would occur. But why would such a scenario be favored for d-bands in comparison to the s-band (or sp) good metals we considered previously?

The answer, quite simply, is that itinerant ferromagnets strongly favor narrow bands and a large DOS at E_f. Flipping a spin-down electron to spin-up increases $n_\uparrow$ and also increases the kinetic energy. For a wideband metal, this is a significant energy penalty but less so in a narrowband metal (e.g., see the discussion regarding the electron kinetic energy at the end of the introduction to this chapter). However, even in a narrow d-band metal this would be an insurmountable energy increase were it not for exchange. If the exchange splitting (Δ, referred to previously and shown in Figure 3.9) is large enough, it is possible to overcome the kinetic energy cost resulting in $n_\uparrow - n_\downarrow > 0$. The resultant exchange splitting is related to the exchange integral, which is typically denoted as J in the modern literature. In various simplified models of itinerant ferromagnetism, analysis leads to expressions for the requirement of obtaining a finite magnetization. In the Stoner model, the following expression is obtained for the total energy change in producing a finite M:

$$\delta E = \frac{1}{4}\left[\frac{1}{N(E_f)} - J\right]M^2 \tag{3.8}$$

Here, $N(E_f)$ is the DOS at the Fermi level. To obtain a negative δE, which favors magnetism, we must have $JN(E_f) > 1$, which is the so-called Stoner criterion (Blundell 2001). In an even simpler model, which treats the bands as having a constant DOS as a function of energy, the result $J > W/5$ is obtained where W is the bandwidth. Surprisingly, using experimental estimates for J and W correctly predicts ferromagnetism in Co and Ni and it is only slightly off for Fe (Harrison 1980). These simple models leave out many details but nonetheless indicate how magnetism may be obtained in metals and highlight how *d*-bands crossing E_f can almost be considered as a necessary (but not sufficient) condition for magnetism in transition metals.

Magneto-optics has a long history beginning in the nineteenth century with Faraday. It took quite some time to understand how light couples to spins. The bottom line is that spin–orbit coupling is required. Were it not for this, magnetism arising from the spin degree of freedom would exist independent of optical phenomena. For example, optical measurements would not be able to distinguish between the ferromagnetic and paramagnetic phases. Fortunately, this is not the case. Despite the long history, obtaining a foundational understanding took quite some time with particularly important theoretical and experimental work in the 1950s, 1960s, and 1970s (Argyres 1955, Bennett and Stern 1965, Erskine and Stern 1973, Wang and Callaway 1974, Erskine and Stern 1975, Singh et al. 1975). We will not go into the theoretical details, nor consider the experimental details as discussed elsewhere in this book. We just mention that for metals, the majority of magneto-optical experiments are performed in what is known as the polar-Kerr configuration where measurements are in a reflection geometry with the magnetic field oriented perpendicular to the surface (the corresponding transmission experiment is referred to as the polar-Faraday configuration) (Chapter 13). In the following, our goal is to simply describe how magnetism appears in the optical response functions (Ebert 1996).

Up to this point we have been treating the optical conductivity as a scalar consisting of real and imaginary terms—that is, $\sigma(\omega) = \sigma_1(\omega) + i\sigma_2(\omega)$. This is a useful starting point and is valid for cubic systems. More generally, response functions are second rank tensors (Chapter 1). Writing the tensor components in matrix form, we can easily express the optical conductivity for a non-ferromagnetic cubic metal. It consists of three equivalent on-diagonal terms with zero for all of the off-diagonal components. Anisotropy would change this, which is exactly the case for a ferromagnet because of the aforementioned spin–orbit coupling. It can be imagined in the following way: In a material with a cubic crystal lattice in the high-temperature paramagnetic state, the optical conductivity tensor would be diagonal. Upon cooling into the ferromagnetic phase in the presence of a magnetic field, there could be changes in the diagonal terms, along with the emergence of off-diagonal terms related to the reduction in symmetry as the spins align and couple to the orbital moment. For a metal with magnetization parallel to the [001] direction, it can be shown that the conductivity tensor would take the following form (Ebert 1996):

$$\underline{\sigma} = \begin{pmatrix} \sigma_{xx} & \sigma_{xy} & 0 \\ -\sigma_{xy} & \sigma_{xy} & 0 \\ 0 & 0 & \sigma_{zz} \end{pmatrix} \tag{3.9}$$

For magneto-optical Kerr effects, it is the off-diagonal components that are crucial in obtaining a measurable signal (Chapter 13). As before, the dielectric response can be expressed in terms of the conductivity as $\underline{\varepsilon}(\omega) = \varepsilon_\infty + i\sigma(\omega)/\varepsilon_0\omega$.

The diagonal and off-diagonal components consist of real and imaginary terms where, as usual, the diagonal real part σ_{xx}^1 describes the absorption. In contrast, for the off-diagonal term, the imaginary part σ_{xy}^2 describes the absorption (note: in the present context, the subscripts 1 and 2 refer to the real and imaginary parts of the conductivity, respectively). This physical interpretation is most insightful when considering the power absorbed by circularly polarized radiation: σ_{xx}^1 measures the sum of the absorption of left and right circularly polarized radiation while σ_{xy}^2 measures the difference in absorption.

Figure 3.10 depicts the optical conductivity as a function of photon energy for a ferromagnet in the vicinity of interband transitions (Ebert 1996). Right- and left-handed circularly

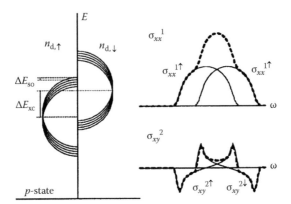

Figure 3.10 Schematic of how magnetism manifests in the optical conductivity from Ebert (1996). The left side is similar to Figure 3.9 but also shows that spin–orbit coupling can contribute to band splitting. More importantly, were it not for spin–orbit coupling, magnetism would not manifest in optical measurements. The upper right depicts how band splitting due to magnetism appears in the real part of the on-diagonal component of the conductivity tensor σ_{xx}^1. The contributions from the spin-up and spin-down bands are additive. In contrast, as shown in the lower right, for the imaginary part of the off-diagonal conductivity σ_{xy}^2, the difference between spin-up and spin-down bands is obtained and can lead to considerable structure in the Kerr rotation spectrum. (Reprinted with permission from Ebert, H., *Rep. Prog. Phys.*, 59, 1665, 1996.)

polarized radiation couple to different spin populations (Chapter 13). As the upper right panel shows, σ_{xx}^1 measures the overall absorption and even with splitting due to exchange and spin–orbit effects can appear relatively featureless and is always positive. In contrast, since σ_{xy}^2 measures the difference, the features can be better resolved and negative values can occur. While σ_{xy}^2 is a useful spectroscopic quantity for magneto-optics, it is not easily obtained from experimental measurements. Just as it can be a complicated matter to obtain σ_{xx}^1 from reflectivity measurements, the determination of σ_{xy}^2 is nontrivial.

For magneto-optic measurements, changes in the polarization state are the measurable quantity. In particular, the ellipticity η and rotation θ of a reflected beam in comparison to the incident beam are measured. These quantities are, respectively, related to the relative difference in absorption between left and right circularly polarized light and the change in phase between right- and left-handed light. Care must be taken to obtain changes, which are related to magnetism as opposed to nonmagnetic. The Kerr rotation can be written as $\phi(\omega) = \theta(\omega) + i\eta(\omega)$ (as in Equation 13.7 of Chapter 13). In terms of the optical conductivity tensor, we have (Chapter 13 gives a similar equation for ϕ in terms of $\varepsilon(\omega)$)

$$\phi \approx \frac{\sigma_{xy}}{\sigma_{xx}\sqrt{1 + i\sigma_{xx}/\omega\varepsilon_o}} \tag{3.10}$$

It is evident from this equation that the diagonal and off-diagonal terms contribute to the Kerr rotation, which is a complication. From careful measurements of the reflectivity and the Kerr rotation, it is possible to determine the off-diagonal tensor components (Chapter 13). In any case, just as the reflectivity as a function of photon energy provides insight into the properties of a metal, so does the Kerr rotation spectrum. Figure 3.11 shows an example of the Kerr rotation spectrum for Ni (Oppeneer et al. 1992). Depending on the photon energy, the rotation can be positive or negative and the spectrum shows some similarities to the schematic σ_{xy}^2

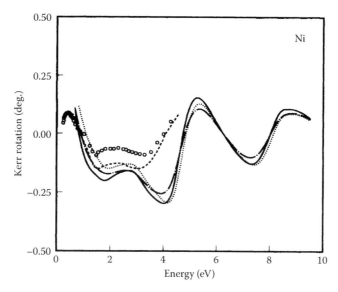

Figure 3.11 Experimental measurement (dashed line and circles) and various theoretical calculations of the Kerr rotation spectrum of nickel. Note that both σ_{xx}^{1} and σ_{xy}^{2} contribute to the response. The calculations are in reasonable agreement with experiments. It is only recently that it is has been possible to carry out such complex calculations using realistic band structures to compare with experiments. (Reproduced with premission from Oppeneer, P.M. et al., *Phys. Rev. B*, 45, 10924, 1992. Copyright 1992 by the American Physical Society.)

spectrum shown in the lower right of Figure 3.10. It is only recently that reasonably accurate calculations of the Kerr rotation have been possible. As Equation 3.10 highlights, this is (at least in part) because the real and imaginary terms of the diagonal and off-diagonal components of the conductivity tensor must be calculated.

The previous discussion provided a brief introduction into how magnetism manifests in optics. Magneto-optical studies are an extremely active area of optics, which include spintronics, nanomagnetism, and surface magnetism. We briefly mention an extremely interesting area of research that has developed since 1996. This is the field of femtomagnetism (Beaurepaire et al. 1996). Just as short optical pulses can be used in pump-probe geometry to measure electron–phonon coupling in good metals, they can be used for dynamic Kerr rotation measurements (Chapter 13). What was particularly surprising from initial measurements was the observation of demagnetization on a sub-picosecond timescale. This was quite unexpected and remains an active area of research as a complete understanding of the microscopic dynamics associated with ultrafast demagnetization remains elusive.

Obtaining a convincing demonstration that femtosecond Kerr rotation measurements measure an ultrafast decrease in M is not simple. The reason for this is highlighted by Equation 3.10. Clearly, photo-induced changes in the Kerr rotation $\Delta\phi$ can arise from changes in the diagonal and off-diagonal components of the conductivity tensor. Thus, photoexcitation will—as described in the section on good metals—lead to changes in occupancy, which could appear in $\Delta\phi$. However, careful comprehensive measurements of the photo-induced changes in reflectivity and Kerr rotation have provided clear evidence that the magnetization can be manipulated on sub-picosecond timescales (Chapter 13). For example, the photo-induced changes (Guidoni et al. 2002) in the Faraday rotation ($\Delta\Theta/\Theta$) and $\Delta q/q$, which are related to the change in magnetization, were measured on $CoPt_3$ (see Figure 13.11 in Chapter 13). This work helped to clarify the nature of quasiparticle and magnetization dynamics in ferromagnetic metals, as a detailed analysis revealed that the dynamics were consistent with changes in M. We refer the reader to Guidoni et al. (2002) for details and to Zhang et al. (2009), Bigot et al. (2009) for recent overviews of theoretical and experimental aspects of femtomagnetism. Chapter 13 also covers much of this material in detail.

3.4 NARROWBAND METALS—TRANSITION METAL OXIDES

In the previous two sections, we considered conventional wideband metals and transition metals, where electrons from d-orbitals play a consequential role in determining the electronic properties. In this section, we consider "bad metals," which are, perhaps, the most interesting case and certainly define the cutting edge in condensed matter physics on the experimental and theoretical fronts (Imada et al. 1998, Fazekas 1999, Tokura 2000, Cooper 2001, Dagotto 2003, Basov and Timusk 2005). Our emphasis will be on TMOs, which exhibit a truly remarkable diversity of phenomena (Imada et al. 1998, Dagotto 2003). Referring back to the rightmost panel of Figure 3.1, we have a simplistic scenario where narrow d-bands dominate the physics. This is too simplistic a picture. For example, in the large majority of TMOs, the transition metal atoms are octahedrally or tetrahedrally coordinated by oxygen atoms. This reduction in spherical symmetry via the crystal field lifts the fivefold degeneracy of the d-orbitals in a predictable manner, which is covered in detail in numerous texts (Tokura 2000, Dagotto 2003) and highlights that any complete description of the physics of TMOs must include hybridization between the transition metal d-orbitals and oxygen p-orbitals. Nonetheless, the concept of a narrow band ($\sim 1\,eV$) is valid. In the schematic view of Figure 3.1, it would appear that, in the case where E_f crosses within the band, a metallic state would result. There are TMOs that are highly conductive, with ReO_3 exhibiting the largest conductivity of any TMO. Indeed, ReO_3 can be considered as a good metal whose properties are, to a large extent, determined from the band structure (Tsuda et al. 2000).

However, what we have in mind in this section are TMOs, which violate the predictions of band theory, and where fierce competition between electron localization and itinerancy dominates the physics. Historically, NiO is such an example where, having an odd number of electrons, band theory predicts a metallic state. In reality, NiO is an insulator, which results from on-site Coulomb repulsion (U) dominating over the ability of an electron to hop from site to site, which is directly related to the bandwidth (W). Materials that are insulating due to correlation are termed Mott–Hubbard (MH) insulators.

Of course, we would not be considering MH insulators in this chapter if they were always insulators. For the present purpose, the most important aspect of MH insulators is that they can be tuned toward metallicity. Chemical substitution is the most common approach and can lead to two main effects. First is doping. For example, for the high-temperature superconducting cuprates, the undoped parent compounds are MH insulators. La_2CuO_4, when doped with Sr (i.e., $La_{2-x}Sr_xCuO_4$), becomes metallic and, with decreasing temperature, superconductive. Doping introduces holes into an otherwise filled band, with a maximum superconducting transition temperature occurring for $x \sim 0.16$. A second example is the manganites whose undoped parent compound—$LaMnO_3$—is also an MH insulator. Doping with divalent substituents such as Sr or Ca results in hole doping of the relevant band, as in the case of the cuprates. In the case of the manganites, this causes (for optimal doping) a transition from a paramagnetic semiconductor to a ferromagnetic metal with decreasing temperature. In the vicinity of this transition temperature, the application of a magnetic field results in a large change in the conductivity. The term "colossal magnetoresistance" was coined to describe this magnetic field-driven insulator to metal transition (Jin et al. 1994, Tokura 2000).

A second effect associated with chemical substitution is tuning of the bandwidth, where changes in the cation size can either enhance or decrease orbital overlap, thus tuning the competition between the bandwidth and on-site Coulomb repulsion. For example, in the case of the manganites Sr or Ca substitution (i.e., $LaMnO_3 \rightarrow La_{1-x}Re_xMnO_3$, Re = Sr, Ca) results in hole doping, but Ca, being smaller than Sr (more precisely, the cation Ca^{2+} radius is smaller than that of Sr^{2+}), has a more pronounced effect on the Mn–O–Mn orbital overlap (see Hwang et al. 1995 for a discussion of the tolerance factor), yielding a lower transition temperature. It is also important to keep in mind a deleterious effect arising from chemical substitution. This is due to the inevitable introduction of disorder due to non-stoichiometry, which complicates the physics and can mask effects associated with band filling or bandwidth control. Nonetheless, since the discovery of the high-T_C superconductors, the progress in controlled synthesis, characterization, and the theoretical underpinning of electron correlations in TMOs has been enormous. These advances, to date, have not led to a complete understanding of high-T_C superconductivity nor have they provided quantitative guidelines in terms of structure–function relationships. It remains a grand challenge to achieve predictive capabilities when it comes to TMOs and other correlated electron materials.

A minimal model that provides considerable insight into the role of electronic correlations is the single orbital Hubbard model. This model can also provide insight into the sensitivity of optical spectroscopy to the effects of correlation. We provide a simple description and refer the reader to several sources, ranging from introductory to advanced (Ashcroft and Mermin 1976, Fazekas 1999, Duan and Guojun 2005). The Hubbard Hamiltonian describes the competition between itinerancy and localization in terms of the bandwidth (W) and on-site Coulomb repulsion (U) parameters that we previously introduced. The right panel in Figure 3.1 would correspond to the situation where $U/W < 1$ for a metallic state, which, from the perspective of optics, would yield a Drude-like response. However, when U is comparable to or larger than the bandwidth W, a splitting of the "band" would occur, resulting in a MH insulating state as indicated in Figure 3.12a. The value of U is the energy cost to place a second electron (of opposite spin) onto an orbital at a given site. For large enough U/W, we would then expect that the ability of electrons to hop from site to site would be energetically costly and yield no Drude-like response. A metal-to-insulator transition would manifest as the disappearance of the Drude response. We also note that as a function of increasing U/W, localization is favored and a band picture in terms of Bloch electrons loses relevance. It becomes more practical to consider a real-space picture in terms of local on-site or inter-site transitions.

A simple real-space cartoon is given in Figure 3.12c for the case of one electron per lattice site, where each site contains one orbital (this situation is termed half-filling since each orbital can hold two electrons of opposite spin). In this cartoon, $U/W \gg 1$ and each electron is confined to a single site. The energy cost is too great for an electron to hop freely about the lattice. However, for photons with energy $\sim U$ (typically on the order of an electron volt), inter-site transitions are possible, leading to a fraction of sites that are empty and a corresponding fraction of sites with double occupancy (Figure 3.12d). Thus, we could imagine that when tuning U/W from less than one to greater than one, we would see a shift of spectral weight from the low-energy Drude-like response (often called a coherent response in the literature) to higher energies, reminiscent of an interband transition. As the examples we show in the following highlight, this is what is observed in numerous real materials. Before moving on to experimental examples, we will first discuss complications that arise and discuss the concept of spectral weight transfer in more detail.

While the Hubbard model provides insight into correlated electron materials, some of the limitations should be mentioned since, while this model is fascinating, real materials are all the more so for reasons that are worth pondering experimentally and theoretically. For example, the single-orbital Hubbard model reduces long-range Coulomb effects to a single-site effect and neglects multi-orbital effects that are important, if not crucial, to consider in the majority of real materials. This includes the oxygen orbitals, which in some cases lead to an insulating state that is better described as a charge transfer insulator as opposed to a

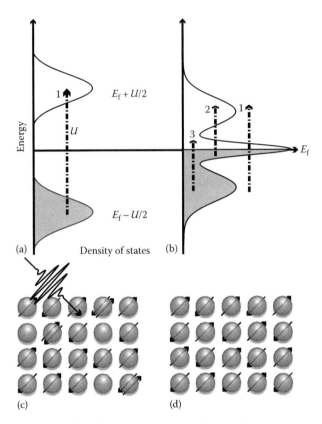

Figure 3.12 Schematic Hubbard picture. (a) For a MH insulator there are upper and lower Hubbard bands separated by an on-site Coulomb repulsion U. At half-filling, optical excitation can lead to a fraction of sites that are empty and a corresponding fraction that are now doubly occupied (see c → d) and indicated in (a) with the vertical arrow labeled 1. (b) A metal–insulator transition caused by doping, bandwidth control, or perhaps photoexcitation can result in a metallic state with a coherent Drude response and residual MH spectral weight. In (b), the arrow labeled 1 is as in (a), while 2 indicates a transition from the Drude-like carriers to the upper Hubbard band and 3 indicates a transition from the lower Hubbard band to the coherent Drude peak.

MH insulator (Fazekas 1999, Duan and Guojun 2005). It is also important to realize that our introductory discussion leaves out many details. For example, a metal–insulator transition is not an all-or-nothing effect where a Drude-like response simply vanishes upon becoming an insulator. In fact, for bad metals (V_2O_3 as discussed in the following being a prime example) the metallic state may be strongly correlated and remnant spectral weight associated with transitions between the upper and lower Hubbard bands could be present. This would be in addition to the coherent Drude response due to transitions between the Hubbard band and the coherent quasiparticle peak (Figure 3.12b).

It is also important to consider other degrees of freedom, such as phonons and spins. The energetics is such that these effects are more pronounced in narrowband materials (e.g., the effect of exchange as mentioned in the section on transition metals). Taken together, effects associated with the spin, lattice, charge, and orbital degrees of freedom lead to numerous phenomena including magnetic ordering, charge and orbital ordering, polaron formation, structural phase transitions, and nanoscale electronic phase separation (Dagotto 2003, 2005). These are just a few of the interesting phenomena occurring in TMOs, all of which can be studied with the optical techniques described in this book.

In addition to the advances in materials synthesis and characterization using optical spectroscopy (and a host of other techniques such as neutron scattering, transport, photoemission, and x-ray absorption), there have been dramatic advances in theory and computation. Of particular interest is the combination of local density approximation (LDA) techniques for band structure calculations with dynamical mean field theory (DMFT) for handling correlations. While the details of LDA + DMFT take us far beyond the scope of this chapter, we mention that modern computational techniques are beginning to enable realistic calculations of the response of correlated electron materials, including determination of the optical conductivity for quantitative comparison with experiments (Kotliar and Vollhardt 2004).

The intuitive idea that there can be a shift in spectral weight from low energies to high energies due to correlation-induced localization means that optical spectroscopy is a powerful probe that provides an empirical characterization of the strength of correlations. That is, optical conductivity measurements can help classify a material as weakly or strongly correlated (Dressel and Grüner 2002, Basov and Timusk 2005, Basov et al. 2010). This idea is obtained from well-known sum rules for the optical conductivity (Chapter 1). For example, the integral of the real part of the optical conductivity rigorously satisfies

$$\int_{0}^{\infty} \sigma_1(\omega)d\omega = \frac{\pi}{2}\omega_p^2 = \frac{ne^2\pi}{2\varepsilon_0 m} \tag{3.11}$$

which simply gives the density of electrons contributing to electromagnetic absorption (when all frequencies are considered, all of the electrons participate). This does not provide insight into electronic correlations. However, restricted sum rules, where the integral is limited to below the onset of interband transitions, can provide insight into correlation. Such an integral defines the spectral weight, N_{eff}:

$$N_{eff}(\omega) = \int_{0}^{\omega} \sigma_1(\omega')d\omega' \propto K_{exp} \tag{3.12}$$

This is related to the effective number of carriers contributing to electromagnetic absorption for frequencies less than ω and to the kinetic energy of the carriers (Basov et al. 2010). The basic idea is that the need to integrate to very high photon energies to exhaust the spectral weight is tantamount to localization, in line with the predictions of the Hubbard model. That is, spectral weight at low frequencies is linked to mobile carriers, while spectral weight at higher frequencies is related to carrier localization. A comparison of the experimentally determined kinetic energy, K_{exp}, using Equation 3.12 with the expectation from band theory, K_{band}, provides a metric for the strength of the correlations. For example, $K_{exp}/K_{band} \sim 1$ signifies weak correlations, while $K_{exp}/K_{band} \ll 1$ would be representative of a strongly correlated electron material. Further details can be found in the references (Rozenberg et al. 1996, Basov et al. 2010). We note that unambiguous determination of K_{exp} from experiment is difficult because of the upper limit in the integration along with experimental uncertainties. We also mention that there are other important sum rules to extract physics from the optical conductivity. For example, the Ferrell–Tinkham–Glover sum rule for superconductors quantifies how spectral weight from the Drude-response shifts to the zero-frequency delta-function condensate response in the superconducting phase (Dressel and Grüner 2002).

The idea of dynamic spectral weight transfer is also of interest in the context of ultrafast optical studies of correlated electron materials (Averitt et al. 2001, Averitt and Taylor 2002, Hilton et al. 2006). As described for the Hubbard model, the electronic properties depend on the occupancy. Optical excitation changes the occupancy, as in the case of excitation across a MH gap. For short pulse excitation, this can initiate a change in the electronic structure, resulting in a temporally evolving spectral weight. By resolving the dynamics of the spectral weight transfer, it is possible to gain insight into how the microscopic degrees of freedom contribute to the properties of a complex material. This is similar to measuring electron–phonon coupling in good metals as mentioned earlier. In such cases, the pump pulse is fairly gentle (e.g., <0.1 mJ/cm^2 pump fluence) and the goal is to investigate coupling between different degrees of freedom within a given phase.

However, given that bad metals can teeter on the edge of phase stability, there has been recent interest in using femtosecond pulses as an external control parameter to induce a phase transition. In fact, the study of photo-induced phase transitions (PIPT) is an emerging area of research, with many classes of materials under investigation (Nasu 2004). Correlated electron materials are of interest precisely because of the dramatic changes in electronic structure that can result from changes in occupancy. In this context, many interesting questions can be posed. For example, can "photodoping" collapse a MH gap and, if so, what is the time scale? Similarly, dramatic changes in the lattice can occur with changes in orbital occupancy, leading to dramatic effects such as orbital ordering. As such, photoexcitation can relax the need for a coherent lattice distortion (e.g., the Jahn-Teller distortion in the manganites) (Rini et al. 2007) and initiate, in principle, a domino-like effect resulting in a change of phase.

Given our brief introduction to correlated electron materials and the role of time-integrated and time-resolved optical spectroscopy in their study, we will now provide several examples to highlight our previous discussion. The vanadates are strongly correlated electron materials that have been extensively studied for 50 years. VO_2 and V_2O_3 are the most thoroughly investigated of the vanadates, in terms of elucidating the role of correlations in driving the insulator-to-metal transition. These materials continue to be extensively studied and we recommend reading several papers from the research literature on vanadates to serve as a starting point to learn about correlated electron materials (Feinleib et al. 1968, Choi et al. 1996, McWhan and Remeika 1970, Castellani et al. 1978, Mott 1990, Limelette et al. 2003, Cavalleri et al. 2005, Pfalzer et al. 2006, Qazilbash et al. 2007, 2008, Baldassarre et al. 2008). Briefly, at low temperature, V_2O_3 is an antiferromagnetic insulator, transitioning to a correlated metallic state at approximately 140 K. V_2O_3 is of historical importance in optical spectroscopy, as it is one of the early examples showing the redistribution of spectral weight (Rozenberg et al. 1996). Figure 3.13 shows experimental data depicting the redistribution of spectral weight in the correlated metallic phase. The mid-IR peak at ~0.5 eV arises from transitions between the Drude-like peak and the upper and lower Hubbard bands (see Figure 3.12b and the arrows labeled 2 and 3). There is an increase in the mid-infrared spectral weight when the temperature is reduced from 300 to 170 K, which is close to the MH insulating state. The inset shows the difference in the conductivity for these two temperatures, which better highlights that the mid-IR peak corresponds to $U/2$, as expected for a MH material. Early theoretical work described the properties of V_2O_3 in terms of a single orbital MH scenario. More recently, it has been realized that a multi-orbital description is needed, though clearly MH physics is the key to understanding the metal–insulator transition and the properties of the bad metallic state. In the metallic state, the DC conductivity is approximately 2000 $(\Omega\,cm)^{-1}$, which is quite low compared to a good metal with a conductivity on the order of 10^5 $(\Omega\,cm)^{-1}$.

As a second example, we briefly consider the manganites, which were mentioned earlier. The bandwidth of manganites can be sensitively controlled. The Mn–O–Mn bond angle depends on the sizes of the rare earth and dopant ions. As a function of decreasing ion radius,

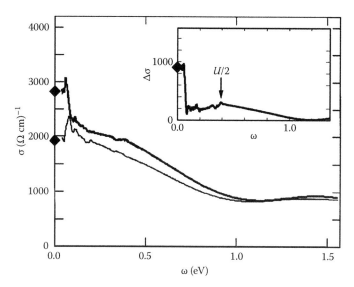

Figure 3.13 Optical conductivity in the metallic state of V_2O_3. Upper line-170K, lower line-300K, and the inset shows the difference between the two, highlighting the development of an MH gap upon approaching the insulating phase. (Reprinted with permission from Rozenberg, M.J. et al., *Phys. Rev. B*, 54, 8452, 1996. Copyright 1996 by the American Physical Society.)

the Mn–O–Mn bond angle decreases with a gradual structural change from cubic to rhombohedral to, eventually, orthorhombic. The observed properties of manganites show a strong correlation to this bond angle (Hwang et al. 1995). For example, $La_{0.7}Sr_{0.3}MnO_3$ is classified as an intermediate bandwidth material exhibiting a transition ($T_C \sim 260\,K$) from ferromagnetic metal to paramagnetic semiconductor. In contrast, the narrow bandwidth manganite $Pr_{0.6}Ca_{0.4}MnO_3$ does not exhibit metallic behavior at any temperature, instead entering a charge-ordered phase with decreasing temperature, though it is very sensitive to external perturbations. These properties manifest in dramatic fashion in optical spectroscopy with significant redistribution of spectral weight from the far-infrared through the visible (Cooper 2001).

Hole doping of $LaMnO_3$ creates mobile carriers, which (for appropriate doping in intermediate bandwidth manganites) leads to incoherent hopping of Jahn–Teller polarons in the paramagnetic phase, crossing over to coherent transport in the low-temperature ferromagnetic metallic state. This manifests in the optical conductivity as shown in Figure 3.14 for $La_{0.825}Sr_{0.175}MnO_3$ ($T_C = 283\,K$). The optical conductivity shows a redshift of an incoherent peak of approximately 1 eV at 293 K to lower energies with the clear onset of a Drude response below 155 K (Takenaka et al. 1999). This data was obtained from reflection measurements on a cleaved single crystal, yielding a considerably larger Drude spectral weight extending to higher energies than previously obtained on polished samples. Figure 3.14 shows that even on a pristine cleaved crystal the incoherent response persists well into the ferromagnetic phase, suggestive of residual polaronic effects, which may be strongly influenced by the orbital degrees of freedom.

Contrasting with this are narrower bandwidth manganites that are not metallic at any temperature (in the absence of an external perturbation). For example, with decreasing temperature, $Pr_{0.6}Ca_{0.4}MnO_3$ transitions from a paramagnetic semiconductor to a charge-ordered insulator ($T_{CO} = 235\,K$). However, an applied magnetic field "melts" the charge order, with a Drude-like peak emerging in σ_1 between 6 and 7 Tesla (Okimoto et al. 1999). This highlights the sensitivity of the optical and electronic properties of manganites resulting from nearly

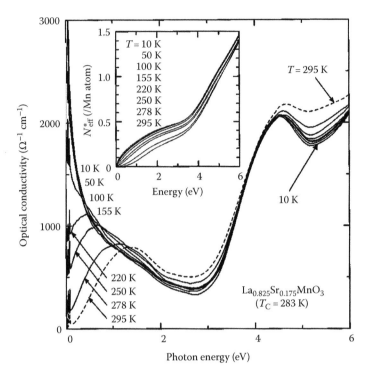

Figure 3.14 σ_1 as a function of photon energy for various temperatures measured on a cleaved single crystal of $La_{0.825}Sr_{0.175}MnO_3$. The polaron peak at approximately 1 eV gradually redshifts with decreasing temperature, transforming into a Drude peak. The inset shows the integrated spectral weight as a function of energy for various temperatures. (Reprinted with permission from Takenaka, K. et al., *Phys. Rev. B*, 60, 13011, 1999. Copyright 1999 by the American Physical Society.)

degenerate ground states with differing order parameters. There is no doubt that these bad metals are extremely interesting from a fundamental point of view.

There have also been interesting time-resolved studies on manganites and related materials with a focus on probing the quasiparticle dynamics within a given phase. Photoexcitation with a pump pulse results in a dynamic redistribution of spectral weight whose subsequent temporal evolution is monitored with a probe pulse. The timescales over which this occurs provides information about which degrees of freedom are involved in the dynamic spectral weight transfer. In the perovskite manganites, optical-pump terahertz-probe measurements in the ferromagnetic metallic phase revealed a two-exponential decrease in the optical conductivity (Averitt et al. 2001, Averitt and Taylor 2002). A short ~1 picosecond response is associated with electron–phonon equilibration while the longer (>10 ps) relaxation is due to spin-lattice thermalization. We refer the reader to the references for further details (Averitt et al. 2001, Averitt and Taylor 2002).

There is a vast literature on the electrodynamics of superconductors (Tinkham 1996, Basov and Timusk 2005). Here we just give a simple example in terms of the two-fluid model as a starting point to appreciate how optical spectroscopy can be used to investigate the electrodynamic response of superconductors, especially in the far-infrared. It is also important to learn about the Mattis–Bardeen description of the electrodynamic response in conventional Bardeen Cooper Schrieffer (BCS) superconductors as described in the references (Tinkham 1996, Dressel and Grüner 2002).

In contrast to conventional narrow-gap superconductors, the far-infrared electrodynamics of high-T_C cuprates is quite different (Brorson et al. 1996, Basov and Timusk 2005).

Most notably, the superconducting gap has a d-wave symmetry and peaks around 30 THz, such that the typical Mattis–Bardeen BCS gap structure in $\sigma_1(\omega)$ is not observed in the 1–3 THz range. However, as with the BCS superconductors, a strong $1/\omega$ dependence from the inductive condensate response is quite prominent in $\sigma_2(\omega)$ below T_C. A model that describes the electrodynamics of superconductors at frequencies far below the gap is the *two-fluid model*. It is useful for approximating the response of high-T_C cuprates in the range up to ≈ 3 THz. The two fluids in this model refer to thermally excited quasiparticles and the superconducting condensate. With decreasing temperature, more quasiparticles join the condensate. Thus, the quasiparticle density $n_N(T)$ and the condensate density $n_{SC}(T)$ are temperature dependent. The sum of quasiparticle and condensate densities is constant with temperature: $n_N(T) + n_{SC}(T) = n$, where n is the normal state carrier density above T_C. This can also be written as $X_N(T) + X_{SC}(T) = 1$, where $X_N(T) \equiv n_N(T)/n$ is the quasiparticle fraction and $X_{SC}(T) \equiv n_{SC}(T)/n$ is the condensate fraction. The conductivity for the two-fluid model is given as follows:

$$\sigma(\omega) = \frac{n_N(T)e^2}{m^*}\frac{1}{1/\tau(T) - i\omega} + \frac{n_{SC}(T)e^2}{m^*}\left[\pi\delta(\omega) + \frac{i}{\omega}\right] \tag{3.13}$$

$$= \frac{\varepsilon_0\omega_p^2}{1/\tau(T) - i\omega}X_N(T) + \frac{1}{\mu_0\lambda_L^2(0)}\left[\pi\delta(\omega) + \frac{i}{\omega}\right]X_{SC}(T) \tag{3.14}$$

The first term is just the Drude response with a temperature-dependent carrier density and scattering time. The second term describes the condensate response with the first term in brackets being a δ-function at zero frequency that describes the infinite DC conductivity of the superconducting condensate. The second term in brackets is a purely imaginary, inductive response (obtained by letting $\tau \to \infty$ in the Drude model). We can thus see that upon condensation of quasiparticles into the superfluid, spectral weight is transferred from the Drude peak to the zero-frequency superconducting peak. In Equation 3.14, the condensate portion is rewritten in terms of the London penetration depth:

$$\lambda_L^2(0) = \frac{m^*}{\mu_0 ne^2} \tag{3.15}$$

which describes how far a DC magnetic field penetrates into the superconductor at zero temperature. The above equations highlight an important aspect about the electrodynamics of superconductors: from a simultaneous measurement of $\sigma_1(\omega)$ and $\sigma_2(\omega)$, the quasiparticle and condensate fractions can be quantitatively determined. The temperature dependence of these fractions can provide essential information about the superconducting state to distinguish, for example, isotropic s-wave from anisotropic d-wave gap symmetries (Brorson et al. 1996).

Figure 3.15 shows a calculation using the two-fluid model for various values of X_N and X_{SC} (keeping $X_N + X_{SC} = 1$). For the calculations, $\tau = 150$ fs (note, in real materials τ is itself temperature dependent), $n = 10^{19}$ cm^{-3}, and $m^* = m_0$, giving $\sigma_{DC} = 10^4$ Ω^{-1}cm^{-1} for $X_N = 1$. The real part $\sigma_1(\omega)$ is shown as a solid line, the total imaginary conductivity $\sigma_2(\omega)$ is plotted as a thick dashed line (this contains the Drude-like fraction and a condensate contribution, as can be determined from Equations 3.13 and 3.14). The Drude contribution to $\sigma_2(\omega)$ is plotted as a dotted line. A pure Drude response results for $X_{SC} = 0$ (Figure 3.16a). With $X_{SC} = 0.2$, a clear $1/\omega$ dependence is evident, although the relative Drude contribution to $\sigma_2(\omega)$ is still quite large but becomes much smaller at $X_{SC} = 0.4$. Finally, for $X_{SC} = 1$ as shown in Figure 3.15d, the response at terahertz (THz) frequencies entirely arises from the superconducting condensate.

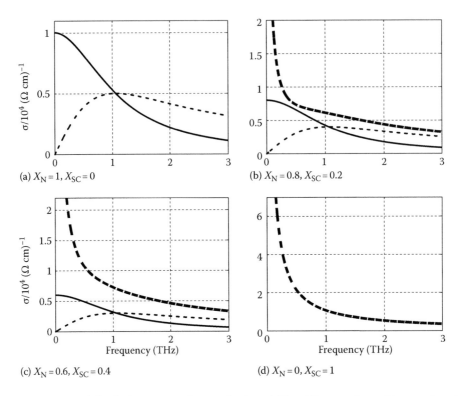

Figure 3.15 Model calculation using the two-fluid model for different values of the quasiparticle and condensate carrier fractions. X_N is the quasiparticle fraction and X_{SC} is the condensate fraction.

Of course this idealized situation is not practically realized, though as Figure 3.15 reveals, it provides a good starting point. Specifically, Figure 3.16 shows the results of THz time-domain spectroscopy (THz-TDS) measurements (Chapter 11) on a near optimally doped film of $YBa_2Cu_3O_7$ ($T_C = 89\,K$) expitaxially grown on <100> MgO using pulsed laser deposition. Figure 3.16a shows $\sigma_2(\omega)$ at $T = 60\,K$ while panel (b) shows $\sigma_2(\omega)$ at 95K. The insets show the corresponding real part $\sigma_1(\omega)$. The thick solid lines are the experimental data and the dashed lines are fits using the two-fluid model. In (a), the Drude contribution to $\sigma_2(\omega)$ is plotted as a thin solid line with the condensate fraction plotted as a thin dotted line. The two-fluid model provides a reasonable fit to the data. In Figure 3.16c, $X_N(T)$ and $X_{SC}(T)$ are shown as extracted from fits to the experimental data. The relation $X_N(T) + X_{SC}(T) = 1$ is seen to be accurate to within 10%. At the lowest temperature measured, $X_N(T) = 0.1$, which is still a substantial quasiparticle fraction. In Figure 3.16d, τ is plotted as a function of temperature. Below T_C, there is a dramatic increase in the scattering time until ~50 K where it saturates at 400 fs due to impurity scattering. This increase in τ results in a substantial narrowing of the Drude peak (inset to Figure 3.16a). However, in even the best films, τ is still much shorter than in the highest quality single crystals, where τ is so long that the Drude peak lies at microwave frequencies.

Despite the subtleties of the cuprate properties, the two-fluid model is still a reasonable starting point for understanding their THz response well below the gap, and accordingly for insight into the nonequilibrium dynamics of superconductors. Further discussion is given in Chapter 11, where the two-fluid model is applied to THz-TDS measurements on a non-cuprate superconductor, MgB_2. The two-fluid model also suggests how we can fruitfully apply optical-pump THz-probe spectroscopy to superconductors. A short near-infrared pulse breaks Cooper pairs, thereby reducing the condensate fraction and increasing the quasiparticle

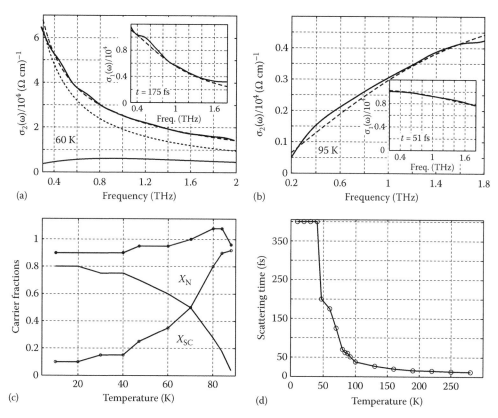

Figure 3.16 THz-TDS measurements on YBa$_2$Cu$_3$O$_7$ with two-fluid model fits. (a) Imaginary conductivity at 60 K, with real part shown in inset. (b) Imaginary conductivity at 95 K, with real part shown in inset. (c) X_N and X_{SC} as a function of temperature as determined from fits. (d) Carrier scattering time as a function of temperature as determined from fits.

fraction. Subsequently, the excess photoexcited quasiparticles will recombine to the condensate. This process can be monitored in exquisite detail using THz pulses, allowing for the simultaneous measurement of the dynamical evolution of quasiparticle and condensate densities, that is, of their time-dependent change $\Delta n_N(t)$ and $\Delta n_{SC}(t)$.

3.5 SUMMARY

We hope that this introductory overview has provided some insight into the role of optical spectroscopy in the study of metals, be they good, poor, or bad. A great deal has been learned, but vast areas remain unexplored and offer fertile ground for new discoveries.

ACKNOWLEDGMENTS

I would like to acknowledge DOE-BES for support under contract DE-FG02-09ER46643.

REFERENCES

Allen, P. B. 1987. Theory of thermal relaxation of electrons in metals. *Phys. Rev. Lett.*, 59, 1460–1463.
Argyres, P. N. 1955. Theory of the Faraday and Kerr effects in ferromagnetics. *Phys. Rev.*, 97, 334.
Ashcroft, N. W. and Mermin, N. D. 1976. *Solid State Physics*, Belmont, CA, Brook Cole.

Averitt, R. D., Lobad, A. I., Kwon, C., Trugman, S. A., Thorsmolle, V. K., and Taylor, A. J. 2001. Ultrafast conductivity dynamics in colossal magnetoresistance manganites. *Phys. Rev. Lett.*, 87, 017401–017404.

Averitt, R. D. and Taylor, A. J. 2002. Ultrafast optical and far-infrared quasiparticle dynamics in correlated electron materials. *J. Phys.: Condens. Matter*, 14, R1357– R1390.

Averitt, R. D., Westcott, S. L., and Halas, N. J. 1999. Ultrafast optical properties of gold nanoshells. *J. Opt. Soc. Am. B*, 16, 1814.

Baldassarre, L., Perucchi, A., Nicoletti, D., Toschi, A., Sangiovanni, G., Held, K., Capone, M., Ortolani, M., Malavasi, L., Marsi, M., Metcalf, P., Postorino, P., and Lupi, S. 2008. Quasiparticle evolution and pseudogap formation in V_2O_3: An infrared spectroscopy study. *Phys. Rev. B*, 77, 113107.

Barnes, W. L., Dereux, A., and Ebbesen, T. W. 2003. Surface plasmon subwavelength optics. *Nature*, 424, 824–830.

Basov, D. N., Averitt, R. D., van der Marel, D., Dressel, M., and Haule, K. 2010. Electrodynamics of correlated electron materials. *Rev. Mod. Phys.*, in press (2011).

Basov, D. N. and Timusk, T. 2005. Electrodynamics of high-T-c superconductors. *Rev. Mod. Phys.*, 77, 721–779.

Beaurepaire, E., Merle, J.-C., Daunois, A., and Bigot, J.-Y. 1996. Ultrafast spin dynamics in ferromagnetic nickel. *Phys. Rev. Lett.*, 76, 4250.

Bennett, H. S. and Stern, E. A. 1965. Faraday effect in solids. *Phys. Rev.*, 137, A 448.

Bigot, J.-Y., Vomir, M., and Beaurepaire, E. 2009. Coherent ultrafast magnetism induced by femtosecond laser pulses. *Nat. Phys.*, 5, 515.

Blundell, S. 2001. *Magnetism in Condensed Matter*, Oxford, U.K., Oxford University Press.

Brorson, S. D., Buhleier, R., Trofimov, I. E., White, J. O., Ludwig, C., Balakirev, F. F., Habermeier, H.-U., and Kuhl, J. 1996. Electrodynamics of high-temperature superconductors investigated with coherent terahertz pulse spectroscopy. *J. Opt. Soc. Am. B*, 13, 1979–1993.

Brorson, S. D., Kazeroonian, A., Moodera, J. S., Face, D. W., Cheng, T. K., Ippen, E. P., Dresselhaus, M. S., and Dresselhaus, G. 1990. Femtosecond room-temperature measurement of the electron-phonon coupling constant γ in metallic superconductors. *Phys. Rev. Lett.*, 64, 2172–2175.

Burns, G. 1985. *Solid State Physics*, Boston, MA, Academic Press.

Cardona, M. 1969. *Modulation Spectroscopy*, New York, Academic Press.

Castellani, C., Natoli, C. R., and Ranninger, J. 1978. Magnetic structure of V_2O_3 in the insulating phase. *Phys. Rev B*, 18, 4945.

Cavalleri, A., Rini, M., Chong, H. H. W., Fourmaux, S., Glover, T. E., Heimann, P. A., Keiffer, J. C., and Schoenlein, R. W. 2005. Band-selective measurements of electron dynamics in VO_2 using femtosecond near-edge x-ray absorption. *Phys. Rev. Lett.*, 95, 067405.

Choi, H. S., Ahn, J. S., Jung, J. H., and Noh, T. W. 1996. Mid-infrared properties of a VO_2 film near the metal-insulator transition. *Phys. Rev. B*, 54, 4621.

Cooper, S. L. 2001. Optical spectroscopic studies of metal insulator transistions in perovskite-related oxides. *Struct. Bonding*, 98, 161–218.

Dagotto, E. 2003. *Nanoscale Phase Separation and Colossal Magnetoresistaince: The Physics of Manganites and Related Compounds*, Berlin, Germany, Springer.

Dagotto, E. 2005. Complexity in strongly correlated electronic systems. *Science*, 309, 257–262.

Degiorgi, L. 1999. The electrodynamics of heavy fermion compounds. *Rev. Mod. Phys.*, 71, 687.

Dressel, M. and Grüner, G. 2002. *Electrodynamics of Solids*, New York, Cambridge University Press.

Driscoll, T., Kim, H. T., Chae, B. G., Kim, B. J., Lee, Y. W., Jokerst, N. M., Palit, S., Smith, D. R., Di Ventra, M., and Basov, D. N. 2009. Memory metamaterials. *Science*, 325, 1518.

Duan, F. and Guojun, J. 2005. *Introduction to Condensed Matter Physics*, Volume 1, Singapore, World Scientific.

Ebert, H. 1996. Magneto-optical effects in transition metal systems. *Rep. Prog. Phys.*, 59, 1665.

Eesley, G. L. 1983. Observation of non-equilibrium heating in copper. *Phys. Rev. Lett.*, 51, 2140–2143.

Erskine, J. L. and Stern, E. A. 1973. Magneto-optic Kerr effect in Ni, Co, and Fe. *Phys. Rev. Lett.*, 30, 1329.

Erskine, J. L. and Stern, E. A. 1975. Calculation of the M_{23} magneto-optical absorption spectrum of ferromagnetic nickel. *Phys. Rev. B*, 12, 5016.

Fazekas, P. 1999. *Lecture Notes on Electron Correlation and Magnetism*, Singapore, World Scientific.

Feinleib, J., Scouler, W. J., and Ferretti, A. 1968. Optical properties of the metal ReO_3 from 0.1 to 22 eV. *Phys. Rev.*, 165, 765–774.

Fujimoto, J. G., Liu, J. M., Ippen, E. P., and Bloembergen, N. 1984. Femtosecond laser interaction with metallic tungsten and nonequilibrium electron and lattice temperatures. *Phys. Rev. Lett.*, 53, 1837–40.

Groeneveld, R. H. M., Sprik, R., and Lagendijk, A. 1995. Femtosecond spectroscopy of electron-electron and electron-phonon energy relaxation in Ag and Au. *Phys. Rev. B*, 51, 11433–11445.

Guidoni, L., Beaurepaire, E., and Bigot, J.-Y. 2002. Magneto-optics in the ultrafast regime: Thermalization of spin populations in ferromagnetic films. *Phys. Rev. Lett.*, 89, 017401.

Harrison, W. A. 1980. *Electronic Structure and the Properties of Solids the Physics of the Chemical Bond*, San Francisco, CA, W.H. Freeman.

Hilton, D. J., Prasankumar, R. P., Trugman, S. A., Taylor, A. J., and Averitt, R. D. 2006. On photo-induced phenomena in complex materials: Probing quasiparticle dynamics using infrared and far-infrared pulses. *J. Phys. Soc. Jpn.*, 75, 011006/1-13.

Hurd, C. M. 1975. *Electrons in Metals*, New York, John Wiley & Sons.

Hwang, H. Y., Cheong, S.-W., Radaaelli, P. G., Marezio, M., and Batlogg, B. 1995. Lattice effects on the magnetoresistance in doped $LaMnO_3$. *Phys. Rev. Lett.*, 75, 914.

Imada, M., Fujimori, A., and Tokura, Y. 1998. Metal-insulator transitions. *Rev. Mod. Phys.*, 70, 1039–1263.

Jin, S., Tiefel, T. H., McCormack, M., Fastnacht, R. A., Ramesh, R., and Chen, L. H. 1994. Thousand fold change in resistivity in magnetoresistive La-Ca-Mn-O films. *Science*, 264, 413–15.

Johnson, P. B. and Christy, R. W. 1972. Optical constants of the nobel metals. *Phys. Rev. B*, 6, 4370.

Kotliar, G. and Vollhardt, D. 2004. Strongly correlated material: Insights from dynamical mean-field theory. *Phys. Today*, 57, 53.

Krebig, U. and Vollmer, M. 1995. *Optical Properties of Metal Clusters*, Berlin, Germany, Springer.

Limelette, P., Georges, A., Jerome, D., Wzietek, P., Metcalf, P., and Honig, J. M. 2003. Universality and critical behavior at the Mott transition. *Science*, 302, 89.

Maier, S. A. 2007. *Plasmonics: Fundamentals and Applications*, Berlin, Germany, Springer.

Marder, M. P. 2000. *Condensed Matter Physics*, New York, John Wiley & Sons.

McWhan, D. B. and Remeika, J. P. 1970. Metal-insulator transition in $(V_{1-x}Cr_x)2O3$. *Phys. Rev. B*, 2, 3734.

Mott, N. F. 1990. *Metal Insulator Transitions*, 2nd edn., London, U.K., Taylor & Francis.

Nasu, K. 2004. *Photoinduced Phase Transition*, Singapore, World Scientific.

Okimoto, Y., Tomioka, Y., and Al, E. 1999. Optical study of $Pr_{1-x}Ca_xMnO_3$ ($x = 0.4$) in a magnetic field: Variation of electronic structure with charge orderin and disordering phase transitions. *Phys. Rev. B*, 59, 7401.

Oppeneer, P. M., Maurer, T., Sticht, J., and Kubler, J. 1992. Ab initio caclulated magneto-optical Kerr effect of ferromagnetic metals: Fe and Ni. *Phys. Rev. B*, 45, 10924.

Pfalzer, P., Obermeier, G., Klemm, M., and Horn, S. 2006. Structural precursor to the metal-insulator transition in V_2O_3. *Phys. Rev. B*, 73, 144106.

Qazilbash, M. M., Brehm, M., Chae, B.-G., Ho, P.-C., Andreev, G. O., Kim, B.-J., Yun, S. J., Balatsky, A. V., Maple, M. B., Keilmann, F., Kim, H.-T., and Basov, D. N. 2007. Mott transition in VO_2 revealed by infrared spectroscopy and nano-imaging. *Science*, 318, 1750.

Qazilbash, M. M., Schafgans, A. A., Burch, K. S., Yun, S. J., Chae, B. G., Kim, B. J., Kim, H. T., and Basov, D. N. 2008. Electrodynamics of the vanadium oxides VO_2 and V_2O_3. *Phys. Rev. B*, 77, 115121.

Rakic, A. D. 1995. Algorithm for the determination of intrinsic optical constants of metal films: Application to aluminum. *Appl. Opt.*, 34, 4755.

Rini, M., Tobey, R., Dean, N., Itatani, J., Tomioka, Y., Tokura, Y., Schoenlein, R. W., and Cavalleri, A. 2007. Control of the electronic phase of a manganite by mode-selective vibrational excitation. *Nature*, 449, 72.

Rosei, R. 1972. Thermomodulation spectra of Al, Au, and Cu. *Phys. Rev. B*, 5, 3883.

Rozenberg, M. J., Kotliar, G., and Kajueter, H. 1996. Transfer of spectral weight in spectroscopies of correlated electron systems. *Phys. Rev. B*, 54, 8452.

Schoenlein, R. W., Lin, W. Z., Fujimoto, J. G., and Eesley, G. L. 1987. Femtosecond studies of nonequilibrium electronic processes in metals. *Phys. Rev. Lett.*, 58, 1680–1683.

Singh, M., Wang, C. S., and Callaway, J. 1975. Spin-orbit coupling, Fermi surface, and optical conductivity of ferromagnetic iron. *Phys. Rev. B*, 11, 287.

Solymar, L. and Shamonina, E. 2009. *Waves in Metamaterials*, Oxford, U.K., Oxford University Press.

Sun, C. K., Vallee, F., Acioli, L., Ippen, E. P., and Fujimoto, J. G. 1993. Femtosecond investigation of electron thermalization in gold. *Phys. Rev. B*, 48, 12365–8.

Takenaka, K., Sawkai, Y., and Sugai, S. 1999. Incoherent-to-coherent crossover of optical spectra in $La_{0.825}Sr_{0.175}MnO_3$: Temperature-dependent reflectivity spectra on cleaved surfaces. *Phys. Rev. B*, 60, 13011.

Tinkham, M. 1996. *Introduction to Superconductivity*, New York, McGraw-Hill.

Tokura, Y. (ed.) 2000. *Colossal magnetoresistive oxides*, Amsterdam, the Netherlands, Gordon and Breach.

Tsuda, N., Nasu, K., Fujimori, A., and Siratori, K. 2000. *Electronic Conduction in Oxides*, Berlin, Germany, Springer.

Wang, C. S. and Callaway, J. 1974. Band structure of nickel: Spin-orbit coupling, the Fermi Surface, and the optical conductivity. *Phys. Rev. B*, 9, 4897.

Wooten, F. 1972. *Optical Properties of Solids*, New York, Academic Press.

Zhang, G. P., Hubner, W., Lefkidis, G., Bai, Y., and George, T. F. 2009. Paradigm of the time-resolved magneto-optical Kerr effect for femtosecond magnetism. *Nat. Phys.*, 5, 499.

PART II

LINEAR OPTICAL SPECTROSCOPY

Chapter 4

METHODS FOR OBTAINING THE OPTICAL CONSTANTS OF A MATERIAL

Hidekazu Okamura

CONTENTS

4.1 OVERVIEW OF THE EXPERIMENTAL TECHNIQUES

4.1.1 Introduction

Optical constants such as the refractive index n, dielectric constant (electrical permittivity) ε, absorption coefficient α, and optical (AC) conductivity σ are characteristic physical quantities related to the properties of a material, as discussed in Chapter 1. Despite the term "constant," they may be strongly varying functions of the frequency, ω. There are probably two main cases where one would like to measure and evaluate the optical constants of a material. In the first case, one is interested in the material's optical properties for optical applications such as lenses, optical filters and fibers, and lasers. Then the optical constants of the material are very important. In the second case, one is more interested in the microscopic properties of a material, such as the electronic structure, chemical bonding, and phonons/molecular vibrations. The optical constants of a material provide rich information about them (specific examples are discussed in Section 4.6).

There are various methods to experimentally evaluate the optical constants of a material. An optical constant is experimentally obtained by analyzing optical spectra, such as the reflectance $R(\omega)$ and transmittance $T(\omega)$, measured for the material of interest. Assuming for simplicity normal incidence in the Fresnel's equations discussed in Chapter 1, $R(\omega)$ and $T(\omega)$ of a material are expressed in terms of the complex refractive index $\hat{n}(\omega) = n_1(\omega) + in_2(\omega)$ as

$$R = \frac{(n_1-1)^2 + n_2^2}{(n_1+1)^2 + n_2^2}, \quad T = 1 - R = \frac{4n_1}{(n_1+1)^2 + n_2^2}. \tag{4.1}$$

(ω dependencies are omitted.) If a material is transparent at ω and if both $R(\omega)$ and $T(\omega)$ can be measured, one can derive $n_1(\omega)$ and $n_2(\omega)$ by solving the above two equations for them (two equations for two unknowns). One can then obtain other complex optical constants such as $\hat{\varepsilon}(\omega)$ and $\hat{\sigma}(\omega)$ as discussed in Chapter 1. (In practice, multiple internal reflections at the sample surfaces may need to be considered.) This method is not applicable in an opaque spectral range since $T(\omega)$ cannot be measured there. Therefore, here we will mainly discuss reflection-based methods for deriving the optical constants. These methods can derive both $n_1(\omega)$ and $n_2(\omega)$ based on reflection measurements only and are more generally applicable to various materials.

4.1.2 Near-Normal Incidence and Grazing Angle Incidence Configurations

Two optical reflection configurations are mainly used to obtain the complex optical constants, *near-normal incidence* and *grazing angle incidence*. In the former case, shown in Figure 4.1a, the intensity spectrum of the reflected light, $I(\omega)$, and that of the incident light, $I_0(\omega)$, are measured to obtain the reflectance $R(\omega) = I(\omega)/I_0(\omega)$. Alternatively, the spectrum of a reference material, $I_r(\omega)$, with known reflectance, $R_r(\omega)$, is measured in addition to the sample to obtain $R(\omega) = [I(\omega)/I_r(\omega)]R_r(\omega)$. The obtained $R(\omega)$ may be analyzed by either Kramers–Kronig (K-K) analysis (Section 4.5.1) or spectral fitting (Section 4.5.2). In the K-K analysis, the K-K relations discussed in Chapter 1 are used to derive the complex optical constants from $R(\omega)$. The complex reflectivity of an electromagnetic wave can be expressed as follows (Wooten 1972, Dressel and Gruner 2002):

$$\hat{r}(\omega) = r(\omega)\exp[i\phi(\omega)]. \tag{4.2}$$

Here, $r(\omega)$ is the reflectivity of the electric field amplitude, related to the measured reflectance $R(\omega)$ by $r(\omega) = \sqrt{R(\omega)}$. $\phi(\omega)$ is the phase shift upon reflection. $r(\omega)$ and $\phi(\omega)$ are not independent but are related through the K-K relation (Chapter 1, Wooten 1972):

$$\phi(\omega) = -\frac{2\omega}{\pi} P \int_0^{+\infty} \frac{\ln r(\omega')}{\omega'^2 - \omega^2} \, d\omega', \tag{4.3}$$

where P denotes the Cauchy principal value. One can therefore calculate $\phi(\omega)$ from the measured $R(\omega)$ and derive all the other complex optical constants from them (Wooten 1972, Burns 1990). In contrast to the K-K analysis, the spectral fitting analysis relies on specific optical models. Namely, a theoretical $R(\omega)$ is calculated based on the Drude–Lorentz model discussed in Chapter 1, and the parameters involved, such as the peak energy, damping rate, and plasma frequency, are adjusted to fit the measured $R(\omega)$. The K-K technique is more generally applicable, while spectral fitting is useful only when distinct spectral features due to free carriers, phonons, and interband transitions are clearly identified in $R(\omega)$.

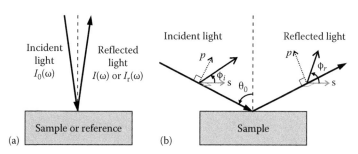

Figure 4.1   Optical reflectivity at (a) near-normal incidence and (b) grazing-angle incidence. In (a), $I_0(\omega)$, $I(\omega)$, and $I_r(\omega)$ indicate the spectra of the incident light, reflected light from the sample, and reflected light from the reference, respectively. In (b), the thick, open arrows indicate the polarization vectors, and p and s denote their components along the plane of incidence and the sample surface, respectively. θ_0 is the angle of incidence and ϕ_i and ϕ_r are the polarization directions. In reality, the reflected light may become elliptically polarized, but for simplicity, it is drawn as linearly polarized.

Grazing angle incidence is widely used with the ellipsometry technique for deriving the optical constants. In this technique, as illustrated in Figure 4.1b, a linearly polarized incident beam is reflected off the sample surface with the angle of incidence set to Brewster's angle, which is generally ~70° (Chapter 1). Then, the ratio of complex reflectivity (Equation 4.2) between p and s polarization components is evaluated (Fujiwara 2007)

$$\hat{\rho} \equiv \frac{\hat{r}_p}{\hat{r}_s} = \tan \psi \cdot \exp(i\Delta). \qquad (4.4)$$

Here
   $\tan \psi$ is the ratio of field amplitude reflectivity
   Δ is the difference in the phase shift between the p and s polarizations

In practice, $\tan \psi$ and Δ can be derived by measuring the reflected light intensity, rather than the reflectance, as a function of polarization direction. The complex optical constants can be calculated from the obtained $\tan \psi$ and Δ (Fujiwara 2007).

The advantages and disadvantages of K-K analysis at normal incidence and ellipsometry at grazing angle incidence are compared in the following:

1. *Spectral range requirement*: The K-K method requires $R(\omega)$ data over a wide frequency range, since the integration in K-K relations runs from $\omega = 0$ to ∞. Even if one is only interested in the infrared (IR), $R(\omega)$ generally has to be measured up to the visible or ultraviolet (UV). In contrast, ellipsometry can give the complex optical constants from data over a narrow spectral range, even at a single frequency, without using K-K analysis.
2. *Experimental setup*: The K-K method generally has a simpler setup—it utilizes a rather simple near-normal-incidence reflection configuration, while ellipsometry requires additional optical elements such as polarizers, wavelength plates, and modulators.
3. *Measurement procedure and time*: In the K-K method, to determine $R(\omega)$, not only the sample but also a reference (or incident light spectrum) need to be measured. With ellipsometry, in contrast, only the sample needs to be measured, as a function of polarization direction.
4. *Spatial resolution*: The K-K method at near-normal incidence generally gives much higher spatial resolution. $R(\omega)$ of a material only weakly depends on the incident angle near normal incidence (Chapter 1). This enables focusing of the light beam with a spot diameter of ~1 mm or smaller in both the visible and the mid-infrared (MIR). (A microscope may be used to further reduce the spot size, as discussed later.) In ellipsometry, in contrast, a collimated beam is ideal, because the signal intensity more strongly depends on the incident angle, especially in p polarization. Combined with the large angle of incidence, the spot size in an ellipsometry experiment is generally much larger.
5. *Sensitivity*: For highly reflective metals and thin films/layers, ellipsometry is generally more sensitive for measuring their optical constants. At normal incidence, it is not easy of such metals is very close to 1 and that of a thin film is very low due to the small interaction volume.

One can see that these two methods have different advantages and disadvantages. Roughly speaking, the K-K method at normal incidence is simpler in terms of instrumentation and data analysis, provided the sample of interest is sufficiently thick and its reflectance is not too close to 1. Ellipsometry at grazing angle incidence is most useful when one needs to deal with thin films and layers or when one needs to quickly measure the complex optical constants over a certain spectral range, provided the sample has a sufficiently large area.

In the following sections, the near-normal incidence method is discussed in detail. The experimental techniques in the visible–UV and IR regions are discussed in Sections 4.2 through 4.4: Section 4.2 describes visible–UV spectroscopy, Section 4.3 discusses Fourier transform infrared spectroscopy (FT-IR), and Section 4.4 gives some examples of actual experimental apparatuses. Data analysis and examples of measured optical constants are discussed in Sections 4.5 and 4.6. For detailed discussions on ellipsometry, there are many good references. For example, Fujiwara, (2007) offers an excellent introduction to the principles of spectroscopic ellipsometry. Tompkins and Haber (2006) is a more expanded handbook of ellipsometry, with various applications described in detail. Schubert, (2009) focuses on the IR applications of ellipsometry.

4.2 VISIBLE–UV SPECTROSCOPY

Evaluation of the complex optical constants in the UV and visible region is important for the characterization of band structures in bulk materials such as semiconductors, insulators, and metals. For example, the ω-dependent complex dielectric function $\hat{\varepsilon}(\omega)$ and the optical conductivity $\hat{\sigma}(\omega)$ have been measured for typical semiconductors such as Si and GaAs (Chapter 2, Philipp and Ehrenreich 1963), insulators such as $BaTiO_3$ and $SrTiO_3$ (Cardona 1965), and metals such as Au, Ag, Cu, and Al (Chapter 3, Ehrenreich and Philipp 1962, Wooten 1972) and have been compared with band calculations. In addition, knowledge of the optical constants is important for the characterization of materials to be used for practical optical device applications. Visible–UV spectroscopy with K-K analysis is a powerful technique for obtaining the complex optical constants.

4.2.1 Visible–UV Spectrometers and Their Components

4.2.1.1 Overview

A spectrometer for measuring $R(\omega)$ in the visible and UV consists of a light source, a monochromator or polychromator, and a detector. Figure 4.2 depicts the components of a typical visible–UV spectrometer and how they are used to measure $R(\omega)$ in two different configurations. Optical elements such as mirrors and lenses are used to manipulate the light beams. Monochromators and polychromators are instruments that disperse the different wavelength

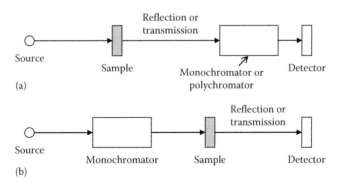

Figure 4.2 Schematic diagrams showing the components and operation of a spectrometer. In (a), the light from the source is sent to the sample, and then the reflected or transmitted light is wavelength dispersed by a monochromator or a polychromator and recorded by a detector. In (b), the light from the source is first dispersed by a monochromator, and the resulting monochromatic light is sent to the sample and recorded by a detector. As discussed later, a polarizer may be placed before or after the sample (not shown in the figure) to study anisotropic optical responses.

components of the light emitted by the source. A monochromator outputs a particular wavelength component from the input light, while a polychromator outputs different wavelength components simultaneously. In most modern visible–UV monochromators and polychromators, the wavelength dispersion is created by a diffraction grating. Various types of visible–UV spectrometers, which contain the above optical components and a control computer, are commercially available. One may also design and construct an efficient spectrometer with relatively lower cost. In the following, we briefly summarize the key components used in spectrometers.

4.2.1.2 Light Sources

A broadband, "white" source is needed to cover the wide spectral range required for the reflection K-K method. In the UV, common sources are the Xe arc lamp and the D_2 (deuterium) lamp. They are discharge lamps, which utilize light emission from excited states of Xe and D_2. A Xe arc lamp has a continuous emission spectrum over a wide wavelength range of $\lambda = 200–2000$ nm and nicely covers the entire UV and visible ranges. On the other hand, a D_2 lamp emits over a narrower wavelength range of $\lambda = 120–600$ nm* but generally has a longer lifetime than a Xe lamp. Since UV radiation below 200 nm (6.2 eV) is strongly absorbed by air, a D_2 lamp is generally used above 200 nm for reflection K-K studies. In the visible, the halogen lamp, which utilizes the black body radiation from a tungsten filament at ~3000 K, is the most common source. A hybrid D_2/halogen lamp, where both D_2 and halogen lamps are built into a single package with a single output beam, is also available to cover the entire visible–UV range. Some commercial sources also have an integrated fiber optic output, enabling a portable white light fiber source that can be directly introduced into the optical system.

Since an actual light source has a finite size, its image projected on the sample surface also has a finite size. To accurately measure the reflectance, it is important to keep the beam diameter on the sample sufficiently small. If not, an aperture should be placed in front of the source to reduce its effective size, in addition to the focusing optics. (The required beam size depends on the size of the sample, but a diameter of 1 mm is usually sufficient.)

4.2.1.3 Monochromators

Figure 4.3a illustrates the operation principle of a Czerny–Turner-type grating monochromator. (Many other types are available, but the fundamental principle of operation is essentially the same.) The incident light is focused onto the entrance slit S_1 and then the input beam is collimated by the concave mirror M_1 and diffracted by the grating G. The diffracted beam is

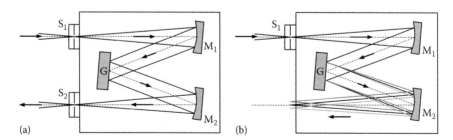

(a)   (b)

Figure 4.3 Illustration of a Czerny–Turner-type monochromator (a) and polychromator (b). S_1 and S_2 indicate entrance and exit slits, M_1 and M_2 denote collimating and focusing mirrors, and G denotes a diffraction grating, respectively.

---

* The long wavelength cutoff varies among different commercial D_2 lamps. The intensity is much weaker above 400 nm than below 400 nm. The short wavelength cutoff depends on the window material, which is ~120 nm for MgF_2 and ~160 nm for fused silica.

focused onto the exit slit S_2 by the concave mirror M_2. A particular wavelength component is extracted through the exit slit by rotating the grating. A spectrum is obtained by recording the output light intensity while the wavelength is scanned by rotating the grating.

The spectral resolution of a monochromator is roughly given by the spectral bandwidth of the output light, which is determined by the opening width of the slits and the dispersion created by the grating, usually expressed in nm/mm or Å/mm (bandwidth per 1 mm slit width). The dispersion depends on the groove density of the grating, the focal length of the concave mirrors, and the angle at which the grating is used. For reflection K-K studies where a high resolution is not required, a groove density of 300–600 mm^{-1} and a focal length of 10–25 cm are commonly used.

The spectral range covered by a monochromator depends on the groove density, coating material, and blaze angle of the grating (if it is a blazed type) (Hecht 2001). To cover a wider spectral range, some monochromators are equipped with two or three gratings with different groove densities and/or coatings, which are mounted on a rotating table and can be switched with one another. Only one grating is used at a time, unlike in a double- and triple-grating monochromator used for high-resolution applications such as Raman spectroscopy (Chapter 6). The wavelength of a monochromator should be calibrated with a standard source such as a low-pressure Hg lamp, which emits discrete lines with known wavelengths.

4.2.1.4 Polychromators

Figure 4.3b illustrates a polychromator, which is also of the Czerny–Turner type. The difference from the monochromator in Figure 4.3a is that a polychromator does not have an exit slit. Therefore, a wide wavelength range can be output and detected simultaneously by a multichannel (multielement) detector such as a photodiode array or a charge-coupled device (CCD) detector. Thanks to recent technological advancements in micro-fabrication, very compact, "mobile" polychromators with a built-in multichannel detector have become available. They may allow measurement of a spectrum in the entire visible–UV range (200–800 nm) in a matter of seconds. This is very advantageous for the K-K method, especially when other experimental parameters such as temperature, magnetic field, and external pressure are varied. The wavelength of a polychromator can also be calibrated as discussed for a monochromator.

4.2.1.5 Mirrors

To propagate and manipulate the light beam (focusing, collimating, etc.), various types of lenses and mirrors are used (Hecht 2001). One should be careful in using lenses since most lenses have chromatic aberration when used over the entire visible–UV range. This can cause a spatial distribution of different wavelength components, which may result in an erroneous spectrum. For example, a fused silica lens designed to have a focal length of $f = 150$ mm at $\lambda = 550$ nm has $f = 136$ mm at $\lambda = 250$ mm. Achromatic lenses that have little chromatic aberration are available; however, they usually cover the visible range only or the UV range only.

By using a mirror, one can avoid chromatic aberrations. Commonly used focusing/collimating mirrors include spherical, off-axis parabolic (paraboloidal), and off-axis ellipsoidal mirrors. A spherical mirror has the advantage of being inexpensive, but the monochromatic aberrations may become large for light rays far off the optical axis. If such aberrations are significant, for example, the light beam may not be well focused on the sample surface. To reduce this effect, one may use, for example, an aperture to select the light near the optical axis only. A parabolic mirror, in contrast, ideally has no monochromatic aberrations; it can focus parallel light rays into a point or collimate the light from a point source into parallel rays. One should also be careful with the surface flatness of a mirror for UV applications. The flatness is usually expressed relative to the wavelength of a He–Ne laser, $\lambda = 632.8$ nm. A mirror with a flatness of λ would be good enough for the long wavelength IR range but not for the short wavelength UV range because Mie scattering due to the surface roughness would significantly reduce the light intensity in the UV. (This also applies to UV windows and lenses.)

One may also choose from various surface materials for mirrors. To cover a wide spectral range in the visible–UV region, aluminum is the most common material. The reflectance of a good aluminum surface is >0.9 throughout the visible–UV range, but there is a weak dip centered at ~800 nm (Chapter 3). Uncoated aluminum may be gradually oxidized, causing its reflectance to decrease with time. Protective coatings are used to avoid this, but one should use them carefully since they may introduce some spectral structure.

4.2.1.6 Detectors

Common detectors used in the visible–UV range include Si photodiodes, photomultiplier tubes (PMTs), and CCD detectors (more detail on these detectors is provided in Chapter 5). Since the signal intensity is relatively high in reflectance experiments, photodiodes and uncooled CCD detectors generally provide enough sensitivity and signal-to-noise ratio (SNR). (If not, a PMT or a Peltier-cooled CCD detector may be used.) For multichannel detection with a polychromator, photodiode array and CCD detectors are often used, as already discussed above. Due to the high signal intensity in reflectance studies, one should always ensure that the detector is not saturated and that there is linearity between the light intensity and the output electrical signal. The light intensity entering the detector should be kept low enough to avoid saturation-related errors.

4.2.1.7 Polarizers

Measurements of $R(\omega)$ with different polarization states of light are often useful in studying the electronic states of materials, in particular those with an anisotropic crystal structure. (A specific example is given in Section 4.3.3.6) Although both circular and linear polarizations can be used, here we discuss the linear case only. Circular polarization is useful in studying, for example, anisotropic atomic orbitals and chiral molecules in chemistry and biology. Linear polarizers based on birefringent crystals such as calcite are commonly used in the UV and visible ranges and those based on polymer sheets are also used in the visible (Hecht 2001). Note that the mirrors in the optical path may have different reflectances with p and s polarizations (Section 4.1 and Chapter 1) since the mirrors in a reflectance apparatus may have a large angle of incidence (Figures 4.4 and 4.9). To record the polarization dependence of $R(\omega)$, therefore, it is generally better to fix the polarizer direction and to rotate the sample, rather than to rotate the polarizer and to fix the sample direction.

4.2.2 Optical Layouts, Reference Material, and Optical Alignment

As already stated, to perform a K-K analysis, $R(\omega)$ of a sample must be measured either relative to a reference material with known reflectance or relative to the spectrum of incident light. Examples of typical optical layouts for these two methods are illustrated in Figure 4.4. In the "reference method" shown in Figure 4.4a, the same optical layout is used to measure the sample and reference by moving the holder on which they are both mounted.

In the "no-reference" method of Figure 4.4b and c, optical components are also moved to measure the spectrum with and without the sample. Although this method is desirable in the sense that it does not require a reference, it is not technically easy, as discussed later.

In the reference method of Figure 4.4a, the sample holder must be moved precisely, for example, by a micrometer-driven manipulator, to keep the surfaces of the sample and the reference aligned in the same plane. If not, the reflected beams from the sample and reference will travel different optical paths, which may cause a significant error in the measured $R(\omega)$. Figure 4.4d illustrates an example of a sample holder, which is machined from a copper block.

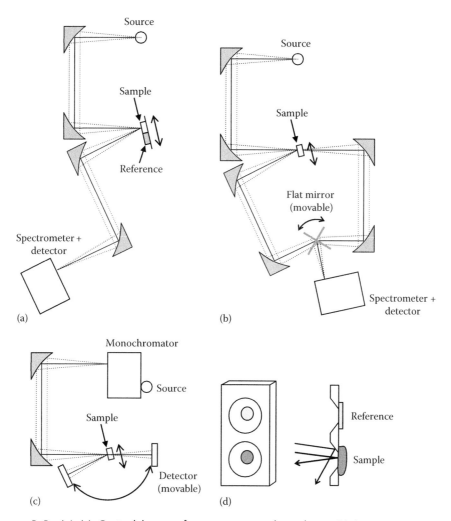

Figure 4.4 (a)–(c) Optical layouts for a near-normal incidence $R(\omega)$ measurement in the visible–UV region. In (a), a reference material with a known $R(\omega)$ is used, but in (b) and (c) no reference is used. Two different optical paths to record the spectra with and without the sample, switched by turning a flat mirror in (b) and by moving the detector in (c) are instead used. (d) An example of a sample holder and its cross section. See text for more details.

This holder keeps the sample and reference surfaces along the same plane and also keeps their active areas the same. Incident light that is not on the sample will be rejected from the correct optical path as illustrated. Good heat conduction and cooling of the sample is ensured by mounting it with conducting paste.

As reference materials, Ag and Al mirrors are often used in the visible and UV, respectively. As discussed in Chapter 3, a fresh, uncoated surface of Ag has a reflectance very close to 1 in the visible ($\lambda > 400\,\text{nm}$), and Al has a high (~0.9), almost constant reflectance in the UV (200–400 nm). One may easily deposit films of Ag and Al on a glass substrate by vacuum evaporation or use commercially available mirrors. It is important to use a freshly or recently deposited film since the reflectance of Ag and Al mirrors gradually decreases with time due to oxidization. One should not use coated mirrors as a reference since the coating may cause spectral structure, even though they are more durable. In any case, one should accurately know or estimate the reflectance of the reference, since an uncertainty in it directly leads to that in the measured $R(\omega)$ of the sample. One should always remember that the reflectance of a specific Ag or Al mirror may be substantially different from the standard data in the published literature.

In the no-reference method of Figure 4.4b and c, two different optical paths are used to record spectra with and without the sample. Needless to say, the two optical channels must be aligned as symmetrically as possible to ensure that the two beams experience identical optical paths. This may sound easy but can be quite difficult in practice. Although the configuration of Figure 4.4c is simpler than that in Figure 4.4b, the former would be difficult if the beam divergence is large compared with the detector size. (See also the discussion in Section 4.4.2.) Nevertheless, the no-reference method, once carefully and successfully implemented, has the obvious advantage of not requiring a reference. This method is also important in vacuum UV (VUV), as discussed in Sections 4.2.4 and 4.4.2.

Finally, methods for aligning a spectrometer should be mentioned here. Roughly speaking, there are two main steps in aligning a spectrometer. First, one follows the visible light from the source by eye and adjusts optical components so that the light beam from the source correctly reaches the sample and then the detector. This can be done while wearing UV protection glasses, by using a sheet of paper to follow the light beam along the optical path. As the second step, one tries to maximize the detector signal by fine-adjusting the angles and/or positions of the optical components and the sample. It is important to do a visual alignment before such a signal-based alignment, since the latter alone might result in serious errors, such as picking up signal from the sample holder rather than the sample and reference.

4.2.3 Use of a Microscope in Visible–UV Spectroscopy

If the sample to be studied by reflectance is large and has a flat surface, the measurement is relatively straightforward. However, it happens often that only a small sample is available for the material of interest or that the sample has many grains or inhomogeneities, so one needs to perform a spatially resolved measurement on the sample. In such cases, it is common to use an optical microscope to focus the light beam to a small spot. The optical layout of a microscope used for a reflectance measurement is very similar to that used for micro-photoluminescence spectroscopy (Chapter 16), so it is not discussed in detail here.

4.2.4 Vacuum UV Region

At wavelengths shorter than 200 nm, the entire optical system must be kept in vacuum due to strong absorption of light by air, hence the term vacuum UV (VUV). The spectral range covered by a commercial, broadband source is only above 120 nm (Section 4.2.1). For wavelengths shorter than 120 nm, synchrotron radiation (SR) is the only broadband source available.* SR refers to the electromagnetic waves radiated by a beam of high-energy electrons under a magnetic field. SR is a very bright, completely white source whose spectrum extends from the IR up to x-ray, providing an indispensable spectroscopic tool for various fields. Many SR facilities in the world have VUV beam lines open to general users, and one may submit a proposal to obtain beam time. For $R(\omega)$ measurements in the VUV, the no-reference method discussed in Section 4.2.2 is used due to a lack of a good reference material. An actual example of a SR-based VUV spectroscopy facility is discussed in Section 4.4.2.

---

* Before SR became available as a VUV source, special discharge sources were used with an evacuated sample chamber to produce discrete spectral lines. The reflectance of a sample was measured at these discrete wavelengths, and then the data points were interpolated to produce a continuous $R(\omega)$.

4.3 FOURIER TRANSFORM INFRARED SPECTROSCOPY (FT-IR)

4.3.1 Overview

The optical constants of a material in the IR range are important in many fields of basic and industrial research. For example, there are various important materials used in IR-based technologies such as remote sensing, thermography, and night vision. In addition, rich information can be obtained about the molecular vibrations in organic materials by IR spectroscopy. The natural (eigen-) frequencies of various molecules and chemical bonds appear as sharp absorption lines in the range 500–3500 cm⁻¹.* They may be regarded as the "fingerprints" of the molecules and are very useful for the analysis of various materials and devices, ranging from basic science and industrial research to areas as diverse as archaeological and forensic investigations. In particular, the range 1000–2000 cm⁻¹ contains a large number of important natural frequencies and is therefore referred to as the "fingerprint region." These characteristic frequencies of various molecules have been thoroughly studied and listed in tables (Socrates 2004). In solid state physics and materials science, on the other hand, a wide range of low-energy excitations may be studied in the IR region, as summarized in Figure 4.5. Common examples are optical excitations related to a semiconducting band gap (Yu and Cardona 2001), a superconducting state (Basov and Timusk 2005), or a charge density wave state (Sacchetti et al. 2007). Optical phonons may also exhibit strong structure in the far IR (Chapter 2). Furthermore, the dynamics of free charge carriers, either electrons or holes, give rise to the so-called Drude spectral component (Chapters 1 and 3).

Today the most common technique for performing IR spectroscopy is Fourier transform infrared spectroscopy (FT-IR), which relies on an interferometer rather than a grating-based spectrometer. In FT-IR, an interference pattern of the source is measured and Fourier transformed to obtain the spectrum. This interferometer-based operation offers many advantages, as discussed later. In fact, it is the development of the FT-IR technique that has made IR spectroscopy so popular for the characterization and analysis of modern materials.

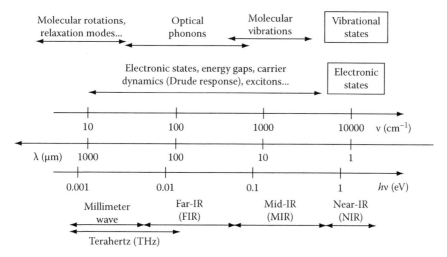

Figure 4.5 Summary of various phenomena, which have their characteristic frequencies (energies) in the IR range, indicated relative to the wave number (ν), wavelength (λ), and photon energy (hν). Shown at the bottom are the technical terms specifying particular wave number ranges, frequently used in the IR literature.

---

* In the FT-IR literature, the frequency of light is usually indicated in units of cm⁻¹, i.e., the wave number. For example, a wave number of 10,000 cm⁻¹ corresponds to a wavelength of 1 μm, 1,000 cm⁻¹ to 10 μm, etc.

Below we will discuss only the essential principles and instrumentation of FT-IR; for a more detailed account, there are many good references. For example, the textbooks by Smith (1995) and Stuart (2004) give a good introduction for those who are new to the FT-IR technique. The textbook by Griffiths and De Haseth (2005) provides thorough mathematical descriptions of the principles, data processing, and analysis of FT-IR, as well as a comprehensive description of the latest instrumentations and applications.

4.3.2 Principles and Advantages of FT-IR

As sketched in Figure 4.6a, a typical FT-IR spectrometer consists of a source, a Michelson interferometer with a beam splitter (BS), and a detector. The sample is usually placed between the BS and the detector in both reflection and transmission configurations but is not indicated here for simplicity. The light from the source, with a spectrum $S(\lambda)$, is split into two beams by the BS. The two beams are reflected by the two mirrors, recombined by the BS, and directed to the detector. One of the two mirrors can be moved, and the detector output is recorded as a function of the mirror position, defined as $x/2$ so that x gives the optical path difference of the two beams. The resulting interference pattern $I(x)$, called the interferogram, is Fourier transformed with a computer to obtain the spectrum $S(\lambda)$. The fact that $I(x)$ is the Fourier

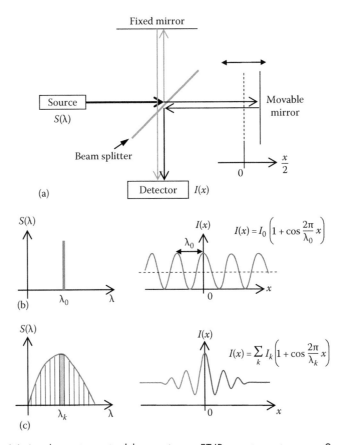

Figure 4.6 (a) A schematic optical layout in an FT-IR spectrometer. $x = 0$ corresponds to the zero path difference between the two beams. The source spectrum $S(\lambda)$ and the resulting interferogram $I(x)$ are sketched for (b) monochromatic and (c) broadband sources. See text for more details.

transform of $S(\lambda)$ can be easily seen as follows. First, assume a monochromatic source with a wavelength of λ_0 and intensity I_0, as shown in Figure 4.6b. Then, as x is varied, $I(x)$ will oscillate between a maximum value of $2I_0$ and zero due to the interference between the two beams, with the form

$$I(x) = I_0\left(1 + \cos\frac{2\pi}{\lambda_0}x\right),\tag{4.5}$$

which is drawn in the graph of Figure 4.6b. If the source is broadband, its spectrum may be regarded as a collection of many wavelength components λ_k, as shown in Figure 4.6c. Then $I(x)$ will be a superposition of Equation 4.5 with different wavelengths λ_k and amplitudes I_k, expressed as

$$I(x) = \sum_k I_k\left(1 + \cos\frac{2\pi}{\lambda_k}x\right).\tag{4.6}$$

Here, the amplitude I_k of a particular λ_k component in Equation 4.6 should be proportional to $S(\lambda_k)$, according to the superposition principle of linear systems. Then Equation 4.6 may be rewritten as

$$I(x) \propto \sum_k S(\lambda_k)\left(1 + \cos\frac{2\pi}{\lambda_k}x\right).\tag{4.7}$$

In Equation 4.7, the summation from the constant part only gives a constant but that from the cosine part is a Fourier series. In the limit of a large number of λ_ks, (4.7) may be expressed as follows:

$$I(x) \propto \frac{1}{L}\int_0^\infty S(\lambda)\left(1 + \cos\frac{2\pi}{\lambda}x\right)d\lambda,\tag{4.8}$$

where the constant L has dimensions of length. Equation 4.8 shows that the cosine part of $I(x)$ is indeed the Fourier transform of $S(\lambda)$.

There are three major advantages of an FT-IR spectrometer over a grating-based IR spectrometer (Griffiths and De Haseth 2005):

1. *Multiplex advantage*: An FT-IR spectrometer detects the entire spectrum at once. In contrast, a grating monochromator, as discussed in Section 4.1, detects only a narrow spectral width at once and requires a long time to scan the wavelength. This multiplex advantage of FT-IR enables much faster acquisition of a spectrum and hence a much higher SNR.*
2. *Throughput advantage*: An FT-IR spectrometer has a high optical throughput since it operates without a slit and hence utilizes the entire input light beam. This results in a high SNR in the obtained spectrum. A grating spectrometer, in contrast, relies on slits for its operation and hence utilizes only a small portion of the input beam.

---

* A grating polychromator with a multichannel IR detector would have a similar advantage, but currently, such instruments are available only above 4000 cm^{-1} (with an InGaAs array detector).

3. *Accuracy advantage*: In an FT-IR spectrometer, the wavelength in a measured spectrum is measured very accurately since it is determined by the position of the moving mirror. The mirror position can be accurately measured relative to the wavelength of a He–Ne laser by simultaneously recording the interference fringe of the laser and the interferogram. Hence, an FT-IR does not require spectral calibration with a standard source, unlike a grating spectrometer.

In most FT-IR instruments, a "rapid scan" mode of operation is employed, where the mirror is moved back and forth rapidly and repeatedly (on an air-bearing or mechanical linear motion stage, with a typical speed of $1 \sim 20$ mm/s) while $I(x)$ is recorded. This enables one to accumulate a large number of interferograms quickly, further enhancing the high SNR given by (1) and (2) above.

In practice, $I(x)$ can be recorded only over a finite optical path difference although the integration in (4.8) runs to ∞. The maximum path difference, x_m, gives the spectral resolution in the obtained spectrum as $\Delta v = 1/x_m$, where $v = 1/\lambda$ is the wave number. (The same is true for terahertz time-domain spectroscopy (THz-TDS) [Chapter 11].) For example, a spectral resolution of 2 cm^{-1}, which is a typical value for studying molecular vibrations, would require $x_m = 0.5$ cm, and a mirror displacement of 0.25 cm from zero path difference. In addition, $I(x)$ is measured discretely with an interval Δx, which defines the spectral bandwidth: the largest wave number in an obtained spectrum is given as $v_m = 1/(2\Delta x)$.* Equivalently, Δx should be set to $\lambda_m/2$ or smaller where $\lambda_m = 1/v_m$ is the shortest wavelength to be measured.

Before performing the Fourier transform of $I(x)$, the finite range of x should be extended to ∞. In addition, although $I(x)$ is ideally symmetric about $x = 0$, it is generally not so in actual experiments. An FT-IR software uses a procedure called apodization to extend the x range to ∞ and another called phase correction to deal with an asymmetric interferogram. The details of these procedures (Griffiths and De Haseth 2005) are not discussed here, but even if a user is unfamiliar with them, the FT-IR software generally takes care of them very well. The numerical Fourier transform is performed using an algorithm called the fast Fourier transform (FFT) (Cooley and Tukey 1965), which greatly helped the development of the FT-IR technique. With a modern PC, the FFT of an interferogram, including the apodization and phase correction, can be done in a few seconds (unless the spectral resolution is set to be very high).

4.3.3 FT-IR Spectrometer and Its Components

The specifications of an FT-IR instrument, such as the spectral range and resolution, vary widely among different commercial instruments. An FT-IR instrument is either vacuum pumped or nitrogen (or dry air) purged, to avoid the strong absorption of IR radiation by molecular vibrations of H_2O and CO_2 in air. A vacuum-pumped FT-IR is more expensive and massive than a purged one, all else being equal, but the former generally has more accuracy and stability. There are two modes of operation for scanning the mirror, namely rapid scan and step scan. The rapid scan mode, already discussed above, is employed by most commercial FT-IR instruments.[†]

The spectral range covered by an FT-IR instrument depends on the specific combination of optical components used. Typical spectral ranges covered by various sources, BSs, and detectors are summarized in Table 4.1. Examples of measured spectra with several typical

---

* Here the factor of 2 comes from the sampling theorem of discrete signal processing (Griffiths and De Haseth 2005) and is usually taken into account within the control software of an FT-IR.

† In the step scan mode, the mirror is moved step-by-step, and the signal is accumulated at each mirror position (Griffiths and De Haseth 2005). This mode may be useful when the signal is weak. Signals in reflectance studies are relatively strong, so the rapid scan is usually more advantageous.

TABLE 4.1

Spectral Ranges Covered by Various IR Sources, Beam Splitters, Detectors, and Optical Window Materials

Sources	Spectral Range (cm^{-1})	Comments
Globar lamp	100–6,000	
Halogen lamp	3,000–25,000	
Mercury lamp	10–700	Water-cooled, high-pressure type. Brighter than a globar below 100 cm^{-1}.

Detectors	Spectral Range (cm^{-1})	Comments
TGS (tryglycine sulphate) DTGS (deuterated TGS)	50–10,000	Spectral range depends on the window used (PE for FIR, KRS-5 for MIR, etc.).
Si bolometer	10–700	Liquid helium cooled.
MCT (HgCdTe)	450–8,000 (Broadband type)	Liquid nitrogen cooled. Various types with different spectral ranges and sensitivities are available.
InGaAs photodiode	5,000–11,000	Visible-enhanced type (up to 20,000 cm^{-1}) is also available.
Si photodiode	9,000–50,000	

BSs	Spectral Range (cm^{-1})	Comments
Ge film on KBr (Ge/KBr)	400–7,000	KBr is highly hygroscopic.
Quartz (coated)	3,000–25,000	Spectral range depends on the coating.
CaF$_2$ (coated)	3,000–15,000	Spectral range depends on the coating.
Mylar (polyethylene terephthalate) with thickness d	50–700 (multilayer) 100–550 ($d = 6\,\mu m$) 30–120 ($d = 23\,\mu m$)	"Multilayer" refers to a 6 μm thick Mylar BS coated with a Ge film.

Windows	Spectral Range (cm^{-1})	Max. Transmission, Comments
Quartz (SiO$_2$)	3,000–60,000 (160 nm)	~90%
BaF$_2$	900–50,000	~90%
KRS-5 (mixed TlBr–TlI)	250–16,000	~75%, orange color to the eye.
NaCl	700–UV	~90%, hygroscopic (but less than KBr).
KBr	400–UV	~90%, highly hygroscopic, soft material.
Polyethylene Polypropylene	Below 700 (also visible range for transparent PE and PP)	Transmission depends on the thickness. White, transparent, and black PEs with various thicknesses are used depending on the desired mechanical strength and spectral range.
Diamond (type IIa)	Throughout FIR–UV	Two-phonon absorption at 2000–3500 cm^{-1}.

combinations of these optical components are shown in Figure 4.7. Traditionally, the terms "far-infrared" (FIR), "mid-infrared" (MIR), and "near-infrared" (NIR) have been used to distinguish different portions in the IR range, as indicated at the bottom of Figure 4.5. There is no strict definition of the three regions, and here we will use them as follows: By MIR, we mean a range 500–7000 cm^{-1}, which is typically covered by a combination of a Ge-coated KBr BS and an HgCdTe detector. The MIR range is the most commonly used one for FT-IR, since it contains the fingerprint region. By FIR, we mean the region below 700 cm^{-1}, typically covered by a liquid He-cooled Si bolometer and a Mylar BS. By NIR, we mean the region above 6000 cm^{-1}, covered by coated quartz or CaF$_2$ BS and a photodiode detector.

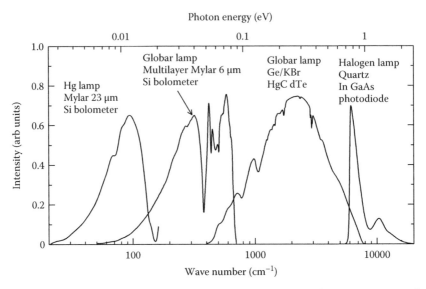

Figure 4.7 Output spectra obtained with several combinations of sources, BSs, and detectors. They were recorded using the apparatus shown in Figure 4.8a, with a gold or silver mirror placed at the sample position. (H. Okamura, unpublished data.)

4.3.3.1 Light Sources

A globar lamp, namely a rod made of SiC heated to ~1400 K, is a common source of black body radiation covering the MIR and FIR ranges (100–6000 cm⁻¹). It is inexpensive, has a long lifetime, and is very stable in its emission intensity. A mercury lamp is brighter than a globar lamp below 100 cm⁻¹ due to additional radiation from the plasma inside the lamp. Finally, a halogen lamp, which is already discussed in Section 4.2.1, is a common broadband source for the NIR region, covering a range above 3000 cm⁻¹.

4.3.3.2 Beam Splitters

The BS is an important component in an FT-IR spectrometer. A Ge-coated KBr (Ge/KBr) BS generally covers 400–7000 cm⁻¹ and is the most widely used BS in the MIR region. KBr is an extremely hygroscopic (moisture absorbing) material, so extreme care must be taken in its handling. In the FIR, Mylar* films with various thicknesses between 3 and 50 μm have been used as BSs. The transmission spectrum of a single layer Mylar BS has a periodic variation with wave number due to internal reflections, so an appropriate thickness should be chosen, depending on the desired spectral region. For example, Mylar BSs with thicknesses of 6.25, 12.5, 25, and 50 μm cover 80–450, 40–220, 20–100, and 12–50 cm⁻¹, respectively (Griffiths and De Haseth 2005). More recently, multilayer (Ge-coated) Mylar BSs, which cover a wide range of 50–700 cm⁻¹, have become available. In addition to Mylar films, wire grid polarizers have been used as BSs for the low-frequency range below 100 cm⁻¹. (An interferometer based on such wire grid BSs is called a Martin-Pupplet interferometer [Griffiths and De Haseth 2005].) In the NIR, CaF_2 and quartz (SiO_2) with appropriate coatings have been commonly used as BSs, whose valid spectral ranges are determined by the coatings.

---

* "Mylar" is a registered trade name for polyethylene terephthalate, also known as PET.

4.3.3.3 Optical Windows

Optical windows are needed between a high vacuum region containing the sample, such as a cryostat or sample chamber, and a low-vacuum or ambient-pressure region, such as the spectrometer or beam transport line. Common IR window materials are listed in Table 4.1. In the MIR region, a BaF_2 plate is a common and convenient window material with a high transmission of about 0.9 from $900\,cm^{-1}$ up to the UV. If the region below $900\,cm^{-1}$ is needed, KRS-5, a mixed crystal of TlBr and TlI (thallium iodide), is a popular window material that may be used down to ~$250\,cm^{-1}$ although the transmission is lower than BaF_2, about 0.7, and extends only to ~$18,000\,cm^{-1}$ (550 nm). NaCl and KBr also have good optical properties as IR windows, with high transmission ($T \sim 0.9$) from 700 and $400\,cm^{-1}$ to the UV, respectively. However, they are both soft and hygroscopic, so care must be taken in their handling. In the NIR, quartz windows are widely used, with high transmission ($T > 0.9$) from $3000\,cm^{-1}$ to the UV.

In the FIR, a thin plate of polyethylene (PE) and polypropylene (PP) can be used as a window below $700\,cm^{-1}$. Since these are soft materials, a PE or PP window is deformed by the pressure difference when one side is evacuated while the other side is at ambient pressure. Nevertheless, it can withstand the pressure unless the active area of the window is too large. For example, a transparent plastic folder (e.g., the kind sold at stationery stores) can be cut into a disk, sealed with an O-ring, and used as a FIR window for a cryostat. It can withstand the pressure difference, even with an active diameter as large as 4 cm. Interference fringes may arise, however, in measured spectra from multiple internal reflections. One may try different thicknesses to reduce interference. (Of course, it should be thick enough to withstand the 1 atm pressure.) When both sides of a window are in vacuum (high vacuum and low vacuum, for example, as is the case between the FT-IR and the sample chamber in Figure 4.9a), a thin PE or PP film, such as the one used for wrapping food, may be used. Obviously, it cannot withstand the 1 atm pressure, so evacuation and venting should be done simultaneously on both sides of the window.

Finally, although very expensive, diamond has good optical properties as an IR window. High-quality diamonds with a very low density of impurities, classified as type IIa, are transparent from the FIR to the UV ($E_g = 5.5\,eV$), except for the 2000–$3500\,cm^{-1}$ region where broad absorption bands due to two-phonon absorption are observed. Diamond plates of diameters up to 100 mm and thicknesses up to 2 mm have been produced with chemical vapor deposition (Dore et al. 1998) and have been commercially available as IR windows. The transmittance of a diamond plate with polished faces is ~0.7 (the refractive index is 2.4) in the transparent range and ~0.4 in the 2000–$2300\,cm^{-1}$ range for a thickness of 0.5 mm (Dore et al. 1998).

4.3.3.4 Mirrors

As in the visible–UV case discussed in Section 4.2, mirrors are used for handling the beams in IR spectroscopy. Parabolic, ellipsoidal, and spherical mirrors are used with a gold or aluminum coating. A gold coating has the obvious advantage of being impervious to oxidization but is more expensive. Aluminum coatings can be used as less expensive alternatives but their reflectance gradually decreases with time due to oxidization. Surface protection coatings are available for both gold and aluminum, but one should be careful in their use since they may introduce spectral structures in the IR. The requirement for the surface flatness is not as stringent as in UV applications since the wavelength is much longer in the IR.

4.3.3.5 Detectors

A wide variety of IR detectors are available (Griffiths and De Haseth 2005). Regarding the MIR range, a tryglycine sulfate (TGS) or deuterated TGS (DTGS) pyroelectric detector is a standard detector that is initially installed in a commercial FT-IR spectrometer. It operates without cooling and has wide dynamic and spectral ranges. Another common, more sensitive detector in the MIR is a liquid nitrogen-cooled HgCdTe (MCT) photoconductive detector. MCT detectors with various spectral ranges, element sizes, and sensitivity are available. A broadband-type MCT can cover a wide spectral range of 450–10,000 cm^{-1}, which is very useful for the reflectance K-K studies. Since MCT detectors have a high sensitivity, care must be taken about saturation effects. Similar to the visible–UV case in Section 4.2.1, linearity between the light intensity input and the signal intensity output from the detector must be ensured. To do so, the intensity of the light and also the gain of the preamplifier must be kept sufficiently low. The saturation of an MCT typically appears as a strong signal below the valid wave number range specified by the manufacturer.

For the FIR region, a TGS detector is a common detector for applications that do not require high sensitivity. As a more sensitive detector, a liquid He-cooled Si bolometer is widely used. In the NIR region, a Si photodiode covers the range from ~9000 cm^{-1} to the visible–UV. InGaAs photodiodes with enhanced sensitivity in the visible range are useful for covering a wide spectral range of 6,000–20,000 cm^{-1}.

4.3.3.6 Polarizers

Polarization-dependent $R(\omega)$ studies in the IR range are often useful in studying materials with anisotropic crystal structure and electronic states. For example, high-T_c superconductors such as $La_{1-x}Sr_xCuO_4$ have highly anisotropic crystal structures containing CuO_2 planes. As a result, their IR reflectance with in-plane polarization (along the CuO_2 plane) is significantly different from that with out-of-plane polarization (Basov and Timusk 2005). A commonly used linear polarizer in the IR is the wire grid polarizer, which is a grid of parallel conducting wires (Hecht 2001). In the FIR and THz range, free-standing wire grid polarizers are often used, while in the MIR and NIR, a wire grid formed on a substrate such as KRS-5, BaF_2, and quartz is used. As already discussed in Section 4.2.1, to analyze the polarization dependent $R(\omega)$ data, different reflectances of the mirrors with s and p polarizations should be taken into account. Circular polarization studies in the IR range (Griffiths and De Haseth 2005) are not discussed here.

4.3.4 Optical Layouts, Reference Material, and Optical Alignments

A commercial FT-IR spectrometer usually contains a sample compartment designed for transmission measurement, as in the example discussed in Section 4.1. To measure $R(\omega)$, therefore, an optical insertion for a reflectance measurement must be placed there. Such an insertion is usually available as an optional accessory from the manufacturer or it can also be designed and built by the user. For low-temperature measurements, a cryostat with optical windows may be inserted into the sample compartment. Alternatively, one may extract the IR beam to an external sample chamber, as shown in the example of Section 4.1. The cost for constructing such an external sample chamber tends to be higher but it provides more flexibility in the experiment, such as performing an in situ evaporation of a reference gold film onto the sample surface. (This is discussed in more detail in Section 4.3.6.). In addition, an external sample chamber enables the use of a cryogen-free, closed-cycle refrigerator cryostat. Since this type of refrigerator causes mechanical vibrations, it is difficult to use with the internal sample compartment of an FT-IR.

Regarding the reference material used for IR reflectance studies, a gold mirror is most appropriate due to its high reflectance (Chapter 3), usually assumed to be 1, and durability. One can easily deposit a film of gold on a glass substrate by vacuum evaporation or may use a commercial uncoated mirror. Coated gold mirrors should not be used as a reference since they may cause spectral structures in the IR. A sample holder of the type described in Figure 4.4d may also be used in the IR range.

The optical alignment for IR spectroscopy can be performed in a similar fashion to that discussed in Section 4.2.2 for visible–UV spectroscopy. Namely, one does the visual alignment first and then maximizes the detector signal by finely adjusting the optical components. Here, the interferometer in a FT-IR is usually already aligned by the manufacturer,* so the user only needs to align the other components. When the source used is a globar (which does not emit strongly in the visible) or when the BS is a Ge/KBr or a Ge-coated Mylar (not transparent in the visible), the visible output from the interferometer is very weak and a visual alignment is difficult. In these cases, one may first use a halogen lamp and a quartz or (uncoated) Mylar BS, which gives a visible output beam, for a visual alignment. After this, one may switch the source and/or the BS and can complete the alignment by maximizing the detector signal.

4.3.5 Sample Surface Conditions (Natural vs. Polished)

In the discussions regarding $R(\omega)$ in this Chapter, a specular (i.e., mirror-like) reflection at a perfectly flat and smooth sample surface has been assumed. In reality, such a case is rare except for, for example, commercially available wafers and substrates. Actual samples may need mechanical polishing to have a flat and shiny (smooth) surface. Even after polishing, the obtained surface may still deviate from a perfect plane and/or may contain irregular voids. Then, the magnitude and/or spectral shape of the measured $R(\omega)$ may be significantly altered from the intrinsic $R(\omega)$ due to diffraction and scattering of light by the surface. Therefore, careful consideration and preparation of the sample surface is crucial for reflectance studies.

If a sample of interest has a natural surface, either as grown or cleaved, it is desired to measure it without polishing. This is because mechanical polishing may damage the sample surface and suppress the physical properties of interest. If natural surfaces are available but only in small dimensions, a microscope may be used as discussed in Sections 4.2.3 and 4.3.8. For example, $R(\omega)$ of the magnetoresistive compound $La_{1-x}Sr_xMnO_3$ (Takenaka et al. 1999) and the valence transition compound $YbInCu_4$ (Okamura et al. 2007b) were measured with micro-FT-IR on small cleaved surfaces. They showed quite different $R(\omega)$ between polished and cleaved samples, which was attributed to damage due to polishing. If a large natural surface is available but is not very flat, the gold evaporation technique discussed in Section 4.3.6 may be used to measure its $R(\omega)$.

If a sample only has very rough surfaces, polishing is required to measure its $R(\omega)$.[†] A surface obtained by polishing should not only be shiny, but also as flat as possible. A polishing method used by the author is illustrated in Figure 4.8. First, the sample is mounted in an acrylic resin block (~3 cm diameter), as in Figure 4.8a and b. A commercial, fast-cure (<10 min) acrylic mounting kit is used, where the cured resin may be easily removed from the cup. Then, as in Figure 4.8c, the bottom of the resin is polished against sandpaper adhered on a glass plate, either by hand or by a polishing machine with rotating arm. This method is quite effective in maintaining a flat sample surface. (If the sample is held directly by hand, the edge of the polished surface would be notably rounded.) Sandpaper with large particles (~80 μm) is used first and then those with finer particles, down to 5–10 μm, are used in a step-by-step manner. As a final step, a polishing cloth is used in place of sandpaper, with 1 μm diameter abrasive powders (Al_2O_3 or SiC) dispersed in water. A soft polishing cloth is used for ductile materials (metals and intermetallics), while a harder one is used for brittle ones

---

* The interferogram intensity may decrease after changing BS's, but most modern FT-IR instruments automatically align the interferometer to maximize the signal intensity.

[†] A technique called "diffuse reflectance" is also used to evaluate a non-specular (non-reflecting) surface (Griffiths and De Haseth 2005). This may be useful when a measurement without polishing is required, for example, in a manufacturing process.

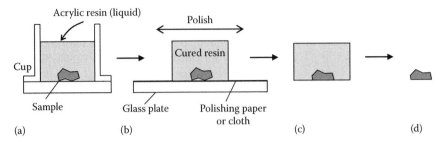

Figure 4.8 Illustration of polishing procedure discussed in text. (a) A sample is placed at the bottom of a cup and acrylic resin is cast in the cup. (b) The sample and the resin are polished together against a polishing paper or cloth adhered on a glass plate. (c) and (d) The resulting sample with flat and shinny surface is removed from resin by immersing it in acetone. See text for more details.

(oxides, ceramics, and semiconductors).* While more sophisticated and expensive polishing machines and jigs are available, the simple method described here has been quite effective in most cases. If the polished surface still has voids or is still not sufficiently flat for some reason, its $R(\omega)$ may be measured with the gold-coating method discussed in Section 4.3.6 to correct for the remaining irregularity.

4.3.6 Reflectance of Samples with Irregular Shapes

As discussed in the last section, it is sometimes necessary to measure $R(\omega)$ on an irregular surface. A powerful and convenient means of obtaining $R(\omega)$ corrected for an irregular surface is gold coating of the sample by evaporation, performed in situ in the sample chamber during the $R(\omega)$ measurement (Homes et al. 1993). When ideally done, this technique creates a gold reference that has exactly the same irregular shape as the sample. Hence, by dividing the reflected spectrum of the sample before coating by that after coating, one obtains a $R(\omega)$ spectrum that has been corrected for the irregular surface. The evaporation setup may be constructed easily by combining a tungsten filament, an electrical feedthrough, and a motion feedthrough, the last of which is used to move the setup in and out of the optical path. There are certain requirements for this method to work correctly (Homes et al. 1993), but we have found this to be a simple yet powerful means of correcting for irregular surfaces, especially in the FIR range. Since this method is performed in vacuum, less expensive silver may be used instead of gold. In most cases, the gold or silver coating on the sample may be removed by adhesive tape, without damaging the sample.

4.3.7 Cryostats Used with FT-IR

There are two types of cryostats that can reach 10 K or below and are used with FT-IR experiments: a liquid He continuous flow cryostat and a closed-cycle refrigerator. The former requires liquid He but does not cause mechanical vibrations and is also more compact. The latter is cryogen free, but the refrigerator head is relatively large and causes mechanical vibrations. The vibrations may affect the operation of the interferometer. Therefore, in the internal sample compartment of an FT-IR, the former type of cryostat should be used. The cryostat should be equipped with appropriate optical windows to keep a high vacuum ($\sim 10^{-7}$ Torr or

---

* One may choose from a variety of polishing cloths intended for different material hardnesses. Polishing with sandpaper may be done more effectively under water flow, provided the sample is not damaged by water. The surface condition of the sample should be checked with a microscope at each step.

below for measurements in the MIR), as opposed to the low vacuum or ambient pressure in the FT-IR spectrometer. A high vacuum is needed to prevent ice formation on the sample surface. Even a thin layer of ice absorbs IR radiation very strongly around $3200\,cm^{-1}$ due to OH stretching vibrations, which may cause structure in the measured $R(\omega)$. If an external sample chamber is to be used, on the other hand, either type of cryostat may be used. When using a refrigerator, one may reduce the amount of vibration reaching the detector and the chamber by using a bellow to connect the refrigerator head to the chamber.

4.3.8 FT-IR Combined with a Microscope

In the section on visible–UV spectroscopy, the advantages of a microscope-based apparatus for small samples and spatially resolved measurements are already discussed. These advantages also apply to IR spectroscopy. Accordingly, micro-FT-IR instruments, where a microscope designed for the IR is combined with an FT-IR, are commercially available and are widely used. (Griffiths and De Haseth 2005, Griffiths 2009). An IR microscope is usually equipped with reflective objectives known as Schwarzschild mirrors* to avoid chromatic aberrations, which could be significant with a lens-based objective in the IR range. In addition, to reduce the beam diameter on the sample, an IR microscope has a small aperture in the optical path to reduce the effective source size. Since a globar is not a bright source, such an aperture greatly reduces the signal intensity. The smallest beam diameter at the sample with an acceptable SNR in the MIR is generally ~$50\,\mu m$ in the fingerprint region (8–10 times λ). This sets the spatial resolution with a regular MCT detector. Recently, MCT array detectors have become available, which have dramatically improved the performance of micro-FT-IR in terms of the spatial resolution, SNR, and measurement time. They enable, for example, the 2D mapping of molecular vibrational states in a sample and are very useful for various applications (Griffiths 2009).

Another direction toward improving the spatial resolution of micro-FT-IR is the use of SR as a highly bright IR source. Since SR is emitted from an electron beam, as already discussed in Section 4.2.4, its effective source size and beam divergence are both very small. Accordingly, the brightness (the number of photons emitted into unit solid angle per unit source area) of an IR SR source is generally two to three orders of magnitude greater than that of a globar. By using IR SR with an IR microscope, it is possible to obtain a beam diameter of the order of a wavelength, without using any aperture unlike the case discussed above (Carr 2001). This has enabled many interesting applications, such as IR studies under high pressure using a diamond anvil cell (Sacchetti et al. 2007, Matsunami et al. 2009). Nearly 20 SR facilities in the world have IR beam lines open to general users, and one may obtain beam time by submitting a proposal. For the latest status of SR-based micro-FT-IR techniques, see (Predoi-Cross and Billinghurst 2010).

4.4 EXAMPLES OF WORKING APPARATUSES

4.4.1 An IR–Visible–UV Spectroscopy Apparatus

As an example of a working apparatus, Figure 4.9 shows the top view of an IR–visible–UV spectroscopy apparatus used at Kobe University. An FT-IR and a grating polychromator are used with various combinations of sources, detectors and optical components to cover from the FIR to the UV ($\nu = 60$–$50,000\,cm^{-1}$, $\lambda = 0.2$–$150\,\mu m$, $h\nu = 0.008$–$6.2\,eV$). A custom-made,

---

* A Schwarzschild mirror consists of a pair of concave and convex mirrors that are nearly concentric. Schwarzschild mirrors are also widely referred to as Cassegrain mirrors. A Schwarzschild mirror is a special case of a Cassegrain mirror (Griffiths 2009).

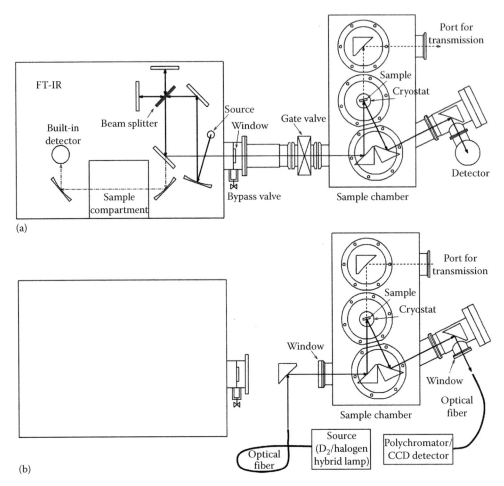

Figure 4.9 Schematic top view of an IR–visible–UV spectroscopy apparatus used at Kobe University. (a) shows the IR–visible (50–18,000 cm⁻¹) setup, which consists of an FT-IR and a vacuum sample chamber containing focusing/collecting optics. The beam path under reflection configuration is shown by the solid lines (the detector is moved for a transmission study). The internal sample compartment and detector of the FT-IR are used only for the interferometer alignment. (b) shows the visible–UV (200–800 nm) configuration, which uses the same sample chamber as in (a) but with a different source and spectrometer. See text for more details.

vacuum-pumped sample chamber, which contains focusing and collecting optics, is used for all the spectral ranges. (The spectra shown in Figure 4.7 were taken with this apparatus, using the reflected light from an Au or Ag mirror). Two cryostats and a heater probe are used to cover a temperature (T) range of 3.3–700 K. A closed-cycle refrigerator is used for T = 7–300 K. It is inserted into the vacuum chamber from its bottom, where a welded bellow is used to connect it to the chamber to minimize the amount of mechanical vibration reaching from the cryostat head to the chamber and detector. (In addition, the chamber and the cryostat are mounted on separate supports.) A liquid He continuous flow cryostat is used to reach T below 7 K (down to 3.3 K), which is inserted into the chamber from its top. A heater probe is used to reach T = 300–700 K, which is inserted into the chamber from its side. In addition, as discussed in Section 4.3.6, an evaporation setup can be inserted into the chamber for in situ coating of the sample with gold or silver, to measure $R(\omega)$ of a sample with an irregular surface. The sample chamber is evacuated by a 300 L/s turbo molecular pump, with a typical pressure between 6×10^{-8} and 3×10^{-7} Torr during a measurement. Since the entire chamber is in high vacuum, no additional window is needed for the cryostat.

For the IR–visible range (60–20,000 cm^{-1}), a FT-IR is used as shown in Figure 4.9a. A window is used between the low vacuum (~4 Torr) of the FT-IR and the high vacuum of the sample chamber and beam transport pipe. The detectors used in this range are a liquid He-cooled Si bolometer (50–700 cm^{-1}), a liquid N$_2$-cooled MCT (450–10,000 cm^{-1}, 2 mm element size), and an InGaAs photodiode (6,000–18,000 cm^{-1}, 2 mm element size). For the visible–UV range (200–800 nm), the configuration shown in Figure 4.9b is used with the same sample chamber. (The beam transport pipe used for IR–visible is removable.) A polychromator with a built-in CCD detector (1 nm resolution) and a hybrid halogen/D$_2$ lamp are used to measure the entire 200–800 nm range at once, which is extremely useful for the K-K method.

4.4.2 A Vacuum–UV Spectroscopy Apparatus

As already discussed in Section 4.2.4, one needs to use an SR source to record $R(\omega)$ over a wide wavelength range in the VUV. As an example, the beam line BL7B of the UVSOR Facility, Institute for Molecular Science (Okazaki, Japan) is schematically described in Figure 4.10a (Fukui et al. 2001). This beam line covers not only the VUV, but also the entire UV, visible, and some of the near-IR range, 40–1000 nm or 1.2–33 eV. Examples of obtained source spectra, measured without a sample in the optical path, are shown in Figure 4.10b. In addition to the wide spectral range, one can measure $R(\omega)$ without a reference material, as explained below. The SR emitted from a 750 MeV electron beam passing through a bending magnet is collected and dispersed by a grating monochromator. Then, as sketched in Figure 4.10a, the

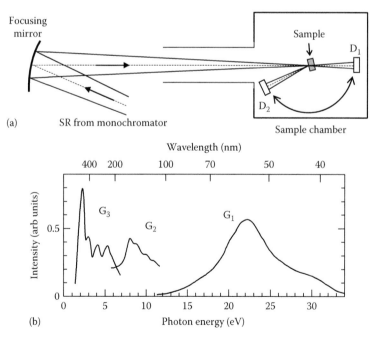

Figure 4.10 (a) Schematic top view of the VUV apparatus based on the SR at beam line BL7B of UVSOR Facility, Institute for Molecular Science (Okazaki, Japan). (Modified from Fukui, K. et al., *Nucl. Instrum. Methods Phys. Res. A*, 467, 601, 2001.) Only a downstream portion of the beam line is shown. D$_1$ and D$_2$ indicate the two detector positions discussed in the text. (b) The source spectra $I_0(\omega)$ measured with three gratings having 1200 (G$_1$), 600 (G$_2$), and 300 (G$_3$) grooves/mm.

output beam from the exit slit is focused onto the sample. The sample reflection spectrum $I(\omega)$ and the source spectrum $I_0(\omega)$ can be measured simply by moving the detector between the two positions D_1 and D_2 in Figure 4.10a. The detector is a $10 \times 10\,mm^2$ Si photodiode designed as an x-ray detector, which can also operate as a VUV detector. Thanks to the small beam divergence of SR, this detector can collect the light without additional optics, unlike the configuration of Figure 4.4b. When recording $I(\omega)$ the sample is placed as in Figure 4.10a, but when recording $I_0(\omega)$ it is moved vertically and out of the optical path. The sample and detector positions can be controlled from outside the chamber using motion feedthroughs. The monochromator is equipped with three interchangeable gratings (300, 600, and 1200 grooves/mm) to cover 40–1000 nm, as shown by the $I_0(\omega)$ spectra in Figure 4.10b.

4.5 DATA ANALYSES FOR OBTAINING OPTICAL CONSTANTS FROM MEASURED REFLECTANCE SPECTRA

As already mentioned, there are two main methods of analyzing a measured near-normal incidence $R(\omega)$ to obtain the optical constants. The K-K method relies on numerical integration of the K-K relations with the measured $R(\omega)$, which also requires extrapolations of the measured $R(\omega)$ toward $\omega = 0$ and ∞. The spectral fitting analysis, on the other hand, relies on a parameter fitting of the measured $R(\omega)$ based on the Drude–Lorentz model (Chapter 1). These methods are outlined in the following, and actual examples are given in Section 4.6.

4.5.1 Numerical Evaluation of the K-K Relations

4.5.1.1 Calculation of the Integrals in the K-K Relations

As already shown by Equation 4.3, the reflectivity of the electric field amplitude $r(\omega)$ and the phase shift $\phi(\omega)$ are related to each other through the K-K relation as follows:

$$\phi(\omega) = -\frac{2\omega}{\pi} P \int_0^{+\infty} \frac{\ln r(\omega')}{\omega'^2 - \omega^2}\, d\omega', \qquad (4.3)$$

where
    P denotes the Cauchy principal value
    $r(\omega)$ is the square root of the experimentally measured reflectance $R(\omega)$

Although the integrand in Equation 4.3 diverges at $\omega' = \omega$, it has been shown (Cardona and Greenaway 1964, Wooten 1972) that this integral can be transformed into another form

$$\phi(\omega) = -\frac{\omega}{\pi} \int_0^{+\infty} \frac{\ln r(\omega') - \ln r(\omega)}{\omega'^2 - \omega^2}\, d\omega'. \qquad (4.9)$$

The integrand in Equation 4.9 is now finite and continuous at $\omega' = \omega$, so that the integration can be performed numerically with a computer.

4.5.1.2 Extrapolations Needed for Performing the Integrals in K-K Relations

Since $R(\omega)$ can be measured only over a finite frequency range, it has to be extrapolated at both high- and low-frequency regions to carry out the integration in Equation 4.9 from $\omega = 0$ to ∞. For the low-energy extrapolations, there are mainly two different forms used, namely the so-called Hagen–Rubens function for metals and a constant for insulators. The Hagen–Rubens function, which is an approximate expression of $R(\omega)$ for a Drude metal at low frequencies, is expressed as follows (Wooten 1972, Burns 1990):

$$R_{HR}(\omega) = 1 - \frac{2}{n_1(\omega)} = 1 - \sqrt{\frac{8\varepsilon_0 \omega}{\sigma_{dc}}}, \qquad (4.10)$$

where
$n_1(\omega)$ is the real part of the complex refractive index
σ_{dc} is the dc conductivity of the material in $\Omega^{-1}\,m^{-1}$
ε_0 is the vacuum permittivity*

This expression is valid when $\omega \ll 1/\tau \ll \omega_p$, where $\tau(1/\tau)$ is the scattering time (damping rate) of the electrons and ω_p is the plasma frequency, defined in Chapters 1 and 3. Note that it depends only on σ_{dc} and does not contain any adjustable parameters. Since the original work by Hagen and Rubens reported in 1904, it has been shown that this expression agrees well with experiment for many metals (Burns 1990). In practice, however, the condition $\omega \ll 1/\tau \ll \omega_p$ may not be met for a compound under study and $R_{HR}(\omega)$ may not connect to the measured $R(\omega)$. It may also happen that σ_{dc} is simply unknown and Equation 4.10 cannot be used. In such a case, a similar form

$$R_{HR}(\omega) = 1 - a\sqrt{\omega} \qquad (4.11)$$

is used, where the parameter a is adjusted to smoothly connect $R_{HR}(\omega)$ and the measured $R(\omega)$.

In the high-frequency range, extrapolations of a ω^{-n} form have been used by many workers. Cardona and Greenway (1964) suggested $n = 4$ in the high-frequency limit. $n = 2, 3$, and combinations of these have been also used. In Section 4.6, we will discuss the effects of different high-frequency extrapolations on the obtained $\sigma(\omega)$ using actual $R(\omega)$ data.

4.5.2 Spectral Fitting

When a measured $R(\omega)$ has distinct features, such as a Drude reflection with a clear cutoff (plasma edge) or a phonon peak, spectral fitting may be performed to obtain the optical constants without K-K analysis. According to the Lorentz model discussed in Chapter 1, the complex dielectric function $\tilde{\varepsilon} = \varepsilon_1 + i\varepsilon_2$ due to bound electrons is expressed as follows:

$$\varepsilon_1 = 1 + \omega_p^2 \cdot \frac{\omega_0^2 - \omega^2}{(\omega_0^2 - \omega^2)^2 + \gamma^2\omega^2}, \quad \varepsilon_2 = \omega_p^2 \cdot \frac{\gamma\omega}{(\omega_0^2 - \omega^2)^2 + \gamma^2\omega^2}. \qquad (4.12)$$

where
ω_0 is the eigen frequency
γ is the damping rate
$\omega_p^2 = ne^2/\varepsilon_0 m$ is the plasma frequency (Chapters 1 and 3)

---

* R_{HR} is better known as $R_{HR}(\omega) = 1 - [2\omega/(\pi\sigma_{dc})]^{1/2}$ in CGS units, where σ_{dc} is in s^{-1}. See the Appendix in Jackson (1975) for more information about this unit conversion.

For simplicity, only the *relative* values of ε_1 and $\varepsilon_2 (\varepsilon_1/\varepsilon_0$ and $\varepsilon_2/\varepsilon_0)$ are indicated in Equation 4.12 and in the following discussion. The Drude model for free carriers is obtained by setting $\omega_0 = 0$ in Equation 4.12:

$$\varepsilon_1 = 1 - \frac{\omega_p^2}{\omega^2 + \gamma^2}, \quad \varepsilon_2 = \frac{\omega_p^2 \gamma}{\omega(\omega^2 + \gamma^2)}, \tag{4.13}$$

Hence the total dielectric function is expressed as follows:

$$\varepsilon_1 = \varepsilon_\infty - \frac{\omega_p^2}{\omega^2 + \gamma^2} + \sum_j^N \frac{\omega_{p,j}^2(\omega_{0,j}^2 - \omega^2)}{(\omega_{0,j}^2 - \omega^2)^2 + \gamma_j^2 \omega^2}, \quad \varepsilon_2 = \frac{\omega_p^2 \gamma}{\omega(\omega^2 + \gamma^2)} + \sum_j^N \frac{\omega_{p,j}^2 \gamma_j \omega}{(\omega_{0,j}^2 - \omega^2)^2 + \gamma_j^2 \omega^2}, \tag{4.14}$$

where N is the number of Lorentz oscillators needed. Initially, N may be set to the number of peaks identified in $R(\omega)$ but may be increased or decreased depending on the quality of the resulting fits. (The factor of 1 in the expression of ε_1 in Equation 4.13 has been replaced by ε_∞ in Equation 4.14 to represent the polarization caused by higher frequency interband transitions, as discussed in Chapter 1.) One Drude term is assumed in Equation 4.14, but may be dropped for an insulator, or two of them may be used if necessary. In the fitting, for a set of parameters in Equation 4.14, the resulting ε_1 and ε_2 are used to calculate $R(\omega)$ using the results of Chapter 1. Then the parameters are adjusted until the agreement between the calculated and measured $R(\omega)$ spectra becomes satisfactory. This is usually done with a least squares fitting method. Commercial data analysis software may contain such a fitting feature.*

It should be carefully examined whether or not the result of a given spectral fit is physically meaningful. For example, the spectral shape of a Drude reflection due to free carriers in the measured $R(\omega)$ may not agree exactly with that given by a simple Drude model. This may happen because carriers with different effective masses have contributed or because the Drude component has a strong overlap with an interband transition component. In such a case, it may be necessary to use two Drude components or a combination of a Drude and a Lorentz component to have a satisfactory fit. Apparently, the quality of a fitting becomes better as more spectral components are used, but at the same time, the arbitrariness among the fitting parameters increases. Two different sets of fitting parameters may yield equally good fits to a measured $R(\omega)$. Examples of spectral fitting on measured $R(\omega)$ are given in Section 4.6.

4.6 EXAMPLES OF MEASURED SPECTRA AND THEIR DATA ANALYSIS

In this section, examples of measured $R(\omega)$ data and their analyses based on K-K and spectral fitting analyses are presented for Ag, InSb, $Tl_2Mn_2O_7$, and $YbAl_3$. The data presented here were all measured using the FT-IR-based apparatus in Figure 4.9a below 2.3 eV (18,000 cm^{-1}) and using the SR-based apparatus in Figure 4.10a from 2.3 to 33 eV. (The visible–UV apparatus shown in Figure 4.9b was constructed more recently and not used for this data.) $\sigma(\omega)$ will refer to the real part of the complex conductivity unless noted, and the dielectric functions $\varepsilon_1(\omega)$ and $\varepsilon_2(\omega)$ will refer to those normalized relative to ε_0.

---

* Dedicated software for spectral fitting called "RefFIT" has been developed and made available free of charge by Alexey Kuzmenko (http://optics.unige.ch/alexey/reffit.html). It can perform both spectral fitting and K-K analyses under various circumstances (e.g., reflectance and transmittance of bulk and film samples under normal and grazing incidence). See also Kuzmenko (2005).

4.6.1 Ag

We first consider Ag to illustrate how the techniques discussed above can be applied to this well-known and familiar metal. The optical properties of Ag (Ehrenreich and Philipp 1962, Wooten 1972) have been already discussed to some depth in Chapter 3, so we will emphasize the analysis of $R(\omega)$. Figure 4.11a shows the near-normal incidence $R(\omega)$ of a ~0.2-µm-thick Ag film, formed on a glass substrate by vacuum evaporation, at room temperature. As discussed in Chapter 3, the sharp drop of $R(\omega)$ at ~3.9 eV (315 nm) corresponds to the plasma frequency, which is reduced from the "bare" value of 9 eV given by the simple Drude model. This reduction is caused by Coulomb screening due to polarization of bound electrons,* or equivalently, by a large value of ε_∞ due to interband transitions, as illustrated in Figure 3.2a of Chapter 3. K-K analysis of $R(\omega)$ was performed with the Hagen–Rubens extrapolation (Equation 4.11) below 1.7 eV and a ω^{-4} extrapolation above 30 eV. The obtained $\varepsilon_1(\omega)$, $\varepsilon_2(\omega)$, and $\sigma(\omega)$ are shown in Figure 4.11b and c. $\varepsilon_1(\omega)$ shows a sharp peak near 3.9 eV, and $\varepsilon_2(\omega)$ and $\sigma(\omega)$ also show a steep rise near 3.9 eV, which are due to interband transitions from the $4d$ states located about 4 eV below the Fermi level (E_F), as sketched in Figure 3.1a of Chapter 3, to the unoccupied states above E_F (Wooten 1972). Note that $\varepsilon_1(\omega)$ crosses zero from negative to positive

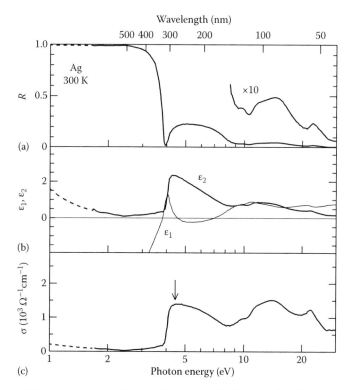

Figure 4.11 (a) Near-normal incidence reflectance (R) of an Ag film formed on a glass substrate, in a logarithmic scale. (b) $\varepsilon_1(\omega)$ and $\varepsilon_2(\omega)$ and (c) $\sigma(\omega)$ obtained with K-K analysis of $R(\omega)$. The downward arrow indicates the contribution of $4d$ interband transitions discussed in the text. The dotted portions of the spectra are in the extrapolated range. (H. Okamura, unpublished data.)

---

* A plasma oscillation of free electrons causes an internal field $\vec{E}_{int}$. $\vec{E}_{int}$ then polarizes the bound electrons, which in turn create an opposite internal field and partly cancel $\vec{E}_{int}$. This reduces the restoring (Coulomb) force of the plasma oscillation, and hence ω_p.

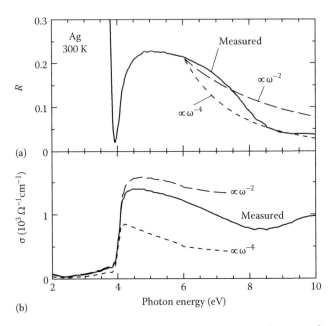

Figure 4.12   Comparison of different high-energy extrapolations for $R(\omega)$ of Ag. (a) The measured $R(\omega)$ below 10 eV and extrapolation functions ω^{-2} and ω^{-4} connected at 6 eV. (b) The resulting $\sigma(\omega)$ spectra. (H. Okamura, unpublished data.)

values at 3.9 eV. This type of zero crossing in $\varepsilon_1(\omega)$ is a well-known condition for a plasma frequency, that is, $\varepsilon_1 = 0$ and $\varepsilon_2 \ll 1$ (Wooten 1972). This interband transition also leads to the strong peak in $\sigma(\omega)$, shown by the arrow in Figure 4.11c.

If one uses a Xe lamp or D_2 lamp in the air rather than a SR source, the short-wavelength limit is usually 200 nm (6.2 eV), as discussed in Section 4.2.1. Therefore, let us see what kind of spectra result from measuring $R(\omega)$ below 6 eV only, with extrapolations above 6 eV. To do so, as shown in Figure 4.12a, extrapolations of ω^{-2} and ω^{-4} forms were connected to $R(\omega)$ at 6 eV. $\sigma(\omega)$ spectra obtained with their K-K analysis are shown in Figure 4.12b. The ω^{-4} extrapolation underestimates $R(\omega)$ above 6 eV, while the ω^{-2} extrapolation overestimates it. Accordingly, $\sigma(\omega)$ spectra obtained with these extrapolations are notably different from that given with the full $R(\omega)$ to 30 eV. The spectral shapes of the three $\sigma(\omega)$ spectra are nevertheless similar to one another. Therefore, if the spectral shape is the main interest, measurement up to 6 eV does not cause a significant problem. But if the spectral weight (intensity) in $\sigma(\omega)$ is also important for, for example, an analysis based on the optical sum rule, the uncertainty arising from the limited spectral range could be significant. Similar results are often obtained for other systems: namely, limited spectral range and different high-energy extrapolations affect the spectral intensity more strongly than the spectral shape in the K-K obtained $\sigma(\omega)$.

4.6.2 InSb

As a second example, we shall consider InSb as an example where both $R(\omega)$ and $T(\omega)$ can be measured and where both phonon and Drude spectral components can be studied. InSb is a well-known narrow-gap semiconductor with the zinc-blend crystal structure, a direct band gap of $E_g = 0.17$ eV at 300 K, and an extremely small electron effective mass of $m^* = 0.015\,m_0$ (Yu and Cardona 2001).

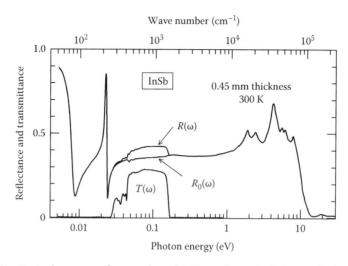

Figure 4.13 Optical spectra of an undoped InSb wafer with 0.45 mm thickness and both faces polished. $R(\omega)$ and $T(\omega)$ are the measured reflectance and transmittance, and $R_0(\omega)$ is the reflectance of a single surface derived by analyzing $R(\omega)$ and $T(\omega)$, as discussed in the text. Note that $R_0(\omega)$ and $R(\omega)$ are identical to each other when $T(\omega) \sim 0$. (H. Okamura, unpublished data.)

Figure 4.13 shows $R(\omega)$ and $T(\omega)$ of an InSb plate, which was cut from an undoped wafer of 0.45 mm thickness. Both surfaces of the sample were polished. $R(\omega)$ shows different structures in different spectral regions, and $T(\omega)$ is nonzero only at 0.03–0.17 eV. The rise of $R(\omega)$ at $\hbar\omega < 10$ meV is due to the Drude response of thermally generated free carriers. The sharp feature in $R(\omega)$ near 25 meV is due to an optical phonon. A similar phonon peak in $R(\omega)$ is commonly observed for semiconductors with the zinc-blend crystal structure (Yu and Cardona 2001). Since the Drude and phonon structures in $R(\omega)$ are clearly identified, the spectral fitting analysis discussed in Section 4.5.2 is very useful to analyze them, as we will see below.

The solid curves in Figure 4.14a and b show $R(\omega)$ of InSb at $T = 160$ and 400 K plotted in cm^{-1} since these units are more common for phonon studies. At 160 K a Drude component is not seen in the measured range but at 400 K it is quite pronounced. This large difference results from the strongly T-dependent carrier density caused by both the small band gap and the small m^*.* The broken curves in Figure 4.14a and b show the result of spectral fitting, which was performed with a Drude term and a Lorentz term in Equation 4.14 with six adjustable parameters (ω_p and γ for the Drude, ω_0, ω_p, and γ for the Lorentz, and ε_∞). The quality of the fit is quite good, especially for the 160 K data (the fitted and measured data almost overlap). The parameters obtained from the fitting are summarized in Table 4.2, together with the calculated carrier density n and dc conductivity σ_{dc}.

Table 4.2 shows that, with increasing T, n and σ_{dc} increase rapidly from 1.7×10^{16} cm^{-3} and 110 Ω^{-1} cm^{-1} at 300 K to 3.7×10^{17} cm^{-3} and 710 Ω^{-1} cm^{-1} at 600 K. $\varepsilon_1(\omega)$ obtained with fitting on the 400 K data is shown in Figure 4.14c, together with its Drude and Lorentz components $\varepsilon_D(\omega)$ and $\varepsilon_L(\omega)$. Here, $\varepsilon_D(\varepsilon_L)$ was plotted with the parameters given by the fitting but without the Lorentz (Drude) term in Equation 4.14. A zero crossing of $\varepsilon_D(\omega)$ from negative to positive values is observed at $\omega_p^* \sim 165$ cm^{-1}. ω_p^* is a "screened" plasma frequency, which is much smaller than $\omega_p \sim 630$ cm^{-1} given by the fitting. This is due to the large value of $\varepsilon_\infty \sim 15$, which is similar to the case of Ag (Section 4.6.1 and Chapter 3). In fact, ω_p^* is close to $\omega_p/\sqrt{\varepsilon_\infty}$ at all

---

* The small m^* of InSb also leads to a large cyclotron frequency and very strong magneto-optical effects in the FIR range (Lax and Wright 1960).

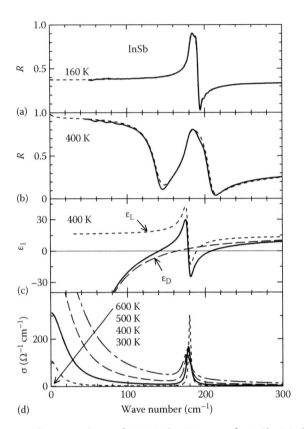

Figure 4.14 Spectral fitting analyses of $R(\omega)$ in the FIR range for InSb. (a)–(b) Measured $R(\omega)$ (solid curve) and the fitting result (broken curve) of InSb at 160 and 400 K. The resolution was 2 cm^{-1} in (a) and 4 cm^{-1} in (b). (c) Solid curve shows $\varepsilon_1(\omega)$ given by the fitting, and ε_L and ε_D show the phonon (Lorentz oscillator) and free carrier (Drude) contributions. (d) $\sigma(\omega)$ spectra at 300–600 K given by the fitting. (H. Okamura, unpublished data.)

temperatures in Table 4.2. A similar zero crossing of $\varepsilon_L(\omega)$ in Figure 4.14c corresponds to the longitudinal optical (LO) phonon frequency (Chapter 2, Yu and Cardona 2001).* On the other hand, ω_0 given by $\varepsilon_L(\omega)$ corresponds to the transverse optical (TO) phonon frequency. Figure 4.14d shows $\sigma(\omega)$ of InSb at different temperatures. Unlike in $R(\omega)$ and $\varepsilon_1(\omega)$, the Drude and Lorentz components are well separated in $\sigma(\omega)$.

Returning to Figure 4.13, above the phonon energy, there is a spectral region where $T(\omega)$ is relatively high. Clearly, the upper end of this transmissive region corresponds to the band gap of InSb. In this region, as discussed in Chapter 1, internal and multiple reflections from the sample surfaces may affect $R(\omega)$ and $T(\omega)$. In fact, $R(\omega)$ in Figure 4.13 exhibits a dome-shaped extra intensity at this region due to internal reflections. Each of $R(\omega)$ and $T(\omega)$ in the presence of internal reflections may be expressed in terms of the single surface reflectance, $R_0(\omega)$, and the absorption coefficient, $\alpha(\omega)$, yielding two equations (Heavens 1991). Hence, $R_0(\omega)$ and $\alpha(\omega)$ can be obtained by solving the two equations for them and by using measured $R(\omega)$ and $T(\omega)$ data, without using a K-K analysis. The obtained $R_0(\omega)$ and $\alpha(\omega)$ are indicated in Figures 4.13 and 4.15, respectively. The dome shape seen in $R(\omega)$ has been indeed removed, and $R_0(\omega)$ is

---

* Note that the total $\varepsilon_1(\omega)$ in Figure 4.14c has two zero crossings from negative to positive, at frequencies different from ω_p^* and ω_L. They represent two normal modes of the coupled plasmon-LO phonon oscillation (Yu and Cardona 2001).

TABLE 4.2

Parameters Obtained by the Spectral Fitting of $R(\omega)$ Shown in Figure 4.14

| | Parameters Obtained from Fitting | | | | | | | | |
| | Drude | | Lorentz | | | | | | |
Temperature (K)	ω_p (cm^{-1})	γ (cm^{-1})	ω_0 (cm^{-1})	γ (cm^{-1})	ε_∞	ω_p^{*a} (cm^{-1})	ω_p^{*b} (cm^{-1})	$\sigma_{dc}^{\ c}$ (Ω^{-1} cm^{-1})	n^d (cm^{-3})
160			181.5	1.6	15.3				
300	272.6	11.5	179.9	3.1	14.4	71.8	70.9	110	1.7×10^{16}
400	629.5	21.0	178.3	6.2	14.4	165.9	164.5	310	8.8×10^{16}
600	1288	38.9	175.7	12.2	14.9	333.7	331.5	710	3.7×10^{17}

[a] Calculated as $\omega_p^* = \omega_p / \sqrt{\varepsilon_\infty}$ where ω_p and ε_∞ are those given by the fitting.

[b] Obtained from the zero crossing in $\varepsilon_D(\omega)$, which is the Drude contribution to $\varepsilon_1(\omega)$.

[c] Calculated from the Drude parameters ω_p and γ given by the fitting.

[d] Calculated from ω_p assuming $m^* = 0.015 m_0$ and using $\omega_p^2 = (2\pi \cdot \nu_p)^2 = (ne^2)/(\varepsilon_0 m^*)$ (in SI units). Note that $\nu_p = 29.95$ cm^{-1} for $n = 10^{16}$ cm^{-3} and $m^* = m_0$.

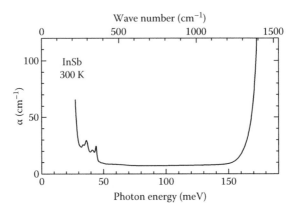

Figure 4.15 Absorption coefficient $\alpha(\omega)$ of InSb in the band gap region derived from $R(\omega)$ and $T(\omega)$, without using a K-K analysis. (H. Okamura, unpublished data.)

identical to $R(\omega)$ outside the transmissive region as expected. $\alpha(\omega)$ is small ($\sim 8\,\mathrm{cm}^{-1}$) and constant in the transmissive region but steeply rises in the vicinity of the band gap and phonon energies. It is possible, of course, to use the K-K method to evaluate $\alpha(\omega)$ from the measured $R_0(\omega)$ only. (The signal due to internal reflections may be suppressed with a wedged sample or by scratching the rear surface.) However, empirically speaking, a reflection K-K analysis on such a transmissive region tends to be less accurate. For example, unphysical results of negative $\alpha(\omega)$ and $\sigma(\omega)$ are sometimes obtained, presumably due to errors in the numerical integrations of K-K relations. Hence, to accurately measure $\alpha(\omega)$ in a transmissive spectral range, it is important to measure not only $R(\omega)$ but also $T(\omega)$.*

Above the band gap region, $R(\omega)$ in Figure 4.13 shows an almost constant value of ~ 0.4 below 1.5 eV. It then exhibits strong structure above 1.5 eV due to interband transitions, which are related to particular portions of the band structure with high density of states (Philipp and Ehrenreich 1963). $R(\omega)$ is fairly high (~ 0.5) throughout the visible–UV range, causing a polished surface of InSb to have a metallic, silvery look to the eye, although InSb is not a metal. The high $R(\omega)$ in this spectral range shows that the corresponding interband transitions have strong oscillator strength, which is also responsible for the large value of $\varepsilon_\infty \sim 15$ discussed above. In Figure 4.16, we show $\varepsilon_1(\omega)$ and $\varepsilon_2(\omega)$ in the visible–UV range. They were calculated with K-K analysis of $R(\omega)$ [$R_0(\omega)$ is inserted for the transmissive range], but again different extrapolation schemes were tried. In Figure 4.16a, the full $R_0(\omega)$ is used for K-K analysis, with a Hagen–Rubens extrapolation (Equation 4.11) below 6 meV and a ω^{-4} extrapolation above 30 eV. In Figure 4.16b through d, $R(\omega)$ at different ranges are used with extrapolations: 1–30 eV in (b), 1–10 eV in (c), and 1–6 eV in (d). Below and above these ranges, a constant and a ω^{-4} extrapolation were used, respectively. The difference between (a) and (b) is negligible. This shows that, for a semiconductor like InSb, the range below E_g does not have to be measured if one is only interested in the region above E_g. Thus, the differences among (b)–(d) are due to different high-energy extrapolations. These data show that $\varepsilon_2(\omega)$ is not affected very much by the high-energy extrapolations. In contrast, $\varepsilon_1(\omega)$ is noticeably affected, in particular its magnitude at 2.5–3.5 eV range, where $\varepsilon_1(\omega)$ is positive in (a)–(c) but negative in (d). Namely, although the spectral shape of $\varepsilon_1(\omega)$ is not significantly affected by different extrapolations, the magnitude of $\varepsilon_1(\omega)$ is more strongly affected, which is similar to the case of Ag in Section 4.6.1.

---

* Note that it is also possible to derive $\alpha(\omega)$ by measuring $T(\omega)$ only, if two samples with different thicknesses are available.

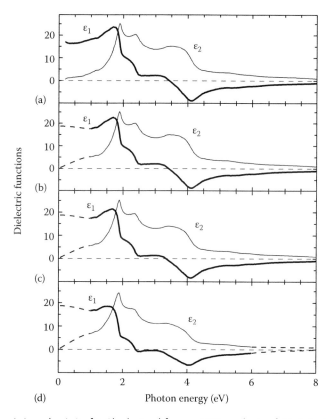

Figure 4.16 $\varepsilon_1(\omega)$ and $\varepsilon_2(\omega)$ of InSb derived from a K-K analysis of $R(\omega)$. In (a), the full $R_0(\omega)$ spectrum of Figure 4.13 was used. The data in (b)–(d) were obtained from narrower ranges of $R(\omega)$, namely, (b) 1–30 eV, (c) 1–10 eV, and (d) 1–6 eV. The dashed portions of the spectra are in the extrapolated region. (H. Okamura, unpublished data.)

4.6.3 $Tl_2Mn_2O_7$

We will describe the optical properties of $Tl_2Mn_2O_7$ (Okamura et al. 2001) as an example of an oxide compound showing a metal–nonmetal transition, where the optical sum rule discussed in Chapter 1 proves very useful. $Tl_2Mn_2O_7$ has the so-called pyrochlore crystal structure and undergoes a ferromagnetic (FM) transition at a Curie temperature (T_C) of 120–140 K (Shimakawa et al. 1996). The electrical resistivity (ρ) decreases dramatically upon cooling through T_C, and a large magnetoresistance has been observed near T_C. Figure 4.17 shows the measured $R(\omega)$ of $Tl_2Mn_2O_7$ at $T = 40$–295 K and $\sigma(\omega)$ derived from a K-K analysis of $R(\omega)$ in the absence of a magnetic field. The spectra at 295 K clearly indicate that $Tl_2Mn_2O_7$ is an insulator, with low $\sigma(\omega)$ and clear phonon peaks in the FIR range. From the onset of $\sigma(\omega)$, as indicated by the dotted line and the upward arrow in Figure 4.17b, the band gap can be estimated to be ~1.6 eV. Note the tail of $\sigma(\omega)$ below 1.6 eV, indicated by the downward arrow in Figure 4.17b. Although $\sigma(\omega)$ should drop to zero exactly at the band gap energy for an ideal insulator (if not for exciton states), that is rarely observed in actual experimental data. The T dependence of $R(\omega)$ and $\sigma(\omega)$ below 0.5 eV are also shown in Figure 4.17. It is seen that $R(\omega)$ increases rapidly and $\sigma(\omega)$ develops a rising component toward zero energy as T is lowered. These are Drude components due to free carriers and demonstrate that the FM transition in $Tl_2Mn_2O_7$ is also an insulator–metal transition. According to the optical sum rule discussed in

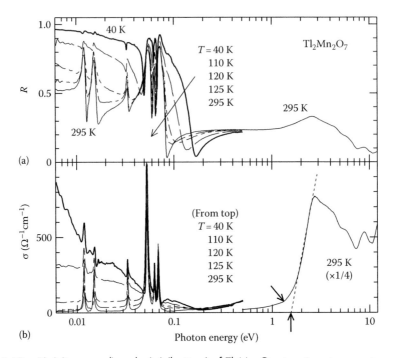

Figure 4.17 $R(\omega)$ (top panel) and $\sigma(\omega)$ (bottom) of $Tl_2Mn_2O_7$ at various temperatures (T). Low T data are shown only below 0.5 eV. In (b), the dotted line and the upward arrow indicate the estimated band gap energy (1.6 eV) and the downward arrow indicates the tail of the absorption edge, as discussed in the text. (Reprinted from Okamura, H. et al., *Phys. Rev. B*, 64, 180409(R)-1, 2001. With permission.)

Chapter 1, the area of a spectral component in $\sigma(\omega)$ is related to the effective carrier density N_{eff}^* by Wooten 1972 and Dressel and Gruner 2002:

$$N_{eff} = \frac{n}{m^*} = \frac{2m_0}{\pi e^2} \cdot \frac{1}{4\pi\varepsilon_0} \int \sigma(\omega)d\omega, \qquad (4.15)$$

where n and m^* are the density and effective mass (in units of the rest electron mass m_0), respectively. Figure 4.18a shows plots of N_{eff} calculated with this formula for the Drude component in $\sigma(\omega)$ as a function of T. It is seen that N_{eff} is low, on the order of 10^{-2} per $Tl_2Mn_2O_7$ formula unit ($\sim 10^{19}$ cm^{-3}) even in the metallic state below T_C, as compared to ~ 1 per formula unit in a good metal such as Ag. It is also seen that N_{eff} decreases steeply upon cooling through T_c. For comparison, the measured magnetization (M) of the same sample is also plotted in Figure 4.18a. M and N_{eff} have very similar T dependences, suggesting that the metallic state is closely related to the magnetic order. $R(\omega)$ and $\sigma(\omega)$ show large changes also with applied magnetic field at temperatures near T_c (Okamura et al. 2001, not shown here). In Figure 4.18b, the values of N_{eff} obtained from data measured at various values of T and B are plotted versus M. They scale beautifully as $N_{eff} \propto M^2$ over the measured T (40–295 K) and B (0–6 T) ranges. Unfortunately, the microscopic origin of this scaling has not been clarified yet. Nevertheless, it is interesting to note that other magnetoresistive compounds such as perovskite manganites (Takenaka et al. 2000) and EuB_6 (Broderick et al. 2002), have shown a similar M^2 scaling with N_{eff} in their optical response.

In addition to the reflectance study discussed here, the carrier dynamics in $Tl_2Mn_2O_7$ have also been studied more directly in the time domain (Prasankumar et al. 2005) using ultrafast

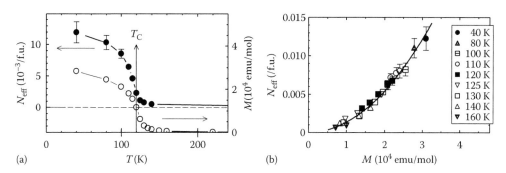

Figure 4.18 Effective carrier density (N_{eff}) in $Tl_2Mn_2O_7$, obtained using the optical sum rule, plotted as a function of (a) temperature (T) and (b) magnetization (M). In (a), M is also plotted versus T for comparison. In (b), N_{eff} obtained at various values of T and external magnetic fields up to 6 T are plotted together. The solid curve is a guide to the eye, showing a M^2 dependence of N_{eff}. (Reprinted from Okamura, H. et al., *Phys. Rev. B*, 64, 180409(R)-1, 2001. With permission.)

pump-probe spectroscopy (Chapter 9). The reflectance data discussed here was critical in understanding the time-resolved studies, in which it was shown that carrier relaxation was strongly influenced by spin disorder at temperatures near T_C. This is only one example showing that the time-integrated optical data described in this chapter is often essential in understanding the results of time and spatially resolved experiments (Parts III and IV in this book) on complex materials; the next section describes another example of this.

4.6.4 YbAl₃

As a final example, we will describe the optical spectra of YbAl$_3$ (Okamura et al. 2004), where the so-called "extended Drude" analysis of the optical spectra proves very useful. YbAl$_3$ is one of the so-called "heavy electron" compounds (Degiorgi 1999, Lawrence 2008). In a heavy electron compound, typically a Ce- or Yb-based intermetallic compound, the hybridization between the mobile conduction (c) electrons and the otherwise localized 4f electrons results in a duality between the localized and delocalized characteristics. Such a duality leads to many interesting physical properties, including the formation of heavy electrons at low temperatures (T). The mass enhancement is due to the strong correlations between 4f electrons, and observed as, for example, a large T coefficient (γ) in the electronic specific heat (Lawrence 2008). YbAl$_3$ has a γ of 58 mJ/K^2 mol as compared to 4 mJ/K^2 mol for LuAl$_3$, which is a nonmagnetic (filled 4f shell) reference material for YbAl$_3$. Namely, there is a factor of ~15 mass enhancement for YbAl$_3$ over LuAl$_3$.

Figure 4.19a shows $R(\omega)$ of YbAl$_3$ at 295 K, measured from 7 meV to 30 eV. At 295 K, $R(\omega)$ is relatively high over the entire IR range, but it also has a broad, marked dip centered at ~0.25 eV. The measured $R(\omega)$ of nonmagnetic LuAl$_3$ is also shown, but it does not have such a dip. Hence, this dip should be a characteristic feature of the electronic structure caused by the unpaired f electrons. Figure 4.19b shows $R(\omega)$ spectra measured at $T = 8$–690 K. The above-mentioned dip exhibits quite a strong T dependence and becomes deeper and wider with decreasing T. $\sigma(\omega)$ spectra derived from $R(\omega)$ with K-K analysis are shown in Figure 4.19c. The dip in $R(\omega)$ gives rise to a marked mid-IR peak centered at 0.25 eV in $\sigma(\omega)$. Such a mid-IR peak has been observed for various Ce and Yb compounds (Okamura et al. 2007a and the cited references). It has been attributed to optical excitations related with the c-f hybridized states near the Fermi level (E_F). In Figure 4.19c, the Drude component in $\sigma(\omega)$ is seen to lose spectral weight with decreasing T. Note, however, that YbAl$_3$ is a good metal, and the measured σ_{DC} keeps *increasing* with decreasing T, exceeding 1×10^6 Ω^{-1} cm^{-1} below 10 K.

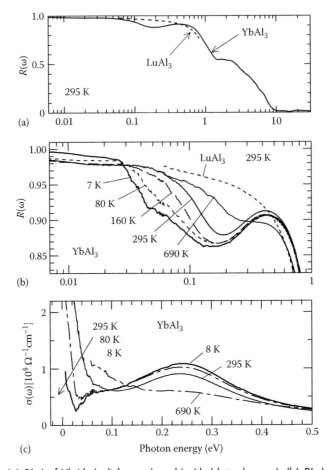

Figure 4.19 (a) $R(\omega)$ of YbAl$_3$ (solid curve) and LuAl$_3$ (dotted curve). (b) $R(\omega)$ of YbAl$_3$ at various temperatures, and that of LuAl$_3$ at 295 K. (c) $\sigma(\omega)$ obtained with K-K analysis. (Reprinted from Okamura, H. et al., *J. Phys. Soc. Jpn.*, *73*, 2045, 2004. With permission.)

Such a large difference between σ_{DC} and $\sigma(\omega)$ is due to the development of an extremely narrow Drude peak and is a characteristic feature observed for many heavy electron metals (Degiorgi 1999). It results from the formation of a narrow, δ-function-like density of states at E_F due to c-f hybridization. The rise of $\sigma(\omega)$ observed below 30 meV at $T = 8$ K, seen in Figure 4.19, is the tail of this Drude peak.

Figure 4.20 shows the ω-dependent scattering rate $1/\tau(\omega)$ and effective mass $m^*(\omega)$ in YbAl$_3$, derived using the "extended Drude model" (Chapter 1, Puchkov et al. 1996). They were calculated from the real $[\sigma_1(\omega)]$ and imaginary $[\sigma_2(\omega)]$ parts of the complex conductivity, obtained from the measured $R(\omega)$ with K-K analysis, as follows:

$$\frac{1}{\tau}(\omega) = \frac{ne^2}{m} \cdot \frac{\sigma_1(\omega)}{\sigma_1(\omega)^2 + \sigma_2(\omega)^2}, \quad \frac{m^*}{m_0}(\omega) = -\frac{ne^2}{m} \cdot \frac{1}{\omega} \cdot \frac{\sigma_1(\omega)}{\sigma_1(\omega)^2 + \sigma_2(\omega)^2}. \quad (4.16)$$

In Figure 4.20a, as T is lowered from 295 K, $1/\tau(\omega)$ shows overall increases, but, below 80 K, where the coherent heavy electron band is formed, $1/\tau(\omega)$ decreases toward zero photon energy and develops a ω^2 dependence, which is a Fermi liquid property. In addition, $m^*(\omega)$ is seen to increase with decreasing T in Figure 4.20b. These results demonstrate that a Fermi

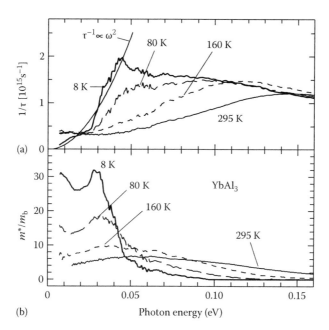

Figure 4.20 Frequency-dependent scattering rate $(1/\tau)$ and effective mass (m^*/m_b) of $YbAl_3$, derived using the extended Drude analysis as discussed in the text. (Reprinted from Okamura, H. et al., *J. Phys. Soc. Jpn.*, 73, 2045, 2004. With permission.)

liquid with enhanced m^* is indeed formed at low T. Similar results have been observed for other heavy electron compounds, such as $CePd_3$ (Webb et al. 1986) and $YbRh_2Si_2$ (Kimura et al. 2006). The extended Drude scheme has been also used extensively for strongly correlated d electron systems, such as the high-temperature cuprate superconductors (Puchkov et al. 1996, Basov and Timusk 2005).

Finally, the dynamics of heavy electrons in $YbAl_3$ have also been explored using ultrafast pump-probe spectroscopy (Demsar et al. 2009, Chapter 8). The increase of the carrier scattering time in the heavy electron state at low T was directly demonstrated in the time domain. As in the case of $Tl_2Mn_2O_7$ discussed earlier, the time-integrated (frequency domain) optical measurements discussed here were important in analyzing the time-resolved data.

ACKNOWLEDGMENTS

The author would like to thank all the colleagues and former students who have contributed to the development of experimental methods and results discussed here. The unpublished data on InSb shown in Figures 4.13 through 4.15 were measured as a part of undergraduate research projects with Keigo Nishi and Satoshi Ishida. The numerical K-K analyses were performed by a computer code provided by Dr. Shin-ichi Kimura. The spectral fittings presented here were performed by the RefFIT software developed by Dr. Alexey Kuzmenko.

REFERENCES

Basov, D. N. and T. Timusk. 2005. Electrodynamics of high-T_c superconductors. *Rev. Mod. Phys.* 77: 721–779.

Broderick, S., B. Ruzicka, L. Degiorgi, H. R. Ott, J. L. Sarrao, and Z. Fisk. 2002. Scaling between magnetization and Drude weight in EuB_6. *Phys. Rev. B* 65: 121102(R)-1–121102(R)-4.

Burns, G. 1990. *Solid State Physics*. Academic Press, San Diego, CA.

Cardona, M. 1965. Optical properties and band structure of $SrTiO_3$ and $BaTiO_3$. *Phys. Rev.* 140: A651–A655.

Cardona, M. and D. L. Greenaway. 1964. Optical properties and band structure of group IV–VI and group V materials. *Phys. Rev.* 133: A1685–A1697.

Carr, G. L. 2001. Resolution limits for infrared microspectroscopy explored with synchrotron radiation. *Rev. Sci. Instrum.* 72: 1613–1619.

Cooley J. W. and J. W. Tukey. 1965. An algorithm for the machine calculation of complex Fourier series. *Math. Comput.* 19: 297–301.

Degiorgi, L. 1999. The electrodynamic response of heavy-electron compounds. *Rev. Mod. Phys.* 71: 687–734.

Demsar, J., V. V. Kabanov, A. S. Alexandrov et al. 2009. Hot electron relaxation in the heavy-fermion $Yb_{1-x}Lu_xAl_3$ compound using femtosecond optical pump-probe spectroscopy. *Phys. Rev. B* 80: 085121-1–085121-6.

Dore, P., A. Nucara, D. Cannavo et al. 1998. Infrared properties of chemical-vapor deposition polycrystalline diamond windows. *Appl. Opt.* 37: 5731–5736.

Dressel, M. and G. Gruner. 2002. *Electrodynamics of Solids: Optical Properties of Electrons in Matter*. Cambridge University Press, Cambridge, U.K.

Ehrenreich, H. and H. R. Philipp. 1962. Optical properties of Ag and Cu. *Phys. Rev.* 128: 1622–1629.

Fujiwara, H. 2007. *Spectroscopic Ellipsometry: Principles and Applications*. Wiley, Chichester, U.K.

Fukui, K., H. Miura, H. Nakagawa et al. 2001. Performance of IR-VUV normal incidence monochromator beamline at UVSOR. *Nucl. Instrum. Methods Phys. Res. A* 467–468: 601–604.

Griffiths, P. R. 2009. Infrared and Raman instrumentation for mapping and imaging. In *Infrared and Raman Spectroscopic Imaging*, ed. R. Salzer and H. W. Siesler, pp. 3–62. Wiley-VCH, Weinheim, Germany.

Griffiths, P. and J. A. De Haseth. 2005. *Fourier Transform Infrared Spectrometry*. Wiley-Interscience, Hoboken, NJ.

Heavens, O. S. 1991. *Optical Properties of Thin Solid Films*. Dover, New York.

Hecht, E. 2001. *Optics* (4th edn.). Addison Wesley, San Francisco, CA.

Homes, C. C., M. Reedyk, D. A. Cradles et al. 1993. Technique for measuring the reflectance of irregular, submillimeter-sized samples. *Appl. Opt.* 32: 2976–2983.

Jackson, J. D. 1975. *Classical Electrodynamics*. John Wiley & Sons, New York.

Kimura, S., J. Sichelschmidt, J. Ferstl et al. 2006. Optical observation of non-Fermi-liquid behavior in the heavy fermion state of $YbRh_2Si_2$. *Phys. Rev. B* 74: 132408-1–132408-5.

Kuzmenko, A. B. 2005. Kramers–Kronig constrained variational analysis of optical spectra. *Rev. Sci. Instrum.* 76: 083108.

Lawrence, J. 2008. Intermediate valence metals. *Mod. Phys. Lett. B* 22: 1273.

Lax, B. and G. B. Wright. 1960. Magnetoplasma reflection in solids. *Phys. Rev. Lett.* 4: 16–18.

Matsunami, M., H. Okamura, A. Ochiai, and T. Nanba. 2009. Pressure tuning of an ionic insulator into a heavy electron metal: An infrared study of YbS. *Phys. Rev. Lett.* 103: 237302-1–237302-4.

Okamura, H., T. Koretsune, M. Matsunami et al. 2001. Charge dynamics in the colossal magnetoresistance pyrochlore $Tl_2Mn_2O_7$. *Phys. Rev. B* 64: 180409(R)-1–180409(R)-4.

Okamura, H., T. Michizawa, T. Nanba, and T. Ebihara. 2004. Pseudogap formation and heavy-carrier dynamics in intermediate-valence $YbAl_3$. *J. Phys. Soc. Jpn.* 73: 2045–2048.

Okamura, H., T. Watanabe, M. Matsunami et al. 2007a. Universal scaling in the dynamical conductivity of heavy fermion Ce and Yb Compounds. *J. Phys. Soc. Jpn.* 76: 023703-1–023703-5.

Okamura, H., T. Michizawa, T. Nanba, and T. Ebihara. 2007b. Infrared study of the valence transition compound $YbInCu_4$ using cleaved surfaces. *Phys. Rev. B* 75: 041101-1–041101-4.

Philipp, H. R. and H. Ehrenreich. 1963. Optical properties of semiconductors. *Phys. Rev.* 129: 1550–1560.

Prasankumar, R. P., H. Okamura, H. Imai et al. 2005. Coupled charge-spin dynamics of the magnetoresistive pyrochlore $Tl_2Mn_2O_7$ probed using ultrafast midinfrared spectroscopy. *Phys. Rev. Lett.* 95: 267404-1–267404-4.

Predoi-Cross, A. and B. E. Billinghurst (ed). 2010. *5th International Workshop on Infrared Microscopy and Spectroscopy with Accelerator Based Sources*, Baniff, Canada. AIP Conference Proceedings 1214.

Puchkov, A. V., D. N. Basov, and T. Timusk. 1996. The pseudogap state in high-T_c superconductors: An infrared study. *J. Phys.: Condens. Matter* 8: 10049–10082.

Sacchetti, A., E. Arcangeletti, A. Perucchi et al. 2007. Pressure dependence of the charge-density-wave gap in rare earth tritellurides. *Phys. Rev. Lett.* 98: 026401-1–026401-4

Schubert, M. 2009. *Infrared Ellipsometry on Semiconductor Layer Structures: Phonons, Plasmons, and Polaritons.* Springer, New York.

Shimakawa, Y., Y. Kubo, and T. Manako. 1996. Giant magnetoresistance in $Tl_2Mn_2O_7$ with the pyrochlore structure. *Nature* 379: 53–55.

Smith, B. C. 1995. *Fundamentals of Fourier Transform Infrared Spectroscopy.* CRC Press, Boca Raton, FL.

Socrates, G. 2004. *Infrared and Raman Characteristic Group Frequencies: Tables and Charts.* Wiley, Chichester, U.K.

Stuart, B. H. 2004. *Infrared Spectroscopy: Fundamentals and Applications.* Wiley, Chichester, U.K.

Takenaka, K., K. Iida, Y. Sawaki et al. 1999. Optical reflectivity spectra measured on cleaved surfaces of $La_{1-x}Sr_xMnO_3$: Evidence against extremely small Drude weight. *J. Phys. Soc. Jpn.* 68: 1828–1831.

Takenaka, K., Y. Sawaki, R. Shiozaki et al. 2000. Electronic structure of the double exchange ferromagnet $La_{0.825}Sr_{0.175}MnO_3$ studied by optical reflectivity. *Phys. Rev. B* 62: 864–867.

Tompkins, H. and E. A. Haber. 2006. *Handbook of Ellipsometry* (*Materials Science and Process Technology*). William Andrew, Norwich, NY.

Webb, B. C., A. J. Sievers, and T. Mihalisin. 1986. Observation of an energy- and temperature-dependent carrier mass for mixed-valence $CePd_3$. *Phys. Rev. Lett.* 57: 1951–1954.

Wooten, F. 1972. *Optical Properties of Solids.* Academic Press, New York.

Yu, P. Y. and M. Cardona. 2001. *Fundamentals of Semiconductors: Physics and Materials Properties.* Springer, Berlin, Germany.

Chapter 5

METHODS FOR OBTAINING THE OPTICAL RESPONSE AFTER CW EXCITATION

Sajan Saini

CONTENTS

Optically active materials absorb radiant energy and return to thermal equilibrium by releasing this absorbed energy, primarily, through a radiative de-excitation. This process of excitation by higher energy photons, followed by the emission of lower energy photons, after a characteristic time, is referred to as photoluminescence. Fundamentally, all materials have absorption spectra, a range of wavelengths or photon energies for which a finite portion of the electromagnetic spectrum is absorbed (e.g., iron atom impurities within the insulator silica (SiO_2:Fe) absorb visible light). Fundamentally, all materials also have emission spectra, with an emission peak that shifts as a function of the temperature (e.g., blackbody radiation from metals at elevated temperatures). Optically active materials are the subset class of materials that couple an absorption transition to an emission transition, efficiently (without dissipating excessive energy into heat) and on a short time scale (ms to sub-ns). SiO_2 doped with rare-earth atoms such as ytterbium, erbium (Er), or thulium is an optically active material, absorbing visible or infrared (IR) light and emitting lower energy IR light. Silicon (Si) and gallium arsenide (GaAs) semiconductors are also optically active materials, absorbing visible light and emitting IR light (albeit Si does this less efficiently than GaAs).

This chapter examines three spectroscopic techniques for measuring the atomic or electronic energy transitions involved in (1) absorbing and emitting light, (2) the efficiency of absorbing or emitting light, (3) mechanisms of energy transfer, and (4) characterization of absorption by a nonoptical detection technique, namely, measuring the concentration of photo-generated electronic carriers. The three techniques we consider in this chapter—Photoluminescence (PL), Photoluminescence Excitation Spectroscopy (PLE), and Photocurrent Spectroscopy (PC)—are measured under continuous wavelength (CW) excitation, i.e., steady-state illumination by an optical power source (time-resolved PL is discussed in Chapter 12).

In a typical PL experiment, one unique wavelength (the output of a laser) is used to excite a specific energy transition, and the emitted spectrum of light is observed by a photodetector as a plot, or "scan," of light intensity versus emission wavelength. It is important to appreciate that there may be multiple emission wavelengths, each having a spectral linewidth. Radiative de-excitation conventionally implies these emitted spectra occur at longer wavelengths (i.e., lower photon energies) relative to the excitation wavelength. However, under high-power excitation, cooperative phenomena between excited energy levels in the sample may result in "upconverting" energy transitions (see Section 5.2), resulting in emission spectra at relatively shorter wavelengths.

In a PLE experiment, a spectrum of excitation wavelengths is available to excite the sample, and spectral scanning tunes a transmission passband (of selected spectral width) across a range of this spectrum. During this scan of excitation wavelengths, a detector monitors the emission of light, at a fixed wavelength, and produces a plot of emitted light intensity versus the excitation wavelength. For a PLE experiment, a priori knowledge of emission spectra, particularly PL peaks, is required, in order to select an emission wavelength for the photodetector to monitor.

In a PC experiment, a transmission passband is used once again to scan across a spectrum of excitation wavelengths. This time, however, there is no photodetector used to measure the emission of light. Instead, an ammeter is used to measure the amount of total current (comprising the photocurrent generated by the optical excitation and the "dark" current due to a sample's conductivity at thermal equilibrium) passing through the sample. A plot of total current versus excitation wavelength summarizes the number of mobile carriers (electrons and holes), generated by absorption at a given wavelength that drift to the electrical contacts.

Section 5.1 introduces readers to the details for constructing a CW excitation spectroscopy experiment, addressing pertinent experimental issues involving sample excitation and signal collection. Sections 5.2 through 5.4 review case studies for the three spectroscopy techniques, demonstrating for readers the essential aspects of data analysis from the optical response after CW excitation and what we can conclude about the atomic or electronic energy models for an optically active material.

5.1 SPECTROSCOPY LINK: EXPERIMENTAL SETUP FOR MEASURING AN OPTICAL RESPONSE

Figure 5.1 is a schematic sketch of an optical link (a link is defined as a fully determined experimental layout) for CW excitation spectroscopy. The link comprises several components: an excitation light source (a laser or a lamp), spectral or spatial manipulation of the source beam (by means of a monochromator, filter, reflective optics and/or mechanical chopper), absorption of the beam by a sample (housed in a room- or low-temperature sample holder), and the measurement of emitted light (with a monochromator and photodetector) or generated photocurrent (with an ammeter). While a spectroscopy experiment does not need to be set up on a vibration-damping table, it is conventional practice to construct this optical link on an optical table or breadboard to easily align the light source, sample, and photodetector. We review the role and operating guidelines for each component in this section.

5.1.1 Optical Excitation Source

For PL experiments, an excitation wavelength lying anywhere within the absorption spectrum will be sufficient to optically excite the sample and generate an emission spectrum; a narrow

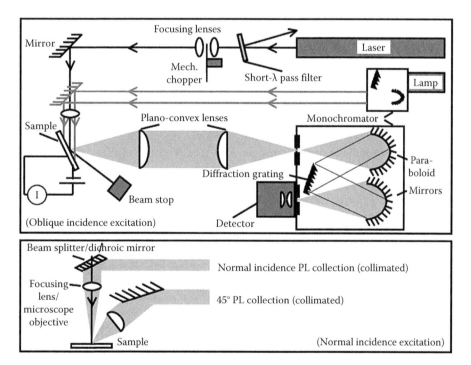

Figure 5.1 Schematic diagram of an optical link for a spectroscopy experiment. PL relies on a laser for optical excitation of the sample, while PLE and PC rely on a lamp and source monochromator for spectrally selective optical excitation. The optical axis refers to the trajectory of the excitation beam (arrow lines) and collimated emission (gray beam). PL and PLE measure light emission from the sample, while PC measures photogenerated current in the sample. The top and bottom schematics show the sample at oblique and normal incidence, respectively, to the excitation beam; for normal incidence excitation, PL emission can also be collected at normal incidence (using a 50/50 beam splitter (or dichroic mirror) to separate the PL beam path from the excitation beam path) or at 45°.

TABLE 5.1

Primary Excitation Wavelengths (Most Commonly Used Wavelength Shown in Bold), Power Range, and Beam Quality (Beam Size × Beam Divergence) for Some CW Gas Lasers and a Solid-State Laser Diode

	Excitation Wavelengths (nm)	Power (mW)	Beam Diameter × Beam Divergence
Ar-ion laser[a]	**488**, 514.5	10–2,000	1.5 mm × 0.5 mrad
HeNe[b]	**633**, 612	1.5–35	0.77 mm × 1.0 mrad
Pulsed Nd:YAG[c]	**1,064**, 946, 532[e]	30 mJ/160 ns pulse (10 kHz repetition)	4 mm × 10 mrad
InGaAs LD[d]	980	100–20,000	0.6 mm × 174 mrad

Note: A pulsed Nd:YAG laser has been included as well.

[a] Sample Argon laser data sheet. http://www.coherent.com/downloads/Innova300Datasheet.pdf

[b] Sample Helium–Neon laser data sheet. http://www.edmundoptics.com/onlinecatalog/DisplayProduct.cfm?productid=2550

[c] Sample Nd:YAG laser data sheet. http://www.photonix.com/PDF/DM.pdf

[d] Sample IR laser diode data sheet. http://www.lasermate.com/LD980-100A.html

[e] The 532 nm emission from Nd:YAG lasers is generated by frequency doubling of the 1064 nm emission.

spectral linewidth laser is a suitable choice for this excitation source, offering several advantageous properties. The coherent nature of a laser concentrates its optical power within a narrow beam size, ensuring a high photon flux density and high rate of optical excitation (see Section 5.2 for a description of the excitation rate equation). Lasers originating from an atomic transition have a mm–scale beam diameter (see Table 5.1) and an angular divergence on the order of 1–10 mrad. The narrow divergence of an Ar ion laser thus gives the experimenter the freedom to place their laser within a working distance of 1–1.5 m from the sample holder, and retain a beam spot size of ≤1 mm on the sample surface (in particular, by refocusing the beam with a plano-convex lens, as discussed in Section 5.1.2).

Laboratory lasers are available in three principal spectral ranges of interest: ultraviolet (UV), visible, and infrared (IR). Table 5.1 lists the performance parameters for some commonly available gas cavity lasers and a semiconductor-state laser diode. The beam diameter and divergence have been stated as a product, referred to as beam quality, that describes the spatial extent of the beam (see "optical throughput" definition in Section 5.1.5).

The experimenter should take extra care to be aware of the full emission spectrum of the laser and the transmission spectrum of the selected dielectric mirrors. Ar-ion lasers, for example, have six visible emission lines (between wavelengths 457.9 and 528.7 nm) and one IR emission line (at wavelength 1092.3 nm). The highest intensity emission lines occur at 514.5 nm (green) and 488.0 nm (blue-green), which can be selectively reflected by dielectric mirrors (provided by the vendor), to achieve lasing. Thus, Ar-ion lasers are referred to as multiline lasers. However, these mirrors are unable to completely remove the IR emission line. If the experimenter is looking for PL from a sample in the IR spectrum, this spontaneous emission line will illuminate the sample (along with the desired excitation wavelength) and a portion of its scattered reflection will be collected by collimating optics (see Section 5.1.5) and routed into the photodetector, resulting in the observation of an artifact PL spectrum with a peak wavelength of 1092.3 nm. For the Ar-ion laser, this IR emission line can be removed by placing a short-wavelength pass filter (with a 3 dB transmission point in the 650–850 nm range) in front of the laser head output. Care should be taken to place this filter at a slight tilt with respect to the optical axis (Figure 5.1). This tilt ensures that the portion of laser light that reflects from the filter will not re-enter the laser cavity, and rather will illuminate the outer casing of the laser head. Reflecting a portion of laser light back into the cavity results in degradation, after long time periods, in the power versus current performance of the laser.

Gas lasers consume a considerable amount of electrical power and dissipate considerable heat within the laser head when operating at output powers in the 0.1–5 W range; hence water cooling is required to cool both the power supply and laser head. This cooling is achieved by means of a closed loop water line, circulating filtered water through the power supply, laser head, and a water chiller (in a series connection). An interlock within the laser controller will shut down current injection, and thus continued heat dissipation, if water flow decreases below a threshold value that is measured in units of gallons per minute. Because deionized water reacts aggressively with the (typically copper) tubing of the water line, and tap water results in mineral deposits, filtered water is exclusively recommended for the closed loop.

While there has been a recent proliferation of low-power air-cooled Ar-ion and HeCd lasers (output power ≤0.1 W), for the study of low-efficiency light emission from thin film samples, high-power lasers remain a necessary tool in a spectroscopy link.

Semiconductor laser diodes are driven by the electron–hole recombination of a forward biased current, supplied by an external current source referred to as a laser driver (LD). These lasers are typically cooled by the thermoelectric effect, controlled by an external thermoelectric controller (TEC). Like air-cooled gas lasers, these lasers are essentially "turn-key" systems with minimal to no maintenance required. It is good practice with laser diodes to power the TEC before the LD and to restrict the LD current to the linear range of the optical power versus electrical current performance plot (commonly called the "LI" curve), supplied by the vendor.

The experimenter should occasionally measure the output optical power of a laser with an independent power meter, to confirm that the output power is in close agreement with the measure of the internal power sensor (for gas lasers) or predicted LI curve (for laser diodes). Care should be taken to note the power meter's saturation specification, and restrict input power (using neutral density filters) to be 5–10× below this limit.

For PLE and PC experiments, a lamp offers a practical solution for excitation wavelength-dependent measurements over a broad spectral range. The excitation spectrum of a xenon (Xe) or tungsten (W) lamp closely approximates a blackbody (Drozdowicz et al. 2004), with typical maximum output optical power of 450 W. Transmitting the lamp output through a monochromator diffraction grating (defined by its grating groove density ρ_{groove} and blaze wavelength λ_{blaze}, Lerner and Thevenon 1996, in Section 5.1.5) results in a transmission passband (of linewidth $\Delta\lambda$) that selects a collection of excitation wavelengths. $\Delta\lambda$ can be tuned across most of the excitation spectrum, as determined by λ_{blaze} (Section 5.1.5). A typical tuning range is 233–700 nm, and the typical passband values are $1 < \Delta\lambda < 10$ nm.

While the isotropic emission of a lamp filament results in a >10× loss in wallplug efficiency (Drozdowicz et al. 2004), its broadband spectrum makes for a unique optical excitation source for PLE and PC. A condenser lens is added to optimize the extraction of a lamp's light, and the excessive heat dissipation of the filament implies that electrical input power must be increased at a moderate rate, to mitigate the stress of thermal cycling on the condenser lens.

This section concludes with a brief note about eye safety (MIT Radiation Protection Committee 2008), an issue of paramount importance when working with lasers in a CW experiment. While both CW and pulsed lasers are used in spectroscopy experiments, the high intensity of pulsed lasers is more dangerous to the human eye. That being said, CW lasers are dangerous to work with as well, in particular when operating at UV and IR wavelengths, portions of the spectrum the eye cannot see. We won't go into the details of how an eye injury develops from exposure to laser light, other than to simply note that prolonged exposure (on the order of <1 s) to a coherent source of radiation results in the boiling of blood and rupture of capillary eye vessels. Whenever possible, it is preferable to work with a laser emitting at a visible wavelength, one that can be rapidly detected and responded to by the human eye (i.e., by blinking and looking away), in an accidental situation when laser light scatters off the optical table and into an experimenter or bystander's eyes. When operating a laser, the experimenter should take great care to wear protective eyewear (laser goggles) that attenuates light intensity at the laser wavelength by several orders of magnitude. Bystanders can be protected from inadvertent laser exposure by drawing an opaque curtain about the optical table. If a curtain is not available, a warning light must be installed above the door to the lab to alert incoming colleagues of the laser's operation. If a colleague must work in the same lab on an independent experiment, they must wear laser goggles as well.

Laser goggles are specified by their optical density (OD) and spectral range of attenuation. Within the spectral range of attenuation, laser goggles diminish the intensity of transmitted light by a factor of 10^{OD}. OD = 7–9 is a safe margin of eye protection for 1–5 W lasers. When aligning a laser beam onto a sample (Section 5.1.3), it helps to be able to observe a visible wavelength beam spot, scattered from the surface of a piece of paper (UV, visible excitation line) or IR card (IR excitation line). For goggles attenuating UV or IR light, the OD in the visible range may be low enough to observe this visible beam spot. For goggles attenuating visible light, the visible beam spot will not be observable. Most experimenters respond to this problem by taking off their goggles and "sneaking a peak" at the visible beam spot. A safer and more effective solution is the purchase of additional laser goggles with an OD = 2 in the visible, permitting the eye to briefly observe the beam spot at a lower and less harmful intensity.

5.1.2 Optical Path of Excitation Beam

The excitation beam in Figure 5.1 must be directed across the working space of the optical table and to the sample holder, either by free space or by fiber-optic propagation. The conventional practice in spectroscopy is to work with free space optics, where the excitation beam is steered through turns by reflective mirrors. Mirrors are a near ideal optical component, as the weak "skin depth" penetration of light into their metallic surface (Omar 1994) implies minimal sampling of the metal's polarizability; the refractive index of aluminum-coated mirrors is largely constant across the UV–visible–IR spectrum (Chapter 3).

When working with laser beams that are invisible to the human eye ($\lambda > 800$ nm) modern practices of the last 30 years have equally relied on the use of multimode optical fiber. A well-designed spectroscopy link that routes IR laser light via optical fiber into and out of its components reduces the risk of laser light scattering into the experimenter's eye. It is recommended that the experimenter work with (1) a large core diameter multimode fiber ($d_{core} > 500$ μm) that has (2) a large numerical aperture (Saleh and Teich 1991) NA $\equiv \sin \theta_a > 0.35$, corresponding to an acceptance angle of $\theta_a > 20.5°$ from normal incidence, and (3) a polished tip. Light can easily be coupled into a large core diameter multimode fiber on an optical breadboard surface, using a mechanical xy-positioning stage and focusing the laser beam onto its polished tip (Senior 1985) with a plano-convex lens or microscope objective. In contrast, coupling into single mode fiber requires the fine-tuning of a piezoelectric actuated xy-positioning stage. Single mode optical fiber is useful for low signal dispersion over hundreds of kilometers, but it offers no advantage in a spectroscopic link with an optical path length of 1–2 m. Multimode fiber should be the exclusive choice for a spectroscopy experimenter.

The input and output end of a fiber must be smoothly cleaved, in order to minimize scattering of inserted or extracted light; light is coupled by means of a plano-convex lens or microscope objective with matching NA (Section 5.1.5). The insertion and extraction loss of optical power should be accounted for in the power budget for the excitation beam, before it reaches the sample. For custom fiber lengths, it is recommended that the experimenter acquires an optical fiber cleaving and polishing kit.

Lastly, when purchasing commercial optical fiber for IR applications, the experimenter should take care to purchase "low OH" fiber,* containing a reduced concentration of water moisture—the historic culprit responsible for IR absorption at $\lambda \sim 1.4\,\mu m$. If the excitation beam is a UV laser, attenuation loss in commercial optical fiber can be 40–50× higher than at visible wavelengths.* Thus, it is largely recommended to work with free space optics in this case, and conventional mirrors should be replaced with UV-enhanced (e.g., MgF_2-coated aluminum) mirrors.[†]

A mechanical chopper is another optical element, frequently inserted in the excitation beam path, for triggering optical excitation of the sample: the interruption of a CW beam by the periodic 50/50 duty cycle slots of a chopper blade[‡] results in a periodic modulation of the excitation intensity, PL emission, and photodetector current; a lock-in amplifier filters background noise frequencies (primarily room lighting) from the triggered photodetector current and enables a high signal-to-noise ratio (SNR) (see dark current versus photocurrent discussion in Section 5.1.4). Chopper blades come in two common configurations, with 6/5 or 30/25 slot windows (corresponding to modulation frequencies of 4–400 Hz or 400 Hz–3.7 kHz, respectively). For chopper frequencies below 50 Hz, a mechanical instability in the chopper motor imparts a tumbling motion to the rotating chopper blade; this tumbling motion contributes an additional intensity modulation, appearing as an artifact in time-resolved PL measurements. Thus, whenever possible, the experimenter should try to operate the chopper at higher rotation speeds.

As a side-note, 0.1–1 ms lifetime measurements are routinely done with a mechanical chopper and triggered oscilloscope. When measuring with chopper frequencies below 50 Hz, it takes an appreciable time for the chopper blade to fully occlude the laser beam, and this imparts a linear skew in the oscilloscope's PL versus time measurement. This temporal skew can be minimized by using two plano-convex lenses to converge and re-collimate the laser beam, while positioning the chopper blade at the narrow beam waist produced between the two lenses.

Before the excitation beam illuminates the sample, it is advantageous to focus the beam spot on the sample surface (in order to increase the excitation intensity) using a plano-convex lens on a z-positioning stage that allows fine adjustment of the focal distance. Common optics lenses are made of fused silica, a synthetic form of quartz (crystalline SiO_2); thus, when working with a UV laser beam, high-crystallinity fused silica lenses that are rated for UV photon energies should be acquired. Visible and UV-rated lenses will normally also have antireflective coatings to minimize back reflections.

As a safety protocol, when designing an optical beam path with a laser, the experimenter must keep track of reflected beam segments at the interface with each optical component, thereby minimizing the amount of scattered laser light that leaves the optical table or breadboard. Lastly, it is highly recommended to place a beam stop in front of an active laser to safely block its beam, when the experimenter is fine-tuning any component along the optical beam path.

---

* Sample low OH fiber data sheet. http://www.thorlabs.com/catalogPages/v20/890.pdf
[†] Sample UV Mirror data sheet. http://www.thorlabs.com/catalogPages/v20/681.pdf
[‡] Sample chopper data sheet. http://www.thinksrs.com/downloads/PDFs/Catalog/SR540c.pdf

5.1.3 Sample Mounting and Temperature Control

The sample mount or "holder" is ideally an *xyz*-positioning stage that holds a backing plate or a similar nonluminescing surface (e.g., a glass slide), on which the sample is mounted with nonluminescing glue. For low temperature measurements between 4 K and room temperature, a commercial cryostat can be purchased in either a continuous flow CF* or closed cycle CC† design. A CF cryostat requires the continuous flow of the cryogens liquid helium (L-He), or liquid nitrogen (LN2). These cryogens are extracted from their storage dewars via a transfer line, that is typically pumped on by a rotary pump. While LN2 (enables cooling between 77 K–room temperature) is a relatively low cost consumable, L-He (enables cooling between 4 K–room temperature) is not. A CC cryostat recirculates its cryogen (either LN2 or L-He) and thus uses up less of this consumable. More recently, CC cryostats have become available that require no cryogen at all, and achieve low temperature cooling with the use of He gas (which is considerably less costly that L-He). While CC cryostats can be more costly as an upfront expense in comparison to the CF design, over the long term, they will contribute less expense to cryogen consumption.

All cryostats are essentially dewars (with quartz windows allowing for line of sight access to the sample for excitation and PL emission collection) whose vacuum is maintained by a mechanical and turbo pump combination. Temperatures intermediate to a liquid cryogen temperature are achieved by means of resistive heating of the sample holder, using a PID controller feedback loop.

Typical adhesives used to mount a sample onto a sample holder are rubber cement, silver paste, or GE varnish. The experimenter should always check for background PL from the adhesive to ensure there is no artifact peak that overlaps with the PL spectrum of the sample.

There are two basic orientations for the sample, with respect to the excitation beam (see Figure 5.1): normal incidence and oblique incidence. In normal incidence, a portion of the excitation beam reflects back into the direction from where it was incident, and PL emission (emitted isotropically in all directions from the sample surface) is collected either (1) by a collimating lens, at an angle with respect to normal incidence (typically 45°); or (2) from normal incidence, using a very large NA lens (e.g., a microscope objective lens) to both tightly focus the excitation beam and optimally collimate the divergent PL, and including a 50/50 beam splitter or dichroic mirror to spatially separate the excitation beam from the PL spectra. It is typically recommended to focus the excitation beam onto the sample surface, to increase photon flux density (and optical excitation rate W_p, as described in Section 5.2). The small focal length of a microscope objective has an unmatched ability to focus; however, the resulting proximity of the objective to the sample blocks collection of PL emission by a separate collimating lens. Thus, when using a microscope objective, sample PL must be collected and collimated in normal incidence (see lower half of Figure 5.1). In oblique incidence, the reflected excitation beam is blocked and the sample surface is oriented toward a collimating lens. The collimating lens is a plano-convex lens that directs the PL emission into the monochromator. In practice, it is easier to collect PL emission for oblique incidence excitation, where the beam spot has a larger footprint. However, this larger footprint also results in a decrease in photon flux density and rate of optical excitation (see Section 5.2). For both orientations, a portion of the non-specular reflection, or scattered light, from the excitation beam is inevitably collected by the collimating lens. This scattered beam can be filtered out by a long wavelength pass filter, placed before the monochromator entrance slit.

---

* Sample continuous flow cryostat. http://www.oxford-instruments.com/products/low-temperature/opticaland-spectroscopy/optistatcf/Documents/OptistatCF-liquid-helium-continuous-flow-optical-spectroscopy-cryostat-product-guide.pdf

† Sample closed cycle cryostat. http://www.janis.com/products/productsoverview/4KCryocooler/SampleInVacuum SHI-4SeriesClosedCycleRefrigerators.aspx

For absorption cross-sectional measurements (Section 5.2), it is necessary to know the size of the excitation beam spot and this can be measured by placing a razor blade (attached to an *xy*-positioning stage with micrometer graduated scale) at the sample location, with an optical meter head just behind the blade. By advancing the razor blade through 5–10 μm intervals, the unoccluded beam power is measured by the optical meter. Plotting this power measurement versus relative blade position and taking the derivative of this plot will give the cross-sectional profile of the excitation beam, in one dimension. A Gaussian fit estimates the beam diameter as twice the standard deviation of this profile. The most accurate beam spot measurement is done at normal incidence; for oblique incidence, the experimenter should estimate the angle of incidence and correct a normal incidence beam spot with the appropriate geometric conversion factor.

5.1.4 Photocurrent/Photoconductivity Measurements

In semiconductors and thin films containing semiconductor nanostructures, absorption of the excitation beam at a given wavelength results in the generation of an electron–hole pair. Section 5.4 discusses in detail the formation physics of these carriers. Here, we will concern ourselves with the basic experimental issues to consider for measuring this excess charge density as a photocurrent. In PC spectroscopy, the optical link terminates at the sample, so there is no further optical path to manage. The experimenter's principal concern is to extract the photo-generated carriers out of the sample; when studying thin films, care should be taken to deposit the film on a high-resistance substrate that does not absorb the excitation light and generate excess carriers (e.g., quartz or sapphire are common resistive substrates). This ensures the measured photocurrent to be entirely attributed to the sample film.

Current and photocurrent measurements are typically done using a two-probe measurement technique (Figure 5.2a), comprising two metal contacts to the surface of a sample that is deposited on a highly resistive substrate. The flow of current through the sample (total resistance R_s) represents current flow through an equivalent circuit with two parallel resistances, representative of the thin film (R_{film}) and the substrate (R_{sub}), hence $1/R_s = 1/R_{film} + 1/R_{sub}$; for a highly resistive substrate, $R_s \simeq R_{film}$.

The schematic in Figure 5.2a shows a current I (or current density J) to flow exclusively through a thin film (i.e., assume a highly resistive substrate), parallel to the length l (separating the two probe contacts) and orthogonal to the cross-sectional area $A = wt$ (w, t are the sample width and thickness, respectively). The conductivity σ of the thin film relates $J \equiv \sigma E$ to the electric field E between the two probes (which can be expressed in

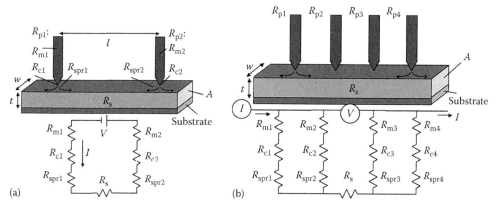

(a) (b)

Figure 5.2 (a) Schematic of geometry (with cross-sectional area A) for two-probe PC measurement and equivalent resistive circuit. (b) Schematic of geometry for four-probe PC measurement and equivalent resistive circuit.

terms of the applied voltage as $E = V_A/l$). Substituting these identities leads to the well known Ohm's Law $V_A = IR_s$, where the sample resistance is defined as $R_s \equiv l/\sigma A = \rho\, l/A$ (and $\rho = \sigma^{-1}$ is the thin film resistivity). For a large sample, $l \gg t$, $w \gg t$, and it is more useful to rewrite the resistance as $R = (\rho/t)(l/w) = R_{sheet}\, l/w$, where $R_{sheet} \equiv \rho/t$, termed the sheet resistance, is interpreted as a resistance per aspect ratio for the sample. For square samples, $l = w$ and $R_s = R_{sheet}$, and hence the resistivity (and photoconductivity) can be calculated directly from a measure of R, provided t is known. In practice, $R_s = R_{sheet}/F$, where F is a unitless correction factor, defined below.

The simplest way to form metal contacts is cold-weld by using a metal that is soft at room temperature, such as indium (In). A cold-weld is formed by pressing the soft metal and sample surfaces together, under moderate pressure (i.e., force applied by tweezers) at room temperature. The sample can be slightly heated on a hot plate as well, in order to reflow the In and ensure a microscopically smooth interface between metal and sample surfaces. Figure 5.2a depicts the total resistance in a two-probe measurement as the series sum $R_{p1} + R_s + R_{p2}$; $R_{p1} + R_{sheet}/F + R_{p2}$. $F = F_1 F_2 F_3$ is a correction factor that accounts for (1) finite sample thickness ($0 \le F_1 \le 1$); (2) finite sample lateral dimension, perpendicular to the direction of current flow ($0 \le F_2 \le 1$); and (3) the placement of the probes at a finite distance from the sample edge ($0 \le F_3 \le 1$). For thin films, with large in-plane dimensions, we can approximate $F_2 = F_3 = 1$ and $F_1 = \pi/\ln(2)$; it is recommended practice to place the metal contacts >5 mm from the sample edges.

The resistance of a probe 1 (R_{p1}) has three contributions: resistance from the metal tip of the probe (R_{m1}), contact resistance between the tip and the sample (R_{c1}), and spreading resistance of the current from this contact point (R_{spr1}). Most semiconductor films form a thin oxidized surface coating, referred to as a native oxide (Ghandhi 1994), from exposure to moisture at standard room temperature and pressure conditions. In addition, if the sample has not been cleaned with solvents (such as acetone and methanol), a microscopic coating of grease from casual handling may be present as well. These two coatings help contribute to R_{c1}. Lastly, a rough microscopic interface between the probe's metal contact and sample surface contributes to R_{spr1}.

In a two-probe measurement, current is measured as $I = V_A/(R_{p1} + R_{sheet}/F + R_{p2})$. In the absence of an excitation beam (and rigorously, any source of illumination) on the sample, this measurement gives a dark current $I_0 \equiv I$(no excitation), representative of the sample's inherent conductivity. In the presence of an excitation beam, the conductivity increases due to carrier photo-generation (see Section 5.4 for details), R_{sheet} decreases, and the measured current $I_{ph} \equiv I$(excitation), will be greater than I_0. Rigorously, a photocurrent is the current extracted from a rectifying diode device, where light absorption in the depletion zone leads to excess carrier generation. Colloquially, experimenters refer to the current I_{ph} from a photo-excited material as a photocurrent, leading to the description of such a measurement as photocurrent spectroscopy; this convention is adopted here as well. For an infinitesimally weak excitation beam, we can relate the differential increase in current dI to a differential decrease in the sample's sheet resistance dR_{sheet}, by the relation $dI = -V_A/(R_{p1} + R_{sheet}/F + R_{p2})^2 (dR_{sheet}/F)$. If R_{p1}, R_{p2} become prohibitively large, it may become difficult to differentiate between I_{ph} and I_0 and the signal-to-noise ratio $\mathrm{SNR} \equiv I_{ph}/I_0 = \left(I_0 + \int dI\right)/I_0$ will be low. It is imperative, therefore, to minimize R_{p1}, R_{p2} (Schroder 2006).

Alternatively, gold (Au) or aluminum (Al) may provide more suitable Ohmic contacts than In. Au and Al are typically deposited by lithographic patterning and thermal evaporation (Ghandhi 1994). The increased thermal energy of Au or Al atoms during deposition results in the formation of a stable contact with low R_{c1}, R_{spr1}.

If the sample is a semiconductor, the choice of metal results in an alignment of Fermi levels that forms an Ohmic contact (bidirectional flow of carriers, see Figure 5.3a inset) or Schottky

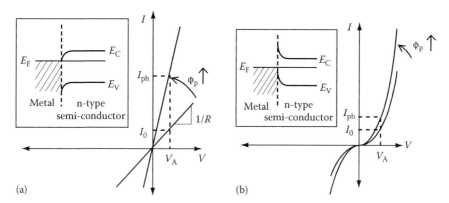

Figure 5.3 (a) Inset: schematic of energy-band diagram for an Ohmic contact. Main plot: Ohmic I–V curve slope under no optical excitation (lower slope plot) and under excitation (larger slope plot). (b) Inset: schematic of an energy-band diagram for a Schottky contact. Main plot: hybrid linear-rectifying I–V curve under no optical excitation (lower curve plot) and under excitation (larger curve plot). Note that both contacts in (b) are Schottky contacts). For both types of contacts, increasing the excitation flux density ϕ_p results in a photocurrent contribution to the dark current (for a given bias voltage V_A).

contact (thermionic emission of electrons in forward bias, see Figure 5.3b, inset) (Pierret 1996). If a dark current and photocurrent are both measured at a common potential bias V_A, then the Ohmic or Schottky nature of the contact is of little concern.

In the case of Ohmic contacts, the excitation beam (with flux density ϕ_p, as defined in Section 5.2) generates carriers and decreases R_{sheet}: the slope of the I–V curve increases in magnitude, for forward and reverse bias (Figure 5.3a). Inserting the identity R_{sheet} into the above expression for dI and assuming $R_{p1} \approx R_{p2} \approx 0$, it can readily be shown that $dI/I_0 = d\sigma/\sigma_0$. This relation makes explicit the intuitive trade-off interpreted from an Ohmic I–V curve: a larger V_A may help the experimenter resolve a larger $\Delta I = \int dI = I_{ph} - I_0$, but at the price of increasing the background dark current I_0. Thus, a high SNR is only realized by (1) a high-intensity excitation beam $\left(\text{large valued } \Delta\sigma = \int d\sigma \right)$ and (2) a low dark conductivity (small valued σ_0). While (1) is something the researcher can experimentally increase, (2) is an intrinsic property of the sample under study.

In the case of Schottky contacts, the excitation beam (flux density ϕ_p) generates carriers in the sample and decreases the *differential* resistance dR_s/dV: the curvature of the I–V curve increases in magnitude, for forward and reverse bias (Figure 5.3b). The I–V curve depicted has a hybrid mix of linear and rectifying characteristics, which is the result of two Schottky barriers forming a mirror-image bending of energy bands (Figure 5.3b, inset), at either contact interface. In a rectifying Schottky device—comprising a Schottky contact to the semiconductor at one end and an Ohmic contact at the other end—carrier generation in the Schottky depletion zone region results in immediate charge separation and inhibits recombination, thereby contributing to an increased reverse bias current (Pierret 1996). However, in the case of two Schottky contacts, the photogenerated excess electrons encounter a conduction band potential barrier at the opposing-end contact, and recombine with trapped holes. As a result, (1) there is no photocurrent at zero bias; (2) the change in current with photogeneration qualitatively resembles the Ohmic case (more positive current in forward bias, more negative current in reverse bias); and (3) the I–V relationship has a superlinear functional dependence, due to the exponential dependence of current on voltage through a Schottky barrier.

Four-probe contacts (Figure 5.2b) offer a more precise measurement of resistivity than two-probe contacts, because this technique can directly determine R_s, independent of the influence of probe resistances. In a four-probe contact, a constant current source is used to send an applied current I_A from probe 1, through the sample and into probe 4. This current results in a voltage drop $V_s = I_A R_s$, where R_s is the sample (sheet) resistance between probes 2 and 3. Connecting a voltmeter, with high internal resistance, R_V, to probes 2 and 3, implies that the resistance sum $R_{p1} + R_{p2} + R_V$ will be in parallel with R_s. If $R_V \gg R_{p1}, R_{p2}$ (which is usually the case); then the voltage measured by the voltmeter will be, to an excellent approximation, $V_V \simeq V_s$. In principle, R_s can be determined by this measure of V_s, for a given I_A. In practice, the presence of voltage offsets V_{os} (Keithley 1986)—due to thermal electromotive forces generated at relay or connector contacts, or inherent to a voltmeter—contribute an additional voltage drop, so that $V_V = V_s + V_{os}$. However, reversing the direction of the applied current I_A will result in a measured voltage $V_V = -V_s + V_{os}$; thus, subtracting these two voltmeter measurements and dividing by two will give an accurate measure of V_s. Permuting the flow of I_A through all combinations of adjacent probes, and measuring the voltage after reversing the direction of I_A, results in a total of eight measurements, from which V_s can be derived to very good accuracy. Lastly, Van der Pauw contacts (van der Pauw 1958) are an oft-cited four-probe contact geometry, specifically for Hall mobility measurements, that easily lead to the above mentioned optimized values for F_1, F_2, F_3.

5.1.5 Photoluminescence Collimation and Emission Monochromator

As mentioned in Section 5.1.3, the (mixed polarization) excitation beam is focused to a spot on the sample surface (normal or oblique incidence) and in most cases, PL emission emerges isotropically (to an excellent approximation) and with mixed polarization from this area of illumination. There are exceptional cases to consider, such as quantum wells that selectively absorb a transverse magnetic (TM) polarized excitation beam at oblique incidence (resulting in a quantum well transition with an electric field dipole moment that is perpendicular to the sample surface); the luminescence is also TM-polarized and optimally extracted from a cleaved facet edge (Franz et al. 2008). Similarly, if a nanowire is excited by a beam with polarization parallel to the long axis of the nanowire, the resulting PL is similarly polarized (Wang et al. 2001). In addition, PL emission from a microcavity will have a more directional far-field profile (Schubert and Poate 1995).

In all cases, the emerging PL has some degree of divergence; efficient collection and spectral analysis is a straightforward example of geometrical optic design, with a few subtle points to keep in mind. The discussion in this section adapts several explanations laid out in extensive and more thorough detail in (Drozdowicz et al. 2004); the experimenter is recommended to consult this reference and the online (Lerner and Thevenon 1996) for a more comprehensive discussion.

The goal is to collect as large a solid angle of PL intensity from the emitting beam spot as possible and to direct it into a monochromator that is attached to a photodetector. A monochromator in turn enables the photodetector to spectrally resolve this PL, i.e., measure the emitted light intensity as a function of emission wavelength. Figure 5.4a shows a schematic diagram of isotropic PL emission from a beam spot, approximated as a point source. The divergent PL rays are collected and converged by a plano-convex "collimating" lens, placed at a distance of the focal length f, into a collimated beam of parallel rays. The collimating lens is specified by two parameters: focal distance f and lens diameter D. Together, these two parameters define an acceptance angle θ that corresponds to the collected solid angle of PL emission. This solid angle of collected emission, also referred to as the light-gathering power of the collimating lens, is quantified by the numerical aperture $NA = \sin \theta \simeq D/(2f)$ or F-number $F/\# = 1/(2NA) \simeq f/D$, where the approximate equalities are valid for small values of θ (the paraxial approximation). Hence, a higher NA or lower F-number collects more PL;

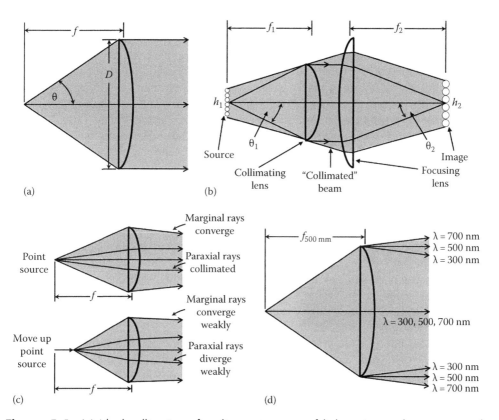

Figure 5.4 (a) Ideal collimation of a divergent source of light, using a plano-convex collimating lens. (b) Cross-sectional schematic depicting the PL emission from a finite-sized beam spot, collimated into a nearly parallel beam. The beam is converged by a second plano-convex focusing lens into an image on a monochromator entrance slit. (c) The effect of lens spherical aberration on collimation; moving the collimating lens closer to the point source results in a more collimated beam. (d) Chromatic aberration results in longer wavelengths of light being less collimated than shorter wavelengths of light. (Adapted from Drozdowicz, Z. et al., *The Book of Photon Tools* (*Newport Oriel Catalog*), Newport Oriel, Stratford, U.K., 2004.)

camera enthusiasts will be familiar with the *F*-number notation, where $F/1$ implies $f = D$, $F/2$ implies $f = 2D$, etc. Concave spherical or paraboloid mirrors (used inside monochromators) also collect and focus light, and are similarly defined by an $F/\#$.

The point source approximation for PL emission is reasonably valid for laser excitation (spot diameter ~ 500 µm), but not for lamp excitation (spot diameter ~ 0.5 cm). Figure 5.4b schematically shows a beam spot with finite extent that cannot be ideally collimated into a parallel beam; this beam is ultimately converged by a "focusing lens" onto the monochromator entrance slit, forming an image with magnification $m = (F/\#_1)/(F/\#_2)$ ($F/\#_{1,2}$ refers to the collimating and focusing lenses, respectively).

The spherical aberration of real lenses also results in an excessive convergence of point source rays from the edges of the collimating lens (Figure 5.4c, top); this results in coupling loss when light enters the monochromator. In practice, it is found that moving the collimating lens slightly closer to the PL emission beam spot results in a more collimated beam (with minor divergence in "paraxial" rays, located closer to the optical axis) and higher throughput of PL emission to the photodetector (in Figure 5.4c, bottom, the point source has been moved up to the lens).

Lastly, the refractive index of fused silica varies with wavelength, and thus collimating lenses are unable to converge longer wavelengths as effectively as shorter wavelengths (see Figure 5.4d). The index variation of fused silica weakly increases from the UV to the IR, corresponding with a more pronounced decrease in light-gathering power. This chromatic aberration is most significant with a broadband light source; thus, while the effect limits the throughput efficiency of a broadband lamp through an excitation monochromator (Section 5.1.1), it does not significantly influence the collection of PL emission (typically a narrow spectrum existing wholly within the UV, visible or IR range). The experimenter only needs to remember to re-optimize the distance between beam spot and collimating lens (so that a maximum solid angle of PL is directed into the photodetector), when switching over from the detection of IR PL emission to visible PL emission, or vice versa.

We now turn our attention to the inner workings of a monochromator, and the essential role it plays prior to the detection of PL emission. Photodetectors are fundamentally "color-blind" devices (we will discuss the exception of the charge coupled device [CCD], later), where the measured photocurrent can be attributed to any photon with energy lying within the detector's absorption spectrum. Thus, an additional device is required, performing the essential function of a prism: to split the mixed colors of an emission spectrum into an angular distribution, such that different wavelengths (colors), *disperse* into different angular directions. A monochromator accomplishes this by means of a diffraction grating, and can be finely calibrated to relate an angle of grating rotation to an emission wavelength, enabling one to attribute the steady-state intensity measured by a photodetector at a given grating angle to a particular emission wavelength.

Figure 5.5a shows an overhead schematic view of a single grating monochromator, in a Czerny–Turner geometry (Lerner and Thevenon 1996). PL emission (A) converges into the entrance slit (B) and diverges, reflects from a parabolic mirror (C), is collimated, and is incident on a diffraction grating (D). Figure 5.5b shows the magnified cross section of a diffraction grating to be a stepped surface (referred to as rulings or grooves), spaced a period a apart. Two parallel rays A and B (for monochromatic light with wavelength λ), incident at angle θ with respect to the grating normal, reflect non-specularly from adjacent grooves, resulting in reflected rays A' and B' at angle ϕ with respect to the normal. If the extra path difference between rays A' and B' is an integer number of wavelengths, the grating equation

$$a\sin\theta + a\sin\phi = m\lambda, \quad m = \pm 1, \pm 2, \ldots \tag{5.1}$$

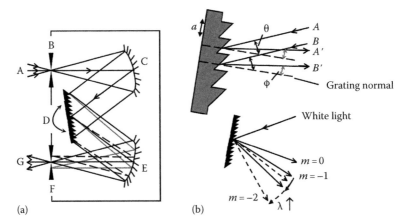

Figure 5.5 (a) Schematic diagram of Czerny–Turner monochromator. (b) A diffraction grating disperses polychromatic light into different angles and diffraction orders m. (Adapted from Drozdowicz, Z. et al., *The Book of Photon Tools* (*Newport Oriel Catalog*), Newport Oriel, Stratford, U.K., 2004.)

determines, for a given λ, θ, ϕ, which interference peaks add constructively, in clockwise (positive m) or counterclockwise (negative m) directions. The lower part of Figure 5.5b schematically illustrates the dispersion, or angular splitting of different wavelengths of reflected light, from an incident polychromatic source: the $m = \pm1$ wavelength modes disperse within a contiguous range of angles (the longer the wavelength for this mode, the larger the angle ϕ), before an overlap occurs with $m = \pm2$ wavelength modes.

Spectrometry, defined as the measurement (by a photodetector) of polychromatic light after transmission through a monochromator, primarily focuses on the measurement of light intensity diffracted into the $m = \pm1$ modes, because the facet angle of the grating grooves can be patterned to diffract an optimal fraction of intensity into the lowest order mode. Because this optimization happens for a unique "blazing" angle, an associated blaze wavelength (e.g., $\lambda_{\text{blaze}} = 350\,\text{nm}$) implies the efficiency of dispersion will decrease for wavelengths shorter and longer than λ_{blaze}. Hence, a grating imparts a transmission window to a monochromator (Lerner and Thevenon 1996), and the rule of thumb approximating its 3 dB passband is from $(2/3)\lambda_{\text{blaze}}$ to $2\lambda_{\text{blaze}}$ (in this example, $233 < \lambda < 700\,\text{nm}$). Commercial gratings today are manufactured by mechanical or holographic patterning of grooves. Typical values for the grating groove density (g/mm) are 600 or 1200 g/mm. λ_{blaze} is typically in the 250–1000 nm range for most UV–visible–IR spectroscopy experiments.

In combination with the narrow entrance slit of the monochromator, a diffraction grating exerts a remarkable influence on the spectral resolution of a PL experiment. Differentiating the grating equation with respect to λ, for a fixed angle θ, gives $d\phi/d\lambda = m/a \cos\phi$. $d\phi/d\lambda$ is the angular dispersion or rate of change in diffraction angle, per wavelength. The relation confirms that angular dispersion increases with diffraction mode m, grating density $1/a$, and diffraction angle ϕ. Figure 5.5a shows the diffracted beam to reflect from a second paraboloid mirror (E), with curvature identical to the first mirror (C), and converge at an exit slit (F), placed at a focal distance f. The displacement x of a given ray on the flat plane of the exit slit is expressed in the paraxial approximation as $x = f\phi$. From this relation, a linear dispersion relation (at the exit slit) is found to be

$$\frac{dx}{d\lambda} = f\frac{d\phi}{d\lambda} = \frac{fm}{a\cos\phi} \tag{5.2}$$

The reciprocal of this relation, $d\lambda/dx$, has an important interpretation: for a slit width Δx, the wavelength instrument resolution of a PL spectrum will be $\Delta\lambda_{\text{instr}} = \Delta x \cdot d\lambda/dx$, where $\Delta\lambda_{\text{instr}}$ represents the instrument linewidth of the PL peak (i.e., the spectral range for which the PL intensity is greater than half of its peak intensity). Section 5.2 will explain the natural forms of spectral broadening that contribute to the linewidth of a PL peak. To quantitatively measure these natural linewidths, it is imperative to reduce the slit size of the monochromator sufficiently, so that $\Delta\lambda_{\text{instr}}$ is less than the natural linewidth. Inspection of $d\lambda/dx$ reveals that a higher grating density $(1/a)$ and longer focal length (f) give a higher wavelength resolution. Typical focal length values for commercial monochromators are 0.25, 0.5, and 1 m.

While the derivation of $\Delta\lambda_{\text{instr}}$ suggests only the monochromator exit slit has to be adjusted, this is a misconception. Both entrance and exit slits must be the same size and be adjusted in tandem. This is because the symmetric design of a monochromator implies that the image of a finite sized beam spot at the exit slit will have approximately the same size as the image at the entrance slit. This symmetric design ensures optimal throughput of optical power through the monochromator, as well. The principle is more fully grasped by defining the optical throughput, or geometrical extent, of a beam as $G \equiv A\Omega$, where A is the area of a source of light (e.g., a PL beam spot in Figure 5.6a) and Ω is the solid angle subtended by the aperture of an optical system (e.g., for a collimating lens, $G_c = A\Omega_c$ in Figure 5.6a). Figure 5.6b shows the finite size of

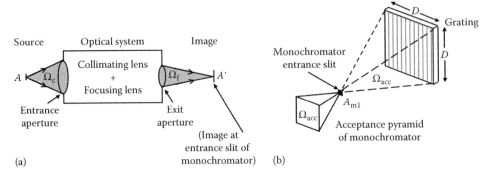

Figure 5.6 (a) Modeling the collimating and focusing lens as an optical system: a PL beam spot of size A has a solid angle Ω_c of its emission collected into a collimating lens ("entrance aperture"), and a focusing lens with solid angle Ω_f focuses the emission light into an image of size A' ("exit aperture"). (b) The acceptance pyramid (approximated as an acceptance cone of solid angle Ω_{acc}) of a finite slit size A_{m1} projects beam spot image A' onto a diffraction grating (for simplicity, an intermediate collimating paraboloid mirror with focal length f is omitted). The finite $D \times D$ size of the grating determines Ω_{acc}. (Adapted from Lerner, J.M. and Thevenon, A., *The Optics of Spectroscopy*, Horiba Scientific website, Metuchen, http://www.horiba.com/us/en/scientific/products/optics-tutorial//?Ovly=1, 1996.)

the diffraction grating (for simplicity, the first concave mirror is omitted) corresponding to an "acceptance pyramid", i.e., the portion of the solid angles outside the monochromator entrance slit that direct light onto the grating. PL emission incident from angles outside of this acceptance pyramid will be directed to areas surrounding the grating, resulting in background stray light that contributes a uniform noise at all wavelengths, in the photodetector. If the grating is approximated as a circle with the same area, the grating distance to the entrance slit defines an acceptance solid angle Ω_{acc}. An efficient PL design matches the optical throughput of a focusing lens, $G_f = A'\Omega_f$ (A' is the area of the beam spot image at the entrance slit and Ω_f is the solid angle subtended by the focusing lens, see Figure 5.6a), to the optical throughput of the monochromator entrance slit, $G_{m1} = A_{m1}\Omega_{acc}$ (A_{m1} is the area of the entrance slit, see Figure 5.6b). In this way, the grating size and distance to the entrance slit dictate the $F/\#$ of the focusing lens. If the image beam spot is desired to be the same magnification as the object beam spot, the $F/\#$ of the collimating lens must be identical as well. Matching optical throughput at the monochromator exit slit G_{m2} (not shown) requires $G_{m2} = G_{m1}$. Since the diffraction grating is placed symmetrically with respect to both entrance and exit slits, Ω_{acc} must be the same for the exit slit, which implies $A_{m1} = A_{m2}$. This means that if the entrance slit is fully opened (to collect the optimal PL intensity from the focusing lens) and the exit slit is independently narrowed (to achieve high spectral resolution), the mismatch $G_{m2} < G_{m1}$ will result in excess stray light within the monochromator, resulting in an increased background noise for the photodetector. As a rule of thumb, there exists a fundamental trade-off between collected PL power and spectral resolution of the measured spectrum: decreasing slit size (or increasing the optical path length of the spectrometer) reduces $\Delta\lambda_{instr}$ to an amount below a PL peak's natural linewidth, but this also reduces the PL intensity that reaches the photodetector. By defining an F-number $F/\#_{m1}$ for the entrance slit (where D is the height and width of a square diffraction grating, and f is the focal distance of the first parabolic mirror), the optical throughput from the focusing lens ($F/\#_f$) into the entrance slit can be quantified by the ratio $F = (F/\#_f)/(F/\#_{m1})$ for $F/\#_f < F/\#_{m1}$ and $F = 1$ otherwise.

The optical throughput of the acceptance pyramid (or cone) implies an additional rule of thumb for PL detection: optimize the filling of Ω_{acc} with PL emission, while passing as much of the beam spot image through the entrance slit as possible. A high-resolution measurement (with narrow monochromator slits) results in a small Ω_{acc}, which can be matched by a focusing lens with a large F-number ($F/\#_f$). However, it is desirable to have a small F-number for the collimating lens ($F/\#_c$), in order to collect as large a solid angle of PL from the beam spot as possible. A mismatch in F-number for these two lenses will result in a magnification of the beam spot, at the entrance slit, of $m = (F/\#_f)/(F/\#_c)$. If the magnified beam spot image (of area $A' = m^2 A$) is larger than the narrow entrance slit area (A_{ml}), this occlusion of a portion of the image results in loss of optical power transmitted to the photodetector, and is called the vignetting loss factor (defined as $V = A_{ml}/A'$ for $A' > A_{ml}$, $V = 1$ otherwise).

The optical power transmitted to the photodetector is summarized as $P_{exit} = P_{in} V F \eta_g R^n$, where P_{in} is the incident power at the monochromator entrance slit, η_g is the spectral efficiency of the diffraction grating (determined by the passband of λ_{blaze}), R is the mirror reflectivity, and n is the number of mirrors within the monochromator.

When a given wavelength λ, from the dispersion of the $m = 1$ mode, is selected by the exit slit, stray light within the monochromator (originating from higher order modes of λ and other wavelengths comprising the PL spectrum) couples into the acceptance pyramid of the exit slit. A double monochromator (comprising two serial diffraction gratings) helps further minimize the coupling of this stray light; it is a more costly solution, and for most applications a single monochromator serves adequately well.

This section has presented an extensive discussion on the collection and spectral manipulation of light, and not without good reason. The experimenter should appreciate that the monochromator represents the central component of a spectroscopy link: it is the tool whose fundamental physics enables spectral manipulation of an excitation beam and spectral analysis of an emission beam. While spectroscopy is not possible without the essential contributions of a light source and photodetector, it is the function of the monochromator that helps the experimenter visualize transitions between the energy levels of a sample.

5.1.6 Photoluminescence Measurement with a Photodetector

A spectral signal diverges from the exit slit of a monochromator and is refocused by lenses onto a photodetector. Photodetectors are typically compound semiconductor materials that absorb light for photon energies above their electronic band gap. The material may be processed into (1) a photomultiplier tube (e.g., InGaAs PMT), (2) a photodiode (e.g., Ge junction photodiode), (3) a photodiode array or charge coupled device (e.g., Si CCD). All device types are subjected to an external bias voltage, resulting in an electric field that sweeps excess photogenerated carriers out of the region of absorption and into an external electrical circuit. A voltage drop across a series resistor gives a measure of the resulting total current.

A PMT is an evacuated glass tube—formally called a "vacuum tube"—comprising a photocathode (coated onto the transmission window), a serial array of metal plates (referred to as dynodes), and an anode electrical contact. Incident light enters the PMT through the transmission window on the top (head-on design) or the side of the tube (end-on design) and excites free carriers in the photocathode via the photoelectric effect. An external voltage bias accelerates the photoelectrons toward the first dynode plate, and impact excitation with successive plates in the dynode array results in a multiplication of secondary electron emission and an amplified output photocurrent at the anode. The PMT design offers high current amplification with minimal addition of noise, making this aging relic of mid-twentieth century electronics a competitive contemporary option to photodiodes (Hamamatsu 1999).

The photodiode comprises a semiconductor p–n junction, doped with a high concentration positive carrier p+-type region and a moderate concentration negative carrier n-type region. The interface between these doped regions undergoes electrochemical exchange and expands into a finite depletion zone with a built-in electric field. Reverse biasing the photodiode (commonly referred to as the photoconductive mode of operation) extends the depletion zone and predominantly increases the electric field. Incident photons typically penetrate the thin p^+-type region and are absorbed within the depletion zone, generating electron–hole pairs that are charge separated by the electric field and extracted as photocurrent. Photodiodes are compact, proving advantageous for immersion into a cryogen (specifically, LN2) for low temperature operation (thereby reducing dark current). Avalanche photodiodes are operated at a high reverse bias, resulting in carrier multiplication and increased photocurrent, due to inelastic scattering between carriers and the crystal lattice (Pierret 1996).

Both classes of photodetectors are specified by two key parameters: (1) quantum efficiency (or, equivalently, responsivity) and (2) noise. The internal quantum efficiency $\eta_{int}(\lambda)$ measures the number of electron–hole pairs generated per absorption of each photon, and its wavelength dependence results from the photocathode/photodiode material's absorption spectrum. The internal quantum efficiency, technically, is a spectral density variable, with units of efficiency per wavelength. The external efficiency of the photodetector is defined as $\eta_{ext}(\lambda) = \eta_{extr}\eta_{int}(\lambda)$ $\eta_{couple}(\lambda)$, where $\eta_{couple}(\lambda)$ is the (wavelength dependent) efficiency of coupling PL emission into the detector (accounting for throughput efficiency of the photodetector, including any refractive index mismatch between the photocathode/photodiode material, the detector's glass window, and ambient air), and η_{extr} is the extraction efficiency of the photocurrent (accounting for carrier recombination mechanisms and the ohmic quality of metal contacts). These additional efficiency factors are fractional, and thus $\eta_{ext} < \eta_{int}$. The spectral responsivity of the photodetector is then defined as $R(\lambda) \equiv n_{ext}(\lambda)\, q\lambda/hc$ and is typically quoted in units of milliamps (of extracted photocurrent) per watt (of incident optical power) per nm (of wavelength). The experimenter should take note that commonly, when reporting quantum efficiency and responsivity, commercial vendors don't bother to acknowledge $\eta_{ext}(\lambda)$, $R(\lambda)$ as spectral densities: a typical specification plot will label $R(\lambda)$ with units of mA/W, and $n_{ext}(\lambda)$ as a unitless fraction (or its percentage equivalent). Rigorously, the responsivity of a detector (in units of mA/W), at a given PL emission wavelength λ, is measured as $R \equiv R(\lambda) \cdot \Delta\lambda_{instr}$. Vendors thus report responsivity and quantum efficiency specifications, for a given $\Delta\lambda_{instr}$.

Noise in a photodetector represents an ever-present background of electrical current (under external voltage bias), and it is against this nonzero magnitude that a photocurrent must distinguish itself, as quantified by its signal-to-noise ratio $\mathrm{SNR} = I_{ph}/I_0$ (I_{ph} is the photocurrent and I_0 is the dark or noise current, as first mentioned in Section 5.1.4). The higher the SNR, the "cleaner" or "less bumpy" a plot of photocurrent signal versus wavelength will be and the more confidently an experimenter can interpret features of the PL spectrum.

Shot noise represents the random arrival of photons and random generation of carriers within a PMT or photodiode. However, the most important contribution to noise current is the generation of carriers by nonradiative excitation mechanisms; this contribution is referred to as the dark current, measured from biased PMTs and photodiodes in the absence of light. In PMTs, the primary contributor to dark current is thermionic emission (Hamamatsu 1999) of electrons from the photocathode. In photodiodes, the primary contributor to dark current is thermal generation (Pierret 1996) of carriers (also referred to as a "leakage" current in reverse biased diodes). Cooling a PMT with a water line, the flow of cold nitrogen gas (collected from the evaporation of an LN2 dewar), or a thermoelectric cooler reduces the dark current appreciably. A photodiode can be cooled by the thermoelectric effect or by direct immersion (due to its small form factor) into a dewar-like receptacle containing LN2. Cooling from immersion in LN2 (77 K temperature) results in a large quench of dark current and high SNR performance. Table 5.2 presents a summary of performance specifications for several commercially available photodetectors.

There is no single detector with high responsivity across the entire UV–IR spectrum. Thus, careful thought must be given to the spectrum under study (UV, visible, or near-IR) and, for an a priori estimate of PL intensity, what level of dark current will accommodate an acceptable signal-to-noise ratio (e.g., SNR > 10 is a recommended lower limit).

TABLE 5.2

Performance Specifications (Responsivity R, External Quantum Efficiency η, Dark Current I_0, Gain G) for Two Typical Photodetectors

Detector[a]	R (mA/W) or η_{ext} (%)[b]	Responsivity Range (nm)	I_0 (nA)	Gain
InGaAs PMT	$R \approx 20$, η;2	900–1600	40 ($T = -60°C$)	10^6 ($V_A = -800\,V$)
Si CCD	20–95	300–1000	2–4 e⁻/pixel/h ($T = -120°C$)	<1 e⁻/digital count

Note: For the Si CCD data, e⁻ is an abbreviation for electron.

[a] Sample photomultiplier tube detector data sheet. http://jp.hamamatsu.com/products/sensoretd/pd002/pd395/H10330-75/index_en.html

[b] Sample CCD detector data sheet. http://www.horiba.com/scientific/products/optical-spectroscopy/detectors/multi-channel/ccds/details/symphony-ccd-detectors-214/

The third and final specification for a photodetector is its gain G, which scales with the applied reverse bias voltage V_A in a power law behavior (gain versus voltage graphs are commonly presented as log-log plots, where the power law determines the straight line slope). The photodetector gain increases output current $I_{gain} = G \cdot I_{ph}$ to values that can be appreciably measured as an output voltage drop, $V_{out} = I_{gain}R$ with high SNR (R is an output series resistor with adjustable values of $1\,M\Omega$–$100\,k\Omega$). This resistor, along with a collection of intrinsic capacitance contributions (from the dynode array of a PMT or the depletion zone of a photodiode), forms a resistive–capacitive RC time delay for the measured voltage signal. (Section 5.1.7 addresses the trade-off implications of this time delay.)

A photodiode array and CCD are examples of "color-sensitive" photodetectors; for brevity, this review will focus on the CCD, a favored choice amongst contemporary experimenters. A CCD is a lithographically patterned one- or two-dimensional micron-scale array of photodiodes. Fundamentally, a CCD can temporarily store the photogenerated charge from a given photodiode CCD element, or pixel, and shift it after a characteristic time, to an adjacent pixel. Eventually, charge from every pixel can be shifted to the boundaries of the CCD and measured as a digital count. This "shift-register" design (Howell 2006) makes it possible for a CCD pixel array to capture the light intensity pattern of an image, leading to the invention of the digital camera. For photodetection, the advantage of the CCD is the ability to simultaneously detect several wavelengths of light, where a given wavelength λ (and a narrow spectral bin about this value, determined by $\Delta\lambda_{instr}$) is dispersed by the monochromator grating onto a given row of pixels, as described by the linear dispersion formula from Section 5.1.5. Whereas the PMT and traditional photodiode are "single-channel" devices—measuring the PL intensity at one emission wavelength, or "channel," at a time—the CCD is a multichannel device that measures PL intensity at multiple emission wavelengths, simultaneously. A CCD is a two-dimensional array of pixels, and in the direction parallel to the monochromator grating groves, there is no dispersion, and thus the same wavelength λ is incident on a row of m pixels (e.g., $m = 256$). It is the collective count of carriers from this row of pixels that is the digital count analogous to $I_{ph}(\lambda)$. Thus, while a gain per pixel element can be specified for the CCD (see Table 5.2), this specification does not adequately quantify the output digital count of PL photons.

While the multichannel design offers a significantly reduced acquisition time for PL data, compared to the single-channel photodiode, the reduced optical throughput per CCD pixel row may require a counteracting (but not necessarily equivalent) increase in acquisition time, for achieving the same SNR per wavelength channel, as the single channel device. The practical constraints against opting for a CCD photodetector are cost, and the requirement of multiple diffraction gratings. CCDs must be calibrated with a monochromator that is designed to direct the diffraction grating's $m = 1$ mode of dispersed PL emission, in its entirety, onto the exit slit.

Differing angular orientations for each wavelength channel in this dispersed mode result in each channel's incidence on a unique CCD element. With a traditional PMT or photodiode, the choice of grating density has no bearing on the spectral range of PL detection. For a CCD, the increased dispersion of a higher grating density still results in higher spectral resolution; but the blocking of highly disperse angles by the exit slit results in a reduced spectral range of PL detection. Thus, for a CCD measurement across a broad spectral range, either a lower density grating must be installed, or several scans must be taken with the high density grating and stitched together to form a composite broad PL spectrum measurement. Analogous to the photodiode, the tiny CCD form factor ensures that its pixel array can be immersed in LN2 and have a low dark count.

Careful investigation by the experimenter into photodetector responsivity, dark current, gain and response time specifications will impress upon them the essential trade-offs in detector choice. In general, the CCD and PMT offer the highest sensitivity (in that order), while the fast response time of the PMT enables ns-scale lifetime measurements; the PMT is found to be the workhorse detector in most spectroscopy labs. Water cooling and LN2 cooling are essential to resolve the weak SNR of PL from thin film samples. For bulk samples or high brilliance quantum well samples, the PL may be large enough so that an uncooled photodiode is sufficient. Lastly, multiple photodetectors may be required in order to observe PL throughout the UV-to-visible or visible-to-IR spectral ranges; it is good practice to select the responsivity of the photodetectors to have some range of spectral overlap, so that PL in these neighboring spectral regions can be renormalized with respect to one another, in case an experiment requires some relative comparison of PL peak intensities.

5.1.7 Comments Concerning Data Acquisition

In PL and PLE measurements, the PL emission from the sample can either be collected with CW excitation, or with a mechanically chopped, periodic excitation. In the latter case, a lock-in amplifier is required to acquire the photodetector's current signal. The lock-in amplifier must be triggered with respect to a mechanical chopper, to identify the photocurrent signal as measured current during the excitation portion of the chopper's duty cycle. Frequency space Fourier-transform filtering within the lock-in amplifier removes background ambient noise (the most ubiquitous source being 60 Hz room lighting) and delivers a high SNR electrical signal to a computer data acquisition to generate an intensity versus wavelength plot. The program must also communication with the diffraction grafting controller for the monochromator so that it can accurately assign a wavelength to the measured intensity. The lock-in amplifier can time-average the collected photodetector signal over chopper excitation cycles by selection of a longer lock-in time constant. The lock-in amplifier and oscilloscope measure photocurrent as a voltage drop across a series resistor; thus, a larger resistance R results in a larger voltage drop.

For lifetime measurements on an oscilloscope, increasing R also increases the RC time delay, and a long RC delay makes it difficult to temporally resolve the lifetime decay of PL intensity. For an a priori estimate of the PL transition's natural lifetime, it is recommended to select an RC delay that is 10 times shorter; if this reduction in R degrades the SNR considerably, one should try countering with a longer oscilloscope time average. Lastly, when measuring the lifetime decay, one must slow down the chopper rate sufficiently so as to ensure that the PL saturates to its maximum intensity and decays to zero intensity, during the chopper's excitation and non-excitation duty cycle, respectively.

5.2 PHOTOLUMINESCENCE SPECTROSCOPY

Steady-state photoluminescence (PL) spectroscopy is the study of the emission spectrum of light from a sample (PL intensity as a function of wavelength), with a functional dependence on two parameters that can be changed at will in the experiment: sample temperature and

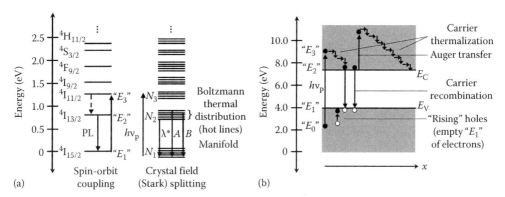

Figure 5.7 Energy scales (relative values) depicting two examples of electron transitions. (a) Atomic transition involving the collective 4f-electrons of Er-doped SiO_2, due to spin-orbit splitting. Left: the ground state can be optically pumped (by photons with energy $h\nu_p$) as a three-level system; nonradiative decay (phonon creation) dominates $E_3 \rightarrow E_2$, whereas the larger $E_2 \rightarrow E_1$ transition is dominated by radiative decay. Right: an anisotropic nearest-neighbor coordination of Si and O atoms results in a Stark effect crystal field splitting of spin-orbit levels into a multiplet manifold, populated according to a Boltzmann thermal probability distribution. Peak PL emission occurs for $\lambda_{PL} = 1.54\,\mu m$. Emission from higher $I_{13/2}$ manifold levels results in hot line transitions A and B. (b) Electronic transition involving electron–hole recombination in the semiconductor ZnO ($E_G = 3.22\,eV$). x represents position in the sample. Carriers can be optically pumped from the valence to the conduction band, approximated as a four-level system. "E_0" valence electrons that absorb a pump photon are promoted to the conduction band, leaving behind a hole that rapidly "rises" to the top of the valence band, thereby depleting the "E_1"energy level of electrons. Carrier thermalization from inelastic scattering decreases the excited electrons' conduction band energy from "E_3" to "E_2". Radiative or nonradiative recombination (e.g., Auger) results in the "E_2" $\rightarrow$ "E_1" transition.

optical excitation power. Figure 5.7 shows schematic energy scales representing atomic and electronic transitions in a PL experiment. A laser with peak frequency ν_p and power $P_p = h\nu_p\phi_p A$ (h is Planck's constant, ϕ_p is the photon flux density, measured as number of photons per cross-sectional area per unit time, and A is the excitation beam spot) is used to excite an electron from an atomic (or electronic) ground state to an excited state. The rate of this excitation is quantified as $W_p = \phi_p\sigma_p$ (units of s^{-1}), where $\sigma_p = \sigma_p(\nu)$ is the absorption cross section that quantifies the oscillator strength (Saleh and Teich 1991), coupling the ground state to the excited state (units of cm^2). As a result of this excitation, a *nonequilibrium* concentration of excited state atoms, or conduction electrons and valence holes, is created within the portion of the sample illuminated by the beam spot. Radiative and nonradiative processes proceed to de-excite and return the concentration of excited atoms or excess carriers to their ground state (Chapter 2). In order to experimentally resolve the luminescence spectrum from that of the optical pump, a pump energy $h\nu_p$ is selected that is considerably greater than the peak PL emission energy. It is worth noting that in some materials, upconverting interactions between energy states (Miniscalco 1991) or two-photon absorption (Yan-hong et al. 2006) lead to the occupation of higher energy states and PL emission at an energy greater than $h\nu_p$.

In atomic and quantized energy transitions, discrete three- or four-level models (Saleh and Teich 1991, see also Figure 5.7) accurately describe the dynamics relating the excitation and emission transitions. For simplicity, the three-level model will be discussed here, wherein the PL emission of interest corresponds to a transition from an excited state energy level E_2 to a ground state energy level E_1, and the optical pump photons of energy $h\nu_p$ are resonantly absorbed and excite the atom or quantum well/dot from E_1 to a higher excited state E_3.

The excited atom/quantum dot de-excites from E_3 by a variety of competing processes: radiative and nonradiative de-excitation from $E_3 \rightarrow E_1$, or radiative and nonradiative de-excitation from $E_3 \rightarrow E_2$. In a solid or soft matter sample, amorphous or crystal lattice vibrations correspond to a phonon density of states (Chapter 2) and $h\nu_{ph}^*$ is defined as the phonon energy with the highest density of states. If the energy of $h\nu_{ph}^*$ closely matches the energy difference $E_3 - E_2$, one (or two) phonons can easily be generated as a by-product of an $E_3 \rightarrow E_2$ nonradiative transition R_{32}. This nonradiative transition will dominate the de-excitation of the energy level E_3 and lead to an efficient populating of E_2, in turn leading to an $E_2 \rightarrow E_1$ transition and radiative PL emission.

The $E_2 \rightarrow E_1$ transition is due to a combination of nonradiative processes (including phonon generation), and radiative de-excitation, quantified by the rates $1/\tau_{nr}$ and $1/\tau_r$, respectively. If the defect concentration is low in the sample of interest (typically the case), phonon generation is an important contributor to $1/\tau_{nr}$ (Pankove 1975). When the difference $E_2 - E_1 \gg h\nu_{ph}^*$, multiple phonons (a low probability many-body process) must be generated as a by-product of an $E_2 \rightarrow E_1$ nonradiative transition; thus, the radiative transition rate $1/\tau_r$ will dominate and PL emission will be efficiently observed. Light emission from this $E_2 \rightarrow E_1$ transition will be observed with a peak wavelength λ_{PL} (corresponding to $h\nu_{PL} = E_2 - E_1$).

The dynamics of excitation and radiative versus nonradiative de-excitation for an atom or quantized nanostructure are summarized by the population rate differential equations

$$\frac{dN_1}{dt} = -W_p N_1 + \frac{N_2}{\tau_{nr}} + \frac{N_2}{\tau_r}, \quad \frac{dN_2}{dt} = R_{32} N_3 - \frac{N_2}{\tau_{nr}} - \frac{N_2}{\tau_r}, \quad \frac{dN_3}{dt} = W_p N_1 - R_{32} N_3 \quad (5.3)$$

where N_1, N_2, N_3 are the concentration of atoms or quantum dots in the ground state, first excited state, and second excited state, respectively. The decay terms N_2/τ_{nr}, N_2/τ_r can be summed together, provided a total rate of de-excitation is defined: $1/\tau \equiv 1/\tau_{nr} + 1/\tau_r$. In steady-state ($dN_1/dt = dN_2/dt = dN_3/dt = 0$), and the third equation's equality (i.e., $R_{32}N_3 = W_p N_1$) can be substituted into the second equation. More generally, assuming $N_3 \approx 0$ for all times, $N_1 + N_2 \approx N$ and $dN_1/dt = -dN_2/dt$; thus, analysis of the second equation is sufficient. The steady-state solution for N_2 is found to be $N_2 = N(W_p/(W_p + 1/\tau))$.

For an electronic energy transition, such as the electron–hole recombination in semiconductors, a continuum of valence and conduction band states are separated by a band gap, which is somewhat analogous to a four-level model (Figure 5.7b). Valence electrons are excited from below the valence band E_V, to available unoccupied states above the conduction band E_C. In the conduction band, picosecond time-scale electron-electron and electron-phonon scattering events (Pankove 1975, Chapter 8) result in rapid thermalization and relaxation to energy levels close to E_C. Recombination of these excess conduction band electrons with excess valence band holes results in an $E_C \rightarrow E_V$ transition and radiative PL emission.

Photogenerated excess electron and hole carriers ($\Delta n = \Delta p$) recombine by radiative or nonradiative processes, and the carrier densities are modeled by identical rate equations:

$$\frac{dn}{dt} = G - \frac{n}{\tau_r} - \left(\frac{n}{\tau_{nr}}\right) \qquad\qquad \frac{dp}{dt} = G - \frac{p}{\tau_r} - \left(\frac{p}{\tau_{nr}}\right)$$

$$= G - Bnp - (R_d + R_{sr} + R_A) \qquad\qquad = G - Bnp - (R_d + R_{sr} + R_A)$$

$$R_d = \Delta n \frac{n_0 + p_0}{n_0 \tau_h + p_0 \tau_e}, \quad R_{sr} = \frac{A_{sr}}{V_{sr}} \cdot \Delta n \frac{n_0 + p_0}{n_0/v_h + p_0/v_e}, \quad R_A = C_n n^2 p + C_p np^2 \text{ (for } \Delta n = \Delta p)$$

$$(5.4)$$

For a given excitation frequency v, $G = \phi_p(v)\alpha(v)$ is the rate of carrier density photo-generation, where $\alpha(v)$ is the absorption coefficient, defined as $\alpha(v) \equiv \sigma_p(v)\rho(v)f_a(v)$); $\rho(v) = (4\pi/h^2)(2m_r)^{3/2}\sqrt{hv - E_G}$ is the optical joint or reduced density of states (Coldren and Corzine 1995) (m_r is the reduced mass, defined as $1/m_r \equiv (1/m_e^{eff}) + (1/m_h^{eff})$, where m_e^{eff}, m_h^{eff} are the effective masses of conduction and valence band carriers, respectively); and $f_a(v) \equiv f_n(E_1)$ $f_p(E_2) = f(E_1)(1 - f(E_2))$ is the probability for an absorption transition ($f(E) \equiv 1/(1 + e^{(E-E_F)/k_BT})$ is the Fermi occupancy function (Saleh and Teich 1991)). $f_n(E_1) \equiv f(E_1)$ represents the probability of electron occupancy in valence band energy level E_1, and $f_p(E_2) \equiv 1 - f(E_2)$ represents the probability of hole occupancy in conduction band energy level E_2, where $hv = E_2 - E_1$.

Note that while $1/\tau_r$, $1/\tau_{nr}$ are recombination rates (with units of s^{-1}), Bnp, R_d, R_{sr}, R_A are carrier density recombination rates (with units of $cm^{-3}s^{-1}$). Carrier density recombination is dependent on the probability of conduction electrons forming a spatial overlap with valence holes, and thus depends on the product of their concentrations, n, p (Coldren and Corzine 1995). The bimolecular recombination coefficient B is an empirical measure of this spatial overlap and gates the radiative carrier density recombination rates n/τ_r, p/τ_r. Then n/τ_{nr}, p/τ_{nr} represent the total nonradiative carrier density recombination rates, including trap state processes such as defect R_d (point defects, impurities, and dislocations) or surface-mediated R_{sr} (polycrystalline grain boundary and surface edges of a microscale patterned structure) recombination, and Auger recombination R_A involving a second conduction electron (term with coefficient C_n) or valence hole (term with coefficient C_p). Both R_d and R_{sr} mechanisms contribute sub-band gap states that enable recombination by means of multiparticle phonon generation; whereas the low density of point defect dangling bonds contributes discrete energy levels, the high density of surface defect dangling bonds contributes a miniband of energy levels. Auger recombination is the dominant nonradiative process in low defect materials, at high excitation powers, or (as seen in Section 5.2.3) at high temperatures—the latter two scenarios occurring when the carrier concentration is large. Auger recombination occurs when an electron or hole collides with a like second carrier, transferring the electron–hole recombination energy to the second carrier (Pankove 1975) and increasing its kinetic energy (Figure 5.7b and Chapter 8). In the (low carrier injection limit) expressions for R_d and R_{sr}, τ_e, τ_h are carrier lifetimes, proportional (using the electron as an example) to $1/\tau_e \propto \sigma_{t,e}N_t$, where N_t is the defect trap density and $\sigma_{t,e}$ is the carrier capture cross section. v_e, v_h are carrier velocities that are a combination of drift and diffusion components. A_{sr}, V_{sr} are the surface area and volume of the region where carrier recombination is occurring. In practice, $\Delta n \gg n_0$ and $\Delta p \gg p_0$ (n_0, p_0 are the electron and hole equilibrium carrier concentrations), thus $n \approx p$ and the two carrier rate equations can reduce to one rate equation for n, with terms proportional to n, n^2, and n^3.

Close inspection of the carrier rate equations shows them to be similar to the rate equations for atomic systems. Thus, the simpler atomic model will be invoked without loss of generality, to illustrate properties applicable to both atomic and semiconductor transitions.

Using the atomic model as a simple case study, the PL intensity I_{PL} is expressed as a quantity that is proportional to N_2, by the relation

$$I_{PL} = \frac{hv_{PL}}{A}\left(\frac{N_2}{\tau_r}V\right) \tag{5.5}$$

where V is the volume of the sample that is excited by the pump beam and the quantity in parentheses is the number of photons emitted per second. (Note: in an actual PL experiment, the experimentally collected signal corresponds to $I_{PL}A\eta_{coll}$, where η_{coll} is an empirically determined efficiency quantifying the fraction of PL emission collected by the combination of collimating and focusing lenses, monochromator, and photodetector (Sections 5.1.5 and 5.1.6).

More complex energy transfer processes, such as upconversion (Desurvire 1994) in the Erbium atom and Auger recombination in bulk and quantum confined semiconductors, add an additional decay term of the form CN_2^2 (C is a proportionality constant) to the dN_2/dt equation. These more complex processes are typically dependent on high excited state concentrations, and thus, for lower optical excitation powers, can be effectively neglected. For low excitation powers, $W_p \ll 1/\tau$, the steady-state expression for N_2 can be simplified to $N_2 \approx \tau W_p N$. This leads to the well-known relation $I_{PL} \propto (\tau/\tau_r)W_p N = \eta W_p N$, where η is the internal quantum efficiency, representing the fraction of total decay transitions that are radiative (Saleh and Teich 1991):

$$\eta = \frac{1/\tau_r}{1/\tau_{nr} + 1/\tau_r} = \frac{1 + \tau_r}{1/\tau} = \frac{\tau}{\tau_r} \tag{5.6}$$

5.2.1 Photoluminescence Linewidth

Figure 5.8 shows the PL emission from an atomic transition and an electronic transition. The atomic transition is from an Erbium (Er) atom $4f$-shell de-excitation, occurring between spin-orbit degeneracy lifted levels (see Figure 5.7a, from Saini 2004), and the electronic transition is from recombination of an excess electron and hole in zinc oxide (ZnO) (see Figure 5.7b, from Zhang et al. 2006). Figure 5.8 clearly demonstrates that the PL emission is not at a unique wavelength in either system, but rather over a range, or *spectrum*, of wavelengths. The width of this spectrum, at the full-width-half-maximum (FWHM) of a given PL peak, is called the *linewidth* of the emission peak. The *broadening*, or finite size of a PL linewidth, occurs for three principal reasons in solid-state materials.

5.2.1.1 Heisenberg Broadening

Fundamentally, the Heisenberg uncertainty principle implies an uncertainty in the energy of E_2 (or E_C), corresponding to the uncertainty in the lifetime of this excited state. The total rate of de-excitation corresponds to a characteristic lifetime τ for the excited state, and an exponential temporal probability distribution $P(t) = (1/\tau)e^{-t/\tau}$ (Bain and Engelhardt 1992) determines that some $E_2 \to E_1$ transitions occur after a shorter lifetime, some after a longer lifetime.

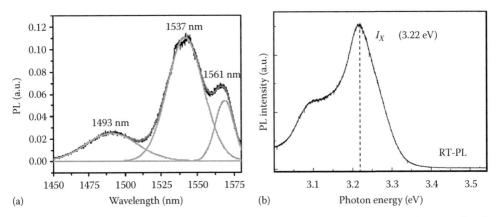

Figure 5.8 Room temperature PL examples. (a) An atomic transition in SiO₂:Er (Reprinted with permission from Saini, S., Gain efficient waveguide optical amplifiers for Si microphotonics, PhD thesis, Massachusetts Institute of Technology, Cambridge, MA, 2004). (b) Electron–hole recombination in ZnO. Reprinted from *J. Lumin.*, 119–120, Zhang, Y.J. et al., Temperature-enhanced ultraviolet emission in ZnO thin film, 242, 2006. Copyright 2006, with permission from Elsevier.

The mean lifetime of this distribution is $\bar{t} = \tau$, and the standard deviation is $\Delta t \equiv \sqrt{\overline{(t-\bar{t})^2}} = \tau$. Heisenberg's principle then shows $\Delta\nu_{HB} = (1/h)\Delta E \sim 1/\Delta t = 1/\tau$. This corresponds to a linewidth of $\Delta\lambda_{HB}$, where $\Delta\lambda_{HB}$ for the excited state lifetime of SiO_2:Er ($\tau \sim 10\,ms$ lifetime, from Polman et al. 1990) will be much smaller than $\Delta\lambda_{HB}$ for the excited state lifetime of ZnO ($\tau \sim 0.5\,ns$ lifetime, from Cho et al. 2006).

5.2.1.2 Thermal (Homogeneous) Broadening

At a given temperature, thermal energy manifests as vibrational or rotational motion within solid, soft, and liquid state materials. This steady-state vibration influences the energy levels E_2, E_1 of the atoms or quantum well/dots, resulting in a distribution of emission energies over time and a broadening of the PL peak. Because this broadening effect is contributed to by each emission site in the sample, it is referred to as homogeneous broadening; i.e., the linewidth broadening will be the same, no matter how small the beam spot and where the sample PL is collected from.

For electron–hole recombination in a semiconductor, an entirely different mechanism dominates as the source of thermal broadening. The conduction band is in thermal contact with a heat reservoir at temperature T, resulting in a Boltzmann-like occupancy probability $f_n(E) \simeq e^{-(E-E_F)/k_BT}$ (derived from the Fermi function for electrons, in the case $E - E_F > 3k_BT$, from Coldren and Corzine 1995). The electron density of states in the conduction band is $\rho_C(E) = (4\pi/h^3)\left(2m_e^{eff}\right)^{3/2}\sqrt{E-E_C}$ (Saleh and Teich 1991), where m_e^{eff} is the effective electron mass at energies near E_C. As mentioned earlier, optically excited excess conduction band electrons equilibrate within ~1 ps (Chapter 8) and form an electron density distribution $n(E) = f_n(E)\rho_C(E)$. A similar argument applies to the excess holes in the valence band, leading to (for $E_F - E > 3k_BT$) the hole-density distribution $p(E) = f_p(E)\rho_V(E)$ $\left(f_p(E) \simeq e^{-(E_F-E)/k_BT}\right.$ $\rho_V(E) = (4\pi/h^3)\left(2m_h^{eff}\right)^{3/2}\sqrt{E_V-E}\right)$. These two carrier distributions are plotted in Figure 5.9; the peak PL intensity corresponds to the energy transition $E_2 \to E_1$, for which $n(E_2)$ and $p(E_1)$

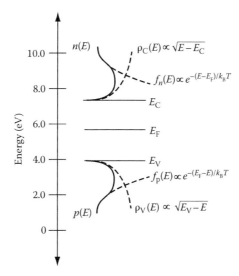

Figure 5.9 In semiconductors, the population of electron (and hole) carriers, $n(E)$ and $p(E)$, as a function of energy above (below) the conduction (valence) band, implies a thermal broadening of the PL emission linewidth, on the order of $\sim1.8k_BT$. The Fermi level, E_F, is drawn near mid-band gap in this schematic.

have their largest values. This is found to be $E_2 = E_C + (k_B T/2)$, $E_1 = E_v - (k_B T/2)$, resulting in a peak emission at $v = (1/h)(E_2 - E_1) = (1/h)(E_G + k_B T)$, where E_G is the semiconductor band gap. Since the oscillator strength for the transition coupling conduction to valence band electrons is effectively the same for energies within $k_B T$ of the band edge, the FWHM of the carrier distribution has a one-to-one correspondence with the thermal broadening of the PL peak and is found to be $h\Delta v_{TB} \sim 1.8 k_B T$. Since the Boltzmann distribution applies identically to every electron–hole recombination site, thermal broadening in semiconductors is another example of homogeneous linewidth broadening.

5.2.1.3 Inhomogeneous Broadening

Lastly, a disordered material (such as the amorphous host surrounding Er in Er-doped SiO_2, or amorphous Si) exhibits a variation of E_2 and E_1 energy levels per unit volume of the sample. As a result, the emission linewidth for Er-doped SiO_2 is considerably broader ($\Delta\lambda_{IB} = 3\,nm$ linewidth in Figure 5.10a) than Er-doped single crystal Si ($\Delta\lambda_{IB} < 2\,nm$ (Kimerling et al. 1997), where Figure 5.10b shows $\Delta\lambda_{instr} \approx 6\,nm$-limited linewidth). This inhomogeneous variation in $E_2 - E_1$ thus leads to inhomogeneous linewidth broadening; this linewidth cannot be a priori predicted and is experimentally cataloged. To unambiguously resolve this effect and separate it from thermal broadening, it is best to measure the experimental PL linewidth at low temperatures.

5.2.2 Spectral Features: Hot Lines and Phonon Replicas

For solid and soft matter, PL from an emission site occurs within a medium that conducts thermal energy by phonon propagation. Two examples are considered here: (1) the $4f$-shell transition within the erbium atom, doped in a silicon host (Si:Er); and (2) the electron–hole recombination of an exciton that binds to an iodine atom, doped in a silver bromide host (AgBr:I). The source of the host medium's thermal energy is the heat reservoir the sample is in equilibrium with (e.g., $T = 300\,K$ at room temperature, $T < 300\,K$ within a cryostat) and, as previously mentioned, available phonon energies are determined by the phonon density of

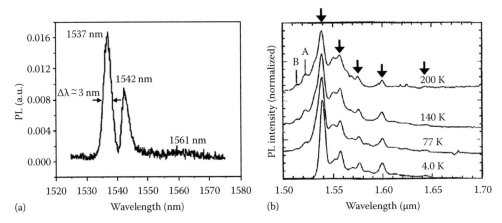

Figure 5.10 (a) Low temperature PL spectra for SiO_2:Er at $T = 4.0\,K$ reveals an inhomogeneous linewidth of ~3 nm for the 1537 nm PL peak. (Reprinted with permission from Saini, S., Gain efficient waveguide optical amplifiers for Si microphotonics, PhD thesis, Massachusetts Institute of Technology, Cambridge, MA, 2004.) (b) PL spectra for Si:Er at several temperatures. Arrows have been overlaid, identifying the five dominant radiative $I_{13/2} \rightarrow I_{15/2}$ manifold transitions. Hot lines PL peaks A and B increase in intensity in the PL spectrum of Si:Er, as temperature increases. (Reprinted and adapted with permission from Kimerling, L.C. et al., *Solid State Phys.*, 50, 333, 1997.)

states. Aside from thermal broadening, the interaction of thermal energy with an optical center can give rise to unique spectral features, such as hot-lines and phonon replicas.

When Er is doped within a solid-state material, such as silicon (or silicon oxide), the angstrom length scale proximity of electropositive Si and oxygen (O) atoms strips away three outer electrons, resulting in Er^{3+}, with a $4f^{11}$ electron multiplet that exhibits degeneracy splitting (Figure 5.7a), due to spin-orbit coupling (Dieke 1968). The lowest energy multiplet states, $I_{15/2}$ and $I_{13/2}$, represent the ground and first excited states for a long, ms time-scale infrared radiative transition ($\lambda = 1.54 \mu m$). The asymmetric distribution of nearest neighbor Si and O atoms contributes a nonzero local electric field about each Er atom, resulting in a Stark effect (interaction of an external electric field with the Er $4f^{11}$ charge distribution, quantified by a linear combination of charge dipole moments, see Griffiths 2004) that lifts the multiplet degeneracy and splits the collective energy states of the $4f$-electrons further, into a *manifold* of energy levels. This Stark splitting, due to an asymmetric crystal field, results in five radiative transitions from the bottom of the $I_{13/2}$ manifold to energy levels within the $I_{15/2}$ manifold (Kimerling et al. 1997). These transitions correspond to the five principal peaks resolved in the $T = 4.0 K$ PL spectrum for Si:Er in Figure 5.10b. At higher temperatures ($T = 77 - 200 K$), thermal energy promotes the excited $4f$-electrons into higher energy levels of the $I_{13/2}$ manifold, with a Boltzmann probability distribution $P_{13/2}(E) \propto e^{-(E-E_{13/2}^*)/k_B T}$ (Reif 1965) ($E_{13/2}^*$ is the energy of the lowest $I_{13/2}$ manifold level). A radiative transition to the bottom of the $I_{15/2}$ manifold results in the emergence of two new higher energy emission lines, labeled A and B in the $T = 200 K$ spectrum (see schematic transitions in Figure 5.7a and corresponding PL peaks in Figure 5.10b). In an atomic transition, where all Er atoms are excited and therefore the $I_{15/2}$ manifold is unoccupied, N_2 and the PL intensity for the A and B emission lines are proportional to $P_{13/2}(E)$. Thus, the PL intensity ratio $I(\lambda_{A,B})/I(\lambda = 1.54 \mu m)$ is proportional to $e^{-(E_{A,B}-E_{13/2}^*)/k_B T}$; an Arrhenius plot finds good agreement for $E_{A,B} - E_{13/2}^*$, with the energy difference directly calculated from the spectral location of the A, B emission lines and the peak $\lambda = 1.54 \mu m$ emission line (Kimerling et al. 1997). The A, B emission lines form as a result of the thermal energy $k_B T$ driving the population of collective $4f$-electrons to higher $I_{13/2}$ manifold levels. The A, B emission lines thus increase in intensity with temperature, and are referred to as hot lines in the PL spectrum.

Whereas hot lines spectrally resolve higher manifold level transitions of an emission site, phonon replicas spectrally resolve the vibrational modes, or phonon distribution, of the host medium. In the case of an exciton within the AgBr:I system, an electron–hole pair is bound to an isoelectronic I atom site; before undergoing recombination, the exciton has a strong interaction probability with the longitudinal optical phonon.

There are two distinct modes of phonon propagation, acoustic and optical phonons, uniquely identified by their dispersion relations (see Chapter 2 for a detailed discussion). Both of these modes further differentiate as longitudinal acoustic and optical (atoms displace along the direction of wave propagation), LA and LO, phonons; or transverse acoustic and optical (atoms displace orthogonal to the direction of wave propagation), TA and TO, phonons. Historically, optical phonons were so named because of the experimental ease with which these modes are excited in ionic solids, by incident infrared light, due to strong spatial overlap of electric field between an electromagnetic wave and an LO phonon (Dean and Herbert 1979).

In ionic and, to a lesser extent, covalent bond crystals, a strong interaction between photons and phonons (Chapter 2) influences the emission of light. Recombination of a longer lifetime (bound) exciton, corresponding to a high degree of charge localization (Haynes 1960), results in a higher probability of coupling exciton energy to the creation of one or more LO phonons and subsequent emission of a red-shifted photon. Figure 5.11a, upper half, shows a schematic representation of PL spectra from such an exciton at $T = 0 K$. Consider the first PL

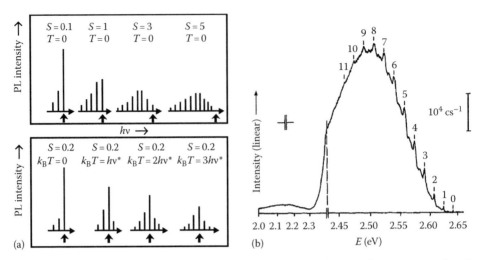

Figure 5.11 (a) Top: schematic representation of PL emission lines in the presence of S phonons, at $T = 0$ K. Bottom: for a given number S phonons, the appearance of anti-Stokes phonons for increasing temperature. The vertical arrows identify the no-phonon lines. (Adapted with permission from Dean, P.J. and Herbert, D.C., Bound excitons in semiconductors, in *Excitons* (Vol. 14 of Topics in Current Physics), ed. K. Cho, pp. 55–181, Springer-Verlag, Berlin, Germany, 1979.) (b) A celebrated phonon replica example: excitonic emission from AgBr:I, revealing the presence of eleven phonon lines. (Reprinted with permission from Czaja, W. and Baldereschi, A., *J. Phys. C: Solid State Phys.*, 12, 405, 1979.)

spectrum, labeled $S = 0.1$, $T = 0$: the emission line corresponding to direct recombination of the exciton (at energy $E_{exc} \equiv h\nu_0$) has the highest intensity and is referred to as the no-phonon or zero-phonon line. An exciton that recombines by emitting one LO phonon (with energy $h\nu_{LO}$) will have less intense PL at emission energy $E_{exc} - h\nu_{LO}$, and is referred to as a one-phonon line. Similarly, there is an even lower intensity two-phonon line at emission energy $E_{exc} - 2h\nu_{LO}$. Depending on the chemistry of the host medium, several phonon interactions may be preferred, resulting in a two-phonon or three-phonon line with stronger PL intensity than the no-phonon line. This influence of the host's chemistry on exciton-phonon interactions is phenomenologically represented by the Huang–Rhys factor, S, equal to the average number of phonons involved. The Poisson probability function $P(h\nu_0 - nh\nu_{LO}) = e^{-S}S^n/n!$, where n is the number of phonons, describes an envelope function for the collection of red-shifted emission lines (Dean and Herbert 1979). This collection of Stokes-like emission lines is called satellite lines, the phonon sideband, or phonon replicas. The literature describes phonon replicas as replicating the no-phonon line; hence, the no-phonon line is also referred to as the principal exciton luminescence line. The subsequent PL spectra in the upper half of Figure 5.11a show that, for a given temperature, a larger value of S results in a broader sideband. Furthermore, peak intensity is now at one of the multi-phonon lines, and no longer the no-phonon line. The lower half of Figure 5.11a demonstrates how increasing the temperature (for a fixed value of S, i.e., for one unique sample) results in the creation of a second phonon sideband, representing blueshift emission lines. These anti-Stokes emission lines originate from the LO phonon energy (from the increasingly vibrating lattice) combining with E_{exc}. The experimenter may recognize the formation of such phonon-mediated Stokes and anti-Stokes lines to simply be an example of Raman scattering (Chapter 6), in this case occurring with photons generated within the sample from PL emission (and not from an external light source). While the LO phonon is the dominant phonon mode that PL emission couples to, it should be noted that the

TA, LA, and TO phonons can also be coupled to, giving rise to additional phonon replicas. It is of interest to note that the LO phonons primarily generated from this interaction proceed to dissipate by coupling to pairs of lower energy acoustic phonons (Dean and Herbert 1979). Lastly, phonon replicas occur in both PL and absorption spectra, but with the phonon side-bands reflected about the no-phonon line.

As the schematic examples in Figure 5.11a suggest, strong phonon coupling can have a considerable influence on PL spectral lineshape. Figure 5.11b shows low temperature PL from AgBr:I (Czaja and Baldereschi 1979), a system with very strong exciton–phonon coupling (Huang–Rhys factor of $S = 9.6$); the large number of Stokes lines dramatically demonstrates how in extreme cases, the peak of an experimentally measured spectrum can be considerably removed from the no-phonon line (in this case, over 140 meV apart). The no-phonon line in this example lies well outside the PL linewidth, and contributes negligible photon emission. Thus, phonon replicas result in a reduction of the no-phonon line intensity and potentially drastic alterations of the PL lineshape. Care must be taken to look for the telltale peaks of replica phonon lines (labeled by their phonon number in Figure 5.11b), in long lifetime systems such as Type II quantum dots or nanocrystals.

5.2.3 Photoluminescence versus Temperature

De-excitation mechanisms for atomic transitions versus electron–hole recombination have distinctly different temperature dependence; we examine the simpler atomic case first. The temperature of a system can only have an effect on the decay rate, in the rate equations at the beginning of Section 5.2. For an atomic transition, the radiative rate of decay, described by Fermi's Golden Rule, has a functional dependence on the transition matrix element, reduced density of states, and optical density of states (Coldren and Corzine 1995). None of these factors has an appreciable dependence on temperature, and thus the radiative rate of decay, $1/\tau_r$, is approximately independent of T. In general, nonradiative decay mechanisms involve coupling to a continuum of energy states (such as sub-band gap defect states or phonon modes, see Pankove 1975). An energy mismatch or band structure bending (in the presence of trapped charge carriers) will pose an activation energy barrier, E^*, to this recombination mechanism. Provided thermal energy can be supplied to surmount E^*, the nonradiative decay mechanism can proceed; thus, $1/\tau_{nr}$ is almost always a function of T, with the Boltzmann factor form $1/\tau_{nr} = (1/\tau'_{nr})e^{-E^*/k_B T}$, where $1/\tau'_{nr}$ is the upper limit of the nonradiative decay rate (i.e., in the presence of no activation barrier). Note that the nonradiative de-excitation rate decreases with lower temperature. Inserting this form into the rate equations, one can solve for N_2, for a given excitation power and temperature. The influence of temperature can be more easily assessed by analyzing the low excitation power expression for PL intensity, $I_{PL} \propto \eta W_p N$. Recalling the identity for η (Section 5.2), the internal quantum efficiency can be rewritten as $\eta = 1/(1 + \tau_r/\tau'_{nr} \cdot e^{-E^*/k_B T})$. Inserting this identity into the expression for I_{PL} results in an exponential dependence on T. Noting that $\lim_{T \to 0} \eta = 1$, the PL intensity can be expressed as

$$I_{PL} = \frac{I(0)}{1 + \tau_r/\tau'_{nr} \cdot e^{-E^*/k_B T}} \tag{5.7}$$

where, $I(0)$ is the PL intensity in the limit of $T \to 0$. Figure 5.12a shows an example of integrated PL intensity versus temperature from silicon nanocrystals (Dal Negro et al. 2006). The dominant emission mechanism from these nanocrystals is understood to be a HOMO-LUMO transition: a dangling Si covalent bond at the interface between the nanocrystal and the host medium contributes to the formation of molecular orbital states. PL emission then originates

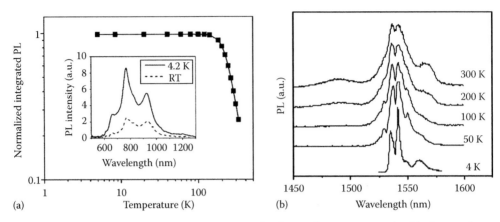

Figure 5.12 (a) Intensity versus temperature plot for np-Si emission, embedded in a silicon rich nitride host. (Reprinted with permission from Dal Negro, L. et al., *Appl. Phys. Lett.*, 88, 233109, 2006. American Institute of Physics.) (b) PL emission versus temperature for SiO$_2$:Er. (Reprinted with permission from Saini, S., Gain efficient waveguide optical amplifiers for Si microphotonics, PhD thesis, Massachusetts Institute of Technology, Cambridge, MA, 2004.)

from an electronic transition between the Highest Occupied Molecular Orbital (HOMO) and Lowest Unoccupied Molecular Orbital (LUMO).

Figure 5.12b shows for lower PL temperatures, the reduction in thermal vibrations eliminates thermal broadening: a $T = 4.0$ K PL spectrum of SiO$_2$:Er resolves the true extent of inhomogeneous broadening, as compared to its $T \simeq 300$ K room temperature PL spectrum. This increase in spectral resolution helps in accurately identifying peak emission wavelength(s).

In semiconductors, for carrier recombination at moderate excitation powers, such that the electron and hole concentrations n, p are close to their equilibrium values, the influence of temperature is approximated by the Boltzmann-like relations $n = N_C e^{-(E_C - E_F)/k_B T}$, $p = N_V e^{-(E_F - E_V)/k_B T}$ $\left(\text{where } N_C = 2 \left(2\pi m_e^{\text{eff}} k_B T / h^2\right)^{3/2} \text{ and } N_V = 2 \left(2\pi m_h^{\text{eff}} k_B T / h^2\right)^{3/2}\right)$. The dependence of radiative and nonradiative carrier density recombination rates on n, p implies a temperature influence. For the case of Auger-dominated nonradiative recombination in n-type semiconductors, an Auger lifetime $1/\tau_A = R_A/n$ can be solved (accounting for momentum conservation) to be (Pankove 1975)

$$\frac{1}{\tau_A} \propto \left(\frac{k_B T}{E_G}\right)^{3/2} \exp\left(\frac{-E_G}{k_B T} \cdot \frac{1 + 2M}{1 + M}\right) \tag{5.8}$$

where $M = m_e^{\text{eff}} / m_h^{\text{eff}}$ if $m_e^{\text{eff}} < m_h^{\text{eff}}$ or $M = m_h^{\text{eff}} / m_e^{\text{eff}}$ if $m_e^{\text{eff}} > m_h^{\text{eff}}$.

Experimental measurements of the total semiconductor lifetime τ (Tsang et al. 1968, Pankove 1975), compared with theoretical values for radiative versus Auger-dominated nonradiative lifetimes reveal that for higher temperatures, the Auger recombination rate will exceed the radiative recombination rate and result in a net reduction in η and I_{PL}. In samples where point defects dominate nonradiative recombination, an Arrhenius activation energy typically governs the dependence of I_{PL} on T (Yoon 1998).

In semiconductors, the position of electronic energy levels is also sensitive to contraction of the crystal lattice as the temperature is reduced. This dependence results in an increase of the band gap for lower temperatures, as described by $E_G(T) = E_G(0) - AT^2/(T + B)$ (A and B

are proportionality constants, see Pankove 1975). This increase in E_G for lower T results in a blueshift of the PL intensity. Simultaneously, the decrease in temperature also reduces the occupancy of higher energy conduction or valence band states by electron or holes, decreasing the thermal broadening. Since thermal broadening dominates semiconductor PL spectra, linewidth narrowing is observed for lower temperatures as well (e.g., refer to a GaAs electroluminescence study in Carr 1965, where such a spectrum is equivalent to PL, originating from the carrier recombination processes mentioned above). It is worthwhile to note that this blueshift and linewidth narrowing for lower T is also observed for exciton emission in semiconductors (Yoon 1998); however, exciton linewidth narrowing is attributed to a reduction in inelastic scattering processes with phonons and ionized impurities.

5.2.4 Photoluminescence versus Optical Pump Power

The steady-state solution for N_2 (Section 5.2) that in the limiting case of large excitation powers ($W_p \to \infty$), the PL intensity will saturate to a constant value ($I_{PL} \infty N/\tau_r$). However, de-excitation processes that are proportional to the population of excited atoms (or carriers), such as cooperative upconversion (Desurvire 1994) in a rare-earth atomic transition and radiative or Auger recombination in a semiconductor electronic transition, will have a parasitic quenching effect on the population of excited states (carriers), i.e., the rate of de-excitation increases with the degree of optical pumping. Figure 5.13a shows, as an example, a plot of PL intensity versus excitation power for an atomic transition in Er-doped alumina (Al_2O_3:Er). In the presence of a higher concentration of upconverting Er atoms, the PL intensity will saturate to a lower value, relative to the ideal prediction (dotted lines), with increasing excitation powers.

As a general rule of thumb, it is recommended to perform PL experiments with as low an excitation power as possible, so that lifetime measurements (a topic for Chapter 14) are negligibly influenced by such excited concentration-dependent rates. However, when collecting PL from thin film or low-yield nanostructure samples, the experimenter may find it necessary to increase the excitation power, in order to collect PL intensity with an adequate SNR. In such situations, it is up to the experimenter to negotiate an acceptable trade-off between higher excitation power versus onset of parasitic quenching.

High power excitation is typically used to probe emission from the higher excited states of saturable samples, such as long lifetime atoms (Miniscalco 1991) or semiconductor nanostructures (Prasankumar et al. 2008). In atomic systems, at high rates of optical excitation, the ground state E_1 is depleted (such samples are referred to as saturable absorbers, i.e., the ground state absorption goes to zero) and the first excited state E_2 remains occupied long enough so that a second absorption event (referred to as excited state absorption, ESA) promotes the electron to a higher excited state $E_n(n = 4, 5, \ldots)$ within the atom. PL emission and lifetime studies from the higher excited state ($E_n \to E_2$, or $E_n \to E_1$) are easily observed. It should be noted, that ESA can only occur if the pump photon energy $h\nu_p$ resonantly matches the energy difference $E_n - E_2$.

In semiconductor nanostructures (with a low absolute number of energy states), at high excitation powers, the rate of carrier promotion to $E_n(n = 3, 4, \ldots)$ can exceed the $E_2 \to E_1$ decay rate, such that in steady-state, all the E_2 states are filled with carriers. In this situation, further optical pumping will result in an increased E_n population and increased $E_n \to E_1$ radiative transitions. Thus, high optical excitation rates can very effectively resolve PL peaks at wavelengths corresponding to $h\nu = E_n - E_1$ and thus help determine the higher energy levels E_n.

Certain carrier-based phenomena can have unanticipated effects under increased excitation power. For example, the PL spectrum of type II band offset nanostructures (Chapter 2) exhibits a blueshift in peak wavelength with increased excitation power. Figure 5.13b shows a one-dimensional energy-band diagram of a ZnTe/ZnSe quantum dot. For this system, steady-state

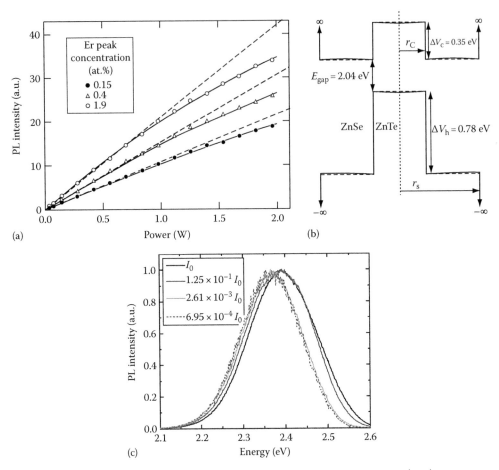

Figure 5.13 (a) PL intensity versus excitation power for Al_2O_3:Er. (Reprinted with permission from Snoeks, E. et al., *J. Appl. Phys.*, 73, 8179, 1993. American Institute of Physics.) (b) One-dimensional energy-band diagram (dotted line) of a ZnTe quantum dot in ZnSe host. Formation of an electric dipole, due to spatially separated carriers, results in band structure bending (solid line). (c) Blueshift in ZnTe exciton PL emission, for increasing excitation power. (b and c, Reprinted with permission from Shuvayev, V.A. et al., *Phys. Rev. B.*, 79, 115307, 2009. Copyright 2009 American Physical Society.)

excitation results in excess holes trapped within the ZnTe region. The positive charge of these trapped holes attracts excess electrons to the ZnTe/ZnSe barrier. While a fraction of the carriers recombine in steady state, the remaining carriers form an extended electric dipole, resulting in a bending of the heterostructure bands. The band bending results in an increase of the energy level difference between electron states confined near the heterointerface and hole states confined within the quantum dot. Figure 5.13c shows a corresponding increase in PL peak energy, for increasing excitation power.

Lastly, the absorption cross section for the excitation wavelength, σ_p, can be determined by a power-dependent experiment. While time-dependent PL analysis is discussed more thoroughly in Chapter 12, the dynamic population dependence on a rise and decay time will be briefly summarized here (it is assumed, for low excitation powers, parasitic quenching effects are negligible). For simplicity, once again the atomic transition model is used without loss of generality (for semiconductors). Labeling the steady-state solution for N_2 (derived at the start of Section 5.2) as N_2^{max}, the time-dependent solution to dN_2/dt assumes an initial

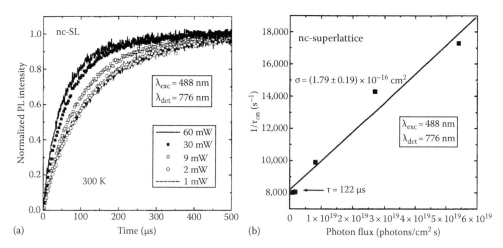

Figure 5.14 (a) PL intensity versus time (rise time) of np-Si (embedded within a SiO_2 host), for different excitation powers. (b) A plot of $1/\tau_{rise}$ versus photon flux of excitation beam yields the absorption cross section from the slope (vertical intercept is $1/\tau$). (Reprinted with permission from Priolo, F. et al., *J. Appl. Phys.*, 89, 264, 2001. American Institute of Physics.)

condition $N_2(t=0) = 0$ and a final condition $N_2(t \to \infty) = N_2^{max}$. The solution is found to be $N_2(t) = N_2^{max}(1 - e^{-t/\tau_{rise}})$, where the rise time is defined by $1/\tau_{rise} \equiv W_p + (1/\tau) = \phi_p\sigma_p + (1/\tau)$ (τ is the net lifetime consisting of radiative and nonradiative decay, from the beginning of Section 5.2). Similarly, for an initial condition $N_2(t=0) = N_2^{max}$ and final condition $N_2(t \to \infty) = 0$, $N_2(t) = N_2^{max}e^{-t/\tau_{decay}}$, where the decay time is defined by $1/\tau_{decay} \equiv 1/\tau$. In a PL measurement, when the excitation beam is chopped at a slow enough rate so that the N_2 population in the sample approaches its steady-state value during the illumination cycle (and approaches zero during the blocked cycle), a complete rise or decay of time-varying PL (Figure 5.14a) can be collected by an oscilloscope. Note that, according to their definitions, $\tau_{rise} < \tau_{decay}$. Both the characteristic times are measured by an exponential fit (or a linear fit on a semilog plot). For different excitation powers, a plot of $1/\tau_{rise}$ versus ϕ_p (Figure 5.14b) will produce a straight line with slope σ_p and vertical intercept $1/\tau$. The simultaneous fitting of τ_{decay} during the decay cycle of the time-varying PL scan gives several independent measures (one for each excitation power ϕ_p) of τ. Thus, the $1/\tau_{rise}$ versus ϕ_p plot can be fitted by a linear regression with a fixed intercept value of $1/\tau$, thereby resulting in a more accurate fit for σ_p.

These analytical techniques offer a basic approach for quantitative interpretation of PL data, collected from a CW (i.e., steady-state) experiment. In Sections 5.3 and 5.4, the focus turns toward two complementary techniques that help to understand the excitation and carrier generation characteristics of samples, using CW spectroscopy.

5.3 PHOTOLUMINESCENCE EXCITATION SPECTROSCOPY

Whereas PL relies on exciting a sample at one excitation wavelength, and observing an emission spectrum, Photoluminescence Excitation spectroscopy (PLE) is the complementary experiment. In PLE, a sample is excited by scanning across a spectrum of excitation wavelengths, while observing light emission at one wavelength (typically the peak of a PL spectrum). Xenon or tungsten lamps are frequently employed as excitation sources that approximate a blackbody spectrum (excepting a collection of intensity spikes in the xenon spectrum). An excitation monochromator, placed between the excitation source and the sample, selects the

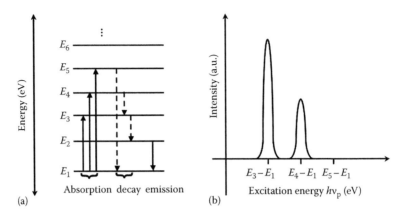

Figure 5.15 (a) Schematic of energy levels within a sample. In a PLE scan, the PL emission from the $E_2 \rightarrow E_1$ transition is monitored, while scanning through pump energies $h\nu_p$ that initiate absorption from E_1 to higher energy excited states. The excited state populations decay (either radiatively or nonradiatively) and some of these decay transitions couple, with varying efficiency or transition rates, to the first excited state E_2. (b) Schematic of an imaginary PLE spectrum, corresponding to the coupling of higher energy excited states with E_2 from (a).

pump photon wavelength or energy $h\nu_p$ (with a transmission passband $\Delta\nu$, also determined by the monochromator) from the near-blackbody spectrum.

PLE reveals the coupling of high-energy states to lower energy states. The energy level schematic in Figure 5.15a demonstrates an imaginary system in which light can be absorbed from a ground state to four possible excited energy states. Suppose the $E_2 \rightarrow E_1$ energy transition, corresponding to PL emission, is of particular interest to us; determining if the higher energy states E_3, E_4, E_5 couple to the first excited state E_2 would help in developing a model that describes the atomic or electronic origin of these states and the mechanism of energy transfer to E_2. A PLE spectrum is plotted as intensity of emission (the $E_2 \rightarrow E_1$ transition) as a function of the excitation or pump energy $h\nu_p$. At $h\nu_p = E_3 - E_1$, $E_4 - E_1$, $E_5 - E_1$, a pump photon is resonantly absorbed and promotes the system from the ground state to the corresponding excited state. The observed intensity of $E_2 \rightarrow E_1$ emission then gives a measure of the efficiency with which the higher excited states (E_3, E_4, E_5) couple to the first excited state (E_2). Suppose, in the PLE spectrum, there is a high peak intensity at $h\nu_p = E_3 - E_1$, no peak intensity at $h\nu_p = E_5 - E_1$, and relatively weak peak intensity at $h\nu_p = E_4 - E_1$ (Figure 5.15b). This corresponds to a strong to moderate coupling transition between $E_3 \rightarrow E_2$ and $E_4 \rightarrow E_2$, but not between $E_5 \rightarrow E_2$. An unknown decay mechanism couples very strongly to the energy state E_5, and the relatively weaker PLE intensity corresponding to the $E_4 \rightarrow E_2$ transition (versus $E_3 \rightarrow E_2$) suggests other competing decay processes preferentially couple to the E_4 state (as compared to the E_3 state), reducing the efficiency of energy transfer to E_2.

In comparing the PLE intensity at $h\nu_p = E_4 - E_1$ to $h\nu_p = E_3 - E_1$, it is essential to normalize the collected luminescence flux density $\phi_{PL}(h\nu)$, with respect to the incident excitation flux density $\phi_p(h\nu)$, in order to ensure that the number of pump photons per second is the same at all pump wavelengths. For a PL photodetector with responsivity $R(\nu)$, $\phi_{PL}(h\nu) \propto I_{ph}(\nu)/(h\nu \cdot R(\nu))$, where $I_{ph}(\nu)$ is the measured photocurrent (Section 5.1.6). For a reference photodiode (that measures the excitation intensity) with responsivity $R_{ref}(\nu)$, $\phi_p(h\nu) \propto I_{ref}(\nu)/(h\nu \cdot R_{ref}(\nu))$, where $I_{ref}(\nu)$ is the measured photocurrent. The near-blackbody spectrum of the excitation source (Section 5.1.1) is a nonuniform intensity profile $I_p(h\nu)$; dividing this profile by $h\nu$ yields a similarly nonuniform pump photon flux density profile $\phi_p(h\nu)$. In the linear regime of optical excitation, the luminescence intensity ($I_{PL}(\nu) = h\nu \cdot \phi_{PL}(\nu)$) is directly proportional to the optical pump rate W_p (Section 5.2), and thus the nonuniform spectral profile of ϕ_p can be removed from the PLE spectrum by dividing $\phi_{PL}(h\nu)$ by $\phi_p(h\nu)$; thus, $I_{PLE}(\nu) \propto \phi_{PL}(\nu)/\phi_p(\nu)$. The use of an incoherent light source—namely, a lamp, instead of a laser—ensures a low enough ϕ_p such that the optical excitation can be suitably approximated in the linear regime, for most samples. Experimentally, ϕ_p is sampled by the reference photodiode after the excitation monochromator. It is very important that the excitation spectrum be sampled *after* the excitation monochromator, and not from between the lamp and monochromator, as the grating transmission and chromatic lens aberration alter the transmitted profile of the excitation spectrum (Section 5.1.5).

5.3.1 Photoluminescence Excitation Probe of Energy Levels and Coupling Mechanisms

In complex soft and solid-state materials, a combination of energy states and generation/recombination processes—originating from diverse physical origins—may couple or compete with one another, resulting in an unpredictable relation between excitation and emission mechanisms. PLE helps identify some of these processes, as shown in the case study of a nanocomposite: amorphous silicon nanoparticles (np-Si) embedded in a silicon nitride host.

Excitation at 3.82 eV results in a PL spectrum (Yu et al. 2007) that can be fitted by two overlapping Gaussian peaks (a dominant peak at 2.45 and weaker peak at 2.9 eV). PLE spectra are collected at these two emission energies (Figure 5.16). Figure 5.16a shows two excitation bands (monitored at 2.9 eV emission) with peaks at 3.75 and 4.5 eV, and the 4.5 eV band dominates I_{PLE}; the inset figure resolves a PL spectrum (excited at 4.5 eV) with the PL peak at 2.45 eV almost entirely quenched. Figure 5.16b shows three I_{PLE} excitation bands (monitored at 2.45 eV emission), with the lower excitation energy bands at 3.5 and 4.3 eV.

Figure 5.17a overlays the PL spectrum (excited at 3.82 eV) over the measured absorption spectra of this sample, showing a large Stokes shift of ~1 eV. The gradually rising tail of the absorption spectrum between 3 and 4 eV suggests the origin of this absorption to be host band-tail states that contribute to sub-band gap absorption in silicon nitride (E_G for stoichiometric

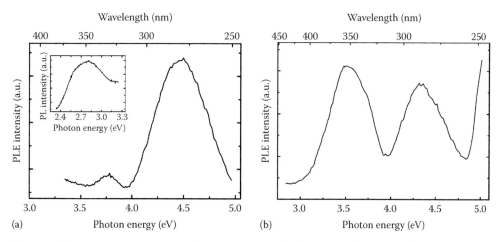

Figure 5.16 PLE measurements for nanostructured silicon nitride (embedded with np-Si). (a) PLE at 2.9 eV emission energy resolves two excitation bands (3.75, 4.5 eV); 4.5 eV is the dominant excitation band. Inset: excitation at 4.5 eV results in a PL peak only at 2.9 eV (the 2.45 eV PL peak is no longer present). (b) PLE at 2.45 eV emission energy resolves two excitation bands (3.5, 4.3 eV) of comparable intensity. (With kind permission from Speringer Science + Business Media: *Eur. Phys. J.B*, Excitonic photoluminescene characteristics of amorphous silicon nanoparticles embedded in silicon nitride film, 57, 53, 2007.)

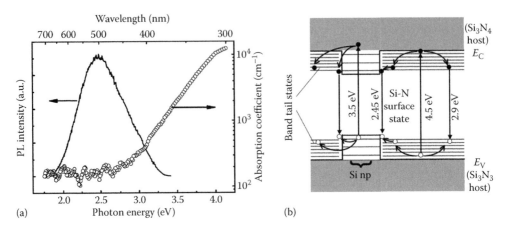

Figure 5.17 (a) PL (excited at 3.82 eV) and absorption spectra for the nanostructured silicon nitride. (Reprinted with permission from Yu, W. et al., *Eur. Phys. J. B*, 57, 53, 2007.) (b) A proposed energy-state model, from the interpretation of PL, PLE, and absorption data. Carriers are excited within both the np-Si and the Si_3N_4 host matrix, and diffuse/thermalize with varying efficiencies to either the nanoparticle–host interface (Si–N bond) or bottom of the host's extended states band tail.

Si_3N_4 reportedly varies between 4 and 5 eV, see Deshpande et al. 1995); carrier thermalization would explain the large Stokes shift between emission and absorption. The np-Si states are attributed to the 3.5 eV absorption band, and citing prior investigations (Noma et al. 2001), the 2.45 eV emission is attributed to a Si–N bond on the np-Si surface. Absorption of light at 3.5 eV is proposed to excite an exciton within np-Si that transfers its energy to the Si–N surface state.

The composite model for energy states suggested by this PLE data is summarized in Figure 5.17b: incident light can be absorbed either by (1) quantized-to-continuum state transitions in np-Si (3.5, 3.75 eV excitation bands), with a weak absorption cross section, or (2) inter-band absorption between the extended states or band tail states of the silicon nitride host medium (4.3, 4.5 eV excitation bands), with a strong absorption cross section. In the case of (1), most of the photogenerated carriers are trapped by np-Si and recombine via the Si–N surface state (2.45 eV emission); a small fraction of these carriers diffuse into the host medium, settling via thermalization to the bottom of the host's band tail states, and then recombine (2.9 eV emission). Carriers generated from 3.75 eV excitation have a higher continuum energy state and higher probability to settle into the host's band tail states. In the case of (2), photogenerated carriers are mostly captured by np-Si and recombine via the Si–N surface state (2.45 eV emission); a small fraction of the carriers settle via thermalization to the bottom of the host's band tail states and recombine (2.9 eV emission). Carriers generated by absorption at 4.3 eV appear to get trapped by np-Si more effectively.

In this manner, the position and relative intensity of PLE excitation bands prove to be highly effective in developing a comprehensive energy model for complex systems.

5.3.2 Photoluminescence Excitation as a Surrogate for Absorption

Quantum dots (QDs) and quantum wells (QWs) are efficient light emitters for room temperature lasers and candidates for intersubband absorption detectors in mid-to-far infrared sensing (Chapter 2). To model the energy states of a quantum dot, the positions of bound conduction and valence energy levels must be accurately identified, usually by means of intensity-dependent PL (Section 5.2.4), absorption, or PLE spectra.

Absorption measurements are typically done by measuring the attenuation of an incident continuous spectrum (with intensity I_0), after transmission through the sample ($I \cong T^2 \cdot I_0 e^{-\alpha d}$, where T is the power transmittance at the interface between air and the sample, d is the film thickness, and α is the absorption coefficient, as shown in Coldren and Corzine 1995). The $10^{17}\,cm^{-3}$ or lower density of quantum dots formed within thin films deposited by molecular beam epitaxy (Gu et al. 2005) potentially results in samples with a very weak attenuation ratio, I/I_0, that lies within the noise floor of the detector being used. In the case that a thick film can be formed (such as QDs formed by colloidal synthesis (Norris and Bawendi 1996) that are embedded in a host matrix), or the QD density can be increased by a factor of 10^2–10^4, an attenuation ratio can be resolved and α can be calculated with little noise (Figure 5.18a). Here too, a second complication exists: the size distribution of nucleated QDs results in severe inhomogeneous broadening of the absorption spectra, making it difficult to identify absorption peaks that correspond to excitation from E_1 to higher energy bound states.

A low intensity PL spectrum of the QDs (Figure 5.18a) typically resolves emission from only ground state recombination (Chapter 2): upon excitation to higher energy bound or extended states, carriers rapidly de-excite (via Auger mediated recombination) to the conduction and valence bound states, $E_{1,C}$ and $E_{1,V}$, respectively. Radiative emission is thus only observed from the $E_{1,C} \rightarrow E_{1,V}$ transition. The 56 meV linewidth of this PL peak is representative of inhomogeneous broadening. In a PLE measurement (Figure 5.18b), a narrow emission

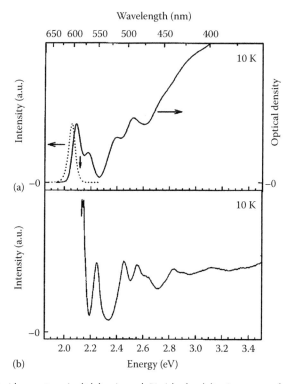

Figure 5.18 (a) Absorption (solid line) and PL (dashed line) spectra for colloidally synthesized (~28-Å-radius) QDs. In luminescence, the sample was excited at 2.655 eV (467.0 nm). The downward arrow marks the emission wavelength monitored for PLE. (b) PLE scan for these QDs. (Reprinted with permission from Norris, D.J. and Bawendi, M.G., *Phys. Rev. B*, 53, 16338, 1996. Copyright 1996 American Physical Society.)

passband (~8 meV, centered about the energy identified by a downward arrow in Figure 5.18a) observes luminescence from a subset of the optically excited QD population, namely, QDs whose size corresponds to an emission transition within the passband. Thus, the PLE spectrum is equivalent to sampling the absorption spectrum of this sub-set population, resulting in reduced inhomogeneous broadening. The higher resolution plot of Figure 5.18b enables the fitting of absorption peaks (corresponding to excitation from E_1 to higher energy bound states). The direct correlation between PLE and absorption spectra is only true if all of the absorption/excitation transitions result in electron–hole recombination with similar quantum efficiency. If the higher energy states involved in this absorption/excitation are also bound states of the QD, this is a reasonable assumption to make; one exception is in colloidally synthesized QDs with no capping layer or surface treatment, in which most of the photoexcited electrons and holes would get trapped in surface states before recombining.

Using PLE as a stand-in for absorption spectra is especially effective in thin film samples because the PL photodetector may be cooled (Section 5.1.6) to achieve a high SNR. In contrast, the detector used in absorption spectra measurements is overwhelmed by a large background photocurrent, due to the transmitted continuous spectrum. This large background current represents a large background noise floor, resulting in a low SNR.

5.4 PHOTOCURRENT SPECTROSCOPY

Photocurrent spectroscopy (PCS) measures the photoconductivity of a semiconductor sample, as a function of wavelength. The spectroscopic link from Figure 5.1 contains a lamp and spectrometer upstream from the sample (similar to PLE), but with no spectrometer or detector downstream from the sample (dissimilar to both PL and PLE). Instead, a metal two- or four-probe contact measures the sample photocurrent or photoconductivity, respectively. While a two-probe contact is sufficient to qualitatively measure a photocurrent and thus determine excitation energies for which mobile carriers are photogenerated, the four-probe contact (see Section 5.1.4) allows for a precise measure of the illuminated sample's sheet resistance, thus allowing one to determine its photoconductivity. PCS as a technique is largely applicable to semiconductors, since PCS requires the measurement of an observable conductivity, whereas PL and PLE are applicable to both semiconductors and insulators.

When the sample is excited at a particular wavelength, photogenerated carriers lead to a non-equilibrium increase in electron ($n = \Delta n + n_0$) and hole ($p = \Delta p + p_0$) concentrations (n_0, p_0 are the equilibrium concentrations of electrons and holes, respectively). The excess concentration for electrons is initially the same as that for holes ($\Delta n = \Delta p$), as the two carriers are photogenerated in pairs. Under no illumination, the equilibrium concentrations of electrons and holes contribute to an equilibrium conductivity $\sigma_0 \equiv n_0 e \mu_e + p_0 e \mu_h$ (where e is the electron charge and μ_e, μ_h are the electron and hole mobilities, respectively). Under steady-state excitation, the solution of the rate equation (see the beginning of Section 5.2) for n (and thus p) leads to an increased conductivity $\sigma = n e \mu_e + p e \mu_h$, which leads to an increased photocurrent density ($J_{ph} = \sigma E$, where E is the electric field) and photocurrent I_{ph} ($I_{ph} \equiv J_{ph} \cdot A$, where A is the cross section of the sample orthogonal to the current flow, depicted in Figure 5.2). Under no excitation, the experimenter measures the dark current $I_0 = J_0 \cdot A = \sigma_0 E \cdot A$; under steady-state excitation, the photocurrent $I_{ph} = J_{ph} \cdot A = \sigma E \cdot A$ is measured (see Section 5.1.4 for measurement details).

5.4.1 Photocurrent versus Absorption and Photoluminescence

Similar to PLE, the horizontal axis of a PC spectrum represents the energy of excitation (or absorption). Figure 5.19 contrasts (a) the absorption spectrum with (b) the PC spectrum for a heterostructure comprising InAs QDs arrays, sandwiched between spacer layers of GaAs

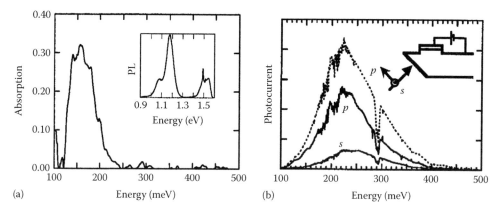

Figure 5.19 (a) Absorption (main panel) and PL spectra (inset) for InAs QD heterostructure. (b) PC spectrum from the same structure, with schematic depiction of the excitation beam orientation and charge collection (inset). (Reprinted with permission from Sauvage, S. et al., *Appl. Phys. Lett.*, 78, 2327, 2001; Schroder, D., *Semiconductor Material and Device Characterization*, John Wiley & Sons, Inc. (Wiley-IEEE), New York, 2006.)

(Sauvage et al. 2001). A PL spectrum (Figure 5.19a, inset) shows two interband emission peaks, at 1.18 and ~1.1 eV, from the QDs (and a 1.5 eV emission peak attributed to GaAs). In contrast, the absorption spectrum in the main panel of Figure 5.19a reveals intersubband absorption peaks, corresponding to promotion of bound electrons intentionally doped into the lowest conduction band energy levels of the InAs QDs to the extended states of the GaAs spacer layers. It is these intersubband transitions that are of interest for infrared detector devices, and the PC in Figure 5.19b reveals preferential absorption of p-polarized light over s-polarized light. While this polarization-preference can also be observed in an absorption experiment, comparison with PCS helps the experimenter determine if p-polarized photogenerated carriers are more readily extracted than s-polarized photogenerated carriers; this is found not to be the case. A blueshift of 65 meV between the absorption peak and the PC peak suggests the influence of QDs as traps that capture lower kinetic energy conduction electrons from the extended states of GaAs.

5.4.2 Junction Photocurrent Spectroscopy and Responsivity Characterization

As emphasized in Section 5.1.4, PCS requires processing of the sample into a primitive circuit element—the resistor—by formation of metal contacts. Junction photocurrent spectroscopy (JPCS) is an example of taking the next step and processing an electronic device. Figure 5.20a illustrates a p–i–n structure approach: the thin film sample is deposited onto a p-doped substrate (e.g., Si:B), where the assumption is made that the film represents an undoped, intrinsic region. Cladding the film with an n-type material (Si:P) results in a depletion zone with a built-in electric field that sweeps the photogenerated carriers in the intrinsic region in opposite directions (electrons and holes drift toward n- or p-type regions, respectively). Reverse-biasing the device increases the carrier drift velocity and drives the I–V curve into the current saturation regime (Pierret 1996). In thermal equilibrium, under no illumination, the saturation current for a semiconductor p–n or p–i–n junction is $I_{sat} = eA((D_e/L_e)n_{p,0} + (D_h/L_h)p_{n,0})$, where e is the Coulomb charge, D_e, D_h are the carrier diffusion coefficients, L_e, L_h are the minority carrier diffusion lengths, and $n_{p,0}$, $p_{n,0}$ are the minority carrier concentrations. Steady-state illumination increases $n_{p,0}$, $p_{n,0}$, leading to an increase in I_{sat}. JCPS measures the photocurrent

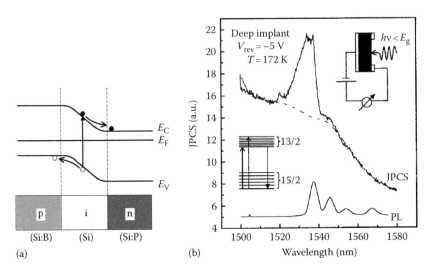

Figure 5.20 (a) Schematic illustration of p–i–n homostructure sample for JPCS and accompanying energy-band diagram. (b) *I–V* of a JCPS measurement for a sample of Si:Er, under steady-state excitation; the PL spectrum is shown for contrast. (Reprinted from *Physica B*, 273–274, Chen, T.D. et al., 322, Copyright (1999), with permission from Elsevier.)

I_{ph} versus excitation energy, typically under reverse potential bias (Figure 5.20b), so that $I_{ph} \simeq I_{sat}$ (excitation). Similar to the PLE requirement to normalize measured emission by the excitation flux density ϕ_p, I_{ph} must also be renormalized, to allow photocurrent comparison at different excitation energies, for uniform ϕ_p.

JCPS is the preferential technique for characterizing materials for potential use as photodetectors. Photodetector studies characterize pn or p–i–n devices by their total responsivity R_T $\left(\text{or spectral responsivity R}(v), \text{ where } R_T \equiv \int R(v)dv\right)$ at a given reverse potential bias, measuring the photocurrent generated per total optical power excitation P_T (or $P(\lambda)$). Responsivity is reported in units of amps/watts (or A/(W · Hz)) and determined by a fit of the linear regime in a plot of I_{ph} versus P_T, for a given reverse potential bias. The responsivity is proportional to the external efficiency of the photodetector. If the internal quantum efficiency is to be estimated, care must be taken to use the transmitted value of power $\int T(\lambda)P(\lambda)d\lambda$, where $T(\lambda)$ is the transmittance from air into the sample, for wavelength λ.

In general, since there may be unknown efficiencies within the sample for current extraction, it is best to simply report the external quantum efficiency from R_T (or $R(v)$).

While PCS and JPCS rely on two-probe contacts for measurement of photocurrent, as mentioned at the start of Section 5.4, photoconductivity measurements rely on the more accurate four-probe contact geometry. Combined with a Hall effect measurement of mobility, photoconductivity measurements allow the experimenter to determine the concentration of excess photogenerated carriers and develop more accurate differential equation models for electron/hole generation and recombination.

This chapter has detailed three workhorse spectroscopy techniques for analyzing the optical response of a sample material under CW excitation. The techniques probe a sample's steady-state optical and electrical properties, representing the first line of characterization for optically active materials. With these spectroscopic analyses, the experimenter can design more specific optoelectronic experiments, building from the broad collection of topics presented in other chapters of this book.

REFERENCES

Bain, L.J. and Engelhardt, M. 1992. *Introduction to Probability and Mathematical Statistics.* Boston, MA: PWS-KENT Publishing Company.

Carr, W.N. 1965. Characteristics of a GaAs spontaneous infrared source with 40 percent efficiency. *IEEE Trans. Electron Dev.* ED-12: 531–535.

Chen, T.D., Platero, M., Lipson, M., Palm, J., Michel, J., and Kimerling, L.C. 1999. The temperature dependence of radiative and nonradiative processes at Er-O centers in Si. *Physica B* 273–274: 322–325.

Cho, S., Kim, S.I., Kim, Y.H. et al. 2006. Effects of growth temperature on exciton lifetime and structural properties of ZnO films on sapphire substrate. *Phys. Stat. Solidi A* 203: 3699–3704.

Coldren, L.A. and S.W. Corzine. 1995. *Diode Lasers and Photonic Integrated Circuits.* New York: John Wiley & Sons (Wiley-Interscience).

Czaja, W. and Baldereschi, A. 1979. The isoelectronic trap iodine in AgBr. *J. Phys. C: Solid State Phys.* 12: 405–424.

Dal Negro, L., Yi, J.H., Michel, J., Kimerling, L.C., Chang, T.-W.F., Sukhovatkin, V., and Sargent, E.H. 2006. Light emission efficiency and dynamics in silicon-rich silicon nitride films. *Appl. Phys. Lett.* 88: 233109.

Dean, P.J. and Herbert, D.C. 1979. Bound excitons in semiconductors. In *Excitons* (Vol. 14 of Topics in Current Physics), ed. K. Cho, pp. 55–181. Berlin, Germany: Springer-Verlag.

Deshpande, S.V., Gulari, E., Brown, S. et al. 1995. Optical properties of silicon nitride films deposited by hot filament chemical vapor deposition. *J. Appl. Phys.* 77: 6534–6541.

Desurvire, E. 1994. *Erbium-Doped Fiber Amplifiers: Principles and Applications.* New York: John Wiley & Sons (Wiley-Interscience).

Dieke, G.H. 1968. *Spectra and Energy Levels of Rare Earth Ions in Crystals.* New York: John Wiley & Sons.

Drozdowicz, Z. et al. 2004. *The Book of Photon Tools (Newport Oriel Catalog).* Stratford, U.K.: Newport Oriel.

Franz, K.J., Charles, W.O., Shen, A. et al. 2008. ZnCdSe/ZnCdMgSe quantum cascade electroluminescence. *Appl. Phys. Lett.* 92: 121105.

Ghandhi, S. 1994. *VLSI Fabrication Principles: Silicon and Gallium Arsenide.* New York: John Wiley & Sons (Wiley-Interscience).

Griffiths, D. 2004. *Introduction to Quantum Mechanics.* San Francisco, CA: Benjamin Cummings.

Gu, Y., Kuskovsky, I.L., van der Voort, M. et al. 2005. Zn-Se-Te multilayers with submonolayer quantities of Te: Type-II quantum structures and isoelectronic centers. *Phys. Rev. B* 71: 045340.

Hamamatsu Corp. 1999. *Photomultiplier Tubes: Basics and Applications.* Hamamatsu, Japan: Hamamatsu Photonics.

Haynes, J.R. 1960. Experimental proof of the existence of a new electronic complex in silicon. *Phys. Rev. Lett.* 4: 361–363.

Howell, S. 2006. *Handbook of CCD Astronomy.* Cambridge, U.K.: Cambridge University Press.

Keithley Instruments, Inc. 1986. *Model 7065 Hall Effect Card Instruction Manual* (http://www.keithley.co.uk/data?asset=1118).

Kimerling, L.C., Kolenbrander, K.D., Michel, J. et al. 1997. Light emission from silicon. *Solid State Phys.* 50: 333–381.

Lerner, J.M. and Thevenon A. 1996. *The Optics of Spectroscopy.* Horiba Scientific website (http://www.horiba.com/us/en/scientific/products/optics-tutorial//?Ovly=1)

Miniscalco, W.J. 1991. Erbium-doped glasses for fiber amplifiers at 1500 nm. *J. Lightwave Technol.* 9: 234–250.

MIT Radiation Protection Committee. 2008. *Massachusetts Institute of Technology Radiation Protection Office Laser Safety Program,* 8th edn. MIT Environment, Health and Safety Office (http://web.mit.edu/environment/pdf/Laser_Safety.pdf)

Noma, T., Seol, K.S., Kato, H. et al. 2001. Origin of photoluminescence around 2.6–2.9 eV in silicon oxynitride. *Appl. Phys. Lett.* 79: 1995–1997.

Norris, D.J. and Bawendi, M.G. 1996. Measurement and assignment of the size-dependent optical spectrum in CdSe quantum dots. *Phys. Rev. B* 53: 16338–16346.

Omar, M.A. 1994. *Elementary Solid State Physics: Principles and Applications*. Reading, MA: Addison-Wesley Publishing Company.

Pankove, J. 1975. *Optical Processes in Semiconductors*. New York: Dover Publications.

Pierret, R. 1996. *Semiconductor Device Fundamentals*. Reading, MA: Addison-Wesley Publishing Company.

Polman, A., Lidgard, A., Jacobson, D.C. et al. 1990. 1.54 µm room-temperature luminescence of MeV erbium-implanted silica glass. *Appl. Phys. Lett.* 57: 2859–2861.

Prasankumar, R.P., Attaluri, R.S., Averitt, R.D. et al. 2008. Ultrafast carrier dynamics in an InAs/InGaAs quantum dots-in-a-well heterostructure. *Opt. Exp.* 16: 1165–1173.

Priolo, F., Franzo, G., Pacifici, D., Vinciguerra, V. et al. 2001. Role of the energy transfer in the optical properties of undoped and Er-doped interacting Si nanocrystals. *J. Appl. Phys.* 89: 264–272.

Reif, F. 1965. *Fundamentals of Statistical and Thermal Physics*. New York: McGraw-Hill.

Saini, S. 2004. Gain efficient waveguide optical amplifiers for Si microphotonics. Ph.D. thesis, Massachusetts Institute of Technology, Cambridge, MA.

Saleh, B.E.A. and Teich M.C. 1991. *Fundamentals of Photonics*. New York: John Wiley & Sons (Wiley-Interscience).

Sauvage, S., Boucaud, P., Brunhes, T. et al. 2001. Midinfrared absorption and photocurrent spectroscopy of InAs/GaAs self-assembled quantum dots. *Appl. Phys. Lett.* 78: 2327–2329.

Schroder, D. 2006. *Semiconductor Material and Device Characterization*. New York: John Wiley & Sons (Wiley-IEEE).

Schubert, E.F. and Poate, J.M. 1995. Er doped Si/SiO_2 microcavities. *Proc. SPIE* 2397: 495–514.

Senior, J. 1985. *Optical Fiber Communications: Principles and Practice*. Harlow: Prentice Hall Financial Times.

Shuvayev, V.A., Kuskovsky, I.L., Deych, L.I., Gu, Y., Gong, Y., Neumark, G.F., Tamargo, M.C., and Lisyansky, A.A. 2009. Dynamics of the radiative recombination in cylindrical nanostructures with type-II band alignment. *Phys. Rev. B* 79: 115307.

Snoeks, E., van den Hoven, G.N., and Polman, A. 1993. Optical doping of soda-lime-silicate glass with erbium by ion implantation. *J. Appl. Phys.* 73: 8179–8183.

Tsang, J.C., Dean, P.J., and Landsberg, P.T. 1968. Concentration quenching of luminescence by donors or acceptors in gallium phosphide and the impurity-band Auger model. *Phys. Rev.* 173: 814–823.

van der Pauw, L.J. 1958. A Method of measuring specific resistivity and hall effect of discs of arbitrary shape. *Philips Res. Rep.* 13: 1–9.

Wang, J., Gudiksen, M.S., Duan, X. et al. 2001. Highly polarized photoluminescence and photodetection from single indium phosphide nanowires. *Science* 293: 1455–1457.

Yan-hong, Y., Run-cai, M., and Bao-ying, L. 2006. Van Hove singularities and nonlinear photoluminescence in multiwalled carbon nanotubes. *Optoelectron. Lett.* 2: 186–188.

Yoon, I.T. 1998. Temperature dependence of free-exciton luminescence of undoped $In_{0.5}Ga_{0.5}P$ layers grown by liquid phase epitaxy. *J. Mater. Sci. Lett.* 17: 2043–2045.

Yu, W., Zhang, J.Y., Ding, W.G. et al. 2007. Excitonic photoluminescence characteristics of amorphous silicon nanoparticles embedded in silicon nitride film. *Eur. Phys. J. B* 57: 53–56.

Zhang, Y.J., Xu, C.S., Liu, Y.C. et al. 2006. Temperature-enhanced ultraviolet emission in ZnO thin film. *J. Lumin.* 119–120: 242–247.

Chapter 6

RAMAN SCATTERING AS A TOOL FOR STUDYING COMPLEX MATERIALS

S. Lance Cooper, Peter Abbamonte, Nadya Mason,
C.S. Snow, Minjung Kim, Harini Barath,
John F. Karpus, Cesar E. Chialvo, James P. Reed,
Young-Il Joe, Xiaoqian Chen, Diego Casa, and Y. Gan

CONTENTS

6.1 INTRODUCTION

Inelastic light (Raman) scattering has proven to be a versatile and powerful technique for studying a diverse range of materials, including semiconductors (Menendez 2000), semiconductor heterostructures (Gammon 2000), carbon-based materials (Dresselhaus et al. 2000), magnetic systems (Hayes and Loudon 1978, Cottam and Lockwood 1986), dilute magnetic semiconductors (Ramdas and Rodriguez 1991), low-dimensional spin systems (Lemmens et al. 2003), high-T_C superconductors (Thomsen and Cardona 1989, Thomsen 1991, Cardona 2000, Cooper 2001a, Devereaux and Hackl 2007), complex oxides (Scott 1974, Podobedov and Weber 2000, Cooper 2001b), and rare earth and actinide materials (Zirngiebl and Guntherodt 1991). The efficacy of this technique arises from several characteristics, which are discussed in greater detail in this chapter: (1) inelastic light scattering can provide important symmetry, energy, and lifetime information regarding a remarkable range of excitations in materials, including one- and two-phonon excitations (Pinczuk and Burstein 1975, Klein 1982), one- and two-magnon excitations (Cottam and Lockwood 1986, Lemmens et al. 2003), single-particle electron and plasmon modes (Klein 1975), crystal-electric-field excitations (Zirngiebl and Guntherodt 1991, Schaak 2000), spin-flip excitations (Ramdas and Rodriguez 1991, Snow et al. 2001), electromagnon modes (Cazayous et al. 2008), charge-density-wave phase and amplitude modes (Sooryakumar and Klein 1981, Tutto and Zawadowski 1992, Snow et al. 2002), and superconducting gap excitations (Thomsen and Cardona 1989, Thomsen 1991, Cardona 2000, Cooper 2001a, Devereaux and Hackl 2007); (2) because inelastic light scattering typically employs visible lasers as excitation sources, it lends itself to spectroscopic studies of materials and phase transitions under "extreme" conditions: for example, in high-magnetic fields of small-bore magnets that are often difficult to access using other spectroscopic techniques (Fleury et al. 1971, Sooryakumar and Klein 1981, Cottam and Lockwood 1986, Ruf et al. 1988, Ramdas and Rodriguez 1991, Blumberg et al. 1997, Nyhus et al. 1997, Karpus et al. 2004, Kim et al. 2008), under high-pressure conditions that are more difficult to study with other spectroscopies (Adler et al. 1994, Kremer et al. 2000, Loa et al. 2001,

Snow et al. 2002, 2003, Goncharov and Struzhkin 2003) and with fast time-resolution, which is readily obtainable using pico- and femtosecond pulsed lasers and pump-probe techniques (Chapter 14) (Takada et al. 2005); (3) using resonance (Cardona 1982), near-field (Verma et al. 2006), and surface-enhanced (Talley et al. 2005) Raman scattering methods to significantly enhance Raman scattering intensities, Raman scattering can be used to study very small samples, microstructures, and even nanometer-scale materials; and (4) Raman scattering with high-energy photons, that is, so-called inelastic x-ray scattering (IXS), can be used to probe the momentum dependence of various excitations, particularly collective modes like phonons (Burkel et al. 1991) plasmons (Schülke et al. 1998) and very recently, magnons (Hill et al. 2008, Schlappa et al. 2009). Additional information about the properties of some of these excitations can be found in other chapters of this book, particularly Chapters 2 (phonons and magnons) and 3 (plasmons).

The purpose of this chapter is to provide guidance to researchers who want to develop Raman scattering as a capability in their laboratories, with an emphasis on providing instruction in some of the more advanced methods useful in the study of complex materials, including high-pressure and high-magnetic-field methods and inelastic x-ray scattering. This chapter also provides a number of examples—aimed at the novice in Raman scattering methods—that illustrate how electronic, phononic, and magnetic excitations are identified and studied in materials, and how temperature-, magnetic-field-, and pressure-dependent Raman scattering methods can be used to study the properties of, and phase transitions in, materials. Because the examples shown cannot provide an exhaustive discussion of the broad capabilities of and applications for Raman scattering, we also provide extensive references to help direct those interested in a wider range of examples of Raman scattering in materials.

The organization of the remainder of this chapter is as follows: In Section 6.2, we provide enough theoretical background to enable researchers to understand the illustrative examples provided later in the chapter. In Section 6.3, we provide detailed descriptions of Raman scattering experimental methods in general, and of high-pressure and high-magnetic-field methods in particular, with the goal of giving interested researchers guidance toward developing a Raman scattering system capable of temperature-, pressure-, and high-magnetic-field-dependent measurements. In Section 6.4, we provide example applications of the Raman scattering method, both to illustrate the powerful capabilities of Raman scattering for the study of materials and to demonstrate how one analyzes and interprets Raman scattering data from complex materials.

6.2 RAMAN SCATTERING FUNDAMENTALS

There are numerous detailed derivations of the Raman scattering cross-section $d^2\sigma/d\Omega d\omega_s$ in the literature (Hayes and Loudon 1978, Cardona 1982, Cottam and Lockwood 1986). We emphasize the classical derivation of the Raman scattering cross section to provide the reader a more intuitive understanding of the various significant contributions to the Raman scattering intensity. Detailed microscopic derivations of the Raman scattering cross section can be found in Hayes and Loudon (1978) and Devereaux and Hackl (2007), and the cross section for inelastic x-ray scattering (IXS) is derived in Sinha (2001).

In the classical description of the Raman scattering process, one considers an electric field vector $E_1(r, t)$—with polarization ε_1, wave vector k_1, and angular frequency ω_1—that is incident on a material. This incident field induces a polarization $P(r, t)$ in the material

$$P_i(r,t) = \varepsilon_o \sum_j \chi_{ij} E_{j1}(r,t), \qquad (6.1)$$

where

χ_{ij} is a second-rank tensor component describing the linear electric susceptibility of the material
ε_o is the permittivity of free space (Chapter 1)

Excitations in the material cause fluctuations in the induced polarization $\boldsymbol{P}$—which are reflected in the susceptibility tensor χ—resulting in a scattered electric field $\boldsymbol{E}_S(\boldsymbol{r}, t)$. Although the scattered light will, in general, be unpolarized and have a continuum of scattered frequencies and scattered wave vectors, the specific design of the Raman scattering apparatus (see Section 6.3) allows one to detect the intensity of scattered light having a specific angular frequency ω_S, a particular wave vector $\boldsymbol{k}_S$, and a specific polarization ε_S. The inelastic light-scattering cross section resulting from induced polarization fluctuations includes two contributions: (1) the Stokes component, in which the incident light field $\boldsymbol{E}_I(\boldsymbol{r}, t)$ creates an excitation of frequency, ω, and wave vector, $\boldsymbol{q}$, causing a reduction in the frequency of the scattered field $\boldsymbol{E}_S(\boldsymbol{r}, t)$ given by $\omega_S = \omega_I - \omega$; and (b) the anti-Stokes component, in which the scattered field gains energy from thermal excitations in the sample, $\omega_S = \omega_I + \omega$. In scattering processes having time-reversal symmetry (i.e., nonmagnetic excitations) and no resonant enhancements, the principle of detailed balance relates the Stokes and anti-Stokes intensities (Hayes and Loudon 1978):

$$I_{AS}(\omega) = I_S(\omega)\left(\frac{\omega_I + \omega}{\omega_I - \omega}\right)^2 e^{-\hbar\omega/k_B T}, \tag{6.2}$$

where

ω_I is the incident light frequency
ω is the frequency of the excitation
T is the temperature

The relationship between the Stokes' and anti-Stokes' Raman intensities therefore provides a method for determining the "actual" temperature of the sample in the scattering region, for example, allowing one to account for such important effects as laser heating of the sample.

6.2.1 Kinematical Constraint in Raman Scattering

In an ideal crystal with full translational symmetry, kinematical constraints imposed by wave vector conservation dictate the following constraint on the Raman scattering process, $\boldsymbol{q} = \boldsymbol{k}_I - \boldsymbol{k}_S$, where $\boldsymbol{k}_I$ and $\boldsymbol{k}_S$ are the wave vectors of the incident and scattered photons, respectively, and $\boldsymbol{q}$ is the wave vector of the excitation. In typical light scattering experiments—with incident and scattered photons in the visible frequency range ($\omega_{I,S} \approx 10^{14-15}$ Hz) and excitation frequencies in the far- to near-infrared range ($\omega \approx 10^{11-13}$ Hz) (which covers the frequency range in which optical phonon, magnon, and intraband electronic excitations are typically observed in solids, as will be discussed below)—the conditions $\omega \ll \omega_I$ and $|\boldsymbol{k}_I| \approx |\boldsymbol{k}_S|$ apply, and therefore the wave-vector-conservation condition can, to a good approximation, be written $|\boldsymbol{q}| = 2|\boldsymbol{k}_I|\sin(\theta/2)$, where θ is the angle between the incident and scattered light directions. Consequently, the excitation wave vectors typically probed in light-scattering experiments have the following range of magnitudes, $0 < |\boldsymbol{q}| < 3 \times 10^{-3}$ Å^{-1} (Hayes and Loudon 1978), which is generally several orders of magnitude smaller than the size of the Brillouin zone boundary, $|\boldsymbol{k}_{ZB}| \approx 2\pi/a \approx 1$ Å^{-1}, where a is the lattice parameter of the crystal. Importantly, therefore, Raman scattering measurements generally probe only excitations very near the Brillouin zone center; that is, at $|\boldsymbol{q}| \approx 0$. This is a severe limitation compared, for example, to neutron scattering measurements, which can probe phonon and magnetic excitations throughout the Brillouin zone. However, this limitation is circumvented under certain circumstances, such as when two-particle (e.g., two-phonon and two-magnon) excitations of equal and opposite momenta, q and $-q$, are excited by the incident light, and when wave vector conservation is lost in amorphous and disordered materials (Brodsky 1975, Klein et al. 1978). More recently, inelastic x-ray scattering (IXS)—which employs photons with wavelengths $\lambda \sim 1$ Å—has been developed as a method for probing wave vectors throughout the Brillouin zone. Thus, IXS provides a method—similar but complementary to neutron scattering techniques—for studying the dispersion of collective excitations in materials.

6.2.2 Raman Scattering Cross Section

The spectrum of scattered light in the Raman scattering process is given by the differential scattering cross-section $d^2\sigma/d\Omega d\omega_S$, which reflects the fraction of photons inelastically scattered into the differential solid angle $d\Omega$ with a scattered frequency in the range ω_S to $\omega_S + d\omega_S$. In the classical derivation, the differential scattering cross section for light polarized in the direction ε_S is related to the power spectrum of polarization fluctuations (Hayes and Loudon 1978, Cardona 1982, Cottam and Lockwood 1986):

$$\frac{d^2\sigma}{d\Omega d\omega_S} = \frac{\omega_I \omega_S^3 V}{(4\pi\varepsilon_o)^2 c^4} \frac{n_S}{n_I} \frac{1}{|E_I|^2} \left\langle \varepsilon_S \cdot P_S^*(k_S, \omega_S) \varepsilon_S \cdot P_S(k_S, \omega_S) \right\rangle, \tag{6.3}$$

where
    V is the volume of the sample that contributes to scattering ("scattering volume")
    n_I and n_S are the indices of refraction of the incident and scattered light *inside the sample*, respectively
    $\langle \cdots \rangle$ represents the power spectrum of polarization fluctuations induced by phononic, magnetic, or electronic excitations in the material

6.2.2.1 Nonmagnetic Excitations

In the case of nonmagnetic excitations, the relationship between the induced polarization and the incident electric field is given by Equation 6.1, and an excitation in the material is generally represented by a dynamical variable, $X(r, t)$, which can be associated, for example, with deviations of atomic position or charge density from equilibrium. The effect of these fluctuations on the induced polarization in the material is obtained by first expanding χ in powers of the dynamical variable, $X(r, t)$, around the static susceptibility, χ_o:

$$\chi_{ij} = (\chi_{ij})_o + \sum_k \left(\frac{\partial \chi_{ij}}{\partial X_k}\right)_o X_k + \frac{1}{2} \sum_{k,m} \left(\frac{\partial^2 \chi_{ij}}{\partial X_k \partial X_m}\right)_o X_k X_m + \cdots. \tag{6.4}$$

Ignoring the first term in this expansion—which is associated with elastic scattering—and considering only the first-order inelastic contribution to the power spectrum for polarization fluctuations in Equation 6.3, the differential scattering cross section can be written:

$$\frac{d^2\sigma}{d\Omega d\omega_S} = \frac{\omega_I \omega_S^3 V}{(4\pi)^2 c^4} \frac{n_S}{n_I} \left| \varepsilon_S \cdot \frac{d\chi(\omega)}{dX} \cdot \varepsilon_I \right|^2 \langle X(q, \omega) X^*(q, \omega) \rangle. \tag{6.5}$$

There are several noteworthy features in this expression for the scattering cross section:

Dependence on the susceptibility derivative—The scattering cross section in Equation 6.5 depends on the susceptibility derivative, $d\chi/dX$; thus, the Raman intensity of an excitation provides a measure of how strongly that excitation modulates the susceptibility.

Dependence on scattering geometry—Equation 6.5 shows that, by varying the incident and scattered polarizations, ε_I and ε_S, one can couple to different components of the susceptibility derivative tensor, $d\chi/dX$. This is a powerful feature of Raman scattering that enables one to identify the symmetries of excitations studied, and will be discussed further in Section 6.2.3.

Correlation and Raman response functions—Equation 6.5 also shows that the Raman scattering cross section for Stokes scattering is related to the correlation function associated with the dynamical variable, $\langle X(q,\omega)X^*(q,\omega)\rangle$. This correlation function is related to the imaginary part of the Raman scattering response function, Im $\chi(q,\omega)$, by the fluctuation-dissipation theorem (Hayes and Loudon 1978):

$$\langle X(q,\omega)X^*(q,\omega)\rangle = S(q,\omega) = \frac{\hbar}{\pi}[n(\omega)+1]\,\text{Im}\,\chi(q,\omega), \qquad (6.6)$$

where $[n(\omega) + 1]$ is the Bose–Einstein thermal factor and $n(\omega) = [\exp(\hbar\omega/k_BT) - 1]^{-1}$.

Finally, it should be noted that the Raman scattering cross section in Equation 6.5 only describes the radiated intensity inside the scattering medium. Because the Raman scattering intensity is measured outside the material, the differential scattering cross section is actually related to the measured Raman scattering intensity by several additional contributions that depend upon the optical response of the material. Consequently, when interpreting measured Raman intensity variations as functions of doping, pressure, and magnetic field, changes in the material's optical properties must also be accounted for. For details, see (Reznik et al. 1993) or (Cooper 2001a).

6.2.2.2 Magnetic Excitations

In magnetic materials, the susceptibility in Equation 6.1 can be dependent on the magnetization M, that is, $P = \chi(M)\cdot E_I$, via magnetostrictive and magnetic exchange effects (Moriya 1967, Cottam and Lockwood 1986). The electric susceptibility of a magnetic material has the symmetry property, $\chi_{ij}(M) = \chi_{ji}(-M)$ because of the axial nature of the magnetization. Further, in materials with no linear absorption, the susceptibility also satisfies the relation, $\chi_{ij}(M) = \chi_{ij}^*(M)$. Consequently, an expansion of the susceptibility similar to that leading to Equation 6.4 results in the following linear contribution to the induced polarization:

$$P^{(1)} = i\varepsilon_o G M \times E_I, \qquad (6.7)$$

where $iG = \partial\chi_{ij}(M)/\partial M_k$, and G is a magneto-optical constant. The resulting differential cross section obtained by inserting this polarization in Equation 6.3 is given by (Cottam and Lockwood 1986, Ramdas and Rodriguez 1991):

$$\frac{d^2\sigma}{d\Omega d\omega_S} \sim (\varepsilon_S \times \varepsilon_I)_i(\varepsilon_S \times \varepsilon_I)_j \langle M_i M_j^*\rangle. \qquad (6.8)$$

Equation 6.8 illustrates that the light-scattering cross section is sensitive to the correlation function associated with fluctuations in the magnetization, and that the associated scattering response is observed in the "depolarized" scattering geometry, that is, in the $(\varepsilon_I \perp \varepsilon_S)$ geometry. Examples of such scattering are spin-flip (see Section 6.4.2.3) and magnon Raman scattering (see Section 6.4.5).

6.2.3 Raman Scattering Selection Rules

The symmetry selection rules for the Raman scattering process are contained in the factor $|\varepsilon_S \cdot (d\chi(\omega)/dX) \cdot \varepsilon_I|$ in Equation 6.5, and are governed by the requirement (Neumann's principle) that the physical properties of a crystal—and in particular the relationship between the incident field, induced polarization, and Raman susceptibility tensor—must be invariant under all the symmetry operations of the crystal's point group (i.e., the set of symmetry operations, such as rotations and reflections, that leave a point in the crystal fixed while moving each atom of the crystal to the position of another atom of the same type). This requirement imposes two important symmetry constraints on the Raman process. First, because both the incident and scattered polarization vectors ε_I and ε_S transform like polar vectors—or, more formally, according to the irreducible representation (i.e., a representation of the group that describes the transformation properties of the group—e.g., as a matrix—and cannot be reduced to a series of other representations) Γ_{PV} of the crystal's point group—the only susceptibility derivatives $d\chi/dX$ that are nonzero in a Raman scattering measurement are those with symmetries Γ_x contained in the decomposition of $\Gamma_{PV}^* \otimes \Gamma_{PV}$ (Hayes and Loudon 1978).

For example, for the cubic O_h point group associated with the hexaborides (Section 6.4.2) and the cubic spinels (Section 6.4.3), the irreducible representation T_{1u} transforms like a polar vector; consequently, the allowed Raman symmetries associated with this point group are given by $\Gamma_{Raman}(O_h) = T_{1u} \otimes T_{1u} = A_{1g} + E_g + T_{1g} + T_{2g}$. The Raman tensors associated with these symmetries are shown in Table 6.1. As another example, for the tetragonal D_{4h} point group associated with $Ca_3Ru_2O_7$ (Section 6.4.5), the irreducible representations A_{2u} and E_u transform like polar vectors. The Raman-allowed symmetries for the D_{4h} point group are therefore given by $\Gamma_{Raman}(D_{4h}) = A_{2u} \otimes A_{2u} + E_u \otimes E_u = A_{1g} + A_{2g} + B_{1g} + B_{2g} + E_g$ (see Table 6.1). Note in Table 6.1 that A and B are associated with singly degenerate representations, E is associated with a doubly degenerate representation, and T is associated with a triply degenerate representation. Second, the application of Neumann's principle to the physical quantities involved in the Raman scattering process places constraints on the components of the different Raman tensors, requiring some elements to be zero and requiring others to be related. A complete listing of the allowed Raman tensors for the 32 crystal point groups can be found in Poulet and Mathieu (1976) and Hayes and Loudon (1978), but some examples of the symmetries of the Raman tensors are given in Table 6.1.

The relationship $|\varepsilon_S \cdot (d\chi(\omega)/dX) \cdot \varepsilon_I|$ in Equation 6.5 also shows that one can experimentally couple to different components of the susceptibility tensor—and thereby identify the symmetry of a given excitation—by varying the polarizations of the incident and scattered light relative to the material's crystalline axes. The nomenclature often used to indicate the geometry of a particular Raman scattering experiment is $k_I(\varepsilon_I, \varepsilon_S)k_S$, where k_I and k_S are the wave vectors of the incident and scattered photons, respectively, and ε_I and ε_S are the polarizations of the incident and scattered photons, respectively. For example, $z(x, y)\bar{z}$, indicates a "true backscattering" and "depolarized" scattering geometry in which the incident and scattered wave vectors are directed along the z- and $\bar{z}$-axes of the crystal, respectively, and the incident and scattered photons are polarized along the x- and y-axes, respectively. In the case of EuB_6 (Section 6.4.2) and the cubic spinels (Section 6.4.3), it is easily shown by calculating $(\varepsilon_S \cdot \Gamma_x \cdot \varepsilon_I)$—where Γ_x are the Raman tensors associated with the O_h point group—that the following Raman-allowed excitation symmetries can be accessed in the following scattering geometries:

TABLE 6.1
Allowed Raman Tensors for Cubic (O_h) and Orthorhombic (D_{4h}) Point Groups

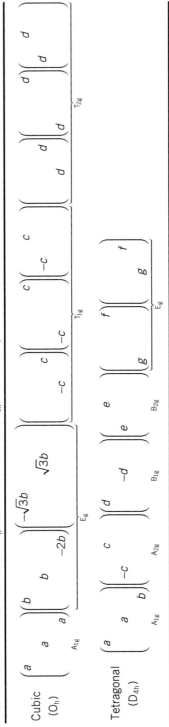

Note: Nonzero elements are shown. Missing elements are zero.

$z(x, x)\overline{z}$: $A_{1g} + E_g$, $z(x, y)\overline{z}$: $T_{1g} + T_{2g}$, $z(x', x')\overline{z} = z(x + y, x + y)\overline{z}$: $A_{1g} + (1/4)E_g + T_{2g}$, and $z(x', y')$ $\overline{z} = z(x + y, x - y)\overline{z}$: $(3/4)E_g + T_{1g}$, where $x = (1,0,0)$, $y = (0,1,0)$, and $z = (0,0,1)$, and where the prefactors indicate the relative intensities of the mode expected in that particular scattering geometry. In practice, the excitation symmetries of different excitations are verified by varying the scattering geometries in the Raman experiment and observing the intensities of the excitations in each geometry. For a description of how these selection rules are used in IXS, see Abbamonte et al. (1999).

6.2.4 Raman-Allowed Phonons and the Correlation Method

In the absence of disorder, the Raman scattering process only couples to a subset of the $q = 0$ phonons, the so-called Raman-allowed modes (Chapter 2). The number and symmetry of $q = 0$ normal modes can be determined using group theoretical methods, and here we describe the "correlation method" for identifying the symmetry of $q = 0$ phonons in a crystal. The "correlation method" (Fateley et al. 1972) involves the following steps: (1) First, for each atom A in the crystal, determine the site symmetry—which describes the subset of symmetry operations the atom satisfies within the larger space group of symmetries satisfied by the Bravais unit cell—using standard crystallographic handbooks (Wyckoff 1964, Henry and Lonsdale 1965); (2) next, for each site group found in (1), identify (using standard character tables from any group theory text, for example, Fateley et al. (1972) or Tinkham (1986)) the irreducible representations of the site group that transform like simple translation vectors $(T_x, T_y,$ and $T_z)$—we are concerned with these representations only, because lattice vibrations involve displacements that transform like translation vectors; and (3) finally, expand—using correlation tables provided in standard group theory textbooks such as Fateley et al. (1972) or Tinkham (1986)—the irreducible representations identified in (2) into the irreducible representations of the full space group, that is, $\Gamma_{phonon} \in \Gamma^{site}(T_x, T_y, T_z) = \sum_R C_R \Gamma_R^{space}$, where Γ^{site} and Γ^{space} are the irreducible representations of the site and space groups, respectively. The symmetries of "Raman-allowed" phonons in which a particular atom A is involved are simply those irreducible representations identified in step (3) that transform like the polarizability tensor α; these representations can be identified using standard character tables. The tensor character of the Raman-allowed modes reflects the two-photon Raman scattering process and the tensor character of $|\varepsilon_S \cdot (d\chi(\omega)/dX) \cdot \varepsilon_I|$ in the Raman scattering cross section (see Section 6.2.2). "Infrared-allowed" modes, on the other hand, are represented by irreducible representations that transform like translation vectors T, reflecting the character of the dipole interaction Hamiltonian for infrared absorption. Consequently, in crystals with a center of inversion symmetry, Raman-allowed phonons can be observed with Raman scattering but not with infrared absorption measurements, while the converse is true for infrared-allowed modes. Finally, modes that are not Raman- or infrared-allowed are called "silent," and must be observed using other techniques, such as neutron scattering. Some examples of the application of the correlation method are described in Section 6.4.

6.3 EXPERIMENTAL DETAILS

The essential experimental components for performing a Raman scattering experiment include an excitation source for generating the incident light, optical elements for directing the incident light to the sample and collecting the scattered light, polarization optics for generating and analyzing polarized incident and scattered light, a spectrometer for filtering the incident light beam and spatially dispersing the scattered light into a flat image on the detector plane, and a detector for counting the scattered photons. Additionally, the typical Raman

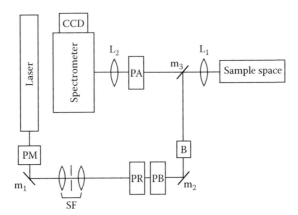

Figure 6.1 Schematic illustration of the Raman scattering system described in Section 6.3.1, showing the prism monochromator (PM), spatial filter (SF), polarization rotator (PR), polarizing beamsplitter (PB), Berek compensator (B), polarization analyzer (PA), lenses (L), mirrors (m), and CCD camera.

system includes equipment for controlling the environment of the sample, including the sample temperature, and less commonly, the pressure, magnetic field, and/or electric field applied to the sample. A schematic of a typical Raman system is illustrated in Figure 6.1. Following is a brief description of the different components of a typical Raman system.

6.3.1 Design of the Basic Raman Scattering System

6.3.1.1 Excitation Sources

Since its development in the 1960s, the laser has been the primary excitation source used in most Raman scattering applications, primarily because of its high power and collimated beam, which can be easily directed to and tightly focused on the sample, even in complicated experimental arrangements. There are numerous laser light sources available for spectroscopy applications (Asher and Bormett 2000). Water-cooled continuous-wave (CW) gas ion lasers such as argon-ion and krypton-ion lasers have been traditionally used as excitation sources. The benefits of gas lasers are that they generally emit light with a stable output, offer high incident powers, provide numerous discrete excitation wavelengths from which to select, and have narrow bandwidths (~1–10 GHz), which is essential for measuring excitations with very small Raman shifts. For example, the krypton-ion laser emits a number of discrete lines with relatively high power (~50–500 mW) in the wavelength range 476–752 nm, while the argon-ion laser emits several discrete lines in the wavelength range 458–542 nm. Helium–neon (633 nm) and helium–cadmium (325 and 442 nm) are also commonly used gas lasers for Raman spectroscopy. A gas laser of sufficient power can also be used to excite (pump) a dye laser—which employs an organic dye as the lasing medium—from which an essentially continuous range of excitation wavelengths can be selected within a narrow "band" of wavelengths. Different wavelength band ranges can be accessed by changing the dye used in the dye laser. The drawbacks of gas lasers include their relatively high cost, bulky size, need for water cooling, and undesired emission of plasma discharge lines that accompany the coherent emission lines.

Semiconductor lasers have become increasingly popular for use in Raman experiments because they are compact, relatively inexpensive, and air-cooled, and do not produce undesirable "plasma lines." However, semiconductor lasers generally have significantly lower output powers than gas lasers (~10–100 mW), they only produce a single excitation wavelength, and they generally have substantially broader bandwidths (~1 THz) than gas lasers, which can inhibit the measurement of excitations with very small Raman shifts. Some common visible-range semiconductor laser materials—and their associated, composition-dependent wavelength ranges—are $Al_{1-x}Ga_xAs$ (780–880 nm), $Al_xGa_yIn_{1-x-y}P$ (630–680 nm), GaSe (590 nm), GaN (400 nm), and ZnS (330 nm).

6.3.1.2 Optical Components

The optical components typically used to direct the incident light to the sample and collect the scattered light are shown in Figure 6.1. A *prism monochromator (PM)* can be used to prevent the plasma discharge lines emitted by a gas laser from entering the spectrometer or exciting the sample. The PM uses a prism to refract the incident light of the laser beam through an exit hole, thereby blocking unwanted emission lines. Alternatively, a holographic "notch" or "pass-band" filter—which transmits the desired emission line of the laser, but rejects light away from the fundamental emission line—can be used (Asher and Bormett 2000). *Polarizers and polarization rotators* (PR) are used to generate and orient the polarization of the incident light in order to couple to particular components of the Raman susceptibility tensor, as discussed in Section 6.2.3. Incident light from laser sources is usually in a well-defined linear polarization state, so generally only a polarization rotator—which can continuously rotate the polarization of a linearly polarized input beam—is needed to orient the polarization along a preferred crystallographic direction of the sample. *Polarization beamsplitters* (PB) are often used in conjunction with polarization rotators to create a clean polarized beam in the incident plane. These beamsplitters—which consist of two right-angle prisms that are glued together—transmit the *p*-polarized component of the light but reflect the *s*-component of light at roughly 90° to the incident beam direction. The ratio of transmitted to reflected components can be greater than 2000:1. A *beam expander and spatial filter* (SF) is sometimes useful between the prism monochromator and polarization rotator to help achieve a uniform Gaussian beam that can be more tightly focused on the sample. A λ/4-plate or *Berek polarization compensator* (B) can be used in the "incident" side of the optical chain to convert linear polarized light to circularly polarized light. The use of circularly polarized light is particularly important in Raman measurements in high-magnetic fields, since linearly polarized light experiences a Faraday rotation in the sample that can complicate the interpretation of magnetic-field-dependent Raman intensity data. Another benefit of a Berek compensator is that it can be used as a variable waveplate, providing λ/4 or λ/2 retardation at any wavelength between 200 and 1600 nm. The final component of the "input optics" of a Raman system is a *focusing lens* (L1), which is generally chosen to be a convex lens that can focus light onto the sample. Using a commercial camera lens for the focusing lens, a laser spot size of ~10–50 μm can be obtained. However, by using a microscope objective as the focusing lens, laser spot sizes of ≤1 μm can be obtained for the study of small samples. Note that Raman experiments on opaque solids are often performed in either a "true backscattering" geometry (Figure 6.1) or a "pseudo-backscattering" geometry, in which the wave vector of the incident light has a small angle to the sample surface normal. In this configuration, the large index of refraction of opaque materials will refract the light inside the sample so that the wave vector of light is nearly perpendicular to the surface normal. The *collection lens* collects the scattered light from the sample and directs a collimated beam toward the spectrometer. This lens is generally a camera lens or high-quality achromatic doublet, chosen to minimize spherical and chromatic aberrations. In the schematic system shown in Figure 6.1, the focusing lens L1 also serves as the collection lens. The *spectrometer focusing lens* (L2) focuses the light collected from the sample onto the spectrometer entrance slit. This lens should be chosen to match the *f*-number of the spectrometer to avoid underfilling (reducing attainable resolution) or overfilling (reducing signal throughput) the spectrometer gratings. A *polarization analyzer* (PA) is used to select the polarization of the scattered beam. By controlling the polarization states of both the incident and scattered light, specific components of the Raman susceptibility tensor can be selected and excitation symmetries can be identified. A polarization analyzer generally consists of a $\lambda/4$ retarder plate and linear polarizer to select a particular polarization direction, followed by a $\lambda/2$ retarder plate to rotate the selected polarization to the direction that maximizes the transmission efficiency of the spectrometer gratings.

6.3.1.3 Spectrometers

The most common system used for spectrally dispersing the scattered light in Raman scattering experiments is a spectrometer. The function of the spectrometer is to separate the relatively weak inelastically scattered light that one would like to detect from the unwanted—but significantly more intense (by $\sim 10^{11}$–10^{14} orders of magnitude)—specularly reflected incident light that is elastically scattered by the sample. A double-grating (two-stage) spectrometer is normally sufficient for measuring inelastically scattered components of the light with energy shifts in excess of roughly $50\,\text{cm}^{-1}$ ($\sim 6\,\text{meV}$). However, many modern Raman scattering systems employ triple-grating (three-stage) spectrometers, which can enable the detection of inelastically scattered light with energy shifts lower than $8\,\text{cm}^{-1}$ ($\sim 1\,\text{meV}$).

Three-stage spectrometers can be configured either in an "additive" arrangement—in which all three gratings spatially disperse the light in the same direction—or in a "subtractive" arrangement—in which the first two gratings spatially disperse the light in opposite directions. In the latter configuration, the first two stages of the spectrometer serve as a non-dispersing filter of the elastically scattered light, while the third grating of the spectrometer disperses the light onto the image plane of the detector (Kuzmany 1998, Asher and Bormett 2000). A more detailed discussion of spectrometers and spectrometer performance can be found in Chapters 4 and 5.

The energy scale typically used when plotting Raman scattering (and infrared) results is the "cm^{-1}," which has the following conversions to other energy units [$c = 3 \times 10^{10}$ cm/sec; $h = 6.63 \times 10^{-34}$ Joules · sec; and $k_B = 1.38 \times 10^{-23}$ Joules/K]:

$$0.03\,\text{THz} = 1\,\text{cm}^{-1}(\text{more generally}, \nu[\text{THz}] = c\bar{\nu}[\text{cm}^{-1}] = 0.03\bar{\nu}[\text{cm}^{-1}]);$$

$$1\,\text{meV} = 8.06\,\text{cm}^{-1}(E[\text{meV}] = hc\bar{\nu}[\text{cm}^{-1}] = 0.124\bar{\nu}[\text{cm}^{-1}]);$$

and

$$1.44\,\text{K} = 1\,\text{cm}^{-1}\left(T[\text{K}] = \frac{hc}{k_B}\bar{\nu}[\text{cm}^{-1}] = 1.44\bar{\nu}[\text{cm}^{-1}]\right).$$

6.3.1.4 Detectors

The most common detectors used in Raman scattering experiments are photomultiplier tube (PMT) and charge-coupled-device (CCD) detectors (Chapter 5). Photomultiplier tubes contain a photocathode with a photoemissive surface, followed by an electron multiplier consisting of a series of electrodes (dynodes), each of which is kept at a more positive potential than its predecessor in the "dynode chain." A photon hitting the PMT photocathode causes an electron to be emitted; this electron is accelerated to the first of the dynodes, resulting in the emission of more electrons, which are further amplified by the dynode chain, generating a large current on the order of 10^6 electrons.

A more efficient, multichannel, method for light detection in Raman scattering involves a CCD detector (Janesick 2001). A CCD is an analog solid-state image sensor consisting of a 2D array of pixels typically consisting of doped silicon. The dimensions of each pixel—typically on the order of $20 \times 20 \mu m^2$—define the smallest linear dispersion that can be detected, and therefore provide a limit on the spectrometer resolution. Photons striking the pixels with energies greater than the band gap of silicon generate electron–hole pairs, causing the accumulation of an electric charge in a pixel that is proportional to the number of photons striking that pixel. After exposing the array of pixels to light for a period of time, the charge is transferred to the neighboring pixel in a "shift register" operation; this process is repeated until the charge is transferred to the last pixel in the array, which transfers its charge to a charge amplifier that converts the charge to a voltage. A CCD detector provides several benefits in comparison to PMT detectors in Raman scattering measurements, including the ability to record the entire scattered light spectrum at once, rather than sequentially; greater quantum efficiency (70%–90%) than a PMT detector (<30%); and relatively low noise. CCD detectors also have a fairly broad spectral range, 300–1000 nm.

6.3.1.5 Calibration

To obtain the Raman response function from the signal recorded using the most commonly used Raman system—that is, a spectrometer used in conjunction with a CCD detector—two corrections to the data are important. The first concerns the energy axis calibration: spectrometers disperse the scattered signal across the CCD detector as a linear function of wavelength, but results are typically plotted as a function of energy. The CCD pixel number to wavelength conversion is readily obtained by placing a calibrated lamp source—for example, neon or krypton lamps, which emit known spectra of sharp, well-characterized emission lines—at the entrance slit of the spectrometer. By measuring the emission of such a calibrated source, every pixel position on the CCD can be converted to a wavelength value for the incident photons. A second calibration associated with the Raman intensity (y-axis) is also necessary to account for the wavelength dependence of the collection optics and spectrometer. For the intensity calibration, a light source that has been calibrated by the manufacturer (the source's "calibration profile") is placed in front of the collection optics, and the measured profile of the light source is obtained at the detector. The "system response function" (SRF) is obtained by dividing the measured profile (measured in wavelength units) of the calibration source by its calibration profile. The Raman response function (RRF) from a sample can then be obtained by dividing the measured signal from a sample by the system response function. Note also that when converting the Raman spectrum to energy units, one also needs to account for the fact that, while the wavelength interval does not depend on pixel position, the energy interval does, according to $\Delta\omega \propto \Delta\lambda/\lambda^2$; consequently, the scattered response must be weighted by an extra factor of λ^2 to account for the variation of the energy interval $\Delta\omega$ across the CCD detector. In sum, the Raman response function in energy units is obtained from the measured spectrum (MS) using the following correction, RRF = (MS/SRF) * λ^2.

6.3.2 High-Pressure Methods

High-pressure Raman and x-ray measurements have proven to be very effective methods for studying phononic, electronic, and magnetic excitations in complex materials, ranging from cuprates (Goncharov and Struzhkin 2003), ferrites (Adler et al. 1994), manganites (Loa et al. 2001, Postorino et al. 2002), ruthenates (Snow et al. 2002, Snow 2003, Gupta et al. 2006, Karpus et al. 2006), vanadates (Kremer et al. 2000), pyrochlores (Saha et al. 2009), charge-density-wave systems (McWhan et al. 1980, Snow 2003, Snow et al. 2003, Lavagnini et al. 2008), and spin-Peierls systems (Goni et al. 1996). In the following, we discuss some

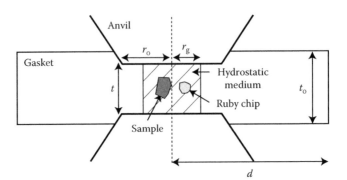

Figure 6.2 Illustration of anvil cell (AC) described in Section 6.3.2.

of the essential features of an anvil cell (AC), schematically illustrated in Figure 6.2, which is the most common pressure transmitting device used for Raman scattering (Dunstan and Scherrer 1988).

6.3.2.1 Anvils

An anvil cell contains a pair of *anvils*, which exert pressure on the sample (indirectly through some *pressure transmitting medium*, generally [Section 6.3.2.3]) and serve as the window through which the incident and scattered photons travel to and from the sample. The three most common anvil materials used in optical experiments are diamond (Eremets 1996), sapphire (Takano and Wakatsuki 1991), and moissanite, that is, $6H$-SiC (Xu and Mao 2000, Snow 2003). Several characteristics of anvils need to be considered when performing Raman experiments:

Hardness of the anvils: The hardness of the anvil material governs, in part, the highest accessible pressures for different possible anvil materials. The highest reported pressures have been obtained with diamond anvils (~2000 kbar = 200 GPa), but pressures exceeding 300 kbar have been attained with moissanite anvils (Xu and Mao 2000) and pressures close to 150 kbar have been achieved using sapphire anvils (Patselov et al. 1989).

Optical properties of the anvils: In addition to generating the pressure in an anvil cell, the anvils serve as the window through which the incident and scattered photons are transmitted. Consequently, it is crucial to consider the optical properties of the anvil material when performing high-pressure optical measurements. The optical transmissions of diamond, sapphire, and moissanite are relatively high throughout the visible wavelength range: between 400 and 800 nm, the transmission of diamond varies between 60% and 75%, the transmission of sapphire is roughly 80%, and the transmission of moissanite varies between 80% and 95%, although the specific details of the transmission will depend somewhat on the impurity content of the material used (Snow 2003). Of additional concern, strong fluorescence from the anvils—which primarily arises from impurities in the anvil material—can easily overwhelm the weak Raman signal from the sample; consequently, it is important to minimize anvil fluorescence as much as possible. This can be achieved by using a different excitation wavelength, or by using higher quality—and therefore more expensive—anvil materials (Adams and Sharma 1977). In particular, synthetic anvils are generally more chemically pure and can exhibit substantially lower background fluorescence than natural anvil materials (Liu and Vohra 1996). Finally, the anvils have their own Raman signatures, associated with vibrational modes in the anvil material, which can also overwhelm Raman modes from the sample. The main Raman modes from different anvil materials are as follows: diamond, ~1330 cm^{-1}; (Goncharov et al. 1997) sapphire, 378, 418, 432, 451, 578, 645, and 751 cm^{-1}; (Kadleikova et al. 2001) and moissanite, 156, 245, 273, 511, 770, 792, and 966 cm^{-1} (Snow 2003).

Design and mounting of the anvils: Diamond anvils used for high-pressure research are typically cut with 16 facets, a large back surface of the anvil (the "table"), and a small front, or working, surface of the anvil (the "culet") (Snow 2003). To attain the highest pressures and to avoid breaking the anvil, it is important to have a small bevel around the culet (Moss and Goettel 1987). However, this design is not suitable for anvils made from softer materials such as moissanite and sapphire. In fact, cracks in sapphire anvils tend to originate at the facets or other sharp edges (Patselov et al. 1989); consequently, sapphire and moissanite anvil designs ideally involve conical shapes with no faceting, and beveling of the table and culet edges (Snow 2003).

6.3.2.2 Gaskets

Anvil cells also incorporate metal gaskets (see Figure 6.2), which not only contain the fluid that serves as the medium for transmitting quasi-hydrostatic pressure to the sample, but also provide important massive support to the edges of the anvil, allowing higher pressures to be generated. Dunstan and Spain (1989) studied the plastic deformation of a thin cylindrical gasket with a sample hole when compressed between two anvils, and showed that higher hydrostatic pressures are achievable for gaskets that have (1) a higher compressive strength, (2) a smaller thickness (t) relative to the anvil culet radius (r_o), that is, a smaller t/r_o, and (3) a larger coefficient of friction (Dunstan and Spain 1989, Snow 2003).

The massive support of the anvils provided by the gasket is improved if the gasket is "pre-indented" prior to a high-pressure experiment, which involves placing a flat gasket between the anvils and pressing the anvils together to bring the center of the gasket to the desired final thickness t (Figure 6.2). Once the gasket is compressed to the desired thickness, a hole for the sample is drilled in the center of the gasket using an electric discharge machine (EDM). Dunstan and Spain (1989) suggest that the gasket parameters (Figure 6.2) have the approximate ratio $(t:r_g:r_o:t_o:d_o) = (1:3:10:10:100)$. Eremets (1996) and Snow (2003) list common gasket materials. BeCu is probably the best gasket material at low temperatures, as this material has a high thermal conductivity, becomes harder and more plastic at low temperatures, and has a low electrical conductivity. On the other hand, while steels and polymers can be suitable for room temperature pressure measurements, they are less suitable for low-temperature measurements because their plasticity and impact toughness decrease at low temperatures.

6.3.2.3 Hydrostatic Pressure Media

The purpose of the hydrostatic medium is to uniformly distribute the pressure around the sample. It is important to note, however, that no pressure medium provides completely hydrostatic pressure transfer from the anvils, and all hydrostatic media undergo liquid-to-solid phase transitions at sufficiently low temperature and high pressure. Jayaraman (1983) and Snow (2003) list a number of common quasi-hydrostatic pressure media used in high-pressure Raman experiments. More extensive plots of the pressure–temperature phase diagrams for these media can be found in Eremets (1996). Of particular note, helium and argon liquid provide fairly hydrostatic pressure transfer over a wide range of temperature and pressure: Burnett et al. (1990) found that helium provides hydrostatic pressure transfer even at low temperatures, while argon becomes increasingly non-hydrostatic when increasing pressure at low temperatures. However, studies by Venkateswaran et al. (1986) and Snow et al. (2002, 2003) suggest that the non-hydrostatic behavior of argon is fairly small (~1% of the pressure applied) at low temperatures for pressures <100 kbar. While helium provides better hydrostatic pressure transfer than other pressure media, it is generally more complicated to "load" (i.e., to introduce into the anvil cell) liquid helium than other pressure media because of its lower condensation temperature (4 K).

6.3.3 High-Magnetic-Field Methods

Numerous studies have demonstrated the benefits of high-magnetic-field Raman scattering studies of excitations in complex materials, such as charge-density-wave superconductors (Sooryakumar and Klein 1981), high-T_C superconductors (Blumberg et al. 1997, Qazilbash et al. 2005), dilute magnetic semiconductors (Ramdas and Rodriguez 1991), ferromagnetic/polaronic materials (Nyhus et al. 1997, Snow et al. 2001, Rho et al. 2002), colossal magnetoresistance materials (Liu et al. 1998, Romero et al. 1998), antiferromagnetic materials (Fleury et al. 1971, Benfatto et al. 2006), and charge- and orbital-ordering systems (Karpus et al. 2004, 2006, Kim et al. 2008).

A common method for generating a high-magnetic field in Raman scattering measurements is to use a split-coil superconducting magnet cryostat with optical access, which consists of a superconducting solenoid for generating the high-magnetic field, an inner sample space that positions the sample in the center of the magnet coils, a helium reservoir that provides cooling for both the sample space and the superconducting solenoid, and a liquid nitrogen jacket and vacuum space that helps thermally insulate the cold inner chambers from the outside environment. These magnet systems provide optical access to the sample in several directions, allowing high-magnetic-field (typically up to ~11 T) measurements to be readily made in both Faraday ($k \| H$) and Voigt ($k \perp H$) geometries, where k is the wave vector of the incident light and H is the magnetic field direction. A drawback of split coil magnets is that they require a relatively large separation between the sample and outside collection optics, which increases the f-number (typically $f/2.4 - f/5.9$) and reduces the scattered light collected from the sample. Another limitation is that the cryogenics for the superconducting solenoid and the sample space are closely coupled in these systems, making it more difficult to perform field-dependent measurements at high temperatures. One can separate the magnet and sample cooling in high-field Raman scattering by using an open bore magnet into which a separate optical cryostat is inserted. This arrangement not only separates the magnet and sample cryogenics, but also provides greater flexibility for placing optics close to the sample space, allowing f-numbers as low as $f/1.4$ to be attained. A limitation of this configuration is that it provides optical access in the Faraday ($k \| H$), but not the Voigt ($k \perp H$), geometry. However, a Voigt ($k \perp H$) geometry can be readily obtained using an open bore magnet by mounting the sample "sideways" in the magnet and inserting a 45° mirror near the sample (Karpus et al. 2004, 2006).

6.3.4 Inelastic X-Ray Scattering

Inelastic x-ray scattering (IXS or "x-ray Raman scattering") involves a process analogous to optical Raman scattering, but employs photons with wavelengths in the x-ray regime, $\lambda \sim 1$ Å. Consequently, this method generally involves the excitation of deep core electrons and can probe wave vectors throughout the Brillouin zone. Thus, this method is capable of measuring the full q- and ω-dependence of the Raman response function, Im $\chi(q, \omega)$, of collective excitations such as phonons (Burkel et al. 1991), plasmons (Schülke et al. 1988), and magnons (Hill et al. 2008, Schlappa et al. 2009). Additionally, if an excitation is sampled over a sufficiently broad range of momentum and energy, then its temporal response (i.e., the "propagator") can be reconstructed—via a Kramers–Kronig transformation—with angstrom spatial resolution and attosecond time resolution; application of this method to the charge propagator in graphite is discussed in Section 6.4.6 (Abbamonte et al. 2004, 2008).

Inelastic x-ray scattering experiments follow the same principles as Raman experiments, though the details of the experimental setup are quite different (Figure 6.3). The photon source in IXS is not a laser, but an undulator insertion device at a third-generation

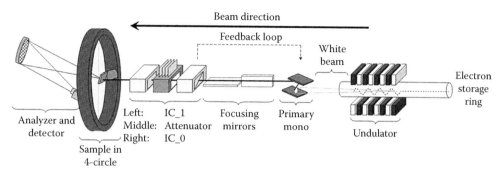

Figure 6.3 Schematic of a moderate energy resolution inelastic x-ray scattering setup, such as that at Sector 9 at the Advanced Photon Source. A synchrotron beam passes through an undulator, creating a broadband beam of x-rays. The beam then passes through a crystal monochromator and is focused on the sample by a pair of grazing-incidence mirrors. The beam position and intensity is monitored by a series of split ion chambers, which serve a real-time feedback loop that compensates for drifts. The sample lies in the center of a goniometer that allows adjustment of three Euler angles. The scattered x-rays are energy-analyzed with a backscattering analyzer.

synchrotron facility. Some examples of such facilities are the Advanced Photon Source at Argonne National Laboratory, the European Synchrotron Radiation Facility in Grenoble, France, or SPring-8 in Hyogo, Japan. The light emitted by a hard x-ray undulator has a bandwidth of typically 50–100 eV, so the emission must be made monochromatic. This is normally accomplished with a Si or diamond crystal monochromator. Because the integrated power of an undulator source is extremely high, the crystals are usually water- or liquid nitrogen-cooled. High-energy resolution (~1 meV) can be achieved with combinations of multiple, nested crystals (Bartels 1983).

IXS experiments may be done in either reflection or transmission geometry, and the scattered light is analyzed with a spherical, diced, backscattering crystal, which is mounted on a rotatable detector arm. This analyzer back-reflects the x-rays onto a detector located near the sample. The advantage of the backscattering geometry is that, at this angle, Bragg's law has its minimum dispersion, so the effects of aberrations and other misalignments on the overall energy resolution are minimized. The energy transfer or "Raman shift" can either be tuned by rotating the incident monochromator or the analyzer. The momentum transfer is adjusted via a coordinated rotation of the analyzer arm and the sample.

6.4 ILLUSTRATIVE CASE STUDIES OF RAMAN SCATTERING IN NOVEL MATERIALS

6.4.1 Raman Scattering from Phonons in Graphene

6.4.1.1 Graphene and Graphite: Introduction

Graphite is a semimetal composed of weakly coupled planes of carbon. The carbon in each plane is arranged in a honeycomb lattice, with four atoms per unit cell; the atoms hybridize to form sp^2 bonds, which cause π-bonds to form above and below the carbon plane. Incident photon energies in the visible frequency range are resonant with these π states, resulting in several strong Raman phonon and defect-related modes in graphite (Ferrari 2007). Raman spectroscopy has thus proved an excellent method of characterizing the structure of graphite-based materials, in addition to providing spectroscopic information about electron–phonon coupling (Dresselhaus et al. 2000). Further, the recent discovery of isolated single sheets of graphite, or graphene, has generated additional excitement about using Raman scattering to study these materials (Geim 2009). Graphene is one atomic layer thick and has unusual electronic properties, particularly an electronic spectrum akin to that of Dirac fermions (Geim 2009). Raman spectroscopy in graphene has proven valuable for (1) efficiently and noninvasively determining the layer thickness (Ferrari et al. 2006); (2) determining stress and crystal orientation (Huang et al. 2009); (3) monitoring doping levels (Das et al. 2008); and (4) examining quantum electrodynamics-like effects, such as the breakdown of the Born–Oppenheimer approximation in Dirac fermion–phonon interactions (Pisana et al. 2007). A case study of Raman scattering from graphene serves as a nice illustration of phonon Raman scattering, defect-induced scattering, and resonance enhancement effects.

6.4.1.2 Phonon Raman Scattering in Graphene

The correlation method described in Section 6.2.4 can be used to identify the Raman-allowed phonons in graphene. Graphene has the point group D_{6h}, with the C ions having site symmetries D_{3h}. Standard character tables identify A_2'' and E' as the irreducible representations of D_{3h} that transform like translation vectors. Using standard correlation tables, one can see that A_2'' correlates with the following irreducible representations of the full D_{6h} point group, $A_2'' \Rightarrow A_{2u}$ and B_{2g}, while E' correlates with the following representations of the full D_{6h} point group, $E' \Rightarrow E_{1u}$ and E_{2g}. Consequently, the $q = 0$ phonon modes of graphene are given by $\Gamma_{\text{phonons}}^{\text{graphene}} = A_{2u} + B_{2g} + E_{1u} + E_{2g}$. Of these representations, only E_{2g} transforms like a polarizability tensor according to the character table for the D_{6h} point group. Thus, there should be one doubly degenerate $q = 0$ Raman-allowed phonon mode in graphene, $\Gamma_{\text{Raman}}^{\text{graphene}} = E_{2g}$, which involves motions of the carbon ions within the graphene plane. The A_{2u} and E_{1u} modes transform like translation vectors, and hence are infrared allowed modes, while B_{2g} is a "silent" mode.

Figure 6.4 shows Raman scattering spectra for disordered (top) and clean (bottom) single-layer graphene, obtained using 488 nm incident light from an argon laser on samples that were mechanically exfoliated onto Si/SiO$_2$ substrates. The "clean" graphene sample (bottom plot, Figure 6.4)—which was treated to a hydrogen cleaning process before measurement—exhibits

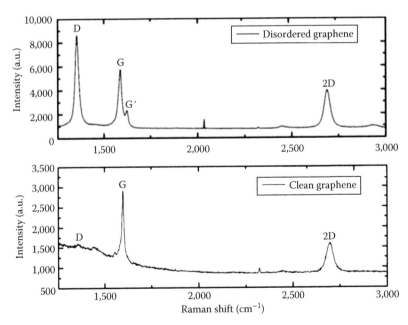

Figure 6.4 Phonon Raman scattering spectra of single-layer graphene at $T = 300\,$K for disordered (top) and clean (bottom) samples, obtained with linearly polarized incident light but unpolarized collected light.

the expected E_{2g} symmetry "G band" at 1560 cm^{-1}. The E_{2g} symmetry of this mode is evident from the appearance of this mode in both "parallel" ($\varepsilon_I \| \varepsilon_S$) and "crossed" ($\varepsilon_I \perp \varepsilon_S$) scattering geometries, consistent with the symmetry of the E_{2g} Raman tensor (Section 6.2.3). On the other hand, in the highly disordered sample (top plot, Figure 6.4)—which was not cleaned and had visible tears on the surface—two additional peaks are observed: first, a strong disorder-induced mode ("D" mode) is apparent near ~1360 cm^{-1}; this defect-induced mode is associated with the disorder-induced breakdown of the kinematical constraint discussed in Section 6.2.1, and arises from phonon scattering from the TO branch near the K point of the Brillouin zone (Reich and Thomsen 2004). Interestingly, the energy of the D mode is dependent upon the incident wavelength; this has been attributed to a "double-resonant Raman process" in which the incident photon first resonantly excites an electron between the π bands of graphene, and then a phonon resonantly scatters the electron between the π bands. Second, another disorder-induced peak (G′)—with origins similar to the D peak, but associated with phonon scattering from the LO branch near the Γ point—is also evident near 1622 cm^{-1} in the disordered sample. Finally, both the clean and disordered graphene samples show a second harmonic of the D peak near 2700 cm^{-1} (2D)—which also arises from double resonance electron–phonon scattering (Ferrari et al. 2006).

6.4.2 Raman Scattering from Phonons, Plasmons, and Polarons in EuB$_6$

A nice demonstration of the range of excitations and phases that can be effectively explored using Raman scattering is found in studies of EuB$_6$, which is a low-carrier-density magnetic semiconductor with a cubic (O_h^1) crystal structure that exhibits a transition from paramagnetic semiconductor to ferromagnetic metal phases near $T_C = 12$ K (Snow et al. 2001).

6.4.2.1 Phonon Raman Scattering in EuB$_6$

Hexaborides (RB$_6$) such as EuB$_6$ crystallize in the cubic space group O_h^1, with the Eu (R) ions having a site symmetry O_h and the B ions having a site symmetry C_{4v} (inset of Figure 6.5). The Eu ions sit at a site of inversion symmetry in the crystal, and thus are not involved in Raman-active modes. This is confirmed using the correlation method (Section 6.2.4): the $q = 0$ phonon modes of the Eu ions in EuB$_6$ are $\Gamma_{phonons}^{Eu} = T_{1u}$, which transforms like a translation vector in the O_h point group, so the Eu ions are involved in infrared-active, but not Raman-active, phonon modes. On the other hand, for the B ions in EuB$_6$, the irreducible representations A_1 and E of the C_{4v} site group transform like translation vectors according to the character tables. Using standard correlation tables, one sees that A_1 correlates with the following irreducible representations of the full O_h point group, $A_1 \Rightarrow A_{1g}$, E_g, and T_{1u}, while E correlates with the following representations of the full O_h point group, $E \Rightarrow T_{1g}$, T_{2g}, T_{1u}, and T_{2u}. Consequently, the $q = 0$ phonon modes of the B ions in EuB$_6$ are given by $\Gamma_{phonons}^{B} = A_{1g} + E_g + T_{1g} + T_{2g} + 2T_{1u} + T_{2u}$. Of these representations, only A_{1g}, E_g, and T_{2g} transform like polarizability tensors according to the character tables for the O_h point group. Therefore, we expect three Raman-allowed phonon modes for EuB$_6$, with symmetries $\Gamma_{Raman} = A_{1g} + E_g + T_{2g}$, all of which involve displacements of the boron ions. The triply degenerate T_{1u} mode is infrared-allowed and the other modes are "silent."

The number and symmetry of Raman-allowed modes in EuB$_6$ is experimentally confirmed in Figure 6.5, which shows the three Raman-allowed phonons of EuB$_6$ at energies of 762 cm^{-1} (T_{2g}), 1098 cm^{-1} (E_g), and 1238 cm^{-1} (A_{1g}) (Nyhus et al. 1997, Snow et al. 2001). As illustrated in Figure 6.5, the phonon symmetries can be verified by examining the phonon intensities in the (x',x') and (x',y') scattering geometries discussed in Section 6.2.3. Note that the 5% shift of these modes to higher energy in ^{10}B-substituted EuB$_6$ (Figure 6.6), consistent with the mass ratio $\sqrt{11/10}$, confirms that these modes are associated with B vibrations.

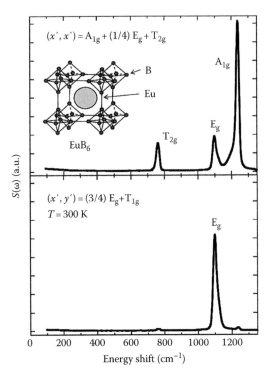

Figure 6.5   Phonon Raman scattering spectra of EuB_6 at $T = 300\,K$ and in (top) $(x', x') = A_{1g} + (1/4)E_g + T_{2g}$ and (bottom) $(x', y') = (3/4)E_g + T_{1g}$ scattering geometries. (After Snow, C.S. et al., *Phys. Rev. B*, 64, 174412, 2001.)

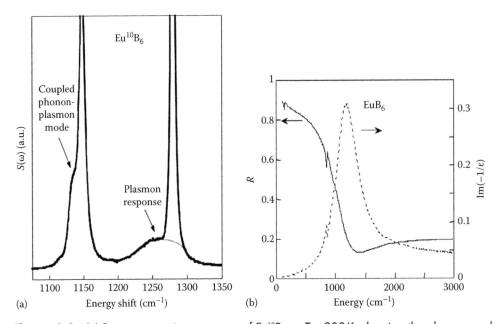

Figure 6.6   (a) Raman scattering spectrum of $Eu^{10}B_6$ at $T = 300\,K$, showing the plasmon and coupled plasmon–phonon modes. The dashed line is a guide to the eye. (b) Room temperature reflectance and loss function, $Im(-1/\varepsilon)$, data for EuB_6, illustrating the plasmon response obtained from optical reflectivity data. (After Snow, C.S. et al., *Phys. Rev. B*, 64, 174412, 2001.)

6.4.2.2 Electron and Plasmon Raman Scattering in EuB$_6$

Inelastic light scattering from itinerant electrons can also be observed in doped materials via light scattering from Fermi surface fluctuations that are induced by electronic excitations at the Fermi surface (Kosztin and Zawadowski 1991, Strohm and Cardona 1997, Cooper 2001a, Devereaux and Hackl 2007). In materials without low energy, interband transitions, or substantial Fermi surface anisotropy, these mass fluctuations are dominated by isotropic fluctuations, corresponding to ordinary charge density fluctuations. Electronic Raman scattering from these isotropic fluctuations provides a measure of the $q = 0$ density-density correlation function $\langle \rho(\boldsymbol{q})\rho^*(\boldsymbol{q}) \rangle$, which is related to the loss function $\mathrm{Im}(-1/\varepsilon(\boldsymbol{q}, \omega))$ (Pines 1963) (Chapter 1) and has a Raman scattering cross section given by (Klein 1975, Hayes and Loudon 1978)

$$\frac{\mathrm{d}^2\sigma}{\mathrm{d}\Omega \mathrm{d}\omega_S} \sim (1 + n(\omega)) r_o^2 \frac{\omega_S V}{\omega_I} (\boldsymbol{\varepsilon}_I \cdot \boldsymbol{\varepsilon}_S)^2 \frac{\hbar q^2}{(4\pi e)^2} \mathrm{Im}\left[\frac{-1}{\varepsilon_o(q, \omega)} \right], \qquad (6.9)$$

where

    $(1 + n(\omega))$ is the thermal factor
    $r_o = e^2/mc^2$ is the classical electron radius
    $\boldsymbol{\varepsilon}_I$ and $\boldsymbol{\varepsilon}_S$ are the incident and scattered photon polarizations
    $\varepsilon_o(q, \omega)$ is the dielectric function

6.4.2.2.1 Plasmon Scattering

At high energies, the loss function in Equation 6.9 is dominated by the plasmon response. In metals with large free carrier densities, bulk plasmon excitations are far too high in energy to be observed by conventional Raman scattering; for example, the plasmon energies $\hbar\omega_p(=4\pi\hbar ne^2/m)$ for some alkali metals are ~8 eV for Li, 6 eV for Na, and 4 eV for K. However, in doped semiconductors and semimetals such as EuB$_6$, the carrier densities can be sufficiently low (i.e., with $\hbar\omega_p < 0.2$ eV) to observe the plasmon response in the range of energy shifts typically probed in Raman scattering, that is, 8–2000 cm^{-1} (i.e., 1–250 meV). For example, Figure 6.6a shows the plasmon response near 1200 cm^{-1} in the Raman scattering spectrum of EuB$_6$; the position of the plasmon peak corresponds very well to the measured loss function of EuB$_6$, shown for comparison in Figure 6.6b. Note that Equation 6.9 predicts that the plasmon response should be dominant in "polarized" scattering geometries, that is, those for which $\boldsymbol{\varepsilon}_I\|\boldsymbol{\varepsilon}_S$, which is consistent with observations (Klein 1975). Similar plasmon excitations have also been observed in Raman scattering measurements of other low-carrier-density materials, including n-type GaAs (Mooradian and McWhorter 1967) and n-type InSb (Patel and Slusher 1968). It should also be noted that in materials with plasmon energies close to phonon energies, a strong interaction can result between the plasmon and longitudinal-optical (LO) phonons, because plasmon oscillations generate a longitudinal electric field; this can lead to phonon–plasmon mode mixing and mutual repulsion of phonon and plasmon energies (Klein 1975) an example of which is also observed in EuB$_6$ (Figure 6.6a).

6.4.2.2.2 Free Electron Scattering

In addition to the higher-energy plasmon and phonon modes, Figure 6.7 shows that EuB$_6$ also exhibits a broad low-energy Raman scattering response that is associated with scattering from itinerant electrons. To understand the basis for this interpretation—and the origins of this electronic scattering response—first note that the low energy contribution to the loss function $\mathrm{Im}(-1/\varepsilon(\boldsymbol{q}, \omega))$ in Equation 6.9 arises from single-particle electronic excitations, which should exhibit a linear-in-ω frequency dependence and a high-energy cutoff near qv_F, where q is the electron momentum

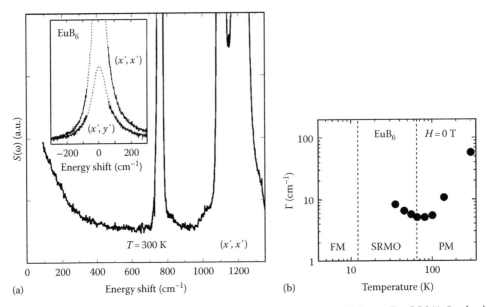

Figure 6.7 (a) Diffusive electronic Raman scattering response in EuB_6 at $T = 300\,K$. Dashed lines in inset show fits to the data using the collision-dominated scattering response expression in Equation 6.10. (b) Temperature dependence of the scattering rate Γ for the collision-dominated scattering response, obtained from fits to the EuB_6 spectra using Equation 6.10. The scattering rate shows a crossover from the paramagnetic (PM) regime to the short-range magnetic order (SRMO) regime. (After Snow, C.S. et al., *Phys. Rev. B*, 64, 174412, 2001.)

and v_F is the Fermi velocity of the electrons (Pines 1963). However, because they involve a change in charge density, isotropic density fluctuations are strongly screened by the long-range Coulomb interaction, which pushes much of the spectral weight associated with the density fluctuations up to the plasma frequency; consequently, one does not typically observe a strong scattering response associated with charge density fluctuations in conventional metals.

To understand the strong electronic scattering response observed at low energies in EuB_6 (Figure 6.7a) and other complex materials, first note that materials with low-energy interband transitions or with complex (e.g., anisotropic or multi-sheeted) Fermi surfaces also exhibit non-isotropic contributions to the mass density fluctuations. Because these non-isotropic mass fluctuations do not involve a net change in the charge density, they are not screened by the long-range Coulomb interaction, and consequently, their light scattering intensities can be significant (Kosztin and Zawadowski 1991, Cooper 2001a, Devereaux and Hackl 2007). Kosztin and Zawadowski (1991) also pointed out that strong intraband electronic scattering in complex materials requires impurity scattering, inelastic scattering, or strong electron–electron interactions to provide momentum conservation in the scattering process. Thus, the Raman scattering response associated with non-isotropic density fluctuations is not limited by momentum conservation to energies less than qv_F.

Notably, the low-energy electronic scattering response observed in EuB_6 (Figure 6.7a) exhibits a distinctive "simple relaxational" scattering intensity that decreases with frequency according to $S(\omega) \sim 1/\omega$. This electronic Raman scattering response was first explained by Ipatova et al. (1981) when analyzing "intervalley" electronic Raman scattering in doped semiconductors such as Ge and Si. Ipatova et al. pointed out that the electronic mean free path ℓ in many materials is shorter than the optical penetration depth δ, that is, $\ell < \delta \sim q^{-1}$ (or equivalently, the carrier scattering rate $\Gamma > qv_F$). Under these conditions, the electronic Raman scattering cross section is given by a "collision-dominated" scattering response (Ipatova et al. 1981, Zawadowski and Cardona 1990):

$$\frac{d^2\sigma}{d\Omega d\omega_S} \sim (1 + n(\omega))|\gamma_L|^2 \frac{\omega \Gamma_L}{\omega^2 + \Gamma_L^2}, \tag{6.10}$$

where

γ_L is the Raman scattering vertex associated with a particular symmetry (i.e., irreducible representation of the material's point group)

Γ_L is the carrier scattering rate associated with electrons located at regions of the Fermi surface having symmetry L

Consequently, by measuring the electronic scattering response associated with different symmetries, that is, by varying ε_I and ε_S appropriately, one can in principle isolate the carrier scattering rates associated with different portions of the Fermi surface (Zawadowski and Cardona 1990, Opel et al. 2000).

The collision-dominated scattering response describes very well the low-energy electronic scattering observed in EuB_6—as illustrated by the dashed-line fit to the data in Figure 6.7 (Nyhus et al. 1997, Snow et al. 2001)—and also has been used to describe electronic Raman scattering observed in a variety of other complex materials, including $Sr_{1-x}La_xTiO_3$ (Katsufuji and Tokura 1994), the manganese perovskites (Liu et al. 1998, Yoon et al. 1998), the nickel perovskites (Yamamoto et al. 1998), EuO (Snow et al. 2001, Rho et al. 2002), the high-T_C cuprates (Cooper 2001, Devereaux and Hackl 2007), spin-ladder compounds (Blumberg et al. 2002), and layered ruthenates (Kirillov et al. 1995, Yamanaka et al. 1996). In many materials, the scattering rate Γ in collision-dominated scattering is governed by electronic scattering from static impurities (Zawadowski and Cardona 1990, Katsufuji and Tokura 1994). However, in ferromagnetic semiconductors exhibiting strong magnetoresistance, such as EuB_6 (Nyhus et al. 1997, Snow et al. 2001), EuO (Snow et al. 2001, Rho et al. 2002), and the manganese perovskites (Liu et al. 1998, Yoon et al. 1998), the electronic scattering rate appears to be dominated by spin-fluctuation scattering over a substantial temperature range near the Curie temperature, T_C. This is evidenced, for example, by the scaling of the scattering rate with the magnetic susceptibility, $\Gamma \propto T\chi$, for $T < 100\,K$ in EuB_6 (Figure 6.7) and by the strong magnetic field-dependence of the collision-dominated Raman scattering intensity in EuB_6 (Nyhus et al. 1997, Snow et al. 2001), EuO (Snow et al. 2001, Rho et al. 2002), and the manganese perovskites (Liu et al. 1998). These results demonstrate that electronic Raman scattering can be used as a contactless method for probing magnetotransport in magnetic materials.

6.4.2.3 Spin-Flip Raman Scattering from Magnetic Polarons in EuB_6

Figure 6.8 shows that the transition from the paramagnetic semiconducting/semimetal phase to the ferromagnetic metal phase in EuB_6 is clearly evident in the electronic Raman scattering response as a change from a collision-dominated scattering response above T_C to a flat "continuum" of electronic scattering, similar to that observed in correlated metals such as the high-T_C cuprates, ruthenates, and titanates (Cooper 2001a, Devereaux and Hackl 2007). This crossover behavior is consistent with the observation in optical reflectivity of an abrupt shift of the plasma frequency to higher energies upon cooling below T_C in EuB_6 (Degiorgi et al. 1997). However, the Raman spectrum of EuB_6 reveals a subtlety in the phase behavior between the collision-dominated regime of the paramagnetic phase and the ferromagnetic metal phase below T_C: in the narrow temperature range $T_C < T < T^* \sim 50\,K$, a broad inelastic peak develops between 50 and 100 cm^{-1} (Figure 6.8); this inelastic peak only appears in "depolarized"

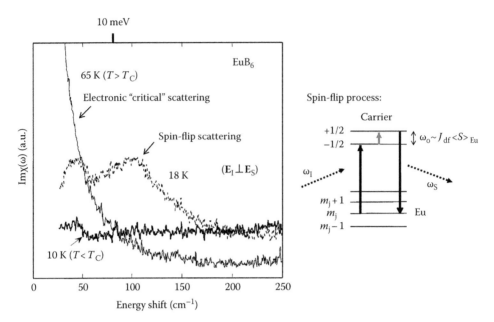

Figure 6.8 Crossover from the collision-dominated Raman response in the paramagnetic phase (65 K) to the flat electronic continuum Raman response in the ferromagnetic metal phase (10 K), via an intermediate spin-flip Raman response (18 K) indicative of the formation of spin polarons. The spin-flip response is seen only in the "depolarized" ($\mathbf{E}_I \perp \mathbf{E}_S$) scattering geometry, as described in Section 6.4.2.3. The illustration at right shows the spin-flip Raman process described in Section 6.4.2.3. (After Snow, C.S. et al., *Phys. Rev. B*, 64, 174412, 2001.)

($\mathbf{\varepsilon}_I \perp \mathbf{\varepsilon}_S$) scattering geometries, indicating that the symmetry of this excitation transforms like the totally antisymmetric representation of the EuB_6 point group, T_{1g}. Because the antisymmetric representation T_{1g} has the transformation properties of an axial vector—like spin and magnetization—this distinctive scattering response can be associated with spin-flip Raman scattering associated with the formation of magnetic polarons (Ramdas and Rodriguez 1991, Nyhus et al. 1997, Snow et al. 2001), which are ferromagnetically polarized clusters of spins to which electrons can become bound via spin exchange. Spin-flip Raman scattering due to the presence of magnetic polarons has been observed in dilute magnetic semiconductors such as $Cd_{1-x}Mn_xSe$ and $Cd_{1-x}Mn_xTe$ (Dietl et al. 1991, Ramdas and Rodriguez 1991) and in dense magnetic materials such as EuB_6 and EuO (Nyhus et al. 1997, Snow et al. 2001, Rho et al. 2002). The spin-flip Raman scattering process is depicted for the specific case of EuB_6 in Figure 6.8: the effective magnetic field of the polarized Eu^{2+} spins splits the spin-up and spin-down levels of the bound electron by an amount $\sim J_{df} \langle S \rangle_{Eu}$, where J_{df} is the exchange coupling constant between the d- and f-electron spins, and $\langle S \rangle_{Eu}$ is the thermally averaged value of the net Eu^{2+} spins inside the magnetic polaron, and Raman scattering can detect the transition of the bound electron between spin-up and spin-down configurations. As illustrated in Figure 6.8, the spin-flip Raman process involves, first, a photon-induced electric-dipole transition between the local moment and carrier states, followed by a d-f exchange-induced spin-flip of the excited electron, and concluding with another dipole transition from the carrier state back to the local moment state (Ramdas and Rodriguez 1991). As described in Section 6.2.2, the characteristic (totally antisymmetric) symmetry of these spin-flip excitations originates from the form of the polarization induced by the magnetic excitation, $P_S = iGM \times E_I$ (Hayes and Loudon 1975).

The spin-flip Raman energy ω_o is related to the net spin—and therefore the size—of the magnetic polarons via the relationship $\omega_o \sim J_{df}\langle S \rangle_{Eu}$, while the linewidth of the spin-flip response reflects the distribution of polaron sizes in the material; consequently, spin-flip Raman scattering provides a direct means by which the evolution of magnetic polarons in EuB_6 and other magnetic systems can be studied as functions of temperature, magnetic field, etc. (Snow et al. 2001).

Among the significant information provided by temperature- and field-dependent spin-flip Raman scattering measurements of EuB_6 are (Nyhus et al. 1997, Snow et al. 2001): (1) the paramagnetic-to-ferromagnetic transition is preceded by the nucleation below T^* of local ferromagnetic clusters (Figure 6.9a); as summarized in Figure 6.9c, the spin-flip Raman energy ω_o increases systematically with decreasing temperature according to the functional form $\omega_o \sim \bar{x}\alpha N_o(7/2)B_{7/2}(g\mu_B H/k_B T^*) + g^*\mu_B H$ (Ramdas and Rodriguez 1991, Snow et al. 2001),

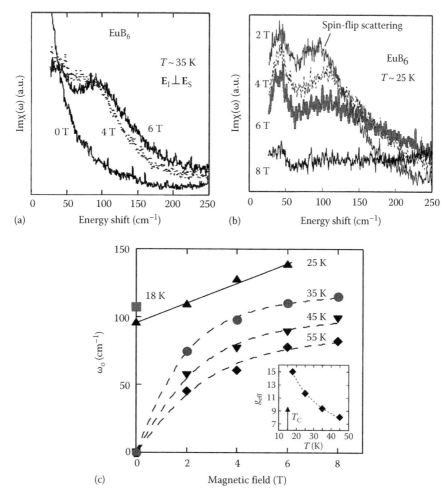

Figure 6.9 (a) Field-dependent evolution of spin-flip Raman scattering for $T = 35\,K$ ($\gg T_C$) in EuB_6, indicating the field-dependent nucleation of magnetic polarons. (b) Field-dependent disappearance of the spin-flip Raman response for $T = 25\,K$ ($>T_C$), demonstrating field-induced "melting" of magnetic polarons into a homogeneous FM phase in EuB_6. (c) Plot of the spin-flip Raman scattering energy ω_o as functions of field and temperature, showing the fits to the Brillouin function $\omega_o \sim B_{7/2}(g\mu_B H/k_B T^*)$ (dashed lines). (After Nyhus, P. et al., *Phys. Rev. B*, 56, 2717, 1997; Snow, C.S. et al., *Phys. Rev. B*, 64, 174412, 2001.)

where $\bar{x}$ is the number of Eu^{2+} moments contributing to the magnetization, αN_0 is the exchange constant, $B_{7/2}$ is the Brillouin function for $J = 7/2$, g is the Lande g-factor, g^* is the intrinsic g-factor, and T^* is the temperature associated with the onset of magnetic correlations. (2) Magnetic field-dependent measurements of the spin-flip Raman energy show that both the polaron nucleation temperature $T^*(H)$ and the polaron percolation temperature $T_C(H)$ increase with increasing magnetic field. This field-dependence demonstrates that the magnetic cluster sizes increase with decreasing temperature, and that a magnetic field can induce a percolation transition of the magnetic clusters into a uniform ferromagnetic phase; this transition causes the spin-flip scattering response to disappear (Figure 6.9b), as electrons bound to the magnetic polarons become delocalized. (3) The effective g-factor (g_{eff})—inferred from the magnetic field dependence of the spin-flip Raman energy, $g_{eff} = \hbar\Delta\omega_0/\mu_B\Delta H$—is more than an order of magnitude larger than the Lande g-factor for Eu^{2+} (inset of Figure 6.9c), illustrating that the effects of an applied magnetic field are dramatically enhanced by the d-f exchange interaction and magnetic polaron formation in EuB_6.

6.4.3 Raman Scattering Studies of Phonons and Structural Phase Transitions in Mn_3O_4

Because it is a sensitive probe of $q = 0$ phonons, Raman scattering can also provide important information regarding structural phases and phase transitions. A good example of this is observed in Mn_3O_4, which has the spinel structure, AB_2O_4, where A is generally a tetrahedrally coordinated metal ion and B is generally an octahedrally coordinated metal ion.

6.4.3.1 Raman Scattering from Phonons in Cubic Spinels

Cubic spinels AB_2O_4 crystallize in the O_h^7 symmorphic space group (Figure 6.10a), where the tetrahedral A ions have a site symmetry T_d, the octahedral B ions have the site symmetry D_{3d}, and the O ions have the site symmetry C_{3v}. To identify the $q = 0$ normal and Raman-allowed modes in this material, first consider the A ions: the irreducible representation T_2 transforms like a translation vector in the T_d site group. T_2 correlates in the following way with irreducible representations of the full O_h point group, $T_2 \Rightarrow T_{2g}$ and T_{1u}. Similarly, for the B-site ions, the representations A_{2u} and E_u transform like translation vectors in the group D_{3d}, which have the following correlations with the representations in O_h: $A_{2u} \Rightarrow A_{2u}$ and T_{1u}; $E_u \Rightarrow E_u$, T_{1u}, and T_{2u}. Finally, for the O ions, the representations A_1 and E transform like translation vectors in the C_{3v} group, which have the following correlations with the representations in O_h: $A_1 \Rightarrow A_{1g}$, T_{2g}, A_{2u}, and T_{1u}; $E \Rightarrow E_g$, T_{1g}, T_{2g}, E_u, T_{1u}, and T_{2u}. Consequently, the $q = 0$ phonon modes of the cubic spinels are given by $\Gamma^{spinel}_{phonons} = A_{1g} + E_g + 3T_{2g} + T_{1g} + 2A_{2u} + 2E_u + 5T_{1u} + 2T_{2u}$. Of these representations, only A_{1g}, E_g, and T_{2g} transform like polarizability tensors according to the

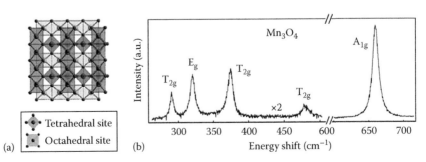

Figure 6.10 (a) Structure of cubic spinels, AB_2O_4, where A = tetrahedral site ion, B = octahedral site ion, and small circles are O ions. (b) Room temperature Raman spectrum of Mn_3O_4.

character tables for the O_h point group. Consequently, one expects five Raman-allowed phonon modes in cubic spinels, $\Gamma_{Raman} = A_{1g} + E_g + 3T_{2g}$; the A_{1g} and E_g modes involve only the O ions, while one of the T_{2g} modes involves the tetrahedral (A) site ion. Note also that there are 5 T_{1u} modes, which transform like translation vectors and so are infrared-allowed modes.

Figure 6.10b shows the room temperature Raman spectrum of Mn_3O_4, which displays the five Raman-active modes expected in cubic spinels. Mn_3O_4 actually has a slight tetragonal distortion, but this distortion does not strongly influence the Raman-active phonons, which exhibit the number and symmetries expected for cubic spinels (Malavasi et al. 2002).

6.4.3.2 Raman Scattering Study of Structural Phase Transitions in Mn_3O_4

Mn_3O_4 (A = Mn^{2+}, B = Mn^{3+}) is an interesting illustration of the benefits of Raman scattering because of its diverse magnetic, orbital, and structural transitions, which can be carefully studied via the Raman-allowed phonons: Below $T_C = 43$ K, the Mn spins ferrimagnetically order, with the net spin of the octahedrally coordinated Mn^{3+} spins oriented antiparallel to the [110] direction of the tetrahedrally coordinated Mn^{2+} spins, and with pairs of Mn^{3+} spins canted from the [$\bar{1}\,\bar{1}0$] direction. Below $T_2 = 33$ K, Mn_3O_4 has a "commensurate" spin structure in which the spin ordering has a periodicity that is double the periodicity of the atomic lattice (Jensen and Nielsen 1974). Figure 6.11b shows the temperature dependence of the ~295 cm^{-1} T_{2g} mode—which is a triply degenerate mode associated with Mn-O bond-stretching vibrations of the tetrahedral site ions—for light polarized along the [$1\bar{1}0$] crystallographic direction of Mn_3O_4. As a function of decreasing temperature above 33 K, the T_{2g} phonon frequency increases and the phonon linewidth decreases (inset, Figure 6.11b), consistent with anharmonic (two-phonon decay) effects, which contribute a temperature-dependent phonon frequency, $\omega(T) = \omega_o - A\Delta_1(T)$, and linewidth, $\Gamma(T) = B\Delta_1(T)$, where ω_o

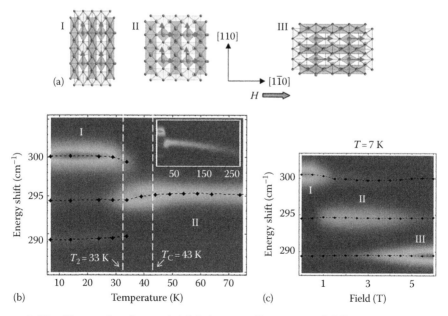

Figure 6.11 (See color insert.) (a) Schematic illustrations of different structural phases exhibited by Mn_3O_4. (b) Temperature dependence of the T_{2g} symmetry Mn-O stretch mode in Mn_3O_4, showing splitting of the mode degeneracy below tetragonal (II) to monoclinic (I) transition. The inset shows an expanded view from 10 to 300 K. (b) Magnetic field dependence of the T_{2g} symmetry Mn-O stretch mode at $T = 7$ K in Mn_3O_4, showing the field-induced phase changes for **H**||[$1\bar{1}0$]. (After Kim, M. et al., *Phys. Rev. Lett.*, 104, 136402, 2010.)

is the bare phonon frequency, A and B are constants, and $\Delta_1(T) = 1 + 2/(\exp(\hbar\omega_0/2k_B T) - 1)$ (Mihailovic et al. 1993). However, below 33 K, the degeneracy of the ~295 cm^{-1} T$_{2g}$ phonon in Mn$_3$O$_4$ splits into two modes at ~290 and ~300 cm^{-1}; this mode-splitting reflects a tetragonal-to-monoclinic structural distortion that expands the Mn^{2+}–O^{2-} bond length in the easy-axis ([110]) direction (Figure 6.11a)—decreasing the energy for Mn–O vibrations in the [110] direction—while contracting the Mn^{2+}–O^{2-} bond length along [1T0], increasing the energy for Mn–O vibrations in the [1T0] direction. The larger Raman intensity for the ~300 cm^{-1} mode in Figure 6.11b shows that the contracted Mn^{2+}–O^{2-} bonds are oriented in the direction of the incident light polarization, that is, along [1T0] (Kim et al. 2010).

Significantly, one can use the Raman spectroscopic signatures of the different structural phases to map the magnetic-field- or pressure-dependent structural phase diagrams, and to obtain microscopic details of field- and pressure-dependent phase changes. For example, Figure 6.11c shows that, at $T = 7$ K, an applied magnetic field *transverse* to the ferrimagnetic Mn moments, that is, **H**‖[1T̄0], induces a monoclinic distortion along the [1T̄0] direction for high (>4T) transverse magnetic fields (Figure 6.11a); this is evidenced by the larger Raman intensity for the ~290 cm^{-1} mode at $H > 4T$, which shows that the expanded Mn^{2+}–O^{2-} bonds have been oriented along the incident light polarization direction by the applied field. However, in the intermediate field range $1T < H < 4T$, the structure reverts to its tetragonal configuration—as indicated by the presence of the ~295 cm^{-1} "tetragonal" mode—reflecting the structural response of the material to the competition between the internal field—which tends to align the Mn moments along [110]—and the applied transverse field—which tends to force the moments along [1T̄0] (Kim et al. 2010).

6.4.4 Raman Scattering from Phonons and Charge-Density-Wave Modes in 1T-TiSe$_2$

Raman scattering is also an excellent probe of electronic phase transitions—particularly those involving a strong coupling to the lattice—such as those observed in charge- and/or orbital-ordering materials like the nickelates (Blumberg et al. 1998, Yamamoto et al. 1998), manganites (Yamamoto et al. 1999, Abrashev et al. 2001, Gupta et al. 2002, Kim et al. 2008), and ruthenates (Snow et al. 2002, Karpus et al. 2004, 2006) or those found in charge-density-wave (CDW) materials like 1T-TiSe$_2$ (Holy et al. 1977, Snow et al. 2003), 2H-TaSe$_2$, (Holy et al. 1976) 2H-NbSe$_2$ (Tsang et al. 1976), K$_{0.3}$MoO$_3$ (Travaglini et al. 1983), and K$_2$SeO$_4$ (Lee et al. 1988). A nice illustration of the efficacy of temperature- and pressure-dependent Raman scattering in such "charge-ordered" materials is found in Raman scattering studies of the charge-density-wave material 1T-TiSe$_2$, (Snow et al. 2003) in which a semimetal or small-gap semiconductor in the normal state develops a commensurate CDW, that is, a periodic modulation of the charge density that differs from the lattice periodicity—with a simple $2a_0 \times 2a_0 \times 2c_0$ superlattice structure (Figure 6.12b)—below a second-order phase transition near $T_{CDW} \sim 200$ K (DiSalvo et al. 1976).

6.4.4.1 Phonon Raman Scattering in 1T-TiSe$_2$

1T-TiSe$_2$ crystallizes in the D$_{3d}^3$ symmorphic space group, with the Ti ions having a site symmetry D$_{3d}$ and the Se ions having a site symmetry C$_{3v}$ (Figure 6.12b). To identify the $q = 0$ normal and Raman-allowed modes in this material, first consider the Ti ions: the irreducible representations A$_{2u}$ and E$_u$ transform like translation vectors in the D$_{3d}$ site group. Thus, because the site symmetry of the Ti ions has the full point group symmetry of the crystal, the $q = 0$ phonon modes involving the Ti ions have symmetries $\Gamma_{phonons}^{Ti} = A_{2u} + E_u$, which are infrared-active.

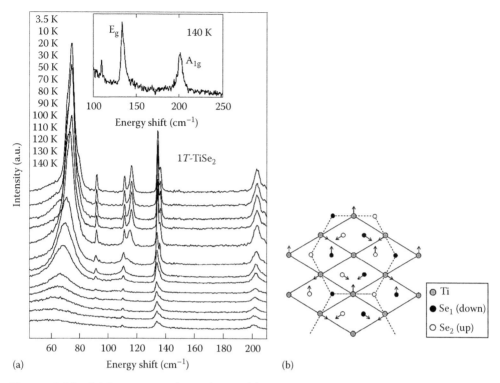

Figure 6.12 (a) Temperature dependence of the Raman spectra of 1T-TiSe$_2$. The inset shows an expanded view of the Raman-active phonons in TiSe$_2$ at 140 K. (b) Structure of 1T-TiSe$_2$, showing the displacement pattern responsible for the charge-density-wave (CDW) state. Dashed lines illustrate the Wigner–Seitz cell for the CDW superlattice. (After Snow, C.S. et al., *Phys. Rev. Lett.*, 91, 136402, 2003.)

On the other hand, for the Se ions in 1T-TiSe$_2$, the irreducible representations A$_1$ and E of the C$_{3v}$ site group transform like translation vectors according to the character tables. Using standard correlation tables, A$_1$ correlates with the following irreducible representations of the full D$_{3d}$ point group, A$_1 \Rightarrow$ A$_{1g}$ and A$_{2u}$, while E correlates with the following representations of the full D$_{3d}$ point group, E $\Rightarrow$ E$_g$ and E$_u$. Consequently, the $q = 0$ phonon modes of the Se ions in 1T-TiSe$_2$ are given by $\Gamma^{\mathrm{Se}}_{\mathrm{phonons}} =$ A$_{1g}$ + E$_g$ + A$_{2u}$ + E$_u$. Of these representations, only A$_{1g}$ and E$_g$ transform like polarizability tensors according to the character tables for the O$_h$ point group. Consequently, one expects two Raman-allowed phonon modes in the normal state of 1T-TiSe$_2$, $\Gamma_{\mathrm{Raman}} =$ A$_{1g}$ + E$_g$, both of which involve displacements of the Se ions. The A$_{2u}$ and E$_u$ modes, on the other hand, transform like translation vectors, and hence are infrared-allowed modes. The number and symmetry of Raman-allowed modes in the normal state of 1T-TiSe$_2$ is confirmed in the inset of Figure 6.12a, which shows the E$_g$-symmetry phonon at 134 cm^{-1} and the A$_{1g}$-symmetry phonon at 201 cm^{-1} (Holy et al. 1977, Sugai et al. 1980, Snow et al. 2003). The symmetries of these phonons are confirmed by observing these modes in the following scattering geometries, $z(x, x)\bar{z}$: A$_{1g}$ + E$_g$ and $z(x, y)\bar{z}$: E$_g$; this correspondence is obtained by calculating $(\varepsilon_S \cdot \Gamma_X \cdot \varepsilon_I)$ using the Raman tensors Γ_X associated with the D$_{3d}$ point group (Section 6.2.3).

6.4.4.2 Raman Scattering from Charge-Density-Wave Modes in 1T-TiSe$_2$
In addition to the two "normal state" phonons observed above $T = 200$ K, Figure 6.12a shows that the transition to the CDW state in 1T-TiSe$_2$ results in the appearance of several

new modes, the most prominent of which include an E_g-symmetry mode near 74 cm^{-1} and an A_{1g}-symmetry mode near 116 cm^{-1} (Holy et al. 1976, Snow et al. 2003). These new modes are "charge-density-wave" modes associated with phonons from the A, L, and M points of the original $a_0 \times a_0 \times c_0$ lattice that are "folded" into the zone center (Γ point) when the $2a_0 \times 2a_0 \times 2c_0$ superlattice forms. The atomic displacements that freeze in as a result of the CDW distortion, forming the superlattice structure in 1T-TiSe$_2$, are illustrated in Figure 6.12b. The identification of these modes as CDW modes is supported by their "soft mode" behavior: as $T \rightarrow T_{CDW}$ with increasing temperature, these modes exhibit temperature-dependent frequencies given by $\omega_o \sim (1 - T/T_{CDW})^\gamma$, where γ is a scaling parameter that is significantly lower in TiSe$_2$ than the mean-field value of $\gamma = 0.5$ (Barath et al. 2008). For a review of Raman scattering studies of soft mode behavior in various materials, see Scott (1974).

One can use the correlation method to determine the number and symmetry of the new modes expected in the CDW state of 1T-TiSe$_2$. The space group for the CDW superlattice structure of 1T-TiSe$_2$ is the nonsymmorphic space group D_{3d}^4, and the atoms in TiSe$_2$ have the following site symmetries in the Wigner–Seitz cell of the superlattice structure (dashed lines in Figure 6.12b): the Ti atoms at the center of the Wigner–Seitz cell have the site symmetry D_3, the Ti atoms at the edge of the cell have the site symmetry C_2, the Se atoms inside the cell have site symmetry C_1, and the Se atoms at the corners of the cell have site symmetry C_3. One can identify the new Raman-allowed modes in the superlattice structure by identifying the irreducible representations that transform like translation vectors in each of these site groups, then by determining the correlations of these irreducible representations with the irreducible representations of the D_{3d} point group of the superlattice structure. For example, the irreducible representations that transform like translation vectors in the D_3 site group of the center Ti atoms are A_2 and E. These irreducible representations have the following correlations with the irreducible representations of the full D_{3d} point group: $A_2 \Rightarrow A_{2g}$ and A_{2u}, and $E \Rightarrow E_g$ and E_u. Repeating this procedure for all the inequivalent sites in the superlattice of 1T-TiSe$_2$ in the CDW state leads to the following $q = 0$ normal modes of the CDW superlattice: $\Gamma_{superlattice} = 5A_{1g} + 7A_{2g} + 12E_g + 5A_{1u} + 7A_{2u} + 12E_u$. Of these, only the 5 A_{1g} and 12 E_g modes transform like polarizabilities and are therefore Raman-allowed phonons; hence, one expects $4A_{1g} + 11E_g$ new Raman-allowed modes in the CDW state of 1T-TiSe$_2$. Figure 6.12a shows that only about five new modes are actually observed in the CDW phase of 1T-TiSe$_2$; the Raman intensities of the other CDW modes are presumably too weak to be observed under these experimental conditions.

6.4.4.3 Pressure-Dependent Raman Scattering in 1T-TiSe₂

Low temperature pressure- and magnetic-field tuning of materials through "quantum ($T \sim 0$) phase transitions" has become an important avenue of study in condensed matter physics, because of interest in novel phases and phase transitions governed—or at least influenced— by quantum fluctuations (Sachdev 1999, Millis 2002). Pressure-tuned Raman scattering is a particularly powerful means of studying the evolution of excitations through such quantum phase transitions, primarily because visible lasers make performing low-temperature and high-pressure Raman measurements relatively simple compared to other spectroscopies.

Studies of the low-temperature, pressure-tuned "quantum melting" of the CDW in 1T-TiSe$_2$ provide a simple illustration of pressure-tuned Raman scattering. Figure 6.13a shows the pressure dependence of both the ~135 cm^{-1} E_g-symmetry phonon and the ~115 cm^{-1} A_{1g}-symmetry CDW mode at $T = 3.5$ K. The E_g phonon shows a small decrease in intensity and a systematic increase in energy (hardening) with increasing pressure; from the pressure dependence of the phonon frequency, one can estimate this phonon mode's "Gruneisen parameter" $\gamma = K_o(d(\ln \omega_o)/dP)$—where K_o is the bulk modulus of the material and ω_o is the mode frequency—which

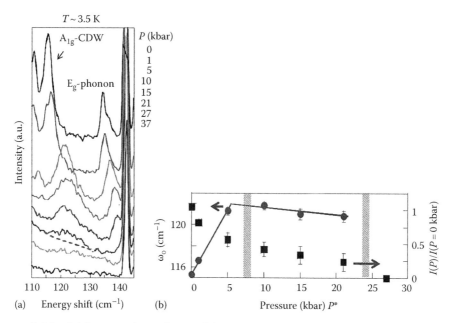

Figure 6.13   (a) Pressure dependence of the A_{1g}-symmetry CDW mode and the E_g-symmetry phonon mode at $T = 3.5\,K$ in 1T-TiSe$_2$. (b) Summary of the pressure dependence of the A_{1g} CDW mode intensity (squares) and frequency (circles) at $T = 3.5\,K$ in 1T-TiSe$_2$, showing a "crystalline" CDW regime below 10 kbar, a "soft" CDW regime between 10 and 25 kbar and pressure-tuned melting of the CDW state above $P^* \sim 25$ kbar. (After Snow, C.S. et al., *Phys. Rev. Lett.*, 91, 136402, 2003.)

provides information regarding the electron–phonon coupling associated with particular phonon modes (Goncharov et al. 2001).

The pressure dependence of the A_{1g} CDW mode in 1T-TiSe$_2$ is more interesting. For example, the intensity of this mode disappears above a critical pressure $P^* \sim 25$ kbar: Because the CDW mode reflects fluctuations of the CDW state, the intensity of this mode serves as an order parameter of the CDW state, and the disappearance of this mode above P^* suggests that the CDW phase in 1T-TiSe$_2$ collapses above roughly 25 kbar. Additionally, there is important information in the pressure-dependence of the mode frequencies, which provides details regarding the stiffness of the atomic and CDW lattices. In particular, the frequency of the A_{1g} CDW mode has a complex pressure dependence (Figure 6.13b): The frequency of the CDW mode increases linearly with increasing pressure below ~10 kbar; however, the CDW mode frequency becomes pressure independent for $10\,\text{kbar} < P < 25\,\text{kbar}$, suggesting that the charge density wave becomes incommensurate with the lattice in this pressure range (Snow et al. 2003). A detailed discussion of the pressure-induced phases of 1T-TiSe$_2$ is beyond the scope of this review, but these results show how pressure- (or field-) dependent Raman studies at low temperatures can be used to explore microscopic details of "quantum" ($T \sim 0$) phase transitions.

6.4.5 Raman Scattering from Magnons in Ca$_3$Ru$_2$O$_7$

6.4.5.1 Magnon Raman Scattering: Introduction

Raman scattering is also an excellent technique for studying magnetic excitations—including one- and two-magnon excitations—and microscopic details of magnetic phases in materials (Cottam and Lockwood 1986, Ramdas and Rodriguez 2001, Lemmens et al. 2003). To understand how light couples to magnetic excitations, first recall from Section 6.2.2 that the electric susceptibility in Equation 6.4 is spin-dependent in magnetic materials, and hence can be expanded in powers of the spin operator S (Moriya 1967, Cottam and Lockwood 1986):

$$\chi_{ij}(r) = (\chi_{ij})_0 + \sum_k K_{ijk}(r)S_k^r + \sum_{k,m} G_{ijkm}(r)S_k^r S_m^r + \sum_\delta \sum_{k,m} H_{ijkm}(r,\delta)S_k^r S_m^{r+\delta} + \cdots, \quad (6.11)$$

where the first term is the susceptibility in the absence of magnetic excitations, S^r is the spin operator at site r, and the magneto-optical coefficients K and G are tensors related to magnetic circular birefringence and magnetic linear birefringence, respectively; the second and third terms in Equation 6.11 are primarily responsible for one-magnon scattering. On the other hand, the magneto-optical coefficient H is a tensor involving spin operators at different sites, and so the fourth term in Equation 6.11 is primarily responsible for two-magnon scattering (Hayes and Loudon 1978, Cottam and Lockwood 1986). Microscopically, in one-magnon light-scattering, the photon couples to the spin via the spin–orbit interaction, $\lambda L \cdot S$ (Fleury and Loudon 1968, Cottam and Lockwood 1986). For example, consider the simple case of a magnetic ion having a ground state with $L = 0$, as shown in Figure 6.14a: the spin–orbit interaction splits the excited $L = 1$ state into different spin components, resulting in finite

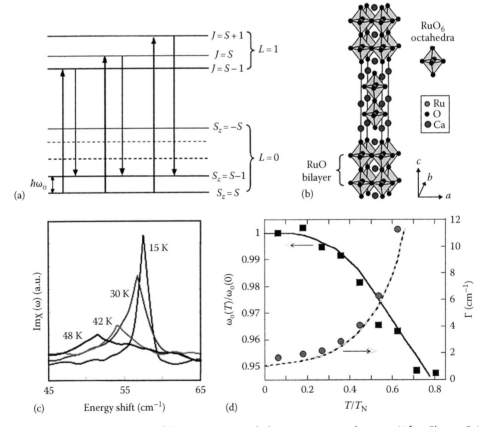

Figure 6.14 (a) Illustration of the one-magnon light scattering mechanism. (After Fleury, P.A. and Loudon, R., *Phys. Rev.*, 166, 514, 1968.) (b) Illustration of the room temperature $Ca_3Ru_2O_7$ structure, showing pairs of RuO layers (bilayers). (c) Temperature dependence of the magnon scattering in $Ca_3Ru_2O_7$ ($T_N = 56\,K$). (d) Magnon scattering normalized frequency, $\omega_o(T)/\omega_o(0)$ (squares), and linewidth, Γ (circles), as a function of normalized temperature, T/T_N, in $Ca_3Ru_2O_7$. (After Karpus, J.F. et al., *Phys. Rev. B*, 73, 134407, 2006.)

dipole transition probabilities for the pairs of transitions shown in Figure 6.14a. The net result of the transitions in Figure 6.14a on the orbital and spin states is $\Delta L = 0$ and $\Delta S = -1$, which corresponds to the excitation of a magnon. A more detailed discussion of the mechanism for the scattering of light by magnons is beyond the scope of this introductory review, but a good description is provided in Cottam and Lockwood (1986).

There is a large literature on one- and two-magnon Raman scattering in simple antiferromagnets like FeF_2 (Fleury and Loudon 1968), MnF_2 (Cottam and Lockwood 1985), $KNiF_3$ (Lyons and Fleury 1982), and $RbCoF_3$ (Nouet et al. 1973); canted antiferromagnets like NiF_2 (Fleury et al. 1971); metamagnetic antiferromagnets like $FeBr_2$ (Psaltakis et al. 1984) and $FeCl_2$ (Lockwood et al. 1982); ferrimagnets such as $RbNiF_3$ (Fleury et al. 1969); and ferromagnets (using Brillouin scattering) such as $CrBr_3$ (Sandercock 1974).

6.4.5.2 Magnon Scattering in Ca₃Ru₂O₇

As a simple illustration of the detailed information that can be gained from temperature-, field-, and pressure-dependent light scattering studies of magnons, we consider here the material $Ca_3Ru_2O_7$ (Figure 6.14b), which is a bilayered, Ruddlesden–Popper phase material (Cao et al. 2004, Yoshida et al. 2005) that exhibits a transition from a paramagnet to an A-type antiferromagnet (ferromagnetically aligned spins within the bilayers, antiferromagnetic alignment of adjacent bilayers) below $T_N = 56\,K$, and a structural transition—involving a compression of the RuO_6 octahedra—below $T_o \sim 48\,K$ (Cao et al. 2004, Yoshida et al. 2005, Karpus et al. 2006). $Ca_3Ru_2O_7$ has an orthorhombic (D_{2h}) structure at room temperature due to the RuO_6 octahedral tilts and rotations (Yoshida et al. 2005), which contributes to the complex phonon Raman spectra observed in this material compared to tetragonal $Sr_3Ru_2O_7$, for example (Iliev et al. 2005); however, for the purpose of discussing the magnon excitations in this material, it is sufficient to index the symmetries to the tetragonal (D_{4h}) point group.

Figure 6.14c shows that, below $T_N = 56\,K$, a narrow peak evolves in the Raman scattering spectrum of $Ca_3Ru_2O_7$ near $56\,cm^{-1}$. This mode can be identified as magnon scattering associated with scattering from $q = 0$ spin waves in a two-sublattice antiferromagnet—with frequencies given by (Cottam and Lockwood 1986)

$$\omega^{\pm} = \left[\alpha J_{\perp} + \beta J_{\parallel} + g\mu_B H_A \right] \pm g\mu_B H = \Delta_{AF} \pm g\mu_B H,$$

where

$J_{\perp}$ and $J_{\parallel}$ are the inter- and intra-bilayer exchange coupling parameters, respectively
H_A is the anisotropy field of the material
Δ_{AF} is the zero-field magnon energy
α and β are exchange parameters
H is the applied magnetic field

In view of the A-type antiferromagnetism in $Ca_3Ru_2O_7$, this doubly degenerate magnon can be thought of as involving ferromagnetic spins precessing in opposite senses on adjacent RuO bilayers. The identification of this $56\,cm^{-1}$ mode as a magnon is aided by its following characteristics (Karpus et al. 2006): (1) the magnon mode frequency ω_o increases with decreasing temperature below T_N (squares, Figure 6.14d), as observed in numerous other antiferromagnets, reflecting the increase in $\langle S_z \rangle$ with decreasing temperature; (2) the magnon mode linewidth—which is primarily associated with magnon–magnon interactions (Cottam and Lockwood 1986)—increases dramatically with increasing temperature below T_N (circles, Figure 6.14d), as expected theoretically and from previous experiments; (3) the magnon is observed in the "crossed" (or "depolarized") ($= z(x, y)\bar{z}$) scattering geometry

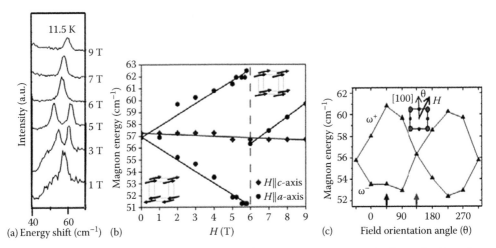

(a) Energy shift (cm^{-1}) (b) H (T) (c) Field orientation angle (θ)

Figure 6.15 (a) Magnetic field dependence of the magnon in $Ca_3Ru_2O_7$ for **H**∥[110] and at $T = 11.5\,K$. (b) Summary of the magnon energy in $Ca_3Ru_2O_7$ for **H**∥[001] (diamonds) and **H**∥[110] (circles). (c) Dependence of the magnon energy in $Ca_3Ru_2O_7$ on the in-plane magnetic field orientation with $H = 4\,T$, with $\theta = 0$ corresponding to **H**∥[100]. (After Karpus, J.F. et al., *Phys. Rev. B*, 73, 134407, 2006.)

(Section 6.2.3), which is consistent with theoretical expectations that magnon scattering involves off-diagonal terms in the Raman scattering tensors (Fleury and Loudon 1968, Cottam and Lockwood 1986); and (4) the Zeeman splitting, $\Delta E = g\mu_B H$ ($g = 2$), of this mode in the presence of a magnetic field (**H**∥ab-plane), as shown in Figure 6.15a and b, is consistent with a lifting of the degeneracy of the ferromagnetic spins that precess in opposite senses on adjacent RuO bilayers.

6.4.5.3 Pressure and Field Dependence of Magnon Scattering in $Ca_3Ru_2O_7$

Substantial microscopic information can be obtained regarding the magnetic phases and magnetic exchange parameters of materials from the field and pressure dependence of magnon scattering. In $Ca_3Ru_2O_7$, for example, the magnon mode energy exhibits no field dependence for **H**∥c-axis (diamonds in Figure 6.15b). However, the magnon mode exhibits a splitting for **H**∥ab-plane, which exhibits an angular dependence given by $\omega^\pm(\theta) = \Delta_{AF} \pm g\mu_B H \cos\theta$ (triangles in Figure 6.15c). The observation of a magnon mode splitting only for **H**∥ab-plane implies that the Ru-spins are aligned in the plane; further, the observation that the magnon mode splitting is maximum for H∥[110] suggests that [110] is the "easy-axis" direction in $Ca_3Ru_2O_7$, that is, the Ru spins are oriented at 45° to the [100] direction below T_N. Figures 6.15a and b also show that the split magnon modes collapse into a single mode at $H_c = 5.9\,T$, reflecting a metamagnetic transition from A-type antiferromagnetic alignment of adjacent RuO bilayers to (forced) ferromagnetic alignment of the bilayers. From the critical field, one can estimate an interbilayer exchange energy—roughly, the exchange energy gained by the system when adjacent RuO bilayers switch between AF and FM alignments—of $\alpha J_\perp = g\mu_B H_c \sim 5.5\,cm^{-1}$ (~0.7 meV).

As a characteristic excitation of the A-type antiferromagnetic ground state, the magnon mode in $Ca_3Ru_2O_7$ can also be usefully studied as a function of pressure to explore the pressure-dependent phases of this material. Figure 6.16 shows that the magnon energy in $Ca_3Ru_2O_7$ decreases slightly with increasing pressure, reflecting a slight decrease in Δ_{AF} and, hence, the anisotropy field H_A, with pressure. More significantly, the magnon mode disappears near $P \sim 46\,kbar$, suggesting that the antiferromagnetic state of $Ca_3Ru_2O_7$ collapses above this pressure. For $Ca_3Ru_2O_7$, the pressure dependences of phonon modes (Snow et al. 2002) and

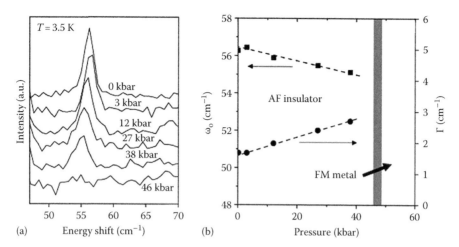

Figure 6.16 (a) Pressure dependence of the magnon mode in $Ca_3Ru_2O_7$ at T = 3.5 K. (b) Summary of the magnon mode frequency (ω_o) and linewidth (Γ) in $Ca_3Ru_2O_7$, and evidence for a pressure-induced phase change above 46 kbar where no magnon scattering is observed. (After Snow, C.S. et al., *Phys. Rev. Lett.*, 89, 226401, 2002; Karpus, J.F. et al., *Phys. Rev. B*, 73, 134407, 2006.)

magnetic properties (Yoshida et al. 2008) support the interpretation of a pressure-dependent magnetic phase change in $Ca_3Ru_2O_7$ near P ~ 40 kbar.

6.4.6 Inelastic X-Ray Scattering from Plasmons in Graphite

As discussed earlier, inelastic x-ray scattering (IXS) is a direct probe of the collective charge dynamics of a material. Specifically, under nonresonant conditions, the scattered intensity is proportional to Im[$\chi(\mathbf{q}, \omega)$], where χ is the charge propagator. $\chi(\mathbf{q}, \omega)$ is the Fourier transform of $\chi(\mathbf{x}, t)$, which physically represents the probability that a disturbance in the electron density at the origin at t = 0 will propagate to some location, $\mathbf{x}$, at some later time, t. We illustrate here the capabilities of IXS with a study of graphite, which is a quasi-2D material with an unusual Fermi surface consisting of small, nested pockets. The charge dynamics of graphite are innately interesting from the standpoint of understanding screening in low-density and low-dimensional systems.

In Figure 6.17a, we show two IXS spectra from a single crystal of graphite, taken at Sector 9 at the Advanced Photon Source. Graphite exhibits two plasmons, which arise from collective oscillations of electrons in the π and $\sigma-\pi$ bands (Schülke et al. 1988, Hiraoka et al. 2005). These excitations are the primary means by which graphite screens charge and are the largest contributors to the charge propagator of the system. If these excitations are sampled over a sufficiently broad range of momentum and energy, it is possible to completely reconstruct the charge propagator (Abbamonte et al. 2004, 2008), allowing electron dynamics to be imaged in real time. Shown in Figure 6.17b is a time slice of the propagator at t = 400 as. The space and time resolutions are determined by the range of ω and $\mathbf{q}$ surveyed in the experiment, which for x-rays can be extremely large (up to ω ~ 1 keV and $\mathbf{q}$ ~ 10 Å^{-1}). For the image in Figure 6.17, the spatial and temporal resolutions were $\Delta\mathbf{x}$ = 0.2 Å and Δt = 10.3 as (1.03×10^{-17} s), respectively. Visible in the propagator shown in Figure 6.17b is a disturbance with sixfold rotational symmetry, which arises because of scattering of the π plasmon off the underlying honeycomb lattice. This data is preliminary and is still being analyzed, but it provides fundamental information like the effective fine structure constant for graphite.

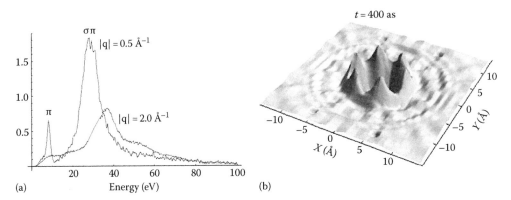

Figure 6.17   (See color insert.) (a) Raw IXS spectra taken from single crystal graphite for two values of the in-plane transferred momentum (shown). (b) Time slice of the density propagator at $t = 400$ as, showing hexagonal symmetry from the underlying honeycomb lattice.

6.5 SUMMARY

In this review, we have attempted to provide enough information to help researchers set up a Raman scattering system and employ more advanced capabilities such as inelastic x-ray scattering and high-pressure and high-magnetic-field Raman scattering. We have also provided a broad range of examples, not only to give researchers a sense of the diverse range of materials' phenomena and phases that can be effectively studied using Raman scattering, but also to provide guidance to researchers in interpreting Raman scattering from different excitations in a range of materials. Given the great breadth of materials applications for which Raman scattering has proven to be useful, we are aware that this review can only scratch the surface of possible applications for Raman scattering in materials science. For instance, space limitations have prevented more detailed discussion of Raman scattering from semiconductor materials (Menendez 2000) and heterostructures (Gammon 2000), carbon-based materials (Dresselhaus et al. 2000), magnetic materials (Hayes and Loudon 1978, Cottam and Lockwood 1986), and superconductors (Thomsen and Cardona 1989, Cardona 2000, Devereaux and Hackl 2007). Additionally, many important techniques related to or based on Raman scattering—some of which are discussed elsewhere in this book—have not been extensively covered in this review, including Brillouin scattering (Sandercock 1974), time-resolved Raman scattering (Chapter 14) (Takada et al. 2005), resonance Raman scattering (Cardona 1982), near-field Raman scattering (Verma et al. 2006), and surface-enhanced Raman scattering (Talley et al. 2005). Yet, in spite of its admittedly narrow scope, we hope that this review provides at least a glimpse of the great power and usefulness of Raman scattering for studying diverse phenomena and materials, and offers some incentive for interested researchers to start developing this powerful tool in their own laboratories.

ACKNOWLEDGMENTS

This material is based on work supported by the U.S. Department of Energy, Division of Materials Sciences, under Award No. DE-FG02-07ER46453, through the Frederick Seitz Materials Research Laboratory at the University of Illinois at Urbana-Champaign, and by the National Science Foundation under Grant NSF DMR 08-56321. Use of the Advanced Photon Source was supported by DOE Contract DE-AC02-O6CH11357.

REFERENCES

Abbamonte, P., Burns, C.A., Isaacs, E.D. et al. 1999. Resonant inelastic x-ray scattering from valence excitations in insulating copper oxides. *Phys. Rev. Lett.* 83: 860.

Abbamonte, P., Finkelstein, K.D., Collins, M.D., and Gruner, S.M. 2004. Imaging density disturbances in water with a 41.3-attosecond time resolution. *Phys. Rev. Lett.* 92: 237401.

Abbamonte, P., Graber, T., Reed, J. et al. 2008. Dynamical reconstruction of the exciton in LiF with inelastic x-ray scattering. *Proc. Natl. Acad. Sci. U.S.A.* 105: 12159.

Abrashev, M.V., Backstrom, J., Borjesson, L., Pissas, M., Koley, N., and Illiev, M.N. 2001. Raman spectroscopy of the charge- and orbital-ordered state in $La_{0.5}Ca_{0.5}MnO_3$. *Phys. Rev. B* 64: 144429.

Adams, D.M. and Sharma, S.K. 1977. Selection of diamonds for infrared and Raman spectroscopy, *J. Phys. E: Sci. Instrum.* 10: 680.

Adler, P., Goncharov, A.F., Syassen, K., and Schoenherr, E. 1994. Optical reflectivity and Raman spectra of Sr_2FeO_4 under pressure. *Phys. Rev. B* 50: 11396–11402.

Asher, S.A. and Bormett, R. 2000. Raman instrumentation. *Raman Scattering in Materials Science*, eds. W.H. Weber and R. Merlin, p. 35. Berlin, Germany: Springer.

Barath, H., Kim, M., Karpus, J.F. et al. 2008. Quantum and classical mode softening near the charge-density-wave—Superconductor transition of Cu_xTiSe_2. *Phys. Rev. Lett.* 100: 106402.

Bartels, W.J. 1983. Characterization of thin layers on perfect crystals with a multipurpose high resolution x-ray diffractometer. *J. Vac. Sci. Technol. B* 1: 338–345.

Benfatto, L., Neto, M.B.S., Gozar, A., Dennis, B.S., Blumberg, G., Miller, L.L., Komiya, S., and Ando, Y. 2006. Field dependence of the magnetic spectrum in anisotropic and Dzyaloshinskii-Moriya antiferromagnets. II. Raman spectroscopy. *Phys. Rev. B* 74: 024416.

Blumberg, G., Kang, M.S., and Klein, M.V. 1997. Electronic Raman scattering of overdoped $Tl_2Ba_2CuO_{6+d}$ in high magnetic fields. *Phys. Rev. Lett.* 78: 2461–2464.

Blumberg, G., Klein, M.V., and Cheong, S.-W. 1998. Charge and spin dynamics of an ordered stripe phase in $La_{12/3}Sr_{1/3}NiO_4$ investigated by Raman spectroscopy. *Phys. Rev. Lett.* 80: 564.

Blumberg, G., Littlewood, P., Gozar, A., Dennis, B.S., Motoyama, N., Eisaki, H., and Uchida, S. 2002. Sliding density wave in $Sr_{14}Cu_{24}O_{41}$ ladder compounds. *Science* 297: 584–587.

Brodsky, M.H. 1975. Raman scattering in amorphous semiconductors. In *Topics in Applied Physics*, vol. 8, ed. M. Cardona, p. 205. Berlin, Germany: Springer-Verlag.

Burkel, E. 1991. Inelastic scattering of x-rays with very high energy resolution. In *Springer Tracts in Modern Physics*, vol. 125. Berlin, Germany: Springer-Verlag.

Burnett, J.H., Cheong, H.M., and Paul, W. 1990. The inert gases Ar, Xe, and He as cryogenic pressure media. *Rev. Sci. Instrum.* 61: 3904–3905.

Cao, G., Balicas, L., Lin, X.N. et al. 2004. Field-tuned collapse of an orbitally ordered and spin polarized state: Colossal magnetoresistance in the bilayer ruthenate $Ca_3Ru_2O_7$. *Phys. Rev. B* 69: 014404.

Cardona, M. 1982. Resonance phenomena. In *Light Scattering in Solids II*, eds. M. Cardona and G. Guntherodt. Berlin, Germany: Springer-Verlag.

Cardona, M. 2000. Raman scattering in high T_c superconductors: phonons, electrons, and magnons. *Raman Scattering in Materials Science*, eds. W.H. Weber and R. Merlin, p. 151. Berlin, Germany: Springer.

Cazayous, M., Gallai, Y., Sacuto A., de Sousa, R., Lebeugle, D., and Colson, D. 2008. Possible observation of cycloidal electromagnons in $BiFeO_3$. *Phys. Rev. Lett.* 101: 037601.

Cooper, S.L. 2001a. Electronic and magnetic Raman scattering studies of the high T_c cuprates. In *Handbook on the Physics and Chemistry of Rare Earths 31*, eds. K.A. Gschneidner and M.B. Maple, pp. 509–562. Amsterdam, the Netherlands: Elsevier Science.

Cooper, S.L. 2001b. Optical spectroscopic studies of metal-insulator transitions in perovskite-related oxides. *Struct. Bonding* 98: 161–219.

Cottam, M.G. and Lockwood, D.J. 1985. One-magnon Raman scattering in a S = 5/2 antiferromagnet. *Phys. Rev. B* 31: 641.

Cottam, M.G. and Lockwood, D.J. 1986. *Light Scattering in Magnetic Solids*. New York: Wiley.

Cuk, T., Struzhkin, V., Devereaux, T.P., Goncharov, A., Kendziora, C., Eisaki, H., Mao, H., and Shen, Z.-X. 2008. Uncovering a pressure-tuned electronic transition in $Bi_{1.98}Sr_{2.06}Y_{0.68}Cu_2O_{8+\delta}$ using Raman scattering and x-ray diffraction. *Phys. Rev. Lett.* 100: 217003.

Das, A., Pisana, S., Chakraborty, B. et al. 2008. Monitoring dopants by Raman scattering in an electrochemically top-gated graphene transistor. *Nat. Nanotechnol.* 3: 210–215.

Degiorgi, L., Felder, E., Ott, H.R., Sarrao, J.L., and Fisk, Z. 1997. Low temperature anomalies and ferromagnetism of EuB_6. *Phys. Rev. Lett.* 79: 5134–5137.

Devereaux, T.P. and Hackl, R. 2007. Inelastic light scattering from correlated electrons. *Rev. Mod. Phys.* 79: 175–233.

Dietl, T., Sawicki, M., Dahl, M. et al. 1991. Spin-flip scattering near the metal-to-insulator transition in $Cd_{0.95}Mn_{0.05}Se:In$. *Phys. Rev. B* 43: 3154.

DiSalvo, F.J., Moncton, D.E., and Waszczak, J.V. 1976. Electronic properties and superlattice formation in the semimetal $TiSe_2$. *Phys. Rev. B* 14: 4321.

Dresselhaus, M.S., Pimenta, M.A., Eklunk, P., and Dresselhaus, G. 2000. Raman scattering in fullerenes and related carbon-based materials. In *Raman Scattering in Materials Science*, eds. W.H. Weber and R. Merlin, p. 314. Berlin, Germany: Springer.

Dunstan, D.J. and Scherrer, W. 1988. Miniature cryogenic diamond-anvil high-pressure cell. *Rev. Sci. Instrum.* 59: 3789.

Dunstan, D.J. and Spain, I.L. 1989. The technology of diamond anvil high pressure cells: 1. Principles, design, and construction. *J. Phys. E: Sci. Instrum.* 22: 913–923.

Eremets, M.I. 1996. *High Pressure Experimental Methods*. Oxford, U.K.: Oxford University Press.

Fateley, W.G., Dollish, F.R., McDevitt, N.T., and Bentley, F.F. 1972. *Infrared and Raman Selection Rules for Molecular and Lattice Vibrations: The Correlation Method*. New York: Wiley-Interscience.

Ferrari, A.C. 2007. Raman spectroscopy of graphene and graphite: Disorder, electron-phonon coupling, doping and nonadiabatic effects. *Sol. St. Commun.* 143: 47–57.

Ferrari, A.C., Meyer, J.C., Scardaci, V. et al. 2006. Raman spectrum of graphene and graphene layers. *Phys. Rev. Lett.* 97: 187401.

Fleury, P.A. and Loudon, R. 1968. Scattering of light by one- and two-magnon excitations. *Phys. Rev.* 166: 514.

Fleury, P.A., Loudon, R., and Walker, L.R. 1971. Magnetic field dependence of magnon frequencies in canted spin antiferromagnet NiF_2. *J. Appl. Phys.* 42: 1649.

Fleury, P.A., Worlock, J.M., and Guggenheim, H.J. 1969. Light scattering from phonons and magnons in $RbNiF_3$. *Phys. Rev.* 185: 738.

Gammon, D. 2000. Raman scattering in semiconductor heterostructures. In *Raman Scattering in Materials Science*, eds. W.H. Weber and R. Merlin, p. 109. Berlin, Germany: Springer.

Geim, A.K. 2009. Graphene: Status and prospects. *Science* 324: 1530–1534.

Goncharov, A.F., Hemley, R.J., Mao, H., and Shu, J. 1997. New high-pressure excitations in para-hydrogen. *Phys. Rev. Lett.* 80: 101–104.

Goncharov, A.F., and Struzhkin, V.V. 2003. Raman spectroscopy of metals, high-temperature superconductors and related materials under high pressure. *J. Raman Spectrosc.* 34: 532.

Goncharov, A.F., Struzhkin, V.V., Gregoryanz, E., Hu J., Hemley, R.J., Mao, H., Lapertot, G., Bud'ko, S.L., and Canfield, P.C. 2001. Raman spectrum and lattice parameters of MgB_2 as a function of pressure. *Phys. Rev. B* 64: 100509.

Goni, A.R., Zhou, T., Schwarz, U., Kremer, R.K., and Syassen, K. 1996. Pressure-temperature phase diagram of the spin-Peierls compound $CuGeO_3$. *Phys. Rev. Lett.* 77: 1079–1083.

Gupta, R., Venketeswara Pai, G., Sood, A.K., Ramakrishnan, T.V., and Rao, C.N.R. 2002. Raman scattering in charge-ordered $Pr_{0.63}Ca_{0.37}MnO_3$: Anomalous temperature dependence of linewidth. *Europhys. Lett.* 58: 778.

Gupta, R., Kim, M., Barath, H., Cooper, S.L., and Cao, G. 2006. Field- and pressure-induced phases in $Sr_4Ru_3O_{10}$: A spectroscopic investigation. *Phys. Rev. Lett.* 96: 067004.

Hayes, W. and Loudon, R. 1978. *Scattering of Light by Crystals*. New York: Wiley.

Henry, N.F.M. and Lonsdale, K. 1965. *International Tables for X-Ray Crystallography*, vol. 1. Birmingham, U.K.: Kynoch.

Hill, J.P., Blumberg, G., Kim, Y.J. et al. 2008. Observation of a 500 meV collective mode in $La_{2-x}Sr_xCuO_4$ and Nd_2CuO_4 using resonant inelastic x-ray scattering. *Phys. Rev. Lett.* 100: 97001.

Hiraoka, N., Ishii, H., Jarrige, I., and Cai, Y.Q. 2005. Inelastic x-ray scattering studies of low-energy charge excitations in graphite. *Phys. Rev. B* 72: 75103.

Holt, M., Zschack, P., Hong, H., Chou, M.Y., and Chiang, T.-C. 2001. X-ray studies of phonon softening in $TiSe_2$. *Phys. Rev. Lett.* 86: 3799.

Holy, J.A., Klein, M.V., McMillan, W.L., and Meyer, S.F. 1976. Raman-active lattice vibrations of commensurate superlattice in $2H$-$TaSe_2$. *Phys. Rev. Lett.* 37: 1145–1148.

Holy, J.A., Woo, K.C., Klein, M.V., and Brown, F.C. 1977. Raman and infrared studies of superlattice formation in $TiSe_2$. *Phys. Rev. B* 16: 3628.

Huang, M.Y., Yan, H.G., Chen, C.Y., Song, D.H., Heinz, T.F., and Hone, J. 2009. Phonon softening and crystallographic orientation of strained graphene studied by Raman spectroscopy. *Proc. Natl. Acad. Sci.* U.S.A. 106: 7304–7308.

Iliev, M.N., Jandl, S., Popov, V.N., Litvinchuk, A.P., Cmaidalka, J., Meng, R.L., and Meen, J. 2005. Raman spectroscopy of $Ca_3Ru_2O_7$: Phonon line assignment and electron scattering. *Phys. Rev. B* 71: 214305.

Ipatova, I., Subashiev, A.V., and Voitenko, V.A. 1981. Electron light scattering from doped silicon. *Sol. St. Commun.* 37: 893.

Janesick, J.R. 2001. *Scientific Charge-Coupled Devices*, SPIE Press Monograph vol. PM83. Washington, DC: SPIE Press.

Jayaraman, A. 1983. Diamond anvil cells and high pressure physical investigations. *Rev. Mod. Phys.* 55: 65.

Jensen, G.B. and Nielsen, O.V. 1974. Magnetic structure of Mn_3O_4 between 47K and Neel point 41K. *J. Phys. C* 7: 409.

Kadleikova, M., Berza, J., and Vesely, M. 2001. Raman spectra synthetic sapphire. *Microelectronics Journal* 32: 955–958.

Karpus, J.F. 2007. Raman scattering studies of the orbital, magnetic, and conducting phases in double-layer ruthenates. PhD thesis, University of Illinois at Urbana-Champaign, Urbana, IL.

Karpus, J.F., Gupta, R., Barath, H., Cooper, S.L., and Cao, G. 2004. Field-induced orbital and magnetic phases in $Ca_3Ru_2O_7$. *Phys. Rev. Lett.* 93: 167205.

Karpus, J.F., Snow, C.S., Gupta, R., Barath, H., Cooper, S.L., and Cao, G. 2006. Spectroscopic study of the field- and pressure-induced phases of the bilayered ruthenate $Ca_3Ru_2O_7$. *Phys. Rev. B* 73: 134407.

Katsufuji, T. and Tokura, Y. 1994. Electronic Raman scattering in filling-controlled metals: $Sr_{1-x}La_xTiO_3$. *Phys. Rev. B* 49: 4372.

Kim, M., Barath, H., Cooper, S.L., Abbamonte, P., Fradkin, E., Rubhausen, M., Zhang, C.L., and Cheong, S.-W. 2008. Raman scattering studies of the temperature- and field-induced melting of charge order in $La_xPr_yCa_{1-x-y}MnO_3$. *Phys. Rev. B* 77: 1344111.

Kim, M., Chen, X.M., Joe, Y.I., Fradkin, E., Abbamonte, P., and Cooper, S.L. 2010. Mapping the magneto-structural quantum phases of Mn_3O_4. *Phys. Rev. Lett.* 104: 136402.

Kirillov, D., Suzuki, Y., Antognazza, L., Char, K., Bozovic, I., and Geballe, T.H. 1995. Phonon anomalies at the magnetic phase transition in $SrRuO_3$. *Phys. Rev. B* 51: 12825–12828.

Klein, M.V. 1975. Electronic Raman scattering. In *Topics in Applied Physics*, vol. 8., ed. M. Cardona., p. 147. Berlin, Germany: Springer-Verlag.

Klein, M.V. 1982. Raman studies of phonon anomalies in transition-metal compounds. In *Light Scattering in Solids III,* eds. M. Cardona and G. Guntherodt. Berlin, Germany: Springer-Verlag.

Klein, M.V., Holy, J.A., and Williams, W.S. 1978. Raman scattering induced by carbon vacancies in TiC_x. *Phys. Rev. B* 17: 1546–1556.

Kosztin, J. and Zawadowski, A. 1991. Violation of the f-sum rule for Raman scattering in metals. *Sol. St. Commun.* 78: 1029.

Kremer, R.K., Loa, I., Razavi, F.S., and Syassen, K. 2000. Effect of pressure on the magnetic phase transition in α'-NaV_2O_5. *Solid State Commun.* 113: 217–220.

Kuzmany, H. 1998. *Solid State Spectroscopy: An Introduction.* Berlin, Germany: Springer.

Lavagnini, M., Baldini, M., Sacchetti, A., Di Castro, D., Delley, B., Monnier, R., Chu, J.-H., Ru, N., Fisher, I.R., Postorino, P., and Degiorgi, L. 2008. Evidence for coupling between charge density waves and phonons in two-dimensional rare-earth tritellurides. *Phys. Rev. B* 78: 201101.

Lee, W.K., Cummins, H.Z., Pick, R.M., and Dreyfus, C. 1988. Amplitude mode in K_2SeO_4: Temperature dependence of the Raman cross section. *Phys. Rev. B* 37: 6442–6445.

Lemmens, P., Guntherodt, G., and Gros, C. 2003. Magnetic light scattering in low dimensional quantum spin systems. *Phys. Rep.* 375: 1–103.

Liu, J. and Vohra, Y.K. 1996. Flourescence emission from high purity synthetic diamond anvil to 370 GPa. *Appl. Phys. Lett.* 68: 2049.

Liu, H.L., Yoon, S., Cooper, S.L., Cheong, S.-W., Han, P.D., and Payne, D.A. 1998. Probing anisotropic magnetotransport in manganese perovskites using Raman spectroscopy. *Phys. Rev. B* 58: 10115–10118.

Loa, I., Adler, P., Grzechnik, A., Syassen, K., Schwarz, U., Hanfland, M., Rozenberg, G. Kh., Gorodetsky, P., and Pasternak, M.P. 2001. Pressure-induced quenching of the Jahn-Teller distortion and the insulator-to-metal transition in $LaMnO_3$. *Phys. Rev. Lett.* 87: 125501.

Lockwood, D.J., Mischler, G., Zwick, A., Johnstone, I.W., Psaltakis, G.C., Cottam, M.G., Legrand, S., and Leotin, J. 1982. Excitations in $Fe_{1-x}Co_xCl_2$—A randomly mixed antiferromagnet with competing spin anisotropies. *J. Phys. C* 15: 2973.

Lyons, K.B. and Fleury, P.A. 1982. Magnetic energy fluctuations—Observations by light scattering. *Phys. Rev. Lett.* 48: 202–205.

Malavasi, L., Galinetto, P., Mossati, M.C., Azzoni, C.B., and Flor, G. 2002. Raman spectroscopy of AMn_2O_4 (A=Mn, Mg, and Zn) spinels. *Phys. Chem. Chem. Phys.* 4: 3876–3880.

Marinopoulos, A.G., Reining, L., Rubio, A., and Olevano, V. 2004. Ab initio study of the optical absorption and wave-vector-dependent dielectric response of graphite. *Phys. Rev. B* 69: 245419.

Martin, R.M. and Falicov, L.M. 1975. Resonant Raman scattering. In *Topics in Applied Physics*, vol. 8, ed. M. Cardona, p. 79. Berlin, Germany: Springer-Verlag.

Massey, M.J., Chen, N.H., Allen, J.W., and Merlin, R. 1990. Pressure-dependence of 2-magnon Raman scattering in NiO. *Phys. Rev. B* 42: 8776–8779.

McWhan, D.B., Fleming, R.M., Moncton, D.E., and DiSalvo, F.J. 1980. Reentrant lock-in transition of the charge density wave in $2H$-$TaSe_2$ at high pressure. *Phys. Rev. Lett.* 45: 269.

Menendez, J. 2000. Characterization of bulk semiconductors using Raman spectroscopy. *Raman Scattering in Materials Science*, eds. W.H. Weber and R. Merlin, p. 55. Berlin, Germany: Springer.

Mihailovic, D., McCarty, K.F., and Ginley, D.S. 1993. Anharmonic effects and the two-particle continuum in the Raman spectra of $YBa_2Cu_3O_{6.9}$, $TlBa_2CaCu_2O_7$, and $Tl_2Ba_2CaCu_2O_8$. *Phys. Rev. B* 47: 8910–8916.

Millis, A.J. 2002. Whither correlated electron theory? *Physica B* 312–313: 1–6.

Mooradian, A. and McWhorter, A.L. 1967. Polarization and intensity of Raman scattering from plasmons and phonons in gallium arsenide. *Phys. Rev. Lett.* 19: 849.

Moriya, T. 1967. Theory of light scattering by magnetic crystals. *J. Phys. Soc. Jpn.* 23: 490.

Moss, W.C. and Goettel, K.A. 1987. Finite element design of diamond anvils. *Appl. Phys. Lett.* 50: 25–27.

Nouet, J., Toms, J.J., and Scott, J.F. 1973. Light scattering from magnons and excitons in $RbCoF_3$. *Phys. Rev. B* 7: 4874.

Nyhus, P., Yoon, S., Kauffman, M., Cooper, S.L., Fisk, Z., and Sarrao, J. 1997. Spectroscopic study of bound magnetic polaron formation and the metal-semiconductor transition in EuB_6. *Phys. Rev. B* 56: 2717–2721.

Opel, M., Nemetschek, R., Hoffman, C. et al. 2000. Carrier relaxation, pseudogap, and superconducting gap in high T_c cuprates: A Raman scattering study. *Phys. Rev. B* 61: 9752–9774.

Patel, C.K.N. and Slusher, R.E. 1968. Light scattering by plasmons and Landau levels of electron gas in InAs. *Phys. Rev.* 167: 413.

Patselov, A.M. Demchuk, K.M., and Starostin, A.A. 1989. Generation of 15-GPa pressure by means of a sapphire cell. *Pri. Tek. Eksp.* 6: 159–163.

Pinczuk, A. and Burstein, E. 1975. Fundamentals of inelastic light scattering in semiconductors and insulators. In *Topics in Applied Physics*, vol. 8, ed. M. Cardona, p. 23. Berlin, Germany: Springer-Verlag.

Pines, D. 1963. *Elementary Excitations in Solids*. Reading, MA: Benjamin/Cummings.

Pisana, S., Lazzeri, M., Casiraghi, C. et al. 2007. Breakdown of the adiabatic Born-Oppenheimer approximation in graphene. *Nat. Mater.* 6: 198–201.

Podobedov, V.B. and Weber, A. 2000. Raman scattering in perovskite manganites. *Raman Scattering in Materials Science*, eds. W.H. Weber and R. Merlin, p. 448. Berlin, Germany: Springer.

Postorino, P., Congeduti, A., Degiorgi E., Itie, J.P., and Munsch, P., 2002. High-pressure behavior of $La_xSr_{2-x}MnO_4$ layered manganites investigated by Raman spectroscopy and x-ray diffraction. *Phys. Rev. B* 65: 224102.

Poulet, H. and Mathieu, J.P. 1976. *Vibration Spectra and Symmetry of Crystals*. Paris: Gordon & Breach.

Psaltakis, G.C., Mischler, G. Lockwood, D.J., Cottam, M.G., Zwick, A., and Legrand, S. 1984. One-magnon and 2-magnon excitations in $FeBr_2$. *J. Phys. C* 17: 1735.

Qazilbash, M.M., Koitzsch, A., Dennis, B.S., Gozar, A., Balci, H., Kendziora, C.A., Greene, R.L., and Blumberg, G. 2005. Evolution of superconductivity in electron-doped cuprates: Magneto-Raman spectroscopy. *Phys. Rev. B* 72: 214510.

Ramdas, A.K. and Rodriguez, S. 1991. Light scattering in diluted magnetic semiconductors. In *Topics in Applied Physics 68*, eds. M. Cardona and G. Guntherodt. Berlin, Germany: Springer-Verlag.

Reich, S. and Thomsen, C. 2004. Raman spectroscopy of graphite. *Philos. Trans. R. Soc. Lond. A* 362: 2271.

Reznik, D., Cooper, S.L., Klein, M.V., Lee, W.C., Ginsberg, D.M., Maksimov, A.A., Puchkov, A.V., Tartakovskii, I.I., and Cheong, S.-W. 1993. Plane-polarized Raman continuum in the insulating and superconducting layered cuprates. *Phys. Rev. B* 48: 7624.

Rho, H., Snow, C.S., Cooper, S.L., Fisk, Z., Comment, A., and Ansermet, J.-Ph. 2002. Evolution of magnetic polarons and spin-carrier interactions through the metal-insulator transition in $Eu_{1-x}Gd_xO$. *Phys. Rev. Lett.* 88: 127401.

Romero, D.B., Podobedov, V.B., Weber, A., Rice, J.P., Mitchell, J.F., Sharma, R.P., and Drew, H.D. 1998. Polarons in the layered perovskite manganite $La_{1.2}Sr_{1.8}Mn_2O_7$. *Phys. Rev. B* 58: 14737–14740.

Rosenblum, S.S. and Merlin, R. 1999. Resonant two-magnon Raman scattering at high pressures in the layered antiferromagnetic $NiPS_3$. *Phys. Rev. B* 59: 6317–6320.

Ruf, T., Thomsen, C., Liu, R., and Cardona, M. 1988. Raman-study of the phonon anomaly in single-crystal $YBa_2Cu_3O_{7-}$ in the presence of a magnetic field. *Phys. Rev. B.* 38: 11985–11987.

Sachdev, S. 1999. Quantum phase transitions. Cambridge, U.K.: Cambridge University Press.

Saha, S., Muthu, D.V.S., Singh, S. et al. 2009. Low temperature and high pressure Raman and x-ray studies of pyrochlore $Tb_2Ti_2O_7$: Phonon anomalies and possible phase transition. *Phys. Rev. B* 79: 134112.

Sandercock, J.R. 1974. Light scattering study of ferromagnet $CrBr_3$. *Sol. St. Commun.* 15: 1715.

Schaak, G. 2000. Raman scattering by crystal-field excitations. In *Light Scattering in Solids VII*, eds. M. Cardona and G. Guntherodt. Berlin, Germany: Springer-Verlag.

Schlappa, J., Schmitt, T., Vernay, F. et al. 2009. Collective magnetic excitations in the spin ladder $Sr_{14}Cu_{24}O_{41}$ measured using high-resolution resonant inelastic x-ray scattering. *Phys. Rev. Lett.* 103: 47401.

Schülke, W., Bonse, U., Nagasawa, N., Kaprolat, A., and Berthold, A. 1988. Interband transitions and core excitation in highly oriented pyrolytic graphite studied by inelastic synchrotron x-ray scattering. *Phys. Rev. B* 38: 2112.

Scott, J.F. 1974. Soft mode spectroscopy: Experimental studies of structural phase transitions. *Rev. Mod. Phys.* 46: 83–128.

Sinha, S.K. 2001. Theory of inelastic x-ray scattering from condensed matter. *J. Phys.: Condens. Mater.* 13: 7511–7523.

Snow, C.S. 2003. Probing the dynamics of pressure- and magnetic-field-tuned transitions in strongly correlated electron systems: Raman scattering studies. PhD thesis, University of Illinois at Urbana-Champaign, Urbana, IL.

Snow, C.S., Cooper, S.L., Cao, G., Crow, J.E., Nakatsuji, S., and Maeno, Y. 2002. Pressure-tuned collapse of the Mott-like state in $Ca_{n+1}Ru_nO_{3n+1}$ (n = 1,2): Raman spectroscopic studies. *Phys. Rev. Lett.* 89: 226401.

Snow, C.S., Cooper, S.L., Young, D.P., Fisk, Z., Comment, A., and Ansermet, J-P. 2001. Magnetic polarons and the metal-semiconductor transitions in $(Eu,La)B_6$ and EuO: Raman scattering studies. *Phys. Rev. B* 64: 174412.

Snow, C.S., Karpus, J.F., Cooper, S.L., Kidd, T.E., and Chiang, T.-C. 2003. Quantum melting of the charge density wave state in *1T*-$TiSe_2$. *Phys. Rev. Lett.* 91: 136402.

Sooryakumar, R. and Klein, M.V. 1981. Raman scattering from superconducting gap excitations in the presence of a magnetic field. *Phys. Rev. B* 23: 3213–3221.

Strohm, T. and Cardona, M. 1997. Electronic Raman scattering in $YBa_2Cu_3O_7$ and other supercon-ducting cuprates. *Phys. Rev. B* 55: 12725.

Sugai, S., Murase, K., Uchida, S., and Tanaka, S. 1980. Raman studies of lattice dynamics in *1T*-$TiSe_2$. *Sol. St. Commun.* 35: 433.

Takada, Y., Koichi, I., and Hiroo, H. 2005. Time-resolved Raman spectroscopy. *Catal. Catal.* 47: 341–345.

Takano, K.J. and Wakatsuki, M. 1991. An optical high pressure cell with spherical sapphire anvils. *Rev. Sci. Instrum.* 62: 1576–1580.

Talley, C.E., Jackson, J.B., Oubre, C. et al. 2005. Surface-enhanced Raman scattering from indi-vidual Au nanoparticles and nanoparticle dimer substrates. *Nanoletters* 5: 1569–1574.

Thomsen, C. 1991. Light scattering in high-T_c superconductors. In *Topics in Applied Physics 68*, eds. M. Cardona and G. Guntherodt. Berlin, Germany: Springer-Verlag.

Thomsen, C. and Cardona, M. 1989. Raman scattering in high T_c superconductors. *Physical Properties of the High T_c Superconductors I*, ed. D.M. Ginsberg, pp. 409–507. Singapore: World Scientific.

Tinkham, M. 1986. *Group Theory and Quantum Mechanics*. New York: McGraw-Hill.

Travaglini, G., Morke, I., and Wachter, P. 1983. CDW evidence in one-dimensional $K_{0.3}MoO_3$ by means of Raman scattering. *Sol. St. Commun.* 45: 289–292.

Tsang, J.C., Smith, J.E., and Shafer, M.W. 1976. Raman spectroscopy of soft modes at the charge-density-wave phase transition in *2H*-$NbSe_2$. *Phys. Rev. Lett.* 37: 1407.

Tutto, I. and Zawasowski, A. 1992. Theory of Raman scattering of superconducting amplitude modes in charge-density-wave superconductors. *Phys. Rev. B* 45: 4842–4854.

Venkateswaran, U., Chandrasekhar, M., Chandrasekhar, H.R., Vojak, B.A., Chambers, F.A., and Meese, J.M. 1986. High pressure studies of $GaAs$-$Ga_{1-x}Al_xAs$ quantum wells of widths 26 Å to 150 Å. *Phys. Rev. B* 33: 8416–8423.

Verma, P., Yamada, K., Watanabe, H., Inouye, Y., and Kawata, S. 2006. Near-field Raman scatter-ing investigation of tip effects on C_{60} molecules. *Phys. Rev. B* 73: 045416.

Wyckoff, R.W.C. 1964. *Crystal Structures*, vols. 1 and 2. New York: Wiley-Interscience.

Xu, J. and Mao, H. 2000. Moissanite: A window for high-pressure experiments. *Science* 290: 783.

Yamamoto, K., Katsufuji, T., Tanabe, T., and Tokura, Y. 1998. Raman scattering of the charge-spin stripes in $La_{1.67}Sr_{0.33}NiO_4$. *Phys. Rev. Lett.* 80: 1493.

Yamamoto, K., Kimura, T., Ishikawa, T., Katsufuji, T., and Tokura, Y. 1999. Probing charge/orbital correlation in $La_{1.2}Sr_{1.8}Mn_2O_7$ by Raman spectroscopy. *J. Phys. Soc. Jpn.* 68: 2538–2541.

Yamanaka, A., Asayama, N., Sasada, M., Inoue, K., Udagawa, M., Nishizaki, S., Maeno, Y., and Fujita, T. 1996. Electronic Raman scattering spectra of Sr_2RuO_4 single crystals. *Physica C* 263: 516–518.

Yoon, S., Liu, H.L., Schollerer, G. et al. 1998. Raman and optical spectroscopic studies of small-to-large polaron crossover in the perovskite manganese oxides. *Phys. Rev. B* 58: 2795.

Yoshida, Y., Ikeda, S.-I., Matsuhata, H., Shirakawa, N., Lee, C.H., and Katano, S. 2005. Crystal and magnetic structure of $Ca_3Ru_2O_7$. *Phys. Rev. B* 72: 054412.

Yoshida, Y., Ikeda, S.I., Shirakawa, N., Hedo, M., and Uwatoko, Y. 2008. Magnetic properties of $Ca_3Ru_2O_7$ under uniaxial pressures. *J. Phys. Soc. Jpn.* 77: 093702.

Zawadowski, A. and Cardona, M. 1990. Theory of Raman scattering on normal metals with impu-rities. *Phys. Rev. B* 42: 10732.

Zirngiebl, E. and Guntherodt, G. 1991. Light scattering in rare earth and actinide intermetallic compounds. In *Topics in Applied Physics 68*, eds. M. Cardona and G. Guntherodt. Berlin, Germany: Springer-Verlag.

PART III

TIME-RESOLVED OPTICAL SPECTROSCOPY

Chapter 7

ULTRASHORT PULSE GENERATION AND MEASUREMENT

Andrew Kowalevicz

CONTENTS

7.1 INTRODUCTION

The high-speed photographic work of Harold Edgerton allowed for the study of nature on a timescale never before possible. The ability to "freeze" time allowed for a completely new way to study macroscopic phenomena. A similar but more profound paradigm shift in scientific study occurred with the advent of mode-locked lasers. In this case, the laser pulses themselves are used as "high-speed cameras" to gather information about carrier relaxation dynamics (Chemla and Shah, 2000) and other molecular processes.

As the field of ultrafast lasers has matured, more reliable and shorter pulses, with broader spectral content, have been generated with increased stability and usability. Initially, mode-locked lasers produced pulses in the nanosecond range (Gurs, 1964; Statz and Tang, 1964). Since then, directly generated pulse durations have fallen approximately six orders of magnitude to just under two optical cycles at 800 nm or approximately 5 fs in duration (Morgner et al., 1999; Sutter et al., 1999). Dye lasers, which were difficult and somewhat dangerous to operate, have given way to solid-state gain media. This progression in ultrafast technology has led to increasingly widespread use of femtosecond sources for optical communication (Nuss et al., 1996), terahertz generation (Mittleman et al., 1996), and frequency metrology (Cundiff et al., 2001; Jones et al., 2000).

Applications have become more diverse and the development of specialized sources has become more critical in order to meet the specifications of the experiment. Besides the development of commercial systems that can routinely generate sub-10 fs pulses, laboratory-based sources have produced pulses shorter than 1 fs (Baltuska et al., 2003; Hentschel et al., 2001; Paul et al., 2001). In addition, Kerr-lens mode-locked (KLM) laser oscillators have produced spectra that span a full octave (Ell et al., 2001) while amplifiers have been able to generate peak intensities that rival star cores (Mourou and Umstader, 2002).

The purpose of this chapter is to provide its reader with a firm understanding of the guiding principles behind the theory and practice of ultrashort pulse generation and measurement. Entire books have been devoted to even subsets of the topics above. Therefore, a successful reading of this chapter should be expected to provide those new to ultrafast optics with the capability to "hit the ground running," as it were, in the race for further scientific exploration. More specifically, an understanding of femtosecond technology will be vital for a full and complete utilization of the information provided in subsequent chapters. A plethora of manuscripts elucidate the finer points beyond that which the scope of these few pages will allow; several excellent references will also be mentioned in the pages to come and a summary of suggested references will be given at the conclusion of this chapter.

7.2 THE MODE-LOCKED LASER

In order to produce a train of pulses in the time domain, an optical spectrum with a bandwidth determined by the temporal duration of the pulse to be generated is required. The method proven to produce such pulses is described as mode locking. The term "mode locking" comes from early work on the subject (DiDomenico, 1964; Yariv, 1964) and refers to the process whereby the adjacent longitudinal modes of a resonator are locked in phase to each other.

Figure 7.1 provides a representation of the allowed frequencies of an oscillator within an arbitrary gain bandwidth. In reality, the broad gain bandwidths available in modern day sources support hundreds of thousands of longitudinal modes or more. Each distinct mode is a solution to the Helmholtz equation that describes the laser cavity. The fact that destructive interference results for any frequency that does not reproduce itself after each roundtrip dictates that the modes of the cavity (eigenmodes) are standing waves. As a direct result, the allowed frequencies are integer multiples of the cavity repetition rate. For a linear cavity, the nth mode can be expressed in the frequency domain as $\nu_n = n(c/2d)$, with an intermode frequency spacing of $\Delta\nu = c/2d$, where d is the single pass optical path of the linear resonator and c is the speed of light.

In general, the output of a laser will be a superposition of several modes around the gain maximum with random phases. This results in continuous wave (CW) emission with approximately constant output power. The mathematical representation of the steady-state output electric field that repeats itself with a roundtrip period of $T_R = 2d/c$ can be expressed as follows:

$$E(t) = \sum_n E_n e^{i((\omega_0 + n\Delta\omega)t + \phi_n)} \tag{7.1}$$

where
    ω_0 is the center frequency
    $\Delta\omega$ is $2\pi\Delta\nu = \pi c/d$
    E_n and ϕ_n are the amplitude and phase of the nth mode

In the special case where the laser is mode locked, the output is typically described by the slowly varying envelope approximation (SVEA). The SVEA assumes that the pulse envelope changes slowly compared to the fast modulation of the carrier wave, allowing us to express Equation 7.1 as follows:

$$E(t) = A(t)e^{i\omega_0 t} \tag{7.2}$$

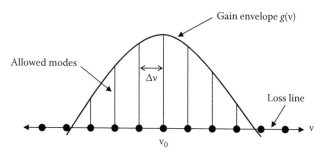

Figure 7.1 Gain envelope and allowed longitudinal modes, with appropriate mode spacing in a laser resonator.

where

$$A(t) = \sum_n E_n e^{i(n\Delta\omega t + \phi_n)} \tag{7.3}$$

When the modes that comprise the oscillation are assumed to have the same phase, i.e., when the laser is mode locked, a dramatic increase in peak intensity can be achieved compared to CW operation. Consider, as an example, a laser oscillating with $N = 20$ longitudinal modes, each of unit amplitude ($E_n = 1$). For the non-mode-locked case, each of the modes has a random phase. In such a case, the instantaneous electric field and intensity envelope of the CW output are shown in Figure 7.2a and c and plotted over 2.5 round-trips. The output intensity fluctuates randomly, staying close to an average value of ~20. However, if the laser modes all have the same phase (ϕ_n = constant), i.e., they are mode locked, then the electric field can be simplified to

$$E(t) = \sum_{n=0}^{N-1} E_n e^{i(\omega_0 + n\omega)t} \propto \frac{e^{iN\omega t} - 1}{e^{i\omega t} - 1} e^{i\omega_0 t} \tag{7.4}$$

giving a time-dependent intensity of

$$I(t) \propto |E(t)|^2 \propto \frac{\sin(N\omega t/2)^2}{\sin(\omega t/2)^2} \tag{7.5}$$

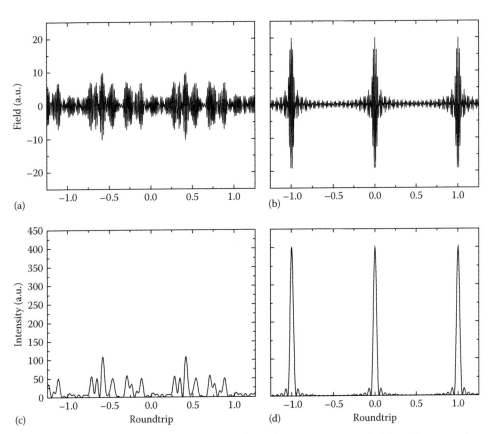

Figure 7.2   (a) Electric field and (c) intensity for a laser operating with 20 modes in random phase. By locking the phases of the laser together, pulses can be produced that significantly increase the peak values of the (b) electric field and (d) intensity.

Plots showing the corresponding mode-locked electric field and intensity envelope are also shown in Figure 7.2b and d respectively. By plotting both CW and mode-locked results on the same vertical axis, it becomes clear that phase locking adjacent longitudinal modes dramatically enhances the peak power. As the number of locked modes increases, the pulse duration, τ_p, is decreased and can be approximated by $\tau_p = T_R/N$. Since, to a good approximation, the peak intensity is N^2 times the average intensity of the N modes that are locked together, mode locking enables very high peak-power pulses to be generated from low average power systems.

7.2.1 Methods of Mode Locking

Locking the phases of the longitudinal modes of a laser may be accomplished either actively or passively. The early work of Haus has shown that analytic explanations of active (Haus, 1975a) and passive mode locking (Haus, 1975b,c) are more easily understood in the time domain; therefore, we will emphasize this view.

7.2.1.1 Active Mode Locking

Active mode locking was the first of the two methods to be demonstrated experimentally (Gurs, 1964; Gurs and Muller, 1963). This technique operates on the principle of time-dependent loss, irrespective of the specific mechanism used for implementation. It can be understood by considering a laser with many modes oscillating simultaneously. When a loss modulator (typically an acousto-optic modulator) is inserted into the cavity, we may lock the modes by modulating the loss at a period equal to the cavity round-trip time T_R. If we keep the loss of the cavity higher than the gain, except for a brief interval of time every T_R, then in general, no single mode can oscillate. This results because the window of net gain is open too briefly for the resonator to build up to oscillation.

The exception to this general case occurs if a superposition of the modes constructs a pulse that arrives at the modulator just before it opens and passes through just before it closes as shown in Figure 7.3. On each pass the modulator would clip off the leading and trailing tails of the pulse. The rest of the pulse would continue to pass unattenuated thus reinforcing its mode-locked oscillation.

In the specific case where $f_{modulator} = 2L/c = 1/T_R$, in the frequency domain, the shutter modulates the modes of the laser at the intermode frequency spacing. If we consider an initial center mode frequency of f_0, we find the production of phase-coherent sidebands at $f_0 \pm f_{modulator}$. Because the modulation frequency was chosen to be the cavity round-trip frequency, these bands sit on top of or are resonant with adjacent axial modes. Therefore, these sidebands will tend to injection lock the axial mode with which they are in resonance. As a result, the intracavity modulator will couple together, or mode lock, each axial mode

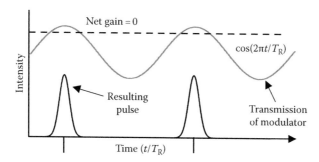

Figure 7.3 Representation of pulse generation by active mode locking in the region where net gain >0.

to its neighboring modes. It is important to emphasize that this modulation does not occur for just the center mode frequency, but for all modes. Thus, it becomes plain to see how each mode is phase locked to its neighbor, and so on, until all modes are locked together. The theory developed by Haus (1975a) predicts that Gaussian pulse shapes result from active mode locking.

In the interest of completeness, it should also be noted that active mode locking can be achieved through phase modulation (Kuizenga and Siegman, 1970), but in either case, there are several drawbacks to active mode locking. One is the complexity of control and timing electronics. Also, as the pulse becomes shorter and shorter, the strength of the pulse short-ening mechanism is reduced because the pulse length and the window of net gain (through the open "shutter") reach approximately the same duration. The physical limitations on the speed of the mechanical device (typically an electro-optic or acousto-optic modulator), in addition to the spectral narrowing effects of the gain medium (Siegman, 1986), mean that active mode locking is not suitable for the production of pulses shorter than about 500 fs in duration.

7.2.1.2 Passive Mode Locking

Passive mode locking is the most common method used to generate ultrashort pulses. Since it uses the pulse to modulate itself, it offers two advantages over active schemes: (1) it requires no external synchronization and (2) the response of the modulator can be extremely fast. In passive mode-locking schemes, the active modulator is replaced by some type of saturable absorber. A saturable absorber is a nonlinear device that has an intensity-dependent loss. It has constant absorption for low incident light intensities that saturates and decreases to lower absorption values at high intensities. Depending on their recovery time with respect to the pulse duration, saturable absorbers can be classified as either slow or fast. In contrast to active mode locking, passive mode locking leads to hyperbolic secant temporal pulse shapes (Haus, 1975b).

Slow saturable absorbers have recovery times that are long compared to the pulse duration. These materials are typically semiconductors, with relaxation times that range from 1 ps to 1 ns. These absorbers can respond to the leading edge of the pulse but require a second pulse-shaping mechanism, typically gain saturation, to close the window of net gain and thus form ultrashort pulses (Figure 7.4). Pulse durations as short as 27 fs have been reported from a dye laser using slow saturable absorber mode locking (Valdmanis et al., 1985).

Fast saturable absorbers, on the other hand, can recover their initial absorption almost instantaneously and therefore are able to shape both the leading and trailing ends of the pulse (Figure 7.5). Since there is no naturally occurring "fast" saturable absorber for pulses that are

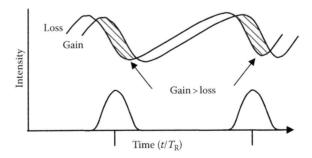

Figure 7.4 Schematic representation of pulse generation using a slow saturable absorber. The window of net gain is opened by the intensity-dependent loss of the saturable absorber and closed by gain saturation.

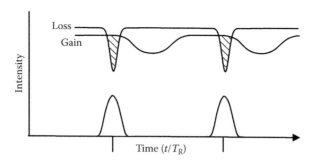

Figure 7.5 In fast saturable absorber mode locking, the absorber itself is able to both open and close the window of net gain because its response time can be treated as nearly instantaneous.

on the order of a few femtoseconds, researchers have discovered an artificial fast saturable absorber that results from the optical Kerr effect (Ippen et al., 1989; Spence et al., 1991).

7.2.2 Pulse Generation by Kerr-Lens Mode Locking

The discovery and development of an artificial fast saturable absorber in the form of KLM has revolutionized ultrafast optics and the study of ultrafast phenomena; this is by far the most widely used method for generating ultrashort laser pulses today. In order to routinely produce few-cycle laser pulses (which are now commercially available) directly from a laser oscillator, one requires a thorough understanding of the theoretical and practical underpinnings. Even at an elementary level, it is vital to be aware of the interplay between the pulse-shaping mechanisms of dispersion, self-amplitude modulation (SAM), and self-phase modulation (SPM) (both resulting from the Kerr effect), along with the cavity design considerations that make KLM possible.

7.2.2.1 Self-Amplitude Modulation and Kerr-Lens Mode Locking

The optical Kerr effect is the result of an optical nonlinearity, related to the third order susceptibility, which causes the refractive index to depend on intensity. In the presence of a strong electric field, the resultant polarization can be expressed as follows:

$$P = \varepsilon_0 (\chi^{(1)} E + \chi^{(2)} E^2 + \chi^{(3)} E^3 + \cdots) \tag{7.6}$$

While the second-order term (proportional to E^2) vanishes in centrosymmetric media, the third-order term ($\propto E^3$) is always present. Using the constitutive relationship, $D = \varepsilon_0 E + P$, we find that an intensity-dependent index of refraction results:

$$n = n_0 + n_{2I} I \tag{7.7}$$

Here

    n_0 is the linear index of refraction

    $n_{2I} = 2n_{2E}/\varepsilon_0 c n_0$ is the nonlinear index of refraction

    n_{2E} is the coefficient of the nonlinear index

It is related to the third-order susceptibility by $n_{2E} = 3\chi^{(3)}/8n_0$. It should be noted that n_{2E} is normally provided in electrostatic units (esu) and n_{2I} in cm²/W. The pulse envelope is normalized such that $I = \frac{1}{2} c \varepsilon_0 n_0 |E|^2$. We assume that the beam is linearly polarized, so that only one component of the $\chi^{(3)}$ tensor contributes to the refractive index.

Because the Kerr effect is nonresonant, it responds on extremely short timescales of only a few femtoseconds (Owyoung et al., 1972). Nevertheless, to realize the goal of truly fast saturable absorber action, this effect must be utilized to give intensity-dependent loss. This goal can be achieved in practice when the nonlinear index of refraction in the laser cavity causes a change in the beam parameters that allows pulses to have higher gain than a CW beam would. Using the Kerr nonlinearity, SAM can be achieved through self-focusing when utilized in conjunction with an intracavity aperture as described in the following.

The transverse profile of an optical pulse generated in a laser resonator generally has a Gaussian intensity distribution. The high-intensity center of the beam will thus experience a larger index of refraction (for n_2 positive) in comparison to the low-intensity wings, which causes the beam to self-focus when propagating through a Kerr medium. If we approximate the intensity distribution as parabolic and the nonlinear index change as small, we find the focal length of the Kerr medium to be $f = w^2/4n_{21}I_0L$, where w is the beam radius, I_0 the peak radial intensity, and L the length of the Kerr medium. Obviously, the focusing caused by the Kerr effect, called a Kerr lens, is small for CW light and increases in magnitude for higher peak intensities (such as those found in short pulse operation).

Figure 7.6 illustrates two mode-locking schemes that create intensity-dependent loss using a Kerr lens and an aperture. In the figures, the laser cavities consist of two planar high reflecting end mirrors and curved folding mirrors (M1 and M2) around the laser gain medium. Hard-aperture mode locking (Figure 7.6a) uses a physical slit to block CW components of light. Soft-aperture mode locking, which is the more typical contemporary approach, (Figure 7.6b) uses the gain medium as the aperture. This scheme is realized when the pump mode within the gain medium is smaller than the CW laser mode. In this case, higher-intensity pulses get focused more tightly by the Kerr lens. Because pulses have better overlap with the pumped region, they experience more gain than their CW counterpart. Under these conditions, nature favors a pulsed operating regime. The Kerr lens was first used in 1991 to self mode lock a laser (Spence et al., 1991), after which it was given the name Kerr-lens mode locking (KLM).

7.2.2.2 Self-Phase Modulation
The second important pulse-shaping mechanism that results from the Kerr effect is SPM. Whereas self-focusing results from variations in the spatial intensity of the pulse, SPM results from the time-varying intensity of the pulse envelope. The total nonlinear phase shift that results from a pulse propagating through a medium with an intensity-dependent index is

$$\Delta\varphi_{nl}(t) = -k_0 n_{21} L I(t) = -\delta A_{\text{eff}} I(t) \tag{7.8}$$

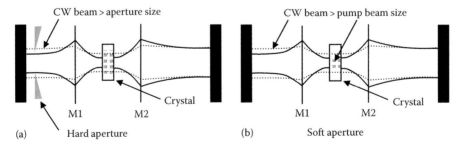

(a) Hard aperture   (b) Soft aperture

Figure 7.6 Comparison between the cavity setup for (a) hard-aperture mode locking, in which the beam is obstructed by a slit, and (b) soft-aperture mode locking, where the pumped region creates an intensity-dependent loss.

where $k_0 = 2\pi/\lambda$, L is the length of the Kerr medium, and $\delta = k_0 n_{21} L / A_{eff}$ is the SPM coefficient, with A_{eff} being the effective cross-sectional area in the medium. By taking the time derivative of the phase shift, we are left with the instantaneous frequency shift

$$\Delta\omega(t) = \frac{d}{dt}\Delta\varphi_{nl}(t) = -\delta A_{eff}\frac{dI(t)}{dt} \tag{7.9}$$

Equation 7.9 implies that the frequency content will vary across the pulse as the intensity envelope changes, an effect called "chirp." As a point of clarification, a frequency chirp on a pulse results not only from SPM but any time there is a medium with a frequency-dependent index of refraction. A schematic representation of the effects of SPM with positive n_2 is shown in Figure 7.7. Figure 7.7a shows the initial pulse with constant phase (unchirped). The instantaneous frequency shift, resulting from the pulse envelope passing through a material, is shown in Figure 7.7b. The constant spectral phase or uniform frequency across the pulse is modified due to the nonlinear interaction with the Kerr medium (Figure 7.7c). In this example, the SPM has caused a down-chirp or redshift in instantaneous frequency on the leading edge of the pulse where $dI(t)/dt$ is positive. An up-chirp or blueshift on the trailing edge of the pulse results where $dI(t)/dt$ is negative (note the negative sign in Equation 7.9).

The effect of SPM is therefore to change the frequency content of the pulse. Even though there are additional frequency components on the leading and trailing ends of the pulse, the pulse envelope has remained unchanged. It is at this point that dispersion becomes an extremely important pulse-shaping mechanism. We will see in the following section that as the pulse propagates through a Kerr medium, the frequency-dependent index can serve to either compress or broaden the pulse temporally.

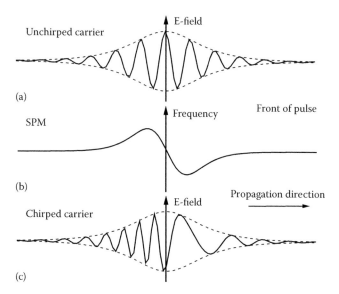

Figure 7.7 Frequency generation by SPM: (a) the initial unchirped electric field and envelope, (b) the frequency shift that is induced due to the effects of SPM, and (c) the resultant pulse after SPM modifies the spectral content and generates a chirped carrier wave.

7.2.2.3 Dispersion

The optical properties of a transparent medium are primarily determined by the index of refraction. The fact that materials are dispersive (i.e., the index is a function of frequency) has serious repercussions for the propagation of ultrashort pulses. Since pulses are composed of a range of optical frequencies, dispersion will cause different frequencies within the pulse wave packet to propagate at different speeds. As we will see, unless care is taken to recompress the spectral content of the pulse by carefully balancing the dispersive effects, the intensity envelope of the pulse will broaden in time from its transform-limited duration. A transform-limited pulse is the shortest pulse that the spectral bandwidth permits and results when all frequency components in the pulse are overlapped temporally such that the pulse has a flat spectral phase.

Let us consider an ultrashort pulse, described by the SVEA (Equation 7.2). In this case, the spectral phase evolves according to

$$\varphi(d,\omega) = \frac{\omega n(\omega)L}{c} \tag{7.10}$$

with L the length of propagation and $n(\omega)$ the frequency-dependent index. We may expand the phase around the center frequency ω_0 to second order, giving

$$\varphi(\omega) \cong \varphi_0 + (\omega - \omega_0)(\partial\varphi/\partial\omega)_{\omega_0} + \frac{(\omega - \omega_0)^2}{2}(\partial^2\varphi/\partial\omega^2)_{\omega_0} \tag{7.11}$$

In Equation 7.11 we may define the group delay (GD), $T_g = (\partial\varphi/\partial\omega)_{\omega_0}$ and the group delay dispersion, GDD $= (\partial^2\varphi/\partial\omega^2)_{\omega_0}$.

The physical significance of the GD can be understood by considering that the carrier wave propagates at the phase velocity, $v_p(\omega_0) = c/n(\omega_0)$, while the intensity envelope propagates at the group velocity, $v_g(\omega_0) = L/T_g$. This difference in propagation speeds leads to a difference in the arrival time between the carrier envelope and the frequency components that compose it. This effect alone does not cause a spreading of the pulse but an offset between the carrier and envelope phase (Brabec and Krausz, 2000; Paulus et al., 2001), which will be further explained in the next section. Pulse broadening (or compression) is governed by the GDD, which is the variation of the GD between the different frequency components that compose the pulse.

The index of refraction of many materials can be determined by Sellmeier coefficients. Therefore, it is straightforward to calculate the expected GD and GDD for a given optical path, $P = nL$. This can be done for either frequency or wavelength:

$$T_g(\omega) = \frac{P}{c} + \frac{\omega}{c}\frac{dP}{d\omega} \Leftrightarrow T_g(\lambda) = \frac{P}{c} - \frac{\lambda}{c}\frac{dP}{d\lambda} \tag{7.12}$$

$$\text{GDD}(\omega) = \frac{1}{c}\left(2\frac{dP}{d\omega} + \omega\frac{d^2P}{d\omega^2}\right) \Leftrightarrow \text{GDD}(\lambda) = \frac{\lambda^3}{2\pi c^2}\frac{d^2P}{d\lambda^2} \tag{7.13}$$

The GDD is then a measure of how strongly the material influences the spread of the spectral components in the pulse. Propagation through a material with positive dispersion, $(d^2n/d\lambda^2) > 0$, causes the long wavelength components to outrun the shorter wavelengths and bunch at the front end of the pulse; negative dispersion has the opposite effect. If we now revisit the frequency-generating effect of the intensity-dependent nonlinear index, the importance of the balance between SPM and dispersion becomes apparent.

The effect of SPM in Figure 7.7 was to generate a frequency profile in the pulse similar to positive dispersion. If left uncorrected, the pulse envelope would begin to spread due to dispersive effects. However, if the additional components could be redistributed by propagation in a region of negative dispersion, the shorter wavelengths could catch up to the longer wavelengths, and the spectral phase (instantaneous frequency) of the pulse could be made flat (unchirped). Since SPM generated new frequency components at the leading and trailing end of the pulse, its spectral content (bandwidth) has been broadened. Dispersion compensation would lead to pulse compression, resulting in a pulse with a shorter duration than the original pulse. Although beyond the scope of this work, a careful balance between SPM and negative dispersion can lead to a soliton (Nijhof et al., 1997), a pulse that propagates over long distances without changing its shape (Kamiya et al., 1999).

The pulse broadening due to a wavelength-independent GDD (no higher order dispersion) can be exactly quantified for a Gaussian pulse (Marcuse, 1980). In this case, the intensity envelope, originally at a transform-limited full-width at half maximum (FWHM) duration of τ_p, stretches to

$$\tau'_p = \tau_p \sqrt{1 + \left(4 \cdot \ln(2) \frac{GDD}{\tau_p}\right)^2} \tag{7.14}$$

For extremely short pulses, on the order of several optical cycles, higher order dispersion terms become important. In this case, third-order dispersion, $TOD = (d^3\varphi/d\omega^3)_{\omega_0}$, and fourth-order dispersion, $FOD = (d^4\varphi/d\omega^4)_{\omega_0}$ must also be considered.

7.2.2.4 Carrier-Envelope Phase

Although not strictly related to mode locking, an elementary understanding of the carrier-envelope phase (CEP) is important when femtosecond pulses are used for measurement and characterization. An ultrashort pulse can be mathematically described by an intensity envelope, which propagates at the group velocity, riding along on top of a sinusoidally varying carrier wave that propagates at the phase velocity. In general (except when propagating in vacuum), the group and phase velocities are different, so the relative position of the peak in the envelope changes with respect to its carrier as the pulse propagates. If we were to sample the electric field of a free-running mode-locked laser just after the output coupler, we would find that the waveform would evolve between successive pulses. The observed change would be due to the relative position of the carrier within its envelope, which would limit the peak electric field to the envelope value at that point.

The CEP, φ_{ce}, is the phase difference between the peak in the electric field and the peak in the electric field envelope, with the latter being converted into a phase value. The effect that the CEP has on the electric field is shown in Figure 7.8a for a few-cycle laser pulse and (b) for a many-cycle pulse with $\varphi_{ce} = 0, \pi$, which are the phases that result in the maximum and minimum peak electric fields. Most experiments utilize pulses of more than a few optical cycles, and, therefore, the change from pulse to pulse of the peak electric field does not significantly vary because the magnitude of the field envelope does not vary much over one optical cycle. On the other hand, as shorter and shorter pulses are generated, the SVEA breaks down and the difference in the peak electric field from one pulse to the next can be considerable. For experiments that rely upon the electric field instead of the intensity, such as efficient

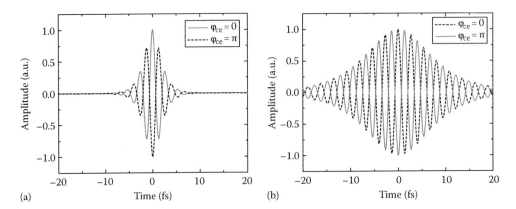

Figure 7.8 The effect of CEP on (a) a few-cycle pulse in comparison to (b) a longer pulse.

high-harmonic generation (Judson and Rabitz, 1992) and coherent control of atomic/molecular systems (Baltuska et al., 2003), controlling the CEP is critical.

7.2.2.5 Master Equation of Mode Locking

Previously, we introduced several of the major pulse-shaping components: SPM, SAM, and dispersion. By using mathematical representations for these processes, along with gain and loss, we can examine the interplay among them in a unifying equation for ultrashort pulses. The purpose of this section is not to gain a deep comprehension of the origin and implications of the master equation but rather to provide a basis to link laboratory-based observations of ultrashort pulse generation to a mathematical formalism so that the reader might gain additional insight and understanding. The master equation of mode locking is a time-dependent treatment of pulse evolution in the steady state. It relates $A(t)$, the temporal pulse shape whose magnitude squared is normalized to give the power, before and after a single round-trip through the resonator.

The various effects that influence the pulse are represented mathematically in Figure 7.9. Starting at the top and proceeding clockwise, there is linear loss and a phase shift, dispersion (GDD), SAM (saturable absorption) and SPM, and frequency-dependent gain. The linear loss, l, accounts for any loss not associated with saturable absorption such as scatter, output coupling, low reflectivity mirrors, etc. The phase-shift term $(i\Psi)$ represented in the figure results from the mismatch between the group and phase velocity and contributes to a change in the CEP. It will not be considered here because, as mentioned in the CEP section, it does not affect the pulse shape. For an in-depth derivation of these quantities, the

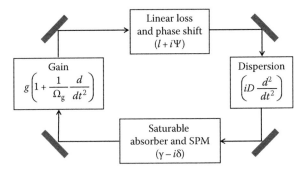

Figure 7.9 Pulse-shaping mechanisms considered in Haus' master equation for mode locking.

interested reader should consult Haus (1975b) and Haus et al. (1991). By assuming small changes, the separate effects become additive and the steady-state master equation with GDD and SPM becomes

$$A(t) - A(t + T_R) = \left[g\left(1 + \frac{1}{\Omega_g}\frac{d}{dt^2}\right) - l + iD\frac{d}{dt^2} + (\gamma - i\delta)|A(t)|^2 \right] A(t) = 0 \qquad (7.15)$$

In the equation, the first term is the gain with associated filtering due to finite bandwidth (approximated as parabolic), the SPM coefficient is given by Equation 7.8 as $\delta = k_0 n_{21} L/A_{\text{eff}}$, and the saturable absorber action is represented by γ. Also, in a medium of length L and index n, with a propagation constant ($k = k_0 n$) whose second derivative is k'', the dispersion parameter is $D = (\frac{1}{2})k''L$. The exact solution to the full master equation has a hyperbolic secant form. The solution was originally recognized by Martinez et al. (1984):

$$A(t) = A_0 \operatorname{sech}^{(1-i\beta)}\left(\frac{t}{\tau}\right) \qquad (7.16)$$

where
  β is the chirp parameter
  τ is the pulse width
  A_0 is the amplitude

A further examination of the results will give some intuition into the effects of GDD and SPM on laser performance. By substituting Equation 7.16 back into Equation 7.15, two complex equations result. By grouping the real and imaginary parts, one may obtain normalized equations for the pulse width (τ_n), bandwidth ($\Delta\omega_n$), chirp parameter (β), and stability ($l - g$) as shown in Equations 7.17 through 7.20, respectively:

$$\tau_n = \frac{D_n\beta^2 - 3\beta - 2D_n}{\delta} \qquad (7.17)$$

$$\Delta\omega_n = \frac{\sqrt{1+\beta^2}}{\tau_n} \qquad (7.18)$$

$$\beta = -\frac{3}{2}\left(\frac{\delta D_n - \gamma}{\delta + \gamma D_n}\right) - \operatorname{sign}(\delta + \gamma D_n)\sqrt{\left(\frac{3}{2}\frac{\delta D_n - \gamma}{\delta + \gamma D_n}\right)^2 + 2} \qquad (7.19)$$

$$l - g = (1 - \beta^2) - 2\beta D_n > 0 \qquad (7.20)$$

Equations 7.17 through 7.20 can be evaluated to give an overview of the important effects of system parameters on mode locking. The results are plotted as a function of normalized dispersion D_n. The family of curves represents different values of the SPM parameter, δ, for a fixed SAM coefficient, γ.

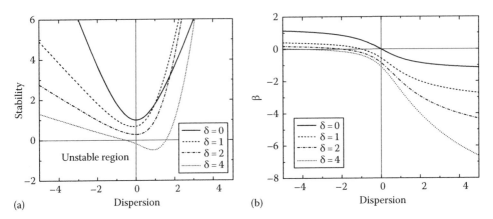

Figure 7.10 (a) Plots of stability (Equation 7.20) and (b) chirp (Equation 7.19) over several orders of SPM as a function of normalized dispersion. (From Haus, H.A. et al., *J. Opt. Soc. Am. B,* 8(10), 2068, 1991. With permission.)

Figure 7.10a shows that the solutions for the master equation are not always stable. The stability is plotted in terms of $l − g$, so that a necessary, but not sufficient condition for stability is that the intracavity loss should exceed the gain before and after the pulse to prevent perturbations from noise. The figure illustrates that dispersion plays an important role in pulse generation as the value of δ is increased. When the magnitude of SPM is large, a larger magnitude of dispersion (positive or negative) is required to maintain stable mode-locked operation.

Figure 7.10b plots the chirp parameter, β, which exhibits two distinct regions of operation separated by $D_n \sim 0$. In the negative dispersion region ($D_n < 0$), the balance between SPM, which generates a positive chirp, and negative GDD leads to soliton-like pulse shaping. In this regime where the chirp parameter is approximately zero, pulses have nearly transform-limited durations. In the positive dispersion regime ($D_n > 0$), there is no longer a balance in net dispersion. Both SPM and GDD contribute to a positive chirp, and pulses are always longer than their transform-limited durations.

The normalized bandwidth for mode-locked pulses is plotted in Figure 7.11a. As we have seen, SPM generates additional frequency components on the leading and trailing ends of

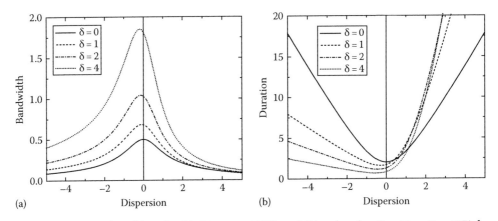

Figure 7.11 (a) Plot of bandwidth (Equation 7.18) and (b) pulse duration (Equation 7.7) for several SPM parameter values as a function of normalized dispersion. (From Haus, H.A. et al., *IEEE J. Quantum Electron.,* 28, 2086, 1992. With permission.)

the pulse. It makes sense then that as the order of δ increases, the normalized bandwidth also increases. The asymmetry observed in the plots that becomes more pronounced for larger δ is a direct result of the interplay between dispersion, SPM, bandwidth, and pulse duration. To understand the relationship among these four quantities, the results depicted in Figure 7.11a and b should be considered together.

The increase in bandwidth shown in Figure 7.11a is associated with a decrease in pulse duration represented in Figure 7.11b, because the two quantities are related by a Fourier transform (larger bandwidth → shorter pulses). In the absence of SPM (δ = 0), there are no additional frequency components created to modify the chirp of the initial pulse. The shortest pulses are observed when $D_n = 0$, because the net dispersion stays zero and the pulse remains transform limited (chirp free). The pulse duration increases symmetrically for positive and negative values of dispersion because the magnitude of dispersion affects the pulse chirp equivalently.

As δ becomes increasingly positive, the additional bandwidth allows for a shorter pulse to be generated when the additional frequency components are overlapped in time. The symmetry in bandwidth and pulse duration is broken because SPM generates a positive chirp (Equation 7.9). The addition of positive dispersion reinforces the positive chirp and causes the pulse to expand rapidly. On the other hand, the chirp due to SPM can be compensated with negative dispersion. Under these conditions, the chirp is removed and the laser is said to be operating in the soliton-like pulse-shaping regime where the pulse duration is minimized. Soliton-like pulse shaping has been shown to reduce the pulse duration by a factor of ~2.5 as compared to lasers without SPM and GDD (Haus et al., 1991). A final feature that should be pointed out is that even though pulses with more SPM produce shorter pulses for net negative GDD, their pulse durations increase more rapidly for positive GDD because their degree of chirp is increased.

The master equation with GDD and SPM is valid as long as the changes in the pulse during the round-trip are small and should reasonably describe pulses down to approximately 10 fs. For shorter pulses, the intensity envelope of the pulse is changed significantly by the individual elements in the cavity and necessitates an approach called dispersion-managed mode locking (Chen et al., 1999). Nevertheless, the results of the master equation given above are pertinent to most ultrashort pulse lasers and serve as an excellent basis for an intuitive understanding of mode-locking dynamics.

7.2.3 The Kerr-Lens Mode-Locked Oscillator

The realization of mode-locked laser operation requires careful design and alignment of the laser cavity. While two- and three-mirror resonator configurations have been analyzed (Haus et al., 1992; Kalashnikov et al., 1997), the four-mirror resonator design is the standard in KLM lasers. Its layout is relatively simple and has been extensively examined to describe stability and sensitivity to misalignment as well as to optimize performance (Brabec et al., 1992; Magni et al., 1993, 1995). This section will provide the practical tools and techniques to understand, if not build, a femtosecond laser.

7.2.3.1 Laser Cavity Layout and Stability

The basic four-mirror laser cavity, shown in Figure 7.12, consists of two curved folding mirrors (M1, M2), separated by a distance *d*, that focus the beam into the Brewster cut gain medium. The incidence angle is set for Brewster's angle to minimize the reflection loss. Moreover, the reduced loss for transverse magnetic field (TM) waves determines the polarization for the resonator. The folding mirrors can be oriented to reflect the beams on the same or opposite sides of the crystal, resulting in an X or Z folded cavity, respectively (Figure 7.12

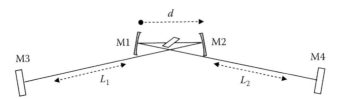

Figure 7.12 Standard X-folded four-mirror laser cavity.

depicts an X folded cavity). Since the gain medium is usually optically pumped through the folding mirrors, they typically have coatings designed to transmit the pump wavelength, while maintaining high reflectivity over the bandwidth of the gain material. Arm 1, of length L_1, is terminated by a planar high reflecting mirror (M3). Arm 2, of length L_2, uses a planar output coupling mirror (M4) to retroreflect the beam. The output coupler is a partially reflecting mirror used to transmit a fraction (~1%–20%) of the intracavity power.

In order for the cavity to behave as a stable resonator, the spatial electric field profile must repeat itself after every round-trip. This general stability condition can be met by Hermite–Gaussian functions. For the specific case of stable, mode-locked operation, the lowest order or fundamental Gaussian solution is required. In this case, an *ABCD* matrix analysis of the *q*-parameter can determine the stability requirements. The cavity is considered to be stable when the beam waist has a finite size.

Knowledge of the *q*-parameter provides full information about the spot size and curvature of the beam. Specifically, the *q*-parameter is a complex quantity defined as follows:

$$q(z) = z + i\frac{2\pi w_0^2}{\lambda} = z + i2z_R = z + ib \tag{7.21}$$

Figure 7.13 gives a graphical representation of the quantities expressed in Equation 7.21, where z is the axial distance from the beam waist (its narrowest point), z_R is the Rayleigh length, the distance of propagation over which the area of the beam increases by a factor of two from its minimum area, and the confocal parameter, $b = 2z_R$. The inverse of the *q*-parameter is also useful and is defined as follows:

$$\frac{1}{q(z)} = \frac{1}{R(z)} - i\frac{\lambda}{\pi w^2(z)} \tag{7.22}$$

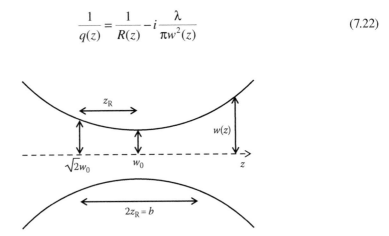

Figure 7.13 Schematic representation of a Gaussian beam with key parameters illustrated.

where

$$w(z)^2 = w_0^2 \left[1 + \left(\frac{\lambda z}{\pi w_0^2} \right)^2 \right] \tag{7.23}$$

Here, $w(z)$ is the spot size, defined by the radius where the intensity drops by $1/e^2$, and

$$R(z) = z \left[1 + \left(\frac{\pi w_0^2}{\lambda z} \right)^2 \right] \tag{7.24}$$

is the radius of curvature of the beam. The minimum spot size, found when $z = 0$ ($R = \infty$), is w_0.

As a Gaussian beam propagates through an optical system, the transformation of its q-parameter is governed by the well-established formalism of $ABCD$ matrices. Specifically, it undergoes a bilinear transformation given by the $ABCD$ law (Kogelnik and Li, 1966):

$$q' = \frac{Aq + B}{Cq + D} \tag{7.25}$$

Here A, B, C, and D are the elements of the final transformation matrix, T, that governs the complete optical system. The T matrix for an arbitrary array of different optical elements, such as displacements and lenses, can be calculated by multiplying the representative matrices for each element along the direction of propagation. For example, if a Gaussian beam started at $z = 0$, propagated a distance z_1 and then propagated through a lens of focal length f, then $T = M_f \cdot M_{z_1}$, where M_{z_1} and M_f are the $ABCD$ matrices that represent free-space propagation and the lens, respectively. By this method, the Gaussian beam parameters can be determined at any location in the cavity and the stability criteria can be established.

Let us take a linear resonator as an example; in this case, we will use a simplified version of the four-mirror cavity shown in Figure 7.11 but without the gain medium. Figure 7.14 shows the "unfolded" cavity with the curved mirrors represented as thin lenses with $f = R/2$. We assume the beam makes one round-trip through the laser cavity. In this specific case, we have chosen to start and end at the position midway between the curved mirrors, but any location can be chosen. Using this representation along with Table 7.1, which lists some useful $ABCD$ matrices for calculations such as ours, determining the final transformation matrix is straightforward.

In order for the beam to be maintained within the resonator, it must reproduce itself upon returning to the same location in the cavity. Therefore, the mathematical representation for the round-trip propagation can be expressed as follows:

$$q' = \frac{Aq + B}{Cq + D} = q \tag{7.26}$$

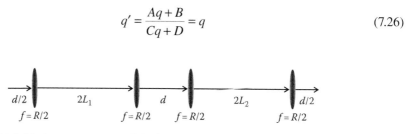

Figure 7.14 Unfolded representation of the four-mirror laser cavity that can be used for $ABCD$ analysis.

TABLE 7.1
ABCD Matrices for Commonly Encountered Optical Elements

Optical system	Transfer matrix	Optical system	Transfer matrix
Freespace propagation	$\begin{pmatrix} 1 & d \\ 0 & 1 \end{pmatrix}$	Distance–lens combination	$\begin{pmatrix} 1 & d \\ \frac{-1}{f} & 1-\frac{d}{f} \end{pmatrix}$
Thin lens	$\begin{pmatrix} 1 & 0 \\ \frac{-1}{f} & 1 \end{pmatrix}$	Dielectric ($n_1 = 1$)	$\begin{pmatrix} 1 & \frac{d}{n} \\ 0 & 1 \end{pmatrix}$

From Equation 7.26, the general stability criteria can be established. For a periodic optical array to be stable, the trace of the *ABCD* matrix ($A + D$) must obey the inequality (Kogelnik and Li, 1966):

$$-1 < \frac{1}{2}(A+D) < 1 \tag{7.27}$$

Beyond establishing stability requirements, Equation 7.26 allows us to determine z and b by solving the quadratic equation for the q-parameter directly. We may also solve the equation as being quadratic in ($1/q$), with solution

$$\frac{1}{q} = \frac{D-A}{2B} \pm i\frac{\sqrt{4-(A+D)^2}}{2B} \tag{7.28}$$

By equating the terms with Equation 7.22, we can solve for R(z), which is just equivalent to the real term in the solution, or $w(z)$, the spot size, which is

$$w^2 = \frac{\lambda}{\pi}\frac{2|B|}{\sqrt{4-(A+D)^2}} \tag{7.29}$$

The absolute value of B is used to assure that the sizes are positive.

We have presented a simple four-mirror cavity, but there are more complicated resonator structures than the one discussed above. In particular, some cavities contain multiple folding mirrors or secondary foci. Nevertheless, in every case, these unfolded resonators are equivalent to a periodic sequence of identical optical systems and therefore allow the *ABCD* matrix formalism to be used to calculate the mode parameters of the resonator. Using this technique, let us gain some insight into how the stability regions change for different cavity parameters.

We will examine the spot sizes on the end mirrors in the section on Kerr-lens sensitivity, but for now let us consider the beam waist, w_0, between the two curved folding mirrors. Let us evaluate several similar laser cavities emitting at 800 nm, with the focal lengths of M1 and M2

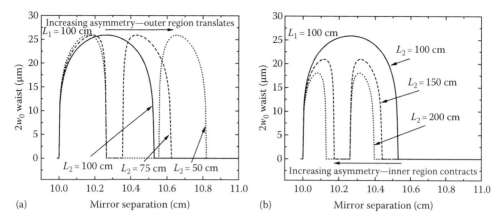

Figure 7.15   Plot shows the changes that occur in the size and position of the stability region of a four-mirror laser cavity as the arm length symmetry is broken by (a) decreasing the length of second arm and (b) increasing the length of the second arm.

equal to 5 cm. Figure 7.15 shows the results for lasers with $L_1 = 100$ cm and $L_2 = 50, 75, 100, 150,$ and 200 cm as a function of folding mirror separation. The condition for stability established by Equation 7.27 still governs, but, more practically, we see that the cavity becomes a stable resonator when the curved mirror separation, $d \geq f_1 + f_2$, with f_1 and f_2 being the focal lengths of the folding mirrors. For a symmetric cavity, where $L_1 = L_2$, there is a single stability region. When the lengths of the arms become unequal, the stability region splits into two. The inner stability region, as it is commonly called, is the region on the left that is obtained for lower d values. The region on the right is known as the outer stability region. Figure 7.15a shows the beam waist for several cavities when L_2 is shorter than L_1. The location of the inner stability region is fixed and the outer stability region shifts away from the inner region when the arm length asymmetry increases. Even though the region shifts, the approximate width and maximum size of the beam waist is maintained. On the other hand, when L_2 is longer than L_1, it is the outer stability region whose inner edge stays fixed. Figure 7.15b shows that both regions shrink in size as the length of the long arm increases, with the inner edge of the outer region staying fixed in position. With all other things being equal, the size of the stability regions is set by the length of the longer arm.

7.2.3.2 Astigmatic Correction

The treatment of cavity stability for the linear resonator neglected the effects of beam astigmatism. Because the focused beam is incident on the crystal at an angle, it becomes astigmatic. This leads to a beam waist size difference between the sagittal (perpendicular to the plane of incidence) and tangential (parallel to the plane of incidence) planes. There is, therefore, an independent stability region for each of the two orthogonal beam axes. This can lead to reduced output power and Kerr-lens strength resulting from poor mode matching between the pump and laser modes, resulting in overall reduced laser performance.

Since the curved mirrors also introduce astigmatism due to off-axis illumination, the deleterious effects of the Brewster-cut crystal can be compensated. It has been shown (Kogelnik et al., 1972) that for a crystal of index n and thickness t, with a mirror of focal length f, the correct compensation angle, θ, can be determined from

$$f\left(\frac{1}{\cos\theta} - \cos\theta\right) = t\frac{\sqrt{1+n^2}}{n^2}\left(1 - \frac{1}{n^2}\right) \tag{7.30}$$

To be clear, Equation 7.30 gives the angle of incidence required for astigmatic correction from a single mirror. In most practical examples, the compensation is done by a pair of mirrors with the same radius of curvature $R = 2f$. In this case, the equation can be simplified to give the angle of incidence necessary for an equal correction from both mirrors:

$$\theta_{1/2} = \arccos\left[\sqrt{1+\left(\frac{Nt}{2R}\right)^2} - \left(\frac{Nt}{2R}\right)\right] \tag{7.31}$$

with

$$N = \frac{\sqrt{1+n^2}}{n^2}\left(1-\frac{1}{n^2}\right) \tag{7.32}$$

7.2.3.3 Kerr-Lens Sensitivity

We saw in Section 7.2.2 that fast saturable absorber mode locking can be achieved by introducing an intensity-dependent loss with either a hard or soft aperture. Cavity alignment can influence the magnitude of the loss and contribute to the resulting mode-locking strength. The optimum alignment for mode locking can be found by analyzing a quantity that is sensitive to Kerr-lensing as a function of relevant cavity parameters (Bouma et al., 1997; Cerullo et al., 1994b; Magni et al., 1995). A common method is to calculate the Kerr-lens sensitivity, δ, which represents the change in beam spot, w, with respect to power, P, in the form

$$\delta = \left(\frac{1}{2w}\frac{dw}{dP}\right)_{P=0} \tag{7.33}$$

To first order, the variations of the cavity losses with power are proportional to δ; therefore, KLM is only possible if an aperture of a suitable size is placed in a position corresponding to negative values of δ. In this situation, higher-intensity pulses will be decreased in radius and have higher transmission through the aperture.

The Kerr-lens sensitivity is usually calculated as a function of crystal position and mirror separation, but the arm length ratio also influences the magnitude of δ. Symmetric cavities have maximum Kerr-lens sensitivity when the laser is operated near the center of the single stability region (curved mirror separation midway between the confocal and concentric positions (Figure 7.15a), with the crystal positioned midway between the folding mirrors. The modulation depth at this operating point is large enough to obtain self-starting KLM (Cerullo et al., 1994a; Magni et al., 1995). Because the arm lengths need to be precisely matched to prevent the stability region from bifurcating, symmetric cavities are particularly susceptible to environmental effects.

As a result of the instability associated with symmetric cavities, asymmetric cavities are more typically used for KLM lasers. As was discussed in the previous section, unequal resonator arms lead to a bistable cavity. A calculation of the transverse Kerr-lens sensitivity reveals that in such a setup, there are two configurations that optimize δ. The first occurs when the laser is operated on the outer edge of the inner stability region, with the crystal positioned slightly closer to M1. In the second configuration, the crystal is slightly closer to M2, with the laser operating on the inner edge of the outer stability region. Although both regions lead to KLM, the second is more typically used because of its low sensitivity to misalignment (Cerullo et al., 1994b). For best KLM performance, the arm length ratios should range between 4:3 and 2:1 (Sutter, 2000).

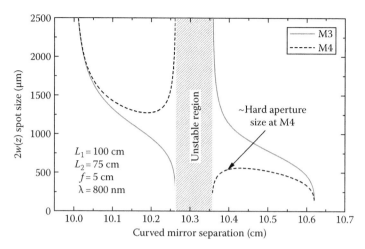

Figure 7.16 Plot of laser spot sizes on each of the end mirrors of a four-mirror cavity as a function of curved mirror separation. Results show that M4 would be an appropriate place to locate a hard aperture for mode locking.

For calculations of δ, it is necessary to consider which position in the cavity is most appropriate for the desired mode-locking mechanism. For hard-aperture KLM, (Figure 7.16b), δ is most often calculated near an end mirror because this is the most convenient position for a slit to be placed. Figure 7.16 shows the relative spot sizes on end mirrors M3 and M4 for the four-mirror cavity in Figure 7.12 as a function of the curved mirror separation. Since the spot size on the short arm of the oscillator is reduced in pulsed operation, a slit slightly smaller than the CW mode shown in the figure at the inner edge of the outer stability region will increase the KLM strength. Previous work (Magni et al., 1995) has shown that astigmatism causes the Kerr-lens sensitivity to increase in the tangential plane, while decreasing in the sagittal plane. For this reason, hard-aperture KLM is usually implemented by using a vertically oriented slit to cut the beam in the tangential plane.

For soft-aperture KLM, which was represented schematically in Figure 7.16a, the Kerr-lens sensitivity must be summed over the length of the crystal. A direct integration along the length of the crystal (Bouma et al., 1997) and a split-step method (Grace et al., 2000; Penzkofer et al., 1996) have been used to calculate δ for soft-aperture KLM. In practice, one typically achieves soft-aperture KLM by setting the pump mode size to be slightly smaller than the CW resonator mode inside the gain medium while the laser is operating on the inner edge of the outer stability region. This can be accomplished by changing the focal length of the pump lens or adjusting the length of the long arm of the laser cavity.

7.2.3.4 Dispersion Compensation

Effective dispersion management is one of the greatest challenges in the generation of ultrashort pulses. As discussed previously, dispersion is caused by a frequency-dependent GD, T_g. Since any wave packet can be described as a superposition of narrow band components, each traveling at their own group velocity, a frequency-dependent group velocity will lead to a change in the pulse shape. For the majority of solids, liquids, and gases in the visible to near-infrared (IR) wavelength range, the group velocity increases with increasing wavelength, so the GD increases with frequency. Since the GDD is defined as follows:

$$\text{GDD}(\omega) = \left(\frac{dT_g(\omega)}{d\omega} \right)_{\omega_c} = \left(\frac{d^2\varphi}{d\omega^2} \right)_{\omega_c} \tag{7.34}$$

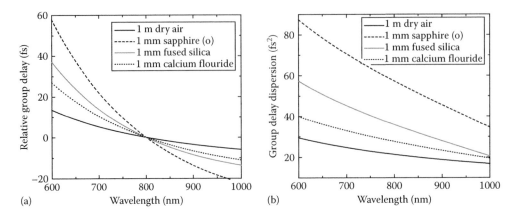

Figure 7.17 Using Equations 7.12 and 7.13, the (a) GD and (b) GDD can be found for common materials. The relative GD had been set to zero at 800 nm for the figure shown; this does not affect the calculation of GDD.

where ω_c is the center frequency, the GDD will be positive in this case. Positive GDD is also referred to as normal dispersion. Normal dispersion causes the red components to propagate faster than the blue components.

In general, the dispersion introduced by an optical element is measured with white-light interferometry (Diddams and Diels, 1996; Naganuma et al., 1990); however, by using Equations 7.12 and 7.13, T_g and the GDD can be calculated in the near-IR for some common materials used in femtosecond laser cavities. To get an idea of how much dispersion is present in the cavity of an ultrafast oscillator, let us consider a typical source. The most common femtosecond lasers have a round-trip cavity length of approximately 3 m (operating at 100 MHz) and use Ti:Al$_2$O$_3$ as a gain medium. In our calculation, the gain crystal is 3 mm long and the laser uses dielectric or metal mirrors that we will assume, for now, contribute negligible positive or negative dispersion. The total round-trip dispersion of the cavity, obtained from Figure 7.17, would be approximately 410 fs^2, with 350 fs^2 coming from the crystal and the remainder from air. Even though the dispersion of air is significantly lower than the solid-state material, its overall contribution to the total dispersion is significant because the beam propagates through 3 m of air as opposed to only 6 mm of the crystal. In order to produce femtosecond pulses, negative dispersion must therefore be introduced to balance the positive dispersion of the cavity.

There are three common methods for generating negative or "anomalous" dispersion, as it is sometimes called: (1) prismatic compensation, (2) dispersion compensating mirrors, and (3) grating compression. Gratings can generate large values of dispersion and are typically used in laser amplifiers or for external cavity compensation but are generally too lossy for intracavity dispersion management. Therefore, we will focus on the first two methods in our discussion of dispersion compensation within a mode-locked laser oscillator.

7.2.3.5 Prism Compressor

A schematic representation of prisms configured to introduce tunable dispersion is shown in Figure 7.18. Prism compressors, because of their low cost and flexibility, are the most common technique to introduce negative dispersion. While they are discussed here for intracavity usage, they are commonly used extracavity to compensate for dispersion that results from optical elements in an experimental setup. The four-prism sequence was originally introduced by Fork et al. (1984). This symmetric configuration of prisms allows for

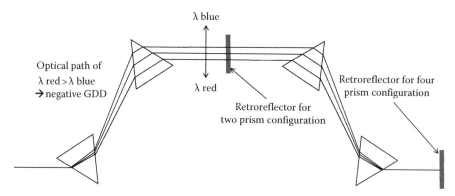

λ blue

Optical path of
λ red > λ blue
→ negative GDD

λ red

Retroreflector for
two prism configuration

Retroreflector for four
prism configuration

Figure 7.18 Schematic representation of a four-prism compressor that is capable of introducing negative GDD. Often, a mirror is placed after a two-prism sequence to retroreflect the beam and simplify alignment.

the beam to be spatially dispersed, compressed, and recombined in a single pass. A more simple design, with the second pair of prisms replaced by a retroreflecting mirror, is now the standard configuration.

The prisms in the compressor are rotated 180° with respect to each other so that their bases remain parallel. Properly aligned prisms allow the beam to propagate parallel to their bases. The apex angle is cut for a specific center wavelength so that the beam may enter and leave the prism at Brewster's angle. This reduces loss for incident p-polarized light, and since the Fresnel loss near Brewster's angle is not strongly angle dependent, low loss can be maintained across a wide bandwidth without the need for an antireflection coating.

The prism compressor generates negative dispersion by introducing a wavelength-dependent path length. Equation 7.13 can then be used to calculate the resulting GDD. The fully generalized solution has been calculated elsewhere (Sherriff, 1998), but if we limit ourselves to rays incident at Brewster's angle only ($\lambda_{\text{Brewster}}$), the solutions for both GDD and third-order dispersion, TOD (Sutter, 2000), become significantly simpler:

$$\text{GDD}(\lambda_{\text{Brewster}}) \equiv \frac{\lambda_{\text{Brewster}}^3}{2\pi c^2} P'' \tag{7.35}$$

$$\text{TOD}(\lambda_{\text{Brewster}}) \equiv -\frac{\lambda_{\text{Brewster}}^4}{4\pi^2 c^3}\left(3P'' + \lambda_{\text{B}} P'''\right) \tag{7.36}$$

Here, the derivatives of the optical path with respect to wavelength at Brewster's angle are

$$P'' = 2\left[n'' + \left(2n - n^{-3}\right)(n')^2\right]\frac{2n}{n^2 + 1} H - 4(n')^2\left(L + \frac{n^2 - 1}{n^2 + 1}H\right)$$

$$P''' = \left[6(n')^3\left(n^{-6} + n^{-4} - 2n^{-2} + 4n^2\right) + 12n'n''\left(2n - n^{-3}\right) + 2n'''\right]\frac{2n}{n^2 + 1}H \tag{7.37}$$

$$+ 12\left[\left(n^{-3} - 2n\right)(n')^3 - n'n''\right]\left(L + \frac{n^2 - 1}{n^2 + 1}H\right)$$

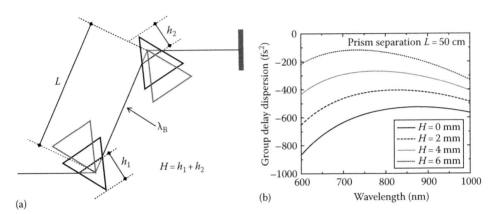

Figure 7.19 (a) Schematic representation of the beam path through a two-prism compressor and (b) the resulting GDD observed for a single pass through the compressor for several different prism insertion values.

To finely tune the dispersion, one or both of the prisms are mounted on translation stages that allow them to be inserted into the beam (introducing positive material dispersion). In Equation 7.37, L is the apex to apex distance when the wavelength $\lambda_{\text{Brewster}}$ just barely passes through both prisms at their minimum insertion distance and H is the sum of the length of insertion of both prisms into the beam. Figure 7.19a gives a schematic representation of the variables L and H with (b) showing the dispersion obtained in the near-IR for a pair of fused-silica prisms cut at 69.1°(Brewster's angle for $\lambda_{\text{Brewster}} = 800\,\text{nm}$). Note that the equations above and Figure 7.19 give the dispersion for a single pass through the prism sequence, the resulting dispersion should be doubled for its return path or if the beam propagates along a single pass through a four-prism sequence.

Generating sub-10 fs pulses using prisms is difficult because there is no way to compensate for GDD and TOD or higher order dispersion independently. Changing the prism material offers some flexibility to compensate GDD and TOD simultaneously but only over limited wavelength ranges. Even with these limitations, prismatic compensation alone has produced pulses as short as 8.5 fs (Zhou et al., 1994). However, more sophisticated dispersion management is required to reliably produce pulses directly from an oscillator with sub-10 fs durations.

7.2.3.6 Dispersion Compensating Mirrors

As the transform-limited pulse duration becomes shorter, the effect of higher order dispersion becomes greater. For example, in order to approximately double the duration of an unchirped, 20 fs Gaussian-shaped pulse a GDD = 400 fs² or a TOD = 8000 fs³ would be required. By comparison, the duration of a 10 fs pulse would be approximately doubled for a GDD = 100 fs² or a TOD = 1000 fs³. While we have seen that a prism pair can compensate for the GDD quite well, only higher order dispersion correction can enable the generation of few-cycle pulses. This can be realized by customizing the multilayered coatings on dielectric mirrors, a technology called double-chirped mirrors (DCMs) that has allowed for few-cycle pulses to be generated in a number of laser systems including $Ti:Al_2O_3$ (Morgner et al., 1999), Cr:LiCAF (Wagenblast et al., 2002), Cr:Forsterite (Chudoba et al., 2001), and Cr:YAG (Ripin et al., 2002) (Figure 7.20).

A standard dielectric mirror, called a Bragg mirror, consists of alternating pairs of high- and low-index material. Each layer has an optical thickness, $\lambda_B/4$, where λ_B is the Bragg wavelength at which the mirror is designed to have its highest reflectivity. If we assume that

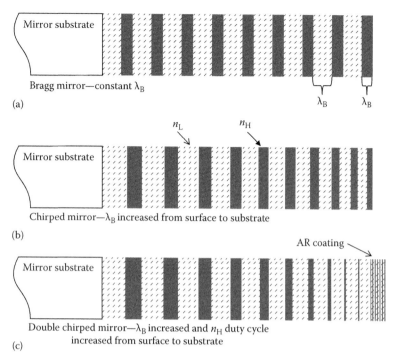

Figure 7.20 Mirror designs for ultrashort pulse generation have progressed from the (a) standard Bragg stack, to the (b) first dispersion-compensating mirror in the form of a chirped mirror, to the (c) DCM that can produce high reflectivity and negative dispersion with reduced ripples in the GDD.

the refractive index of the coating materials is wavelength independent, then the reflectivity bandwidth of the mirrors can be calculated from the Fresnel reflectance, r, resulting from the high (n_H) and low (n_L) index material:

$$\frac{\Delta\omega}{\omega_B} = \frac{4}{\pi}\arcsin\left(\frac{n_H - n_L}{n_H + n_L}\right) = \frac{4}{\pi}\arcsin(r) \tag{7.38}$$

For typical low (MgF$_2$ [$n = 1.37$] and SiO$_2$ [$n = 1.45$]) and high (Ta$_2$O$_5$ [$n = 2.1$] and TiO$_2$ [$n = 2.35$]) index materials, the fractional bandwidth is in the range of one-quarter to one-third of the center frequency. While this bandwidth is fairly broad, it does not reach the octave spanning regime that requires a fractional bandwidth of 2/3. Moreover, while the contribution to GDD is small near the center of the reflectivity bandwidth, it grows rapidly at the edges, contributing significant positive (negative) dispersion for wavelengths shorter (longer) than λ_B.

An improvement over the simple Bragg stack was first suggested in the design of "chirped mirrors"(Szipöcs et al., 1994). The new design mimicked the chirped fiber Bragg grating, which was already well developed. By slowly decreasing the Bragg wavelength of the coatings as the layers were deposited, several improvements were realized. First, by slowly varying the Bragg wavelength, the high-reflectivity bandwidth of the mirror could be extended simply by adding more layers. At the same time, because the design deposits the thickest layers starting at the mirror substrate, the short wavelengths are reflected first and allow the long wavelengths to penetrate deeper into the mirror. Since the average GD increases with increasing wavelength, the mirror is able to generate negative GDD.

Despite successfully producing, for the first time, sub-10 fs pulses directly from a cavity without intracavity prisms (Stingl et al., 1995), chirped mirrors were limited in their performance. They exhibited strong GDD oscillations, especially as they were extended to operate over larger wavelength ranges. Two sources for the observed oscillations were discovered. The first issue was that the Fresnel reflection at the air interface created additional dispersion in a Gires–Tournois-like fashion when it interfered with reflections from deeper layers. The second problem was discovered to be essentially an impedance mismatch that resulted from the thicknesses of the high- and low-index layers.

In order to correct for the impedance mismatch and eliminate the unwanted reflections that generated the oscillations, a new design has been developed that relies upon smoothly varying the thickness of the high-index layer from almost zero thickness near the air interface to a full quarter wave near the substrate. At the same time, the Bragg wavelength of the combined high-/low-index unit cell is increased as a function of depth, as before. Since both λ_B and the relative thickness of the high- and low-index layers of these mirrors vary systematically, they are named double chirped mirrors (DCMs). In an additional effort to provide better impedance matching, an antireflective coating is added to the double-chirped dielectric stack and the entire structure is further optimized. This combination of improvements allows DCMs to provide smoother and more precise dispersion compensation over a wider wavelength range, while still maintaining a high reflectivity (Figure 7.21).

DCM designs allow for prismless cavities to be designed. In general, such designs are more environmentally stable because, as opposed to cavities with prisms, a slight change in beam position has no effect on the overall dispersion operating point. On the other hand, DCMs have been designed to work in conjunction with intracavity prism compressors. The inclusion of the prisms can relax the design requirements of the mirrors, while still providing the ability to fine-tune the dispersion for optimal pulse duration (Morgner et al., 1999). Utilizing a single mirror design to provide reflectivity and dispersion compensation for bandwidths in excess of an octave results in dispersion oscillations that are difficult to suppress. In order to overcome

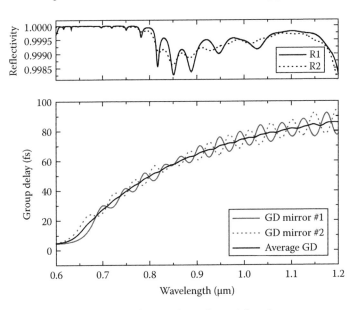

Figure 7.21 Complementary DCM design that allowed for dispersion compensation and high reflectivity over an optical octave. By designing the mirror pairs with oscillations that are out of phase with each other, the average GDD can be made smooth over a wide wavelength range. (Data provided by F. X. Kaertner.)

this challenge, complimentary pair DCMs have been fabricated so that the oscillations from one mirror are out of phase with the other. When the dispersion oscillations from all of the cavity elements are added, they cancel to produce a smoothly varying GD across the entire wavelength range. Such mirrors were used to generate the shortest fully characterized pulses (5 fs) directly from a laser oscillator (Ell et al., 2001).

7.3 AMPLIFICATION AND TUNING OF ULTRASHORT PULSES

Thus far, we have restricted our discussion to the requirements and challenges behind the generation of ultrashort pulses from a laser oscillator. However, the focus of this section of the book is the use of femtosecond pulses to investigate ultrafast processes in materials for which it is important to have the right tool (i.e., producing ultrashort pulses at an appropriate wavelength and intensity) for a given application. As the reliability and availability of femtosecond oscillators has become more widespread, the demand for higher intensities and access to other wavelength ranges has kept pace. In the section to follow, we will review techniques for femtosecond pulse amplification and the nonlinear optical systems that allow new wavelengths to be generated from existing sources.

7.3.1 Amplification of Femtosecond Pulses

The short pulse duration and reasonable average power available from femtosecond oscillators mean that pulses with peak powers in the hundreds of kilowatts and intensities in the MW/cm^2 range for unfocused beams have become commonplace. As such, the direct amplification of such short pulses fails due to a number of nonlinear effects that introduce nonlinear spectral and temporal distortions. A more serious limitation results from the potential damage or destruction of the gain medium or other optical elements due to the realization of extremely high intensities.

A laser gain medium cannot continue to provide a fixed gain for an arbitrarily high input power because this would require an arbitrary amount of energy to be available. As a result, when a high-power pulse is injected into a laser amplifier, the leading edge of the pulse is preferentially amplified, which leads to the observation of self-steepening as shown in Figure 7.22a. In addition, because the gain is changing as the pulse is propagating, the imaginary part of the index of refraction creates a nonlinear shift in the phase of the pulse, as does SPM. The overall pulse shape is modified by the combination of edge steepening due to gain saturation and the pulse distortion resulting from dispersion acting on the induced phase modulation. The result is that the original pulse shape and duration become unrecoverable.

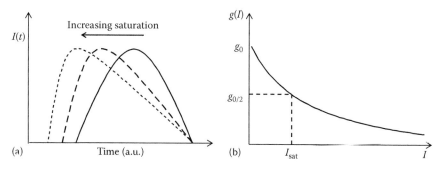

Figure 7.22 (a) Self-steepening effect that gain saturation can have on the pulse envelope when the leading edge is preferentially amplified. (b) Representation of gain saturation. In order to achieve a high overall gain factor, it is important to keep the intensity for the initial passes of amplification well below I_{sat}.

An equally or more serious problem that limits the achievable gain in standard amplification techniques is self-focusing. For very high intensities, the same Kerr nonlinearity that is exploited for KLM leads to catastrophic beam behavior. Because the pulse power is well above the critical power for self-focusing, the beam either collapses and generates filamentation in air or focuses within and destroys the gain material directly. The solution to these obstacles was proposed in 1985 (Strickland and Mourou, 1985) and is known as chirped pulse amplification (CPA).

7.3.1.1 Chirped Pulse Amplification

CPA mitigates the deleterious effects associated with extremely high intensities by stretching the pulse before amplification. This technique, which has revolutionized ultrafast science and technology, is a multistep process. First, the pulse to be amplified is temporally broadened, typically using diffraction gratings, to increase its duration by ~10^3–10^4. This significant increase in pulse width reduces the intensity in the gain medium, which allows operation below the damage threshold and avoids untenable nonlinear effects as discussed above. Thus, the pulse energy can be increased over a number of passes through the amplifying medium. Finally, the pulse is recompressed by a dispersion line (that usually also uses gratings). The pulse compressor would have the opposite sign and magnitude of GDD from that which was used to stretch the pulse, along with enough additional compensation to correct for the dispersion of the gain medium. If the gain bandwidth is not exceeded by the pulse, and the amplifier is not saturated, then the original pulse duration may be recovered (Diels and Rudolph, 2006).

The goal of amplification is not just to increase the energy of the pulse per se, but to increase the intensity that can be achieved. As a result, well-designed amplifiers are expected to achieve high overall gain, while at the same time, maintain excellent mode quality. In order to achieve a high gain factor, it is important to keep the initial pulse intensity low. As shown in Figure 7.22b, when the pulse intensity is much lower than the saturation intensity, I_{sat}, the gain can operate in the small signal regime where the gain coefficient is highest (close to g_0). In a typical setup, the pulse is made to undergo a number of passes through the gain medium, such that the intensity eventually reaches near saturation levels before being dumped from the amplifier. This approach maximizes energy extraction efficiency and enhances pulse-to-pulse stability. There are two major amplifier configurations that utilize CPA to increase the energy of femtosecond pulses: the multipass amplifier and the regenerative amplifier.

7.3.1.2 Multipass Amplifier

Instead of repetitively passing the beam through the gain medium along the same optical path in a scheme similar to oscillators, the multipass amplifier takes a different approach. It makes use of a bow-tie or other similar geometric beam configuration to allow the pulse to pass through the gain medium multiple times along different paths. Depending on the repetition rate and the number of passes, a multipass amplifier can be CW or synchronously pumped. In either case, however, care must be taken to ensure that the profile of the pump beam does not negatively affect the mode quality of the pulse being amplified.

Multipass amplifiers are commonly, although not exclusively, used to give an additional stage of amplification to the high-energy output of a regenerative amplifier. In this case, no control electronics or switching is required at the input of the amplifier. If, on the other hand, the multipass amplifier is used to increase the pulse energy directly from an oscillator, a Pockels cell can be used in conjunction with a polarizing beamsplitter (PBS) as shown in Figure 7.23. A Pockels cell is an electro-optic device that is capable of functioning as a quarter- or half-waveplate when energized by a high-voltage pulse. In this particular case, the input beam is ordinarily cross-polarized with the PBS so that pulses are not transmitted into the

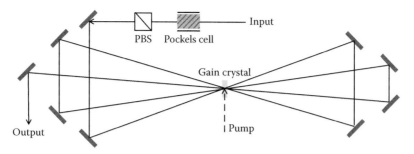

Figure 7.23   Setup of a multipass amplifier in a bow-tie-like configuration.

multipass setup. To switch a pulse into the amplifier, the Pockels cell is activated to function as a half-wave plate, which rotates the polarization of the incoming pulse 90°, allowing it to be transmitted through the PBS and the rest of the amplifier.

Precise alignment and an increased number of optical elements are required for a multipass amplifier. The rigid geometric design means that the number of passes cannot be easily changed. In addition, practical issues restrict the amplification to no more than 4–8 passes. This constraint limits the overall gain of the system, even though this amplifier is typically operated near the damage threshold. For higher gain efficiency, one has to make use of the more complex regenerative amplifier. On the other hand, the multipass amplifier is the only technique capable of producing pulse energies above 50 mJ (Rulliere, 2005) because of the damage threshold of the intracavity elements, such as the Pockels cells, used in regenerative amplifiers. Commercially available systems operating at 10 Hz repetition rate can generate 1 J pulses with durations shorter than 25 fs, which equates to peak powers in excess of 10 TW.

7.3.1.3 Regenerative Amplifier

Regenerative amplifiers trade off the technological simplicity of multipass amplifiers for generally higher gain and superior beam quality. These systems require high-speed, high-voltage electronics and precise timing to pick a temporally dispersed pulse from the oscillator chain. The amplifier section is basically a modified laser cavity whose alignment is often confirmed by operating it as a laser without a seed pulse. The injected pulse makes multiple trips, ranging up to around 50, through the gain medium that has been inverted by a separate pump laser. After the required number of passes, which is controlled by the electronics and set by the user, the pulse is dumped from the cavity and is restored to its original pulse duration by a grating compressor setup. Precise timing and a high extinction ratio for injection/dumping optics are required to avoid pre- or post-pulses from entering/leaving the cavity and to maintain high-fidelity amplification.

To get a better idea of how pulse picking, injection, and dumping take place, let us consider a representative setup as shown in Figure 7.24. The amplifier must be able to effectively select a single pulse from the pulse train of an oscillator. To achieve this, the thin film polarizer (TFP) reflects an incoming pulse, with vertical polarization, into the cavity. The optical pulse passes through the Pockels cell. High-speed electronics apply a high voltage to the electro-optic material within the cell to allow it to function as a quarter-wave plate, which leaves the pulse circularly polarized. After retroreflecting from the end mirror, the second pass through the Pockels cell, which is still biased, rotates the polarization to horizontal. The pulse is transmitted through the PBS/TFP and is injected into the cavity.

While the pulse is being amplified, the Pockels cell is kept unbiased. In the passive state, the Pockels cell does not affect the polarization. This configuration prevents pulses from entering or leaving the cavity. The pulses from the oscillator are reflected out of the cavity by the TFP on

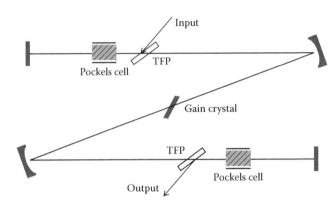

Figure 7.24 Representation of a regenerative amplifier that uses Pockels cells and TFPs to switch the pulse in and out of the cavity using high-speed, high-voltage electronics.

their return pass and the pulse being amplified is transmitted on both passes. After a number of round-trips, the fully amplified pulse is dumped from the amplifying section by the second Pockels cell (some configuration use a single cell for both injection and dumping). The energized cell rotates the polarization state from horizontal to vertical through two quarter-wave retardations so that the pulse reflects from the TFP and is dumped from the cavity.

The example presented above outlines the key operating principles of a regenerative amplifier, but there are several variations possible depending, most notably, on the repetition rate of the system. The system above represented an amplifier that operates in the 1–10 kHz range. At this pulse-repetition rate, Q-switched pumps are used to invert the gain medium and the overall gain is commonly on the order of 10^6. Higher repetition-rate systems that operate up to around 250 kHz are also commonly used. For these systems, a CW-pump source is used and the electro-optic Pockel cells and TFPs are replaced by acousto-optic devices that can operate at much higher frequencies. Except for the need to match the cavity length to that of the oscillator, the concept of operation remains the same. For more specific information on these high-repetition-rate systems, the interested reader should refer to Norris (1992). Commercially available regenerative amplifiers can now produce milijoule pulse energies at kilohertz repetition rates or microjoule pulses at hundreds of kilohertz.

7.3.2 Wavelength Tunability through Parametric Processes

As will be seen in subsequent chapters, the generation of femtosecond pulses over a broad wavelength range is often required to characterize ultrafast processes in complex materials. While commercial Ti:Al$_2$O$_3$ systems now offer continuous tuning from 680–1080 nm, it is often necessary to access a much broader spectral range. An intense pulse can be focused into a transparent material to produce a broad bandwidth, "white" light supercontinuum with a bandwidth exceeding 1 μm (Alfano, 2006). Unfortunately, the intensity profile is often highly structured, and even when parameters have been controlled to provide a smooth spectrum (Ranka et al., 2000), the power spectral density is often not sufficient to provide a pulse with enough energy over a desired wavelength range to perform a measurement. An alternative is to use multiple laser systems, each based on different solid-state materials, but the added complexity and cost as well as the existence of wavelength gaps make this undesirable in most cases. Having a single source capable of producing femtosecond pulses that can be continuously tuned from the UV to the IR is advantageous and often necessary. This can only be achieved by relying on nonlinear processes in optical crystals to convert existing frequencies to desired frequencies.

7.3.2.1 Nonlinear Optics

Equation 7.6 demonstrates that in the presence of a strong electric field, a nonlinear polarization response is induced and new frequencies are created. Whereas SPM and self-focusing are the result of the third-order susceptibility, $\chi^{(3)}$, in this case we only need to consider effects associated with the second-order susceptibility. Crystals with non-centrosymmetric symmetry produce a nonzero $\chi^{(2)}$ in which the second-order polarization can be expressed in general as follows:

$$P_i = \sum_{jk} \chi_{ijk}^{(2)} E_j E_k \tag{7.39}$$

All second-order processes relate the interaction of three photons such that $\omega_1 + \omega_2 = \omega_3$, which is just a restatement of the conservation of energy. If we explicitly express Equation 7.39, six terms result that represent several different processes:

$$P(2\omega_1) = \chi^{(2)} E_1^2 \qquad \text{Second Harmonic Generation of } \omega_1$$

$$P(2\omega_2) = \chi^{(2)} E_2^2 \qquad \text{Second Harmonic Generation of } \omega_2$$

$$P(\omega_1 + \omega_2) = 2\chi^{(2)} E_1 E_2 \qquad \text{Sum Frequency Generation} \tag{7.40}$$

$$P(\omega_1 - \omega_2) = 2\chi^{(2)} E_1 E_2^* \qquad \text{Difference Frequency Generation}$$

$$P(0) = 2\chi^{(2)} (E_1 E_1^* + E_2 E_2^*) \qquad \text{Optical Rectification}$$

The processes of sum frequency generation (SFG), different frequency generation (DFG), and second-harmonic generation (SHG) (which is just SFG with two photons that have the same energy) form the basis for continuous wavelength tunability. Even higher frequency ranges can be reached through third and fourth harmonic generation by cascading a combination of the lower order processes of SFG and SHG. In addition, optical rectification, although not discussed in detail here, can be used to generate far-IR terahertz pulses (Nahata et al., 1996; Rice et al., 1994) (Chapter 11).

In SFG, phase matching is necessary to efficiently convert light between the three wavelengths. The nonlinear process is said to be phase matched when

$$\vec{k}_1 + \vec{k}_2 = \vec{k}_3 \tag{7.41}$$

where $\vec{k}_i$ is the wave vector for a particular wave. Essentially, the incident wave and generated wave need to have the same group velocity as they propagate through the nonlinear optical crystal for efficient conversion. If we assume that the beams propagate in a collinear fashion, the phase-matching condition can be rewritten as a scalar:

$$\frac{n_1}{\lambda_1} + \frac{n_2}{\lambda_2} = \frac{n_3}{\lambda_3} \text{ or } n_1\omega_1 + n_2\omega_2 = n_3\omega_3 \tag{7.42}$$

For a real crystal with an isotropic index of refraction, this condition cannot be satisfied. However, in uniaxial and biaxial crystals the index depends upon the polarization of the incident light and the direction of propagation through the crystal, which can be utilized to

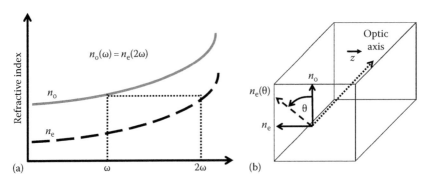

Figure 7.25 (a) Type I phase matching for a uniaxial crystal where the ordinary index of refraction at ω is matched to the extraordinary index at 2ω. (b) Schematic of a uniaxial crystal showing the optic axis and the polarization directions that coincide with the ordinary and extraordinary principal axes.

efficiently phase match light conversion between different wavelengths. Non-isotropic crystals that are commonly used because they can satisfy phase-matching conditions and possess a large $\chi^{(2)}$ are β-borate (BBO), lithium triborate (LBO), or potassium titanyl phosphate (KTP).

Figure 7.25a shows the index versus frequency for a negative uniaxial crystal, where the extraordinary index is smaller than the ordinary index. The ordinary and extraordinary indices apply to any wave polarized along the principal axes of the crystal, n_o or n_e, respectively. When a wave is polarized along neither principal axis, it experiences an index given by

$$\frac{1}{n_e(\theta)^2} = \frac{\sin^2\theta}{\bar{n}_e^2} + \frac{\cos^2\theta}{n_o^2} \tag{7.43}$$

where
   θ is the angle taken with respect to the n_o direction
   $\bar{n}_e$ is the principal value of the extraordinary index (Figure 7.25b)

If we want to phase match for SHG, for example, where $\omega_{SHG} = 2\omega_1$, then we need $n_e(2\omega, \theta) = n_o(\omega)$. This phase matching approach, where both input beams are polarized along the same direction, is known as Type I phase matching.

In general, solutions do not always exist, but, by cutting the crystals along appropriate directions, phase matching can sometimes be achieved among the different polarization components of λ_1, λ_2, and λ_3 by using Type II phase matching, where the fundamental frequencies are aligned to axes with different refractive indices. In addition, rotating the crystal angle with respect to the incident optical beam allows for wavelength tuning, because as the observed index along the direction of propagation changes, different wavelengths may become phase matched. For the interested reader, a more in-depth and complete discussion of the various phase-matching approaches can be found elsewhere (Bloembergen, 1996; Shen, 2002).

Even when the phase-matching condition is met, however, there are limitations on the conversion efficiency. While one might expect that a thick crystal would be required to achieve optimum conversion, the practical length is bounded by GDD and beam walk-off. GDD affects the propagation velocity of the beams within the crystal and limits the range over which they overlap spatially. Since dispersion also affects the temporal spread of a pulse, the intensity-dependent conversion efficiency is reduced as the beams propagate deeper into the crystal and their temporal overlap is reduced. Finally, when the k-vector of the extraordinary wave is not aligned with a principal axis of the crystal, the Poynting vector and the propagation direction do not overlap, leading to beam walk-off, which is another mechanism that reduces the spatial overlap and limits the useful crystal thickness. In practice, the conversion efficiency in SFG is limited to ~25%–50%.

7.3.2.2 Optical Parametric Oscillators and Amplifiers

Due to the high intensities achieved by femtosecond laser pulses, the nonlinear processes described above can still lead to high conversion efficiency despite the rigid requirements for phase matching and limitations imposed by dispersion and beam walk-off. In order to obtain broad and continuously tunable femtosecond sources, optical parametric oscillators (OPOs) and optical parametric amplifiers (OPA) are used. In both OPOs and OPAs, there are three propagating waves, the signal, the idler, and the pump. Through the parametric process of difference-frequency generation, the pump photon is split into a signal photon and an idler photon with $\omega_p > \omega_s > \omega_i$.

In practice, an OPA makes use of a high-energy input pulse that is split into three unequal parts. A small portion of the pump is focused into a material in order to generate a white light supercontinuum (used as a "seed" for the amplifier). In the preamplifier stage, the second part of the pump, with slightly higher power, is overlapped with the continuum inside a nonlinear crystal. The wavelength of the signal and idler are selected by tuning the crystal angle; the phase-matching relationship causes one wavelength to be efficiently converted for a given angle. The amplification process efficiently transfers energy from the pump to the signal and idler, which can be enhanced (Rulliere, 2005) by a factor of ~10^6. The final stage is the power amplifier stage, where the majority of the initial pump power is used to amplify the signal and idler in a second pass through a nonlinear crystal (often the same crystal) with appropriate phase matching. The overall conversion efficiency from pump to signal and idler can be >30%.

The OPO works in a similar fashion as the OPA, with some important differences. Because OPO are not seeded with amplified pulses, the output is built up over several pump cycles and requires that the pump and the signal/idler pulses are temporally overlapped (synchronous pumping) to exceed the oscillation threshold. In the OPO, the initial high-energy pump photons undergo a spontaneous down conversion process whereby two lower energy photons (one each for the signal and idler) are created. Depending on the cavity mirrors, either the signal (singly resonant) or the signal and the idler (doubly resonant) will travel through the cavity and be time aligned within the crystal along with the next pump pulse from the seed oscillator. Power continues to be transferred from the pump pulse to the signal (or signal and idler) through DFG on each successive pass. Power is transmitted from the cavity through a partially reflecting output coupling mirror. The OPO functions much in the same way as a laser, except that the gain is not provided by an inverted medium but by a nonlinear frequency conversion process.

While both OPOs and OPAs typically deliver broadband wavelength tunability ranging from approximately 1.0 to 1.6 and 1.6 to 2.6 μm for the signal and idler, respectively (without additional mixing stages), there are several other aspects of their operation that make one or the other preferable, depending on the application. OPAs generate high-energy output pulses at low repetition rates from amplified pump pulses with energies ~1 μJ–1 mJ. Conversely, OPOs produce low-energy output pulses at highrepetition rates from nanojoule energy oscillator pumps. Finally, the output from OPOs generally has higher spatial and coherence properties with lower noise than OPAs.

7.4 CHARACTERIZATION OF ULTRASHORT PULSES

The characterization of ultrashort pulses has made significant strides as the technology itself has enabled the generation of successively shorter pulse durations. As early as 1965, pulses were generated which were too short to be measured by conventional electronic techniques

(DeMaria et al., 1966). The measurement of any transient event requires an even shorter event, which makes the measurement of ultrashort pulse durations difficult since they are the shortest known man-made event. As a result, techniques have been developed to use the pulse to measure itself.

In this section, we will consider a total of four pulse-characterization techniques. All have advantages and disadvantages related to the information desired as well as the over-all complexity and cost of the measurement setup. The autocorrelation has been the standard technique for measuring pulse durations, but techniques such as frequency-resolved optical gating (FROG) and spectral phase interferometry for direct electric field reconstruction (SPIDER) have recently become more widespread because they can provide additional information about the pulse. The newest technique we will review, multiphoton intrapulse interference phase scan (MIIPS), is not only able to measure pulse properties but to shape the pulse for a transform-limited duration or to imprint certain specific phase characteristics.

7.4.1 Autocorrelation

The autocorrelation was the first successful approach for characterizing ultrashort pulses. There are two basic autocorrelation techniques: the intensity autocorrelation and the fringe-resolved or interferometric autocorrelation (IAC). It should be pointed that neither are field correlation techniques that can be used to calculate the spectrum of the light source (as in an FTIR measurement [Chapter 4]). Instead, the intensity correlation and IAC are second-order autocorrelations that obtain information about the profile of the pulse, from which one can estimate the pulse duration. Both approaches use a BS to split the pulse to be measured into two, after which one pulse is delayed with respect to the other in a Michelson interferometer. The pulses are then recombined inside a nonlinear crystal, generating a second-harmonic signal that can be detected as the delay is swept, yielding the autocorrelation.

7.4.1.1 Intensity Autocorrelation

Figure 7.26 shows a schematic representation of a typical intensity autocorrelator. It is basically a noncollinear Michelson interferometer that uses a single BS. One of the arms contains mirrors mounted on a translation stage or a cube retroreflector, that is configured to vary the delay time, τ. After the beam is divided and the pulse replicas propagate through their respective arms, they are focused into a nonlinear crystal (a photodiode with a bandgap of higher energy than the pulse photons may also be used [Ranka et al., 1997]) by a lens. If dispersion is a concern, a curved mirror can be used as the focusing element.

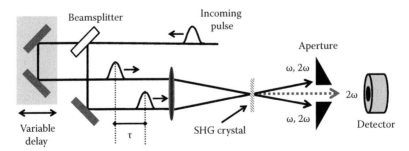

Figure 7.26 Experimental representation of an intensity autocorrelator. The relative delay between the pulses can be controlled by the variable delay stage. The single beam fundamental and second-harmonic signals are blocked by an aperture to provide a background-free measurement of the pulse duration.

Each beam will independently generate an SHG signal. However, phase-matching requirements, discussed in the previous section, dictate that only the signal that results from the interaction between the two pulses will propagate along the optic axis, which is the angle bisector of the incoming beams (Figure 7.26). Therefore, when aligning an intensity autocorrelator one can align the two beams and crystal such that each beam generates an SHG signal, and then vary the time delay until an SHG signal is observed along the optic axis. In this way, the signal generated is background free and can be expressed as follows:

$$I_{SHG}(t,\tau) \propto |A(t)A(t-\tau)|^2 \propto I(t)I(t-\tau) \tag{7.44}$$

where $A(t)$ is the complex envelope of the field. Because the detector is too slow to capture the quickly oscillating signal in time, what will be measured is

$$I_{AC}(\tau) \propto \int_{\infty}^{\infty} I(t)I(t-\tau)dt \tag{7.45}$$

By scanning the delay between the two pulses, the intensity envelope can be captured (Figure 7.27).

Regardless of the actual pulse shape, the intensity autocorrelation will be symmetric because clearly, under proper experimental conditions, $I_{AC}(\tau) = I_{AC}(-\tau)$. Nevertheless, if a pulse shape is known (or, more likely, assumed), the estimated pulse width can be extracted from the data by using a deconvolution factor. The Gaussian and hyperbolic secant are the two most commonly assumed shapes for femtosecond pulses. For these two shapes, the ratios of the autocorrelation-measured FWHM pulse widths to the actual pulse widths are $\tau_{ac}/\tau_p = 1.414$ and 1.543 for the Gaussian and sech-shaped pulses, respectively. Because of its relative simplicity, the intensity autocorrelation does not contain information about the electric field since it cannot measure the temporal phase of the pulse.

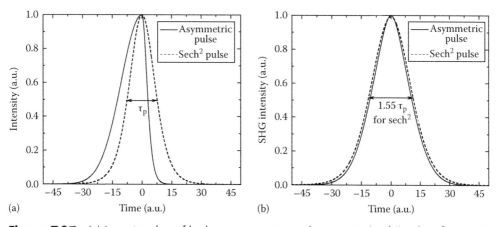

(a)   Time (a.u.)

(b)   Time (a.u.)

Figure 7.27   (a) Intensity plots of both an asymmetric and symmetric (sech²) pulse of approximately the same pulse duration. (b) Intensity autocorrelation of the two pulses. Because the autocorrelator always provides a symmetric output, the actual shape of the pulse is lost and the measurement of the two different pulses becomes indistinguishable.

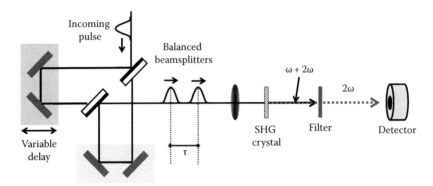

Figure 7.28 Schematic representation of an interferometric or collinear autocorrelator. In order to get accurate measurements, two counter-oriented beam splitters must be used to balance the phase between the arms and make them appear identical.

7.4.1.2 Interferometric Autocorrelation

Figure 7.28 shows a schematic representation of an interferometric autocorrelator. It is very similar to the experimental setup for an intensity autocorrelator, except that the beams are collinearly incident on the second-harmonic crystal (again, the focusing lens may be replaced by a curved mirror to eliminate any additional dispersion). The output from the crystal must be filtered to remove the fundamental wavelength so that only the SHG is detected. Special care must be taken to ensure that the beams follow identical paths, and the dispersion in both arms should be balanced. In addition, two counter-oriented beamsplitters should be used to balance the phase acquired on reflection, ensuring that there is no phase errors imparted to the two pulses by the experimental setup.

If the autocorrelator is designed as described above, the total field that results after the pulses are recombined will be

$$E_{TOT}(t,\tau) = A(t+\tau)e^{iw_c(t+\tau)}e^{i\varphi_{ce}} + A(t)e^{i\omega_c t}e^{i\varphi_{ce}} \tag{7.46}$$

where
  $A(t)$ is the complex amplitude
  ω_c is the carrier frequency
  φ_{ce} is the CEP

As previously discussed, the SHG electric field will be proportional to $E(t, \tau)^2$, and the SHG intensity will be proportional to the magnitude of the SHG field squared. However, as before, the slow detector will measure

$$I_{LAC}(\tau) = \int_{-\infty}^{\infty} \left| \left(A(t+\tau)e^{i\omega_c(t+\tau)}e^{i\varphi_{ce}} + A(t)e^{i\omega_c t}e^{i\varphi_{ce}} \right)^2 \right|^2 dt \tag{7.47}$$

After the terms in Equation 7.47 are evaluated, they may be grouped and expressed as follows:

1. A constant background term that remains when the pulses no longer overlap in time

$$I_{BG} = \int_{-\infty}^{\infty} I^2(t) + I^2(t+\tau)dt \tag{7.48}$$

2. The intensity autocorrelation term, which is essentially the same as Equation 7.45

$$I_{AC}(\tau) = 4 \int_{-\infty}^{\infty} I(t)I(t-\tau)dt \tag{7.49}$$

3. Coherence term of $E(t)$, oscillating at the fundamental ω_c

$$I_{\omega_c} = 4 \int_{-\infty}^{\infty} \text{Re}\left[\left(I(t) + I(t+\tau) \right) A(t) A*(t+\tau) e^{i\omega_c \tau} \right] dt \tag{7.50}$$

4. The interferogram of the second harmonic, oscillating at $2\omega_c$

$$I_{2\omega_c} = 2 \int_{-\infty}^{\infty} \text{Re}\left[A^2(t) \left(A*(t+\tau) \right)^2 e^{i2\omega_c \tau} \right] dt \tag{7.51}$$

In the calculation above, the φ_{ce} term drops out because it is identical in both pulses, demonstrating that autocorrelation techniques cannot measure the absolute value of φ_{ce} (nor can any of the techniques discussed in this chapter). However, methods for measuring the absolute value of φ_{ce} have been recently developed, and although outside the scope of this book, are discussed in Paulus et al. (2001). Summing Equations 7.48 through 7.51 gives the full expression for the detected autocorrelation trace:

$$I_{LAC}(\tau) = I_{BG} + I_{AC} + I_{\omega_c} + I_{2\omega_c} \tag{7.52}$$

Figure 7.29a shows an IAC for a short, unchirped pulse of a few cycles. Based on the figure, there are several values that serve as useful comparisons to confirm proper autocorrelator alignment in practice. Notably, the peak of the signal at zero delay has an 8:1 ratio with the

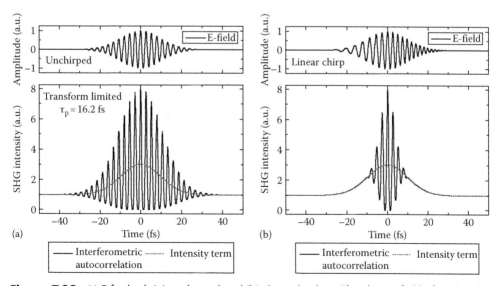

Figure 7.29 IAC for both (a) unchirped and (b) chirped pulses. The electric field of each pulse is shown in its own plot at the top and the corresponding intensity autocorrelation term is shown as the average of the fringe trace. For the same pulse duration, the IAC is able to distinguish between unchirped and chirped pulses.

detected signal level at long delay times. The base of the measured IAC signal should touch zero. If the delay arm were scanned fast enough, the interference terms in the IAC would average to zero and the resulting trace would appear as the intensity autocorrelation sitting on top of a background of unity. In this case, the peak to baseline ratio would be 3:1 when measured from the zero signal level. The background term does not appear in the noncollinear intensity autocorrelation setup since the SHG signal generated by a single beam propagates along a different direction. This is due to the fact that the phase-matching condition for propagation along the optic axis is only met when two beams are present.

As with the intensity autocorrelation, the fundamental pulse shape cannot be determined directly from the IAC. Nevertheless, the IAC does provide the capability to quantitatively measure the linear chirp associated with a pulse, thus providing a representation of the electric field. Figure 7.29b shows a pulse with linear chirp, which is clearly distinguishable in shape from a transform-limited pulse. While beyond the scope of this chapter, the interested reader should consult Diels and Rudolph (2006) for an in-depth discussion on this topic. The IAC can give a qualitative assessment as to the presence or absence of phase perturbations, but the reconstruction of amplitude and phase information cannot be performed directly. It has been shown, however, that the pulse shape and phase can be found by iterative fitting if the pulse spectrum is known (Naganuma et al., 1989). Unfortunately, due to the fact that the algorithm does not always converge, this method has generally been abandoned in favor of FROG, SPIDER, and other more advanced techniques if knowledge of the electric field is important.

7.4.2 Frequency Resolved Optical Gating

FROG was born from the need for a method to reconstruct the electric field not only of ultrashort laser pulses but also of short bursts of light with phases that may not be as well behaved. In order to retrieve both the amplitude and phase of a pulse, more than just time domain information is needed. For that reason, the FROG trace is basically a spectrogram that is displayed in the hybrid time-frequency domain. A more familiar example of a spectrogram is the musical score, where time is on the horizontal axis, frequency along the vertical axis, and intensity along the top. With respect to FROG, the mathematical representation of such a trace is

$$\sum_{E}(\omega, \tau) \equiv \left| \int_{-\infty}^{\infty} E(t)g(t - \tau)e^{-i\omega t}dt \right|^2 \tag{7.53}$$

where
  $E(t)$ is the electric field we are interested in determining
  $g(t - \tau)$ is a yet-to-be-defined variable gate function

Without a second pulse (known as the gate), the equation for Σ_E when only one pulse propagates through the setup would be reduced to the equation for power spectral density and yield only the non-time-resolved spectrum of the pulse. On the other hand, the gating function makes it possible to measure the spectrum of the pulse at a given delay. In this way, Σ_E represents the set of spectra taken over a number of time slices through $E(t)$.

The basic FROG setup is very similar to the intensity autocorrelation. In fact, for SHG FROG, the only difference is that the photodetector is replaced by a spectrometer as shown in Figure 7.30. There are several variations of FROG that principally differ in the type of gating that is used: polarization gating (PG), second harmonic (SHG), third harmonic (THG),

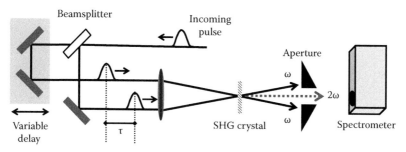

Figure 7.30 FROG setup is similar to the intensity autocorrelator, except that the detector is replaced by a spectrometer.

transient grating (TG), and self-diffraction (SD). In all cases, a time-delayed replica of the initial pulse is used to perform the gating function in conjunction with a nonlinear process. Depending on the process, $g(t - \tau)$ will take several functional forms. For example, in the case of PG FROG, $g(t - \tau) = |E(t - \tau)|^2$ and for SHG FROG, $g(t - \tau) = E(t - \tau)$.

Even though we cannot know the exact gating function without knowing $E(t)$, we do know the relationship between the gating function and the electric field of the pulse (by knowing the type of FROG we are using). If we define $E_{\text{sig}}(t, \tau) = E(t)g(t - \tau)$, and substitute it into Equation 7.53, we are left with

$$I_{\text{FROG}}(\omega, \tau) \propto \left| \int_{-\infty}^{\infty} E_{\text{sig}}(t, \tau)e^{-i\omega t} dt \right|^2 \tag{7.54}$$

Unfortunately, Equation 7.54 is an example of a 1D phase-retrieval problem, which does not have a unique solution. However, even if we define $E_{\text{sig}}(t, \tau)$ as the Fourier transform of a new function $\hat{E}_{\text{sig}}(t, \Omega)$, i.e.,

$$E_{\text{sig}}(t, \tau) = \int_{-\infty}^{\infty} \hat{E}_{\text{sig}}(t, \Omega)e^{-i\Omega\tau} d\Omega \tag{7.55}$$

we can still maintain that $\hat{E}_{\text{sig}}(t, 0) \propto E(t)$. Because

$$
\begin{aligned}
\hat{E}_{\text{sig}}(t, 0) &= \int_{-\infty}^{\infty} E_{\text{sig}}(t, \tau)e^{i(0)t} d\tau \\[2mm]
&= \int_{-\infty}^{\infty} E(t)g(t - \tau)e^{i(0)t} d\tau \\[2mm]
&= E(t)\int_{-\infty}^{\infty} g(t - \tau) d\tau \\[2mm]
&= E(t)\int_{-\infty}^{\infty} g(\tau') d\tau' \qquad \leftarrow \tau' = t - \tau \\[2mm]
&\propto E(t)
\end{aligned}
\tag{7.56}
$$

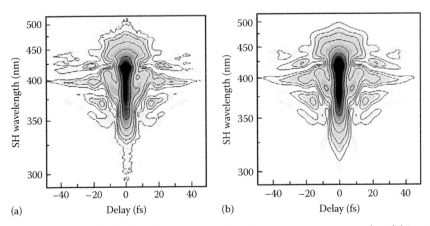

Figure 7.31 Results of SHG FROG characterization showing (a) experimental and (b) retrieved traces for a 6.5 fs compressed pulse. (Reprinted from Trebino, R., *Frequency-Resolved Optical Gating: The Measurement of Ultrashort Laser Pulses,* Kluwer Academic Publishers, Norwell, MA, 2000. With permission.)

Then, by substituting Equation 7.55 into Equation 7.54, we obtain

$$I_{\text{FROG}}(\omega,\tau) \propto \left| \int_{-\infty}^{\infty} \int_{-\infty}^{\infty} \hat{E}_{\text{sig}}(t,\Omega) e^{-i\omega t} e^{-i\Omega\tau} dt d\Omega \right|^2 \tag{7.57}$$

Whereas the 1D phase-retrieval problem does not have a unique solution, the 2D phase-retrieval problem (Equation 7.57) does. By using an algorithm that makes use of the nonlinear optical constraint, that $E_{\text{sig}}(t, \tau) \propto E(t)g(t - \tau)$, and the so-called data constraint defined by Equation 7.54, an essentially unique solution for $E(t)$ can almost be guaranteed. Using this algorithm, the intensity and phase of the pulse electric field in both the time and frequency domains can be recovered. In addition, as a feedback mechanism to verify that the recovered $E(t)$ is correct, the algorithm reconstructs the FROG trace and compares it to the experimentally measured trace. Figure 7.31 shows the typical output format for a FROG measurement. For more information on FROG and related pulse-characterization techniques, the interested reader should refer to Trebino (2000).

7.4.3 Spectral Phase Interferometry for Direct Electric Field Reconstruction

The capability to reconstruct the electric field and thus the pulse shape does not require an iterative fitting approach. An alternative technique, SPIDER, enables the pulse shape to be recovered from an interferogram and the pulse spectrum. The experimental setup has no moving parts and the acquired data is only a 1D scan from a spectrometer. The algorithm calculates two Fourier transforms and extracts the spectral phase from the oscillatory term in the interferogram. As a result, SPIDER can retrieve the characteristics of a single pulse (for amplified pulses) and has demonstrated an update rate of greater than 1 kHz.

SPIDER is based on a technique called spectral interferometry (SI) that had been proposed for pulse characterization in the 1970s (Froehly et al., 1973). The basic idea is to start with two pulses: a reference pulse, $E_{\text{ref}}(t)$, with a well-known electric field and the pulse that needs to be characterized, $E_{unk}(t)$, with unknown characteristics. $E_{unk}(t)$ is then delayed with respect to $E_{ref}(t)$ by a time τ and the two are overlapped on a spectrometer after passing through a nonlinear optical crystal. Under such conditions, the spectrometer would measure

$$S_{SI}(\omega) = \left| \int_{-\infty}^{\infty} \left[E_{ref}(t) + E_{unk}(t-\tau) \right] e^{-i\omega t} dt \right|^2 = \left| E_{ref}(\omega) + E_{unk}(\omega) e^{-i\omega\tau} \right|^2$$

$$= \left[S_{ref}(\omega) + S_{unk}(\omega) \right] + S_-(\omega) e^{i\omega\tau} + S_+(\omega) e^{-i\omega\tau}$$

$$= S_{DC}(\omega) + S_-(\omega) e^{i\omega\tau} + S_+(\omega) e^{-i\omega\tau} \tag{7.58}$$

where $S_-(\omega) = E_{ref}(\omega) E_{unk}^*(\omega)$ and $S_+(\omega) = E_{ref}^*(\omega) E_{unk}(\omega)$ (Figure 7.32).

By inverse Fourier transforming Equation 7.58 to the pseudo-time domain, three signals will be represented along the time axis, $S_{DC}(t)$, $S_-(t+\tau)$, and $S_+(t-\tau)$, centered about $t = 0$, $-\tau$, and τ, respectively, where $S(t) = \Im^{-1}\{S(\omega)\} = S_{DC}(t) + S_-(t+\tau) + S_+(t-\tau)$. If τ is sufficiently long, the three signals will be well separated and $S_+(t-\tau)$ (or $S_-(t+\tau)$) can be filtered using a fourth-order super-Gaussian centered at $t = \tau$ with a temporal width of τ. At this point, $S_+(t-\tau)$ can be Fourier transformed back to the frequency domain and the spectral phase of the isolated signal can be expressed as follows:

$$\Phi_+(\omega) = \arg\left[S_+(\omega) e^{-i\omega\tau} \right] = \varphi_{unk}(\omega) - \varphi_{ref}(\omega) - \omega\tau \tag{7.59}$$

From the total phase of Equation 7.59, the known phase of the reference pulse can be subtracted off, along with the linear term proportional to the known delay τ, leaving us with the spectral phase of the unknown pulse. $E_{unk}(t)$ is then uniquely determined by combining the magnitude $|E(\omega)|$, from a separate measurement of the spectrum, with the spectral phase, $\varphi_{unk}(\omega)$, and Fourier transforming back to the time domain.

The process of recovering an unknown pulse shape if one has a well-referenced pulse is straightforward. The unfortunate reality is that no such pulse exists for measuring femtosecond pulses. Using the concept of spectral shearing interferometry (SSI), in conjunction with the basic SI concept from above, Iaconis and Walmsley (Iaconis and Walmsley, 1998)

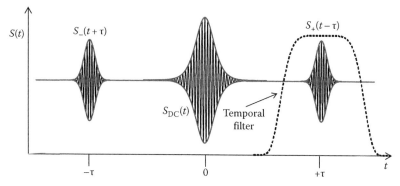

Figure 7.32 Pseudo-time domain representation of the three Fourier transformed terms from Equation 7.59 and the temporal filtering required in SI.

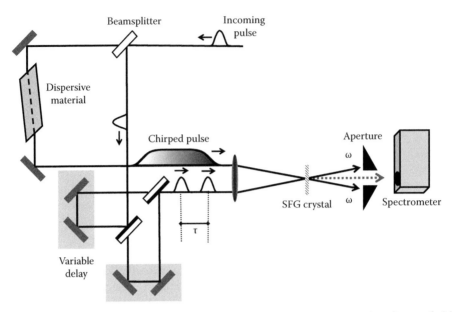

Figure 7.33 Experimental SPIDER setup that uses SSI to reconstruct the electric field of the pulse.

invented SPIDER. The experimental layout of SPIDER is shown in Figure 7.33. Instead of using a reference and an unknown pulse, three copies of the input pulse are produced. Two of the pulses are passed through a delay line to introduce a lag time τ between them. The third pulse is temporally broadened, typically by propagation through a known thickness of glass or by a grating stretcher, by a factor typically greater than 10^3. It is this stretched pulse that is successively combined with the other two copies of the input through a noncollinear SFG process within a nonlinear crystal. Spectral shearing is introduced because each of the two input pulses is overlapped with a different frequency component of the stretched pulse in the SFG process. The upshift is assumed to be quasi-monochromatic (bandwidth < 1 nm) due to the high temporal broadening of the stretched pulse. In addition, SPIDER is able to provide a broader phase matching bandwidth for a given crystal thickness than is typically possible for SHG-based characterization techniques. This advantage is made possible by using Type II phase matching with the input pulses aligned to the broadband axis of the crystal and the stretched pulse aligned to the narrow bandwidth axis. The output is measured on a spectrometer.

To obtain the spectral phase, we follow the same general approach as previously. After the Fourier transform to the pseudo-time domain, temporal filtering, and Fourier transforming back to the frequency domain, the total phase for the SPIDER technique, analogous to Equation 7.59, is

$$\Phi(\omega) = \arg\left[S_+(\omega)e^{-i\omega\tau} \right] = \varphi(\omega + \omega_s + \delta\omega) - \varphi(\omega + \omega_s) - \omega\tau \tag{7.60}$$

where
  ω is the fundamental frequency
  ω_s is the upshift frequency
  $\delta\omega$ is the shear

Since $\delta\omega$ is known to be small compared to the center frequency of the pulses, the first two terms may be rewritten as a derivative. The derivative, in this case, is equivalent to the GD $(T_g = (\partial\varphi/\partial\omega)_{\omega_0})$. The total phase becomes

$$\Phi(\omega) = \delta\omega\left(\frac{\varphi(\omega + \omega_s + \Omega) - \varphi(\omega + \omega_s)}{\delta\omega}\right) - \omega\tau = \delta\omega\frac{d\varphi(\omega)}{d\omega} - \omega\tau = \delta\omega\tau_g - \omega\tau \qquad (7.61)$$

The linear phase term can be subtracted off because the delay, τ, is known, leaving

$$\Phi(\omega) = \delta\omega\frac{d\varphi(\omega)}{d\omega}$$

$$\Downarrow \qquad\qquad\qquad (7.62)$$

$$\varphi(\omega) = \frac{1}{\delta\omega}\int_0^\omega \Phi(\omega')d\omega'$$

Thus, the spectral phase can be determined from a measurement of the GD, which is what SPIDER actually measures. While the algorithm contains no phase ambiguities or the need for iterations, the accuracy of the measurement is highly dependent on a precise knowledge of τ. For an error in the delay, $\Delta\tau$, the recovered spectral phase error becomes $\omega^2\Delta\tau/2\delta\omega$. For a more complete description of pulse characterization using SPIDER, the interested reader should consult the chapter by I. A. Walmsley in Kaertner (2004).

7.4.4 Multiphoton Intrapulse Interference Phase Scan

The previous techniques obtain varying levels of information about the pulse duration and/ or the electric field. While all three methods have seen widespread use, each is restricted to only measurement capabilities. None offer the capability to correct the spectral phase to either generate a transform-limited pulse or produce a pulse of a particular shape. The natural progression from measurement to control is made possible by a technique called MIIPS.

The MIIPS experimental setup is straightforward and composed of three parts: (1) a femtosecond pulse shaper, (2) a second-harmonic crystal, and (3) a spectrometer. The pulse to be measured is injected into the shaper, which is shown schematically in Figure 7.34. A dispersive element such as a grating (or prism) is used to spectrally separate the input beam. The dispersed beam then propagates to a lens (or mirror), which is spatially separated from the grating by the focal length, f. A spatial light modulator (SLM) is placed in the Fourier plane of the lens, which is an additional distance f away. In this 4f-configuration, the spectral components of the pulse are spread across the elements of the SLM, which can impart a

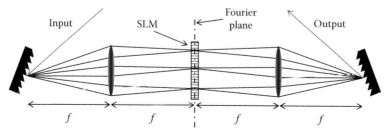

Figure 7.34 Setup for pulse shaper using a 4-f configuration with an SLM in the Fourier plane.

frequency-dependent phase shift. The beam is then recombined after propagating through the inverse optical system. Because this 4-f configuration has an SLM in the Fourier plane, it can change the phase of a particular component of the pulse spectrum and be used for femtosecond pulse shaping (Weiner et al., 1990). The output beam from the device then passes through a second-harmonic crystal and onto a spectrometer for analysis by the MIIPS algorithm.

The fundamental process behind the MIIPS characterization scheme is multiphoton intrapulse interference (MII) (Lozovoy et al., 2003). As the name suggests, MII refers to the phase-dependent interference between different frequency components within a single pulse that affects the efficiency of a multiphoton process. Based on this concept, the amplitude of the two-photon effective electric field at $2\omega_0$ can be expressed as follows:

$$E^{(2)}(2\omega_0) \propto \int_{-\infty}^{\infty} E(\omega_0 + \Omega)E(\omega_0 - \Omega)d\Omega \tag{7.63}$$

where $\Omega = \omega_0 - \omega$ represents the spectral detuning from a given frequency ω_0. To make the importance of the spectral phase, $\varphi(\omega)$, apparent, $E(\omega)$ can be represented as a product of its magnitude and phase explicitly, such that Equation 7.63 becomes

$$E^{(2)}(2\omega_0) \propto \int_{-\infty}^{\infty} \left|E(\omega_0 + \Omega)\right|\left|E(\omega_0 - \Omega)\right| \cdot e^{i\{\varphi(\omega_0 + \Omega) + \varphi(\omega_0 - \Omega)\}}d\Omega \tag{7.64}$$

In this latter form, we can see that a maximum signal is observed when the phase term, $\varphi(\omega_0 + \Omega) + \varphi(\omega_0 - \Omega)$ equals zero. Thus, the largest second-harmonic signal at a given wavelength results when there is flat (zero) spectral phase across the entire fundamental pulse spectrum.

Instead of attempting to monitor the total second-harmonic intensity while using the pulse shaper to sort through an infinite set of possible phase corrections to determine the global maximum signal, a more facile method is made possible if we assume that the phase distortions in the pulse are continuous. This assumption is generally acceptable since physical processes typically lead to continuous phase modulations and allows us to Taylor expand the phase locally around ω_0. However, it should be pointed out for completeness that, if the pulse has a pathological spectral phase with discontinuities, MIIPS may fail, like other pulse characterization methods, to measure them accurately. In the vast majority of laser-produced pulses, this expansion is valid and the phase term can be expressed as follows:

$$\varphi(\omega_0 + \Omega) + \varphi(\omega_0 - \Omega) = 2\varphi(\omega_0) + \varphi''(\omega_0)\Omega^2 + \cdots + \frac{2}{(2n)!}\varphi^{2n\prime}(\omega_0)\Omega^{2n} \tag{7.65}$$

If we consider terms up to only the second order in the expansion, we find that the local SHG signal, $I_{SHG}(2\omega_{Local})$, is optimized when $\varphi''(\omega_{Local}) = 0$. Therefore, we can use the pulse shaper to introduce a known correction, $f''(\omega_{Local})$, to the unknown second-order phase distortion (the distortion of the pulse itself), $\phi''(\omega_{Local})$, such that $\varphi''(\omega_{Local}) = \phi''(\omega_{Local}) + f''(\omega_{Local}) = 0$. By this process, $\phi''(\omega_{Local}) = -f''(\omega_{Local})$ and the local spectral phase of the pulse we are measuring, $\phi(\omega_{Local})$, can be recovered.

To determine the spectral phase over the full frequency range (not just for ω_{Local}) of the pulse, MIIPS relies upon the characterization of a sequence of spectra, in a similar fashion to FROG. For MIIPS, however, instead of a temporal scan, a series of phase scans are applied to the pulse to create what is known as the MIIPS trace. The process of phase recovery and

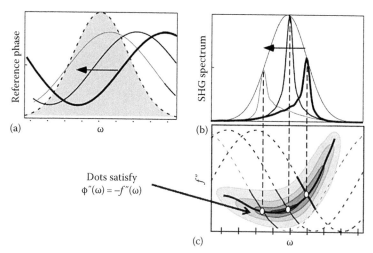

Figure 7.35 Schematic representation of MIIPS algorithm. (a) The reference phase is scanned across the spectrum of the pulse on the SLM (the shaded area represents the pulse spectrum). (b) An SHG spectrum is captured for each reference position of the phase scan. (c) Using the spectral curves, the locations of maximum SHG, where $\phi''(\omega) = -f''(\omega)$ are determined.

correction is shown schematically in Figure 7.35a through c. A sinusoidal phase variation of the form $f(\delta, \omega) = \alpha \sin(\gamma\omega - \delta)$, where α is the amplitude of the phase variation (typically around 1.5π), γ is an estimate of the pulse duration, and δ is the phase factor that is varied from 0 to 4π over the scan, is applied over the spectrum of the pulse using the SLM (Figure 7.35a). For each value of δ, the SHG spectrum is captured on the spectrometer (Figure 7.35b). By creating a density plot, $I_{SHG}(\delta, \omega)$, of the intensity data as a function of wavelength and reference phase position, the condition, $\delta_{max}(\omega)$, where the maximum SHG signal is obtained, can be determined (Figure 7.35c). Using these values, the second-order phase distortion can be evaluated from $\phi''(\omega) = -f''(\omega) = \alpha\gamma^2\sin(\gamma\omega - \delta_{max}(\omega))$. It is then straightforward to compute a double integration in the frequency domain to recover the spectral phase of the pulse $\phi(\omega)$. The SHG intensity and MIIPS are not affected by the choice of integration constants; in the calculation, the relative phase, ϕ^0, and linear term, ϕ^1, are assumed zero, like in all other pulse-characterization methods (Xu et al., 2006).

Although the phase is retrieved in a single scan, higher accuracy can be accomplished if subsequent iterations are performed, as represented in Figure 7.36. In each step, the phase correction from the previous measurement is implemented so that the absolute accuracy of MIIPS improves as the phase distortions diminish. MIIPS is versatile with the capability to compensate for very large phase distortions, on the order of hundreds of radians, or very small phase distortions, on the order of ~0.01 rad, with the accuracy and precision to compensate the entire phase distortion to values under 0.1 rad over the entire spectrum of the pulse (Lozovoy et al., 2003). While not as fast as some implementations of FROG, SPIDER, or IAC, MIIPS measures and typically compresses pulses to within 1% of the transformed-limited duration within 1 min. MIIPS has demonstrated the capability to compensate for phase distortions of an ultrabroad laser spectrum and generate pulses as short as 4.3 fs (Coello et al., 2008).

While the explanation of MIIPS was given assuming that the nonlinear crystal was placed directly after the pulse shaper, such a configuration is not required. There are numerous applications where it is desirable to have a transform-limited pulse (or some other known pulse form, which MIIPS is also capable of producing) at a particular location within an optical system. In such cases, since the pulse characterization is performed at the position of the

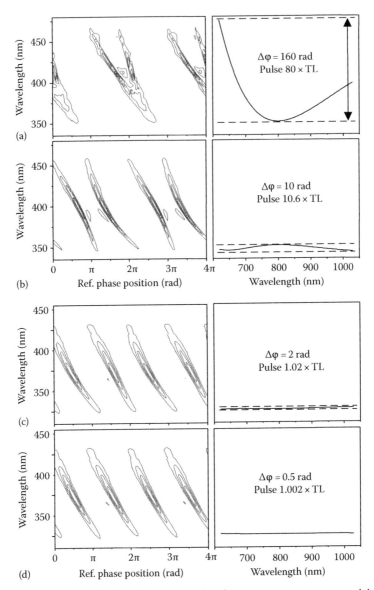

Figure 7.36 Repetitive MIIPS algorithm corrects the phase over successive scans (a) through (d). The left panel shows the MIIPS trace, while the right panel shows the spectral phase. The phase error is reduced from 160 to 0.5 rad and the pulse duration is reduced from 80 times the transform limit (TL) to 1.004 times TL. The final pulse duration was 4.4 fs.

second-harmonic crystal, MIIPS can be used to generate a pulse with known characteristics at an arbitrary location in an experimental setup as long as the beam is accessible for analysis at that location. In this way, MIIPS has been used to compensate for microscope objectives to achieve increased contrast in biomedical imaging applications (Laura et al., 2006). In addition, by placing the pulse shaper before the input to a laser amplifier, with the crystal and spectrometer at the output of the amplifier, transform-limited pulses with the full output power of the amplifier can be realized (Pestov et al., 2009). In the latter configuration, the loss associated with the pulse shaper (typically tens of percent) is taken before the amplification process and allows the amplifier gain to make up for the insertion loss.

7.5 TITANIUM SAPPHIRE LASER

Throughout the course of the chapter, we have laid out a basic theoretical framework to explain the generation of ultrashort pulses. We have, in addition, reviewed fundamental calculations for the practical design and layout of a femtosecond laser resonator. The information was provided in a very general way so that it would remain widely applicable and be able to establish a firm foundation for the understanding of femtosecond technology and measurement. Sometimes, however, specific information about common practice and "how-tos" can be at least as important as or even more important than abstract theories and equations. Given that, along with the practical focus of this book, we would be remiss if we did not spend at least a few paragraphs discussing the KLM $Ti:Al_2O_3$ (titanium sapphire) laser. Much of the following discussion, however, is applicable to nearly any solid-state mode-locked laser system.

7.5.1 $Ti:Al_2O_3$ as a Gain Medium

There are many favorable characteristics that make $Ti:Al_2O_3$ the most common, most successful, and most widely used femtosecond laser. One attractive feature is that its absorption cross section peaks around 500 nm, which makes it easily accessible to a number of pump sources such as argon ion lasers and frequency-doubled solid-state lasers based on Neodymium-doped gain media. $Ti:Al_2O_3$ also has a reasonably small Stokes shift (quantum defect), which allows for good optical-to-optical conversion efficiency and a reduced heat load. The thermal conductivity is excellent at room temperature and further improves with cryogenic cooling. Despite these qualities, probably the most impressive characteristic of $Ti:Al_2O_3$ is its gain bandwidth, which spans from ~600 to 1100 nm and enables it to support few-cycle optical pulses.

7.5.2 Typical $Ti:Al_2O_3$ Lasers

The versatility of the gain medium has supported truly amazing demonstrations of pulse duration and energy, as mentioned in the introduction. Furthermore, $Ti:Al_2O_3$ based laser systems have taken on variations in size and configuration to meet a wide variety of applications. Nevertheless, rather than concentrate on the exotic, let us consider common characteristics that are widely available commercially or in standard home-built laser systems.

In a typical $Ti:Al_2O_3$-based laser, gain crystals with lengths ranging from about 2 mm to 1 cm are used. Standard prism pairs (often made from the materials depicted in Figure 7.17) can be used for intracavity dispersion compensation to generate sub-100 fs pulses. Shorter durations are likely if the gain crystal is kept short and the dispersion is well compensated. These systems are regularly pumped with 5 W green lasers and generate output powers in the range of 500 mW. The pulse-repetition rate is set by the cavity length (each pulse traverses the cavity once before exiting), but standard systems operate at 80–100 MHz, which results in pulse energies in the low nanojoule range. The pump threshold of these systems depends heavily on spot size, power absorbed, and optical loss (including output coupling percentage), but generally 2–3 W is required in order to achieve mode-locked operation.

7.5.3 Initiation of Mode Locking

Mode locking is initiated by one of several common techniques, which are all designed to produce a "noise" spike in the CW operation of the laser that will, for reasons discussed earlier, develop into a pulse. One common technique is to tap on the optics table in the vicinity of the laser.

A better method is to use a dispersion compensating prism to start the mode locking. Because the prism is already mounted on a translation stage for dispersion tuning, rapidly de-inserting the prism and slowly returning it to its set position often initiates pulse generation. Finally, tapping or translating an end mirror mounted on a translation stage is another effective method to start the mode-locking process.

Several characterization devices are normally used to provide helpful information related to the initiation and continued stable operation of a mode-locked oscillator. The first such device is a fast photodiode (typical response time ~1 ns), which is monitored on an oscilloscope or spectrum analyzer. The purpose of the photodiode is not to monitor the pulse duration, which we already know is too short for direct electrical measurement, but to monitor the pulse train. While attempting to initiate mode locking, observing the output from the photodiode in either the time or frequency domain will help determine how close the laser is to commencing pulsed operation.

The second useful tool for initiating and monitoring the operation of a mode-locked laser is a spectrometer. If the laser alignment and dispersion are favorable for mode-locked operation, the CW spectrum can be observed to be unstable or "jumpy." Since the switch from CW to pulsed operation requires an increase in bandwidth, monitoring the output spectrum while trying to initiate mode locking can give insight as to when the transition is beginning to occur.

7.5.4 Undesirable Regimes of Operation

Even though a laser may initially generate single pulses with good pulse-to-pulse repeatability, environmental changes may cause slight misalignments or may directly affect the long-term stability of the laser. For this reason, it is common practice to utilize a fast photodiode and spectrometer, as mentioned above for pulse initiation, to also perform long-term monitoring of the pulse train. In doing so, one can determine if the laser has slipped into an undesired multiple-pulse or Q-switched mode-locked regime of operation. A passively mode-locked laser within the appropriate subset of operating parameters will support the production of a single pulse at the fundamental repetition rate of the cavity. When the conditions are less than optimal, more than one pulse per round-trip of the cavity can be produced. A photodiode can determine if there are multiple pulses circulating within the cavity, but its limited bandwidth will support detection only if the pulses are separated by more than around 100 ps. There are, however, physical effects that allow pulses to form and remain in a stable configuration even closer together. As a result, without the help of a spectrometer, single-pulsed operation cannot easily be confirmed. If, however, a laser is operating in a multiple-pulse fashion with temporally close pulses, a strong modulation will be observed on the optical spectrum due to pulse interference effects, as shown in Figure 7.37a.

There are many conditions that can lead to multiple pulsing, but in most cases, the phenomenon can be removed by optimizing the configuration of the laser cavity. In this case, one should confirm good overlap between the laser mode and the pump beam within the gain medium by optimizing the CW output power. In addition, the user should validate that the laser is operating on the inner edge of the outer stability region where the Kerr-lens sensitivity is strongest. A slight adjustment of the crystal position or of the dispersion operating point may also assist in eliminating pulse breakup. Finally, reducing the pump power can also help in achieving single-pulse operation.

Even if the laser is "single pulsing," large variations in output intensity or power with time are possible. If the change in pulse-to-pulse intensity is greater than a few percent, it is likely that the laser is Q-switched mode locking. In this operating regime, the pulse train undergoes periodic modulation as represented in Figure 7.37b. In practice, the oscilloscope trace

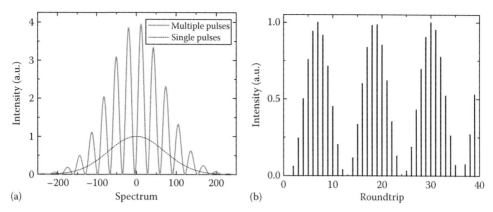

(a) Spectrum   (b) Roundtrip

Figure 7.37 (a) Representation of the modulated spectrum that results when a laser is producing two pulses that are too close temporally to be resolved by a fast photodiode. (b) An example of the large variations in pulse-to-pulse intensity that are obtained when a laser is Q-switched mode locking.

from the fast photodiode will appear unstable due to the combination of ineffective triggering and the display anomalies that result from large pulse intensity variations. The suppression of Q-switched mode locking can be achieved by changing the pump power (typically lowering it), the dispersion compensation, and/or minor cavity realignment.

For a more technical discussion of multiple pulsing and Q-switched mode locking, the interested reader should consider Hönninger et al. (1999) and/or Schibli et al. (2000).

ACKNOWLEDGMENTS

The author would like to thank Franz Kaertner for his assistance with several topics covered in this chapter and also for providing the data for the complementary octave-spanning DCMs. Special thanks are also extended to Rick Trebino for assistance with FROG and the figures that were adapted from his website. A grateful acknowledgement should also be made to Marcos Dantus for helpful discussions and assistance with composing the MIIPS section of the chapter. Thanks are also due to Idan Mandelbaum and D. Ahmasi Harris for helpful discussions. Finally, the author is indebted to Louise Sengupta and BAE Systems, Technology Solutions, who provided support for this effort.

REFERENCES

Alfano, R. R. (2006) *The Supercontinuum Laser Source: Fundamentals with Updated References*, New York, Springer.

Baltuska, A., Udem, T., Uiberacker, M., Hentschel, M., Goulielmakis, E., Gohle, C., Holzwarth, R., Yakovlev, V. S., Scrinzi, A., Hansch, T. W., and Krausz, F. (2003) Attosecond control of electronic processes by intense light fields. *Nature*, 421, 611–615.

Bloembergen, N. (1996) *Nonlinear Optics*, River Edge, NJ, World Scientific Publishing Company.

Bouma, B. E., Ramaswamy, M., and Fujimoto, J. G. (1997) Compact resonator designs for mode-locked solid-state lasers. *Appl. Phys. B*, B65, 213–215.

Brabec, T. and Krausz, F. (2000) Intense few-cycle laser fields: Frontiers of nonlinear optics. *Rev. Mod. Phys.*, 72, 545–591.

Brabec, T., Spielmann, C., Curley, P. F., and Krausz, F. (1992) Kerr lens modelocking. *Opt. Lett.*, 17, 1292–1294.

Cerullo, G., De Silvestri, S., and Magni, V. (1994a) Self-starting Kerr lens mode-locking of a Ti:sapphire laser. *Opt. Lett.*, 19, 1040–1042.

Cerullo, G., De Silvestri, S., Magni, V., and Pallaro, L. (1994b) Resonators for Kerr-lens mode-locked femtosecond Ti:sapphire lasers. *Opt. Lett.*, 19, 807–809.

Chemla, D. S. and Shah, J. (2000) Ultrafast dynamics of many-body processes and fundamental quantum mechanical phenomena in semiconductors. *Proc. Natl. Acad. Sci.*, U.S.A. 97, 2437–2444.

Chen, Y., Kartner, F. X., Morgner, U., Cho, S. H., Haus, H. A., Ippen, E. P., and Fujimoto, J. G. (1999) Dispersion-managed mode locking. *J. Opt. Soc. Am. B*, 16, 1999–2004.

Chudoba, C., Fujimoto, J. G., Ippen, E. P., Haus, H. A., Morgner, U., Kärtner, F. X., Scheuer, V., Angelow, G., and Tschudi, T. (2001) All-solid-state Cr:forsterite laser generating 14-fs pulses at 1.3 um. *Opt. Lett.*, 26, 292–294.

Coello, Y., Lozovoy, V. V., Gunaratne, T. C., Xu, B., Borukhovich, I., Tseng, C.-H., Weinacht, T., and Dantus, M. (2008) Interference without an interferometer: A different approach to measuring, compressing, and shaping ultrashort laser pulses. *J. Opt. Soc. Am. B*, 25, A140–A150.

Cundiff, S. T., Ye, J., and Hall, J. L. (2001) Optical frequency synthesis based on mode-locked lasers. *Rev. Sci. Instrum.*, 72, 3749–3771.

Demaria, A. J., Stetser, D. A., and Heynau, H. (1966) Self mode-locking of lasers with saturable absorbers. *Appl. Phys. Lett.*, 8, 174–176.

Diddams, S. and Diels, J.-C. (1996) Dispersion measurements with white-light interferometry. *J. Opt. Soc. Am. B*, 13, 1120–1129.

Didomenico Jr., M. (1964) Small-signal analysis of internal (coupling-type) modulation of lasers. *J. Appl. Phys.*, 35, 2870–2876.

Diels, J. C. and Rudolph, W. (2006) *Ultrashort Laser Pulse Phenomena*, Burlington, MA, Elsevier.

Ell, R., Morgner, U., Kaertner, F. X., Fujimoto, J. G., Ippen, E. P., Scheuer, V., Angelow, G., and Tschudi, T. (2001) Generation of 5 fs pulses and octave-spanning spectra directly from a Ti:sapphire laser. *Opt. Lett.*, 26, 373–375.

Fork, R. L., Martinez, O. E., and Gordon, J. P. (1984) Negative dispersion using a pair of prisms. *Opt. Lett.*, 9, 150–152.

Froehly, C., Lacourt, A., and Vienot, J. C. (1973) Notions de reponse impulsionelle et de fonction de tranfert temporelles des pupilles opticques, justifications experimentales et applications. *Nouv. Rev. Opt.*, 4.

Grace, E. J., Ritsataki, A., French, P. M. W., and New, G. H. C. (2000) New optimization criteria for slit-apertured and gain-apertured KLM all-solid-state lasers. *Opt. Commun.*, 183, 249–264.

Gurs, K. (1964) Beats and Modulation in optical ruby lasers. In Grivet, P. and Bloembergen, N. (Eds.) *Quantum Electronics III*, New York, Columbia University Press.

Gurs, K. and Muller, R. (1963) Breitband-modulation durch Steuerung der emission eines optischen masers (Auskopple-modulation). *Phys. Lett.*, 5, 179–181.

Haus, H. A. (1975a) Theory of forced mode-locking. *IEEE J. Quantum Electron.*, QE11, 323–330.

Haus, H. A. (1975b) Theory of mode-locking with a fast saturable absorber. *J. Appl. Phys.*, 46, 3049–3058.

Haus, H. A. (1975c) Theory of mode-locking with a slow saturable absorber. *IEEE J. Quantum Electron.*, 11, 736–746.

Haus, H. A., Fujimoto, J. G., and Ippen, E. P. (1991) Structures for additive pulse modelocking. *J. Opt. Soc. Am. B*, 8(10), 2068–2076.

Haus, H. A., Fujimoto, J. G., and Ippen, E. P. (1992) Analytic theory of additive pulse and Kerr lens mode locking. *IEEE J. Quantum Electron.*, 28, 2086–2096.

Hentschel, M., Kienberger, R., Spielmann, C., Reider, G. A., Milosevic, N., Brabec, T., Corkum, P. B., Heinzmann, U., Drescher, M., and Krausz, F. (2001) Attosecond metrology. *Nature*, 414, 509–513.

Hönninger, C., Paschotta, R., Morier-Genoud, F., Moser, M., and Keller, U. (1999) Q-switching stability limits of continuous-wave passive mode locking. *J. Opt. Soc. Am. B*, 16, 46–56.

Iaconis, C. and Walmsley, I. A. (1998) Spectral phase interferometry for direct electric-field reconstruction of ultrashort optical pulses. *Opt. Lett.*, 23, 792–794.

Ippen, E. P., Haus, H. A., and Liu, L. Y. (1989) Additive pulse modelocking. *J. Opt. Soc. Am. B*, 6, 1736–1745.

Jones, D. J., Diddams, S. A., Ranka, J. K., Stentz, A., Windeler, R. S., Hall, J. L., and Cundiff, S. I. (2000) Carrier-envelope phase control of femtosecond mode-locked lasers and direct optical frequency synthesis. *Science*, 288, 635–639.

Judson, R. S. and Rabitz, H. (1992) Teaching lasers to control molecules. *Phys. Rev. Lett.*, 68, 1500.

Kaertner, F. X. (2004) *Few-Cycle Laser Pulse Generation and Its Applications*, Berlin, Germany, Springer.

Kalashnikov, V. I., Kalosha, V. P., Poloyko, I. G., and Mikhailov, V. P. (1997) Optimal resonators for self-mode locking of continuous-wave solid-state lasers. *J. Opt. Soc. Am. B*, 14, 964–969.

Kamiya, F. S. T., Wada, O., and Yamjima, H. (1999) *Femtosecond Technology: From Basic Research to Application Prospects*, Berlin, Germany, Springer.

Kogelnik, H. and Li, T. (1966) Laser beams and resonators. *Appl. Opt.*, 5, 1550–1567.

Kogelnik, H. W., Ippen, E. P., Dienes, A., and Shank, C. V. (1972) Astigmatically compensated cavities for cw dye lasers. *IEEE J. Quantum Electron.*, 8, 373–379.

Kuizenga, D. and Siegman, A. E. (1970) FM and AM mode locking of the homogeneous laser—Part I: Theory. *IEEE J. Quantum Electron.*, 6, 694–708.

Laura, T. S., Janelle, C. S., and Marcos, D. (2006) Advantages of ultrashort phase-shaped pulses for selective two-photon activation and biomedical imaging. *Nanomedicine*, 2, 177–181.

Lozovoy, V. V., Pastirk, I., Walowicz, K. A., and Dantus, M. (2003) Multiphoton intrapulse interference. II. Control of two- and three-photon laser induced fluorescence with shaped pulses. *J. Chem. Phys.*, 118, 3187–3196.

Magni, V., Cerullo, G., and De Silvestri, S. (1993) ABCD matrix analysis of propagation of Gaussian beams through Kerr media. *Opt. Commun.*, 96, 348–355.

Magni, V., Cerullo, G., De Silvestri, S., and Monguzzi, A. (1995) Astigmatism in Gaussian-beam self-focusing and in resonators for Kerr-lens mode locking. *J. Opt. S. Am B*, 12, 476–485.

Marcuse, D. (1980) Pulse distortion in single-mode fibers. *Appl. Opt.*, 19, 1653–1660.

Martinez, O. E., Fork, R. L., and Gordon, J. P. (1984) Theory of passively modelocked lasers including self-phase modulation and group-velocity dispersion. *Opt. Lett.*, 9, 156–158.

Mittleman, D. M., Jacobsen, R. H., and Nuss, M. C. (1996) T-ray imaging. *IEEE J. Sel. Top. Quantum Electron.*, 2, 679–692.

Morgner, U., Kaertner, F. X., Cho, S. H., Haus, H. A., Fujimoto, J. G., Ippen, E. P., Scheuer, V., Angelow, G., and Tschudi, T. (1999) Sub-two cycle pulses from a Kerr-lens modelocked Ti:sapphire laser. *Opt. Lett.*, 24, 411–413.

Mourou, G. and Umstader, D. (2002) Extreme light. *Sci. Am.*, 286, 81–86.

Naganuma, K., Mogi, K., and Yamada, H. (1989) General method for ultrashort light pulse chirp measurement. *IEEE J. Quantum Electron.*, 25, 1225–1233.

Naganuma, K., Mogi, K., and Yamada, H. (1990) Group-delay measurement using the Fourier transform of an interferometric cross correlation generated by white light. *Opt. Lett.*, 15, 393–395.

Nahata, A., Weling, A. S., and Heinz, T. F. (1996) A wideband coherent terahertz spectroscopy system using optical rectification and electro-optic sampling. *Appl. Phys. Lett.*, 69, 2321–2323.

Nijhof, J. H. B., Doran, N. J., Forysiak, W., and Knox, F. M. (1997) Stable soliton-like propagation in dispersion-managed systems with net anomalous, zero, and normal dispersion. *Electron. Lett.*, 33, 1726–1727.

Norris, T. B. (1992) Femtosecond pulse amplification at 250 kHz with a Ti:sapphire regenerative amplifier and application to continuum generation. *Opt. Lett.*, 17, 1009–1011.

Nuss, M. C., Knox, W. H., and Koren, U. (1996) Scalable 32 channel chirped-pulse WDM source. *Electron. Lett.*, 32, 1311–1312.

Owyoung, A., Hellwarth, R. W., and George, N. (1972) Intensity-induced changes in optical polarization in glasses. *Phys. Rev. B*, 5, 628–633.

Paul, P. M., Toma, E. S., Breger, P., Mullot, G., Auge, F., Balcou, P., Muller, H. G., and Agostini, P. (2001) Observation of a train of attosecond pulses from high harmonic generation. *Science*, 292, 1689–1692.

Paulus, G. G., Grasbon, F., Walther, H., Villoresi, P., Nisoli, M., Stagira, S., Priori, E., and De Silvestri, S. (2001) Absolute-phase phenomena in photoionization with few-cycle laser pulses. *Nature*, 414, 182–184.

Penzkofer, A., Wittmann, M., Lorenz, M., Siegert, E., and Macnamara, S. (1996) Kerr lens effects in a folded-cavity four-mirror linear resonator. *Opt. Quantum Electron.*, 28, 423–442.

Pestov, D., Lozovoy, V. V., and Dantus, M. (2009) Multiple independent comb shaping (MICS): Phase-only generation of optical pulse sequences. *Opt. Express*, 17, 14351–14361.

Ranka, J. K., Gaeta, A. L., Baltuska, A., Pshenichnikov, M. S., and Wiersma, D. A. (1997) Autocorrelation measurement of 6-fs pulses based on the two-photon-induced photocurrent in a GaAsP photodiode. *Opt. Lett.*, 22, 1344–1346.

Ranka, J. K., Windeler, R. S., and Stentz, A. (2000) Visible continuum generation in air-silica microstructure optical fibers with anomalous dispersion at 800 nm. *Opt. Lett.*, 25, 25–27.

Rice, A., Jin, Y., Ma, X. F., Zhang, X. C., Bliss, D., Larkin, J., and Alexander, M. (1994) Terahertz optical rectification from <110> zinc-blende crystals. *Appl. Phys. Lett.*, 64, 1324–1326.

Ripin, D. J., Chudoba, C., Gopinath, J. T., Fujimoto, J. G., Ippen, E. P., Morgner, U., Kärtner, F. X., Scheuer, V., Angelow, G., and Tschudi, T. (2002) Generation of 20-fs pulses by a prismless Cr4+:YAG laser. *Opt. Lett.*, 27, 61–63.

Rulliere, C. (2005) *Femtosecond Laser Pulses*, New York, Springer.

Schibli, T. R., Thoen, E. R., Kärtner, F. X., and Ippen, E. P. (2000) Suppression of Q-switched mode locking and break-up into multiple pulses by inverse saturable absorption. *Appl. Phys. B Lasers Opt.*, 70, S41–S49.

Shen, Y. R. (2002) *The Principles of Nonlinear Optics*, Hoboken, NJ, John Wiley & Sons, Inc.

Sherriff, R. E. (1998) Analytic expressions for group-delay dispersion and cubic dispersion in arbitrary prism sequences. *J. Opt. Soc. Am. B*, 15, 1224–1230.

Siegman, A. E. (1986) *Lasers*, Mill Valley, CA, University Science Books.

Spence, D. E., Kean, P. N., and Sibbett, W. (1991) 60-fsec pulse generation from a self-mode-locked Ti:Sapphire laser. *Opt. Lett.*, 16, 42–44.

Statz, H. and Tang, C. L. (1964) Zeeman effect and nonlinear interactions between oscillating laser modes. In Grivet, P. and Bloembergen, N. (Eds.) *Quantum Electronics III*, New York, Columbia University Press.

Stingl, A., Lenzner, M., Spielmann, C., Krausz, F., and Szipöcs, R. (1995) Sub-10-fs mirror-dispersion-controlled Ti:sapphire laser. *Opt. Lett.*, 20, 602–604.

Strickland, D. and Mourou, G. (1985) Compression of amplified chirped optical pulses. *Opt. Commun.*, 55, 447–449.

Sutter, D. H. (2000) New frontiers of ultrashort pulse generation. *Physics*. Zurich, Technical University of Switzerland.

Sutter, D. H., Steinmeyer, G., Gallmann, L., Matuschek, N., Morier-Genoud, F., Keller, U., Scheuer, V., Angelow, G., and Tschudi, T. (1999) Semiconductor saturable-absorber mirror-assisted Kerr-lens mode-locked Ti:sapphire laser producing pulses in the two-cycle regime. *Opt. Lett.*, 24, 631–633.

Szipöcs, R., Ferencz, K., Spielmann, C., and Krausz, F. (1994) Chirped multilayer coatings for broadband dispersion control in femtosecond lasers. *Opt. Lett.*, 19, 201–203.

Trebino, R. (2000) *Frequency-Resolved Optical Gating: The Measurement of Ultrashort Laser Pulses*, Norwell, MA, Kluwer Academic Publishers.

Valdmanis, J. A., Fork, R. L., and Gordon, J. P. (1985) Generation of optical pulses as short as 27 fs directly from a laser balancing self-phase modulation, group-velocity dispersion, saturable absorber, and saturable gain. *Opt. Lett.*, 10, 131–133.

Wagenblast, P. C., Morgner, U., Grawert, F., Schibli, T. R., Käärtner, F. X., Scheuer, V., Angelow, G., and Lederer, M. J. (2002) Generation of sub-10-fs pulses from a Kerr-lens mode-locked Cr3+:LiCAF laser oscillator by use of third-order dispersion-compensating double-chirped mirrors. *Opt. Lett.*, 27, 1726–1728.

Weiner, A. M., Leaird, D. E., Patel, J. S., and Wullert, J. R. (1990) Programmable femtosecond pulse shaping by use of a multielement liquid-crystal phase modulator. *Opt. Lett.*, 15, 326–328.

Xu, B., Gunn, J. M., Cruz, J. M. D., Lozovoy, V. V., and Dantus, M. (2006) Quantitative investigation of the multiphoton intrapulse interference phase scan method for simultaneous phase measurement and compensation of femtosecond laser pulses. *J. Opt. Soc. Am. B*, 23, 750–759.

Yariv, A. (1964) Internal modulation in multimode laser oscillations. *J. Appl. Phys.*, 36, 388–391.

Zhou, J., Taft, G., Huang, C. P., Murnane, M. M., Kapteyn, H. C., and Christov, I. P. (1994) Pulse evolution in a broad-bandwidth Ti:sapphire laser. *Opt. Lett.*, 19, 1149–1151.

ADDITIONAL READING

Boyd, R. W., 2003. *Nonlinear Optics*. San Diego, CA: Academic Press.

Diels, J.-C. and Rudolph, W., 2006. *Ultrashort Laser Pulse Phenomena*. Burlington, MA: Academic Press.

Kärtner, F. X., 2004. *Few-Cycle Laser Pulse Generation and Its Applications*. Berlin, Germany: Springer-Verlag.

Rulliere, C., 2005. *Femtosecond Laser Pulses: Principles and Experiments*. New York: Springer.

Siegman, A. E., 1986. *Lasers*. Sausalito, CA: University Science Books.

Svelto, O., 1998. *Principles of Lasers*. New York: Springer.

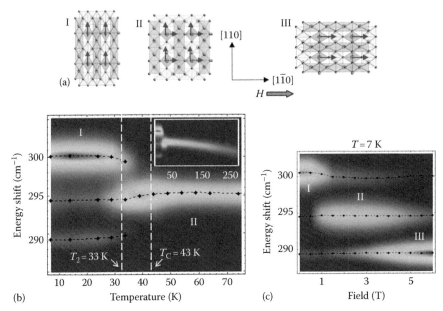

Figure 6.11 (a) Schematic illustrations of different structural phases exhibited by Mn_3O_4. (b) Temperature dependence of the T_{2g} symmetry Mn-O stretch mode in Mn_3O_4, showing splitting of the mode degeneracy below tetragonal (II) to monoclinic (I) transition. The inset shows an expanded view from 10 to 300 K. (b) Magnetic field dependence of the T_{2g} symmetry Mn-O stretch mode at $T = 7$ K in Mn_3O_4, showing the field-induced phase changes for **H**‖[1$\bar{1}$0]. (After Kim, M. et al., *Phys. Rev. Lett.*, 104, 136402, 2010.)

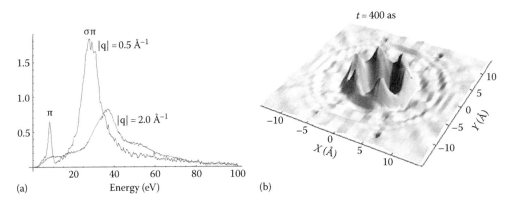

Figure 6.17 (a) Raw IXS spectra taken from single crystal graphite for two values of the in-plane transferred momentum (shown). (b) Time slice of the density propagator at $t = 400$ as, showing hexagonal symmetry from the underlying honeycomb lattice.

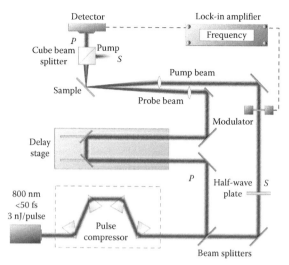

Figure 9.1 Time-resolved differential reflection spectrometer using degenerate pump–probe spectroscopy. The system can be similarly configured for time-resolved differential transmission, with the collection optics and detection relocated behind the sample. The polarization of the pump beam is rotated by 90° using a half-wave plate. After either reflection (as shown) or transmission (not shown), the probe beam can be isolated from the pump beam using a polarizing cube beam splitter. If using a balanced detector, a small portion of the probe beam right after the pulse compressor is split off and used as the detector reference input (not shown). The second detector and the balanced detector circuitry use this reference to cancel any power fluctuations in the laser power from the differential change to the probe transmission or reflectivity.

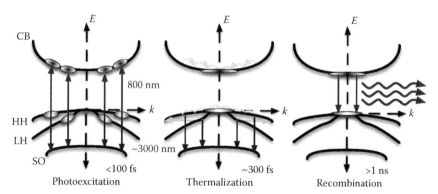

Figure 9.5 Optical pump, mid-infrared probe measurements on a III–V semiconductor (left). The mid-infrared probe pulse is used to isolate the hole subband dynamics from the conduction band dynamics. Initial photoexcitation is determined by the photon energies present in the pump pulse (red). Mid-infrared probe pulses (blue) promote holes from the upper valance subbands to the split-off subband. The wavelength of this probe pulse determines the region in k-space probed in this experiment. (middle) Energy thermalization is the result of inter- and intrasubband carrier–carrier scattering (yellow) and phonon scattering between subbands (green). (right) The system returns to equilibrium using electron–hole recombination, which results in the emission of a photon (red), carrier trapping (not shown), or thermal diffusion to the substrate (not shown). CB, conduction band; HH, heavy-hole subband; LH, light-hole subband; and SO, split-off–hole subband. (From Hilton, D.J. and Tang, C.L., *Phys. Rev. Lett*, 89, 146601, 2002; Scholz, R., *J. Appl. Phys*, 77, 3219, 1995; Braunstein, R. and Kane, E., *J. Phys. Chem. Solids*, 23, 1423, 1962; Braunstein, R. and Magid, L., *Phys. Rev*, 111, 480, 1958.)

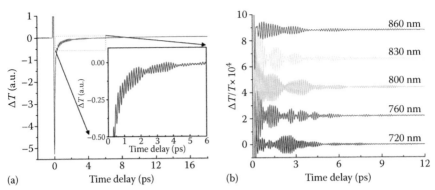

Figure 9.9 (a) The time-resolved differential transmission (ΔT) of an ensemble of carbon nanotubes in solution using a *degenerate* differential transmission measurement at $\lambda = 800\,nm$. The periodic modulation resulting from the coherent modulation of the nanotube diameter is highlighted in the inset. (b) The extracted coherent phonon oscillations in the ensemble as a function of wavelength generated by removing the electronic component using a form similar to Equation 9.10. The presence of multiple phonon frequencies in these data results from the existence of multiple chiral species of nanotubes (n, m) in this HiP$_{CO}$ nanotube sample. (Reprinted with permission from Lim, Y.-S., Yee, K.-J., Kim, J.-H., Haroz, E., Shaver, J., Kono, J., Doorn, S.K., Hauge, R., and Smalley, R., Coherent lattice vibrations in single-walled carbon nanotubes, *Nano Lett.*, 6, 2696–2700, 2006. Copyright 2006 American Chemical Society.)

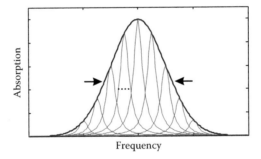

Figure 10.1 Schematic of an inhomogeneously broadened absorption line. The absorption profile (thick line) is actually due to the sum of many narrower lines (thin lines) at different center frequencies (the center frequencies are continuously distributed; only a few are shown). The inhomogeneous (arrows) and homogeneous (dotted line) line widths are indicated. The oscillators with a given resonance frequency are known as a "frequency group."

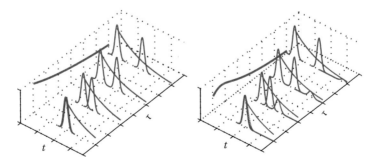

Figure 10.4 Sketch for two-pulse excitation showing excitation pulses (solid lines), coherences (dashed lines), signals (dotted lines), the time-integrated signal (thick solid line), and decays of the polarization in individual frequency groups (thin solid lines) for a homogeneously broadened system (left) and an inhomogeneously broadened system (right). The delay, τ, between the two excitation pulses increases from front to back.

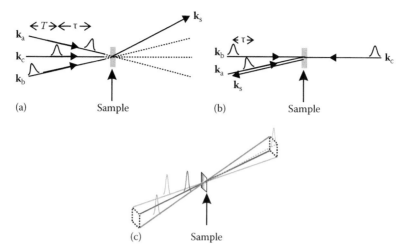

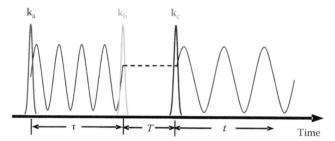

Figure 10.6 3P-TFWM configurations: (a) coplanar, (b) phase conjugate, and (c) box.

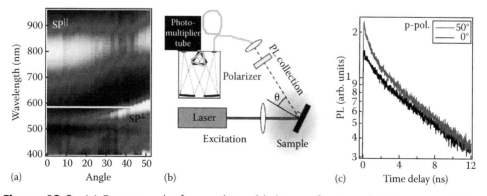

Figure 10.9 The pulse sequence for 2DFTS and phase evolution. The initial pulse, $\mathbf{k}_a$, excites an initial coherence that evolves during a time period τ. The second pulse, $\mathbf{k}_b$, stores the phase of the initial coherence in a population state. The third pulse, $\mathbf{k}_c$, generates the coherence that radiates during time period t. The overall phase of the radiating coherence is determined by the phase evolution during time period t. By taking a two-dimensional Fourier transform of the signal with respect to τ and t, the frequencies of both the initial coherence and the emitting coherence can be determined. For uncoupled resonances, these two frequencies will always be the same, whereas they can be different if two resonances are coupled, for example, the two transitions of a three-level system.

Figure 12.8 (a) Extinction plot for *p*-polarized light as a function of incidence angle (low extinction: dark; high extinction: yellow). Indicated are the in-plane SP∥ and out-of-plane SP⊥ resonance and the center PL wavelength of the NCs (yellow line). (b) Schematic of the angle and polarization-resolved PL dynamics setup. (c) *p*-polarized PL decay dynamics collected at 0° and 50°. (Reprinted from Wang, Y. et al., *Phys. Rev. Lett.*, 102, 163001, 2009.)

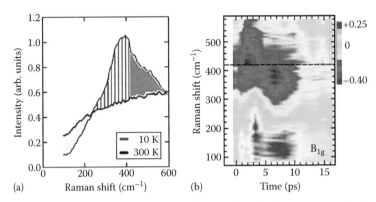

Figure 14.9 (a) Steady-state Raman spectra of Bi-2212 in the B_{1g} geometry (in Porto notation $z(xy)z$, where z is perpendicular to the CuO planes) at 10 and at 300 K. A gap opens below 250 cm^{-1} (blue area) and a pair-breaking peak appears around 420 cm^{-1} (red area). (b) Temporal evolution of the time-resolved Raman difference spectra at an equilibrium temperature of 10 K in the B1g geometry. The figure represents a contour plot consisting of 12 Raman difference spectra for different delay times. The dashed line separates two energy regions of the pair-breaking peak that reveal different characteristic behaviors. The intensity changes are color coded, demonstrating the transfer of spectral weight from high to low energies after 1 ps. (Reprinted with permission from Saichu, R.P., Mahns, I., Goos, A., Binder, S., May, P., Singer, S.G., Schulz, B., Rusydi, A., Unterhinninghofen, J., Manske, D., Guptasarma, P., Williamsen, M.S., and Rbhausen, M., *Phys. Rev. Lett.*, 102, 177004. Copyright 2009 by the American Physical Society.)

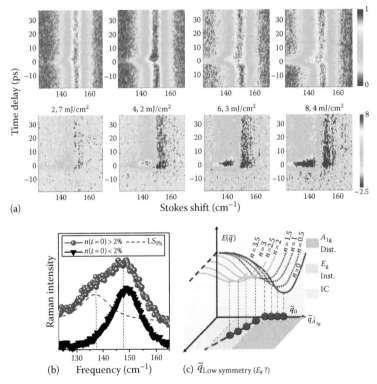

Figure 14.11 Photoinduced low-symmetry phase in antimony single crystal. (a) False color plot of the Raman response of the Ag phonon at different times after photoexcitation for various excitation densities (in the lower panel the Raman response at negative times has been subtracted). (b) The reduction of symmetry due to the photoinduced phase transition leads to the appearance of an additional phonon mode on the low-energy side of the Ag mode for excitation densities larger than 2%. (c) Cartoon of the free energy curves as a function of the A_{1g} and E_g lattice distortion. The location of minimum free energy is indicated in the bottom plane of the sketch. (Adapted with permission from Fausti D. et al., *Phys. Rev B*, 80, 161207, 2009.)

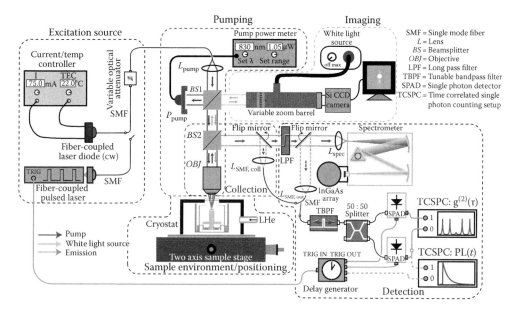

Figure 16.2 Schematic of a typical micro-PL setup for investigating a sample in a cryogenic environment. SMF = single-mode fiber, *BS* = beam splitter, *L* = lens, *OBJ* = objective, LPF = long-pass filter, TBPF = tunable bandpass filter, SPAD = single-photon avalanche photodiode, TCSPC = time-correlated single-photon counting.

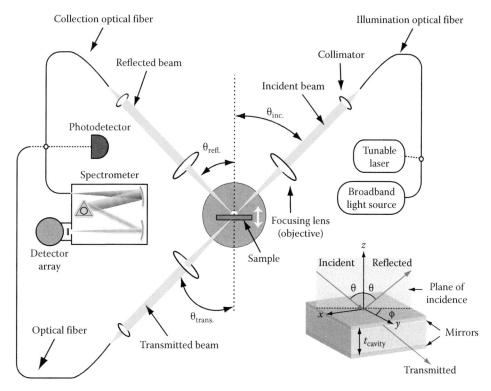

Figure 16.15 Angle-resolved reflectivity or transmission setup. The goniometer consists of two arms that rotate around a common axis, so that the angle between the two may be precisely adjusted. Collimating and focusing optics, as well as collection and illumination fiber optics shown, are mounted on the rotating arms. The bottom right image shows the orientation of the beam on a planar microcavity sample.

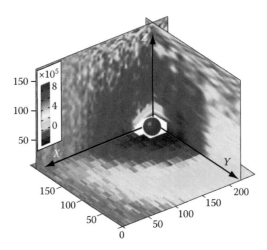

Figure 17.15 3D rendering of the near-field coupling between a Si tip and a fluorescent sphere. The white cutout region indicates the topographical volume traced by the AFM, while the false-color data represents the fluorescence count rate, as indicated by the inset (in counts/s). The spatial dimensions are in units of nanometers. (Reprinted with permission from Mangum, B.D. et al., *Nano Lett.*, 9, 3440, 2009.)

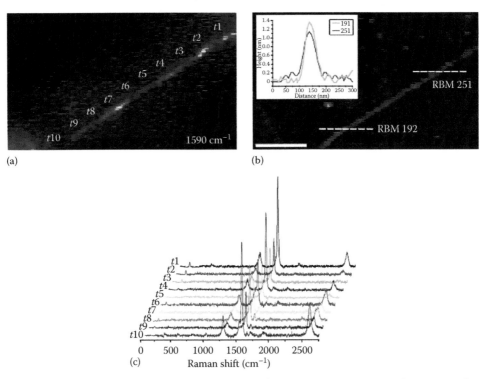

Figure 17.18 TERS image of an SWCNT. Near-field Raman image (a) and corresponding topography image (b) of an isolated SWCNT, where the optical resolution was determined to be 40 nm. Also shown are a series of tip-enhanced Raman spectra (c) acquired along the length of the SWNT. From the recorded spectra, two resonant RBM modes are detected: one at 251 cm⁻¹ corresponding to a semiconducting chirality and the second recorded from the lower section of the SWCNT at 192 cm⁻¹, corresponding to a metallic chirality. The inset of (b) displays two cross-sectional profiles acquired from both the upper and lower sections, respectively, revealing that the expected diameter change occurs as the SWCNT undergoes the transition from a semiconducting to metallic chirality. Scale bar denotes 200 nm and is valid for both (a) and (b). (Reprinted with permission from Anderson, N. et al., *Nano Lett.*, 7, 577, 2007.)

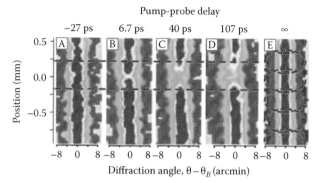

Figure 18.1 Ultrafast XRD experiments on a Ge film at different time delays. The region of the sample photoexcited by the pump pulse is indicated by the dotted lines in (A)–(D). The image at infinite time delays (E) depicts six single-shot damage regions, clearly showing that the diffraction signal has almost completely recovered after several seconds. (From Siders, C. et al., Detection of nonthermal melting by ultrafast x-ray diffraction, *Science*, 286, 1340–1342, 1999. Reprinted with permission from AAAS.)

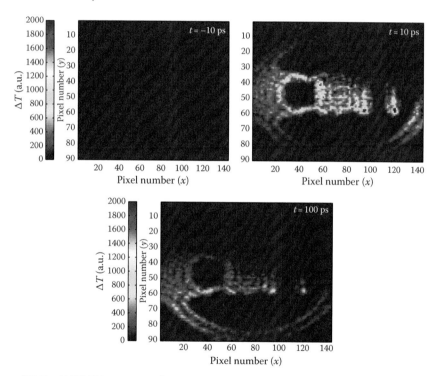

Figure 18.9 UOWFM images of a patterned amorphous Si film at different time delays between the pump and probe pulses. (From Prasankumar, R.P., Ku, Z., Gin, A.V., Upadhya, P.C., Brueck, S.R.J., and Taylor, A.J., Ultrafast optical wide field microscopy, in *Conference on Lasers and Electro-Optics*, Baltimore, MD, 2009. With permission.)

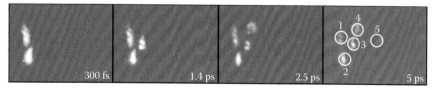

Figure 18.10 Time-resolved PL images of CdSSe nanobelts as a function of time delay. (Reprinted with permission from Gundlach, L. and Piotrowiak, P., Ultrafast spatially resolved carrier dynamics in single CdSSe nanobelts, *J. Phys. Chem. C.*, 113, 12162–12166, 2009. Copyright 2009 American Chemical Society.)

Chapter 8

CARRIER DYNAMICS IN BULK SEMICONDUCTORS AND METALS AFTER ULTRASHORT PULSE EXCITATION

Jure Demsar and Thomas Dekorsy

CONTENTS

The power of time-resolved optical spectroscopy is best illustrated using simple model systems. This chapter deals with the basics of carrier dynamics in bulk semiconductors and metals after ultrashort excitation; brief introductions to these subjects are given in Chapters 2 and 3, respectively. The discussion here will serve as a basis for understanding dynamics in more complex systems (some of which are discussed here as well as in other chapters). These systems were carefully investigated by ultrafast optical pump-probe methods, which provided the first insight into ultrafast scattering events and carrier relaxation processes.

8.1 ULTRAFAST CARRIER DYNAMICS IN SEMICONDUCTORS

Semiconductors are the most important materials for today's electronic and optoelectronic devices. As device dimensions continuously shrink to the nanometer scale, quantum mechanical effects become important when the device dimensions reach the de Broglie wavelength of electrons. In addition to the important implications of reduced spatial dimensions, the timescale at which fundamental interactions in semiconductors happen becomes increasingly important. One example is a high-frequency field-effect transistor; the ultimate speed limit of a transistor is the transit time an electron needs to travel the distance from the source electrode to the drain electrode. Assuming a distance between source and drain electrodes of 100 nm and a drift velocity of electrons of 10^7 cm/s, this transit time is only 1 picosecond (ps). The corresponding frequency is 1 terahertz (THz), which is two orders of magnitude higher than standard present-day transistors in silicon technology operate. This example illustrates that processes on a femtosecond (fs) timescale become relevant for the performance of such a device. The only method for studying electronic processes on such a timescale is by means of time-resolved optical experiments using ultrashort laser pulses, as is described in detail in Part III of this book. These experiments rely on an ultrashort laser pulse for triggering an event that is recorded in the time domain with a time-delayed optical probe pulse. The purpose of this chapter is to establish the most important processes that happen on this timescale. We will not discuss in detail the so-called *coherent regime*, where the optical excitation of the material has to be treated as a polarization rather than in terms of excited charge carriers (Meier 2007). This regime is discussed in more detail in Chapter 10. Since semiconductors are complicated many-body systems with many degrees of freedom strongly depending on each other, we will treat different processes separately and ask the reader to bear in mind that several of these processes occur on the same timescale. In order to bring some logical order into the many processes involved we will order the different processes along their "typical" timescale. However, these timescales are not universal and one will find specific materials in which certain processes are faster and more important than in other materials. Finding out which process dominates under certain circumstances in a given material, e.g., under reduced dimensions, is one of the fascinating topics of ultrafast spectroscopy in present-day research. Table 8.1 summarizes the hierarchy of the different regimes as a function of time including the most relevant interactions. We suggest that the reader return to this table while studying this chapter.

For further reading on this topic we suggest the following books:

- W.T. Wenckebach, Wiley 1999, for a theoretical treatment of electron–phonon coupling, charge transport, and carrier scattering in semiconductors.
- C. Klingshirn, Springer 2006, for a complete overview of semiconductor optics with emphasis on excitons and polaritons.
- W. Schäfer and M. Wegener, Springer 2002, for a cohesive presentation of modern semiconductor physics written jointly by a theoretician and an experimentalist (Schäfer 2002).
- J. Shah, Springer 1999, for a complete survey of experiments on ultrafast dynamics in semiconductors and semiconductor nanostructures.

- T. Meier, P. Thomas, and S. Koch, Springer 2007, for a theoretical treatment of ultrafast dynamics in semiconductors with emphasis on the coherent regime.
- F.T. Vasko and A.V. Kuznetsov, Springer 1998, for an emphasis on the electronic and optical properties of semiconductor heterostructures including tunneling phenomena, intersubband transitions, and ultrafast phenomena.

8.1.1 The Hierarchy of Relaxation Processes

We make the following simplified gedanken experiment: at time zero, we put an electron into the conduction band of a semiconductor, with a given excess energy E_{ex} above the minimum of the conduction band and a given momentum in reciprocal space according to the band structure. Now we ask the question, "what will happen to this single electron?" The electron will emit this excess energy to the crystal lattice by emission of phonons. The associated scattering time depends on the material, i.e., its band structure, the type of available phonon modes, and the interaction strength of different phonons. These mechanisms will be discussed in the sections on intravalley carrier relaxation, i.e., when the final state of the relaxation process is in the same valley of the conduction band, and intervalley scattering, i.e., when the final state of the scattering process is a different local minimum of the conduction band.

TABLE 8.1

Overview of Different Temporal Regimes of a Semiconductor after Excitation with an Ultrashort Laser Pulse

Regime and Timescale	Relevant Interaction	Remarks
Optical excitation 10–100 fs	Optical dipole moment of interband transitions (or intersubband transitions in quantum structures)	The duration of the excitation pulse is closely related to the energetic width of the excited carrier distribution
Coherent regime 10 fs to several 100 fs	Interaction of optical polarization with electronic states, decay of coherent polarization through scattering processes	Description in terms of coherent wave functions, coherent polarizations, and optical Bloch equations, observation of coherent wavepacket dynamics (quantum beats, Bloch oscillations)
Nonthermal regime <200 fs	Carrier–carrier scattering	Carrier distribution cannot be assigned a temperature, occurrence of a spectral hole
Thermalized or hot-carrier regime 100 fs to ps	Carrier–carrier scattering Carrier–phonon interactions	Carrier distribution is defined by a temperature; temperatures of electron and hole distributions may be different and are larger than the lattice temperature; temperatures of electron and hole distributions equilibrate with the lattice temperatures
Recombination or isothermal regime ps to µs	Radiative or non-radiative recombination Carrier trapping	Temperatures of carrier distributions and lattice are equal; timescale strongly depends on the material, e.g., direct or indirect semiconductor, density of defects, and quality of surfaces

Note: The discrimination between the regimes and the given time intervals are not strict, since all processes are related to an exponential decay of the investigated quantity (e.g., coherent polarization, excess energy, temperature, and carrier density) and several processes occur simultaneously.

The same statements describe holes in the valence bands of a semiconductor. In this simplified gedanken experiment, important interactions are neglected: for example, a semiconductor is a many-body system with up to 10^{18} free electrons or holes per cm^3 depending on the doping level or optical excitation, but here the interaction among those carriers is neglected. These effects will be also discussed when we talk about the description of charge carriers with a distribution function. In addition, the "single electron at time zero" will be created in a real experiment by an optical pulse, which always creates an electron–hole *pair*. The electron–hole pair is coupled by the Coulomb interaction, which also has to be considered. In the case of many electron–hole pairs the Coulomb interactions will be reduced since electric fields are screened by electron–hole pairs. This effect gives rise to a strong dependence of Coulomb interactions on carrier density.

8.1.2 Intravalley Carrier Relaxation

The possible channels for energy relaxation of an electron in a single valley of the conduction band of a semiconductor are sketched in Figure 8.1. For the case of holes in the valence band the description of energy relaxation and electron–phonon interaction is very similar, though the scattering times may vary due to differences in effective masses of holes or different symmetries of hole wave functions compared to electronic wave functions (Yu 1996).

The electron can either emit (or absorb at finite lattice temperatures) an acoustic or optical phonon under energy and momentum conservation. Hence, the phonon dispersion discussed in Chapter 2 has to be taken into account to describe a final state that lies on the electronic dispersion curve. If the excess energy of the electron is smaller than the optical phonon energy E_{TO} or E_{LO}, only emission of acoustic phonons is possible. The relaxation rates for acoustic phonon emission are determined by the acoustic deformation potential, which describes the change in the eigenstate of the electron due to the deformation associated with an acoustic phonon. This interaction strength is proportional to the wave vector (q) of the acoustic phonon involved. Intuitively this can be understood, since q approaching zero represents a pure translation of the whole crystal lattice. The relaxation rates for optical phonons strongly depend on the type of semiconductor involved. In *nonpolar* semiconductors which do not possess a TO–LO splitting at $q = 0$ (e.g., Si or Ge [Fox 2001]) the optical deformation potential defines the interaction strength (Yu 1996). This interaction is approximately independent of q. In polar semiconductors which possess a TO–LO splitting at $q = 0$ (e.g., GaAs, InP, CdSe, GaN) the interaction of the polar longitudinal lattice vibrations with electrons and holes is much stronger than the

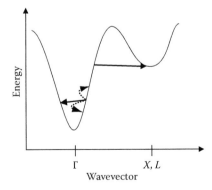

Figure 8.1 Schematic band diagram of the conduction band of a semiconductor. The arrows indicate different electron–phonon scattering processes (solid arrows: acoustic phonons with small energy transfer, dashed arrows: optical phonons with large energy transfer).

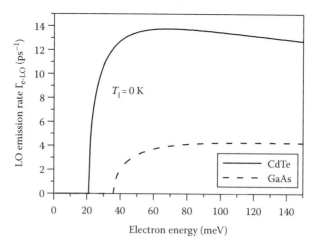

Figure 8.2 Phonon emission rate for electron–longitudinal optical phonon scattering in GaAs (dashed line) and CdTe (solid line) versus electron energy relative to the top of the conduction band at a lattice temperature of 0 K. The difference in the onset of phonon emission stems from the different longitudinal optical phonon energies in the two materials.

optical deformation potential interaction. This is clear since LO phonons are accompanied by a lattice polarization that strongly interacts with the dielectric displacement of charge carriers. The smaller the wave vector of the involved phonon, the more extended is the phonon polarization and the stronger the interaction with charge carriers. Hence, this interaction strength scales with $1/q$, and thus the intravalley relaxation in polar semiconductors is extremely fast, with phonon emission rates in the range of $1–10\,ps^{-1}$. This is shown in Figure 8.2 for two different polar semiconductors (GaAs and CdTe) versus the excess energy of the electron. The interaction of longitudinal optical phonons in a polar semiconductor is often referred to as the Fröhlich interaction. The Fröhlich scattering rate is proportional to the difference of the inverse high-frequency dielectric constant ε_∞ ("high" refers to frequencies above the highest phonon frequencies) and the static dielectric constant ε_{static} ("static" refers to zero frequency, i.e., below the lowest phonon frequencies): $(1/\varepsilon_\infty - 1/\varepsilon_{static})$. The evaluation of this formula results in a factor of 4 difference in the scattering rates between GaAs and CdTe (Shah 1999).

Up to this point, we have only discussed the process of phonon emission. At higher temperatures of the semiconductor the phonon occupation increases and phonon absorption achieves higher probabilities.

8.1.3 Intervalley Scattering and Umklapp Processes

In the previous section, only scattering within one valley of the band structure was considered. Now we expand this picture to the band structure of the entire Brillouin zone, which may contain different local minima in different directions of k-space (Chapter 2). Through interaction with a phonon with a sufficiently large wave vector the electron may be scattered from one valley into another. This valley may either lie in the same Brillouin zone as the original valley (intervalley scattering) or in an adjacent Brillouin zone (Umklapp process). These different processes are sketched in Figure 8.3. Since these scattering processes require the interaction of a large q phonon, they are dominated by acoustic phonon scattering. It should be noted that since only one central valley exists as compared to many side valleys, and the density of states in side valleys is typically larger (i.e., the effective mass of the side valleys is larger), the scattering from the central valley into a side valley is typically an order of magnitude faster than the reverse process.

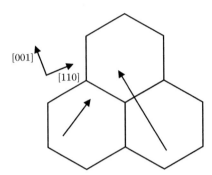

Figure 8.3 Schematic sketch of different electron–phonon scattering processes within one Brillouin zone (left arrow) and between adjacent Brillouin zones (right arrow). Each hexagon represents a cross section through the Brillouin zone of a model semiconductor.

8.1.4 Dynamics of the Electron (Hole) Distribution Function

The dynamics described so far have been restricted to the dynamics of a single electron. In a real experiment one has to deal with electron densities in the large range of some 10^{14}–10^{21} cm^{-3}. Hence a description in a single-electron picture is incomplete. The electrons (holes) are described in terms of a distribution function defined by a temperature, i.e., a Maxwell–Boltzmann or Fermi–Dirac function (Shah 1999). One should always bear in mind that a distribution function is only well defined in thermal equilibrium, which is definitely not the case following ultrashort optical excitation. Since the electrons (holes) are typically excited with some excess energy above the conduction band (below the valence band), the temperature of the distribution function will be higher than the temperature of the unexcited crystal lattice. Hence electron and hole distributions are characterized by two different temperatures, which are both higher than the lattice temperature. This is the regime of hot carriers. At very short timescales, more electrons "sit" at higher energies in the conduction band. Such a carrier distribution cannot be described by a distribution function. This regime has been coined the nonthermal regime. Experimentally, the nonthermal regime has been searched for intensively. The difficulty in observing a nonthermal distribution is (1) the ultrafast equilibration of the internal energy of the carriers through electron–electron and electron–hole scattering and (2) the fast polar-optical phonon emission rates associated with the Fröhlich interaction. Successful observation of a nonthermal carrier distribution could only be made with a time resolution better than 100 fs at low carrier densities (Leitenstorfer 1996).

The description of carrier dynamics in terms of distribution functions is also important in the case of nonequilibrium transport under an applied electric field. In equilibrium transport the electron distribution function can be described as a distribution that is displaced in reciprocal space into the field direction. However, under nonequilibrium conditions electrons may be accelerated to energies such that intervalley scattering occurs. Then, the electron distribution is expressed by the occupancy of different valleys and the scattering rates back and forth between these valleys (Wysin 1988). This is described in more detail in a following section.

8.1.5 Changes of the Optical Properties of an Excited Semiconductor

The excitation of electron–hole pairs influences the optical properties of the semiconductor in several ways. A typical experiment to probe these changes on a fs timescale is a pump-probe scheme (Chapter 9), where the optical changes induced by the photoexcited carriers are spectrally resolved by a broadband probe pulse via transmission changes in a thin semiconductor film (Collet 1994, Hunsche 1993). The absorption of the semiconductor can be described as $\alpha(h\nu) = \alpha_{ex}(h\nu) + \alpha_{in}(h\nu)$, with α_{ex} the excitonic contribution to the absorption due to the Coulomb attraction of electrons and holes, α_{in} the interband absorption between valence and conduction bands, and $h\nu$ the photon energy. The excitonic contribution α_{ex} gives rise to one or more sharp absorption lines with transition energies a few 10 meV to a few meV below the energy of the band gap (Chapter 2). It should also be noted that the interband absorption is also influenced by Coulomb effects. α_{in} is given as $\alpha_{in} = \alpha_0 \times CEF(h\nu,\rho) \times (h\nu - E_G)^{1/2} \times \{1 - f_e(\Delta E_e) - f_h(\Delta E_h)\}$ with $CEF(h\nu,\rho)$ the Coulomb enhancement factor, E_G the band gap, and f_e (f_h) the distribution function of electrons (holes), indicating the fraction of occupied states at a given excess energy ΔE_i (calculated from the photon energy $h\nu$, the gap energy E_G, and the band dispersion).

For the unperturbed case, where no electron–hole pairs are excited, i.e., $f_e = f_h = 0$, this equation represents the standard optical absorption of an undoped semiconductor, with a typical square-root dependence above the band gap. Upon optical excitation the Coulomb interaction of excitons is screened depending on the excitation density, giving rise to a reduction of α_{ex} corresponding to an increase of transmission. The interband contribution of the absorption is reduced due to the bleaching of the transition, when states in the valence band are not occupied (holes) and states in the conduction band are occupied (electrons), i.e., $1 - f_e(h\nu) - f_h(h\nu) < 1$. In addition, the unperturbed band gap of the semiconductor shrinks in the presence of electron–hole pairs due to their Coulomb interaction. The amount of shrinkage strongly depends on the carrier density (Banyai 1986). This effect is known as band gap renormalization. The Coulomb enhancement factor CEF is also reduced as shown by a comparison of experiment and theory in a thin GaAs film (Collet 1994, Hunsche 1993). In lower dimensional semiconductors, the effects of changes in the Coulomb interaction are typically enhanced compared to the three-dimensional case of a bulk semiconductor.

8.1.6 Static Electric Fields: Velocity Overshoot

In most electronic semiconductor devices carrier dynamics are triggered by electric fields. As an example of important device physics, we discuss the effect of velocity overshoot, which gives rise to transient carrier velocities exceeding the equilibrium drift velocity of a given material by up to an order of magnitude. Hence, this effect can be exploited in high-frequency electronics. In addition, velocity overshoot is a source for electromagnetic radiation at terahertz frequencies. Velocity overshoot is closely related to the Gunn effect, which is the radiation source in every microwave oven.

Let us start again in the simple single electron picture. Assume an electron is at the bottom of the conduction band at time zero in the presence of a static electric field. This scenario can be easily obtained by excitation of a direct semiconductor with photon energies closely above the band gap energy. The electron will be accelerated according to the Newtonian theorem $\partial p / \partial t = eF$, i.e., the electron momentum will increase linearly in time and its energy will increase according to the band structure. Depending on the strength of the electric field the electron will acquire enough energy to (1) emit an optical phonon or (2) scatter into a side valley through intervalley scattering. Interactions with acoustic phonons can be neglected since the Fröhlich interaction is much stronger than the electron–acoustic phonon interaction in polar semiconductors. Before such a scattering process occurs, the velocity of the electron may exceed the static drift velocity in equilibrium. This effect is denoted as velocity overshoot. In the case of intervalley scattering the electron will be strongly decelerated since the effective masses are larger in the side valleys than in the central valley. This deceleration is the origin of electromagnetic radiation at terahertz frequencies (Leitenstorfer 2000). The dynamics of electrons at different electric field strengths in GaAs are excellently obtained from Monte

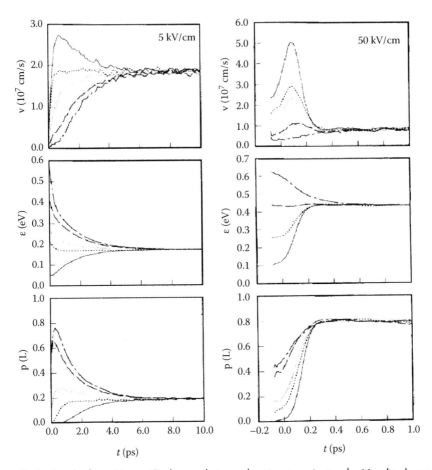

Figure 8.4 Results from Monte Carlo simulations showing transient velocities (top), average energies of the electronic distribution function (middle), and fractional L-valley occupation (bottom) for 5 kV/cm (left) and 50 kV/cm (right) electric fields at different excitation energies. The different curves correspond to photon energies of 1.5 eV (solid line), 1.7, 1.8, 2.0, and 2.2 eV (dashed-solid line), respectively. Please notice the change in the timescale for the two field strengths. (Reproduced from Wysin, G.M. et al., *Phys. Rev. B*, 38, 12514, 1988. With permission. Copyright 1988 by the American Physical Society.)

Carlo simulations (Wysin 1988). Figure 8.4 shows transient velocities and side-valley occupation and excess energies of the electrons for different static electric field strengths and optical excitation energies (i.e., different initial excess energies) as obtained from Monte Carlo simulations. Studying the details of this figure is very rewarding for a basic understanding of high-field transport in semiconductors.

8.1.7 Spatiotemporal Dynamics: Surface Field Screening and the Photo-Dember Effect

Surface field screening and the photo-Dember effect are not covered in most standard textbooks on semiconductor physics, but they have gained importance in the context of the generation of terahertz electromagnetic waves from optically excited semiconductors.

When a semiconductor surface is excited by an ultrashort laser pulse with a photon energy larger than the band gap, the excited carrier density decreases away from the surface according to the Lambert–Beer law $I(z) = I_0 \exp(-\alpha z)$, with α the absorption coefficient, I_0 the incident intensity, and z the coordinate perpendicular to the excited surface (Meier 2007). A typical value for α to remember is $\alpha = 10^4\,\text{cm}^{-1}$ for GaAs excited closely above the band gap at room temperature. This *spatial* carrier distribution will diffuse into the bulk of the semiconductor along its gradient with the ambipolar diffusion D_{amb} coefficient (which we will define later). The associated current density of this diffusion current is $j_D = -q D_{\text{amb}}\, dn/dx$.

In almost all semiconductors a static built-in electric field exists at the surface. This surface field is set up by charges at the surface which result, for example, from dangling bonds, i.e., unsaturated bonds of surface atoms. Further descriptions of these surface fields can be found under the terms Fermi-level pinning or surface band bending (Darling 1991). Such a surface field strongly modifies the spatial carrier distribution: electrons and holes will drift along the electric field setting up drift currents with current densities $j_{e,h} = q\mu_{e,h} n_{e,h} F$, with $\mu_{e,h}$ the mobilities of electrons and holes, respectively, $n_{e,h}$ their densities, and F the field strength. Since the charge q has opposite signs for electrons and holes, the electric field tends to separate electrons and holes until the driving field is screened. The time for screening the electric field strongly depends on the optically excited carrier concentration and may range from sub-ps to several ps. Since the initial separation of electrons and holes by the surface field occurs on sub-ps timescales, a dynamic polarization is set up, which is the origin of electromagnetic radiation in the terahertz frequency range from the surface of semiconductors. Figure 8.5 shows electron and hole distribution functions at different times after the optical excitation at zero time delay (Dekorsy 1993). The separation of electrons and holes is clearly visible.

Let us now consider a semiconductor without a built-in electric field, which can be achieved by chemical treatment of the surface (Darling 1991). In this case only diffusion of electrons and holes away from the surface is expected. In general, the diffusion coefficient is different for electrons and holes, i.e., D_e does not equal D_h. The diffusion coefficient is related to the carrier mobility via the Einstein equation $D_i/\mu_i = k_b T/q$, i.e., a higher carrier mobility implies a larger diffusion coefficient. For GaAs at low doping concentration ($10^{16}\,\text{cm}^{-3}$), $\mu_e = 8500\,\text{cm}^2\,\text{V}^{-1}\,\text{s}^{-1}$, $\mu_h = 400\,\text{cm}^2\,\text{V}^{-1}\,\text{s}^{-1}$, $D_e = 200\,\text{cm}^2/\text{s}$, and $D_h = 10\,\text{cm}^2/\text{s}$. Hence, the electrons will diffuse much faster away from the surface than the holes. This difference in the diffusion currents builds up a net polarization between electrons and holes. The resulting electric field couples the further

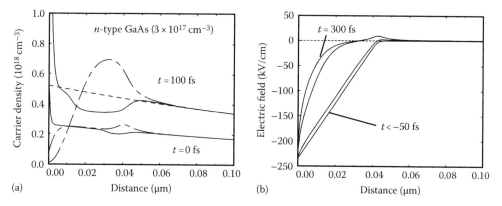

Figure 8.5 (a) Transient carrier distributions of electrons (dashed-dotted lines) and holes (solid lines) in n-doped GaAs at zero time delay and a time delay of 100 fs when excited with a 50 fs laser pulse. The dotted line represents the carrier distribution as it would result without carrier transport after the excitation pulse. (b) Transient electric fields at different time delays due to screening by charge carriers (highest field at negative time delays before the excitation pulse <50, 0, 100, and 300 fs). (Adapted with permission from Dekorsy, T., Pfeifer, T., Kütt, W., and Kurz, H., *Phys. Rev. B, 47*, 3842, 1993. Copyright 1993 by the American Physical Society.)

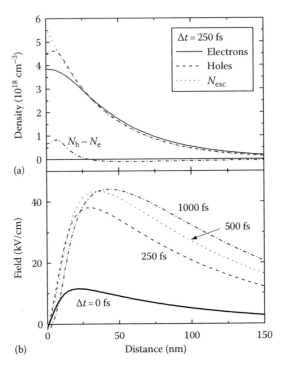

Figure 8.6 Photo-Dember effect in tellurium. (a) Spatial carrier distribution perpendicular to the excited surface 250 fs after the excitation pulse. (b) Electric field distribution for different times after the excitation pulse. (Adapted with permission from Dekorsy, 1966. Copyright 1996 by the American Physical Society.)

diffusion of electrons and holes, which diffuse concomitantly into the bulk. The heavier holes thus hinder the diffusion of the faster electrons. This results in much slower diffusion, which is governed by the smaller diffusion coefficient. That is the reason why for steady-state carrier transport the diffusion of electrons *and* holes is described by a single diffusion coefficient, the so-called ambipolar diffusion $D_{amb} = 2D_eD_h/(D_e + D_h)$ for the case of equal electron and hole densities (Seeger 2004). For $D_e \gg D_h$ the ambipolar diffusion coefficient is only a factor of 2 larger than the hole diffusion coefficient. The build-up of the electric field that couples the diffusion of electrons and holes is known as the photo-Dember effect (Dember 1931). Figure 8.6 shows the build-up of the photo-Dember electric field and the associated spatial carrier densities at different times following pulsed excitation of a highly absorbing semiconductor ($\alpha^{-1} = 40$ nm). Terahertz emission from the transient photo-Dember effect has been observed for a variety of semiconductors (Ascazubi 2006, Dekorsy 1996, Gu 2002).

8.1.8 Impulsive Generation of Coherent Phonons and Plasmons

When the duration of the laser pulse is shorter than the typical period of a phonon mode, e.g., 113 fs for LO phonons in GaAs, the coherent excitation of phonons in the excited volume may result, where all of the atoms vibrate in phase. There exist different mechanisms for exciting coherent optical phonon modes in semiconductors. These mechanisms are based either on (1) impulsive stimulated Raman scattering (ISRS) (Yan 1985), (2) displacive excitation of coherent phonons (Zeiger 1992), or (3) coupling of polar phonons to ultrafast electric field changes as described in the previous section (Cho 1990, Dekorsy 1996). The coherent excitation of these modes appears as an oscillatory modulation of the reflectivity or transmission change in a pump-probe experiment, or may be detected through the emitted terahertz radiation. If the semiconductor is doped or—as is mostly the case—charge carriers are simultaneously excited through the optical excitation pulse, a coupled plasmon–phonon mode is coherently excited (Cho 1996, Kersting 1997). A review about coherently excited phonon modes can be found in (Dekorsy 2000).

8.1.9 Recombination of Carriers

The previous sections dealt with the processes following optical excitation on timescales ranging from sub-ps to several ps. The ultimate return of the excited semiconductor into its unexcited equilibrium state occurs through the recombination of charge carriers (in the case of applied electric fields the carriers may also leave the optically excited volume through transport processes). One distinguishes between radiative and non-radiative recombination. Radiative recombination, occurring when a photon is emitted when an electron and hole recombine, is usually the dominant recombination mechanism in high-purity direct semiconductors, with associated time constants in the ns range. In indirect semiconductors radiative recombination is only possible via phonon-assisted processes in order to maintain momentum conservation. This process is less probable and has a time constant in the μs range (depending also on the lattice temperature).

Recombination rates in both direct and indirect semiconductors are strongly influenced by defects and doping of the crystal, which may dominate over the intrinsic radiative recombination. Figure 8.7 gives a schematic representation of different recombination channels.

Another recombination channel is Auger recombination, which becomes relevant at higher excitation levels and may dominate the recombination dynamics at carrier densities larger than $10^{20}\,\mathrm{cm^{-3}}$ (Klingshirn 2006). In the Auger recombination process an excited electron (hole) gives its energy to another electron (hole) exciting the second electron (hole) higher into the conduction (valence) band while the first electron (hole) recombines with a hole (electron). The energy of the excited electron (hole) is transferred to the lattice via electron–phonon (hole–phonon) interaction.

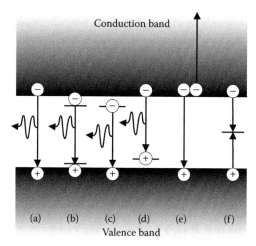

Figure 8.7 Schematic sketch of different recombination processes in bulk semiconductors: (a) band-to-band or direct recombination, (b) exciton recombination, (c) donator–band recombination, (d) acceptor–band recombination, (e) Auger recombination, (f) recombination via deep levels. (a) through (d) are radiative recombination processes, where a photon is emitted. (e) and (f) are non-radiative recombination channels.

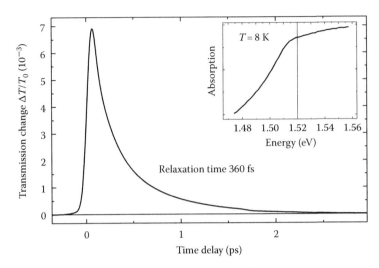

Figure 8.8 Transient transmission change through a thin low temperature–grown GaAs film, showing a carrier trapping time of 360 fs. The inset shows the static absorption spectrum of the sample close to the band gap (at low lattice temperature), which does not exhibit sharp excitonic absorption lines and defect-induced extended absorption below the band gap. (Adapted from Dekorsy, T., Cho, G.C., in *Light Scattering in Solid VIII*, eds. M. Cardona and G. Giintherodt, Springer, Topics in Applied Physics 76: 169, 2000. With permission. Copyright 2000 by the American Physical Society.)

8.1.10 Defect-Induced Ultrafast Trapping

For many ultrafast or THz optoelectronic devices, a sub-ps carrier lifetime is required. Some applications for such materials are as the active zones in a semiconductor saturable absorber (Keller 1996) or in a photoconductive terahertz antenna (Brown 1995). Ideally the carrier lifetime can be tailored by processes that are compatible with semiconductor technology. One prominent example is low temperature–grown GaAs (LT-GaAs). LT-GaAs is grown by molecular beam epitaxy (MBE) at substrate temperatures (e.g., 200°C), much lower than the temperature used for high-quality crystal growth (e.g., 600°C). In addition As overpressure is used in the MBE chamber. Depending on the pressure and temperature, excess As is incorporated into the crystal at an amount of 1%–2%. In a post-growth high-temperature annealing step the crystal quality is improved and the amount As defects can be tailored through precipitation of As clusters (Nolte 1999). This material exhibits a very high electrical breakdown field and ultrashort carrier lifetimes in the range of several hundreds of fs (Segschneider 1997). The short electron lifetime is achieved through ultrafast trapping of electrons in As_{Ga}^+ antisite defects. Figure 8.8 shows transient transmission changes from such an LT-GaAs film with a carrier lifetime of only 360 fs. Another way to tailor the carrier lifetime into the sub-ps regime is the implantation of ions into the semiconductor (Krotkus 1995). Both LT-GaAs and ion-implanted semiconductors are frequently used as terahertz emitters and detectors in photoconductive antenna structures due to their short carrier lifetimes (Chapter 11).

8.2 ULTRAFAST CARRIER DYNAMICS IN METALS

The successful utilization of ultrafast optical techniques to study various phenomena in semiconductors, as discussed in Section 8.1, is based largely on their well-known electronic band structure and on the extreme purity of the available samples (impurity concentrations as low as 10^{13} cm^{-3} can be achieved). In the last 20 years, however, we have experienced an increased attention by the condensed matter physics community to ultrafast phenomena in strongly correlated electron systems, driven by heightened interest in these technologically-important systems. The discoveries led to significant progress in the understanding of these materials even though their electronic structure is complicated and in general not agreed upon. Systematic studies implementing pump-probe techniques on high-temperature superconductors (Demsar 1999a, Kabanov 1999), colossal magnetoresistive compounds (Averitt 2001, Ogasawara 2001, 2005), or, for example, low-dimensional charge density waves (Demsar 1999b, Perfetti 2006) have shown that these techniques can yield new, important, and complementary information (to the more conventional time-averaging frequency-domain spectroscopies [Part II of this book]) on their low-energy electronic structure, as well as on the interaction strengths between various degrees of freedom (electronic, lattice, spin, orbital). Moreover, recent studies have shown that these techniques can be used not only to probe relaxation phenomena near thermodynamic equilibrium, but also to drive and track the transitions between different electronic or structural phases on an fs timescale. Examples of such studies in manganites (Rini 2007), high temperature superconductors (Kusar 2008), low-dimensional charge density waves (Schmitt 2008, Tomeljak 2009), and VO_2 (Cavalleri 2001, Kübler 2007), are opening up new possibilities for technological applications.

Many of these correlated electron systems are metallic in their normal (high-temperature) state (here the word metallic is being used in the most general sense possible), but possess various types of ordering phenomena at low temperatures like superconductivity, charge ordering, spin and charge density waves, and heavy electron behavior. Therefore, understanding the relaxation phenomena in metals is of utmost importance for understanding these processes in their low-temperature phases. For such materials the models developed for simple metals can often present a good starting point for the description of relaxation phenomena in their normal states.

The aim of this section is therefore to present a basic description of the relaxation processes in metals. While most of the discussion will involve dynamics in noble metals like Au, Ag, and Cu, the underlying phenomena should present a starting point for the understanding of the ultrafast phenomena also in strongly correlated electron systems, at least in their normal (metallic) state.

In the sections to follow we will first briefly discuss the typical timescales in question, considering only the electronic and lattice degrees of freedom. Then we continue with a discussion of the two-temperature model (TTM), the model that is most commonly used to interpret relaxation phenomena in metals. The basic premise of this model is the assumption that electron–electron (e–e) scattering proceeds on a much shorter timescale than electron–phonon (e–ph) scattering, resulting in the description of the relaxation process as electron–phonon thermalization, where the electrons quickly thermalize with each other, to a temperature much higher than the initial temperature, and subsequently thermalize with the lattice. Within this model, the electron–phonon thermalization rate is proportional to the dimensionless electron–phonon coupling constant (λ), the important parameter in the theory of superconductivity. As we will show in the following sections, this model has several shortcomings. For example, the basic assumption of electron–electron thermalization being faster than electron–phonon thermalization is found to fail at low temperatures. Therefore care must be taken before using any of the developed theoretical models to interpret the data.

As for the experimental techniques to be used in studying relaxation phenomena in metals, the most common technique is optical pump-probe spectroscopy (Chapter 9); however, recently time-resolved (and angle-resolved) photoemission studies are also being performed, where the time evolution of the electronic occupation in the vicinity of the Fermi level can be studied (Perfetti 2007). We should also mention time-resolved two-photon photoemission experiments, where the lifetime of an electron as a function of energy can be studied (Schmuttenmaer 1994)—see Part V for a more detailed discussion of time-resolved photoemission experiments.

8.2.1 Typical Timescales

Similar to the discussion of relaxation processes in semiconductors, the so-called *coherent regime* will not be discussed here—some recent developments in this field are discussed by Petek (1997). We are interested in what happens after photoexcitation, when high-energy electron–hole pairs are created, and how the absorbed energy is distributed within the electronic subsystem and further transferred to other degrees of freedom, typically phonons. Furthermore, in these experiments we are studying the electron and phonon population densities, or their respective distribution functions (thermal or nonthermal—see Section 8.1.4). We should further note that almost exclusively* one measures the electronic response (electronic distribution function), and based on these measurements, infers what is happening with the phonon subsystem or, alternatively, treats the phonon subsystem as an (infinite) thermal bath.

In ferromagnetic metals, the spin order is an additional factor that must be considered. Here scattering events not only modify the energy and momentum of carriers, but can also induce spin-flips, thereby affecting the net magnetization. The demagnetization dynamics, typically studied by the time-resolved magneto-optical Kerr effect (TR-MOKE), has been found to take place on the sub-ps timescale (Beaurepaire 1996, Lisowski 2005) and has been a topic of intense research over the last decade or so (Beaurepaire 1996, Bigot 2009, Carpene 2008, Lisowski 2005). In the following we will focus on the electron–electron and electron–phonon processes and invite the reader to consult Chapter 13 for details on magnetization dynamics.

According to the Drude theory (Chapters 1 and 3), momentum scattering processes occur on a timescale of ~1–100 fs, leading to an (incoherent) population of hot (highly nonthermal) electrons and holes. The lifetime of hot electrons as a function of energy after photoexcitation of a typical metal like copper can be measured by time-resolved two-photon photoemission (see Figure 8.9), and is found to be in qualitative agreement with the Fermi liquid[†] result (Pines 1996)

$$\tau_{e-e} = \frac{128}{\pi^2 \sqrt{3}\omega_p} \left(\frac{E_F}{E - E_F} \right)^2 \tag{8.1}$$

where
  E_F is the Fermi energy
  ω_p the bare plasma frequency

In a typical metal (see Figure 8.9) $e-e$ scattering times are ~10 fs at energies of several eV above E_F and diverge toward E_F. Based on these results, $e-e$ thermalization to a Fermi–Dirac distribution with electronic temperature T_e (much higher than the base temperature) is expected to occur on a typical timescale of ~100 fs at high excitation densities,[‡] presumably before significant energy transfer can occur to the phonon bath. Since the heat capacity of the electronic system is much smaller than that of the lattice (at temperatures above ~10 K), extremely high electronic temperatures, T_e, can be achieved. The $e-e$ thermalization time can be quantitatively measured using time-resolved photoemission by monitoring the time evolution of the occupation density near E_F.

---

* Dynamics of the lattice temperature can be in principle studied by time-resolved Raman spectroscopy by probing at the same time the Stokes and anti-Stokes spectra, and determining the transient lattice temperature by following their ratio as a function of time—see Chapter 14.

[†] Fermi liquid theory (see, e.g., Kittel 1986) is a phenomenological theory introduced by Landau, who showed that many properties of an interacting fermion system are similar to those of a noninteracting one (Fermi gas).

[‡] At low excitation densities, where the resulting T_e is just slightly higher than the base temperature, $e-e$ thermalization times can be substantially longer, as discussed later in the text.

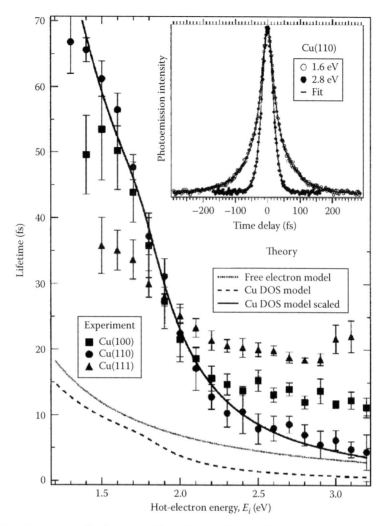

Figure 8.9 Experimentally determined hot-electron lifetimes for Cu(100), Cu(110), and Cu(111) surfaces obtained by time-resolved two-photon photoemission. Calculated lifetimes using different models (the Fermi-liquid result is shown with a dotted line) are also shown (see Ogawa 1997 for details). The inset shows the two-pulse correlation measurements for 1.6 and 2.8 eV hot electrons at the Cu(110) surface, from which the hot-electron lifetimes are deduced. (Reproduced from Ogawa, S. et al., *Phys. Rev. B*, 55, 10869, 1997. With permission. Copyright 1997 by the American Physical Society.)

The experimental results qualitatively corroborate this picture. One should note, however, that even in noble metals like gold it can take ~800 fs before thermalization is achieved (Fann 1992a,b), while at earlier times the distribution is characterized by tails expanding far beyond k_BT_e. The main reason for slow $e–e$ thermalization in metals is Pauli blocking, i.e., electron–electron scattering into states below the (initial) Fermi level is strongly reduced due to the small fraction of unoccupied states to which electrons can be scattered. Electron–phonon scattering proceeds in parallel with $e–e$ thermalization, with typical timescales ranging from 10 fs to 1 ps. While perhaps in good metals the $e–e$ thermalization and $e–ph$ relaxation timescales could be well decoupled (at least at high excitation densities), one can expect that this may not be the case for correlated electron systems like high-temperature superconductors. In semiconductors, on the other hand, $e–e$ thermalization is typically much faster than $e–ph$ scattering (Section 8.1.4 and Table 8.1), since the large number of available states in the conduction band minimizes Pauli blocking; however, at extremely low excitation densities ($\sim 10^{15}$ cm^{-3}) in semiconductors $e–e$ thermalization also becomes slower than $e–ph$ energy transfer (Leitenstorfer 1996).

8.2.2 Femtosecond Thermomodulation Spectroscopy

Numerous studies of photoexcited carrier relaxation dynamics in noble metals have been performed already in the 1980s and 1990s. The first time-resolved experiments using ps and sub-ps laser pulses were dubbed fs thermomodulation experiments (Brorson 1990a), since they were considered to be an extension of conventional thermomodulation spectroscopy. Conventional thermomodulation spectroscopy is a method that is used for band structure investigations in metals (Kittel 1986). Since interband transitions are usually masked by free electrons up to the plasma energy (4–10 eV [Kittel 1986] [Chapter 3]) in conventional optical reflectivity or transmission experiments performed on metals, modulation spectroscopy (measuring the photon energy dependence of the derivatives of the optical constants by varying temperature, current, etc.) is ideal for studying the band structure of metals. Thermomodulation spectroscopy involves periodically perturbing the sample's temperature and measuring changes in the optical absorption spectrum occurring in phase with the perturbation. Since only the changes in optical constants are measured, modulation spectroscopy is very sensitive to critical points in the band structure. For example, the presence of narrow d-bands below the Fermi energy gives rise to the large joint density of states (JDOS) for optical transitions from the d-band to E_F (see Figure 8.10, also Figure 3.1 and associated discussion in Chapter 3).

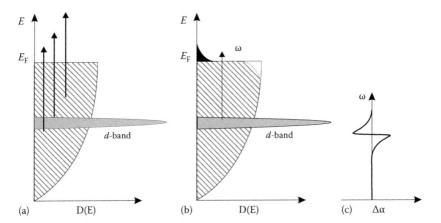

Figure 8.10 Femtosecond thermomodulation mechanism: (a) photoexcitation, where the arrows represent possible electronic transitions from occupied (hatched) to unoccupied electronic states, results in (b) Fermi-level smearing, thereby opening some states and blocking others for optical transitions. When the change in absorption ($\Delta\alpha$) is measured as a function of the probe photon energy ω one observes (c) a derivative-like feature at the probe energy corresponding to the energy separation between the low-lying d band and E_F. The time evolution of $\Delta\alpha(\omega)$ is measured by delaying the probe with respect to the photoexcitation pulse.

Generally the change in temperature modifies the reflectivity of the metal in two ways:

1. The increase in temperature causes a broadening of the occupation of states near E_F (Fermi level smearing), thereby blocking some states and opening others for optical transitions. This modifies the sample's absorption, and therefore also transmission and reflectivity, which are directly measured. This purely electronic effect is usually the most pronounced, and is expected to be dominant on the fs/ps timescale.
2. The temperature change causes strain in the sample due to thermal expansion, which can lead to shifts of the electronic energy bands, thereby influencing optical properties. The rise time of this effect depends on the electron–phonon coupling, and can be on the order of 100 fs. However, because this is a lattice effect, the relaxation timescale is typically on the order of tens of ps or more, determined by the phonon escape time. Therefore, nanosecond (ns) dynamics in time-resolved experiments are usually attributed to lattice heating effects.

In the case of fs thermomodulation experiments performed with fs laser pulses, most often only fast fs/ps transient changes are analyzed. Therefore, only the Fermi level smearing effect on a sample's optical properties is usually considered.

Figure 8.10 presents the idea behind fs thermomodulation. Here the high-intensity pump pulse excites electrons from occupied to unoccupied states. This is followed by initial electron–electron thermalization. The electron–electron scattering time is $\tau_{e-e} \sim \hbar E_F / 2\pi E^2$— see Equation 8.1, where E is the energy measured from the Fermi energy E_F, and is on the order of several tens of fs. Since this is believed to be fast compared to the pulse duration, one can assume the process to be instantaneous. As τ_{e-e} is much faster than the period of a typical phonon vibration, electrons are decoupled from the lattice, and the electronic system can be described with the electronic temperature T_e, which differs from the lattice temperature T_l. Because the heat capacity of the electron gas is much smaller than the heat capacity of the lattice, T_e can be much higher than T_l. In experiments performed with high-photoexcitation density pulses, T_e can reach several thousand Kelvin above T_l.

Fermi level smearing affects the reflectivity of the delayed probe pulse. In the first approximation it affects the probe optical transition probability only if the initial or final states for the optical transition lie near E_F. The amplitude of the reflectivity change therefore strongly depends on the probe photon energy (Figure 8.10c). In the case of Cu, for example, the most pronounced feature appears near 2.15 eV, which corresponds to the d-band to E_F transition. The strong dependence of the photoinduced signal amplitude on the wavelength of the probe pulse made these materials especially interesting for studying energy relaxation processes, and the experiments gave new insight into nonequilibrium phenomena in metals.

The first room-temperature ps thermomodulation experiments on Cu were performed by Eesley in 1983 (Eesley 1983). He used two synchronously pumped dye lasers, one tunable between 2.03 and 2.17 eV (610–572 nm) as a probe and the other one at 1.92 eV (645 nm) as a pump. He observed a fast pulsewidth-limited initial signal that changed sign at the probe photon energy near 2.15 eV (Figure 8.11a), in accordance with conventional thermomodulation measurements, as shown in the inset to Figure 8.11a. The initial fast signal was followed by a slowly decaying signal that did not change sign as the probe photon energy was changed and was attributed to lattice effects, with typical timescales being determined by heat diffusion. Similar results were found by Fujimoto et al. (Fujimoto 1984) in tungsten. The first measurements of the electronic relaxation time were done by Schoenlein et al. in Au (Schoenlein 1987) and Elsayed-Ali et al. in Cu (Elsayed-Ali 1987), where 60 fs CPM (colliding pulse mode-locked) lasers were used as a probe. In Schoenlein's experiment, high-intensity pulses from a CPM laser were used to produce a 60 fs broadband continuum, enabling measurements of

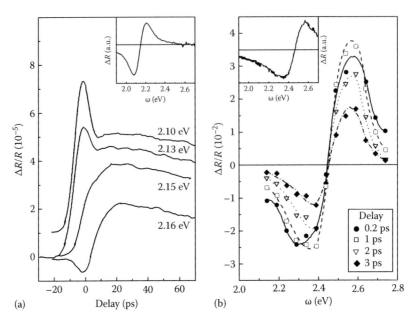

Figure 8.11 (a) Picosecond thermomodulation response in Cu. (Adapted from Eesley, G.L., *Phys. Rev. Lett.*, 51, 2140, 1983. With permission.) By tuning the probe laser energy around 2.15 eV the sign of the fast component changes according to Figure 9.10. (b) Transient reflectivity change in Au as a function of probe photon energy at various time delays after photoexcitation. (Adapted from Schoenlein, R.W., *Phys. Rev. Lett.*, 58, 1680, 1987. With permission.) Insets: the corresponding thermomodulation spectra of Cu and Au at 300 K. (Adapted from Brorson, S.D., Femtosecond thermomodulation measurements of transport and relaxation in metals and super-conductors, thesis, Research Laboratory of Electronics, Massachusetts Institute of Technology, Cambridge, MA, 1990b. With permission), where thermomodulation was achieved by heating the sample with current.

the electron relaxation with <100 fs resolution over a broad energy range (Figure 8.11b). The experiment showed the electron decay time to be on the order of 2–3 ps.

It should be noted, however, that the derivative-like feature observed in fs and conventional thermomodulation is much broader than what one might expect on the basis of the simple model, which suggests that the width of the feature should be on the order of $k_B T_e$, where k_B is the Boltzmann constant. However, the fact that neither d-band nor p-like states near E_F are dispersionless should lead to a broadening of the observed feature.

By measuring the temporal dependence of the reflectivity change on fs/ps timescales one can experimentally determine the relaxation time(s) of the photoexcited electronic system. In general, two effects should be considered: first, the thermalization due to electron–electron scattering and energy relaxation via electron–phonon scattering, and second, the transport of hot electrons out of the probed volume. The latter can be neglected in materials with very short electron mean free paths (like many of the correlated electron systems mentioned above), but should be important in the noble metals, where the electron mean free paths even at room temperature are on the order of 100 nm, which is much longer than the optical penetration depth (in metals typically 10 nm in the visible range). However, even in metals transport effects can, in principle, be avoided by performing experiments (in the optical pump-probe configuration) on thin films on dielectric substrates, where the film thickness is smaller or similar to the optical penetration depth. For the sake of simplicity, we will first focus on the description of relaxation processes in the absence of transport effects, leaving the discussion of (ballistic) transport to Section 8.2.5.

8.2.3 The Two-Temperature Model

The original work on what is commonly referred to as the two-temperature model (TTM) (Anisimov 1974) is by Kaganov, Lifshitz, and Tanatarov back in 1956 (Kaganov 1956), when considering deviations from Ohm's law at high currents. Given the fast timescales in question and lack of short pulsed laser sources at the time, the authors argued that these processes were unlikely to be ever experimentally measured. Motivated by the experimental data from the first ps relaxation dynamics studies in metals, energy relaxation through electron–phonon scattering was theoretically revisited by Allen in 1987 (Allen 1987). In particular, he made the explicit connection between the measured relaxation time at high temperatures with the dimensionless electron–phonon coupling constant, λ, an important parameter in the theory of superconductivity.

The main assumptions used in this model are that (1) the $e-e$ thermalization time is much faster than the $e-ph$ relaxation time; (2) $e-e$ (Coulomb) and $ph-ph$ (anharmonic) collisions are active only in keeping the distribution functions in quasi-equilibrium at their respective temperatures (in this case the $e-e$ and $ph-ph$ collision integrals are equal to zero); (3) diffusion driven by spatial inhomogeneities is negligible, which is a good approximation when the mean free path is short; (4) acceleration due to external or internal fields is negligible; and (5) no other collision processes are important. Then, the system of electrons and phonons is described with thermal distribution functions f_k and n_q that are determined by T_e and T_l respectively, both of which vary with time.

The collision integrals describing the time evolution of f_k and n_q can be approximated by

$$\frac{\partial f_k}{\partial t} = -\frac{2\pi}{\hbar N_c} \sum_q |M_{kk'}|^2 \left\{ f_k(1-f_{k'})\left[(n_q+1)\delta(\varepsilon_k-\varepsilon_{k'}-\hbar\omega_q)+n_q\delta(\varepsilon_k-\varepsilon_{k'}+\hbar\omega_q)\right] \right.$$

$$\left. -(1-f_k)f_{k'}\left[(n_q+1)\delta(\varepsilon_k-\varepsilon_{k'}+\hbar\omega_q)+n_q\delta(\varepsilon_k-\varepsilon_{k'}-\hbar\omega_q)\right] \right\}$$

$$\frac{\partial n_q}{\partial t} = -\frac{4\pi}{\hbar N_c} \sum_k |M_{kk'}|^2 f_k(1-f_{k'})\left[n_q\delta(\varepsilon_k-\varepsilon_{k'}+\hbar\omega_q)-(n_q+1)\delta(\varepsilon_k-\varepsilon_{k'}-\hbar\omega_q)\right] \quad (8.2)$$

where

$\varepsilon_k, \varepsilon_{k'}$ and $\hbar\omega_q$ are the energies of electrons (with momentum k, k') and phonons (with momentum q), respectively

N_c is the number of the unit cells in the sample

$M_{kk'}$ is the electron–phonon matrix element normalized to the unit cell with magnitude $(E_F\hbar\omega_D)^{1/2}$

ω_D is the Debye frequency (Kittel 1986)

The additional factor of 2 in the second equation accounts for electron spin degeneracy. If f_k and n_q are known at a time $t = 0$, and the above assumptions stand, then the two equations determine their temporal evolution.

Allen calculated (Allen 1987) the rate of energy exchange in the high temperature limit, where T_l is larger than the Debye temperature* (Θ_D). Above Θ_D the lattice energy is $E_l \approx 3N_ck_BT_l$.

---

* The Debye model is used for estimating the phonon contribution to the specific heat in a solid (see Kittel 1986). It treats the vibrations of the lattice as phonons in a box. It correctly reproduces the T^3 temperature dependence of the specific heat at low temperatures. The Debye temperature is a temperature above which the phonon specific heat is nearly constant, approaching the classical value of $3N_ck_B$. The Debye frequency, ω_D, is related to the Debye temperature as $\hbar\omega_D = k_B\Theta_D$.

The energy in the electronic subsystem after photoexcitation and e–e thermalization to a temperature T_e, where $T_e > T_1$, is equal to* $E_e \approx E_0 + \gamma_{Somm}(T_e - T_1)^2/2$, where E_0 is the energy prior to photoexcitation, $\gamma_{Somm} = \pi^2 D(E_F)k_B^2/3$ is the Sommerfeld constant, and $D(E_F)$ is the electron density of states at the Fermi energy. A simple expression for the rate of change of the electronic temperature was derived (Allen 1987):

$$\frac{\partial T_e}{\partial t} = \gamma_T(T_1 - T_e); \quad \gamma_T = \frac{3\hbar\lambda\langle\omega^2\rangle}{\pi k_B T_e}\left(1 - \frac{\hbar^2\langle\omega^4\rangle}{12\langle\omega^2\rangle k_B^2 T_e T_1} + \cdots\right). \tag{8.3}$$

Here $\lambda\langle\omega^n\rangle = 2\int_0^\infty \left[\alpha^2 F(\Omega)/\Omega\right]\Omega^n \, d\Omega$, with $\alpha^2 F(\Omega)$ being the product of the electron–phonon coupling strength α^2 and the phonon density of states $F(\Omega)$, both functions of phonon energy, Ω, and $\langle\omega^2\rangle$ is the mean square phonon frequency. Assuming that the fs thermomodulation picture stands, and that the photoinduced reflectivity is indeed proportional to the difference of the electronic and lattice temperatures, $\Delta R/R(t) \propto (T_e - T_1)$, Allen was able to extract the values of the average electron–phonon coupling constant λ from the published data in Au (Schoenlein 1987), Cu (Elsayed-Ali 1987), and W (Fujimoto 1984). The obtained values were in very good agreement with λs extracted from resistivity and neutron scattering data (Allen 1987).

8.2.3.1 Determination of the e–ph Coupling Constant λ in Superconductors

Based on Allen's relation between the measured $(e$–$ph)$ relaxation and the dimensionless e–ph coupling constant λ, the important parameter in the theory of superconductivity, several studies of various metals exhibiting superconductivity at low temperatures were carried out. Figure 8.12 presents the induced reflectivity traces[†] in various metallic superconductors measured by Brorson et al. (Brorson 1990c). Analysis of the measured relaxation times with Equation 8.3, using the published values of the Sommerfeld constant to determine the initial electronic temperature (note that the e–ph thermalization rate γ_T is proportional to the final electronic temperature—see Equation 8.3) enabled the extraction of $\lambda\langle\omega^2\rangle$. Assuming a Debye spectrum for phonons, i.e., taking the electron–phonon coupling strength α^2 to be independent of the phonon frequency, values of λ (see the table in Figure 8.12) were obtained, in excellent agreement with literature values. Unfortunately, the authors did not measure the excitation intensity dependence of the relaxation time γ_T^{-1}, which should, according to Equation 8.3, increase with excitation fluence due to the fact that $\gamma_T^{-1} \propto T_e$ and the electronic temperature after photoexcitation is $T_e = \sqrt{T_i^2 + 2U_i/\gamma}$, where U_1 is the absorbed energy density.

While the TTM is often used to quantitatively interpret the data and to extract microscopic quantities like λ, excitation-dependent studies are only rarely performed to prove that one is indeed in the appropriate limiting case to be able to utilize Equation 8.3. Only at very high excitation densities the dependence of the relaxation time on fluence, which follows the TTM prediction $\gamma_T^{-1}(U_i) \propto \sqrt{T_i^2 + 2U_i/\gamma}$, has been observed in thin Au films (Elsayed-Ali 1993).

---

* In simple metals the electronic specific heat increases linearly with temperature as $C_e(T) = \gamma_{Somm}T$.
† Several metals did not show fast electronic dynamics at the measured probe photon energy (1.98 eV); therefore a thin ~40 Å Cu film was deposited on top of them and served as a "detector" of the electronic temperature in the underlying film, assuming that electrons in both have the same temperature (Brorson 1990c).

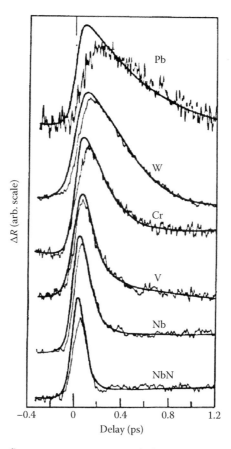

Figure 8.12 Induced reflectivity transients recorded in various metallic superconductors at room temperature, together with numerical fits to the data. Below, the extracted values of $\lambda \langle \omega^2 \rangle$ and their decomposition are compared to the literature values extracted by other techniques. (Reproduced from Brorson, S.D. et al., *Phys. Rev. Lett.*, 64, 2172, 1990c. With permission. Copyright 1990 by the American Physical Society.)

	$T_e(0)$ (K)	$\lambda \langle \omega \rangle^2$ (exp.) (meV^2)	$\langle \omega \rangle^2$ (meV^2)	λ_{exp}	λ_{lit}
Cu	590	29 ± 4	377	0.08 ± 0.01	0.10
Au	650	23 ± 4	178	0.13 ± 0.02	0.15
Cr	716	128 ± 15	987	0.13 ± 0.02	
W	1200	112 ± 15	425	0.26 ± 0.04	0.26
V	700	280 ± 20	352	0.8 ± 0.06	0.82
Nb	790	320 ± 30	275	1.16 ± 0.11	1.04
Ti	820	350 ± 30	601	0.58 ± 0.05	0.54
Pb	570	45 ± 5	31	1.45 ± 0.16	1.55
NbN	1070	640 ± 40	673	0.95 ± 0.06	1.46
V$_3$Ga	1110	370 ± 60	448	0.83 ± 0.13	1.12

Brorson et al. performed a similar type of measurement on a series of high-temperature cuprate superconductors (Brorson 1990a), where, assuming the Debye model for calculating $\langle\omega^2\rangle$, they determined the value of the e–ph coupling constant to be on the order of 1, suggesting strong e–ph coupling. Recently, the first time-resolved photoemission experiments have been performed on the cuprate superconductor $Bi_2Sr_2CaCu_2O_{8+\delta}$ (Perfetti 2007), where spectral changes in the vicinity of the Fermi level are studied as a function of time delay after photoexcitation (see Figure 18.8 in Chapter 18 and associated discussion). Fitting the electronic distribution function with the Fermi distribution function enabled the authors to trace the time evolution of the electronic temperature. From these results, similar values for $\lambda\langle\omega^2\rangle$ are obtained, as in the original report using all-optical methods (Brorson 1990a). However, assuming the Einstein model for the phonon distribution function, which gives a much higher value of the mean square phonon frequency $\langle\omega^2\rangle$, a much lower value of the e–ph coupling constant has been obtained. In both cases, however, no excitation fluence dependence of the dynamics has been reported.

8.2.3.2 Temperature Dependence of the e–ph Thermalization Time within the TTM

As shown above, the approximate solution of the TTM for temperatures above the Debye temperature has a particularly simple form, where the relaxation rate is proportional to the dimensionless electron–phonon coupling constant, and inversely proportional to the temperature of the electronic subsystem following the initial e–e thermalization. Here we review how the e–ph thermalization time should depend on temperature in the low excitation regime, where the electronic temperature after photoexcitation is only slightly increased.

Under the assumptions discussed above, e–ph thermalization can be described by the two coupled differential equations that describe the time evolution of the electronic (T_e) and lattice (T_l) temperatures (Anisimov 1974):

$$C_e(T_e)\frac{\partial T_e}{\partial t} = -g(T_e - T_l) + S(t)$$

$$C_i\frac{\partial T_l}{\partial t} = g(T_e - T_l)$$

(8.4)

where
$C_e(T_e)$ and $C_l(T_l)$ are the respective heat capacities of the electrons and lattice
$g\ (=g(T_l))$ is the e–ph coupling function*
$S(t)$ describes the absorbed energy from the photoexcitation pulse

In the weak perturbation limit, when the change in the electronic temperature is small compared to the initial temperature $T_e - T_l \ll T_l$, the relaxation of the electronic temperature is exponential, with the e–ph thermalization time given by

$$\tau_{e-ph} = \frac{1}{g}\frac{C_e C_l}{C_e + C_l}.$$

(8.5)

Here $g(T_l)$ is in the linear response limit ($T_e - T_l \ll T_l$), particularly straightforward for the case of simple metals, where the electron bandwidth is much larger than the Debye temperature, Θ_D.

---

* The electron–phonon coupling function, $g(T)$, should not be confused with the (dimensionless) electron–phonon coupling constant, λ. However, at high temperatures $g(T) \to g_\infty$, which is proportional to λ.

In this case, the Debye model for the electron–phonon coupling function can be used and $g(T) = dG(T)/dT$, where (Groeneveld 1995, Kaganov 1956)

$$G(T) = 4g_\infty \left(\frac{T}{\Theta_D}\right)^5 \int\limits_0^{\Theta_D/T} \frac{x^4\,dx}{e^x - 1}. \qquad (8.6)$$

Given that, the T-dependence of τ_{e-ph} (Equation 8.5) is completely determined by $g_\infty(\propto\lambda)$, Θ_D, and $C_e(T)$. In simple metals, the lattice specific heat C_l is a factor of 100 larger than the electronic specific heat $(C_l \gg C_e)$ over the temperature range of interest (typically 10–300 K). Therefore, the $e-ph$ thermalization time τ_{e-ph} can be approximated by:

$$\tau_{e-ph} \approx \frac{C_e}{g(T_i)} = \frac{\gamma T_i}{g(T_i)}. \qquad (8.7)$$

Here the electronic specific heat $C_e = \gamma T_l$ has been substituted explicitly in the model equations. For $T_l \ll \Theta_D$ the function $g(T_l)$ varies as T_l^4, while for $T_l \geq \Theta_D$, $g(T_l)$ becomes constant (g_∞). At $T_l \geq \Theta_D$ Equation 8.7 becomes equal to the weak perturbation limit of Equation 8.3, where $T_e - T_l \ll T_l$ and $T_e \approx T_l$. In simple metals, g_∞ is typically $\approx 10^{17}$ W/m^{3}K, e.g., for Cu, $g(300\,\mathrm{K}) = 1 \times 10^{17}$ W/m^{3}K (Groeneveld 1995). In the low perturbation limit, the TTM suggests two limiting cases for the temperature dependence of the $e-ph$ thermalization time (Ahn 2004, Demsar 2003a, Groeneveld 1995):

$$\tau_{e-ph} \approx T_i^{-3} \text{ at } \quad T_i \leq \frac{\Theta_D}{5}$$
$$\tau_{e-ph} \approx T_i \text{ at } \quad T_i \geq \Theta_D \qquad (8.8)$$

Figure 8.13 shows the predicted temperature dependence of τ_{e-ph} for metallic LuAgCu$_4$ $(\Theta_D = 280\,\mathrm{K})$ (Demsar 2003a); similar observations follow for most of the metals studied thus far (Groeneveld 1995). While the agreement at high temperatures is reasonable (at least qualitatively), no upturn in relaxation time was observed at low temperatures. As will be discussed in Section 8.2.4, this result can be attributed to the failure of the TTM at low temperatures, where the underlying assumption of $e-e$ thermalization being much faster than $e-ph$ relaxation was shown to fail.

As far as the excitation dependence of the recovery time is concerned, no analysis for the low temperature regime has been made thus far. On the other hand, at temperatures above Θ_D the model predicts that τ_{e-ph} should increase with fluence, proportional to the electronic temperature immediately after $e-e$ thermalization (see Equation 8.3).

Apart from the temperature dependence of the $e-ph$ relaxation time, the T-dependence of the photoinduced changes in the complex dielectric constant can also be derived for the TTM (Hase 2005). The underlying assumption is that the changes in the reflectivity (due to the changes in the dielectric constant) at optical frequencies, B, are proportional to the photoinduced change in the $e-h$ density. Assuming fast $e-e$ thermalization (the basic assumption of the TTM), the induced change in reflectivity B is proportional to the photoinduced quasiparticle density in the energy range of $k_B T_e$ around the Fermi energy. In the limit where $k_B T_e \ll E_F$, which is usually the case here, the number density of the electron–hole pairs n in the Landau Fermi liquid is exactly proportional to the temperature, $n \propto T_e$. If after photoexcitation all the

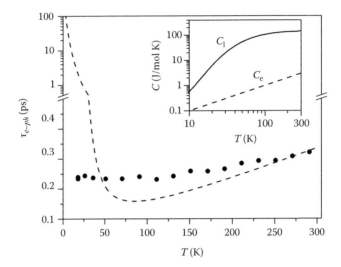

Figure 8.13 The expected T-dependence of the e–ph thermalization time in the TTM in the weak perturbation regime compared to the data on LuAgCu$_4$ (Demsar 2003a). The inset shows the T-dependence of the electronic and lattice specific heat of LuAgCu$_4$. As can be seen at high temperatures, there is a qualitative agreement with the model (the relaxation time increases with temperature); however, no upturn in τ_{e-ph} is observed at low temperatures. (Adapted from Demsar, J. et al., *Phys. Rev. Lett.*, 91, 027401, 2003a. With permission. Copyright 2003 by the American Physical Society.)

absorbed energy initially goes in the electronic subsystem and the electronic subsystem can be described by an increased electronic temperature $T_e(>T_1)$, one obtains

$$B \propto n_p = n_{T_e} - n_{T_i} \propto T_e - T_i,$$ (8.9)

where n_{T_i} is the quasiparticle density at the initial temperature T_1, while n_{T_e} is the quasiparticle density after electrons have thermalized to the temperature T_e (before e–ph thermalization has taken place). Taking into account that the electronic specific heat $C_e = \gamma T_e$, and using the energy conservation law, it follows that

$$B \propto T_e - T_1 = \sqrt{T_1^2 + \frac{2U_1}{\gamma}} - T_1,$$ (8.10)

where U_1 is the photoexcitation energy intensity.

There are several important implications of Equation 8.10. First, the amplitude of the transient is maximum at low temperatures and decreases as the sample temperature increases. Second, at low temperatures the model predicts sub-linear excitation fluence dependence, while at high temperatures the dependence becomes more and more linear. Unfortunately, there are only few temperature- and fluence-dependent studies of the photoinduced signal amplitude in metals (Demsar 2006, Hase 2005), none of which were performed on noble metals. The experiments on Zn, however, show a rather good agreement with the TTM prediction for high temperatures (see Figure 8.14).

In Figure 8.14 we see that the model reproduces the observed temperature dependence reasonably well at high temperatures and that B shows saturation below 50 K, contrary to the model prediction. As we shall discuss in Section 8.2.4, this discrepancy can be attributed to the failure of the TTM at low temperatures.

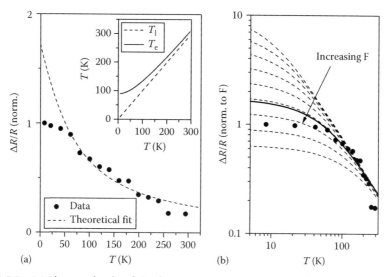

Figure 8.14 (a) The amplitude of the fast electronic transient as a function of temperature in Zn in the weak perturbation limit (after Hase 2005). The inset shows the corresponding initial (T_i, dashed line) and final (T_e, solid line) electronic temperatures as a function of temperature. (b) The expected T-dependence of the amplitude (within the TTM) as a function of excitation fluence (F), where at low temperatures sub-linear behavior is expected, while at high temperatures the dependence is linear. (Adapted from Hase, M. et al., *Phys. Rev. B*, 71, 184301, 2005. With permission. Copyright 2005 by the American Physical Society.)

8.2.4 Beyond the Two-Temperature Model

As shown in Section 8.2.3, numerous studies of carrier relaxation dynamics in metals are in reasonable agreement with the TTM: (1) the relaxation time at high temperatures increases roughly linearly with temperature (Groeneveld 1995), (2) the excitation fluence-dependent measurements on noble metals show at high enough excitation densities an increase of the relaxation time with fluence (Elsayed-Ali 1993). On the other hand, the prediction of the TTM that the relaxation time should (in the weak perturbation limit) at low temperatures increase upon cooling was never observed.

Figure 8.15 reveals the T-dependence of the relaxation time in Ag and Au (Groeneveld 1992), showing very similar behavior to the dependence observed in LuAgCu$_4$ (Figure 8.13). The relaxation time, while at high temperatures exhibiting the trend expected from the TTM, was in general longer than the TTM prediction. Moreover, no measurable change in relaxation time upon increasing excitation density was observed. At about the same time, the first reports of time-resolved photoemission data became available (Fann 1992a,b), revealing that, even at room temperature, the $e-e$ thermalization time is as long as 800 fs. This led Groeneveld et al. to suggest that the origin of the disagreement is in the fact that the main assumption of the TTM, that $e-e$ thermalization is much faster than $e-ph$ relaxation, fails in the limit of low temperatures and low excitation densities. They developed the so-called *nonthermal electron model* (NEM) (Groeneveld 1995). By numerically solving the Boltzmann equations for electrons with $e-e$ and $e-ph$ scattering, starting from an initial nonthermal electron distribution created by a laser pulse and assuming phonons to have a thermal distribution with a time-independent T, they were able to account for the data down to about 50 K (Groeneveld 1995). Within this model, the effectively slower $e-ph$ relaxation time with respect to the prediction of the TTM (Figure 8.15) can be qualitatively explained, assuming that on the timescale of the $e-ph$ thermalization time the electrons have a nonthermal distribution. The energy relaxation

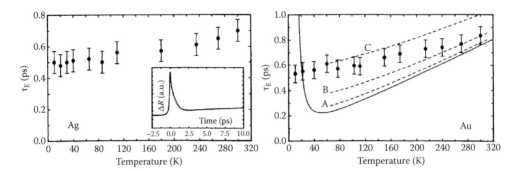

Figure 8.15 The T-dependence of the e–ph relaxation time in Ag (left) and Au (right), extracted from the recovery time of the fast transient (inset left). The TTM simulations are shown in the right panel: the weak perturbation limit is shown as a solid line, while the dashed lines correspond to general TTM solutions with the absorbed energy densities increasing from A to C. (Reproduced with permission from Groeneveld, R.H.M., Sprik, R., and Lagendijk, A., *Phys. Rev. B*, 45, 5079, 1992. Copyright 1992 by the American Physical Society.)

to the phonon subsystem proceeds via spontaneous phonon emission. Since the average electron energy in the distribution of excess quasiparticles is (much) larger than the width of the phonon spectrum ($\sim k_B T$) each electron has a constant phonon emission rate. Since the density of nonequilibrium electrons is roughly doubled after each e–e scattering event (a high-energy electron kicks another electron out of the Fermi sea) the effective e–ph energy relaxation rate increases with time. Therefore, the incomplete e–e thermalization results in a longer e–ph thermalization time with respect to the TTM, as experimentally observed. Since the Fermi liquid theory predicts that in the weak perturbation limit, the e–e thermalization time, τ_{e-e}, should follow a T^{-2} dependence, it is also clear that the discrepancy between the TTM and experiment should be more pronounced at low temperatures.

While Groeneveld et al. proposed the essential idea of a nonthermal electron system, their analysis did not capture the low-T region, where the most striking difference between the TTM ($\tau \propto T^{-3}$) and the experimental data (τ almost T independent) occurs (Figures 8.13 and 8.15). This temperature range was discussed in detail by Ahn (2004), who performed extensive numerical simulations of the coupled Boltzmann equations in the low-temperature regime. Indeed, the analysis suggests that the nonthermal electron distribution results in faster and less-T-dependent relaxation behavior at low T than the thermal distribution.

The physical argument for this behavior is depicted in Figure 8.16: If the electrons have a thermal distribution at T which is just slightly higher than the equilibrium one (solid and dashed lines in Figure 8.16a, respectively), the e–ph scattering important for the relaxation happens within an energy range $k_B T$ from E_F. The relaxation rate depends on the phase space available for relaxation, which is in this case limited to phonons within $\omega < k_B T$. Using the Debye model, where the phonon density of states is $D_p(\omega) \sim \omega^2$, the relaxation rate τ^{-1} is proportional to T^3. In contrast, if the electrons do not possess a thermal distribution (solid line in Figure 8.16b), then e–ph relaxation occurs over an energy range on the order of the Debye energy ($k_B \Theta_D$), making the e–ph relaxation faster and less T dependent. This implies that the e–ph relaxation should be faster and less T dependent in the case of a nonthermal electron distribution, as indeed experimentally observed.

The above-mentioned numerical results point to the importance of including e–e relaxation processes in the analysis, as opposed to assuming that e–e thermalization is nearly instantaneous. However, from these simulations of the time-dependent Boltzmann equation alone it is hard to determine the e–ph coupling strength in terms of the dimensionless e–ph coupling function λ (this is probably one of the main motivations for such experiments), since they require a

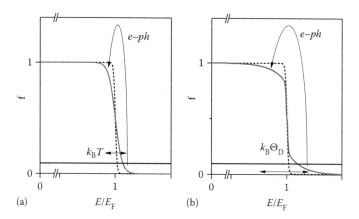

Figure 8.16 Schematic pictures explaining the different relaxation dynamics for (a) thermal and (b) nonthermal electron distributions at low temperatures. The lines represent the nonequilibrium (solid) and thermal (dashed) electron distributions for both cases. (Adapted from Ahn, K.H. et al., *Phys. Rev. B*, 69, 045114, 2004. With permission. Copyright 2004 by the American Physical Society.)

number of input parameters, which might be unknown a priori. Recently, relaxation processes in metals have been theoretically reexamined by Kabanov and Alexandrov (Kabanov 2008). An analytical approach to this problem was developed, where the integral Boltzmann equation was reduced to a differential Schrödinger-type equation in an auxiliary space (reciprocal to energy). For the limit of weak excitation densities, assuming that the phonon system is an infinite thermal bath (a reasonable approximation in the very weak excitation regime and at temperatures where the electronic specific heat is considerably lower than the lattice specific heat—see Figure 8.13), the exact analytical time-dependent electron distributions, when both $e–e$ and $e–ph$ relaxation were taken into consideration, were derived (Kabanov 2008). The results of this analysis can be summarized as follows.

8.2.4.1 The T-Dependence of the e–e Thermalization Time
The T-dependence of the $e–e$ thermalization time is found to have the form expected from the Fermi liquid theory ($1/T^2$)

$$\tau_{e-e} = \frac{2}{\pi^2 K (k_B T)^2}, \tag{8.11}$$

with $K \approx \pi \mu_c^2 / 2\hbar E_F$, and μ_c being the Coulomb pseudo-potential. Importantly, the characteristic $e–e$ relaxation time is quite long due to Pauli blocking. Using realistic values of $\mu_c = 1$ and $E_F = 5 – 10\,\text{eV}$, one obtains $\tau_{e-e} \approx 0.6 – 1.2\,\text{ps}$ at room temperature, consistent with the time-resolved photoemission results on Au (Fann 1992a, b). As the temperature is reduced, τ_{e-e} increases further as $1/T^2$.

8.2.4.2 The T-Dependence of the (Pure) e–ph Thermalization Time
Here, analytical results for the high-temperature ($T > \Theta_D$) and low-temperature limits $T \ll \Theta_D$ are obtained. In the high-temperature regime, the relaxation rate is, up to a numerical factor on the order of 1, given by $\tau_{e-ph} = 2k_B T / \pi \hbar \lambda \langle \omega^2 \rangle$, which is nearly identical to the TTM result, $\tau_{TTM} = \pi k_B T_e / 3\hbar \lambda \langle \omega^2 \rangle$—see Equation 8.3. Interestingly, the relaxation rate was found to be almost independent of the particular shape of the nonequilibrium distribution, and in the linear limit, excitation fluence independent.

For the low-temperature limit, two cases are considered: poor metals and clean metals (Chapter 3). The definition of poor/clean refers to the low-energy asymptotes of the Eliashberg function, $\alpha^2 F(\Omega) = \lambda n(\omega/\omega_D)^n/2$, where n depends on the impurity concentration, disorder, and sample dimension. In clean metals $n = 2$, while in disordered metals (or thin metallic films) it is reduced to $n = 1$ due to phonon trapping (Belitz 1995, Kabanov 2008). In the theoretical analysis both extreme limits were analyzed; however, in any real system one would expect to see some intermediate behavior. For both cases the asymptotic behavior of the photoexcited carrier density $n(t)$ was found to follow

$$n(t) \propto \frac{e^{-t/\tau_{e-ph}}}{t}, \tag{8.12}$$

with the e–ph thermalization time, τ_{e-ph}, exhibiting the following T dependencies (again the time evolution of $n(t)$ is widely independent of the initial electron distribution function):

$$\tau_{e-ph}^{poor} = \frac{2\hbar^2\omega_D}{\pi^3\lambda(k_B T)^2} \quad \text{(poor metals)}$$
$$\tau_{e-ph}^{clean} = \frac{1.76\hbar^3\omega_D^2}{3\pi^3\lambda(k_B T)^3} \quad \text{(clean metals)} \tag{8.13}$$

There are two important conclusions that can be drawn from the above analysis. First of all, from the comparison of τ_{e-e} with the low-temperature limit of τ_{e-ph} (e.g., for poor metals) one obtains

$$\frac{\tau_{e-e}}{\tau_{e-ph}^{poor}}(T \ll \Theta_D) = \frac{E_F}{\hbar\omega_D}\frac{2\lambda}{\mu_c^2}, \tag{8.14}$$

implying that in the low-temperature limit e–e thermalization is faster than e–ph relaxation (TTM) only in the limit of extremely weak e–ph coupling ($\lambda \ll 1$). Second, comparing the two timescales in the high-temperature limit one finds

$$\tau_{e-e} > \tau_{e-ph} \quad \text{for} \quad T < \Theta_D \left(\frac{E_F}{2\pi\hbar\omega_D}\right)^{1/3}. \tag{8.15}$$

It follows that in good metals with large E_F and low Θ_D as well as in advanced materials, e.g., cuprate superconductors, with high Θ_D and small E_F, e–ph relaxation is faster than e–e thermalization up to very high temperatures; e.g., for a good (clean) metal with $E_F = 5\,\text{eV}$ and $\hbar\omega_D = 20\,\text{meV}$ this temperature is in the range of 1000–2000 K, while for an advanced metal this T corresponds to Θ_D (which is also in the range of 500–1000 K).

Combining both e–e and e–ph processes, the relaxation time is found to be dependent on the ratio of the e–e and e–ph relaxation times (Kabanov 2008). In the limit of $\tau_{e-e} \ll \tau_{e-ph}$ (as in the case of the TTM) the relaxation time is determined by the bare τ_{e-ph}. In the opposite case, $\tau_{e-e} > \tau_{e-ph}$ (which is probably more appropriate for materials with reasonable e–ph coupling), the exact relaxation time is $8\tau_{e-ph}/5$. Therefore, one may underestimate the e–ph coupling constant by about a factor of two using the TTM analysis.

Figure 8.17 presents the temperature dependences of the various limiting cases discussed above, for a metal with $\Theta_D = 280\,\text{K}$, $\lambda = 0.1$, $E_F = 5\,\text{eV}$.

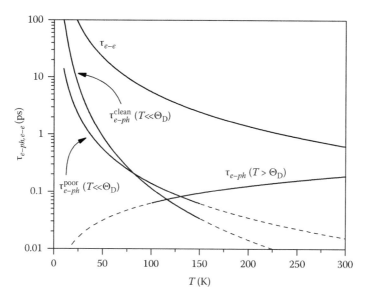

Figure 8.17 The T-dependence of the e–e and e–ph relaxation times in the weak perturbation limit, obtained from the analytical solutions (see Equations 8.11 and 8.13) for both high- and low-temperature limits (the solutions are plotted over the entire temperature range). Here parameters for a typical metal with $\Theta_D = 280\,K$, $\lambda = 0.1$, and $E_F = 5\,eV$ are used.

The main result of the above analysis is that the e–e thermalization time due to the Pauli exclusion principle is, in most of the experimentally accessible range, slower than the e–ph relaxation time. Indeed, there are many observations (apart from the discussed time-resolved photoemission results [Fann 1992a]) that seem to corroborate this result. Various studies of relaxation processes in superconductors have shown that, following photoexcitation with a 50 fs optical pulse, it takes several ps before superconductivity is suppressed (e.g., in MgB$_2$ [Demsar 2003b] this timescale is ~10 ps at low temperatures and excitation densities [Figure 8.18], and in a cuprate superconductor like La$_{2-x}$Sr$_x$CuO$_4$ this timescale is ~1–2 ps [Kusar 2008]). Moreover, the absorbed optical energy required to destroy superconductivity was found to be over an order of magnitude higher than the superconducting condensation energy in La$_{2-x}$Sr$_x$CuO$_4$ (Kusar 2008), while one would expect these energies to be comparable if, on a fast timescale, the energy remains within the electronic subsystem. In fact, it follows from the analysis of these data that in the initial process of relaxation, e–ph processes are dominant and superconductivity is suppressed only after high-frequency phonons are absorbed, breaking "Cooper pairs" in the process.

On the other hand, analytical results on the relaxation dynamics in metals suggest (see Figure 8.17) that, even in the case of a nonthermal electron distribution, the e–ph relaxation time should increase at low temperatures as $1/T^3$ or $1/T^2$, for the clean and pure metal cases, respectively. This conclusion is clearly at odds with the experimental results. One possible reason for this inconsistency could be that the weak excitation limit is experimentally not realized at low temperatures. Alternatively, it could also be that some underlying assumptions, e.g., particle–hole symmetry, are not fulfilled. Clearly further systematic studies are required to resolve these issues.

8.2.5 Ballistic Transport of Hot Carriers

In the previous sections we have discussed relaxation processes where optically thin samples were considered; i.e., no transport of hot carriers out of the probed volume was considered.

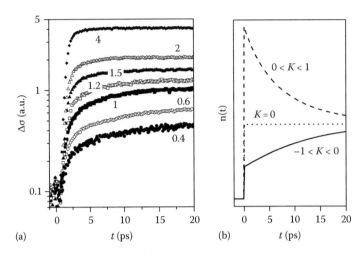

Figure 8.18 Dynamics of superconducting state suppression in MgB$_2$ by means of time-resolved optical pump–THz probe spectroscopy. (a) The superconducting "pair breaking" dynamics at 7 K taken at various excitation densities in units of μJ/cm^2 (all traces shifted vertically by 0.1); at low-excitation densities the typical timescale is ~10 ps. The data are fit by the analytical solution of the Rothwarf–Taylor model (see Demsar 2003b for details). (b) Rothwarf–Taylor model solutions for different initial conditions, where 0 < K < 1 corresponds to the situation where following photoexcitation e–e scattering is dominant, while in the case of –1 < K < 0 following photoexcitation e–ph scattering is dominant and high-frequency phonons subsequently break Cooper pairs, consistent with the experimental data. (Adapted from Demsar, J. et al., *Phys. Rev. Lett.*, 91, 267002, 2003b. With permission. Copyright 2003 by the American Physical Society.)

In the case of materials with a very short mean free path (momentum relaxation length), l_p, transport may not play an important role even when experiments on bulk samples are performed. However, in clean metals, the mean free path even at room temperature can exceed the optical penetration depth (~10 nm) by an order of magnitude. Given the fact that transport of hot electrons presents a parallel relaxation channel to, for example, e–ph relaxation, this process needs to be taken into account when performing experiments on optically thick samples.

Considering the transport of high-energy carriers out of the probed volume, one has to bear in mind that the motion of an individual electron is a random walk, where the electrons move with a velocity comparable to the Fermi velocity v_F. In the limit of lengths longer than l_p, such random walk behavior is averaged and the electron motion is subject to the diffusion equation. However, on a length scale shorter than l_p the electrons move ballistically with ~v_F. In metals, the electron–electron scattering length l_{e-e} was found to be $l_{e-e} \propto (E - E_F)^2$ and was calculated for Au (Krolikowski 1970) to be 80 nm for 1 eV electrons. On the other hand, the electron–phonon scattering length l_{e-ph} is usually inferred from conductivity data. Using the Drude relaxation times one obtains $l_{e-ph} \approx 50$ nm for Au at room temperature. When compared to the characteristic length—optical skin depth, which is on the order of 10 nm in metals—we can see that in pump-probe optical experiments one is probing ballistic, rather than diffusive, transport of electrons.

Femtosecond electronic heat transport in thin Au films was experimentally studied by Brorson et al. in 1987 (Brorson 1987) (see Figure 8.19), when two types of pump-probe experiments were performed on films of different thicknesses on sapphire substrates. First, the differential reflectivity of the probe at the back surface of the film was measured as the sample was pumped with a 100 fs pulse from the front. By measuring the delay of the rising edge as a function of the film thickness they found that it increases linearly with the film thickness, yielding an energy transport velocity of ~10^8 m/s, which is on the same order of magnitude

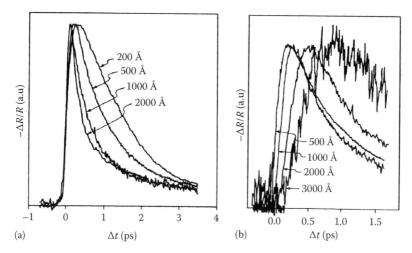

Figure 8.19   Femtosecond relaxation dynamics in thin gold films. (a) Dynamics of the induced reflectivity change as a function of film thickness, when the reflectivity change was measured at the front of the film (where photoexcitation took place). (b) The same measurements, performed in the configuration when the sample was photoexcited from the front and the induced reflectivity change was measured from the substrate side. (Reproduced from Brorson, S.D. et al., *Phys. Rev. Lett.*, 59, 1962, 1987. With permission. Copyright 1987 by the American Physical Society.)

as v_F (1.4 × 10^8 m/s in Au). The "front pump–front probe" relaxation time measurements revealed that the relaxation time was indeed much shorter in thicker films.

These data can be understood in the following way. When the sample thickness is large in comparison to the optical skin depth, the transport and the energy relaxation occur simultaneously. In this case, the photoinduced reflectivity decay is very fast since two competing processes remove energy from the probed region of the sample. Conversely, when the sample length is decreased to be comparable to the optical penetration depth, less transport occurs and the photoinduced reflectivity dynamics are primarily due to the energy relaxation. Indeed, from the thickness dependence of the relaxation dynamics, the authors concluded that the ballistic transport of hot electrons with v_F takes place in the optically thick samples. Similar results were later obtained also by other groups using optical pump-probe spectroscopy (Hohlfeld 2000, Juhasz 1993), as well as two-photon photoemission (Knoesel 1998, Lisowski 2004).

Recently, ballistic transport and its influence on relaxation dynamics was considered theoretically when discussing the anomalous T-dependence of the relaxation dynamics in a mixed valence heavy fermion compound YbAl$_3$ (Demsar 2009)—see also Chapter 4. There, the relaxation time increases upon cooling from room temperature for this entire class of materials (Demsar 2006); however, this increase in relaxation time is truncated at some temperature T^* (Figure 8.20a). Below this temperature the relaxation time is roughly T-independent, and the functional form or relaxation is found to change. Measurements of the de Haas-van Alphen* (dHvA) effect in YbAl$_3$ indicate a large mean free path, $l \approx 120$ nm, several times larger than the optical penetration depth $l_{opt}(l_{opt}(\mathrm{YbAl}_3) \approx 22$ m). From the T-dependence of resistivity, which is a measure of the momentum scattering rate, it follows that $l > l_{opt}$ up to ≈50 K, which coincides with the temperature where the relaxation time becomes T independent. The characteristic relaxation time at low temperature was ~3 ps, which is comparable

---

* The dHvA effect is the oscillation of the magnetic moment in a metal as a function of magnetic field. It is a powerful method of determining the topology of the Fermi surface, the cyclotron mass, and the lifetime of the conduction electrons or the electron mean free path (see, e.g., Kittel 1986).

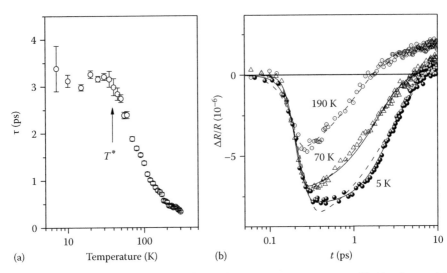

(a) Temperature (K)   (b) t (ps)

Figure 8.20 (a) The T-dependence of the electronic recovery time in YbAl$_3$ obtained by a single exponential fit to the photoinduced reflectivity data. (b) The data recorded on YbAl$_3$ at selected temperatures, fit with a single exponential decay (dashed line) and with Equation 8.19 (solid line), where $l_{opt}/v_F \approx 1.33$ ps. At low temperatures, the fit is entirely governed by the term l_{opt}/v_F (with $\tau > 5$ ps), while above 50 K the relaxation is best fit with a single exponential decay. (Adapted from Demsar, J. et al., *Phys. Rev. B*, 80, 085121, 2009. With permission. Copyright 2009 by the American Physical Society.)

to the rough estimate for the timescale of ballistic transport of electrons out of the probed volume, $l_{opt}/v_F \approx 0.5$ ps.

To describe the relaxation process for such a case, where, in addition to the relaxation process with time constant τ, ballistic transport of hot electrons takes place, the Boltzmann kinetic equation was used with the collision integral in the τ-approximation (Kabanov 2008),

$$\frac{\partial f}{\partial t} + v_F \cos(\theta)\frac{\partial f}{\partial x} = -\frac{f}{\tau(\varepsilon,\theta)}. \tag{8.16}$$

Here $f(t, x, \varepsilon, \theta)$ is the nonequilibrium correction to the equilibrium distribution function, which depends on time t, distance from the surface x, energy ε relative to the Fermi energy, and the angle θ between the velocity and the transport direction, x (the relaxation time τ in general depends on the energy and the angle). Assuming the electron–hole symmetry is preserved (no electric field) and that the Fermi energy is large compared to the photon energy, one can neglect the dependence of the speed of hot electrons and holes on their relative energy, i.e., $v \approx v_F$. Then, Equation 8.16, supplemented by the initial condition $f(t = 0, x, \varepsilon, \theta) = F(\varepsilon, x)$, describing the initial distribution of the hot quasiparticles after photoexcitation, has the solution:

$$f(t,x,\varepsilon,\theta) = \exp\left(\frac{-t}{\tau(\varepsilon,\theta)}\right)F(\varepsilon, x - v_F \cos(\theta,t)). \tag{8.17}$$

The spatial and temporal distribution of the electron–hole pair density is found by integrating

$$n(x,t) = \int_0^\infty d\varepsilon \int_0^\pi f(t,x,\varepsilon,\theta)\sin(\theta)\,d\theta. \tag{8.18}$$

For comparison with the experimental results, one can assume that τ is energy- and angle-independent, $\tau(\varepsilon, \theta) = \tau$, and take the Gaussian form of the energy-integrated excitation profile at $t = 0$, $\int_0^\infty F(\varepsilon, x)\,d\varepsilon = (2/\sqrt{\pi})\exp(-x^2/l_{opt}^2)$. Integrating Equation 8.18 with the Gaussian probe profile, $\exp(-x^2/l_{opt}^2)$, and normalizing to 1 at $t = 0$ one obtains (using the generally accepted assumption that $\Delta R/R$ is proportional to the number of photoinduced carriers within l_{opt})

$$\frac{\Delta R}{R} \propto n(t) = \sqrt{\frac{\pi}{2}}\frac{l_{opt}}{v_F t}e^{-t/\tau}erf\left(\frac{v_F t}{\sqrt{2}l_{opt}}\right). \tag{8.19}$$

Equation 8.19 describes the recovery dynamics of the photoinduced carrier density where, in addition to intrinsic relaxation processes (in simple metals this would be e–ph thermalization), hot electrons ballistically move out of the probed volume. On a short timescale, $t \le l_{opt}/v_F$, Equation 8.19 reduces to

$$n(t) = \exp\left(\frac{-t}{\tau} - \frac{v_F^2 t^2}{2l_{opt}^2}\right), \tag{8.20}$$

while for $t \gg l_{opt}/v_F$ one obtains

$$n(t) = \frac{l_{opt}}{v_F t}e^{-t/\tau}. \tag{8.21}$$

Figure 8.20b presents the photoinduced reflectivity traces in $YbAl_3$ at a few selected temperatures above and below T^*. Clearly the relaxation dynamics are non-exponential at very low temperatures. The relaxation dynamics is fit with both a single exponential decay (dashed) and with Equation 8.19 (solid lines).

At $T < T^*$ the model given by Equation 8.19 clearly describes the experimental data much better than a single exponential decay. In fact, the fit is mainly governed by the l_{opt}/v_F ratio (best fit gives $l_{opt}/v_F = 1.33 \pm 0.05$ ps in good agreement with the simple estimate given above) while τ can be anywhere between 5 and 100 ps. The situation is reversed for $T > T^*$, where the fit is entirely governed by τ. Since the mean free path quickly decreases below l_{opt} upon increasing T, e.g., in $YbAl_3$ this should happen at ≈ 50 K, the ballistic transport becomes ineffective for $T > T^*$. Correspondingly, the fit to the 70 K data in Figure 8.20b with fixed $l_{opt}/v_F = 1.33$ becomes inadequate. In fact, the 70 K data are much better fit with the single exponential decay (dashed). Thus, T^* can be thought of as the temperature for $l \approx l_{opt}$. Below T^* the relaxation is dominated by ballistic transport, while above this temperature, quasiparticle relaxation is governed by the specific relaxation process involved (in this particular case this is attributed to relaxation across the indirect hybridization gap [for details see Demsar 2006, 2009]).

ACKNOWLEDGMENTS

We would like to acknowledge valuable discussions with Alfred Leitenstorfer and Viktor V. Kabanov. We also thank our many collaborators on time-resolved studies of semiconductors, metals, and correlated electron systems reviewed here, in particular K.H. Ahn,

A.S. Alexandrov, R.D. Averitt, H.J. Bakker, G.C. Cho, M. Hase, M.J. Graf, V.V. Kabanov, M. Kitajima, H. Kurz, D. Mihailovic, J.L. Sarrao, A. J. Taylor, S.A. Trugman, and many more. Preparation of this review was in part supported by Sofja Kovalevskaja Award from the Humboldt Foundation. We would also like to thank the authors of various articles for the permission to reproduce their figures in the preceding text.

REFERENCES

(Ahn 2004) Ahn, K.H., Graf, M.J., Trugman, S.A. et al. 2004. Ultrafast quasiparticle relaxation dynamics in normal metals and heavy-fermion materials. *Phys. Rev. B* 69: 045114.

(Allen 1987) Allen, P.B. 1987. Theory of thermal relaxation of electrons in metals. *Phys. Rev. Lett.* 59: 1460.

(Anisimov 1974) Anisimov, S.I., Kapeliovitch, B.L., and Perel'man, T.L. 1974. Electron-emission from surface of metals induced by ultrashort laser pulses. *Sov. Phys-JETP* 39: 375.

(Ascazubi 2006) Ascazubi, R., Wilke, I., Kim, K.J., and Dutta, P. 2006. Terahertz emission from Ga1-xInxSb. *Phys. Rev. B* 74: 075323.

(Averitt 2001) Averitt, R.D., Lobad, A.I., Kwon, C. et al. 2001. Ultrafast conductivity dynamics in Colossal magnetoresistance manganites. *Phys. Rev. Lett.* 87: 017401.

(Banyai 1986) Banyai, L. and Koch, S.W. 1986. A simple theory for the effects of plasma screening on the optical-spectra of highly excited semiconductors. *Z. Phys. B* 63: 2283.

(Beaurepaire 1996) Beaurepaire, E., Merle, J.-C., Daunois, A., and Bigot, J.-Y. 1996. Ultrafast spin dynamics in ferromagnetic nickel. *Phys. Rev. Lett.* 76: 4250.

(Belitz 1995) Belitz, D. and Wybourne, M.N. 1995. Eliashberg function of amorphous metals, *Phys. Rev. B* 51: 689.

(Bigot 2009) Bigot, J-Y., Vomir, M., and Beaurepaire, E. 2009. Coherent ultrafast magnetism induced by femtosecond laser pulses. *Nat. Phys.* 5: 515 and references therein.

(Brorson 1987) Brorson, S.D., Fujimoto, J.G., and Ippen, E.P. 1987. Femtosecond electronic heat-transport dynamics in thin gold films. *Phys. Rev. Lett.* 59: 1962.

(Brorson 1990a) Brorson, S.D., Kazeroonian, A., Face, D.W. et al. 1990. Femtosecond thermo-modulation study of high-Tc superconductors. *Solid. State. Commun.* 74: 1305.

(Brorson 1990b) Brorson, S.D. 1990. Femtosecond thermomodulation measurements of transport and relaxation in metals and superconductors, thesis, Research Laboratory of Electronics, Massachusetts Institute of Technology, Cambridge, MA.

(Brorson 1990c) Brorson, S.D. et al., 1990. Femtosecond room-temperature measurement of the electron-phonon coupling constant l in metallic superconductors. *Phys. Rev. Lett.* 64: 2172.

(Brown 1995) Brown, E.R., McIntosh, K.A., Nichols, K.B., and Dennis, C.L. 1995. Photomixing up to 3.8-thz in low-temperature-grown GaAs. *Appl. Phys. Lett.* 66: 285.

(Carpene 2008) Carpene, E., Mancini, E., Dallera, C. et al. 2008. Dynamics of electron-magnon interaction and ultrafast demagnetization in thin iron films. *Phys. Rev. B* 78: 174422.

(Cavalleri 2001) Cavalleri, A., Toth, C., Siders, C.W. et al. 2001. Femtosecond structural dynamics in VO_2 during an ultrafast solid-solid phase transition. *Phys. Rev. Lett.* 87: 237401.

(Cho 1990) Cho, G.C., Kütt, W., and Kurz, H. 1990. Subpicosecond time-resolved coherent-phonon oscillations in GaAs. *Phys. Rev. Lett.* 65: 764.

(Cho 1996) Cho, G.C., Dekorsy, T., Bakker, H.J., Hövel, R., and Kurz, H. 1996. Generation and relaxation of coherent majority plasmons. *Phys. Rev. Lett.* 77: 4062.

(Collet 1994) Collet, J.H., Hunsche, S., Heesel, H., and Kurz, H. 1994. Influence of electron-hole correlations on the absorption of GaAs in the presence of nonthermalized carriers. *Phys. Rev. B* 50: 10649.

(Darling 1991) Darling, R.B. 1991. Defect-state occupation, fermi-level pinning, and illumination effects on free semiconductor surfaces. *Phys. Rev. B* 43: 4071.

(Dekorsy 1993) Dekorsy, T., Pfeifer, T., Kütt, W., and Kurz, H. 1993. Subpicosecond carrier transport in GaAs surface-space-charge fields. *Phys. Rev. B* 47: 3842.

(Dekorsy 1996) Dekorsy, T., Auer, H., Bakker, H.J., Roskos, H.G., and Kurz, H. 1996. THz electromagnetic emission by coherent infrared-active phonons. *Phys. Rev. B* 53: 4005.

(Dekorsy 2000) Dekorsy, T. and Cho, G.C. 2000, in *Light Scattering in Solids VIII*, eds. M. Cardona and G. Güntherodt, Springer, Berlin, Germany, Topics in Applied Physics 76: 169.

(Dember 1931) Dember, H. 1931. A photoelectrical-motor energy in copper-oxide crystals. *Phys. Z.* 32: 554.

(Demsar 1999a) Demsar, J., Podobnik, B., Kabanov, V.V., Wolf, Th., and Mihailovic, D. 1999. Superconducting Gap D_c, the Pseudogap D^p, and Pair Fluctuations above T_c in overdoped $Y_{1-x}Ca_xBa_2Cu_3O_{7-d}$ from femtosecond time-domain spectroscopy. *Phys. Rev. Lett.* 82: 4918.

(Demsar 1999b) Demsar, J., Biljakovic, K., and Mihailovic, D. 1999. Single particle and collective excitations in the one-dimensional charge density wave solid $K_{0.3}MoO_3$ probed in real time by femtosecond spectroscopy. *Phys. Rev. Lett.* 83: 800.

(Demsar 2003a) Demsar, J., Averitt, R.D., Ahn, K.H. et al. 2003. Quasiparticle relaxation dynamics in heavy Fermion compounds. *Phys. Rev. Lett.* 91: 027401.

(Demsar 2003b) Demsar, J., Averitt, R.D., Taylor, A.J. et al. 2003. Pair-breaking and superconducting state recovery dynamics in MgB_2. *Phys. Rev. Lett.* 91: 267002.

(Demsar 2006) Demsar, J., Sarrao, J.L., and Taylor, A.J. 2006. Dynamics of photoexcited quasiparticles in heavy electron compounds. *J. Phys. Condens. Mat.* 18: R281.

(Demsar 2009) Demsar, J., Kabanov, V.V., Alexandrov, A.S. et al. 2009. Hot electron relaxation in the heavy-fermion $Yb_{1-x}Lu_xAl_3$ compound using femtosecond optical pump-probe spectroscopy. *Phys. Rev. B* 80: 085121.

(Eesley 1983) Eesley, G.L. 1983. Observation of non-equilibrium heating in copper. *Phys. Rev. Lett.* 51: 2140.

(Elsayed-Ali 1987) Elsayed-Ali, H.E., Norris, T.B., Pessot, M.A., and Mourou, G.A. 1987. Time-resolved observation of electron-phonon relaxation in copper. *Phys. Rev. Lett.* 58: 1212.

(Elsayed-Ali 1993) Elsayed-Ali, H.E. and Juhasz, T. 1993. Femtosecond time-resolved thermomodulation of thin gold-films with different crystal-structures. *Phys. Rev. B* 47: 13599.

(Fann 1992a) Fann, W.S., Storz, R., Tom, H.W.K., and Bokor, J. 1992. Direct measurement of nonequilibrium electron-energy distributions in subpicosecond laser-heated gold-films. *Phys. Rev. Lett.* 68: 2834.

(Fann 1992b) Fann, W.S., Storz, R., Tom, H.W.K., and Bokor, J. 1992. Electron thermalization in gold. *Phys. Rev. B* 46: 13592.

(Fox 2001) Fox, M. 2001. *Optical Properties of Solids*. Oxford University Press, Oxford, U.K.

(Fujimoto 1984) Fujimoto, J.G., Liu, J.M., Ippen, E.P., and Bloembergen, N. 1984. Femtosecond laser interaction with metallic tungsten and nonequilibrium electron and lattice temperatures. *Phys. Rev. Lett.* 53: 1837.

(Groeneveld 1992) Groeneveld, R.H.M., Sprik, R., and Lagendijk, A. 1992. Effect of a nonthermal electron-distribution on the electron-phonon energy relaxation process in noble-metals. *Phys. Rev. B* 45: 5079.

(Groeneveld 1995) Groeneveld, R.H.M., Sprik, R., and Lagendijk, A. 1995. Femtosecond spectroscopy of electron-electron and electron-phonon energy relaxation in Ag and Au. *Phys. Rev. B* 51: 11433, and references therein.

(Gu 2002) Gu, P., Tani, M., Kono, S., Sakai, K., and Zhang, X.-C. 2002. Study of terahertz radiation from InAs and InSb. *J. Appl. Phys.* 91: 5533.

(Hase 2005) Hase, M., Ishioka, K., Demsar, J., Ushida, K., and Kitajima, M. 2005. Ultrafast dynamics of coherent optical phonons and nonequilibrium electrons in transition metals. *Phys. Rev. B* 71: 184301.

(Hohlfeld 2000) Hohlfeld, J., Wellershoff, S.S., Gudde, J. et al. 2000. Electron and lattice dynamics following optical excitation of metals. *Chem. Phys.* 251: 237.

(Hunsche 1993) Hunsche, S., Heesel, H., Ewertz, A., Kurz, H., and Collet, J.H. 1993. Spectral-hole burning and carrier thermalization in GaAs at room-temperature. *Phys. Rev. B* 48: 17818.

(Juhasz 1993) Juhasz, T., Elsayedali, H.E., Smith, G.O., Suarez, C., and Bron, W.E. 1993. Direct measurements of the transport of nonequilibrium electrons in gold-films with different crystal-structures. *Phys. Rev. B* 48: 15488.

(Kabanov 1999) Kabanov, V.V., Demsar, J., Podobnik, B., and Mihailovic, D. 1999. Quasiparticle relaxation dynamics in superconductors with different gap structures: Theory and experiments on $YBa_2Cu_3O_{7-2}d$. *Phys. Rev. B.* 59: 1497.

(Kabanov 2008) Kabanov, V.V. and Alexandrov, A.S. 2008. Electron relaxation in metals: Theory and exact analytical solutions. *Phys. Rev. B* 78: 174514.

(Kaganov 1956) Kaganov, M.I., Lifshitz, I.M., and Tanatarov, L.V. 1957. Relaxation between electrons and lattice. *Zh. Exsp. Theor. Fiz.* 31: 232 [*Sov. Phys.* JETP 4: 173 (1957)].

(Keller 1996) Keller, U., Weingarten, K.J., Kärtner, F.X. et al. 1996. Semiconductor saturable absorber mirrors (SESAM's) for femtosecond to nanosecond pulse generation in solid-state lasers. *IEEE J. Sel. Top. Quant. Electron.* 2: 435.

(Kersting 1997) Kersting, R., Unterrainer, K., Strasser, G., Kauffmann, H.F., and Gornik, E. 1997. Few-cycle THz emission from cold plasma oscillations. *Phys. Rev. Lett.* 79: 3038.

(Kittel 1986) Kittel, C. 1986. *Introduction to Solid State Physics.* John Wiley & Sons, Inc., New York.

(Klingshirn 2006) Klingshirn, C.F. 2006. *Semiconductor Optics*, 3rd edn. Springer, Berlin, Germany.

(Knoesel 1998) Knoesel, E., Hotzel, A., and Wolf, M. 1998. Ultrafast dynamics of hot electrons and holes in copper: Excitation, energy relaxation, and transport effects. *Phys. Rev. B* 57: 12812.

(Krolikowski 1970) Krolikowski, W.F. and Spicer, W.E. 1970. Photoemission studies of the noble metals. II. Gold. *Phys. Rev. B* 1: 478.

(Krotkus 1995) Krotkus, A., Marcinkevicius, S., Jasinski, J. et al. 1995. Picosecond carrier lifetime in GaAs implanted with high-doses of as ions—An alternative material to low-temperature GaAs for optoelectronic applications. *Appl. Phys. Lett.* 66: 3304.

(Kübler 2007) Kübler, C., Ehrke, H., Huber, R. et al. 2007. Coherent structural dynamics and electronic correlations during an ultrafast insulator-to-metal phase transition in VO_2. *Phys. Rev. Lett.* 99: 116401.

(Kusar 2008) Kusar, P., Kabanov, V.V., Demsar, J. et al. 2008. Controlled vaporization of the superconducting condensate in cuprate superconductors by femtosecond photoexcitation. *Phys. Rev. Lett.* 101: 227001.

(Leitenstorfer 1996) Leitenstorfer, A., Fürst, C., Laubereau, A. et al. 1996. Femtosecond carrier dynamics in GaAs far from equilibrium. *Phys. Rev. Lett.* 76: 1545.

(Leitenstorfer 2000) Leitenstorfer, A., Hunsche, S., Shah, J., Nuss, M.C., and Knox, W.H. 2000. Femtosecond high-field transport in compound semiconductors. *Phys. Rev. B* 61: 16642.

(Lisowski 2004) Lisowski, M., Loukakos, P.A., Bovensiepen, U. et al. 2004. Ultra-fast dynamics of electron thermalization, cooling and transport effects in Ru(001). *Appl. Phys. A* 78: 165.

(Lisowski 2005) Lisowski, M., Loukakos, P.A., Melnikov, A. et al. 2005. Femtosecond electron and spin dynamics in Gd(0001) studied by time-resolved photoemission and magneto-optics. *Phys. Rev. Lett.* 95: 137402.

(Meier 2007) Meier, T., Thomas, P., and Koch S.W. 2007. *Coherent Semiconductor Optics: From Basic Concepts to Nanostructure Applications.* Springer, Berlin, Germany.

(Nolte 1999) Nolte, D.D. 1999. Semi-insulating semiconductor heterostructures: Optoelectronic properties and applications. *J. Appl. Phys.* 85: 6259.

(Ogasawara 2001) Ogasawara, T., Kimura, T., Ishikawa, T., Kuwata-Gonokami, M., and Tokura, Y. 2001. Dynamics of photoinduced melting of charge/orbital order in a layered manganite $La_{0.5}Sr_{1.5}MnO_4$. *Phys. Rev. B* 63: 113105.

(Ogasawara 2005) Ogasawara, T., Ohgushi, K., Tomioka, Y. et al. 2005. General features of photoinduced spin dynamics in ferromagnetic and ferrimagnetic compounds. *Phys. Rev. Lett.* 94: 087202.

(Ogawa 1997) Ogawa, S., Nagano, H., and Petek, H. 1997. Hot-electron dynamics at Cu(100), Cu(110), and Cu(111) surfaces: Comparison of experiment with Fermi-liquid theory. *Phys. Rev. B.* 55: 10869.

(Perfetti 2006) Perfetti, L., Loukakos, P.A., Lisowski, M. et al. 2006. Time evolution of the electronic structure of $1T$-TaS_2 through the insulator-metal transition. *Phys. Rev. Lett.* 97: 067402.

(Perfetti 2007) Perfetti, L., Loukakos, P.A., Lisowski, M. et al. 2007. Ultrafast electron relaxation in superconducting $Bi_2Sr_2CaCu_2O_8^+$ delta by time-resolved photoelectron spectroscopy. *Phys. Rev. Lett.* 99: 197001.

(Petek 1997) Petek, H. and Ogawa, S. 1997. Femtosecond time-resolved two-photon photoemission studies of electron dynamics in metals. *Prog. Surf. Sci.* 56: 239.

(Pines 1996) Pines, D. and Nozieres, P. 1996. *Theory of Quantum Liquids*. Benjamin, New York.

(Rini 2007) Rini, M., Tobey, R., Dean, N. et al. 2007. Control of the electronic phase of a manganite by mode-selective vibrational excitation. *Nature* 449: 7158.

(Schäfer 2002) Schäfer W. and Wegener M. 2002. *Semiconductor Optics and Transport Phenomena*. Springer, Berlin, Germany.

(Schmitt 2008) Schmitt, F., Kirchmann, P.S., Bovensiepen, U. et al., 2008. Transient electronic structure and melting of a charge density wave in TbTe3., *Science* 321: 1649.

(Schmuttenmaer 1994) Schmuttenmaer, C.A., Aeschlimann, M., Elsayedali, H.E. et al. 1994. Time resolved two photon photoemission from Cu(100): Energy dependence of electron relaxation. *Phys. Rev. B* 50: 8957.

(Schoenlein 1987) Schoenlein, R.W., Lin, W.Z., Fujimoto, J.G., and Eesley, G.L. 1987. Femtosecond studies of nonequilibrium electronic processes in metals. *Phys. Rev. Lett.* 58: 1680.

(Seeger 2004) Seeger, K.H. 2004. *Semiconductor Physics: An Introduction*, 9th edn. Springer, Berlin, Germany.

(Segschneider 1997) Segschneider, G., Dekorsy, T., Kurz, H., Hey, R., and Ploog, K. 1997. Energy resolved ultrafast relaxation dynamics close to the band edge of low-temperature grown GaAs. *Appl. Phys. Lett.* 71: 2779.

(Shah 1999) Shah, J. 1999. *Ultrafast Spectroscopy of Semiconductors and Semiconductor Nanostructures*, 2nd edn. Springer, Berlin, Germany.

(Tomeljak 2009) Tomeljak, A., Schäfer, H., Städter, D. et al. 2009. Dynamics of photoinduced charge-density-wave to metal phase transition in $K_{0.3}MoO_3$. *Phys. Rev. Lett.* 102: 066404.

(Vasko 1998) Vasko F.T. and Kuznetsov, A.V. 1998. *Electronic States and Optical Transitions in Semiconductor Heterostructures*. Springer, New York.

(Wenckebach 1999) Wenckebach, W. T. 1999. *Essentials of Semiconductor Physics*. John Wiley & Sons, New York.

(Wysin 1988) Wysin, G.M., Smith, D.L., and Redondo, A. 1988. Picosecond response of photo-excited GaAs in a uniform electric-field by Monte Carlo dynamics. *Phys. Rev. B* 38: 12514.

(Yan 1985) Yan, Y.X., Gamble, E.B., and Nelson, K.A. 1985. Impulsive stimulated scattering—General importance in femtosecond laser-pulse interactions with matter, and spectroscopic applications. *J. Chem. Phys.* 83: 5391.

(Yu 1996) Yu, P.Y. and Cardona, M. 1996. *Fundamentals of Semiconductors*. Springer, Berlin, Germany.

(Zeiger 1992) Zeiger, H.J., Vidal, J., Cheng, T.K. et al. 1992. Theory for displacive excitation of coherent phonons. *Phys. Rev. B* 45: 768.

Chapter 9

ULTRAFAST PUMP–PROBE SPECTROSCOPY

David J. Hilton

CONTENTS

9.1 INTRODUCTION

The origin of the observed electronic and optical properties of materials (resistance, imped-ance, dielectric constant, and refractive index) is a complex interplay between lattice, elec-tronic, and spin degrees of freedom that generally occur on a timescale that ranges from 10^{-18} to 10^{-9} s (or longer). Standard electrical characterization techniques can be used to determine the complex dielectric function of the material. These measurements typically use network analyzers, cavity resonance, capacitance bridge circuits, and a wide variety of other methods; a broad review of these experimental tools can be found in Horowitz and Hill (1989). These electrical measurements are limited to a maximum bandwidth determined by the inverse of the electronic response time (RC time constant) of the electrical components used. Electronic signals faster than this response time are quickly damped by the measurement apparatus and cannot be detected. Current electrical techniques, therefore, are limited to measurements with timescales >0.1 ns (10 GHz). To overcome this response time limitation and access the relevant timescales ($\ll 10^{-9}$ s) in complex materials, alternate *nonelectrical* methods are needed.

Experimental methods to measure the dynamic properties of materials on a femtosecond tim-escale have transformed our understanding of their electronic and optical properties. Pump–probe spectroscopic techniques are perhaps the simplest ultrafast optical measurement techniques. Here, a high-intensity optical pump pulse perturbs the system from equilibrium and a time-delayed weak probe pulse measures the photoinduced change in either the transmission or reflection of the sample at that delay. Pump–probe measurements use sampling techniques to overcome electronic detection limitations and recover the dynamics of electronic and optical properties on femtosec-ond or picosecond timescales. Pump–probe spectroscopy can be used to determine the nonequi-librium dynamics of electron–electron interactions (Groeneveld et al., 1995), examine coupling between electron, phonon, and spin subsystems (Groeneveld et al., 1995; Beaurepaire et al., 1996), time-resolve photoinduced phase transitions (Ogawa et al., 2000; Cavalleri et al., 2001, 2004; Nasu, 2004; Rini et al., 2005; Yonemitsu and Nasu, 2008), and elucidate the electronic structure of complex materials (Demsar et al., 2002, 2003) (see Chapter 8).

The rapid development of ultrafast laser sources over the past two decades has driven advances in the field of ultrafast pump–probe spectroscopy. Commercially available titanium:sapphire (Ti:Al$_2$O$_3$ or Ti:sapphire) lasers routinely generate sub-50 fs pulses (1 fs = 10^{-15} s) with a wide range of available powers (Chapter 7) (Backus et al., 1998). *Custom-designed* laser systems that have achieved pulse durations of a few femtoseconds at optical and infrared frequencies have been reported in the literature (Sander et al., 2009). The energy per pulse, E, of these systems spans the range of a few picojoules (Bartels et al., 1999) to joules (Miller et al., 2004). Optical parametric generation and amplification techniques (Chapter 7) can be used to tune ultrafast laser wavelengths to the visible, infrared, and terahertz ranges

(Tang and Cheng, 1995; Zhang et al., 1995; Mittleman, 2003). Using high harmonic generation techniques (Chapter 7 and Part V) (Christov et al., 1997), ultraviolet and x-ray wavelengths can be generated using the output of a high-power Ti:sapphire laser. This source diversity permits the study of materials over a very wide range of excitation conditions.

This chapter focuses on time-resolved differential transmission and differential reflection measurements in condensed matter systems. This general picture also applies to all of the other time-resolved techniques described in Part II of this book, which all employ similar experimental configurations and operate on similar principles.

9.2 ULTRAFAST PUMP–PROBE TECHNIQUES

Optical measurements currently span timescales from *static* measurements in a Fourier transform (FT) infrared spectrometer (Chapter 4) (Padilla et al., 2004) to femtosecond (10^{-15} s) (Sander et al., 2009) and attosecond (1 as = 10^{-18} s) measurements (Christov et al., 1997; Goulielmakis et al., 2008). Pump–probe spectroscopy measures the transient optical properties of materials with a temporal resolution that greatly exceeds the electrical bandwidth limitation of the system electronics. To overcome the response time limitations of the electronics, pump–probe spectroscopic techniques employ a *sampling* technique to reconstruct the signal. The full ultrafast time domain signal is not measured at once, but is reconstructed by probing the pump-induced change in the system reflectivity or transmission at a given time delay using a second ultrafast pulse and repeating the measurement multiple times for multiple time delays. Each individual data sample is the *average* signal over many reflected or transmitted pulses at a given time delay, thus eliminating the detector response time as a limitation of the temporal resolution. In an ultrafast pump–probe spectrometer, the minimum time resolution is limited by the pump and probe pulse durations and the interval between the measured time delays, which can be as small as 0.1 fs for a piezoelectric transducer with a $d = 30$ nm spatial resolution. In this way, the temporal resolution can be ≤ 50 fs (depending on the laser source and stage chosen) while using a detector with a response time of $\gg 0.1$ ns.

Time-resolved differential transmission $\Delta T/T$ and differential reflectivity $\Delta R/R$ measurements are possible with these techniques. These two measurements together permit the determination of the changes to the full *complex* refractive index or, equivalently, the complex dielectric constant, complex conductivity, and complex susceptibility (see Sections 9.5.4 and 9.5.5). The following section discusses the general design of this experiment, design decisions, common enhancements, and ways to estimate the performance of a system from its design.

9.3 EXPERIMENT DESIGN

A diagram of a degenerate,* noncollinear pump–probe spectrometer is shown in Figure 9.1. The output of the laser propagates through a two- or four-prism sequence to provide the necessary anomalous dispersion to compensate for the normal dispersion of the optics in the system (prism compressors are discussed in further detail in Chapter 7). This ensures that the temporal widths of the optical pump and probe pulses at the sample position are a minimum. This laser beam is then split into two components using a beam splitter. The transmitted pump beam receives the majority of the optical power and is used to optically perturb the sample from equilibrium. The reflected beam becomes the probe beam and is used to sample the electronic configuration of the sample at a fixed time delay determined by the difference between the pump and probe path lengths.

---

* A *degenerate* pump–probe spectrometer uses pump and probe pulses of the same wavelength. A *nondegenerate* pump–probe spectrometer uses different pump and probe wavelengths.

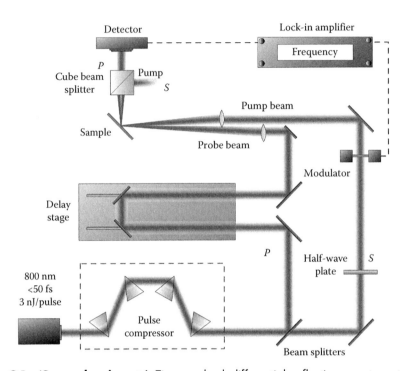

Figure 9.1 (See color insert.) Time-resolved differential reflection spectrometer using degenerate pump–probe spectroscopy. The system can be similarly configured for time-resolved differential transmission, with the collection optics and detection relocated behind the sample. The polarization of the pump beam is rotated by 90° using a half-wave plate. After either reflection (as shown) or transmission (not shown), the probe beam can be isolated from the pump beam using a polarizing cube beam splitter. If using a balanced detector, a small portion of the probe beam right after the pulse compressor is split off and used as the detector reference input (not shown). The second detector and the balanced detector circuitry use this reference to cancel any power fluctuations in the laser power from the differential change to the probe transmission or reflectivity.

The *pump beam* passes through a broadband half-wave plate to rotate its polarization ($\hat{s}$ after the half-wave plate) so that it is perpendicular to the probe beam polarization ($\hat{p}$). An optical chopper provides a known frequency modulation to the pump beam ($f_c \approx 1\,\text{kHz}$ when using a mechanical chopper) that is used as the reference input to a lock-in amplifier. A lens or off-axis parabolic mirror (to minimize dispersion or nonlinear effects) focuses this pump beam onto the surface of the sample to be studied. This lens usually has a shorter focal length than the lens used to focus the probe beam to ensure that the diameter of the focused pump spot is larger than that of the probe spot.

The relative power of the *probe beam* should be much less than the pump beam (ideally a pump/probe power ratio >10:1) so that the probe merely *samples* the material properties ($\Delta R/R$ or $\Delta T/T$) at each time delay without significantly altering them. After the beam splitter, the probe beam travels through a variable length delay stage to control the optical path length and, therefore, the relative timing of the pump and probe pulse incidence on the sample. This stage is the origin of the temporal delay and of the ultrafast resolution of this system; its minimum step size, along with the laser pulse width, restricts the temporal resolution of the system. This probe beam is then focused onto the sample position so that it is spatially overlapped with the focused pump beam spot. The focused spot size of the probe beam should be smaller than that of the pump (typically a pump/probe ratio >2:1) so that the probe measures an area of the photoexcited sample that is approximately uniformly excited. This is achieved by using a smaller $f/\#$ lens, which has a smaller diffraction-limited spot size (Hecht, 1987). If both the pump and the probe beam diameters are the same before the lens, then the probe lens focal length should be longer than the pump lens focal length by the ratio of the desired spot sizes (i.e., if $f_{probe} = N \times f_{pump}$, then the pump spot radius will be approximately N times the probe radius).

In *degenerate* pump–probe spectroscopy, both the pump and the probe wavelengths are the same. This makes isolating the transmitted or reflected probe beam from the scattered pump beam after the sample difficult and leads to a significant background signal on the detector. The chopped pump beam modulates the reflectivity of the sample at the chopper frequency f_c, which then modulates the transmitted or reflected probe beam power at that chopper frequency (to be discussed further in Figure 9.4). As a result, both the pump and probe beams have a component at the chopper frequency, but only the modulated probe beam is dependent on the pump-induced change in the material properties. This is different from *nondegenerate* pump–probe spectroscopy, where a wavelength filter that blocks the pump beam and transmits the probe wavelength reduces this background.

In a *degenerate* experiment, both the scattered pump beam and the transmitted or reflected probe beam are incident on the detector and the lock-in amplifier is unable to distinguish between them. The polarization optics (half-wave plate and polarizing beam splitter) are critical components in this experiment needed to separate the pump and probe beams, as the scattered pump beam will otherwise saturate the lock-in amplifier if it is incident on the detector. The pump beam polarization is rotated by 90° by the half-wave plate ($\hat{p} \rightarrow \hat{s}$) before incidence on the sample. After either reflection (as shown in Figure 9.1) or transmission through the sample, a polarizing cube beam splitter or broadband polarizer splits the pump polarization ($\hat{s}$) from the perpendicularly polarized probe beam ($\hat{p}$). The probe beam is then focused onto the detector while the pump beam travels a different physical path and is blocked.

After the polarizing beam splitter, the probe beam is focused onto a detector and converted to an electrical signal. As discussed earlier, it is not necessary for the detector to have the time resolution to measure the full ultrafast signal (i.e., a detector with tera- or petahertz bandwidth). This detector must, instead, be capable of resolving the signal at the chopper frequency. The electrical signal at the chopper frequency is then measured using a lock-in amplifier to isolate the changes in the probe beam transmission or reflection at each value of the pump–probe delay. The temporal resolution instead comes from the relative optical path length difference between the pump and the probe pulses. The full ultrafast signal results from averaging repeated measurements of the photoinduced change to the probe signal over the full range of pump–probe delays. Also, observing the measured traces while performing several time delay scans can reveal long-term pump-induced changes in the sample properties (i.e., if the amplitude and/or shape of the traces changes significantly from scan to scan), which are typically undesirable; reducing the pump fluence and/or changing the wavelength often helps remove these effects.

In nondegenerate pump–probe spectroscopy, the pump and the probe wavelengths are different. The previous example uses a broadband Ti:sapphire laser oscillator (high repetition rate, low pulse energy) (Spence et al., 1991; Bartels et al., 1999), but other sources (Chapter 7) can include an amplified Ti:sapphire laser (Backus et al., 1998), optical parametric amplifiers (OPAs) (Zhang et al., 1995), or optical parametric oscillators (OPOs) (Tang and Cheng, 1995). In the case of nondegenerate pump–probe spectroscopy, the probe beam would have one of these frequency conversion devices to tune the probe wavelength from the source laser wavelength.

The polarization-based pump–probe design described in Figure 9.1 is not appropriate for pump–probe experiments in waveguides as the external laser fields (pump and probe) must couple to an available mode of the waveguide structure. The supported modes of waveguide structures strongly depend on the excitation wavelengths, the polarization state, the geometry of the waveguide structure, and the material properties of the waveguide; this is discussed in further detail by Yariv (1997) for common rectangular and cylindrical waveguide geometries. It is unlikely that both the pump and the probe fields in an apparatus like the one shown in Figure 9.1 would simultaneously be supported modes of the cavity with different polarizations and incidence angles. As a result, it is necessary to eliminate the polarization optics in Figure 9.1 and use a beam splitter to recombine the pump and probe beams for collinear incidence into the waveguide structure. To isolate the transmitted or reflected probe signal from the pump beam (the main purpose of the polarization optics in Figure 9.1), a number of different strategies can be adopted. For example, a nondegenerate pump–probe experiment can use wavelength-specific optical components (band pass filter, color filter, etc.) to isolate the pump from the probe signal (Sanders et al., 1994). Heterodyne detection is an alternate strategy, which uses three beams (pump, probe, and reference) in an approximately 1:0.2:1 power ratio (Hall et al., 1992; Borri et al., 1999). The reference (f_{ref}) and probe (f_{probe}) pulses are each modulated using a high-frequency acousto-optic modulator at slightly different frequencies such that their difference is $\Delta f = (f_{\text{reference}} - f_{\text{probe}}) \approx 1\,\text{MHz}$ (where $\Delta f \ll f_{\text{reference}}$ and $\Delta f \ll f_{\text{probe}}$), while the pump beam is modulated using an optical chopper at $f_{\text{pump}} \approx 1\,\text{kHz}$. An RF spectrum analyzer with a bandwidth greater than the optical chopper frequency isolates the difference between the probe and the reference beam at $\approx 1\,\text{MHz}$ to isolate the component of the signal due to the probe transmission or reflection from the sample. This filtered output is then detected using a lock-in amplifier using the optical chopper reference (1 kHz) to isolate the pump-induced change to the probe beam. It is still necessary to reduce or eliminate the pump and reference beam powers incident on the detector to avoid saturation of the detector with the pump, as in the design of Figure 9.1.

9.3.1 Design Criteria

General issues to consider in a robust pump–probe experiment design include the following:

Photon energy: The photon energy (laser wavelength) determines the electronic transitions excited by the pump pulse and examined by the probe pulse. In semiconductors, an above-band-gap photon creates an electron–hole pair in the semiconductor and increases the conductivity of the material, while at photon energies below the band-gap, free carrier absorption (Chapter 2) generally dominates the response (Blakemore, 1981). In semiconducting systems, the band gap is the relevant energy scale for the photon energy, as band-to-band absorption is typically stronger than free carrier absorption. In metals, free carrier absorption of electrons at the Fermi surface is the dominant absorption mechanism (Groeneveld et al., 1995). Further discussion of this can be found in Chapters 2, 3, and 8.

Excitation fluence: The pulse energy per unit area defines the *fluence* **F** and determines the local heating and density of quasiparticles photo-generated by the absorbed pump pulse. This fluence must be large enough to produce a measurable change with a signal-to-noise ratio greater than 1, which sets a lower limit on the fluence needed. Each sample may also require a different absorbed fluence at each temperature, pump and probe polarization, and at each pump and probe wavelength, depending on the material's properties (Hilton et al., 2006). Higher pump pulse fluences can result, in many cases, in a larger change to the material's properties and a proportionate increase in the change of the probe pulse transmission or reflection. The exception to this is a system that undergoes a photoinduced phase transition with a threshold fluence to trigger a transition to an alternate phase (Nasu, 2004; Yonemitsu and Nasu, 2008). If the experiment is intended to measure the dynamics of a particular phase (Goodenough, 1971), this determines the upper limit on the allowed fluence, as a higher fluence would trigger a transition to the alternate order (Hilton et al., 2006). Finally, the material damage threshold should also be considered when photo-exciting a sample at high fluences.

Temporal resolution: Temporal sensitivity is primarily determined by the pump and probe pulse widths and the spatial resolution of the delay generator (usually a mechanical translation stage, as discussed earlier). This will determine the fastest dynamics that the pump–probe spectrometer is capable of resolving. Because of the presence of dispersive optics in the beam paths, the shortest temporal resolution requires the use of dispersion compensation to minimize the pulse width at the sample position.

Frequency bandwidth: An ultrafast optical pulse necessarily has a finite bandwidth that can be calculated from the time domain waveform using an Fourier transform (FT) (Goodman, 1996). This distribution of pump wavelengths excites a distribution of initial states, while the distribution of probe wavelengths measures the electronic properties of the material over a range of states. This can result in a balance between the system temporal response and the range of states excited.

High sensitivity: Ultrafast pump–probe spectrometers can measure relative changes in transmission or reflectivity of 10^{-5}–10^{-7} using high-repetition-rate lasers, high frequency modulation, lock-in detection, and high-sensitivity avalanche photodiodes. The use of an optical chopper and a lock-in amplifier eliminates noise that is not within the pass band of the lock-in, reducing the overall system noise. Higher-repetition-rate lasers take advantage of the enhancement of the signal-to-noise ratio by increased counting statistics of the measured signal (signal to noise ratio $\propto \sqrt{N}$, where N is the number of samples taken). Higher-repetition-rate lasers and higher frequency chopping will generally improve the overall signal-to-noise ratio. Currently available high-repetition-rate ultrafast lasers are limited to lower pulse energies (a few nanojoules per pulse) and would not be suitable in experiments where a high optical fluence is required. Mechanical choppers are typically limited to a few kilohertz modulation frequencies while acousto-optic modulators can be used to increase the modulation frequency to megahertz frequencies (Demsar et al., 2003, 2006).

Each of these restrictions dictates the laser choice, pulse width, and the electronics needed to perform the ultrafast pump–probe experiment. The competition between many of these design criteria requires compromises in the experiment design. With modern Ti:sapphire technology, it is not currently possible, for example, to use a high-repetition-rate (>1 MHz) Ti:sapphire laser with a millijoule pulse energy.

9.3.2 Laser Choice

The choice of the laser system is dependent on each of the design criteria given earlier. Ultrafast pump–probe spectrometers typically use a Ti:sapphire gain medium, which has replaced previous generations of dye lasers due to Ti:sapphire's superior performance and reliability (see Chapter 7 for more detail). The available gain bandwidth, $\Delta\lambda = 680$–1100 nm (Rapoport and Khattak, 1988), of Ti:sapphire is sufficient to produce pulses as short as a few femtoseconds (Rapoport and Khattak, 1988; Sander et al., 2009). Ti:sapphire's absorption band begins at 490 nm and extends into the visible range. As a result, common commercially available pump lasers include frequency-doubled Nd:VO$_4$ and Nd:YAG (either continuous wave, Q-switched, or flashlamp-pumped) lasers that emit at 1064 nm (doubled to 532 nm to match the absorption spectra of Ti:sappire). These pump sources result in available Ti:sapphire output powers from a few picojoules to several millijoules per pulse.

A Ti:sapphire laser has an output spectrum typically centered at 800 nm ($h\nu = 1.55$ eV = 12,500 cm^{-1}) with a spectral bandwidth determined by the pulse width of the system. This range of photon energies must correspond to absorption in the material under study. An experiment in GaAs would satisfy this requirement, as its band gap is $E_g = 1.42$ eV at room temperature and an $h\nu = 1.55$ eV photon can create electrons in the conduction band and holes in both the heavy and light-hole subbands (Blakemore, 1981). In metals, heating of the electrons on the Fermi surface by free carrier absorption of the photons can also result in a differential change to the sample reflection or transmission (Groeneveld et al., 1995).

The high available power of Ti:sapphire lasers permits the extra-cavity operation of parametric oscillators (Tang and Cheng, 1995), amplifiers (Zhang et al., 1995), and high harmonic generation experiments (Christov et al., 1997) to tune the pulse wavelength over a broad range. This greatly simplifies parametric device design and permits the generation of ultrashort pulses from the x-ray (Christov et al., 1997), ultraviolet (Ringling et al., 1993; Reed et al., 1995), visible (Boyd, 1991), infrared (Bartels et al., 1999), and terahertz (Mittleman, 2003) ranges. Optical pump, terahertz probe experiments will be covered in greater detail in Chapter 11.

Mode-locked Erbium-doped fiber lasers ($\lambda = 1550$ nm) have recently emerged as an alternate laser medium for femtosecond-pulsed experiments (Tamura et al., 1993). The benefits of a fiber laser include increased portability, reliability, and pointing stability as well as lower noise operation, but such lasers are not currently capable of as high power or broad bandwidth (i.e., short pulses) an operation when compared to Ti:sapphire. Currently, typical powers average a few hundred milliwatts with pulse durations of 100 fs (Hoffmann et al., 2008; Tauser et al., 2008). Nonetheless, these have seen numerous applications in pump–probe spectroscopy, ultrafast terahertz spectroscopy (Azad et al., 2008), and, when frequency doubled to $\lambda = 775$ nm, as the seed lasers for amplified Ti:sapphire systems (Hoffmann et al., 2008).

9.3.3 Pulse Energy

Currently, commercially available Ti:sapphire lasers can have pulse energies from nanojoules to millijoules. For example, a typical laser oscillator at 80 MHz has an average power of 1 W and a pulse energy of $E = 1.25$ nJ per pulse. In contrast, an amplified Ti:sapphire laser operating at 1 kHz with 3.5 W of average power has $E = 3.5$ mJ per pulse, which is an amplification of five orders of magnitude. The robustness of the Ti:sapphire gain medium and the extensive development of this medium over the past two decades has resulted in the wide range of choices in pulse energy.

The two-temperature model is commonly employed to simulate the interaction of a pulsed laser with a sample, allowing one to estimate the resulting rise in sample temperature (Groeneveld et al., 1995). The model used here is one of two interacting systems. The first subsystem is the ensemble of electrons in the material. The electrons are (internally) in equilibrium and characterized by a temperature T_E. Since the external laser field usually interacts solely with this system, initially all energy deposited is placed in this subsystem. The second subsystem is the phonons (lattice) of the material, which has also established a thermal equilibrium temperature T_L that can differ from the electronic subsystem. In the limit of small temperature changes to the electron subsystem, the transfer of energy from electrons to phonons is a linear function of the temperature difference $H(T_E, T_L) \approx g_\infty(T_E - T_L)$, where g_∞ is the electron–phonon coupling constant. The two-temperature model is described by the following coupled differential equations:

$$C_E \frac{\partial T_E}{\partial t} = -g_\infty(T_E - T_L) + P(t)$$

$$(9.1)$$

$$C_L \frac{\partial T_L}{\partial t} = +g_\infty(T_E - T_L)$$

where
$C_E(T)$ and $C_L(T)$ are the temperature-dependent specific heats of electrons and phonons
$P(t)$ is the time-dependent laser pumping

A numerical solution of these equations will estimate the temperature rise of the sample in response to photoexcitation. Further detail on this model is included in Chapter 8.

The design of the system must account for the temperature rise in both the electronic and phonon subsystems. If the goal is to determine the near-equilibrium properties of the material, then the pump fluence should be chosen to minimize $\Delta T/T$ or $\Delta R/R$ while still achieving an acceptable signal-to-noise ratio. Alternately, if the experiment is to study a photoinduced phase transition (Nasu, 2004), then the large temperature change induced by a high fluence laser system may be required. The wide diversity of Ti:sapphire laser systems permits these kind of design decisions.

9.3.4 Repetition Rate

The repetition rate of Ti:sapphire laser systems currently varies from as low as a few hertz to 2 GHz (Bartels et al., 1999), depending on the design of the system and the required performance. As was stated earlier, a higher repetition rate is desirable since it increases the number of experiments performed for each second of acquisition time and improves the statistics of the measurement. The choice of pulse energy, however, may conflict with the choice of repetition rate. Generally, the energy per pulse is higher with lower-repetition-rate systems, requiring the use of these systems when high pulse energy is desired. A low-fluence study would measure the near-equilibrium properties of a material (Cavalleri et al., 2004) while a high-fluence study could, instead, trigger a photoinduced transition to an alternate phase (Nasu, 2004; Hilton et al., 2006). As a result, repetition rate and pulse energy are design trade-offs in an ultrafast pump–probe spectrometer.

9.3.5 Temporal Resolution

The required pulse width is governed by two competing factors: temporal resolution and spectral resolution. The temporal resolution of a system needs to be better than the shortest temporal feature to accurately recover the various lifetimes present in a relaxation process. As an example, in this section we will assume that the measured relaxation can be modeled with a double exponential decay (time constants τ_1 and τ_2). We define τ_1 to be the shorter of the lifetimes for convenience ($\tau_1 < \tau_2$). This is a model similar to the thermalization and recombination dynamics of a semiconductor after above-band-gap photoexcitation with a femtosecond source (Tommasi et al., 1996; Ganikhanov et al., 1999). The thermalization lifetime (τ_1) is approximately 1 ps, while the recombination lifetime (τ_2) is typically hundreds of picoseconds or nanoseconds. We also assume that the excitation process is "instantaneous" (faster than the temporal resolution of this experiment) and neglect any finite rise time of the signal. The time domain signal, $g_0(t)$, is given by

$$g_0(t) = u(t)\left[A_1 \exp\left(\frac{-t}{\tau_1}\right) - A_2 \exp\left(\frac{-t}{\tau_2}\right) \right] \tag{9.2}$$

where
  $u(t)$ is a step function that has $u = 0$ for $t < 0$ and $u = 1$ for $t \geq 0$
  A_1 and A_2 are the relative amplitudes of the two temporal components

This signal could represent the differential transmission or reflection depending on the experimental configuration, as the time domain analysis proposed here is identical for either case. Alternate decay dynamics, such as a power law decay:

$$g_0(t) = u(t)[A_1 t^{-n} + A_2 t^{-m}] \tag{9.3}$$

would result from an analysis similar to that outlined here; the numerical form of Equation 9.10 (shown later) would, however, be different.

9.3.5.1 Pump Pulse Width

The laser source must be capable of resolving τ_1 (the shorter lifetime). The temporal profile of a typical mode-locked laser pulse is Gaussian and described by a pulse width, σ, and a central frequency, ν_0, which is near the peak Ti:sapphire pulse spectrum (800 nm = 375 THz) (Chapter 7):

$$p_{\text{pump}}(t) = p_{\text{probe}}(t) = B \exp\left(\frac{-t^2}{\sigma^2}\right)\sin(2\pi\nu_0 t) \tag{9.4}$$

This laser source will generate the pump (p_{pump}) and probe pulses (p_{probe}) in this example (degenerate pump–probe spectroscopy), as each would be derived from the same laser source. In a nondegenerate pump–probe experiment or in an optical pump, terahertz probe experiment, a portion of this laser would be frequency shifted using an OPA or OPO or other nonlinear optical device.

Equation 9.2 assumes that the rise time of the ideal signal (g_0) is "instantaneous." Since the pump pulse width σ is not instantaneous in practice, there will be a *finite* rise time to a differential transmission or reflection signal that is typically limited by the pump pulse width σ. Fundamentally, this represents the response to the system when the pump pulse energy is delivered to the sample over the pulse temporal profile $p(t)$. This delivery is not instantaneous; therefore, the actual material response is no longer instantaneous. In this case, the differential transmission or reflection signal measures a rise time that results from the experimental apparatus and not the material property being studied.

Mathematically, the actual material dynamics are a convolution integral of the model $g_0(t)$, given by Equation 9.2, and the pump pulse profile $p_{\text{pump}}(t)$, given by Equation 9.4:

$$g_1(t) = \int_{Y=-\infty}^{\infty} g_0(Y)p_{\text{pump}}(Y-t)dY \tag{9.5}$$

Figure 9.2 shows the finite rise time that results from the finite pump pulse width. Figure 9.2a shows the "instantaneous" response function (solid) and the pump pulse (dashed). Figure 9.2b again shows the pump pulse profile and the actual system dynamics due to the finite pulse width of the pump beam (Equation 9.5).

9.3.5.2 Probe Pulse

The finite width of the *probe pulse* also results in a modification of the detected signal from the pump-induced dynamics in the sample. Each individual pulse samples the change in reflectivity or transmission of the material over the probe pulse width. This corresponds to a second convolution integral of the induced differential reflection or transmission signal $g_1(t)$, and the probe pulse $p_{\text{probe}}(t)$:

$$g_2(t) = \int_{X=-\infty}^{\infty} g_1(X)p_{\text{probe}}(X-t)dX$$

$$= \int_{X=-\infty}^{\infty}\left[\int_{Y=-\infty}^{\infty} g_0(Y)p_{\text{pump}}(Y-X)dY\right]p_{\text{probe}}(X-t)dX$$

$$= \int_{Y=-\infty}^{\infty} g_0(Y)\left[\int_{X=-\infty}^{\infty} p_{\text{pump}}(Y-X)p_{\text{probe}}(X-t)dX\right]dY \tag{9.6}$$

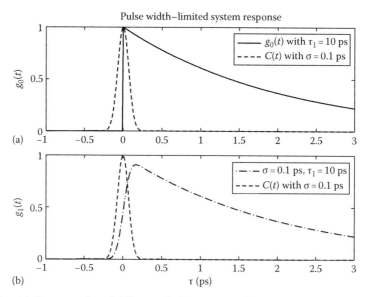

Figure 9.2 (a) The normalized differential change $g_0(\tau)$ for a δ-function input (i.e., σ is negligible). For a sample that has an instantaneous response (Equation 9.2), the actual rise time would be the cross correlation width. A longer rise time than the cross correlation width–limited rise time would indicate that the system response at the probed state is not instantaneous. An example of this can be found in Cavalleri et al. (2004), where the finite response time is attributed to the induced structural change in the sample after photoexcitation. (b) The dynamics of the induced system differential absorption or refractive index (dash-dot) when excited by a finite pump pulse of width σ. The system rise time is limited by the width of the pump pulse and not by a system response time.

The order of integration can be exchanged in Equation 9.6 to generate an autocorrelation integral $C(t)$, which is the temporal overlap of the pump pulse profile with the probe pulse profile:

$$C(\tau) = \int\limits_{Z=-\infty}^{\infty} p_{\text{pump}}(Z)p_{\text{probe}}(Z-\tau)dZ$$

$$= \int\limits_{Z=-\infty}^{\infty} B\exp\left(\frac{-Z^2}{\sigma^2}\right)B\exp\left(\frac{-(Z-\tau)^2}{\sigma^2}\right)dZ$$

$$= B^2\exp\left(\frac{-\tau^2}{2\sigma^2}\right) = C_0\exp\left(\frac{-\tau^2}{w^2}\right) \tag{9.7}$$

This autocorrelation can be independently measured using second harmonic generation in a thin nonlinear material (Sacks et al., 2001) or using two-photon absorption in a semiconductor diode (Ranka et al., 1997) (Chapter 7). For the Gaussian pulses described in Equation 9.4, the autocorrelation function is a Gaussian with a width $w = \sigma\sqrt{2}$, where σ is the temporal width of the pump and probe pulses.

Since the pump and probe autocorrelation function $C(\tau)$ can be independently measured, the actual probed signal can be written as Equation 9.8:

$$g_2(t) = \int_{Y=-\infty}^{\infty} g_0(Y)C(Y-t)dY \tag{9.8}$$

An instantaneous signal (Equation 9.2) excited and probed by Gaussian pulses would result in a detected signal $g_2(t)$, given by

$$g_2(t) = C_0 \int_{Y=0}^{\infty} \left[A_1 \exp\left(\frac{-Y}{\tau_1}\right) + A_2 \exp\left(\frac{-Y}{\tau_2}\right) \right] \exp\left(\frac{-(Y-t)^2}{w^2}\right) dY \tag{9.9}$$

This integral can be written using the error function integral erf(x), which is tabulated in many sources (Abramowitz and Stegun, 1965) and commonly available in many mathematical packages:

$$g_2(t) = \left\{ D_1 \exp\left(\frac{-t}{\tau_1}\right) \left[1 + \mathrm{erf}\left(\frac{w}{2\tau_1} - \frac{t}{w}\right) \right] \right\} + \left\{ D_2 \exp\left(\frac{-t}{\tau_2}\right) \left[1 + \mathrm{erf}\left(\frac{w}{2\tau_2} - \frac{t}{w}\right) \right] \right\} \tag{9.10}$$

The previous example assumes the presence of two lifetimes (τ_1 and τ_2); this can be extended to include multiple relaxation processes by introducing additional terms into Equation 9.2. This functional form can be fitted to the measured data to extract the relative amplitudes, $D_1 = C_0 A_1$ and $D_2 = C_0 A_2$, and time constants for each of the two lifetime components for a known cross correlation width $w = \sigma\sqrt{2}$.

9.3.5.3 Finite System Rise Time

Figure 9.2b shows a simulated pump–probe scan where the rise time in Equation 9.10 is determined by the cross correlation width only. A rise time that is significantly longer than w indicates that the system response to the pump pulse is not "instantaneous." A finite response time can be important in several cases (Cavalleri et al., 2004) and can be accounted for in the above analysis by multiplying Equation 9.2 by the expression $[1 - \exp(-t/\tau_R)]$:

$$g_0(t) = u(t)\left(1 - \exp\left(\frac{-t}{\tau_R}\right)\right) \left[A_1 \exp\left(\frac{-t}{\tau_1}\right) + A_2 \exp\left(\frac{-t}{\tau_2}\right) \right] \tag{9.11}$$

This finite response time suggests that the probed state is not populated "instantaneously" but requires a finite lifetime either due to an indirect excitation mechanism or other excitation bottleneck.

Figure 9.3 shows the effect of the finite rise time in Equation 9.11 on the final differential transmission or reflection signal. Figure 9.3a shows the differential transmission assuming a pump and probe pulse with a cross correlation width $w = 0.144\,\mathrm{ps}$, with a recovery lifetime $\tau_1 = 10\,\mathrm{ps}$, as described by Equation 9.2. The product of these two components (finite rise time and recovery) is shown on Figure 9.3b as a solid line, while the autocorrelation $C(\tau)$ is plotted in a dashed line. This shows that the rise time τ_R delays the formation of the peak differential signal and reduces its peak magnitude compared to the "instantaneous" response function magnitude.

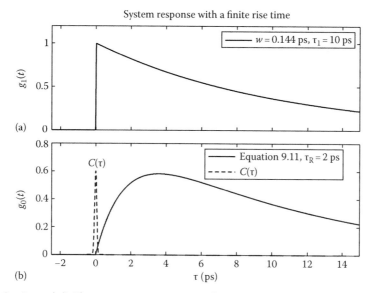

Figure 9.3 (a, solid) The instantaneous material response assuming a pump and probe pulse described by a cross correlation width $w = 0.144$ ps (Equation 9.2). (b, solid) The system response when a finite material response delays the formation of the excited state (Equation 9.11). (b, dashed) The pump–probe pulse autocorrelation trace with $w = 0.144$ ps ($\sigma = 0.1$ ps). The rise time observed in these (simulated) data is significantly longer than the pump–probe autocorrelation width, which indicates the existence of a nonnegligible rise time τ_R in the materials response.

9.3.6 System Temporal Resolution

9.3.6.1 Temporal Resolution (Pulse Width-Limited)

It is clear from Equation 9.10 that lifetimes τ_i substantially shorter than the cross correlation width $w = \sigma\sqrt{2}$ cannot be resolved by a pump–probe experiment. The cross correlation width limits the temporal resolution of the experiment. For $\tau_1 \approx w$, τ_1 may or may not be resolvable depending on the system's signal-to-noise ratio (Burr and Tang, 1999; Hilton and Tang, 2002).

9.3.6.2 Temporal Resolution (Delay Stage-Limited)

The temporal resolution can also be limited by the mechanical hardware used for the delay generation. A mechanical stage with N passes of the laser beam in a medium of refractive index n that has a minimum step size δd results in a temporal resolution limit given by

$$\delta t_{\text{stage}} = N \times \delta d \times \frac{c}{n} \tag{9.12}$$

For stages operating in air or vacuum, $n = 1$, while c is the speed of light in vacuum. A stage that has a laser beam pass down to the moving platform and back (as shown in Figure 9.1) is a "double-pass" stage with $N = 2$. Additional round trip passes ($N = 4, 6, 8, \ldots$) will decrease the temporal resolution while increasing the temporal window size. The actual temporal resolution is the maximum of the cross correlation width, w, and the resolution limit imposed by the stage, δt_{stage}. The stage resolution is, however, typically not the limit in many experiments as commercially available stages have sub-femtosecond resolution.

9.3.6.3 Temporal Window

The resolution of the longer lifetime τ_2 is limited by the length of the delay stage d. This must be longer than the physical distance traveled by light in the medium of refractive index, n, in the lifetime τ_2:

$$d = \frac{n\tau_2}{Nc} \tag{9.13}$$

Adding additional round-trip passes ($N = 4, 6, 8,...$) to a double-pass stage can increase the temporal window of the measurement without requiring new hardware, at the potential expense of the temporal resolution (δt_{stage}) as discussed earlier.

9.3.6.4 Spectral Resolution

The above analysis shows that the system pulse width should be minimized for the best temporal resolution. An ultrashort pulse necessarily has a broad spectrum due to the FT relationship between the temporal and spectral profiles. The spectrum of the Gaussian pulse $p_{\text{pump}}(t)$ described in Equation 9.4 is given by $P(\nu)$:

$$P(\nu) = A \exp\left[-\pi^2\sigma^2(\nu - \nu_0)^2\right] \tag{9.14}$$

obtained by using the mathematical definition of the FT from Goodman (1996) to calculate $P(\nu)$ from Equation 9.4. The spectral width (full width at half maximum) of this pulse is therefore inversely proportional to the pulse temporal width, $\Delta\nu = (\sigma\pi\sqrt{\ln 2})^{-1} \approx 0.38\sigma^{-1}$. Thus, a shorter transform-limited optical pulse has more frequency components. This potentially results in more available optical transitions that are energetically allowed. The broad spectrum of an ultrashort pulse may complicate some experiments, such as photo-exciting one of two closely spaced spectral features that are each narrower than the spectrum of the pulse itself (Walmsley et al., 1988). The pulse bandwidth $\Delta\nu$ can be spectrally narrowed to match this feature spacing, but the system will then have a longer pulse width σ.

9.3.7 Dispersion Compensation

The four-prism sequence in Figure 9.1 is required in femtosecond systems to minimize the pulse width at the sample position and achieve the minimum time resolution. A femtosecond laser system necessarily consists of a wide bandwidth that is inversely proportional to the temporal width (i.e., $\sigma = 100\,\text{fs}$ has a bandwidth of $\Delta\lambda \sim 10\,\text{nm}$). Propagation of this pulse through a dispersive media $n(\nu)$ necessarily leads to a broadening of a transform-limited pulse width, σ_0, compared to the input pulse. The *second-order dispersion* of a material (β_2) is the second derivative of n with respect to frequency ν. After propagation through a dispersive media of physical length z, either the pump or the probe pulse width will be $\sigma = \sigma_0\left(1 + z^2/L_D^2\right)^{1/2}$, where $L_d = \sigma_0^2 \times |\beta_2|^{-1}$ is the dispersion length (Agrawal, 2001). The actual pulse width, $\sigma(z)$, at a distance z within a dispersive media is necessarily longer than the transform limited pulse width σ_0 in a dispersive medium regardless of the sign of β_2. Achieving the minimum temporal resolution therefore requires that the experiment design use reflective or low-dispersion optical elements (minimize β_2 to maximize L_D), minimize the thickness of these dispersive elements (minimize z), or compensate any normal dispersion with optical elements with anomalous dispersion of equal magnitude (minimize β_2).

The four equilateral–prism sequence (or its two-prism equivalent [Chapter 7]) shown in Figure 9.1 was first proposed by Fork et al., as a method of generating lossless and anomalous

($\beta_2 < 0$) dispersion to overcome the normal dispersion ($\beta_2 > 0$) of the optical elements (Fork et al., 1984). The effective dispersion, D, of this four-prism sequence results from the frequency-dependent optical path and is given by

$$D = \frac{\lambda}{cd} \frac{d^2 P}{d\lambda} \tag{9.15}$$

where
　c is the speed of light
　d is the separation of the first and second prisms
　λ is the central wavelength of the pulse
　P is the total path length as a function of wavelength (Fork et al., 1984)

The correct anomalous (negative) dispersion can be introduced into the cavity to balance the normal (positive) dispersion of the optical components by choosing the separation of the prisms in this sequence, d. The resulting total system dispersion is the sum of the positive dispersion from the system optics and the negative dispersion from the prism. The total accumulated dispersion *at the sample position* is therefore zero, which results in a dispersion length, L_D, that is infinite and a minimum pulse width $\sigma(z) = \sigma_0$. Further details on prism compression and other dispersion compensation methods can be found in Chapter 7.

9.3.8 Detection

The choice of detector is primarily determined by the following three criteria:

Response time: The temporal resolution of the system results from the path length difference between the pump and the probe pulses and not from the response time of the detector. The required detector response time is, instead, given by the chopper modulation frequency of the pump and/or probe beams. For example, a chopper frequency of $f_c = 500\,\text{Hz}$ would require a detector response time of $\tau_{\text{det}} = 2\,\text{ms}$, which is easily achievable with most semiconductor photodiodes, avalanche photodiodes, or photomultiplier (PMT) detectors.

Wavelength: The detector must have a response at the *probe* wavelength to generate a measurable signal. In semiconductor and avalanche photodiodes, the probe photon energy must be larger than the band gap of the semiconductor material. Long wavelength detectors are frequently cooled to minimize the detection of thermal blackbody radiation that would otherwise be a large source of noise.

Noise equivalent power (NEP): The NEP is the minimum optical power needed to produce a detector signal with a signal-to-noise ratio of 1 (Boyd, 1983). This is the smallest signal that can be recovered from the system and is an important figure of merit to minimize when choosing a detector (or pair of detectors in a balance configuration).

9.3.8.1 Lock-in Detection

The ultrafast time resolution of a pump–probe spectrometer results from repeated measurements of the change to the probe pulse transmission or reflectivity over a range of pump–probe delays, τ. A lock-in detection scheme with a low noise detector is a common combination to reduce system noise (Groeneveld et al., 1995), although boxcar averaging is also an alternative appropriate for low-repetition-rate laser systems to increase counting statistics (this alternative is discussed in a later section). In a lock-in-based pump–probe experiment, the power of the pump beam is modulated "on" and "off" at a known frequency f_c ($f_c = 0.500\,\text{kHz}$ in the example shown on the top of Figure 9.4, where "on" = 1 and "off" = 0). If the pump laser has a repetition rate f_{laser}, and is chopped with a *synchronized* chopper (e.g., $f_c = f_{\text{laser}}/2$), then "on" will represent

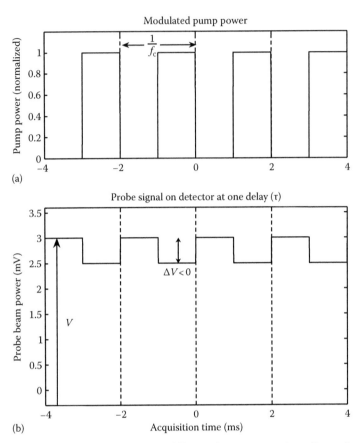

Figure 9.4 A simulation of a time-resolved differential transmission (or reflection) experiment at one value of the pump–probe delay t using lock-in detection. (a) The pump beam (solid) is modulated "on" and "off" at the chopper frequency f_c, which is 500 Hz in this example. (b) The modulated probe signal at f_c. The full detector signal in this example is $V = 3$ mV absent the pump beam. In this example, the pump power decreases the sample transmission (reflection) as shown in the solid curve. The phase of the lock-in is established by using the pump beam to determine the correct sign of the differential change. The $\Delta T/T(\tau)$ or $\Delta R/R(\tau)$ is the ratio of $\Delta V/V$ at this value of the pump–probe delay τ. The full ultrafast dynamics result from repeating this measurement for multiple values of τ.

one pulse transmitted by the chopper and "off" will be one pulse blocked by the chopper. The chopper is not synchronized to the laser repetition rate when using high-repetition-rate lasers, in which suitable modulators and lock-in amplifiers at those frequencies do not exist. In this case, "on" will transmit many optical pulses and "off" will block many optical pulses.

In both cases, the probe beam is not chopped.* Because of the modulation of the pump beam, there will be twice as many probe pulses as pump pulses for a 50:50 chopper duty cycle (50% "on" and 50% "off"). One-half of the probe pulses will be incident on the sample without a pump pulse to perturb their optical properties and the other half will be incident on the photoexcited sample. The modulated probe signal at the chopper frequency f_c will therefore be proportional to the *change* to the material properties due to the photoexcitation; this is shown on the bottom of Figure 9.4.

---

* Both the pump and the probe can be modulated at different frequencies f_{c_1} and f_{c_2} to remove the effects of background scattered pump light if the polarization or wavelength filter technique is insufficient. The lock-in reference should then be the sum of these two frequencies $f_c = f_{c_1} + f_{c_2}$.

To correctly measure the sign of the differential transmission or reflection, the phase of the lock-in should be set by temporarily redirecting a portion of the pump beam into the detector while blocking the probe beam and maximizing the in-phase component using the lock-in phase adjustment. Once this is done, a positive lock-in voltage ΔV corresponds to an increase in the sample transmission (reflection) and a negative signal to a decrease in transmission (reflection). The lock-in time constant $\tau_{\text{lock-in}}$ should be 2× to 3× the acquisition time to allow the lock-in sufficient time to stabilize; a $\tau_{\text{lock-in}} = 300\,\text{ms}$ time constant, for example, would permit an ~1 s acquisition time per point. The change in voltage *at the chopper frequency* is $\Delta V = -0.5\,\text{mV}$ in the previous example. To determine the *differential* change of the sample property, the measured ΔV should be divided by the full voltage of the detector absent the pump beam V; this is $V = 3\,\text{mV}$ in the previous example. The ratio ($\Delta V/V = -16.7\%$ in this example) at this delay τ is either $\Delta T/T(\tau)$ or $\Delta R/R(\tau)$ depending on the experiment configuration; typical values are often much smaller than this example (i.e., $\Delta R/R \sim 10^{-4}$–10^{-6}) (Groeneveld et al., 1995). The full ultrafast temporal dynamics, $\Delta T/T(\tau)$ or $\Delta R/R(\tau)$, result from repeating this measurement for multiple values of the pump–probe delay τ.

9.3.8.2 Boxcar Integrators

Boxcar integrators are an alternative to lock-in detection, used when the repetition rate of the laser is low ($f_{\text{laser}} \ll 1\,\text{kHz}$). A lock-in-based detection scheme inherently reduces the counting statistics of the detection since the pump pulse is modulated at the chopper frequency to modulate the sample reflectivity. The bandwidth filter that the lock-in time constant provides enhances the system's signal-to-noise ratio by rejecting all signals within one inverse time constant ($\Delta f \approx \pm 1/\tau_{\text{lock-in}}$) of the chopper frequency to reduce the overall noise. The overall system's signal-to-noise ratio is proportional to the square root of the number of pulses measured, signal to noise ratio $\sim \sqrt{N}$, which should, ideally, be maximized to improve system performance. The gain in signal-to-noise ratio from the lock-in filter Δf is at least partially offset by the reduction in the counting statistics N. When the repetition rate of the laser is sufficiently low, the signal-to-noise ratio no longer benefits from the lock-in detection scheme. This is a particularly important issue in very high–pulse energy experiments, as high–pulse energy laser sources frequently employ low-repetition-rate pump lasers (currently $E \geq 10\,\text{mJ}$ per pulse with $f < 1\,\text{kHz}$).

A boxcar integrator also compares the detector signal when the sample is excited by a pump pulse to that of the unexcited sample. The pump power is modulated on and off using either a shutter or chopper. The lock-in amplifier returns at most one data point per lock-in time constant; in contrast, the signal is recorded over the entire "on" window of the boxcar to increase the number of data points acquired N. Since there is no frequency filter Δf here, the measured boxcar signal contains additional sources of noise that would have otherwise been rejected by the lock-in. Nevertheless, a boxcar integrator's improved counting statistics can overcome the broader range of noise sources in certain cases.

9.4 RELAXATION DYNAMICS

9.4.1 Semiconductors

In semiconductors, there exists an energy gap E_g at the Fermi surface, which limits the maximum wavelength $\lambda \leq hc/E_g$, where h is Planck's constant and c is the speed of light, that can be absorbed by a band-to-band transition. A general picture of the band structure of a group IV or III–V semiconductor in the diamond or zinc blende band structure is shown in Figure 9.5. Wavelengths shorter than the cutoff wavelength have photon energies sufficient to promote an electron from the valence subbands into the conduction band. Common bulk group IV and III–V semiconductors, including silicon and gallium arsenide, have degenerate valence

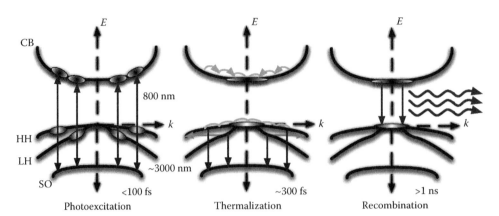

Figure 9.5   (See color insert.) Optical pump, mid-infrared probe measurements on a III–V semiconductor (left). The mid-infrared probe pulse is used to isolate the hole subband dynamics from the conduction band dynamics. Initial photoexcitation is determined by the photon energies present in the pump pulse (red). Mid-infrared probe pulses (blue) promote holes from the upper valance subbands to the split-off subband. The wavelength of this probe pulse determines the region in *k*-space probed in this experiment. (middle) Energy thermalization is the result of inter- and intrasubband carrier–carrier scattering (yellow) and phonon scattering between subbands (green). (right) The system returns to equilibrium using electron–hole recombination, which results in the emission of a photon (red), carrier trapping (not shown), or thermal diffusion to the substrate (not shown). CB, conduction band; HH, heavy-hole subband; LH, light-hole subband; and SO, split-off-hole subband. (From Hilton, D.J. and Tang, C.L., *Phys. Rev. Lett,* 89, 146601, 2002; Scholz, R., *J. Appl. Phys,* 77, 3219, 1995; Braunstein, R. and Kane, E., *J. Phys. Chem. Solids,* 23, 1423, 1962; Braunstein, R. and Magid, L., *Phys. Rev,* 111, 480, 1958.)

subbands near the zone center (known as heavy and light-hole subbands), so optical excitation to the conduction subband creates populations of holes in both the heavy and light-hole subbands. Shorter pump wavelengths can also excite electrons to the conduction subband from lower-energy subbands, like the split-off subband ($\lambda \leq hc/(E_g + \Delta_0)$). This will result in a population of holes in the split-off subband along with the heavy and light-hole subbands (Burr and Tang, 1999).

The relaxation of a semiconductor after photoexcitation follows the general pathway shown in Figure 9.5 (further detail can also be found in Chapter 8). Initially, the electron distribution in the conduction (upper) band and the hole distribution in the valence (lower) subbands are strongly in nonequilibrium and are not described by Fermi–Dirac or Boltzmann statistics or by a thermodynamic temperature. We assume here that the excitation pulse is "short" compared to any of the relaxation lifetimes (electron–electron, electron–hole, and electron–phonon) within the band structure. After excitation, carrier–carrier (electron–electron, electron–hole, and hole–hole) scattering establishes a Fermi–Dirac/Boltzmann distribution within each subband, described by different temperatures (T_{CB}, T_{HH}, T_{LH}, and T_{SO}). Further redistribution of carriers between subbands occurs, including phonon-assisted SO-to-LH scattering (if applicable) and LO phonon–assisted LH-to-HH scattering (Scholz, 1995). Typical lifetimes for electron and hole thermalization are approximately 0.5–1 ps in semiconductors (Tommasi et al., 1996; Ganikhanov et al., 1999). Electron–hole recombination, which results in the emission of photons at the transition energy, occurs on a nanosecond timescale in high-quality semiconductors (Rossi and Kuhn, 2002). Excess energy deposited by the laser then diffuses to the substrate and surrounding environment with a microsecond to second lifetime, depending on the experimental configuration.

Free carrier absorption is also a possible absorption mechanism in semiconductors (Blakemore, 1981) (Chapter 2). This is particularly significant in degenerately doped semiconductors, where the Fermi level crosses one of the energy subbands, or when using a long-wavelength probe beam. In the latter case, the optical pump pulse is still above the band gap and generates electron–hole pairs in the conduction and valance subbands. If the long-wavelength probe pulse is insufficient to excite a band-to-band transition, then intraband free carrier absorption is the available absorption mechanism. In this case, similar relaxation dynamics governed by carrier–carrier and electron–phonon interactions will thermalize carriers within the affected subband, inter-subband scattering and electron–phonon interactions will reestablish equilibrium within the sample, and diffusion to the substrate will thermalize the entire system.

9.4.2 Metals

In metals, no energy gap exists at the Fermi level and thus there is no minimum wavelength for absorption. Free carrier absorption is therefore the dominant absorption mechanism in metals at long wavelengths (Groeneveld et al., 1995). The following general picture describes the relaxation pathways taken by photo-excited carriers as they return to equilibrium in a metallic (gapless) system (Chapter 8). The initial excitation pulse generates a nonequilibrium (i.e., non-Fermi–Dirac) distribution of electrons at the Fermi surface, which is, by definition, within one of the energy bands. Electron–electron scattering establishes a new Fermi–Dirac distribution on a sub-picosecond timescale. This new distribution is described by an electron temperature T_E, which differs from the lattice temperature T_L and the initial system temperature. Electron–phonon scattering establishes an equilibrium between lattice and phonons $(T_E = T_L)$, at a temperature that is elevated compared to the initial temperature, on a timescale of a few to hundreds of picoseconds, depending on the material. The final relaxation phase has the excess laser-deposited energy diffuse into the substrate and surrounding environment over a timescale of milliseconds to seconds, depending on the experimental configuration.

In metals, band-to-band transitions, as described earlier in semiconductors, are still possible if the photon energy is sufficient to promote an electron into a higher-order subband (Groeneveld et al., 1995). These dynamics would be similar to the band-to-band pump–probe experiment described in the previous section.

9.5 DIFFERENTIAL REFLECTIVITY AND TRANSMISSION

The magnitude of the differential transmission or differential reflection at any time t depends on the initial and final wave functions of the probed transition, the joint density of states, and the electron occupancy of the initial, f_i, and final, f_f, states. The probed transition quasimomentum k is determined by the probe photon energy using

$$E_f(k) - E_i(k) = \hbar\omega_{probe} \tag{9.16}$$

where

    $E_i(k)$ is the energy dispersion relation (conduction, heavy hole, light hole, or split-off hole for group III–V or IV semiconductors) function for the initial subband

    $E_f(k)$ is the corresponding energy dispersion relation for the final state

    $\hbar\omega_{probe}$ is the probe photon energy

This probe photon energy, therefore, determines the region of the Brillouin zone that the differential reflection or differential transmission measures, while the pump photon energy determines the region of the band structure initially perturbed from equilibrium. The magnitude and sign of the differential reflection and transmission depends on three main factors: the wave function parity, the joint density of states, and the occupancy functions of the probed states.

9.5.1 Wave Function Parity

The required initial and final wave function parity is restricted by the conservation of angular momentum of the initial and final quantum mechanical states after absorption of the optical photon. Group III–V (i.e., GaAs, InAs, GaN) or group IV (i.e., Si, Ge) semiconductors primarily have wave functions near the Brillouin zone center ($k \simeq 0$) with $|P\rangle$ like symmetry in the valence subbands and $|S\rangle$ -like symmetry in the conduction band (Taylor et al., 1985). Materials with transition metals as atomic components would potentially need to employ $|D\rangle$-like wave functions (Chapter 3), while heavy fermion systems have partially occupied $4f$ and $5f$ electronic states at the Fermi level (Sarrao, 2002; Demsar et al., 2003). In chemical systems, appropriate singlet and triplet wave functions would be needed to describe the initial and final states (Lakowicz, 1999).

This section assumes the sample under study is a semiconductor with initial and final states of $|S\rangle$ and $|P\rangle$ symmetry, respectively (i.e., Group IV and Group III–V semiconductors). We will use the electric dipole approximation $\hat{\vec{D}} = e\hat{\vec{R}}$ to model the interaction of matter with the electric field of the probe beam in the semiclassical limit, with e the electron charge and $\hat{\vec{R}}$ the position operator vector. The differential transmission is proportional to the dipole matrix element M_{if} from the initial to the final state, which is given by

$$M_{if} = \left\langle u_i \left| \hat{D}_e \right| u_f \right\rangle \tag{9.17}$$

where $|u_i\rangle$ and $|u_f\rangle$ are the initial and final wave function states, respectively. Only components of $\hat{\vec{D}}$ in the direction of polarization of the electric field ($\vec{E} = E_0\hat{e}$) contribute to this matrix element ($\hat{\vec{D}} \cdot \hat{e} = \hat{D}_e$).

If we assume a linearly polarized probe pulse ($\hat{e} = \hat{x}$), the dipole vector operator becomes $\hat{D}_e = e\hat{X}$. This has a nonzero value only for transitions with $\Delta\ell = \pm1$ (i.e., $|S\rangle \to |P\rangle$ and $|P\rangle \to |S\rangle$ for semiconductors like GaAs), and forbids transitions with $\Delta\ell \neq \pm1$ (i.e., $|S\rangle \to |S\rangle$, $|P\rangle \to |P\rangle$, $|S\rangle \to |D\rangle$, etc.). Higher-order interactions (quadruple, magnetic dipole, etc.) relax some of these parity requirements but are generally weaker interactions than the dipole interaction (Dui et al., 1978).

The wave function $\psi_i(\mathbf{r})$ of electrons and holes in a periodic lattice can be factored into an atomic-like component $u_i(\mathbf{r})$ and a periodic component of the form in Equation 9.18:

$$\psi_i(\vec{r}) = u_i(\vec{r})\exp(i\vec{k} \cdot \vec{r}) \tag{9.18}$$

The potential $V(\mathbf{r})$ has the symmetry of the space group of the crystalline lattice (i.e., $V(\mathbf{r}) = V(\mathbf{r} + \mathbf{r}_0)$ where $\mathbf{r}_0$ is a lattice translation vector). Substituting Equation 9.18 into the time-independent Schrödinger equation results in an equation for the atomic-like u_i:

$$(H_0 + H_1)u_i = E_i u_i \tag{9.19}$$

where E_i is the eigen energy $H_0 = p^2/2m + V(r)$ and $H_1(\vec{k}) = (h\vec{k} \cdot \vec{p})/m + (h^2 k^2)/2m$. The H_0 component of the Hamiltonian can be exactly solved at $k = 0$ to find wave functions (the spin-degenerate $|S\rangle$, $|P\rangle$, $|P_u\rangle$, and $|P_z\rangle$ wave functions), while H_1 can be solved using perturbation theory in the $|S\rangle$ and $|P\rangle$ basis. Because of the $\vec{k} \cdot \vec{p}$ interaction term, the atomic-like component of the electron or hole wave functions away from the Brillouin zone center (i.e., away from $k = 0$) are an admixture of the $|S\rangle$ and $|P\rangle$ states (Luttinger and Kohn, 1955; Braunstein and Magid, 1958; Braunstein and Kane, 1962). The full wave functions resulting from this calculation for a group III–V or group IV material can be found in Taylor et al. (1985).

The main consequence of this state mixing is that inter-valence subband transitions that will be dipole forbidden at $k = 0$ (i.e., $\Delta \ell = 0$ for $|P\rangle \rightarrow |P\rangle$) will be permitted for finite k and, thus, the relevant dipole matrix element, M_{if}, is from this initial mixed state to a final mixed state and will be nonzero. Optical absorption that directly results from this mixing of $|S\rangle$ and $|P\rangle$ states has been previously observed in p-type gallium arsenide in the middle infrared (mid-IR) range (Braunstein and Magid, 1958). The admixture of states in the valence subband permits direct transitions from the split-off subband to either the heavy or light-hole subbands with photons in the mid-IR range ($\lambda \sim 3.6\,\mu m$ in gallium arsenide with a spin–orbit splitting of $\Delta_0 = 0.341\,eV$ (Blakemore, 1981)) or in the far-IR range between the light- and heavy-hole subbands (Braunstein and Magid, 1958). Additionally, these direct transitions from the split-off subband to the light and heavy-hole subbands have been used to isolate the hole dynamics from the electron dynamics in ultrafast pump–probe experiments (Tommasi et al., 1996; Burr and Tang, 1999; Ganikhanov et al., 1999; Hilton and Tang, 2002).

As will be discussed further in Chapter 13, the use of circularly polarized pump and probe fields ($\hat{\sigma}_\pm = (\hat{x} \pm i\hat{y})/\sqrt{2}$) permits the isolation and study of one specific angular momentum transition (either $\Delta m_\ell = +1$ for $\hat{\sigma}_-$ or $\Delta m_\ell = -1$ for $\hat{\sigma}_+$). In materials with significant spin–orbit coupling, this all-optical technique can study the interaction of orbital and spin angular momentum in the magnetic properties of materials using time-resolved magneto-optical Kerr or Faraday rotation.

9.5.2 Joint Density of States

Ultrashort optical pulses necessarily have a finite bandwidth that results in excitation from a finite range of initial states near the point of excitation in k-space into a finite range of states in the final subband, also near k. The joint density of states is the density of available transitions from this initial range of states to the corresponding range of final subband states and is given by

$$\rho_J(\hbar\omega_{probe}) = \frac{k^2}{\hbar^2} \frac{1}{|\partial E_i /\partial k - \partial E_f /\partial k|} \tag{9.20}$$

where the energies of the initial and final subbands of the probed transition are given by $E_i(k)$ and $E_f(k)$, respectively (Zory, 1993). The presence of nearly parallel subbands results in a large joint density of states (a van Hove singularity) and a proportionate increase in the differential reflection or transmission (Hilton and Tang, 2002).

9.5.3 State Occupancy

The Pauli exclusion principle forbids multiple occupancy of a quantum state by spin-½ particles. A band-to-band optical transition is permitted when the initial state has an electron (with an occupancy, $f_i = 1$) that can be promoted into an upper state that does not (with an occupancy, $f_f = 0$). In contrast, if the upper state is already populated ($f_f = 1$), no transition is permitted and the absorption of the material decreases. Similarly, if the initial state is empty ($f_i = 0$), the probe beam photon is not absorbed. The differential transmission is therefore proportional to the product $f_i(t) \times [1 - f_f(t)]$ of these occupancies that represents the required electron configuration for optical absorption of the probe energy. Partial occupancy of states, $0 \leq f \leq 1$, reflects the probability that a particular state is occupied at a given instant within the ensemble of carriers in the sample.

The optical pump pulse promotes electrons from the initial state to the final state, resulting in a vacancy in the initial subband as compared to the equilibrium occupancy. Additionally, this will increase the occupancy of the final state compared to the equilibrium value. In both cases, the primary effect of this is to permit (forbid) optical transitions that would have been otherwise forbidden (permitted) near equilibrium depending on the probe wavelength chosen. This resulting change to the state occupancies $f_i(t)$ and $f_f(t)$ would appear as a transient absorption (transient bleaching) of the sample at this photon energy that can be measured in a pump–probe configuration, either by examining changes to the material transmission or reflection.

The relaxation dynamics can reveal the internal dynamics of the full band structure examined by the probe pulse through these differential transmission and reflection measurements. The initial carrier distribution resulting from direct excitation into an upper state will rapidly thermalize within the subband before electron–hole recombination (Chapter 8) (Groeneveld et al., 1995; Ganikhanov et al., 1999). This upper state can also be occupied through indirect excitation to a second state and subsequent carrier–carrier or carrier–phonon scattering into this state (Ridley, 2000). In either case, the state occupancies will be time-dependent functions whose timescale(s) are dependent on the various intra- and inter-subband relaxation rates. A full time- and wavelength-dependent pump–probe scan with sufficient temporal resolution would then reveal the intra- and inter-subband relaxation dynamics over a range of k in the Brillouin zone.

9.5.4 Differential Transmission

The differential transmission $\Delta T/T(\hbar\omega_{\text{probe}}, t)$ is dependent on the initial and final wave function parities (i.e., the dipole matrix element M_{if}), the joint density of states (ρ_J), the electron occupancy of the initial subband (f_i), and requires the absence of an electron in the final subband ($1 - f_f$). We assume here that the sample is antireflection coated for the probe wavelength to eliminate any Fresnel reflection due to the index contrast between the sample and the surrounding medium. The change to the absorption under these conditions is given by

$$\Delta\alpha(\hbar\omega_{\text{probe}}) = |M_{if}|^2 \times \rho_J(\hbar\omega_{\text{probe}}) \times \{f_i(t)[1 - f_f(t)]\} \tag{9.21}$$

The resulting differential transmission for a thin film of thickness d under these conditions is given by (Ganikhanov et al., 1999)

$$\frac{\Delta T}{T}(\hbar\omega_{\text{probe}}, t) \approx -\Delta\alpha \times d_{\text{sample}}$$

$$= -d_{\text{sample}} \times \left| \left\langle u_i \left| \hat{\vec{D}} \cdot \hat{e} \right| u_f \right\rangle \right|^2 \rho_J(\hbar\omega_{\text{probe}}) f_i(t)[1 - f_f(t)] \tag{9.22}$$

The negative sign indicates that an increase in absorption results in a *decrease* in transmission, as expected. This formula is valid for film thicknesses d_{sample}, which are much less than the e^{-1} absorption length for this material (i.e., in the limit of small changes to the absorption). In most cases, the time dependence of the differential transmission signal originates in the time-dependent occupancy of the initial $f_i(t)$ and final $f_f(t)$ subbands probed by $\hbar\omega_{probe}$.

9.5.5 Differential Reflection

Causality requires an integral transform relationship between the real, $n_{2r}(\nu)$, and imaginary, $n_{2i}(\nu)$, parts of the complex refractive index, $\tilde{n}_2(\nu) = n_{2r}(\nu) + in_{2i}(\nu)$, that is commonly called the Kramers–Kronig relation (Born and Wolf, 1999) (Chapter 1). Thus, an induced change in absorption, $\Delta\alpha = 2\pi\Delta n_{2i}/\lambda$, is *necessarily* associated with a differential refractive index Δn_{2r} over an extended wavelength range. Determination of the real part of the refractive index for all wavelengths would therefore permit the calculation of the imaginary part of the refractive index. In metallic or other high loss samples where the unperturbed absorption α is large ($\alpha \times d_{sample} \gg 1$), it may not be possible to achieve a measurable differential transmission. In this case, reflection measurements offer an alternate method of recovering the dielectric constants.

The following calculation assumes that the sample surface is well polished to minimize any scattering losses of the pump and probe beams. The electric field reflection coefficient r of the sample is determined by the input polarization state (if either field is not normally incident), the angle of incidence (θ_1) and the refractive index of the material ($\tilde{n}_2$) and its surrounding medium ($\tilde{n}_1$) by the familiar Fresnel reflection formula (Born and Wolf, 1999). If we assume that the probe beam is normally incident ($\theta_1 \approx 0°$), then the *field* reflection coefficient (for either $\hat{s}$ or $\hat{p}$ polarized light) is given by

$$r = \frac{\tilde{n}_2 - \tilde{n}_1}{\tilde{n}_1 + \tilde{n}_2} \tag{9.23}$$

The *intensity* reflection coefficient is typically measured in pump–probe experiments and is the square modulus of the field reflectivity $R = |r|^2$. The optical pump pulse perturbs the complex refractive index of the sample ($n_{2r} \rightarrow n_{2r} + \Delta n_{2r}$) and, thus, changes the sample reflectivity ($R \rightarrow R + \Delta R$). In the limit of a small change to the real refractive index ($|\Delta n_{2r}| \ll |n_{2r}|$), the differential reflectivity is given by (Ganikhanov et al., 1999)

$$\frac{\Delta R}{R}(\hbar\omega_{probe}, t) \cong \frac{4\Delta n_{2r}(\hbar\omega_{probe}, t)}{n_{2r}^2(\hbar\omega_{probe}) - 1} \tag{9.24}$$

In samples with weak or no absorption before the pump pulse perturbation, simultaneous measurements of the time-dependent differential reflection ($\Delta n_{2r} = \text{Re}[\Delta\tilde{n}_2]$) *and* the time-dependent differential transmission ($\Delta n_{2i} = \text{Im}[\Delta\tilde{n}_2] = \lambda\Delta\alpha(2\pi)^{-1}$) will recover the full time-dependent change to the complex index of refraction ($\Delta\tilde{n}_2 = \Delta n_{2r} + \Delta n_{2i}$) of the material. From this information, the time-dependent state occupancies $f_i(t)$ and $f_f(t)$ can be determined with a temporal resolution of approximately the pump–probe cross correlation width w.

9.5.6 Coherent Artifact

The above analysis neglects the *coherent* interaction between the pump and probe pulses within the sample (Walmsley et al., 1988). A full theoretical treatment of this artifact can be found in Yan et al. (1989) and Luo et al. (2009); the basic effect is summarized here. The interaction of the pump and the probe pulses results in the formation of a periodic modulation of the complex index of refraction and the formation of a transient diffraction grating.

This phenomenon is commonly referred to as the "coherent artifact" and appears in pump–probe scans within the timescale of the cross correlation of the pump and probe pulses (i.e., w). This transient diffraction grating diffracts a portion of the pump beam into the direction of the probe beam and results in an additional component in the differential reflection or transmission signal not described by the above formula for $\Delta T/T$ and $\Delta R/R$ (Palfrey and Heinz, 1985).

This coherent artifact is a significant complication when the goal of the experiment is to determine lifetimes that are similar to the cross correlation width w of the pump and probe pulses. The use of cross-polarized laser beams, as was proposed in the previous section, will reduce or eliminate this coherent artifact due to the polarization specificity of the interference process (Saeta et al., 1991; Groeneveld et al., 1995); an alternate strategy based on a variation of the pump and probe angles of incidence is discussed in Luo et al. (2009). Different pump and probe wavelengths would also eliminate this artifact if the detector chosen is insensitive to the pump photon energy diffracted into the probe beam direction or if a wavelength-selective filter is placed before the detector (Hilton and Tang, 2002). It is also possible to exploit this coherent artifact to determine the location of the zero delay point between the pump and probe pulses (Baumberg et al., 1994).

9.6 ALTERNATE PUMP–PROBE TECHNIQUES

A major limitation of traditional pump–probe experiments is the substantial acquisition time needed to recover the full temporal dynamics. This is a consequence of the sampling techniques used to measure the differential transmission or reflection of a sample over a short temporal window (i.e., the probe pulse width) and recover the full signal by scanning the pump–probe delay over the full duration of the signal. As a consequence, the *effective* bandwidth of this pump–probe technique (the speed at which data can be acquired) is much less that the actual resolution bandwidth (the minimum temporal feature that can be resolved). For example, a typical pump–probe scan takes 15–30 min to recover data over a window of several hundred picoseconds with femtosecond resolution (effective bandwidth = $(30\,\text{min})^{-1}$ = 0.56 mHz), whereas the system has an ultimate temporal resolution governed by the pulse width (*resolution-limited bandwidth* ~ $(100\,\text{fs})^{-1}$ = 10 THz for a 100 fs laser system). This is a bandwidth reduction of 16 orders of magnitude.

This section focuses on alternate experimental configurations that can improve upon the effective bandwidth of the acquisition system. The development of alternate sampling techniques would assist the development of a wider array of technological applications including communications systems capable of terahertz frequency communication bandwidths (Foster et al., 2008). These alternate techniques can be naturally divided into two groups: those with designs that implement a *faster modulation* of the pump–probe delay and those that implement true single-shot detection of the probe signal through *time-to-frequency mapping*.

9.6.1 Faster Path Length Modulation

The simplest modification to the experimental setup to increase the effective bandwidth from an ~1 mHz to 10–20 Hz (three orders of magnitude) replaces the lock-in amplifier, chopper, and computer-controlled delay stage in Figure 9.1 with an analog data acquisition board, function generator, and a delay-line shaker. This shaker is a retroreflector mirror whose path length changes sinusoidally (at a modulation frequency f_M) with a driving sinusoidal voltage. The shaker converts the femtosecond time domain signal into a temporally scaled signal that is now a few hundred milliseconds long (determined by the inverse shaker frequency). This is still a reconstructed signal sampled from the repeated pump–probe experiment with a variable delay between the pump and probe pulses. This method merely speeds up the acquisition time and allows the entire temporally scaled signal to be viewed on an oscillo-scope or acquired using a data acquisition board. This method is also commonly employed to measure the pulse width σ in an autocorrelator using either two-photon absorption (Ranka et al., 1997) or second harmonic generation (Diels et al., 1985).

In Figure 9.1, the pump and probe delay are held fixed and a single data point is taken from the lock-in amplifier corresponding to this value of the pump–probe pulse delay τ. The entire signal using that configuration is reconstructed from a series of such measurements at differ-ent pump–probe delays, but each individual measurement is an average power measurement of the probe beam at that value of delay and is not a time-dependent signal. In the shaker con-figuration, the femtosecond signal is now scaled in time (i.e., $m \times \tau = t$ where m is the scaling factor, τ is the femtosecond timescale, and t is the millisecond timescale) to a millisecond tim-escale signal with a period of the inverse shaker frequency. The time-dependent signal from the detector is recorded directly in the time domain using an analog data acquisition board or an oscilloscope. Calibration of the temporal scaling factor m is best accomplished experi-mentally by introducing a known delay τ_0 into either the pump or the probe beam path ($m \times (\tau + \tau_0) = t + t_0$) and measuring the resulting time shift of the millisecond signal $t_0 = m \times \tau_0$.

The introduction of the shaker increases the effective bandwidth of the pump–probe exper-iment, but requires changes to the electrical equipment. In Figure 9.1, the *pump* beam is chopped at a known frequency and a lock-in detector detects only this frequency in the *probe* beam signal, which results from the interaction between the pump and probe pulses in the sample. Fluctuations in system power or any other portion of this signal not at this chopper frequency are rejected by the lock-in bandwidth filter (the bandwidth is proportional to the inverse time constant) of the lock-in amplifier. Since the signal on the detector in a shaker-based system is a millisecond time domain signal, multiple frequencies are still present and all are needed to fully recover the signal. The bandwidth of a lock-in would need to be wide enough to cover the millisecond time domain signal, negating the advantage of the lock-in amplifier. As a result, the use of a shaker is incompatible with lock-in detection schemes.

The removal of the lock-in amplifier results in an experiment that is now sensitive to "slow" fluctuations in laser power that would otherwise have been rejected by the lock-in bandwidth filter. One method of overcoming slow fluctuations is the acquisition and averaging of the signal over many shaker periods. Unfortunately, this does so at the expense of some of the effective bandwidth gain achieved by employing the shaker. A second method would measure the average power of the laser source at the same time the femtosecond signal is acquired using a balanced detector, which will divide out the power fluctuations (in a linear approximation) and their effect on the pump–probe signal. A balanced detector is a pair of detectors that record both the reflected probe beam and a small portion of the pump beam power (using a beam splitter before the sam-ple). In a commercially available balanced detector, these two signals are mathematically divided by the associated detector circuit to produce a single signal that is the probe beam normalized by the fluctuations in the pump power. A balanced detector is also commonly added to the pump–probe configuration shown in Figure 9.1 to remove the effects of fluctuations from the experi-ment, although the lock-in amplifier already removes fluctuations faster than its time constant.

Using a path length shaker will not recover the maximum resolution-limited bandwidth (i.e., $(100\,\text{fs})^{-1} = 10\,\text{THz}$) available in a pump–probe apparatus due to the mechanical limita-tions of rapid physical path modulation (a few Hertz). Additionally, this design requires the use of a data acquisition board and cannot easily incorporate a lock-in amplifier, likely reduc-ing the overall signal-to-noise ratio of the experiment. A shaker-based pump–probe tech-nique can, however, be an advantage when there is limited availability of acquisition time (Crooker, 2002), a desire to monitor the full temporal signal during acquisition in quasi-real time (Chapter 11), or a need to monitor changes to the sample differential transmission or reflection on the timescale of a few seconds to minutes (Lim et al., 2006).

An additional limitation of this technique results from the limited physical length of path length modulation. The overall temporal window of the experiment design in Figure 9.1 is limited by the length of the computer-controlled translation stage, which can be as long as 1 m ($\tau_2 \approx 7$ ns for a double-pass stage) for commercially available stages. In contrast, shaker path length modulation is a few centimeters (1 cm → $\tau_2 \approx 66$ ps), which limits the maximum temporal window the system can recover. As the motion of the shaker is sinusoidal, the actual useful range over which dynamics can be resolved is further restricted to the motion of the stage when it is away from the end points of its motion and the displacement is a linear function of acquisition time.

9.6.2 Asynchronous Optical Sampling (ASOPS)

This technique employs two separate femtosecond mode-locked lasers with different repetition rates f_1 and f_2 respectively, to generate the pump and probe pulses (Bartels et al., 2006, 2007). The difference in repetition rates is the source of the optical delay in this case, as opposed to the use of a mechanical path modulation. If two pulses are perfectly in phase at $t = 0$, the Nth pair of emitted pulses have a relative delay given by $\Delta T = N \times (f_1 - f_2)^{-1}$, as shown in Figure 9.6. This repetition rate difference, $f_M = (f_1 - f_2)$, can be larger (kilohertz to megahertz) than is possible using path length modulation (a few Hertz) with a mechanical delay stage.

The alignment of a pair of mode-locked lasers at *asynchronous frequencies* is a significant optical and electrical experimental challenge (Bartels et al., 2006, 2007). The repetition rate of any mode-locked laser is determined by the round-trip time for light propagation in the laser cavity (Chapter 7). A laser operating at $f_1 = 80$ MHz, for example, would have a round-trip time of $\tau = 12.5$ ns and a full cavity round-trip length of approximately $L = 3.75$ m,

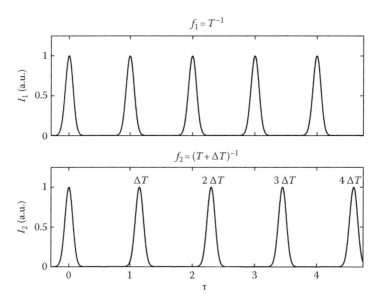

Figure 9.6 Pulse train from a pair of asynchronously mode-locked lasers. The higher-repetition-rate laser is shown on the top axis, while the lower axis shows the pulse train of a lower-repetition-rate laser. These lasers are temporally in phase at $\tau = 0$ but build up a relative delay ΔT between the pulse trains on each round trip of the higher-repetition-rate laser. This frequency difference f_M is then used as the delay generation mechanism in an ASOPS-based experiment. (From Bartels, A. et al., *Opt. Express*, 14, 430, 2006, Bartels, A. et al., *Rev. Sci. Instrum.*, 78, 035107, 2007.)

assuming a medium of index $n = 1$ inside a single-pass cavity. Small changes in this cavity length cause small changes to the repetition rate and can be due to mechanical instability in the laser mirror positions or changes to the refractive index of the laser gain medium due to either thermal fluctuations or laser wavelength drift. In a repetition-rate-stabilized laser, one cavity mirror is mounted on a piezoelectric transducer to generate variation in the round-trip length. The repetition rate of the laser is monitored, and a feedback circuit controls this mirror position to stabilize the cavity length to ensure that the repetition rate f_1 remains fixed throughout the pump–probe experiment.

A second cavity length-stabilized mode-locked laser generates the probe pulse. Since the desired repetition rate, $f_2 = f_1 + f_M$, is different from that of the pump laser, the cavity length of this laser is set to be a different length L_2 than the pump laser. The repetition rate of the pump laser, f_1, is then used to stabilize the (different) repetition rate of the probe laser, f_2. The electronic feedback circuit of this is more complex than the stabilization of a single laser and is outlined in detail in Bartels et al. (2007). The pump laser repetition rate f_1 is added to an external fixed reference frequency f_M to generate an output signal at the sum frequency $f_1 + f_M$. This sum frequency signal is then used to stabilize the repetition rate of the probe laser to $f_2 = f_1 + f_M$ using the piezoelectric transducer–mounted mirror. Figure 9.6 shows the buildup in relative delay, $N\Delta T$, between pulses in an asynchronously locked pump–probe experiment.

The pump–probe signal is now a temporally scaled signal at the difference frequency f_M as the relative path length between the two pulses changes by a fixed amount ΔT with each successive pulse generated. The repetition rate difference f_M can be stabilized at a much higher frequency than a mechanical shaker can operate. As a result, the effective bandwidth of this configuration can be significantly higher than either the delay stage or the shaker configuration, although this is still significantly lesser than the resolution bandwidth (10 THz) of the system. A secondary benefit of ASOPS is the elimination of the moving optical rail or shaker, which are sensitive to misalignment and result in a more mechanically robust system that improves the system's signal-to-noise ratio (Bartels et al., 2006, 2007).

Again, the ASOPS technique is not compatible with lock-in detection. The signal is a temporally scaled version of the femtosecond time domain signal and the lock-in time constant would need to be longer than this signal to avoid distorting the temporal signal. The advantages (rapid acquisition) and disadvantages (reduced signal-to-noise ratio) are qualitatively the same as in the shaker system. The distinction here is that the ASOPS technique can produce a faster effective variation in pump and probe pulse delay than a mechanical shaker can (i.e., a higher f_M). The final acquired signal can be the average of many different pump–probe scans as the effective bandwidth f_M results in $N = f_M^{-1}$ pump–probe experiments in each second; this can be used to improve the system's signal-to-noise ratio by improving the overall counting statistics. The scan length (τ_2 in Equation 9.13) is also potentially longer and limited only by the pulse repetition rate.

9.6.3 Electrically Controlled Optical Sampling

As in the ASOPS technique, the electrically controlled optical sampling (ECOPS) technique requires the use of two independent laser systems (Tauser et al., 2008). The difference here is that the two laser systems are first length stabilized to the *same* repetition rate ($f_1 = f_2$) and the relative timing between these two pulse trains is controlled with a piezoelectric transducer to produce an electrically controlled variable delay between these two pulses. This relative phase can be scanned over a wide temporal window and at a fast (effective bandwidth $\approx$ 100 Hz) rate. Again, the lack of a moving delay line (either a shaker or computer-controlled stage) to generate optical delay is a significant advantage, as this increases the reliability of the experimental apparatus.

9.6.4 Streak Camera

One traditional method of rapidly acquiring a picosecond time domain signal over a wide temporal window employs a streak camera. This is a sampling technique that does not record a true "single-shot" pump–probe scan, but records a signal that is reconstructed from multiple repeated experiments. The optical field is incident on a photocathode that emits electrons synchronized to the optical field dynamics through the photoelectric effect. These electrons then pass through a second time-dependent bias field $V_\perp(t)$, perpendicular to the propagation. Electrons arriving at $t = t_0$ are deflected spatially by the perpendicular bias field $V_\perp(t_0)$, while later electrons (arriving at $t = t_1$) are deflected at a different angle by the different perpendicular bias voltage existing at that time, $V_\perp(t_1)$. The exact time-to-space mapping is dependent on the speed of electrons in the vacuum tube (determined by the accelerating potential) and the time dependence of the perpendicular voltage $V_\perp(t)$ (usually linear). Since the sweep voltage rate is not synchronized to the electron beam dynamics, the resulting image is a reconstruction of the pump–probe dynamics from multiple repeated traces and not a true single-shot detection technique; alternate strategies for true single-shot detection are, however, possible (Murnane et al., 1990).

The spatially mapped signal is incident on a phosphor screen in which each pixel element corresponds to a small window δt of the original time domain signal. The minimum temporal resolution δt is limited by the time-to-space mapping function and the minimum feature size the camera can resolve, while the temporal window (τ_2 in Equation 9.13) is similarly limited by the camera and phosphor screen size.

9.6.5 Time-Frequency Mapping Techniques

The techniques discussed earlier improve upon the millihertz effective bandwidth detection of the traditional pump–probe spectroscopic techniques, but none approach the resolution-limited bandwidth of a typical femtosecond spectrometer. A broad range of communications applications would benefit from techniques to recover femtosecond dynamics in a single-shot measurement, which would dramatically expand the available bandwidth of communications systems (Foster et al., 2008). Additionally, the experimental capacity to measure femtosecond dynamics in a "single shot" would permit the measurement of the femtosecond dynamics of non-repeatable processes that damage the sample or otherwise irreversibly alter it. Each of the techniques outlined earlier are not capable of true *single pulse recovery*, as the gains in effective bandwidth result from the faster modulation of the pump–probe beam delay. Each recovered signal is reconstructed from a series of sampled experiments that are repeated multiple times. A true single-shot measurement, therefore, will require an alternate experimental method to convert the time domain signal into either a spatial domain signal or into the frequency domain (Dorrer et al., 1999; Oba et al., 1999; Azaña et al., 2004; Solli et al., 2007).

Among the more promising recent designs was the one proposed by Foster et al. (2008). This "ultrafast oscilloscope" demonstrated a single-shot resolution of 220 fs with a wide temporal window of 100 ps. The parametric "time-lens" described in this work temporally lengthens the pulse by using temporal phase shifting of a single pulse and not high-repetition-rate sampling; this makes this a true *single-shot* recovery technique. The four-wave mixing process employed for the time-lens is a third-order ($\chi^{(3)}$) process and does not require a centro-symmetric medium (Boyd, 1991). As a result, this ultrafast oscilloscope is constructed using standard complementary metal oxide semiconductor (CMOS) silicon fabrication techniques, commonly used in the manufacture of a broad range of semiconductor devices, and a 1550 nm mode-locked fiber laser. The use of CMOS is a significant advantage due to the widespread availability and sophistication of CMOS manufacturing techniques.

A conventional (spatial) lens generates a 2D FT of an image, $E_o(\xi, \eta)$, at the focal plane in the paraxial limit $E_i(x, y) = FT[E_o(\xi, \eta)]$, where (ξ, η) is the object coordinate system and (x, y) is the image coordinate system located at a distance f from the lens. It does this by adding a quadratic phase $\exp[i2\pi n(\xi^2 + \eta^2)/\lambda f]$ to the wave at each point in the object plane (ξ, η), where f is the focal length of the lens, n is the refractive index of the lens material, and λ is the wavelength. The main effect of the lens is to map each *spatial frequency* of the input field $E_o(\xi, \eta)$ onto a *spatial position* (x, y) in the focal plane of the lens to generate an image that is proportional to the 2D spatial FT of the object $(E_i(x, y) = FT[E_o(\xi, \eta)])$ (Hecht, 1987; Goodman, 1996).

The technique used by Foster et al. (2008) is the *time domain* analog to the *spatial domain* action of a lens described in the last paragraph. This technique maps the ultrafast time domain signal into the spectral domain using a *time lens* in much the same manner that a (conventional) lens maps the 2D spatial image onto its 2D spatial frequency signal in the focal plane. A complete treatment of temporal lensing can be found in Chapter 3 of Yariv (1997); a brief summary is presented here. The ultrafast signal is first propagated through a length L_A of optical fiber with a second-order dispersion β_{2A}, which provides a group delay dispersion $D_A = \beta_{2A} \times L_A$ that will temporally stretch the signal in time. Next, the temporal lens adds a quadratic phase to the time domain signal $\exp[i\omega_0 t^2/2f_T]$, where f_T is the temporal focal length of the lens. The temporal lens is generated using four-wave mixing (Chapter 10) in silicon waveguides designed to have a certain temporal focal length f_T, and is discussed in further detail by Salem et al. (2009). A second fiber with group delay dispersion $D_B = \beta_{2B} \times L_B$ is then used to generate the scaled spectrum that is proportional to the time domain waveform depending on the length of this second fiber, L_B, and its dispersion, β_{2B}. The magnification $M = D_B/D_A$ of the signal in time can be controlled by varying the relative dispersion of fiber A and fiber B to balance the minimum resolvable time step δt with the acquisition window τ_2. This spectrum can then be recorded using a spectrometer and CCD camera in a single shot or integrated in time to improve the system signal-to-noise ratio.

This method operates on a single ultrafast time domain signal, in contrast to the other methods described in this chapter that employ sampling to reconstruct the femtosecond dynamics from multiple pump and probe pulses. This kind of true single-shot detection scheme has numerous applications to conventional pump–probe experiments described in this chapter as well as to high-bandwidth telecommunications.

9.7 EXAMPLES

This section will highlight a few selected pump–probe experiments to demonstrate some of the more commonly occurring situations and materials physics to be recovered using this technique. Examples will include nondegenerate pump–probe spectroscopy in semiconductors, free carrier absorption in metals, the formation of coherent phonons, and the study of photoinduced insulator-to-metal phase transitions.

9.7.1 Time-Resolved Degenerate Differential Reflection Spectroscopy in Metals

Direct time domain observation of the differential reflectivity of a metal can isolate the processes of electron–electron scattering and the electron–phonon interaction (g_∞ in Equation 9.1) that occur on a sub-picosecond timescale (Chapter 8). The first experimental investigations of the femtosecond dynamics in gold and silver thin films were performed by Groeneveld et al. (1995) and used an experimental configuration similar to that in Figure 9.1. The pump and probe pulses originate from a colliding-pulse mode-locked dye laser (degenerate pump–probe spectroscopy) and have a pulse duration $\sigma = 150\,fs$ corresponding to a cross correlation width, $w = 214\,fs$. This cross correlation width is sufficient to recover the electron–phonon thermalization lifetime $\tau_l \approx 700\,\mathrm{fs} \ll w$ in gold, and this pioneering experiment was the first to recover the electron–phonon coupling lifetime using pump–probe spectroscopy. The electron–electron, τ_{ee}, and electron–phonon scattering, τ_{ep}, lifetimes directly influence the complex conductivity in metals and semiconductors (Ridley, 2000) while the electron–phonon constant, g_∞, is an important parameter in the Bardeen–Cooper–Schrieffer (BCS) theory of superconductors (Poole et al., 2007). Ultrafast pump–probe spectroscopy provides a direct, noncontact method to study these systems on a femtosecond timescale that electronic techniques cannot.

In gold, there exist no energetically accessible band-to-band transitions at the pump and probe energies ($h\nu_0 = 2.0\,\mathrm{eV}$). As a result, free carrier absorption of carriers within $h\nu_0$ of the Fermi surface is the primary absorption mechanism. The femtosecond pump pulse excites these carriers into a nonequilibrium carrier distribution above the Fermi surface. These nonequilibrium electrons thermalize through electron–electron scattering ($\tau_{ee} \sim 100\,\mathrm{fs}$). This establishes the appropriate statistics needed to define a new electron temperature $T_E(t)$, which is higher than the initial temperature $T_E(0)$ and higher than the lattice temperature $T_L(t)$. Electron–phonon scattering then equilibrates $T_E(t)$ with $T_L(t)$ on a timescale called the electron–phonon lifetime, as discussed in further detail by Groeneveld et al. (1995) and in Chapter 8.

The probe pulse samples the changes to the state occupancy $f_i(t)$ in the conduction subband by probing the transition between the initial and final states with a time-delayed ultrashort pulse. Before excitation, this subband-to-subband transition was allowed, as the initial state was occupied by the electron ($f_i = 1$) while the final state was empty ($f_f = 0$) (i.e., $f_i(t) \times [1 - f_f(t)] = 1$); this photon would have been absorbed or reflected by the sample. After the pump pulse, the electron occupancy in the final state increases while the initial state occupancy decreases, resulting in a time-dependent decrease in $f_i(t) \times [1 - f_f(t)] < 1$; the probability of absorption decreases and consequently the reflection increases. In the limit where the pump pulse induces a small change to $f_i(t)$, the electron occupancy in the upper state is small ($1 - f_f(t) \approx 1$). In this case, the measured temporal dynamics reflect the time-dependent state occupancy $f_i(t)$ near the Fermi surface and, therefore, the scattering dynamics that thermalize electrons and phonons (see Equations 9.22 and 9.24).

Figure 9.7 shows the differential reflectivity of gold after photoexcitation of carriers at the Fermi surface. The initial increase in ΔR is due to the coherent artifact, caused by the

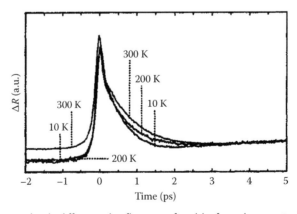

Figure 9.7 Time-resolved, differential reflection of gold after photoexcitation at 2.0 eV with 0.5 J cm^{-2}. The initial spike in ΔR is a coherent artifact that results from the interference of the pump and probe pulses in the sample (both are in the same linear polarization state). The sub-picosecond recovery dynamics were fit with a model similar to Equation 9.10 to extract an electron–phonon thermalization lifetime $\tau_1 \approx 700\,\mathrm{fs}$, and an electron–phonon coupling constant $g_\infty = 3.0 \pm 0.5 \times 10^{16}\,\mathrm{W\,m^{-3}\,K^{-1}}$. (Reprinted with permission from Groeneveld, R.H.M., Sprik, R., and Lagendijk, A., *Phys. Rev. B*, 51, 11433, 1995. Copyright 1995 by the American Physical Society.)

formation of an interference grating by the pump pulse and the diffraction of a component of this energy into the direction of the probe pulse. The small offset (compared to the temperature before $t = 0$) is the long lifetime ($\tau_2 \gg 5\,$ps) that results from the rise in system temperature after absorbing the pump fluence. This additional energy will diffuse to the surrounding environment or substrate on a microsecond timescale, which is longer than the time between probe pulses. This is discussed in the manuscript in further detail as well as in Chapter 8. The dynamics of electron–electron scattering are at least partially obscured by the coherent artifact that appears within the first ~200 fs after pump pulse excitation. Neglecting this coherent artifact, these data can be fitted using a model similar to Equation 9.10 to extract the observed sub-picosecond relaxation lifetime τ_{ep} and the electron–phonon coupling constant g_∞, using purely optical, noncontact methods. This technique is now a well-established method for determining this constant in a wide variety of materials (Hilton et al., 2006) (Chapter 8).

9.7.2 Nondegenerate Pump–Probe Spectroscopy in Semiconductors

Gallium arsenide has been studied extensively using pump–probe techniques due to the availability of high-quality samples, numerous optoelectronic applications, which use GaAs, and its compatibility with Ti:sapphire laser emission ($E_G = 1.42\,$eV in GaAs at room temperature compared to a central photon energy, $h\nu_0 = 1.55\,$eV in Ti:sapphire) (Portella et al., 1992; Tommasi et al., 1996; Rossi and Kuhn, 2002). The conventional *band-insulator* model describes the dynamics of carriers in semiconductors (Chapter 2).

This section focuses on *nondegenerate* pump–probe experiments used to isolate hole subband dynamics independent of the conduction subband dynamics (Ganikhanov et al., 1999; Hilton and Tang, 2002). The pump source was a low-pulse energy Ti:sapphire laser oscillator ($\lambda = 800\,$nm with a pump pulse energy of $E \sim 2\,$nJ) focused to a diameter of approximately 100 µm ($F = 5\,$µJ cm^{-2}). The probe used the tunable idler from a synchronously pumped femtosecond OPO ($\lambda \approx 3000\,$nm) based on periodically poled lithium niobate (Tang and Cheng, 1995). The polarization optics shown in Figure 9.1 are not needed, as the pump (800 nm) and probe beams (~3000 nm) can be isolated using a color filter. The significant difference between the pump and probe wavelengths also suppresses the coherent artifact in this configuration. While a component of the pump wavelength is likely diffracted into the direction of the probe beam, the wavelength filter used to isolate these two beams will block this and prevent its detection (Hilton and Tang, 2002).

The femtosecond pump pulse generates electrons in the conduction subband and holes in both the heavy- and light-hole subbands. Undoped gallium arsenide is transparent to mid-IR wavelengths at equilibrium (Blakemore, 1981). Before excitation, photons with an energy equal to the split-off to heavy-hole or split-off to light-hole transitions ($\Delta_0 = 0.341\,$eV or $\lambda = 3635\,$nm for GaAs) are not absorbed, as the final states for these transitions in the heavy- and light-hole subbands are filled with electrons (Pauli blocking, $f_f = 1$). In the pump–probe experiment, the pump pulse generated holes in the upper valence subbands ($1 - f_f > 0$ for $t > 0$) that allow transitions from the split-off subband to the upper valence subbands that would have otherwise been blocked. This results in transient absorption $\Delta\alpha(t) > 0$ near $\lambda \sim 3000\,$nm whose dynamics reflect the heavy- and light-hole subband thermalization. From the measured differential transmission $\Delta T/T(t)$ an energy relaxation lifetime of $\tau_E = 300\,$fs in GaAs for holes was extracted from these studies using a model similar to Equation 9.10. As in the prior section, there exists a second lifetime τ_2, which is much longer than the measurement window that is determined by electron–hole recombination ($\gg 1\,$ns in GaAs) and thermal diffusion to the substrate and surrounding environment (Ganikhanov et al., 1999).

The amplitude of the differential transmission signal in this experiment is strongly dependent on the complexities of the valence subbands in gallium arsenide. Ordinarily, transitions between the valence subbands would be dipole forbidden, as all three valence subbands have P-like symmetry near $k = 0$. The transition matrix elements M_{ij} in intervalence subband transitions (split-off to either heavy- or light-hole subbands) are nonzero due to the $\vec{k} \cdot \vec{p}$ interaction that mixes the hydrogenic wavefunctions away from $k = 0$ (Taylor, 1985). In addition, due to the strongly non-parabolic light-hole subband dispersion, the joint density of states ρ_J diverges at a probe wavelength that corresponds to a wave vector where the light and split-off hole subbands are parallel ($\lambda_{probe} \sim 4000\,nm$), resulting in a large increase in ΔT (Hilton and Tang, 2002).

Figure 9.8 shows a time-resolved mid-IR differential transmission study of an antireflection-coated GaAs sample after pumping by an $\lambda_0 = 800\,nm$ pump pulse with a duration of $\sigma = 80\,fs$. After the valence-to-conduction band pump pulse photogenerates holes in the heavy- and light-hole subbands, the initial decrease in transmission (increase in absorption) results from the lifting of the Pauli restrictions on inter-valence subband transitions after

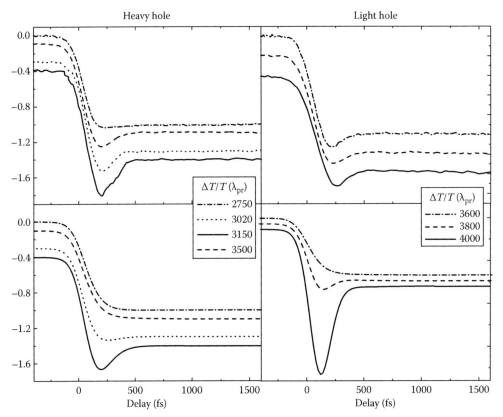

Figure 9.8 Nondegenerate pump–probe spectroscopy of valence subband dynamics in an epitaxially grown, intrinsic gallium arsenide GaAs single crystal. (top left) Experimentally measured $\Delta T(\tau)$ (a.u.) for the heavy-hole subband. (top right) Experimentally measured $\Delta T(\tau)$ (a.u.) for the light-hole subband. (bottom left) A simulation of the numerical Boltzmann equation for the heavy-hole subband. (bottom right) A simulation of the numerical Boltzmann equation for the light-hole subband. The deviation between the simulation and the data (top right) is discussed in Ganikhanov et al. (1999). These transitions from the split-off subband to the upper valance subbands are dipole allowed due to the mixing of the valence subband wavefunctions away from $k \approx 0$ (i.e., the $\vec{k} \cdot \vec{P}$ interaction). (Reprinted with permission from Ganikhanov, F., Burr, K.C., Hilton, D.J., and Tang, C.L., *Phys. Rev. B*, 60, 8890, 1999. Copyright 1999 by the American Physical Society.)

the formation of photogenerated holes. Intervalence subband transitions are not possible in undoped gallium arsenide before the pump pulse generates open states (i.e., holes) in the heavy- and light-hole subbands into which electrons from the split-off subband can be promoted. The range of probe wavelengths in this experiment (2750 nm $\leq \lambda \leq$ 4000 nm) studies transitions from the split-off subband to the upper subbands over a range of quasimomenta (k). This will directly measure the carrier scattering between different k states and, thus, formation of Boltzmann statistics in GaAs from the initial, nonequilibrium carrier distribution produced by the femtosecond 800 nm pump pulse. From these data, an energy relaxation lifetime τ_E of approximately 300 fs (the fast relaxation) was deduced by comparison of these data with a solution to the appropriate nonequilibrium Boltzmann transport equation (shown in Figure 9.8, bottom left and right). The longer lifetime in Figure 9.8 is longer than the measurement window (τ_2) and is determined by electron–hole recombination rates across the band gap and thermal diffusion into the substrate (Ganikhanov et al., 1999).

This experiment highlights one of the important design decisions to consider when building a pump–probe apparatus. The overall system's signal-to-noise ratio can be improved *either* by increasing the signal by increasing the pump power or decreasing the noise by increasing the repetition rate, detector sensitivity, or acquisition time. Despite the low pump fluence (F $\approx$ 5 μJ cm^{-2} at 800 nm), this system still possesses an acceptable signal-to-noise ratio comparable to or better than an amplified system due to the high repetition rate (80 MHz = 8 × 10^7 pulses per second) of both the Ti:sapphire laser and the synchronously pumped OPO, which also operates at the Ti:sapphire repetition rate. A similar experiment using amplified Ti:sapphire (1 kHz = 1 × 10^3 pulses per second) would have a pump fluence of approximately F $\approx$ 1 mJ cm^{-2} at 800 nm and would use an OPA as a probe pulse source to generate pulses near 3000 nm. In the linear absorption limit, the induced change to the material (ΔR or ΔT) would be 5 × 10^{-3} times smaller with the oscillator-based pump source when compared to the amplifier-based pump due to the difference in pump pulse energy. The higher-repetition-rate, however, improves the signal-to-noise ratio (S:N ~ $\sqrt{N}$) of the oscillator system over the comparable amplifier system by $\sqrt{8 \times 10^7 / 1 \times 10^3} \approx 283$ times, assuming otherwise identical data acquisition conditions (i.e., modulation frequency and time constant for lock-in detection, acquisition time, and detector sensitivity). As a result, the enhancement in the system's signal-to-noise ratio due to the improved counting statistics overcomes the smaller differential change to the reflectivity or transmission in the higher-repetition-rate system (by 5 × 10^{-3} × 283 $\approx$ 1.4x in this example). As discussed earlier in this chapter, the overall pump fluence should be small for perturbative experiments to determine the optical properties and rate constants of the material *near equilibrium*, which is easier to achieve in high-repetition-rate systems with lower pulse energies. In other experiments where the goal is to induce large changes to the material (or to trigger a phase transition), a higher fluence may be desirable, which currently requires lower-repetition-rate lasers with higher pulse energies. In *both* fluence limits, comparable signal-to-noise ratios are available from both low- and high-repetition-rate Ti:sapphire-based pump–probe experiments.

9.7.3 Generation of Coherent Phonons in Bulk and Nanoscale Materials

In the preceding examples, the ultrafast pump–probe experiments measured the optical constants of materials (dielectric constant, refractive index, susceptibility) to probe their underlying physics. Femtosecond excitation can also trigger the formation of *coherent phonons* that periodically modulate the crystalline structure, the Bloch states that result from the periodic lattice (Chapter 8), and the optical constants (n_r and α). The modulation of n_r and α results in a change to the differential transmission $\Delta T/T$ (Equation 9.22) and differential reflection $\Delta R/R$

(Equation 9.24). They have been observed in numerous bulk and nanoscale materials (Cheng et al., 1991; Pfeifer et al., 1992; Hase et al., 1996; Wehner et al., 1998; Kasami et al., 2004, 2006). The use of pump–probe techniques to observe coherent phonons in the time domain is an ultrafast optical counterpart of Raman spectroscopy, which determines phonon frequencies by measuring frequency shifts in reflected light in the spectral domain (Ferraro et al., 2003).

These effects are not limited to bulk materials, but can also manifest in nanoscale materials including semiconductor quantum wells (Özgür et al., 2001) and carbon nanotubes (Lim et al., 2006; Song et al., 2008). Carbon nanotubes are a single monolayer of sp^2-bonded graphene that has been rolled into a nanoscale tube. The nanotube diameter is smaller than the coherence length of electrons (Zaric et al., 2004) that leads to a complex electronic structure dominated by many-body (excitonic) effects (Wang et al., 2005). The electronic properties of these tubes depend heavily on the chiral indices n and m that describe the angle between the hexagonal graphene lattice and the nanotube axis. There exist numerous different possible chiral indices, which each result in different electronic states (metal, narrow-band-gap semiconductor, or wide-band-gap semiconductor), different phonon frequencies, and different nanotube radii (Ando, 2006, Avouris et al., 2008).

Ultrafast pump–probe spectroscopic studies of carbon nanotubes have revealed much information about the complex electronic and lattice structure of these 1D conductors. In addition to producing a change to the electronic state of the nanotube, an above-band-gap optical pulse also excited the radial breathing mode (RBM) of the tube, which periodically modulates the nanotube diameter (Avouris et al., 2008; Song et al., 2008). The resulting time-dependent complex refractive index $\Delta \tilde{n}_2(t)$ modulates the transmission or reflection of the sample that can be used to determine the lifetime of these coherent phonons in each chirality of the nanotube. Figure 9.9 shows a time-resolved differential transmission study of an ensemble of suspension of single-wall carbon nanotubes (made using the HiP$_{CO}$ technique) using

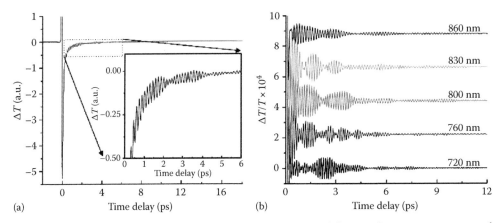

Figure 9.9 (See color insert.) (a) The time-resolved differential transmission (ΔT) of an ensemble of carbon nanotubes in solution using a *degenerate* differential transmission measurement at $\lambda = 800$ nm. The periodic modulation resulting from the coherent modulation of the nanotube diameter is highlighted in the inset. (b) The extracted coherent phonon oscillations in the ensemble as a function of wavelength generated by removing the electronic component using a form similar to Equation 9.10. The presence of multiple phonon frequencies in these data results from the existence of multiple chiral species of nanotubes (n, m) in this HiP$_{CO}$ nanotube sample. (Reprinted with permission from Lim, Y.-S., Yee, K.-J., Kim, J.-H., Haroz, E., Shaver, J., Kono, J., Doorn, S.K., Hauge, R., and Smalley, R., Coherent lattice vibrations in single-walled carbon nanotubes, *Nano Lett.*, 6, 2696–2700, 2006. Copyright 2006 American Chemical Society.)

a degenerate pump–probe experiment similar to Figure 9.1 in transmission (Lim et al., 2006). Figure 9.9a shows the full recovered change to the transmission that includes the electronic and coherent phonon components and highlights the periodic modulation from the nanotube motion of ΔT in the inset.

After removing the electronic component (Equation 9.10), the subtracted dynamics shown in Figure 9.9b consist of a number of different oscillation frequencies and have a pronounced dependence on the pump and probe wavelengths. The multiple oscillation frequencies in this HiP_{CO} nanotube sample result from the presence of multiple chirality nanotubes (n, m), which each have different RBM frequencies and energy gaps E_{11}, as discussed further by Lim et al. (2006). The wavelength dependence reflects the individual energy gaps E_{11} of each of the nanotubes relative to the pump and probe wavelengths. A direct time domain fitting of these dynamics is difficult due to multiple nanotube chiralities that result in multiple different phonon frequencies in this sample. The lifetimes of the RBMs can instead be determined by calculating the FT of the subtracted time domain dynamics, fitting the spectrum line widths, which are inversely proportional to the phonon lifetimes, and central frequencies that are the RBM for the nanotube chiralities present in the HiP_{CO} sample.

9.7.4 Insulator-to-Metal Phase Transitions in Correlated Electronic Materials

Ultrafast spectroscopic techniques are particularly suited to unravel the complex interactions between lattice, spin, and electronic degrees of freedom in correlated electronic materials. These complex interactions are not well described by the conventional band-insulator theory that has been successfully been applied to describe metals and semiconductors (as in the previous examples); this is not sufficient to describe the electronic and optical properties of materials whose interactions are sufficiently large or complicated. Correlated electronic materials possess exotic and technologically relevant properties such as superconductivity (Orenstein and Millis, 2000), colossal magnetoresistance (Averitt et al., 2001), charge ordering (Prasankumar et al., 2007), and insulator-to-metal transitions (Cavalleri et al., 2004). Ultrafast spectroscopic techniques are particularly useful tools for unraveling these complex interactions, as they can examine the timescales and elucidate their influence on the observed materials properties (Hilton et al., 2006).

Vanadium dioxide is a prototype correlated electron material that undergoes an insulator-to-metal transition at 340 K that is accompanied by a structural phase transition from a monoclinic-semiconductor phase to a rutile-metal phase (Goodenough, 1971). Despite nearly five decades of research, the nature of this phase transition is still a significant topic of debate (Goodenough, 1971; Hood and Denatale, 1991; Choi et al., 1996; Cavalleri et al., 2001, 2004; Rini et al., 2005; Qazilbash et al., 2007, 2008; Perucchi et al., 2009; Wei et al., 2009). The open question is the mutual influence of this structural phase transition and the electronic phase change (i.e., does the structural phase transition drive the insulator-to-metal transition, or does the opposite occur?). The *Mott transition model* describes the electronic phase transition due to electron–electron correlations, which then leads to the structural phase transition (Qazilbash et al., 2007), while the *Peierls model* attributes the electronic transition to the changes in crystalline structure within the framework of the band-insulator theory (Cavalleri et al., 2004).

Besides the thermally driven transition described earlier, *high optical fluence* pump–probe spectroscopy can trigger transition from the monoclinic-semiconductor to the rutile-metal phase (Hilton et al., 2007). Optically driven *photoinduced phase transitions* are a novel method of studying the interactions between the multiple degrees of freedom in correlated electronic materials that have been enabled by the development of high–pulse energy

Ti:sapphire lasers (Yonemitsu and Nasu, 2008). Further discussion of photoinduced phase transitions of this type can be found in Nasu (2004).

This section highlights one ultrafast optical study of vanadium dioxide under high optical fluence sufficient to trigger the phase transition starting from the monoclinic-semiconductor phase and ending in the rutile-metal phase (Cavalleri et al., 2004). The significant feature of this time-resolved differential reflection study is the existence of a finite rise time (τ_R in Equation 9.11) for ΔR that is unrelated to the pump and probe pulse cross correlation widths w. Figure 9.10 shows the extracted rise time from differential reflection measurements for a series of experiments that systematically reduced w from 1 ps to 15 fs. It is clear from these fits that the rise time is limited to ≥ 80 fs even when pumped by a pulse that is substantially shorter. The other significant data are the appearance of coherent oscillations in the differential reflectivity that indicate the formation of (displacive) A_g coherent phonons by the optical pump pulse. As a result, the connections between the electronic and lattice degrees of freedom can be studied using ultrafast differential reflection spectroscopy directly in the time domain.

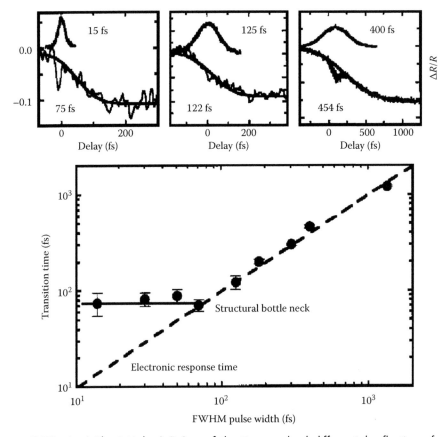

Figure 9.10 (top) The initial ~0.5–1 ps of the time-resolved differential reflection of vanadium dioxide after photoexcitation using an OPA as a pump source. The authors systematically reduced the cross correlation width w from ~1 ps to 15 fs and measured the differential reflection $\Delta R/R$ as a function of delay at a constant pulse energy. When $w < 80$ fs, the rise time of the differential transmission is still 80 fs and indicates the existence of a fundamental limiting timescale in the material τ_R that limits the formation time of the metallic state after photoexcitation. (Reprinted from Cavalleri, A., Dekorsy, T., Chong, H.H.W., Kieffer, J.C., and Schoenlein, R.W., *Phys. Rev. B*, 70, 161102. Copyright 2004 by the American Physical Society.)

The existence of a rise time τ_R in a time-resolved differential transmission or reflection experiment indicates that the probed state (the metallic phase, in this case) forms with a finite response time that is significantly longer than the cross correlation width ($w = \sigma\sqrt{2}$, where σ is the width of the pump or probe pulse). In vanadium dioxide, the finite rise time $\tau_R \sim 80\,fs$ cannot be explained in the Mott transition picture. There, the rise time limited by purely electronic interactions would be substantially faster than τ_R (on the order of 1 fs or less and often termed "instantaneous"). Cavalleri et al. (2004), instead, attribute this rise time to a "structural bottleneck" in the phase transition that prevents the instantaneous formation of the electronic phase. The femtosecond pulse itself can trigger a coherent A_g phonon mode of the monoclinic phase that results in the displacement of atoms from their lattice site. This structural process forms with a finite lifetime that corresponds to, at minimum, one phonon period, in contrast to a purely electronic process that would occur "instantaneously." This structural limiting timescale, extracted from the rise time–limited differential reflection measurements, indicates the influence of the structural transition on the electronic degree of freedom, although recent evidence from spatially localized, time-integrated spectroscopic measurements has also pointed to the influence of electron correlations (Qazilbash et al., 2007).

9.8 SUMMARY

Ultrafast pump–probe spectroscopic techniques are a well-established method for overcoming the limitations of electronic detectors to recover sub-picosecond dynamics in materials. This is accomplished by implementing different sampling techniques to measure femtosecond carrier dynamics in materials, although techniques for true single-shot recovery are now available. The high sensitivity of these techniques permits the resolution of dynamics on the order of a few femtoseconds while measuring photoinduced changes in reflection or transmission as small as 10^{-7}. As new laser technology develops, these capabilities will undoubtedly increase and will contribute to our understanding of condensed matter systems on unprecedented time and energy scales.

ACKNOWLEDGMENTS

I thank Mr. Nathaniel F. Brady (The University of Alabama at Birmingham), Mr. Mark Tolbert (Toptica, Inc.), and Professor Chung Tang (Cornell University) for their assistance with this chapter.

REFERENCES

Abramowitz, M. and Stegun, I.A. (Eds.) (1965) *Handbook of Mathematical Functions: With Formulas, Graphs, and Mathematical Tables*. Dover Publications, New York.

Agrawal, G. (2001) *Nonlinear Fiber Optics*, 3rd edn. Academic Press, San Diego, CA.

Ando, T. (2006) Effects of valley mixing and exchange on excitons in carbon nanotubes with Aharonov-Bohm flux. *J. Phys. Soc. Jpn*, **75**, 024707.

Averitt, R., Lobad, A., Kwon, C., Trugman, S.A., Thorsmølle, V.K., and Taylor, A.J. (2001) Ultrafast conductivity dynamics in colossal magnetoresistance manganites. *Phys. Rev. Lett.*, **87**, 017401.

Avouris, P., Freitag, M., and Perebeinos, V. (2008) Carbon-nanotube photonics and optoelectronics. *Nat. Photon.*, **2**, 341–350.

Azad, A., Prasankumar, R.P., Talbayev, D., Taylor, A.J., Averitt, R.D., Zide, J.M.O., Lu, H., Gossard, A.C., and OHara, J.F. (2008) Carrier dynamics in InGaAs with embedded ErAs nanoislands. *Appl. Phys. Lett*, **93**, 121108.

Azaña, J., Berger, N., Levit, B., and Fischer, B. (2004) Spectral Fraunhofer regime: Time-to-frequency conversion by the action of a single time lens on an optical pulse. *Appl. Opt*, **43**, 483–490.

Backus, S., Durfee, C.G., Murnane, M.M., and Kapteyn, H.C. (1998) High power ultrafast lasers. *Rev. Sci. Instrum.*, **69**, 1207.

Bartels, A., Cerna, R., Kistner, C., Thoma, A., Hudert, F., Janke, C., and Dekorsy, T. (2007) Ultrafast time-domain spectroscopy based on high-speed asynchronous optical sampling. *Rev. Sci. Instrum.*, **78**, 035107.

Bartels, A., Dekorsy, T., and Kurz, H. (1999) Femtosecond Ti: sapphire ring laser with a 2-GHz repetition rate and its application in time-resolved spectroscopy. *Opt. Lett.*, **24**, 996–998.

Bartels, A., Thoma, A., Janke, C., and Dekorsy, T. (2006) High-resolution THz spectrometer with kHz scan rates. *Opt. Express*, **14**, 430–437.

Baumberg, J., Awschalom, D., Samarth, N., Luo, H., and Furdyna, J. (1994) Spin beats and dynamical magnetization in quantum structures. *Phys. Rev. Lett.*, **72**, 717–720.

Beaurepaire, E., Merle, J.-C., Daunois, A., and Bigot, J.-Y. (1996) Ultrafast spin dynamics in ferromagnetic nickel. *Phys. Rev. Lett.*, **76**, 4250–4253.

Blakemore, J.S. (1981) Intrinsic density $n_i(T)$ in GaAs: Deduced from band gap and effective mass parameters and derived independently from Cr acceptor capture and emission coefficients. *J. Appl. Phys.*, **53**, 520–531.

Born, M. and Wolf, E. (1999) *Principles of Optics*. Cambridge University Press, Cambridge, U.K.

Borri, P., Langbein, W., Mørk, J., and Hvam, J.M. (1999) Heterodyne pump-probe and four-wave mixing in semiconductor optical amplifiers using balanced lock-in detection. *Opt. Commun.*, **169**, 314–324.

Boyd, R.W. (1983) *Radiometry and the Detection of Optical Radiation* (Wiley Series in Pure and Applied Optics). Wiley-Interscience, New York.

Boyd, R.W. (1991) *Nonlinear Optics*. Academic Press, San Diego, CA.

Braunstein, R. and Kane, E. (1962) The valence band structure of the III-V compounds. *J. Phys. Chem. Solids*, **23**, 1423–1431.

Braunstein, R. and Magid, L. (1958) Optical absorption in *p*-type gallium arsenide. *Phys. Rev.*, **111**, 480–481.

Burr, K.C. and Tang, C.L. (1999) Femtosecond midinfrared-induced luminescence study of the ultrafast dynamics of split-off holes in GaAs. *Appl. Phys. Lett.*, **74**, 1734–1736.

Cavalleri, A., Dekorsy, T., Chong, H.H.W., Kieffer, J.C., and Schoenlein, R.W. (2004) Evidence for a structurally-driven insulator-to-metal transition in VO_2: A view from the ultrafast timescale. *Phys. Rev. B*, **70**, 161102.

Cavalleri, A., Tòth, C., Siders, C.W., Squier, J.A., Ráksi, F., Forget, P., and Kieffer, J.C. (2001) Femtosecond structural dynamics in VO_2 during an ultrafast solid-solid phase transition. *Phys. Rev. Lett.*, **87**, 237401.

Cheng, T.K., Vidal, J., Zeiger, H.J., Dresselhaus, G., Dresselhaus, M.S., and Ippen, E.P. (1991) Mechanism for displacive excitation of coherent phonons in Sb, Bi, Te, and Ti_2O_3. *Appl. Phys. Lett.*, **59**, 1923.

Choi, H.S., Ahn, J.S., Jung, J.H., Noh, T.W., and Kim, D.H. (1996) Mid-infrared properties of a VO_2 film near the metal-insulator transition. *Phys. Rev. B*, **54**, 4621–4628.

Christov, I.P., Murnane, M.M., and Kapteyn, H.C. (1997) High-harmonic generation of attosecond pulses in the "single-cycle" regime. *Phys. Rev. Lett.*, **78**, 1251–1254.

Crooker, S.A. (2002) Fiber-coupled antennas for ultrafast coherent terahertz spectroscopy in low temperatures and high magnetic fields. *Rev. Sci. Instrum.*, **73**, 3258–3264.

Demsar, J., Averitt, R.D., Ahn, K.H., Graf, M.J., Trugman, S.A., Kabanov, V.V., Sarrao, J.L., and Taylor, A.J. (2003) Quasiparticle relaxation dynamics in heavy fermion compounds. *Phys. Rev. Lett.*, **91**, 027401.

Demsar, J., Forrò, L., Berger, H., and Mihailovic, D. (2002) Femtosecond snapshots of gap-forming charge-density-wave correlations in quasi-two-dimensional dichalcogenides 1T-TaS$_2$ and 2H-TaSe$_2$. *Phys. Rev. B*, **66**, 041101.

Demsar, J., Thorsmølle, V.K., Sarrao, J.L., and Taylor, A.J. (2006) Photoexcited electron dynamics in Kondo insulators and heavy fermions. *Phys. Rev. Lett.*, **96**, 037401.

Diels, J.-C.M., Fontaine, J.J., Mcmichael, I.C., and Simoni, F. (1985) Control and measurement of ultrashort pulse shapes (in amplitude and phase) with femtosecond accuracy. *Appl. Opt.*, **24**, 1270–1282.

Dorrer, C., Beauvoir, B.D., Blanc, C.L., Ranc, S., Rousseau, J.-P., Rousseau, P., and Chambaret, J.-P. (1999) Single-shot real-time characterization of chirped-pulse amplification systems by spectral phase interferometry for direct electric-field reconstruction. *Opt. Lett.*, **24**, 1644–1646.

Dui, B., Cohen-Tannoudji, C., Diu, B., and Laloe, F. (1978) *Quantum Mechanics*, Vol. 1. Wiley-Interscience, New York.

Ferraro, J.R., Nakamoto, K., and Brown, C.W. (2003) *Introductory Raman Spectroscopy*. Academic Press, New York.

Fork, R., Martinez, O., and Gordon, J. (1984) Negative dispersion using pairs of prisms. *Opt. Lett.*, **9**, 150–152.

Foster, M.A., Salem, R., Geraghty, D.F., Turner-Foster, A.C., Lipson, M., and Gaeta, A.L. (2008) Silicon-chip-based ultrafast optical oscilloscope. *Nature*, **456**, 81.

Ganikhanov, F., Burr, K.C., Hilton, D.J., and Tang, C.L. (1999) Femtosecond optical-pulse-induced absorption and refractive-index changes in GaAs in the midinfrared. *Phys. Rev. B*, **60**, 8890–8896.

Goodenough, J.B. (1971) The two components of the crystallographic transition in VO_2. *J. Solid State Chem.*, **3**, 490–500.

Goodman, J.W. (1996) *Introduction to Fourier Optics*. McGraw-Hill Science/Engineering/Math, New York.

Goulielmakis, E., Schultze, M., Hofstetter, M., Yakovlev, V., Gagnon, J., Uiberacker, M., Aquila, A., Gullikson, E., Attwood, D., Kienberger, R., Krausz, F., and Kleineberg, U. (2008) Single-cycle nonlinear optics. *Science*, **320**, 1614.

Groeneveld, R.H.M., Sprik, R., and Lagendijk, A. (1995) Femtosecond spectroscopy of electron-electron and electron-phonon energy relaxation in Ag and Au. *Phys. Rev. B*, **51**, 11433–11445.

Hall, K.L., Lenz, G., Ippen, E.P., and Raybon, G. (1992) Heterodyne pump-probe technique for time-domain studies of optical nonlinearities in waveguides. *Opt. Lett.*, **17**, 874–876.

Hase, M., Mizoguchi, K., Harima, H., Nakashima, S., Tani, M., Sakai, K., and Hangyo, M. (1996) Optical control of coherent optical phonons in bismuth films. *Appl. Phys. Lett.*, **69**, 2474.

Hecht, E. (1987) *Optics*. Addison-Wesley, Reading, MA.

Hilton, D.J., Prasankumar, R.P., Fourmaux, S., Cavalleri, A., Brassard, D., Khakani, M.A.E., Keiffer, J.C., Taylor, A.J., and Averitt, R.D. (2007) Enhanced photosusceptibility near T_c for the light-induced insulator-to-metal phase transition in vanadium dioxide. *Phys. Rev. Lett.*, **99**, 226401.

Hilton, D.J., Prasankumar, R.P., Trugman, S.A., Taylor, A.J., and Averitt, R.D. (2006) On photo-induced phenomena in complex materials: Probing quasiparticle dynamics using infrared and far infrared pulses. *J. Phys. Soc. Jpn.*, **75**, 011006.

Hilton, D.J. and Tang, C.L. (2002) Optical orientation and femtosecond relaxation of spin-polarized holes in GaAs. *Phys. Rev. Lett.*, **89**, 146601.

Hoffmann, M.C., Yeh, K.-L., Hwang, H.Y., Sosnowski, T.S., Prall, B.S., Hebling, J., and Nelson, K.A. (2008) Fiber laser pumped high average power single-cycle terahertz pulse source. *Appl. Phys. Lett.*, **93**, 141107.

Hood, P.J. and Denatale, J.F. (1991) Millimeter-wave dielectric properties of epitaxial vanadium dioxide thin films. *J. Appl. Phys.*, **70**, 376.

Horowitz, P. and Hill, W. (1989) *The Art of Electronics*. Cambridge University Press, Cambridge, U.K.

Kasami, M., Mishina, T., and Nakahara, J. (2004) Femtosecond pump and probe spectroscopy in Bi under high pressure. *Phys. Status Solidi (B)*, **241**, 3113–3116.

Kasami, M., Ogino, T., Mishina, T., Yamamoto, S., and Nakahara, J. (2006) Femtosecond pump–probe studies of phonons and carriers in bismuth under high pressure. *J. Lumin.*, **119–120**, 428–432.

Lakowicz, J.R. (1999) *Principles of Fluorescence Spectroscopy*. Springer, New York.

Lim, Y.-S., Yee, K.-J., Kim, J.-H., Haroz, E., Shaver, J., Kono, J., Doorn, S.K., Hauge, R., and Smalley, R. (2006) Coherent lattice vibrations in single-walled carbon nanotubes. *Nano Lett.*, **6**, 2696–2700.

Luo, C.W., Wang, Y.T., Chen, F.W., Shih, H.C., and Kobayashi, T. (2009) Eliminate coherence spike in reflection-type pump-probe measurements. *Opt. Express*, **17**, 11321–11327.

Luttinger, J. and Kohn, W. (1955) Motion of electrons and holes in perturbed periodic fields. *Phys. Rev.*, **97**, 869–883.

Miller, G.H., Moses, E.I., and Wuest, C.R. (2004) The national ignition facility. *Opt. Eng.*, **43**, 2841.

Mittleman, D. (2003) *Sensing with Terahertz Radiation*. Springer, Berlin, Germany.

Murnane, M.M., Kapteyn, H.C., and Falcone, R.W. (1990) X-ray streak camera with 2 ps response. *Appl. Phys. Lett.*, **56**, 1948.

Nasu, K. (2004) *Photoinduced Phase Transitions*. World Scientific, Hackensack, NJ, p. 345.

Oba, K., Sun, P., Mazurenko, Y., and Fainman, Y. (1999) Femtosecond single-shot correlation system: a time-domain approach. *Appl. Opt.*, **38**, 3810–3817.

Ogawa, Y., Koshihara, S., Koshino, K., Ogawa, T., Urano, C., and Takagi, H. (2000) Dynamical aspects of the photoinduced phase transition in spin-crossover complexes. *Phys. Rev. Lett.*, **84**, 3181–3184.

Orenstein, J. and Millis, A. (2000) Advances in the physics of high-temperature superconductivity. *Science*, **288**, 468–474.

Özgür, Ü., Lee, C.-W., and Everitt, H. (2001) Control of coherent acoustic phonons in semiconductor quantum wells. *Phys. Rev. Lett.*, **86**, 5604–5607.

Padilla, W., Li, Z., Burch, K., Lee, Y., Mikolaitis, K.J., and Basov, D.N. (2004) Broadband multi-interferometer spectroscopy in high magnetic fields: From THz to visible. *Rev. Sci. Instrum.*, **75**, 4710–4717.

Palfrey, S.L. and Heinz, T.F. (1985) Coherent interactions in pump-probe absorption measurements: the effect of phase gratings. *J. Op. Soc. Am. B*, **2**, 674–679.

Perucchi, A., Baldassarre, L., Postorino, P., and Lupi, S. (2009) Optical properties across the insulator to metal transitions in vanadium oxide compounds. *J. Phys. Condens. Matter*, **21**, 323202.

Pfeifer, T., Dekorsy, T., Kiitt, W., and Kurz, H. (1992) Generation mechanism for coherent LO phonons in surface-space-charge fields of III-V-compounds. *Appl. Phys. A*, **55**, 482–488.

Poole, C.P., Farach, H.A., Creswick, R.J., and Prozorov, R. (2007) *Superconductivity*. Academic Press, Amsterdam, the Netherlands.

Portella, M.T., Bigot, J.Y., Schoenlein, R.W., Cunningham, J.E., and Shank, C.V. (1992) l-space carrier dynamics in GaAs. *Appl. Phys. Lett.*, **60**, 2123–2125.

Prasankumar, R.P., Zvyagin, S., Kamenev, K.V., Balakrishnan, G., Paul, D.M., Taylor, A.J., and Averitt, R.D. (2007) Phase inhomogeneities in the charge-orbital-ordered manganite $Nd_{0.5}Sr_{0.5}MnO_3$ revealed through polaron dynamics. *Phys. Rev. B*, **76**, 020402.

Qazilbash, M.M., Brehm, M., Chae, B.-G., Ho, P., Andreev, G., Kim, B.-J., Yun, S., Balatsky, A., Maple, M.B., Keilmann, F., Kim, H.-T., and Basov, D.N. (2007) Mott transition in VO_2 revealed by infrared spectroscopy and nano-imaging. *Science*, **318**, 1750.

Qazilbash, M.M., Schafgans, A.A., Burch, K.S., Yun, S.J., Chae, B.G., Kim, B.J., Kim, H.T., and Basov, D.N. (2008) Electrodynamics of the vanadium oxides VO_2 and V_2O_3. *Phys. Rev. B*, **77**, 115121.

Ranka, J.K., Gaeta, A.L., Baltuska, A., Pshenichnikov, M.S., and Wiersma, D.A. (1997) Autocorrelation measurement of 6-fs pulses based on the two-photon-induced photocurrent in a GaAsP photodiode. *Opt. Lett.*, **22**, 1344–1346.

Rapoport, W.R. and Khattak, C.P. (1988) Titanium sapphire laser characteristics. *Appl. Opt.*, **27**, 2677–2684.

Reed, M.K., Steiner-Shepard, M.K., and Negus, D.K. (1995) Tunable ultraviolet generation using a femtosecond 250 kHz Ti:sapphire regenerative amplifier. *IEEE J. Quantum Electron.*, **31**, 1614–1618.

Ridley, B.K. (2000) *Quantum Processes in Semiconductors*. Oxford University Press, New York.

Ringling, J., Kittelmann, O., Noack, F., Korn, G., and Squier, J. (1993) Tunable femtosecond pulses in the near vacuum ultraviolet generated by frequency conversion of amplified Ti:sapphire laser pulses. *Opt. Lett.*, **18**, 2035–2037.

Rini, M., Cavalleri, A., Schoenlein, R., López, R., Feldman, L.C., Haglund, R.F., Boatner, L., and Haynes, T.E. (2005) Photoinduced phase transition in VO_2 nanocrystals: Ultrafast control of surface-plasmon resonance. *Opt. Lett.*, **30**, 558–560.

Rossi, F. and Kuhn, T. (2002) Theory of ultrafast phenomena in photoexcited semiconductors. *Rev. Mod. Phys.*, **74**, 895–950.

Sacks, Z., Mourou, G., and Danielius, R. (2001) Adjusting pulse-front tilt and pulse duration by use of a single-shot autocorrelator. *Opt. Lett.*, **26**, 462–464.

Saeta, P., Wang, J.-K., Siegal, Y., Bloembergen, N., and Mazur, E. (1991) Ultrafast electronic disordering during femtosecond laser melting of GaAs. *Phys. Rev. Lett.*, **67**, 1023–1026.

Salem, R., Foster, M.A., Turner, A.C., Geraghty, D.F., Lipson, M., and Gaeta, A.L. (2009) Optical time lens based on four-wave mixing on a silicon chip. *Opt. Lett.*, **33**, 1047–1049.

Sander, M.Y., Birge, J., Benedick, A., Crespo, H.M., and Kärtner, F.X. (2009) Dynamics of dispersion managed octave-spanning titanium:sapphire lasers. *J. Opt. Soc. Am. B*, **26**, 743–749.

Sanders, G.D., Sun, C.K., Fujimoto, J.G., Choi, H.K., Wang, C.A., and Stanton, C.J. (1994) Carrier-gain dynamics in $In_xGa_{1-x}As/AL_yGa_{1-y}As$ strained-layer single-quantum-well diode lasers: Comparison of theory and experiment. *Phys. Rev. B*, **50**, 8539–8558.

Sarrao, J. (2002) From $Ce_3Bi_4Pt_3$ to $CeCoIn_5$: 10 years of new materials research at LANL. *Phys. B Condens. Matter*, **318**, 87–91.

Scholz, R. (1995) Hole-phonon scattering rates in gallium arsenide. *J. Appl. Phys.*, **77**, 3219–3231.

Solli, D., Chou, J., and Jalali, B. (2007) Amplified wavelength–time transformation for real-time spectroscopy. *Nat. Photon.*, **2**, 48–51.

Song, D., Wang, F., Dukovic, G., Zheng, M., Semke, E.D., Brus, L.E., and Heinz, T.F. (2008) Direct measurement of the lifetime of optical phonons in single-walled carbon nanotubes. *Phys. Rev. Lett.*, **100**, 225503.

Spence, D.E., Kean, P.N., and Sibbett, W. (1991) 60-fsec pulse generation from a self-mode-locked Ti:sapphire laser. *Opt. Lett.*, **16**, 42–44.

Tamura, K., Ippen, E.P., Haus, H.A., and Nelson, L.E. (1993) 77-fs pulse generation from a stretched-pulse mode-locked all-fiber ring laser. *Opt. Lett.*, **18**, 1989–1082.

Tang, C.L. and Cheng, L.K. (1995) *Fundamentals of Optical Parametric Processes and Oscillations*. Harwood Academic Publishers, Amsterdam, the Netherlands.

Tauser, F., Rausch, C., Posthumus, J.H., and Lison, F. (2008) Electronically controlled optical sampling using 100 MHz repetition rate fiber lasers. *Proc. SPIE*, **6881**, 68810O.

Taylor, A.J., Erskine, D.J., and Tang, C.L. (1985) Ultrafast relaxation dynamics of photoexcited carriers in GaAs and related compounds. *J. Opt. Soc. Am. B*, **2**, 663–673.

Tommasi, R., Langot, P., and Vallée, F. (1996) Femtosecond hole thermalization in bulk GaAs. *Appl. Phys. Lett.*, **66**, 1361–1363.

Walmsley, I., Mitsunaga, M., and Tang, C.L. (1988) Theory of quantum beats in optical transmission-correlation and pump-probe experiments for a general Raman configuration. *Phys. Rev. A*, **38**, 4681–4689.

Wang, F., Dukovic, G., Brus, L., and Heinz, T. (2005) The optical resonances in carbon nanotubes arise from excitons. *Science*, **308**, 838–841.

Wehner, M., Ulm, M., Chemla, D., and Wegener, M. (1998) Coherent Control Of Electron-LO-phonon scattering in bulk GaAs. *Phys. Rev. Lett.*, **80**, 1992–1995.

Wei, J., Wang, Z., Chen, W., and Cobden, D.H. (2009) Metal-insulator transition in vanadium dioxide nanobeams: Probing sub-domain properties of strongly correlated materials. ArXiv/cond-mat, 0904.0596.

Yan, Y.J., Fried, L.E., and Mukamel, S. (1989) Ultrafast pump-probe spectroscopy: Femtosecond dynamics in Liouville space. *J. Phys. Chem.*, **93**, 8149–8162.

Yariv, A. (1997) *Optical Electronics in Modern Communications* (Oxford Series in Electrical and Computer Engineering). Oxford University Press, New York.

Yonemitsu, K. and Nasu, K. (2008) Theory of photoinduced phase transitions in itinerant electron systems. *Phys. Rep.*, **465**, 1–60.

Zaric, S., Ostojic, G.N., Kono, J., Shaver, J., Moore, V.C., Strano, M.S., Hauge, R.H., Smalley, R.E., and Wei, X. (2004) Optical signatures of the Aharonov-Bohm phase in single-walled carbon nanotubes. *Science*, **304**, 1129–1131.

Zhang, J.-Y., Huang, J.Y., and Shen, Y.R. (1995) *Optical Parametric Generation and Amplification* (Laser Science and Technology, Vol. 19). CRC, Australia.

Zory, P.S. (Ed.) (1993) *Quantum Well Lasers*. Academic Press, New York.

Chapter 10

TRANSIENT FOUR-WAVE MIXING

Steven T. Cundiff

CONTENTS

10.1 INTRODUCTION

Until the advent of laser, optical spectroscopy was limited to absorption and emission. These linear techniques often provided incomplete information because of the presence of inhomogeneous broadening, which arises in gases from Doppler shifts and in solids from random crystal fields or structural disorder (Figure 10.1). Lasers provided the high intensity needed to implement nonlinear spectroscopic techniques, which can make measurements "below" the

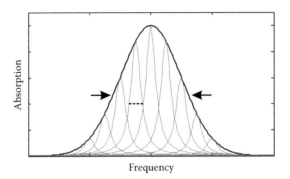

Figure 10.1 (See color insert.) Schematic of an inhomogeneously broadened absorption line. The absorption profile (thick line) is actually due to the sum of many narrower lines (thin lines) at different center frequencies (the center frequencies are continuously distributed; only a few are shown). The inhomogeneous (arrows) and homogeneous (dotted line) line widths are indicated. The oscillators with a given resonance frequency are known as a "frequency group."

inhomogeneous line width and determine the underlying homogeneous line widths. Often the homogeneous width is more interesting because it reflects dynamics such as scattering and/ or radiative decay.

Nonlinear laser spectroscopic techniques for probing below the inhomogeneous width can be implemented in either the frequency or the time domain. In the frequency domain, the main technique is known as spectral hole burning, a phenomena that was observed in the gain medium of a HeNe laser (Bennett 1962) shortly after it was invented. In the time domain, inhomogeneous broadening can be canceled by the formation of an echo, known as a photon echo in analogy to the spin echo used in nuclear magnetic resonance (NMR).

In principle, time and frequency domain techniques provide the same information. However, there are practical reasons for choosing one over the other. From an experimental perspective, generating time delays of greater than a few nanoseconds becomes challenging, as is generating accurate frequency detuning of greater than a few gigahertz. Thus, time domain techniques are typically used to study processes that occur on a sub-nanosecond timescale, whereas the frequency domain is used for slower processes. If processes are occurring on multiple timescales, time domain techniques generally emphasize the fastest process, while frequency domain techniques emphasize the slowest. This difference can lead to apparent discrepancies when comparing results between the two domains. With the development of multidimensional Fourier transform techniques in the optical and infrared regions of the spectrum, the distinction between time and frequency domains is becoming somewhat blurred.

Photon echoes were first demonstrated in cryogenic ruby (Kurnit et al. 1964) using nanosecond pulses from a ruby laser. A photon echo is formed when an inhomogeneously broadened sample is illuminated with a sequence of two excitation pulses and the signal in the appropriate direction is detected. During the time between the pulses, each frequency group accumulates a phase that is proportional to the frequency of the group and the delay; thus, different frequency groups are initially out of phase immediately after the second pulse. During the time after the second pulse, the phase evolution of each group is reversed, and thus all the frequency groups come back into phase to create a macroscopic polarization, which then radiates the "echo" pulse. The delay between the second pulse and the echo is the same as the delay between the excitation pulses. The time integral of the echo decays exponentially with increasing delay between the excitation pulses at the homogeneous dephasing rate;

thus, measuring its time-integrated intensity as a function of delay between excitation pulses allows the homogeneous dephasing rate to be determined.

The term "photon echo" is often used to describe a class of time-domain spectroscopic techniques, regardless of whether or not an echo is actually produced. A plethora of names for the variants, such as reverse photon echo or virtual photon echo, are used. However, all of these techniques are some form of transient four-wave mixing (TFWM); thus, this more general term will be used here.

TFWM is a third-order nonlinear effect, that is, the interaction of the incident excitation pulses in the sample due to the third-order nonlinear susceptibility, $\chi^{(3)}$, gives rise to a nonlinear polarization:

$$P^{NL} \propto \chi^{(3)} E^3 \tag{10.1}$$

where E is the electric field of the incident pulses. The nonlinear polarization radiates to produce a signal field that is detected. If E consists of three separate fields, for example, three pulses, they will interact to produce the signal field. The three incident fields plus the signal field correspond to the four waves in TFWM. Note that in some cases, there may only be two distinct incident fields, but one must act twice; thus, it is still thought of as a four wave process. Spectroscopic information is obtained by measuring the signal field as the properties of the incident pulses are varied. A simple case would be measuring the intensity of the signal field as the delay between the incident pulses is varied, although much more sophisticated methods are often used.

The incident field can be made up of two or three separate pulses with an adjustable time delay between them. Most often, these pulses are replicas of the same pulse that has a carrier frequency ω and complex envelope function $\hat{E}(t)$. Thus, if two excitation pulses are used, the incident field can be written as follows: $E = \hat{E}(t - t_a)e^{i(\mathbf{k}_a x - \omega t)} + \hat{E}(t - t_b)e^{i(\mathbf{k}_b x - \omega t)} + c.c$ where pulse i has wave vector $\mathbf{k}_i$ and arrives at time t_i. Inserting this electric field into Equation 10.1 and choosing the terms that oscillate at ω, that is, dropping third harmonic and rectification terms, and picking those that depend on both pulses results in 48 possible combinations of the two incident pulses. These terms correspond to nonlinear polarizations that radiate in different directions corresponding to combinations of $\mathbf{k}_a$ and $\mathbf{k}_b$. All of these terms can be considered a form of TFWM; however, the terms resulting in a photon echo are in the directions $2\mathbf{k}_a - \mathbf{k}_b$ and $2\mathbf{k}_b - \mathbf{k}_a$. For a two-level system, these two terms are the same except for the time ordering of the pulses, with the signal being emitted in the direction $\mathbf{k}_s = 2\mathbf{k}_b - \mathbf{k}_a$ when the pulse in direction $\mathbf{k}_a$ arrives first. A schematic of this case is shown in Figure 10.2.

10.1.1 Physical Description of TFWM for a Two-Level System

For a homogeneously broadened ensemble of two-level systems, generation of the TFWM signal can be described in a fairly simple physical picture based on inducing a diffraction grating

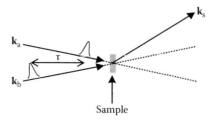

Figure 10.2 Schematic of a two-pulse TFWM experiment.

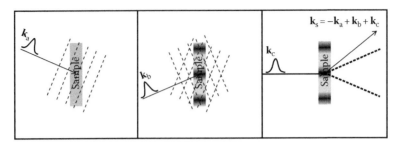

Figure 10.3 Grating picture of TFWM. Left panel: First pulse arrives at the sample; the dashed lines designate phase fronts. The phase is stored in the coherence induced in the sample. Center panel: Second pulse converts coherence to excited state (darker regions) or ground state (lighter regions) population in sample depending on the pulse phase relative to the coherence. Right panel: Third pulse scatters off grating in signal direction.

in the sample. This picture is easier and more general to describe for three pulses. It is readily simplified to describe the two-pulse case. While this picture has significant limitations, it is a useful starting point for discussing more complex situations. It is shown schematically in Figure 10.3. In this "grating" picture, the first pulse puts the two-level systems in a "coherence," that is, a superposition of ground and excited states. The coherence has an initial phase that depends on the spatial location within the sample and the direction of the first pulse, $\mathbf{k}_a$. The phase of the coherence evolves based on the transition frequency of the two-level systems until the second pulse arrives. The second pulse then converts the coherence to either a ground state or excited state population, depending on the relative phase of the coherence with respect to the second pulse. Since there is an angle between the propagation directions of the two pulses, the relative phase will vary with a wave vector corresponding to the vector between $\mathbf{k}_a$ and $\mathbf{k}_b$. If the second pulse is in phase with the coherence established by the first pulse, then the coherence will be converted to an excited state population, whereas if it is out of phase, the coherence will become a ground state population. The absorption is reduced for a system that is in the excited state; thus, the spatially modulated populations correspond to a spatial modulation of the optical properties of the sample. Since the modulation is sinusoidal, it can act as a diffraction grating and diffract a portion of a third beam into the signal direction. For two-pulse TFWM, the second pulse acts as both the second and third pulses in this description; namely, it both forms the grating and diffracts off of it.

It is important to understand that this discussion has only described the excitation sequence that results in a TFWM signal in the signal direction, $-\mathbf{k}_a + \mathbf{k}_b + \mathbf{k}_c$ for three pulses or $2\mathbf{k}_b - \mathbf{k}_a$ for two pulses. Simultaneously, other processes will be occurring, for example, each pulse can create an excited state population on its own. However, these processes do not directly contribute to the quantum pathway that results in a signal in the specified direction. In the presence of excitation-induced effects, such as those that occur in semiconductors, these other pathways may nevertheless modify the signal, for example, by changing its decay rate or emission frequency.

The signal decays exponentially with the dephasing rate as a function of time after the last pulse. In addition, as the delay between the first and second pulses is increased, the strength of the emitted signal will decay due to the decay of the coherence that is established by the first pulse. Thus, the time-integrated intensity of the signal is $I(\tau) \sim e^{-2\gamma_{ph}\tau}$, where γ_{ph} is the dephasing rate and τ is the delay between the pulses. The factor of 2 is due to measuring the intensity. The TFWM signal as a function of τ and "real" time t is sketched in Figure 10.4.

In an inhomogeneously broadened system, each frequency group can be thought of as an independent homogeneously broadened ensemble. While each frequency group produces an

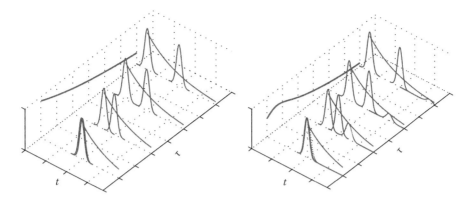

Figure 10.4 (See color insert.) Sketch for two-pulse excitation showing excitation pulses (solid lines), coherences (dashed lines), signals (dotted lines), the time-integrated signal (thick solid line), and decays of the polarization in individual frequency groups (thin solid lines) for a homogeneously broadened system (left) and an inhomogeneously broadened system (right). The delay, τ, between the two excitation pulses increases from front to back.

exponentially decaying signal, they are initially out of phase with each other and thus do not coherently add up to give a net signal. However, during the time after the last pulse, the phase evolution is opposite to that during the time between pulses. Thus, at a time $t = \tau$ after the second pulse, the signals from all the frequency groups come back in phase to produce a net signal pulse, the photon echo. The time-integrated intensity of the signal from an inhomogeneously broadened system is $I(\tau) \sim e^{-4\gamma_{ph}\tau}$ for τ large compared to the inverse inhomogeneous width. The signal decays as a function of τ twice as fast as for the homogeneously broadened case because the echo is effectively sampling the exponential decay as a function of t as it occurs at later and later times. At shorter delays, the signal rises because the echo is truncated due to causality (Yajima and Taira 1979).

These results show that the time-integrated TFWM signal decays exponentially as a function of delay τ at a rate that is proportional to the dephasing rate, that is, inverse of the homogeneous linewidth. In a two-pulse experiment, τ is the delay between the pulses; in a three-pulse experiment, τ is the delay between the first two pulses. Thus, it is possible to determine the homogeneous linewidth from a TFWM measurement using a slow (time integrating) detector, but only if it is known whether the system is homogeneously or inhomogeneously broadened.

10.1.2 Overview of Variants and Extensions of TFWM

The simplest TFWM experiment is the two-pulse configuration with time-integrated detection, as described in the previous section. The configuration of the excitation pulses is the same as for a pump-probe experiment; however, the signal is detected in a background-free direction. However, there are several variants and extensions that can provide greater information than the two-pulse time-integrated configuration. They all come at the price of greater experimental complexity.

The two-pulse excitation configuration can be extended by time resolving the signal or spectrally resolving it. These are often designated as TR-TFWM and SR-TFWM, respectively. For clarity, time-integrated detection can be designated as TI-TFWM although it is often assumed that "TFWM" without a modifier refers to TI-TFWM, as has been done heretofore in this chapter. In a TI-TFWM experiment, the signal is measured as a function of delay between the excitation pulses, whereas in a TR-TFWM experiment it is measured as a function of time after the second pulse for a fixed delay between excitation pulses (of course a series of such measurement may be made for varying delay). It should also be noted that before the use of TR-TFWM became common, TI-TFWM experiments were sometimes described as "time resolved," which simply meant that they were performed in the time domain, as opposed to the frequency domain. TR-TFWM is usually implemented by cross-correlation using a second harmonic crystal (Section 10.3.3) (Schultheis et al. 1985, Webb et al. 1991a, Koch et al. 1992); however, it has also been achieved using a streak camera (Noll et al. 1990). SR-TFWM is implemented by detecting the signal using a spectrometer (Lyssenko et al. 1993). An important use of both TR-TFWM and SR-TFWM is to determine if beats observed in a TI-TFWM signal are due to simple electromagnetic interference from two uncoupled resonances, known as polarization interference, or rather are true quantum beats from a system consisting of two excited states coupled to a single ground state.

Another extension is to use three excitation pulses, rather than two. If the three pulses have wave vectors $\mathbf{k}_a$, $\mathbf{k}_b$, and $\mathbf{k}_c$, the signal is emitted in the direction $\mathbf{k}_s = -\mathbf{k}_a + \mathbf{k}_b + \mathbf{k}_c$. The physical description given in the previous section still applies; however, here pulse $\mathbf{k}_b$ creates the grating and $\mathbf{k}_c$ scatters off of it. Three-pulse TFWM is sometimes designated 3P-TFWM. Time and spectrally resolved versions of it can also be implemented. 3P-TFWM gives the ability to measure spatial and spectral diffusion (discussed in more detail in Section 10.2). Spatial diffusion is measured by varying the angle between the first two pulses. If the first two pulses are held coincident in time, then this configuration is often called a transient grating experiment. Spectral diffusion can be observed by scanning the delay between the first two pulses for various delays between the second and third pulses (Weiner et al. 1985). 3P-TFWM also has the ability to distinguish between homogeneously and inhomogeneously broadened systems. If there is a fixed finite delay between the second and third pulses, scanning the conjugated pulse, $\mathbf{k}_a$, through the second pulse will produce a symmetric trace for a homogeneously broadened system, whereas it will be asymmetric for an inhomogeneously broadened system because a photon echo is only produced when $\mathbf{k}_a$ arrives first.

Either temporally or spectrally resolving the signal results in a signal as a function of two variables, that is, the delay between excitation pulses and the "real" time for TR-TFWM or the emission photon energy for SR-TFWM. A different approach is known as coherent excitation spectroscopy (CES) (Euteneuer et al. 1999). Based on earlier work that showed it was useful to use pulses with differing spectra (Cundiff et al. 1996), CES uses a narrowband first pulse and a broadband second pulse and measures the signal spectrum as a function of the photon energy of the first pulse. The result is a plot in two frequency dimensions that can clearly show coupling between resonances. TR-TFWM and SR-TFWM are discussed in more detail in Section 10.3.

More recently, multidimensional Fourier transform techniques have been developed. These methods were originally developed in NMR (Ernst et al. 1987). They essentially constitute an extension of TFWM, where the signal field is fully measured in phase and amplitude and the phase is correlated with the phase delay between excitation pulses. Two-dimensional Fourier transform spectroscopy (2DFTS) has given significant insight into the nonlinear optical response of semiconductors (Zhang et al. 2007, Cundiff et al. 2009). Multidimensional spectroscopy is discussed in more detail in Section 10.5.

10.1.3 Use of TFWM Spectroscopy in Solids, Semiconductors, and Complex Systems

TFWM spectroscopy has been extensively used to study exciton resonances in semiconductors and semiconductor nanostructures. Most studies were performed on III–V semiconductors because they interact strongly with light. This work was motivated by the realization that TFWM is very sensitive to many-body interactions among the optically created excitations in semiconductors (Chemla and Shah 2001). Most of this work has focused on excitonic resonances. Excitons are electron–hole pairs bound by Coulomb attraction (Chapter 2). Excitons have much longer dephasing times than free electron–hole pairs because they are neutral particles and thus experience weaker interactions with their environment and give a stronger TFWM signal. Many-body effects further enhance the signal.

The dephasing of electron–hole pairs in the presence of a two-dimensional electron gas was studied using TFWM in the quantum Hall effect regime (Fromer et al. 1999). As the magnetic field was increased, a crossover from Markovian to non-Markovian dynamics was observed. The dephasing time underwent jumps at even Landau-level filling factors. The observations were qualitatively reproduced by a model based on scattering by the collective excitations of the two-dimensional electron gas. Further experiments were done using 3P-TFWM, which showed a strong oscillatory off-resonant signal (Dani et al. 2006). By comparison to a microscopic theory, quantitative information about dephasing and interference of many-particle coherences was extracted.

Infrared TFWM experiments have been used to study the ultrafast vibrational dynamics and stability of amorphous silicon (Wells et al. 2002). Deuterated amorphous silicon showed differences in the Si-D stretch vibrations compared to Si-H vibrations in hydrogenated amorphous silicon. Photon-echo measurements of the vibrational dephasing indicated that nonequilibrium transverse optical phonons contributed at low temperatures, while elastic phonon scattering from transverse acoustic phonons contributed at elevated temperatures.

10.2 THEORY

For a simple system, such as an isolated two-level system, it is possible to develop a complete theoretical description of the nonlinear light–matter interaction that gives rise to TFWM signals. In certain limits, such as approximating the excitation pulses as being infinitely short, analytic results can be obtained. For complex systems, especially where strong many-body interactions dominate the optical response, the theory is much more complex (Haug and Koch 2004, Meier et al. 2007) and typically analytically intractable.

To provide insight into TFWM spectroscopy, a perturbation theory treatment for an ensemble of two-level systems will be presented here. The first approach will completely neglect many-body effects but include inhomogeneous broadening and spectral diffusion for three-pulse excitation (Weiner et al. 1985). The second approach will phenomenologically add many-body effects for two-pulse excitation, but neglect inhomogeneous broadening and diffusion. As described in Section 10.3.4, three-pulse experiments provide more information than two-pulse experiments, although they are more complex from both an experimental and theoretical perspective.

Spectral diffusion describes any process by which an oscillator can change its frequency within an inhomogeneous distribution. Such processes occur in a variety of situations, such as excitons migrating between localization sites in a semiconductor nanostructure or molecules undergoing changes in conformation or solvent environment. The spectral shift will be described by a kernel $W(\omega, \omega')$, which is the rate at which oscillators shift from frequency ω' to ω. If the spectral diffusion rate is much larger than the overall decay of the excitation, a quasi-equilibrium distribution will be created after a sufficiently long time so that all the oscillators have undergone multiple frequency shifts. The quasi-equilibrium distribution may be the same as the inhomogeneous distribution, but in general they are not identical. Once established, the quasi-equilibrium distribution decays at the population decay rate. It is also assumed that all coherence is lost in the spectral diffusion process. This assumption is reasonable as each oscillator will acquire an arbitrary phase shift during the diffusion process. As a consequence, there will be no fixed phase relationship between the oscillators that have diffused into a given frequency group.

The analysis is done using the density matrix, ρ. The diagonal elements of the density matrix represent populations, while the off-diagonal elements represent coherences. Since the dipole moment operator only has off-diagonal elements, the electric field of the incident laser pulses couples the diagonal elements with the off-diagonal elements of the density matrix. The induced polarization, which is the driving term in Maxwell's equations and thus produces the signal, is $P = \text{Tr}(\mu \cdot \rho)$ where μ is the dipole moment operator. The equation of motion for the density matrix is

$$i\hbar\dot{\rho} = [H, \rho] + \frac{\partial\rho}{\partial t}\bigg|_{\text{decay}} \qquad (10.2)$$

where H is the Hamiltonian and the decay term includes the effects of spectral diffusion, spontaneous emission, and dephasing due to various scattering processes. The Hamiltonian can be divided into two contributions, $H = H_0 + H_1$, where H_0 describes the energy level structure of the system and H_1 describes the dipolar interaction between the system and light field.

10.2.1 TFWM Signal from an Inhomogeneously Broadened Ensemble of Two-Level Systems

From Equation 10.2, we find the equations of motion for the elements of ρ are

$$\dot{\rho}_{12}(\omega,t) = i\omega\rho_{12}(\omega,t) - i\frac{\mu E(t)}{\hbar}\left(2\rho_{22}(\omega,t) - G(\omega)\right) - \gamma_{\text{ph}}^t\rho_{12}(\omega,t) \qquad (10.3)$$

$$\dot{\rho}_{22}(\omega,t) = i\frac{\mu E(t)}{\hbar}\left(\rho_{12}(\omega,t) - \rho_{21}(\omega,t)\right) - \left(\gamma_{\text{pop}} + \Gamma\right)\rho_{22}(\omega,t) + \int W(\omega,\omega')\rho_{22}(\omega',t)d\omega' \qquad (10.4)$$

where

$E(t)$ is the electric field of the applied pulses
γ_{pop} is the population decay rate of the upper level
Γ is the spectral diffusion rate
$\gamma_{\text{ph}}^t = (1/2)\gamma_{\text{pop}} + \Gamma + \gamma_{\text{ph}}$, γ_{ph} represents pure dephasing processes
$G(\omega)$ is the inhomogeneous distribution of oscillators with $\int G(\omega)d\omega = 1$

Equation 10.3 describes evolution of the coherent superposition state, with the first term resulting in it oscillating at frequency ω, the second term describing the interaction with the incident electric field, and the last term describing dephasing of the coherence. Equation 10.4 describes the evolution of the excited state populations, where the first term describes the interaction with the incident electric field, the second term describes population decay due to spontaneous emission and spectral diffusion, and the last term is a source term due to spectral diffusion from other frequency groups into this one. Equations 10.3 and 10.4 are also known as the optical Bloch equations (OBEs) although typically the OBEs neglect spectral diffusion. In writing these equations, it has been assumed that all coherence is lost in the spectral diffusion process; thus, there is no source term in Equation 10.3. The equation of motion for ρ_{11} is determined by the requirement that population be conserved at each frequency, that is, $\rho_{11}(\omega) + \rho_{22}(\omega) = G(\omega)$. To obtain an analytic solution to Equation 10.4, we assume that the spectral diffusion kernel is independent of initial frequency, that is, $W(\omega, \omega') = \Gamma F(\omega)$ with $\int F(\omega)d\omega = 1$.

To calculate the TFWM signal, we perform a perturbation expansion of the OBEs treating $E(t)$ as a small perturbation. Assuming $\rho_{11}^{(0)}(\omega,t) = G(\omega)$, that is, all the oscillators start in the ground state, and expanding to third order in E yields

$$\rho_{12}^{(3)}(\omega,t) = \frac{-2i\mu^3}{\hbar^3} \int_{-\infty}^{t} \exp[(i\omega - \gamma_{ph}^t)(t-t')]E(t') \int_{-\infty}^{t'} \exp[-(\gamma_{pop} + \Gamma)(t'-t'')]$$

$$\times \{G(\omega)\int_{-\infty}^{t''} (E(t'')E(t''')\exp[(-i\omega - \gamma_{ph}^t)(t''-t''')] + c.c.)dt'''$$

$$+\Gamma F(\omega)\int_{-\infty}^{t''} \exp[-\gamma_{pop}(t''-t''')]\int_{-\infty}^{+\infty} G(\omega')$$

$$\times \int_{-\infty}^{t'''} (E(t''')E(t'''')\exp[(-i\omega' - \gamma_{ph}^t)(t'''-t'''')] + c.c.)dt''''d\omega'dt'''dt''\}dt' \quad (10.5)$$

This result is general within the approximation of being a perturbation expansion. To proceed, we assume that the electric field has the form

$$E(t) = \sum_i \hat{E}(t)\exp[i(\omega_i t - \mathbf{k}_i \cdot \mathbf{x})] \quad (10.6)$$

and make the rotating wave approximation. Additionally, we choose the experimental geometry for 3P-TFWM, namely, that three pulses are incident on the sample, where a pulse with wave vector $\mathbf{k}_1$ arrives at time $t = -\tau$, that with $\mathbf{k}_2$ arrives at $t = 0$, and that with $\mathbf{k}_3$ arrives at $t = T$. We choose the signal direction to be $-\mathbf{k}_1 + \mathbf{k}_2 + \mathbf{k}_3$. Making the approximation that the pulses are infinitely short, that is, $\hat{E}(t) \to \delta(t)$ where $\delta(x)$ is the Kronecker delta function, we can evaluate the integrals in Equation 10.5. Selecting the terms that correspond to $T > 0$ and $t > 0$ and assuming Gaussian profiles for the inhomogeneous profile (center frequency ω_0^{in} and width $\Delta\omega_{in}$) and redistribution kernel (center frequency ω_0^{re} and width $\Delta\omega_{re}$) yields

$$\rho_{12}^{(3)} = -\frac{2i\mu^3}{\hbar} \exp\left[i\omega_l(T+\tau) - i(-i\mathbf{k}_1 + i\mathbf{k}_2 + i\mathbf{k}_2)\cdot\mathbf{x}\right]\exp\left[-\gamma_{ph}^t(t-T+\tau)\right]\Theta(t-T)\Theta(T)\Theta(\tau)$$

$$\times\left\{\exp\left[-(\gamma_{pop}+\Gamma)T\right]\exp\left[i\omega_0^{in}(t-T-\tau)\right]\exp\left[-\frac{\Delta\omega_{in}^2}{4\ln 2}(t-T-\tau)^2\right]\right.$$

$$+\exp\left[-\gamma_{pop}T\right](1-\exp[-\Gamma T])\exp\left[-i\omega_0^{in}\tau + i\omega_0^{re}(t-T)\right]$$

$$\times\exp\left[-\frac{\Delta\omega_{in}^2}{4\ln 2}\tau^2\right]\exp\left[-\frac{\Delta\omega_{in}^2}{4\ln 2}(t-T)^2\right]\right\} \quad (10.7)$$

where $\Theta(x)$ is the Heaviside step function. There are two contributions to this response. The first is a pulse emitted at a time τ after the arrival of the third pulse and corresponds to the classic photon echo. The other is emitted promptly with respect to the third pulse and arises from spectrally diffused excitations. The two contributions correspond to the two terms in curly brackets in Equation 10.7. The temporal width of the echo is determined by the width of the inhomogeneous distribution (broader inhomogeneous distributions give shorter echoes). For times long compared to both the pulse width and inverse inhomogeneous width, the time-integrated photon echo intensity is $I(\tau,T) \propto \exp[-4\gamma'_{ph}\tau - 2(\gamma_{pop} + \Gamma)T]$, allowing easy determination of both the dephasing rate by scanning τ and total population decay rate by scanning T. In each case the other delay is simply held at a fixed value, typically close to zero to have the strongest signal. However, it is best not to set the fixed delay exactly to zero, but rather to one pulse width, because the calculation has assumed strict time ordering and ignored terms that may contribute when the pulses overlap.

Since the system has to be in a coherent superposition state to have a wave function that oscillates as a function of time, the TFWM signal is generated by the induced third-order nonlinear polarization

$$P^{(3)}(t) = \int \mu(\rho_{12}^{(3)}(\omega,t) + c.c)d\omega \qquad (10.8)$$

Thus, from the third-order off-diagonal elements, we can compute the emitted TFWM signal.

An important aspect of the response is the ability to distinguish between inhomogeneously broadened and homogeneously broadened systems. In contrast to the earlier discussion, the signal emitted for a homogeneously broadened system occurs coincidentally with the arrival of the third pulse and is an exponential decay with respect to t. The time-integrated intensity is $I(\tau, T) \alpha \exp[-2\gamma_t\tau - 2\gamma_{sp}T]$. The factor of 2 difference in the decay of the signal with respect to τ makes it important to determine the nature of the broadening when interpreting TFWM signals. In the case of strong inhomogeneous broadening, this determination can be made by comparison to linear spectra. Otherwise, TR-TFWM or SR-TFWM can be used. Understanding the nature of the broadening can also give important physical insight. The temporal behavior of the signal can be exploited to determine the nature of the broadening by performing TR-TFWM measurements.

The contribution due to the spectrally diffused population has a somewhat more complicated behavior. It is only observable for $T > 0$ and $\tau \sim 0$, which provides a characteristic signature of spectral diffusion. If the signal is measured as a function of τ, then for $T = 0$ it is purely an exponential decay; however, for $T > 0$, an added contribution becomes observable at $\tau = 0$. This contribution is Gaussian with respect to t with a width proportional to the inverse inhomogeneous linewidth. It depends on T as $1 - \exp[\Gamma T]$.

To explain the behavior of the signal in the presence of spectral diffusion, the origin of the echo must be more carefully examined. The first pulse excites the oscillators, which subsequently evolve in time and acquire a phase that is frozen by the arrival of the second pulse (as discussed in Section 10.1.1). However, this phase is unique for each frequency group. When the third pulse arrives, it causes the oscillators to resume their temporal evolution; however, they have an initial phase that was acquired in the time interval between the first and second pulses. Due to the opposite sign for the frequency, this initial phase is such that at a time after the arrival of the third pulse that is equal to the interval between the first and second pulses, the net phase for all frequency groups returns to zero and coherent emission is observed. The oscillators are always emitting; however, at times other than the echo time, the emissions are not in phase and thus destructively interfere. The loss of signal due to spectral diffusion for a large delay between the second and third pulses can now be understood by observing that the spectral diffusion results in the oscillator having the wrong frequency during the rephasing process and hence an incorrect net phase. Except, of course, for zero delay between the first and second pulses, in which case no rephasing is required and a strong signal is still emitted, resulting in a spike for $\tau = 0$.

10.2.2 Phenomenological Inclusion of Many-Body Effects

The coherent optical response of a two-level system is described by the OBE. Inclusion of phenomenological corrections to describe many-body effects, such as excitation-induced dephasing (EID), excitation-induced shifts (EIS), and local field correction (LFC), yields a set of modified OBE (MOBE):

$$\dot{\rho}_{22} = -\gamma_{pop}\rho_{22} + \frac{i}{\hbar}\left[\mu(E+LP)\right](\rho_{12} - \rho_{21})$$

$$\dot{\rho}_{12} = -(\gamma_{ph}^t + \gamma'N\rho_{22})\rho_{12} + i(\omega_0 + \omega'N\rho_{22})\rho_{12} + \frac{i}{\hbar}\left[\mu(E+LP)\right](\rho_{22} - \rho_{11})$$

(10.9)

where EID is included through the parameter γ', which causes the dephasing rate to increase for increasing excited state population if $\gamma' > 0$. Similarly, EIS is included through ω', which causes the resonant frequency to increase with increasing excited state population for $\omega' > 0$. Both ω' and γ' can be either positive or negative. The renormalization of the electric field by the induced polarization P due to LFC is controlled by the parameter L. N is the density of oscillators, which in a semiconductor can be set to obtain the correct absorption strength. Here, we have neglected inhomogeneous broadening and spectral diffusion. Including them makes the equations intractable to solve analytically. It is worth mentioning that these equations are not the same as the semiconductor Bloch equations (SBE), which describe the evolution of the polarization and carrier populations in k-space (Haug and Koch 2004). Generally, numerical methods must be used to solve the SBE.

The MOBE can be solved analytically using perturbation theory if the applied electric field is assumed to be Dirac-delta function pulses or in the CW limit. Solving to third order using delta function pulses and selecting the term that radiates in the direction $2\mathbf{k}_2 - \mathbf{k}_1$ yields the third-order polarization:

$$P_{12}^{(3)}(t) = i\frac{\mu^4}{\hbar}\frac{N}{8}E_2^2E_1^*\left\{\left[1 + C\left(1 - e^{-\gamma_{pop}(t-\tau)}\right)\right]e^{\lambda_{ph}^t t}\Theta(t-\tau)\Theta(\tau)\right.$$

$$\left. + C\left(1 - e^{-\gamma_{pop}t}\right)e^{\lambda_{ph}^t(t-2\tau)}\Theta(t)\Theta(-\tau)\right\}e^{i(\omega_0 - \omega_l - \eta N)(t-2\tau)} + c.c \qquad (10.10)$$

where $C = N(\gamma' - i2\eta - i\omega')/\gamma_{pop}$, $\eta = \mu^2 L/\hbar$. The time-integrated signal can be calculated from Equation 10.7, yielding

$$I_s(\tau) = \frac{\mu^8}{\hbar^6}\frac{N^2}{128}\frac{I_2^2 I_1}{\gamma_{ph}^t}\left\{\left(1 + \frac{2\gamma'N}{\gamma_{pop} + 2\gamma_{ph}^t}\right)e^{-2\gamma_{ph}^t\tau}\Theta(\tau)\right.$$

$$\left. + \frac{N^2\left(\gamma'^2 + (2\eta + \omega')^2\right)}{\left(\gamma_{pop} + 2\gamma_{ph}^t\right)\left(\gamma_{pop} + \gamma_{ph}^t\right)}\left(e^{-2\gamma_{ph}^t\tau}\Theta(\tau) + e^{+4\gamma_{ph}^t}\Theta(-\tau)\right)\right\} \qquad (10.11)$$

where I_n is the intensity of pulse n, which shows the presence of a signal for $\tau < 0$ in the presence of EID, EIS, or LFC. The similarity in the effect of all three is evident from the fact that L, γ', and ω' appear together. Indeed, they are virtually inseparable: based on the perturbation result it is impossible to determine which is responsible for a signal at $\tau < 0$. This conclusion is the main point of this calculation. Thus, further information such as SR-TFWM or TR-TFWM is required to determine which effect(s) is(are) responsible for a signal at $\tau < 0$. The signals described by Equations 10.10 and 10.11 are clearly inconsistent with the underlying physics, namely, the decays (or corresponding linewidths) are independent of excitation density, whereas they should be density dependent because of the inclusion of EID and EIS. This discrepancy is an inherent limitation of a perturbation approach. The density-dependent decays can be added in by hand (Wang et al. 1993) or numerical calculations can be performed (Shacklette and Cundiff 2003).

10.3 EXPERIMENTAL DETAILS OF TFWM

TFWM experiments share many details with other time-domain spectroscopic techniques such as transient absorption (also known as pump-probe) (Chapter 9). The following sections give a basic overview of how to build a TFWM setup. They will describe a "typical" setup, but there are always alternate methods and not all of them can be discussed.

To start, a source of appropriate pulses is needed. These are typically produced by a mode-locked laser, possibly followed by amplification stages (Chapter 7). The main characteristics of the pulses are wavelength, duration, pulse energy, and repetition rate. Often, the last two are coupled as typically the available average power is approximately constant; thus, higher pulse energy requires lower repetition rate. However, it may be necessary to lower the repetition rate if the sample has slow relaxation dynamics, otherwise excitations will accumulate and effectively bleach the optical response of the sample, even if the coherent dynamics are occurring on a much shorter timescale. The wavelength needs to match any resonances to be studied. While shorter pulses provide greater temporal resolution and provide greater spectral coverage, they also reduce spectral selectivity. Experiments that include some spectral selections, such as SR-TFWM or 2DFTS can provide spectral selectivity in the detection; however, the presence of many-body effects may mean that the results are different if short, broadband, excitation pulses are used as compared to longer, narrower band pulses.

Clearly the choice of light source will be dictated by the system to be studied. If a resonance with a large dipole moment (resulting in high light absorption) is to be studied, such as excitons in semiconductors, a mode-locked oscillator provides sufficient pulse energy. Semiconductors based on GaAs have their fundamental band gap around 800–900 nm (depending on temperature, structure, and exact material composition), which fall conveniently in the tuning range of femtosecond Ti:sapphire lasers. Wide-gap materials, such as GaN, can be accessed with higher harmonics of a Ti:sapphire laser. Narrower gap materials, such as InAs and self-organized quantum dots based on it, typically require the use of an optical parametric oscillator. Materials with smaller dipole moments, such as colloidal quantum dots, require the use of amplified pulses. Generally, it is desirable to maintain the highest repetition rate possible to obtain the best signal-to-noise ratio.

10.3.1 Basics of Excitation Pulse Generation, Alignment, and Signal Detection

Alignment of the excitation beams in a TFWM experiment presents similar challenges to other transient techniques, namely, overlap between the excitation pulses in space and time must be found. In addition, TFWM presents the challenge of being a "background-free" experiment in that the signal direction is unique and does not match that of any of the excitation pulses. Finally, when aligning a TFWM experiment for the first time, a GaAs/AlGaAs multiple-quantum-well sample with ~10 periods is useful for generating a large signal at 800 nm, if available.

The excitation pulses are produced by using partially reflective beam splitter(s) to split the laser output into two or more arms. A variable delay is produced in one or more arms by simply using the time-of-flight of light,* for example, by mounting a retroreflector on a motorized and computer-controlled translation stage. The retroreflector can be a corner-cube; however, if the range of delay needed is relatively small, a pair of mirrors carefully set at right angles can be sufficient. The beam must be aligned carefully along the axis of motion for the delay stage, otherwise the retroreflected beam will move transversely as the stage is moved. The accuracy of this alignment must be greater if long delay scans are expected. However, it must be remembered that reasonably long scans may be needed to find zero delay, and thus sloppy alignment can impede finding zero delay. Since only the relative delay between the pulses matters, one arm can have a fixed delay. It can be convenient to use a manual translation stage in that arm to allow the position of temporal overlap to be adjusted.

The excitation pulses must spatially overlap at the sample. Typically, it also makes sense to have the beams focus at the sample as well to produce the highest intensity and thus typically the highest nonlinear signal. In TFWM, the spots on the sample are usually the same size. Since the transverse profile of the beams is a Gaussian, the signal will be produced from regions with a range of excitation densities. Having a range of excitation densities is not a problem if the incident powers are kept low enough that the signal scales cubically with excitation intensity (known as the "$\chi^{(3)}$ regime"). A convenient way to produce beams that cross and focus at the same spot is to have parallel beams pass through a single lens equidistant from the center. It is important to initially align the beams very parallel to each other, and making sure that they are parallel to the table is also important. Then simply inserting a lens will cause them to cross and focus at the same place, as long as their divergence angle is small. If the beams have a diameter of a few mm, spacing them by 20 mm and using a 50 mm diameter lens are usually good choices. The focal length of the lens can be chosen to give the desired spot size or angle between beams. Of course, the spacing between the beams can also be adjusted to change the angle between them. Even if the beams are perfectly parallel, it is important to verify that they are crossing at the sample position. For reasonably large spots in the visible (or near-IR with a viewer), it is possible to do so by eye (while following standard eye safety guidelines for working with lasers (Chapter 5)). Alternatively, the sample can be imaged using a CCD camera (or other sensors depending on wavelength), which can give magnification for small spots. If none of these approaches are possible, then the sample can be replaced with an aperture that is approximately the same diameter as the beams, and the power transmitted through the aperture can be optimized for each beam. It helps to mount the input lens on a translation stage that allows the distance from the lens to the sample to be adjusted.

The next step is to find zero delay, where the pulses temporally overlap in the sample. The approximate position of the delay stage can be determined by simply measuring the path lengths of the arms. Rather than explicitly measuring each segment, it is usually sufficient to simply use a piece of string to trace out the path. Additional delay due to passing through optical elements must be taken into consideration, although it is good practice to try using matching optics in the arms to minimize the differential chirp on the pulses. However, simple path length measurement is rarely sufficient to get an "exact" zero delay, but rather it provides an approximate position. In some cases, the approximate position is sufficient to

---

* Remembering the speed of light as 300 μm/ps is useful.

then begin hunting for a TFWM signal by running scans of the translation stage. However, because the TFWM signal can also be hard to locate (since its exact direction is unknown), it typically makes sense to use another experiment to refine the zero delay position. One possibility is to do a transient absorption experiment on the same sample, if it has a reasonably strong transient absorption signal (Chapter 9). This experiment can easily be performed by simply moving the detector into the probe beam. It is particularly convenient if the signal is reasonably long lived as it can then easily be found by setting the "probe" pulse arrival time sufficiently late that it will definitely arrive after the pump, given the uncertainty in the path lengths. It is also possible to determine the zero delay position by performing an autocorrelation, in which the sample is replaced with a second harmonic crystal or a photodiode with a large band gap so that only two-photon absorption occurs (e.g., GaAsP photodiodes work well with Ti:sapphire lasers) (Chapter 7). The latter option is particularly convenient as there is no issue of phase matching or finding the second harmonic signal.

Once spatial and temporal overlaps are achieved, the search for the signal begins. If zero delay is accurately known, then set the delay to be approximately one pulse width, with the appropriate ordering (i.e., make sure that pulse $\mathbf{k}_2$ arrives second by about one pulse width if searching for a signal in direction $2\mathbf{k}_2 - \mathbf{k}_1$), as the signal typically does not peak at zero delay unless the dephasing time is shorter than the pulse width. If the delay is approximately known, and the sample is thought to have a reasonably long dephasing time, the delays can be set to approximately the $1/e$ time. Otherwise, the delay will have to be systematically scanned to find the signal. It is generally best to put a lens after the sample that matches the one before the sample. This second lens recollimates the signal and makes it propagate parallel to the transmitted excitation pulses. Actually, it is best to install this lens prior to inserting the sample. The spatial location of the signal can be estimated from the positions of the transmitted excitation pulses. The specific position will be discussed in the following as it depends on the experimental configuration. To start, it is best to simply put a detector (e.g., a photodiode with integrated amplifier or photomultiplier tube) at the expected position of the signal. Placing an iris in front of the detector helps suppress background light. Initially, the iris can be set to be slightly larger than the beam diameters, and then later it should be adjusted to optimize the signal-to-background ratio. In some experimental geometries, care needs to be taken that the signal beam is not being accidently blocked by, for example, the finite size of a cryostat window or other aperture.

Since TFWM signals are relatively weak, especially before full optimization, it is common to use lock-in detection (Chapter 9). Mechanically chopping one beam is the easiest and usually sufficient. However, the signal-to-noise ratio can be further increased by chopping two beams at different frequencies and detecting the sum or difference frequency. Some mechanical choppers are specifically designed for dual frequency chopping. Even if more complicated detection will be used, for example, time or spectrally resolving the TFWM signal, it is usually best to start by using a simple detector and lock-in detection to optimize the TFWM signal.

The initial signal will typically be several orders of magnitude weaker than when it is fully aligned. Obviously, the first step is simply to vary all alignment knobs and the delay to find the maximum. In most cases, simple maximization is sufficient as the various degrees of freedom are orthogonal. One exception is the parallelism of the beams before the input lens and the position of the lens. If the beams are not focusing at the crossing point, simply adjusting the lens position will result in a reduction of signal because the overlap decreases, despite the fact that the focusing is improved. An iterative procedure is required where the lens is moved a small amount and then the angle between the beams is adjusted. If the signal goes up after both steps, then continue to move the lens in the same direction, performing both steps each time. If not, try going the other way. When moving the lens in either direction results in a net reduction in signal after optimizing the beam pointing, then it is at the optimum position.

Usually the dominant source of "noise" is actually scattering of the excitation pulses off the sample. In addition to using lock-in detection, spatial filtering in the signal beam is the best way to suppress the scattered light. The spatial filtering can simply be some irises placed along the signal beam path or a true spatial filter with a tight focus on a pinhole. In the latter case, care must be taken with day-to-day adjustments. The spatial filter will strongly select a specific signal direction, which in turn sets the wave vectors of the excitation pulses. Drift in the alignment may result in the optimum overlap not corresponding to the direction determined by the spatial filter. Thus, it is best to occasionally remove the pinhole and re-optimize the alignment of the excitation beams, followed by optimization of the pinhole without touching the excitation alignment. It is useful to note the signal strength and all relevant settings (e.g., gain settings and any attenuation) each day to help diagnose if drift is occurring. In some cases, polarization can be used to partially suppress scatter from the excitation pulses. However, the polarization dependence of the signal can be very nontrivial (see Section 10.4.5), so great care must be taken when varying the polarization of the incident beams.

There are a multitude of problems that can arise in a TFWM experiment. A few of the more common ones will be briefly mentioned. A very common problem is the presence of a spurious signal when the pulses temporally overlap on the sample. This signal is due to interference between scattered light from the excitation pulses. Since the paths are typically not phase stable, the signal will fluctuate wildly. While it can be useful for finding zero delay, this signal is typically undesirable. Often it is not a problem once the TFWM is optimized. If it persists, the sample may be scattering light excessively, so trying a different part of the sample may help. If nothing else works, purposely jiggling a mirror, for example, using a piezoelectric transducer or a voice coil, can average out the interference. Spurious signals can also arise due to a combination of scattered light and transient absorption in the sample. These signals are more troublesome as they will persist at long delays. In a three-pulse experiment, they can be subtracted out by blocking the unchopped beam; however, in a two-pulse experiment that is not possible.

10.3.2 Time-Integrated Two-Pulse TFWM

A typical two-pulse TI-TFWM setup is shown in Figure 10.5. In general, the signal can be easily located. If a card is placed after the sample, the signal will be on a line drawn through the two transmitted excitation beams and spaced away from one excitation spot by the same distance as between the two excitation spots.

The two-pulse geometry is not phase matched, that is, $|2\mathbf{k}_2 - \mathbf{k}_1| > |\mathbf{k}_i|$, unless the angle between the pulses is zero, where i is either 1 or 2. However, for a thin sample (such as an epilayer)

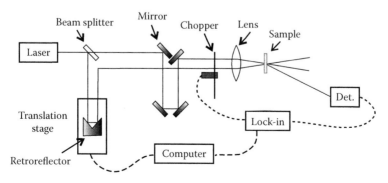

Figure 10.5   Typical two-pulse TI-TFWM setup.

and small angle between the beams, the phase mismatch is negligible. For samples with high index of refraction, such as semiconductors, the internal angle is even smaller due to refraction, and thus the phase mismatch is even further reduced. A detailed discussion of phase matching is given in Chapter 7.

A two-pulse experiment can also be done in reflection. In this case, the signal is located in the same position, but with respect to the reflected excitation beams. For a thin sample, the strength of the reflected signal is comparable to that of the transmitted signal. The reflection geometry can also be used for opaque samples. The reflected signal is due to the same nonlinear polarization as the transmitted signal, although it is not phase matched, and thus contains the same information.

10.3.3 Time and Spectrally Resolved Two-Pulse TFWM

Time or spectrally resolved TFWM have usually been implemented using two-pulse excitation, although they can also be implemented using 3P-TFWM (next section). SR-TFWM is fairly straightforward to implement as the detector is simply replaced with a spectrometer. If a scanning spectrometer is used, lock-in detection is still possible. If a spectrograph with a CCD camera is used instead, then lock-in detection cannot be used. Often some type of automated background subtraction must be implemented, typically using automated shutters to gate one or multiple excitation beams. TR-TFWM is typically implemented using up-conversion in a crystal with second-order nonlinearity (Chapter 7). The reference pulse for up-conversion is generated in a manner similar to the excitation pulses.

Similar information is gained by performing TR-TFWM and SR-TFWM. However, they are not simply related to each other because both measure the intensity and not the full electric field with phase information. Measurements of the full field have been achieved (Chen et al. 1997) using spectral interferometry (Lepetit et al. 1995).

10.3.4 Three-Pulse TFWM

3P-TFWM has several advantages: it can measure population relaxation as well as dephasing; it is sensitive to non-radiative coherences such as Raman or two-quantum coherences; it is sensitive to both spatial and spectral diffusion; and it can be fully phase matched. Of course, these advantages come at the cost of the additional complexity of a third excitation pulse.

Three common 3P-TFWM configurations are shown in Figure 10.6. The planar geometry is a simple extension of two-pulse TFWM that is not phase matched, although the phase mismatch is typically negligible, as for two-pulse TFWM. The second two configurations are both fully phase matched, so they can be used for thick samples.

Finding the signal for a 3P-TFWM experiment is typically no harder than for two-pulse TFWM and can even be easier. For the planar case, it is important to keep the three excitation beams in the same plane so that the signal will be in that plane. The signal spot will be the same distance from $\mathbf{k}_b$ as the distance between the spots for $\mathbf{k}_a$ and $\mathbf{k}_c$ (for beams labeled as in Figure 10.2). For the phase-conjugate geometry, the signal is counter propagating with respect to $\mathbf{k}_a$ and thus a beam splitter is used to separate it. Thus, by simply reflecting $\mathbf{k}_a$ back on itself, the detector can easily be positioned, and spatial filters can even be aligned prior to finding the actual signal. Alignment of the box geometry is relatively simple if care is taken to position the excitation pulses on three corners of a square. The signal is then emitted into the direction corresponding to the fourth corner of the square. Both the phase conjugate and box geometries have the advantage of needing less total angle than the two-pulse or planar three-pulse geometries.

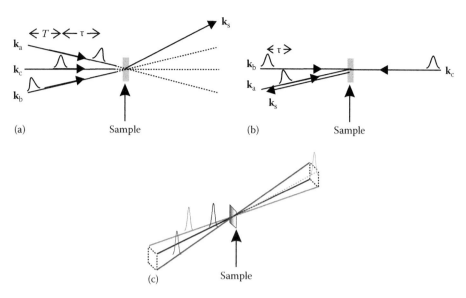

Figure 10.6 (See color insert.) 3P-TFWM configurations: (a) coplanar, (b) phase conjugate, and (c) box.

10.3.5 Co-linear TFWM

The TFWM configurations described so far all rely on wave vector selection to isolate the signal from the incident pulses. While the use of a background free geometry is an advantage of these types of TFWM, it can also be a disadvantage in certain situations. One example is when the sample does not permit the propagation of beams with different wave vectors, for example, a waveguide. Another possibility is if a single quantum emitter is being probed, in which case there is insufficient spatial extent to form a signal beam. By modulating the excitation beams, it is possible to isolate the signal from the excitation because the nonlinear optical response results in the signal having a modulation that is distinct from any of the excitation beams. A very successful technique is to frequency shift the excitation using acousto-optic modulators (Borri et al. 1999). For the pulse train produced by a mode-locked laser, a frequency shift is equivalent to phase modulation. By using appropriate modulation frequencies with respect to the laser repetition rate, it is possible to arrange for the final detection frequency to be in the range of a standard lock-in amplifier.

10.4 CASE STUDY: TFWM FROM EXCITONS IN GaAs QUANTUM WELLS

Direct gap semiconductors have been extensively studied using TFWM. Optical excitation creates electron–hole pairs, which are expected to experience scattering from other carriers, phonons, or disorder, on picosecond to sub-picosecond timescales. Furthermore, optical excitation can create nonequilibrium carrier distributions, which are expected to relax to equilibrium on sub-picosecond timescales (Chapter 8). At low temperature, where phonon scattering is sufficiently suppressed to reveal other processes, the electron–hole pairs can form an exciton, a bound state with a hydrogenic wave function for the relative coordinate (Chapter 2). Excitons result in strong resonant absorption features just below the gap and consequently tend to dominate the nonlinear optical response.

10.4.1 Dephasing

Coherent spectroscopic techniques were initially applied to semiconductors to determine the dephasing time. Much of this work focused on exciton resonances at low temperatures, as they have dephasing times in the picosecond range and large oscillator strengths. By studying the dependence of the dephasing rate on temperature and carrier density, the contributions from phonon scattering and carrier–carrier scattering could be determined.

One of the earliest demonstrations of a TFWM signal from the exciton resonance in a GaAs quantum well by Hegarty et al. (1982), showed that a strong signal could be obtained; however, the dephasing was not determined due to insufficient time resolution. This experiment was 3P-TFWM in the phase conjugate geometry. In subsequent work, an upper limit could be determined from a comparison to calculations (Schultheis et al. 1985). The sample displayed significant inhomogeneous broadening due to disorder; TR-TFWM experiments showed that the signal was a photon echo. The strong TFWM signal obtained in these experiments motivated further studies.

In a subsequent paper, the dephasing time (T_2) of excitons in a single quantum well (Schultheis et al. 1986) was measured (Figure 10.7) to be in the range of a few picoseconds using two-pulse TI-TFWM. The dephasing time matched the absorption linewidth, showing that the sample was homogeneously broadened. By varying the temperature, it was also possible to separate the contribution due to phonons, which was determined to be dominated by acoustic phonon scattering. The linear temperature dependence also supported the idea that the excitons were delocalized.

To characterize the excitonic dephasing due to carrier–carrier scattering, the dephasing time was measured in the presence of excitons injected by a pre-pulse or free electron–hole

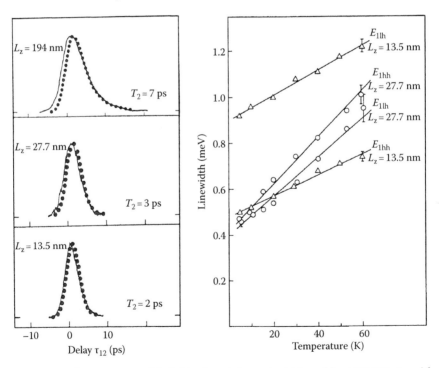

Figure 10.7 (Left) Experimental (solid line) and theoretical (dotted line) TFWM signal from a bulk GaAs layer (top) with thickness L_z and two single quantum wells at 2 K. Fitting the theory to the experiment yields the dephasing time T_2. (Right) Temperature dependence of the homogeneous linewidths for the heavy-hole and light-hole exciton transitions in the single quantum wells. (Reproduced from Schultheis, L. et al., *Phys. Rev. B*, 34, 9027, 1986. With permission. Copyright 1986 by the American Physical Society.)

pairs injected by a CW beam with higher photon energy (Honold et al. 1989). Since excitons are neutral particles they interact with each other through the dipole interaction, whereas the free carriers have direct Coulomb interactions with excitons. Not surprisingly, free electron–hole pairs were much more efficient at dephasing excitons. The presence of dephasing due to exciton–exciton scattering means that there are new terms in the nonlinear response, although it was not recognized at the time.

In parallel with these experimental developments, the theory of the coherent response of semiconductors was being developed (Schmitt-Rink et al. 1985, Lindberg and Koch 1988). These efforts resulted in SBE that have a form similar to the OBE. However, these early results were derived within a mean field (Hartree–Fock)-type approximation, which was later found to be insufficient to describe many of the experimentally observed effects.

10.4.2 Many-Body Effects

The studies presented in the previous section were motivated by the ability of coherent spectroscopy, mainly TFWM, to measure dephasing times. However, anomalies were observed in the signals, the most dramatic of which was the signal for "negative delay" in two-pulse TFWM experiments. As we showed earlier, two-pulse calculations for a two-level system only yield a signal for positive delay, that is, $\mathbf{k}_a$ arriving before $\mathbf{k}_b$. The negative delay signal in semiconductors is due to many-body interactions. However, there are multiple phenomena that can give rise to such a signal.

The initial observations of negative delay signals (Figure 10.8) attributed them to local field effects (Leo et al. 1990a, Wegener et al. 1990). As shown in Equation 10.9, local field effects can easily be incorporated into the OBE. Inclusion of the local field produces a correction to the oscillation frequency, known as the Lorenz–Lorentz shift. At third order, it also produces a TFWM signal for negative delays that rises as $\exp\left[4\gamma_{ph}^t \tau\right]$, which was

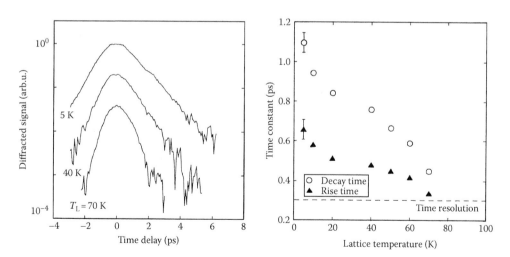

Figure 10.8 (Left) TFWM signal from a GaAs quantum well as a function of temperature, showing signals for negative delays. (Right) The rise and fall times obtained from an exponential fit, showing that they differ by a factor of 2, as expected for a negative delay signal due to local fields. (Reproduced from Leo, K. et al., *Phys. Rev. Lett.*, 65, 1340, 1990a; Leo, K. et al., *Appl. Phys. Lett.*, 57, 19, 1990b. Copyright 1990 by the American Physical Society. Copyright 1990 by American Institute of Physics.)

in agreement with the experiments, which showed that on a log scale, the slopes of the rise and fall differed by a factor of 2.

The generation of a signal in two-pulse TFWM for negative delays due to local fields has a straightforward physical explanation. The incident pulses produce a first-order polarization that radiates in the same direction as the pulse, that is, the well-known free decay (also known as the "free induction decay" from NMR). The re-radiated field can then drive the system on its own. However, this field decays with the dephasing rate and so can persist long after the pulse is gone. Thus, when the $\mathbf{k}_b$ pulse arrives first, the free decay can persist until after the $\mathbf{k}_a$ pulse arrives, thereby producing a TFWM signal. In 3P-TFWM, a "negative" delay corresponds to $\mathbf{k}_a$ arriving last. For nonzero delay between $\mathbf{k}_b$ and $\mathbf{k}_c$, $\mathbf{k}_a$ can arrive second (i.e., between $\mathbf{k}_b$ and $\mathbf{k}_c$), which can result for a signal from a homogeneously broadened system even in the absence of local fields or other many-body effects. A similar physical explanation can be used when $\mathbf{k}_a$ arrives last.

More sophisticated calculations based on the SBE also showed signals for negative delay (Lindberg et al. 1992, Schäfer et al. 1993). Following the previously developed language, these were also called "local field corrections"; however, their origin was not purely due to optical local fields but also due to many-body effects. Unfortunately, this usage has engendered some confusion. The calculations predicted that the TR-TFWM signal would show interesting dynamics, as did the OBE with an LFC. These effects were observed experimentally (Kim et al. 1992, Weiss et al. 1992).

The presence of strong excitation-induced effects was observed in early measurements of exciton dephasing. However, it was not realized at the time that these constituted many-body phenomena that resulted in new coherent signals. The fact that a dephasing rate that depended on the excitation level, known as "excitation-induced dephasing" (EID), would produce a signal was first discussed in the context of the polarization dependence of the signal from bulk GaAs (Wang et al. 1993). Similar effects were observed in quantum wells (Hu et al. 1994). Calculations were made based on the OBE with a correction to include EID (Wang et al. 1994).

While EID was fairly easy to observe, as it was readily apparent in the dependence of TFWM decay on the intensity of the incident beams, an EIS in the exciton resonance frequency can also produce a signal. However, an EIS is not directly apparent in the TFWM signal in the same way as EID. SR-TFWM experiments provided evidence for contributions from EIS (Shacklette and Cundiff 2002). In this work, spectrally resolved measurements were performed to verify that the EIS was present and determine its strength relative to EID.

Physically, the signal produced by EID can be understood in terms of a linewidth grating. Just as for the ordinary TFWM signal, the interference between the polarization produced by the first pulse and the second pulse results in a grating in the excited state population. This excited state grating in turn results in a modulation of the width of the exciton resonance due to the fact that EID means that the dephasing rate, which is just the linewidth, depends on excited state population. Increasing the width of a resonance decreases the absorption at line center, while increasing the absorption in the line wings. Thus, a spatial modulation in the absorption coefficient still occurs, which diffracts the second pulse into the signal direction. A similar explanation for EIS can be given, except the excited state grating results in a spatial modulation of the resonance frequency.

The picture of a linewidth grating does not explain why a signal occurs for negative delay nor does it intuitively explain the slow rise of the signal. These effects can be understood as follows. The first pulse ($\mathbf{k}_b$ in Figure 10.2, which arrives first for negative delay) creates a coherence. This coherence produces a free decay that radiates in the same direction. The second ($\mathbf{k}_a$ in Figure 10.2) pulse interacts with this polarization to produce a spatially modulated excited state population that acts on the coherence induced by the first pulse. In regions of large population, the original coherence begins to decay faster due to EID. After a time period determined by how much the excited state population increases the dephasing rate, there will be a spatial modulation of the coherence, which will change its radiation pattern so that it emits in the signal direction. The first pulse acts twice by both establishing the initial coherence and participating in the formation of the population grating. The second pulse only participates in forming the grating. This process can be thought of as scattering of the initial coherence into the signal direction by the population.

10.4.3 Disorder

Disorder can dramatically change the properties of materials, for example, transforming a conductor into an insulator. If the disorder results in local regions of lower potential energy, it is not surprising that the transport properties will be degraded as charge carriers are trapped. The localization of the wave function is important, as it means that the Bloch wave functions of a perfect lattice do not apply and that the density of states will be modified. Often theoretical approaches assume Bloch wave functions, which will affect their validity in the presence of disorder.

An important, and usually dominant, form of disorder in semiconductor quantum wells is fluctuations in the width of the well. Even in the highest quality samples, fluctuations of one monolayer are inevitable in the transition between the barrier and well materials. In wider regions of the well, the confinement energy will be lower, and the exciton center-of-mass wave function will tend to be spatially localized in the potential minimum. For localized excitons, migration amongst localization sites is an important relaxation mechanism, leading to dephasing and spectral diffusion. At the lowest temperature, migration occurs due to phonon assisted hopping (Takagahara 1985), while at higher temperatures thermal activation to the delocalized states occurs. 3P-TFWM experiments observed the temperature dependence predicted by these mechanisms (Webb et al. 1991b). These measurements also displayed the signatures of spectral diffusion.

Since disorder produces inhomogeneous broadening, the TFWM signal should be a photon echo, which was confirmed by TR-TFWM (Schultheis et al. 1985, Noll et al. 1990, Webb et al. 1991a). However, studies on intensity dependence showed a transition from an echo to a free decay (Webb et al. 1991a). Most likely this occurs due to changing many-body contributions; however, it has not been fully explained.

The presence of a mobility edge close to line center means that the dephasing rate varies significantly across the inhomogeneous distribution (Hegarty and Sturge 1985, Cundiff and Steel 1992). Below the mobility edge, the wave functions are localized, which reduces their interactions with other excitations and hence also reduces their dephasing rate. Above the mobility edge, the wave functions are extended, which increases their interactions with other excitations and hence their dephasing rate. For excitation pulses that cover the entire inhomogeneous distribution, non-exponential decays would be expected as some components decay faster than others (Spivey and Cundiff 2007). This effect was most evident as a change in the beat period for heavy-hole–light-hole beats.

10.4.4 Quantum Beats

When multiple resonances are excited by the incident pulses, beating will occur in the TFWM signal. The beats will occur at the frequency corresponding to the difference between the frequencies of the resonances. Beating can be classified as either originating from simple electromagnetic interference in the light emitted by two uncoupled resonances, often called polarization interference, or originating from quantum beats that are due to coupled resonances where a quantum mechanical oscillation in the wave function will occur. The first observation of beating in the TFWM signal from a semiconductor quantum well was attributed to the presence of distinct regions of the quantum well that differed in thickness by one monolayer (Göbel et al. 1990). In that paper, the beats were described as "quantum beats," although it was actually unclear whether or not the excitons localized in spatially separated areas really constituted a quantum mechanically coupled system.

It was found that both TR-TFWM (Koch et al. 1992) and SR-TFWM (Lyssenko et al. 1993) could distinguish between beats due to polarization interference and true quantum beats. In the presence of inhomogeneous broadening, the correlation between the broadening of the two transitions needs to be taken into account (Cundiff 1994). Two-dimensional Fourier transform spectroscopy (Section 10.5) can readily distinguish between these two cases as well (Zhang et al. 2005).

Beats were also observed between the heavy-hole and light-hole excitons in a GaAs quantum well (Feuerbacher et al. 1990, Leo et al. 1990b). For co-linear excitation, it was clear that these were quantum beats, and the results were in good agreement with theory. For co-circular excitation, the two allowed transitions are from the −3/2 heavy-hole valence band to the −1/2 conduction band and from the −1/2 light-hole valence band to the +1/2 conduction band, which appear to result in a pair of uncoupled transitions as there are no states in common. However, experiments showed that they were indeed coupled, which was attributed to many-body interactions (Chen et al. 1997, Phillips and Wang 1999, Smirl et al. 1999).

10.4.5 Polarization Dependence

Disorder was also shown to significantly affect the polarization dependence of the TFWM signal. It was observed that the signal was much weaker and decayed faster when the excitation pulses were cross-linearly polarized as compared to co-linearly polarized (Cundiff and Steel 1992, Cundiff et al. 1992, Yaffe et al. 1993). A connection between the polarization dependence and the inhomogeneous linewidth was observed, suggesting that the disorder played a role (Bennhardt et al. 1993). Furthermore, time resolving the signal showed that the signal changed from a photon echo for co-polarized excitation to a free decay for cross-polarized excitation (Cundiff et al. 1992). Eventually, these observations were reproduced by calculations that included disorder and correlations beyond the Hartree–Fock approximation (Weiser et al. 2000). Two-dimensional Fourier transform spectroscopy promises further insight into these effects (Kuznetsova et al. 2007).

10.5 MULTIDIMENSIONAL FOURIER TRANSFORM SPECTROSCOPY

In the last few years, a powerful technique has been developed that preserves the best features of TFWM while also providing the information gained by spectrally resolved transient absorption. This technique, known as multidimensional Fourier transform spectroscopy, is inspired by work in NMR (Ernst et al. 1987) and more recently in ultrafast chemistry (Jonas 2003).

Multidimensional Fourier transform spectroscopy is based on 3P-TFWM, but with the enhancement that the phase of the emitted signal is measured and correlated with the phase between the excitation pulses. The correlation is performed by taking a multidimensional Fourier transform. Typically two time dimensions are correlated, in which case the technique is known as two-dimensional Fourier transform spectroscopy (2DFTS). The pulse sequence and phases are shown schematically in Figure 10.9. This enhancement is powerful for several reasons: (1) Measuring the signal phase alone is important as it reveals information about the phenomenological origin of the coherent response. (2) Correlating the phases allows coupling

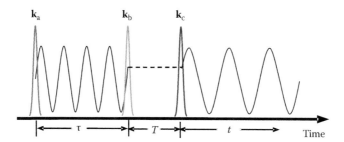

Figure 10.9 (See color insert.) The pulse sequence for 2DFTS and phase evolution. The initial pulse, $\mathbf{k}_a$, excites an initial coherence that evolves during a time period τ. The second pulse, $\mathbf{k}_b$, stores the phase of the initial coherence in a population state. The third pulse, $\mathbf{k}_c$, generates the coherence that radiates during time period t. The overall phase of the radiating coherence is determined by the phase evolution during time period t. By taking a two-dimensional Fourier transform of the signal with respect to τ and t, the frequencies of both the initial coherence and the emitting coherence can be determined. For uncoupled resonances, these two frequencies will always be the same, whereas they can be different if two resonances are coupled, for example, the two transitions of a three-level system.

between transitions to be determined. (3) The various terms that contribute to the nonlinear optical response can be separated and in some cases isolated. (4) The homogeneous and inhomogeneous linewidths can be determined simultaneously and unambiguously. (5) The evolution of non-radiative coherences, such as Raman or two-quantum coherences, can be observed.

The implementation of 2DFTS in the optical regime is challenging because it requires interferometric stability and measurement of the phase and amplitude of the signal (Zhang et al. 2005, Bristow et al. 2009). Interferometric stability can be achieved by active stabilization of the delays, which typically requires having a continuous wave laser traverse the same path as the femtosecond pulses. Measurement of the amplitude and phase of the signal can be done using spectral interferometry (Lepetit et al. 1995).

10.6 SUMMARY

TFWM is a powerful spectroscopic technique for studying complex materials such as semiconductor heterostructures. The technique can give information on dynamics, such as dephasing or energy relaxation. However, it is also very sensitive to many-body interactions among optically created excitations. Careful analysis of the TFWM signal can give insight into fundamental many-body interactions in these materials.

REFERENCES

Bennett, W. R. 1962. Hole burning effects in a He-Ne optical maser. *Phys. Rev.* 126: 580–593.

Bennhardt, D., P. Thomas, R. Eccleston, E. J. Mayer, and J. Kuhl. 1993. Polarization dependence of four-wave-mixing signals in quantum wells. *Phys. Rev. B* 47: 13485–13490.

Borri, P., W. Langbein, J. Mørk, and J. M. Hvam. 1999. Heterodyne pump-probe and four-wave mixing in semiconductor optical amplifiers using balanced lock-in detection. *Opt. Commun.* 169: 317–324.

Bristow, A. D., D. Karaiskaj, X. Dai, T. Zhang, C. Carlsson, K. R. Hagen, R. Jimenez, and S. T. Cundiff. 2009. A versatile ultrastable platform for optical multidimensional Fourier-transform spectroscopy. *Rev. Sci. Instrum.* 80: 073108.

Chemla, D. S. and J. Shah. 2001. Many-body and correlation effects in semiconductors. *Nature* 411: 549–557.

Chen, X., W. J. Walecki, O. Buccafusca, D. N. Fittinghoff, and A. L. Smirl. 1997. Temporally and spectrally resolved amplitude and phase of coherent four-wave-mixing emission from GaAs quantum wells. *Phys. Rev. B* 56: 9738–9743.

Cundiff, S. T. 1994. Effects of correlation between inhomogeneously broadened transitions on quantum beats in transient 4-wave-mixing. *Phys. Rev. A* 49: 3114–3118.

Cundiff, S. T., M. Koch, W. H. Knox, J. Shah, and W. Stolz. 1996. Optical coherence in semiconductors: Strong emission mediated by nondegenerate interactions. *Phys. Rev. Lett.* 77: 1107–1110.

Cundiff, S. T. and D. G. Steel. 1992. Coherent transient spectroscopy of excitons in GaAs-AlGaAs quantum-wells. *IEEE J. Quantum Electron.* 28: 2423–2433.

Cundiff, S. T., H. Wang, and D. G. Steel. 1992. Polarization-dependent picosecond excitonic nonlinearities and the complexities of disorder. *Phys. Rev. B* 46: 7248–7251.

Cundiff, S. T., T. H. Zhang, A. D. Bristow, D. Karaiskaj, and X. Dai. 2009. Optical two-dimensional Fourier transform spectroscopy of semiconductor quantum wells. *Acc. Chem. Res.* 41: 1423–1432.

Dani, K. M., J. Tignon, M. Breit, D. S. Chemla, E. G. Kavousanaki, and I. E. Perakis. 2006. Ultrafast dynamics of coherences in a quantum hall system. *Phys. Rev. Lett.* 97: 057401.

Ernst, R., G. Bodenhausen, and A. Wokaun. 1987. *Principles of Nuclear Magnetic Resonance in One and Two Dimensions*. Oxford, U.K.: Oxford Science Publications.

Euteneuer, A., E. Finger, M. Hofmann, W. Stolz, T. Meier, P. Thomas, S. W. Koch, W. W. Rühle, R. Hey, and K. Ploog. 1999. Coherent excitation spectroscopy on inhomogeneous exciton ensembles. *Phys. Rev. Lett.* 83: 2073–2076.

Feuerbacher, B., J. Kuhl, R. Eccleston, and K. Ploog. 1990. Quantum beats between the light and heavy hole excitons in a quantum well. *Solid State Commun.* 74: 1279–1283.

Fromer, N. A., C. Schüller, D. S. Chemla, T. V. Shahbazyan, I. E. Perakis, K. Maranowski, and A. C. Gossard. 1999. Electronic dephasing in the quantum hall regime. *Phys. Rev. Lett.* 83: 4646–4649.

Göbel, E. O., K. Leo, T. C. Damen, J. Shah, S. Schmitt-Rink, W. Schäfer, J. F. Muller, and K. Köhler. 1990. Quantum beats of excitons in quantum-wells. *Phys. Rev. Lett.* 64: 1801–1804.

Haug, H. and S. W. Koch. 2004. *Quantum Theory of the Optical and Electronic Properties of Semiconductors*. River Edge, NJ: World Scientific.

Hegarty, J. and M. D. Sturge. 1985. Studies of exciton localization in quantum-well structures by nonlinear-optical techniques. *J. Opt. Soc. Am. B* 2: 1143–1154.

Hegarty, J., M. D. Sturge, A. C. Gossard, and W. Wiegmann. 1982. Resonant degenerate 4-wave mixing in GaAs multiquantum well structures. *Appl. Phys. Lett.* 40: 132–134.

Honold, A., L. Schultheis, J. Kuhl, and C. W. Tu. 1989. Collision broadening of two-dimensional excitons in a GaAs single quantum well. *Phys. Rev. B* 40: 6442–6445.

Hu, Y. Z., R. Binder, S. W. Koch, S. T. Cundiff, H. Wang, and D. G. Steel. 1994. Excitation and polarization effects in semiconductor 4-wave-mixing spectroscopy. *Phys. Rev. B* 49: 14382–14386.

Jonas, D. 2003. Two-dimensional femtosecond spectroscopy. *Annu. Rev. Phys. Chem.* 54: 425–463.

Kim, D. S., J. Shah, T. C. Damen, W. Schäfer, F. Jahnke, S. Schmitt-Rink, and K. Köhler. 1992. Unusually slow temporal evolution of femtosecond 4-wave-mixing signals in intrinsic GaAs quantum-wells—Direct evidence for the dominance of interaction effects. *Phys. Rev. Lett.* 69: 2725–2728.

Koch, M., J. Feldmann, G. von Plessen, E. O. Göbel, P. Thomas, and K. Köhler. 1992. Quantum beats versus polarization interference: An experimental distinction. *Phys. Rev. Lett.* 69: 3631–3634.

Kurnit, N. A., S. R. Hartmann, and I. D. Abella. 1964. Observation of photon echo. *Phys. Rev. Lett.* 13: 567–569.

Kuznetsova, I., T. Meier, S. T. Cundiff, and P. Thomas. 2007. Determination of homogeneous and inhomogeneous broadening in semiconductor nanostructures by two-dimensional Fourier-transform optical spectroscopy. *Phys. Rev. B* 76: 153301.

Leo, K., M. Wegener, J. Shah, D. S. Chemla, E. O. Göbel, T. C. Damen, S. Schmitt-Rink, and W. Schäfer. 1990a. Effects of coherent polarization interactions on time-resolved degenerate 4-wave-mixing. *Phys. Rev. Lett.* 65: 1340–1343.

Leo, K., T. C. Damen, J. Shah, E. O. Göbel, and K. Köhler. 1990b. Quantum beats of light hole and heavy hole excitonsin quantum wells. *Appl. Phys. Lett.* 57: 19–21.

Lepetit, L., G. Cheriaux, and M. Joffre. 1995. Linear techniques of phase measurement by femto-second spectral interferometry for applications in spectroscopy. *J. Opt. Soc. Am. B* 12: 2467.

Lindberg, M., R. Binder, and S. W. Koch. 1992. Theory of the semiconductor photon-echo. *Phys. Rev. A* 45: 1865–1875.

Lindberg, M. and S. W. Koch. 1988. Effective Bloch equations for semiconductors. *Phys. Rev. B* 38: 3342–3350.

Lyssenko, V. G., J. Erland, I. Balslev, K. H. Pantke, B. S. Razbirin, and J. M. Hvam. 1993. Nature of nonlinear 4-wavemixing beats in semiconductors. *Phys. Rev. B* 48: 5720–5723.

Meier, T., P. Thomas, and S.W. Koch. 2007. *Coherent Semiconductor Optics.* Berlin, Germany: Springer.

Noll, G., U. Siegner, S. G. Shevel, and E. O. Göbel. 1990. Picosecond stimulated photon echo due to intrinsic excitations in semiconductor mixed crystals. *Phys. Rev. Lett.* 64: 792–795.

Phillips, M. and H. Wang. 1999. Coherent oscillation in four-wave mixing of interacting excitons. *Solid State Commun.* 111: 317–321.

Schäfer, W., F. Jahnke, and S. Schmitt-Rink. 1993. Many-particle effects on transient 4-wave-mixing signals in semiconductors. *Phys. Rev. B* 47: 1217–1220.

Schmitt-Rink, S., D. S. Chemla, and D. A. B. Miller. 1985. Theory of transient excitonic optical nonlinearities in semiconductor quantum-well structures. *Phys. Rev. B* 32: 6601–6609.

Schultheis, L., A. Honold, J. Kuhl, K. Köhler, and C. W. Tu. 1986. Optical dephasing of homoge-neously broadened two-dimensional exciton-transitions in GaAs quantum-wells. *Phys. Rev. B* 34: 9027–9030.

Schultheis, L., M. D. Sturge, and J. Hegarty. 1985. Photon echoes from two-dimensional excitons in GaAs-AlGaAs quantum wells. *Appl. Phys. Lett.* 47: 995–997.

Shacklette, J. M. and S. T. Cundiff. 2002. Role of excitation-induced shift in the coherent optical response of semiconductors. *Phys. Rev. B* 66: 045309.

Shacklette, J. M. and S. T. Cundiff. 2003. Nonperturbative transient four-wave-mixing line shapes due to excitation induced shift and excitation induced dephasing. *J. Opt. Soc. Am. B* 20: 764–769.

Smirl, A. L., M. J. Stevens, X. Chen, and O. Buccafusca. 1999. Heavy-hole and light-hole oscil-lations in the coherent emission from quantum wells: Evidence for exciton-exciton correla-tions. *Phys. Rev. B* 60: 8267–8275.

Spivey, A. G. V. and S. T. Cundiff. 2007. Inhomogeneous dephasing of heavy-hole and light-hole exciton coherences in GaAs quantum wells. *J. Opt. Soc. Am. B* 24: 664–670.

Takagahara, T. 1985. Localization and energy-transfer of quasi-2-dimensional excitons in GaAs-AlAs quantum-well heterostructures. *Phys. Rev. B* 31: 6552–6573.

Wang, H., K. B. Ferrio, D. G. Steel, P. R. Berman, Y. Z. Hu, R. Binder, and S. W. Koch. 1994. Transient 4-wave-mixing line-shapers—Effects of excitation induced dephasing. *Phys. Rev. A* 49: R1551–R1554.

Wang, H. L., K. Ferrio, D. G. Steel, Y. Z. Hu, R. Binder, and S. W. Koch. 1993. Transient nonlinear-optical response from excitation induced dephasing in GaAs. *Phys. Rev. Lett.* 71: 1261–1264.

Webb, M. D., S. T. Cundiff, and D. G. Steel. 1991a. Observation of time-resolved picosecond stimulated photon echoes and free polarization decay in GaAs/AlGaAs multiple quantum wells. *Phys. Rev. Lett.* 66: 934–937.

Webb, M. D., S. T. Cundiff, and D. G. Steel. 1991b. Stimulated-picosecond-photon-echo studies of localized exciton relaxation and dephasing in GaAs/Al$_x$Ga$_{1-x}$As multiple quantum-wells. *Phys. Rev. B* 43: 12658–12661.

Wegener, M., D. S. Chemla, S. Schmitt-Rink, and W. Schäfer. 1990. Line-shape of time-resolved 4-wave-mixing. *Phys. Rev. A* 42: 5675–5683.

Weiner, A. M., S. De Silvestri, and E. P. Ippen. 1985. 3-Pulse scattering for femtosecond dephasing studies—Theory and experiment. *J. Opt. Soc. Am. B* 2: 654–662.

Weiser, S., T. Meier, J. Mobius, A. Euteneuer, E. J. Mayer, W. Stolz, M. Hofmann, W. W. Ruhle, P. Thomas, and S. W. Koch. 2000. Disorder-induced dephasing in semiconductors. *Phys. Rev. B* 61: 13088–13098.

Weiss, S., M. A. Mycek, J. Y. Bigot, S. Schmitt-Rink, and D. S. Chemla. 1992. Collective effects in excitonic free induction decay—Do semiconductors and atoms emit coherent-light in different ways? *Phys. Rev. Lett.* 69: 2685–2688.

Wells, J. P. R., R. E. I. Schropp, L. F. G. van der Meer, and J. I. Dijkhuis. 2002. Ultrafast vibrational dynamics and stability of deuterated amorphous silicon. *Phys. Rev. Lett.* 89: 125504.

Yaffe, H. H., Y. Prior, J. P. Harbison, and L. T. Florez. 1993. Polarization dependence and selection rules of transient four-wave mixing in GaAs quantum-well excitons. *J. Opt. Soc. Am. B* 10: 578–583.

Yajima, T. and Y. Taira. 1979. Spatial optical parametric coupling of picosecond light pulses and transverse relaxation effects in resonant media. *J. Phys. Soc. Jpn.* 47: 1620–1626.

Zhang, T. H., C. N. Borca, X. Q. Li, and S. T. Cundiff. 2005. Optical two-dimensional Fourier transform spectroscopy with active interferometric stabilization. *Opt. Express* 13: 7432–7441.

Zhang, T., I. Kuznetsova, T. Meier, X. Li, R. Mirin, P. Thomas, and S. Cundiff. 2007. Polarization-dependent optical 2D Fourier transform spectroscopy of semiconductors. *Proc. Natl. Acad. Sci. U.S.A.* 104: 14227–14232.

Chapter 11

TIME-DOMAIN AND ULTRAFAST TERAHERTZ SPECTROSCOPY

Robert A. Kaindl

CONTENTS

11.1 INTRODUCTION

Many basic condensed matter properties and phenomena are governed by the spectrum of electronic and vibrational degrees of freedom at low energies, comparable to the Boltzmann thermal energy ($k_B T$). They determine both the material's ground state at finite temperature and its dynamical response to electromagnetic or electronic perturbations. Optical access to these low-energy excitations is the realm of *terahertz* (THz) spectroscopy: an electromagnetic wave of frequency $\nu = 1\,\text{THz} = 10^{12}\,\text{Hz}$ corresponds to a photon energy $h\nu \approx 4.136\,\text{meV}$. As a result, measurement of the THz electromagnetic response in the ≈ 0.1–$50\,\text{THz}$ range yields insight into a particularly relevant spectrum of excitations in complex materials (Dressel and Grüner, 2002; Basov and Timusk, 2005). These include, for example, the conductivity of itinerant charges, plasmons, and polarons, as well as transitions across internal exciton states, quantized levels of nano-confined carriers, and superconducting gaps.

Despite its importance, however, the few-THz regime in the past was difficult to access due to the lack of compact, efficient sources and sensitive detectors. With 1 THz corresponding to a $\lambda = 300\,\mu m$ vacuum wavelength, this region of the electromagnetic spectrum lies in between the more accessible areas of microwave electronics and near-infrared (near-IR) optoelectronics. In what follows, we will not discuss more conventional sources such as blackbody radiation, or large-scale free-electron laser facilities. In this chapter, we will review instead the generation and detection of coherent few-cycle THz pulses based on tabletop ultrafast lasers, and their application to time-domain spectroscopy (TDS) and time-resolved studies of complex materials. We discuss practical aspects of these field-resolved techniques in a compact format along with applications to complex materials studies, and we point to recent books and reviews for more exhaustive treatment of either aspect (Mittleman, 2003; Schmuttenmaer, 2004; Sakai, 2005; Dexheimer, 2007; Reimann, 2007; Lee, 2008; Zhang and Xu, 2009).

The basic scheme of THz time-domain spectroscopy (THz-TDS) is illustrated in Figure 11.1. A femtosecond (fs) laser pulse is employed to generate a coherent few-cycle THz pulse, which interacts with the sample. A second, temporally delayed fs pulse is then used in conjunction with nonlinear optical detection to resolve the THz electric field $E(t)$ directly in the time domain. To determine a material's THz properties, the field $E_{sample}(t)$ transmitted through the sample can be compared with a reference $E_{ref}(t)$ without the sample. As both amplitude and phase of the field are measured, a Fourier transform then provides directly the complex-valued field transmission function $t(\omega) = E_{sample}(\omega)/E_{ref}(\omega)$, and in turn the *complex-valued* dielectric function $\varepsilon(\omega) = \varepsilon_1(\omega) + i\varepsilon_2(\omega)$. This represents a key advantage of THz-TDS: both real and imaginary parts of the dielectric response are obtained on equal footing in the measured frequency range, avoiding the need for Kramers–Kronig transforms that require measurement over a very large spectral range. Other advantages of THz-TDS include its high sensitivity, which results from electric field detection due to the sublinear scaling $E \propto \sqrt{I}$ with intensity and from gated detection that ignores THz fields outside the ultrashort sampling pulse.

A second, important aspect is the ultrafast time resolution itself arising from the pulsed nature of the coherent THz transients. This allows for optical-pump THz-probe experiments, which can detect ultrafast, transient changes of a material's THz dielectric function after

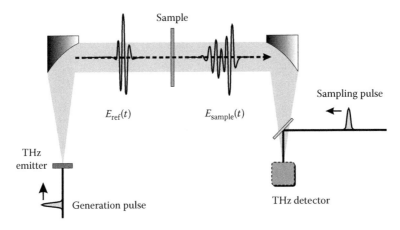

Figure 11.1   Scheme of THz spectroscopy in the time domain.

photoexcitation. Transient THz pulses can follow the dynamics of photoexcited quasiparticles, such as free carriers and excitons in semiconductors, or separate in the time domain charge, lattice, or spin degrees of freedom in correlated electron materials whose cause–effect relationships remain hidden in equilibrium. Finally, we note that intense THz pulses can also be utilized to resonantly excite low-energy excitations. This can provide insight into the coupling of important system excitations which otherwise remain thermally inaccessible.

In the following, Section 11.2 will first discuss the generation of coherent THz pulses and their broadband detection. Next, in Section 11.3 we will review the basics of THz-TDS. This includes the determination of the dielectric function from the measured THz traces for different sample geometries, along with examples. Section 11.4 will describe optical pump-THz probe spectroscopy. Practical aspects of the experimental setup will be described along with recipes for extracting the transient dielectric function. Moreover, we will illustrate case examples of transient conductivity studies of complex materials, including quantum transitions in excitons and charge pairing kinetics in superconductors. Section 11.5 will briefly review the new and rising field of THz excitation studies. The chapter will conclude with a summary and outlook of future opportunities.

11.2 THz GENERATION AND DETECTION

Coherent THz pulses are strongly promoted by the wide availability of near-IR fs lasers—Ti:sapphire or fiber oscillators and amplifiers—used in their generation and detection. The duration of such coherently generated THz pulses often spans only a few optical cycles, analogous to the shortest few-fs pulses generated in the near-IR or visible. A number of different methods exist for THz generation, including difference frequency mixing (DFM), generation from semiconductor surfaces, photoconductive (PC) antennas, or four-wave-mixing in laser-produced plasmas. Detection can likewise proceed along several different routes. Here, we will focus on DFM and electro-optic (EO) sampling in nonlinear optical crystals, providing spectroscopic access to a wide frequency range from roughly 0.1 to 50 THz.

11.2.1 Generation of Coherent THz Pulses

In the following, we will discuss THz generation via DFM. In this scheme, ultrashort pulses in the visible or near-IR are incident on a suitable nonlinear optical medium with a nonzero second-order susceptibility tensor $\chi^{(2)} \neq 0$, typically a thin crystal such as ZnTe or GaSe. The electric field $\mathbf{E}$ of these incoming waves then induces a polarization

$$\mathbf{P} = \varepsilon_0 \chi^{(1)} \mathbf{E} + \varepsilon_0 \chi^{(2)} \mathbf{E}\mathbf{E} + \cdots \quad (11.1)$$

(Chapter 7, [Boyd, 2003]). Let us consider two different frequency components ω_1 and ω_2 of the incident waves. For simplicity, we ignore the vector nature and propagation of the fields, such that the electric field of these two components at the crystal can simply be written as $E(t) = E_1 \cos(\omega_1 t) + E_2 \cos(\omega_2 t)$. The nonlinear polarization in Equation 11.1 then results in the mixed term

$$E_1 \cos(\omega_1 t) \cdot E_2 \cos(\omega_2 t) = \frac{1}{2} E_1 E_2 \left[\cos((\omega_1 + \omega_2)t) + \cos((\omega_2 - \omega_1)t) \right] \quad (11.2)$$

oscillating at the sum and difference frequencies $\omega_1 + \omega_2$ and $\omega_2 - \omega_1$. Accordingly, a THz component is generated at $\omega_{THz} \equiv \omega_2 - \omega_1$ for *each* frequency pair ω_1, ω_2 within the broad spectrum of a single, ultrashort input pulse. This downconversion process is illustrated in Figure 11.2. Note that while $\omega = 2\pi\nu$ designates the angular frequency in the equations, THz spectra are

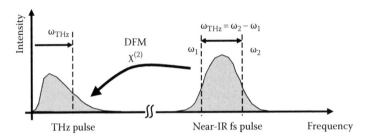

Figure 11.2 Illustration of THz pulse generation via DFM of frequency pairs ω_1 and ω_2 within the broad spectrum of a near-IR fs pulse.

usually plotted as a function of the cycle frequency ν. This process is often called "optical rectification" when the incoming pulses spectrally overlap (or when the two different frequency components are part of the same pulse), leading to a THz spectrum that extends toward zero frequency. In contrast, the term "difference frequency mixing" is more commonly used for generation of THz or mid-IR pulses with a distinct center frequency, resulting, for example, from spectrally separated input pulses. The physical process in both cases is the same.

Practical implementations require more careful consideration of the nonlinear process, including the transmission, dispersion, and symmetry properties of the nonlinear material. Suitable crystals must simultaneously fulfill several unconventional criteria: large transparency both at near-IR pump and THz output wavelengths, high nonlinearity, the capability for phase-matching (Chapter 7), and a large damage threshold. Two-photon absorption should also be small at all interacting wavelengths. Hence, the number of available materials is small. Table 11.1 summarizes several crystals used in THz generation and detection. We will now illustrate practical implementation with two widely used materials—ZnTe and GaSe.

ZnTe is suitable for generation of far-infrared pulses, typically via optical rectification, in a range that spans approximately 0.1–3 THz for crystals that are several hundreds of μm thick. Absorption from phonon difference modes around 1.6 and 3.7 THz quenches the efficiency at high frequencies, while the overall THz dispersion of ZnTe is governed by the TO phonon at 5.3 THz (Gallot et al., 1999; Brazis and Nausewicz, 2008). There are two important aspects illustrated here for ZnTe, which must be generally taken into account for THz generation with nonlinear crystals.

First, the tensor property of the nonlinear susceptibility implies that the nonlinear process will strongly depend on the orientation of the light polarizations with respect to the

TABLE 11.1

Properties of Typical Nonlinear Crystals for THz Generation and EO Detection, with Values from Dmitriev et al. (1997), Nahata et al. (1996), Wu and Zhang (1997b)

Crystal	Structure (Point Group)	Spectral Range in Thick Crystals (THz)	Nonlinear or Electro-Optic Coefficients d, r (pm/V)	Near-IR Group Index n_g ($\lambda_{NIR} =$ 800 nm)	THz Refractive Index n ($\nu = 1$ THz)
ZnTe	Zincblende ($\bar{4}3m$)	0.1–3	$r_{41} = 4$ ($\lambda = 633$ nm)	3.24	3.17
GaP	Zincblende ($\bar{4}3m$)	0.1–7	$r_{41} = 1$ ($\lambda = 633$ nm)	3.55	3.32
GaSe	Hexagonal ($\bar{6}2m$)	14–43	$d_{22} = 54$ ($\lambda = 10.6$ μm)	3.13 (o) 2.86 (e)	3.26

Note: The spectral range is quoted for relatively thick crystals, while ultrathin crystals ($d \approx 10$–30 μm) can cover frequencies up to 100 THz.

crystal axes. ZnTe has a lattice with zincblende symmetry, whose only nonzero elements of the nonlinear tensor are $d_{41} = d_{52} = d_{63}$ (the compact notation $d_{il} \equiv \frac{1}{2}\chi^{(2)}_{ijk}$ for the nonlinear coefficient is generally used with tabulated assignments $l \leftrightarrow jk$). For normal incidence, the crystal should be cut in a <110> orientation, while a <100>-oriented ZnTe crystal does not generate any THz fields. These symmetry aspects, including the dependence of the THz emission on the rotation about the crystal normal, were studied by Chen et al. (2001), who found that an angle of $\approx 54.7°$ between the near-IR polarization and the in-plane [001] axis leads to maximum THz emission from a <110> crystal. This is illustrated along with a typical spectrum in Figure 11.3. In this optimized case, the emitted THz polarization is parallel to that of the incoming beam. In practice, the optimal configuration is often found by rotating the generation and EO detection crystals (care must be taken when using THz polarizers to confirm the orientation, since the detector itself acts as a polarization-sensitive element).

Second, the generated THz field strength in a thick nonlinear medium is greatly enhanced if the THz waves add up coherently as they are emitted throughout the crystal. This "phase matching" over the crystal length (Chapter 7) occurs when the wave vector difference Δk between the three interacting fields becomes negligible, i.e., $\Delta k = (k_2 - k_1) - k_{THz} \approx 0$. With $k \equiv 2\pi/\lambda = \omega\, n(\omega)/c_0$, where $n(\omega)$ is the refractive index and c_0 the light speed in vacuum, we can also write the phase matching condition as

$$n(\omega_{THz})\omega_{THz} = n(\omega_2)\omega_2 - n(\omega_1)\omega_1 \tag{11.3}$$

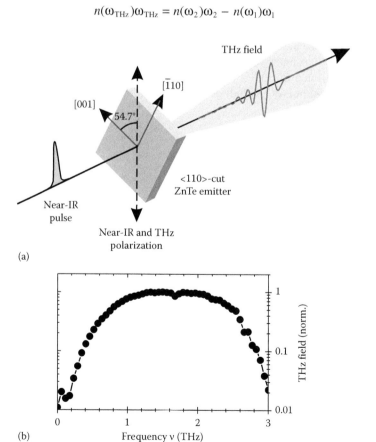

Figure 11.3 (a) Optimal alignment of a <110>-oriented zincblende emitter crystal such as ZnTe for THz generation, following Chen et al. (2001). (b) Typical THz spectrum generated in this configuration with a 500 μm thick <110> ZnTe emitter.

The above expression can be rewritten to show that for sufficiently small ω_{THz}, phase matching in THz generation corresponds to the matching of the group velocity $v_g = (dk(\omega)/d\omega)^{-1}$ of the near-IR pulse with the phase velocity $v_{ph} = c_0/n(\omega_{THz})$ of the coherent THz pulse (Nahata et al., 1996). This condition is intuitively clear: when the near-IR pulse is shorter than a half-cycle of the THz pulse (often the case in practice), it needs to propagate with the peak of the THz cycle to coherently "pile up" the THz electric field during the nonlinear interaction. Phase matching can be achieved in birefringent crystals through a proper selection of beam polarizations and crystal angle (Chapter 7). In fortuitous cases, Equation 11.3 may already be approximately fulfilled, enabling efficient THz generation over significant crystal thicknesses even without the need to resort to birefringent phase matching. This is the case for ZnTe, a major reason for its widespread use.

GaSe (gallium selenide) is a nonlinear crystal suitable for phase-matched generation of THz pulses at much higher frequencies, ranging into the mid-infrared section (Abdullaev et al., 1972; Vodopyanov and Voevodin, 1995). It is characterized by a very large nonlinear coefficient of around 54 pm/V (Table 11.1). GaSe cannot be polished with sufficient quality and is commercially available only with a cleaved surface perpendicular to the [001] direction, although In-doped crystals were reported that can be polished at arbitrary angles (Das et al., 2006). However, even in its naturally cleaved form, GaSe is suitable for phase matching of near-IR pump waves due to small internal phase-matching angles around 10°. Tunable mid-IR/THz pulses can be generated with a complex, cascaded scheme where a parametric amplifier converts near-IR pump pulses into signal and idler pulses in the 1–2 μm range, with subsequent DFM (Bayanov et al., 1994; Fraser et al., 1997; Ehret and Schneider, 1998; Ventalon et al., 2006). A much simpler, phase-matched scheme was demonstrated where THz light fields are generated in GaSe by mixing the components *within* the broad spectrum of very short (sub-20 fs) pulses (Kaindl et al., 1998, 1999). It is illustrated in Figure 11.4a.

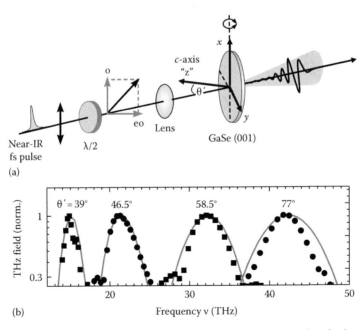

Figure 11.4 (a) Phase matched THz generation in GaSe, via DFM within the broad spectrum of a near-IR pulse, following the scheme of Kaindl et al. (1998). (b) Tunable mid-IR pulses generated in 500 μm thick GaSe, as per the data of Kaindl et al. (1999), using a 2 MHz, 10 nJ cavity-dumped Ti:sapphire oscillator.

The single-step conversion from the near-IR into the THz range yields an inherently simple setup, which is easily switched to optical rectification in crystals such as ZnTe and GaP to access other spectral ranges. Most importantly, however, mixing components within the same pulse spectrum automatically entails a vanishing carrier-envelope offset frequency, i.e., the phase of the generated THz electric field remains the same from pulse to pulse. This is an important prerequisite for THz-TDS.

The beam and crystal orientation shown in Figure 11.4a allows for phase matching of the DFM process. Two types of phase matching are possible in GaSe:

$$\text{type I}: n_o(\omega_{THz})\omega_{THz} = n_{eo}(\omega_2,\theta)\omega_2 - n_o(\omega_1)\omega_1 \tag{11.4a}$$

$$\text{type II}: n_{eo}(\omega_{THz},\theta)\omega_{THz} = n_{eo}(\omega_2,\theta)\omega_2 - n_o(\omega_1)\omega_1 \tag{11.4b}$$

where

n_o and n_{eo} denote the ordinary (o) and extraordinary (eo) refractive indices

θ is the polar angle between the beam direction k inside the crystal and the optical axis z

The latter is related to the external angle θ' (shown in Figure 11.4) via $\theta = \arcsin[\sin(\theta')/n]$. These processes are governed by the effective nonlinearity $d_{ooe} = d_{22}\cos\theta\sin3\phi$ (type I) and $d_{eoe} = d_{22}\cos^2\theta\cos3\phi$ (type II), due to the hexagonal crystal symmetry (Abdullaev et al., 1972). Here, ϕ is the azimuthal angle between the plane of incidence (the k–z plane) and the crystallographic x–z plane. Thus, to achieve type-I phase matching, it is necessary for the incident beam to propagate within the yz-plane of the GaSe crystal as shown in Figure 11.4a (Vodopyanov and Kulevskii, 1995). Rotating the crystal around its z-axis (angle ϕ) in 30° increments switches between type I and type II, and accordingly flips the THz polarization by 90°. Type I phase matching exhibits smaller angles of incidence and larger nonlinearity. A key element is to employ a half-wave plate to adjust the polarization of the input pulses such that it encloses an angle of 45° with respect to the plane of incidence. In this way, a single input beam simultaneously provides the components for both ordinary *and* extraordinary polarized waves (Kaindl et al., 1998). Wavelength tuning follows the phase-matching conditions of Equations 11.4a and b by rotating the crystal via the angle of incidence θ'. Typical spectra are shown in Figure 11.4b for phase-matched DFM of sub-20 fs pulses in a 500 μm thick GaSe crystal. This results in a compact tunable source covering 15–43 THz, with an average power that is increased about hundredfold beyond that from non-phase-matched rectification in GaSe (Kaindl et al., 1999).

In sufficiently thick crystals, the effective interaction length is limited by the temporal separation of the two interacting ordinary and extraordinary polarized near-IR pulses, which travel at different group velocities. The bandwidth of the THz/mid-IR pulse depends on the spectral acceptance of the DFM process: it follows from the interaction length given by the group velocity mismatch between the two near-IR components, and from the THz ↔ near-IR group velocity mismatch over this length. Moreover, group velocity dispersion lengthens the short near-IR pulses during propagation. Typical values in GaSe are ≈100–200 fs/mm group velocity mismatch between the near-IR pulses, and ≈1 ps/mm mismatch between THz and near-IR. This limits the interaction length typically to ≈200 μm for sub-20 fs input pulses, so that even for thicker GaSe crystals fs pulses can be generated in the mid-IR range.

Precise modeling of the generated THz pulses requires a numerical calculation employing coupled nonlinear wave equations for the $\chi^{(2)}$ DFM process (Kaindl et al., 1999). As the THz field is much weaker than the near-IR field, depletion of the near-IR pulse can be neglected. The near-IR pulse at each position s is then determined by linear dispersion from the field

$E_{NIR}(\omega)$ at the crystal entrance face ($s = 0$). It acts as the nonlinear source term for the THz wave equation, enabling calculation of the THz field (Kaindl et al., 1999; Reimann, 2007):

$$E_{THz}(\omega_{THz}, s) \propto d_{eff}(\theta)\omega_{THz}^2 \exp(ik(\omega_{THz})s)$$

$$\times \int E_{NIR}^*(\omega + \omega_{THz}) \cdot E_{NIR}(\omega) \frac{\exp\left[i\Delta k(\omega, \omega_{THz}, \theta)s\right] - 1}{\Delta k(\omega, \omega_{THz}, \theta)(2k(\omega_{THz}) + \Delta k(\omega, \omega_{THz}, \theta))} d\omega \quad (11.5)$$

where the phase mismatch for the case of type I phase matching is defined as

$$\Delta k(\omega, \omega_{THz}, \theta) = \frac{1}{c_0}(n_{eo}(\omega + \omega_{THz}, \theta)(\omega + \omega_{THz}) - n_o(\omega)\omega - n_o(\omega_{THz})\omega_{THz}) \quad (11.6)$$

The refractive indices $n_o(\omega)$ and $n_{eo}(\omega, \theta)$ are determined by the "Sellmeier" equations for GaSe (or whichever material is being simulated), i.e., from model curves fitted to refractive index data (Tropf et al., 1995; Vodopyanov and Kulevskii, 1995; Dmitriev et al., 1997). An example is shown as the lines in Figure 11.4b. The frequency-domain field $E_{THz}(\omega_{THz})$ can be Fourier transformed to obtain the time-domain representation $E(t)$. The calculation includes the full dispersion and birefringent effects of the crystal but ignores the boundary conditions and hence internal multiple reflections (the latter are discussed in Reimann [2007]). In practice, however, only the first pulse replica is normally measured.

Ultrabroadband pulses spanning 30 THz or more into the mid-IR have been generated via non-phase-matched optical rectification of very short, ≈10 fs Ti:sapphire pulses on semiconductor surfaces or nonlinear crystals (Bonvalet et al., 1995; Joffre et al., 1996; Han and Zhang, 1998). Such sources are characterized by very low power, which makes spectroscopic applications difficult. A compromise can be found between the comparatively narrow pulses generated in a thick GaSe crystal described above and the non-phase-matched generation of ultrabroadband THz pulses. For this, phase matching is weakened by employing a thinner GaSe crystal with thickness ranging from approximately 20 to 150 μm (Huber et al., 2000; Kübler et al., 2005). This also increases the transmission below ≈15 THz, which is otherwise limited by phonon absorption. Detuning of the crystal away from perfect phase matching broadens the spectrum at the expense of the generated THz power. Spectra with measurable frequencies extending up to ≈100 THz can be generated, although for practical purposes the signal drops very rapidly above ≈45 THz (Kübler et al., 2005). An example of a broadband THz field from a thin GaSe crystal is shown in Figure 11.5a and d with a spectrum that covers ≈10–20 THz.

To cover the full range from below 1 THz to above 50 THz several nonlinear crystals must be combined, each having respective advantages or disadvantages. ZnTe and GaSe offer high nonlinearity and cover 0.1–3 THz and ≈10–50 THz, respectively. The resulting gap can be partially covered by GaP (see Figure 11.5b and e) with emission up to ≈7 THz, although combined THz generation and detection with GaP is almost one order of magnitude less powerful due to its lower nonlinearity (Table 11.1). Even broader spectra were recently generated by phase-matched DFM of 7 fs pulses in LiIO₃, resulting in ultrabroadband 50–130 THz pulses (Zentgraf et al., 2007). Besides the crystals discussed earlier, THz optical rectification has been demonstrated in a larger range of materials including LiNbO₃ and LiTaO₃ (Auston and Nuss, 1988; Kawase et al., 2001; Yeh et al., 2007; Stepanov et al., 2008), GaAs (Zhang et al., 1992), and periodically poled crystals (Lee et al., 2000; Vodopyanov, 2009). Generation in poled polymers and organic crystals such as 4-dimethylamino-*N*-methylstilbazolium tosylate (DAST) also received much attention due to their large nonlinear coefficients, although the emitted spectra are often quite structured and thus limit spectroscopic applications (Nahata et al., 1995; Han et al., 2000; Brunner et al., 2009).

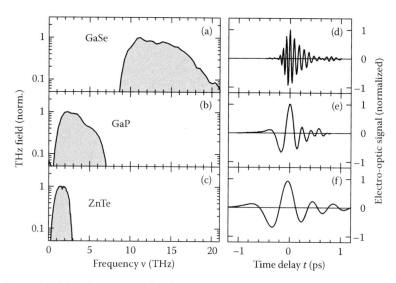

Figure 11.5 (a)–(c) THz spectra of pulses generated in different crystals with the 250 kHz amplifier system described in the text. (d)–(f) Corresponding electric field traces measured via EO sampling. Data correspond to (a),(d) 80 µm thick GaSe emitter and 50 µm GaSe detector, (b),(e) 200 µm GaP emitter and detector, and (c),(f) 500 µm thick ZnTe emitter and detector.

11.2.2 Electro-Optic Sampling

Broadband time-domain detection of the THz fields can be achieved via free-space EO sampling in a nonlinear optical crystal (Wu and Zhang, 1996b; Wu et al., 1996). This scheme relies on the EO Pockels effect, i.e., on the near-IR birefringence induced in a medium by an applied electric field. To detect a freely propagating THz wave, a weak fs near-IR pulse is sent collinearly with the THz pulse through an EO crystal (Wu and Zhang, 1996b). When the near-IR pulse is comparable to or shorter than the half-cycle of the THz field, it can detect the quasi-instantaneous coherent THz field E_{THz} that exists during its short fs duration by measuring the phase retardance due to the THz-induced birefringence. The THz field $E_{\text{THz}}(t)$ is then sampled in the time domain by varying the relative time delay t between the THz field and near-IR probe pulse. Conditions for this include that the crystal be sufficiently thin to avoid temporal walkoff, and that the THz phase remain steady with respect to the probe pulse, as achieved, for example, via optical rectification using the same laser source.

A typical EO sampling arrangement is shown in Figure 11.6a, for the case of a <110>-oriented zincblende-symmetry crystal such as ZnTe or GaP. For maximum phase retardation, the THz field is polarized along [$\bar{1}$10], and the near-IR pulse is linearly polarized either parallel or perpendicular to it (Wu and Zhang, 1996b; Chen et al., 2001; Planken, 2001). After the crystal, a quarter-wave plate is used to impose a circular polarization on the near-IR pulse, whose two components are split apart by a Wollaston polarizer. The intensity difference between the two polarizations is analyzed as the differential photocurrent ΔI from a pair of balanced silicon photodiodes. The birefringence induced by the THz field E_{THz} will lead to a polarization ellipticity, which on the photodiodes results in the relative difference signal (Planken, 2001):

$$\frac{\Delta I}{I} = \frac{\omega_{\text{THz}} n^3}{c} dr E_{\text{THz}} \tag{11.7}$$

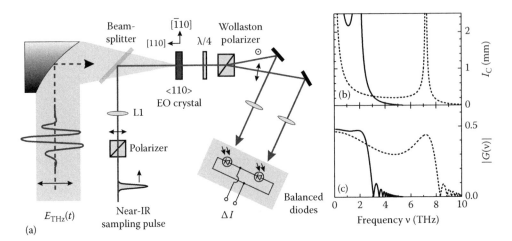

Figure 11.6 (a) Scheme for free-space EO sampling in <110> zincblende crystals such as ZnTe or GaP (Wu and Zhang, 1996b; Chen et al., 2001; Planken, 2001). For broadband THz sampling an achromatic $\lambda/4$ waveplate is used. (b),(c) Coherence length l_C and response function $G(\omega)$ calculated for ZnTe (solid line) and GaP (dashed line) with the models of Wu and Zhang (1997b) and Nahata et al. (1996) for $\lambda_{\rm NIR} = 800\,$nm. In panel (c), the crystal thicknesses are $d = 500\,\mu$m (ZnTe) and $200\,\mu$m (GaP).

In the above equation, n is the near-IR refractive index of the EO crystal, d is the crystal thickness, and r represents the EO coefficient (with typical values given in Table 11.1). The $\lambda/4$ plate results in a linear response $\Delta I \propto E_{\rm THz}$, but other configurations are also used, especially for THz imaging (Lu et al., 1997).

The balanced photodiode circuit used for sensitive EO detection can be easily fabricated by electrically cross-connecting the cathode and anodes of two photodiodes as shown in Figure 11.6a. The difference output ΔI is then amplified in a current preamplifier and measured via phase-sensitive detection with a lock-in amplifier, while the THz generation beam is modulated (usually with a mechanical chopper). With oscillators or high-repetition rate amplifiers, probe current photomodulations $\Delta I/I$ as small as $10^{-8}/\sqrt{\rm Hz}$ can be detected in optimized configurations, close to the shot-noise limit. In the 250 kHz amplifier-based setup discussed later, a signal-to noise (S/N) ratio of 10^4–10^5:1 in the electric field measurement was routinely achieved using a $500\,\mu$m ZnTe emitter and detector. Such sensitivities are even more remarkable when considered in terms of the THz intensity that scales as $I_{\rm THz} \propto E_{\rm THz}^2$.

The EO Pockels effect underlying this detection can also be regarded as a sum-frequency process, where the THz photon mixes with the near-IR photon to generate, in turn, an orthogonally polarized near-IR photon at somewhat higher frequency. This is the reverse of the THz generation process described above. EO sampling is accordingly limited by the same issues, including THz absorption, dispersion, and temporal walkoff between the interacting waves that can lead to cancellation of positive and negative going half-waves. For a near-IR probe pulse much shorter than the THz cycle period, the frequency-dependent EO detector response function can be written as (Wu and Zhang, 1997b):

$$G(\omega) = \frac{2}{n(\omega)+1} \times \frac{e^{i\omega\delta(\omega)}-1}{i\omega\delta(\omega)} \tag{11.8}$$

In the above equation, $\delta(\omega) \equiv [n_{\rm g}(\lambda_{\rm NIR}) - n(\omega)] \times d/c$ is the temporal walkoff between the THz field (phase velocity) and the near-IR pulse (group velocity) after propagating through a crystal of thickness d. Here, $n(\omega)$ is the THz refractive index and $n_g(\lambda_{NIR})$ the near-IR group index at wavelength λ_{NIR}. We can also consider the coherence length l_C for EO sampling, which is the propagation length over which the probe pulse sweeps across one half of the THz cycle period, i.e., across the timespan π/ω (Nahata et al., 1996). Therefore, $l_C(\omega) \equiv (\pi/\omega) \times d/|\delta(\omega)|$, or

$$l_C(\omega) = \frac{\pi c}{\omega \left| n_g(\lambda_{NIR}) - n(\omega) \right|} \tag{11.9}$$

Figure 11.6b shows a calculation of $l_C(\omega)$ for ZnTe (solid line) and GaP (dashed line). The high-frequency peaks correspond to exact matching between near-IR group and THz phase velocities. ZnTe exhibits large values of l_C, rendering it highly suitable for a spectral range below ≈ 3 THz. This is confirmed in Figure 11.6c by the response function $|G(\omega)|$ for a 500 μm thick crystal. This thickness represents a reasonable trade-off between bandwidth and sensitivity. GaP exhibits a broader response (dashed lines in Figure 11.6b and c) due to its higher phonon frequency, but the coherence length motivates a smaller thickness of ≈ 200 μm. Near the optical phonon frequencies, the dispersion of the EO coefficient must also be taken into account, leading to a corrected response $G(\omega) \cdot r(\omega)$ as discussed by Leitenstorfer et al. (1999). Due to the above reasons, and despite the usual literature designation, the measured signals do not exactly represent the THz field. However, most THz experiments measure *relative* changes, referenced to THz spectra measured with the same EO detector, so that the response function drops out of the equation. Nevertheless, the EO crystals should also be kept thin because of THz and near-IR pulse dispersion effects not included in the above analysis (Bakker et al., 1998; Gallot and Grischkowsky, 1999).

The measurement of THz transients in an even larger "ultrabroadband" THz range necessitates very careful consideration of the dispersion behavior in EO sampling. Ultrabroadband THz detection up to 37 THz in the mid-IR range was demonstrated by employing $d \approx 30$ μm thick ZnTe emitters (Wu and Zhang, 1997a; Han and Zhang, 1998). Use of ultrathin ZnTe and GaP crystals demonstrated detectable signals up to 70 THz (Leitenstorfer et al., 1999). One problem with such thin EO crystals is multiple reflections, which severely limit the available detection time window. To counter this, the detector can be glued onto a thick, EO-inactive <100> crystal of the same material for index matching (Wu and Zhang, 1996a,b; Leitenstorfer et al., 1999), with the <110> side facing the incoming beam to minimize dispersion. Fabrication of these detectors is challenging, as the glue layer must be pressed very thin and polishing of ZnTe to around 10 μm thickness with sufficient optical quality is difficult. It was later shown that in thick ZnTe crystals surprisingly smooth and broadband THz detection is possible, comparable in sensitivity to 10 μm thick ZnTe sensors (Kampfrath et al., 2007). However, the overall detection sensitivity is very limited in either case. Since EO sampling can be considered the reverse of THz generation, as mentioned above, birefringent phase matching can be employed to significantly enhance the EO detectivity in a defined spectral region. Birefringent phase-matched EO sampling was demonstrated in GaSe (Kübler et al., 2004; Liu et al., 2004), allowing detection of frequency components up to ≈ 100 THz.

Another interesting development is single-shot EO sampling, which enables detection of the THz field trace without the need to scan the delay t between the THz and probe pulses (Jiang and Zhang, 1998; Shan et al., 2000). For this, the probe pulse is chirped so that time information is mapped onto the spectral domain, which can be combined with frequency-resolved detection of the EO phase retardation. This limits the achievable S/N ratio but is advantageous, for example, for non-repetitive experiments or for THz imaging with CCD sensors. The detectable THz bandwidth in single-shot measurements was initially restricted due to loss of temporal resolution with chirped pulses, but this limitation was recently overcome (Kim et al., 2006).

11.2.3 Photoconductive Antennas and Other Schemes

There are numerous other methods for tabletop coherent THz generation and detection. While a comprehensive discussion is beyond the scope of this text, we briefly review PC antennas, rectification at semiconductor surfaces, and high-field THz generation.

PC antennas are of particular importance and are utilized extensively by researchers both for THz generation and detection. The geometry illustrated in Figure 11.7, also called the Auston switch, consists of a pair of striplines lithographically fabricated on a semiconductor surface (Auston, 1975; Smith et al., 1988). The striplines run to a dipole antenna structure, which, however, is separated at its center by a micron-sized gap. For use as a THz emitter, a voltage is applied across the gap. Photoexcitation induces photocarriers in the semiconducting gap region, switching it into a conducting state. In turn, this results in a fast current pulse $j(t)$ with an ultrafast rise set by the pulse duration and RC time constant of the structure, and a decay time τ_R mostly determined by carrier recombination. The time-dependent current $j(t)$ acts as a fast dipole, emitting radiation with a broadband spectrum that extends to several THz if τ_R is short enough (Smith et al., 1988; Reimann, 2007). Suitable semiconductors with short τ_R are, for example, low-temperature grown (LT) GaAs or radiation-damaged silicon-on-sapphire. For use as a THz detector, the PC structure is connected to a current preamplifier (instead of the bias voltage), which measures the photoexcited carriers accelerated by the transient THz field. Ultrabroadband THz detection with PC antennas with detectable components as high as $\approx$95 THz was demonstrated by operating in an integrating mode independent of τ_R which takes advantage of a fast rise time using 10 fs probe pulses (Ashida, 2008).

Direct comparison between PC detection and EO sampling depends on many variables, and several studies have investigated the trade-offs involved (Brener et al., 1998; Park et al., 1999; Kono et al., 2001). In general, PC antennas are particularly useful when combined with MHz oscillators at nJ pulse energies, which can be tightly focused without damage onto the small gap. In such systems, the near-IR excitation pulse can also be delivered through fibers to the PC antenna, allowing for measurements in high magnetic fields and for compact and flexible THz spectrometers which are commercially available (Crooker, 2002). Fiber coupling, moreover, alleviates the sensitivity of the microscale gap structures to laser beam pointing fluctuations. Recent PC antenna designs have enabled simultaneous polarization analysis of orthogonal field components (Hussain and Andrews, 2008). In the PC emitter, the majority of the THz power is emitted into the substrate side. Collection into this direction is hampered by the total internal

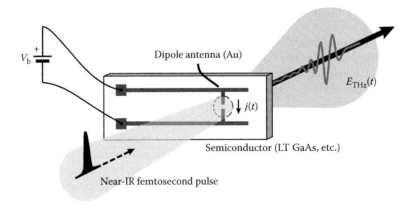

Figure 11.7 Scheme of photoconducting dipole antenna. A THz field is emitted due to the rapidly time-varying current $j(t)$ that flows between the electrode gap during the picosecond lifetime τ_R of photoexcited e-h pairs in a semiconductor film.

reflection within the substrate. As a consequence, hemispherical silicon lenses are often attached to the substrate backside to collimate the THz beams and strongly enhance the emission.

THz generation is also observed from carriers photoexcited on semiconductor surfaces without an externally applied field (Zhang et al., 1990). In this case, THz emission is governed by the built-in surface depletion field (Section 8.1.7). Typical materials are, for example, InP, GaAs, CdTe, InSb, or InAs, with photoexcitation at non-normal incidence in the non-transparent region *above* the bandgap. Three mechanisms can contribute (Greene et al., 1992): (1) acceleration of the photoinjected carriers in the depletion field, (2) DFM via a $\chi^{(2)}$ nonlinearity induced by the depletion field, and (3) DFM from the bulk $\chi^{(2)}$ within the short absorption length. The relative contribution of these effects depends on the intensity, wavelength, semiconductor bandgap, and propagation geometry. At normal incidence, the depletion field points along the propagation direction so that the first effect is suppressed: this enables, for example, broadband optical rectification from GaAs (Rice et al., 1994; Wu and Zhang, 1997a). For non-normal excitation the THz emission from the first effect can be strongly enhanced by magnetic fields (Zhang et al., 1993; Ohtake et al., 2000). Work is progressing on improving these kinds of emitters, including small-bandgap semiconductors such as InAs and InN that are compatible with long-wavelength fiber lasers (Wilke et al., 2008).

High-field THz pulses can be used to photoexcite materials or induce nonlinear changes of the THz conductivity. Generation of such intense fields, in contrast to the weaker THz probe beams, can proceed along different routes. PC antennas can be combined with amplified, mJ-class lasers by using large-aperture structures, e.g., electrodes spaced several millimeters apart with high-voltage (kV) electrical bias, or micron-sized gaps between interdigitated electrodes (Budiarto et al., 1996; Dreyhaupt et al., 2005; Peter et al., 2007). These generally represent sources in the few-THz range with fields up to several 100 kV/cm. Optical rectification in large-aperture crystals such as ZnTe or phase-matched DFM in GaSe with mJ-class pump sources yields intense light fields of several 100 kV/cm (μJ-level energies) in the few-THz range (Siders et al., 1999; Blanchard et al., 2007) and up to 100 MV/cm ($\approx 20 \mu$J) in the mid-IR range (Reimann et al., 2003; Sell et al., 2008). Moreover, very recent work with 10–100 Hz repetition rate TW pump lasers combined with tilted pulse-front phase matching in LiNbO$_3$ enabled generation of few-THz pulses with up to 30 μJ pulse energy (Yeh et al., 2007; Stepanov et al., 2008).

A particularly exciting development concerns generation of intense THz pulses in gases. For this, the strong fundamental and weaker second-harmonic of a mJ-level fs Ti:sapphire amplifier are both focused into air, N$_2$, Ar, or other gases, creating a two-color light field at high intensity around 10^{14} W/cm^2 (Hamster et al., 1993; Cook and Hochstrasser, 2000; Loeffler et al., 2000). This leads to partial ionization of the gas and the emission of intense THz radiation in a broad spectral range, with frequencies up to 75 THz and MV/cm electric field strengths corresponding to μJ pulse energies (Bartel et al., 2005; Kim et al., 2008). While initial explanations were given in terms of four-wave mixing in the gas, the THz emission at high intensities is now understood as resulting from the dynamics of electrons in the plasma after tunnel ionization (Kim et al., 2008; Kim, 2009). Laser-induced plasmas in gases can also be used for THz detection, via electric-field induced second-harmonic generation (Dai et al., 2006). Advantages of these gas-based schemes include the absence of phonon absorption or laser-induced damage at high intensities that plague nonlinear crystals, enabling ultrabroadband THz light fields scalable to dramatically higher intensities.

11.3 THz TIME-DOMAIN SPECTROSCOPY

This section will describe THz-TDS. Measurement of the THz electric field $E(t)$ in the time-domain provides both real and imaginary parts of the dielectric function $\varepsilon(\omega)$. The material response can also be expressed by the optical conductivity $\sigma(\omega) = \sigma_1(\omega) + i\sigma_2(\omega)$, which is useful to understand systems with itinerant charges, and is connected to the dielectric function via $\sigma(\omega) = i\omega\varepsilon_0 [1 - \varepsilon(\omega)]$ (Chapter 1). The conductivity describes the frequency-dependent current response $J(\omega) = \sigma(\omega)E(\omega)$ of the system to an incident transverse electromagnetic field. The real part $\sigma_1(\omega)$ is a measure of the absorbed power density and allows for analysis of oscillator strengths, while the imaginary part $\sigma_2(\omega)$ quantifies the out-of-phase, inductive system response. The dielectric function is also linked to the complex-valued refractive index $n(\omega) \equiv n_1(\omega) + in_2(\omega) = \sqrt{\varepsilon(\omega)}$, with the absorption coefficient obtained as $\alpha(\omega) = 2\omega n_2(\omega)/c$. Of course, all these response functions can be used interchangeably (once both real and imaginary parts are known) and their analytic utility depends on the physics of the material, such as the prevalence of bound versus free charge carriers. In the following, a specific experimental setup is described to illustrate the instrumentation, along with general design considerations. The extraction of the dielectric function from the THz fields will then be explained for different geometries, along with the electrodynamics of superconductors as an example of THz-TDS measurements.

11.3.1 Experimental Setup

We will now illustrate THz-TDS instrumentation with a specific experimental setup (Figure 11.8), as employed in various studies including optical-pump THz-probe measurements (Kaindl et al., 2002, 2003, 2005, 2009; Huber et al., 2006). Our discussion will concentrate on THz-TDS, with the optical-pump THz probe aspects explained in the next section. While the setup can access ≈0.3–20 THz, we will concentrate here on the 3 THz range covered with ZnTe

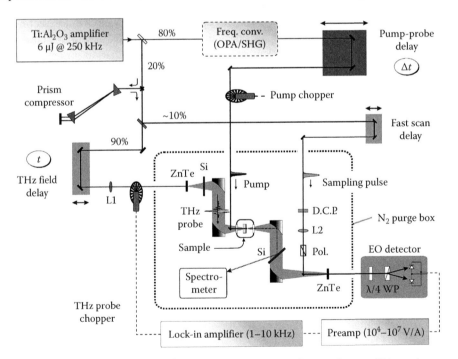

Figure 11.8 Setup for THz time-domain spectroscopy and optical-pump THz-probe experiments. The sample is mounted in a cryostat with mylar windows. Focal lengths are f = 300 mm (L1), 500 mm (L2). Effective focal lengths of the off-axis parabolic mirrors are 60, 120, 100, and 100 mm in sequence after the emitter; the beam deviation at each mirror was 90°, except for the first two which were 45° for better alignment (not shown). DCP: dispersion compensation plate, WP: Wollaston polarizer.

emitters and detectors. At the outset, a commercial Ti:sapphire regenerative amplifier system operating at a 250 kHz repetition rate (Coherent RegA) generates a train of fs pulses centered at 800 nm, with about 6 μJ energy per pulse. The output pulse duration is either ≈50 or 130 fs, depending on the type of compressor, with the shorter pulses needed to reach beyond 3 THz.

Optical rectification and free-space EO sampling is used to generate and detect the transient THz pulses. For this, approximately 20% of the laser output is split off, with the remainder used for photoexcitation as described later. About 150 mW is focused onto a 500 μm thick <110> ZnTe emitter crystal to a 300 μm diameter spot, resulting in the generation of a strongly divergent THz beam. Focusing into the generation crystal is limited by two-photon absorption, leading to free-carrier generation such that the emitted THz power scales less than the expected square power law (and ultimately limited by optical damage). Part of the near-IR generation beam leaks through the emitter, and is therefore blocked with a 1 mm thick high-resistivity Si wafer (transparent to THz radiation) to avoid excitation of the sample or interference with EO detection.

Since the THz beam diverges much more rapidly than the near-IR beam, it is collected and collimated with short focal-length optics, specifically with a gold-coated off-axis parabolic mirror. This may be omitted if the generated THz beam is nearly collinear, e.g., when using some large-aperture emitters. Additional off-axis parabolic mirrors focus the collimated THz beam onto the sample, mounted in a cryostat, and then collimate, and refocus it after the sample onto a 500 μm <110> ZnTe detection crystal. The THz beam path is enclosed and purged with dry N_2 or alternatively other gases to avoid distortion of the pulses by THz absorption from ambient water vapor (Exter et al., 1989).

THz detection via EO sampling follows the scheme described earlier in the context of Figure 11.6. The polarization retardance induced by the THz field $E(t)$ on the near-IR sampling pulse in the EO crystal is sensitively recorded as the difference signal ΔI from the balanced Si photodiodes. The electric field $E(t)$ is then mapped out in the time domain by varying the delay time t between the THz field and near-IR sampling pulse. This is achieved via the computer-controlled motorized delay stage, indicated as "THz field delay" in Figure 11.8. For maximum sensitivity, the EO signal is converted with a transimpedance preamplifier and then measured with a lock-in amplifier, with the near-IR generation beam (before the THz emitter) modulated at several kilohertz with an optical chopper.

A typical field obtained with 500 μm thick <110> ZnTe was shown in Figure 11.5f. In this example, the measurement consisted of three consecutive scans averaged together, with 600 ms accumulation per time step in each scan. The spectrum in Figure 11.5c, obtained via Fourier transform of the time-domain signal, extends from ≈0.3 to 3 THz. The shape of such THz transients is influenced by dispersion and absorption of any material they pass through (including the generation crystal and gas), which causes chirp and ringing of the THz field in the time domain. At low frequencies, the spectra are also curtailed by diffraction (Faure et al., 2004). For optimal sensitivity, the power of the near-IR sampling beam should be maximized while remaining in the linear response regime of the Si photodiodes. This limit is often around ≈5 mW. When comparing the peak THz field to the background noise with the THz beam blocked, an S/N ratio of 10^4–10^5 in the electric field was routinely achieved with the setup of Figure 11.8.

A few additional aspects should also be noted. First, the near-IR pulse duration should be fully compressed at the THz emitter and detector location, although in some cases a slight pre-chirp is necessary to maximize the interaction length (Kaindl et al., 1999). A dispersion compensator (Chapter 7) such as the prism compressor in Figure 11.8 can be used to independently compensate the THz and optical pump sections. Similarly, the optical components in the THz generation and detection arms should be balanced. Fine adjustment of the group velocity dispersion can be achieved with dispersion compensation plates (DCP in Figure 11.8), such as windows, thin crystals, or chirped mirrors (Chapter 7). Second, in the setup of Figure 11.8 a fast galvanometric scanner operating at ≈10–20 Hz was added into the THz probe arm. Such a capability allows the signal to be displayed on an oscilloscope in real time during optical alignments. This is useful since mirror adjustments affect the relative path lengths, and therefore can lead to deceptive amplitude changes of the THz signal during alignment that instead result from changes of the *phase*.

Finally, several issues contribute to the overall S/N performance. Purging can lead to turbulent gas flow and vibrations. Time-dependent variations of the laser power and beam pointing will affect the THz intensity, and pulse duration drifts also couple to the THz spectral bandwidth. Fluctuations during the scan couple into the spectrum, motivating a sufficiently fast scan speed if the low-frequency part of the spectrum is important. Moreover, the residual birefringence of the EO crystal and small differences between the photodiode characteristics lead to a signal offset, which can be canceled by adjusting the λ/4 waveplate (Figure 11.8) and by signal processing. Material imperfections (e.g., microscale stress), in turn, lead to spatial variation of the birefringence, which also couples beam-pointing noise into the EO signal. The ultimate goal is to achieve EO sampling at the shot-noise limit, which further necessitates close matching of the photodiodes and careful consideration of all the sources of electronic noise.

Of course, the setup discussed above is just one example, and many different variations have been demonstrated and published (Mittleman, 2003; Dexheimer, 2007). Amplifier-based setups as in Figure 11.8 are often driven by the need to perform optical-pump THz probe experiments. Oscillator-based systems at MHz repetition rates are optimal for THz-TDS due to lower noise and more compact arrangements. The combination of two oscillators with different repetition rates enables asynchronous EO sampling (Klatt et al., 2009), where the THz field is rapidly scanned without any mechanical delay stage (similar to ASOPS and ECOPS, discussed in Chapter 9). The optical elements vary greatly. Off-axis parabolic mirrors combine short focal lengths, lack of chromatic aberration, and a high reflectivity over extreme bandwidths, but are difficult to align. For narrowband THz pulses, Si, teflon, or polyethylene lenses may be used instead. For the beam combination of THz and near-IR beams before the EO detector a small mirror is shown in Figure 11.8. Many alternative implementations are employed, such as thin pellicles, high-resistivity Si wafers, and ITO films as beamsplitters, or a parabolic mirror with a drilled hole (Bauer et al., 2002). The spatial propagation of THz beams should be considered especially for non-confocal geometries (Gürtler et al., 2000; Côté et al., 2003).

11.3.2 Dielectric Function Analysis: Thin-Film versus Thick Slab Geometries

As mentioned earlier, the dielectric function $\varepsilon(\omega)$ of a material can be determined in THz-TDS by comparing the THz field $E_{sample}(t)$ transmitted through the sample to that of a reference $E_{ref}(t)$ transmitted through air or vacuum. A Fourier transform yields the complex-valued field transmission spectrum $t(\omega) = E_{sample}(\omega)/E_{ref}(\omega)$. With knowledge of the relationship between $t(\omega)$ and $\varepsilon(\omega)$ for a given sample geometry, the dielectric response is obtained.

For the Fourier transform, the time-domain data are taken with a common spacing of time steps δt. The maximum useful frequency is then given by the Nyquist limit $\nu_{max} = \omega_{max}/2\pi = 1/(2\delta t)$, although in practice the scans should be sampled more smoothly. Likewise, the scan length T determines the frequency resolution as $1/T$. The scan length cannot be increased arbitrarily, but is limited by multiple reflections within the THz emitter and detector, the sample, and any other materials in the beam path that lead to undesirable replicas of the THz pulse away from the main peak. This motivates the use of sample substrates, windows, beamsplitters, and other materials in the beam which are either sufficiently thick (usually several hundreds of microns) or very thin ($d < c/2\nu_{max}n$).

In the following section, we will discuss two commonly encountered geometries, that of a thick slab and of a thin multilayered film on a substrate.

11.3.2.1 Thick Slab Geometry

Let us first consider a thick slab of a dielectric material of refractive index $n(\omega) = \sqrt{\varepsilon(\omega)}$ and thickness d. This case is important to evaluate the dielectric properties of bulk materials, and has provided, for example, very useful information on the dispersion of dielectrics or the conductivity of semiconductors (Grischkowsky and Keiding, 1990; Grischkowsky et al., 1990; Jeon and Grischkowsky, 1997; Nashima et al., 2001; Kojima et al., 2003; Azad et al., 2006). The complex-valued field transmission can be written as

$$t(\omega) = \frac{4n(\omega)}{[n(\omega)+1]^2} e^{i\frac{\omega d}{c}[n(\omega)-1]} \tag{11.10}$$

which takes into account the Fresnel reflections at the entrance and exit surface and the dispersive propagation through the bulk material. The factor $n(\omega) - 1$ in the exponential results from the comparison of propagation in the material with pulse propagation in vacuum ($n_{vac} = 1$).

To determine the complex-valued refractive index from $t(\omega)$, the above expression needs to be solved with a numerical method, e.g., the Newton–Raphson iterative algorithm or more elaborate schemes (Duvillaret et al., 1996; Dorney et al., 2001). One should note that the dispersive real part of $n(\omega)$ implies rapid variation of the phase of $t(\omega)$, which quickly reaches values well beyond 2π. Since the Fourier transform intrinsically cannot distinguish between a phase ϕ and $\phi + 2\pi$, the phase needs to be "unwrapped": in the software this is achieved by an algorithm that tests for large phase jumps close to 2π and adds the necessary 2π each time. Moreover, as THz data do not extend to $\omega = 0$, the phase offset at the minimum frequency must be determined through extrapolation of the low-frequency slope of $\phi(\omega)$. These are subtle but crucial points, since otherwise the extracted $n(\omega)$ exhibits unphysical discontinuities and falls short of the actual value. In Equation 11.10, multiple reflections within the thick substrate are ignored. This is often sufficient when the time-domain THz traces are measured in a time window that samples only the first pulse replica. However, numerical techniques exist that allow for reliable extraction of the dielectric response even when including multiple reflections (Duvillaret et al., 1996).

An example is given in Figure 11.9 for a THz-TDS measurement on a $LaAlO_3$ crystal, a material used, for example, as a substrate for high-T_C superconducting films. The measurement shown is for a 478 µm thick wafer at $T = 95 \, K$. The time-domain scans in Figure 11.9a compare $E_{ref}(t)$ without sample (gray line) with $E_{sample}(t)$ after transmission through the sample (black line). A time delay of several picoseconds (ps) is clearly apparent, which reflects the pulse propagation with smaller group velocity in the sample. Moreover, the THz pulse lengthens in time and shows clear ringing, a result of the material dispersion. The phase $\phi(\omega)$ of the complex transmission function $t(\omega)$ is shown in Figure 11.9b: it exhibits many 2π phase jumps, which were corrected by unwrapping before the final analysis. The refractive index was extracted using the numerical algorithm of Duvillaret et al. (1996), with $n_1(\omega)$ and $\alpha(\omega)$ shown in Figure 11.9c and d. While $LaAlO_3$ is useful as a substrate for superconducting films, it also exhibits absorption and twinning of the microscale domains that restrict its use to the few-THz range. Measurement of the substrate at different temperatures was instrumental in the optical-pump THz-probe studies of Bi-2212 described later.

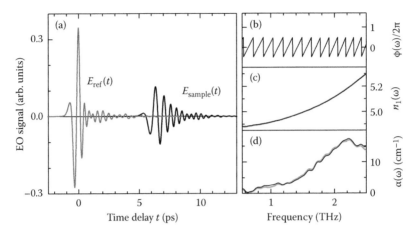

Figure 11.9 Time-domain THz spectroscopy of bulk LaAlO$_3$ at $T = 95$ K. (a) THz fields $E_{ref}(t)$ in vacuum and $E_{sample}(t)$ after transmission through the 478 μm thick wafer (thickness at 300 K). (b) Phase $\phi(\omega)$ before unwrapping. (c),(d) Refractive index and absorption coefficient, comparing the single scan (thin black line) from panel (a) with an average over twelve $E_{sample}(t)/E_{ref}(t)$ pairs (thick gray line).

11.3.2.2 Multilayered Film Geometry

The complex transmission of a layered system can be written as

$$t(\omega) = \frac{4n_S(\omega)}{(S_{11} + S_{12})n_S(\omega) + S_{22} + S_{21}} \cdot \frac{e^{i(\omega/c)[n_S(\omega)-1]d_S}}{n_S(\omega) + 1} \tag{11.11}$$

using a matrix approach, where **S** is a matrix describing the response of the multilayer stack on the substrate (Born and Wolf, 1999). The substrate refractive index and thickness are $n_S(\omega)$ and d_S, respectively. All reflections within the thin stack are included in this expression, while multiple reflections in the thick substrate are again ignored since only the first pulse replica is considered. The matrix **S** can be decomposed into its N constituent layers via $\mathbf{S} \equiv \mathbf{L}_1 \cdot \mathbf{L}_2 \cdot \mathbf{L}_3 \ldots \mathbf{L}_N$ where the response of each single layer is given by a matrix

$$\mathbf{L}_m = \begin{pmatrix} \cos(\beta_m) & -\dfrac{i}{n_m}\sin(\beta_m) \\ -in_m\sin(\beta_m) & \cos(\beta_m) \end{pmatrix} \tag{11.12}$$

where $\beta_m \equiv n_m d_m \omega/c$, with n_m and d_m being the mth layer refractive index and thickness, respectively. This general expression is very useful when evaluating the response of layered systems, such as quantum wells (QWs) and embedded layers. For instance, an analytical expression can be derived for multiple QWs (Kaindl et al., 2009). Also, in systems where the photoexcitation decays exponentially into the material, the above expression with a large number of layers can be used to numerically approximate the refractive index profile. In general, interfaces lead to frequency-dependent phase shifts which must be taken into account in THz-TDS to avoid artifacts and properly evaluate the time-domain signals (Schall and Jepsen, 2000).

In the following, for brevity, we will consider only the simplest case resulting from Equation 11.11, namely, a single layer with refractive index $n(\omega)$ and thickness d. Furthermore, we will assume that the layer is *optically thin*, i.e., $\beta(\omega) = n(\omega)d\omega/c \ll 1$.

In this case, $\mathbf{L}_{11} = \mathbf{L}_{22} \approx 1$, $\mathbf{L}_{12} \approx -[i/n(\omega)]\beta(\omega)$, and $\mathbf{L}_{21} \approx -in(\omega)\beta(\omega)$ such that, taking into account that $\omega d/c \ll 1$, we can write

$$t(\omega) = \frac{4n_S(\omega)}{n_S(\omega)+1-(i\omega d/c)\varepsilon(\omega)} \cdot \frac{e^{i(\omega/c)[n_S(\omega)-1]d_S}}{n_S(\omega)+1} \tag{11.13}$$

where $\varepsilon(\omega) = n(\omega)^2$ is the dielectric function.

From the above expression it is clear that the dielectric function can be obtained from the frequency-dependent, complex THz transmission function $t(\omega)$ if the film thickness and substrate properties are known precisely. Let us call this the "absolute" method. The resulting $\varepsilon(\omega)$ is quite susceptible to the substrate optical pathlength $n_S(\omega) \cdot d_S$, especially at higher frequencies. Due to the exponential function in Equation 11.13, an incorrect d_S can lead to unphysical oscillations of the extracted optical response function. Therefore, a precise determination of $\varepsilon(\omega)$ demands exact measurement of the substrate thickness. Alternatively, knowledge of $\varepsilon(\omega)$ at the high-frequency end of the spectrum (obtained, for example, via FTIR spectroscopy) can be used to determine the thickness with even higher precision.

An alternative is the "relative" measurement, where the properties of the thin film on the substrate are measured with respect to a reference. In particular, one can measure the identical substrate both with and without the film, such that the substrate thickness then drops out of the equation for the relative transmission

$$\frac{t_{\text{bare substrate}}(\omega)}{t_{\text{film}}(\omega)} = 1 - \frac{i\omega d}{(n_S(\omega)+1)c}\varepsilon(\omega) \tag{11.14}$$

The precision of this method can be maximized by taking the two substrate pieces to be compared (with and without film) from the same wafer or removing the film after measurement to expose the bare substrate (Corson, 2000). Similarly, thin-film THz properties have been measured with very high precision by coating only half a substrate with the film, and then quickly vibrating the sample between the coated and uncoated halves accompanied with synchronous lock-in detection (Ferguson and Zhang, 2002).

Relative measurements can also be useful when *changes* of the dielectric response functions should be determined, for example, by changing the sample temperature across a phase transition (as treated in the following text), or by variation of a magnetic field, pressure, etc. The dependence of the substrate optical path length $n_S(\omega) \cdot d_S$ on temperature, magnetic field, etc. in this case should be characterized and included in the equation for maximum precision. Likewise, optical-pump THz-probe spectroscopy, described later, also involves a comparison between equilibrium and photoexcited states, and thus yields high precision. The majority of THz studies were carried out in the transmission mode discussed above, in part because in THz reflection studies the phase becomes extremely sensitive to the sample position. However, schemes for THz time-domain reflection and ellipsometric studies have been devised and experimentally applied (Khazan et al., 2001; Nagashima and Hangyo, 2001; Pashkin et al., 2003; Jepsen and Fischer, 2005; Matsumoto et al., 2009; Matsuoka et al., 2009).

11.3.3 Real and Imaginary Conductivity of Superconductors

In this section, we will illustrate the use of THz time-domain spectroscopy in the case of the electrodynamics of superconductors. Long before THz-TDS, far-IR spectroscopy already played an important role in characterizing conventional *s*-wave superconductors governed by the BCS phonon-mediated pairing mechanism (Glover and Tinkham, 1957; Tinkham,

1996). In 1986, superconductivity at much higher transition temperatures T_C was discovered by K. Müller and J. Bednorz in the cuprates, whose pairing mechanism remains a puzzle to this day (Bednorz and Müller, 1986). Broadband FTIR measurements are of great importance in studying these materials (Basov and Timusk, 2005). THz-TDS has also been applied to investigate conventional superconductors and high-T_C materials (Nuss et al., 1991; Jaekel et al., 1994; Corson et al., 1999, 2000; Wilke et al., 2000). THz electrodynamics represents a powerful contactless probe of low-energy excitations in superconductors including their pairing gap 2Δ and collective superfluid response.

A very interesting test for the limits of phonon-mediated superconductivity is MgB_2, whose superconductivity was discovered in 2001 with a surprisingly high T_C up to 39 K (Nagamatsu et al., 2001), much larger than previously thought possible within conventional BCS. Due to the ease of processing, MgB_2 is also of much applied interest. Figures 11.10 and 11.11 show the results of THz-TDS of MgB_2 with the above setup (Kaindl et al., 2002). The samples investigated were 100–200 nm thick MgB_2 films with T_C between 30.5 and 34 K, fabricated on sapphire substrates by annealing of boron films in the presence of bulk MgB_2 and Mg vapor.

Electric field traces of THz pulses transmitted through an MgB_2 film are shown in Figure 11.10a for temperatures just above T_C (dashed line, $T = 40$ K), and after cooling far below the transition (solid line, $T \approx 6$ K). Clearly, the THz field increases in amplitude *and* shifts in phase as the material transitions from normal to superconducting. Power transmission spectra $T(\omega) = |t(\omega)|^2$ are obtained via Fourier transform of the electric field traces, and are shown in Figure 11.10b normalized to the transmission just above T_C (which changes little above the transition). A dramatic change in $T(\omega)$ is clearly observed below T_C. At 6 K (dots in Figure 11.10b) a more than twofold increase in transmission occurs around 1.7 THz, accompanied by a decrease below 1 THz. This transmission peak shifts to lower frequency with increasing temperature.

More detailed physical insight can be derived from understanding the changes of the complex-valued optical conductivity $\sigma(\omega)$ of the superconducting film that underlies the observed THz field reshaping in Figure 11.10a. For this, we solve the above-discussed thin-film formula for $t(\omega)$ in the superconducting and normal states to obtain the conductivity $\sigma(\omega)$ relative to

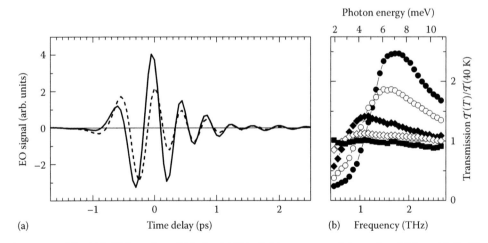

(a) Time delay (ps)   (b) Frequency (THz)

Figure 11.10 THz-TDS studies of MgB_2. (a) THz fields transmitted through a 100 nm MgB_2 film at $T = 6$ K (solid line) and 40 K (dashed line). (b) Power transmission $T(\omega)$ normalized to 40 K, as obtained from the transients below T_C at $T = 6$ K (dots), 20 K (circles), 27 K (solid diamonds), 30 K (open diamonds), and above T_C for $T = 33$ K (squares). (Adapted from Kaindl, R.A., Carnahan, M.A., Orenstein, J., Chemla, D.S., Christen, H.M., Zhai, H.Y., Paranthaman, M., and Lowndes, D.H., *Phys. Rev. Lett.*, 88, 027003, 2002. Copyright 2002 by the American Physical Society.)

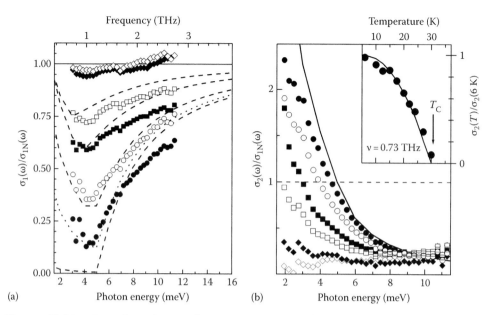

Figure 11.11 Optical conductivity from THz-TDS studies of MgB$_2$. (a) $\sigma_1(\omega)$ normalized to the 40 K normal-state value σ_{1N} for T = 6 K (dots), 17.5 K (circles), 24 K (solid squares), 27 K (open squares), 30 K (solid diamonds), and 50 K (open diamonds). Dashed and dotted lines: Mattis-Bardeen model for T_C = 30 K, $2\Delta_0$ = 5 meV, and T = 6, 12, 17.5, 24, and 27 K (bottom to top). (b) Corresponding normalized imaginary part $\sigma_2(\omega)/\sigma_{1N}(\omega)$. Solid line: MB model at 6 K. Inset: Normalized temperature dependence of $\sigma_2(0.73\,\text{THz})$ (dots) compared to the Mattis-Bardeen result (solid line). (Adapted with permission from Kaindl, R.A., Carnahan, M.A., Orenstein, J., Chemla, D.S., Christen, H.M., Zhai, H.Y., Paranthaman, M., and Lowndes, D.H., *Phys. Rev. Lett.*, 88, 027003, 2002. Copyright 2002 by the American Physical Society.)

its normal-state value. Figure 11.11 shows the resulting normalized real and imaginary parts. A strong depletion in the real part $\sigma_1(\omega)/\sigma_N(\omega)$ occurs below T_C. At the lowest temperature, an absorption onset is evident around ≈5 meV, which represents the signature of the superconducting gap. In turn, the imaginary part $\sigma_2(\omega)/\sigma_N(\omega)$ exhibits the buildup of a component that strongly increases with decreasing frequency. This $1/\omega$-like response is the hallmark of the superconducting state, since it corresponds to the purely inductive motion of the superfluid being accelerated by the electromagnetic THz field.

The complex-valued conductivity of superconductors is, as described in more detail in Chapter 3, often described by the well-known "two-fluid" model (Waldram, 1996),

$$\sigma(\omega) = \rho_{QP}\,\frac{1}{1/\tau - i\omega} + \rho_S\left\{\pi\delta(\omega) + \frac{i}{\omega}\right\} \tag{11.15}$$

where ρ_{QP} and ρ_S are the quasiparticle and superfluid spectral weights, corresponding to carrier densities n_{QP}, n_S via $\rho_i = n_i\,e^2/m^*$ (m^*: effective mass). The first term is the Drude response of unpaired quasiparticles with scattering rate $1/\tau$ (Chapters 1 and 3). The second term represents the electromagnetic response of the Cooper pair condensate: its δ-function in the real part σ_1 results from the infinite DC conductivity of the superfluid, while the imaginary part $\sigma_2 \propto 1/\omega$ is the inductive response of dissipationless supercurrents.

The two-fluid model works well for frequencies far below the gap 2Δ. The spectra in Figure 11.11 approach the two-fluid model at their low-frequency end, but generally require the inclusion of electromagnetic transitions across the superconducting gap. A corresponding theory of THz electrodynamics within the microscopic BCS theory was formulated by Mattis and Bardeen (1958). Fits with this model are shown in Figure 11.11 (lines) and faithfully represent both $\sigma_1(\omega)$ and $\sigma_2(\omega)$. The steep onset in σ_1 at low temperatures reflects the fact that Cooper pair breaking is possible only for photon energies that exceed 2Δ. At higher temperatures, an additional Drude-like low-frequency conductivity occurs, explained by thermally excited unpaired quasiparticles similar to the ρ_{QP} term in the two-fluid model.

Importantly, the THz data for MgB_2 are only reproduced if a low-temperature gap value $2\Delta_0 = 5\,\text{meV}$ is assumed, almost a factor of two smaller than the BCS prediction $2\Delta_0 = 3.5\,k_B$ $T_C \approx 9\,\text{meV}$. This small gap is also observed in other MgB_2 studies. In contrast to surface-sensitive probes such as scanning tunneling microscopy, the THz field penetrates into the bulk. This shows that the condensate persists up to the large transport-derived T_C (inset, Figure 11.11b) and confirms the presence of the small gap throughout the film, ruling out a scenario where the T_C used in the theory comparison is skewed by percolative transport. Theory explains the novel physics in MgB_2 via two bands whose coupling leads to the unusual phenomenon of two different superconducting gaps with a single T_C, where the smaller gap dominates the THz conductivity (Liu et al., 2001; Choi et al., 2002).

Several THz-TDS experiments have been carried out also on high-T_C cuprates, accessing excitations of particular relevance to their physics (Nuss et al., 1991; Jaekel et al., 1994; Corson et al., 1999, 2000; Wilke et al., 2000). The THz range is above that measurable by microwave studies and difficult to access with conventional FTIR. Corson et al. (1999) investigated the THz transmission of 40–65 nm thick films of the high-T_C superconductor $Bi_2Sr_2CaCu_2O_{8+\delta}$ (Bi-2212), with key results shown in Figure 11.12. The THz-TDS spectrometer was based on a 90 MHz Ti:sapphire oscillator, whose 100 fs output pulses were focused onto ≈10 μm sized gaps of PC THz emitters and detectors (Corson, 2000). The spectral coverage extended from ≈0.1 to 1 THz, and the optics and sample size were optimized for the resulting large foci, with up to 2 cm beam waist at 0.1 THz. This range lies almost 100 times below the d-wave gap maximum. The two-fluid model is thus generally a good

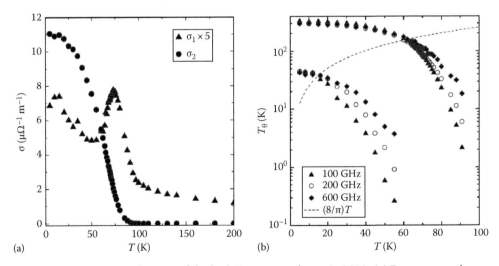

(a)

(b)

Figure 11.12 THz conductivity of the high-T_C superconductor Bi-2212. (a) Temperature dependence of the real and imaginary conductivity at 0.1 THz for a sample with $T_C = 74\,\text{K}$. (b) Phase-stiffness temperature T_θ obtained from σ_2 at three frequencies. Results are for a sample with $T_C = 33\,\text{K}$ (left side) and $T_C = 74\,\text{K}$ (right side). Dashed: Kosterlitz-Thouless criterion $T_\theta = (8/\pi)\,T$. (Reprinted by permission from Macmillan Publishers Ltd. (Corson, J., Mallozi, R., Orenstein, J., Eckstein, J.N., and Bozovic, I. Vanishing of phase coherence in underdoped $Bi_2Sr_2CaCu_2O_{8+\delta}$, *Nature*, 398, 221–223), Copyright 1999.)

approximation for high-T_C cuprates in this range, although for Bi-2212 deviations occur in σ_1 at the low-frequency end due to nanoscale inhomogeneities in the condensate density (Corson et al., 2000). The optical conductivity of Bi-2212 with $T_C = 74\,K$ is shown in Figure 11.12a at 0.1 THz. A peak is observed around T_C in the real part σ_1, explained by the competition between a decreasing quasiparticle density and reduced scattering rate $1/\tau$ with falling temperature (Nuss et al., 1991). The imaginary part σ_2 (dots, Figure 11.12a), in turn, represents an excellent probe of the superconducting order parameter—it is dramatically enhanced by the onset of superconductivity below T_C.

An important observation, however, is that σ_2 clearly persists above T_C and becomes unobservable only at much higher temperatures of $\approx 100\,K$. While static measurements show the onset of the infinite DC conductivity only at T_C, it is smeared out when viewed on short timescales with the rapidly varying THz electric field. The underlying reason is revealed by analysis of a larger set of data and comparison with theory (Corson et al., 1999). Superconductivity requires not only the binding of charges into Cooper pairs (with excitation gap 2Δ) but also a long-range *phase coherence* of the pairs. The condensate weight ρ_S can be understood as "stiffness," i.e., rigidity toward perturbations, of the quantum phase. Superconductivity disappears when the phase stiffness energy $k_B T_\theta = (\hbar^2 d/e^2) \times \rho_S$ becomes comparable to that of thermal excitations ($k_B T$), where d is the Bi-2212 interlayer spacing. In conventional BCS, however, the binding energy 2Δ is much smaller than $k_B T_\theta$ so that Cooper pair breakup instead governs the overall thermal transition into the normal state. A very different physics is given by the "Kosterlitz–Thouless" model, where thermal excitations generate *vortices* in the quantum phase corresponding to circulating currents (Corson et al., 1999). When the phase stiffness energy is small enough for vortex proliferation, long-range phase coherence and thus superconductivity is lost while pairing remains.

To test this model, Figure 11.12b shows the phase stiffness temperature from data at different frequencies from 0.1 to 0.6 THz, and for two different underdoped samples with $T_C = 33$ and 74 K. The temperature is defined as $T_\theta(\omega) = -(\hbar^2 d/e^2) \times i\omega\sigma_2(\omega)/k_B$, using the relationship between σ_2 and ρ_S from the two-fluid model. The underlying idea is that a sufficiently high-frequency (THz) probe cannot distinguish between an order parameter that is macroscopically coherent or fluctuating. At lower probe frequencies, however, the fluctuations will become apparent as $T_\theta(\omega)$ drops below the superconducting value extrapolated from higher-frequency data. For a macroscopically coherent state $T_\theta(\omega)$ should be the same regardless of frequency. Indeed, Figure 11.12b shows that at low temperatures the T_θ values at different frequencies are commensurate, while above a certain temperature they begin to deviate. Kosterlitz–Thouless theory predicts that at a given temperature T phase coherence is lost when the T_θ (linked to the pair density) drops below a value $T_{KT} = (8/\pi)T$. This criterion is shown in Figure 11.12b as the dashed line, and it falls right on top of the points where the data fan out. The above THz-TDS measurements thus reveal how the phase coherence vanishes in Bi-2212, persisting as short-range phase fluctuations in a significant temperature range above T_{KT} until all coherence is lost.

These examples illustrate the importance of obtaining the full complex-valued response function, to access the different physical information encoded in its real versus imaginary parts.

11.4 OPTICAL PUMP-THz PROBE SPECTROSCOPY

Optical-pump THz-probe experiments determine transient changes of the THz dielectric response. THz probes provide the necessary spectral selectivity in ultrafast studies to resonantly reveal optical coherences, transient correlations and phase transitions, or relaxation processes with meV-scale signatures. Conversely, adding temporal resolution to THz spectroscopy conveys the potential to distinguish spectral features and processes hidden in time-averaged equilibrium. A comprehensive review is beyond the scope of this section, but we note some examples. One area that was extensively studied concerns collective excitations, e.g., the Drude response (Chapters 1 and 3) in photoexcited semiconductors, nanowire surface plasmons, or the non-Drude conductance in disordered nanocrystalline materials (Beard et al., 2000; Huber et al., 2001; Prasankumar et al., 2005b; Cooke et al., 2006; Parkinson et al., 2007). Semiconductor quantum-size effects also lead to new optical transitions, e.g., interlevel and intersubband (IS) transitions in nanocrystals and QWs (Chapter 2). Carrier dynamics, stimulated emission, and transient coherence in nanomaterials were studied via THz and mid-IR probes after optical or electric pumping (Prasankumar et al., 2005b; Wang et al., 2006; Kroll et al., 2007; George et al., 2008; Choi et al., 2009). Moreover, intra-excitonic resonances enable entirely new ways to investigate electron–hole (*e-h*) pairs (Groeneveld and Grischkowsky, 1994; Kaindl et al., 2003; Kubouchi et al., 2005; Hendry et al., 2007). In cuprate superconductors, manganites, and other correlated materials ultrafast THz studies can discern basic interactions and phase transitions via thermally inaccessible perturbations of the correlated ground state (Kaindl et al., 2000, 2005; Averitt et al., 2001; Demsar et al., 2003; Prasankumar et al., 2005a; Hilton et al., 2006).

In the following, we will first describe the practical aspects of optical-pump THz probe experiments, along with the scheme for evaluating the measured signals. To illustrate the technique, we will then discuss two examples yielding insight into Cooper pair dynamics and the transient THz conductivity of photoexcited *e-h* gases.

11.4.1 Experimental Considerations

For a typical experimental scheme, we refer back to the setup in Figure 11.8. While a small fraction of the laser output is used to generate and probe the THz pulses, the major amount is available for photoexcitation of the sample. In the measurement, the optical excitation beam is chopped, while the synchronous change of the EO signal is detected with the lock-in amplifier. This yields the *photoinduced change* of the electric field $\Delta E(t)$, for a given pump-probe delay Δt between the optical excitation and THz probe pulses; conceptually, this is similar to a pump-probe experiment at optical frequencies (Chapter 9). By evaluating $\Delta E(t)$ along with the field $E(t)$ transmitted in equilibrium, we can determine the corresponding photoinduced *change* in the THz dielectric function, $\Delta \varepsilon(\omega)$. The pump-probe delay is stepped through different values Δt with a motorized stage. This gives access to the transient change of the THz response as it evolves on an fs to ns timescale after photoexcitation.

The use of amplified laser systems in these experiments is driven by the need to provide intense, tunable excitation pulses to cover the large probe spots. The long THz wavelengths imply large focus diameters at the sample, and the pump spot must accordingly provide a flat excitation profile across this area. In the setup of Figure 11.8, the probe diameter is ≈1 mm (FWHM) at 1.5 THz. This diameter is a typical value for THz foci but is strongly frequency-dependent, scaling approximately inversely with the THz frequency. To avoid low-frequency artifacts, a metal aperture can be placed over the sample to confine the transmitted THz field to the photoexcited region. In addition to direct photoexcitation with the 800 nm output of the Ti:sapphire amplifier, broadly tunable pump pulses ($\lambda \approx 0.4$–20 µm) can be generated with an optical parametric amplifier coupled to sum and/or DFM stages. In the above setup a two-lens telescope (not shown in Figure 11.8) was used to adjust the pump spot diameter for different wavelengths. Moreover, for selective excitation of narrow resonances, a pulse shaper or spectral filters can be employed to control the shape of the pump spectrum. Overlap with sample resonances, such as narrow exciton lines, can then be verified by spectrally resolving the transmitted pump pulse (Figure 11.8) while optimizing its center wavelength and bandwidth.

The choice of materials such as sample substrates or cryostat windows is limited by the need to transmit both at the THz and photoexcitation wavelengths. Useful materials include crystals such as MgO, SiC, sapphire, quartz, or diamond, or in some cases plastics such as picarin or polymethylpentene. Two-photon absorption and birefringence must be avoided in materials used as substrates, as well as materials with high refractive index where multiple reflections within the substrate will lead to recurring excitation of the sample. As cryostat windows, we have employed 25–50 μm thick Mylar foils cemented to an aluminum ring, which are transparent in the ≈1–3 THz range as well as the near-IR. The ultimate substrate and window material for ultrabroadband THz applications is diamond, available as polished synthetic CVD-grown discs, which is transparent from the far-IR to the ultraviolet (UV) range.

Many different types of setups have been employed for optical-pump THz probe spectroscopy. The 250 kHz setup detailed here provides intense and tunable pump pulses while simultaneously allowing for very sensitive THz detection. Ti:sapphire and fiber oscillators at MHz repetition rates with few-fs pulses can also be employed, with limited pump fluence and tunability but much easier access to ultrabroadband THz detection. Most optical-pump THz-probe setups are based on mJ-scale Ti:sapphire amplifiers at ≈1–3 kHz repetition rate, which enables very high excitation fluence at the expense of lower S/N ratio. However, using the much more stable, unamplified oscillator pulses for EO sampling was shown to significantly increase the THz detection sensitivity (Reimann et al., 2003).

A related technique providing insight into dynamics is *THz emission spectroscopy*, in which an fs pulse is used to prepare a coherent superposition of several quantum states. THz emission from IS quantum beats, for instance, provided insight into the decay of IS coherence and plasmon excitations (Planken et al., 1992; Roskos et al., 1992; Bonvalet et al., 1996; Bratschitsch et al., 2000). THz emission was also observed from coherent phonons (Dekorsy et al., 1996). Ultrafast transport phenomena are also linked to THz emission, which can be a probe of their dynamics, for example, in the case of Bloch oscillations in superlattices (Waschke et al., 1993; Dekorsy et al., 1995) or coherently controlled currents in bulk and nanoscale semiconductors (Côté et al., 1999; Newson et al., 2008).

11.4.2 Transient Dielectric Functions

After photoexcitation, the THz dielectric response of the sample transiently changes to a modified value $\varepsilon(\omega) + \Delta\varepsilon(\omega)$, where $\varepsilon(\omega)$ denotes the equilibrium dielectric function. In the experiment, the induced change $\Delta\varepsilon(\omega)$ is determined for each fixed pump-probe time delay Δt between the arrival of the pump and probe pulses on the sample. Here we describe how it can be obtained from the measured THz fields.

First, the THz reference probe field $E(t)$ transmitted through the sample in equilibrium is measured by chopping the THz generation beam, with the pump beam blocked (identical to THz-TDS, described in Section 11.3). The pump-induced change $\Delta E(t)$ is then recorded for each pump-probe delay Δt by chopping the pump beam and measuring with a lock-in amplifier the corresponding modulated EO signal. In both cases, the delay t is scanned. The full measurement over a range of delays Δt thus involves a two-dimensional (2D) dataset $E(t, \Delta t)$ as shown in Figure 11.13a. Since such scans can take significant amounts of time, up to hours for very small signals, careful consideration of the time window and number of steps is often necessary. Fourier transformation of these traces provides the corresponding frequency-domain fields $E(\omega)$ and $\Delta E(\omega)$, at each pump-probe delay Δt. As photoexcitation can result in a change of both the THz field amplitude *and* phase, 1D scans obtained, for example, by sitting at the peak of the electric field often yield misleading kinetics. Detection of the full 2D electric field change is thus important to analyze the transient dielectric response, and 1D scanning should

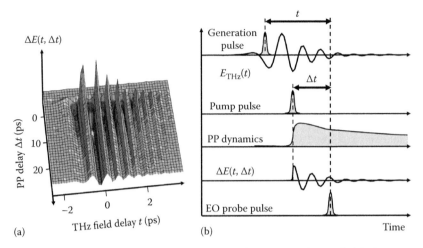

Figure 11.13 (a) Typical, experimentally measured transient electric field change $\Delta E(t, \Delta t)$ as a function of the THz field delay t and the pump-probe delay Δt. The field is added onto the equilibrium field $E(t)$ to obtain the transient response. The specific dataset shown represents the response of Bi-2212 at $T = 6\,\mathrm{K}$ after near-IR photoexcitation, discussed later in this chapter. (b) Definition of time delays as discussed in the text.

only be employed as a general overview of the timescales involved, or if phase shifts of $\Delta E(t)$ are absent with varying pump-probe delays Δt.

As discussed previously, in equilibrium the experimentally accessible complex transmission coefficient $t(\omega) = E(\omega)/E_{\mathrm{ref}}(\omega)$ is directly linked to the static dielectric function $\varepsilon(\omega)$ for a given sample geometry. Likewise, in the photoexcited state a modified dielectric function $\varepsilon(\omega) + \Delta\varepsilon(\omega)$ can be defined and linked to the modified complex transmission

$$t^*(\omega) = \frac{E(\omega) + \Delta E(\omega)}{E_{\mathrm{ref}}(\omega)} \tag{11.16}$$

The ratio of these two transmission coefficients is then given by

$$\frac{t(\omega)}{t^*(\omega)} = \frac{E(\omega)}{E(\omega) + \Delta E(\omega)} \tag{11.17}$$

By inserting the specific expression for the complex transmission for a given sample geometry, we can obtain the change of the dielectric response as a function of the above field ratio. In many cases, this necessitates a numerical solution, which can be obtained quickly with a computer. A very common case is that of a thin film on a substrate, for which using Equations 11.13 and 11.17 we obtain the analytical expression for the pump-induced *change* of the THz dielectric function:

$$\Delta\varepsilon(\omega) = \frac{\Delta E(\omega)}{E(\omega) + \Delta E(\omega)} \left(\frac{-ic}{\omega d} (1 + n_{\mathrm{S}}) - \varepsilon(\omega) \right) \tag{11.18}$$

When recording the 2D THz scans, the way the time delays are scanned is very important to avoid artifacts in the evaluation. The relative timing is illustrated in Figure 11.13b. The pump pulse causes a rapid change of the THz dielectric response of the sample around zero delay that decays with time. Accordingly, ΔE is nonzero for only the fraction of the THz field that hits the sample *after* arrival of the pump pulse. Scanning the *EO sampling* delay with respect to the pump pulse to measure ΔE would thus result in significant artifacts, particularly for small values of Δt or samples with significant transmission changes on a sub-ps timescale, since each part of the field trace would see a different point in time after excitation. Instead, by keeping this delay Δt between the pump and EO sampling pulse *fixed* the field change ΔE always samples the system at a specific time after excitation. The time delay t is then varied by changing the timepoint of THz *generation* with respect to the fixed EO sampling pulses. The change $\Delta E(t)$ is thus recorded for different times across the THz transient $E(t)$ while keeping the pump-probe delay Δt fixed. This scanning sequence is implemented through the location of the motorized stages as shown in Figure 11.8, i.e., the THz field is scanned via the THz *generation* arm rather than the EO sampling arm. More details, including a mathematical description of this issue, are given in Kindt and Schmuttenmaer (1999) and Beard et al. (2000).

In the above example, we have implicitly assumed that there is no pulse propagation between the sample and EO sampling crystal. This assumption is reasonable as long as the propagation is dispersionless, i.e., it does not significantly reshape the temporal profile of the THz electric field. Otherwise, the THz field (and hence the change ΔE) sampled at a specific time t will be affected by the sample dynamics at different Δt as they mix into the temporal profile during propagation. To avoid such issues, the optical components between the sample and EO detector must be carefully chosen to minimize pulse dispersion. For larger THz bandwidths, reflective optics for beam steering and materials such as silicon, Mylar, or thin pellicles as windows and beamsplitters should be used. A particularly important aspect is the sample orientation. For a thin film on a substrate, the film side should be oriented toward the EO detector to avoid dispersion of the THz pulse after traversing the photoexcited medium. In some cases, when the dynamics around time zero ($\Delta t \approx 0$) must be determined down to the time-resolution limit, the dispersion of the EO detector crystal should also be taken into account. The THz signals can then be corrected for this effect numerically but this necessitates a very dense mesh size of the 2D dataset, limiting the range of Δt. Nonequilibrium response functions such as $\Delta\varepsilon(\omega, \Delta t)$ can be extracted from the THz data, but the physical significance as a change of the equilibrium dielectric response is well defined only when the dynamics is slow compared to the THz cycle period. Time-energy limitations apply in THz spectroscopy (Kindt and Schmuttenmaer, 1999; Nemec et al., 2005), and must be taken into account as in other time-resolved experiments (Chapter 9).

11.4.3 Dynamics of Photoexcited *e-h* Pairs: Intra-Excitonic Spectroscopy

In the following, we will illustrate the application of optical-pump THz-probe spectroscopy with a study of photoexcited *e-h* pairs in semiconductors. The binding of *e-h* pairs into *excitons* via the Coulomb interaction is relevant both for optoelectronic applications and basic physics. Excitons are characterized by a center-of-mass momentum K that describes their pair motion, and additionally by a relative momentum connected to the pair's internal quantum state (Chapter 2). In the past, excitons were extensively studied by visible or near-IR optical techniques operating near the band gap (Shah, 1999). Near-IR absorption and photoluminescence (PL) correspond to *e-h* pair creation or annihilation (Chapter 5) with important limitations. For example, PL is often restricted by momentum conservation to $K \approx 0$ and depends in a complex way on the distribution function. Also, determination of *absolute* pair densities from PL is exceedingly difficult due to uncertainties on the collection efficiency and dipole moment.

Fundamentally different access to excitons is offered by *intra-excitonic* absorption, corresponding to transitions between different low-energy exciton quantum states, i.e., between the 1s exciton ground state and higher relative-momentum states. With typical exciton binding energies between 1 and 100 meV, intra-excitonic transitions occur at THz frequencies. Analogous to atomic absorption spectroscopy, intra-excitonic absorption probes electromagnetic transitions of *existing* excitons rather than pair generation and annihilation. By gauging the cross section of these transitions, absolute exciton densities can be determined. Although initial studies were scarce, rapid advances in ultrafast THz technology have rendered intra-excitonic spectroscopy a powerful tool for investigating the low-energy structure and dynamics of excitons (Timusk, 1976; Groeneveld and Grischkowsky, 1994; Cerne et al., 1996; Kira et al., 2001; Jörger et al., 2003; Kaindl et al., 2003, 2009; Kubouchi et al., 2005; Huber et al., 2006; Hendry et al., 2007; Ideguchi et al., 2008; Lloyd-Hughes et al., 2008).

The result of optical-pump THz-probe studies of the transient THz conductivity of photo-excited *e-h* gases in GaAs QWs are shown in Figure 11.14 (Kaindl et al., 2003, 2009). These experiments were performed on a stack of ten 14 nm wide undoped GaAs QWs separated by 10 nm wide $Al_{0.3}Ga_{0.7}As$ barriers. At low temperatures, the near-IR spectrum (not shown) is dominated by a sharp absorption peak of 1s heavy-hole excitons at 1.54 eV, followed by additional lines and a broadband continuum at higher energies. To avoid THz artifacts from photoexcited carriers, the substrate was removed by selective etching after attaching the sample to MgO. The experiment was conducted with the above 250 kHz setup (Figure 11.8) using 500 μm thick ZnTe crystals and with the near-IR pump pulses spectrally narrowed to 2 meV for selective excitation of either the 1s exciton or continuum.

Figure 11.14a shows the THz response of unbound *e-h* pairs after *nonresonant* excitation into the band-to-band continuum, measured at room temperature to ensure that the *e-h* pairs remain fully ionized. The pump-induced field change $\Delta E(t)$ (solid line) resembles the reference THz pulse $E(t)$ (dashed) with a phase shift. This points to a spectrally broadband

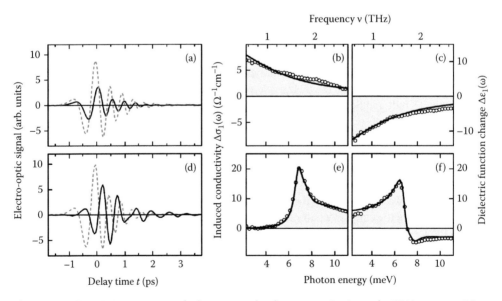

Figure 11.14 THz response of photoexcited *e-h* pairs in GaAs multi-QWs at $\Delta t = 10$ ps. (a) THz reference field E (dashed) and pump-induced change ΔE (solid line, ×50) after excitation 100 meV above the bandgap at $T = 300$ K. (b),(c) Spectra $\Delta\sigma_1$ and $\Delta\varepsilon_1$ (dots) corresponding to panel (a), along with a Drude model (solid line) with $n_{eh} = 2 \times 10^{10}$ cm^{-2} and $\Gamma_D = 4.8$ meV. (d–f) THz response after excitation at the 1s heavy-hole line, at $T_L = 6$ K. Solid lines in panels (e),(f): intra-excitonic model with $n_X = 2.7 \times 10^{10}$ cm^{-2}. (Reprinted from Kaindl, R.A., Hägele, D., Carnahan, M.A., and Chemla, D.S., *Phys. Rev. B*, 79, 045320, 2009. Copyright 2009 by the American Physical Society.)

response, which is confirmed in the corresponding THz spectra in Figure 11.14b and c (dots). The dielectric change is decomposed as (Chapter 1)

$$\Delta\varepsilon(\omega) = \Delta\varepsilon_1(\omega) + \frac{i}{\varepsilon_0\omega}\Delta\sigma_1(\omega) \tag{11.19}$$

where the induced conductivity $\Delta\sigma_1(\omega)$ represents the absorbed power density and $\Delta\varepsilon_1(\omega)$ measures the out-of-phase response. The THz spectra were computed from the time-domain traces as discussed above, taking into account the QW layer sequence. The large low-frequency conductivity $\Delta\sigma_1$ in Figure 11.14b underscores the conducting nature of the unbound pairs, while the dispersive $\Delta\varepsilon_1 < 0$ apparent in Figure 11.14c is characteristic of a zero-frequency, Drude-like oscillator. The response is indeed well described by a pure Drude response (Chapters 1 and 3)

$$\Delta\varepsilon(\omega) = \frac{n_{\mathrm{eh}}e^2}{d_{\mathrm{W}}\varepsilon_0 m^*} \times \frac{-1}{(\omega^2 + i\omega\Gamma_{\mathrm{D}})} \tag{11.20}$$

where
   n_{eh} is the *e-h* pair 2D sheet density
   d_{W} is the QW width
   Γ_{D} denotes the Drude scattering rate
   m^* is the reduced effective mass of the *e-h* pairs

The Drude fit (solid lines in Figures 11.14b and c) also closely agrees with the density estimated from the excitation fluence.

   A completely different behavior is observed after *resonant* excitation of the 1s exciton at low temperature, as shown in Figures 11.14d and f. The pump-induced change of the THz field in Figure 11.14d displays a phase shift that strongly varies across the time delay *t*, indicating a complex spectral shape. This is confirmed by the spectra in Figure 11.14e and f. They reveal an induced conductivity $\Delta\sigma_1$ dominated by a distinct, asymmetric peak around $h\nu \approx 7\,\mathrm{meV}$, with an oscillatory response in $\Delta\varepsilon_1$ around the same energy. This response is explained by *intra-excitonic* transitions. The 7 meV peak arises from 1s → 2p exciton level transitions, in agreement with known binding energies. This low-energy oscillator (absent in equilibrium) occurs in the transparent region of the semiconductor far below its bandgap. The vanishing conductivity $\Delta\sigma_1$ toward low frequencies results from the insulating nature of the charge-neutral excitons. Moreover, at low frequencies the sign of $\Delta\varepsilon_1$ is opposite to that of the Drude response. These differences illustrate how insulating (exciton) and conducting (free carrier) phases can be discriminated in the THz regime, and they again underscore the importance of measuring both real and imaginary parts of the response.

   For a quantitative description, calculations of the THz dielectric function of intra-excitonic transitions in 2D *e-h* gases are shown in Figure 11.14e and f as solid lines. They reproduce the shape of the experimental data extremely well and confirm that the peak stems from the 1s–2p transition, while the shoulder is explained by broadened transitions into higher bound *p* levels and the continuum. Importantly, absolute *e-h* pair densities can be obtained (Kaindl et al., 2009). In this respect, we note that both the Drude and intra-excitonic models fulfill the so-called partial oscillator strength sum rule (Chapter 1)

$$\int_0^\infty \sigma_1(\omega)d\omega = n_{\mathrm{3D}}\frac{\pi e^2}{2m^*} \tag{11.21}$$

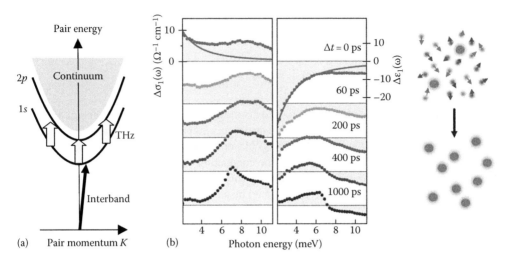

Figure 11.15 (a) Electron–hole pair energy dispersion versus center-of-mass momentum K. Arrows: near-IR exciton generation, and 1s–2p intra-excitonic transitions. (b) THz response $\Delta\sigma_1(\omega)$ and $\Delta\varepsilon_1(\omega)$ showing exciton formation in GaAs QWs at $T = 6\,K$ after nonresonant excitation. Solid line at $\Delta t = 0$ ps: Drude model. The cartoon illustrates the formation process. (Adapted by permission from Macmillan Publishers Ltd. (Kaindl, R.A., Carnahan, M.A., Hägele, D., Lövenich, R., and Chemla, D.S., Ultrafast terahertz probes of transient conducting and insulating phases in an electron-hole gas, *Nature*, 423, 734–738). Copyright 2003.)

within a parabolic band approximation. Here, the 3D carrier density n_{3D} is obtained from the exciton or free carrier 2D sheet density by dividing through the QW width d_W. Equation 11.21 links the total pair density to the integral over $\sigma_1(\omega)$ to below the onset of interband transitions, i.e., to the intraband spectral weight. This underscores the capability of THz spectroscopy as a gauge of absolute carrier densities.

Intra-excitonic transitions measure changes between *relative* momentum states and therefore can detect excitons well outside the optically accessible K, as illustrated in Figure 11.15a. This sensitivity to the *total* exciton density renders the transient THz conductivity an ideal tool to investigate exciton formation and ionization (Kaindl et al., 2003, 2009). Figure 11.15b shows the THz response after nonresonant excitation above the band gap at $T = 6\,K$. At 0 ps, the response is broad with a large low-frequency conductivity, and $\Delta\varepsilon_1(\omega)$ mostly follows the inductive signature of the Drude model (solid line, Figure 11.15b). Thus, the photoexcited QWs are initially dominated by a conducting gas of unbound *e-h* pairs. Already at this time delay, however, $\Delta\sigma_1(\omega)$ deviates significantly from the Drude response and shows instead a distinct exciton peak. This signifies the rapid appearance of excitonic *e-h* correlations at low lattice temperatures. The peak increases in spectral weight and sharpens as the system evolves over several 100 ps, while the Drude components decay. Figure 11.15b thus reveals the formation of excitons from a gas of unbound *e-h* pairs, with two different timescales. The immediate appearance of an exciton peak points to rapid formation via optical phonon and Coulomb interactions. The overall transformation into a pure exciton gas, however, proceeds much slower within several 100 ps, linked to a permanent release of the excess binding energy into the acoustic phonon bath. This is reflected in the decay of the Drude-like inductive system response at 1 THz and a crossover to $\Delta\varepsilon_1 > 0$ as the remaining *e-h* pairs bind into excitons. A quantitative description of the mixed THz response with both intra-excitonic and Drude-like features was obtained with a two-component model (Kaindl et al., 2009), enabling exciton and free-carrier fractions to be inferred. This yields quantitative insight into the dynamics and thermodynamics of exciton ionization.

Intra-excitonic spectroscopy has a broad range of applications. The intra-excitonic response of a high-density exciton gas revealed strong renormalization of the binding energy and a crossover into a purely conducting state (Huber et al., 2005). Measurements of exciton magneto-conductivity up to 6.5 T showed the splitting of the intra-excitonic transition into $1s$–$2p^+$ and $1s$–$2p^-$ peaks with different magnetic quantum number (Lloyd-Hughes et al., 2008). In contrast to absorption, THz *gain* from inverted exciton populations was not initially observed until a recent study investigating the transient low-energy response in Cu_2O (Huber et al., 2006). Access to the low-energy structure and dynamics of excitons can also be of fundamental relevance to the study of low-temperature collective phenomena, such as the elusive exciton Bose–Einstein condensate (Butov et al., 2002; Snoke et al., 2002; Kasprzak et al., 2006). Theory predicts strong changes of the THz response during exciton BEC (Johnsen and Kavoulakis, 2001). Candidate excitons are ideally optically "dark" with vanishing *interband* dipole moment, so that they recombine slowly and cool to low temperature. A key contender is Cu_2O whose $1s$ exciton is interband dipole forbidden, with PL studies yielding conflicting interpretations (Jang and Wolfe, 2006). In contrast, intra-excitonic THz transitions are independent of these symmetry restrictions. Several recent studies probed the mid-IR $1s$–$2p$ transitions in Cu_2O (see e.g., Karpinska et al., 2005; Kubouchi et al., 2005; Ideguchi et al., 2008). The $1s$ exciton is split via exchange interactions into a para-exciton and a threefold degenerate ortho-exciton at 12 meV higher energy. Kubouchi et al. (2005) studied transient mid-IR absorption in Cu_2O after two-photon orthoexciton generation. They observed the buildup of para-excitons within ≈ 100 ps via a 31 THz (129 meV) intra-excitonic resonance, accompanied by the decay of a 28 THz ortho-exciton peak. This ortho-para-interconversion was attributed to electron spin exchange. Intra-excitonic studies thus give new insight into low-energy quantum states inaccessible with near-IR interband probes.

11.4.4 Systems with a Correlated Ground State: Cooper Pair Dynamics

We have discussed earlier semiconductor experiments where the THz-active carriers are introduced by photoexcitation, forming a transient many-particle gas that interacts and decays. In contrast, THz pulses can also interrogate the dynamics of systems whose *ground state* itself is correlated, where ultrafast perturbation generates an excited state whose relaxation back to equilibrium may give insight into the interactions underlying the equilibrium phase. Of particular interest are superconductors, characterized at low temperatures by the quantum coherence of the Cooper pair condensate. Understanding quasiparticle interactions is particularly important for the high-T_C cuprates, which exhibit a complex phase diagram with chemical doping and whose charge pairing mechanism remains unresolved. Seeking to understand quasiparticle interactions in cuprates, numerous experiments were carried out that study fs optical reflectivity changes in the visible or near-IR (Han et al., 1990; Stevens et al., 1997; Gedik et al., 2005). These ≈ 1–2 eV probe energies, however, are far above the fundamental, low-energy excitations, such as the superconducting gap and Cooper pair response discussed earlier. In contrast, ultrafast THz studies can directly couple to these excitations, providing a particularly insightful approach to measure quasiparticle interactions hidden in linear spectroscopy. Several such studies were reported on conventional superconductors and cuprates (Federici et al., 1992; Feenstra et al., 1997; Carr et al., 2000; Kaindl et al., 2000, 2005; Averitt et al., 2001).

Results from the first optical-pump THz-probe study of a high-T_C superconductor are shown in Figure 11.16. THz pulses were generated and detected electro-optically using a 1 kHz Ti:sapphire amplifier producing 1 mJ, 150 fs pulses. The experiments were carried out

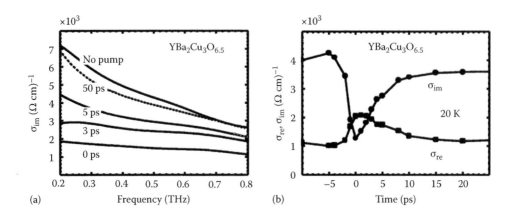

Figure 11.16 Optical-pump THz probe spectroscopy of underdoped $YBa_2Cu_3O_{6.5}$ (T_C = 50 K). (a) THz spectra of the imaginary conductivity σ_{im} [i.e., $\sigma_2(\omega)$] in the photoexcited state at different time delays. (b) Dynamics of the real and imaginary parts of the conductivity at 0.5 THz. (Reprinted from Averitt, R.D., Rodriguez, G., Lobad, A.I., Siders, J.L. W., Trugman, S.A., and Taylor, A.J., *Phys. Rev. B*, 63, 140502, 2001. Copyright 2001 by the American Physical Society.)

on near-optimally doped and underdoped $YBa_2Cu_3O_7\text{-}\delta$ (YBCO), grown as c-axis oriented thin films via rf magnetron sputtering on THz-transmissive MgO (Averitt et al., 2001). The equilibrium spectra of such films were discussed in Chapter 3 (see Figure 3.16), showing an imaginary conductivity $\sigma_2(\omega)$ that evolves from a flat spectrum to an approximately $1/\omega$ shape as the temperature drops below T_C. The THz response discussed in the following is for a 300 nm thick, underdoped YBCO film, with $\delta = 0.5$ and $T_C = 50$ K. Figure 11.16a shows a spectrum at $T = 10$ K in equilibrium before excitation ("no pump"). The complex-valued conductivity in this THz spectral range, far below the superconducting gap, can be understood within the framework of the previously discussed two-fluid model (Equation 11.15). At low frequencies, σ_2 is governed by the condensate, while σ_1 (not shown) reflects the contribution of unpaired quasiparticles.

After photoexcitation the imaginary conductivity $\sigma_2(\omega)$ (designated as σ_{im}) is strongly reduced, as shown in Figure 11.16a within the measured 0.2–0.8 THz range. It approaches the normal state response directly after excitation ($\Delta t = 0$ ps), which demonstrates the reduction of ρ_S, i.e., the breakup of the Cooper pair condensate. At later times, σ_2 increases toward its low-temperature equilibrium value, thereby directly following the reestablishment of the coherent superfluid. The experimental traces were measured after excitation with 150 fs, 1.5 eV pulses with 60 μJ energy, corresponding to the case of high excitation fluence with an initially photoexcited quasiparticle density around 4×10^{19} cm^{-3}. The dynamics are further illustrated by the fixed-frequency, time-dependent cuts shown in Figure 11.16b at 0.5 THz: the strong reduction of σ_2 (σ_{im}) is accompanied by an increase of σ_1 (σ_{re}), as expected from the increase of ρ_{QP} due to the generation of excess quasiparticles from Cooper pair breakup. Above T_C the dynamics were much faster, associated with thermalization dynamics analogous to that in metals (Chapter 8). The decay time τ_σ was also analyzed: in optimally doped YBCO it increased from a low-temperature value of 1.5–3.5 ps near T_C. This was interpreted as resulting from a dependence inverse to the gap size, $\tau_\sigma \propto 1/\Delta(T)$, reminiscent of the behavior in BCS superconductors. In the underdoped case, however, no temperature dependence was found, pointing to a predominant influence of the pseudogap (see following text) on the decay time. The quick, albeit incomplete recovery on a ≈5 ps timescale in both cases also strongly contrasts with the much slower, nanosecond (ns) kinetics observed in conventional s-wave BCS superconductors (Carr et al., 2000).

Optical-pump THz-probe experiments were also carried out to detect the transient THz conductivity in another prototypical high-T_C superconductor, Bi-2212 (Kaindl et al., 2005). That study investigated 62 nm thick films with near-optimal doping and $T_C \approx 88$ K, grown by molecular beam epitaxy on $LaAlO_3$ substrates (Eckstein and Bozovic, 1995). THz generation and EO detection with high sensitivity was achieved with 500 µm ZnTe using the above-described 250 kHz setup (Figure 11.8). Near-IR excitation at 1.55 eV resulted in a reduction of σ_2 accompanied by an increase of σ_1, with the signs and spectral shapes of these changes clearly demonstrating Cooper pair breakup and excess quasiparticles (Kaindl et al., 2005). Two-fluid model fits revealed that the spectral weight was fully conserved in this process, i.e., $\Delta\rho_S = -\Delta\rho_{QP}$. A central aspect of the study was a detailed investigation of the relaxation kinetics of the condensate, which was strongly non-exponential. A strikingly simple description of the kinetics was obtained by plotting the conductivity change on a "reciprocal" scale, i.e., as $1/\Delta\sigma_2$, as shown in Figure 11.17 (as a function of the delay time designated as t). At low temperatures, this type of plot reveals a linear time dependence (Figure 11.17a), which is the well-known hallmark of *bimolecular* kinetics. This can be understood by writing the kinetics as $1/\Delta\sigma_2(t) = r \cdot t + 1/\Delta\sigma_2(0)$, with slope r, which after taking the time derivative results in $d(\Delta\sigma_2)/dt = -r(\Delta\sigma_2)^2$. Since $\Delta\sigma_2$ is proportional to the density of broken Cooper pairs (confirmed by two-fluid fits to the spectra), the natural explanation of these dynamics is the pairwise interaction of quasiparticles as they recombine into Cooper pairs. The bimolecular scenario is further confirmed by the density dependence (Figure 11.17a), showing the persistence of the linear bimolecular shape at different densities.

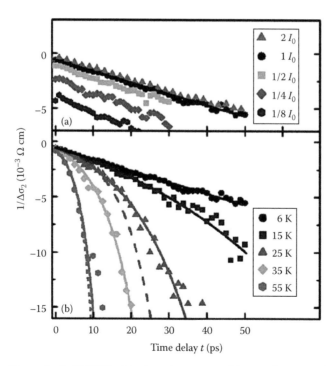

Figure 11.17 Dynamics in Bi-2212 of the imaginary part of the conductivity, plotted on a reciprocal scale at 1.33 THz probe frequency. (a) Intensity dependence at $T = 6$ K. (b) Temperature dependence at intensity $I_0 \equiv 0.7$ µJ/cm². Solid lines: fits with the model of Equation 11.22. Dashed lines: single exponential decay. (Reprinted from Kaindl, R.A., Carnahan, M.A., Chemla, D.S., Oh, S., and Eckstein, J.N., *Phys. Rev. B*, 72, 060510(R), 2005. Copyright 2005 by the American Physical Society.)

Various theories of quasiparticle "recombination," i.e., Cooper pair formation, have been formulated (see Nicol and Carbotte, 2003), with the Rothwarf–Taylor rate equation being particularly simple and insightful (Rothwarf and Taylor, 1967; Twerenbold, 1986):

$$\frac{d}{dt}n^* = -R(n^*)^2 - 2Rn_T n^* + 2\tau_B^{-1}N_\omega^* \tag{11.22}$$

where

n^* is the photoexcited quasiparticle density
n_T is the thermal equilibrium density
R is the recombination coefficient

Cooper pair formation can occur either via mutual interaction of two excess quasiparticles via the $(n^*)^2$ term or by recombination of a thermal quasiparticle with a photoexcited one ($n_T n^*$ term). The third term describes Cooper pair breaking at rate $2\tau_B^{-1}$ via absorption of nonequilibrium bosons (density N_ω^*) described by an additional equation (not shown): it leads to a slow, typically ns exponential decay in conventional superconductors via "phonon trapping," where phonons emitted during Cooper pair formation are repeatedly reabsorbed (Rothwarf and Taylor, 1967). This was observed in the condensate THz dynamics in conventional superconductors (Carr et al., 2000; Demsar et al., 2003). The much faster bimolecular kinetics in Bi-2212 points to a subordinate role of phonon trapping and may be linked to recombination from the anti-nodes of the d-wave gap (Kaindl et al., 2005). The quadratic term in Equation 11.22 dominates for low temperatures ($n^* \gg n_T$), which explains the bimolecular kinetics. At high temperatures, n_T becomes large so that the second term in Equation 11.22 dominates the kinetics and leads to single-exponential kinetics. Indeed, as evidenced in Figure 11.17b, with increasing temperature the decay exhibits a crossover from a linear bimolecular shape ($T = 6\,\text{K}$) to one closely described by single-exponential kinetics (dashed line, $T = 55\,\text{K}$). All intermediate cases are well described by Equation 11.22 by variation of n_T (solid lines in Figure 11.17b). The kinetics differs from that observed in YBCO, which may be linked to nanoscale electronic inhomogeneity in Bi-2212 that relaxes momentum restrictions for scattering. Importantly, in equilibrium spectroscopy the Cooper pair formation rate is hidden: even for the fastest rate ($T = 55\,\text{K}$) it contributes a broadening $\hbar/\tau = 0.2\,\text{meV}$, well below the line widths in linear THz and photoemission spectroscopy.

While the transient THz spectra in the above experiments followed the condensate dynamics, they cannot reveal changes to the superconductor's electronic gap at higher photon energies. This is of great importance in cuprates due to the "pseudogap" that mimics the superconducting gap in energy but exists up to much higher temperatures T^* (Basov and Timusk, 2005). To access its dynamics, a first study detected ultrafast mid-IR reflectivity changes in YBCO using tunable probe pulses between ≈14 and 43 THz generated in GaSe (Kaindl et al., 2000). In optimally doped YBCO, the response was consistent with the above-discussed Cooper pair breakup and re-pairing observed around 1 THz, showing a transient modulation of the gap on a ≈5 ps timescale. In contrast, underdoped YBCO ($T_C = 68\,\text{K}$) revealed complex gap dynamics with *two* constituents: (1) a ≈5 ps component below T_C due to the superconducting gap dynamics as in optimally doped YBCO and (2) a faster ≈700 fs component surviving up to $T^* \approx 160\,\text{K}$, which reveals the dynamics of the pseudogap correlations. The ultrafast study thus enabled the identification and separation of these two gap components along the time axis, while they remain indistinguishably superimposed in linear spectra. A proportionality between the ultrafast signals and strength of anti-ferromagnetic resonances was also observed, providing support for strong coupling of carriers and spins.

11.5 TERAHERTZ EXCITATION

In addition to detection of low-energy dynamics, THz and mid-IR sources scaled to high pulse energies—and thus very high THz electric field strengths up to the MV/cm range—are of increasing interest in materials studies. Their applications can include, for example, (1) resonant excitation of low-energy vibrational or electronic modes, (2) studies of high-field charge transport, or (3) investigations of a system's higher-order nonlinearities via all-THz wave mixing. To illustrate these new opportunities we will list a few examples in the following.

Among the most intriguing applications of intense THz pulses to complex materials is the coherent excitation of vibrational modes, to study the coupling of specific lattice distortions with a material's *electronic ground state*. Such an experiment was reported in (Rini et al., 2007), which investigated the ultrafast dynamics of the magnetoresistive manganite $Pr_{0.7}Ca_{0.3}MnO_3$ (PCMO) after direct vibrational excitation with a 17 THz pulse. The PCMO unit cell, shown in Figure 11.18a, exhibits strong orthorhombic distortion with a large Mn–O–Mn bond angle. The conductivity is strongly influenced by the distortion, since the charge itinerancy depends on Mn–Mn 3d electron transfer via the intermediate O-2p atomic orbitals, and thus on the orbital overlap. A prominent vibrational mode around 17 THz ($h\nu \approx 71$ meV,

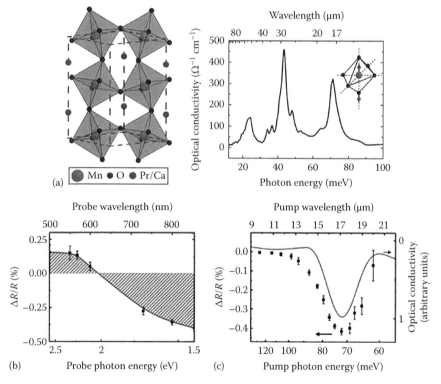

Figure 11.18 (a) Unit cell of $Pr_{0.7}Ca_{0.3}MnO_3$ (left side) and low-temperature optical conductivity (right side) with major optically active phonons. Inset: atomic displacement of the 17 THz (≈ 71 meV) mode. (b) Transient reflectivity change 1 ps after resonant THz vibrational excitation (≈ 1 mJ/cm^2 fluence, $T = 30$ K). (c) Maximum signal amplitude at an 800 nm probe wavelength as a function of the pump wavelength (dots), along with the sample conductivity (solid line) convolved with the pulse bandwidth. (Reprinted by permission from Macmillan Publishers Ltd. (Rini, M., Tobey, R., Dean, N., Itatani, J., Tomioka, Y., Tokura, Y., Schoenlein, R.W., and Cavalleri, A. Control of the electronic phase of a manganite by mode-selective vibrational excitation, *Nature*, 449, 72–74), Copyright 2007.)

see the right side of Figure 11.18a) modulates the Mn–O distance and was resonantly excited in the experiments of (Rini et al., 2007) by intense, μJ mid-IR pulses generated via DFM in GaSe. Figure 11.18b shows the resulting low-temperature reflectivity changes $\Delta R/R$ in the visible range, indicating a spectral weight shift to lower photon energies that represents the switchover into a *conducting* phase from the insulating ground state. This agrees with optical changes during a phase transition induced by magnetic fields. These changes can be linked to the collapse of a 0.3 eV gap in the insulating electronic ground state of PCMO. This occurs on an extremely fast timescale after vibrational excitation, with a sub-300 fs rise time, and persists in a metastable state for several hundreds of ps. The signal amplitude strongly depends on resonant excitation of the 71 meV mode as shown in Figure 11.18c. A strong threshold dependence on pump fluence was observed, indicating a vibrationally driven phase transition. Sample resistivity measurements upon vibrational excitation showed a dramatic resistance drop, corresponding to an increase of conductivity of almost five orders of magnitude. This first study elucidates how specific lattice distortions influence the electronic ground state of the manganite PCMO. Mode-selective vibrational excitation was also applied to a different manganite, $La_{1/2}Sr_{3/2}MnO_4$ (Tobey et al., 2008), and is of interest for a large range of correlated materials.

Intense THz fields can also be used to coherently drive *electronic* polarizations in materials, with several studies investigating intersubband (IS) transitions in quantum wells as a model system. Resonant THz driving was shown to generate IS sidebands on the near-IR interband transitions and to "undress" collective electron coupling (Craig et al., 1996; Carter et al., 2005). THz *Rabi oscillations* constitute a quantum optical effect where the Bloch vector of a two-level system is driven coherently between its coupled states. A vivid example of direct field-resolved measurements in the Rabi oscillation regime was given by Luo et al. (2004), which investigated a 24 THz, ($n = 1$) ↔ ($n = 2$) IS transition in n-type GaAs multi-QWs. Detection of the re-emitted fields from the THz-driven sample via EO sampling (using a 10 μm thick ZnTe crystal) provided a time-domain picture of the crossover from absorption (weak field limit) to stimulated emission (high THz fields). This experiment showed Rabi oscillations between the coupled subbands and deviations linked to Coulomb-mediated collective coupling between different QW layers. Rabi oscillations with high-field THz pulses were also observed in other systems, including 1s–2p transitions of donor impurities in GaAs and intra-excitonic transitions (Cole et al., 2001; Leinss et al., 2008). Moreover, high-field THz pulses enabled observation of IS gain without inversion (Frogley et al., 2006). Superradiant decay of optically inverted impurity transitions was also observed after THz excitation (Gaal et al., 2006). Extension of coherent population transfer schemes may enable unprecedented control of low-energy populations and polarizations in quantum structures (Sherwin et al., 1999; Batista, 2006; Paspalakis et al., 2006).

We should also mention very recent developments. First, the high-field THz-driven acceleration of polarons—electrons coupled to a surrounding lattice distortion—was studied in GaAs (Gaal et al., 2007). Using a mid-IR probe this revealed impulsive phonon emission once the polaron accelerated to a kinetic energy exceeding one LO phonon. The electron was found to oscillate around the polaron's center of mass at the phonon frequency, a new quantum kinetic effect relevant to high-field transport in nanostructures (Gaal et al., 2007). Second, THz-pump THz-probe spectroscopy with μJ pulses in the few-THz range was recently applied to materials studies, revealing carrier heating and impact ionization in InSb in 100 kV/cm electric fields (Hoffmann et al., 2009). Finally, coherent 2D Fourier transform spectroscopy using nonlinear wave mixing was extensively applied in ultrafast studies in the visible to mid-IR range (Hochstrasser, 2007). A scheme was recently demonstrated in solids that combines *collinear* wave mixing with direct EO field detection, enabling 2D THz spectroscopy of higher-order nonlinearities

(Kuehn et al., 2009). This vastly extends, for example, the previous capabilities of mid-IR four-wave-mixing of IS transitions (Kaindl et al., 2001). These novel all-THz schemes promise completely new insight into coherent and incoherent low-energy dynamics in complex materials.

11.6 SUMMARY AND OUTLOOK

In this chapter, we presented an introduction into the techniques of THz-TDS and ultrafast THz spectroscopy as applied to condensed matter studies. Generation and detection of broadband coherent THz fields with fs lasers was reviewed, with particular emphasis on the widely applied methods of DFM and EO sampling. THz-TDS takes advantage of the availability of both amplitude and phase that results from time-resolved detection of the electric field $E(t)$. We discussed how the complex-valued dielectric function is obtained from THz-TDS scans for a given sample geometry. Availability of both the real and imaginary parts of the response function places strict limits on theoretical models, important for discerning the materials physics. In superconductors, for example, this reveals quasiparticle and condensate densities and the excitation gap 2Δ of the correlated charge pairs. The short THz pulse duration can, moreover, be exploited in optical-pump THz-probe spectroscopy to follow the dynamics of low-energy excitations. For illustration, a setup operating at a 250 kHz repetition rate was described in detail, along with methods for extracting the dielectric function change from the signals. We gave examples of the application of this method to detection of Coulomb-bound e-h pairs via intra-excitonic transitions and to Cooper pair formation in high-T_C cuprates. Recent studies using resonant low-energy excitation with intense THz pulses were also discussed.

The ubiquitous nature of low-energy excitations in complex materials coupled with the fast-paced development of THz technology opens many pathways for future exploration. The selective excitation of low-energy modes with intense THz pulses and ultrabroadband THz detection of transient dielectric changes are of particular interest. Combining both offers many opportunities to unravel the coupling of low-energy modes in strongly correlated electron systems. Resonant vibrational excitation, for instance, may provide insight into electron–phonon coupling in high-T_C cuprates. Photoinduced phase transitions between metal–insulator or ferromagnetic–paramagnetic states are also of great interest and may be manipulated with THz pulses. Intra-excitonic spectroscopy, as discussed above, represents a powerful approach for exciton studies, where low-energy detection of optically dark excitons is of increasing interest. Moreover, degenerate THz wave mixing could find applications to probe transient THz coherences and energy transfer, for example, in nanostructured materials.

Beyond ensemble measurements, probing of the THz near field for sensitive studies of single nanostructures promises unprecedented insight into charge conductance and low-energy excitations at the nanoscale. Such experiments may ultimately follow the spatiotemporal dynamics of charges as they transfer between heterointerfaces and along networks of quasi-1D and 0D structures in novel energy materials. Despite initial work, the combination of near-field and THz spectroscopy is in its infancy and will profit greatly from technology improvements. The combination of THz optics and nano-plasmonics provides important impetus for these endeavors.

Ultimately, limitations of THz methods must also be taken into account in a critical comparison with competing low-energy probes such as Raman, neutron, or photoelectron spectroscopies. Mutual combination, however, offers interesting opportunities. For instance, ultrafast photoemission could be combined with selective THz excitation to track the momentum-resolved electronic structure dynamics linked to specific vibrational or electronic modes. Applications thus abound, and as in the past we can expect that new opportunities will be strongly driven by further rapid THz technology developments. For studies of complex materials at the forefront of contemporary condensed matter physics, coherent time-domain and ultrafast THz spectroscopy thus promises continued, unique insights into low-energy excitations essential to basic physics and applications.

ACKNOWLEDGMENTS

I would like to thank my collaborators on ultrafast THz and IR studies, especially D. S. Chemla, T. Elsaesser, M. Woerner, M. A. Carnahan, H. Y. Choi, and many others. I am also grateful to those who granted permission for reproduction of their figures. The preparation of this chapter was supported by the Office of Basic Energy Sciences, Materials Sciences and Engineering Division, of the U.S. Department of Energy under Contract No. DE-AC02-05CH11231.

REFERENCES

Abdullaev, G. B., Kulevskii, L. A., Prokhorov, A. M., Savel'ev, A. D., Salaev, E. Y., and Smirnov, V. V. (1972) GaSe, a new effective material for nonlinear optics. *JETP Lett.*, 16, 90–92.

Ashida, M. (2008) Ultra-broadband terahertz wave detection using photoconductive antennas. *Jpn. J. Appl. Phys.*, 47, 8221–8225.

Auston, D. H. (1975) Picosecond optoelectronic switching and gating in silicon. *Appl. Phys. Lett.*, 26, 101–103.

Auston, D. H. and Nuss, M. C. (1988) Electrooptic generation and detection of femtosecond electrical transients. *IEEE J. Quantum Electron.*, 24, 184–196.

Averitt, R. D., Rodriguez, G., Lobad, A. I., Siders, J. L. W., Trugman, S. A., and Taylor, A. J. (2001) Nonequilibrium superconductivity and quasiparticle dynamics in $YBa_2Cu_3O_{7-\delta}$. *Phys. Rev. B*, 63, 140502.

Azad, A. K., Han, J., and Zhang, W. (2006) Terahertz dielectric properties of high-resistivity single crystal ZnO. *Appl. Phys. Lett.*, 88, 021103.

Bakker, H. J., Cho, G. C., Kurz, H., Wu, Q., and Zhang, X.-C. (1998) Distortion of terahertz pulses in electro-optic sampling. *J. Opt. Soc. Am. B*, 15, 1795–1801.

Bartel, T., Gaal, P., Reimann, K., Woerner, M., and Elsaesser, T. (2005) Generation of single-cycle THz transients with high electric-field amplitudes. *Opt. Lett.*, 30, 2805–2807.

Basov, D. N. and Timusk, T. (2005) Electrodynamics of high-T_C superconductors. *Rev. Mod. Phys.*, 77, 721–779.

Batista, A. A. (2006) Pulse-driven interwell carrier transfer in n-type doped asymmetric double quantum wells. *Phys. Rev. B*, 73, 075305.

Bauer, T., Kolb, J. S., Loffler, T., Mohler, E., Roskos, H. G., and Pernisz, U. C. (2002) Indium–tin-oxide-coated glass as dichroic mirror for far-infrared electromagnetic radiation. *J. Appl. Phys.*, 92, 2210–2212.

Bayanov, I. M., Danielus, R., Heinz, P., and Seilmeier, A. (1994) Intense subpicosecond pulses tunable between 4 μm and 20 μm. *Opt. Commun.*, 113, 99–104.

Beard, M. C., Turner, G. M., and Schmuttenmaer, C. A. (2000) Transient photoconductivity in GaAs as measured by time-resolved terahertz spectroscopy. *Phys. Rev. B*, 62, 15764–15777.

Bednorz, J. G. and Müller, K. A. (1986) Possible high T_C superconductivity in the Ba-La-Cu-O system. *Z. Phys. B*, 64, 189–93.

Blanchard, F., Razzari, L., Bandulet, H. C., Sharma, G., Morandotti, R., Kieffer, J. C., Ozaki, T. et al. (2007) Generation of 1.5 μJ single-cycle terahertz pulses by optical rectification from a large aperture ZnTe crystal. *Opt. Express*, 15, 13212–13220.

Bonvalet, A., Joffre, M., Martin, J. L., and Migus, A. (1995) Generation of ultrabroadband femtosecond pulses in the mid-infrared by optical rectification of 15 fs light pulses at 100 MHz repetition rate. *Appl. Phys. Lett.*, 67, 2907–2909.

Bonvalet, A., Nagle, J., Berger, V., Migus, A., Martin, J.-L., and Joffre, M. (1996) Femtosecond infrared emission resulting from coherent charge oscillations in quantum wells. *Phys. Rev. Lett.*, 76, 4392–4395.

Born, M. and Wolf, E. (1999) *Principles of Optics*, Cambridge, U.K.: Cambridge University Press.

Boyd, R. W. (2003) *Nonlinear Optics*, San Diego, CA: Academic Press.

Bratschitsch, R., Müller, T., Kersting, R., Strasser, G., and Unterrainer, K. (2000) Coherent terahertz emission from optically pumped intersubband plasmons in parabolic quantum wells. *Appl. Phys. Lett.*, 76, 3501–3503.

Brazis, R. and Nausewicz, D. (2008) Far-infrared photon absorption by phonons in ZnTe crystals. *Opt. Mater.*, 30, 789–791.

Brener, Y. C., Lopata, J., Wynn, J., Pfeiffer, L., Stark, J. B., Wu, Q., Zhang, X.-C., and Federici, J. F. (1998) Coherent terahertz radiation detection: Direct comparison between free-space electro-optic sampling and antenna detection. *Appl. Phys. Lett.*, 73, 444–446.

Brunner, F. D. J., Schneider, A., and Günter, P. (2009) Velocity-matched terahertz generation by optical rectification in an organic nonlinear optical crystal using a Ti:sapphire laser. *Appl. Phys. Lett.*, 94, 061119.

Budiarto, E., Margolies, J., Jeong, S., Son, J., and Bokor, J. (1996) High-intensity terahertz pulses at 1-kHz repetition rate. *IEEE J. Quantum Electron.*, 32, 1839–1846.

Butov, L. V., Lai, C. W., Ivanov, A. L., Gossard, A. C., and Chemla, D. S. (2002) Towards Bose–Einstein condensation of excitons in potential traps. *Nature*, 417, 47–52.

Carr, G. L., Lobo, R., Laveigne, J., Reitze, D. H., and Tanner, D. B. (2000) Exploring the dynamics of superconductors by time-resolved far-infrared spectroscopy. *Phys. Rev. Lett.*, 85, 3001–3004.

Carter, S. G., Ciulin, V., Hanson, M., Huntington, A. S., Wang, S., Gossard, A. C., Coldren, L. A., and Sherwin, M. S. (2005) Terahertz-optical mixing in undoped and doped GaAs quantum wells from excitonic to electronic intersubband transitions. *Phys. Rev. B*, 72, 155309.

Cerne, J., Kono, J., Sherwin, M. S., Sundaram, M., Gossard, A. C., and Bauer, G. E. W. (1996) Terahertz dynamics of excitons in GaAs/AlGaAs quantum wells. *Phys. Rev. Lett.*, 77, 1131–1134.

Chen, Q., Tani, M., Jiang, Z., and Zhang, X.-C. (2001) Electro-optic transceivers for terahertz-wave applications. *J. Opt. Soc. Am. B*, 18, 823–831.

Choi, H., Borondics, F., Siegel, D. A., Zhou, S. Y., Martin, M. C., Lanzara, A., and Kaindl, R. A. (2009) Broadband electromagnetic response and ultrafast dynamics of few-layer epitaxial graphene. *Appl. Phys. Lett.*, 94, 172102.

Choi, H. J., Roundy, D., Sun, H., Cohen, M. L., and Louie, S. G. (2002) The origin of the anomalous superconducting properties of MgB_2. *Nature*, 418, 758–760.

Cole, B. E., Williams, J. B., King, B. T., Sherwin, M. S., and Stanley, C. R. (2001) Coherent manipulation of semiconductor quantum bits with terahertz radiation. *Nature*, 410, 60–63.

Cook, D. J. and Hochstrasser, R. M. (2000) Intense terahertz pulses by four-wave rectification in air. *Opt. Lett.*, 25, 1210–1212.

Cooke, D. G., Macdonald, A. N., Hryciw, A., Wang, J., Li, Q., Meldrum, A., and Hegmann, F. A. (2006) Transient terahertz conductivity in photoexcited silicon nanocrystal films. *Phys. Rev. B*, 73, 193311.

Corson, J. (2000) Advances in terahertz spectroscopy of high-Tc superconductors. Berkeley, CA, U. C. Berkeley PhD thesis.

Corson, J., Mallozi, R., Orenstein, J., Eckstein, J. N., and Bozovic, I. (1999) Vanishing of phase coherence in underdoped $Bi_2Sr_2CaCu_2O_{8+\delta}$. *Nature*, 398, 221–223.

Corson, J., Orenstein, J., Oh, S., O'Donnell, J., and Eckstein, J. N. (2000) Nodal quasiparticle lifetime in the superconducting state of $Bi_2Sr_2CaCu_2O_{8+\delta}$. *Phys. Rev. Lett.*, 85, 2569.

Côté, D., Fraser, J. M., Decamp, M. F., Bucksbaum, P. H., and van Driel, H. M. (1999) THz emission from coherently controlled photocurrents in GaAs. *Appl. Phys. Lett.*, 75, 3959–3961.

Côté, D., Sipe, J. E., and van Driel, H. M. (2003) Simple method for calculating the propagation of terahertz radiation in experimental geometries. *J. Opt. Soc. Am. B*, 20, 1374–1385.

Craig, K., Galdrikian, B., Heyman, J. N., Markelz, A. G., Williams, J. B., Sherwin, M. S., Campman, K., Hopkins, P. F., and Gossard, A. C. (1996) Undressing a collective intersubband excitation in a quantum well. *Phys. Rev. Lett.*, 76, 2382–2385.

Crooker, S. A. (2002) Fiber-coupled antennas for ultrafast coherent terahertz spectroscopy in low temperatures and high magnetic fields. *Rev. Sci. Instrum.*, 73, 3258–3264.

Dai, J., Xie, X., and Zhang, X.-C. (2006) Detection of broadband terahertz waves with a laser-induced plasma in gases. *Phys. Rev. Lett.*, 97, 103903.

Das, S., Ghosh, C., Gangopadhyay, S., Chatterjee, U., Bhar, C. B., Voevodin, V. G., and Voevodina, O. G. (2006) Tunable coherent infrared source from 5–16 μm based on difference frequency mixing in an indium-doped GaSe crystal. *J. Opt. Soc. Am. B*, 23, 282–288.

Dekorsy, T., Auer, H., Bakker, H. J., Roskos, H. G., and Kurz, H. (1996) THz electromagnetic emission by coherent infrared-active phonons. *Phys. Rev. B*, 53, 4005–4014.

Dekorsy, T., Ott, R., Kurz, H., and Köhler, K. (1995) Bloch oscillations at room-temperature. *Phys. Rev. B*, 51, 17275–17278.

Demsar, J., Averitt, R. D., Taylor, A. J., Demsar, J., Averitt, R. D., Taylor, A. J., Kabanov, V. V., Kang, W. N., Kim, H. J., Choi, E. M., and Lee, S. I. (2003) Pair-breaking and superconducting state recovery dynamics in MgB_2. *Phys. Rev. Lett.*, 91, 267002–267004.

Dexheimer, S. L. (Ed.) (2007) *Terahertz Spectroscopy: Principles and Applications*, Boca Raton, FL, CRC Press.

Dmitriev, V. G., Gurzadyan, G. G., and Nikogosyan, D. N. (1997) *Handbook of Nonlinear Optical Crystals*, Heidelberg, Germany: Springer Verlag.

Dorney, T. D., Baraniuk, R. G., and Mittleman, D. M. (2001) Material parameter estimation with terahertz time-domain spectroscopy. *J. Opt. Soc. Am. A*, 18, 1562–1570.

Dressel, M. and Grüner, G. (2002) *Electrodynamics of Solids*, Cambridge, U.K.: Cambridge University Press.

Dreyhaupt, A., Winnerl, S., Dekorsy, T., and Helm, M. (2005) High-intensity terahertz radiation from a microstructured large-area photoconductor. *Appl. Phys. Lett.*, 86, 121114.

Duvillaret, L., Garet, F., and Coutaz, J. L. (1996) A reliable method for extraction of material parameters in terahertz time-domain spectroscopy. *IEEE J. Sel. Top. Quantum Electron.*, 2, 739.

Eckstein, J. N. and Bozovic, I. (1995) High-temperature superconducting multilayers and heterostructures grown by atomic layer-by-layer molecular beam epitaxy. *Ann. Rev. Mater. Sci.*, 25, 679.

Ehret, S. and Schneider, H. (1998) Generation of subpicosecond infrared pulses tunable between 5.2 and 18 μm at a repetition rate of 76 MHz. *Appl. Phys. B*, 66, 27–30.

Exter, M. V., Fattinger, C., and Grischkowsky, D. (1989) Terahertz time-domain spectroscopy of water vapor. *Optics Lett.*, 14, 1128–1130.

Faure, J., Van Tilborg, J., Kaindl, R. A., and Leemans, W. P. (2004) Modelling laser-based table-top THz sources: Optical rectification, propagation and electro-optic sampling. *Opt. Quantum Electron.*, 36, 681–697.

Federici, J. F., Greene, B. I., Saeta, P. N., Dykaar, D. R., Sharifi, F., and Dynes, R. C. (1992) Direct picosecond measurement of photoinduced Cooper-pair breaking in lead. *Phys. Rev. B*, 46, 11153–11156.

Feenstra, B. J., Schützmann, J., van der Marel, D., Pérez Pinaya, R., and Decroux, M. (1997) Nonequilibrium superconductivity and quasiparticle dynamics studied by photoinduced activation of mm-wave absorption. *Phys. Rev. Lett.*, 79, 4890.

Ferguson, B. and Zhang, X.-C. (2002) Materials for terahertz science and technology. *Nat. Mater.*, 1, 26–33.

Fraser, J. M., Wang, D., Haché, A., Allan, G. R., and van Driel, H. M. (1997) Generation of high-repetition rate femtosecond pulses from 8 to 18 μm. *Appl. Opt.*, 36, 5044–5047.

Frogley, M. D., Dynes, J. F., Beck, M., Faist, J., and Phillips, C. C. (2006) Gain without inversion in semiconductor nanostructures. *Nat. Mater.*, 5, 175–178.

Gaal, P., Kuehn, W., Reimann, K., Woerner, M., Elsaesser, T., and Hey, R. (2007) Internal motions of a quasiparticle governing its ultrafast nonlinear response. *Nature*, 450, 1210–1213.

Gaal, P., Reimann, K., Woerner, M., Elsaesser, T., Hey, R., and Ploog, K. H. (2006) Nonlinear terahertz response of n-type GaAs. *Phys. Rev. Lett.*, 96, 187402.

Gallot, G. and Grischkowsky, D. (1999) Electro-optic detection of terahertz radiation. *J. Opt. Soc. Am. B*, 16, 1204–1212.

Gallot, G., Zhang, J., Mcgowan, R. W., Jeon, T.-I., and Grischkowsky, D. (1999) Measurements of the THz absorption and dispersion of ZnTe and their relevance to the electro-optic detection of THz radiation. *Appl. Phys. Lett.*, 74, 3450–3452.

Gedik, N., Langer, M., Orenstein, J., Ono, S., Abe, Y., and Ando, Y. (2005) Abrupt transition in quasiparticle dynamics at optimal doping in a cuprate superconductor system. *Phys. Rev. Lett.*, 95, 117005.

George, P. A., Strait, J., Dawlaty, J., Shivaraman, S., Chandrashekhar, M., Rana, F., and Spencer, M. (2008) Ultrafast optical-pump terahertz-probe spectroscopy of the carrier relaxation and recombination dynamics in epitaxial graphene. *Nano Lett.*, 8(12), 4248–4251.

Glover, R. E. and Tinkham, M. (1957) Conductivity of superconducting films for photon energies between 0.3 and $40kT_c$. *Phys. Rev.*, 108, 243–256.

Greene, B. I., Saeta, P. N., Dykaar, D. R., Schmitt-Rink, S., and Chuang, S.-L. (1992) Far-infrared light generation at semiconductor surfaces and its spectroscopic applications. *IEEE J. Quantum Electron.*, 28, 2302–2312.

Grischkowsky, D. and Keiding, S. (1990) THz time-domain spectroscopy of high T_C substrates. *Appl. Phys. Lett.*, 57, 1055–1057.

Grischkowsky, D., Keiding, S., Exter, M. V., and Fattinger, C. (1990) Far-infrared time-domain spectroscopy with terahertz beams of dielectrics and semiconductors. *J. Opt. Soc. Am. B*, 7, 2006–2015.

Groeneveld, R. H. M. and Grischkowsky, D. (1994) Picosecond time-resolved far-infrared experiments on carriers and excitons in GaAs-AlGaAs multiple quantum wells. *J. Opt. Soc. Am. B*, 11, 2502.

Gürtler, A., Winnewisser, C., Helm, H., and Jepsen, P. U. (2000) Terahertz pulse propagation in the near field and the far field. *J. Opt. Soc. Am. B*, 17, 74–83.

Hamster, H., Sullivan, A., Gordon, S., White, W., and Falcone, R. W. (1993) Subpicosecond, electromagnetic pulses from intense laser-plasma interaction. *Phys. Rev. Lett.*, 71, 2725.

Han, P. Y., Tani, M., Pan, F., and Zhang, X.-C. (2000) Use of organic crystal DAST for terahertz beam applications. *Opt. Lett.*, 25, 675–677.

Han, S. G., Vardeny, Z. V., Wong, K. S., Symko, O. G., and Koren, G. (1990) Femtosecond optical detection of quasiparticle dynamics in high-T_C $YBa_2Cu_3O_{7-\delta}$ superconducting thin films. *Phys. Rev. Lett.*, 65, 2708.

Han, P. Y. and Zhang, X.-C. (1998) Coherent broadband mid infrared terahertz beam sensors. *Appl. Phys. Lett.*, 73, 3049–3051.

Hendry, E., Koeberg, M., and Bonn, M. (2007) Exciton and electron-hole plasma formation dynamics in ZnO. *Phys. Rev. B*, 76, 045214.

Hilton, D. J., Prasankumar, R. P., Trugman, S. A., Taylor, A. J., and Averitt, R. D. (2006) On photoinduced phenomena in complex materials: Probing quasiparticle dynamics using infrared and far-infrared pulses. *J. Phys. Soc. Jpn.*, 75, 011006.

Hochstrasser, R. M. (2007) Multidimensional ultrafast spectroscopy. *Proc. Natl. Acad. Sci.*, 104, 14189.

Hoffmann, M. C., Hebling, J., Hwang, H. Y., Yeh, K.-L., and Nelson, K. A. (2009) Impact ionization in InSb probed by terahertz pump-terahertz probe spectroscopy. *Phys. Rev. B*, 79, 161201.

Huber, R., Brodschelm, A., Tauser, F., and Leitenstorfer, A. (2000) Generation and field-resolved detection of femtosecond electromagnetic pulses tunable up to 41 THz. *Appl. Phys. Lett.*, 76, 3191–3193.

Huber, R., Kaindl, R. A., Schmid, B. A., and Chemla, D. S. (2005) Broadband terahertz study of excitonic resonances in the high-density regime in GaAs/Al_xGa_{1-x}As quantum wells. *Phys. Rev. B*, 72, 161314.

Huber, R., Schmid, B. A., Shen, Y. R., Chemla, D. S., and Kaindl, R. A. (2006) Stimulated terahertz emission from intra-excitonic transitions in Cu_2O. *Phys. Rev. Lett.*, 96, 017402.

Huber, R., Tauser, F., Brodschelm, A., Bichler, M., Abstreiter, G., and Leitenstorfer, A. (2001) How many-particle interactions develop after ultrafast excitation of an electron-hole plasma. *Nature*, 414, 286–289.

Hussain, A. and Andrews, S. R. (2008) Ultrabroadband polarization analysis of terahertz pulses. *Opt. Express*, 16, 7251–7257.

Ideguchi, T., Yoshioka, K., Mysyrowicz, A., and Kuwata-Gonokami, M. (2008) Coherent quantum control of excitons at ultracold and high density in Cu_2O with phase manipulated pulses. *Phys. Rev. Lett.*, 100, 233001.

Jaekel, C., Waschke, C., Roskos, H. G., Kurz, H., Prusseit, W., and Kinder, W. (1994) Surface resistance and penetration depth of $YBa_2Cu_3O_{7-\delta}$ thin films on silicon at ultrahigh frequencies. *Appl. Phys. Lett.*, 64, 3326–3328.

Jang, J. I. and Wolfe, J. P. (2006) Relaxation of stress-split orthoexcitons in Cu_2O. *Phys. Rev. B*, 73, 075207.

Jeon, T.-I. and Grischkowsky, D. (1997) Nature of conduction in doped silicon. *Phys. Rev. Lett.*, 78, 1106–1109.

Jepsen, P. U. and Fischer, B. M. (2005) Dynamic range in terahertz time-domain transmission and reflection spectroscopy. *Opt. Lett.*, 30, 29–31.

Jiang, Z. and Zhang, X.-C. (1998) Electro-optic measurement of THz field pulses with a chirped optical beam. *Appl. Phys. Lett.*, 72, 1945–1947.

Joffre, M., Bonvalet, A., Migus, A., and Martin, J. L. (1996) Femtosecond diffracting Fourier-transform infrared interferometer. *Opt. Lett.*, 21, 964–966.

Johnsen, K. and Kavoulakis, G. M. (2001) Probing Bose-Einstein condensation of excitons with electromagnetic radiation. *Phys. Rev. Lett.*, 86, 858–861.

Jörger, M., Tsitsishvili, E., Fleck, T. and Klingshirn, C. (2003) Infrared absorption by excitons in Cu_2O. *Phys. Stat. Sol. (b)*, 238, 470–473.

Kaindl, R. A., Carnahan, M. A., Chemla, D. S., Oh, S., and Eckstein, J. N. (2005) Dynamics of Cooper pair formation in $Bi_2Sr_2CaCu_2O_{8+\delta}$. *Phys. Rev. B*, 72, 060510(R).

Kaindl, R. A., Carnahan, M. A., Hägele, D., Lövenich, R., and Chemla, D. S. (2003) Ultrafast terahertz probes of transient conducting and insulating phases in an electron-hole gas. *Nature*, 423, 734–738.

Kaindl, R. A., Carnahan, M. A., Orenstein, J., Chemla, D. S., Christen, H. M., Zhai, H. Y., Paranthaman, M., and Lowndes, D. H. (2002) Far-infrared optical conductivity gap in superconducting MgB_2 films. *Phys. Rev. Lett.*, 88, 027003.

Kaindl, R. A., Eickemeyer, F., Woerner, M., and Elsaesser, T. (1999) Broadband phase-matched difference frequency mixing of femtosecond pulses in GaSe: Experiment and theory. *Appl. Phys. Lett.*, 75, 1060–1062.

Kaindl, R. A., Hägele, D., Carnahan, M. A., and Chemla, D. S. (2009) Transient terahertz spectroscopy of excitons and unbound carriers in quasi-two-dimensional electron-hole gases. *Phys. Rev. B*, 79, 045320.

Kaindl, R. A., Reimann, K., Woerner, M., Elsaesser, T., Hey, R., and Ploog, K. H. (2001) Homogeneous broadening and excitation-induced dephasing of intersubband transitions in a quasi-two-dimensional electron gas. *Phys. Rev. B*, 63, 161308.

Kaindl, R. A., Smith, D. C., Joschko, M., Hasselbeck, M. P., Woerner, M., and Elsaesser, T. (1998) Femtosecond infrared pulses tunable from 9 to 18 mu m at an 88-MHz repetition rate. *Opt. Lett.*, 23, 861–863.

Kaindl, R., Woerner, M., Elsaesser, T., Smith, D. C., Ryan, J. F., Farnan, G. A., Mccurry, M. P., and Walmsley, D. G. (2000) Ultrafast mid-infrared response of $YBa_2Cu_3O_7$. *Science*, 287, 470–473.

Kampfrath, T., Noetzold, J., and Wolf, M. (2007) Sampling of broadband terahertz pulses with thick electro-optic crystals. *Appl. Phys. Lett.*, 90, 231113.

Karpinska, K., Mostovoy, M., van der Vegte, M. A., Revcolevschi, A., and van Loosdrecht, P. H. M. (2005) Decay and coherence of two-photon excited yellow orthoexcitons in Cu_2O. *Phys. Rev. B*, 72, 155201.

Kasprzak, J., Richard, M., Kundermann, S., Baas, A., Jeambrun, P., Keeling, J. M. J., Marchetti, F. M. et al. (2006) Bose–Einstein condensation of exciton polaritons. *Nature*, 443, 409–414.

Kawase, K., Shikata, J., Imai, K., and Ito, H. (2001) Transform-limited, narrow-linewidth, terahertz-wave parametric generator. *Appl. Phys. Lett.*, 78, 2819–2821.

Khazan, M., Meissner, R., and Wilke, I. (2001) Convertible transmission-reflection time-domain terahertz spectrometer. *Rev. Sci. Instrum.*, 72, 3427–3430.

Kim, K.-Y. (2009) Generation of coherent terahertz radiation in ultrafast laser-gas interactions. *Phys. Plasmas*, 16, 056706.

Kim, K. Y., Taylor, A. J., Glownia, J. H., and Rodriguez, G. (2008) Coherent control of terahertz supercontinuum generation in ultrafast laser-gas interactions. *Nat. Photonics*, 2, 605–609.

Kim, K. Y., Yellampalle, B., Rodriguez, G., Averitt, R. D., Taylor, A. J., and Glownia, J. H. (2006) Single-shot, interferometric, high-resolution, terahertz field diagnostic. *Appl. Phys. Lett.*, 88, 041123.

Kindt, J. T. and Schmuttenmaer, C. A. (1999) Theory for determination of the low-frequency time-dependent response function in liquids using time-resolved terahertz pulse spectroscopy *J. Chem. Phys.*, 110, 8589–8596.

Kira, M., Hoyer, W., Stroucken, T., and Koch, S. W. (2001) Exciton formation in semiconductors and the influence of a photonic environment. *Phys. Rev. Lett.*, 87, 176401.

Klatt, G., Gebs, R., Janke, C., Dekorsy, T., and Bartels, A. (2009) Rapid-scanning terahertz precision spectrometer with more than 6 THz spectral coverage. *Opt. Express*, 17, 22847–22854.

Kojima, S., Tsumura, N., Takeda, M. W., and Nishizawa, S. (2003) Far-infrared phonon-polariton dispersion probed by terahertz time-domain spectroscopy. *Phys. Rev. B*, 67, 035102.

Kono, S., Tani, M., and Sakai, K. (2001) Ultrabroadband photoconductive detection: Comparison with free-space electro-optic sampling. *Appl. Phys. Lett.*, 79, 898–900.

Kroll, J., Darmo, J., Dhillon, S. S., Marcadet, X., Calligaro, M., Sirtori, C., and Unterrainer, K. (2007) Phase-resolved measurements of stimulated emission in a laser. *Nature*, 449, 698–701.

Kübler, C., Huber, R., and Leitenstorfer, A. (2005) Ultrabroadband terahertz pulses: Generation and field-resolved detection. *Semicond. Sci. Technol.*, 20, S128–S133.

Kübler, C., Huber, R., Tubel, S., and Leitenstorfer, A. (2004) Ultrabroadband detection of multi-terahertz field transients with GaSe electro-optic sensors: Approaching the near infrared. *Appl. Phys. Lett.*, 85, 3360–3362.

Kubouchi, M., Yoshioka, K., Shimano, R., Mysyrowicz, A., and Kuwata-Gonokami, M. (2005) Study of orthoexciton-to-paraexciton conversion in Cu_2O by excitonic lyman spectroscopy. *Phys. Rev. Lett.*, 94, 016403.

Kuehn, W., Reimann, K., Woerner, M., and Elsaesser, T. (2009) Phase-resolved two-dimensional spectroscopy based on collinear n-wave mixing in the ultrafast time domain. *J. Chem. Phys.*, 130, 164503.

Lee, Y.-S. (2008) *Principles of Terahertz Science and Technology*, New York: Springer.

Lee, Y.-S., Meade, T., Perlin, V., Winful, H., Norris, T. B., and Galvanauskas, A. (2000) Generation of narrow-band terahertz radiation via optical rectification of femtosecond pulses in periodically poled lithium niobate. *Appl. Phys. Lett.*, 76, 2505–2507.

Leinss, S., Kampfrath, T., V. Volkmann, K., Wolf, M., Steiner, J. T., Kira, M., Koch, S. W., Leitenstorfer, A., and Huber, R. (2008) Terahertz coherent control of optically dark paraexcitons in Cu_2O. *Phys. Rev. Lett.*, 101, 246401.

Leitenstorfer, A., Hunsche, S., Shah, J., Nuss, M. C., and Knox, W. H. (1999) Detectors and sources for ultrabroadband electro optic sampling: experiment and theory. *Appl. Phys. Lett.*, 74, 1516–1518.

Liu, A. Y., Mazin, I. I., and Kortus, J. (2001) Beyond Eliashberg superconductivity in MgB_2: Anharmonicity, two-phonon scattering, and multiple gaps. *Phys. Rev. Lett.*, 87, 087005.

Liu, K., Xu, J., and Zhang, X. C. (2004) GaSe crystals for broadband terahertz wave detection. *Appl. Phys. Lett.*, 85, 863–865.

Lloyd-Hughes, J., Beere, H. E., Ritchie, D. A., and Johnston, M. B. (2008) Terahertz magnetoconductivity of excitons and electrons in quantum cascade structures. *Phys. Rev. B*, 77, 125322.

Loeffler, T., Jacob, F., and Roskos, H. G. (2000) Generation of terahertz pulses by photoionization of electrically biased air. *Appl. Phys. Lett.*, 77, 453–455.

Lu, Z. G., Campbell, P., and Zhang, X.-C. (1997) Free-space electro-optic sampling with a high-repetition-rate regenerative amplified laser. *Appl. Phys. Lett.*, 71, 593.

Luo, C. W., Reimann, K., Woerner, M., Elsaesser, T., Hey, R., and Ploog, K. H. (2004) Phase-resolved nonlinear response of a two-dimensional electron gas under femtosecond intersubband excitation. *Phys. Rev. Lett.*, 92, 047402.

Matsumoto, N., Fujii, T., Kageyama, K., Takagi, H., Nagashima, T., and Hangyo, M. (2009) Measurement of the soft-mode dispersion in $SrTiO_3$ by terahertz time-domain spectroscopic ellipsometry. *Jpn. J. Appl. Phys.*, 48, 09KC11.

Matsuoka, T., Fujimoto, T., Tanaka, K., Miyasaka, S., Tajima, S., Fujii, K., Suzuki, M., and Tonouchi, M. (2009) Terahertz time-domain reflection spectroscopy for high T_c superconducting cuprates. *Physica C*, 469, 982–984.

Mattis, D. C. and Bardeen, J. (1958) Theory of the anomalous skin effect in normal and superconducting metals. *Phys. Rev.*, 111, 412–417.

Mittleman, D. M. (Ed.) (2003) *Sensing with Terahertz Radiation*, Berlin, Germany: Springer.

Nagamatsu, J., Nakagawa, N., Muranaka, T., Zenitani, Y., and Akimitsu, J. (2001) Superconductivity at 39 K in magnesium diboride. *Nature*, 410, 63–64.

Nagashima, T. and Hangyo, M. (2001) Measurement of complex optical constants of a highly doped Si wafer using terahertz ellipsometry. *Appl. Phys. Lett.*, 79, 3917–3919.

Nahata, A., Auston, D. H., Wu, C., and Yardley, J. T. (1995) Generation of terahertz radiation from a poled polymer. *App. Phys. Lett.*, 67, 1358–1360.

Nahata, A., Weling, A. S., and Heinz, T. F. (1996) A wideband coherent terahertz spectroscopy system using optical rectification and electro-optic sampling. *Appl. Phys. Lett.*, 69, 2321–2323.

Nashima, S., Morikawa, O., Takata, K., and Hangyo, M. (2001) Temperature dependence of optical and electronic properties of moderately doped silicon at terahertz frequencies. *J. Appl. Phys.*, 90, 837–842.

Nemec, H., Kadlec, F., Surendran, S., Kuzel, P., and Jungwirth, P. (2005) Ultrafast far-infrared dynamics probed by terahertz pulses: A frequency domain approach. I. Model systems. *J. Chem. Phys.*, 122, 104503.

Newson, R. W., Ménard, J.-M., Sames, C., Betz, M., and van Driel, H. M. (2008) Coherently controlled ballistic charge currents injected in single-walled carbon nanotubes and graphite. *Nano Lett.*, 8, 1586–1589.

Nicol, E. J. and Carbotte, J. P. (2003) Comparison of s- and d-wave gap symmetry in nonequilibrium superconductivity. *Phys. Rev. B*, 67, 214506.

Nuss, M. C., Mankiewich, P. M., O'malley, M. L., Westerwick, E. H., and Littlewood, P. B. (1991) Dynamic conductivity and coherence peak in $YBa_2Cu_3O_7$ superconductors. *Phys. Rev. Lett.*, 66, 3305–3308.

Ohtake, H., Ono, S., Sakai, M., Liu, Z., Tsukamoto, T., and Sarukura, N. (2000) Saturation of THz-radiation power from femtosecond-laser-irradiated InAs in a high magnetic field. *Appl. Phys. Lett.*, 76, 1398–1400.

Park, S.-G., Melloch, M. R., and Weiner, A. M. (1999) Analysis of terahertz waveforms measured by photoconductive and electrooptic sampling. *IEEE J. Quantum Electron.*, 35, 810–819.

Parkinson, P., Lloyd-Hughes, J., Gao, Q., Tan, H. H., Jagadish, C., Johnston, M. B., and Herz, L. M. (2007) Transient terahertz conductivity of GaAs nanowires. *Nano Lett.*, 7, 2162–2165.

Pashkin, A., Kempa, M., Němec, H., Kadlec, F., and Kužel, P. (2003) Phase-sensitive time-domain terahertz reflection spectroscopy. *Rev. Sci. Instrum.*, 74, 4711–4717.

Paspalakis, E., Tsaousidou, M., and Terzis, A. F. (2006) Coherent manipulation of a strongly driven semiconductor quantum well. *Phys. Rev. B*, 73, 125344.

Peter, F., Winnerl, S., Nitsche, S., Dreyhaupt, A., Schneider, H., and Helm, M. (2007) Coherent terahertz detection with a large-area photoconductive antenna. *Appl. Phys. Lett.*, 91, 081109.

Planken, P. C. M. (2001) Measurement and calculation of the orientation dependence of terahertz pulse detection in ZnTe. *J. Opt. Soc. Am. B*, 18, 313–317.

Planken, P. C. M., Nuss, M. C., Brener, I., Goossen, K. W., Luo, M. S. C., Chuang, S.-L., and Pfeiffer, L. (1992) Terahertz emission in single quantum wells after coherent optical excitation of light hole and heavy hole excitons. *Phys. Rev. Lett.*, 69, 3800–3803.

Prasankumar, R. P., Okamura, H., Imai, H., Shimakawa, Y., Kubo, Y., Trugman, S. A., Taylor, A. J., and Averitt, R. D. (2005a) Coupled charge-spin dynamics of the magnetoresistive pyrochlore $Tl_2Mn_2O_7$ probed using ultrafast midinfrared spectroscopy. *Phys. Rev. Lett.*, 95, 267404.

Prasankumar, R. P., Scopatz, A., Hilton, D. J., Taylor, A. J., Averitt, R. D., Zide, J. M., and Gossard, A. C. (2005b) Carrier dynamics in self-assembled ErAs nanoislands embedded in GaAs measured by optical-pump terahertz-probe spectroscopy. *Appl. Phys. Lett.*, 86, 201107.

Reimann, K. (2007) Table-top sources of ultrashort THz pulses. *Rep. Prog. Phys.*, 70, 1597–1632.

Reimann, K., Smith, R. P., Weiner, A. M., Elsaesser, T., and Woerner, M. (2003) Direct field-resolved detection of terahertz transients with amplitudes of megavolts per centimeter. *Opt. Lett.*, 28, 471–473.

Rice, A., Jin, Y., Ma, X. F., Zhang, X.-C., Bliss, D., Larkin, J., and Alexander, M. (1994) Terahertz optical rectification from <110> zinc-blende crystals. *Appl. Phys. Lett.*, 64, 1324–1326.

Rini, M., Tobey, R., Dean, N., Itatani, J., Tomioka, Y., Tokura, Y., Schoenlein, R. W., and Cavalleri, A. (2007) Control of the electronic phase of a manganite by mode-selective vibrational excitation. *Nature*, 449, 72–74.

Roskos, H. G., Nuss, M. C., Shah, J., Leo, K., Miller, D. A. B., Fox, A. M., Schmitt-Rink, S., and Köhler, K. (1992) Coherent submillimeter-wave emission from charge oscillations in a double-well potential. *Phys. Rev. Lett.*, 68, 2216–2219.

Rothwarf, A. and Taylor, B. N. (1967) Measurement of recombination lifetimes in superconductors. *Phys. Rev. Lett.*, 3, 27–30.

Sakai, K. (Ed.) (2005) *Terahertz Optoelectronics*, Berlin, Germany: Springer.

Schall, M. and Jepsen, P. U. (2000) Photoexcited GaAs surfaces studied by transient terahertz time-domain spectroscopy. *Opt. Lett.*, 25, 13–15.

Schmuttenmaer, C. A. (2004) Exploring dynamics in the far-infrared with terahertz spectroscopy. *Chem. Rev.*, 104, 1759–1779.

Sell, A., Leitenstorfer, A., and Huber, R. (2008) Phase-locked generation and field-resolved detection of widely tunable terahertz pulses with amplitudes exceeding 100 MV/cm. *Opt. Lett.*, 33, 2767–2769.

Shah, J. (1999) *Ultrafast Spectroscopy of Semiconductors and Semiconductor Nanostructures*, Berlin, Germany, Springer Verlag.

Shan, J., Weling, A. S., Knoesel, E., Bartels, L., Bonn, M., Nahata, A., Reider, G. A., and Heinz, T. F. (2000) Single-shot measurement of terahertz electromagnetic pulses by use of electro-optic sampling. *Opt. Lett.*, 25, 426–428.

Sherwin, M. S., Imamoglu, A., and Montroy, T. (1999) Quantum computation with quantum dots and terahertz cavity quantum electrodynamics. *Phys. Rev. A*, 60, 3508–3514.

Siders, C. W., Siders, J. L. W., Taylor, A. J., Park, S.-G., Melloch, M. R., and Weiner, A. M. (1999) Generation and characterization of terahertz pulse trains from biased, large-aperture photoconductors. *Opt. Lett.*, 24, 241–243.

Smith, P. R., Auston, D. H., and Nuss, M. C. (1988) Subpicosecond photoconducting dipole antennas. *IEEE J. Quantum Electron.*, 24, 255–260.

Snoke, D., Denev, S., Liu, Y., Pfeiffer, L., and West, K. (2002) Long-range transport in excitonic dark states in coupled quantum wells. *Nature*, 418, 754.

Stepanov, A. G., Bonacina, L., Chekalin, S. V., and Wolf, J.-P. (2008) Generation of 30 µJ single-cycle terahertz pulses at 100 Hz repetition rate by optical rectification. *Opt. Lett.*, 33, 2497–2499.

Stevens, C. J., Smith, D., Chen, C., Ryan, J. F., Podobnik, B., Mihailovic, D., Wagner, G. A., and Evetts, J. E. (1997) Evidence for two-component high-temperature superconductivity in the femtosecond optical response of $YBa_2Cu_3O_7.\delta$. *Phys. Rev. Lett.*, 78, 2212.

Timusk, T. (1976) Far-infrared absorption study of exciton ionization in germanium. *Phys. Rev. B*, 13, 3511–3514.

Tinkham, M. (1996) *Introduction to Superconductivity*, New York: McGraw-Hill.

Tobey, R. I., Prabhakaran, D., Boothroyd, A. T., and Cavalleri, A. (2008) Ultrafast electronic phase transition in $La_{1/2}Sr_{3/2}MnO_4$ by coherent vibrational excitation: Evidence for nonthermal melting of orbital order. *Phys. Rev. Lett.*, 101, 197404.

Tropf, W. J., Thomas, M. E., and Harris, T. J. (1995) Properties of crystals and glasses. In Bass, M. (Ed.) *Optical Society of America Handbook of Optics*. 2nd edn., New York: McGraw Hill.

Twerenbold, D. (1986) Nonequilibrium model of the superconducting tunneling junction x-ray detector. *Phys. Rev. B*, 34, 7748.

Ventalon, C., Fraser, J. M., Likforman, J.-P., Villeneuve, D. M., Corkum, P., and Joffre, M. (2006) Generation and complete characterization of intense mid-infrared ultrashort pulses. *J. Opt. Soc. Am. B*, 23, 332–340.

Vodopyanov, K. L. (2009) Terahertz-wave generation with periodically inverted gallium arsenide. *Laser Phys.*, 19, 305–321.

Vodopyanov, K. L. and Kulevskii, L. A. (1995) New dispersion relationships for GaSe in the 0.65–18 µm spectral region. *Opt. Commun.*, 118, 375–378.

Vodopyanov, K. L. and Voevodin, V. G. (1995) 2.8 µm laser pumped type I and type II travelling-wave optical parametric generator in GaSe. *Opt. Commun.*, 114, 333–335.

Waldram, J. R. (1996) *Superconductivity of Metals and Cuprates*, Bristol, U.K.: Institute of Physics Publishing.

Wang, F., Shan, J., Islam, M. A., Herman, I. P., Bonn, M., and Heinz, T. F. (2006) Exciton polarizability in semiconductor nanocrystals. *Nat. Mater.* 5, 861–864.

Waschke, C., Roskos, H. G., Schwedler, R., Leo, K., Kurz, H., and Köhler, K. (1993) Coherent submillimeter-wave emission from Bloch oscillations in a semiconductor superlattice. *Phys. Rev. Lett.*, 70, 3319–3322.

Wilke, I., Ascazubi, R., Lu, H., and Schaff, W. J. (2008) Terahertz emission from silicon and magnesium doped indium nitride. *Appl. Phys. Lett.*, 93, 221113.

Wilke, I., Khazan, M., Rieck, C. T., Kuzel, P., Kaiser, T., Jaekel, C., and Kurz, H. (2000) Terahertz surface resistance of high temperature superconducting thin films. *J. Appl. Phys.*, 87, 2984–2988.

Wu, Q., Litz, M., and Zhang, X. C. (1996) Broadband detection capability of ZnTe electro-optic field detectors. *Appl. Phys. Lett.*, 68, 2924–2926.

Wu, Q. and Zhang, X.-C. (1996a) Design and characterization of traveling-wave electrooptic terahertz sensors. *IEEE J. Quantum Electron.*, 2, 693–700.

Wu, Q. and Zhang, X.-C. (1996b) Ultrafast electro-optic field sensors. *Appl. Phys. Lett.*, 68, 1604–1606.

Wu, Q. and Zhang, X.-C. (1997a) Free-space electro-optics sampling of mid-infrared pulses. *Appl. Phys. Lett.*, 71, 1285–1286.

Wu, Q. and Zhang, X. C. (1997b) 7 terahertz broadband GaP electro-optic sensor. *Appl. Phys. Lett.*, 70, 1784–1786.

Yeh, K. L., Hoffmann, M. C., Hebling, J. and Nelson, K. A. (2007) Generation of 10 µJ ultrashort terahertz pulses by optical rectification. *App. Phys. Lett.*, 90, 171121.

Zentgraf, T., Huber, R., Nielsen, N. C., Chemla, D. S., and Kaindl, R. A. (2007) Ultrabroadband 50–130 THz pulses generated via phase-matched difference frequency mixing in $LiIO_3$. *Opt. Express*, 15, 5775–5781.

Zhang, X. C., Hu, B. B., Darrow, J. T., and Auston, D. H. (1990) Generation of femtosecond electromagnetic pulses from semiconductor surfaces. *Appl. Phys. Lett.*, 56, 1011–1013.

Zhang, X.-C., Jin, Y., Kingsley, L. E., and Weiner, M. (1993) Influence of electric and magnetic fields on THz radiation. *Appl. Phys. Lett.*, 62, 2477–2479.

Zhang, X. C., Jin, Y., Yang, K., and Schowalter, L. J. (1992) Resonant nonlinear susceptibility near the GaAs band gap. *Phys. Rev. Lett.*, 69, 2303–2306.

Zhang, X.-C. and Xu, J. (2009) *Introduction to THz Wave Photonics*, New York: Springer.

Chapter 12

TIME-RESOLVED PHOTOLUMINESCENCE SPECTROSCOPY

Marc Achermann

CONTENTS

12.1 INTRODUCTION

Modern research on photoluminescence (PL) phenomena dates back to the middle and end of the nineteenth century, when J. Herschel, A. E. Becquerel, and G. G. Stokes published their reports on fluorescent and phosphorescent materials (Herschel 1845; Stokes 1852; Becquerel 1867). Since then, the origin of PL has been largely understood, and PL studies have thus provided important insight into fundamental material properties. More recently, with the advent of fast detectors and ultrafast laser sources, *time-resolved* photoluminescence (TRPL) techniques have emerged as useful tools to understand a wealth of dynamic processes that include charge and energy relaxation, recombination, and transfer. In this chapter, I will introduce four TRPL techniques that offer time resolution in the pico- and femtosecond time range, followed by a few comments on the interpretation and analysis of TRPL measurements. I will finish with three research examples, in which TRPL data allowed us to understand ultrafast charge carrier and energy processes in single-component and hybrid nanostructures.

12.2 TECHNIQUES

The time evolution of luminescence signals can be determined by a variety of different measurement techniques. However, the general concept of all TRPL techniques is similar; the fluorophores of interest are excited with a pulsed- or an on/off-modulated light source, and the emission is subsequently recorded while the excitation source is off. Optical excitation with high repetition rates is advantageous, since it increases count rates and allows averaging and therefore enhances the signal-to-noise ratio for a given data acquisition time. An upper limit to the repetition rate is set by the minimum off time between subsequent excitation pulses that should be longer than the relaxation time of the excited fluorophore to its ground state. This condition ensures that excitation of the unrelaxed system to higher energy levels is avoided, simplifying data interpretation. Another general consideration for all TRPL techniques is that the TRPL signal is proportional to the *radiative* decay rate of the emitter. Therefore, it is inherently easier to measure ultrafast PL dynamics of emitters with a high radiative decay rate, because more photons are emitted within a given time window.

A rather slow, but simple and cheap TRPL technique with microseconds or slower time resolution is based on an on/off-modulated light source that excites the fluorophores, a fast photodiode that detects the PL, and a fast oscilloscope that records the photodetector's response. If internal/electronic modulation of the light source is not possible, external modulation can be achieved with a mechanical chopper or an acousto-optic modulator that operates at maximum repetition rates in the kHz and the MHz ranges, respectively. Small area, fast photodetectors with a bandwidth in the GHz range are readily available at low cost. Biased versions are preferable because their better linearity increases the reliable dynamic range of the measured PL dynamics.

More sophisticated techniques are necessary to perform TRPL measurements with better time resolution. In the following sections, I introduce four *time-domain* TRPL techniques with time resolution in the femtosecond and picosecond range that differ in their performance (time resolution and sensitivity), experimental complexity, and budget requirements. As a general rule, time resolution comes at the expense of sensitivity and can be overcome to some extent by expensive equipment. All of these techniques rely on optical excitation with short laser pulses that are most commonly produced by mode-locked lasers (femtosecond pulses) or fast diode lasers (picosecond pulses). A short pulse excitation source is necessary because the overall time resolution is determined by the temporal response of the TRPL detection system as well as by the temporal width of the excitation pulse. The four TRPL techniques that are discussed here are based on time-correlated single photon counting (TCSPC), streak cameras, photoluminescence upconversion (PLU), and optical Kerr gating (OKG). A discussion on frequency-domain techniques that operate on longer time scales can be found in Lakowicz (2006). Before going into the details of the four TRPL techniques, the following table presents a brief comparison of their key parameters. The indicated numbers have to be understood as general guidelines and will change over time as equipment evolves.

12.2.1 Time-Correlated Single Photon Counting

TCSPC is one of the most widely used techniques for TRPL because of its versatility, low complexity, and cost-effectiveness. TCSPC modules are commercially available and come either as a computer card or a stand-alone unit with a USB computer connection. As highlighted here, TCSPC modules are used for emission lifetime measurements; however, they can also be employed in fluorescence correlation spectroscopy. In TCSPC measurements, the fluorophores are excited with a periodic light source. After each excitation pulse, the time delay between a single emitted photon and a reference trigger pulse is recorded, and a histogram of the time delays is generated (Figure 12.1a). This histogram of number of occurrences versus time delay then reconstructs the PL dynamics. Because of the counting concept, TCSPC is a technique that is inherently highly linear over many orders of magnitude. The statistical nature of the TCSPC concept requires multiple, repetitive excitations of the fluorophores to obtain PL decay traces with statistical significance; single shot experiments are meaningless.

More specifically, after pulsed laser excitation of the fluorophores their emission is collected by a lens system and spectrally filtered to suppress scattered light from the excitation pulse and to obtain wavelength-dependent PL dynamics. Spectral filtering is achieved either by a monochromator or by band pass filters for high or low spectral resolution, respectively. The PL is then detected by a fast and sensitive photodetector, such as a photomultiplier tube (PMT) or a single photon avalanche photodiode (SPAD). The requirements on the detectors are moderate to high sensitivity in the spectral range of interest (at least a few percent detection probability), low dark count rate (preferably below 100 Hz), fast signal rise time (less than a few hundred ps), and small transit time spread (TTS). The TTS is the variation in the time delay between the absorption of a photon by the detector and the generation of an electrical pulse at the detector's output and is caused by the amplification process. Typically, a small TTS is in the tens of ps range. In single photon counting techniques, TTS is the main parameter that determines the time resolution of the detection system. Both PMTs and SPADs are

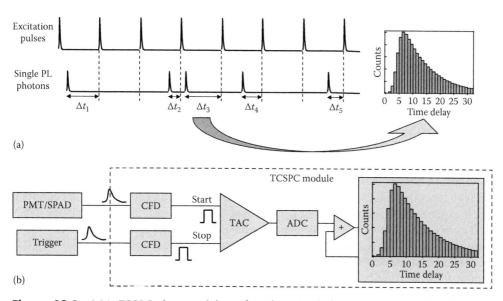

Figure 12.1 (a) In TCSPC, the time delays of single emitted photons are recorded and used to generate a histogram. (b) Schematic of the functioning of a TCSPC module; see text for details.

available with a broad range of specifications; therefore, the choice for a specific detector strongly depends on the application and its requirements. In general, PMTs have larger detection areas and often smaller TTS but are less sensitive and more expensive than SPADs.

Both detector types convert a detected single photon into an electrical pulse that is fed into the TCSPC module. Most TCSPC modules operate in a reversed start/stop configuration (Figure 12.1b), that is, the single photon detector pulse starts a timer that is stopped by a subsequent trigger pulse provided by the laser source. Trigger pulses have to be synchronized to the excitation laser pulse train (most often at the same repetition rate). If the excitation pulses are generated by a diode laser, the laser driver often provides electrical trigger pulses. In case of excitation by a passively mode-locked laser, trigger pulses can be generated by a fast photodiode that detects a small fraction of the excitation pulse train. In a first processing step, the TCSPC module suppresses electrical noise in both the trigger and the single photon signal channel by a fast discriminator that triggers only when the leading edge of the electrical pulse surpasses a threshold value. Instead of using a simple *constant* threshold value, better time resolution is obtained by a constant *fraction* discriminator (CFD). The CFD triggers at a constant fraction of the pulse maxima of electrical pulses that are above a lower threshold level. This CFD concept eliminates timing jitter caused by amplitude variations that arise from the randomness of the detector's amplification mechanism. The CFD starts a time-to-amplitude converter (TAC) that is stopped by the reference trigger pulse. The TAC often consists of a capacitor that is charged during the time between the single photon pulse and the trigger pulse. Finally, an analog-to-digital converter (ADC) digitizes the analog capacitor voltage and converts it into time delays that are binned to create a histogram representing the PL decay dynamics.

As indicated by the name, the TCSPC technique is based on the detection of a *single* photon per excitation pulse. Two or more detected photons per excitation pulse distort the statistics, because only the temporal information of the first photon enters the histogram, rendering the recorded dynamics faster than actual. Since spontaneous emission of photons is a random process that follows a Poisson distribution, one detected photon in average actually means that the photodetector registers zero or one photon with a probability of 37% ($1/e$) each and two or more photons with a 26% probability. Hence, in this case it is almost as likely to detect a single photon as multiple photons, which is the situation that should be avoided. Therefore, it is generally recommended to keep the detection rate approximately one order of magnitude below the repetition rate to keep two or more photon events negligible (e.g., by reducing the PL intensity with neutral density filters in front of the detector). In this case, the average photon count is ~0.1 per excitation pulse, and the ratio between multiple and single photon events is around 5% based on Poisson statistics. Besides the upper limit imposed by the repetition rate, most detectors also restrict the maximum count rate to avoid overexposure.

The limitation on the detected photon rate has several consequences. To obtain good signal-to-noise ratios within reasonable collection times, it is required to use lasers that operate with repetition rates in the high kHz and MHz range. As an example, let us consider a single exponential PL decay with decay rate Γ that we want to measure with a relative statistical error of less than $\sigma = 10\%$. Such an error requires the acquisition of at least $1/\sigma^2 = 100$ signal counts per data point (a data point corresponds to a time bin of temporal width Δt). If we want to measure the PL decay over two orders of magnitude, the maximum of the exponential decay needs to be 100 times larger than the minimum count number, that is, 10,000 counts per time bin in our example. Then the required photon count rate k_{PL} is given by

$$k_{PL}T = \frac{10,000}{\Delta t \Gamma}. \tag{12.1}$$

Equation 12.1 indicates that the total number of counts acquired within the collection time T (left side) is equal to the time-integrated exponential decay (right side). Considering a typical order of magnitude acquisition time of $T = 100\,s$ and a reasonable temporal mapping of the PL dynamics ($\Delta t\Gamma \leq 0.1$), we obtain $k_{PL} = 1\,kHz$. Consequently, a laser repetition rate of 10 kHz is required.

As a side note, the imposed less than unity photon detection probability (~10% per excitation pulse) is the reason for the reverse start/stop concept, because the timer is only started when a photon is detected and not for every reference trigger pulse. Ultimately, the limitation on the detected photon rate also imposes an upper limit to the dark count rate of the detector. Further information about the TCSPC technique can be found in Becker (2005).

12.2.2 Streak Cameras

The general purpose of a streak camera is to convert the temporal dependence of an optical signal into a spatial profile that is recorded with an optical camera. Originally, streak cameras used fast-moving or rotating components, such as a recording film or mirrors and prisms that deflect the optical signal. The spatial location where the photons hit the recording element then provides the temporal information. These mechanical streak cameras achieved time resolutions in the low nanosecond range. Nowadays, photoelectronic streak cameras are primarily used. These cameras, to be discussed in this section, achieve typical time resolutions of a few ps.

The concept of a streak camera used for TRPL measurements is schematically shown in Figure 12.2 (an interactive Java applet of the mode of operation of a streak camera can be found on http://learn.hamamatsu.com/tutorials/streakcamera). Photons from the emitting sample that impinge on a streak camera are absorbed by a photocathode (PC) that produces electrons through the photoelectric effect. The electrons are accelerated in a cathode ray tube toward a microchannel plate (MCP). On their trajectories, they pass through a pair of sweep electrodes (SEs) that generate an electric field perpendicular to the main propagation direction of the photoelectrons. The applied sweep voltage is synchronized to the PL excitation pulse and ramped up at high speed (can be as fast as a few hundred ps), resulting in a photoelectron deflection that depends on the moment when they pass through the SEs. Hence, the spatial location where the electrons hit the MCP contains the temporal information of the photons that produced the photoelectrons. The MCP consists of a slab of highly resistive material that is perforated with a dense array of parallel pores with ~10 μm diameters. The pores are oriented at a small off-normal angle and a strong electric field is applied normal to the MCP. Hence, when the electrons pass through the MCP, they are accelerated by the

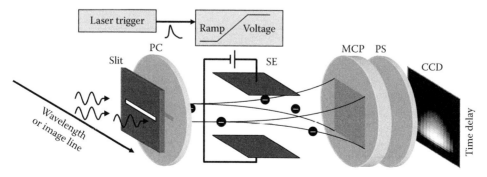

Figure 12.2 Schematic of a streak camera with the following components: PC, SEs, MCP, PS, and CCD.

electric field and hit the sidewalls of the pores, because of their off-normal angle. In this process, the photoelectrons are multiplied several thousand times, similar to the avalanche process in PMTs. The deflected and amplified photoelectrons finally hit a phosphor screen (PS) that converts them back into light, which is recorded most often by a digital charge-coupled device (CCD) camera.

Streak cameras are available in a broad range of specifications. Unlike the TCSPC technique that requires a repetitive signal to build up a histogram with statistical significance, streak cameras operate both at high repetition rates and in the single shot or at low repetition rate regime. In the latter regime sub-ps time resolution is achievable. However, operation at high repetition rates is more common, limiting the time resolution to a few ps, because of timing jitter between individual sweeps. Besides the high temporal resolution, the main advantage of a streak camera is the possibility to make use of the second spatial coordinate that allows parallel data acquisition. This is accomplished by an entrance slit in front of the PC (Figure 12.2) that is perpendicular to the sweep electric field direction and imaged onto the recording device perpendicular to the time axis. In microscopy, the PL dynamics of a line segment of an image can be recorded simultaneously and turned into an image with a time and a spatial axis. In spectroscopy, the PL is spectrally resolved with a spectrometer and then detected and temporally resolved by the streak camera. The resulting image displays PL dynamics as a function of wavelength.

In comparison with the TCSPC technique, streak cameras provide an order of magnitude better time resolution and allow parallel processing for microscopy and spectroscopy purposes. However, to make full use of these advantages, strong PL signals are required. For example, highest time resolution is only obtained in single sweep mode that requires a strong enough PL signal that can be temporally resolved with a single excitation/sweep process. Another disadvantage is that the specifications of streak cameras often allow less flexibility in their mode of operation. For example, specific modules only operate either at low or high frequencies and have either high time resolution with a small temporal sweep range or vice versa. Finally, streak cameras tend to be more expensive than TCSPC systems (see Table 12.1).

12.2.3 Ultrafast Upconversion

While TCSPC and streak camera systems are commercially available, the PLU technique (Shah 1988) relies on home-built setups. Due to its complexity and lower sensitivity, PLU is only recommended when ultrafast time resolution in the femtosecond range is required. In PLU, the sample is excited with an ultrafast laser pulse and its PL is frequency-mixed (gated) with a second synchronous laser pulse. Since this technique is all-optical, time resolutions

TABLE 12.1

	TCSPC	Streak Camera	PLU	OKG
Time resolution	30 ps	1 ps	100 fs	100 fs
Maximum time range	1 μs	1 ms	1 ns	1 ns
Sensitivity	High	High	Medium	Medium
Commercially available	Yes	Yes	No	No
Laser costs	$20k[a]	$80k[b]	$80k–$300k[b,c]	$300k[c]
TRPL setup costs	$30k–$60k	$150k–$250k	$50k	$50k

[a] Picosecond diode laser.
[b] Femtosecond Ti:sapphire oscillator.
[c] Femtosecond Ti:sapphire amplifier.

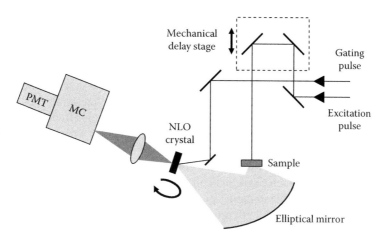

Figure 12.3 Schematic of a PL upconversion setup with PL collection in a forward-scattering configuration that is appropriate for dilute concentrations of fluorophores in transparent host materials or on transparent substrates. MC: monochromator, PMT: photomultiplier tube, NLO crystal: nonlinear optical crystal.

around 100 fs are achievable, limited mainly by the temporal widths of the excitation and gating laser pulses and the spectral bandwidth of the frequency-mixing process. The latter occurs in a nonlinear optical crystal and requires sufficiently strong PL signals and gating pulses with high peak intensities. In general, PL upconversion is less sensitive than TCSPC and streak camera approaches, because it relies on a nonlinear process.

The basic outline of a PLU setup is schematically depicted in Figure 12.3. A fluorescent sample is excited by a train of ultrafast laser pulses that is commonly provided by an ultrafast laser oscillator or amplifier (Ti:sapphire lasers are the most common). The wavelength of the excitation pulse is often modified by nonlinear optical processes to fall into the absorption band of the fluorophores (e.g., by second harmonic generation or by parametric amplification in optical parametric oscillators or amplifiers [Chapter 7]). The PL is collected and refocused onto a nonlinear optical crystal, preferably by reflective, aberration-free optics to avoid pulse spreading through dispersion (Chapter 7), thereby improving time resolution and upconversion efficiencies. A synchronous high-power gating pulse that is provided by the same laser source (most often at the fundamental laser wavelength) is sent through a mechanical delay stage to adjust the time delay with respect to the excitation pulse and then focused and spatially overlapped with the PL on the nonlinear optical crystal. Because of the nonlinearity of the crystal, the fraction of the PL, $PL(t)$, that temporally overlaps with the gating pulse mixes with it and produces an optical signal at the sum or difference frequency. The gated $PLU(t)$ signal is proportional to the peak intensity of the gating pulse I_G and under optimal phase matching conditions (explained in the following), it is given by (Shah 1988):

$$PLU(t) \propto PL(t)d_{\text{eff}}^2 I_G L^2, \tag{12.2}$$

where

L is the thickness of the nonlinear crystal

d_{eff} is its effective nonlinear coefficient that is proportional to the second-order susceptibility $\chi^{(2)}$

The frequency-mixed signal is then spatially and spectrally filtered to reduce the background signal from the unconverted PL and gating pulse and detected by a PMT. Since PMTs are available with excellent performance in the blue and ultraviolet spectral range, sum frequency rather than difference frequency generation is used, leading to the term "PL upconversion." Typical nonlinear optical crystals that are used for PLU measurements are beta-barium borate (BBO), lithium triborate (LBO), and potassium dihydrogen phosphate (KDP). These crystals have a moderately high nonlinearity, wide transparency range into the ultraviolet that is important for sum-frequency generation, and good chemical and mechanical properties.

A fundamental requirement for the PL upconversion effect to be efficient is the so-called phase matching condition (discussed in more detail in Chapter 7), when the wave vectors of the gate and the PL photons, $\vec{k}_G$ and $\vec{k}_{PL}$, respectively, add up to the wave vector of the sum-frequency photon: $\vec{k}_{uPL} = \vec{k}_{PL} + \vec{k}_G$ (momentum conservation). Because of the birefringence of nonlinear optical crystals, such phase matching is obtained for a specific orientation of the crystal with regard to the PL and gating pulse polarization and direction. Phase matching is only satisfied in a small wavelength range for a given crystal orientation; therefore, spectrally resolved PLU measurements require rotation of the nonlinear crystal (this requirement has been softened in the PLU setup of Schanz et al. 2001). The spectral selectivity can also be a limiting factor for the time resolution when using ultrashort gating pulses with broad optical spectra. Moreover, the phase matching condition is only valid within a finite acceptance angle of the incoming beams; therefore, the phase matching acceptance angle sets a lower limit on the focal length of the lenses that focus the PL and the gating pulses (Shah 1988).

As indicated by Equation 12.2, the upconverted signal can be significantly enhanced by increasing the thickness of the nonlinear crystal, because of its quadratic dependence on L. However, increasing the crystal thickness also enhances spectral dispersion effects, making Equation 12.2 invalid. Specifically, second-order dispersion causes a temporal broadening of the gating pulse and a temporal walk-off between PL and gating pulses (group velocity mismatch) (Chapter 7). Both effects reduce the upconversion efficiency and the time resolution.

12.2.4 Ultrafast Optical Kerr Gating

The ultrafast OKG technique is conceptually very similar to the previously introduced PLU technique (Duguay and Hansen 1969; Kinoshita et al. 2000; Takeda et al. 2000). Both techniques are all-optical, take advantage of an ultrafast optical nonlinearity, and require femtosecond laser systems to take full advantage of the technique. In addition, both are generally less sensitive than TRPL techniques based on TCSPC and streak cameras and are not commercially available. The main difference between the OKG and the PLU technique is the type of optical nonlinearity used. As indicated by the name, the nonlinearity in the OKG method is provided by the optical Kerr effect, which refers to a change in the refractive index $\Delta n(I) = n_2 I$ induced by an optical beam of intensity I in a medium with a second-order nonlinear refractive index n_2 (more detail is given in Chapter 7).

A conceptual setup of the OKG technique is depicted in Figure 12.4. After ultrafast excitation, the collected PL is passed through a polarizer that most often is oriented parallel to the polarization of the excitation pulse. The linear polarized PL is then focused onto a Kerr medium, where the PL is overlapped with a time-delayed gating pulse. As in the PLU technique, excitation and gating pulse have to be synchronized, and they often have different wavelengths that are obtained through nonlinear optical processes. The high-intensity gating pulse, which is polarized at 45° with respect to the PL polarizer, changes the refractive index of the Kerr medium along its polarization direction, causing a transient birefringence. The phase change Φ induced by the birefringence depends on the peak intensity of the gating pulse and the strength of the Kerr nonlinearity and is given by $\Phi = k\Delta nL$, in which k is the wave vector and L is the thickness of the Kerr medium. Since the polarization of the gating

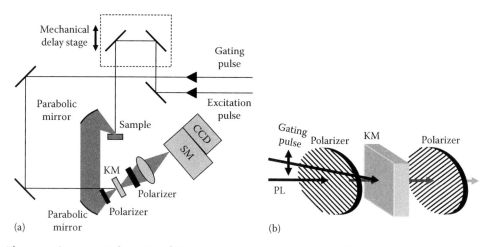

Figure 12.4 (a) Schematic of the Kerr gating technique for TRPL measurements in a back-scattering configuration. KM: Kerr medium: SM: spectrometer; CCD: charge-coupled device. (b) Configuration of the orientation of the two cross-polarizers and the polarization of the gating pulse. Since the emission of many fluorophores is partially polarized along the polarization of the excitation pulse, it is recommended to rotate the polarization of the excitation or gating pulses such that they are at 45° with respect to each other.

pulse and the PL are not parallel but rotated by 45°, the transient birefringence causes a transient change of the PL polarization. The PL is then sent through a second polarizer that is cross-polarized with the first one and, therefore, only transmits the time-gated part of the PL. The gated emission $PL_{OKG}(t)$ is given by Takeda et al. (2000):

$$PL_{OKG}(t) = PL(t)\sin^2\left(\frac{\Phi}{2}\right) \approx PL(t)\frac{k^2 L^2 n_2^2 I_G^2}{4}, \qquad (12.3)$$

with all quantities defined as in Equation 12.2. It is noteworthy that PL_{OKG} in Equation 12.3 is proportional to the square of I_G, in contrast to the PLU signal that increases linearly with I_G (Equation 12.2). Therefore, the OKG technique generally requires high gating pulse intensities provided by amplified lasers, while the PLU technique also works with non-amplified ultrafast laser pulses. By changing the time delay between excitation and gating pulses (like in the PLU technique in Figure 12.3), the PL dynamics can be recorded. The main advantage of the OKG method is its spectral insensitivity that facilitates the recording of time-resolved PL spectra, unlike the PLU technique that requires adjustment of the crystal orientation to optimize phase matching while tuning the wavelength. As with the PLU technique, the time resolution and gating efficiency of the OKG method can be improved by using aberration-free optics to collect and refocus the emission onto the Kerr medium.

The most common Kerr media for femtosecond PL measurements are high refractive index glasses, because of their moderately high third-order nonlinear susceptibility and instantaneous response (certain liquids or solutions have higher nonlinearities at the expense of longer response times that are caused by rotational relaxation). A large number of different glasses have been tested by Nakamura and Kanematsu (2004) for their usage as a Kerr medium in TRPL experiments. High OKG efficiency requires a high Kerr nonlinearity ($PL_{OKG} \propto n_2^2$, from Equation 12.3) that often comes at the expense of strong dispersion. As already mentioned in the previous section, second-order dispersion reduces the gating efficiency and the time resolution due to temporal broadening of the gating pulse and temporal walk-off between gating pulse and PL.

12.3 INTERPRETATION OF MEASUREMENTS

TRPL techniques are commonly used to study excited state lifetimes, deduce non-radiative and radiative decay rates, and determine spectral dynamics. Here, I will discuss a few phenomena that are typically investigated with TRPL techniques and provide some guidelines on how to analyze and interpret the acquired TRPL data.

12.3.1 TRPL Dynamics of Resonantly Excited Identical Emitters

Let us consider identical emitters that do not undergo rotational diffusion (described in detail in the following) and that are resonantly excited, such that relaxation processes between absorptive and emissive states are not relevant or occur on much shorter time scales than the time resolution of the TRPL technique. In this simple case, the time-resolved PL power is the convolution of the instrument response function, $\eta(t)$, and a single exponential decay function:

$$PL_{\mathrm{IE}}(t) = \eta(t) * \Gamma_{\mathrm{rad}}e^{-\Gamma t}, \tag{12.4}$$

in which $\eta(t)$ has units of energy and includes the overall collection efficiency, instrument sensitivity, and the temporal response function of the experimental setup. Γ is the total decay rate of the measured excited state, equal to the sum of radiative and non-radiative decay rates, Γ_{rad} and Γ_{nr}, respectively. The pre-factor Γ_{rad} in Equation 12.4 results from energy conservation arguments and ensures that the energy emitted by a single excited fluorophore in the absence of non-radiative decay channels is a single photon independent of the radiative decay rate. Equation 12.4 tells us that any radiative transition results in a TRPL signal at *short time delays*, even if the emission quantum yield (QY) is small, as long as the time resolution of the applied TRPL technique is shorter than the PL decay time. This is different from continuous wave (*cw*) PL techniques that tend to be insensitive to transitions with low QYs and high non-radiative rates that are related by

$$QY = \frac{\Gamma_{\mathrm{rad}}}{\Gamma} = \frac{\Gamma_{\mathrm{rad}}}{\Gamma_{\mathrm{rad}} + \Gamma_{\mathrm{nr}}}. \tag{12.5}$$

Equation 12.5 directly results from the definition of the QY, that is, the ratio of the time-integrated exponential decay in Equation 12.4 with and without non-radiative decay channels. The effect of different radiative and non-radiative rates on TRPL and *cw* PL measurements is demonstrated in Figure 12.5. It can be seen that transitions with different radiative rates (e.g., different emission dipole moments) are easily distinguishable in TRPL (Figure 12.5a) but identical in *cw* experiments (Figure 12.5b). Moreover, non-radiative decay channels do not affect the TRPL signal at zero time delay but only the decay dynamics. Hence, from TRPL measurements and Equations 12.5 and 12.4, we can deduce Γ_{rad} and Γ_{nr} and obtain information about excited state lifetimes, radiative transition rates, non-radiative decay channels, etc.

It is noteworthy that the decay function of TRPL measurements can deviate from a single exponential decay if the emission dipoles rotate during the decay time of the excited fluorophores. Imagine an excited emission dipole that is oriented along the *x*-direction and detected in the *z*-direction. If the molecule undergoes rotational diffusion within the lifetime of the excitation, the emission dipole will have one-third probability to be oriented along each of the major axes. Since dipoles oriented along the *z*-axis do not emit into the *z*-direction of the detector, the PL signal decays to two-thirds of its original value in the absence of other decay mechanisms, with a decay time equal to the angular correlation time. Rotational diffusion can be studied using cross-polarized detection schemes that are commonly used to investigate

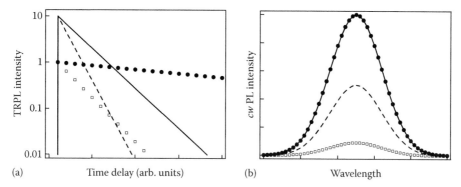

Figure 12.5 Comparison of TRPL (a) and *cw* PL (b) measurements of transitions with 100% QY and different radiative rates Γ_{high} and Γ_{low} (solid line and filled circles) and the same two transitions with an additional non-radiative decay rate $\Gamma_{non\text{-}rad}$ that results in QYs of 50% and 9% (dashed line and open squares). In this example, the rates were the same in both figures, and chosen such that $\Gamma_{high} = 10 \times \Gamma_{low}$ and $\Gamma_{non\text{-}rad} = \Gamma_{high}$.

fluorescence anisotropy effects. It can be omitted under specific conditions for the polarization direction of the excitation beam and the polarizer in front of the TRPL detector. In case of parallel absorption and emission dipoles, an angle of 54.7° (the "magic angle," given by $\cos^2\theta = 1/3$) between excitation and emission detection polarizations eliminates dynamic signatures in the TRPL signal caused by rotational diffusion that occurs for most molecules and small particles in solutions (Lakowicz 2006).

12.3.2 General Considerations in TRPL Measurements

The TRPL signal often contains more information and is more complicated than what has been discussed in the previous section. Here, I will introduce only a few of the additional phenomena that can be studied with TRPL techniques. For example, when fluorophores are excited with high-energy photons, the excitation can relax through several intermediate energy levels to the ground state. If the intermediate transitions are optically active and the TRPL time resolution is shorter than the relaxation time, the relaxation dynamics can be followed by spectrally resolved TRPL techniques. Even if the intermediate transitions are not optically active, relaxation to the lowest radiative transition can be observed as a delayed onset of the TRPL signal. In the example section, I will discuss such a case in more detail. It is worth noting that it is usually not possible to study relaxation dynamics with *cw* techniques.

Another frequently encountered situation is that TRPL decays are not well fitted by a single exponential function, as introduced in the previous section, but are better approximated with multi-exponential or stretched exponential functions. The origins of non-exponential PL decays are manifold, including distributions of dipole moments of nonidentical emitters, variations in the environment of the emitters, etc. The result is that the measured PL decay, *TRPL(t)*, is governed by a distribution of lifetimes $\rho(\tau)$:

$$TRPL(t) \propto \int \rho(\tau)e^{-t/\tau}\mathrm{d}\tau. \qquad (12.6)$$

If the distribution function is known, the decay dynamics can be reconstructed and an averaged or weighted lifetime can be determined. In other cases, a stretched exponential function $e^{-(t/\tau)^\beta}$ was found to be a good phenomenological approximation for the decay in disordered systems.

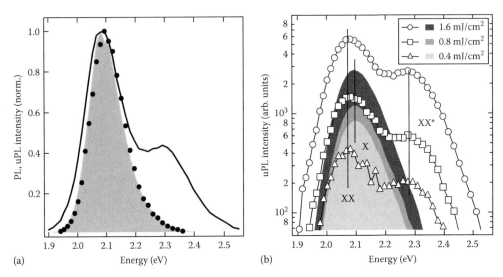

Figure 12.6 (a) Normalized *cw* (shaded area) and time-resolved PLU spectra measured at time delays of 1 ps (solid line) and 200 ps (solid circles). The *cw* and the PLU spectra at 200 ps originate from single exciton emission, whereas the PLU spectrum at early time delays displays a redshifted main peak due to excitons and neutral biexcitons and a high-energy peak caused by charged biexcitons. (b) Because of the large difference between exciton and biexciton decay rates, the exciton contribution (shaded areas) can be removed from the PLU spectra, resulting in pure multiexciton spectra (markers) that allow one to determine the energies of the neutral and charged biexcitons. The measured spectra are obtained for different excitation pulse intensities. (Reprinted from Achermann M. et al., *Phys. Rev. B*, 68, 245302, 2003a.)

The stretching coefficient β is between 0 and 1, with $\beta = 1$ giving the single exponential response. The mean lifetime $\langle \tau \rangle$ is given by

$$\langle \tau \rangle = \int e^{-(t/\tau)^\beta} \, dt = \frac{\tau}{\beta} \Gamma\left(\frac{1}{\beta}\right), \tag{12.7}$$

in which $\Gamma(x) = \int_0^\infty t^{x-1} e^{-t} dt$ is the gamma function. The shortest mean lifetime is obtained with $\beta = 1$ and becomes longer for $\beta < 1$ when the decay is temporally stretched.

In the special case of spectrally overlapping transitions with distinctly different transition rates, the distribution of lifetimes is a series of delta functions, and the TRPL signal can be well described by a multi-exponential decay. In this case, TRPL is very powerful, since it can disentangle the different components and associate lifetimes with each individual transition. Such a case is demonstrated in Figure 12.6, where the time-resolved spectra of excitons and biexcitons in CdSe nanocrystals (NCs) are shown. The lifetimes of biexcitons in such systems are in the picosecond range, limited by non-radiative Auger recombination, while excitons decay on a nanosecond time scale. Because of this distinct difference in decay dynamics, we can resolve the small spectral shift between the inhomogeneously broadened exciton and biexciton spectra (Achermann et al. 2003a).

12.3.3 Energy Transfer Dynamics

Other dynamic processes that have been successfully studied with TRPL techniques include energy transfer (ET) phenomena. An important reason for this preference is that

TRPL techniques are sensitive to excited states, but not to charged species. Measurements done by techniques such as transient absorption spectroscopy (Chapter 9) can be difficult to analyze because they contain information about excited states that give rise to ET *and* charged species that can occur in charge transfer processes. Among the different kinds of ET phenomena, fluorescence resonance energy transfer (FRET) (Forster 1948), which relies on near-field dipole–dipole interactions, is the most widely studied ET process and has been observed in many disciplines including physics, chemistry, and biology. In FRET, an excited donor (D) molecule with dipole moment μ_D non-radiatively transfers its energy to a proximal acceptor (A) molecule with dipole moment μ_A with a characteristic ET rate, Γ_{ET} that is given by

$$\Gamma_{ET} = \frac{2\pi}{\hbar} \frac{\mu_D^2 \mu_A^2 \kappa^2}{n^4 r_{DA}^6} \Theta,$$ (12.8)

where

n is the medium's refractive index

r_{DA} is the D–A separation

κ^2 is a factor that takes into account the relative orientation of the donor and acceptor dipole directions (for averaged random orientations, $\kappa^2 = 2/3$)

Θ is the spectral overlap integral between normalized donor emission and acceptor absorption line shapes and demonstrates the resonant nature of FRET

The strong D–A distance dependence is characteristic for near-field interactions between point dipoles and makes FRET-based techniques sensitive to distances of a few nanometers. In a classical picture, FRET can be understood as ET between two coupled oscillators (donor and acceptor), while in quantum electrodynamics FRET is described by the emission of a virtual photon from the excited donor and its subsequent absorption by the acceptor molecule (Juzeliunas and Andrews 1994a,b). Once the energy is transferred to the acceptor molecule, the excitation normally relaxes to a lower energy level within the acceptor molecule, inhibiting energy back transfer to the donor molecule (assuming that the energy was transferred from the lowest excited energy level of the donor).

In TRPL measurements, FRET manifests itself as an accelerated decay of the donor emission dynamics in the presence of the acceptor compared to the donor decay in the absence of the acceptor. If the acceptor dipole has an emissive transition at lower energies, FRET also results in a spectrally redshifted and time-delayed acceptor emission. Considering single donor to single acceptor FRET and assuming that the relaxation from absorptive to emissive transitions is much faster than any other dynamics in the system, the PL dynamics of the donor and acceptor emission can be described by the following coupled rate equations of the donor and acceptor population probabilities N_D and N_A, respectively:

$$\frac{\partial N_D}{\partial t} = -\Gamma_D N_D - \Gamma_{ET} N_D \quad \text{and} \quad \frac{\partial N_A}{\partial t} = \Gamma_{ET} N_D - \Gamma_A N_A,$$ (12.9)

in which Γ_D and Γ_A are the donor and acceptor decay rates in the absence of FRET. The solutions to Equations 12.9 are

$$N_D(t) = N_{D,0} e^{-(\Gamma_D + \Gamma_{ET})t} \quad \text{and} \quad N_A(t) = N_{D,0} \frac{\Gamma_{ET}}{\Gamma_D + \Gamma_{ET} - \Gamma_A} \left(e^{-\Gamma_A t} - e^{-(\Gamma_D + \Gamma_{ET})t} \right),$$ (12.10)

for initial conditions $N_D(0) = N_{D,0}$ and $N_A(0) = 0$. The donor PL dynamics is simply a single exponential decay that is accelerated in the presence of a proximal acceptor. Comparing donor decay dynamics in the presence and absence of acceptors then allows one to determine the FRET rate. However, in certain systems other processes (e.g., charge transfer to the acceptor) can result in an accelerated donor decay rate as well. In these cases, the acceptor emission dynamics has to be analyzed to characterize the FRET process. It is worth distinguishing two cases for the acceptor dynamics that are both governed by Equations 12.10. For $\Gamma_D + \Gamma_{ET} > \Gamma_A$, the rise of the acceptor PL signal is given by $\Gamma_D + \Gamma_{ET}$ and the decay by Γ_A; the opposite is true for $\Gamma_D + \Gamma_{ET} < \Gamma_A$. Hence, depending on the relative magnitudes of these rates, the FRET rate can be obtained by analyzing the delayed rise or the prolonged decay of the acceptor emission.

The strong D–A distance dependence of FRET indicated in Equation 12.8 is the reason why FRET is a widely used tool to determine nanometer scale distances. The characteristic length scale for the FRET process is the so-called Förster radius, r_0, that is, the D–A distance for which the FRET rate is equal to the donor decay rate. For typical molecular donor–acceptor pairs r_0 is in the 3–5 nm range, while in semiconductor NC FRET systems r_0 is larger than 10 nm (Crooker et al. 2002; Achermann et al. 2003b). In *cw* experiments, the donor emission intensity changes dramatically for D–A separations around r_0, and hence, *cw* techniques are well suited to determine D–A distances in the close vicinity of r_0. In contrast, TRPL techniques can use the FRET process to accurately determine D–A separations over a larger range, in particular, toward shorter separations. Considering dye molecules with lifetimes of a few nanoseconds and a TRPL technique with a time resolution of a few picoseconds (streak cameras), one can determine D–A separations as small as $\sim r_0/3$.

12.4 EXAMPLES OF TRPL STUDIES

In the following paragraphs, I will provide a few research examples of how TRPL can be used to study dynamic processes in nanoscale structures. Specifically, I will discuss ultrafast relaxation of hot electron–hole (e–h) pairs in CdSe nanorods using PLU techniques, radiative rate engineering in metal–semiconductor nanostructures investigated by TCSPC techniques, and FRET dynamics in hybrid semiconductor nanostructures, also measured by TCSPC techniques.

12.4.1 Ultrafast Relaxation Dynamics in Semiconductor Nanocrystals

As an illustrative example of using TRPL techniques to reveal ultrafast relaxation dynamics, I review our recent work on carrier cooling in semiconductor quantum rods (QRs) (Achermann et al. 2006a). Quantization of electronic and phonon energies and large surface-to-volume ratios significantly modify energy relaxation mechanisms in nanoscale semiconductors compared to bulk materials (Chapter 2). In the case of ultrasmall semiconductor NCs (Alivisatos 1996), strong quantum confinement leads to greatly enhanced carrier–carrier interactions that open new NC-specific energy relaxation and recombination channels (Chepic et al. 1990; Efros et al. 1995; Klimov and McBranch 1998; Guyot-Sionnest et al. 1999; Klimov et al. 2000; Mohamed et al. 2001; Htoon et al. 2003; Wang et al. 2003). For example, enhanced carrier–carrier interactions in NCs lead to large non-radiative Auger recombination rates (Chepic et al. 1990; Klimov et al. 2000; Xu et al. 2002; Achermann et al. 2003a; Htoon et al. 2003; Wang et al. 2003). In this process, the e–h recombination energy is not emitted as a photon but is transferred to a third carrier that is re-excited to a higher-energy state (inset of Figure 12.7a). Auger recombination leads to heating of the e–h system (i.e., an increase of the average e–h pair energy) that can, in

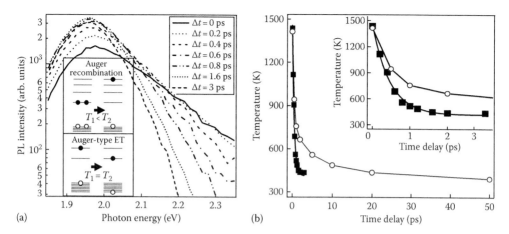

Figure 12.7 (a) TRPL spectra taken at different time delays Δt after excitation. Inset: Schematic of Auger-type e–h energy transfer (ET) that does not change the average energy per e–h pair (i.e., carrier temperature), in contrast to Auger recombination that increases the total energy of carriers in an NC by approximately E_g, therefore heating the electronic system. (b) Carrier temperature dynamics extracted from the TRPL spectra measured at n_{eh} = 5.5 × 10^{18} cm^{-3} (solid squares) and n_{eh} = 2.2 × 10^{19} cm^{-3} (open circles). The inset is a blowup of the carrier temperature dynamics during the first 3 ps. (Reprinted from Achermann, M. et al., *Nat. Phys.*, 2, 557, 2006a.)

principle, affect (slow down) carrier relaxation dynamics. This behavior is in contrast to e–h ET, which does not change the total energy of the e–h pair but only leads to the redistribution of energy between the electron and the hole (inset of Figure 12.7a).

For our study of carrier relaxation in strongly confined one-dimensional (1D) electronic systems, we used highly monodisperse, colloidal CdSe QRs prepared as hexane solutions (Murray et al. 1993; Manna et al. 2000; Peng and Peng 2001). We studied a series of samples with the same QR diameter of 4.6 nm and various lengths from 22 to 44 nm. The samples were excited by frequency-doubled (400 nm wavelength), 100 fs pulses from an amplified Ti:sapphire laser operating at a 100 kHz repetition rate. The ultrafast carrier relaxation dynamics was recorded by the previously mentioned femtosecond PLU technique (Shah 1988) with a time resolution of approximately 300 fs. All measurements were conducted at room temperature.

Figure 12.7a displays typical PLU spectra of 29 nm long QRs taken at different time delays (Δt) after excitation for an initial e–h pair density, n_{eh}, of ~5.5 × 10^{18} cm^{-3}, corresponding to two to three e–h pairs per QR on average. The PLU spectra are divided by the linear absorption spectra to account for the spectral dependence of the joint e–h density of states. In contrast to hot PL spectra of spherical NCs (Xu et al. 2002; Achermann et al. 2003a) that show well-separated emission peaks arising from distinct quantized states (Figure 12.6), the TRPL spectra of QRs consist of a single peak with an extended, exponential, high-energy tail, which reflects the distribution of charge carriers over the dense spectrum of high-energy, QR states. We use the decay of this tail as a measure of the instantaneous carrier temperature T_e. No nonthermal features are observed in the PLU spectra, indicating that carrier thermalization occurs on time scales shorter than our temporal resolution (~300 fs). The progressive increase of the high-energy slope of the PLU spectra with time in Figure 12.7a reflects carrier cooling dynamics.

The carrier temperature T_e is obtained by fitting the high-energy tails of the PLU spectra with exp($-\hbar\omega/kT_e$) (k is Boltzmann's constant) and plotted as a function of time in Figure 12.7b (solid squares). At pump levels below ~5.5 × 10^{18} cm^{-3}, the temperature relaxation (time constant is 0.5 ps) does not show a significant dependence on either rod length or pump level, indicating that high-carrier-density effects play a minor role. In this regime of low pump levels, carrier cooling in QRs is bulk-like and has been explained in terms of strong coupling between the e–h and the longitudinal optical (LO)-phonon subsystems that are in equilibrium with each other and cool together via interactions with acoustic phonons (Klimov et al. 1995) (Chapter 8). At high excitation densities, the energy relaxation changes significantly. In Figure 12.7b, we compare the temperature dynamics in 29 nm long QRs at $n_{ch} = 5.5 \times 10^{18} \, cm^{-2}$ and $n_{ch} = 2.2 \times 10^{19} \, cm^{-2}$. In contrast to the low density case, we find that the cooling process at higher carrier densities is not terminated after a few picoseconds, but instead persists up to tens of picoseconds. As we showed in Achermann et al. (2006a), the slowdown of the temperature decay is caused by carrier heating through Auger recombination that is efficient enough at high excitation densities to compete with the cooling process. More specifically, in the high excitation density regime, the carrier temperature is determined by the balance between carrier cooling through interactions with phonons and energy inflow produced by Auger recombination. This approximate equilibrium results in a cooling dynamics that is controlled by the carrier recombination process. This peculiarity has never been observed previously either in bulk or low-dimensional materials. In addition, because of the difference in scaling of Auger recombination rates with respect to carrier density in zero-dimensional (0D) and 1D semiconductors (Htoon et al. 2003) and the direct correlation between energy relaxation and recombination dynamics, we found that the carrier cooling behavior is significantly different in short and long QRs. More specifically, the temperature difference between electron and lattice temperatures increases linearly with excited carrier density in long rods (1D Auger recombination) and quadratically in short rods (0D Auger recombination). Further details on this interesting behavior can be found in Achermann et al. (2006a).

12.4.2 Radiative Rate Enhancements in Hybrid Metal–Semiconductor Nanostructures

As mentioned earlier, TRPL measurements are well suited for distinguishing and quantifying non-radiative and radiative relaxation pathways. Both of these processes are affected by the local environment of the emitter, as has been illustrated recently by our work on interactions between surface plasmons (SPs) in metal nanostructures and emission dipoles in semiconductor NCs (Wang et al. 2009). More specifically, we showed that the PL decay is accelerated when detected at a wavelength, angle, and polarization that correspond to a maximum in extinction measurements. Since the extinction spectra are dominated by SP effects, the accelerated PL decay is attributed to an increased radiative rate caused by local SP-induced field enhancements in the vicinity of the metal nanostructures. Our findings indicate that NC–SP interactions can result in directionally enhanced emission that can be beneficial for lighting, optical sensing, and microscopy applications.

Our samples for this study consisted of square arrays of gold nanodisks with nominal dimensions of 170 nm diameter, 300 nm center-to-center distance, and 70 nm height. The disk arrays were produced on a 100 nm thick SiN membrane that was suspended on a silicon wafer (Xiao et al. 2008). We characterized the Au disk arrays with angle and polarization-resolved transmission measurements using a tungsten halogen light source that was lightly focused onto the sample. From the spectra measured through the SiN membrane with and without the disk array, $I(\theta,\lambda)$ and $I_0(\theta,\lambda)$, respectively, we calculate the extinction $Q = -\ln[I(\theta, \lambda)/I_0(\theta, \lambda)]$ as a function of angle θ, wavelength λ, and p- and s-polarizations. The extinction maps feature a SP resonance at ~800 nm that is associated with the in-plane SP^ resonance (Figure 12.8a). The extinction resonance in the range 550–600 nm and for $\theta > 30°$ is related to the out-of-plane

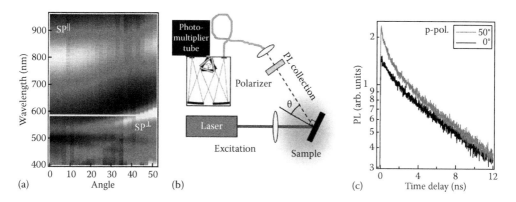

Figure 12.8 (See color insert.) (a) Extinction plot for *p*-polarized light as a function of incidence angle (low extinction: dark; high extinction: yellow). Indicated are the in-plane SP$^\parallel$ and out-of-plane SP$^\perp$ resonance and the center PL wavelength of the NCs (yellow line). (b) Schematic of the angle and polarization-resolved PL dynamics setup. (c) *p*-polarized PL decay dynamics collected at 0° and 50°. (Reprinted from Wang, Y. et al., *Phys. Rev. Lett.*, 102, 163001, 2009.)

SP$^\perp$ resonance, because it is only active for *p*-polarized light at off-normal angles (Figure 12.8a) and completely absent in *s*-polarized extinction maps (not shown here).

Here, we focus on the interaction between excited NCs and the SP$^\perp$ resonance. As dipole emitters, we chose CdSe/ZnS core/shell NCs with an emission centered around 585 nm that overlaps with the SP$^\perp$ resonance (Figure 12.8a). We spin coated NCs onto the gold nanodisk arrays and obtained an approximate monolayer of NCs with randomly oriented emission dipoles that cover the Au disks and the substrate between the disks. The NCs were excited at a fixed off-normal angle with a pulsed diode laser emitting ~50 ps pulses at 407 nm and 10 MHz repetition rate (Figure 12.8b). The emission of the NCs that is affected by the orientation of their two-dimensional (2D) emission dipoles and their local environment is collected at variable angle and polarization (Figure 12.8b). The PL is detected at λ = 585 nm with a PMT and analyzed with a TCSPC system with ~70 ps time resolution, yielding time-, angle-, and polarization-resolved PL dynamics.

We find that for *p*-polarized PL, the decay dynamics at θ = 50° shows a significant initial acceleration that is absent at a θ = 0° detection angle (Figure 12.8c). In contrast, the *s*-polarized PL dynamics is essentially independent of the detection angle and identical to the *p*-polarized PL at θ = 0° (Wang et al. 2009). From extinction measurements, we know that the SP$^\perp$ resonance is excited very efficiently at θ = 50° and λ = 585 nm with *p*-polarized light. At this wavelength, no SP resonance can be excited at normal incidence or at any angle with *s*-polarized light. Hence, there is a direct correlation between the extinction and TRPL measurements, and we conclude that the accelerated PL dynamics is due to interactions between the NC dipole emitters and the SP$^\perp$ resonance.

For a quantitative analysis, we modeled the PL decay dynamics detected at θ = 0° with the sum of two exponential decays with decay rates $\Gamma_m = \Gamma_{m,nr} + \Gamma_{rad}$ and $\Gamma_{sub} = \Gamma_{nr} + \Gamma_{rad}$ that describe the emission dynamics of NCs that are located on the Au disks, Γ_m, and on the substrate between the Au disks, Γ_{sub}. Here, Γ_{nr} and $\Gamma_{m,nr}$ are the non-radiative decay rates that include charge trapping, FRET between NCs, and metal quenching in the case of $\Gamma_{m,nr}$. From a fit with a double exponential function, we obtain $\Gamma_m = 0.7$ ns^{-1} and $\Gamma_{sub} = 0.1$ ns^{-1}. At θ = 50° detection angle, the PL decay that originates from NCs on Au disks is accelerated because of the interaction with the SP$^\perp$ resonance. Hence, the fast decay rate Γ_m changes with detection angle and we determine a decay rate difference of $\Delta\Gamma_m \approx 0.8$ ns^{-1} between Γ_m at 0° and 50°

detection angles. From the definition of Γ_m given earlier, we know that $\Delta\Gamma_m$ is caused by a change of the radiative or the non-radiative rate. Let us assume here that $\Delta\Gamma_m$ is associated with an SP-induced change of the *radiative* rate and *not* related to SP-induced non-radiative energy dissipation. We will provide evidence for this assumption in the following. From $\Delta\Gamma_m$, we can calculate a radiative rate enhancement of $\Gamma_{rad}^{SP}/\Gamma_{rad} = (\Delta\Gamma_m/\Gamma_{rad})+1 \approx 11$' in which $\Gamma_{rad} = 0.08\,ns^{-1}$ is the reference radiative decay rate of NCs on the SiN substrate that was determined from the decay dynamics at long time delays (Crooker et al. 2003). Conceptually, the radiative decay rates of NCs with or without a metal nanostructure in their proximity, Γ_{rad}^{SP} and Γ_{rad}, respectively, are related by the field enhancement factor F, defined by the ratio of the projected local electric field onto the dipole direction with and without a metal nanostructure (Carminati et al. 1998; Shimizu et al. 2002; Farahani et al. 2005):

$$\Gamma_{rad}^{SP} = |F|\Gamma_{rad}. \tag{12.11}$$

Hence, the measured radiative rate enhancement of ~11 is associated with a field enhancement $F \sim 3.3$.

The earlier assertion of a radiative rate enhancement is supported by the following reasoning. The time-resolved PL amplitude of an emitter at zero time delay, PL_0, is proportional to the radiative decay rate through energy conservation (Equation 12.4). Therefore, an increase of the radiative rate results in an increase of PL_0. In contrast, PL_0 remains unchanged for modifications of non-radiative decay rates (Figure 12.5). When comparing the non-normalized data taken at $50°$ for s- and p-polarizations (Wang et al. 2009), we find that PL_0 is significantly larger for p-polarized than for s-polarized emission. The same measurements of NCs on a glass slide or an unpatterned Au film did not show any significant polarization variations in the zero time delay amplitude or the decay dynamics. Likewise, in Figure 12.8c we clearly see that PL_0 depends on the detection angle and is larger at higher angles. Such a behavior cannot be explained by non-radiative effects but has to be attributed to radiative rate changes.

12.4.3 Energy Transfer Dynamics in Assemblies of Semiconductor Nanostructures

While most FRET studies are based on individual donor and acceptor molecules (Lakowicz 2006), I want to present here a special case of FRET involving an extended 2D donor system that, as a result of its dimensionality, affects the FRET process (Achermann et al. 2004). The main motivation for this ET study was the possibility to excite semiconductor NCs by means of ET from a proximal semiconductor quantum well (QW) in order to use NCs in light-emitting devices. Since electrical pumping of NCs is difficult (largely due to the presence of an insulating organic capping layer on the NCs), we suggested the indirect injection of e–h pairs by non-contact, non-radiative FRET from a proximal QW that can be pumped electrically (Achermann et al. 2004; Achermann et al. 2006b). As ET relies on Coulomb interactions rather than a direct wave function overlap, it is not significantly inhibited by the NC capping layer.

The investigated structure consists of a 3 nm thick InGaN QW, on top of which is assembled a close-packed monolayer of highly monodisperse NCs using the Langmuir–Blodgett technique. The NCs contain a CdSe core (radius = 1.9 nm) overcoated with a shell of ZnS (~0.6 nm thickness), followed by a layer of ~1.1 nm long organic molecules. These NCs show efficient emission centered near 575 nm and a structured absorption spectrum with the lowest absorption maximum of the 1S state at ~560 nm (Figure 12.9a). QW samples were grown on

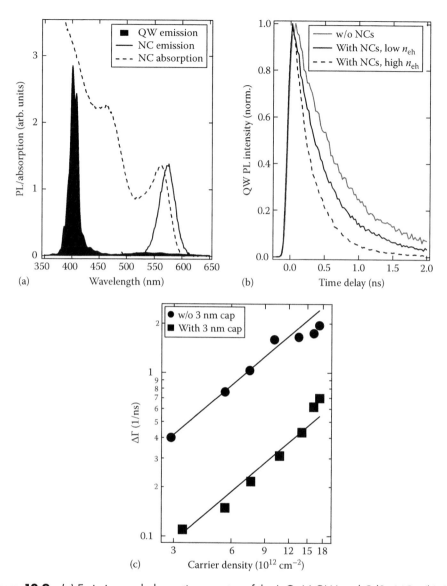

Figure 12.9 (a) Emission and absorption spectra of the InGaN QW and CdSe NCs. (b) QW PL decay without NCs and with NCs for different excitation densities. (c) Energy transfer rate as a function of carrier density in the QW for QWs with and without a 3 nm thick cap layer. (Reprinted from Achermann, M. et al., *Nature*, 429, 642, 2004.)

sapphire substrates by metal–organic chemical-vapor deposition and either terminated with a 3 nm GaN top barrier (capped QW) or left uncapped (Koleske et al. 2002). The In concentration in the QWs was 5%–10%, which corresponds to an emission wavelength of ~400 nm (Figure 12.9a). This wavelength is in the range of strong NC absorption, which provides strong coupling of QW excitations to the absorption dipole of NCs, and should allow efficient FRET. To study ET dynamics, we measured spectrally resolved TRPL from the QW and the NCs using a TCSPC system that provides ~30 ps time resolution. The hybrid QW-NC structures were excited at 266 nm by 200 fs pulses of the frequency-tripled output of an amplified Ti:sapphire laser.

The interactions between the QW and the NC monolayer can be described in terms of a resonant Förster-type ET. Considering that the QW excitations are unbound e–h pairs with density n_{eh}, we determined the FRET rate Γ_{ET} to be (Kos et al. 2005):

$$\Gamma_{ET} \propto \frac{\mu_{NC}^2 \mu_{QW}^2 n_{NC} n_{eh} N_{NC}}{d^4},$$ (12.12)

where

μ_{NC} and μ_{QW} are the transition dipole moments for the NC and the QW, respectively
n_{NC} is the surface density of NCs
N_{NC} is the NC density of states at the QW emission energy
d is the separation between the centers of the QW and the NC monolayer

It is worth emphasizing that the dimensionality of the QW results in a d^{-4} distance dependence in Equation 12.12, in contrast to the r^{-6} dependence of Equation 12.8. Moreover, the FRET rate is proportional to the unbound e–h pair density and, therefore, to the excitation intensity, which is not the case for most other types of FRET.

In Figure 12.9b, we show comparative TRPL measurements for hybrid QW-NC structures and isolated QWs. We observe that the presence of the NC layer adjacent to the QW significantly alters the QW PL dynamics. Namely, the QW PL decay becomes faster in the presence of NCs, indicating an additional relaxation channel for QW excitations, which we assign to QW-NC ET. This NC-induced change in QW dynamics becomes more pronounced with increasing carrier density. To quantify the increase, in Figure 12.9c we plot the additional initial decay rate $\Delta\Gamma = \Gamma_{QW\ with\ NC} - \Gamma_{QW\ without\ NC}$ as a function of n_{eh} for structures based on uncapped and capped QWs. We observe that in both cases the $\Delta\Gamma$ growth is linear with n_{eh}, but absolute values of $\Delta\Gamma$ are approximately 4.4 times greater for the uncapped QWs compared to QWs with a top barrier. Both of these observations are consistent with the fact that the additional decay rate $\Delta\Gamma$ is due to QW-to-NC ET. Equation 12.12 predicts that the ET rate should increase linearly with n_{eh}, which is exactly the dependence observed experimentally. Furthermore, the increase in the transfer rate in the case of the uncapped QW is consistent with its strong dependence on the ET distance ($\Gamma_{ET} \sim d^{-4}$). From the geometrical parameters of our system (d is 8.1 and 5.1 nm for capped and uncapped QWs, respectively), we estimate that the d dependence should result in an increased ET rate by a factor of 5.5, which agrees well with the factor of 4.4 observed experimentally.

Further evidence for efficient QW-NC ET is provided by the analysis of the PL from the NC layer. The energy outflow from the QW should result in a corresponding increase in the emission of the NCs. In our experimental configuration, however, optical pumping directly excites not only the QW but also the NCs. Therefore, to extract the ET-induced increase in the NC PL, we perform a side-by-side comparison of PL data for hybrid QW-NC structures and a NC Langmuir–Blodgett monolayer assembled on a glass slide. One such set of data, plotted as temporally integrated NC PL intensity versus pump fluence, is displayed in Figure 12.10a. To account for the difference in the NC packing densities for Langmuir–Blodgett films assembled on the QW and the glass slide, we introduce a constant scaling factor, which allows us to match the PL intensities detected from QW-NC and glass/NC samples at low pump powers, for which ET from the QW is negligible. The data indicate that at low pump fluences, both types of samples show a similar PL pump dependence. However, the two traces show distinctly different behavior at higher pump fluences, for which ET starts to play a significant role. Whereas emission from the isolated NC layer saturates at ~20 µJ cm^{-2}, the NC PL in the hybrid structure shows a steady growth until ~80 µJ cm^{-2}. As a result of this delayed

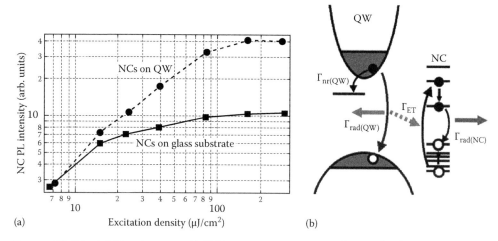

Figure 12.10 (a) Time-integrated NC PL intensity versus pump fluence for the NC monolayer assembled on a glass substrate and on top of a capped QW. (b) Carrier relaxation and energy-transfer processes in the hybrid QW-NC structure. The QW-NC energy transfer competes with radiative and non-radiative decay processes in the QW. High-energy excitations created in the NC through energy transfer rapidly relax to the NC band edge, which prevents transfer back to the QW. (Reprinted from Achermann, M. et al., *Nature*, 429, 642, 2004.)

saturation, the maximum NC PL intensity achievable with the QW-NC structure is four times greater than the PL for the NC monolayer on the glass slide. All of these results indicate a strong additional energy inflow into the NCs as a result of ET from the QW.

The efficiency of non-radiative QW-NC ET (η_{ET}) can be estimated from the expression $\eta_{ET} = \Gamma_{ET}/(\Gamma_{ET} + \Gamma)$, in which $\Gamma = \Gamma_{rad} + \Gamma_{nr}$ is the relaxation time of QW excitations due to both radiative and non-radiative processes (Figure 12.10b). Our experimental results for the uncapped sample indicate that $1/\Gamma \approx 0.6$ ns and $1/\Gamma_{ET} \approx 0.5$ ns (for $n_{eh} = 1.8 \times 10^{13}$ cm^{-2}), which yields η_{ET} as high as 55%. It is interesting that despite the additional step in the ET process, the PL QY of the hybrid QW-NC device (QY$_{QW/NC}$) can be greater than the original QY of the QW (Achermann et al. 2004). QY$_{QW/NC}$ can be estimated from the expression QY$_{QW/NC}$ = QY$_{NC}/(1 + \Gamma/\Gamma_{ET})^{-1}$. This expression indicates that if $\Gamma_{ET} \gg \Gamma$, the QY of the hybrid structure approaches that of NCs. This conclusion further means that even the use of an InGaN QW with poor room-temperature QY can produce highly efficient hybrid devices.

REFERENCES

Achermann M., Bartko A. P., Hollingsworth J. A., and Klimov V. I. 2006a. The effect of Auger heating on intraband carrier relaxation in semiconductor quantum rods. *Nature Physics* 2: 557–561.

Achermann M., Hollingsworth J. A., and Klimov V. I. 2003a. Multiexcitons confined within a sub-excitonic volume: Spectroscopic and dynamical signatures of neutral and charged biexcitons in ultrasmall semiconductor nanocrystals. *Physical Review B* 68: 245302.

Achermann M., Petruska M. A., Crooker S. A., and Klimov V. I. 2003b. Picosecond energy transfer in quantum dot Langmuir-Blodgett nanoassemblies. *Journal of Physical Chemistry B* 107: 13782–13787.

Achermann M., Petruska M. A., Koleske D. D., Crawford M. H., and Klimov V. I. 2006b. Nanocrystal-based light-emitting diodes utilizing high-efficiency nonradiative energy transfer for color conversion. *Nano Letters* 6: 1396–1400.

Achermann M., Petruska M. A., Kos S., Smith D. L., Koleske D. D., and Klimov V. I. 2004. Energy-transfer pumping of semiconductor nanocrystals using an epitaxial quantum well. *Nature* 429: 642–646.

Alivisatos A. P. 1996. Semiconductor clusters, nanocrystals, and quantum dots. *Science* 271: 933–937.

Becker W. 2005. *Advanced Time-Correlated Single Photon Counting Techniques*. Berlin, Germany: Springer.

Becquerel A. E. 1867. *La lumière, ses causes et ses effets*. Paris, France: Didot.

Carminati R., Neito-Vesperinas M., and Greffet J. J. 1998. Reciprocity of evanescent electromagnetic waves. *Journal of the Optical Society of America A—Optics Image Science and Vision* 15: 706–712.

Chepic D. I., Efros A. L., Ekimov A. I., Vanov M. G., Kharchenko V. A., Kudriavtsev I. A., and Yazeva T. V. 1990. Auger ionization of semiconductor quantum drops in a glass matrix. *Journal of Luminescence* 47: 113–127.

Crooker S. A., Barrick T., Hollingsworth J. A., and Klimov V. I. 2003. Multiple temperature regimes of radiative decay in CdSe nanocrystal quantum dots: Intrinsic limits to the dark-exciton lifetime. *Applied Physics Letters* 82: 2793–2795.

Crooker S. A., Hollingsworth J. A., Tretiak S., and Klimov V. I. 2002. Spectrally resolved dynamics of energy transfer in quantum-dot assemblies: Towards engineered energy flows in artificial materials. *Physical Review Letters* 89: 186802.

Duguay M. A. and Hansen J. W. 1969. An ultrafast light gate. *Applied Physics Letters* 15: 192–194.

Efros A. L., Kharchenko V. A., and Rosen M. 1995. Breaking the phonon bottleneck in nanometer quantum dots: Role of Auger-like processes. *Solid State Communications* 93: 281–284.

Farahani J. N., Pohl D. W., Eisler H. J., and Hecht B. 2005. Single quantum dot coupled to a scanning optical antenna: A tunable superemitter. *Physical Review Letters* 95: 017402.

Forster T. 1948. Zwischenmolekulare Energiewanderung Und Fluoreszenz. *Annalen Der Physik* 2: 55–75.

Guyot-Sionnest P., Shim M., Matranga C., and Hines M. 1999. Intraband relaxation in CdSe quantum dots. *Physical Review B* 60: R2181–R2184.

Herschel J. F. W. 1845. On a case of superficial colour presented by a homogeneous liquid internally colourless. *Philosophical Transactions* 135: 143–145.

Htoon H., Hollingsworth J. A., Dickerson R., and Klimov V. I. 2003. Effect of zero- to one-dimensional transformation on multiparticle Auger recombination in semiconductor quantum rods. *Physical Review Letters* 91: 227401.

Juzeliunas G. and Andrews D. L. 1994a. Quantum electrodynamics of resonant energy-transfer in condensed matter. *Physical Review B* 49: 8751–8763.

Juzeliunas G. and Andrews D. L. 1994b. Quantum electrodynamics of resonant energy-transfer in condensed matter. 2. Dynamical aspects. *Physical Review B* 50: 13371–13378.

Kinoshita S., Ozawa H., Kanematsu Y., Tanaka I., Sugimoto N., and Fujiwara S. 2000. Efficient optical Kerr shutter for femtosecond time-resolved luminescence spectroscopy. *Review of Scientific Instruments* 71: 3317–3322.

Klimov V., Bolivar P. H., and Kurz H. 1995. Hot-phonon effects in femtosecond luminescence spectra of electron-hole plasmas in CdS. *Physical Review B* 52: 4728–4731.

Klimov V. I. and McBranch D. W. 1998. Femtosecond 1P-to-1S electron relaxation in strongly confined semiconductor nanocrystals. *Physical Review Letters* 80: 4028–4031.

Klimov V. I., Mikhailovsky A. A., McBranch D. W., Leatherdale C. A., and Bawendi M. G. 2000. Quantization of multiparticle Auger rates in semiconductor quantum dots. *Science* 287: 1011–1013.

Koleske D. D., Fischer A. J., Allerman A. A., Mitchell C. C., Cross K. C., Kurtz S. R., Figiel J. J., Fullmer K. W., and Breiland W. G. 2002. Improved brightness of 380 nm GaN light emitting diodes through intentional delay of the nucleation island coalescence. *Applied Physics Letters* 81: 1940–1942.

Kos S., Achermann M., Klimov V. I., and Smith D. L. 2005. Different regimes of Forster-type energy transfer between an epitaxial quantum well and a proximal monolayer of semiconductor nanocrystals. *Physical Review B* 71: 205309.

Lakowicz J. R. 2006. *Principles of Fluorescence Spectroscopy*. New York: Springer.

Manna L., Scher E. C., and Alivisatos A. P. 2000. Synthesis of soluble and processable rod-, arrow-, teardrop-, and tetrapod-shaped CdSe nanocrystals. *Journal of the American Chemical Society* 122: 12700–12706.

Mohamed M. B., Burda C., and El-Sayed M. A. 2001. Shape dependent ultrafast relaxation dynamics of CdSe nanocrystals: Nanorods vs nanodots. *Nano Letters* 1: 589–593.

Murray C. B., Norris D. J., and Bawendi M. G. 1993. Synthesis and characterization of nearly monodisperse CdE (E = S, Se, Te) semiconductor nanocrystallites. *Journal of the American Chemical Society* 115: 8706–8715.

Nakamura R. and Kanematsu Y. 2004. Femtosecond spectral snapshots based on electronic optical Kerr effect. *Review of Scientific Instruments* 75: 636–644.

Peng Z. A. and Peng X. G. 2001. Formation of high-quality CdTe, CdSe, and CdS nanocrystals using CdO as precursor. *Journal of the American Chemical Society* 123: 183–184.

Schanz R., Kovalenko S. A., Kharlanov V., and Ernsting N. P. 2001. Broad-band fluorescence upconversion for femtosecond spectroscopy. *Applied Physics Letters* 79: 566–568.

Shah J. 1988. Ultrafast luminescence spectroscopy using sum frequency generation. *IEEE Journal of Quantum Electronics* 24: 276–288.

Shimizu K. T., Woo W. K., Fisher B. R., Eisler H. J., and Bawendi M. G. 2002. Surface-enhanced emission from single semiconductor nanocrystals. *Physical Review Letters* 89: 117401.

Stokes G. G. 1852. On the change of refrangibility of light. *Philosophical Transactions* 142: 463–562.

Takeda J., Nakajima K., Kurita S., Tomimoto S., Saito S., and Suemoto T. 2000. Time-resolved luminescence spectroscopy by the optical Kerr-gate method applicable to ultrafast relaxation processes. *Physical Review B* 62: 10083–10087.

Wang L. W., Califano M., Zunger A., and Franceschetti A. 2003. Pseudopotential theory of Auger processes in CdSe quantum dots. *Physical Review Letters* 91: 056404.

Wang Y., Yang T., Tuominen M., and Achermann M. 2009. Radiative rate enhancements in hybrid metal-semiconductor nanostructures. *Physical Review Letters* 102: 163001.

Xiao Q. J., Yang T. Y., Ursache A., and Tuominen M. T. 2008. Clusters of interacting single domain Co nanomagnets for multistate perpendicular magnetic media applications. *Journal of Applied Physics* 103: 07C521.

Xu S., Mikhailovsky A. A., Hollingsworth J. A., and Klimov V. I. 2002. Hole intraband relaxation in strongly confined quantum dots: Revisiting the "phonon bottleneck" problem. *Physical Review B* 65: 53191–53195.

Chapter 13

TIME-RESOLVED MAGNETO-OPTICAL SPECTROSCOPY

Jigang Wang

CONTENTS

13.1 INTRODUCTION

13.1.1 Motivation: Spin Dynamics and Ultrafast Magnetism

There has always been a great interest in exploring and understanding spin-related phenomena in materials at ultrafast time scales, from both fundamental scientific and technological points of view. On one hand, it is of intrinsic scientific interest to understand transient magnetic phenomena and collective spin systems that have been driven far from equilibrium. On the other hand, the emerging "ultrafast frontier" for spins is driven by technological demands for ever-increasing writing and reading speed for magnetic storage and computation devices.

One of the most exciting, yet puzzling, questions in materials science and condensed matter physics today is whether one can detect, understand, and control macroscopic spin orderings in highly nonequilibrium, nonthermal states at femtosecond time scales. Such processes are at least 1000 times faster than those of the traditional thermal magnetic processes (Zvezdin and Kotov, 1997; Sugano and Kojima, 2000; Zhang et al., 2002), which set the limit of the magnetic switching time to 100 ps–10 ns, as used in the modern magneto-optical (MO) recording industry (Zvezdin and Kotov, 1997). Recently, there is growing evidence that, at the femtosecond time scale, coherent modification of magnetism is feasible, which involves direct coupling between the light field and spins. Compelling evidence for such coherent ultrafast magnetism has been obtained from photoinduced demagnetization in nickel films (Bigot et al., 2009), spin precession in orthoferrite $DyFeO_3$ (Kimel et al., 2005), and spin reorientation in the ferromagnetic semiconductor GaMnAs (Kapetanakis et al., 2009; Wang et al., 2009). However, exactly how laser pulses can substantially modify the collective spin ordering in the coherent regime or induce even a complete spin reversal or a magnetic phase transition at ultrafast time scales remains controversial. Some of the most challenging issues in this temporal regime are the highly nonequilibrium processes and photoexcited coherences involved, since the relevant characteristic time scales are comparable to or even shorter than one oscillation cycle of phonons and magnons, spin-dependent scattering, and dephasing times. These raise some fundamental questions deep into the cornerstones of our current understanding of magnetism and phase transitions, for example, microscopic origin of angular momentum conservation and the validity of the thermodynamic description of magnetism at these extremely short time scales.

13.1.2 Ultrafast Magneto-Optical Spectroscopy

MO spectroscopy is a powerful probe for investigating spin-related phenomena in condensed matter. A variety of linear and nonlinear MO techniques, such as magneto-optical Kerr effect (MOKE)/Faraday spectroscopy, magnetic Raman scattering, magnetically induced second harmonic generation (MSHG), and spin-resolved photoemission, have provided extensive information and new insights into collective excitations and diverse phenomena, including magnons in magnetically ordered crystals as well as high-T_c superconductors, magnetic polarons and excitons in II–VI paramagnetic semiconductors (Sugano and Kojima, 2000).

MO spectroscopy can do even more when combined with femtosecond laser pulses, providing direct time-domain information about the magnetic properties of excited states with high temporal resolution, which are often hidden in static MO and transport measurements. Photoexcitation of a magnetic system with ultrashort laser pulses can strongly alter thermodynamic equilibrium among the constituent degrees of freedom (charge carrier, spin, and lattice), triggering a variety of dynamical processes. Subsequently, time-resolved MO techniques are capable of directly revealing the spin and charge fluctuations associated with these dynamic processes, instead of time-averaged mean field values obtained from those static measurements. The additional temporal dimension obtained from ultrafast MO spectroscopy provides a natural way to differentiate various correlation effects in the time domain. In addition, ultrafast photoexcitation opens up fascinating opportunities to control the magnetic properties of materials on an extremely fast time scale and may create new transient spin states that are inaccessible via thermal equilibrium transitions.

To achieve these goals, several types of ultrafast MO schemes have been developed, including time-resolved linear MO effects (MOKE in reflection [Wang et al., 2006] or Faraday effect in transmission [Crooker et al., 1996]), ultrafast spin-resolved photoemission spectroscopy (Scholl et al., 1997; Melnikov et al., 2003), and time-resolved magnetic SHG spectroscopy (see, e.g., Hohlfeld et al., 1997). Particularly, it is now recognized that time-resolved MOKE/Faraday spectroscopy is extremely relevant—if operated with proper care—for accessing genuine spin dynamics, even at femtosecond time scales (see, e.g., Zhang et al., 2002, 2009; Koopmans et al., 2003; Wang et al., 2006). These provide essential contributions to our understanding of fundamental transient magnetic phenomena, and have clear implications for future high-speed, multifunctional MO device technology.

13.1.3 Scope of the Chapter

It should be noted that research in ultrafast magnetism is still in a very vigorous phase, and many important experimental as well as theoretical questions remain open. Therefore, this chapter is not meant to summarize all the important experimental and theoretical progresses; it rather focuses on different implementations of ultrafast MOKE/Faraday spectroscopy techniques in condensed matter systems. This chapter is organized as follows: Section 13.2 first discusses in detail several linear MO effects, such as the MOKE, the Faraday effect, and magnetic circular dichroism (MCD), which will be used to determine the time-dependent magnetization later in this chapter. It is important for the reader to understand the general concepts as well as intricacies involved in time-resolved MO measurements, which are subsequently presented in this section. Some other commonly used ultrafast MO spectroscopy techniques developed in the past decade are also briefly reviewed. Section 13.3 discusses the experimental layouts and detection schemes that allow us to obtain MO signals with high signal-to-noise ratio. The data acquisition and recovery methods using short laser pulses from either a MHz repetition rate laser oscillator or kHz repetition rate laser amplifier are described in detail. In addition, strategies for carrying out high-fidelity measurements—to access genuine spin dynamics at femtosecond time scales—are presented. In Section 13.4, ultrafast spin dynamics and transient magnetic phenomena are discussed in various nonmagnetic and magnetically ordered condensed matter systems, which are roughly divided into four categories: (1) nonmagnetic semiconductors, (2) ferromagnetic metals, (3) ferromagnetic semiconductors, and (4) antiferromagnetic insulators. Finally, conclusions and a future outlook are given in Section 13.5.

13.2 GENERAL CONSIDERATIONS IN ULTRAFAST MAGNETO-OPTICAL MEASUREMENTS

The basic concepts and intricacies of time-resolved MO spectroscopy are presented, after a brief summary of the linear MO response.

13.2.1 Linear Magneto-Optics

13.2.1.1 Local Dielectric Response

The fundamental equation for describing the propagation of electromagnetic waves in condensed media, in the absence of external charge or current, is given by

$$\nabla \times \nabla \times \vec{E} = -\frac{1}{c^2} \frac{\partial^2 \vec{D}}{\partial^2 t}, \tag{13.1}$$

where the electrical displacement $\vec{D}$ is related to the electric field $\vec{E}$ and polarization $\vec{P}$ by the constitutive relation $\vec{D} = \vec{E} + 4\pi\vec{P}$. Here $\vec{E}, \vec{D}, \vec{P}$ are macroscopic variables defined over volumes with dimensions "small" as compared to the wavelength, and "large" as compared to the atomic dimensions. It is often useful to consider the local response to a monochromatic excitation ω. Thus we obtain (Chapter 1)

$$\vec{P}(\omega) = \chi(\omega)\vec{E}(\omega), \tag{13.2}$$

where $\chi(\omega) = \int_{-\infty}^{\infty} dt \chi(t) e^{-i\omega t}$, and, hence,

$$\vec{D}(\omega) = [1 + 4\pi\chi(\omega)], \quad \vec{E}(\omega) = \varepsilon(\omega)\vec{E}(\omega). \tag{13.3}$$

This defines the dielectric constant ε, which fully describes the linear optical response of the medium (Chapter 1). Note that since $\vec{E}, \vec{D}, \vec{P}$ are polar vectors, ε is a second-rank polar tensor in general.

In the presence of a uniform magnetization **M** along the z-axis, the local dielectric tensor ε of an isotropic medium can be expressed as (Chapter 3)

$$\begin{pmatrix} \varepsilon_{xx} & \varepsilon_{xy} & 0 \\ -\varepsilon_{yx} & \varepsilon_{xx} & 0 \\ 0 & 0 & \varepsilon_{xx} \end{pmatrix}, \tag{13.4}$$

where the complex elements ε_{ij} are $\varepsilon_{ij} = \varepsilon'_{ij} + i\varepsilon''_{ij}$. The diagonal elements transform symmetrically under time reversal, that is, there is no sign change if the magnetization changes to the opposite direction. This describes normal linear optical behavior, which is associated with the normal complex refractive index $n = n_1 + in_2$, where n_1 and n_2 are real and imaginary parts. The magnetization breaks the time-reversal symmetry of the crystal, resulting in nonzero off-diagonal components of the dielectric tensor (Equation 13.4). These off-diagonal components transform antisymmetrically upon a magnetization flip, which accounts for MO properties. Neglecting higher order magnetization terms, the off-diagonal elements are proportional to the magnetization to the first order, that is, $\varepsilon_{xy}(M, \omega) = -\varepsilon_{yx}(-M, \omega) \propto M$. Then we can define the MO constant $\tilde{Q}$ as follows:

$$\tilde{Q} = q' + iq'' = -i\frac{\varepsilon_{xy}}{\varepsilon_{xx}}. \tag{13.5}$$

The appearance of these off-diagonal elements implies different microscopic responses to right (RCP) and left circularly polarized (LCP) light, two normal modes of the dielectric matrix (Equation 13.4). Consequently, the two circularly polarized light waves will propagate with different dielectric constants, ε_+ and ε_-, and hence different complex refractive indices n_+, k_+ and n_-, k_-. These quantities relate to the elements of the dielectric tensor matrix Equation 13.4 via

$$\varepsilon_\pm = \varepsilon_{xx} \pm i\varepsilon_{xy} = (n_\pm + ik_\pm)^2. \tag{13.6}$$

The difference of ε_+ and ε_-, directly related to ε_{xy}, which leads to different phase velocities and/or absorption for the two normal modes, is the origin of polarization rotation and ellipticity, respectively. When a linearly polarized light beam is reflected from the sample surface, Kerr rotation is observed as a rotation of the polarization plane of the light. The MOKE ellipticity is observed when linearly polarized light becomes elliptically polarized, which is also often referred to as MCD, as discussed in further detail in Sugano and Kojima (2000) and Zvezdin and Kotov (1997).

Generally, we can define a complex MOKE angle, called the Voigt vector,

$$\tilde{\Theta}_k = \theta_k + i\eta_k, \tag{13.7}$$

whose real part θ_k and imaginary part η_k correspond to the Kerr rotation and ellipticity, respectively. The Faraday rotation θ_F and ellipticity η_F can also be described accordingly. These quantities directly relate to the optical and MO properties of a magnetic sample, that is, the diagonal and off-diagonal elements of the dielectric tensor (13.4).

Next we consider linearly polarized light with E field $[E_x \ 0 \ 0]$ reflected from the sample surface. Here, we assume that the light propagation is in the z-direction, that is, parallel to the sample normal and the magnetization, called the polar MOKE geometry. The complex reflection coefficients for right- and left-circularly polarized light are r_+ and r_-, which can be calculated using the Fresnel equations (Chapter 1) as follows:

$$r_+ = r_x + ir_y = \frac{(n_+ + ik_+) - 1}{(n_+ + ik_+) + 1}, \tag{13.8}$$

$$r_- = r_x - ir_y = \frac{(n_- + ik_-) - 1}{(n_- + ik_-) + 1}. \tag{13.9}$$

The reflected light will be given by $[r_x E_x \ \ r_y E_x \ \ 0]$, and then the complex MOKE angle can be expressed as

$$\tilde{\Theta}_k = \theta_k + i\eta_k = \frac{r_x E_x}{r_y E_x}. \tag{13.10}$$

Using the definitions in Equations 13.4, 13.6, 13.8, 13.9, and 13.10, the MOKE angle and ellipticity can be expressed as

$$\theta_k = F_2(n_1, n_2)\varepsilon'_{xy} + F_1(n_1, n_2)\varepsilon''_{xy} \tag{13.11}$$

and

$$\eta_k = F_1(n_1, n_2)\varepsilon'_{xy} - F_2(n_1, n_2)\varepsilon''_{xy}, \tag{13.12}$$

respectively. Here the elements ε'_{xy} and ε''_{xy} are the real and imaginary components of ε_{xy}. F_1 and F_2 are prefactors as functions of n_1 and n_2:

$$F_1(n_1, n_2) = \frac{A}{A^2 + B^2}, \quad F_2(n_1, n_2) = \frac{B}{A^2 + B^2}, \tag{13.13}$$

where $A = n_1(n_1^2 - 3n_2^2 - 1)$ and $B = n_2(-n_2^2 + 3n_1^2 - 1)$.

For a transparent sample $n_1 \gg n_2$, the complex polar Kerr angle in Equations 13.11 and 13.12 can be simplified as

$$\tilde{\Theta}_k = \theta_k + i\eta_k = \frac{i\varepsilon_{xy}}{\sqrt{\varepsilon_{xx}(\varepsilon_{xx} - 1)}} = \frac{-\varepsilon''_{xy}}{n(n^2 - 1)} + i\frac{-\varepsilon'_{xy}}{n(n^2 - 1)}. \tag{13.14}$$

This formula indicates that the Kerr rotation and ellipticity are proportional to the sample magnetization **M**, according to Equation 13.5. By measuring the MOKE rotation angle or MOKE ellipticity (MCD), one can monitor magnetization changes in the sample under study. Here θ_k and η_k usually do not exceed one degree since the ratio between the diagonal and nondiagonal elements of the susceptibility tensor is typically of the order of 0.01.

For arbitrary incidence angles, the complex MOKE angle for linearly polarized light with electric field $[E_x \ \ 0 \ \ 0]$ can be expressed as

$$\tilde{\Theta}_k = \frac{\cos\theta_0}{\cos(\theta_1 + \theta_0)} \times \left[\frac{-i\varepsilon_{xy}}{\sqrt{\varepsilon_{xx}(\varepsilon_{xx} - 1)}}\right]. \tag{13.15}$$

Here θ_0 and θ_1 are the incidence angle and refractive angle, respectively (You and Shin, 1996). In this expression, it can be seen that the second factor inside the bracket is identical to Equation 13.14 for normal incidence.

There are two other MOKE geometries besides the polar MOKE geometry discussed earlier: when the magnetization is (a) parallel to the surface and in the plane of incidence or (b) parallel to the surface and perpendicular to the plane of incidence. Then MOKE is called (a) the longitudinal Kerr effect or (b) the transverse Kerr effect, respectively (see Zvezdin and Kotov, 1997). The signals from these effects are normally at least one order of magnitude smaller than those of the polar MOKE, and more details can be found, for example, in Zvezdin and Kotov (1997) and You and Shin (1996).

Finally we briefly mention without a detailed derivation that in transmission measurements of transparent samples ($n_1 \gg n_2$), the Faraday rotation and ellipticity (measured per unit length because of the transmission geometry) are given by (Zvezdin and Kotov, 1997)

$$\theta_F + i\eta_F = \frac{\omega}{2\pi}\frac{-\varepsilon''_{xy}}{n} + i\frac{\omega}{2\pi}\frac{-\varepsilon'_{xy}}{n}. \tag{13.16}$$

We can conclude that Faraday rotation and ellipticity are also proportional to the sample magnetization **M**, and correspond to the MOKE ellipticity and rotation, respectively, when one compares their dependence on the real and imaginary parts of ε_{xy} in Equation 13.14.

13.2.1.2 Microscopic Description

Next, we discuss the microscopic physics and models beneath the linear MO response. The microscopic interactions that can significantly contribute to ε_{xy} and hence the magnitudes and features of linear MO spectra are those that can cause spin-dependent energy splitting, such as spin–orbit or spin exchange interactions. These interactions remove the inherent degeneracy to multiple spin states of electronic levels in solids. For example, the spin–orbit interaction couples the electron spin and its charge motion, which poses a correlation between the electronic (orbital) and spin degrees of freedom of the electronic wave functions. This gives rise to different transition strengths for the photons with left and right angular momentum. Therefore, the MOKE/Faraday signals, being sensitive to the difference between the RCP and LCP light transitions, are directly related to the splitting of those involved energy levels. This indicates that the MO spectra provide not only information about the transition strength and joint density of states of the coupling levels, which can be obtained by normal optical measurements, but also about the spin–orbit coupling and spin polarization of states participating in the MO transitions. Consequently, (1) the MO spectra are sensitive to, for example, d states in the transition-metal ions and f states in rare earth ions, respectively, because of the strong spin–orbit coupling of these states; (2) the MO response shows the characteristics of the electronic states involved. For instance, this makes d states distinguishable from s or p states.

To substantiate the above conclusions, we give the following microscopic expressions for the off-diagonal elements of the dielectric tensor using the Kubo formula (Bennett and Stern, 1964):

$$\varepsilon'_{xy} = C \times \sum_{gn} \int_{BZ} dk^3 \frac{[f_{g,k}(1 - f_{n,k})]\omega_{gn,k}(\omega^2_{gn,k} - \omega^2)\left(\left|P^+_{gn}\right|^2 - \left|P^-_{gn}\right|^2\right)}{\omega^2[(\omega^2_{gn,k} - \omega^2)^2 + (2\omega\gamma_{gn,k})^2]}, \tag{13.17}$$

$$\varepsilon'_{xy} = C \times \sum_{gn} \int_{BZ} dk^3 \frac{2 \times [f_{g,k}(1 - f_{n,k})]\omega_{gn,k}\gamma_{gn,k}\left(\left|P^+_{gn}\right|^2 - \left|P^-_{gn}\right|^2\right)}{\omega[(\omega^2_{gn,k} - \omega^2)^2 + (2\omega\gamma_{gn,k})^2]}, \tag{13.18}$$

where the Fermi–Dirac distribution function is $f_{n,k} = 1/[1 + \exp((\hbar\omega_{gn,k} - E_F)/k_BT)]$. Here C is a constant independent of frequency, g and n are the index for the Bloch states—the wave functions of electrons placed in a periodic potential of a solid—which exhibit a momentum (k) dependent transition energy $\omega_{gn,k}$ and broadening $\gamma_{gn,k}$. The summation is over the entire Brillouin zone (BZ) and all the Bloch bands. The transition dipole matrix elements for RCP and LCP light are given by

$$\left|P^\pm_{gn}\right| = \left|\left\langle \psi_{g,k} \left| p_\pm \right| \psi_{n,k} \right\rangle\right|, \tag{13.19}$$

where the momentum operator $p_\pm = i\hbar(\nabla_x \pm i\nabla_y)$ couples the Bloch states with wave function $\psi_{n,k}(\psi_{g,k})$. As we can see here, the subtraction of transition matrix elements in Equations 13.17 and 13.18 are nonzero only if there is spin-dependent energy splitting of the states involved.

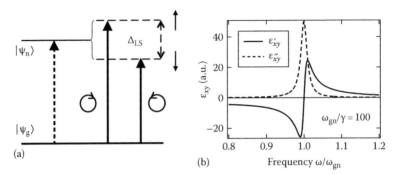

Figure 13.1 (a) Energy scheme for double transition with the excited state split by the spin–orbit interaction Δ_{LS}. MO transitions due to RCP and LCP light are shown. (b) ε'_{xy} and ε''_{xy} calculated for the ratio of transition energy and broadening $\omega_{gn,k}/\gamma_{gn,k} = 100$.

We conclude this section with an example. We consider an electronic state split by the spin–orbit interaction (Δ_{LS}). The allowed MO transitions due to RCP and LCP light are denoted by arrows, as shown in Figure 13.1a, with equal optical dipole transition matrix element p. Thus

$$\left| P_{gn}^{+} \right|^2 - \left| P_{gn}^{-} \right|^2 \propto p^2 \times \Delta_{LS}. \tag{13.20}$$

The real and imaginary parts of the off-diagonal matrix elements, which are plotted in Figure 13.1b, were derived using Equations 13.17 and 13.18. Note that ε'_{xy} and ε''_{xy} exhibit dispersive and dissipative line shapes, respectively.

13.2.2 Ultrafast MOKE/Faraday Spectroscopy

13.2.2.1 Transient MOKE/Faraday Signals: What Is Measured?

In this section, we will address the pump-induced MO signals, obtained in time-resolved MOKE/Faraday measurements. The basis of the static MO effect is the proportionality between the complex Voigt vector in Equation 13.7 and the static magnetization as shown in Equations 13.14 through 13.16. In the dynamic case, pumping a magnetic system with ultrashort laser pulses can drive the system strongly out of thermodynamic equilibrium, and subsequent probe pulses measure the time-dependent MO signals associated with the build-up or decay of the transient magnetic states. Therefore, we can write the transient MO signals phenomenologically by relating the complex Voigt vector to time-dependent magnetization and Fresnel coefficients associated with the transient states of materials:

$$\tilde{\Theta}(t) = N(t) + \sum_{i=x,y,z} \tilde{F}_i(t) M_i(t), \tag{13.21}$$

where $N(t)$ is the nonmagnetic contribution that is symmetric upon magnetization reversal, for example, from nonlinear optical effects such as state/band filling, pump-induced optical anisotropy and two-photon absorption. Such nonmagnetic contributions to the transient MO signals can be eliminated by measuring difference of the photoinduced MO signals under two opposite magnetization directions, that is, $\Delta\tilde{\Theta}(t) = 1/2[\tilde{\Theta}(M(t)) - \tilde{\Theta}(-M(t))]$. We will neglect $N(t)$ in the following discussion for simplicity. In addition, it is highly relevant for many experimental conditions to consider normal incidence, that is, polar MOKE where only M_z gives a nonzero contribution in Equation 13.21. Therefore we can simplify Equation 13.21 for the time-dependent MO signals to

$$\tilde{\Theta}(t) = \tilde{F}(t) \cdot M(t), \tag{13.22}$$

where $\tilde{F}(t)$ is an effective Fresnel coefficient under the polar MOKE condition. If the complex prefactor $\tilde{F}$ is mostly time independent, that is, $\tilde{F} = \tilde{F}_c$, we can further simplify Equation 13.21 as $\tilde{\Theta} = \tilde{F}_c \cdot M(t)$. Hence, the pump-induced change is given by $\Delta\tilde{\Theta} = \tilde{F}_c \cdot \Delta\vec{M}(t)$. Given these conditions, the following relation holds:

$$\frac{\Delta\tilde{\Theta}(t)}{\tilde{\Theta}_0} = \frac{\Delta M(t)}{M_0}, \tag{13.23}$$

which indicates that a proportionality exists between the normalized induced MO signal and the normalized magnetization change. Therefore the proportionality between the complex MOKE vector and magnetization holds, similar to the static case (Equations 13.14 through 13.16). This is the ideal case for induced MO signals, which indicates that the ultrafast MO measurements reveal the genuine time-dependent magnetization changes.

13.2.2.2 Ultrafast Magneto-Optics: Magnetism vs. Optics

The interpretation of the transient MO signal in terms of spin dynamics is not always trivial as discussed in the previous section, especially at femtosecond time scales and under high pump fluences. Generally speaking, since the pump creates a transient carrier population, leading to a redistribution of the oscillator strength of the coupling states seen by the probe, one may imagine that the induced MO signal is not necessarily related to the magnetization change. In such a case, the proportionality between the complex Voigt vector and magnetization, the basis of linear MO effects, is broken. In this case, the prefactor $\tilde{F}$ in Equation 13.22 is time-dependent, that is, $\tilde{F} = \tilde{F}(t)$, and hence Equation 13.23 does not hold, since the pump-induced change

$$\Delta\tilde{\Theta}(t) = \tilde{F}_0 \cdot \Delta M(t) + \Delta\tilde{F}(t) \cdot M_0, \tag{13.24}$$

and

$$\frac{\Delta\tilde{\Theta}}{\tilde{\Theta}_0} = \frac{\Delta M(t)}{M_0} + \frac{\Delta\tilde{F}(t)}{\tilde{F}_0}. \tag{13.25}$$

Therefore, the induced MO signals no longer reflect genuine spin dynamics.

Indeed, some doubts about femtosecond spin dynamics derived from time-resolved MO measurements were raised (Koopmans et al., 2000b). By comparing simultaneously detected Kerr rotation and ellipticity in Ni films, these authors found evidence that immediately after pump excitation, so-called dichroic bleaching dominated the MO response and thus no information about fast magnetization dynamics could be obtained from time-resolved MOKE data below 0.5 ps. This view was further supported by magnetic SHG measurements performed on Ni films (Regensburger et al., 2000). Moreover, in ferromagnetic GaMnAs under relatively strong excitation with a 3.1 eV pump (Kojima et al., 2003), the nonmagnetic contribution persists even at hundreds of picoseconds after photoexcitation. These results indicate that proper experimental design with controlled measurements should be implemented together

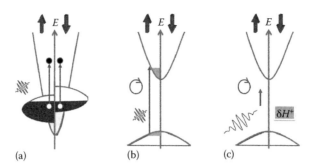

Figure 13.2 Schematics of ultrafast excitation. (a) Transient electronic heating in a ferromagnetic metal with spin-dependent band structures due to ferromagnetic exchange interaction between spin-up and spin-down bands. Note there is no spin selective excitation in this case. (b) Photoexcited spin-polarized carriers in a semiconductor, due to the selection rules for interband optical transitions. (c) Photomagnetic field pulse via coherent virtual excitation using largely below band gap pulses in an insulator.

with ultrafast MO measurements in order to prevent nonmagnetic contributions from contaminating the transient MO signals. Some criteria and strategies for these will be presented in Section 13.3.3.

13.2.2.3 Ultrafast Excitation Schemes
Selectively pumping a magnetic system can make spins "responsive," exhibiting desired dynamics and sensitivity at ultrafast time scales. The main ultrafast laser perturbation schemes of interest are illustrated in Figure 13.2: (1) transient electronic heating, (2) photodoping of spin-polarized carriers, and (3) virtual photomagnetic field pulse.

The first excitation scheme, transient electronic heating, illustrated in Figure 13.2a, plays a major role in describing ultrafast spin dynamics in ferromagnetic metals (Chapter 3). The laser energy is first deposited into the charge carriers via absorption of the pump pulses, followed by thermalization of the high-energy electron population (~1 eV of excess energy) with a characteristic time scale of tens of femtoseconds (Chapter 8). This quasithermal equilibrium electron distribution can be very hot, characterized by a transient electronic temperature much higher than that of the lattice. Moreover, because of the large Fermi sea of charge carriers, largely quenched orbital moments, and mixing of the optically coupled bands in metals, the effects from carrier spin or density changes are mostly negligible. Consequently, the main effect of photoexcitation is heating of the preexisting charge carrier population, which partially led to ultrafast demagnetization in ferromagnetic Ni (Beaurepaire et al., 1996), one of the pioneering experiments in the field.

Furthermore, this excitation scheme has been shown to trigger magnetization rotation followed by collective spin precession in many magnetically ordered crystals, including ferromagnets (Kampen et al., 2002; Qi et al., 2007; Wang et al., 2007a; Hashimoto et al., 2008) and antiferromagnets (Kimel et al., 2004). Transient electronic heating leads to heat transfer to the lattice, which thermally alters the magnetic anisotropy fields of the material. Such a phonon-assisted process can lead to an effective thermal magnetic field pulse $\vec{H}_{\mathrm{eff}}(t)$. The equation of motion for the total spin $\vec{S}$ is given by the Landau-Liftshitz-Gilbert equation (Kapetanakis et al., 2009):

$$\frac{\partial}{\partial t}\vec{S} = -\gamma\vec{S}\times\vec{H}_{\mathrm{eff}}(t) + \frac{\alpha}{S}\vec{S}\times\frac{\partial}{\partial t}\vec{S}, \tag{13.26}$$

which describes Larmor precession and damping related to the zero-momentum magnon, which represents the fundamental quantum of the collective spin precession (Chapter 2). Here γ is the gyromagnetic ratio $g\mu_0\mu_B/\hbar$ and α is the Gilbert coefficient.

In contrast to the first scheme, spin-polarized carriers can be created in nonmagnetic or magnetic semiconductors by a circularly polarized pump pulse close to the Brillouin zone center (Chapter 2), illustrated in Figure 13.2b. Such optical orientation is very efficient in semiconductors, thanks to the interband optical transition selection rules resulting from the spin–orbit coupling of the atomic-like states near the zone center. This spin-polarized excitation scheme has provided essential contributions to our understanding of spin relaxation, spin coherence effects, and dephasing phenomena in semiconductors and their quantum-confined structures, a critical stimulus for the emerging field of semiconductor spintronics (Awschalom et al., 2002).

The final perturbation scheme that is discussed in this chapter explores the coherent nature of photoexcitation as illustrated in Figure 13.2c. Such an effect is proposed in semiconductors where coherent virtual excitation and polarization dephasing have been shown to induce spin rotation (Chovan et al., 2006), and demonstrated in some antiferromagnetic (Kimel et al., 2005) and ferrimagnetic insulators (Hansteen et al., 2006) termed as the inverse Faraday effect. There high intensity, mostly nonresonant circularly polarized driving pulses have been used to induce an effective external magnetic field:

$$\delta H^{\pm} \propto \vec{E}(\omega) \times \vec{E}^{*}(\omega), \tag{13.27}$$

where

$\vec{E}(\omega)[\vec{E}^{*}(\omega)]$ is the electric field of the light wave (its complex conjugate)

$\pm$ indicates photoexcitation by RCP or LCP light to induce magnetic fields of opposite sign

The excitation pulses used normally have small photon energy below the band gap, and thereby the effect mostly exists only during the pulse duration (since there are no real carriers created), in contrast to the two prior schemes. For instance, at a high pump fluence of about $500\,mJ/cm^2$, the effect of a 200 fs laser pulse is shown to be equivalent to a magnetic field pulse of 5 T on a magnetic system with strong spin–orbit coupling, such as rare-earth orthoferrite $DyFeO_3$ (Kimel et al., 2005). The microscopic origin of the photoinduced effective magnetic field is from the angular momentum of the pump photons and strong spin–orbit coupling of the materials.

Applying these three excitation schemes to magnetic and nonmagnetic materials opens up opportunities to study an array of fascinating transient magnetic phenomena complementary to each other, as discussed in Section 13.4.

13.2.2.4 Probe Photon Energy

There are at least three reasons to carefully choose the probe energy in ultrafast MO measurements:

First, according to Equations 13.14 through 13.18, the MO spectra reflect the real and imaginary parts of off-diagonal and diagonal tensor elements, which could be highly structured because of multiple electronic transitions involved. Note that in Equations 13.17 and 13.18 ε'_{xy} and ε''_{xy} are proportional to the oscillator strength and spin–orbit coupling, which will be enhanced at some particular spectral regions that are noticeable in the static MOKE/Faraday spectra. Carefully tuning the probe photon energy to match those resonant enhancements in the MO spectra will significantly improve the signal-to-noise ratio in the ultrafast MO measurements. For example, in the case of ferromagnetic semiconductors such as GaMnAs and InMnAs, the MOKE spectra exhibit enhancement at ~1.5 eV due in part to the Mn d transitions, as discussed in the examples in Section 13.4.3. Some other examples are from II –VI and III–V semiconductors, which exhibit large MOKE/Faraday signals near their band gaps due to the excitonic enhancement (Crooker et al., 1996).

Second, the ability to access different spectral regions allows one to selectively probe those electronic transitions of interest. A specific example is charge-transfer-type excitations from the O_{2p} to Mn e_g and t_{2g} bands in the Kerr spectra of colossal magnetoresistive manganites (Sugano and Kojima, 2000).

Third, tuning the probe energy allows one to search for a photon energy range where the MO response is governed by pure magnetization dynamics. For example, a recent first-principle theoretical calculation of the coherent optical and magnetic response of ferromagnetic Ni to femtosecond laser excitation reveals that for photon energies smaller than 2 eV, the MOKE response is dominated by the magnetization dynamics, even during the pulse duration (Zhang et al., 2009). This provides an effective approach that circumvents the artifacts associated with the ultrafast MO measurements discussed in Section 13.2.2.2.

13.2.2.5 The Laser Source: kHz-Laser Amplifier vs. MHz-Laser Oscillator

Ultrafast MO spectroscopy setups are normally built using two categories of laser systems: (1) an oscillator producing a few nanojoules (nJ) of energy per pulse at a 30–70 MHz repetition rate that permits for high sensitivity, weak perturbation measurements over a relatively narrow spectral range centered in the visible and near-infrared (NIR)*; (2) a 1–250 kHz repetition rate amplifier producing μJ to mJ pulses that permits measurements of strongly excited samples with broadband tunability from the visible/NIR to the mid-infrared (MIR) and far-infrared (FIR). A detailed description of the physical principles and operation of these laser systems can be found in Chapter 7. Next we will focus on how to choose the right laser system, driven by the experimental considerations, material systems, and physics of interest.

The MHz-oscillator setup allows for high-sensitivity, high-speed measurements of the transient MO response as well as video rate MO spatial imaging of a given sample, especially with the use of multi-kHz (~50 kHz) frequency modulation using, for example, a photoelastic modulator (PEM) (Section 13.3.1.3). Such an experimental scheme is extremely suitable for measurements of single/few spins in nanostructures and for photoinduced spin dynamics in more "weakly correlated" systems, such as semiconductors and traditional ferromagnetic metals. However, the relatively weak pump fluence, short pulse-to-pulse separation (~ns), and relatively narrow frequency tuning range of the MHz setup (particularly in the MIR and FIR ranges) may pose some serious limitations when applied to study those "strongly correlated" systems as well as those magnetic phenomena associated with phase transitions and critical behaviors.

Compared to the widely used MHz setup, the kHz- and multi-kHz amplifier setups can solve some of the limitations discussed earlier. Such a setup, especially those using optimized mJ pulses at a few kHz repetition rate, can drive independently tunable optical parametric amplifiers (OPAs) that generate intense femtosecond pulses tunable from the FIR to the UV (Chapter 7). These amplified setups can: (1) strongly perturb the magnetic states or even induce magnetic phase transitions at sub-100 fs time scales; (2) resonantly drive low energy elementary excitations; (3) provide more strategies to eliminate the nonlinear optical effects that could contaminate ultrafast MO signals, for example, by carefully tuning the pump/probe energies; (4) substantially minimize or even eliminate electronic or lattice heating, using below-gap coherent virtual excitation. Therefore, these ultrafast MO spectroscopy setups are more suitable for investigating strongly driven transient magnetic phase transition in more "strongly correlated" electron systems. Some examples can be found in Tobey et al. (2008)

---

* Some high repetition rate lasers with broadband tuning capability, such as optical parametric oscillators, may extend the spectral range probed. However, these lasers produce much less pulse energy compared to that of kHz amplifiers and have not been widely used in ultrafast MO measurements.

and Kise et al. (2000). However, because of large pulse-to-pulse fluctuations and the low repetition rate that restricts the use of high frequency modulation, one challenge in developing such a kHz laser-based time-resolved MO setup (particularly at 1 kHz) is to have a suitable detection scheme with good signal-to-noise ratio. Some experimental implementations and schemes are described in detail in Section 13.3.

13.2.3 Other Commonly Used Ultrafast Spin-Resolved Techniques

Time-resolved MSHG spectroscopy is another powerful ultrafast MO spectroscopy technique (Hohlfeld et al., 1997; Zhang et al., 2002). In general, the nonlinear polarization $P(2\omega)$, responsible for generating a second-harmonic field when used as a source term in the wave equation (2.1), can be written as $P_i(2\omega) = \chi^2_{ijk} E_j E_k$, where χ^2_{ijk} is the second-order optical susceptibility tensor (Chapter 7). It can be decomposed into even (nonmagnetic) and odd (magnetic) components by differentiating the source for breaking the spatial inversion symmetry (Zhang et al., 2002). MSHG techniques allow one to separate these two contributions by carefully measuring second-harmonic intensities and phases for opposite magnetization directions I_+ and I_-, respectively. These can be understood by examining the following analytical expressions for I_+ and I_-, where φ is the relative phase between even and odd components:

$$I_+ \propto \left| \chi^{(2)}_{\text{even}} + \chi^{(2)}_{\text{odd}} e^{i\phi} \right|^2, \tag{13.28}$$

$$I_- \propto \left| \chi^{(2)}_{\text{even}} - \chi^{(2)}_{\text{odd}} e^{i\varphi} \right|^2. \tag{13.29}$$

A more detailed analysis of extracting the phases and reconstructing magnetization induced second-harmonic components can be found in Conrad et al. (2001). One should note that, in time-resolved MSHG, it is also not trivial to distinguish between ultrafast magnetization components and pure nonlinear optical effects, as discussed in Regensburger et al. (2000).

Ultrafast spin-resolved photoemission is another fast growing technique that has attracted some current attention (Scholl et al., 1997; Stamm et al., 2007). Static photoemission spectroscopy has been refined significantly during the past decade, driven in part to the intense research into high-T_c superconductivity. This has proven to be an extremely powerful probe for measuring the single-particle band structure and Fermi surface. The setup can be modified to include time resolution using a pump–probe configuration and spin-resolved capability with a spin analyzer to differentiate the spin states of the photoelectrons. Note that this technique is more relevant for measuring the transient spin polarization of the states near the Fermi surface, instead of the total net magnetization.

There exist some other highly relevant time-resolved techniques, including time-resolved x-ray MCD for recording transient changes of both spin and orbital angular momentum, spin-flip Raman scattering, and polarization-resolved photoluminescence. There are some practical limitations of these techniques such as temporal resolution and the involvement of large light sources, for example, synchrotron facilities. Finally, we should be aware that all these techniques can provide complementary information to the results obtained from ultrafast MOKE/Faraday measurements, but they exhibit their own intricacies and cannot replace the ultrafast MO techniques.

13.3 EXPERIMENTAL IMPLEMENTATION

In this section, we discuss in detail the experimental apparatus and data acquisition methods used in developing ultrafast MOKE/Faraday spectroscopy. It should be straightforward to implement these schemes experimentally. It will be also of great help when setting up an ultrafast MO experiment to reproduce some experimental results in commercially available samples that have been well studied using ultrafast MO techniques, for example, ferromagnetic metal Ni, GaAs, and ZnSe semiconductors and their heterostructures (Beaurepaire et al., 1996; Crooker et al, 1996; Kikkawa and Awschlom 1998). Some of the benchmark results in these systems are discussed in Sections 13.4.1 and 13.4.2.

13.3.1 Basic Techniques

13.3.1.1 Pump and Probe Setup

In ultrafast MOKE/Faraday spectroscopy, a pump–probe configuration (Chapter 9), combined with femtosecond laser pulses, is used to obtain temporal resolution, and a static magnetic field is desirable to control the initial magnetic state of the sample. A schematic of a time-resolved polar MOKE/Faraday experimental setup is depicted in Figure 13.3. An output ultrashort laser pulse tuned to an appropriate wavelength is used as the pump. A time-delayed, linearly polarized probe with energy typically a small fraction (<10%) of the pump energy passes though a mechanical delay line. A polarizer is used to set the probe polarization. A half- and quarter-waveplate can be used as an analyzer for rotation angle and ellipticity measurements, respectively. A quarter- or half-waveplate can also be used to control the polarization of the pump. The sample is mounted in a magnet or cryostat, which controls the temperature to as low as 1.2 K and the magnetic field as high as 10 T depending on the experimental requirements. The reflected (transmitted) probe passes though a polarization-sensitive element used as an analyzer. Changes in the polarization plane, either rotation or ellipticity, of the reflected (transmitted) probe pulse are detected by the optical polarization bridge or polarization modulation techniques, as discussed in the next sections. The transient

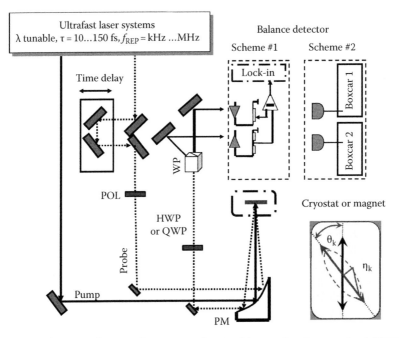

Figure 13.3 Schematic diagram for a typical optical layout for time-resolved MOKE spectroscopy. Both the complex polarization rotation and the ellipticity of the MOKE signals can be acquired (bottom right). Ultrafast Faraday signals can be obtained in a similar layout in transmission geometry. POL, polarizer; PM, parabolic mirror; HWP/QWP, half (quarter) wave plate: WP, Wollaston prism. Two balanced detection schemes are illustrated.

MOKE/Faraday signals are normally defined by measuring the photoinduced MO signals as $\Delta\tilde{\Theta}(t) = 1/2[\tilde{\Theta}(M(t)) - \tilde{\Theta}(-M(t))]$ to eliminate any nonmagnetic contributions, as in Equation 13.21. The pump and probe pulses can be widely tunable from the FIR to the UV using appropriate laser systems (Section 13.2.2.5). For example, the ultrafast spectroscopy setup can employ two NIR OPAs, pumped by a 1 kHz Ti:sapphire amplifier centered at 800 nm wavelength. This allows for independently tunable pump and probe photons from 0.5–20 μm (Tobey et al., 2008; Wang et al., 2010).

13.3.1.2 Optical Polarization Bridge

The simplest scheme for polarization analysis consists of two polarizers placed before and after the sample with roughly perpendicular optical axes. However, compared to an optical polarization bridge that can analyze the full polarization state of the light, it shows lower sensitivity and higher nonmagnetic background, which is particularly damaging for the kHz setup.

The polarization balance bridge represents a major scheme for full polarization analysis. The basic optical setup for such a scheme is illustrated in the detection part of Figure 13.3. A Wollaston prism serves as a polarization-dependent beam splitter, which splits the linearly polarized probe beam into two spatially separated components with polarization directions orthogonal to each other, each with ~50% of the original beam intensity. The two beams are sent to two identical photodiodes, which provide an input to preamplifiers. Then the signal is fed into a differential amplifier. A lock-in amplifier (scheme #1) measures the signal from the output of the differential amplifier or two boxcar integrators (scheme #2) record the two polarization states simultaneously.

We can calculate the differential signals using the Jones matrix method (Born and Wolf, 1999). In this notation, the input E field (s-polarized) and complex reflection from the sample are

$$\begin{pmatrix} E_s \\ E_p \end{pmatrix} = \begin{pmatrix} 1 \\ 0 \end{pmatrix}, \quad R = r_s \begin{pmatrix} 1 & -\tilde{\Theta} \\ \tilde{\Theta} & \lambda \end{pmatrix}, \quad (13.30)$$

respectively. Here $\lambda = r_p/r_s$ and $r_s(r_p)$ are the complex reflection coefficients. $\tilde{\Theta}$ is the complex Voigt vector for s-polarized input light, as defined in Equation 13.7. The MO signal, detected as the difference of the s- and p-polarized light in Figure 13.3, is calculated by

$$I_{\text{MOKE}} = I_p - I_s = \frac{|r_s|^2}{2} \left[\left| (1 \quad 0) \begin{pmatrix} 1 & 1 \\ 1 & -1 \end{pmatrix} \begin{pmatrix} 1 \\ \tilde{\Theta} \end{pmatrix} \right|^2 - \left| (0 \quad 1) \begin{pmatrix} 1 & 1 \\ 1 & -1 \end{pmatrix} \begin{pmatrix} 1 \\ \tilde{\Theta} \end{pmatrix} \right|^2 \right] = 2R\theta, \quad (13.31)$$

where we neglect second order terms in $\tilde{\Theta}$ and $R = |r_s|^2$ (for s-polarized input light). This allows for background-free detection of the MOKE rotation. The MOKE ellipticity can be obtained by replacing the half waveplate with a quarter waveplate that has its axis oriented at 45° with respect to the x-axis (the horizontal axis). Following a similar procedure as for Equation 13.31, the calculated MOKE ellipticity is

$$I_\eta = \frac{R}{4} \left[\left| (1 \quad 0) \begin{pmatrix} 1+i & 1-i \\ 1-i & 1+i \end{pmatrix} \begin{pmatrix} 1 \\ \tilde{\Theta} \end{pmatrix} \right|^2 - \left| (0 \quad 1) \begin{pmatrix} 1+i & 1-i \\ 1-i & 1+i \end{pmatrix} \begin{pmatrix} 1 \\ \tilde{\Theta} \end{pmatrix} \right|^2 \right] = 2R\eta. \quad (13.32)$$

From Equations 13.31 and 13.32, the measured time-dependent rotation $\Delta I_\theta(t)$ and ellipticity $\Delta I_\eta(t)$ are

$$\Delta I_\theta(t) = 2R_0\Delta\theta(t) + 2\Delta R(t)\theta_0, \tag{13.33}$$

$$\Delta I_\eta(t) = 2R_0\Delta\eta(t) + 2\Delta R(t)\eta_0. \tag{13.34}$$

In many realistic cases, the first terms in Equations 13.33 and 13.34 are much bigger than the second terms, therefore transient MOKE contrasts from these measurements directly probe magnetization dynamics. Finally, following the deviations given earlier, one can verify that, in the condition of $\alpha_{HWP} = 1/2 \arctan(\theta_0/2)$, the induced MOKE rotation change yields

$$\Delta I_\theta(t) = 2R_0\Delta\theta(t), \tag{13.35}$$

which gives an exact account of magnetization dynamics without any nonmagnetic contributions.

13.3.1.3 Polarization Modulation

The basic function of a PEM is to directly modulate the relative phase velocity of one optical axis to another at a frequency of ~50 kHz, allowing high precision light polarization measurements of various types in the MHz-oscillator setup (Koopmans et al., 2000a; Sugano et al., 2000). The basic working principle of a PEM is to use birefringence induced by stressing or compressing the material, causing different linearly polarized light beams to have slightly different speeds when passing through the material, to modulate the polarization of an incident light beam. A piezoelectric transducer is attached to the transparent sample and controlled by a driving circuit to provide phase modulation of light. An intuitive way for measuring the MCD (ellipticity) signal is by using quarter wave retardation. In one cycle of quarter wave modulation, the PEM modulates the incident linearly polarized beam (45° with respect to the optical axis of the PEM) from left circularly polarized light to right circularly polarized light. Then, sending the signal from the balanced detection setup to the lock-in for demodulation at the fundamental frequency ($1f$) measures the MCD signal from the sample.

In general, the lock-in signals at the fundamental ($1f$) and second harmonic ($2f$) of the modulation frequency are proportional to the induced ellipticity and rotation, respectively, and a typical configuration is shown in Figure 13.4. The setup consists of two polarizers placed before the PEM and after the sample. The first one sets the initial input polarization to be 45° with respect to the s and p polarizations and the other one is an analyzer set parallel

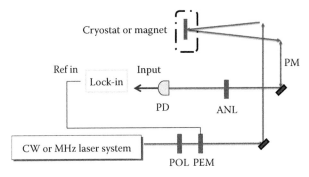

Figure 13.4 Schematic diagram for the polarization modulation scheme. PEM, photoelastic modulator; POL, polarizer; ANL, analyzer; PD, photodiode. PM, parabolic mirror.

to the horizontal p-axis. The following derivations apply for either oblique or normal incidence of the probe light, and the detected magnetization can be either out-of-plane or in-plane depending on the MOKE configuration, as discussed in Section 13.2.1.1. Following the Jones matrix analysis for a PEM whose optical axis is parallel to the s-axis,

$$P(t) = \begin{pmatrix} 1 & 0 \\ 0 & e^{iA(t')} \end{pmatrix}, \tag{13.36}$$

where $A(t') = A_0 \cos(\Omega_{PEM} t')$ is the PEM modulation at $\Omega_{PEM} \sim 50\,\text{kHz}$. Following similar deviations as those in Equations 13.31 and 13.32, one can get the DC and AC responses at $1f$ and $2f$ driving frequency:

$$I_{dc} = R/2, \tag{13.37}$$

$$I_{1f} = J_1(A_0)R\eta, \tag{13.38}$$

$$I_{2f} = J_2(A_0)R\theta, \tag{13.39}$$

which are direct measures of the optical reflectivity and MO responses. Here J_n is the nth order Bessel function.

13.3.2 Magneto-Optical Signal Recovery

Next we will discuss in detail some methods that allow one to extract the magnetization or other relevant physical quantities out of the raw data obtained.

13.3.2.1 Schemes Based on a kHz-laser Amplifier

In this setup, two boxcar integrators (Chapter 9) with two separate identical photodiodes are used with an optical layout similar to the second detection scheme illustrated in Figure 13.3. The detection scheme, following Wang et al. (2005a, 2007, 2008), is shown in Figure 13.5. It is suitable for the detection of short low-duty-cycle pulses at few kHz, which greatly reduces pulse-to-pulse fluctuations from the amplified system. It also has the unique advantage of extracting the polarization and reflectivity changes simultaneously from the same set of raw data (the latter as in a pump–probe measurement [Chapter 9]), via further computation with some algorithms shown in Figure 13.5.

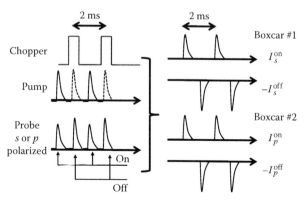

Figure 13.5 The principle of simultaneously detecting photoinduced MOKE rotation (ellipticity) signals and differential reflectivity using two boxcar integrators.

In order to implement this scheme, the pump is modulated by an optical chopper at a frequency of half the laser repetition rate, that is, every other pump pulse is blocked. We record the intensity of the reflected probe pulses, both the s- and p-components, as a function of time delay and magnetic field. Two boxcars, locked to the laser reference and synchronized with each laser shot, are used to simultaneously detect two different polarization states of the probe with and without pump for the adjacent pulses. As shown in Figure 13.5, after recording four sets of raw data at each time delay in one scan, I_s^{on}, I_s^{off}, I_p^{on}, and I_p^{off}, which are the s-polarized probe and p-polarized probe with pump and without pump, respectively, the Kerr rotation angle can be expressed in the following way, according to Equation 13.31:

$$\theta_k = \frac{I_p - I_s}{2(I_p + I_s)}.$$

(13.40)

The analytical expression for the photoinduced Kerr rotation is obtained by taking the first derivative of Equation 13.40 in the small perturbation limit, that is,

$$\Delta\theta_k(t) = \frac{(I_s^{off}\Delta I_p - I_p^{off}\Delta I_s)}{(I_s^{off} + I_p^{off})^2},$$

(13.41)

where $\Delta I_p = I_p^{on} - I_p^{off}$ and $\Delta I_s = I_s^{on} - I_s^{off}$. An analytical expression for the ellipticity change can be obtained in a similar way, after replacing I_s and I_p by two circularly polarized components I_+ and I_-, respectively. The differential reflectivity can be also derived from the raw data, that is,

$$\frac{\Delta R}{R} = \frac{(\Delta I_p + \Delta I_s)}{(I_s^{off} + I_p^{off})}.$$

(13.42)

With a typical kHz amplifier system, the smallest Kerr angle and differential reflectivity that can be measured are, respectively, 2×10^{-5} rad and 10^{-5}, recording 5000 pulses at each time delay.

13.3.2.2 Schemes Based on MHz-Laser Oscillator

Next we will discuss in detail a double-modulation detection scheme, following Koopmans, et al. (2000a) using a PEM for high-frequency polarization modulation and a mechanical chopper for low-frequency modulation. This scheme is particularly relevant for the MHz-oscillator setup, which allows one to simultaneously detect the pump-induced MOKE rotation and ellipticity, as well as the static MOKE rotation and ellipticity (Figure 13.6).

The double modulation is implemented using a mechanical chopper for the pump with $f_M \sim 100$ Hz and a PEM for the probe $\Omega_{PEM} \sim 50$ kHz, respectively ($\Omega_{PEM} \gg f_M$). Two lock-in amplifiers in series, which are fed by the reflected or transmitted probe light, are locked to the two reference frequencies. The analog output from the first lock-in amplifier, referenced to the PEM frequency, is related to static reflection, MOKE ellipticity and rotation for the DC, Ω_{PEM}, and $2\Omega_{PEM}$ components, respectively, according to Equations 13.37 through 13.39. The analog output of the first lock-in is fed to the input of the second lock-in, which is referenced to f_M. When the time constant τ_{lc} of the first lock-in amplifier is set as $1/f_M \gg \tau_{lc} \gg 1/\Omega_{PEM}$,

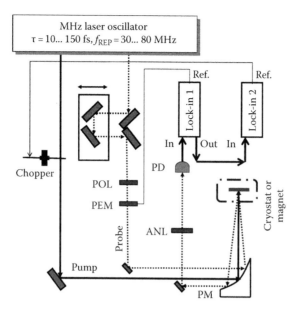

Figure 13.6 Schematic diagram for the double modulation scheme relevant for MHz oscillator. PEM, photoelastic modulator; POL, polarizer; ANL, analyzer; PD, photodiode; PM, parabolic mirror.

the pump-induced change in MOKE rotation and ellipticity can be measured by the second lock-in amplifier and is given by

$$\frac{\Delta I_{1F}(t)}{I_{dc}} = 2J_1(A_0)\Delta\eta(t), \tag{13.43}$$

$$\frac{\Delta I_{2F}(t)}{I_{dc}} = 2J_1(A_0)\Delta\theta(t), \tag{13.44}$$

where A_0 is as defined in Section 13.3.1.3. In deriving Equations 13.43 and 13.44, we assume $\Delta\tilde{\Theta}(t)/\tilde{\Theta}_0 \gg \Delta R(t)/R_0$ which normally holds in practice. We shall discuss more of this assumption next.

13.3.3 High-Fidelity Measurements: Access to Genuine Spin Dynamics

In this section, the criteria for properly analyzing the obtained transient MO signals and the strategies for establishing a correlation with pure magnetization dynamics are given.

13.3.3.1 Dichroic Bleaching and Nondegenerate Pump/Probe

One of the main issues in ultrafast MO techniques is the existence of nonlinear optical effects that can contaminate transient MO signals at fast time scales. For example, dichroic bleaching (Koopmans et al., 2007) is commonly seen in the degenerate pump–probe geometry with similar pump and probe photon energies, used by many prior experiments (Hohlfeld et al., 1997; Scholl et al., 1997; Koopmans et al., 2000; Regensburger et al., 2000; Melnikov et al., 2003; Bigot et al., 2009). Generally, the dichroic bleaching can be understood as follows. Upon photoexcitation the pump creates a transient carrier population that leads to a redistribution of the oscillator strength of the coupling states seen by the probe. While the whole

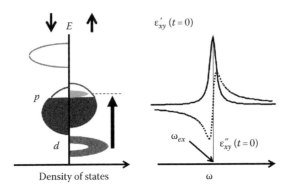

Figure 13.7 (a) Schematic of spin-dependent electronic states of a model ferromagnetic system exhibiting strong p–d exchange coupling. The preferred photoexcitation is in the majority spin band (right side). (b) The transient MO response, probed near excitation photon energy ω_p at zero time delay $t = 0$, is determined by ε'_{xy} and ε''_{xy} as depicted, which exhibits significant contrast from nonmagnetic contributions such as transient dichroic bleaching.

photoexcitation process may conserve the total spin, there could still be substantial induced MO signals that are not necessarily related to the magnetization change. In such a case, the proportionality between the complex Voigt vector and magnetization, the basis of linear MO effects, is broken down. Consequently, it takes some time (tens of femtoseconds or more) for the temporal profile of transient MO signals to reflect genuine magnetization dynamics (Koopmans et al., 2000).

One intuitive example is illustrated in Figure 13.7a for a ferromagnet with p–d exchange interactions, where localized spins (d-like bands) are coupled to carrier spins (p-like bands). A model ferromagnetic system falling into this category is III–V ferromagnetic semiconductors such as (Ga,Mn)As (Ohno, 1998). The pump laser excitation energy ω_p can be chosen to excite carriers *mostly* in the majority spin band from the d states to the top of the p states, because of the energy difference of the spin-split bands as illustrated in Figure 13.7. At early pump–probe delays near $t = 0$ fs, there are characteristic changes in the off-diagonal components of the dielectric tensor, illustrated in Figure 13.7b. This originates from Equations 13.17 and 13.18 because of the perturbation in the Fermi–Dirac distribution function. Such dichroic bleaching could lead to substantial MO signals around ω_p, as seen in Equations 13.17 and 13.18, which are seen by probe photons with similar energy. This intuitive example clearly shows how dichroic bleaching can break the proportionality between the complex MO vector and magnetism shown in Equation 13.25. In addition, the dichroic bleaching can also influence transient MO signals in nonmagnetic semiconductors, but much less considered there due in part to the fact that it is the photoexcited, spin-polarized carriers measured rather than the macroscopic magnetization (or spin-split bands).

The nondegenerate (or two-color) pump–probe geometry using unequal photon energies could provide a much-improved geometry for ultrafast MO measurements. In the past few years, this spectroscopy technique has been shown to be extremely productive in probing genuine femtosecond spin dynamics in magnetic materials. In the two-color method of transient MOKE spectroscopy, one carefully detunes the probe to couple to energy levels *far away* from the quasi-Fermi level of optical excitation, which can very effectively avoid nonlinear optical effects. For instance, in the case of dichroic bleaching (Figure 13.7), the redistribution of the oscillator strength of the coupling states is substantially reduced if the probe is moved higher than ω_p. For instance, some two-color experimental geometries, excitation schemes, and powers suitable for studying femtosecond spin dynamics in semiconductors are documented in a series of publications (Wang et al., 2005a, 2007, 2008a, 2009).

13.3.3.2 Rotation vs. Ellipticity

A particularly important practical criterion for attributing the transient MO signals to genuine magnetization dynamics is that the relative variation of Kerr rotation and ellipticity should have similar temporal profiles, as discussed in the following. Separating the real and imaginary parts of the complex MO response $\Delta\tilde{\Theta}$ in Equation 13.25 using the definition (13.7), we have

$$\frac{\Delta\theta}{\theta_0} = \frac{\Delta M(t)}{M_0} + \frac{\Delta F_1(t)}{F_{10}}, \tag{13.45}$$

$$\frac{\Delta\eta}{\eta_0} = \frac{\Delta M(t)}{M_0} + \frac{\Delta F_2(t)}{F_{20}}, \tag{13.46}$$

where $\tilde{F}(t) = F_1(t) + iF_2(t)$ to include the time dependence of the effective Fresnel coefficients. From these equations, we obtain the following practically applicable equation:

$$\frac{\Delta\tilde{\Theta}(t)}{\tilde{\Theta}_0} = \frac{\Delta M(t)}{M_0} \Leftrightarrow \frac{\Delta\theta(t)}{\theta_0} = \frac{\Delta\eta(t)}{\eta_0}. \tag{13.47}$$

This relation indicates that transient MO responses reflect the genuine magnetization dynamics if there is no discrepancy between the temporal profiles of the rotation angle and ellipticity, and vice versa (Koopmans et al., 2000; Wang et al., 2009). Indeed, in most prior experiments, this equation has been used to assign genuine magnetization dynamics in time-resolved MO measurement. For example, by comparing the simultaneously detected Kerr rotation and ellipticity in Ni films (Koopmans et al., 2002), evidence was found that immediately after pump excitation, Equation 13.47 did not hold and dichroic bleaching dominated the MO response. Thus, no information about fast magnetization dynamics could be obtained from time-resolved MOKE data below 0.5 ps. Only after that, the temporal profiles of Kerr angle and ellipticity began to overlap, reflecting the genuine magnetization dynamics.

We have the following two additional comments: (1) please note although Equation 13.47 is extremely useful and practically applicable in many cases, it may break down in some rare cases, for example, $\Delta F_1(t)/F_{10} = \Delta F_2(t)/F_{20} \gg \Delta M(t)/M_0$; (2) strictly speaking, the overlapping of Kerr angle and ellipticity in Equation 13.47 should be replaced by the real and imaginary parts of the MO constants defined by Equation 13.5, that is, $\Delta q''(t)/q_0' = \Delta q''(t)/q_0'$. We shall discuss some details next.

13.3.3.3 Off-Diagonal vs. Diagonal Elements in the Dielectric Tensor

Carefully separating the dynamics of the diagonal and off-diagonal elements of the time-dependent dielectric tensor will allow one to distinguish between the contributions of spin and charge dynamics in transient MO signals. This can be achieved, for example, by simultaneously measuring the differential transmission $\Delta T/T_0$, the differential reflection $\Delta R/R_0$, and the relative change in the real and imaginary components of $\Delta\tilde{\Theta}/\tilde{\Theta}_0$, following the procedures described in Guidoni et al. (2002 and references therein) and Sections 13.3.1 and 13.3.2 (as well as Chapter 9). These measurements provide essential insight into the exact nature of time-resolved MO signals and their connection to the genuine magnetization. Therefore, the following relation for the complex MO constants defined in Equation 13.5 can be used as another criterion for assigning genuine magnetization dynamics in ultrafast MO measurements:

$$\frac{\Delta q'(t)}{q_0'} = \frac{\Delta q_0''(t)}{q_0''} \gg \frac{\Delta\varepsilon_{xx}''(t)}{\varepsilon_{xx}''}, \quad \frac{\Delta\varepsilon_{xx}'(t)}{\varepsilon_{xx}'}. \tag{13.48}$$

This indicates that the relative variation of the nondiagonal elements is much larger than that of the diagonal elements of the dielectric tensor. Therefore, we can propose the following practical relationship to justify the magnetic origin of transient MO signals:

$$\frac{\Delta\theta(t)}{\theta_0} = \frac{\Delta\eta(t)}{\eta_0} \gg \frac{\Delta R(t)}{R_0}, \quad \frac{\Delta T(t)}{T_0}. \tag{13.49}$$

13.3.4 More Strategies

Taking advantage of the different dependence on external parameters for magnetic and non-magnetic contributions to the transient MO response may help identify genuine magnetic signals. For example, magnetic phases are usually sensitive functions of the temperature and magnetic field, especially near critical points, while the nonlinear optical responses may only have a relatively weak temperature dependence, for example, in the case of ferromagnetic metals such as Ni. This may prove to be a practical way for separating magnetic and non-magnetic contributions in time-resolved MO measurements. For example, a scheme exploiting the different temperature dependences of magnetic and nonmagnetic contributions to the transient MO signals has been demonstrated in studying ultrafast demagnetization dynamics in ferromagnetic Ni (Koopmans et al., 2002).

In addition, the recent development of first principles theoretical simulations has shown that the correspondence between magnetic and nonmagnetic contributions in the photoinduced femtosecond MO response sensitively depends on the incident photon energy (Zhang et al., 2009). It is established that the MO response in nickel reflects the magnetic response for probe photon energies below 2 eV, but not above 2 eV. Therefore, the probe photon energy dependence, together with complementary theoretical simulations, constitutes an important approach for future femtosecond magnetism research.

13.4 ULTRAFAST MAGNETO-OPTICAL KERR/FARADAY SPECTROSCOPY STUDIES OF CONDENSED MATTER SYSTEMS

This section focuses on different implementations of the ultrafast MO spectroscopy schemes discussed earlier in condensed matter physics and materials science.

13.4.1 Nonmagnetic Semiconductors

Recently, there has been much interest in spin-related phenomena in semiconductors. This is partially due to the expectation that the integration of "spintronic" devices with conventional semiconductor devices may be able to integrate information storage and processing capabilities into a single device (Awschalom et al., 2002). In addition to its technological importance, spin coherence and spin transport in semiconductors as well as their nanostructures are also of great interest to fundamental scientific research (Chapter 2).

In these materials, the ground states are not magnetically ordered without applying a magnetic field. However, thanks to the interband optical transition selection rules, spin-polarized carriers can be created in semiconductors by a circularly polarized ultrafast pump pulse (Chapter 2), that is, the second excitation scheme illustrated in Figure 13.2b. Dynamic evolution of the spin polarization can then be detected by a delayed probe pulse via transient MOKE/Faraday spectroscopy. This approach has provided fundamental insight into spin relaxation, spin coherent effects, and dephasing phenomena in semiconductors and their nanostructures. For example, two of the striking early observations made using time-resolved

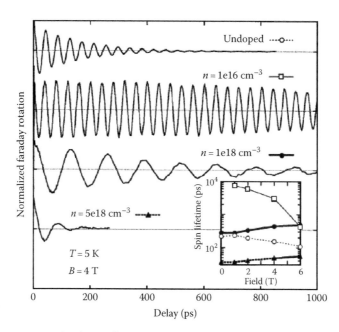

Figure 13.8 Time-resolved Faraday rotation spectroscopy for undoped and *n*-type GaAs under a magnetic field of 4T. Significant variation of the spin dephasing time as function of doping density is clearly shown in the time domain. The extracted spin dephasing times are summarized in the inset. Importantly, very light *n*-doping can increase spin lifetimes more than two order of magnitudes at zero field compared to other conditions. (Reprinted from Kikkawa, J.M. and Awschlom, D.D., *Phys. Rev. Lett.*, 80, 4313, 1998. With permission. Copyright 1998 by the American Physical Society.)

MO spectroscopy are: (1) the electronic spin polarization memory can be stored in lightly *n*-doped semiconductors for nanoseconds, even longer than the electron–hole recombination lifetime (Kikkawa and Awschlom, 1998); (2) spin coherence can be both stored and transported at room temperature over macroscopic lengths of ~100 μm (Kikkawa and Awschlom, 1999). These seminal experiments represented a critical stimulus for the emergence of semiconductor spintronics (Wolf et al., 2001).

One example is reproduced in Figure 13.8 (Kikkawa and Awschlom, 1998) for time-resolved Faraday rotation measurements in the *n*-doped semiconductor GaAs. The experiment is done in the Voigt geometry, where the magnetic field is applied parallel to the sample surface. A net electron spin polarization leads to rotation of the polarization plane of the transmitted probe beam, measured by the polarization bridge technique described previously. Significant oscillations in the temporal profile of the Faraday rotation under a magnetic field clearly show the magnetic origin of the transient MO response. It can be assigned to the Larmor precession of electron spins excited by a normally incident, circularly polarized pump pulse, as shown in Figure 13.2b. Most intriguingly, the oscillatory temporal evolution persists more than 10ns at a certain doping density, even longer than the electron–hole recombination time (~1ns). This indicates that controlled doping of the electron gas is primarily eliminating holes from the system through interband recombination, hence terminating the highly efficient electron–hole spin scattering process. It should be noted that time-resolved Faraday/Kerr spectroscopy is essential for monitoring this process, as the phenomenon occurs for electrons above the Fermi energy, which is hidden in other measurements of spin relaxation, such as time-resolved PL, which probes electrons near the Brillouin zone center.

More recent experiments have shown that manipulation of electron spin coherence can be pushed to the femtosecond time scale in two-dimensional (2D) semiconductor heterostructures

(Gupta et al., 2001) and to the few or single spin level in zero-dimensional semiconductor quantum dots (Hanson et al., 2008). It should also be pointed out here that ultrafast manipulation of spin degrees of freedom using time-resolved MO techniques is essential to achieve fast (above terahertz speeds) semiconductor spintronics, which should be crucial for the development of quantum information and communication using the spin degree of freedom in semiconductors. Finally, recent experimental advances in spin dynamics of semiconductors have also brought up a wealth of emerging concepts, for example, the existence of spin Coulomb drag (Weber et al., 2005) and a persistent spin helix current in a 2D electron gas (Koralek et al., 2009).

13.4.2 Ferromagnetic Metals

Ultrashort laser-excited ferromagnetic metals exhibit intriguing magnetization dynamics, that is, ultrafast demagnetization in Ni (Beaurepaire et al., 1996) and ultrafast magnetization precession by thermally induced reorientation of the magnetic easy axis (Kampen et al., 2002). In particular, the discovery of ultrafast demagnetization at subpicosecond time scales has attracted intense attention in the community in the past decade and is considered to be a seminal experiment in the research field of ultrafast magnetization dynamics. In addition to basic scientific interest, femtosecond demagnetization led to an ultrafast scheme for MO data writing at speeds much faster than those of the traditional thermal-magnetic processes used in modern MO technology.

Thus far, a large number of ultrafast MO studies have been performed to investigate exactly how a laser pulse can effectively change the magnetic moment at femtosecond time scales. These studies in metals have provided essential insight on the microscopic channels of energy and angular momentum transfer for photoexcited collective spins in highly nonequilibrium states, one of the most outstanding issues in modern condensed matter and materials physics. In this section, we will briefly review some of the main ultrafast MO studies performed on metallic magnets and discuss proper interpretation of the transient MO signals.

Figure 13.9a shows simplified physical scenarios for the ultrafast demagnetization of metallic magnets, shown together with that of insulating systems for comparison (Figure 13.9b). Because of the large Fermi sea and strong quenching of orbital moments in $3d$ metals, the main effect of ultrafast excitation in these systems has been electronic heating, that is, the first excitation scheme in Figure 13.2a. The established MO spectroscopy results have shown that after femtosecond laser excitation of itinerant ferromagnets, the pump-induced magnetization decrease mostly follows the thermalization of the hot electron population with a characteristic time scale of 100 fs (Figure 13.9a). On the other hand, in insulating magnets, a much slower demagnetization process is observed with a characteristic time scale of hundreds of picoseconds (Figure 13.9b).

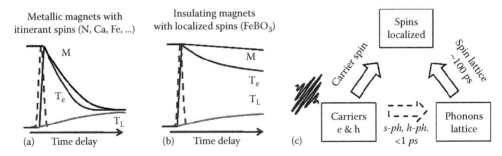

Figure 13.9 A schematic overview of the demagnetization dynamics in metallic (a) and insulating (b) magnets, respectively. (c) Three reservoirs (carriers, spins, lattice) taking part in the light-induced demagnetization process.

The microscopic mechanism underlying ultrafast demagnetization involves three reservoirs illustrated in Figure 13.9c, which we denote as carriers, spins, and lattice. Energy and angular momentum can be exchanged between them: the mechanisms and time scales for the carrier–lattice and spin–lattice couplings are quite well understood (especially the former, Chapter 8). In the standard process of laser-induced demagnetization, as used in MO recording (Sugano et al., 2002), the light pulse excites carriers, which then equilibrate with the lattice on a time scale of picoseconds. The heat deposited in the lattice is then transferred into the spin system through the spin–lattice interaction. Heating the spins leads to an increase in spin fluctuations and results in demagnetization, occurring on the time scale of the spin–lattice relaxation time τ_{sl}, which is on the order of 100 ps (Wang et al., 2005a). However, in order to explain the fast, femtosecond demagnetization in metals, the existence of direct coupling between the carrier and spin reservoirs has to be invoked and microscopic interactions beyond phonon-mediated ones have to be involved, that is, the spin–orbit and spin exchange interaction. In contrast, for insulating magnets, the spin–lattice mechanism is believed to underlie the relatively slow, picosecond demagnetization process seen there. Because of the localized character of the magnetic moments and absence of intrinsic doping carriers in insulating magnets, the primary effect of a femtosecond pump pulse is lattice heating, which is followed by heating of the collective spin system via magnon generation due to the spin–lattice interaction (Kimel et al., 2002).

Ultrafast demagnetization at femtosecond time scales in a metallic magnet was first reported by Bigot's group in 1996 (Beaurepaire et al., 1996). Using ultrafast MOKE to study 22 nm thick Ni films, the authors observed an ultrafast (within 2 ps) reduction in MOKE signals with no applied magnetic field, making this a measure of time-dependent remanent magnetization, which was interpreted as photoinduced demagnetization (left panel, Figure 13.10; Beaurepaire et al., 1996). In a more extreme case, intense laser pulses were shown to increase the electron and spin temperature even above the Curie temperature, driving a ferromagnetic to paramagnetic phase transition on a femtosecond time scale. The right panel of Figure 13.10 shows an example of complete quenching of ferromagnetism induced by laser pulses; it occurred within 500 fs in CoPt$_3$ (Beaurepaire et al., 1998). Finally, it should be mentioned that, in the case of

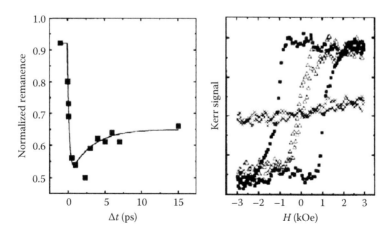

Figure 13.10 Femtosecond demagnetization in ferromagnetic metals Ni and CoPt$_3$, respectively. Left panel: the normalized remnant magnetization for a Ni film decreased within the first picosecond after photoexcitation. Right panel: in CoPt$_3$, ferromagnetic hysteresis loops at different time delays (squares: no pump; triangles: −1 ps; cross: 630 fs) show that a ferromagnet was converted to its paramagnetic state after ~500 fs. (Reprinted from Beaurepaire, E. et al., *Phys. Rev. Lett.*, 76, 4250, 1996; Beaurepaire, E. et al., *Phys. Rev. B*, 58, 12134, 1998. With permission. Copyright 1998 by the American Physical Society.)

complete quenching of the magnetic order occurs, the interpretation is rather convincing since the material switches to its opposite magnetic state when it is kept in a static magnetic field with an amplitude slightly less than the coercive field, for example, the magnetic loop at $-1\,ps$ exhibits the same amplitude but zero coercivity as compared to that of the 630 fs trace. The coercive field here defines the strength of the applied magnetic field required to reduce the magnetization of $CoPt_3$ to zero when sweeping the magnetic field back from saturation. Therefore, in the ferromagnetic metal $CoPt_3$, the evidence from ultrafast MOKE measurements indicates that the intrinsic magnetization can indeed be switched off on a femtosecond time scale.

In the initial investigations on demagnetization in nickel, a phenomenological three-temperature model was used to fit the experimental data (Beaurepaire et al., 1996). This is an extension of the two-temperature model that describes the thermal relaxation of electrons in metals (Chapter 8). As illustrated in Figure 13.9c, the three reservoirs were described by respective temperatures, and the demagnetization resulted from the changes in spin temperature. The equations of the heat flow were of the simple form:

$$C_e(T_e)\frac{dT_e}{dt} = -\gamma_{el}(T_e - T_l) - \gamma_{es}(T_e - T_s) + p, \tag{13.50}$$

$$C_s(T_s)\cdot\frac{dT_s}{dt} = -\gamma_{se}(T_s - T_e) - \gamma_{sl}(T_s - T_l), \tag{13.51}$$

$$C_l(T_l)\cdot\frac{dT_l}{dt} = -\gamma_{ls}(T_l - T_s) - \gamma_{le}(T_l - T_e), \tag{13.52}$$

where
   e, s, and l denote the electronic, spin, and lattice reservoirs, with corresponding specific heat coefficients $C_{e,s,l}$ and coupling constants γ_{ij} $(=\gamma_{ji})$
   p is the laser pumping rate

This three-temperature model has been shown to account for some experimental observations. However, it should be stressed, that although the value of the direct carrier–spin interaction constant γ_{es} could be fit to the experimental data, this phenomenological model does not account for the microscopic coupling mechanism. The other serious limitation of this approach is that the description is solely in terms of the energy transfer, neglecting the fact that the angular momentum also has to be exchanged between the reservoirs.

The initial observation of ultrafast demagnetization in Ni left many open issues including the exact time scale and size for the fast reduction of the collective magnetization and the microscopic mechanism for angular momentum conservation. Even more fundamentally, the interpretation of the observed transient MO signals in terms of magnetization dynamics, especially on a femtosecond time scale, remains nontrivial and needs justification, motivating many further experimental investigations. Indeed, in Ni films, as discussed earlier, it was established that dichroic bleaching dominated the MO response below 500 fs, long after electron thermalization within tens of femtoseconds, and, thus, no information about fast demagnetization could be obtained from the ultrafast MOKE data at these time scales (Koopmans et al., 2000). Only after around 500 fs, the temporal profiles of Kerr angle and ellipticity begin to match, thus reflecting the genuine magnetization dynamics.

In order to further understand ultrafast demagnetization in ferromagnetic metals, a series of time- and spectrally resolved experiments in another model ferromagnetic metal system, $CoPt_3$, has been further performed (Guidoni et al., 2002). With temporal resolution of 20 fs shorter than electron thermalization time and simultaneous detection of various quantities (reflection, transmission, MOKE, and Faraday rotation), the authors distinguished between the contributions of the spin and charge populations in the MO signals by carefully separating the dynamics of (1) the rotation and ellipticity and (2) the diagonal and nondiagonal elements of the time-dependent dielectric tensor. One example is reproduced in Figure 13.11: (1) there is a clear discrepancy in the relative changes of ellipticity and rotation, which has been shown to be due to electron thermalization within 50 fs (Figure 13.11a), instead of dichroic bleaching; (2) after electron thermalization, the dynamics of the real and imaginary parts of the Voigt vector are identical and the relative variation of the nondiagonal elements was 10 times larger than that of the diagonal elements of the tensor (Figure 13.11b and c). These indicate that spin dynamics dominate the MO signal immediately after electron thermalization, as discussed in Sections 13.3.3.2 and 13.3.3.3, and genuine demagnetization therefore appears in $CoPt_3$ at such

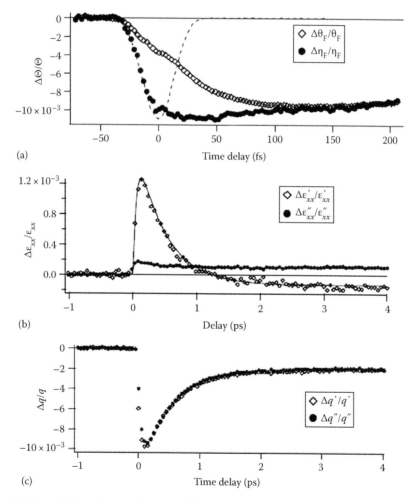

Figure 13.11 (a) The relative changes of ellipticity and rotation as a function of time for $CoPt_3$ samples. These quantities exhibit the same dynamics after the electronic thermalization time of ~50 fs, indicating that ultrafast spin dynamics dominates the transient MO signal afterwards. (b and c) Temporal evolution of the relative variations of the real and the imaginary parts of the MO constant $q' + iq'' = \varepsilon_{xy}/\varepsilon_{xx}$ (b) and diagonal elements of the dielectric tensor (c). (Reprinted from Guidoni, L. et al., *Phys. Rev. Lett.*, 89, 017401, 2002. With permission. Copyright 2002 by the American Physical Society.)

time scales; (3) Moreover, there was also no transient spectral feature in photoexcited $CoPt_3$ related to dichroic bleaching (Bigot et al., 2004). The lack of dichroic bleaching in $CoPt_3$ indicates that detailed transient MO signals sensitively depend on the specific material system and probe photon energy used (Section 13.2.2.4). These measurements serve as a basis for the extraction and interpretation of ultrafast MO and optical signals in metals and constitute a convincing proof for femtosecond demagnetization in laser-excited ferromagnetic metals.

In parallel with these experimental efforts, a microscopic understanding of ultrafast demagnetization and angular momentum transfer to the collective spins has been actively pursued. Theoretical investigations explaining femtosecond demagnetization have been proposed. However, many existing theories have their own deficiencies and the microscopic understanding is not complete yet. For instance, a purely electronic excitation mechanism, that is, the Stoner excitation, proposed by Schroll et al., is not compatible with the conservation of angular momentum (Scholl et al., 1997; Melnikov et al., 2003). Zhang et al. showed that ultrafast demagnetization is the result of a cooperative action between the laser field and spin–orbit coupling (Zhang and Hübner, 2000). However, considering the small number of available photons and the longer time scale observed for demagnetization as compared to the pulse duration, such a mechanism cannot be the sole factor, even if this light-assisted mechanism is highly relevant during the pulse duration, as suggested in recent experiments (Bigot et al., 2009).

Finally, there have been remarkable recent experimental and theoretical advances toward a microscopic understanding of ultrafast demagnetization in metals. However, some of the key issues are still under intense debate. On one hand, new probes such as time-resolved x-ray MCD reveal that the measured signals (Stamm et al., 2007), which are proportional to the sum of the spin and orbital momenta, are almost completely quenched within ~120 fs. This indicates that the spin–lattice interaction, instead of the spin–orbit interaction, is likely to be the dominant pathway for angular momentum conservation in ultrafast demagnetization, that is, the total magnetization of spins and lattice is conserved. However, the temporal resolution at the current stage of ultrafast x-ray technology, limited to ~100 fs, prevents one from reaching an unambiguous conclusion. Some recent theoretical proposals (Koopmans et al., 2005) support angular momentum conservation from the spin–lattice interaction, although it was initially thought too weak to account for the substantial demagnetization at femtosecond time scales. On the other hand, however, Bigot et al. found evidence for a coherent magnetic response during ultrafast demagnetization and therefore proposed a mechanism of dynamically enhanced spin–orbit coupling in the coherent regime (Bigot et al., 2009). This implies the important role of spin–orbit coupling in angular momentum conservation. Some further investigations are highly needed to reconcile some of the outstanding concerns posed in the studies mentioned earlier. In addition, a recent first-principles theoretical calculation (Zhang et al., 2009) of the coherent optical and magnetic response of ferromagnetic metal Ni to femtosecond laser pulses shows that there exists a regime of photon energies (<2 eV) where the MOKE response is dominated by magnetization dynamics, even during the pulse duration. This lays down a solid foundation for the interpretation of the results obtained from ultrafast MO techniques in nickel. These new developments may finally provide us a complete picture for the long-debated puzzle of ultrafast demagnetization.

13.4.3 Ferromagnetic Semiconductors

Magnetic materials displaying carrier-mediated spin–spin exchange interaction are ideal for nonthermal, potentially ultrafast spin manipulation. Prominent examples of such materials are transition-metal-doped (III,Mn)V ferromagnetic semiconductors such as GaMnAs (Ohno, 1998). These materials can serve as a "model" system of interacting spins with "intermediate" complexity, linking the weakly correlated semiconductors and traditional ferromagnetic metals and the more strongly correlated spin systems such as manganites. The exchange interaction between carriers (holes from Mn doping) and spins (Mn ions) in these materials make the magnetic properties, for example, the Curie temperature (T_c) and magnetization M, sensitive functions of the hole density, polarization, distribution, and wave function (Koshihara et al., 1997; Ohno et al., 2000; Wang et al., 2007). This platform provides more "knobs" to investigate photoinduced spin dynamics, for example, not only electronic heating (Figure 13.2a) but also spin-selective excitation and increases in the charge carrier density are relevant (Figure 13.2b). Since the discovery of carrier-mediated ferromagnetism in (III,Mn)V semiconductors, there has been intense interest in exploring the dynamics of nonthermal collective spin states and their correlation on photoexcited carriers in these systems. The underlying physics is rarely addressed on other platforms, such as metallic magnets or nonmagnetic semiconductors.

In addition, (III,Mn)V ferromagnetic semiconductors have served as a very productive playground in developing two-color time-resolved MO spectroscopy techniques to probe *genuine* femtosecond spin dynamics (Wang et al., 2005, 2006, 2007, 2008, 2009). As discussed in Section 13.3.3.1, the two-color method should be very effective in avoiding nonlinear optical effects such as dichroic bleaching in these material systems. Different geometries and excitation powers have been tested in this system, and several schemes and excitation regimes suitable for studying femtosecond spin dynamics have been successfully identified. Next we will discuss three transient magnetic phenomena observed in (III,Mn)V ferromagnetic semiconductors obtained by implementing ultrafast two-color MOKE spectroscopy in these systems and discuss the interpretation of the obtained MO signals.

1. *Ultrafast demagnetization*: One of the main difficulties in understanding ultrafast demagnetization in metallic ferromagnets discussed in Section 13.4.2 is due to the fact that in these materials the distinction between the "carriers" and "spins" is subtle, as the itinerant electrons contribute to both electronic transport and magnetism. In turn, direct microscopic coupling between the electronic degree of freedom and the collective spins, leading to extremely fast demagnetization on a femtosecond time scale, is not well-understood and represents a serious impediment to create a transparent model for ultrafast demagnetization in metals. Carrier-induced ferromagnetism in (III,Mn) V semiconductors provides an interesting alternative for studying ultrafast magnetization dynamics. Unlike in ferromagnetic metals, there is a clear distinction between mobile carriers (holes, p-like band) and localized spins (Mn ions, d-like band), and macroscopic magnetic order is realized through their strong coupling (through the p–d exchange interaction). The sharp separation between the mobile carriers and localized spins makes such a distinction between carrier and spin populations more natural. Therefore, here, demagnetization can be started by the exchange of energy and angular momentum between two-spin ensembles coupled by the spin exchange interaction, an intuitive starting point. At long times, it can also involve the process of spin–lattice relaxation, in which the spins are scattered by fluctuations of the crystal fields produced by phonons. In addition, since two separate issues of current interest, that is, hole relaxation and collective magnetization dynamics, fall into one platform, such as GaMnAs and InMnAs, their correlation, which is rarely addressed elsewhere, can be investigated.

Figure 13.12 shows two-color MOKE experiments in InMnAs (Wang et al., 2005a) and GaMnAs (Wang et al., 2008) to study ultrafast demagnetization in these systems. The two-color spectroscopy setup used consisted of an OPA pumped by a Ti:Sapphire-based regenerative

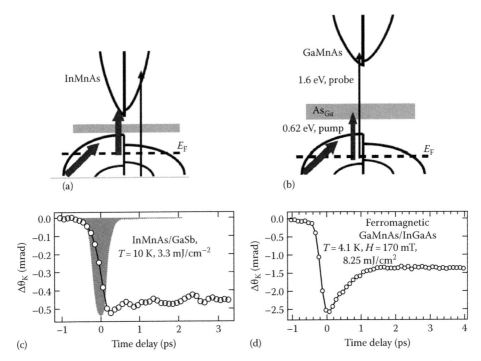

Figure 13.12 (a and b) Schematics of band dispersion in InMnAs (a) and GaMnAs (b) and optical transitions involved in two-color MOKE experiments. Here the conduction band (CB) and valence band (VB) are spin-split below the Curie temperature. The thick gray bars illustrate antisite states. In InMnAs, the transient carriers are created through direct interband and intervalence band transitions at 0.62 eV (thick solid arrows, pump) and the magnetization changes are probed via Kerr rotation by reflected, time-delayed pulses at 1.6 eV (thin arrows, probe). In GaMnAs, below band gap pump pulses at 0.62 eV (thick solid arrows, pump) create holes in the VB via transitions from below the Fermi level to the midgap states (e.g., an AsGa antisite defect band centered ~0.5 eV above the top of the VB), and excite the preexisting hole population by intervalence band transitions. The magnetization changes are probed by 1.6 eV photons (thin arrows, probe). (c and d) Photoinduced two-color MOKE dynamics for (c) InMnAs and (d) GaMnAs, respectively. The shaded area in (c) is the pump–probe autocorrelation. (Reprinted from Wang, J. et al., *Phys. Rev. Lett.*, 95, 167401, 2005b; Wang, J. et al., *Phys. Rev. B*, 77, 235308, 2008. With permission.)

amplifier. The large pump fluences used, on the order of mJ/cm², ensured the study of demagnetization dynamics in these systems, instead of other subtle transient magnetic effects that are manifested at low-pump fluences (Wang et al., 2007, 2009). In the case of InMnAs, as shown in Figure 13.12a and c, the OPA pump beam, which was linearly polarized and tuned to 2 μm, excited transient carriers near the band edge of InMnAs, and a very small fraction of the regenerative amplifier beam (775 nm) was used as a probe; the high photon energy of the probe has been shown to very efficiently diminish dichroic bleaching effects due to the pump excited carriers. The negative sign of $\Delta\theta_k$ at early times clearly indicates transient demagnetization, which can be attributed to the spin-flip scattering of Mn ions with carriers, with the subsequent spin relaxation of carriers occurring through the spin-orbit interaction. The whole process can be envisioned as the reverse of the Overhauser effect (Wang et al., 2005a) (which refers to the transfer of electron spin polarization to nuclear spins); the excited carriers become dynamically polarized at the expense of the localized spins, and the dissipation of magnetization occurs through spin relaxation in the carrier system. Therefore, exploiting the simplicity of carrier–spin coupling in the ferromagnetic semiconductors allows one to provide this physically transparent picture and also put the idea of spin-flip scattering in the demagnetization on a much stronger footing. In addition, the traditional pathway of the demagnetization due to spin–lattice interaction connects to the long time dynamics on a ~100 ps time scale, as shown in (Wang et al., 2005a).

Wang et al. (2005) and Cywiński and Sham (2007) formulated a theory for femtosecond demagnetization in magnetic semiconductors or more general *sp-d* ferromagnets (strongly coupled localized spins and itinerant carrier spins) based on the ideas given earlier. The essence of this theory is to consider the probability of a spin-flip of the Mn spins. In the one band model, it is given by

$$W_{m,m'} = \frac{2\pi}{\hbar}\frac{\beta^2}{4}\sum_{s,s'}\int\frac{d^3k}{(2\pi)^3}\int\frac{d^3q}{(2\pi)^3}\,f_{s'}\!\left(\vec{k}\right)\!\left(1-f_s\!\left(\vec{q}\right)\right)\times\delta\!\left(\varepsilon_s\!\left(\vec{q}\right)-\varepsilon_{s'}\!\left(\vec{k}\right)\pm\delta\right)\!\left|\left\langle\,ms\left|\hat{S}_\pm\hat{s}_\mp\right|m's'\,\right\rangle\right|^2,$$

$$(13.53)$$

where
  δ is the mean-field splitting of the Mn spin levels
  $m = -5/2 \ldots 5/2$
  s is the hole spin

The occupation functions are thermal, with a high effective temperature T_h mimicking the highly nonequilibrium state of holes. Consequently, the dynamics of Mn spins are described by a simple rate equation for diagonal elements of the localized spin density matrix, and the dynamics of the average hole spin is governed by $ds/dt = -\gamma\cdot dM/dt - 1/\tau_s(s - s_{eq}(M, T_h))$, where M is the average Mn spin, γ is the ratio of Mn to hole density, τ_s is the spin-relaxation time, and s_{eq} is the instantaneous equilibrium value of the average hole spin. This equation captures the dynamic polarization of holes. In the calculations $T_h = 1000$ K, which gives an energy spread of the holes comparable to the exchange splitting of the band (~100 meV for typical ferromagnetic semiconductors). Using $\tau_s = 10$ fs, a hole concentration of $p = 4 \times 10^{20}$/cm^3, and a band mass $m_h = 0.5m_0$, a typical value of magnetization drop of 10% within 200 fs is obtained under the experimental conditions. This result clearly shows that femtosecond demagnetization in these materials is possible, and the good agreement with experiments supports the underlying physical picture.

On the other hand, in the case of GaMnAs, shown in Figure 13.12b and d, the probe couples to electronic transitions for which the initial states are in the valence band, close to the Fermi level. These are exactly the states whose population is strongly disturbed by the pump, which leads to dichroic bleaching as discussed in Sections 13.3.3.1 and 13.3.3.2. This gives rise to the observation for GaMnAs, shown in Figure 13.12d, of initial prominent overshoot behavior within the first 1 ps, which is not seen in InMnAs. Actually, this distinct behavior in GaMnAs from dichroic bleaching can be used to understanding the ultrafast relaxation dynamics of holes in (III,Mn)V semiconductors. One of the sources for dichroic bleaching, as illustrated in Figure 13.7, is a simple blocking of transitions by transient holes added to the valence band, which diminishes the absorption for both circular polarizations. Consequently, the partial recovery of the MOKE signal with a time constant of ~1 ps can be attributed to the energy relaxation of highly excited holes. It is important to note that much less is known experimentally about the ultrafast dynamics of holes than that of electrons, especially in a regime of strong excitation. This example shows that dichroic bleaching in ultrafast MOKE may provide a sensitive scheme for probing hole relaxation dynamics, particularly for highly excited regimes that are difficult to access by many other experimental techniques.

2. *Ultrafast enhancement of ferromagnetism*: There has been long and intense interest in searching for possibilities of ultrafast enhancement of collective magnetic order via photoexcitation. Such photoexcitation would lead to fascinating opportunities for establishing a transient cooperative phase from an uncorrelated ground state and for determining the relevant time scales for the buildup of order parameters. However, most prior experiments in magnetically ordered materials only show ultrafast demagnetization and quenching dynamics, due in part to the dominant role of laser-induced electronic heating. The hole-mediated ferromagnetism in (III, Mn)V ferromagnetic semiconductors offers unique opportunities and flexibility for nonthermal control of magnetism, as illustrated in Figure 13.13a. The first time-resolved experiments in (III, Mn)V systems to show transient photoinduced magnetization and measure the time scale for the enhancement of collective order were reported in Wang et al. (2005). Figure 13.13b illustrates that femtosecond pump pulses create a large density of holes in the valence band of GaMnAs. The linearly polarized pump beam was at 3.1 eV, with peak fluences of $10\,\mu J/cm^2$. A small fraction of the fundamental beam at 1.55 eV was used as a probe, detecting magnetization via the polar MOKE angle using the optical polarization bridge and data acquisition schemes described in Section 13.3. The low pump peak fluences and the high pump photon energy have been shown to effectively minimize the influence of spurious effects on the magnetic signals. The "magnetic origin" of the transient MOKE response is confirmed by separate measurements showing the overlap of the pump-induced rotation

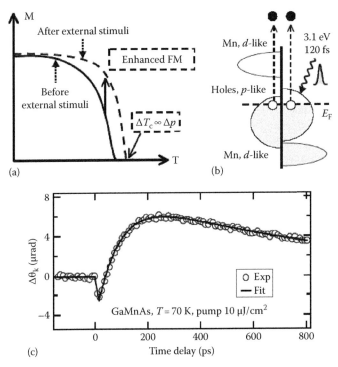

Figure 13.13 (a) Illustration of hole-density-tuning effects via external stimuli in GaMnAs seen in the static experiments. FM: ferromagnetism. Δp is the hole density change. (b) Schematic diagram of the spin-dependent density of states in GaMnAs. Femtosecond pump pulses create a transient population of holes in the valence band. (c) Time-resolved MOKE dynamics at 70 K and under 1.0 T field. Transient enhancement of magnetization, with ~100 ps rise time, is clearly seen after the initial fast demagnetization. (Reprinted from Wang, J. et al., *Phys. Rev. Lett.*, 98, 217401, 2007a. With permission.)

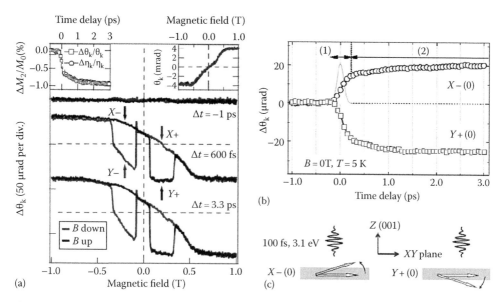

Figure 13.14 (a) Photoinduced femtosecond four-state magnetic hysteresis in ferromagnetic GaMnAs. Inset (left): temporal profiles of normalized Kerr rotation and ellipticity angle changes at 1.0 T. Inset (right): static magnetization curve at 5 K. (b) Photoinduced femtosecond magnetization rotation for two in-plane memory states shown together with the pump–probe cross correlation (shaded). The opposite, out-of-plane M rotations for the X − (0) and Y + (0) initial states are illustrated in (c). (Reprinted from Wang, J. et al., *Appl. Phys. Lett.*, 94, 021101, 2009. With permission.)

and ellipticity through the entire time scan range, for example, as shown in the inset of Figure 13.14a. The authors observed evidence of the ultrafast enhancement of ferromagnetism in GaMnAs, as shown in Figure 13.13c. Here the MOKE data clearly shows a distinct magnetization enhancement occurring on a tens of picosecond time scale ($\Delta\theta_k > 0$), after by an initial subpicosecond demagnetization ($\Delta\theta_k < 0$).

The observed ultrafast photoenhanced ferromagnetism can be understood via the transient hole–Mn interaction via the H_{p-d} exchange, as follows. At early pump–probe delays ($t \sim 0$ fs), the ultrashort laser pulses generate a nonequilibrium distribution of spin-unpolarized electron–hole pairs in GaMnAs under a finite external B field. During the first picosecond ($t < 1$ ps), the photoexcited hot holes will experience efficient spin-flip scattering with the localized Mn moments, manifested as a subpicosecond demagnetization component. This results from the off-diagonal elements of the exchange Hamiltonian H_{p-d}, which cause the spin polarization of the Mn ions to transfer to the holes within several hundred femtoseconds, similar to the femtosecond demagnetization described earlier (Wang et al., 2005a, 2008). Meanwhile, the hot hole distribution quickly loses its energy via carrier–phonon scattering (optical phonon energy ~36 meV), resulting in a rapid termination of demagnetization (within the first picosecond). At longer pump–probe delays of $t > 1$ ps, the photoexcited, thermalized holes, settling down in the spin-split bands (Figure 13.13b), can now participate in the process of hole-mediated ferromagnetic ordering. These extra holes enhance the Mn–Mn exchange correlation and polarize Mn spins via the mean-field (diagonal) elements of the exchange Hamiltonian H_{p-d}, thereby increasing the macroscopic magnetization. In addition, note the photoexcited transient electrons have little effect on the magnetization dynamics described here, because of the short free electron lifetime (<1 ps) (Wang et al., 2005a) in (III,Mn)V materials and their weak coupling to Mn ions.

3. *Photoinduced femtosecond collective spin reorientation*: The strong coupling between carriers (holes) and Mn ions, for example, on the order of 1 eV in GaMnAs, could imply that there is a nonthermal femtosecond cooperative magnetic response to photoexcited carriers. Figure 13.14 shows an example of photoinduced femtosecond spin rotation, which developed during the pulse duration and showed four-state magnetic memory behavior in GaMnAs (Wang et al., 2009). Figure 13.14a shows the magnetic field scan traces of the photoinduced MOKE signal $\Delta\theta_k$ at three time delays. A mere 600 fs after photoexcitation, a clear photoinduced four-state magnetic hysteresis is observed, with four abrupt switchings when sweeping the magnetic field, arising as a direct consequence of magnetic memory effects. The magnetic memory here means that the spin states at a given magnetic field depend on the field scanning history, as discussed in more details in Wang et al. (2009) and Liu et al. (2005). The photoinduced dynamics of the two zero-B field memory states, shown in Figure 13.14b, indicates a photoinduced femtosecond magnetic rotation. More intriguingly, M in the $X - (0)$ and $Y + (0)$ initial states rotates to different z-axis directions, as illustrated in Figure 13.14c, although they are symmetric in the magnetic ground state. The opposite signs of the photoinduced signals are responsible for the four-state magnetic switching. Note that such dynamics, together with the magnetic field dependence, clearly excludes any nonmagnetic origin of the photoinduced MOKE responses in these experiments.

The possibility for the existence of an early nonequilibrium, nonthermal femtosecond regime for collective magnetization rotation in (III,Mn)V semiconductors is consistent with the theoretical proposal in Chovan et al. (2006). Together with simulations, the results clearly show that the observed femtosecond spin dynamics are governed by the dynamics of both coherent and nonthermal photoexcited holes (Kapetanakis et al., 2009). Following the nonequilibrium theory of ultrafast magnetization reorientation developed in Kapetanakis et al. (2009), one can distinguish between thermal, nonthermal, and coherent photoexcited carrier contributions to the mean field equation of motion:

$$\frac{\partial \vec{S}}{\partial t} = -\gamma \vec{S} \times H^{\text{th}} - \frac{\beta}{V} \sum_k \vec{S} \times \Delta s_{\vec{k}}^h + \frac{\alpha}{S} \vec{S} \times \frac{\partial \vec{S}}{\partial t}, \tag{13.54}$$

where
   γ is the gyromagnetic ratio
   α is the Gilbert damping rate

and the photoinduced hole spin is

$$s_{\vec{k}}^h = \sum_{nn'} s_{knn'}^h \left\langle \hat{h}_{-\vec{k}n}^+ \hat{h}_{-\vec{k}n'} \right\rangle, \tag{13.55}$$

which is excited by the nonlinear optical contributions including phase space filling, coupling to the intervalence band coherences, and transient changes in the hole states due to interactions with the light-induced magnetization change (Kapetanakis et al., 2009). In this framework, the femtosecond spin rotation results can be understood as photoexcitation of a pulsed hole spin component perpendicular to the magnetization vector in the ground state, or photoexcitation of an effective magnetic field pulse not parallel to the light propagation direction, as discussed in Kapetanakis et al. (2009). Consequently, ΔS can develop during the optical excitation, due to interactions with the coherent electron–hair pair spin. On the other hand, the thermal contribution in (13.54) is described by $H^{th} = -\partial E_h/\partial \vec{S}$, the thermal hole anisotropy field. Here E_h is the total energy of thermal holes.

The last two examples clearly show how ultrafast MO techniques can reveal new fundamental collective magnetic processes on ultrafast time scales beyond demagnetization and identify the critical roles of spin coherence and the spin–carrier correlations in these fundamental photoinduced cooperative and nonthermal behaviors.

13.4.4 Antiferromagnetic Insulators

Generally speaking, virtual hopping among localized spin states in conventional magnetic insulators leads to an antiferromagnetic superexchange interaction, while interacting itinerant spins in magnetic metals favor a ferromagnetic exchange splitting of spin up and spin down energy bands (Chikazumi and Graham, 1997). As a result of these fundamental differences in magnetic states and transport properties, different dynamic magnetization processes occur after femtosecond laser excitation, as already shown in Figure 13.9 for the case of the ultrafast demagnetization.

Insulating magnets are much less studied in the context of ultrafast magnetization dynamics, but there has been growing interest in this type of material in the past 5 years, thanks to the recent advances described in Kimel et al. (2005) and Hansteen et al. (2006). There are two types of laser excitation schemes demonstrated. First, ultrafast excitation in antiferromagnetic insulators can lead to electronic or phonon heating similar to those of their metallic counterparts, as illustrated in Figure 13.9b. This has given rise to some interesting transient magnetic phenomena, for example, demagnetization of the collective spin system in $BFeO_3$ via magnon generation on hundreds of picoseconds time scale, ultrafast generation of 100s GHz antiferromagnetic magnon precession by thermally induced reorientation of the magnetic easy axis, and the photoinduced antiferromagnetic to ferromagnetic phase transition in FeTh. Second, the perturbation of particular interest in antiferromagnetic insulators is the use of a photo-magnetic pulse via coherent virtual excitations, the third excitation scheme illustrated in Figure 13.2c. One of these coherent driving mechanisms is the inverse Faraday effect, as illustrated in Figure 13.15a, which has been recently observed in antiferromagnetic

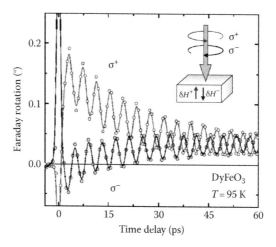

Figure 13.15 Photoinduced MO Faraday rotation in $DyFeO_3$ after a circularly polarized pump pulse. The femtosecond effective magnetic field pulses δB are generated through the inverse Faraday effect by right-handed and left-handed circularly polarized pumps, respectively. (Reprinted from Kimel, A.V. et al., *Nature*, 435, 655, 2005. With permission.)

materials (Kimel et al, 2005). Here high-intensity circularly polarized laser pulses can generate an effective nonthermal B field in the largely transparent region, which can be viewed as a reverse process of the Faraday effect, that is, the polarization rotation of polarized light under static magnetic fields. Right- and left-handed waves will induce effective magnetic fields of opposite sign. Since the mechanism is coherent, the appearance of δB is quasiinstantaneous, limited only by the laser pulse duration, and thereby fast, potentially femtosecond spin rotation should be expected. This is in strong contrast to the previously held picture of optically excited spin precession in ferromagnetic materials, in which the photoexcitation thermally altered the anisotropy fields of the materials via heating of the lattice.

One seminal experiment for the demonstration of the inverse Faraday effect is reproduced in Figure 13.15 for the rare-earth orthoferrite $DyFeO_3$. This antiferromagnetic crystal exhibits a giant Faraday effect due to both large spin–orbit coupling and canting antiferromagnetic spins. Ultrafast MO Faraday spectroscopy allows one to coherently excite and subsequently follow the spin dynamics. Figure 13.15 shows the temporal profile of the induced Faraday rotation for two circularly polarized pump pulses of opposite helicities. A distinctive feature is an oscillatory behavior of the MO signals with a frequency of about 200 GHz, which can be assigned to zero momentum antiferromagnetic magnon oscillations. Most intriguingly, the helicity of the pump light controls the phase of the photoinduced magnetization oscillation, which clearly shows the direct coupling of the angular momentum of pump photons and collective spins in $DyFeO_3$. Ultrafast laser excitation here may be thought as turning on an effective magnetic field δB, which has its microscopic origin in the second-order Raman nonlinearity (Shen and Bloembergen, 1960). This femtosecond photomagnetic field then exerts a spin torque on the $\mathbf{M}$ vector, $\Delta\vec{B} \times \mathbf{M}$, and induces magnetization oscillations in the sample. Such optical pulses are shown to be equivalent to 200 fs magnetic field pulses up to 5 T.

Finally, we note that such a coherent, spin selective excitation scheme has also been demonstrated in some ferrimagnetic insulators where two spin sublattices with unequal magnetization couple antiferromagnetically to each other. Particularly, a complete spin reversal has shown to be achieved with a single excitation pulse in the ferrimagnet GdCoFe (Stanciu et al., 2007), which opens an intriguing question as to whether it is possible to achieve a complete spin reversal during the pulse duration.

Microscopic understanding of femtosecond spin excitation in these antiferromagnets and ferrimagnets is still challenging, and several outstanding issues remain open, regarding, for example, the second-order Raman nonlinearities involved and origin of the large size of the photoinduced magnetic effects. More systematic theoretical simulations and finely controlled experimental data are needed to improve our understanding.

13.5 CONCLUSIONS AND OUTLOOK

The ultrafast MO spectroscopy experiments have been shown to be extremely useful for revealing spin dynamics and fundamental transient magnetic phenomena in many materials. Particularly, they greatly advance our knowledge of collective spin systems driven away from equilibrium by providing time-domain information hidden in static measurements. It is clear that proper attention should be given to the earlier mentioned technical details of the experimental setups and the interpretation of the MO signals obtained. In addition, it should be also noted that application of these techniques to advanced materials is still in a very active phase and many exotic properties as well as new concepts regarding ultrafast magnetism are still to come. One of the foreseeable themes in the future research will be the coherent aspects of photoexcited interacting spins. There has been emerging evidence of such coherent ultrafast magnetism in ferromagnetic metals and semiconductors as well as insulating magnets, where the photoinduced cooperative spin dynamics develop even during the photoexcitation.

Opportunities also lie in the following several aspects. First, the development of new ultra-fast laser sources and spectroscopy techniques in the MIR and FIR regions offer opportunities to explore ultrafast magnetic phenomena via exploiting these ultraintense small photons. Compared to the widely used NIR and visible sources, they allow one to strongly perturb the magnetic *ground* states without involving high-energy states. The second opportunity is in the synthesis of complex materials with "responsive" magnetic ground states that show sensitivity to external stimuli. Some examples are those strongly correlated electronic materials such as colossal magnetoresistive manganites, which represent relatively new territory in the context of ultrafast magnetism. Third, it will be of great value in continuing to develop proper theoretical language to describe femtosecond MO signals and exceed the ultrashort time scale limitations of the common thermodynamic descriptions of photo-excited, non-equilibrium magnetism. These will be critical for further refining the ultrafast MO techniques and proper interpretation of obtained results.

ACKNOWLEDGMENTS

We are grateful to I. E. Perakis, D. S. Chemla, J. Kono, L. Cywinski, L. J. Sham, C. J. Stanton, X. Liu and J. K. Furdyna for useful discussions and suggestions. This work has been supported by the Ames Laboratory of the U.S. Department of Energy-Basic Energy Sciences under contract No. DE-AC02-07CH11358.

REFERENCES

Awschalom, D. D., D. Loss, and N. Samarth (Eds.), 2002, *Semiconductor Spintronics and Quantum Computation*, Springer, Berlin, Germany.

Beaurepaire, E., M. Maret, V. Halte, J.-C. Merle, A. Daunois, and J.-Y. Bigot, 1998, Spin dynamics in CoPt3 alloy films: A magnetic phase transition in the femtosecond time scale, *Phys. Rev. B* 58, 12134.

Beaurepaire, E., J.-C. Merle, A. Daunois, and J.-Y. Bigot, 1996, Ultrafast spin dynamics in ferromagnetic nickel, *Phys. Rev. Lett.* 76, 4250.

Bennett, H. S. and E. A. Stern, 1964, Faraday effect in solids, *Phys. Rev.* 137, A448.

Bigot, J.-Y., L. Guidoni, E. Beaurepaire, and P. N. Saeta, 2004, Femtosecond spectrotemporal magneto-optics, *Phys. Rev. Lett.* 93, 077401.

Bigot, J.-Y., M. Vomir, and E. Beaurepaire, 2009, Coherent ultrafast magnetism induced by femtosecond laser pulses, *Nat. Phys.* 5, 515.

Born, M. and E. Wolf, 1999, *Principles of Optics: Electromagnetic Theory of Propagation, Interference and Diffraction of Light*, 7th Edn., University of Cambridge, Cambridge, U.K.

Chikazumi, S. and C. D. Graham, 1997, *Physics of Ferromagnetism*, Oxford University Press, New York.

Chovan, J., E. G. Kavousanaki, and I. E. Perakis, 2006, Ultrafast light-induced magnetization dynamics of ferromagnetic semiconductors, *Phys. Rev. Lett.* 96, 057402.

Conrad, U., J. Güdde, V. Jähnke, and E. Matthias, 2001, Phase effects in magnetic second-harmonic generation on ultrathin Co and Ni films on Cu(001), *Phys. Rev. B* 763, 144417.

Crooker, S. A., J. J. Baumberg, F. Flack, N. Samarth, and D. D. Awschalom, 1996, Terahertz spin precession and coherent transfer of angular momenta in magnetic quantum wells, *Phys. Rev. Lett.* 77, 2814.

Cywiński, L. and L. J. Sham, 2007, Ultrafast demagnetization in the sp-d model: A theoretical study, *Phys. Rev. B*, 76, 045205.

Guidoni, L., E. Beaurepaire, and J.-Y. Bigot, 2002, Magneto-optics in the ultrafast regime: Thermalization of spin populations in ferromagnetic films, *Phys. Rev. Lett.* 89, 017401.

Gupta, J. A., R. Knobel, N. Samarth, and D. D. Awschalom, 2001, Ultrafast manipulation of electron spin coherence, *Science*, 292, 2458.

Hanson, R., V. V. Dobrovitski, A. E. Feiguin, O. Gywat, and D. D. Awschalom, 2008, Coherent dynamics of a single spin interacting with an adjustable spin bath, *Science* 320, 352.

Hansteen, F., A. Kimel, A. Kirilyuk, and T. Rasing, 2006, Nonthermal ultrafast optical control of the magnetization in garnet films, *Phys. Rev. B* 73, 014421.

Hashimoto, Y., S. Kobayashi, and H. Munekata, 2008, Photoinduced precession of magnetization in ferromagnetic (Ga,Mn)As, *Phys. Rev. Lett.* 100, 067202.

Hohlfeld, J., E. Matthias, R. Knorren, and K. H. Bennemann, 1997, Nonequilibrium magnetization dynamics of nickel, *Phys. Rev. Lett.* 78, 4861.

Kampen, M. V., C. Jozsa, J. T. Kohlhepp, P. LeClair, L. Lagae, W. J. De Jonge, and B. Koopmans, 2002, All-optical probe of coherent spin waves, *Phys. Rev. Lett.* 88, 227201.

Kapetanakis, M. D., I. E. Perakis, K. J. Wickey, C. Piermarocchi, and J. Wang, 2009, Femtosecond coherent control of spins in (Ga,Mn)As ferromagnetic semiconductors using light, *Phys. Rev. Lett.* 103, 047404.

Kikkawa, J. M. and D. D. Awschlom, 1998, Resonant spin amplification in n-type GaAs, *Phys. Rev. Lett.* 80, 4313.

Kikkawa, J. M. and D. D. Awschalom, 1999, Lateral drag of spin coherence in gallium arsenide, *Nature* 397, 139.

Kimel, A. V., A. Kirilyuk, A. Tsvetkov, R. V. Pisarev, and Th. Rasing, 2004, Laser-induced ultrafast spin reorientation in the antiferromagnet $TmFeO_3$, *Nature* 429, 850.

Kimel, A. V., A. Kirilyuk, P. A. Usachev, R. V. Pisarev, A. M. Balbashov, and Th. Rasing, 2005, Ultrafast non-thermal control of magnetization by instantaneous photomagnetic pulses, *Nature* 435, 655.

Kimel, A., R. Pisarev, J. Hohlfeld, and T. Rasing, 2002, Ultrafast quenching of the antiferromagnetic order in $FeBO_3$: Direct optical probing of the phonon-magnon coupling, *Phys. Rev. Lett.* 89, 287401.

Kise, T, T. Ogasawara, M. Ashida, Y. Tomioka, Y. Tokura, and M. Kuwata-Gonokami, 2000, Ultrafast spin dynamics and critical behavior in half-metallic ferromagnet: Sr_2FeMoO_6, *Phys. Rev. Lett.* 89, 287401.

Kojima, E., R. Shimano, Y. Hashimoto, S. Katsumoto, Y. Iye, and M. Kuwata-Gonokami, 2003, Observation of the spin-charge thermal isolation of ferromagnetic Ga0.94Mn0.06As by time-resolved magneto-optical measurements, *Phys. Rev. B* 68, 193203.

Koopmans, B., 2003, Laser-induced magnetization dynamics, in *Spin Dynamics in Confined Magnetic Structures II*, B. Hillebrands and K. Ounadjela (Eds.), Springer, Berlin, Germany, p. 253.

Koopmans, B., J. E. M. Haverkort, W. J. M. de Jonge, and G. Karczewski, 2007, Time-resolved magnetization modulation spectroscopy: A new probe of ultrafast spin dynamics, *J. Appl. Phys.* 85, 6763.

Koopmans, B., M. van Kampen, J. T. Kohlhepp, and W. J. M. de Jonge, 2000a, Femtosecond spin dynamics of epitaxial Cu(111)/Ni/Cu wedges, *J. Appl. Phys.* 87, 5070.

Koopmans, B., M. van Kampen, J. T. Kohlhepp, and W. J. M. de Jonge, 2000b, Ultrafast magneto-optics in nickel: Magnetism or optics? *Phys. Rev. Lett.* 85, 844.

Koopmans, B., M. van Kampen, J. T. Kohlhepp, and W. J. M. de Jonge, 2002, Experimental access to femtosecond spin dynamics, *J. Phys. Condens. Matter* 14, 1.

Koopmans, B., J. J. M. Ruigrok, F. D. Longa, and W. J. M. de Jonge, 2005, Unifying ultrafast magnetization dynamics, *Phys. Rev. Lett.* 95, 267207.

Koralek, J. D., C. P. Weber, J. Orenstein, B. A. Bernevig, S.-C. Zhang, S. Mack, and D. D. Awschalom, 2009, Emergence of the persistent spin helix in semiconductor quantum wells, *Nature* 458, 610.

Koshihara, S., A. Oiwa, M. Hirasawa, S. Katsumoto, Y. Iye, C. Urano, H. Takagi, and H. Munekata, 1997, Ferromagnetic order induced by photogenerated carriers in magnetic III–V semiconductor heterostructures of (In,Mn)As/GaSb, *Phys. Rev. Lett.* 78, 4617.

Liu, X., W. L. Lim, L. V. Titova, M. Dobrowolska, J. K. Furdyna, M. Kutrowski, and T. Wojtowicz, 2005, Perpendicular magnetization reversal, magnetic anisotropy, multistep spin switching, and domain nucleation and expansion in $Ga_{1-x}Mn_xAs$ films, *J. Appl. Phys.* 98, 063904.

Melnikov, A., I. Radu, U. Bovensiepen, O. Krupin, K. Starke, E. Matthias, and M. Wolf, 2003, Coherent optical phonons and parametrically coupled magnons induced by femtosecond laser excitation of the Gd(0001) surface, *Phys. Rev. Lett.* 91, 227403.

Ohno, H., 1998, Making non-magnetic semiconductors ferromagnetic, *Science* 281, 951.

Ohno, H., D. Chiba, F. Matsukura, T. Omiya, E. Abe, T. Dietl, Y. Ohno, and K. Ohtani, 2000, Electric-field control of ferromagnetism, *Nature* 408, 944.

Qi J., Y. Xu, N. H. Tolk, X. Liu, J. K. Furdyna, and I. E. Perakis, 2007, Coherent magnetization precession in GaMnAs induced by ultrafast optical excitation, *Appl. Phys. Lett.* 91, 112506.

Regensburger, H., R. Vollmer, and J. Kirschner, 2000, Time-resolved magnetization-induced second-harmonic generation from the Ni(110) surface, *Phys. Rev. B* 61, 14716.

Scholl, A., L. Baumgarten, R. Jacquemin, and W. Eberhardt, 1997, Ultrafast spin dynamics of ferromagnetic thin films observed by fs spin-resolved two-photon photoemission, *Phys. Rev. Lett.* 79, 5146.

Shen, Y. R. and N. Bloembergen, 1960, Interaction between light waves and spin waves, *Phys. Rev.* 143, 372–384.

Stamm, C., T. Kachel, N. Pontius, R. Mitzner, T. Quast, K. Holldack, S. Khan et al., 2007, Femtosecond modification of electron localization and transfer of angular momentum in nickel, *Nat. Mater.* 6, 740.

Stanciu, C. D., F. Hansteen, A. V. Kimel, A. Kirilyuk, A. Tsukamoto, A. Itoh, and Th. Rasing, 2007, All-optical magnetic recording with circularly polarized light, *Phys. Rev. Lett.* 99, 047601.

Sugano, S. and N. Kojima, 2000, *Magneto-Optics*, Springer-Verlag, Berlin, Germany, 2000.

Tobey, R. I., D. Prabhakaran, A. T. Boothroyd, and A. Cavalleri, 2008, Ultrafast electronic phase transition in La½Sr3/2MnO4 by coherent vibrational excitation: Evidence for nonthermal melting of orbital order, *Phys. Rev. Lett.* 101, 197404.

Wang, J., I. Cotoros, D. S. Chemla, X. Liu, J. K. Furdyna, J. Chovan, and I. E. Perakis, 2009, Memory effects in photoinduced femtosecond magnetization rotation in ferromagnetic GaMnAs, *Appl. Phys. Lett.* 94, 021101.

Wang, J., I. Cotoros, K. M. Dani, X. Liu, J. K. Furdyna, and D. S. Chemla, 2007a, Ultrafast enhancement of ferromagnetism via photoexcited holes in GaMnAs, *Phys. Rev. Lett.* 98, 217401.

Wang, J., L. Cywinski, C. Sun, J. Kono, A. Oiwa, H. Munekata, and L. J. Sham, 2008, Femtosecond demagnetization and hot-hole relaxation in ferromagnetic $Ga_{1-x}Mn_xAs$, *Phys. Rev. B* 77, 235308.

Wang, J., M. W. Graham, Y. Ma, G. R. Fleming, and R. A. Kaindl, 2010, Ultrafast spectroscopy of midinfrared internal exciton transitions in separated single-walled carbon nanotubes, *Phys. Rev. Lett.* 104, 177401.

Wang, J., Y. Hashimoto, J. Kono, A. Oiwa, H. Munekata, G. D. Sanders, and C. J. Stanton, 2005a, Propagating coherent acoustic phonon wave packets in $In_xMn_{1-x}As/GaSb$, *Phys. Rev. B*, 72, 153311.

Wang, D. M., Y. H. Ren, X. Liu, J. K. Furdyna, M. Grimsditch, and R. Merlin, 2007b, Light-induced magnetic precession in (Ga,Mn)As slabs: Hybrid standing-wave Damon-Eshbach modes, *Phys. Rev. B* 75, 233308.

Wang, J., C. Sun, Y. Hashimoto, J. Kono, G. A. Khodaparast, Ł. Cywiński, L. J. Sham, G. D. Sanders, C. J. Stanton, and H. Munekata, 2006, Ultrafast magneto-optics in ferromagnetic III–V semiconductors, *J. Phys. Condens. Matter* 18, R501.

Wang, J., C. Sun, J. Kono, A. Oiwa, H. Munekata, Ł. Cywiński, and L. J. Sham, 2005b, Ultrafast quenching of ferromagnetism in InMnAs induced by intense laser irradiation, *Phys. Rev. Lett.* 95, 167401.

Weber, C. P., N. Gedik, J. E. Moore, J. Orenstein, J. Stephens, and D. D. Awschalom, 2005, Observation of spin Coulomb drag in a two-dimensional electron gas, *Nature* 437, 1330.

Wolf, S. A., D. D. Awschalom, R. A. Buhrman, J. M. Daughton, S. von Molnar, M. L. Roukes, A. Y. Chtchelkanova, and D. M. Treger, 2001, Spintronics: A spin-based electronics vision for the future, *Science* 294, 1488.

You, C.-Y. and S.-C. Shin, 1996, Derivation of simplified analytic formulae for magneto-optical Kerr effects, *Appl. Phys. Lett.* 69, 1353.

Zhang, G. P. and W. Hübner, 2000, Laser-induced ultrafast demagnetization in ferromagnetic metals, *Phys. Rev. Lett.* 85, 3025.

Zhang, G, W. Hubner, E. Beaurepaire, and J.-Y. Bigot, 2002, Laser induced ultrafast demagnetization: Femtomagnetism, a new frontier? in *Spin Dynamics in Confined Magnetic Structures I*, B. Hillebrands and K. Ounadjela (Eds.), Springer, Berlin, Germany, p. 245.

Zhang, G. P., W. Hübner, G. Lefkidis, Y. Bai, and T. F. George, 2009, Paradigm of the time-resolved magneto-optical Kerr effect for femtosecond magnetism, *Nat. Phys.* 5, 499.

Zvezdin, K. and V. A. Kotov, 1997, *Modern Magnetooptics and Magnetooptical Materials*, Institute of Physics Publishing, Bristol, U.K.

Chapter 14

TIME-RESOLVED RAMAN SCATTERING

Daniele Fausti and Paul H.M. van Loosdrecht

CONTENTS

14.1 INTRODUCTION

The physical properties of condensed matter often result from an intricate interplay between the electronic, nuclear, orbital, and spin degrees of freedom. This is in particular also true for the transient processes occurring following photoexcitation of materials. Excitation using photons in the visible or ultraviolet part of the electromagnetic spectrum typically leads to the creation of highly excited electron–hole pairs, which on a very fast time scale thermalize among themselves. After these thermalization processes, kinetic energy relaxation takes place through energy transfer to other fundamental excitations of the system. A simple example of this is photoexcitation in semiconductors, where the electron–hole energy relaxation typically takes place through Fröhlich interaction-mediated excitation of longitudinal optical phonons (Chapter 8). Lattice anharmonicity leads to a subsequent decay of the longitudinal optical phonons into lower-energy phonons belonging, for instance, to the acoustical branches.

In more complex solids like the 3d transition metal oxides (Chapter 3), the decay process becomes correspondingly more complex too, involving a variety of other fundamental excitations including not only phonons, magnons, and orbital excitations but also pair-breaking excitations in superconductors, and phasons and amplitudons in charge density wave materials. For modest photoexcitation densities, the process of electron–hole excitation and subsequent relaxation can be considered as a near equilibrium phenomenon. Fundamentally, the materials do not change their nature. The structural and electronic properties remain intact and the only nonequilibrium changes are in the occupation numbers of the various excitations. When the excitation density is high enough, this situation might change, however. There are many examples where optical excitation fundamentally changes the nature of the material by inducing an optically induced phase transition. One of the most common examples might be the phase change materials used in optical data storage where one optically switches between crystalline and amorphous states of matter. In this particular example, nonequilibrium occupation of excited states does not play a role, and the switching process can be understood in thermodynamical terms and occurs on a nanosecond time scale. More intriguing (and faster) are phase transitions, which are driven by nonequilibrium populations. Examples are found in optically induced ultrafast melting (Sokolowski-Tinten et al. 2003), melting of the magnetic ordering (Beaurepaire et al. 1996), insulator to metal transitions (Chollet et al. 2005), paraelectric to ferroelectric transitions (Collet et al. 2003), melting of Peierls order (Karutz et al. 1998), and a structural phase transition in the A7 metals (Fausti et al. 2009). Many of the transient phenomena described earlier may be observed using, for instance, standard transient reflectivity experiments (see Chapter 9). For a detailed understanding of the transient physical processes, however, one needs probes that more specifically address the various excitations and their population dynamics, as well as the symmetry aspects of the materials under study. Nowadays, a large variety of dedicated fast probes are available to help elucidate the intricate details of ultrafast optically induced phenomena in condensed matter, many of which are discussed throughout this book.

14.1.1 Time-Resolved Raman Spectroscopy

Time-resolved Raman spectroscopy (TRRAS) studies transient phenomena induced by an optical pump pulse through the analysis of the time evolution of the spontaneous Raman spectrum (Chapter 6) measured with a second optical probe pulse, which is time delayed with respect to the pump pulse. Though one of the advantages of this technique is that it can individually address particular low-energy excitations such as phonons and magnons, yielding dynamical information on crystal potential and magnetic interactions and the nature of intermediate and metastable transient states, its main strengths are found in its capabilities to address the dynamical population of these excitations, through detailed balance, as well as the overall symmetry of the material, through the selection rules for Raman scattering. While there are alternative techniques for addressing particular excitations, such as time-resolved terahertz (THz) spectroscopy (see Chapter 11), which in centrosymmetric materials is complementary to TRRAS (see also Chapter 6), and the powerful but more elaborate time-resolved x-ray diffraction techniques (see Part V) to address crystal structure and symmetry, there are currently no other techniques which can address the population dynamics in such a direct manner as time-resolved Raman scattering does. In addition to this, TRRAS has the advantage that one does not need high-power probe pulses, which by themselves may induce photo-physical processes, as is, for instance, the case in coherent anti-Stokes Raman scattering experiments, nor does it typically rely on particular coherent excitations created by the pump pulse, as is the case in transient grating (see Chapter 10) and transient coherent phonon generation experiments. Another advantage of TRRAS is that it quite easily covers a wide frequency range, starting from a few wavenumbers ranging up to several thousands of wavenumbers. There are, however, limitations to this, related to the fact that short pulses are used to generate the Raman signal. As the probe pulse shortens, its spectral width increases. For transform-limited Gaussian pulses the relation between the Gaussian widths in the time ($\sigma_t = \Delta t/2\ln(2)$, with Δt the full width at half maximum) and frequency ($\sigma_\omega = \Delta\omega/2\ln(2)$) domains is given by $\sigma_\omega\sigma_t = 1$. In TRRAS, the observation of low-frequency modes is hampered by the elastically scattered probe light. The Raman process is an instantaneous process, which does not allow for filtering out the elastically scattered light in the time domain. In order to suppress the Rayleigh-scattered light, one is thus forced to spectrally filter the scattered probe light. The filtering bandwidth should be at least equal to the bandwidth of the probe pulse; in practice, it is often several times the bandwidth in order to fully suppress the elastically scattered light. At the same time the large bandwidth of ultrashort pulses poses problems in detection of the photoinduced spectral changes. Intrinsically narrow Raman modes are broadened by the experimental resolution (i.e., typically the bandwidth of the probe pulse), and any optically induced spectral line shift should at least be a reasonable fraction (10% or so) of the band-width to be observable. That the situation is not as bad as it sounds is illustrated in Figure 14.1, which shows the experimentally accessible region in the frequency-time bandwidth plane, together with the experimentally most interesting region. For a 1 ps transform limited pulse, the spectral bandwidth is about $3\,\mathrm{cm^{-1}}$, so that one can easily access excitations down to about $20\,\mathrm{cm^{-1}}$ at a resolution, which is not that much worse than the typical resolution of about $1\,\mathrm{cm^{-1}}$ used in most continuous wave Raman experimental setups (Chapter 6).

Though the Raman effect itself has been known for almost a century, applications in the time domain only started in the 1970s. Apart from the need for pulsed light sources, the development of time-resolved Raman has been hindered by the low scattering efficiency of the Raman process itself; typically only one in 10^8–10^{10} incoming photons undergoes inelastic scattering. Early applications of time-resolved Raman spectroscopy therefore made use of tunable pulsed dye lasers to exploit electronic resonances, thereby boosting the scattering efficiency by several orders of magnitude. An early study by Pagsberg et al. (1976) investigated the lifetime of short-lived radicals by monitoring the resonant Raman spectrum of the radicals excited using a flashlight-pumped tunable 600 ns coumarin-102 dye laser. An even earlier application (Laubereau et al. 1972), making use of the sensitivity of time-resolved Raman to population dynamics, focused on the measurement of vibrational lifetimes of molecules in

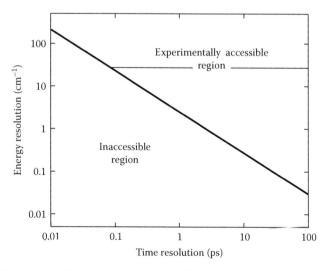

Figure 14.1 Experimentally accessible region in the time-frequency bandwidth plane (upper right part of the graph). The experimentally most interesting region is the one with a resolution of about 30 cm⁻¹ or better (the upper bound is indicated by the horizontal line).

liquids through time-resolved stimulated Raman experiments using a mode-locked amplified picosecond laser system. This experiment provided the first direct Raman determination of the lifetime of vibrational excitations. Nowadays, time-resolved Raman has found wide use in many branches of science (for a recent overview, see for instance reference [ham 2000]). Some recent achievements include studies of real-time protein dynamics (Yamamoto et al. 2000; Balakrishnan et al. 2008), reaction flows (Roy et al. 2008), photoinduced electron and energy transfer phenomena (Fujiwara and Mizutani 2008; Grosserueschkamp et al. 2009), energy dissipation (Iwata and Hamaguchi 1997), spin crossover (Smeigh et al. 2008), and different photo-isomerization processes (Yoshizawa and Kurosawa 1999; Ishii and Hamaguchi 2003; Kukura et al. 2005; Sakamoto et al. 2006; Du et al. 2009). Reviewing the historical approaches to solve different experimental issues and measuring the transient Raman response of molecules and matter in different fields ranging from materials science and chemistry to medical science, molecular biology, and cell biology goes beyond the purpose of this chapter and can be found in various reviews and experimental papers (Boteler and Gupta 1993; Okamoto et al. 1993; Hamaguchi and Gustafson 1994; Zhang and Hamaguchi 1994; Gupta et al. 1995; Uesugi et al. 1997; Kneipp et al. 1999; ham 2000; Everall et al. 2001; Mizutani and Kitagawa 2002; Huang et al. 2005; Lu 2005; Matsubara et al. 2006; Kukura et al. 2007; Pomeroy et al. 2008; Slipchenko et al. 2008; Balakrishnan et al. 2009).

This chapter focuses on time-resolved spontaneous Raman studies of photoinduced dynamics in complex condensed matter. As already sketched, the unique characteristic of Raman spectroscopy is the possibility of measuring in a direct way the amount of energy stored in a particular excitation. The theoretical description of Raman scattering usually proceeds by calculating the dipole moment induced by the incident light (Chapter 6); this is usually done through second-order perturbation theory, considering a classical electromagnetic field as a perturbation. By subsequently making use of the correspondence principle, the intensity of the radiation emitted by the dipole moment is calculated using the classical expression for an oscillating dipole. The full description goes beyond the purpose of this chapter and can be found elsewhere (Chapter 6, or any of the many books on Raman scattering). The aim here is to give an insight into how population dynamics comes into play, giving access, through the measurements of the scattering intensities, to a local "temperature" measurement specific for the excitation investigated. Temperature should not be taken too literally here; what is really measured is the density of excitations, which can only be used to define the temperature in thermodynamic equilibrium. A Jablonski diagram of the Raman scattering process is sketched in Figure 14.2. In the left part (Stokes process), an electron is excited by a photon and this excitation is followed by a simultaneous emission of a lower-energy photon and an excitation into the system, for instance a phonon. Conversely, in the anti-Stokes process the final state differs from the initial one by the absorption of an excitation from the system. The intermediate state (dashed line in Figure 14.2) could be either a real electronic state or a virtual one. In the first case, the scattering process is resonant and consequently called resonant Raman scattering, while in the latter case it is called spontaneous Raman scattering. In the nonresonant case, the intensity of the Stokes (I_s) and anti-Stokes (I_{as}) scattering processes depends on the occupation number n_i, of the excitation investigated as $I_s/I_{as} \sim (n_i + 1)/n_i$. In the resonant case, this ratio has to be corrected by the ratio of the resonant enhancement factors of the Stokes and anti-Stokes processes. At thermal equilibrium, the occupation number depends on the nature of the quasiparticle investigated, usually bosonic excitations like phonons. In the nonresonant case, the ratio between anti-Stokes and Stokes intensities then becomes (Chapter 6)

$$\frac{I_{as}}{I_s} = \left(\frac{\omega + \Omega}{\omega - \Omega}\right)^4 e^{-\hbar\Omega/kT} \tag{14.1}$$

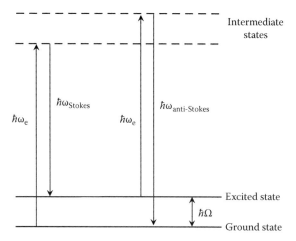

Figure 14.2 Jablonski diagram of the nonresonant Stokes and anti-Stokes Raman scattering processes. Incoming light with frequency ω_e is inelastically scattered through the creation or annihilation of an excitation with energy Ω.

where

> ω and Ω are respectively the frequency of the incoming light and the frequency of the excitation investigated
>
> k is Boltzmann's constant
>
> T is the temperature

This relation is illustrated in Figure 14.3 for several choices of the excitation frequency Ω. It should be noted that the previous expression is valid only in the limit where the Raman tensor for Stokes and anti-Stokes processes is the same. This is usually a valid assumption in transparent crystals, but, in general, it needs to be considered for the particular case investigated.

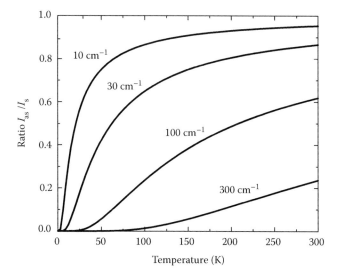

Figure 14.3 Ratio between the Stokes and anti-Stokes scattering intensities as a function of temperature for several fixed excitation energies (indicated in the figure). As the temperature increases, more and more excitations will be thermally excited leading to a much stronger increase of the anti-Stokes scattering intensity (see also Equation 14.4).

In the case of resonant processes, for example, the relative intensity of Stokes and anti-Stokes lines can depend strongly on the nature of the scattering process. There is no a priori relationship between the Stokes and anti-Stokes intensities and a detailed description of the scattering process is required before extracting any statistical information.

A common approach to study the photoinduced population dynamics is to compare the anti-Stokes intensity at a certain delay time after the optical pump pulse with the unperturbed one, rather than to compare the Stokes and anti-Stokes intensities at a given time delay. This method measures the transient population of a particular excitation induced by a pump pulse at $t = 0$. The transient anti-Stokes intensity is then given by $n(t)/n(t < 0) = I_{as}(t)/I_{as}(t < 0)$. The advantage of using this method rather than the comparison between Stokes and anti-Stokes intensities is that resonance effects do not play a role. An additional advantage is that one only needs to record the anti-Stokes part of the spectrum. Still, one has to make the assumption that the anti-Stokes Raman tensor does not depend on the optical excitation, which is usually, but not necessarily always, the case. Though one can also use the Stokes spectrum for the extraction of the population statistics, this is usually less sensitive since there the intensity is proportional to $1 + n$.

Since the Raman process is probing excitations in the neighborhood of zero momentum, it gives a true measure of the population statistics. The decay time measured in TRRAS therefore gives the population decay time τ_{pop} and not the decay time τ_s of a single quantum mechanical state specified by its energy and momentum. It therefore may differ from the single particle lifetime one extracts from the line width of a Raman band in the frequency domain, as well as from the decay time τ_{coh} as observed in transient coherent excitation spectroscopy in the time domain. The latter is determined both by population decay of the excited coherent excitation and by the dephasing processes due to elastic and intraband scattering processes (with dephasing time τ_p), and is determined by $\tau_{coh}^{-1} = \tau_s^{-1} + \tau_p^{-1}$ (Chapter 10).

In principle, one can also use TRRAS to study momentum relaxation, though no experimental study in this direction has been reported so far. Since a photon carries only a very small momentum, the Raman process is essentially restricted to $k \sim 0$ excitations. However, when the scattering process involves multiple particles momentum conservation only requires the sum of the momenta to be zero. In a two-particle process this means that simultaneously two excitations are created with momenta k and $-k$, that is, the Raman spectra will show a band that reflects the joint density of states of the excitations involved, weighted by the k-dependent scattering efficiency. In principle, it should thus be possible to follow the momentum relaxation in time by analyzing the line shape of the multiparticle scattering band. Typical examples are found in multiphonon scattering processes, and in two-magnon scattering in ordered antiferromagnets (Fleury and Loudon 1968) such as MnF_2 (Fleury et al. 1967) and in quantum antiferromagnets such as $CuGeO_3$ (van Loosdrecht et al. 1996).

The following section introduces a few examples of time-resolved Raman spectrometer systems and will go deeper into some of the more technical details. After that, some particular examples of the use of this technique to study complex matter will be reviewed starting with studies on phonon population dynamics and energy relaxation in semiconductors and carbon nanotubes. Subsequently, we will introduce two different prototypical works devoted to study complex matter: a Jones–Peierls system (the A7 semimetals) and high-temperature superconductors ($Bi_2Sr_2CaCu_2O_8$).

14.2 TECHNIQUES AND INSTRUMENTATION

14.2.1 General Considerations

A time-resolved Raman experiment is quite analogous to a conventional pump-probe experiment (Chapter 9). The major difference is that instead of registering the transient intensity of *elastically* scattered light, a TRRAS experiment records the spectral intensity of the

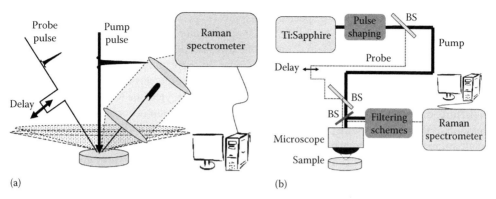

Figure 14.4 (a) Sketch of a Raman pump/probe experiment. The sample is excited by a short, intense pump pulse. The time-delayed Raman response is then measured by a probe pulse at later times. The delay between pump and probe pulses is controlled by a mechanical delay as in standard pump/probe measurements. The dashed lines indicate that, in contrast to elastic pump/probe experiments, the inelastically scattered light is emitted over a large angle. In view of the low efficiency of the Raman process, one therefore typically uses a large numerical aperture lens to collect the scattered light. For this reason, the TRRAS experiments are often performed using microscopes in a back-scattering configuration, allowing usage of large numerical aperture objectives. A scheme utilizing a microscope is depicted in (b), which also includes (schematically) filtering schemes to reject/separate the scattered light from the pump beam as well as a pulse shaping block to tune the resolution of the experiment (see text).

inelastically scattered light in order to detect transient frequency shifts, line width changes, intensity changes, and/or selection rule changes (corresponding to symmetry changes in the material) in the Raman spectrum. A conceptual scheme for an optical pump/Raman probe experiment is depicted in Figure 14.4a. As in standard optical pump/probe experiments, two light pulses are used, the first to excite the sample (pump pulse) and the second to measure its response at later times (probe pulse). Mechanical (or electronic) control of the delay between the pump and the probe pulses is introduced to control the difference in arrival time of the pump and probe pulses at the sample in order to be able to measure the photoinduced changes as a function of the time elapsed after photoexcitation. The spontaneous Raman response of the probe pulses is measured for a fixed time delay by frequency analysis of the Raman scattered light using a monochromator equipped with a charge coupled device (CCD) detector (multichannel detection) or a photomultiplier (single channel detector) (see also Chapter 6). It should be noted that the spontaneous Raman scattering processes preserve the bandwidth of the probe pulses. The time resolution of the experiments is therefore limited by the autocorrelation between the pump and the probe beams. The frequency resolution in a TRRAS experiment is typically limited by the bandwidth of the probe pulses.

Figure 14.4a also shows some of the major conceptual differences between TRRAS experiments and most of the other time-resolved optical techniques. As schematically indicated by the conical-shaped dotted areas, both pump and probe beams produce Raman scattered photons diffused over large solid angles. Clearly, to increase the measurement efficiency, a collecting objective with large numerical aperture is desirable, as is depicted in Figure 14.4a by the high numerical aperture lens collecting the scattered light (gray-shaded cone). In an actual experiment, one usually suppresses part of the elastically scattered light by introducing a beam stop in the detection path, which blocks the direct pump and probe light from entering into the Raman spectrometer.

In some experimental realizations the probe and pump wavelengths are equal. This introduces a potential problem since the pump pulse will also give rise to a Raman response, which spectrally overlaps with the probe spectrum. In typical experiments, the pump response is even stronger than the Raman response of the probe pulse. Later in this section, we will describe some of the different approaches to coupling the diffused light from the probe into the spectrometer, while rejecting the pump-induced response. In particular cases, however, measuring the pump- and probe-induced spectra simultaneously can be quite advantageous. Since Raman scattering is an instantaneous process, the pump response will yield the Raman response of the unperturbed sample as long as the pump-induced changes are not occurring during the pulse duration of the pump pulse. By using pump and probe wavelengths, which are separated by a small amount, sufficient to avoid spectral overlap of the pump and probe Raman spectra, the obtained pump/probe spectrum will simultaneously contain spectrally separated information on both the unperturbed state and the induced transient state, which may be compared directly. As in the comparison between Stokes and anti-Stokes intensities, however, one has to ascertain that resonant effects do not play a significant role here, or otherwise correct for this.

Some of the important characteristics to take into account in the design of a time-resolved Raman spectrometer are listed here:

1. The cross section of spontaneous Raman processes is very low ($<10^{-8}$–10^{-10}) requiring sensitive detection techniques (for general Raman detection methods see Chapter 6).
2. The scattered photons are emitted over large solid angles, which in view of the previous discussion requires implementation of a scheme to collect the probe response over large solid angles.
3. Many relevant excitations in solids are found at low energy, and the frequency of the Raman scattered light will be quite close to the frequency of the excitation laser. Measuring them therefore requires a dedicated scheme to reject the elastic response of both pump and probe beams. This can be achieved partially by introducing beam stops at appropriate places in the setup and partially by using high-resolution spectrometers in the detection path.
4. When equal pump and probe wavelengths are used, one has to suppress the Raman spectrum of the pump light. One of the methods to achieve this is to make use of the polarization properties of the Raman scattered light.
5. The spontaneous Raman scattering processes preserve the bandwidth of the probe pulses, which therefore determines the spectral resolution. The lowest frequency accessible is in principle determined by the tails of the spectral content of the probe pulses, unless the pump and probe pulses have the same wavelength, in which case the sum of the tails of both determines this. It is therefore desirable to introduce a scheme to optimize the bandwidth (spectral resolution) for the particular experiment at hand.

The considerations mentioned earlier are generally valid for all TRRAS schemes, but their implementation in an effective setup should be considered in relation to the specific scientific question of interest. To make this clear let us give a couple of examples. If the materials of interest are characterized by a very low scattering cross section, it will be crucial to choose high repetition rate lasers allowing high average fluence, and a large numerical aperture objective to increase the collection efficiency (Waldermann et al. 2008; Saichu et al. 2009); in case the setup is dedicated to study luminescent or resonant materials, the pump and probe wavelengths should be chosen critically and luminescence rejection schemes should be implemented (Benniston et al. 2000; Song et al. 2008); if the primary interest is the low-energy modes, the spectral content of the pump and probe pulses should be controlled by specific pulse-shaping techniques (Nakajima et al. 2005; Fausti et al. 2009). As usual, a detailed analysis of the perceived experiments should be the starting point of designing a TRRAS experimental setup.

Before entering a more detailed description of some of the experimental approaches in use, a few considerations on the type of laser sources eligible for this kind of experiments need to be made. Given the low scattering efficiency, the average laser power is often critical in the measurements of spontaneous Raman scattering. The number of Raman scattered photons grows linearly with the integrated intensity of the laser source. In this respect, a high average power for the probe beam is desirable. On the other hand, in an ideal pump/probe experiment the intensity of the probe pulses should be low (in the linear regime) in order not to perturb the sample during the probing process. For these reasons, the repetition rate of the laser sources is usually crucial. A high repetition rate allows high average probe power while keeping the energy per pulse relatively low. The trade-off of high repetition rates is that the studies of highly nonlinear regimes are inhibited due to average heating effects of the pump pulses. On the other hand, low repetition rate systems reduce the average probe intensity, facilitating the investigation of a larger dynamical range in the pump excitation density. Considering the case of Ti:Sapphire systems, most of the transient spontaneous Raman experiments reported to date have been performed with high repetition rate lasers: oscillators working at ~80 MHz or cavity-dumped oscillators where the repetition rate is tunable down to a few hundred kilohertz. This requirement is relaxed for studies under resonant conditions, where researchers have succeeded in measuring population dynamics with a regenerative amplified 1 KHz laser (Song et al. 2008). As a final comment regarding the optimal repetition rate, it should be noted that a low repetition rate amplified system allows a large tunability of pump and probe (through the use of optical parametric amplifiers), while oscillators operate in a limited wavelength region (700–1000 nm plus the higher harmonics). This is useful since it allows one to choose the optimal pump wavelength to excite, for instance, a particular electronic state and to choose a probe wavelength, which is tuned to maximize resonance effects. In experiments using different pump and probe wavelengths one has to take care, however, that the penetration depth of the probe light is comparable to or smaller than that of the pump light, since otherwise the probe light might be probing an unperturbed region of the sample.

14.2.2 Time-Resolved Raman Spectrometers

In the detailed description of the time-resolved Raman apparatus, we will focus mostly on modern implementations based on Ti:Sapphire lasers. Nevertheless, the history of time-resolved Raman spectroscopy in condensed matter is older than the Ti:Sapphire laser and it is worth here collecting a set of references to the TRRAS configurations developed before the "Ti:Sapphire era." In the early 1980s the first schemes proposed to measure the transient Raman response of low-energy excitations in condensed matter were proposed by von der Linde et al. (1980). Making use of a system based on a synchronously mode-locked rhodamine 6G (575 nm, 80.6 MHz repetition rate) dye laser, the authors were able to measure directly the time evolution of the optically induced nonequilibrium population of incoherent longitudinal optical phonons in GaAs. A few years later, using a similar system with improved time resolution (and a slit to cut the tails in the pulses' frequency distribution), Kash et al. (1985) were able to temporally resolve the growth of the optically induced nonequilibrium longitudinal optical (LO) phonon population in GaAs induced by kinetic energy decay of optically excited electron–hole pairs. A different scheme with two independently synchronously pumped tunable dye lasers (Oberli et al. 1987; Tatham et al. 1989; Tsen 1992) allowed measurements with different pump and probe colors, but the time resolution was limited by the jitter in the synchronization of the lasers to about 8 ps (Oberli et al. 1987) and 3 ps (Tsen 1992). A few years later Grann et al. (1996) used ultrashort laser pulses generated from a double-jet dye laser synchronously pumped by the second harmonic of a cw mode-locked yttrium aluminum garnet laser to achieve sub-picosecond resolution in TRRAS experiments.

14.2.2.1 Ti:Sapphire-Based Systems

With the discovery of Ti:Sapphire passive mode-locking techniques, most of the dynamical spectroscopic techniques, as described in other chapters, were improved and setups making use of the new laser sources were developed. In contrast, the development of time-resolved Raman spectroscopy based on Ti:Sapphire lasers has been relatively slow, and only in the more recent years various time-resolved Raman experiments in condensed matter have been performed with Ti:Sapphire lasers. This is, in the authors' opinion, due to two main reasons. The first is that passive Kerr lens mode locking works "easily and stably" in the femtosecond regimes yielding high time resolution (<100 fs) measurements with lasers that are omnipresent in optical laboratories (Chapter 7). It is clear though that this is achieved at the cost of frequency resolution, limiting the possibility of studying detailed vibrational dynamics in crystals. Moreover, the most interesting region often is the low-energy region ($E < k_b T \sim$ $100\,cm^{-1}$), which would be hidden under the spectral width of the laser pulses. These observations strongly limit the applicability of the technique in the femtosecond regime. Even though cavities lasing with narrower bandwidth are available, they require active mode-locking schemes and are somewhat less easy to operate. The second, and maybe more practical, reason is that the lasing of most of the Ti:Sapphire passive mode-locked cavities is characterized by the presence of a background luminescence as intense as 10^{-6} with respect to the fundamental laser line (due to the broad Ti:Sapphire luminescence band). Considering that the characteristic scattering cross section of a spontaneous Raman process is on the scale of 10^{-8}–10^{-10}, Ti:Sapphire lasers are not suitable for Raman experiments without proper beam conditioning, requiring additional filtering schemes to suppress the background luminescence of most of the Ti:Sapphire cavity designs.

14.2.2.2 Experimental Realizations

The general layout of a time-resolved Raman experiment based on a single oscillator is depicted in Figure 14.5a and b. The laser beam is divided into two parts and, as in any optical pump/probe experiment, one is used as a pump, while the second one is used to probe the material response at a controllable delay time. Working with the samples in a microscope in back scattering configuration facilitates the use of large numerical aperture objectives (NA > 0.5) necessary to increase the collection efficiency. Pump and probe are combined collinearly through the microscope objective, which also collects the Raman back-scattered light. A beam splitter couples the scattered light into a high-resolution spectrometer. Clearly, both the reflected beams and the inelastic response of the pump beam are also collected by this objective and need to be discriminated by the detection system.

Different schemes have been employed to discriminate between the Raman signals coming from the pump and probe beams. The two common approaches are (1) polarization selection and (2) frequency selection. (1) For single-color time-resolved Raman experiments, the pump and probe beams are cross-polarized, and a polarizing beam splitter (or a beam splitter plus a polarizer) between the microscope and the spectrometer suppresses the scattered light from the pump. The major drawback of this approach is the impossibility to access all possible polarization combinations, that is, all elements of the Raman tensor (Chapter 6) (Letchter et al. 2007; Fausti 2008; Fausti et al. 2009). (2) For a two-color experiment the pump and probe are separated in frequency; therefore, the inelastic response of the pump will appear in a different spectral region and can be removed by a band-pass filter or by selecting the region of interest through the gratings and slits of the spectrometer. The drawback of a single oscillator configuration is that one is forced to either perform single-color experiments or use two-color experiments making use of the various harmonics. In many studies, however, it is desirable to be able to tune the wavelengths independently. This has the advantage that one can more

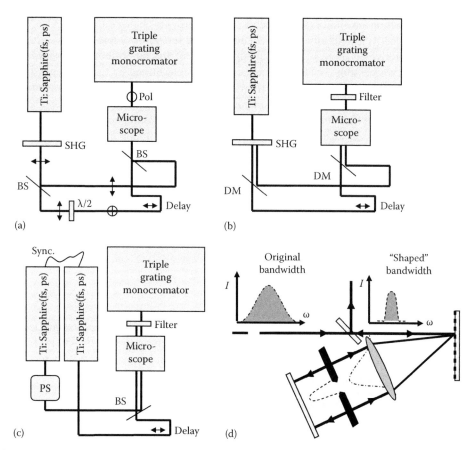

Figure 14.5 Sketches of the two most common schemes implemented to measure TRRAS. The configurations described in (a) and (b) make use of a single oscillator (Machtoub et al. 2003a; Fausti et al. 2009; Saichu et al. 2009). The selection of the probe response is done with polarization (a) and band-pass filters (b). A two-color experiment based on two synchronized oscillators is depicted in (c). The most basic pulse-shaping filter to cut off the tails of the spectral distribution, allowing TRRAS measurements at low frequencies, is shown in (d) (Fausti 2008).

intelligently pump the materials under study by tuning to specific electronic transitions and more efficiently probe by maximizing the Raman response. In addition, it becomes possible to observe the unperturbed and transient spectra at the same time by shifting the probe energy by a few times the frequency resolution with respect to the pump. For these reasons, a setup based on two synchronized oscillators, independently tunable, with the low repetition rate option, would be ideal (Figure 14.5c), with the added advantage that the pump–probe delay time can be controlled by the synchronization electronics. To align an experimental pump–probe setup one usually optimizes on the probe signal. In the case of TRRAS, however, this is usually not the best approach due to the weakness of the scattered Raman signal. Therefore, one usually relies on first optimizing the direct transient reflection signal by recording the transient Rayleigh scattered light. Only after that can one optimize the Raman probe signal itself (which then does not need much tweaking anymore).

14.2.2.3 Controlling the Time and Frequency Resolution Trade-Off

The analysis of low-frequency modes is often of particular interest in studying materials after photoexcitation. For vibrational excitations, these correspond to the external lattice modes yielding information on the crystal structure and (transient) changes therein. Also, single particle and collective excitations in semiconductors and superconductors are typically found in this region. Provided there is sufficient suppression of the pump-induced spectrum, the lowest frequency accessible in TRRAS experiments is determined by the frequency tails of the probe pulse bandwidth. Figure 14.5d shows a filter configuration allowing for continuous trade-off between time and frequency resolutions. The broadband laser beam is dispersed by a grating, collected by a lens and reflected back at the same horizontal angle, but under a small vertical angle, so that the Gaussian intensity profile is restored in the filtered beam. It is now easy to introduce a "soft" slit selecting the spectral bandwidth of the optical pulse leading to a temporally longer but spectrally narrower pulse. In this configuration, the spectral content and the time duration of the probe pulses can be tuned continuously, while at the same time filtering out the long tails and any luminescence coming from the laser cavity. Making use of this optical design in combination with a triple grating monochomator, it has been shown possible to measure excitation spectra down to at least $50\,cm^{-1}$ with a time resolution of about 1 ps (Fausti 2008).

14.3 EXPERIMENTAL STUDIES

14.3.1 Semiconductors

Photoexcitation of a semiconductor using photon energies exceeding the bandgap leads to the creation of a nonequilibrium distribution of excited electrons and holes. After a fast initial charge thermalization process driven by electron–electron interactions, energy relaxation occurs by emission of longitudinal and acoustical phonons, which typically occurs on a picosecond timescale (Chapter 8). In polar semiconductors the dominant process is LO phonon generation (Collins and Yu 1984). This process generates a nonequilibrium phonon distribution, which thermalizes through phonon–phonon interactions, typically also on a picosecond timescale, albeit usually somewhat slower than the electron–phonon relaxation. The first time-resolved Raman experiments in a condensed matter system, performed by von der Linde et al. (1980) studied the relaxation of a nonequilibrium LO phonon population in GaAs generated by energy relaxation of optically created electron–hole pairs by monitoring the transient peak intensity of the LO phonon Raman response at $295\,cm^{-1}$. They used an 80 MHz, 80 mW Rhodamine 6G dye laser operating at 575 nm with a pulse duration of 2.5 ps. The laser pulses were split into two (50:50 ratio) cross-polarized pulses, which were used for pumping and detection using a regular pump–probe scheme. The perpendicular polarization states of the pump and probe pulses were crucial to discriminate between Raman scattered phonons generated by the pump and the probe pulses. They showed that the population relaxation time, due to decay into acoustic phonons is about 7 ps at 77 K. The relatively long excitation pulses in that experiment did not allow observation of LO phonon generation by decay of the optically excited electron–hole pairs. This was done in a later experiment by Kash et al. (1985) using an improved setup with a time resolution better than 1 ps. They were the first to report time-resolved spectra, which are reproduced in Figure 14.6, measuring LO phonon decay times of 4 ps at room temperature and 7.5 ps below 80 K. Their time resolution also allowed a study of the growth of the LO phonon population after photoexcitation, that is, the decay of photoexcited charge carriers into LO phonons, yielding an electron–phonon scattering time of 165 fs. In addition, they observed an appreciable decrease of the scattering efficiency of the LO phonon at high pump excitation densities resulting from screening of the LO phonons by the photoexcited carrier density, in good agreement with earlier excitation power–dependent picosecond excited time–integrated Raman experiments by Collins and Yu (1984). Since the early experiments many other studies focusing on phonon generation and decay have been performed, often using resonant enhancement by tuning probe beam wavelength to an

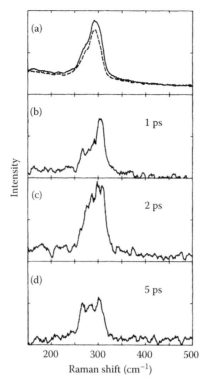

Figure 14.6 (a) Time-resolved anti-Stokes Raman spectrum for room temperature GaAs. The instantaneous carrier concentration induced by the pump laser was about 5×10^{16} cm^{-3}. For the dashed curve the probe preceded the pump by 20 ps; the solid curve is for a probe delay of 1 ps. (b)–(d) Difference spectra at 1, 2, and 5 ps, obtained by subtraction of Raman spectra as in (a). (Reprinted with permissions from Kash, J.A., Tsang, J.C., and Hvam, J.M., *Phys. Rev. Lett.*, 54, 2151. Copyright 1985 by the American Physical Society.)

electronic transition, to improve the signal-to-noise ratio. These studies include work on Si (Letchter et al. 2007) and Ge (Ledgerwood and van Driel 1996), AlGaAs and InGaAs alloys (Kash et al. 1987), GaAs multiple quantum well systems (Tatham et al. 1989; Tsen et al. 1989, 1991), GaN, ZnSe, and GaP (Siegle et al. 1996), InP and InAs (Grann et al. 1996), GaN and AlGaN (Tsen et al. 1999), and InN (Tsen and Ferry 2009).

Apart from monitoring the generation of optical phonons by charge energy relaxation, it is also possible to look at the charge energy decay in a more direct manner by monitoring the electronic Raman scattering spectrum. This approach has been followed by Oberli et al. (1987) who recorded the transient dynamics of the single-particle excitations of the two-dimensional electron–hole plasma in GaAs quantum wells. From their experiments, they were able to extract the transient carrier densities of the electronic subbands as well as the interband scattering rate (~0.5 ns^{-1}) and single-particle lifetimes (~0.3–0.8 ns). Picosecond-excited single-particle spectra in InP and InAs have been reported by Grann et al. (1996), who also, based on previous phonon population decay experiments, suggested a $\tau \sim d^{-10}$ relationship between the LO phonon population relaxation time τ and the bond length d in III–V semiconductors.

14.3.2 Carbon-Based Materials

The decay from optically excited hot charge carriers into optical phonons, and then eventually into acoustic phonons, has also been studied in carbon-based materials. Early

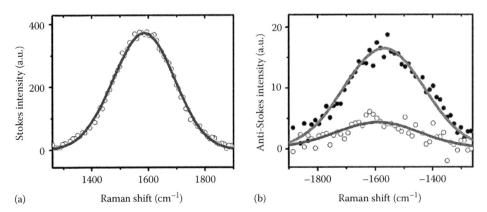

Figure 14.7 Stokes and anti-Stokes Raman spectra of G-mode phonons in single-walled carbon nanotubes measured with femtosecond probe pulses at 2.17 eV. The solid lines are fits to Gaussian functions with a width reflecting that of the probe pulses. (a) Spectrum for Stokes Raman scattering. The signal is nearly independent of the presence or absence of a time-synchronized femtosecond pump beam, since the population factor 1 + n hardly changes. (b) Spectrum for anti-Stokes Raman scattering in the presence (filled circles) and absence (open circles) of pump pulses (1.55 eV, 1 J/m²). The strong effect of the pump beam here is due to the large changes in the anti-Stokes scattering efficiency, which is directly proportional to the induced population, n. (Reprinted with permission from Song, D., Wang, F., Dukovic, G., Zheng, M., Semke, E.D., Brus, L.E., and Heinz, T.F., *Phys. Rev. Lett.*, 100, 225503. Copyright 2008 by the American Physical Society.)

non-time-resolved picosecond work (Tsen et al. 1996) on solid C_{60} used the enhanced anti-Stokes signal, reflecting the phonon occupation, as a measure for the electron–phonon interaction strength. From the observed behavior of the low-frequency A_g mode after photoexcitation it was concluded that vibrations involving C–C stretching couple more strongly to the charge carriers than others, even though the high-frequency pentagonal pinch mode, which also involves C–C stretching, did not show any enhanced photoinduced population. More recently, experiments have been performed on single-walled carbon nanotubes (Kang et al. 2008; Song et al. 2008). Also in the nanotubes, a coupling to the stretch modes has been observed, in particular to the so-called G-mode near 1600 cm⁻¹. An illustration of time-resolved anti-Stokes spectra used to study the G-mode dynamics in carbon nanotubes is shown in Figure 14.7 (Song et al. 2008). For semiconducting nanotubes, the measured lifetime of the G-mode optical phonon was determined to be 1.1 ps (Song et al. 2008). From temperature-dependent measurements, it was concluded that this finite lifetime is due to G-mode decay into two acoustic phonons (Kang et al. 2008), quite similar to what happens in most of the semiconductors discussed earlier. The decay time of the G-mode in metallic nanotubes is found to be similar (0.9 ps) to that of the semiconducting ones (Kang et al. 2008). It is interesting that the corresponding optical phonon in graphite decays a factor of 2 slower, with a decay time of about 2.2 ps (Yan et al. 2009). This is most likely due to the strong coupling of the G-mode to the radial breathing modes in single-walled carbon nanotubes (Gambetta et al. 2006), a process that is absent in graphite. Apart from the apparent population dynamics, the experiments on carbon nanotubes also show a small frequency hardening of the G-mode (see Figure 14.7b) upon increasing pump power. A more detailed time dependence of the corresponding shift in graphite, together with the measured phonon population and derived phonon temperature, is shown in Figure 14.8. The origin of the stiffening is the high electron temperature after photoexcitation, which screens the electron–phonon interaction, leading to the observed frequency hardening (Yan et al. 2009).

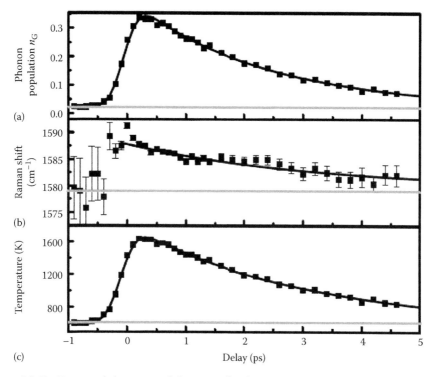

Figure 14.8   Temporal dynamics of the G-mode phonons in graphite as a function of delay time following pump excitation. (a) Experimental anti-Stokes Raman intensity, which is proportional to the phonon mode population. The solid line represents an exponential decay with a time constant of 2.2 ps convoluted with the instrumental response. (b) Measured shifts of the G-mode frequency. The solid line is based on the model of the temperature dependence of the self-energy described in Yan et al. (2009). (c) Temperature of the G-mode phonons inferred from the mode population in (a). At negative delay, the temperature is higher than room temperature. This increased temperature arises from self-pumping by the probe pulse. (Reprinted with permission from Yan, H., Song, D., Mak, K.F., Chatzakis, I., Maultzsch, J., and Heinz, T.F., *Phys. Rev. B*, 80, R121403. Copyright 2009 by the American Physical Society.)

14.3.3 Quasiparticle Dynamics in High-Temperature Superconductors

One of the most intriguing open problems in condensed matter science is the origin of the superconducting properties of the cuprates. It is therefore not surprising that the superconducting phase of high-temperature superconductors (HTSCs) has been widely studied at thermal equilibrium through, for instance, optical conductivity (Basov and Timusk 2005), Raman (Devereaux et al. 1994; Devereaux and Hackl 2007), angle-resolved photoemission (Damascelli et al. 2003), and STM (McElroy et al. 2003) techniques. However, unambiguous answers to the major questions about the physics of HTSC are still to be answered. What is the basic interaction responsible for the binding potential necessary to form Cooper pairs (Monthoux et al. 2007)? What is the nature of the pseudogap phase (Timusk and Statt 1999) where Cooper pairs are probably formed locally in the absence of a macroscopic condensate (Julian and Norman 2007; Hufner et al. 2008)?

Recently, a new approach, based on ultrafast sub-100 fs coherent laser sources, has been developed to investigate the physics of HTSCs making use of their nonequilibrium properties. The basic idea behind these nonequilibrium experiments is to photo-inject an excess of excitations through an ultrashort pump pulse that modify the equilibrium electron distribution on a timescale faster than the thermal heating. Measuring in real time the recombination processes of the photoexcited quasiparticles, the nature of the superconductive gap can be inferred. Various are the studies based on time-resolved reflectivity/transmittivity measurements with femtosecond time resolution (Kabanov et al. 1999; Gedik et al. 2005, 2009; Liu et al. 2008). All these techniques probe the transient optical properties related to the density of the excitations photo-injected into the system, and information on the superconducting gap dynamics can be retrieved only indirectly from the study of the recovery dynamics. In contrast, the unique feature of time-resolved Raman spectroscopy to probe HTSC is the possibility of directly probing the superconducting order parameter and the charge, phonon, and magnon populations as well as their mutual interactions (Tacon et al. 2006).

The basic idea of the time-resolved Raman experiments reported by Machtoub et al. (2003a,b) and Saichu et al. (2009) is the following: an intense short light pulse (pump) perturbs the electronic distribution in the superconducting sample and a second pulse probes the Raman active excitations at a controllable time delay. Both groups used a system based on an 80 MHz Ti:Sapphire laser. Machtoub et al. measured the relaxation times of high-energy excitations in a single-color experiment with cross-polarized second harmonic light (pump and probe at 3.12 eV), while Saichu et al. performed two-color experiments, exciting with the second harmonic (3.44 eV) and probing with the fundamental laser frequency (1.72 eV). The two beams were coupled collinearly into a custom-designed large numerical aperture objective (Schulz et al. 2005), and the scattered light from the probe was frequency selected (Figure 14.9). Machtoub et al. studied the relaxation processes following an O_{2p}-Cu_{3d} charge transfer excitation. The major finding of Machtoub et al. (2003a,b), deduced from a symmetry

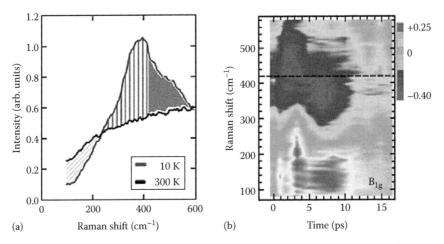

(a)   Raman shift (cm^{-1})   (b)   Time (ps)

Figure 14.9 (See color insert.) (a) Steady-state Raman spectra of Bi-2212 in the B_{1g} geometry (in Porto notation z(xy)z, where z is perpendicular to the CuO planes) at 10 and at 300 K. A gap opens below 250 cm^{-1} (blue area) and a pair-breaking peak appears around 420 cm^{-1} (red area). (b) Temporal evolution of the time-resolved Raman difference spectra at an equilibrium temperature of 10 K in the B1g geometry. The figure represents a contour plot consisting of 12 Raman difference spectra for different delay times. The dashed line separates two energy regions of the pair-breaking peak that reveal different characteristic behaviors. The intensity changes are color coded, demonstrating the transfer of spectral weight from high to low energies after 1 ps. (Reprinted with permission from Saichu, R.P., Mahns, I., Goos, A., Binder, S., May, P., Singer, S.G., Schulz, B., Rusydi, A., Unterhinninghofen, J., Manske, D., Guptasarma, P., Williamsen, M.S., and Rbhausen, M., *Phys. Rev. Lett.*, 102, 177004. Copyright 2009 by the American Physical Society.)

analysis of the scattered Raman signals in different scattering geometries and their time dependence, is that in under-doped cuprates the relaxation times for phonons and magnons are different. The measured phonon and magnon population dynamics are well described by simple rate equations accounting for the relaxation of the photoexcited electronic excitation through non-radiative processes via scattering with phonons and magnons (Machtoub et al. 2003b).

The first direct study of the superconducting order parameter in the superconducting phase has been reported recently by Saichu et al. (2009). The authors studied the superconducting order parameter in slightly over-doped $Bi_2Sr_2CaCu_2O_{8+\delta}$ by the analysis of the transient Raman response of the low-energy electronic excitations (see Figure 14.9). The steady-state Raman spectrum shows two characteristic features below the superconducting transition temperature: the opening of a gap below $250\,cm^{-1}$ leading to a reduction of the scattering in this region, and the appearance of a pair-breaking peak around $420\,cm^{-1}$ (see Figure 14.9a). Photoexcitation by the pump in a time-resolved experiment leads to the partial suppression of the pair-breaking peak, and an enhanced scattering in the gap region, that is, there is a spectral weight transfer from the pair-breaking peak to the sub-gap region upon photoexcitation. The relaxation of this nonequilibrium situation was found to occur in a nontrivial fashion, from which the authors conclude that two relaxation mechanisms should contribute to the recovery of the pair-breaking peak and hence to the recovery of the superconducting condensate. The intensity of the high-energy part of the pair-breaking peak (above the dashed line in Figure 14.9b) shows a fast (2 ps) decrease followed by a slow relaxation (7.4 ps). This decay can be understood by assuming hole–phonon coupling mostly via the in-plane breathing mode. A second type of transient response is found for the low-energy part of the pair-breaking peak, indicative of an important second coupling mechanism. Here, the decrease of intensity only sets in 5 ps after photoexcitation, while the recovery is found to be quite fast (1.4 ps). The authors argue that this observation might be understood in terms of a second coupling mechanism arising from hole–spin coupling. In addition, the charge inhomogeneities characteristic for high-temperature superconductors might play a role here. A more definitive interpretation of the last observation in particular has to await new experiments as well as a better theoretical understanding. These pioneering studies discussed here, in particular, the work by Saichu et al., have demonstrated the possibility of measuring directly and in detail the recombination processes of photoexcited electrons in high-temperature superconductors, providing hope that TRRAS studies in the years to come might shed some light on the nature and origin of the superconductivity in these intriguing and challenging materials.

14.3.4 Ultrafast Symmetry Breaking in a Photo-Stimulated Peierls System

The A7 elemental semimetals like bismuth and antimony have served in the last two decades as the primary playground for studying interactions between ultrafast light pulses and absorbing matter (Cheng et al. 1990; Hase et al. 1998, 2002; Boschetto et al. 2008; Garl et al. 2008). The physics behind the optically induced effects in the A7 semimetals can be sketched in a simple way. While nearly all of the elemental metals crystallize into a cubic or hexagonal closed α-Arsenic structure with rhombohedral space group $R\bar{3}m$, the equilibrium structure of bismuth and antimony sketched in Figure 14.10a may be described as a distorted simple cubic structure, where the (111) planes of atoms have an alternating displacement along the [111] direction. This structural anomaly of the semimetals Bi, Sb, and As has been widely discussed in the past (Peierls 1991) and originates from a strong electron–phonon coupling. In one dimension, this type of distortion is the well-known Peierls distortion. As depicted in Figure 14.10a, the reduction of the crystalline symmetry opens a gap at the Fermi level and thereby reduces the total energy of the populated electronic states. This results in a net energy

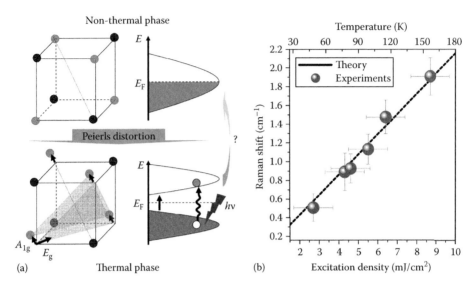

Figure 14.10 Photoinduced low symmetry phase in antimony single crystal. (a) Sketch of the structure of the A7 semimetals and of the idea behind the photoinduced reversed Peierls transition. Photoexcitation reduces the population of the valence band, thereby reducing the electronic energy gain of the Peierls distortion, and driving the system toward the undistorted phase. (b) A_{1g} Phonon frequency in Sb measured 15 ps after photoexcitation. The dashed line is calculated making use of thermal parameters. The bottom scale indicates the energy of the light pulses and the top one the local temperature expected due to heating effects. The agreement between the model and the data indicates that electronic heat diffusion plays a minor role in the relaxation processes (see text). (Adapted with permission from Fausti D. et al., *Phys. Rev B*, 80, 161207, 2009.)

gain compensating the elastic energy cost of the structural distortion. As the temperature is raised, more and more electrons will be excited from the occupied density of states (valence band) into the unoccupied density of states (conduction band) above the gap. A high enough thermal occupation of the valence band leading to a recovery of the undistorted cubic symmetry, however, cannot be reached at any temperature below the melting temperature in Sb or Bi. One can overcome this by using photoexcitation. For sufficiently strong pump pulses, one can in principle deplete the population of the valence band substantially enough to effectively reduce the electronic energy gain of the Peierls distortion, thereby driving the system toward the undistorted phase. Various studies have tried to address this highly nonequilibrium regime but failed to visualize a reversible symmetry change in bismuth and antimony (Sokolowski-Tinten et al. 2003; Fritz et al. 2007).

The strong electron–phonon coupling in the A7 metals leads to a charge density–dependent renormalization of the frequency of the phonon connected to the structural distortion. A simple, phenomenological approach, as suggested by Zeiger et al. (1992), already gives a quite reasonable qualitative description of the phonon softening as a function of the dynamical electronic population $n(t)$ of the conduction band.

Starting from the hypothetical high-symmetry cubic structure, the A7 structure arises from a structural phase transition with the amplitude of the displacement of the atoms along the cube's body diagonal, q (see Figure 14.10a), as the phase transition's order parameter. This corresponds to the atomic displacement pattern associated with the Raman active A_{1g} phonon. In terms of this order parameter, a Landau expansion of the free energy $E(q)$ may be written as $E(q) = -an_0q^2 + bn_0q^4 + \cdots$, where n_0 is the number of unit cells per unit volume. Note that in this expression the temperature dependence of the coefficients in the expansion is not included explicitly, since this is less relevant for the present discussion. It will be assumed that there is a nonzero order parameter, and, hence, that the system is in the distorted A7 phase. The equilibrium position in the distorted phase is obtained by minimizing the free energy with respect to q, and is given by $q_{0S} = \sqrt{a/2b}$. From the equation of motion $\mu n_0 \ddot{q} = -dE/dq$, where μ is the reduced mass of the vibration ($\mu = 1/2mx$, X = Bi, Sb,...), one easily finds the vibrational frequency of the order parameter excitation (the A_{1g} Raman mode) to be $\omega_0 = \sqrt{4a/\mu}$.

Optically induced excitation of n electrons over the Peierls gap results in two additional contributions to the free energy:

1. A term proportional to the number of excited carriers n to account for the electronic energy. This term can be taken as $\propto n\Delta$, where Δ is the Peierls gap.
2. A term accounting for the electron–phonon coupling, which is proportional to the excess charge density and the square of the lattice distortion q. This term may be written as cq^2n, with c the coupling parameter.

Inclusion of these terms in the Landau expansion gives

$$E = (q) = -an_0q^2 + bn_0q^4 + (\Delta + cq^2)n(t) \tag{14.2}$$

The equilibrium position and the frequency of the vibrational mode as a function of the electron population are easily calculated. The new equilibrium position can be derived trivially as $q_n = \sqrt{a/2b - cn/2n_0b} = \sqrt{q_0^2 - (2n_0b/c)n}$. The phonon frequency will also depend on the number of electrons, which within a harmonic approximation can be written as

$$\omega(n) = \sqrt{\omega_0^2 - 4c\frac{n}{n_0}} \tag{14.3}$$

This expression predicts that in the low-density limit $\left(n \ll n_0\omega_0^2/4c\right)$ the phonon frequency depends linearly on the number of excited electrons

$$\omega(n) \approx \omega_0 - \frac{8c}{n_0\omega_0^2}n \tag{14.4}$$

in agreement with earlier predictions and measurements (Fahy and Reis 2004; Murray et al. 2005). TRRAS measurements of the thermalized system (15 ps after photoexcitation) confirm this expectation. The frequency is linear in the applied optical pump excitation density (see Figure 14.10b). Assuming that no substantial heat diffusion has taken place within the first 15 ps, one can convert the applied optical pump excitation density to an expected sample volume temperature using the known thermal parameters. Thus, the calculated temperature dependence of the phonon frequency in Sb corresponds closely to the results obtained from temperature-dependent steady-state Raman spectroscopy (dashed line in Figure 14.10b), leading to the conclusion that all the energy dumped in the sample by the pump pulses relaxes locally by electron–phonon scattering and only at later times the energy diffuses out of the excited volume via electronic and/or vibrational heat transport.

The simple description of photoexcitation of a Peierls system describes the regime of weak excitation quite well, but it fails at high-excitation densities where time-resolved Raman spectroscopy showed the occurrence of a spontaneous symmetry breaking (Fausti 2008; Fausti et al. 2009). As shown in Figure 14.11, for excitation densities exceeding a critical density

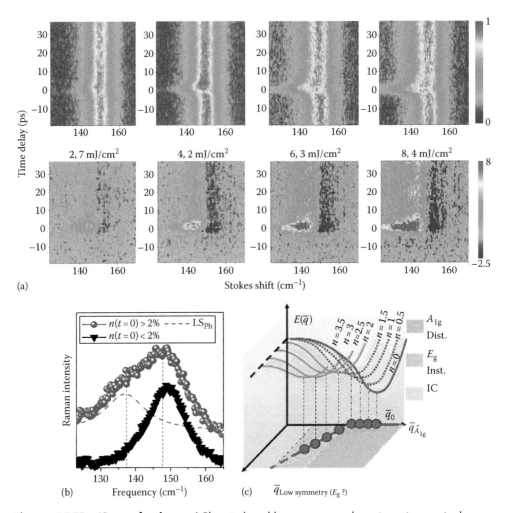

Figure 14.11 (See color insert.) Photoinduced low-symmetry phase in antimony single crystal. (a) False color plot of the Raman response of the Ag phonon at different times after photoexcitation for various excitation densities (in the lower panel the Raman response at negative times has been subtracted). (b) The reduction of symmetry due to the photoinduced phase transition leads to the appearance of an additional phonon mode on the low-energy side of the Ag mode for excitation densities larger than 2%. (c) Cartoon of the free energy curves as a function of the A_{1g} and E_g lattice distortion. The location of minimum free energy is indicated in the bottom plane of the sketch. (Adapted with permission from Fausti D. et al., *Phys. Rev B*, 80, 161207, 2009.)

(5 mJ/cm^2, corresponding to 2 charge excitations per 100 ions) a new low-energy phonon becomes visible in the Raman spectrum for a few picoseconds after laser irradiation, signaling an optically induced phase transition (Figure 14.11a and b). Surprisingly, the induced phase is not the commonly expected high-symmetry cubic phase (in which case there would be no Raman active phonons at all), but rather a phase that has an even lower symmetry than the A7 structure. Based on these observations, the authors suggest that in addition to the A_{1g} distortion present in the A7 phase, the photoinduced phase might be understood in terms of the presence of an additional E_g distortion in the induced phase (see Figure 14.11c), which would be in line with the observation that the observed frequency of the induced optical phonon (Fausti et al. 2009) coincides with a particular zone boundary phonon.

14.4 CONCLUSIONS

The intricate interplay between the various degrees of freedom leads to many remarkable phenomena in complex matter. These include, among others, the formation of charge and spin density waves, static and dynamic Jahn Teller distortions, superconductivity, spin-Peierls states, and multiferroic behavior. Understanding the origins of these remarkable phenomena requires a detailed understanding of the coupling between spin, lattice, charge, and orbital excitations in particular. This is where time-resolved Raman spectroscopy and other time-resolved spectroscopic techniques such as the pump–probe variants of infrared, THz, inelastic electron, and inelastic x-ray spectroscopies can play a crucial role. In semiconductors, time-resolved Raman spectroscopy has greatly helped in understanding charge–lattice coupling phenomena. But, as the examples in the present chapter have shown, the TRRAS technique can be at least as powerful in studies on complex matter, and yield often unique information not easily obtained by other techniques. TRRAS experiments using moderate pump excitation densities, that is, near statistical equilibrium, give direct access to coupling between the initially excited charges and, for instance, the phonon excitations, as has been shown in the case of the carbon-based materials. Excitations far from equilibrium, such as in the superconductors and the A7 metals, are crucial in elucidating the interactions responsible for the formation of a particular ground state. At the same time, TRRAS experiments are powerful in studying the properties of optically induced transient phases, just as normal Raman spectroscopy is for the study of thermodynamically stable states of matter. Given the currently increasing interest in dynamical phenomena in complex matter, together with the rapid recent developments, TRRAS holds great promise for the future. Key areas where experiments as described here might in the authors' view provide crucial insights in the not-too-distant future include the origin of superconductivity in the high T_cs, the nature of electromagnons in the multiferroics, and the nature of photoinduced nonthermal states of matter.

ACKNOWLEDGMENTS

The authors gratefully acknowledge the assistance of Ms. J. de Boer in typesetting and correcting this chapter. This work is part of the research program of the Foundation for Fundamental Research on Matter (FOM), which is financially supported by the Netherlands Organisation for Scientific Research (NWO).

REFERENCES

(2000). *J. Raman Spectros.,* 31(4):229–358.

Balakrishnan, G., Weeks, C. L., Ibrahim, M., Soldatova, A. V., and Spiro, T. G. (2008). *Curr. Opin. Struct. Biol.,* 18:623.

Balakrishnan, G., Zhao, X., Podstawska, E., Proniewicz, L. M., Kincaid, J. R., and Spiro, T. G. (2009). *Biochemistry,* 48:3120.

Basov, D. N. and Timusk, T. (2005). *Rev. Mod. Phys.,* 77:721–779.

Beaurepaire, E., Merle, J.-C., Daunois, A., and Bigot, J.-Y. (1996). *Phys. Rev. Lett.* 76:4250–4253.

Benniston, A. C., Matousek, P., and Parker, A. W. (2000). *J. Raman Spectrosc.,* 31:503.

Boschetto, D., Gamaly, E. G., Rode, A. V., Luther-Davies, B., Glijer, D., Garl, T., Albert, O., Rousse, A., and Etchpare, J. (2008). *Phys. Rev. Lett.,* 100(2):027404.

Boteler, J. M. and Gupta, Y. M. (1993). *Phys. Rev. Lett.,* 71:3497.

Cheng, T. H., Brorson, S. D., Kazeroonian, A., Moodera, J. S., Dresselhaus, G., Dressaulhaus, M. S., and Ippen, E. P. (1990). *Appl. Phys. Lett.,* 57:1004.

Chollet, M., Guerin, L., Uchide, N., Fukaya, S., Shimoda, H., Ishikawa, T., Matsuda, K. et al. (2005). *Science,* 307:86.

Collet, E., Lemee-Cailleau, C. H., Coint, M. B.-L., Cailleau, H., Wulff, M., Luty, T., Koshihara, S. Y. et al. (2003). *Science*, 300:612–615.

Collins, C. L. and Yu, P. Y. (1984). *Phys. Rev. B*, 30:4501.

Damascelli, A., Hussain, Z., and Shen, Z. (2003). *Rev. Mod. Phys.*, 75:473.

Devereaux, T. P., Einzel, D., Stradlober, B., Hackl, R., Leach, D. H., and Neumeier, J. J. (1994). *Phys. Rev. Lett.*, 72:396.

Devereaux, T. P. and Hackl, R. (2007). *Rev. Mod. Phys.*, 79:175.

Du, Y., Xue, J., Li, M., and Phillips, D. L. (2009). *J. Phys. Chem.* A, 113:3344.

Everall, N., Hahn, T., Matousek, P., Parker, A. W., and Towrie, M. (2001). *Appl. Spectrosc.*, 55:1701.

Fahy, S. and Reis, D. A. (2004). *Phys. Rev. Lett.*, 93:109701.

Fausti, D. (2008). Phase transitions and optically induced phenomena in cooperative systems. PhD thesis, University of Groningen, the Netherlands, http://dissertations.ub.rug.nl/faculties/science/2008/d.fausti (accessed 8/26/2010).

Fausti, D., Misochko, O. V., and van Loosdrecht, P. H. M. (2009). *Phys. Rev. B*, 80:161207(R).

Fleury, P. A. and Loudon, R. (1968). *Phys. Rev.*, 166:514.

Fleury, P. A., Porto, S. P. S., and Loudon, R. (1967). *Phys. Rev. Lett.*, 18:658–662.

Fritz, D. M., Reis, D. A., Adams, B., Akre, R. A., Arthur, J., Blome, C., Bucks-baum, P. H. et al. (2007). *Science*, 315:633.

Fujiwara, A. and Mizutani, Y. (2008). *J. Raman Spectrosc.*, 39:1600.

Gambetta, A., Manzoni, C., Menna, E., Meneghetti, M., Cerullo, G., Lanzani, G., Tretiak, S. et al. (2006). *Nat. Phys.*, 2:515.

Garl, T., Gamaly, E. G., Boschetto, D., Rode, A. V., Luther-Davies, B., and Rousse, A. (2008). *Phys. Rev. B*, 78:134302.

Gedik, N., Langner, M., Orenstein, J., Ono, S., Abe, Y., and Ando, Y. (2005). *Phys. Rev. Lett.*, 95:117005.

Gedik, N., Langner, M., Orenstein, J., Ono, S., Abe, Y., and Ando, Y. (2009). *Phys. Rev. B*, 79:224502.

Grann, E. D., Tsen, K. T., and Ferry, D. K. (1996). *Phys. Rev. B*, 53:9847.

Grosserueschkamp, M., Friedrich, M. G., Plum, M., Knoll, W., and Naumann, R. L. C. (2009). *J. Phys. Chem. B*, 113:2492.

Gupta, Y. M., Pangilinan, G. I., Winey, J. M., and Constantinou, C. P. (1995). *Chem. Phys. Lett.*, 232:341–345.

Hamaguchi, H. and Gustafson, T. L. (1994). *Annu. Rev. Phys. Chem.*, 45:593.

Hase, M., Kitajima, M., Nakashima, S., and Mizoguchi, K. (2002). *Phys. Rev. Lett.*, 88:067401.

Hase, M., Mizoguchi, K., and Harima, H. (1998). *Phys. Rev. B*, 58:5448.

Huang, Y.-S., Karashima, T., Yamamoto, M., and Hamaguchi, H. (2005). *Biochemistry*, 44:10009.

Hufner, S., Hossain, M. A., Damascelli, A., and Sawatzky, G. A. (2008). *Rep. Prog. Phys.*, 71:062501.

Ishii, K. and Hamaguchi, H. (2003). *Chem. Phys. Lett.*, 367:672.

Iwata, K. and Hamaguchi, H. (1997). *J. Phys. Chem.*, 101:632–637.

Julian, S. R. and Norman, M. R. T. (2007). *Nature*, 447:537–538.

Kabanov, V., Demsar, J., Podobnik, B., and Mihailovic, D. (1999). *Phys. Rev. B*, 59:1497–1506.

Kang, K., Ozel, T., Cahill, D. G., and Shim, M. (2008). *Nano Lett.*, 8:4642.

Karutz, F. O., von Schütz, J. U., Wachtel, H., and Wolf, H. C. (1998). *Phys. Rev. Lett*, 81:140.

Kash, J. A., Jha, S. S., and Tsang, J. C. (1987). *Phys. Rev. Lett.*, 58:1869.

Kash, J. A., Tsang, J. C., and Hvam, J. M. (1985). *Phys. Rev. Lett.*, 54:2151.

Kneipp, K., Kneipp, H., Itzkan, I., Dasari, R. R., and Feld, M. S. (1999). *Chem. Rev.*, 99:2957–2975.

Kukura, P., McCamant, D. W., and Mathies, R. A. (2007). *Annu. Rev. Phys. Chem*, 58:461.

Kukura, P., McCamant, D. W., Yoon, S., Wandschneider, D. B., and Mathies, R. (2005). *Science*, 310:1006.

Laubereau, A., von der Linde, D., and Kaiser, W. (1972). *Phys. Rev. Lett.*, 28:1162.

Ledgerwood, M. L. and van Driel, H. M. (1996). *Phys. Rev. B*, 54:4926.

Letchter, J. J., Kang, K., Cahill, D. G., and Dlott, D. D. (2007). *Appl. Phys. Lett.*, 90:252104.

Liu, Y. H., Toda, Y., Shimatake, K., Momono, N., Oda, M., and Ido, M. (2008). *Phys. Rev. Lett.*, 101:137003.

Lu, H. P. (2005). *J. Phys. Cond. Mat.*, 17:R333–R355.

Machtoub, L., Machtoub, G. E., Shimoyama, J., Suemoto, T., and Kishio, K. (2003a). *Physica C*, 291294:392396.

Machtoub, L., Machtoub, G. E., Shimoyana, J., Suemoto, T., and Kishio, K. (2003b). *Physica C*, 392–396:291.

Matsubara, E., Inoue, K., and Hanamura, E. (2006). *J. Phys. Soc. Jpn.*, 75:024712.

McElroy, K., Simmonds, R. W., Hoffman, J. E., Lee, D. H., Orenstein, J., Eisaki, H., Uchida, S., and Davis, J. C. (2003). *Nature*, 422:592.

Mizutani, Y. and Kitagawa, T. (2002). *Bull. Chem. Soc. Jpn.*, 75:623.

Monthoux, P., Pines, D., and Lonzarich, G. G. (2007). *Nature*, 450:1177–1183.

Murray, E. D., Wahlstrand, J. K., Fahy, S., and Reis, D. A. (2005). *Phys. Rev. B*, 72:060301(R).

Nakajima, M., Kazumi, K., Isobe, M., Ueda, Y., and Suemoto, T. (2005). *J. Phys. Conf. Ser.*, 21:201.

Oberli, D. Y., Wake, D. R., Klein, M. V., Klem, J., Henderson, T., and Morkoc, H. (1987). *Phys. Rev. Lett.*, 59:696.

Okamoto, H., Nakabayashi, T., and Tasumi, M. (1993). *J. Phys. Chem.*, 39:9873.

Pagsberg, P., Wilbrandt, R., Hansen, K. B., and Weisberg, K. V. (1976). *Chem. Phys. Lett.*, 39:538.

Peierls, R. (1991). *More Surprises in Theoretical Physics*. Princeton University Press, Princeton, NJ.

Pomeroy, J. W., Gkotsis, P., Zhu, M., Leighton, G., Kirby, P., and Kuball, M. (2008). *J. Microelectromech. Syst.*, 17:1315.

Roy, S., Kinnius, P. J., Lucht, R. P., and Gord, J. R. (2008). *Opt. Commun.*, 281:319.

Saichu, R. P., Mahns, I., Goos, A., Binder, S., May, P., Singer, S. G. et al. (2009). *Phys. Rev. Lett.*, 102:177004.

Sakamoto, A., Matsuno, S., and Tasumi, M. (2006). *J. Raman Spectrosc.*, 37:429.

Schulz, B., Bäckström, J., Budelmann, D., Maeser, R., Rubhausen, M., Klein, M. V., Schoeffel, E., Mihill, A., and Yoon, S. (2005). *Rev. Sci. Instr.*, 76:073107.

Siegle, H., Kutzer, V., Hoffmann, A., and Thomson, C. (1996). *Proc. ICPS23*, 23:533.

Slipchenko, M. N., Prince, B. D., Ducatman, S. C., and Stauffer, H. U. (2008). *J. Phys. Chem. A*, 113:135.

Smeigh, A. L., Creelman, M., Mathies, R. A., and McCusker, J. K. (2008). *J. Am. Chem. Soc.*, 130:14105.

Sokolowski-Tinten, K., Blome, C., Blums, J., Cavalleri, A., Dietrich, C., Tarasevitch, A., Uschmann, I. et al. (2003). *Nature*, 422:287.

Song, D., Wang, F., Dukovic, G., Zheng, M., Semke, E. D., Brus, L. E., and Heinz, T. F. (2008). *Phys. Rev. Lett.*, 100:225503.

Tacon, M. L., Sacuto, A., Georges, A., Kotliar, G., Gallas, Y., Colson, D., and Forget, A. (2006). *Nat. Phys.*, 2:537.

Tatham, M. C., Ryan, J. F., and Foxon, C. T. (1989). *Phys. Rev. Lett.*, 63:1637.

Timusk, T. and Statt, B. A. (1999). *Rep. Prog. Phys.*, 62:61–122.

Tsen, K. T. (1992). *Semicond. Sci. Technol.*, 7:B191.

Tsen, K. T. and Ferry, D. K. (2009). *J. Phys. Condens. Matter*, 21:174202.

Tsen, K. T., Ferry, D. K., Goodnick, S. M., Salvador, A., and Morkoc, H. (1999). *Physica B*, 272:406.

Tsen, K. T., Grann, E. D., Guha, S., and Menendez, J. (1996). *Appl. Phys. Lett.*, 68:1051.

Tsen, K. T., Joshi, R. P., Ferry, D. K., and Morkoc, H. (1989). *Phys. Rev. B*, 39:1446.

Tsen, K. T., Wald, K. R., Ruf, T., Yu, P. Y., and Morkoc, H. (1991). *Phys. Rev. Lett.*, 67:2557.

Uesugi, Y., Mizutani, Y., and Kitagawa, T. (1997). *Rev. Sci. Instrum.*, 68:4001.

van Loosdrecht, P. H. M., Boucher, J. P., Martinez, G., Dhalenne, G., and Revcoleschki, A. (1996). *Phys. Rev. Lett.*, 76:311.

von der Linde, D., Kuhl, J., and Klingenberg, H. (1980). *Phys. Rev. Lett.*, 44:1505.

Waldermann, F. C., Sussman, B. J., Nunn, J., Lorenz, V. O., Lee, K. C., Surmacz, K., Lee, K. H. et al. (2008). *Phys. Rev. B: Condens. Matter*, 78:155201.

Yamamoto, K., Mizutani, Y., and Kitagawa, T. (2000). *Biophys. J.*, 79:485.

Yan, H., Song, D., Mak, K. F., Chatzakis, I., Maultzsch, J., and Heinz, T. F. (2009). *Phys. Rev. B*, 80:R121403.

Yoshizawa, M. and Kurosawa, M. (1999). *Phys. Rev. A*, 61:013808.

Zeiger, H. J., Vidal, J., Cheng, T. K., Ippen, E. P., Dresselhaus, G., and Dresselhaus, M. S. (1992). *Phys. Rev. B*, 45:768.

Zhang, X. and Hamaguchi, H. (1994). *Phys. Rev. B*, 50:14718.

PART IV

SPATIALLY-RESOLVED OPTICAL SPECTROSCOPY

Chapter 15

MICROSCOPY

Alexander Neumann, Yuliya Kuznetsova, and Steven R.J. Brueck

CONTENTS

15.1 INTRODUCTION

Microscopy is among the oldest and still the most scientifically important applications of optics with a rich heritage extending hundreds of years. The fundamental physical principles of microscopy were established by Rayleigh, Abbe, and others around the end of the nineteenth century (Born and Wolf 1999). The diffraction limit of approximately one wavelength was established, but as we will see in the following, this is not a simple limit and care must be exercised in its interpretation. Notwithstanding this centuries-old heritage, there are an increasing number of exciting developments in the microscopy of both illuminated (e.g., sensitive to spatial variations in the complex index of refraction) and self-luminous (fluorescent/photoluminescent) objects. Indeed, we are on the verge of enormous breakthroughs in microscopy and ultimately nanoscopy that will extend microscopy's reign as the most important scientific application of optics well into the future and will provide useful competition to other imaging modalities such as electron-beam microscopy and the many variants of atomic force/scanning tunneling microscopy.

So what sets the stage for this explosion in possibilities and improvements? We see at least five major trends in modern microscopy:

1. Developments in lasers for illumination continue with advances in coherence, brightness, ultrafast pulses, etc., making easy and routine experimental arrangements that were previously tour-de-force laboratory triumphs (Chapter 7). Examples are sources for two- and multi-photon microscopies and laser sources optimized for optical coherence tomography of scattering media such as biological tissue and inhomogeneous solids.

2. Advances in the understanding and control of the quantum physics of molecular and solid-state materials. These are particularly critical for advanced fluorescence techniques, such as stimulated emission depletion (STED), ground state depletion (GSD), stochastic optical reconstruction microscopy (STORM), and others, which are briefly reviewed in the following but are mainly outside of the scope of this chapter.

3. Transfer of ideas from both long wavelengths (RF and microwave radar and imaging) and very short wavelengths (lithographic imaging that is being driven by the demands of Moore's law to extreme sub-wavelength regimes.) At the long wavelength end of the spectrum, radar imaging has developed into an all important defense asset, for example, airborne warning and control (AWACs) systems. The technology of AWACs is able to resolve rapidly moving objects in ground clutter at long ranges and has become an integral part of advanced surveillance. In microscopy, similar ideas are the foundation of imaging interferometric microscopy (IIM), reviewed in the following. At the short wavelength end of the spectrum, the silicon-integrated circuit industry, which requires the ability to image and manufacture the smallest features with the greatest density over the largest areas, has both borrowed from old ideas in microscopy (off-axis illumination, immersion, pupil plane filtering) and reintroduced some ideas back into microscopy (computational lithography/microscopy and synthetic aperture techniques). During the production of this book, the silicon industry was manufacturing at the 45 nm node, with gate features of perhaps 20 nm, using a 193 nm photon-based lithography tool—certainly well below the wavelength—and was preparing to introduce the 32 nm node in 2010. An essential aspect of lithography is that the photosensitive material, photoresist, has a highly nonlinear intensity response that offers the possibility of higher spatial harmonics to extend resolution for the binary imaging required for IC manufacturing. This is the basis of double patterning, which is now the leading candidate for the manufacturing of the 32, 22, and 16 nm roadmap nodes (International Technology Roadmap for Semiconductors). On the other hand, once you produce a lithographic

pattern, it is not easily changed except in an overall subtractive (etching) or additive (lift-off) process. In contrast for microscopy, digital manipulation of partial images is readily performed and forms an essential aspect of many of the new techniques.

4. Dramatic and continuing advances in the electronic capture, manipulation, and storage of image information. CCD cameras with up to 500 megapixels (Mpixels) are commercially available, and digital image processing has progressed enormously along with increasing computational capabilities. These advances are, of course, a consequence of the lithographic improvements that allow the definition of ~20 nm-scale features in integrated circuits.

5. Advances in nanophotonics and nanoscale fabrication of novel optical materials (metamaterials and plasmonics) that allow sub-wavelength manipulation of light. Interestingly, as is the case with so many "new" fields, plasmonics is an old field that is being reinvigorated by the fabrication advances and by the antenna ideas of RF and microwave physics transposed to a shorter wavelength regime.

The result is an explosion of new techniques and new possibilities that offer a true revolution in the capabilities of optical microscopy and allow it to extend to the nanoscale, historically the domain of electron-beam techniques. This development is of particular importance for biological studies because of the need for imaging the dynamics of live organisms in water. Semiconductor manufacturing is another important application where the enhanced resolution possible with new techniques will play a significant role. As the benefits of integration and miniaturization are incorporated in photonics developments, microscopy will again have an essential role in characterization and inspection. The future of nanotechnology clearly involves additional integration and hierarchical nanoscale structures that will additionally require diagnostic and characterization tools capable of dealing with vast amounts of nanoscale information and extending beyond serial technologies to true imaging capabilities.

15.2 PARAMETERS THAT DESCRIBE MICROSCOPY

Most generally, the goal of microscopy is to provide full three-dimensional (3D) structural and possibly spectroscopic (chemical) information of complex structures. This is a challenge of a very high order and many different techniques have been developed to attack its parts. These are briefly cataloged in this chapter.

It is useful to briefly define some of the most common specifications of microscope systems to provide a common language (Gu 2000, Török and Kau 2007). A typical microscopy system consists of a light source, an illumination system, a sample arrangement including a 3D scanning capability, a collection system (objective lens and auxiliary optics), and an image recorder, most often today an electronic camera interfaced with a computer or other digital storage device.

Sources range from incoherent, continuous lamps and light emitting diodes to very highly coherent (and therefore bright) lasers and to very high bandwidth, femtosecond-pulse lasers. This represents an enormous diversity of spectral, spatial, and temporal characteristics and many different techniques have been developed to maximize the imaging information content and to take advantage of various source capabilities. The illumination system is characterized by a wavelength (denoted as λ) and numerical aperture that describes the angular distribution of the light incident on the sample plane. It is often desirable to manipulate the angular illumination distribution to enhance certain features of the image and it is important to keep in mind the distinction between NA_{ill}, which is

$$NA_{ill} = \sin(\alpha_{ill}) \qquad (15.1)$$

where α_{ill} is the maximum incidence angle of light that is transmitted through the illumination optics onto the object, and the measurement $NA_{ill-sys}$, which can be any subset of the angles within NA_{ill}. This distinction is often not made explicit in descriptions of microscopy techniques, which can lead to confusion. There is more detailed discussion of this issue in Section 15.3.2.

The collection system consists of an objective lens, again characterized by an NA and auxiliary optics to transport the image to the camera for recording. Additional optics, such as polarizers, wave plates, and prisms for Nomarski optical schemes, and phase plates for phase contrast microscopy, are often incorporated in the optics after the objective lens. The working distance is the distance between the object and the first surface of the objective lens.

There are trade-offs between resolution, depth of field, and working distance that are familiar to anyone who has used a microscope. In general, the ability to resolve small features (see discussion in the following) increases linearly with the NA of the objective lens and inversely with a decrease in the source wavelength. The depth of field decreases as the square of the NA within the paraxial approximation [$\sin \theta_{NA} \sim \theta_{NA}$ where $\theta_{NA} = \sin^{-1}(NA)$]. These are fundamental relationships related to diffraction effects. In general, the working distance and the field of view decrease as the NA is increased, but this is specific to individual lenses and both quantities can be increased for a fixed NA with more complex (and more expensive) objective lenses.

15.3 RESOLUTION

15.3.1 Image Formation

Resolution in an optical image-forming system is a measure of the limits of the system to transfer information about small features in the object to the image. This quantity is very dependent on parameters of the imaging system including, but not restricted to the illumination scheme (including the temporal and spatial coherence of the source), the optical characteristics of the object, the collection optical system including its aberrations (as a function of spatial frequency and position of a feature in the object plane), and the system noise. In general, there is no single parameter that can fully characterize the resolution.

We begin by establishing a qualitative understanding of image formation. The rigorous theory of image formation is quite complex and is given in detail in many places (Goodman 1985, Klein and Furtak 1986, Born and Wolf 1999). We present a much simplified model that gives perhaps a fresh point of view on the imaging system. The essence of this approach is to consider the optical response to a simple amplitude grating and then to recognize that it is possible, by Fourier analysis, to represent an arbitrary object as a superposition of gratings (a Fourier sum). This makes available much of the apparatus of linear systems theory (Goodman 1998). However, it is important to remember that image formation is not a linear process. Light is scattered by the object, transmitted through the optical system, and detected in the image plane by a nonlinear (square law) detector. This nonlinear step has profound implications on the final image and results in much of the complexity but also has many advantages. It is natural to start an investigation of the system with just a simple grating object. Any object can be expressed in a Fourier integral, and this analysis is quite general as long as the nonlinear step at the detection plane is properly taken into account.

First, consider a conventional, single-lens, unity-magnification optical system (Figure 15.1a), with a simple, thin ($\ll$ an optical wavelength) amplitude grating of period d (e.g., a chrome on glass structure) illuminated at normal incidence by coherent light of wavelength λ and a lens with a sufficiently large NA so that the zero-order beam, $E_0 e^{-ik_0 z}$, the symmetrical first-order diffracted beams, $E_1 e^{ik_d x} e^{-i\sqrt{k_0^2 - k_d^2} z}$ and $E_{-1} e^{-ik_d x} e^{-i\sqrt{k_0^2 - k_d^2} z}$, and the symmetrical

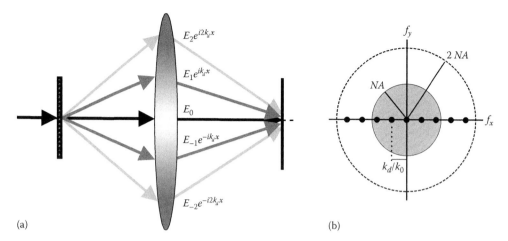

Figure 15.1 Image formation: (a) arrangement and (b) spatial frequency coverage. (For convenience of representation, we show the diffracted beams displaced in the vertical direction and the corresponding spatial frequencies separated in the horizontal direction; effectively the representation is rotated by 90° between parts (a) and (b).)

second-order diffracted beams $E_2 e^{i2k_d x} e^{-i\sqrt{k_0^2 - (2k_d)^2} z}$ and $E_{-2} e^{-i2k_d x} e^{-i\sqrt{k_0^2 - (2k_d)^2} z}$ are collected by the optical system. Here, $k_d \equiv 2\pi/d$ and $k_0 \equiv 2\pi/\lambda$ are the grating and photon wave vectors, respectively, x is the transverse spatial coordinate, and z is the spatial coordinate in the direction of the optical axis. In general, there are higher order diffracted beams, up to the optical transmission limit, $m \leq k_0/k_d$, with m an integer, and evanescent, non-propagating diffracted orders for larger integer values of m. To begin, we use a scalar approximation and ignore the polarization of the incident and scattered beams; we discuss issues with this simplification, which occur at higher NAs and become particularly important for immersion techniques, in the following. Initially, we ignore all z-terms by considering the image in the focal plane $z = 0$; note that there is a set of such planes including the object plane and the conjugate image plane defined by the optical system; depth-of-field issues are discussed in the following.

The image is formed by the interference of the zero-order beam with each of the higher order diffracted beams transmitted through the lens NA. In addition, there are contributions from the interference of each order with itself (constant across the image) and from the interference of different diffracted orders (the nonimaging terms that contribute to a dark-field image). Mathematically, the field at the image plane can be described as follows:

$$E_{\text{total}} = E_0 + E_1 e^{ik_d x} + E_{-1} e^{-ik_d x} + E_2 e^{i2k_d x} + E_{-2} e^{-i2k_d x}$$

$$= E_0 + 2E_1 \cos(k_d x) + 2E_2 \cos(2k_d x) \tag{15.2}$$

where the object is a thin amplitude grating so that $E_i = E_{-i}$, for $i = 1, 2$.

Then the intensity is as follows:

$$I_{\text{total}} = \left| E_0 + 2E_1 \cos(k_d x) + 2E_2 \cos(2k_d x) \right|^2$$

$$= |E_0|^2 + 2|E_1|^2 + 2|E_2|^2 \qquad \text{(baseline)}$$

$$+ 4|E_0 E_1| \cos(k_d x) + 4|E_0 E_2| \cos(2k_d x) \quad \text{(image)}$$

$$+ 4|E_1 E_2| \cos(k_d x) + 2|E_1|^2 \cos(2k_d x) + 4|E_1 E_2| \cos(3k_d x) + 2|E_2|^2 \cos(4k_d x)$$

$$\text{(dark field)} \tag{15.3}$$

This result has been parsed into three categories: baseline or constant terms with no spatial variation, imaging terms corresponding to the interference between the zero and higher order terms, and dark-field terms corresponding to the interference between two nonzero orders. Since the intensity is simply a sum of cosine functions, its two-dimensional (2D) (x, y) Fourier transform is a series of δ-functions along the k_x transform coordinate, for example, the Fourier transform of $\cos(2k_d x) = (e^{2ik_d x} + e^{-2ik_d x})/2$ is just a pair of δ-functions at offset spatial frequencies of $\pm 2k_d$ along the k_x axis. For convenience, we plot this information in dimensionless coordinates, normalized to k_0 ($=2\pi/\lambda$) and denoted by f_x, so that the acceptance cone of the objective lens is a circle of radius NA as shown in Figure 15.1b. The domain of the image terms is restricted to the NA circle, whereas the dark-field terms extend to a radius of $2NA$. Of course, a δ-function is an idealized representation, as limitations of the imaging system (e.g., field of view) will delimit the image and broaden the corresponding Fourier transform. The amplitude of the zero-frequency component, at the origin of the (f_x, f_y) plot, from Equation 15.2, is $|E_0|^2 + 2|E_1|^2 + 2|E_2|^2$. The amplitude of the δ-functions at the fundamental grating frequency, $\pm k_d/k_0$, equal to $4(|E_0 E_1| + |E_1 E_2|)$, has contributions from both imaging and dark-field terms. Similarly, the amplitude of the second-order grating peaks at $\pm 2k_d/k_0$ ($=4|E_0 E_2| + 2|E_1|^2$) contains both imaging and dark-field terms. The higher (third and fourth) order spatial frequency terms arise solely from dark-field interference terms. Note that these terms arise from interference between two waves scattered from the object and that the observed frequencies are, therefore, linear combinations of the scattered frequencies from the object. Dark-field imaging is quite useful in many microscopic applications; additional discussion will be provided in the following. Here, the goal is to recover the scattering coefficients (amplitude and phase) of the object and the dark-field terms will be subtracted to leave the image terms.

It is possible to obtain the higher spatial frequency image terms using a lower (by up to a factor of 2) NA lens along with off-axis illumination. We denote the numerical aperture of the low NA lens by NA_{low} to simplify the bookkeeping. In the case of NA_{low}, the grating period and lens NA are such that only the zero-order (grating transmission) and first-diffracted orders are transmitted through the lens and onto the CCD camera (Figure 15.2a) for normal incidence illumination. In this case, the intensity at the image (camera) plane is as follows:

$$I_{\text{low}} = \left|E_0 + 2E_1 \cos(k_d x)\right|^2 = \left|E_0\right|^2 + 2\left|E_1\right|^2 + 4\left|E_0 E_1\right| \cos(k_d x) + 2\left|E_1\right|^2 \cos(2k_d x) \quad (15.4)$$

The frequency space coverage is shown in part (b) of Figure 15.2. The E_1 beam is inside the lens NA and is transmitted to the image plane; the second-order scattering (E_2 beam) is outside the lens NA and is not captured by the optical system. The two points outside the gray circle represent the final term in Equation 15.4; these are the result of mutual interference between

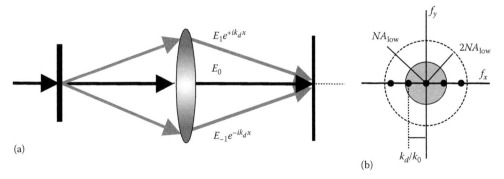

Figure 15.2 Image formation: (a) arrangement with reduced NA and (b) spatial frequency coverage.

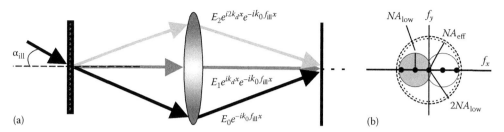

Figure 15.3 (a) Illumination tilt is the key to (b) enhanced frequency coverage.

the complementary first-order fields. The distance between these two orders in frequency space is doubled and corresponding features appear beyond NA_{low}. These high frequency features (part of the "dark field") are a result of the square law detection. The spatial frequency domain of the dark fields always extends over a frequency space with a radius of $2NA_{low}$ (dotted circle in Figure 15.2).

One of the E_2 beams can be transmitted through the objective lens optical system by introducing a tilt of the illumination beam at an angle α_{ill}. Now every scattered plane wave from the object has an additional phase term $e^{-ik_0 \sin \alpha_{ill} x} = e^{-ik_0 f_{ill} x}$, which means that the scattered beams are tilted in accordance with the grating equation, $\sin \theta_{out} = \sin \alpha_{ill} - mk_d/k_0$. Implicit in this result is that some scattered orders may switch from propagating to evanescent (and *vice versa*) depending on the wavelength and the grating period.

Here we choose α_{ill} such that the zero-order beam is near the one edge of the NA_{low} acceptance cone (Figure 15.3a) and the E_2 beam is within the aperture on the opposite side of the acceptance cone.

Now the intensity at the image plane is as follows:

$$I_{high} = |E_0|^2 + |E_1|^2 + |E_2|^2$$

$$+ 2\left[|E_0 E_1| + |E_1 E_2|\right]\cos(k_d x) + 2|E_0 E_2|\cos(2k_d x) \qquad (15.5)$$

The image part (middle term) is now the same as in Equation 15.3 for the larger lens, within a factor of 2. The constant terms and the dark-field terms are different for the two optical systems (normal incidence illumination with NA lens and tilted illumination with NA_{low} lens). In this simple case, the imaging terms are identical to those obtained with the higher NA lens although the dark-field terms are different.

Notice that this is a single sideband system—either (E_0, E_1, E_2) or (E_0, E_{-1}, E_{-2}) are collected. The collected sidebands are shown in the gray-filled circle in Figure 15.3b; the frequencies within the conjugate symmetrical small circle offset in the opposite direction are restored as a result of the square law intensity response.

As will be discussed in more detail in the following, an issue with this configuration is that the image pattern shifts as the object is moved through focus, i.e., this optical configuration is not telecentric. Often this is addressed by illuminating with two mutually incoherent beams from opposite directions, α_{ill} and $-\alpha_{ill}$ to restore telecentricity; this is known as dipole illumination.

The circle of transmitted frequencies (gray-filled circle in Figure 15.3b) is shifted by $\sin \alpha_{ill} = NA_{ill}$, so that the maximum collected frequency corresponds to an effective numerical aperture:

$$NA_{eff} = NA_{low} + \sin \alpha_{ill} = NA_{low} + NA_{ill} \qquad (15.6)$$

In other words, the effective aperture of the imaging system is a sum of both the illumination and collection apertures. Changing the angle of incidence, α_{ill}, shifts the spatial frequency coverage to different spatial frequency ranges. However, the reader should not infer that NA_{eff} corresponds to the same frequency space coverage as a larger lens with an acceptance cone of NA_{eff} and normal incidence illumination. NA_{eff} refers only to the maximal transmitted frequency along the direction of the tilted illumination; the covered area in the transverse direction in spatial frequency space is still constrained by NA_{low}. Full frequency space information (large dotted circle in Figure 15.3b) with radius NA_{eff} can be acquired with additional illumination beam offsets in other azimuthal directions (the azimuth is the angle φ from the x-axis in the x, y plane). As long as there is no coherence between the illumination beams (which would result in complex and undesirable interference effects), these can be applied simultaneously. For the simple grating structure, the optimum direction of the tilt to capture more information is obvious; for arbitrary objects without a preferred direction, annular illumination, with incoherent illumination incident in a ring at a constant tilt angle around the edge of the pupil is often used. Thus, off-axis illumination can be used to collect the diffraction orders that fall outside of the collection NA of the optical system, up to frequencies of $2NA$. Because the offset has to be less than $\sin^{-1} NA$ to allow collection of the zero-order transmission, the offset circles corresponding to the frequency coverage necessarily overlap as shown in Figure 15.3b. This overlap represents unequal weighting of the overlapped spatial frequencies relative to singly covered frequencies that can impact the final image. In addition to varying the azimuthal angle, it is also possible to vary the offset or inclination angle of the illumination beam from the optical axis to cover additional regions of frequency space.

15.3.2 Coherent versus Incoherent Illumination

The imaging system has a unity transfer function for every plane wave within the acceptance cone, or entrance pupil, of the lens. For normal incidence illumination, it follows from Equations 15.2, 15.3, and Figure 15.1 that the scattered electric fields are transmitted through the optical system with unchanged amplitude (and phase in the absence of lens aberrations) until the higher angle, and beyond $\pm\sin^{-1} NA$, plane waves corresponding to scattering from smaller structures are restricted by the entrance pupil (NA) of the system. Therefore, it is evident that the electric field transfer function (ETF) is a rectangular function (rect) in cross section with a cutoff at NA as shown in Figure 15.4a. So, the ETF for an idealized lens and coherent illumination is just a circular (circ) function in two dimensions. Any source could be considered as "perfectly coherent" if its coherence area is larger than the field of view of the objective lens.

Spatially incoherent illumination can be approximated as a superposition of mutually incoherent plane waves with all possible k-vectors filling 2π steradians from normal to grazing incidence at all azimuth angles, which incorporates on-axis as well as all possible off-axis illumination directions at once. For each incident illumination direction, interference between

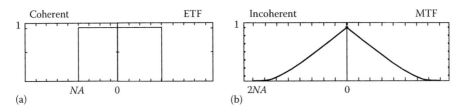

Figure 15.4 Transfer functions: (a) ETF for coherent, normal incidence illumination; (b) MTF for conventional incoherent illumination imaging.

the scattered beams forms an image with an offset spatial frequency response similar to that shown in Figure 15.3. Since the incident plane waves are mutually incoherent, so are the respective scattered beams, and, therefore, each of these images is independent and the final image is a superposition of many incoherently related sub-images (e.g., intensities are added). For incoherent illumination, the optical transfer function for a circular aperture is the auto-correlation of the circular amplitude pupil function, called the modulation transfer function (MTF), and has the following form (Goodman 1998):

$$\mathrm{MTF}(f_i) = \frac{2}{\pi} \left[\cos^{-1}\left(\frac{f_i}{2NA}\right) - \frac{f_i}{2NA}\sqrt{1 - \left(\frac{f_i}{2NA}\right)^2}\; \right] \tag{15.7}$$

where f_i corresponds to the normalized spatial frequency. The MTF falls from 1 to 0 at twice the coherent cutoff (Figure 15.4b). Thus, for incoherent illumination, the spatial bandwidth extends to $2NA$, but with reduced fidelity at high spatial frequencies (Goodman 1998).

Incoherent illumination provides a linear relationship of the spectral intensity from the object to the image in contrast to the coherent case, where the system is linear in electric field up to the square law detection step. Of course, we know from the wave theory of optics that no source can be fully incoherent. No far-field source can be focused to an optical spot smaller than approximately ½ the wavelength; similarly, the coherence area of any far-field source, restricted to propagating waves, is always larger than $\sim\lambda/2$. A necessary condition for Equation 15.7 to apply is that the smallest feature of the object should be larger than the coherence area of the illumination. Nonetheless, Equation 15.7 is a good approximation for many practical cases. The applicability condition is that the angular extent of the incoherent illumination should be greater than the sum of $\sin^{-1}(NA)$ and the angle corresponding to diffraction from the smallest features of the object. There has been much discussion in the optics literature on the practical limits of these ideas (Hopkins 1955, Goodman 1985).

Partial coherence refers to any illumination system between these coherent and incoherent limits. The discussion of partial coherence is often based on the numerical aperture of the combined source, NA_{src}, and illumination system, $NA_{ill-sys}$. For an extended source such as a tungsten filament, $NA_{src} = 1$ is $> NA_{ill-sys}$ and only $NA_{ill-sys}$ impacts the optical performance. For partial and incoherent illumination, there is a continuum of spatial frequencies spanning an extended range of angles incident on the object denoted as $NA_{ill-sys}$ in contrast to the maximum transmitted spatial frequency NA_{ill}. Most often for partially coherent illumination the center of the distribution of incident angles is normal to the object plane and $NA_{ill-sys}$ is equal in this case to NA_{ill} (Figure 15.5a); this is the context for the following discussion on the effects of partial coherence. Figure 15.5b shows the case of tilted illumination at an angle $\alpha_{ill} = \sin^{-1}NA_{ill}$. This case will reduce to coherent illumination as long as the spread in incident spatial frequencies $NA_{ill-sys}$ is close to zero. Other geometries are possible; two that are often used are annular illumination, where there is no illumination at normal incidence, but rather over a range of incident angles (inclinations from α_{min} to α_{max}) and at all azimuths (all φ) (Figure 15.5c), and quadrupole illumination, suitable for rectilinear (x, y) or Manhattan geometry objects as are often used in integrated circuits, where again there is a range of inclinations and φ is constrained to small ranges around $\pi/4$, $3\pi/4$, $5\pi/4$, and $7\pi/4$ (Figure 15.5d). The quadrupole illumination can either be oriented at $\pi/4$ to the (x, y) axes, as shown, to provide a larger transfer function at intermediate frequencies or at $0°$ to the (x, y) axes to provide higher maximum frequencies albeit at a smaller transfer function. The spread in incident spatial frequencies, $NA_{ill-sys}$, defines the degree of spatial coherence of the illumination system. The ratio $\mathcal{R} = NA_{obj}/NA_{ill-sys}$ is often used to facilitate a more quantitative discussion (Goodman 1985).

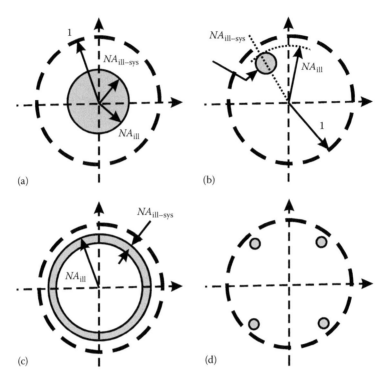

Figure 15.5 Alternate illumination schemes: (a) partial coherent illumination with $NA_{ill-sys}$ = NA_{ill}, (b) tilted partial coherent illumination or tilted coherent illumination if $NA_{ill-sys} \sim 0$, (c) annular illumination, and (d) quadrupole illumination.

Thus, $\mathcal{R} = 1$ corresponds to matching the angular spreads of the illumination and collection optics and $\mathcal{R} = \infty$ corresponds to coherent illumination. There is a subtle difference between $NA_{ill-sys}$ and the NA_{ill} defined earlier in the context of coherent illumination, for example, off-axis coherent illumination means $NA_{ill} > 0$, but $NA_{ill-sys} \sim 0$.

On the other hand, of course, $NA_{ill-sys}$ is necessarily ≤ 1, as the idealized case of incoherent illumination, $\mathcal{R} \rightarrow 0$, includes near-field evanescent waves that are not available in a far-field system. The influence of the degree of partial coherence on the characteristics of the transfer function (Becherer and Parrent 1967) is shown in Figure 15.6; there is little practical difference between the response for $\mathcal{R} \sim 1$, matching the source and objective NAs, and fully incoherent illumination $\mathcal{R} = 0$. In practical terms, using an unfocused laser source provides coherent illumination, while an extended source such as a filament coupled with an appropriate illumination system provides partially coherent or incoherent illumination. In lithography systems, where the limited bandwidth of the optical system requires a highly temporally coherent source (narrow linewidth), great pains are taken to eliminate spatial coherence ($\mathcal{R} > 0$) to avoid speckle effects (interference between directly imaged light and scattered light within the optical system or flare). Note that the transfer functions of coherent and incoherent illuminations are not directly comparable because of the difference in character between the ETF (the square law response is taken after applying the transfer function) and the modulation (intensity) transfer function. For comparison purposes, the apparent transfer function (ATF), which is the ratio of the output versus input intensities of simple grating objects as a function of period, is used.

Figure 15.6 provides important insight into the behavior of optical imaging systems for different illumination conditions. The general trend is that higher values of $\mathcal{R}$ provide a more consistent transfer function across spatial frequencies, but cut off at lower spatial frequencies.

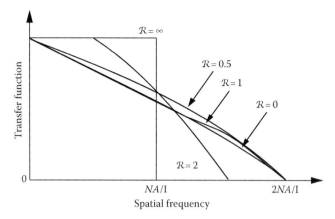

Figure 15.6 ATFs for different degrees of partial coherence. (Reprinted from Becherer, R.J. and Parrent, G.B., *J. Opt. Soc. Am.*, 57, 1479, 1967. With permission.)

The optimal $\mathcal{R}$ is both subjective and dependent on the image pattern. An $\mathcal{R}$ between 0.75 and 0.5 provides a consistently higher image response than incoherent illumination ($\mathcal{R} = 0$), while still providing some response to spatial frequencies as high as $2NA$.

It is important to make a clear distinction between the frequency content within the pupil plane of the objective and that in the image plane. These are related by the square law intensity response of the camera. Representations of the pupil plane for different illuminations are shown in Figure 15.7 along with the frequency responses for a simple grating pattern. This figure is drawn for a somewhat more complex optical system (known as a $4f$ optical system), where the pupil plane is easily identified, still with unity magnification for simplicity of representation. Both coherent (top half of object) and partially coherent illumination (bottom half of object) are shown. The diffraction from the grating immediately gives rise to multiple orders at the object plane. For coherent illumination the 0, ±1, ±2, ±3, and ±4 orders are shown. The period of the grating is chosen so that only the 0 and ±1 order are captured by the objective optical system and appear in the pupil plane. The representation shows, using a gray scale, the amplitudes of the various orders that are focused to Airy disks in the pupil plane; there is a corresponding phase map that is not shown in this simple diagram. At the image plane, the square law intensity response results in an image containing frequencies corresponding to the 0, ±1, and ±2 spatial frequencies of the grating; the Fourier transform of this image is represented in the figure. The ±2 orders arise from the "dark-field" interference of the ±1 orders transmitted through the pupil. The extent of the spatial frequency content of the image extends to $2NA$, while the pupil frequency content is limited to NA. For partially coherent illumination, each of the pupil plane regions is spread out as a result of the range of illumination angles. This results because, in the pupil plane, each plane wave corresponding to a specific coherent illumination angle focuses to a different, shifted Airy disk; the sum of all these Airy disks gives an extended response corresponding to the partial coherence. In the image, each plane wave along with its corresponding diffracted beams image to the same grating features. The figure is drawn for $\mathcal{R} \sim 3$. Notice that a small portion of the ±2 order diffraction information is captured by the optical system and this gives rise to additional higher spatial frequency terms in the image. Finally, for incoherent illumination, the entire pupil is filled with each order, and the roll-off in the MTF is roughly represented by the gray scale of the image Fourier transform. Incoherent illumination has a reduced transfer efficiency at high frequencies. Coherent illumination can be plagued with speckles due to interference associated with reflections from optical surfaces and scattering from optical defects and dust particles. This noise can be partially reduced by decreasing the coherence length and

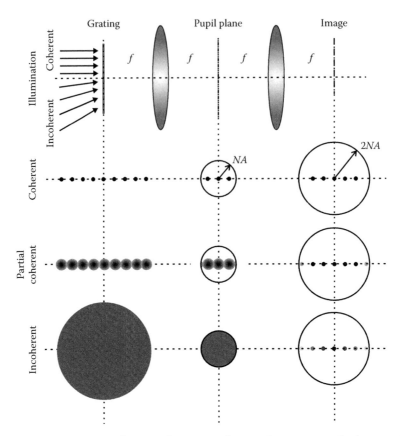

Figure 15.7 Comparison of spectral content of optical system pupil plane and image Fourier plane.

subtracting the background recorded as an image of the illumination beam without an object (Voelz et al. 1997, Kuznetsova et al. 2007).

15.3.3 Definition of Resolution Specific to Optical Configuration

As mentioned earlier, a full discussion of resolution is a complex topic that is not in general amenable to a simple treatment. A resolution criterion is required for estimation of performance of the system and/or for metrology purposes. The achievable resolution is a function of the optical system, including the illumination scheme, and of the details of the object, so a simple definition of resolution applicable across all objects is not possible. Nonetheless, it is useful to consider several of the most established and widely used shorthand characterizations of resolution.

In 1873, Ernst Abbe, then research director of Zeiss Optical Works, understood that the ability of an optical system to resolve small features is directly connected to the ability of the system to transfer high spatial frequency information from the object to the image planes. He was the first to recognize the advantages of the off-axis illumination approach outlined earlier. In the context introduced earlier, the smallest grating period accessible with a given objective corresponds to the case of the zero-order beam incident at one edge of the pupil and the first diffracted order at the opposite side of the pupil (Abbe 1873):

$$d = \left(\frac{\lambda}{2NA} \right); \quad \text{Res} \sim \frac{\lambda}{4NA} \tag{15.8}$$

Here, the resolution is taken as the half pitch of the smallest period grating that can be imaged through the optical system. This criterion (maximum transferred spatial frequency) is perfectly applicable in the case of 1D structures and coherent illumination where there is a sharp cutoff in the band pass function at the edge of the pupil. However, the applicability of this simple criterion becomes more difficult if the transfer function is not constant across spatial frequencies, as in the case of incoherent illumination.

In this chapter, we address a simpler question. First, as have many before us, we restrict the object to a very simple structure, a pair of small ($\ll\lambda$) apertures in a screen for a 2D structure or a pair of small lines for a 1D structure and ask the same question as Rayleigh (1879): What is the smallest distance between a pair of objects where we can identify that there are two separate objects? We assume a perfect optical system without any aberrations and a large signal–noise ratio so no ambiguity is introduced by noise levels. Rayleigh suggested another metric, which states that the smallest resolvable distance between two point sources is the one that brings the maximum of one Airy pattern onto the minimum of the second (Rayleigh 1879). The point spread function (PSF, which is the Fourier transform of the point source spectrum filtered by the optical transfer functions discussed earlier) of a circular pupil aperture is the Airy disk pattern, $[J_1(\pi NAr/\lambda)/(\pi NAr/\lambda)]^2$ where J_1 is the Bessel function of first order and r is the radial coordinate from the center of the spot, the first zero of J_1 is located at $\pi NAr/\lambda = 1.22$, which determines the minimum resolvable separation of geometrical points d, given by the famous Rayleigh resolution criteria:

$$d = 2CD = 0.61\left(\frac{\lambda}{NA}\right); \quad \text{Res} \sim CD \sim 0.31\left(\frac{\lambda}{NA}\right) \tag{15.9}$$

where the CD or critical dimension, a concept borrowed from lithographic terminology, is the smallest resolvable linear dimension or half of the period. Comparing Equations 15.7 and 15.8, the resolution is a linear function of wavelength and inversely proportional to the numerical aperture in both cases. Note that Equation 15.9 was derived for a particular case (incoherent illumination), a fixed pattern (two points of equal intensity), and a certain pupil shape (circular). Thus, it can be only used quantitatively for this particular situation but can serve as a qualitative, not quantitative, reference for comparison of optical systems.

Optical configurations for which the PSFs do not have zeros or obvious minima in the neighborhood of their central maxima (e.g., Gaussian PSFs) require generalization of this approach. A first attempt at such a generalization was undertaken by Rayleigh's contemporary, Sparrow, who reformulated the resolution as the distance for which the ratio of the value at the central dip in the composite intensity distribution to that at the maxima on either side is equal to 0.81. This corresponds to the dip in between the two $[\text{sinc}^2(\pi NAx/\lambda) = [\sin(\pi NAx/\lambda)/(\pi NAx/\lambda)]^2]$ functions in the Rayleigh construct. Sparrow was working with lines in spectrographic applications; the dip is equal to 0.735 for two points (Sparrow 1916). However, Sparrow recognized that even though a generalized Rayleigh criterion is capable of comparing different systems (at least in a qualitative sense), it is not suitable for quantitative metrology purposes because of its very high sensitivity to mutual intensity in the points. Indeed, the criterion should be applied with care even in the case of equal point (line) intensities and cannot be simply extended for more general cases. For example, taking three equal intensity equidistant lines (points) and varying the pupil aperture results in the intensity profiles shown in Figure 15.8. For a sufficiently large NA that passes the spatial frequency corresponding to the spacing, the three lines are resolved. For very small NAs, the three lines collapse into a single feature. However, at intermediate apertures, only two lines are apparent; in other words, while the dip satisfies the generalized Rayleigh condition, the three objects

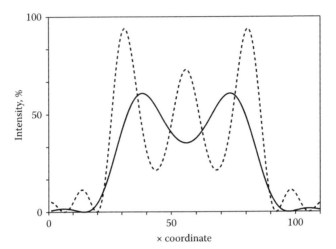

Figure 15.8 Example of a possible misinterpretation in application of the Sparrow resolution criterion (three lines resolved [dotted]; the same pattern after low-pass filtering [solid] shows only two lines with a dip that satisfies the resolution condition).

are represented by the optical system as two apparent objects and the resolution is clearly misrepresented by this criterion.

Sparrow suggested an alternative approach (Sparrow 1916). Two PSFs should be moved toward each other from infinity. At first, a minimum between peaks will develop. Finally, the separation at which the minimum becomes a saddle point should be considered as the resolution limit. This procedure works well even with differences in line/point intensities. However, it is not robust to system noise and is too forgiving for cases with elevated PSF wings. All of these considerations are further complicated by extension to more complex arbitrary objects from simple point–line pairs. Similar issues plague any attempt at a generally useful definition of resolution for arbitrary patterns.

As the *NA* increases beyond the paraxial approximation, polarization effects influence the resolution (Richards and Wolf 1959, Raub et al. 2004, Köklü et al. 2009). The 2D PSF becomes roughly elliptical in the *x-y* plane. The interference of two tilted TE-polarized beams (transverse electric polarization, electric field perpendicular to the plane of incidence defined by the propagation direction and the normal to the image plane) is not influenced by the angle of tilt since the electric field vector remains the same independent of tilt. In contrast, the *E*-vector for TM polarization (magnetic field perpendicular to the plane of incidence) continuously varies in direction with tilt. Both *x*- and *z*-components interfere independently with the corresponding vector components of the second interfering coherent beam with a 180° phase shift between the interference components. At a 45° tilt, the intensities are equal and the resulting contrast is zero, and for larger tilts the contrast is reversed. The resolution is higher (smaller focused spot size) for the TE direction.

15.3.4 Longitudinal Resolution Depends on Illumination (Focusing/Defocusing)

The illumination scheme has a strong influence on the behavior of a defocused image and on the ability to separate features in the *z*-direction. To illustrate, first consider the defocusing associated with a dual sideband (telecentric) coherent optical system (Figure 15.1, Equation 15.2). Equation 15.2 is written at focus (the image plane conjugate to the object plane) where *z* is effectively zero and the propagation terms disappear. Complimentary (conjugate) orders from both sides (+1,–1; +2,–2; etc.), beating with the zero order, form a pair of image gratings. For a thin ($\ll\lambda$) amplitude (e.g., chrome on glass) object, every pair is mutually in phase (in focus at $z = 0$), i.e., the higher order image gratings are in phase with the lowest order grating. Defocusing causes the longitudinal phase terms (with variations in z) to change differently for each spatial frequency. The conjugate terms resulting from E_0E_1 and E_0E_{-1} now have opposite phase shifts, for example, $e^{i\left[k_dx+\left(k_0-\sqrt{k_0^2-k_d^2}\right)z\right]}$ and $e^{-i\left[k_dx-\left(k_0-\sqrt{k_0^2-k_d^2}\right)z\right]}$, and they partially compensate each other. Thus, the resultant intensity for a particular frequency initially reduces, goes to zero, and then reappears at multiple defocus planes. The z-direction periodicity is a function of the transverse period of that frequency component and the full image does not reappear (the different terms do not come into phase again at any focal position away from $z = 0$ except for the limiting case of only a single pair of ±1 diffracted orders). This is easy to show by including the z-dependent terms in Equation 15.2:

$$I_{\text{image}} = 4\left|E_0E_1\right|\cos(k_dx)\cos\left[\left(k_0-\sqrt{k_0^2-k_d^2}\right)z\right]$$

$$+ 4\left|E_0E_2\right|\cos(2k_dx)\cos\left[\left(k_0-\sqrt{k_0^2-(2k_d)^2}\right)z\right] \tag{15.10}$$

Note that the z-directed periodicities are different for the first- and second-order pairs. Hence, for a specific cut through the sample, there is only one conjugate image plane where all frequencies are in the correct phase relationship, i.e., there the image of this part of the object is in focus, at $z = 0$, the plane conjugate to the object plane as defined by the optical system. For other layers of the object, the frequencies have phase shifts with respect to each other that vary with the z-position away from focus and the image is consequently blurred or more severely distorted.

The situation is more severe in the case of single-sideband imaging (Figure 15.2). Here, the grating image shifts continuously to one side with defocus rather than disappearing and reappearing, for example, instead of the separate cosine term, there is a continuous phase shift introduced into the pattern, i.e., instead of $\cos(k_dx)\cos\left[\left(k_0-\sqrt{k_0^2-k_d^2}\right)z\right]$, the variation is $\cos\left(k_dx+\left[\left(k_0-\sqrt{k_0^2-k_d^2}\right)z\right]\right)$. In this sense a single-sideband system has a quasi-infinite depth of focus. Choosing the correct focus even for a thin object becomes challenging in absence of a priori knowledge of the object. In general, there is no possibility of sectioning a thick sample without additional information, which can be provided by using wide spectral range sources including short pulse, short coherence length, tunable lasers (Chapter 7), tomography, and other techniques. Illumination with a short longitudinal coherence length source allows sectioning of the sample up to the scale of the coherence length. Short pulses and illumination with different wavelengths give the same possibility.

For incoherent illumination, which is always a telecentric double-sideband system, the picture is more complex. Here, for every spatial frequency, a continuum of the beating pairs is built by the multiple off-axis illuminations, each pair with its own mutual phase shift (due to the different off-axis angles). The summed amplitude at a given spatial frequency still oscillates with focus with continuously reduced amplitude as shown in Figure 15.9, curves (2–4) (Hopkins 1955, Williams and Becklund 1989). So, if the object is out of focus, the MTF "degenerates" and higher frequencies increasingly disappear as the sample is moved further from the objective focal plane.

Only the region around the conjugate image plane ($z = 0$) is "in focus," and sectioning of the sample is accomplished since only one layer of thickness corresponding to the depth of

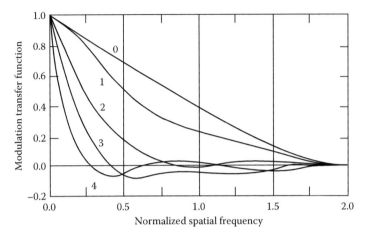

Figure 15.9 MTF curves for object planes in and out of focus. The parameter is the normalized distance from focus in units of the depth of field. (Reprinted from Hopkins, H., *Proc. R. Soc. Lond.*, A231, 91, 1955. With permission.)

focus is visible; other layers almost vanish and don't perturb the focused region. Of course, this assumes that the object is weakly scattering and does not materially impact the propagation of the illumination beam. The depth of focus (Δz) is nearly proportional to the reciprocal of the area of the entrance pupil:

$$\Delta z = \left(\frac{\lambda}{4n\left(1 - \sqrt{1 - NA^2}\right)} \right) \tag{15.11}$$

In a weak scattering limit, so that the propagation of the fundamental beam is not strongly affected by the sample, 3D sectioning is possible by longitudinal scanning of a thick sample in Δz steps.

15.4 ENHANCEMENT TECHNIQUES

Numerous additional techniques have been developed to enhance the images obtained with conventional microscopes. In particular, phase contrast and Nomarski techniques are often employed, particularly for low-contrast biological samples. Phase contrast microscopy was invented by Fritz Zernike (for which he received the 1953 Nobel Prize in Physics) (Zernike 1942). For a sample with variations in thickness and/or in index of refraction, the E_1 and E_{-1} diffracted orders are not equal in phase, and information about the object is encoded in the phase difference between these pairs. This is especially problematic for colorless and transparent biological materials as there is very poor image contrast associated with amplitude variations. Zernike suggested changing the phase of just the zero order, so that the phase difference for a particular pair can be balanced to enhance the contrast of image contours corresponding to a specific annulus in frequency space.

A related technique, phase shift interferometry (PSI), is a contemporary research topic. The gist is that by intentionally changing the zero-order phase in known steps it is possible after some digital analysis of the results to separate the amplitude and phase parts of the electric field in the image. The phase part contains combined information about both the average value of the index of refraction and the longitudinal position of the object in the z-direction.

PSI is ideal for the investigation of reflective surfaces, where the longitudinal position corresponds to the surface itself. This position can be obtained with very high precision of about a few nanometers (Dubois et al. 2001, Pitter et al. 2004) and is the basis of the interferometers used to control stage position in microlithography steppers/scanners. Since this is an interferometric technique, the resolution is not limited by the wavelength but rather by variations in the refractive index in the optical path during the multiple measurements (atmospheric variations) and by the signal–noise levels of the images. Sectioning of a thick sample is not possible by this technique.

Nomarski techniques are closely related to phase contrast microscopy. The basic principle is similar to the phase contrast microscopy technique but the implementation is different. A prism is used to split a polarized illumination beam into two slightly shifted plane waves that pass through the transparent phase object sample and are recombined interferometrically in a second prism before observation. This results in a differentiation of the image wherein regions of rapid phase variation are enhanced. Several texts provide a more detailed description of phase contrast and Nomarski microscopy (Born and Wolf 1999, Gu 2000, Murphy 2001).

15.4.1 Dark-Field Microscopy

As noted earlier, dark-field microscopy refers to the nonimaging terms that arise from interference between scattered waves other than the zero-order transmission (reflection) from the object. In general, there are two approaches to dark-field microscopy. In one approach, a block is used in the objective pupil to eliminate the zero-order light transmitted (reflected) from the object. This, of course, is only possible for coherent and partially coherent illumination; for incoherent illumination, some of the zero-order transmission is necessarily transferred to the image plane. The frequency space content of the image is restricted to $2NA$. Assuming normal incidence illumination, the collected spatial frequency information is the same as that with conventional microscopy; however, dark-field microscopy offers contrast enhancement advantages similar to phase contrast. The resulting square law detection intensity pattern at the image plane is not truly an image, in the sense of a one-to-one correspondence of features of the image with the sample. This approach emphasizes higher frequencies (e.g., edges) and contains frequencies that are not necessarily present in the object (cf. Equation 15.3).

An alternate approach to dark-field microscopy is to illuminate the object at angles beyond the lens NA, similar to the situation depicted in Figure 15.3, but with the tilt increased so that the zero-order transmission is beyond the lens NA. In this case, higher spatial frequency scattered information is collected from the object, but the square law response results in image frequencies again restricted to $2NA$; in other words, the high frequency content of the scattered light is downshifted in the detection process. This approach to microscopy provides an important starting point for the synthetic aperture approaches such as IIM, discussed in the following. Often this illumination is in an annular pattern, similar to the annular illumination discussed earlier, but with all of the inclination angles of the illumination greater than the NA of the collection optical system.

15.4.2 Immersion Microscopy

Immersion techniques to extend the frequency space coverage of microscopy (to $NA \sim 1.4$–1.6, i.c., $\sim\lambda/6$) are well established (Murphy 2001) but remain limited in application as a result of practical issues such as compatibility of immersion fluids with the sample. Traditional approaches with immersion fluids are restricted to NAs of ~ 1.4 by the available immersion fluids and the glass lens materials, as well as the difficulty of fabricating high-NA aberration-corrected optics. Recently, a transmission microscopy approach using opposing immersion lenses and annular illumination has been demonstrated to have a resolution of ~λ/5 (90 nm) (Vainrub et al. 2006), and a commercial product is available (www.cytoviva.com).

15.4.3 Solid Immersion Microscopy

This approach uses a numerical aperture increasing lens (NAIL) placed on the substrate of an object to both illuminate with and collect scattered spatial frequencies at angles beyond the *NA* of the remote objective to reach the full linear systems limits of microscopy (Wu et al. 2000). Full immersion will extend the frequency space coverage to $2n_{sub}/\lambda$. A resolution of λ/9 (145 nm with 1.3 μm illumination) was achieved by the optical setup shown in Figure 15.10 (Ippolito et al. 2000). Spherical aberration is eliminated by the NAIL, but the large *NA* leads to a small field of view and other higher order aberrations.

15.4.4 Near-Field Microscopy

The highest spatial frequency information in the scattered light from the object is contained in the near fields that are localized within distances of order of the wavelength or less from the object and are not accessible with far-field microscopy. The most straightforward approach is to build a small probe that can sample these fields and scan the probe to build up an image of the sample. With the enormous advances in scanning tunneling and atomic force microscopy, this has become both a feasible and a common approach. Relatively recent reviews are available (Courjon 2003, Novotny and Stranick 2006). Both aperture-based and apertureless approaches have been explored. In the aperture-based approaches, a local light source, based for example on an optical fiber that has been stretched to the nanoscale and coated with a metal, leaving only the tip uncoated, is used as the probe. The primary difficulty with this approach is the limited amount of light that can propagate through the cutoff waveguide to the sample. In apertureless schemes, a solid pointed metal probe is used either to scatter the

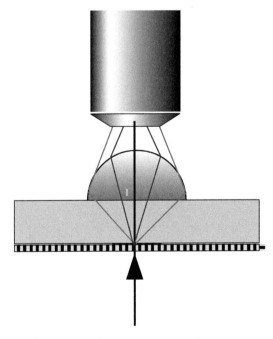

Figure 15.10 High-resolution subsurface microscopy technique.

local field from the sample or to illuminate the sample by exciting surface plasma waves (Chapter 3) that can propagate down the taper to the sample. Detection is by collecting the light scattered by the tip-sample geometry, which varies with the sample optical topography. In all of these approaches, it is important to account for the electromagnetic interaction between the tip and the sample. In effect, the tip perturbs the electromagnetic environment of the sample, changing the scattering/transmission characteristics. Near-field probes are serial scanning techniques, which inherently require a long time to build up an image, what is perhaps their greatest disadvantage. An extensive discussion is given in Chapter 17.

15.5 ADVANCED DIRECTIONS IN MICROSCOPY

Recently, many additional techniques have been introduced to enhance the resolution, either transverse or longitudinal, of conventional microscopy. Many of these rely on the inherent nonlinearity associated with two-photon processes (fluorescence) and with additional non-linearities enabled by saturation and multi-photon excitation. These are somewhat outside of the scope of this chapter and are mentioned for completeness, along with references for further exploration. Codification of many of these techniques has recently been presented (Heintzmann and Ficz 2006, Hell 2007, 2009).

15.5.1 Confocal Microscopy

A technique called confocal imaging was first proposed by P. Nipkow and pioneered by M. Minsky who made the first scanning confocal microscope at Harvard University in 1957 (Wilson 1990). Confocal microscopy is typically applied to fluorescent samples and provides both enhanced depth resolution and improved rejection of scattered light from adjacent objects (Webb 1996, Pawley 2006). In contrast to the imaging techniques described earlier, confocal microscopy uses a point source, typically from a laser, to illuminate a single point on the object. The fluorescence from this point is then passed through a conjugate aperture in the collection system; this discriminates against light emitted from planes of the object that are away from the focus, since they are not reimaged at the conjugate aperture, and dramatically improve the longitudinal resolution. The object is then scanned in all three spatial directions to build up an image. This is a very powerful technique that has found much acceptance, particularly in biological imaging. The method allows scanning a series of thin optical "slices" through the thickness (z-direction) of the specimen. In practice, the best transverse resolution of a confocal microscope is about $0.2\,\mu m$, and the best axial resolution is about $0.5\,\mu m$. Confocal microscopy has also been used for semiconductor diagnostics, particularly for determining the carrier diffusion length in semiconductors, by shifting the collection aperture relative to the input spot and monitoring the spatial decay of the photoluminescence (Fong and Brueck 1992, Fong et al. 1994).

15.5.2 4PI, I5M Microscopy

In conventional microscopy, the longitudinal resolution, along the optical or z-axis is significantly poorer than the transverse resolution, for example, the depth of field is much larger than the transverse resolution. This limitation is addressed by increasing the solid angle (NA of the illumination source). The extreme is coherent illumination with a coherent optical source from both top and bottom sides with high-NA (immersion) lenses, known as 4π microscopy (Nagorni and Hell 1998). To reduce optical scattering effects, two-photon fluorescence is often the observed quantity. Since a coherent laser source is used, this is a confocal configuration, the imaging is point by point, and the sample is scanned across a 3D focal volume. In 1999, interference microscopy (I5M) was demonstrated based on a novel interferometric technique in which the sample is observed and/or illuminated from both sides simultaneously using two opposing objective lenses and an incoherent source of illumination. Separate interference effects in the excitation light and the emitted light give access to higher resolution (~100 nm), and axial information about the sample than can be achieved with conventional wide-field or confocal microscopes (Gustafsson et al. 1999). Since this is a wide-field measurement, scanning is only in the z-direction, but additional artifacts are often present in the image (Bewersdorf et al. 2006).

These schemes are largely based on fluorescence microscopy. Special care has to be taken with biological samples to avoid photobleaching (Davidson 2010) of the chromophore and photoxicity associated with free-radical generation by excited dye molecules (Hopt and Neher 2001). Continuous laser excitation of fluorophores causes either bleaching, i.e. reduction of the fluorescence efficiency, or toxicity (destruction of the biological function) due to continuous and intense illumination. The light intensity and duration must be limited to avoid these effects, which impacts the viewing time and the achievable signal-to-noise ratio.

Some of these problems were solved with multi-photon microscopy (Denk et al. 1990). Multi-photon absorption was predicted by Maria Göppert-Mayer in 1930 (Göppert-Mayer 2009), and a proof-of-principle experiment was performed in the 1960s using continuous-wave laser sources. Multi-photon fluorescence microscopy allows imaging in highly absorbing media, which increases detection sensitivity, image contrast, and enables full-frame video-rate fluorescence lifetime imaging, to reduce considerably the generation of phototoxic products. The resolution in this case is more limited with a given fluorophore as compared to confocal imaging (Hell and Wichman 1992, Rudolph et al. 2003).

Lateral resolution that exceeds the classical diffraction limit by a factor of 2 ($\lambda/3.3$) was achieved by using spatially structured illumination in a wide-field fluorescence microscope by Gustafsson (2000). A sample was illuminated with a series of excitation light patterns, which encoded normally inaccessible high-resolution information into the observed image. Additional nonlinearities associated with saturation extended the resolution to ~$\lambda/10$ (Gustafsson 2005).

15.5.3 Stimulated Emission Depletion and Ground State Depletion

Even more impressive results have been obtained by STED microscopy (Hell et al. 1994). The basic concept is to illuminate a fluorophore with two beams. One is a low-intensity source that excites fluorescence. Since the spot size is limited by diffraction, so is the resolution, as discussed earlier. The second beam is a donut mode at a different frequency that deactivates the fluorescence through stimulated emission at a longer wavelength (to a higher level of the ground state manifold). Once again the definition of the donut hole is limited by diffraction. However, the depletion "saturates," and the fluorescence only survives in regions where the second beam intensity is below a threshold value. Increasing the intensity of the second beam provides a dramatically improved resolution, limited only by the accuracy of the "null" at the center of the donut. In an initial report, resolution of 33 nm ($\lambda/23$) was achieved (Marcus and Hell 2002). This is accomplished by exciting the molecules with a femtosecond pulse and subsequent depletion of the excited state with red-shifted, picosecond-pulsed, counter-propagating, coherent light fields. GSD is a related technique wherein the molecular ground state is shifted to a long-lived "dark" state, for example a triplet, by the saturating donut-shaped pulse. Recently, the resolution was increased to $\lambda/50$ (15–20 nm) (Donnert et al. 2006). In a recent review article, Hell (2007) proposed a revision to the Rayleigh criteria that applies to a number of nonlinear microscopies:

$$R \approx \left(\frac{\lambda}{2NA\sqrt{1 + I_{max}/I_s}} \right) \tag{15.12}$$

where

I_{max} is the maximum intensity of the saturation beam
I_s is the characteristic saturation intensity that quenches the fluorescence

For both STED and GSD, the resolution is ultimately limited by the quality of the null in the donut beam; once the intensity in the null exceeds the saturation intensity, all of the photoluminescence is quenched and no image remains. An example of applying STED to a solid-state microscopy problem, the distribution of color centers in diamond, was presented by Rittweger et al. (2009).

15.5.4 Photoactivated Localization Microscopy and Stochastic Optical Reconstruction Microscopy

Both of these related techniques take advantage of improved localization by fitting the centroid of an isolated PSF rather than resolution or distinguishing between two overlapping PSFs (Betzig et al. 2006, Hess et al. 2006, Rust et al. 2006). The basic idea is to use one illumination source to "turn on" fluorophores that are separated by distances greater than the optical resolution. Then take N number of images of the same fluorophores, by fitting them achieve a centroid definition of $\sim \lambda/2NA\sqrt{N}$, deactivate the fluorophores by some saturation mechanism (perhaps by blinking of quantum dots), and excite another set. Repeat multiple times until a full image is developed. This process is quite lengthy, taking many hours for typical biological entities and is not suitable for real-time observations.

15.5.5 Nanophotonics: Plasmonics, Nanoantennas, and Metamaterials

The extraordinary enhancements observed in surface-enhanced Raman scattering (SERS) with colloidal Ag nanoparticles, up to 10^{15}, provide ample evidence of the strong field enhancements (nanoantennas) that are available with localized surface plasma wave resonances of complex structures (Moskovits 1985, Campion and Kambhampati 1998). This has been an active research field for over 30 years, yet we still do not have reproducible SERS structures that can reliably reproduce these exciting results. The SERS enhancement is a nanoscale electromagnetic effect associated with localized surface plasma waves confined by sub-wavelength composite, metal-dielectric structures; the difficulties have been in fabricating samples and in maintaining their SERS properties under environmental assault (chemical contamination).

Recently, there has been considerable activity on nanoantennas particularly associated with semiconductor lasers for producing intense, near-field, sub-wavelength resolution sources (Cubukcu et al. 2006, Rao et al. 2007). Magnetic storage is a specific, large-scale application that is driving much of this activity since the available storage density is directly related to the available laser spot size (the reason blu-ray disks hold more information than conventional, red-laser-based, DVD optical media). This work has grown out of the observation of enhanced transmission through an array of sub-wavelength apertures in a metal film (Ebbesen et al. 1998) that has led to extensive research efforts at both understanding and applying this phenomenon (Coe et al. 2008). The general picture that has emerged is that the periodic hole array allows coupling (phase matching) between incident radiation and surface plasmon waves that propagate along the metal film (either top or bottom surface). The transmission is also impacted by coupling to the localized modes of the holes. Surface plasma waves inherently have larger wave vectors than the incident plane waves (Chapter 3) and so can be applied to sub-wavelength imaging. Near the cutoff temporal frequency for the surface plasmon waves, this compression can be large and offers the possibility of strongly enhanced resolution (Vedantam et al. 2009). Semiconductor lasers with nanoantennas employ related phenomena to concentrate the laser output in a sub-wavelength spot.

Metamaterials are novel, man-made, sub-wavelength nanostructures that offer optical properties not available from natural materials and have generated enormous interest recently (Engheta and Ziolkowski 2006, Cai and Shalaev 2009). A great deal of attention has been paid to the possibility, and realization, of negative permeability (negative μ) and of related negative-index materials (NIM) with both negative permeability and negative permittivity (Re μ < 0 and Re ε < 0) (Shalaev 2007) (so far to wavelengths as short as the near-IR but not into the visible due to the lossy visible optical properties of metals). Much of this scientific excitement has been driven by the possibility of NIM-based "perfect lenses" made from NIM that operate without any transverse spatial-frequency band pass limitation (Pendry 2000). This improved resolution is necessarily restricted to near-field domains for flat lenses and is limited by materials, fabrication, and impedance-matching constraints (Smith et al. 2003).

Related hyperlenses, which take advantage of non-planar metamaterial/plasmonic variations to generate magnification and thus to convert the evanescent fields at the object to propagating fields at the image, have demonstrated resolution to ~$\lambda/3$ (Lee et al. 2007, Liu et al. 2007) and to $\lambda/7$ (Smolyaninov et al. 2007). The geometric constraints associated with the hyperlens severely restrict the field of view, to date to only a few times the resolution. In the case of a 2D, planar, hyperlens structure (Smolyaninov et al. 2007) the image is necessarily restricted to a 1D line image. To date, these are exciting initial scientific demonstrations but not routine techniques for large-area microscopic investigations.

15.5.6 Digital Holography

Digital holography is a promising method for overcoming the conventional microscopy resolution limit. Digital holography allows reconstruction of both the amplitude and the phase of imaged objects. The scattered light from the sample is mixed in the Fourier plane with a reference wave. Of course, the diffracted orders have to propagate to the detection plane, which limits the resolution to $> \lambda/4$. The resulting hologram is recorded with a CCD camera; then the object wavefront is reconstructed numerically using the Kirchhoff–Fresnel propagation equations (Haddad et al. 1992, Schnars 1994, Schnars and Jüpter 1994, Grilli et al. 2001). Phase-shifting digital holography (PSDH) uses a series of images with a variation of the phase of the reference beam to obtain the complex amplitude at the plane of the CCD (Decker et al. 1978, Yamaguchi and Zhang 1997, Guo and Devaney 2004). PSDH has also been used for 3D microscopy (Zhang and Yamaguchi 1998, Yamaguchi et al. 2001), encryption (Lai and Neifeld 2000), and wavefront reconstruction (Lai et al. 2000).

15.6 SYNTHETIC APERTURE APPROACHES

A significant resolution improvement is obtained using holographic synthetic aperture methods. This is often referred to as "super-resolution," which is a bit of a misnomer. The resolution is indeed better than that provided by the generalized Rayleigh (or any other) criteria, but this is the result of effectively stitching together an effective pupil aperture that is larger than the physical *NA*, as will be described in the following; within this new larger aperture the standard resolution constraints apply. The synthetic aperture can be as large as 2 in air and $2n$ in an immersion medium, in the normalized units ($2\pi/\lambda$) defined earlier, with corresponding Abbe half-pitch limits of $\lambda/4$ and $\lambda/4n$, independent of the *NA* (cf. Equation 15.5).

As an analogue to synthetic aperture radar (SAR), the holographic synthetic aperture methods are based on the generation of a synthetic aperture by combining different sub-images recorded at different camera positions to construct a larger digital hologram (Le Clerc et al. 2001, Massig 2002). The resolution improvement increases with the number of recorded sub-images, as long as each sub-image covers additional (and exclusive) portions of the available spatial frequency space (Françon 1952, Toraldo di Francia 1955, Kartashev 1960, Lohmann and Parish 1964, Lukosz 1967, Toraldo di Francia 1969, Cox and Sheppard 1986, Sun and Leith 1992, Shemer et al. 1999, Zalevsky et al. 1999, Zalevsky and Mendlovic 2002, Zlotnik et al. 2005). The basis of super-resolution is to produce a synthetic enlargement in the system aperture without changing the physical dimensions of the lens or the illumination wavelength (Toraldo di Francia 1955, Sun and Leith 1992, Shemer et al. 1999). Many approaches are based on a certain a priori knowledge about the object, such as its time independence (Françon 1952, Sun and Leith 1992, Shemer et al. 1999), polarization independence (Lohmann and Parish 1964, Zlotnik et al. 2005), and/or wavelength independence (Kartashev 1960) allowing additional information to be accessed. All of these parameters are involved in information capacity theory (Toraldo di Francia 1955, Toraldo di Francia 1969, Cox and Sheppard 1986), which gives an invariance theorem for the number of degrees of freedom of an optical system. This theorem states that it is not the spatial bandwidth but the information capacity of an imaging system that remains constant. Thus, it is possible to extend the spatial bandwidth by encoding or decoding the additional information onto unused parameters of the imaging system. As examples, time independence of the object allows for sequential recording of sub-images with different optical configurations and polarization independence allows for simultaneous recording of sub-images with orthogonal polarizations and complementary frequency space coverage.

15.6.1 Imaging Interferometric Microscopy

IIM allows resolution to the linear systems limit of a half pitch of $\lambda/4$ in air. A related concept had been introduced earlier for lithographic image formation (imaging interferometric lithography or IIL) (Chen and Brueck 1998, Brueck and Chen 1999, Chen and Brueck 1999, Smolev et al. 2006).

A natural approach is to apply some of the techniques invented for higher resolution lithography to high-resolution measurement tools (Schwarz et al. 2003). The goal of IIM is to get high resolution with a relatively low-*NA* microscope objective in order to retain the objective's depth of field, field of view, and working distance. Multiple partial images with off-axis illumination and interferometric optics can be combined to assemble a composite image corresponding to larger frequency space coverage than is available with conventional imaging approaches. As we show in the following, this concept can extend to the linear systems limits of optics, to spatial frequencies of $2k_0$ in air, and $2nk_0$ in a medium of refractive index n (corresponding to the interference between counterpropagating plane waves). In the same sense as the Abbe limit, these limits correspond to the highest spatial frequency terms that can propagate through the respective media. Any higher resolution information corresponds to near fields and demands close proximity ($\ll\lambda$) to the object.

As was shown earlier (Figure 15.3), it is possible to increase the effective *NA* by using off-axis illumination to capture the second-order scattering information, E_2; it was important in that example to pass the E_0 term through the lens as well. This set an upper limit on the off-axis illumination angle. With larger tilt angles of the illumination beam (extreme off-axis illumination, where the extreme refers to the fact that the zero-order beam is outside the lens *NA*), information corresponding to larger scattering angles (smaller features) from the illumination beam is passed through the lens. As discussed earlier, this is one approach to dark-field illumination. In the IIM configuration, a zero-order beam is reintroduced at the image plane using additional optics to provide the appropriate divergence, amplitude, phase, and angle of

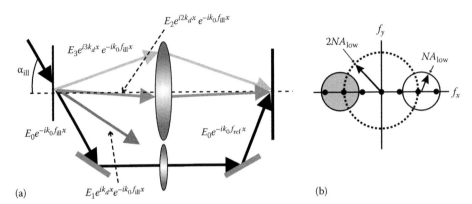

Figure 15.11   (a) High-frequency imaging setup. The extreme off-axis illumination allows the high orders to be transmitted through the lens and interferes with the zero order that is brought around the lens. (b) Frequency space coverage: gray circles = high-frequency images, dashed circle = dark field.

incidence for image formation (Figure 15.11a). Effectively this involves constructing a Mach–Zehnder interferometer around the objective lens. The optics are arranged to provide the same characteristics of the zero-order (reference) beam that would have been available if the lens NA was sufficiently large to collect it directly. If the illumination angle is such that the higher terms E_2 and E_3 are collected and a reference beam is added with additional optics, the intensity in the image plane is as follows:

$$I_{high} = \left| E_0 e^{-ik_0 f_{ref} x} + E_2 e^{i(2k_d - k_0 f_{ill})x} + E_3 e^{i(3k_d - k_0 f_{ill})x} \right|^2$$

$$= \left| E_0 \right|^2 + \left| E_2 \right|^2 + \left| E_3 \right|^2 + 2\left| E_0 E_2 \right| \cos\left\{ [2k_d - k_0(f_{ill} - f_{ref})]x + \phi_2 \right\}$$

$$+ 2\left| E_0 E_3 \right| \cos\left\{ [3k_d - k_0(f_{ill} - f_{ref})]x + \phi_3 \right\} + 2\left| E_1 E_2 \right| \cos\left\{ [k_d - k_0(f_{ill} - f_{ref})]x + \phi_2 - \phi_1 \right\}$$

$$(15.13)$$

where
   $f_{ill} \equiv \sin\alpha_{ill}$ is the normalized frequency shift due to the inclined illumination (high frequencies are shifted by the tilted illumination and are captured by the lens)
   f_{ref} is the normalized frequency shift due to reference beam inclination (which shifts the frequencies back to their original values)

The ϕ's are the respective phases of the diffracted beams. The incident angle of the reference beam is adjusted so that $f_{ref} = f_{ill}$, so that these offsets cancel in the sub-image:

$$I_{high} = \left| E_0 e^{-ik_0 f_{ref} x} + E_2 e^{i(2k_d - k_0 f_{ill})x} + E_3 e^{i(3k_d - k_0 f_{ill})x} \right|^2$$

$$= \left| E_0 \right|^2 + \left| E_2 \right|^2 + \left| E_3 \right|^2 + 2\left| E_2 E_3 \right| \cos(k_d x + \phi_3 - \phi_2)$$

$$+ 2\left| E_0 E_2 \right| \cos(2k_d x + \phi_2) + 2\left| E_0 E_3 \right| \cos(3k_d x + \phi_3)$$

$$(15.14)$$

The corresponding frequencies are shown by the black dots in Figure 15.11b; high frequency space coverage for this case is the gray circle and the matching clear circle resulting from the square law operation, while the dashed circle corresponds to the dark-field region and the black dots to the corresponding frequencies. Notice that as a result of the extreme off-axis illumination, the frequency content now extends beyond the $2NA$ limit set by the optical system. Thus, the synthetic aperture can extend beyond the limits set by traditional optics approaches. However, the low frequencies, and in particular the E_1 term, are missing. This is a high frequency sub-image; combining it appropriately with a normal-incidence illumination, low-frequency image provides a more complete representation of the object.

IIM is based on the incoherent (intensity) addition of several coherent sub-images. Combining the two sub-images from normal incidence and off-axis illumination (Figures 15.2 and 15.11, respectively) results in the following response, which consists of a spatially constant base line, an image part, and a dark field (Equation 15.15). The intensity of the high-frequency image has been multiplied by two to compensate for the reduced intensity due to the collection of only a single sideband. The off-axis dark field is easily measured by blocking the reference beam and is subtracted from the off-axis sub-image before combining with the low-frequency sub-image:

$$I_{\text{total}} = I_{\text{low}} + 2(I_{\text{high}} - I_{\text{high,dark}}) = 3|E_0|^2 + 2|E_1|^2 + 4|E_0 E_1|\cos(k_d x + \phi_1)$$

$$+ \left[4|E_0 E_2|\cos(2k_d x + \phi_2) + 2|E_1|^2 \cos(2k_d x + 2\phi_1)\right]$$

$$+ 4|E_0 E_3|\cos(3k_d x + \phi_3) \tag{15.15}$$

Again, this is a single sideband system; the sideband not collected by the optical system is restored as a result of the square law intensity response. Clearly, in this simple case, it is possible by measuring the spatial frequencies, intensities, and phases, to reconstruct the image (e.g., to determine d, E_0, E_1, E_2, and E_3). However, the reference beam in our experimental setup is transmitted around the lens, so it has an arbitrary amplitude and phase. So more precisely, the intensity distribution of the high-frequency image is (setting $f_{\text{ref}} = f_{\text{ill}}$)

$$I_{\text{high}} = \left|AE_0 e^{-i\phi_{\text{ref}}} e^{-ik_0 f_{\text{ref}} x} + E_2 e^{i(2k_d x + \phi_2)} e^{-ik_0 f_{\text{ill}} x} + E_3 e^{i(3k_d x + \phi_3)} e^{-ik_0 f_{\text{ill}} x}\right|^2 = |AE_0|^2 + |E_2|^2$$

$$+ |E_3|^2 + 2|AE_0 E_2|\cos(2k_d x - \phi_{\text{ref}} + \phi_2) + 2|AE_0 E_3|\cos(3k_d x - \phi_{\text{ref}} + \phi_3) \tag{15.16}$$

where
A is a reference attenuation factor
ϕ_{ref} is the arbitrary phase shift of the reference beam
ϕ_i is the phase of the corresponding high-frequency beam

Using a reference object and adjusting the phase of the reference beam by adjusting the length of the reference optical path, the system can be adjusted in such a way that $\phi_{\text{ref}} = 0$ in order to match the phases (e.g., x-positions) of the low- and high-frequency sub-images. Neutral density filters are used to adjust A so that $|E_0| \sim |E_2|$ in order to have the maximum image contrast (optimum signal–noise ratio). Then the intensities of the sub-images are adjusted electronically by comparing with the reference object. The net result is a resolution corresponding to a larger NA lens. For example, an optical system with an $NA_{\text{low}} = 0.4$ objective acts as a low-pass filter that limits the frequency space with He-Ne laser illumination (633 nm) to a maximum frequency corresponding to a half pitch of ~790 nm ($\lambda/2NA_{\text{low}}$). If the illumination offset is chosen at $2NA$ to provide continuous coverage along the x-axis, the effective $NA_{\text{eff}} = 3NA_{\text{low}}$

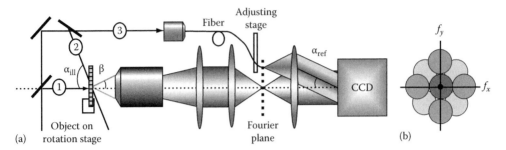

Figure 15.12 (a) Optical arrangement for IIM: $\beta = \sin^{-1}(NA)$, α_{ill} is the incident beam angle of incidence and α_{ref} is the angle of the reference beam onto the image plane. (b) Frequency space coverage for an experiment with five sub-images.

and the half-pitch resolution is decreased to 263 nm. The zero-order transmission is shifted outside of the imaging pupil plane, with the additional interferometric reintroduction of the zero-order beam to the Fourier plane on the low-NA side of the lens in order to restore the original frequencies. A schematic of the experiment is shown in Figure 15.12a. The incident angle of the zero-order α_{ref} is adjusted such that the spectral content is shifted back to the original high frequencies. Changing α_{ill} together with α_{ref} leads to different frequency coverage. A rotation stage can be used to extend the azimuthal coverage across the f_x, f_y plane.

The possible frequency space coverage is shown in Figure 15.12b, where the circle in the middle represents the low-frequency image and the circles on the sides represent high-frequency images. The Fourier transform is symmetric when the object is phase invariant (a thin object $\ll \lambda$). As noted earlier, the imaging is single sideband; the square law (intensity) response of the image formation process restores the conjugate frequency space components, resulting in the two symmetric circles in Figure 15.12b for each sub-image. As before, it is possible to restore telecentricity by combining sub-images from opposing tilt angles in pairs.

In the arrangement of Figure 15.12a, beam 1 (2 and 3 are blocked) is used for on-axis illumination. Beams 2 (extreme off-axis illumination) and 3 (reference), when 1 is blocked, are used for the high-frequency sub-images. It is convenient to use a single-mode fiber for beam 3 to simplify the optical arrangement and provide a clean reference beam. The exit aperture of the fiber is placed in the pupil of the imaging optics; the position of the fiber sets f_{ref} and the path length sets φ_{ref}. There is a requirement that the optical length of path 3 be the same as that of path 2 combined with the collection optical path length to within the source longitudinal coherence length. For the long-coherence-length He-Ne laser used in the demonstration experiments, this requirement is easy to realize. For a shorter coherence length source, care is necessary to ensure that this condition is met. The dark fields and the background/reference images should be subtracted from the stored high-frequency image in order to get the high-frequency image with only the imaging contribution (interference between zero-order and diffraction beams). Thus, the effective numerical aperture is extended to $3NA$.

We use a Manhattan structure (x, y) object (Figure 15.13a) as an example to show the possibility of tiling frequency space and reconstructing an object. The resulting frequency coverage for this structure with lines of width 500 nm is shown in Figure 15.13b. If we use a $NA = 0.4$ objective and a He-Ne laser ($\lambda = 633$ nm), the circles in the center correspond to the spatial frequencies 0.4 and 0.8 normalized to k_0, for coherent and incoherent illumination, respectively (Section 15.3.2), which define the range of possible captured frequencies by conventional illumination. The spatial frequency 0.8 also corresponds to the extent of the dark field. The set of shifted circles of radius $NA_{low} = 0.4$ for the case when $\alpha_{ill} = 53°$, extends the radius to $3NA = 1.2$ in the x- and y-directions or a minimum resolution half pitch of ~260 nm (Schwarz et al. 2003).

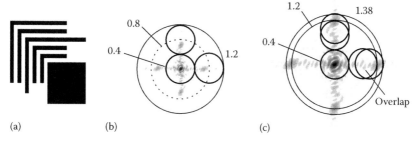

Figure 15.13 (a) Manhattan geometry pattern used for image resolution exploration consisting of five nested "ells" and a large box. (b) Intensity Fourier space components of the pattern, mapped onto the frequency space coverage of the imaging system. Lines and spaces of the "ells" are 500 nm. (c) Lines and spaces of the "ells" are 240 nm.

Additional frequency space coverage is available with a second pair of off-axis sub-images, represented by the outer set of shifted circles (Figure 15.13c), with a larger tilt of the illumination plane wave, approaching grazing incidence. The maximum frequency coverage (from Equation 15.5) extends to $1 + NA = 1.4$. However, in the corresponding experiments, the inclination angle was limited to $80°$ [$\sin(80°) = 0.98$] to the object plane due to the high substrate reflectivity at steeper angles and the increasing wavefront distortion associated with substrate thickness variations. Therefore, the experimental extension was $0.98 + NA = 1.38$, which allowed resolution of a Manhattan structure with lines of width 240 nm using the same $NA = 0.4$ objective (Figure 15.13c). Clearly, the frequency space coverage of the outer circles captures the fundamental frequency components of the image (Schwarz et al. 2003). The problem of overlapping of frequency coverage can be solved by filtering images either optically with apertures in the Fourier plane or electronically by taking Fourier transforms of the sub-images, appropriately filtering, and retransforming.

For some sample structures with a frequency peak at the edge of the collection limit, some higher frequency sidebands of the image are not collected, resulting in distortions, such as extra features and ringing (Gibbs phenomenon) arising from a sharp cutoff in frequency space (Smith 2007). This can be reduced by using apodized filters (Tridhavee et al. 2005) applied in the physical pupil plane or electronically.

The reference beam in the experimental setup is transmitted around the lens, so its amplitude, phase, and position in the pupil have to be adjusted according to the formula (15.12). A reference object is required to assist in setting the reference beam parameters. This object should be as close as possible to the object being imaged. In our initial demonstrations, where we knew a priori the object, we used the known object as a self-reference.

For higher contrast, it is important to choose the object position in such a way that the polarization is parallel to the small image features recorded in each sub-image (TE polarization) (Nesterov and Niziev 2005). This is simple to understand by considering the interference between two plane waves as a function of the angular separation (2θ) between them. For TE polarization, the result is just $|\vec{E}|^2 = |A|^2[1 + \cos(2k\sin\theta x)]$, while for TM polarization it is $|\vec{E}|^2 = |A|^2[1 + \cos(2\theta)\cos(2k\sin\theta x)]$. The visibility [$(I_{max} - I_{min})/(I_{max} + I_{min})$] is unity independent of θ for TE polarization but is reduced at steeper angles for TM polarization, goes through 0 at $\theta = 45°$ and inverts for large angles (switching lines and spaces).

The experimental results with a Manhattan geometry structure (line width 500 nm) and a He-Ne laser used for illumination are shown in Figure 15.14. The low-frequency image after dark-field and background subtraction as well as filtering is shown in Figure 15.14a. The individual nested-"ell" lines are not resolved. The off-axis illumination sub-image of the horizontal features after the same subtraction procedure is shown in Figure 15.14b; the

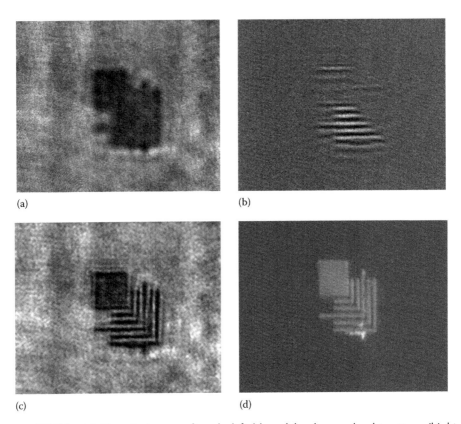

(a)   (b)

(c)   (d)

Figure 15.14 (a) On-axis image after dark-field and background subtraction, (b) high-frequency image of horizontal structures after dark-field and reference image subtraction, (c) filtered reconstructed image, and (d) image taken by a conventional incoherent-illumination microscope with $NA = 0.9$. (Reprinted from Schwarz, C.J. et al., *Opt. Lett.*, 28, 1424, 2003. With permission.)

vertical features are similar. The reconstructed image obtained by adding the three images is shown in Figure 15.14c. The image taken using a conventional microscope with incoherent white light and $NA = 0.9$ is shown in Figure 15.14d. Note that the extra features appearing at the bottom two horizontal lines are due to corrosion and lifting of the chrome features. This demonstrates that IIM is indeed responding to arbitrary structures and is not limited to simple geometries.

A major advantage of IIM compared with its lithographic analogue (IIL) is that the partial images can be electronically manipulated, whereas in the lithography case the images are chemically stored in the photoresist and are not individually accessible. This relates back to the major trends identified in the introduction, particularly the availability of high pixel count digital imaging sensors and the high-speed computation necessary for manipulating the image information.

15.6.2 Development and Advantages of Imaging Interferometric Microscopy

Subsequent to the initial publication of the IIM concept (Schwarz et al. 2003), Alexandrov et al. (2005, 2006) introduced an alternate, but related, concept wherein the images were recorded directly in the Fourier plane. This procedure not only increases the system information capacity for a given camera but also leads to ambiguity of phase determination and possible information loss for highly periodic images due to the restricted dynamic range of the CCD camera. In order to determine the correct phase for image reconstruction it is necessary to have a reference object, which is impossible in this case. In more recent reports (Alexandrov et al. 2007, Hillman et al. 2009), the recording camera is shifted away from the Fourier plane (defocused) to increase the available dynamic range of the recorded intensities. Multiple images (90) were recorded with image rotation in order to extract phase information from overlapped images and the NA was increased from 0.13 to 0.61. An alternate system based on illumination with multiple wavelength sources to cover different regions of frequency space was introduced (Alexandrov and Sampson 2008).

The experiment described by Schwarz et al. (2003) was reproduced at a wavelength of 532 nm by Price et al. (2007) with a 100× objective ($NA = 0.59$) to show an improvement in the effective numerical aperture from 0.59 to 0.78 and by Micó et al. (2007) with $NA_{low} = 0.14$ extended to $NA_{eff} = 0.45$. With a setup similar to that of Schwarz (2003), a vertical-cavity surface-emitting laser (VCSEL) array was used as a source of light for the off-axis illumination, providing a simple optical system that is switched by either turning on individual VCSELs or by relying on their mutual incoherence to record all of the offset images in a single step (Micó et al. 2004, 2006a, b). Micó et al. (2004) used five sources simultaneously to increase the system spatial frequency bandwidth. Recently, Micó et al. (2006a) extended the optical system (discussed in Micó et al. (2004)) for 2D objects. The recording process is by interference of each frequency band with a complementary set of reference plane waves in parallel. The benefit of this system is improved modulation speed, which leads to more rapid image synthesis. Moreover, any desired synthetic coherent transfer function can be realized by changing the electrical drive of the VCSEL array. However, the holograms for the different band passes are incorrectly overlapped since distance between VCSELs is not matched to the optical system, and so the combined image is distorted; this could be resolved by designing the laser array to match specifically to the optical system. Other authors have implemented spatially incoherent illumination sources for a continuum of independent off-axis illuminations, along with a pinhole in the pupil plane for blocking the dark field, to increase the resolution by incoherent-to-coherent conversion (MTF to ETF) with maximum achievable effective aperture of unity (Leith et al. 1987, Leith 1990).

Another method uses a collection of mutually incoherent point sources at different lateral positions to provide tilted, spherical-wave illuminations for the object since every point source gives a spherical wave with a different origin (Micó et al. 2006c). In this case, the input object was illuminated at off-axis illumination angles higher than the NA of the microscope objective (Sheppard and Hegedus 1998, Micó et al. 2006c). In this way a resolution improvement by a factor of 3 was shown to be achievable using off-axis illumination with a maximum illumination angle equal to the NA of the imaging lens. The angle of illumination is limited by $NA_1 + NA_2$, where NA_1 is the numerical aperture of the optical elements between the VCSEL and a sample and NA_2 is the numerical aperture of the imaging system. In later experiments, (Micó et al. 2008b, Granero et al. 2009) a grating near the Fourier plane was moved during image acquisition for phase recovery. In the following papers, the authors discussed the possibility of axial resolution, showing an example of 3D images of swine sperm: (Micó et al. 2008a, c).

Interferometric synthetic aperture microscopy (ISAM) (Tyler et al. 2007, 2008, Davis et al. 2008) is aimed at increasing the 3D volume of the image beyond the traditional depth of field, rather than increasing the transverse resolution. In ISAM, a sample is illuminated by a series of femtosecond laser pulses that make a point-by-point scan of a transverse plane. The scattered field interferes with an original pulse and is dispersed by a grating. The resultant spectral content is used to compute the contribution from various depths. In this way, longitudinal information from the 3D image is extracted without mechanical refocusing. The axial resolution comes from the low coherence length of the short pulse, which only allows interference when the path lengths of the sample and reference beams are matched in length.

15.6.3 Concepts Associated with a Tilted Object Plane

Off-axis illumination allows an increase in NA_{eff} to $(1 + NA)$ with the optical axis being normal to the object plane (position a in Figure 15.15), but there is additional higher frequency (larger angle) scattering to the side of the objective toward the incident, off-axis beam. This information can be captured by tilting the object plane or equivalently, tilting the objective (shown as the offset objective at position b in Figure 15.15) (Kuznetsova et al. 2007, 2008). Notice that, for subimages offset in frequency space by more than $3NA$, there is a separation between the imaging terms and the dark-field terms, which are limited to spatial frequencies less than $2NA$. So, the dark-field terms are easily removed electronically by taking the Fourier transform of the subimage, filtering appropriately in spatial frequency space and transforming back to real space.

In order to obtain an effective aperture for this case, we have to add the object plane tilt angle θ_{tilt} to the angle corresponding to the objective NA in Equation 15.5. The modified formula is as follows:

$$NA_{eff} = \sin(\sin^{-1}(NA) + \theta_{tilt}) + \sin(\alpha_{ill}) \tag{15.17}$$

The highest possible spatial frequencies $(\sim 2/\lambda)$ are captured with a grazing incidence illumination and an object plane tilt to $\pi/2 - \sin^{-1}NA$ $(\sim 66.42°$ for $NA = 0.4)$ (e.g., counterpropagating plane waves), but the constraints of the optical system, both physical and optical (aberrations), restrict the tilt and thus limit the frequency space coverage to slightly less than $2/\lambda$. As a result of the non-paraxial optical system, the extent (minimum to maximum spatial frequency) of the captured frequency range in the direction along the tilt decreases as the tilt increases. In the orthogonal direction, the range is invariant to the tilt, so the covered frequency region becomes elliptical rather than circular.

An example of frequency space coverage using a $NA = 0.4$ objective, tilt $\sim 39°$ and an incident beam angle tilt of $80°$ with respect to the normal to the object plane is shown in Figure 15.16 superimposed on the frequency space intensity plot for an object with 180 nm lines with the same structure as shown in Figure 15.13a. In this case, the second pair of off-axis exposures with the object tilted extends the frequency space coverage out to ~ 1.87. An image restoration procedure is required to adjust the measured spatial frequencies from the laboratory frame to the image frame for full image reconstruction. There is a small gap of $\sim 3\%$ of

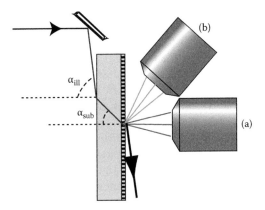

Figure 15.15 Optical arrangement using (a) off-axis illumination and (b) off-axis illumination and tilted object to enhance the frequency space information.

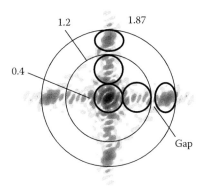

Figure 15.16 Frequency space coverage with a tilted objective.

frequency space between the inner circles and outer ellipses. So, the optimum *NA* to cover the frequency space with two offset images and 39° tilt would be ~0.415, but it is worthwhile investigating other combinations of *NA* and tilting angles for optimum results.

Tilting decreases the field of view from the perspective of geometric optics because only a small portion of the mask is in focus. However, this is a consequence of the optical system and is addressed by the spatial frequency correction as discussed in the following.

15.6.4 Impact of Conical Diffraction on IIM

Our intuition for grating diffraction is built from illumination in a plane that contains the grating wave vector and the surface normal. For example, for normal incidence illumination, the diffraction orders are observed in this plane to the sides of the transmitted (reflected) beam in accordance with the grating equation $\left(\text{e.g.,}\ k_{x,j} = (2\pi/d)j;\ k_y = 0;\ \text{and}\ k_{z,j} = \sqrt{k_0^2 - k_{x,j}^2}\right)$. For a surface normal plane, the variations in $k_{z,j}$ do not affect the image and the diffraction order spots are separated by $L\sin(2\pi j\lambda/d)$. Here, d is the period of the grating, j is the diffraction order, and L is the distance from the object to the image plane. For off-axis illumination in the k_y direction, the diffraction spots are tilted away from the normal direction $\left(k_{x,j} = (2\pi/d)j;\ k_y = k_{y0};\ \text{and}\ k_{z,j} = \sqrt{k_0^2 - k_{x,j}^2 - k_{y0}^2}\right)$ and again the spot positions are not affected by $k_{z,j}$, and the determination of $k_{x,j}$ and k_{y0} from the spot positions is straightforward. However, if the observation plane is tilted away from the object normal into the x, z plane, the spot position becomes dependent on $k_{z,j}$. This is known as conical diffraction since the multiple diffraction spots describe a conic section in the x', z' plane. Since the tilt couples the propagation vector in the z-direction into the observed position in the pupil plane, analysis of this case with a tilted object plane (in the x–z plane) can be extended to the general conical diffraction case.

Using Equation 15.17, we can reset the distorted frequencies:

$$f_x = (f_{xobs} - f_{ref})\sqrt{1 - f_{tilt}^2} + \sqrt{1 - f_y^2 - (f_{xobs} - f_{ref})^2}\, f_{tilt} + f_{ill} \qquad (15.18)$$

where

f_x is the frequency in the x-direction
f_{xobs} is the observed frequency
f_y is the frequency in y-direction
$f_{ill} = \sin(\alpha_{ill})$ is the angle of illumination
$f_{tilt} = \sin(\theta_{tilt})$ is the angle of the plane tilt

The reference beam is adjusted to a known selected frequency f_{sel} of a reference object, and other frequencies should be recalculated from Equation 15.18. Here, α_{ref} can be obtained from Equation 15.18 for the case when $f_{xobs} = f_x = f_{sel}$

$$f_{ref} = \sin\left[\theta_{tilt} - \sin^{-1}(f_{sel} - f_{ill})\right] + f_{sel} \tag{15.19}$$

The need for this frequency correction is a direct consequence of the non-paraxial effects that map the observed spatial frequencies of the diffracted fields from the image spatial frequencies in a nonlinear fashion. Correcting this distortion also provides restoration of the field of view.

This is illustrated in Figure 15.17, which shows the x-offset high-frequency partial images of two adjacent test structures (the one on the left is the object with 180 nm line width, while the one on the right is the 170 nm line width test structure; overall each structure is about 20 half pitch wide (3.6 μm) and the separation is 12 μm). Note that these are the high spatial frequencies in the x-direction, similar to the high y-frequencies of Figure 15.14b but in the orthogonal direction. The full image of this Manhattan structure is available only on combining all of the sub-images. The optical system was adjusted so that for the experimental image the smaller (right-hand) structure was approximately in focus while the larger (left-hand) structure was behind the focal plane due to the tilt and was substantially blurred. The image (Figure 15.17a) was restored using Equation 15.15 and shown in Figure 15.17b, which is in good agreement with a corresponding model (Figure 15.17c and d). The dotted lines and arrows show the significant shifting of the positions of the intense features accompanying the

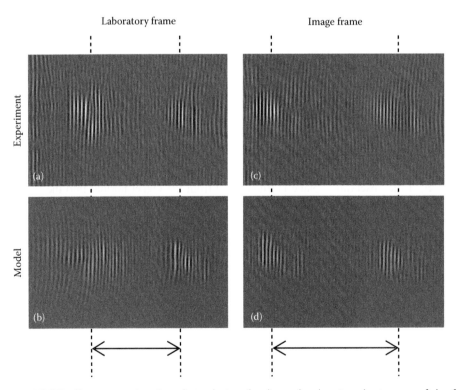

Figure 15.17 Experiment (a, c) and simulation (b, d) results showing the impact of the frequency restoration on the high-frequency partial image. The dotted lines are guides for the eye, showing the significant shift of the out-of-focus object (left) in the laboratory frame. (Reprinted from Kuznetsova, Y. et al., *Opt. Exp.*, 15, 6651, 2007. With permission.)

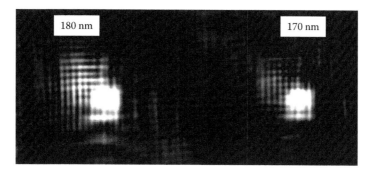

Figure 15.18 Reconstructed image: 180 and 170 nm structures. (Reprinted from Kuznetsova, Y. et al., *Opt. Exp.*, 15, 6651, 2007. With permission.)

transformation from the laboratory frame to the image frame and restoring the overall field of view. This correction requires very precise measurement of all of the relevant angles.

The final reconstructed images are shown in Figure 15.18. As expected, the reconstruction procedure restores both images and clearly improves the field of view that was limited by the tilt of the object plane relative to the focal plane. There are also extra features due to the noise in the system, which is magnified during the restoration procedure. This method requires very precise knowledge of the object tilt and incident illumination offset in order to obtain high-quality, extended-field images.

15.6.5 Structured Illumination for Backward Compatibility with Existing Microscopes

This implementation of IIM required an interferometer around the objective lens and access to the back pupil plane (Figure 15.12a). While this is straightforward in an optics laboratory, it is difficult to retrofit to an existing microscope. A structured illumination approach, in which the object is illuminated by two coherent beams, moves the interferometer to the front of the sample (Neumann et al. 2008a). Another method to obtain the same result is to use a zero-order beam reinjected before the objective using a beamsplitter or other optical element (grating). Effectively, this creates an intermediate spatial frequency, which is reset in the signal processing. An additional advantage of this intermediate frequency technique is that it reduces the pixel count demands on the camera.

15.6.6 Half Solid Immersion IIM

The next contribution to resolution extension was obtained by evanescent illumination from a high-index substrate (total internal reflection [TIR] illumination). This concept has been used extensively in fluorescence measurements to limit the excitation volume of the evanescent field of the exciting light, which is much less than the depth of field of traditional microscopy, providing ~10× enhancement of the axial resolution (Axelrod 2001). A lateral resolution enhancement for fluorescence microscopy was achieved with standing evanescent waves (Cragg and So 2000).

The interferometric reintroduction of a zero-order beam allows capture of the information corresponding to the propagating diffracted waves scattered by the object from the evanescent wave illumination that extends above the substrate surface. Therefore, instead of extreme off-axis illumination in air (Figure 15.15), illumination propagating beyond the TIR angle in the transparent object substrate in the same setup is used (Figure 15.19) (Neumann et al. 2008b). The illumination angle from the substrate α_{ill} is now larger than that for illumination through the back surface of the substrate where the illumination angle is restricted to $\sin^{-1}(1/n_{sub})$

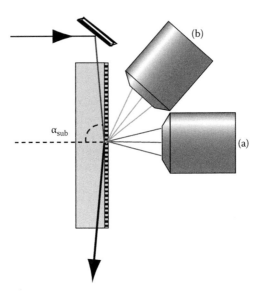

Figure 15.19   Illumination through substrate to enhance spatial frequency coverage.

(Figure 15.15). The evanescent wave associated with the TIR extends beyond the substrate into the sample region. This evanescent wave is scattered by the fine (sub-wavelength) sample structure into propagating waves that provide information on the details of the object at spatial frequencies up to $n_{sub} + NA$ without optical axis tilt and up to $n_{sub} + 1$ with a tilted optical axis.

In this case, the resolution limit depends on the refractive index of the substrate. If a conventional glass substrate with $n = 1.5$ and an $NA = 0.4$ objective are used, the maximum value of the effective NA is $NA_{eff} = 1.9$, which allows resolution of 166 nm half-pitch grating structures with $\lambda = 633$ nm. NA_{eff} is extended to 2.5 for the geometry with a tilted axis, and a corresponding minimum half-pitch resolution is 126 nm.

The frequency space Fourier intensity transform of a Manhattan (x–y geometry) test pattern scaled to different dimensions for a linewidth of 180 nm is shown in Figure 15.20a and for a linewidth of 150 nm is shown in Figure 15.20b. The second pair of circles (Figure 15.20a) is available using evanescent wave illumination extending the frequency space coverage to a radius of $n_{sub}f_{ill} + NA \sim 1.87$ (with $\sin^{-1}f_{ill} = 76°$) without tilt of the microscope optical axis.

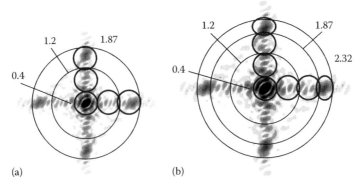

Figure 15.20   Frequency space visualization of IIM: (a) frequency space coverage for the structure with CD = 180 nm, which is resolved for the normal exit configuration of Figure 15.19, (b) frequency space coverage for the structure with CD = 150 nm, which requires the optical axis tilted configuration of Figure 15.19.

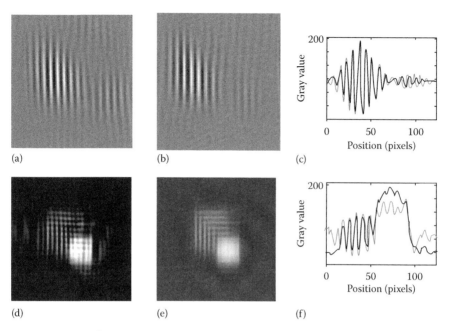

Figure 15.21   IIM of a 150 nm structure using evanescent illumination and a tilted optical system. (a) High-frequency image obtained by evanescent wave illumination of tilted object plane, (b) high-frequency image simulation, (c) comparison of high-frequency structures crosscut with the model, (d) reconstructed image, (e) reconstructed image simulation, and (f) comparison of Manhattan structures crosscut with the model. (Reprinted from Neumann, A. et al., *Opt. Exp.*, 16, 20477, 2008a. With permission.)

The third pair of off-axis sub-images in Figure 15.20b corresponds to the tilted optical axis and extends the frequency space coverage out to $NA_{eff} = n_{sub}f_{ill} + \sin(\theta_{tilt} + \sin^{-1} NA) \sim 2.32$, with illumination angle (in the glass), α_{ill}, of 76° and a tilt angle θ_{tilt} of 35°.

In order to decrease the influence of the camera pixel discretization on this high-frequency image, the reference beam can be adjusted to provide lower intermediate frequencies on the imaging camera, which are then reconstructed computationally. The result of the restored high-frequency image is shown in Figure 15.21a along with the corresponding model, Figure 15.21b, and the crosscut comparison in Figure 15.21c. The experimental reconstructed image from seven sub-images (Figure 15.20b) is demonstrated in Figure 15.21d, the corresponding model in Figure 15.21e, and the crosscuts in Figure 15.21f. The high-frequency image is in very good agreement with the model, but the overall quality of the image (even in the model) is not as well defined as for the 180 nm image.

Scaling of the frequency space coverage to get an equivalent image resolution requires both increasing the high frequency coverage along the principal axes of the Manhattan pattern, which we have accomplished with the off-axis illumination and tilting of the objective and additional coverage away from the principal axes, which we have yet to add with either additional sub-images at different azimuthal angles or a larger *NA* objective. For the smaller pattern, the frequency content spreads away from the major axes and less of the important frequency information is captured in the present configuration. Additionally, Gibb's effects resulting from the sharp cutoff in frequency response in a region with strong spectral content, and the required precision in setting and measuring the tilt and illumination angles make it more difficult to obtain high-quality, extended-field images as the frequency coverage is increased. The noise of the system causes problems for combination of the image from seven (or more) sub-images.

Extension to higher-index materials is straightforward and will provide extensive resolution enhancement. For example, GaP has an index of 3.3 at 633 nm (Nelson and Turner 1968) and for $NA = 0.4$, the spatial bandwidth corresponds to $NA_{eff} \sim 3.7$ without tilt and ~4.3 with tilt. The present results correspond to "half immersion" where the illumination is through the substrate exciting an associated evanescent wave, but the collected scattered light is propagating in air. Full solid immersion leads to further possibilities of resolution extension.

15.6.7 Full Solid Immersion

An important aspect of IIM is that the frequency space is parsed into small, more manageable pieces so that it is not necessary to provide the full NA in a single image. Existing immersion microscopy lens are limited to $NA \sim 1.4$, not only by the available indices of refraction but also by the difficulty of making a single diffraction limited objective at higher NA with acceptable aberrations. With IIM, we can in principle reach an NA of 3.3 at 633 nm. Because each sub-image covers only a portion of frequency space, the demands on the optical system are more easily met.

The full immersion extension is conceptually straightforward, which will further increase the NA_{eff} to $2n$ (= 6.6 for GaP) and the corresponding half-pitch resolution to 48 nm ($\lambda/13$). We envision using a simple set of prisms or gratings to extract and conventional, air-based lenses to capture the information. As is always the case, there is a trade-off between the number of sub-images and the physical NA of the objective lens. This is the same spatial frequency space that is accessed by the solid immersion techniques discussed earlier. Each technique has comparative advantages that need to be further explored.

Using smaller NA objectives in the setup presumes increasing the number of the sub-images to achieve a given resolution, which also leads to added noise and additional experimental complexity. In the case of substrates with higher refractive indexes, it is better to use a higher NA lens to reduce the number of sub-images. In order to cover $NA_{eff} = 6.6$ ($2n$ for GaP), we would need five sub-images in one direction for $NA = 0.9$ or three for $NA = 1.4$. It is possible to increase NA up to 1.4, using a conventional immersion objective lens.

The resolvable dimensions for typical laser source wavelengths are shown in Table 15.1. The columns labeled by n_{max} reflect the largest index transparent material of which we are aware at each wavelength. Interestingly, the usual improvement in resolution with wavelength is modified by the available high-index materials, which have limited transparency regions. Additional materials may further expand the available resolution. At a 193 nm wavelength, the resolution approaches typical SEM resolutions without requiring vacuum and indeed being fully compatible with water immersion. These resolutions are well beyond the current established perceptions of microscopic capabilities and suggest that advances in optical microscopy can yet yield improvements that will have important impacts across a broad swath of science and technology.

The possible increase of NA_{eff} using only low-numerical aperture objective lenses and retaining the depth of field, field of view, and working-distance advantages of the low-numerical

TABLE 15.1

Resolution Limits (Grating Half Pitch) for Various Optical Configurations at Commonly Available Laser Wavelengths (All Dimensions in nm)

λ	IIM (Air; $\lambda/4$)	½-Immersion ($\lambda/2(n + 1)$; $n = 1.5$)	Full Immersion ($\lambda/4n$; $n = 1.5$)	½-Immersion (n_{max})	Full Immersion (n_{max})
1064	266	213	177	116	74 ($n = 3.6$; Si)
633	158	127	106	74	48 ($n = 3.3$; GaP)
488	122	98	81	71	50 ($n = 2.45$; GaN)
193	48	39	32	34	27 ($n = 1.8$; PR)

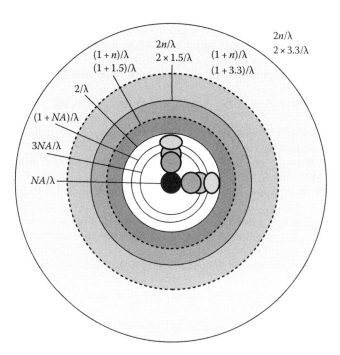

Figure 15.22 Available frequency space coverage for various optical systems.

aperture system is shown in Figure 15.22, drawn for a 0.4 NA system. The dark inner circle corresponds to the NA/λ frequency space coverage associated with coherent, normal-incidence illumination. The successively larger diameter concentric circles correspond to the frequency space coverage of IIM optical configurations: $3NA/\lambda$, which is the coverage for a single offset exposure (Figures 15.12 through 15.14) and $(1 + NA)/\lambda$, which is the coverage including multiple off-axis illumination angles (Figure 15.15), including tilt of the objective (Figures 15.16 through 15.18) that extends the frequency space coverage to $2/\lambda$, independent of the NA of the objective. The two dotted circles at $(1 + n)/\lambda$ show the coverage with half immersion for an index of 1.5 (glass) and 3.3 (GaP) at 633 nm (Figures 15.19 through 15.21) and finally the two solid circles show the full immersion frequency space coverage for the same two indices. As the frequency coverage is extended, the use of higher NA lenses will reduce the number of sub-images required for a more complete coverage of frequency space; the individual sub-image coverage is shown as the small circles for the non-immersion cases. Of course, the required coverage is dependent on the pattern, and there are some applications, for example in metrology for integrated circuits, where coverage of only a subset of the full frequency space is appropriate.

REFERENCES

Abbe, E. 1873. Beiträge zur theorie des mikroskops und der mikroskopischen wahrnehmung. *Arch. Mikrosk. Anat. Entwicklungsmech.* **9**, 413–468. [English translation in SPIE Milestone Series. MS178].

Alexandrov, S. A., T. R. Hillman, T. Gutzler, and D. D. Sampson. 2006. Synthetic aperture Fourier holographic optical microscopy. *Phys. Rev. Lett.* **97**, 168102.

Alexandrov, S. A., T. R. Hillman, T. Gutzler, and D. D. Sampson. 2007. Digital Fourier holography enables wide-field, superresolved, microscopic characterization. *Opt. Photon. News* **18**, 29.

Alexandrov, S. A., T. R. Hillman, and D. D. Sampson. 2005. Spatially resolved Fourier holographic light scattering angular spectroscopy. *Opt. Lett.* **30**, 3305.

Alexandrov, S. A. and D. D. Sampson. 2008. Spatial information transmission beyond a systems diffraction limit using optical spectral encoding of the spatial frequency. *J. Opt. A: Pure Appl. Opt.* **10**, 025304.

Axelrod, D. 2001. Total internal reflection fluorescence microscopy in cell biology. *Traffic* **2**, 764–774.

Becherer, R. J. and G. B. Parrent. 1967. Nonlinearity in optical imaging system. *J. Opt. Soc. Am.* **57**, 1479.

Betzig, E., G. H. Patterson, S. Sougrat et al. 2006. Imaging intracellular fluorescent proteins at nanometer resolution. *Science* **313**, 1642.

Bewersdorf, J., R. Schmidt, and S. W. Hell. 2006. Comparison of I5M and 4Pi-microscopy, *J. Microsc.* **222**, 105.

Born, M. and E. M. Wolf. 1999. *Principles of Optics*. Cambridge, U.K.: Cambridge University Press.

Brueck, S. R. J. and X. Chen. 1999. Experimental comparison of off-axis illumination and imaging interferometric lithography—Approaching the resolution limits of optics. *J. Vac. Sci. Technol.* **B17**, 921–929.

Cai, W. and V. Shalaev. 2009. *Optical Metamaterials: Fundamentals and Applications*. New York: Springer-Verlag.

Campion, A. and P. Kambhampati. 1998. Surface-enhanced Raman scattering. *Chem. Soc. Rev.* **27**, 241–250.

Chen, X. and S. R. J. Brueck. 1998. Imaging interferometric lithography: A wavelength division multiplex approach to extending of optical lithography. *J. Vac. Sci. Technol.* **B16**, 3392–3397.

Chen, X. and S. R. J. Brueck. 1999. Imaging interferometric lithography—Approaching the resolution limits of optics. *Opt. Lett.* **24**, 124–126.

Coe, J. V., J. M. Heer, S. Teeters-Kennedy, H. Tian, and K. R. Rodriquez. 2008. Extraordinary transmission of metal films with arrays of subwavelength holes. *Ann. Rev. Phys. Chem.* **59**, 179–202.

Courjon, D. 2003. *Near-Field Microscopy and Near-Field Optics*. London, U.K.: Imperial College Press.

Cox, I. J. and J. R. Sheppard. 1986. Information capacity and resolution in an optical system. *J. Opt. Soc. Am.* **A3**, 1152–1158.

Cragg, G. E. and P. T. C. So. 2000. Lateral resolution enhancement with standing evanescent waves. *Opt. Lett.* **25**, 46–48.

Cubukcu, E., E. A. Kort, K. B. Crozier, and F. Capasso. 2006. Plasmonic laser antenna. *Appl. Phys. Lett.* **89**, 093120.

Davidson, M. 2010. *Photobleaching*. http://micro.magnet.fsu.edu/primer/java/fluorescence/photobleaching/(accessed 02/11).

Davis, B. J., D. L. Marks, T. S. Ralston, P. S. Carney, and S. A. Boppart. 2008. Interferometric synthetic aperture microscopy: Computed imaging for scanned coherent microscopy. *Sensors* **8**, 3903–3931.

Decker, A., Y. Pao, and P. Claspy. 1978. Electronic heterodyne recording and processing of optical holograms using phase modulated reference waves. *Appl. Opt.* **17**, 917–921.

Denk, W., J. H. Strickler, and W. W. Webb. 1990. Two-photon laser scanning fluorescence microscopy. *Science* **248**, 73.

Donnert, G., J. Keller, R. Medda et al. 2006. Macromolecular-scale resolution in biological fluorescence microscopy. *Proc. Natl. Acad. Sci.* U.S.A. **103**, 11440–11445.

Dubois, A., L. Vabre, and A. C. Boccara. 2001. Sinusoidally phase-modulated interference microscope for high-speed high-resolution topographic imagery. *Opt. Lett.* **26**, 1873–1875.

Ebbesen, T. W., H. J. Lezec, H. F. Ghaemi, T. Thio, and P. A. Wolff. 1998. Extraordinary optical transmission through sub-wavelength hole arrays. *Nature* **391**, 667–669.

Engheta, N. and R. W. Ziolkowski. 2006. *Metamaterials—Physics and Engineering Explorations*. New York: John Wiley & Sons.

Fong, Y. C., E. A. Armour, S. D. Hersee, and S. R. J. Brueck. 1994. A confocal photoluminescence study of MOCVD growth on patterned GaAs substrates. *J. Appl. Phys.* **75**, 3049–3055.

Fong, Y. C. and S. R. J. Brueck. 1992. Confocal photoluminescence: A direct measurement of semiconductor carrier transport properties. *Appl. Phys. Lett.* **61**, 1332–1334.

Françon, M. 1952. Amélioration de la reśolution d'optique. *Nuovo Cimento*. Suppl. **9**, 283–290.

Goodman, J. W. 1985. *Statistical Optics*. New York: John Wiley & Sons.

Goodman, J. W. 1998. *Introduction to Fourier Optics*, 2nd edn. New York: John Wiley & Sons.

Göppert-Mayer, M. 2009 (reprinted) Elementary processes with two quantum transitions, *Annalen der Physik* **18**, 466.

Granero, L., V. Micó, Z. Zalevsky, and J. García. 2009. Superresolution imaging method using phase-shifting digital lensless Fourier holography. *Opt. Express* **17**, 15008–15022.

Grilli, S., P. Ferraro, S. De Incola, A. Finizio, G. Pierattini, and R. Meucci. 2001. Whole optical wavefields reconstruction by digital holography. *Opt. Express* **9**, 294–302.

Gu, M. 2000. *Advanced Optical Imaging Theory*. New York: Springer-Verlag.

Guo, P. and A. J. Devaney. 2004. Digital microscopy using phase shifting digital holography with two reference waves. *Opt. Lett.* **29**, 857–859.

Gustafsson, M. G. L. 2000. Surpassing the lateral resolution limit by a factor of two using structured illumination microscopy. *J. Microsc.* **198**, 82–87.

Gustafsson, M. G. L. 2005. Nonlinear structured-illumination microscopy: Wide-field fluorescence imaging with theoretically unlimited resolution. *Proc Natl. Acad. Sci.* U.S.A. **102**, 13081–13086.

Gustafsson, M. G. L., D. A. Agart, and J. W. Sedat. 1999. I^{5}M: 3D widefield light microscopy with better than 100 nm axial resolution. *J. Microsc.* **195,** 10–16.

Haddad, W. S., D. Cullen, J. C. Solem et al. 1992. Fourier transform holographic microscope. *Appl. Opt.* **31**, 4973–4978.

Heintzmann, R. and G. Ficz. 2006. Breaking the resolution limit in light microscopy. *Brief. Funct. Genom. Proteom.* **5**, 289–301.

Hell, S. W. 2007. Far-field optical nanoscopy (review). *Science* **316**, 1153–1158.

Hell, S. W. 2009. Microscopy and its focal switch. *Nat. Methods* **6**, 24–32.

Hell, S. W. and E. H. K. Stelzer, 1992. Fundamental improvement of resolution with a 4Pi-confocal fluorescence microscope using two-photon excitation. *Opt. Commun.* **93**, 277.

Hell, S. W. and J. Wichmann. 1994. Breaking the diffraction resolution limit by stimulated emission: Stimulated-emission-depletion fluorescence microscopy. *Opt. Lett.* **19**, 7809.

Hess, S. T., T. P. K. Girirajan, and M. D. Mason. 2006. Ultra-high resolution imaging by fluorescence photoactivation localization microscopy. *Biophys. J.* **91**, 4258–4272.

Hillman, T. R., T. Gutzler, S. A. Alexandrov, and D. D. Sampson. 2009. High-resolution, wide-field object reconstruction with synthetic aperture Fourier holographic optical microscopy. *Opt. Express* **17**, 7873.

Hopkins, H. 1955. The frequency response of a defocused optical system. *Proc. R. Soc. Lond.* **A231**, 91.

Hopt, A. and E. Neher. 2001. Highly nonlinear photodamage in two-photon fluorescence microscopy. *Biophys. J.* **80**, 2029.

International Technology Roadmap for Semiconductors (ITRS) at www.public.itrs.net (accessed 11/2009).

Ippolito, S. B., A. K. Swan, B. B. Goldberg, and M. S. Ünlü. 2000. High resolution subsurface microscopy technique. In *Proceedings of LEOS, Annual Meeting*, Vol. 2, 13–16 November, pp. 430–431.

Kartashev, A. I. 1960. Optical system with enhanced resolving power. *Opt. Spectrosc.* **9**, 204–206.

Klein, V. and T. E. Furtak. 1986. *Optics*. New York: John Wiley & Sons.

Köklü, F. H., S. B. Ippolito, B. B. Goldberg, and M. S. Ünlü. 2009. Subsurface microscopy of integrated circuits with angular spectrum and polarization control. *Opt. Lett.* **34**, 1261–1263.

Kuznetsova, Y., A. Neumann, and S. R. J. Brueck. 2007. Imaging interferometric microscopy— Approaching the linear systems limits of optical resolution. *Opt. Express* **15**, 6651–6663.

Kuznetsova, Y., A. Neumann, and S. R. J. Brueck. 2008. Imaging interferometric microscopy. *J. Opt. Soc. Am.* **A25**, 811–822.

Lai, S., B. King, and M. A. Neifeld. 2000. Wave front reconstruction by means of phase-shifting digital in-lineholography. *Opt. Commun.* **173**, 155–160.

Lai, S. and M. A. Neifeld. 2000. Digital wavefront reconstruction and its application to image encryption. *Opt. Commun.* **178**, 283–289.

Le Clerc, F., M. Gross, and L. Collot. 2001. Synthetic aperture experiment in the visible with on-axis digital heterodyne holography. *Opt. Lett.* **26**, 1550–1552.

Lee, H., Z. Liu, Y. Xiong, C. Sun, and X. Zhang. 2007. Development of optical hyperlens for imaging below the diffraction limit. *Opt. Express* **15**, 15886.

Leith, E. N. 1990. Small-aperture, high-resolution, two-channel imaging system. *Opt. Lett.* **15**, 885–887.

Leith, E. N., D. Angell, and C.-P. Kuei. 1987. Superresolution by incoherent-to-coherent conversion. *J. Opt. Soc. Am.* **A4**, 1050–1054.

Liu, Z., H. Lee, Y. Xiong, C. Sun, and X. Zhang. 2007. Far-field optical hyperlens magnifying sub-diffraction-limited objects. *Science* **315**, 1686.

Lohmann, A. W. and D. P. Parish. 1964. Superresolution for nonbirefringent objects. *Appl. Opt.* **3**, 1037–1043.

Lukosz, W. 1967. Optical systems with resolving powers exceeding the classical limit. II. *J. Opt. Soc. Am.* **57**, 932–941.

Marcus, D., S. W. Hell. 2002. Focal spots of size $\lambda/23$ open up far-field florescence microscopy at 33 nm axial resolution. *Phys. Rev. Lett.* **88**, 16390.

Massig, J. H. 2002. Digital off-axis holography with a synthetic aperture. *Opt. Lett.* **27**, 2179–2181.

Micó, V., J. Garcia, and Z. Zalevsky. 2008a. Axial superresolution by synthetic aperture generation. *J. Opt. A: Pure Appl. Opt.* **10**, 125001.

Micó, V., Z. Zalevsky, C. Ferreira, and J. García. 2008b. Superresolution digital holographic microscopy for three dimensional samples. *Opt. Express* **16**, 19260–19270.

Micó, V., Z. Zalevsky, and J. Garcia. 2006a. Superresolution optical system by common-path interferometry. *Opt. Express* **14**, 5168–5177.

Micó, V., Z. Zalevsky, and J. García. 2007. Synthetic aperture microscopy using off-axis illumination and polarization coding. *Opt. Commun.* **276**, 209–217.

Micó, V., Z. Zalevsky, and J. García. 2008c. Common-path phase-shifting digital holographic microscopy: A way to quantitative imaging and superresolution. *Opt. Commun.* **281**, 4273–4281.

Micó, V., Z. Zalevsky, P. Garcia-Martinez, and J. Garcia. 2004. Single-step superresolution by interferometric imaging. *Opt. Express* **12**, 2589–2595.

Micó, V., Z. Zalevsky, P. Garcia-Martinez, and J. Garcia. 2006c. Superresolved imaging in digital holography by superposition of tilted wavefronts. *Appl. Opt.* **45**, 822–826.

Micó, V., Z. Zalevsky, P. García-Martínez, and J. García. 2006c. Synthetic aperture superresolution with multiple off-axis holograms. *J. Opt. Soc. Am.* **A23**, 3162–3170.

Moskovits, M. 1985. Surface-enhanced spectroscopy. *Rev. Mod. Phys.* **57**, 783.

Murphy, D. 2001. *Fundamentals of Light Microscopy and Digital Imaging.* New York: John Wiley & Sons.

Nagorni, M. and S. W. Hell. 1998. 4Pi-confocal microscopy provides three dimensional images of the microtubule network with 100- to 150-nm resolution. *J. Struct. Biol.* **123**, 236–247.

Nelson, D. F. and E. H. Turner. 1968. Electro-optic and piezoelectric coefficients and refractive index of gallium phosphide. *J. Appl. Phys.* **39**, 3337–3343.

Nesterov, A. V. and V. G. Niziev. 2005. Vector solution of diffraction task using the Hertz vector. *Phys. Rev.* **E71**, 046608.

Neumann, A., Y. Kuznetsova, and S. R. J. Brueck. 2008a. Optical resolution below $\lambda/4$ using synthetic aperture microscopy and evanescent-wave illumination. *Opt. Express* **16**, 20477–20485.

Neumann, A., Y. Kusnetsova, and S. R. J. Brueck. 2008b. Structured illumination for the extension of imaging interferometric lithography. *Opt. Express* **16**, 6785–6793.

Novotny, L. and S. J. Stranick. 2006. Near-field optical microscopy and spectroscopy with pointed probes. *Annu. Rev. Phys. Chem.* **57**, 303–331.

Pawley, J. B. 2006. *Handbook of Biological Confocal Microscopy*, 3rd edn. Berlin, Germany: Springer-Verlag.

Pendry, J. B. 2000. Negative refraction makes a perfect lens. *Phys. Rev. Lett.* **85**, 3966.

Pitter, M. C., C. W. See, and M. G. Somekh. 2004. Full-field heterodyne interference microscope with spatially incoherent illumination. *Opt. Lett.* **29**, 1200–1202.

Price, J., P. Bingham, and C. E. Thomas. 2007. Improving resolution in microscopic holography by computationally fusing multiple, obliquely-illuminated object waves in the Fourier domain. *Appl. Opt.* **46**, 827–833.

Rao, Z., L. Hesselink, and J. S. Harris. 2007. High-intensity bowtie-shaped nano-aperture vertical-cavity surface-emitting laser for near-field optics. *Opt. Lett.* **32**, 1995–1997.

Raub, A. K., A. Frauenglass, S. R. J. Brueck et al. 2004. Deep-UV immersion interferometric lithography. *Proc. SPIE* **5377**, 306–318.

Rayleigh, L. 1879. Investigations in optics, with special reference to the spectroscope. *Rayleigh Philos. Mag. Ser.* **8**(49), Part XXXI, 261–274.

Richards, B. and E. Wolf. 1959. Electromagnetic diffraction in optical systems. II. Structure of the image field in an aplanatic system. *Proc. R. Soc. Lond* **A253**, 358–379.

Rittweger, E., K. Y. Han, S. E. Irvine, C. Eggeling, and S. W. Hell. 2009. STED microscopy reveals crystal colour centres with nanometric resolution. *Nat. Photon.* **3**, 144–147.

Rudolph, W., P. Dorn, X. Liu, N. Vretenar, and R. Stock. 2003. Microscopy with femtosecond laser pulses: Applications in engineering, physics and biomedicine. *Appl. Surf. Sci.* **208**, 327–332.

Rust, M. J., M. Bates, and X. Zhuang. 2006. Sub-diffraction-limit imaging by stochastic optical reconstruction microscopy (STORM). *Nat. Methods* **3**, 793–796.

Schnars, U. 1994. Direct phase determination in hologram interferometry with use of digitally recorded holograms. *J. Opt. Soc. Am.* **A11**, 2011–2015.

Schnars, U. and W. P. O. Jüpter. 1994. Direct recording of holograms by a CCD target and numerical reconstruction. *Appl. Opt.* **33**, 179–181.

Schwarz, C. J., Y. Kuznetsova, and S. R. J. Brueck. 2003. Imaging interferometric microscopy. *Opt. Lett.* **28**, 1424–1426.

Shalaev, V. M. 2007. Optical negative-index metamaterials. *Nat. Photon.* **1**, 41.

Shemer, A., D. Mendlovic, Z. Zalevsky, J. Garcia, and P. García-Martínez. 1999. Superresolving optical system with time multiplexing and computer decoding. *Appl. Opt.* **38**, 7245–7251.

Sheppard, C. J. R. and Z. Hegedus. 1998. Resolution for off-axis illumination. *J. Opt. Soc. Am.* **A15**, 622–624.

Smith, S. W. 2007. *The Scientist and Engineer's Guide to Digital Signal Processing*, San Diego: California Technical Publishing, http://www.dspguide.com/ch11/4.htm (accessed 2/28/2011).

Smith, D. R., D. Schurig, M. Rosenbluth, S. Schultz, S. A. Ramakrishna, and J. B. Pendry. 2003. Limitations on subdiffraction imaging with a negative refractive index slab. *Appl. Phys. Lett.* **82**, 156.

Smolev, S., A. Biswas, A. Frauenglass, and S. R. J. Brueck. 2006. 244-nm imaging interferometric lithography test bed. *Proc. SPIE* **6154**, 61542K.

Smolyaninov, I., Y. J. Hung, and C. C. Davis. 2007. Magnifying superlens in the visible frequency range. *Science* **315**, 1699–1701.

Sparrow, C. M. 1916. On spectroscopic resolving power. *Astrophys. J.* **64**, 76–86.

Sun, P. C. and E. N. Leith. 1992. Superresolution by spatial–temporal encoding methods. *Appl. Opt.* **31**, 4857–4862.

Toraldo di Francia, G. 1955. Resolving power and information. *J. Opt. Soc. Am.* **45**, 497–501.

Toraldo di Francia, G. 1969. Degrees of freedom of an image. *J. Opt. Soc. Am.* **59**, 799–804.

Török, P. and F. J. Kau. 2007. *Optical Imaging and Microscopy*. Berlin, Germany: Springer-Verlag.

Tridhavee, T. M., B. Santhanam, and S. R. J. Brueck. 2005. Optimization and apodization of aerial images at high NA for imaging interferometric lithography. *J. Microlith. Microfab. Microsys.* **4**, 023009.

Tyler, S., D. L. Ralston, P. Marks, S. Carney, and S. A. Boppart. 2007. Interferometric synthetic aperture microscopy. *Nat. Phys.* **3**, 129.

Tyler, S., D. L. Ralston, P. Marks, S. Carney, and S. A. Boppar. 2008. Real-time interferometric synthetic aperture microscopy. *Opt. Express* **16**, 2555.

Vainrub, A., O. Pustovyy, and V. Vodyanoy. 2006. Resolution of 90 nm (λ/5) in an optical transmission microscope with an annular condenser. *Opt. Lett.* **31**, 2855–2857.

Vedantam, S., H. Lee, J. Tang, J. Conway, M. Staffaroni, and E. Yablonivitch. 2009. A plasmonic dimple lens for nanoscale focusing of light. *Nano Lett.* **9**, 3447–3452.

Voelz, D. G., K. A. Bush, and P. S. Idell. 1997. Illumination coherence effects in laser-speckle imaging: Modeling and experimental demonstration. *Appl. Opt.* **36**, 1781–1788.

Webb, R. H. 1996. Confocal optical microscopy. *Rep. Prog. Phys.* **59**, 427–471.

Williams, C. S. and O. A. Becklund. 1989. *Introduction to the Optical Transfer Function*. New York: John Wiley & Sons, p. 338.

Wilson, T. 1990. *Confocal Microscopy*. London, U.K.: Academic press.

Wu, Q., L. P. Ghislan, and V. B. Elings. 2000. Imaging with solid immersion lenses, spatial resolution, and applications. *Proc. IEEE* **88**, 1491–1498. www.cytoviva.com (accessed 11/1/2009).

Yamaguchi, I., J. I. Kato, S. Otha, and J. Mizuno. 2001. Image formation in phase-shifting digital holography and applications to microscopy. *Appl. Opt.* **40**, 6177–6186.

Yamaguchi, I. and T. Zhang. 1997. Phase-shifting digital holography. *Opt. Lett.* **22**, 1268–1270.

Zalevsky, Z. and D. Mendlovic. 2002. *Optical Super Resolution*. Berlin, Germany: Springer-Verlag.

Zalevsky, Z., D. Mendlovic, and A. W. Lohmann. 1999. Optical systems with improved resolving power. In *Progress in Optics*, Vol. 15, Ed. E. Wolf, Chap. 4. Amsterdam, the Netherlands: North Holland.

Zernike, F. 1942. Phase contrast, a new method for the microscopic observation of transparent objects. *Physica* **9**, 686–698.

Zhang, T. and I. Yamaguchi. 1998. Three-dimensional microscopy with phase-shifting digital holography. *Opt. Lett.* **23**, 1221–1223.

Zlotnik, A., Z. Zalevsky, and E. Marom. 2005. Superresolution with nonorthogonal polarization coding. *Appl. Opt.* **44**, 3705–3715.

Chapter 16

MICRO-OPTICAL TECHNIQUES

Kartik Srinivasan, Matthew T. Rakher, and Marcelo Davanço

CONTENTS

16.1 INTRODUCTION

This chapter focuses on optical measurements in which spatial resolution on the order of the wavelength of light provides meaningful, distinct information from that which can be obtained through bulk measurements, such as those described in Chapter 6. These techniques include photoluminescence (PL) spectroscopy, photoluminescence excitation (PLE) spectroscopy, electroluminescence (EL), and angle-resolved reflectivity measurements. There are several physical systems for which the ability to perform such measurements with wavelength-scale spatial resolution is needed. For example, it is essential when interrogating nanofabricated photonic devices (Figure 16.1a) such as microcavity lasers (McCall 1992), where the optical cavities are a few micrometers in each planar dimension and are fabricated in arrays with a device-to-device spacing of tens of micrometers. Here, the spatial resolution is needed to distinguish between cavities and between a cavity and unprocessed regions of the chip. A second example is an ensemble of solid-state emitters in or on a substrate (Figure 16.1b), which encompasses structures such as epitaxially grown self-assembled quantum dots (QDs) (Michler 2003) embedded in a semiconductor material, colloidal QDs in solution or deposited on a substrate (Murray 1993, Alivisatos 1996), fluorescent molecules in a host matrix (Moerner 1999), and impurity color centers in a crystal (Gruber 1997). These materials may exhibit a density gradient across the sample, so that spatially resolved measurements can provide an understanding of optical properties as a function of the number of excited emitters, ultimately reaching the single emitter limit in very dilute (≈ 1 emitter per μm^2) regions. Our discussion is restricted to techniques that achieve diffraction-limited spatial resolution through "conventional" methods of high numerical aperture, free space far-field optics. For the near-infrared wavelengths that are our interest, this produces a length scale on the order of a micrometer (hence the title "Micro-optical techniques"). This chapter does not cover some of the important developments in the quest for obtaining better spatial resolution, including near-field scanning optical microscopy (NSOM), the topic of Chapter 17, and so-called super-resolution techniques like stimulated emission depletion microscopy (Hell 2009) (Chapter 15). Such tools can provide a wealth of added information, such as spatial profiles of microcavity modes (Balistreri 1999), or the ability to distinguish between single fluorescent centers within a dense array (Betzig 1993). Nevertheless, as we shall see throughout this chapter,

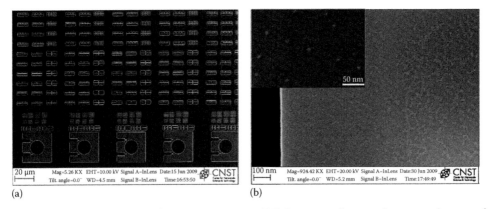

(a)   (b)

Figure 16.1 (a) Scanning electron microscope (SEM) image of a two-dimensional array of nanophotonic devices (waveguides, hole arrays, and microdisk cavities) to be interrogated through micro-PL measurements. The minimum required spatial resolution is that needed to go from device to device on the chip. (b) SEM images of PbS QDs spun on an Si substrate. With sufficient spatial resolution, micro-PL can be used to study QD ensembles of varying density. The inset shows the number of QDs present within an $\approx 260\,nm \times 170\,nm$ field.

wavelength-scale spatial resolution is in many cases preferred, since the improved resolution of a technique like NSOM comes at the cost of increased complexity and sacrifice in collection efficiency. Space constraints also prevent us from addressing promising recent developments in improving PL collection efficiencies, particularly for embedded media like quantum wells or QDs, through solid immersion lenses (Gerardot 2007) or external waveguide probes (Srinivasan 2007).

While this chapter has the general title of "Micro-optical techniques," its primary focus is micro-PL. As discussed in Chapter 6, PL can be loosely defined as a process in which a material absorbs light at one wavelength and emits light at another (usually redshifted) wavelength. For our purposes, micro-PL is a measurement in which the pump beam has been focused down to a micrometer-scale spot on the sample, and the emission from a portion of this region is collected and spectrally resolved. There are several reasons why we have chosen micro-PL as our representative micro-optical technique. The first is that it is one of the most widespread methods of device characterization, providing information about the electronic structure of the material and optical transitions between its states. Next, the experimental apparatus needed for conducting micro-PL measurements is sufficiently general so that PLE, EL, and angle-resolved reflectivity measurements can be incorporated into the same setup if relatively straightforward additions are made. The setup required for micro-Raman spectroscopy (Delhaye 1975), a complementary technique that can provide information about low-frequency transitions within the material (such as vibrational and rotational modes), is conceptually similar, though the specific pieces of equipment needed may differ. Finally, as we use micro-PL routinely within our own laboratory, we have the opportunity to share considerations we made when constructing our setup.

Our goal is to provide the reader with the essential information needed to construct and use a micro-PL setup. The content of this chapter lies somewhere between a set of instructions for a senior undergraduate laboratory and a formal review of research results, and is biased heavily toward the former. The reader should be aware that micro-PL and related optical techniques such as confocal fluorescence microscopy have been used within various research fields for many years, and providing an adequate history of the technique and the scientists who pioneered it would be a significant undertaking that is beyond the scope of what we hope to accomplish. References that may provide the reader with valuable insight include standard optics texts (Hecht 1998), books on optical spectroscopy (Demtroder 1998), Hobbs's book on practical construction of electro-optic systems (Hobbs 2000), and Novotny and Hecht's recent book on nano-optics (Novotny 2006). Several major optics vendors also include "principles of operation" notes in their catalogs, which can be invaluable when determining which optics to purchase. There are also several other review articles on micro-PL, a few of which we cite here (Kasai 1995, Gustafsson 1998, Moerner 2003).

The organization of this chapter is as follows. In Section 16.2, we present a detailed description of a typical micro-PL setup and the process by which it is designed. We have partitioned the apparatus into different subsystems, and for each subsystem, we discuss design choices and trade-offs involved. In Section 16.3, we present examples of data acquired from a micro-PL setup, describe what the measurements tell us about the system, and what further information could be obtained through additional measurements. In Section 16.4, we describe extensions to micro-PL setups that allow for new information to be unlocked. These include PLE spectroscopy, where the excitation wavelength is tuned to provide some insight into the electronic structure of the system; EL measurements, where the excitation channel is electrical rather than optical, and angle-resolved reflectivity, where the resonant optical response of a system can be probed. Finally, in Section 16.5, we include a couple of general lab procedures and rules of thumb that can help beginners when building a new micro-optical measurement setup.

16.2 ANATOMY OF A MICRO-PHOTOLUMINESCENCE SETUP

The core idea of a micro-PL setup is to focus an excitation source down to a wavelength-scale spot on a sample, where it is absorbed and generates emission (typically at a different wavelength) that is directed into a grating spectrometer, through which an emission spectrum can be produced. Spatial resolution is obtained through the limited extent of the pump and collection areas, and PL maps can be generated through translation of the sample and/or excitation beam. If the sample to be interrogated contains specific features to which the pump beam must be aligned, imaging using a white light source and an appropriate camera may be required. The remainder of this section is devoted to providing a detailed, practical description of how these functionalities can be achieved. As is the case whenever one builds optical setups, appropriate attention to safety must be paid, and the user should consult with his/her laser safety officer to ensure compliance with laser safety standards.

A schematic of a typical micro-PL apparatus is shown in Figure 16.2, where a single objective is used to focus the pump beam on the sample and to collimate emission from the sample. For convenience, we have partitioned this setup into several subsystems: (1) excitation source, (2) optics for focusing the excitation beam onto the sample (pumping), (3) optics for PL collection (collection), (4) detection, (5) imaging, and (6) sample environment/positioning. Obviously, this separation is not absolute, and various optical components may serve dual roles, for example, in both sample excitation and collection. Before we proceed to spend time outlining the components of each subsystem, it is worthwhile to first outline a few general considerations that must go into the design of such a measurement apparatus.

These general considerations largely involve some knowledge of the specific physical system(s) to be characterized. Knowledge of the wavelength-dependent absorption and emission cross sections determines the required properties of the excitation source and detector. The required information here can sometimes be quite approximate, and in some cases obtained from literature or from ensemble absorption and/or macro-PL measurements of the system. For example, in a

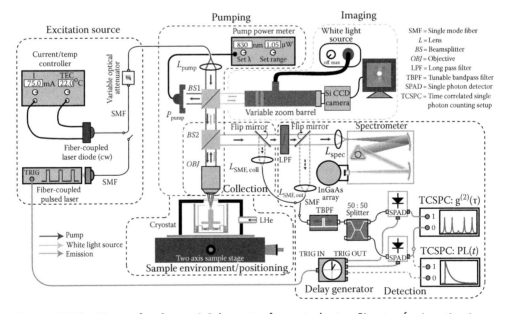

Figure 16.2 (See color insert.) Schematic of a typical micro-PL setup for investigating a sample in a cryogenic environment. SMF = single-mode fiber, *BS* = beam splitter, *L* = lens, *OBJ* = objective, LPF = long-pass filter, TBPF = tunable bandpass filter, SPAD = single-photon avalanche photodiode, TCSPC = time-correlated single-photon counting.

micro-PL measurement of semiconductor QDs, the band gap energy of the host semiconductor will determine the appropriate wavelength for photogeneration of carriers, while the likely emission band determines whether an Si-based or InGaAs-based detector should be used. Optics such as beam splitters, lenses, and objectives are often antireflection coated for use over specific wavelength ranges. If certain optics are to be used in multiple roles (e.g., pump beam focusing, emission collection, and white light imaging), chromatic aberration can be an important factor in deciding what type of optical element to choose. Depending on the characteristics of the material under investigation, control of the polarization of the excitation beam and collected signal can be important, and will have to be taken into account in the choice of optics. The environment in which the sample is to be characterized (e.g., ambient vs. low temperature) can impose a minimum separation between the sample surface and focusing/collection optics, setting limits on the excitation/collection spot and the fraction of emission that will be collected. It can also determine whether pumping and collection will be done on the same side of the sample, or whether backside excitation/collection can be employed. Regardless of the specific implementation, keeping the optical paths as short as possible is generally advisable. Finally, and perhaps most importantly, the user will impose constraints on the setup based upon what is required from the measurement, and the budget available. An experiment requiring limited spectral resolution and moderate collection efficiency from an ensemble of visible wavelength emitters will necessitate a setup that is an order of magnitude less complex (and expensive) than the one needed to produce spectrally resolved time-dependent PL dynamics from single near-infrared emitters. The objectives of the experiment thus set the priority level given to different parts of the setup.

One might envision planning a micro-PL setup by going through the following process:

(1) Determine the general architecture of the setup. What are the most critical aspects of the setup? Will pumping and collection be done from the same side, on opposite sides, or in a noncollinear fashion? How will imaging (if needed) be integrated? The goal of this step is to have a general layout in place for the excitation, collection, and imaging paths. This layout, along with the expected excitation and collection wavelengths, will go a long way toward determining what type of optics (aspheres, achromatic doublets, microscope objectives, etc.) to use.

(2) Based on the system to be investigated and the type of measurement needed, select an excitation source.

(3) Specify the minimum allowable working distance (separation between the sample and the optic's front surface) for the optic that will be focusing light onto the sample. Next, specify the required excitation spot size and determine the optics needed to achieve this.

(4) Specify the required collection angle and, if applicable, collection area. Determine if this is compatible with the minimum working distance allowed by the setup, and with the already specified pump focusing optics, if any optics will be shared. If necessary, consider how additional optics can be added to limit the collection area. If portions of the excitation and collection beam paths are to be shared, decide on what optics will be used to combine/separate them.

(5) Specify if the collected signal is to be spectrally resolved and, if so, the required spectral resolution. Decide on what type of detector is going to be used to measure the collected and spectrally resolved light, and any optics that might be needed to couple into it.

(6) Specify the requirements for sample imaging (resolution, field of view [FOV], and zoom range), and determine the appropriate optics and camera/video system to meet these needs and to integrate the white light beam path with the existing pump/collection optics.

(7) Decide whether translation of the sample or translation of the optics is the preferred approach for performing micro-PL measurements across the sample.

(8) Double-check steps 1–7 and iterate as needed.

As a concrete example, let us review the thought process used to construct our micro-PL setup (Figure 16.2) for studying near-infrared InAs QDs.

(1) Along with low-temperature capabilities, our specific experiments require a system in which additional probes (electrical, near-field optical) can be incorporated. These capabilities are not relevant to this chapter except that they influence the design of the system—in particular, incorporation of such probes is most easily accomplished through pumping and collection from the top surface of the sample. Furthermore, we need to image our samples with micrometer-scale spatial resolution, and this also has to be done from the top. Our priority is collection efficiency, followed by imaging capability and pump efficiency, though there is a minimum requirement for all three.

(2) We use a 830 nm laser diode for excitation, as it can efficiently excite the QD states through absorption in the GaAs and wetting layers that surround the QDs.

(3) The minimum working distance is set by the separation between our cryostat top window and the sample surface, and is ≈15 mm (relatively large due to the introduction of electrical/optical probes), precluding the use of almost any commercially available fluorescence microscope. We require a pump spot size of ≈2 μm, though the collection spot is more important than the pump spot in our measurements. Optics that can achieve this include large diameter aspheres and long working distance microscope objectives.

(4) We want to collect 1300 nm emission from an area of ≈2 μm^2, though there is some flexibility here depending on the exact characteristics of the sample to be studied. While these requirements are in principle compatible with both large diameter aspheres and long working distance microscope objectives, we choose the latter due to better chromatic aberration properties (also needed for the white light imaging). Cube beam splitters or dichroic mirrors will be used to combine/separate the pump and collection beam paths.

(5) The primary detector will be a grating spectrometer equipped with a cooled InGaAs array. Long pass edge filters will remove the residual pump signal, and the collimated collected signal will need to be focused into the spectrometer. For time-resolved measurements, we will use InGaAs photodiodes and single-photon counters with single-mode fiber inputs, requiring a convenient method to switch between detectors. Furthermore, the photodiodes and single-photon counters will require separate optics to focus the collected signal into the single-mode fiber.

(6) Sample imaging with approximately 1 μm resolution is needed since we will be aligning the pump beam to microfabricated devices. Adjustable zoom is needed in situations where near-field probing is combined with micro-PL. These requirements are compatible with an infinity-corrected long working distance microscope objective and zoom barrel system.

(7) Though small adjustments of the pump beam position on the sample provide fine translation capabilities, sample translation is preferred since it requires little realignment of optics, and high-resolution stepping is not required in our measurements.

With the above in mind, we now delve into each of the micro-PL subsystems in detail.

16.2.1 Excitation Source

16.2.1.1 Continuous Wave Sources

We begin by considering continuous wave (cw) excitation sources for use in steady-state PL spectroscopy. The most obvious requirement is that the source be of the appropriate wavelength. For example, consider common III–V semiconductor materials like GaAs and InP, which often serve as a host for light-emitting structures like quantum wells and QDs (Chapter 2), and have their room temperature band gaps at approximately 870 and 930 nm, respectively. By photoexciting GaAs- or InP-based light-emitting structures at photon energies above the band gap, significant absorption of optical energy and production of carriers within the GaAs/InP layers can be achieved. Upon relaxation (e.g., due to phonons), these carriers can then fill the appropriate states of the quantum wells/dots, eventually resulting in recombination and emission. Spectrally resolved PL measurements, such as those discussed in Section 16.4, will not only display the emission due to the quantum wells/dots but can also show emission from recombination at the GaAs band edge and at defect centers in the material.

An inexpensive and compact excitation source for micro-PL measurements on these III–V systems is a semiconductor diode laser (Coldren 1995) emitting at 780 nm or 830 nm. Such lasers, typically edge-emitting Fabry–Perot or distributed feedback structures, can be driven by a current source and stabilized by a temperature controller that regulates the current to a thermoelectric cooler that is often integrated with the laser diode. Threshold currents are typically in the 100 mA range, with output powers of tens or even hundreds of milliwatts available. If wavelengths other than 780 nm or 830 nm are required, semiconductor laser diodes based on the technologically mature GaAs, GaP, and InGaAsP systems provide coverage over the ≈630–1600 nm range, while III-nitride semiconductors can cover green and blue wavelengths. Other choices for visible wavelength excitation include HeNe, Ar ion, and frequency-doubled Nd:YAG lasers, as discussed in Chapter 6. The choice of laser is dictated by issues including power, spectral linewidth, form factor, beam shape, and cost.

A typical edge-emitting diode laser supports emission from multiple cavity modes, and thus does not provide spectrally narrow excitation. In many instances, this is of no concern, and indeed, even an incoherent light-emitting diode can be used as an excitation source if enough output power is available. When spectrally narrow emission is needed, distributed feedback or fiber Bragg grating stabilized lasers, in which a grating is used to select a specific mode of the device, are an option. Diffraction gratings can also be incorporated into external cavity designs, such as the Littrow and Littman–Metcalf geometries (Paschotta 2008). Here, the laser diode has a high reflectivity (HR) coating on one facet and is antireflection-coated on the other facet. The cavity is formed between the HR facet and the grating (Littrow configuration) or between the HR facet and a grating that is followed by an external mirror (Littman–Metcalf). By rotating either the grating (Littrow) or mirror (Littman–Metcalf), the laser wavelength is tuned over the diode's gain bandwidth, and the external cavity geometry typically provides a narrower linewidth (on the order of 1 MHz) than a standard laser diode. As a result, such external cavity diode lasers (ECDLs) can be used in PLE experiments, as discussed in Section 16.3, although the available tuning range (usually a couple of tens of nanometers for wavelengths less than 1 μm and as much as 100 nm for wavelengths near 1.55 μm) may be too narrow, requiring more widely tunable solid-state lasers.

One of the most commonly used tunable solid-state lasers is the Ti:sapphire laser (Paschotta 2008), an optically pumped (in the green) laser that typically provides wavelength coverage between ≈700 nm and 900 nm, though broader wavelength coverage is possible depending on the power requirement and optics used. The laser can also be used as a pump source in different wavelength conversion units, such as an optical parametric oscillator (used to generate longer wavelengths) or a frequency doubler (used to generate shorter wavelengths) (Chapter 7). These features, along with output power levels that can significantly exceed those of tunable diode lasers, make cw Ti:sapphire lasers a versatile source for micro-PL and micro-PLE experiments. Disadvantages include high cost (typically several times more expensive than an ECDL), large size (along with the pump laser can occupy a significant portion of an optical table), and support equipment (often requiring water cooling and significant electrical power consumption). Finally, we note that a host of other cw tunable laser sources have been used in PL experiments, including tunable dye lasers (wavelength coverage in the 550–780 nm range) and color center lasers (various emission bands in the 800 nm to 3 μm region). The development of bright and compact supercontinuum sources (Chapter 7) has made them a potential (quasi-cw) PL spectroscopy source, in which different bandpass filters can be used to select for specific excitation wavelengths.

16.2.1.2 Pulsed Sources

The need for pulsed excitation can arise in a few different situations. While investigation of time-dependent phenomena is the most common (Chapter 12), other instances include steady-state spectroscopy of structures in which thermal dissipation is a problem (e.g., nanofabricated membrane structures, such as the photonic crystal microcavity lasers to be discussed in Section 16.3.1) and material systems in which photobleaching, photodarkening, or blinking occur (Moerner 2003). In such situations, the parameters of the pulsed excitation may be chosen to limit heating or improve signal-to-noise/background levels. Pulsed excitation is also used in lock-in measurements, which can improve the signal-to-noise ratio (SNR) by shifting the detection frequency away from DC. On the other hand, steady-state spectroscopy of thermally stable and photostable structures using pulsed excitation (without lock-in detection) causes a decrease in the number of collected photons, by a fraction approximately equal to the duty cycle (contingent upon the emission lifetime).

The time-dependent measurements considered in this chapter are the PL decay and second-order correlation function, which are commonly performed measurements for single solid-state emitters (ultrafast pump–probe spectroscopy and related techniques are the focus of other chapters within the book in Part III). For these measurements, the key characteristics of the laser source are its wavelength, repetition rate, pulse width, and energy. In general, a PL lifetime measurement is not a single-shot experiment, but rather the average over many experiments in which the sample is excited by a pulse and its PL intensity is detected over a range of times that are delayed with respect to the pulse (Chapter 12). The period (inverse of the repetition rate) must therefore be sufficiently long so that all of the important dynamics occur in the interval between pulses. The pulse width must be narrow enough to occupy a small fraction of the PL decay. The pulse energy is set according to the physics of the system under investigation (e.g., to generate a specific average number of electron–hole pairs per pulse in a III–V QD).

For convenience, we consider two types of pulsed sources, those which are created by modulating an existing cw source and those based on mode-locking. One straightforward and direct way to modulate a cw source is through the energy source that pumps the laser. For example, a semiconductor laser diode can be modulated by switching its current between above threshold and below threshold, creating an "on" and an "off" state, with the pulse width given by the length of time over which the current is kept above threshold (this is called gain switching). Such current modulation of semiconductor lasers is complicated by carrier-dependent gain dynamics (Coldren 1995, Yariv 2007), influencing the achievable pulse shapes and widths and ultimately necessitating the use of external electro-optic modulators in certain applications. For situations in which gain-switched diode laser pulses are acceptable, they can be generated through the current driver for the laser, which will often have a radio frequency (RF) input that can be driven by a general-purpose waveform generator. In this scenario, typical maximum modulation speeds are on the order of 1 MHz, limited by the current source electronics. For generation of pulse widths on the order of 100 ps at repetition rates of tens of MHz, dedicated drivers and laser diodes are commercially available. These systems produce moderate output powers (a few milliwatts of average power), and are a compact and significantly less expensive option than most of the mode-locked lasers to be discussed later. Furthermore, the repetition frequency is often controllable and can be set to a fraction of its maximum value. This can be important when using the source across systems in which carrier dynamics exhibit very different timescales.

Another option is to use an external modulator, which is a gate that blocks the output laser for a desired period of time and at a desired rate. An inexpensive external modulation method is found in a mechanical chopper, which is often used with a lock-in amplifier or as a simple way to reduce heating. The chopper is essentially a very stable rotating fan, with the pulse repetition rate and width being determined by its speed and blade width (typical rates are in the range of

1 Hz to a few kilohertz; tens of kilohertz are possible in some units). To achieve faster repetition rates and narrower pulse widths, acousto-optic modulators (AOMs) or electro-optic modulators (EOMs) can be used (Yariv 2007), providing modulation bandwidths on the order of 100 MHz (AOM) or 10 GHz (EOM). At the fastest modulation speeds, the primary cost involved in an EOM is not in the modulator itself, but rather in the modulator driver, which must produce high-speed electrical pulses. Other than cost, one of the principal challenges in using EOMs for lifetime measurements is in the achievable on/off ratio (the extinction ratio), and its stability over time. Achieving the highest levels of extinction and stability requires precise control of the polarization state of light going into the modulator. Control of the environment (temperature) and feedback on a DC bias applied to the modulator may also be necessary.

Mode-locked lasers constitute a second class of pulsed laser sources for time-dependent PL measurements, with the Ti:sapphire laser being one of the most common choices (Chapter 7). As discussed earlier, cw Ti:sapphire lasers offer the combination of high output powers and broad wavelength tuning ranges. Generation of short pulses is done through mode locking, where the relative phases of the laser cavity's multiple modes are fixed (through introduction of an intra-cavity saturable absorber or through intracavity loss/phase modulation), resulting in a pulse train (Diels 2006, Chapter 7). Mode-locked Ti:sapphire lasers routinely produce sub-100 fs pulse widths at a rate of 80 MHz, which is set by the round-trip time of light in the cavity. Importantly, this repetition rate sets the timescale over which dynamics can be measured—to use an 80 MHz Ti:sapphire laser to study a system with a PL lifetime >10 ns, some form of pulse selection (a "pulse picker") must be used for downsampling. Finally, other tunable sources previously described, such as dye, color center, and rare-earth-doped glass lasers, can also be mode-locked. Rare-earth-doped optical fibers, of crucial importance to optical communications, have also been mode-locked to produce near-single-cycle optical pulses (4.3 fs in Krauss 2009), and their small form factors and relative stability make them attractive alternatives in situations where the accessible wavelength range and power levels are appropriate.

16.2.2 Optics for Focusing the Excitation Beam onto the Sample

This section is devoted to describing how the excitation beam, whose extent is usually a few millimeters, is focused down to a micrometer-scale spot on the sample surface. We discuss shaping to produce a clean Gaussian beam, and focusing of the beam to a desired pump spot size.

16.2.2.1 Beam Shaping

Although certain situations might call for the use of an elliptical or doughnut-shaped beam, in most situations, the desired input into a micro-PL setup is a circular Gaussian beam, which will eventually be collimated to fill the clear aperture of the focusing lens (discussed in more detail in Section 16.2.2.2). The amount of shaping needed depends on the laser source. Lasers based on Fabry–Perot cavities may naturally produce a Gaussian-shaped beam, while edge-emitting semiconductor lasers can produce an elliptical, diverging beam. In other circumstances, the laser output may have a waist significantly larger or smaller than desired. Since semiconductor diode lasers are such an inexpensive and compact option for micro-PL measurements, we will discuss how to circularize their output.

When circularizing an elliptical beam, the first step is to determine its beam shape through beam profiling (Section 16.5.2), or from information provided by the laser diode manufacturer. Two common methods to shape the beam employ either cylindrical lenses or anamorphic prism pairs. A cylindrical lens affects light along only one axis, so that two orthogonal cylindrical lenses, chosen with a focal length ratio equal to the ratio of the beam divergence angles along the axes of the elliptical beam, will produce a collimated circular beam.

An anamorphic prism pair also magnifies a beam along only one axis. The prism pair is placed in the beam path after the laser output has been collimated. The angles of the two prisms relative to the beam propagation direction determine the level of magnification, with ratios of 2–6 being common. An adjustable prism pair allows for circularization of beams of differing eccentricities, while a fixed prism pair can be housed in a compact mount that is usually significantly smaller than the footprint required for a pair of cylindrical lenses, although it should be noted that the input and output beams from the prism pair are not collinear.

Light from the excitation source might also require spatial filtering, in which the goal is to remove unwanted spatial fluctuations. One method of spatial filtering is through a pinhole aperture mounted on a multi-axis translation stage. The basic idea is to focus the laser beam into a pinhole, which is small enough to block high spatial frequency noise components (whose radial distance from the propagation axis is proportional to the spatial frequency), but large enough to transmit the majority of the laser beam power propagating in the spatial mode of interest; a pinhole diameter 1.5 times the expected Gaussian beam diameter is a common choice. The cleaned-up beam can then be re-collimated and introduced into the rest of the micro-PL setup.

An alternate method for spatial filtering is to couple the laser into a single-mode optical fiber. The fiber acts as a modal filter, allowing only its fundamental mode to propagate a significant distance, and the output beam launched from a single-mode fiber closely approximates a Gaussian beam. One note of warning when fiber coupling the laser output is that back reflections from the fiber can be a destabilizing influence on the laser, and are particularly problematic in some cases. A common method to reduce back reflections is to use a fiber that has been terminated in an angled physical contact (APC) connector, where the fiber end face has been polished at an 8° angle. The highest levels of protection from back reflected light are usually accomplished by a Faraday isolator, a nonreciprocal optical element that can produce isolation levels routinely in excess of 30 dB. Many lasers offer an isolator output as an option.

A potential disadvantage of the single-mode fiber approach is that even with relatively good alignment, the coupling efficiency might only be 50%, and depending on the setup stability, it may need to be periodically adjusted. An advantage of the method is that these small misalignments only affect the efficiency with which the output Gaussian beam is launched, and not the spatial profile of the beam. In our laboratory, we generally use fiber-coupled semiconductor diode lasers as PL pump sources. Along with the spatial filtering aspects described earlier, coupling light into a single-mode fiber allows the laser to be separated by a significant distance from the rest of the setup, as light propagates through single-mode fibers with very low loss (<1 dB/km), though if polarization preservation is important, special considerations must be made (e.g., use of polarization maintaining fiber). This is of practical convenience when building the setup, and also makes it convenient to switch between sources by simply disconnecting one fiber and connecting a second (followed by an adjustment of the collimation optics if the excitation wavelength is significantly different). In addition, many vendors offer fiber-pigtailed semiconductor lasers, where the laser-to-fiber coupling has been completed and the system has been packaged so that the alignment remains fixed in place. These sources offer power levels at the milliwatt to tens of milliwatts level, though higher power lasers are available, for example, at 980 nm (the pump band for erbium-doped fiber amplifiers).

16.2.2.2 Beam Focusing

This step involves taking the Gaussian beam produced by the excitation source, and collimating it to an appropriate diameter so that the subsequent focusing optic will produce the desired spot size on the sample surface. Figure 16.3a schematically depicts the block of the overall micro-PL setup (Figure 16.2) devoted to these tasks, while Figure 16.3b through e is schematics of individual beam-shaping operations that may be employed.

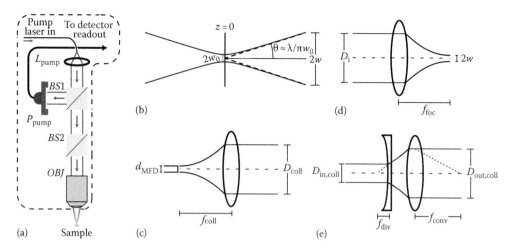

Figure 16.3 (a) Pumping section of a micro-PL setup, taken from Figure 16.2. (b) Gaussian beam schematic, showing the beam waist and divergence angle. (c) Optics for pump collimation from a single-mode fiber, (d) focusing, and (e) beam expansion. In (e), $D_{out,coll}/D_{in,coll} \approx f_{conv}/f_{div}$. The lens shown is meant to be generic and could be an achromatic doublet, aspherical lens, or microscope objective. $BS1$ and $BS2$ are beam splitters used to combine/separate the collection and white light imaging beam paths (not shown).

There are several options for the beam-focusing optic, including microscope objectives, aspheres, and doublets. Singlet lenses are also a possibility, though spherical and chromatic aberration errors usually suggest that one of the aforementioned optics is a better choice. Otherwise, the decision between doublet, asphere, and microscope objective often comes down to the specifics of the experimental layout. If the focusing optic is to be used exclusively with the excitation source, a simple doublet or asphere can work quite well. If the optic is to be used for both excitation and collection, the difference in wavelengths can lead to chromatic aberrations, so that achromatic doublets, achromatic microscope objectives, or low dispersion aspheres are better options. An achromatic doublet consists of a pair of lenses, made of different materials, so that two different wavelengths can be brought into focus in the same plane, providing a level of chromatic correction not available in most single-element optics. If the optic is to be used for white light imaging along with excitation and collection, an apochromatic microscope objective (element OBJ in Figure 16.3) is a versatile option that provides significant levels of spherical and chromatic aberration correction. The price of such optics essentially monotonically increases with the complexity and aberration-correcting ability—apochromatic objectives are multielement, multi-material systems that can significantly outperform achromatic doublets and aspheres but whose cost can exceed them by an order of magnitude.

Along with spot size and chromatic considerations, other factors in choosing an optic include its working distance, depth of focus or confocal parameter (the distance along the propagation axis over which the beam area changes by a factor of 2), and clear aperture (which determines how large an input beam can be accommodated). The working distance is important if the sample to be interrogated must be cooled in a cryostat, or simultaneously electrically or optically contacted with external probes, as in both cases there is a minimum separation that can be achieved between the optic and the sample. In contrast, fluorescence experiments on molecules spun onto a glass slide often place the optic within 1 mm of the sample. As working distance generally decreases with shorter focal lengths and higher resolving powers, there is usually a trade-off to be made between working distance and smallest possible spot size.

As the beam produced by many laser sources can be reasonably approximated as Gaussian, formulas from Gaussian beam optics (Siegman 1986, Yariv 2007) can be used to estimate parameters such as the focused pump spot size. Let us recall some of the important points of Gaussian beam optics, referring to Figure 16.3b. The electric field transverse to the propagation direction ($E(r)$) has the form

$$E(r) = E_0 e^{-r^2/w^2(z)}, \tag{16.1}$$

where
E_0 is the initial value of the field
r is the distance from the center of the beam
$w(z)$ is the radial distance from beam center (at propagation distance z) for which $E = E_0/e$

The beam attains a minimum radius w_0, called the beam waist, at a specific location in the propagation direction ($z = 0$ here), and expands quadratically as a function of z as given by the formula

$$w(z) = w_0 \sqrt{1 + \left(\frac{z}{z_R}\right)^2}. \tag{16.2}$$

The parameter z_R is called the Rayleigh range and is given by $z_R = \pi w_0^2/\lambda$, so that $w = \sqrt{2} w_0$ at $z = \pm z_R$, with $2z_R$ being called the confocal parameter or the depth of focus. For $|z| \gg z_R$, the beam essentially diffracts linearly, at an angle $\theta = \lambda/\pi w_0$ with respect to the propagation axis.

Let us next consider how to collimate a Gaussian beam, taking as an example the beam exiting a single-mode optical fiber (Figure 16.3c). We assume a Gaussian beam with a diameter $2w_0 = d_{MFD}$, where d_{MFD} is the fiber's mode field diameter.* Assuming we place the optic a focal length (f_{coll}) away from the fiber output, and $f_{coll} \gg z_R$, we determine the collimated beam diameter D_{coll} by equating the previously defined diffraction angle θ with $D_{coll}/2f_{coll}$, the expected diffraction angle due to geometric optics in the small angle limit (and equal to the optic's numerical aperture, as we shall discuss later). Doing so yields:

$$D_{coll} = \left(\frac{4\lambda}{\pi}\right)\left(\frac{f_{coll}}{d_{MFD}}\right). \tag{16.3}$$

As a concrete example, single-mode optical fiber designed for 780 nm laser light has $d_{MFD} = 5.6\,\mu m$, so that an $f_{coll} = 15\,mm$ asphere would produce $D_{coll} = 2.7\,mm$.

Now that the beam is collimated, it can be focused down by an optic with focal length f_{foc}, as schematically depicted in Figure 16.3d. The formula to use is the same as the one given earlier (Equation 16.3), but we now calculate the output beam waist given an input collimated beam, that is,

$$2w = \left(\frac{4\lambda}{\pi}\right)\left(\frac{f_{foc}}{D_{coll}}\right). \tag{16.4}$$

---

* The mode field diameter describes the extent of the beam propagating through the optical fiber and, rather than the physical fiber core diameter, is the appropriate parameter to use in calculations involving single-mode fibers. Note that the mode field diameter varies as a function of wavelength.

Combining these equations, we relate the focused spot size to the initial beam waist as

$$2w = \left(\frac{f_{\text{foc}}}{f_{\text{col}}}\right) d_{\text{MFD}}. \tag{16.5}$$

Thus, the minimum achievable spot size is given by the product of the ratio of the focal lengths of the two lenses and the mode field diameter of the initial beam. In our example given earlier, to achieve a spot size of $2w = 1\mu m$, we need $f_{\text{col}}/f_{\text{foc}} \approx 1/5$. In practice, the ability to achieve this (or even smaller spot sizes) can be practically limited by typical focal lengths for the optics, and by the constraint of the minimum working distance needed for the focusing optic. In addition, the collimated beam diameter produced by the first lens must fit within the clear aperture of the focusing lens. Continuing with our example, one might use an $f_{\text{coll}} \approx 25\,mm$ lens to produce an $\approx 4.5\,mm$ collimated beam, which is then focused down by an $f_{\text{foc}} \approx 4.5\,mm$ lens with a clear aperture of $5\,mm$ and working distance of $2.9\,mm$ to a spot diameter of $\approx 1\,\mu m$. If a larger working distance is needed, one option is a large diameter asphere, which can have a much longer focal length and working distance ($>10\,mm$), but also requires a large input beam. A second option is to use a long working distance objective, which combines sub-micrometer resolution capabilities with a working distance of about $20\,mm$ and requires an input beam diameter of about $8\,mm$.

Equation 16.4 tells us that the achievable spot size is inversely proportional to the input collimated beam diameter. While we have assumed that this input beam has been generated by a single optic, it might not be practically possible to use a single element to generate a sufficient diameter to take advantage of the full power of the focusing optic. To do so, one can use multiple lenses to expand the input beam and more completely fill the focusing optic. Such an expander (Figure 16.3e) consists of a pair of lenses, with the laser beam first expanded by a diverging lens (negative focal length) and then collimated by a converging lens (positive focal length). The diverging lens should be chosen to have a clear aperture sufficient to handle the incoming beam (which we have assumed is collimated), and the absolute value of the ratio of the converging lens focal length to the diverging lens focal length should equal the desired beam expansion factor. It is sometimes convenient to adjust the beam expansion ratio. Many optics vendors offer variable beam expanders, where additional lenses have been incorporated in the tube so that the effective focal length of the diverging or converging section of the expander can be varied through path length adjustment, allowing for differing levels of beam expansion.

As a final comment, it is usually important to know the pump power at the sample surface, so that, along with the pump spot area, the delivered pump intensity can be obtained. This intensity is important when comparing a PL spectrum to those produced by another setup. In Figure 16.3a, a detector (P_{pump}) is placed at an open port of beam splitter $BS1$, and by comparing the value measured here with a measurement taken at the sample location, we determine the multiplicative factor to apply to additional pump power measurements (in which the detector remains next to $BS1$). It is important to periodically recheck this factor, and to ensure that the chosen detector is compatible with the pump size at the point at which it is measured.

16.2.3 Optics for Photoluminescence Collection

In this section, we consider the optics needed to collect emission from the sample, schematically depicted in Figure 16.4a. As discussed previously, selection of the optic is strongly influenced by how many roles it has to play in excitation, collection, and white light imaging. That being said, there are still some general comments that can be made. The first is

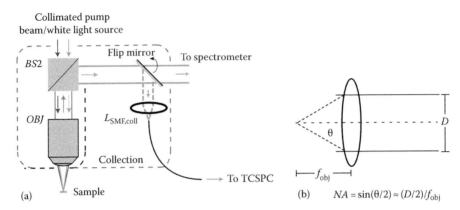

Figure 16.4 (a) Collection block of a micro-PL setup, redisplayed from Figure 16.2. (b) Numerical aperture of a converging lens.

that the numerical aperture (*NA*) of the optic, which is a measure of its acceptance angle (Figure 16.4b), plays a dominant role along with the working distance, which accounts for space constraints in the setup. A high *NA* optic collects light from a significant range of angles and is therefore preferable from a light-gathering perspective, but is often accompanied by a limited working distance. For example, a microscope objective with *NA* = 0.6 (collection angle ≈60°) has a working distance less than 1 mm, while *NA* = 0.4 (collection angle ≈40°) has a working distance of ≈3–4 mm. Aspheres can produce somewhat larger working distances (≈1.8 mm for *NA* = 0.68), and large-diameter (≈25 mm) aspheres that have recently been offered by many companies can combine *NA* ≈ 0.5 with a working distance of around 10 mm.

Long working distance, apochromatic microscope objectives (element *OBJ* in Figure 16.4a) combine the high *NA* and working distance of large-diameter aspheres with chromatic aberration correction. Typical specifications are *NA* = 0.42 for a working distance of 20 mm and *NA* = 0.55 for a working distance of 13 mm. Furthermore, unlike traditional microscope objectives, most long working distance apochromats are infinity corrected, meaning that they transform incoming light rays from the focus into a set of parallel rays that can later be focused to an image plane by a separate tube lens. The importance here is that the region between the objective and tube lens (the tube lens is a zoom barrel in Figure 16.2) is one in which additional optics can be inserted without modification to the objective's working distance. For example, one might consider placing a rotatable polarizer in this section, if polarized emission measurements are needed.

At first glance, it would appear that the sample area over which emission is collected is equal to the pumped area, given by the focused excitation spot size discussed in Section 16.2.2, and indeed this is often the case. There are important exceptions, however. Carrier diffusion in semiconductors can generate an effective pumped area that is larger than the excitation spot. Early experiments in single QD spectroscopy relied on fabrication methods to physically reduce the sample size from which emission could be collected, through etching of sub-micrometer width mesas or the use of sub-micrometer apertures in a deposited metal film.

The area over which emission is collected can also be limited through optics that aperture the collimated signal out of the microscope objective, thereby improving the spatial resolution. A standard way to do this in confocal microscopy is to focus the beam through a pinhole before going into the detector (Chapter 15). A second option, which we employ, is to focus the light into a single-mode optical fiber (element $L_{\mathrm{SMF,coll}}$ in Figure 16.4a), which serves the same function. We calculate the ratio of the spot diameter collected by the fiber (D_{fiber}) to the spot diameter collected by the objective (D_{tot}) as follows. D_{tot} is approximately given by geometric optics as $D_{\text{tot}} = 2NA_{\text{obj}}f_{\text{obj}}$, where NA_{obj} and f_{obj} are the numerical aperture and focal length of the initial collection optic, respectively. Equation 16.4 then gives us the expression for D_{fiber}, so that in total, we have

$$\left(\frac{D_{\text{fiber}}}{D_{\text{tot}}}\right) \approx \frac{(4\lambda/\pi)(f_{\text{fiber}}/d_{\text{MFD}})}{2NA_{\text{obj}}f_{\text{obj}}}, \tag{16.6}$$

where f_{fiber} is the focal length of the lens used to collect light into the single-mode fiber. This ratio is also equal to $(2w_{\text{fiber}}/2w_{\text{pump}})$, the fraction of the pump spot diameter from which light is being collected. Using Equation 16.4 to rewrite the pump spot diameter $2w_{\text{pump}}$ in terms of f_{obj} and D_{i} (the input excitation beam diameter), we have

$$2w_{\text{fiber}} \approx 2w_{\text{pump}}\left(\frac{D_{\text{fiber}}}{D_{\text{tot}}}\right) \approx \frac{(4\lambda/\pi)^2(f_{\text{fiber}}/d_{\text{MFD}})}{2NA_{\text{obj}}D_{\text{i}}}. \tag{16.7}$$

As a concrete example, our experiments involve collection at $\lambda = 1.3\,\mu\text{m}$ into a single-mode fiber with $d_{\text{MFD}} = 7\,\mu\text{m}$, using a long working distance apochromatic objective for which $NA_{\text{obj}} = 0.4$ and $f_{\text{obj}} = 10\,\text{mm}$. If our excitation source collimation optics produce $D_{\text{i}} = 7\,\text{mm}$, we should expect $2w_{\text{pump}} \approx 2.4\,\mu\text{m}$. Let us now assume we want $2w_{\text{fiber}}$ to be reduced with respect to $2w_{\text{pump}}$, bearing in mind the limit (Δr) to which one can expect to resolve point emitters, as determined by diffraction, and given by Novotny (2006) (Chapter 15):

$$\Delta r \approx 0.61\frac{\lambda}{NA} \tag{16.8}$$

For our choice of objective and wavelength, one might try to achieve $2w_{\text{fiber}} = 1.6\,\mu\text{m}$. Equation 16.7 tells us that a $f_{\text{fiber}} = 23\,\text{mm}$ coupling lens is an appropriate choice. While this reduction in collection area relative to pump area can come at the cost of collected photons for certain materials, for isolated single solid-state emitters like self-assembled QDs, the emission should be emanating from much smaller regions, so that the improved spatial resolution has come with no sacrifice in terms of collection efficiency.

Finally, we briefly consider options for combining the pump and collection paths in situations where the experiment calls for it. For example, in Figures 16.3a and 16.4a, elements *BS*1 and *BS*2 depict cube beam splitters used for this purpose. A nonpolarizing cube beam splitter can provide a 50:50 coupling ratio, with a clear aperture that is usually more than 75% of the cube's edge length. Potential drawbacks are that the splitting ratio varies somewhat with wavelength (a pellicle beam splitter provides better chromatic performance), and the specified wavelength range may not cover both the pump and collection beams. In such cases, it is usually advantageous to choose a beam splitter optimized for the collection wavelength, provided that sufficient pump power is available, and the transmission at the pump wavelength is measured so that the pump power delivered to the sample surface is known. Regardless, one disadvantage of the cube beam splitter is in losing half the collected signal. To overcome this, a dichroic beam splitter is a good option. A long wave pass (short wave pass) dichroic transmits the longer (shorter) wavelength and reflects the shorter (longer) wavelength. Since the dichroic reflectivity ($\approx 99\%$) typically exceeds its transmission ($\approx 90\%$), when collection efficiency is at a premium, it is advantageous to make it the reflected signal and use a short wave pass dichroic for redshifted emission. On the other hand, collecting in transmission through a long wave pass dichroic can offer a broader bandwidth. Drawbacks in comparison to a cube beam splitter are that it is not as broadband, and its properties are polarization dependent.

16.2.4 Emission Detection

There are a significant number of options for detecting the PL signal, and our treatment here is just a small sampling of what can be done. We have divided this section into three pieces, the first detailing options for spectrally resolving the PL signal, the second describing the detectors that can be used to detect this spectrally resolved emission to construct a PL spectrum, and the third describing time-dependent measurements that might be done.

16.2.4.1 Spectrally Resolving the Collected Emission

The amount of spectral resolution needed obviously depends on the system under investigation and the specific measurement in question. In some cases, it is adequate to simply remove any residual pump beam, and detect emission over all other wavelengths (within the detector bandwidth). This can be done with a long wave pass interference filter, or if the pump beam is sufficiently spectrally narrow, with a pump rejection filter. If the emission wavelength of interest is well known, a bandpass filter can be used to provide better isolation between the collected signal of interest and background emission (depending on the isolation levels offered, the bandpass filter may be used in combination with an edge-pass filter). A tunable bandpass filter can be used to produce an emission spectrum if the signal past the filter is monitored as a function of filter position. A simple method for creating a tunable bandpass filter is to angle-tune a fixed bandpass filter, which usually is specified to operate at normal incidence. By adjusting the filter angle with respect to the incident beam, the center of the bandpass is shifted to shorter wavelengths. The exact design of the interference filter (e.g., number of cavities within the dielectric stack) will affect the steepness of the bandpass, which in turn will determine the available spectral resolution. A high-resolution filter can be produced by a Fabry–Perot cavity (Siegman 1986), which is an optical resonator consisting of a pair of high reflectivity mirrors. These devices are specified in terms of the free spectral range (FSR), the wavelength separation between modes of the cavity (which show up as transmission peaks in the Fabry–Perot spectrum), and the finesse, the ratio of the free spectral range to the bandwidth of the modes. A scanning Fabry–Perot cavity allows for spectrally resolved measurements because the separation between mirrors can be varied through application of a voltage to a piezoelectric transducer on which one of the mirrors is mounted, thus changing the spectral position of the modes. For spectrally resolved measurements, the criterion needed to use a scanning Fabry–Perot cavity is that the emission window must be narrower than the FSR, and the spectral resolution will be given by the bandwidth of the modes (determined by the reflectivity of the Fabry–Perot mirrors). Scanning Fabry–Perot cavities typically provide much better spectral resolution than either a tunable bandpass interference filter or a grating spectrometer, with picometer level resolution possible over spectrally narrow windows (e.g., FSR = 1 nm and finesse = 1000).

Diffraction gratings are another tool used to provide spectral resolution, as they take an input collimated beam and disperse it, causing a spatial separation of different wavelength components, which then allows for wavelength-resolved detection. A commonly used instrument featuring a diffraction grating is a Czerny–Turner monochromator (Hobbs 2000), shown in Figure 16.5. The Czerny–Turner starts with an entrance slit (S_{ent}), which is placed at the focus of a back collimating mirror (M_{coll}). Light collimated from this mirror is diffracted by the grating (G), and the now spatially separated wavelength components are then refocused

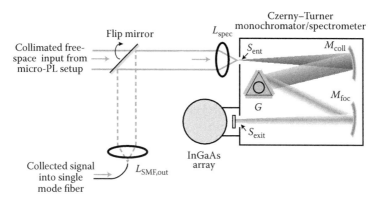

Figure 16.5 Diagram of a Czerny–Turner monochromator, equipped with an InGaAs array detector to produce a grating spectrometer, and with two input paths (collimated free-space beam and single-mode fiber). Wavelength tuning is achieved through rotation of the grating G, with the grating turret holding three different gratings to allow for broad wavelength coverage by switching the gratings.

by a second mirror (M_{foc}) onto an exit slit (S_{exit}), which spatially filters out everything but a small wavelength band, which is detected. Monochromators typically use reflective elements rather than refractive ones, since they tend to be less dispersive. A scanning monochromator produces wavelength-resolved data by recording a detected signal, rotating the grating to vary the wavelength band passing through the exit slit, and repeating; commercially available optical spectrum analyzers common in the optical telecommunications industry are typically scanning monochromators.

The monochromator resolution is set by its linear dispersion and the width of the entrance and exit slits. Dispersion describes how well the monochromator spatially separates two different input wavelengths. It is usually specified in units of nanometer/millimeter and is a function of the monochromator geometry and grating. For example, a 500 mm Czerny–Turner monochromator with a diffraction grating having a groove density of 1000 mm^{-1} may have a dispersion of $\approx$1.5 nm/mm. If incoming light has been well focused into the monochromator (input coupling is described in more detail later), so that the entrance slit width is equal to or smaller than the exit slit width and so that the beam completely fills the grating after being collimated by M_{coll}, the spectral resolution is approximately given by the product of the dispersion and the exit slit width. For example, for an exit slit width of 15 μm, the resolution is (1.5 nm/mm) × (0.015 mm) = 0.0225 nm. In this situation, the entrance slit does not affect the resolution, and the primary reason to reduce its size is to limit the amount of stray light that enters the spectrometer. In practice, it may not be possible to focus the incoming light as tightly as needed. In such situations, the entrance slit plays a role in determining the achievable resolution, as its size determines how well the grating is filled; if underfilled, spectral resolution may be lost. In general, a good procedure to follow is to measure the resolution by recording the spectrum from a narrow, monochromatic source. For example, a laser linewidth of 10 MHz ($\approx$5.6 × 10^{-5} nm at 1300 nm), achievable in an external cavity diode laser, will be below most monochromator resolution capabilities, so that the measured spectrum will essentially be an instrument response function.

A grating spectrometer is an instrument closely related to a monochromator. While a monochromator uses an exit slit and a single channel detector to detect one wavelength at a time, a spectrometer uses a multichannel detector to acquire several wavelengths simultaneously, with each channel of the detector commonly called a pixel (since charge-coupled device (CCD) detectors are often used). The geometry of the Czerny–Turner can remain the same, with the exit slit opened all the way (or at least, to the lateral extent of the multichannel detector). At a fixed grating position, the wavelength span that can be measured is equal to the product of the monochromator's dispersion, the pixel width, and the number of pixels. A 1024 element array with $25\,\mu m$ pixel width will have a span of $(1.5\,nm/mm) \times (0.025\,mm)(1024) = 38.4\,nm$ if attached to the hypothetical spectrometer described earlier, and the best possible spectral resolution, approximately given by the product of the pixel width and the monochromator's dispersion, is $\approx(1.5\,nm/mm) \times (0.025\,mm) \approx 0.0375\,nm$. As was the case with the single channel detector, measurement of the instrument response through a narrow linewidth source is a preferred way to determine the actual resolution of the system.

We next describe the entrance optics needed to take the collected emission (which we assume has been collimated) and feed it into the monochromator. Ideally, the entrance optics will ensure that the grating is perfectly filled (illuminated). If underfilled, loss of spectral resolution may result, while overfilling reduces the detected signal and can lead to unwanted scatter within the system. When ample signal is available, it is common to let the entrance slit S_{ent} do all of the work, in providing an aperture through which a diffracted beam is sent onto M_{coll} and through the rest of the system. As discussed earlier, the entrance slit width will then influence the achievable spectral resolution, and light will be lost as a result of the slit's aperturing effect. In this mode, there is a trade-off between spectral resolution and throughput, and it is therefore not necessarily suited for high-resolution measurements of small optical signals (e.g., single QD spectroscopy). A focusing optic (L_{spec} in Figure 16.5) can be used to circumvent this trade-off.

Monochromators are typically characterized by an f-number, which is a measure of the collection angle of an optic (the higher the f-number, the lower the collection angle). To completely fill the grating, the ratio of the input optic's focal length to its input beam diameter should match this f-number. To be more concrete, let us consider two examples. The first assumes that a long working distance apochromatic objective has been used to produce a collimated collection beam with an 8 mm diameter, which is to be resolved in a f-6.5 monochromator. Matching the spectrometer f-number requires a 52 mm focal length optic. If a shorter focal length optic is used, the collimated beam should still be focused at the entrance slit, but will expand more quickly than is optimal, and will overfill the first mirror (M_{coll}), resulting in a net loss of detected signal. If a longer focal length optic is used, the beam will underfill mirror M_{coll}, and the optimal spectral resolution will not be achieved. As a second example, we consider light that has been collected into a single-mode optical fiber. We expect to first collimate the light coming out of the fiber ($L_{SMF,out}$ in Figure 16.5) before focusing it into the entrance slit. The monochromator f-number of 6.5 corresponds to $NA \approx 0.077$, while the NA of the fiber is typically ≈ 0.14, so as an initial guess, we expect to require a focal length ratio of about 1.8:1 (the NA ratio) for the two lenses. A more precise calculation is as follows: an $f_{SMF,out} = 11\,mm$ optic will produce a 2.6 mm collimated beam if placed at the output of a $7\,\mu m$ mode field diameter fiber operating at $1.3\,\mu m$, so that an $f_{spec} = 17\,mm$ optic is needed to focus the collimated beam into the monochromator. This focal length ratio ($17/11 \approx 1.55:1$) varies from the initial guess because we have used the more precise Gaussian beam optics equation (Equation 16.3) to calculate the collimated beam diameter, rather than the simple ray optics.

16.2.4.2 Detectors for Measuring an Emission Spectrum

Factors to consider in choosing a detector include detection wavelength, sensitivity, single channel versus multichannel capability, and bandwidth. In the visible and short-infrared wavelength region (350–1000 nm), silicon-based detectors are dominant, while in the near-infrared (e.g., 900–1700 nm), InGaAs and Ge detectors are commonly used. Single channel detectors can include p-i-n photodiodes, avalanche photodiodes, photomultiplier tubes (PMTs), and single photon detectors. For photodiodes (often packaged with a transimpedance amplifier to produce an overall photoreceiver unit), the sensitivity will be limited by noise that is often specified as a noise-equivalent power (NEP), which includes contributions from noise sources other than photon shot noise, and is typically written in units of W/Hz$^{1/2}$, so that detection of an incident power equal to the NEP with a 1 s integration time will yield SNR = 1. The minimum detectable signal is obtained for low bandwidths, limiting the ability to achieve ultrasensitive detection with fast time resolution. Higher sensitivity can often be achieved in cooled detectors, where the cooling source is usually a thermoelectric Peltier element or a liquid nitrogen dewar. Such devices also tend to have much lower bandwidths than their uncooled counterparts. Detector noise usually scales with area, so it can be worthwhile to look for units that have as small an active area as possible without sacrificing the detected signal. For example, if the exit slit width is going to be primarily kept at around 15 μm (to achieve good spectral resolution), a small area detector (e.g., diameter ≈ 50 μm) with an appropriate focusing lens will ensure complete detection of the wavelength-resolved signal—there is no benefit to having a larger detector area. Lock-in detection is a technique that can often be used to improve the SNR in single channel detection. By using an optical chopper, for example, to modulate the detected signal, one can shift the detection away from DC to higher frequencies, away from many common noise sources.

Si CCDs and InGaAs photodiode arrays are common multichannel detectors for spectroscopy, and as is the case for single channel units, cooling through a thermoelectric element or liquid nitrogen source is often used to reduce noise. A multichannel detector can provide significant benefit in comparison to a single channel detector in at least two ways. The most obvious is the multiplexed detection, that is, the ability to simultaneously acquire information from multiple wavelength bins. In comparison to the step-and-acquire procedure when scanning the monochromator with a single channel detector, this multiplexed detection provides a time speed-up that is proportional to the array size along the wavelength-dispersed direction, which can often be in the range of 256–1024 elements. Considering that a 10 s integration may be needed to provided an adequate SNR level, the difference between producing one spectrum every 10 s and one every 3 h is considerable. Next, although multichannel detectors have a new noise source (readout noise) in comparison to their single channel counterparts, in some situations, the SNR is not limited by readout noise, but is dark count noise limited. One way to reduce dark count noise is to keep the detector size as small as possible without limiting the number of collected photons. The typical pixel size in a CCD is on the order of 25 μm × 25 μm, which is smaller than what is usually available in a single channel detector, but still large enough to capture the majority of the emission in the corresponding wavelength band, provided that the monochromator and its input optics are properly chosen. Furthermore, two-dimensional arrays such as Si CCDs usually have binning functions, so that multiple pixels in the vertical or horizontal directions can be grouped together. This can then allow the user to select a detector size appropriate to what is produced at the monochromator output.

The SNR available to a multichannel detector is limited by three primary noise sources: (1) photon shot noise, (2) dark count noise, and (3) readout noise. These values can typically be found within the manufacturer's specification sheet. If we call the detected photon flux (in units of counts/s) P_{det}, the dark count rate (also in units of counts/sec) D, and the readout noise R_{rms} (usually provided as a root mean square value in units of counts), the SNR is given by

$$\text{SNR} = \frac{P_{det}t_{int}}{\sqrt{P_{det}t_{int} + Dt_{int} + R_{rms}^2}}, \qquad (16.9)$$

where t_{int} is the integration time. This formula, which implicitly assumes that we have included the detector response (quantum efficiency and gain) into P_{det} and D, shows how a measurement can be noise-limited in different regimes. For a short enough integration time, we expect readout noise to be the dominant contribution, while at a longer integration time, some combination of dark count noise or photon shot noise will dominate. Clearly, the preference is to be photon shot noise limited, where SNR $\approx \sqrt{P_{det}t_{int}}$, and continues to improve with either increasing P_{det} or t_{int}. While shot noise-limited detection is often achievable at visible wavelengths with Si-based detectors, in the near-infrared, it can be harder to achieve in low light level experiments (e.g., single emitter PL measurements), due to lower gain and orders of magnitude higher dark count rates in InGaAs detectors. Particular attention must therefore be paid to efficient photon collection in the construction of near-infrared micro-PL setups. Finally, it should be noted that the pixels in a multichannel detector are typically specified to have a certain full well capacity, which is the maximum number of counts supported by the detector at a given gain setting. The full well capacity limits both the dynamic range of a measurement and the maximum integration time over which an experiment can be conducted. Even though Equation 16.9 indicates that the SNR improves indefinitely with increasing t_{int}, in practice t_{int} will be limited, and the values of P_{det} and D will determine the maximum achievable SNR. Finally, it should be noted that detectors may have additional noise sources not included in Equation 16.9; for example, charge multiplying CCDs have an "excess noise factor" that must be taken into account (Robbins 2003).

16.2.4.3 Detectors for Studying Time-Dependent Phenomena

Until now, we have been concerned with detectors for acquiring a PL spectrum. In such measurements, low noise and single versus multichannel capability are the main concerns, and the detector bandwidth need only be fast enough to respond to the acquisition time, mechanical chopper speed, etc. Of course, the energy level structure mapped out by a PL spectrum only tells part of the story—information about the PL dynamics can also be quite important and is discussed in detail in Chapter 12. The process by which one measures spectrally resolved dynamics is conceptually straightforward. A pulsed laser excites the material under study, and the scanning monochromator is used to select a wavelength band from the resulting emission, which is then studied with a sufficiently high bandwidth detector to obtain the desired temporal resolution. The process is then repeated for a new wavelength band, and after several iterations, a two-dimensional map of PL intensity against wavelength and time is produced.

In considering the detectors to use for such a measurement, we recall that photoreceivers based on p-i-n and avalanche photodiodes are usually specified with a bandwidth and NEP, which correspondingly set the best possible temporal resolution and sensitivity at that temporal resolution. For the low light levels associated with many micro-PL experiments, these detectors are simply not sensitive enough; a gain of 10^7 may be possible, but at a bandwidth of less than 100 Hz. A photomultiplier tube (PMT) is an instrument that can typically provide very high gain (in the 10^7 range) while simultaneously maintaining a temporal resolution in the nanosecond or even sub-nanosecond range, making it a more suitable choice in many situations. Single-photon counters are an important alternative to PMTs for measuring nanosecond-scale dynamics at low light levels. As the name implies, these detectors are sensitive enough to register single-photon events, which is done by biasing an avalanche photodiode above its breakdown voltage (Geiger mode operation), and sub-100 ps resolution can be achieved in some commercially available detectors. The output of the single-photon counter is usually an electronic pulse for each detection event.

Such time-domain measurements are discussed in detail in Chapter 12, so we provide just a few comments here. The first is that if the pulsed laser source has a repetition rate R and pulse width p, and one wants to study the dynamics of a system with a characteristic decay time τ, one must have $p \ll \tau \ll 1/R$, so that each laser pulse initiates a new lifetime trace. The traces are then averaged to generate the final measurement result, with averaging and data acquisition done in different ways depending on the detection equipment used. With standard photodiodes, for example, an oscilloscope or fast data acquisition card that is appropriately synchronized (triggered) to the excitation pulses can be used. This synchronization is produced by an electronic trigger signal that is usually provided as an output from the pulsed laser source. With a single-photon detector, a different set of electronics is required, the sum total of which is usually referred to as a time-correlated single-photon counting (TCSPC) setup (Becker 2005) (Chapter 12). Recently, many commercial vendors have begun to offer TCSPC systems that integrate the needed electronics into a single unit. Included in such units are discriminators that measure the arrival times of the electronic pulses generated by the excitation source and detected photon, a time-to-amplitude converter (TAC) to measure the time difference between them, and an analog-to-digital converter (ADC) to digitize the signal from the TAC and produce a histogram showing the number of detected events as a function of time difference. To reduce noise, it is sometimes appropriate to use the electronic trigger signal from the laser not only on the electronic data acquisition end, but on the detection end itself, to trigger a detection window (essentially, a gate that turns the detector on for time intervals on the order of τ, and at a rate R). Gated detection is commonly used with near-infrared InGaAs single-photon counters (Ribordy 1998), where the avalanche photodiode is periodically biased so that it operates in Geiger (single-photon detection) mode for specified short intervals at a specified repetition rate.

A different approach to lifetime measurements involves a streak camera, again discussed in Chapter 12. The streak camera performs a time-of-flight measurement in which differences in time are mapped to differences in spatial position. This is done in two steps. First, upon illumination, a photocathode generates a number of electrons proportional to the intensity of the incident light. These electrons are fed into a region where a high-speed sweep (synchronized by the excitation laser's trigger signal) is applied between two electrodes. This causes the electrons, which have different arrival times depending on when they were incident on the photocathode, to be mapped to different positions on a phosphor screen. Spectrally resolved measurements can be performed simultaneously by coupling a wavelength-dispersed signal into the streak camera. This will produce a two-dimensional image on the phosphor screen, where time is mapped along one axis and wavelength is mapped along the other. The temporal resolution of streak cameras can be in the sub-picosecond range, exceeding what can be done in most TCSPC measurements, and the sensitivity can be good enough to perform measurements on single emitters (Santori 2002).

Photon correlation measurements are another common time-domain characterization technique used in micro-PL setups. These measurements provide information on the statistics of the fluorescence, which can be used to characterize whether the source is incoherent (e.g., a blackbody radiator), a coherent Poissonian source such as a laser, or a truly nonclassical light source as one might expect for a single emitter (Mandel 1995). The latter point is of particular relevance to micro-PL setups used for single emitter spectroscopy. Depending on the density, there can be many emitters within a diffraction limited spot, and although there can be spectral indications as to whether more than one emitter is producing fluorescence, these indicators usually rely upon arguments related to their inhomogeneous and homogeneous line widths, which can vary from sample to sample. On the other hand, a measurement of photon antibunching (Michler 2003, Mandel 1995) in the second-order correlation function (an intensity correlation measurement) is an unambiguous demonstration of single emitter fluorescence. This normalized intensity correlation function $g^{(2)}(\Delta t)$ relates detection events separated by an interval Δt and is given by

$$g^{(2)}(\Delta t) = \frac{\langle I(t)I(t+\Delta t)\rangle}{\langle I(t)\rangle\langle I(t+\Delta t)\rangle}. \tag{16.10}$$

The brackets denote ensemble averages, which, in a quantum mechanical picture, are expectation values (the expression also invokes normal ordering and time ordering—see Mandel 1995). Antibunching occurs when $g^{(2)}(\Delta t = 0) < g^{(2)}(\Delta t > 0)$, while $g^{(2)}(\Delta t = 0) \geq 1$ is expected for a classical source. Physically, this signifies the fact that in a single optical transition, when the system relaxes from its excited state to its ground state, only one photon can be produced at a time—there should be no chance of detecting a second photon coincident with the detection of a first one. Antibunching measurements have been demonstrated in a number of systems, including single atoms (Kimble 1977), molecules (Basche 1992), colloidal QDs (Michler 2000), impurity color centers (Kurtsiefer 2000), and semiconductor QDs (Michler 2003).

The typical setup for a $g^{(2)}$ measurement is called a Hanbury–Brown and Twiss (HBT) interferometer (depicted in the detection block of Figure 16.2), and involves the use of a 50:50 beam splitter to separate the emitted light into two paths, each of which is fed into a single-photon detector. The outputs of the two single-photon detectors are fed into a TCSPC setup as described earlier, where now the start and stop signals into the TAC are generated by the two detectors, and a histogram of number of coincidence counts as a function of time delay is built up. Unlike the lifetime measurement, we note that a $g^{(2)}(\Delta t)$ measurement can be performed under either continuous wave or pulsed excitation, and conceptually, one might expect to obtain much of the same information from either approach. In practice, signal-to-noise considerations may favor the pulsed approach (particularly if gated detection is applied), though the difference in carrier dynamics between the two excitation methods may produce distinct phenomena.

While $g^{(2)}(\Delta t)$ is an autocorrelation measurement on the emission within a single wavelength band, there can also be benefits to studying the cross-correlation between different wavelength bands. In semiconductor QDs, a spectrum such as that in Figure 16.12a shows the emission from many states, with no information as to the time-ordering of the emission. Cross-correlation measurements provide this information (Aspect 1980), and radiative cascades in QDs have been measured (Moreau 2001). The main difference in the setup required for the cross-correlation measurement is in the spectral filtering. In the autocorrelation measurement, a single spectral filter can be used before the 50:50 beam splitter. In the cross-correlation case, spectral filters must be placed in each arm of the HBT interferometer.

16.2.5 Imaging

In this section, we consider white light imaging, used to resolve the surface of the sample upon which micro-PL measurements are being done (Figure 16.6). For micro-PL conducted in a commercial confocal fluorescence microscope, this type of imaging may be naturally included. In a home-built system, there are many factors involved, starting with an evaluation of the importance of imaging to the experiments at hand. In some situations, such as when performing micro-PL measurements on unprocessed materials, it may be enough to ensure that the pump spot is focused on the sample surface, and a fixed field of view (FOV) with modest resolution will be adequate for this task because the spatial resolution of the measurement will be determined by the pump spot size and how finely the pump beam (or sample) is rastered from point to point. On the other hand, when performing micro-PL on processed devices, it may be necessary to have micrometer-scale spatial resolution in the imaging in order to properly align the pump beam with respect to a fabricated feature. The spatial resolution Δl is basically set by the wavelength and the numerical aperture of the optic, as described in Equation 16.8 ($\Delta l \approx 0.61\lambda/NA$).

In most cases, the imaging path will at least partially coincide with either or both of the pump and collection paths, and because maximizing the collected signal is typically the most important criterion, the choice of optics to image the sample will necessarily be constrained.

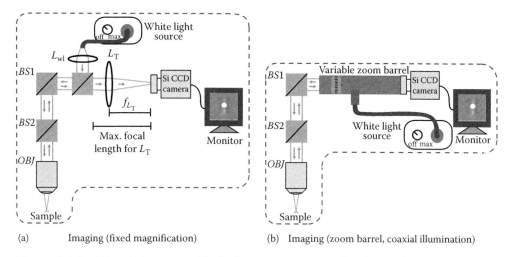

(a)   Imaging (fixed magnification)   (b)   Imaging (zoom barrel, coaxial illumination)

Figure 16.6 White light imaging block of a micro-PL setup: (a) fixed magnification determined by lens L_T; (b) variable magnification using an adjustable zoom barrel with coaxial illumination.

Nevertheless, imaging needs can influence the decision. For example, in the apparatus in Figure 16.2, a single optic (*OBJ*) is used for focusing and collection in the excitation, collection, and imaging paths. If only the first two were required, an achromatic doublet or an asphere might be adequate. However, the need for relatively high-resolution white light imaging suggests using a high *NA* apochromatic optic and, along with the working distance considerations previously discussed, pushes the balance toward a long working distance microscope objective.

Assuming the primary focusing and collection optic is chosen in this way, there is still considerable flexibility in determining the rest of the imaging system. A number of options are available for the illuminating white light source, including bright white LEDs, fiber-optic illuminators that use quartz halogen or metal halide bulbs, and ring lights that provide diffuse lighting. The orientation of the illuminator strongly depends on the properties of the sample and the environment in which it is probed. For opaque materials, reflection of the white light (episcopic mode) is usually done, while in transparent substrates, backside illumination through the sample (diascopic mode) is used. In episcopic mode, there is the possibility of keeping the illuminator off-axis, though the microscope objective and sample environment (e.g., within a vacuum chamber or cryostat) can limit how efficiently this approach can work. Normal incidence illumination, where the white light is sent through the microscope objective using a beam splitter or mirror, is an option in such cases (Figure 16.6). Depending on the characteristics of the white light source and the distance between its output and the microscope objective, focusing the illumination through a condenser lens (L_{wl} in Figure 16.6a) may be useful. The illumination source can affect the material under investigation (e.g., through absorption and subsequent carrier generation), so it is usually prudent to block it during a measurement.

Light reflected off the sample goes back through the objective before proceeding through imaging optics and into a camera, usually a Si CCD. A number of options exist, ranging from inexpensive (<$50) cameras more commonly used for surveillance purposes, to several thousand dollar (or more) imaging cameras with good enough sensitivity to detect single-molecule fluorescence. The chip size (number of pixels), sensitivity, frame rate, and display method are some of the important factors to consider. Along with the objective and imaging optics, the chip size determines the FOV imaged, and in some cases, the pixel size influences the achievable spatial resolution. The rate at which the camera acquires information (its frame rate) can be important in certain situations, though in general micro-PL setups do not require video rate streaming. Typically, the output of the camera (usually a video signal with a standard format like $75\,\Omega$ BNC, composite video, or S-video) is sent to a monitor for display, although such an approach does not allow for image acquisition. This can be done through various computer graphics cards or frame grabbers. In recent years, CCD cameras using the universal serial bus (USB) or Firewire computer interface have become available, allowing for the image to be displayed on a standard computer. This can be advantageous in systems for which a computer is already being used, as image acquisition is then readily available. In other setups, particularly when the imaging is not critical, acquisition is not needed and an inexpensive analog monitor (<$100 for a black-and-white monitor) provides an adequate solution.

In considering image formation on the camera, we focus on the situation in which an infinity-corrected microscope objective has been used. The parallel rays produced by the objective are focused by a lens L_T onto the CCD, as shown in Figure 16.6a. The FOV imaged is given by the ratio of the CCD chip size to the magnification of the composite optical system consisting of the objective and imaging lens. If the objective has a focal length f_{OBJ} and the imaging lens has a focal length f_T, the magnification is given by the ratio $M = f_T/f_{OBJ}$, with the upper bound on M set by the amount of space available (if $f_{OBJ} = 10\,mm$ and $M = 20$ is needed, $f_T = 200\,mm$) and the intensity of white light available. A 1/2″ CCD chip has dimensions of $6.4\,mm \times 4.8\,mm$ (the 1/2″ designation does not refer to the chip size, but is rather a historical artifact), so that a system with 20× magnification will have FOV = $320\,\mu m \times 240\,\mu m$. Pixel size varies, but a typical 1/2″ CCD has 640×480 pixels, so that each pixel is responsible for an area of $0.5\,\mu m \times 0.5\,\mu m$, which is close to the minimum resolvable length scale set by the microscope objective (Δl). For the sharpest imaging, we want to be in the limit where each pixel samples an area much smaller than Δl. To do this, we need a higher level of magnification, while on the other hand, a larger FOV would require lower magnification. An alternate way of thinking of this is in terms of matching the actual CCD pixel size to Δl. In this case, the pixel size is $20\,\mu m \times 20\,\mu m$, and if the microscope objective $NA = 0.4$, $\Delta l = 0.69\,\mu m$ for $\lambda = 550\,nm$. This means we need a magnification of around 30× to have one pixel per Δl. In practice, a higher level of magnification (say 100×) is needed for adequate sampling.

Obviously, the magnification can be adjusted through replacement of the microscope objective and the eyepiece or lens tube, and for fixed FOV imaging this is the most straightforward approach. Since replacement of the objective changes the pump and collection areas, it is probably more desirable to change the eyepiece or lens tube. In some cases, however, it is necessary to switch between different FOVs. An example might be micro-PL measurements on microcavity arrays, where a large FOV is desired when imaging the array of devices, and ascertaining the position of a single device within the large array, but a small FOV is needed when interrogating a specific device. In such circumstances, a zoom system, commonly used in machine vision applications, might be used (Figure 16.6b). The zoom barrel is an adjustable focal length system that provides for a range of possible magnifications, and can usually be outfitted with additional tube adapters to increase the level of magnification possible. As an example, the zoom system we use in our setup has a magnification range between ≈10× and 130× when combined with the 20× microscope objective. The resulting FOV range is $640\,\mu m \times 480\,\mu m - 49\,\mu m \times 37\,\mu m$ for a 1/2″ CCD camera, and as discussed earlier, 130× magnification provides adequate sampling for achieving the highest possible spatial resolution with the chosen objective.

16.2.6 Sample/Objective Positioning

Up to this point, we have discussed topics that fundamentally limit the achievable spatial resolution, such as the pump spot and area from which light is collected, but we have not described how to use this spatial resolution to construct a spatial PL map or, somewhat less ambitiously, how to effectively move from location to location so that the PL can be characterized in different regions. Conceptually, one can envision either rastering the pump and collection paths or moving the sample, and each method has its strengths and disadvantages.

From an optics perspective, moving the sample is easier in that, in principle, it requires no adjustment to the pump or collection optics, provided that sample translation is purely within the plane orthogonal to the optical path, so that it stays in focus at all times. Motorized stages can provide nanometer-scale translation distances, so that the minimum step can be much less than the pump/collection area limited spatial resolution, and the same stages can offer several millimeters of translation, so that large sections of the sample can be covered. These attributes have led us to adopt sample translation as the primary method for producing spatially resolved PL maps in our micro-PL setup. There are some important disadvantages to keep in mind. Though translation stages are relatively stable, they will almost always be less stable than a completely monolithic sample mount. They are also much more costly, particularly if both high-resolution and relatively long translation ranges are needed. Cost and stability become even larger concerns if the sample is to be interrogated in a non-ambient environment such as in a cryostat. In this case, cryogenically compatible motorized stages can be used to translate the sample within the cryostat or, alternately, large platform (and load capacity) stages can move the entire cryostat.

Rastering the pump and collection optics consists of moving the primary focusing/collection optic, and then adjusting the corresponding mirrors to redirect the pump and collection beams through it. To some extent, one might imagine being able to automate such a process through motorized optics mounts, and over relatively small translation ranges (e.g., several micrometers), the amount of adjustment needed is relatively small. Moving over more significant distances does require significant realignment, and can be a tedious process. As mentioned earlier, in our setup, we primarily use sample translation but have also allowed for some amount of pump/collection beam movement, through a stable and relatively high-resolution manual translation stage to position the long working distance objective. It is important to have some ability to adjust the pump and collection beam paths, if only to ensure that the focus is properly set on the sample surface, and small adjustments of the pump beam position can be made through the initial turning mirrors, without having to realign the collection optics.

16.2.7 Alignment and Calibration

Sections 16.5.1 and 16.5.2 provide tips for aligning optical systems, using the example of coupling into a single-mode optical fiber. The alignment of the micro-PL setup roughly follows the same approach, but in multiple stages. A general procedure is as follows: (1) Approximately place the optical components of the micro-PL setup. Check their heights so that they lie along a common optical axis, and position lenses and objectives at the appropriate focal length/working distances. (2) Place the sample at the appropriate working distance with respect to the objective, and approximately align the imaging subsection so that the white light source is focused onto the sample surface and the reflection can be observed in the CCD camera. It is often useful to image the edge of the sample to get the system into focus. (3) Collimate the excitation source, direct it into the objective, and observe the pump spot with the imaging subsection. Small adjustments to the imaging components (e.g., camera/lens position) and the separation between objective and sample surface may be needed. The net result should be that the imaging system simultaneously displays both a tightly focused pump spot and an in-focus image of the sample surface. (4) Try to detect the reflected pump beam in the spectrometer (or whatever detector is being used). This will likely require temporary removal of the pump beam removal filter, in which case attention should be paid to how much power is put into the spectrometer, so as not to damage the detector. Using the methods described in Section 16.5.1 maximize the detected signal. If needed, keep the slit widths open as wide as possible to increase throughput (in the beginning, spectral resolution is not needed). (5) Once the pump beam detection has been maximized, reinsert the pump beam removal filter and try to measure the PL signal. Once any signal has been detected, the input optics to the spectrometer can be adjusted to maximize this signal.

When performing experiments with the micro-PL setup, it is important to have some sense for the source and magnitude of the uncertainties in the measured quantities. Clearly, there are noise sources associated with detection (discussed in Section 16.2.4.2) that are important, but they are not our focus here. Instead, we focus on aspects related to system alignment, which can be significantly reduced through careful calibration. The basic result of a PL measurement is the emission level as a function of wavelength at a given pump power. We therefore focus on how one calibrates emission wavelength, emission level, and pump power.

Wavelength calibration is done through alignment of the grating spectrometer, which is usually a standard procedure performed when first setting up the machine. It typically involves use of a lamp source, such as Ar or Ne, which produces a series of spectral lines at well-known wavelengths that the spectrometer-determined wavelengths are calibrated against—some systems allow for such calibration to be implemented into the operating software. Next, the emission level read by the spectrometer can be calibrated with a laser source and calibrated power meter. The output of the laser source is first measured with the power meter, and then directed into the spectrometer using the input coupling optics as described in Section 16.2.4.1. The integrated photon count rate in the resulting emission spectrum is compared to the value produced by the power meter, providing an estimate of the spectrometer throughput that can be checked against the product of the vendor-specified grating efficiency, mirror reflectivities, and detector quantum efficiency. This throughput should be periodically checked to ensure that the system remains calibrated. As discussed in Section 16.2.4.1, the use of a narrow linewidth source is also important in establishing the system's spectral resolution. Finally, as mentioned in Section 16.2.2, the pump power is typically detected at some reference position, so that the power at the sample surface is determined by placing the power meter at the sample position, reading the power level, and comparing it to the power level when the meter is placed at its reference position. Like the spectrometer calibration, this measurement must be done periodically.

16.3 MEASUREMENTS USING A MICRO-PHOTOLUMINESCENCE SETUP

Having described the construction of a micro-PL setup, we now present a sampling of the data that can be acquired from such a system. In Section 16.3.1, we discuss measurements on microcavity lasers, while in Section 16.3.2, we discuss QD spectroscopy.

16.3.1 Microcavity Lasers

Microfabricated structures often require spatially resolved optical measurements, and microcavity lasers are one example of such a device. The results reviewed in this section were reported by Srinivasan (2003), who fabricated photonic crystal (PC) membrane microcavities in a material containing InAsP/InGaAsP quantum wells on top of an InP sacrificial layer, designed for 1300 nm emission. The devices (Figure 16.7) have in-plane dimensions of ≈8 μm × 11 μm and a thickness of 252 nm, and the spacing between devices is 20 μm.

The devices were tested in a micro-PL setup similar to that depicted in Figure 16.2, where a 20× magnification, long working distance microscope objective (*OBJ*) was used for pumping, collection, and imaging. The sample was kept in atmosphere and pumped with an 830 nm laser diode. To correct for the severe astigmatism of the diode's output, an anamorphic prism pair was used, though the result of the circularization (Figure 16.7c and d) was imperfect.

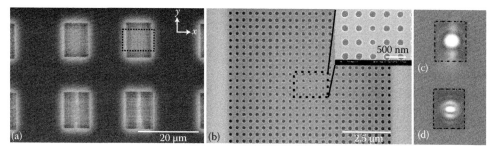

Figure 16.7 (a)–(b) SEM images of PC microcavity laser array. (c)–(d) CCD images of the excitation spot under (c) diffuse and (d) focused pumping conditions, obtained using the zoom barrel imaging subsection of the micro-PL setup described in this chapter. The cavity boundary is shown as a dotted line for reference. (Parts (b)–(d): Reprinted from Srinivasan, K., Barclay, P.E., Painter, O., Chen, J., Cho, A.Y., and Gmachl, C., Experimental demonstration of a high-Q photonic crystal microcavity, *Appl. Phys. Lett.*, 83, 1915. Copyright 2003, American Institute of Physics.)

Heating of the membrane due to absorption of the pump beam limited the maximum pump power that could be used, since the PC cavities are membrane structures with limited thermal contact to the rest of the chip, and would collapse upon absorption of too much average power. This necessitated pulsed excitation, which was accomplished by a pulsed voltage source that drove the laser diode with a pulse width of 10 ns and period of 300 ns. The collected light was separated from the residual pump beam by a free-space edge pass filter, focused into a multimode optical fiber (core diameter = 62.5 μm), and wavelength-resolved using an optical spectrum analyzer (OSA). The OSA was essentially an automated scanning monochromator incorporating a single channel InGaAs detector. The minimum collected power that can be measured with this instrument was around 1 pW, with a spectral resolution as high as ≈0.1 nm.

Figure 16.8a shows a reference PL spectrum collected from an unprocessed portion of the material, under strong cw excitation (≈1 mW), with a wavelength resolution of 10 nm, and a focused pump spot area of 8 μm² (Figure 16.7d). A broad wavelength resolution was chosen to maximize the collected signal because no sharp features were expected. This reference spectrum reveals a few important pieces of information about the material: the peak emission wavelength (1287 nm), the width of the ground-state emission peak (≈140 nm), and the presence of a second (excited) emission peak at 1125 nm. The peak emission wavelength

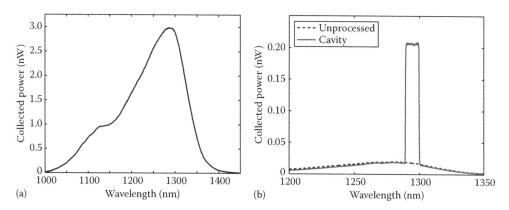

Figure 16.8 (a) PL spectrum from the quantum well material prior to fabrication; (b) PL spectrum from a PC cavity (solid) and an unprocessed region of the chip to the side of the cavity (dashed). Wavelength resolution = 10 nm.

informs the choice of fabricated PC lattice constant (a) and hole radius (r), as simulations of the employed graded lattice geometry specify that $a/\lambda \approx 0.25$ and r/a should vary between 0.23 and 0.31. Since the material emits across a range of wavelengths, cavities of different geometries (and hence different resonant frequencies) can be observed, and it is common practice to check theoretical expectations by confirming the predicted change in cavity emission wavelength as a function of a and r. The spectrum from one cavity (now under pulsed excitation at an average pump power of a few μW) is shown in Figure 16.8b, along with PL from an unprocessed region of the chip to its side. The spectra line up closely, with the cavity spectrum showing a peak at $\lambda \approx 1294$ nm.

Having obtained a broad picture of this device's spectral behavior, we now focus on learning more about the cavity mode emission. By adjusting the wavelength resolution of the OSA (corresponding to a reduction in the monochromator slit width), we find that the observed peak contains a single mode that is quite narrow (inset of Figure 16.9a), and in fact is narrower than the best resolution of the OSA (0.10 nm), provided that heating effects in the membrane are mitigated through the use of a more diffuse pump beam (Figure 16.7c; area ≈ 21 μm^2).

As discussed by Srinivasan (2003), if the emission linewidth is measured at the pump power at which material transparency is reached, an estimate of the cavity quality factor (Q) is obtained (Q is a measure of the photon lifetime in the cavity). The light-in-light-out (L–L) curve (Figure 16.9a through c), given by measuring the emission in the cavity mode as a function of pump power, shows the characteristic turn-on of a laser (Yariv 2007), with the relatively smooth transition (rather than an abrupt kink) indicative of a microcavity, in which the number of modes involved is far less than in a macroscopic device. Extrapolating back from the data at high pump powers gives an estimate of the laser threshold, which in this case is 360 μW. Measurement of the background emission, produced by gain regions external to the cavity, as a function of pump power (Figure 16.9c) indicates (incomplete) gain clamping, with the kink at $\approx$365 μW coinciding with the extrapolated laser threshold value.

The subthreshold spectrum and L–L curve are perhaps the two most common measurements used to characterize microcavity lasers in a micro-PL setup, but there are several other pieces of information that can be obtained. Figure 16.9d shows the emitted power as a function of pump beam position on the cavity, where the position was varied by moving the sample. This measurement provides qualitative understanding of mode localization in such a device, and illustrates why spatial resolution smaller than the cavity size might be useful—we explicitly see the ability to spatially resolve emission with a resolution of $\approx$1 μm, which is ten times smaller than the cavity size. Next, the polarization of the laser emission is determined by placing an adjustable polarizer in the collection path, for example, between *BS2* and *OBJ* in Figure 16.4. Figure 16.9e shows the collected emission as a function of polarizer angle with respect to the cavity's x-axis, confirming that the emission is x-polarized, as expected from calculations.

Another important piece of information, not presented here, is the cavity linewidth as a function of pump power, which can provide information about gain and loss in the laser, but would require a higher spectral resolution device such as a scanning Fabry–Perot cavity, since the OSA resolution is insufficient. Correlation measurements, as discussed in Section 16.2.4.3, provide important information about the onset of lasing in a microcavity (Strauf 2006). Microcavities are typically distinguished by a relatively small number of optical modes, so that the transition between spontaneous emission and lasing in an L–L curve, marked by a pronounced kink in a macroscopic device, is not necessarily readily apparent. Indeed, as the number of optical modes is reduced to unity, this kink is expected to soften and, under certain conditions, disappear. To distinguish between such a device and one operating in the spontaneous emission regime, a measurement of $g^{(2)}(\Delta t)$ can be performed and will show a transition as the emission becomes coherent. Measurements of $g^{(2)}(0)$ as a function of pump power can then be used to establish the laser threshold and probe the behavior of few mode microcavities.

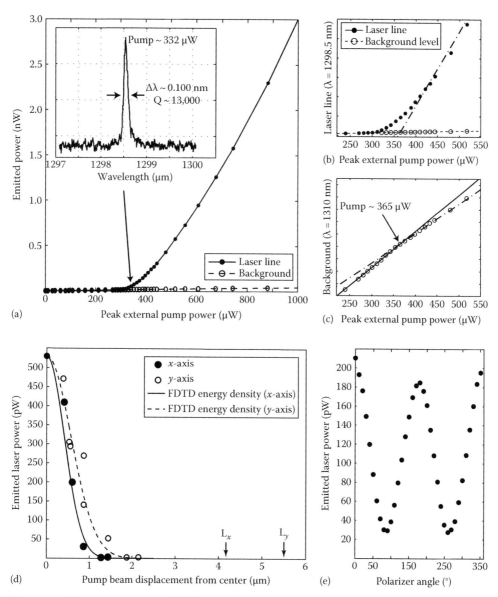

Figure 16.9 (a) L–L curve and subthreshold spectrum (inset) of the PC microcavity pumped, and zoomed-in plots of (b) laser threshold and (c) background emission. (d) Laser power as a function of pump position along the x- and y-axes of the cavity. Finite difference time domain (FDTD)-generated Gaussian fits to the envelope of the electric field energy density of the cavity mode are shown for comparison. L_x and L_y correspond to the physical extent of the PC in the x- and y-directions, respectively. (e) Emitted laser power as a function of polarizer angle with respect to the x-axis of the cavity. (Reprinted from Srinivasan, K., Barclay, P.E., Painter, O., Chen, J., Cho, A.Y., and Gmachl, C., Experimental demonstration of a high-Q photonic crystal microcavity, *Appl. Phys. Lett.*, 83, 1915. Copyright 2003, American Institute of Physics.)

16.3.2 Quantum Dot Spectroscopy

In this section, we discuss micro-PL measurements of self-assembled InAs QDs embedded in GaAs that emit at 1.2–1.3 μm, with PL emission produced from both ensembles of QDs and single QDs. The apparatus used is similar to that described in the previous section and is schematically depicted in Figure 16.2. Our research investigates the interaction of QDs with tightly confined optical fields, and in such experiments, limiting the QD dephasing below some minimum value is required. This is done by cooling the sample to a temperature below 20 K using a liquid helium flow cryostat. The other important modification with respect to the measurements of Section 16.3.1 is the use of a grating spectrometer with a liquid nitrogen–cooled InGaAs array detector, which, as described in Section 16.2.4.2, substantially improves the SNR and measurement time with respect to a scanning monochromator and single channel detector. While the collected power levels in the previous section were in the picowatt to nanowatt range, in these measurements collected signal levels are in the 0.1–10 fW range, necessitating a much more sensitive detector. Finally, we use two different laser sources in our measurements: an 830 nm cw (fiber-coupled) laser diode, and a 780 nm pulsed laser diode with a 50–100 ps pulse width and a 50 MHz repetition rate.

We first consider measurements of QDs that emit at 1.2 μm. These structures, grown by molecular beam epitaxy, have a density gradient along one axis of the wafer (Figure 16.10a). When interrogating a new wafer, our goals are to (1) determine the peak emission

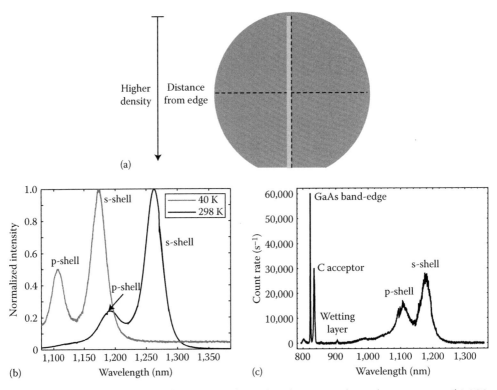

Figure 16.10 (a) Schematic of the QD wafer with a density gradient along one axis; (b) 298 and 40 K PL spectra from a high-density region of the wafer, showing the ground state (s shell) and excited state (p shell) ensemble emission, and an ≈90 nm blueshift between room and low temperature; (c) broad PL spectrum from a high-density region of the wafer at 7 K, showing the GaAs band edge, carbon acceptor peak, and wetting layer emission along with the QD ensemble s shell and p shell peaks.

wavelength of the ground-state ensemble (knowing that the peak position may vary across the wafer), (2) examine the PL as a function of position along the density gradient, ascertaining the portion of the wafer where single QD emission can be observed, and (3) perform additional measurements (pump power–dependent behavior, lifetime, $g^{(2)}(\Delta t)$) on single QDs.

Figure 16.10b shows PL spectra that have been taken from a high-density region of the wafer, at temperatures of 298 K and 40 K. The pump beam diameter is around 5 μm, with an average incident pump power of 60 μW (wavelength = 830 nm), and emission is collected from a 2 μm diameter portion of the excited region with a 1 s integration time. Both PL spectra show two dominant peaks, corresponding to the ground and excited states of the QD ensemble. In comparing room temperature and low temperature results, we note a couple of salient points. First, there is a wavelength blue-shift of about 88 nm, which follows the band gap of the host semiconductor material (Vurgaftman 2001), and can be a qualitative check on the sample temperature (although the cryostat temperature sensor may read a certain value, poor thermal conductivity can sometimes prevent proper cooling of a sample). Next, we see that the linewidth of the peaks does not appreciably narrow at low temperature (the full width at half maximum for the s-shell changes from 33 to 30 nm). This is because the peaks are inhomogeneously broadened due to the size distribution of QDs produced by the Stranski–Krastanov growth process (Michler 2003). In Figure 16.10c, we show another PL spectrum from the sample, now at 7 K, over a much broader wavelength range. By pumping the sample at 780 nm and using an 800 nm edge pass filter to remove residual pump light, we are able to observe other important emission peaks (we also reduce the average pump power to 4 μW in preparation for single QD measurements). At 900 nm, we see emission from the QD wetting layer, a thin region of semiconductor material used to help seed the QD growth. At 833 nm, we see emission from the carbon acceptor impurity state, and at 823 nm, we see emission from the GaAs band edge. The relative strength of these peaks provides information about the material under investigation. For example, the strength of the carbon acceptor peak can be indicative of the growth chamber conditions (how much carbon is in the chamber), while the GaAs band-edge emission can be less prominent in quantum well structures due to rapid relaxation of carriers into the wells.

Next, we look at the PL as a function of position along the direction of the density gradient and near the center of the wafer. Fixing the grating position so that it is centered about the s-shell, we pump relatively weakly (300 nW incident power), to reduce background emission created by QD–wetting layer carrier interactions, and acquire spectra with a 30–60 s integration time. A 0.5 mm step size is used to map PL across the wafer, but sub-micrometer resolution can be utilized if needed. The integrated s-shell emission as a function of position is shown in Figure 16.11a, with spectra at selected positions shown in Figure 16.11b. We observe two dominant trends. First, the integrated emission almost monotonically decreases as we approach the low-density wafer edge, as one would expect, with essentially no emission once we are within 3–4 mm of the edge. Next, the character of the PL spectra begins to change. As we move to lower-density regions, the broad emission curve is punctuated by a multitude of emission spikes, as the QD density becomes sufficiently low so that PL from small numbers of QDs can be resolved, rather than just the inhomogeneously broadened PL from the ensemble. At approximately 5.7 mm from the low-density edge, we see essentially no broad ensemble PL, but instead only a few sharp peaks. These peaks correspond to states of a single QD, as has been described in a multitude of references (Michler 2003). Calculations that account for the shape and composition of the QD can help us distinguish between neutral, charged, and multiple exciton states. Pump power–dependent measurements can also provide useful information, in that the neutral exciton state is generally expected to exhibit emission at pump powers lower than that at which charged states will be present. In addition, the biexciton state

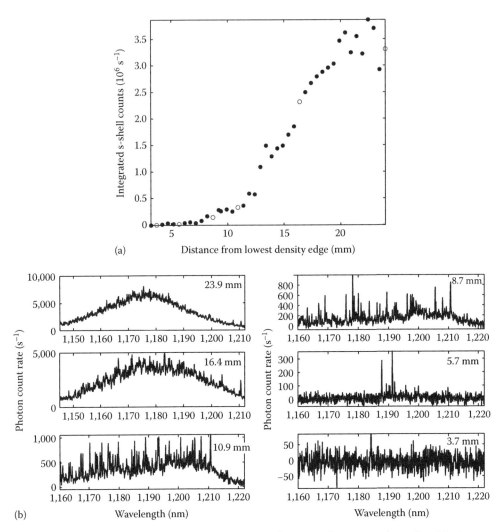

Figure 16.11 (a) Integrated s-shell emission as a function of position along the QD density gradient, taken at 7 K. Open circles denote positions for which full spectra are shown in (b). Single QD emission lines are observed at a distance of ≈4–7 mm from the low-density wafer edge. Average incident pump power = 300 nW.

is expected to have quadratic pump power dependence, in contrast to the linear dependence of the exciton (Michler 2003).

Such identification was presented by Srinivasan (2007) for a different QD epitaxy (a single QD in a well, or DWELL), though the focus there was on improved collection efficiency through use of a fiber taper waveguide probe. Figure 16.12a shows measurements of this DWELL material using the PL setup described earlier for 1.2 μm QDs. As was the case with Srinivasan (2007), excitation at the lowest pump powers shows three emission lines, corresponding to the polarization split neutral exciton states (X_a/X_b) and negatively charged state (X^-) of the QD. As the pump power increases, additional states appear, including the positively charged state (X^+) and biexciton state (2X). At the highest pump levels, a multitude of states are expected, including multiply charged states (e.g., X^{2-}) and QD states that are hybridized with the wetting layer. Precise identification of these spectral lines through PL measurements (and supporting theoretical calculations) alone can be quite difficult. Placing the QD in a field-effect structure to allow for capacitance spectros-copy (Drexler 1994) is a commonly used method to provide a better understanding of the QD energy level structure.

Figure 16.12b shows data from Srinivasan (2007), zooming in on the neutral exciton lines of such a QD and plotting their PL line widths as a function of sample temperature, between 13.8 and 90 K. The broadening of the linewidth corresponds to an increase in the QD dephas-ing, which is expected due to the increased interaction with phonons at higher temperatures; however, the limited spectral resolution of the system used precludes accurate measurements of linewidth below $\approx$25 K. Measurements with better spectral resolution (Bayer 2002) indicate that the broadening is essentially continuous as the temperature is increased above the lowest temperature achieved (2 K), and that the increase is roughly linear with a slope of $\approx$0.4 pm/K for temperatures below 60 K. Figure 16.12c shows the wavelength of the X_b line as a func-tion of sample temperature (starting at 13.8 K), along with least squares fits to the data using the well-known Varshni and Bose–Einstein functional forms (Grilli 1992), which have seen widespread use in investigations of the temperature dependence of direct band gap semicon-ductors (Vurgaftman 2001). The data follow both curves reasonably well, again providing a qualitative check of the sample temperature (the linewidth of the QD lines is a second check). The plateau in the emission wavelength at lowest temperatures seems to fit the Bose–Einstein curve more closely, which has been suggested to be the case at temperatures below 80 K (Grilli 1992).

Finally, in Figure 16.12d, we present a measurement of the PL decay from states of this single QD, using a setup similar to that discussed in Section 16.2.4.3. The sample is pumped

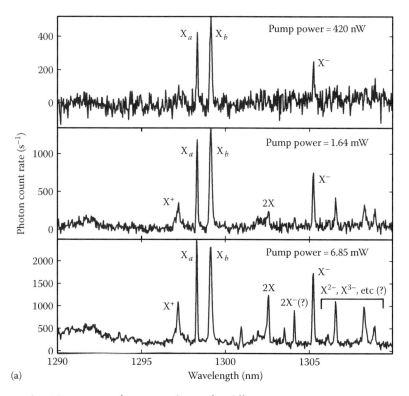

Figure 16.12   (a) PL spectra from a single QD for differing pump powers, with the QD states tentatively identified, taken at 7 K.

(continued)

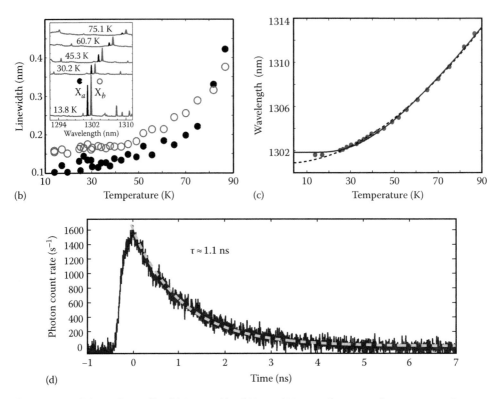

Figure 16.12 (continued) (b) Linewidth of X_a and X_b as a function of temperature (spectra inset). (c) Position of X_b as a function of temperature (dots), with fits to the data using the Varshni (dashed) and Bose–Einstein (solid) functional forms. (d) PL decay and fit for emission from 1294 nm to 1306 nm, under weak excitation so that only X_a, X_b, and X^- are present. (Reprinted with permission from Srinivasan, K., Painter, O., Stintz, A., and Krishna, S., Single quantum dot spectroscopy using a fiber taper waveguide near-field optic, *Appl. Phys. Lett.*, 91, 091102. Copyright 2007, American Institute of Physics.)

with the aforementioned 780 nm pulsed laser, at an average power that excites only the X_a/X_b and X^- states, as in Figure 16.12a. A 12 nm bandpass filter centered at 1300 nm removes emission outside of this wavelength band, and the signal collected by a single-mode optical fiber is sent into an InGaAs single-photon counter operated in gated detection mode with a 20 ns gate width and 10 μs dead time, and whose output is sent to a TCSPC board. The trigger signal of the laser is fed to a digital delay generator, which serves as a master clock for the experiment, with one output triggering the gated detection mode of the single-photon counter, and the other fed to the TCSPC board. The resulting lifetime trace (Figure 16.12d) is fit to a singly decaying exponential (in practice the logarithm of the curve is fit to a linear function), with a lifetime of 1.1 ± 0.1 ns. This decay constant is consistent with the 1 ns lifetime typically measured for InAs QDs (Michler 2003); more precise measurements could be made by using a narrower bandpass filter to study only one QD state at a time, and by improving the collected signal.

16.4 RELATED MICRO-OPTICAL TECHNIQUES

In this section, we describe a few extensions to the micro-PL setup that can equip the user with the ability to probe additional aspects of the material system under investigation. The following discussion is by no means comprehensive, and because the optical techniques involved are largely similar to those involved in micro-PL, less detail is provided.

16.4.1 Photoluminescence Excitation Spectroscopy

The basic measurement performed in PLE spectroscopy is that of the sample's emission spectrum as a function of excitation wavelength (Chapter 6). Until now, we have paid little attention to the excitation wavelength other than to say that it should be at a wavelength for which the sample has adequate absorption to generate the carriers needed to fill the relevant energy states that produce luminescence. There can be more to this story, however, as some materials may have spectrally narrow absorption bands, so that precise spectral alignment is needed, or absorption bands that lead to preferential emission on certain transitions. The InAs QDs described in the previous section provide one example where PLE measurements can be useful. The broad spectrum of Figure 16.10c indicates several luminescence centers other than the ground state of the QD ensemble, including the GaAs band-edge, the C acceptor peak, the QD wetting layer, and the QD ensemble p-shell. The measurements we described so far were at pump wavelengths within the GaAs band gap (780 nm excitation) or at the C acceptor peak (830 nm excitation), so that carriers were generated at much higher energies than the QD transition, and fluorescence at the QD ground state occurs only after carriers relax into the QD states. Pumping at a wavelength closer to resonance with the QD s-shell, for example, at the wetting layer (900 nm) or, preferably, the QD p-shell (1100 nm), can produce different carrier dynamics that will physically influence the behavior of the system. For example, photon correlation measurements have been used to study sub-microsecond correlations in QD samples under above-band and near-resonant pumping (Santori 2004).

PLE measurements have recently been performed by Badolato (2008), in a system consisting of a single InAs QD in a photonic crystal microcavity for which there is only one cavity mode interacting with the QD. Data from this study, reprinted in Figure 16.13, show how the PL intensity in the QD states and cavity mode varies with excitation wavelength and the corresponding photogenerated carrier density. For wavelengths below the GaAs band edge (≈820 nm), PL is collected from the cavity mode* and the X and XX states. At excitation wavelengths between 820 nm and 840 nm, one can find regions in which the X/XX states are bright and the cavity is comparatively dark, and vice versa. The presence of carbon impurities at 833 nm (Figure 16.10c) suggests that excitation of these shallow acceptor states can generate electrons that readily fill the QD states, but do not contribute to emission into the

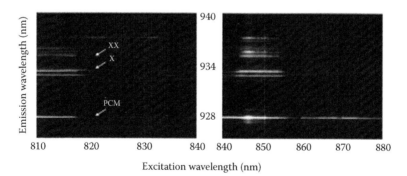

Figure 16.13 PLE spectra of a coupled cavity-QD system when tuning the excitation wavelength over the (a) 810–840 nm region (50 nW pump power), and (b) 840–880 nm region (500 nW pump power). (Reprinted from Badolato, A. et al., *Compt. Rend. Phys.*, 9, 850, 2008. With permission.)

---

* The source of the cavity mode emission when nonresonant with a QD state is an active area of research, and PLE measurements provide information about this, particularly when combined with photon correlation measurements.

cavity mode. Increasing the excitation wavelength to 846 nm produces bright emission from both the cavity and QD states, for reasons currently unknown, and a further increase beyond ≈860 nm (now in the region of the QD wetting layer) produces bright cavity mode emission but no emission from the QD.

From an equipment standpoint, the modifications to the setup in Section 16.2 (Figure 16.2) are straightforward, albeit potentially expensive. The main requirement is a tunable laser; a Ti: sapphire or dye laser will generally provide the broadest wavelength coverage, a necessity if widely spectrally separated absorption centers are to be excited. The spectral resolution of the measurement will be limited by the excitation laser's linewidth. Changing the excitation wavelength may also require adjustment of the filter(s) used to separate the pump from the collected signal. For systems in which laser spontaneous emission is problematic, a tunable bandpass filter may be required at the output of the tunable laser.

16.4.2 Electroluminescence

EL measurements are sometimes used to characterize systems in which electrical injection of carriers into the light-emitting transitions can be achieved. The III–V semiconductors are perhaps the most notable examples of this, and EL measurements are a necessity in the development of technologically important current injection devices, including microcavity lasers (Levi 1992) and single-photon sources (Yuan 2002). Conceptually, EL measurements involve a replacement of the photoexcitation used in PL with current injection. The specifics of the injection scheme (e.g., wire bond vs. electrical probe) and the material design for efficient carrier generation are beyond the scope of this chapter, but in one common situation, an EL sample is made by placing the light-emitting layer in the intrinsic region of a p-i-n diode. Once the applied bias is large enough, a current can pass between the doped layers in the intrinsic region (inset of Figure 16.14a). In this region, carriers can optically recombine, emitting photons that are detected in the same manner as in PL. To modify the setup described in Section 16.2 for EL measurements, in principle all that is needed is to ensure that the beam

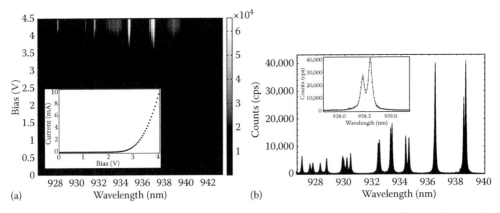

(a)   (b)

Figure 16.14 (a) EL spectrum as a function of applied bias for an electrically gated micropillar cavity with embedded QDs. The dark, sharp lines are cavity modes and the weaker emission between modes is due to the QDs. The onset of luminescence coincides with the current turn-on in the IV curve (see inset). (b) Photoluminescence spectrum of the same device under no applied bias. Inset: A fit to the polarization-split fundamental cavity modes. (Reprinted from Rakher, M.T., Quantum optics with quantum dots in microcavities, PhD thesis, University of California, Santa Barbara, CA, 2008.)

collection optics and the environment housing the sample are compatible with the current injection scheme. For example, electrical testing in a cryogenic probe station usually involves long working distances, so that the apochromatic objectives discussed thus far remain a good choice, and in some cases, even longer working distance optics may be needed. On the other hand, a requirement of good collection efficiency and high numerical aperture optics with shorter working distances may necessitate the experimenter to explore wire-bonded geometries that maintain a smaller form factor.

To illustrate some of these concepts, we consider the EL spectrum from a QD embedded in a microcavity structure, as shown in Figure 16.14 (Rakher 2008). In this structure, a vertical cavity surface emitting laser (Coldren 1995, Yariv 2007) design was used to electrically contact a single micropillar cavity. In such a scheme, the top $GaAs/Al_{0.9}Ga_{0.1}As$ stack was p-doped while the bottom stack was n-doped and the sample was grown on n-type GaAs. The p-type mirror was contacted using a Ti/Pt/Au metallization while the n-type mirror was contacted using an AuGe/Ni/Au metallization after etching. The sample was then wire-bonded and placed in a cryostat at 4 K, a voltage-current source was used to apply an external bias, and the EL was collected as a function of applied bias as shown in Figure 16.14a. The turn-on of luminescence coincides with the turn on of current near 3 V (see inset). EL is detected both from the micropillar cavity modes (narrow, dark lines) and bare QD-related emission (underlying broad emission). This spectrum can be compared to the PL spectrum from the same device under no applied bias, as shown in Figure 16.14b. The broad QD emission is significantly quenched due to the large electric field over the QDs, but emission from the cavity modes remains (the inset of Figure 16.14b shows a fit to the two polarization-split fundamental modes of this cavity).

16.4.3 Angle-Resolved Reflectivity

The final micro-optical technique that we consider is angle-resolved reflectivity (or transmission), where the reflected (transmitted) signal off a sample surface is monitored as a function of angle (Figure 16.15). In comparison to the methods described thus far, it differs in that it probes the resonant response of the system, and although we do not discuss them in this chapter, we note that micro-absorption measurements follow a procedure similar to what we describe. Angle-resolved reflectivity measurements are particularly well suited for characterizing the linear optical properties of planar thin films, yielding information about the electromagnetic waves supported upon illumination at an angle from the surface normal (here, illumination refers to the interrogation beam, and not white light imaging). For example, photonic band gap materials such as thin-film distributed Bragg reflector (DBR) stacks (Yariv 2007) or films of self-assembled photonic crystals (Lopez 2002) may be designed to forbid propagating waves for incidence along certain directions, over wavelength ranges called photonic band gaps. Spectral band gap width and position vary with incidence angle, depending on the sample's structure. In such forbidden bands, the reflectance level is substantial and the transmittance is very low. Outside the band gap, the opposite is generally observed, as propagating waves accessible from outside the structure are available. Tracking band gap position with incidence angle gives insight into the photonic crystal band structure and may help determine physical parameters of the structure (e.g., layer thickness, refractive index, and lattice constants). In spatially inhomogeneous samples, for instance, involving polycrystalline organic films or photonic crystals (Lopez 2002) with micrometer-scale domains, it is important to limit the probed region to dimensions smaller than that of an average sub-domain, to minimize inhomogeneous broadening of spectral features. A similar situation arises when probing fabricated devices with micrometer-scale dimensions.

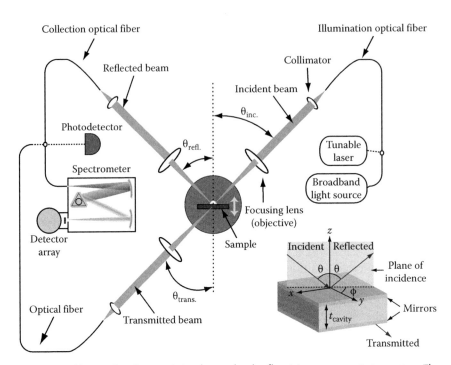

Figure 16.15   (See color insert.) Angle-resolved reflectivity or transmission setup. The goniometer consists of two arms that rotate around a common axis, so that the angle between the two may be precisely adjusted. Collimating and focusing optics, as well as collection and illumination fiber optics shown, are mounted on the rotating arms. The bottom right image shows the orientation of the beam on a planar microcavity sample.

Many of the considerations in an angle-resolved reflectivity measurement are the same as what has been discussed for micro-PL in Section 16.2. A significant difference is that in the micro-PL setup of Figure 16.2, we considered normal incidence excitation and collection, whereas we now want to measure the response as a function of angle (non-normal incidence or collection may be advantageous in micro-PL characterization of certain sample geometries as well). The main element required is a goniometer, which allows precise control of optical beam incidence and collection angles. The goniometer includes one or more arms that rotate around an axis, and are equipped with optical elements for directing, focusing, and collecting optical beams (Figure 16.15). If a single rotating arm is employed, the sample mount must allow for rotation as well, so all angles may be accessed. Multimode optical fibers are generally employed on movable arms, serving as either excitation or collection ports, and can significantly reduce the complexity of the system. Collimation, focusing, and collection of light from and into the fibers are accomplished as discussed in Sections 16.2.2 and 16.2.3. Short focal lengths generally afford tighter focusing, but low NAs lead to illumination with, and collection of, a smaller angular spread, resulting in improved angular resolution. Objectives are often used as focusing and collection lenses, offering small focal lengths with low NAs, as well as aberration correction (Lopez 2002). The total angular range accessed by the apparatus is limited to a range for which the objectives do not touch the sample or each other, so that long working distance objectives are generally advisable. Additional elements such as polarizers and waveplates may be placed in the collimated beam paths on the arms, to control the incident beam polarization and select the collected beam polarization. The incident beam should meet the sample at the goniometer axis, to prevent the illumination area from moving, and to minimize variations in collected power as the incidence angle is changed. A translation stage adjusts the sample position with respect to the rotation axis and the illuminating beam. If the sample is mounted on a rotating stage, its rotation axis must be aligned with that of the goniometer. If two rotating arms are used, the axes must be coincident.

The light source must have a sufficiently broad spectrum to cover the entire range of the dispersion curves. Common sources are tungsten-halogen incandescent lamps, which provide a spectrum ranging from visible to near-infrared wavelengths, and Hg(Xe) arc lamps, which provide UV to near-infrared coverage. Another possibility is fiber supercontinuum sources, which, although pulsed, cover extremely broad wavelength ranges and provide high-power, coherent light beams that are easier to focus. When broadband sources are used, the signal coupled into the collection fiber is brought into a spectrometer, where it is dispersed to reveal the reflectance or transmittance spectra. Alternately, a tunable laser may be used with a photodetector for better spectral resolution, albeit usually at the expense of measurement time and spectral bandwidth.

As an example, we consider measurements of the dispersion characteristics of planar microcavity resonant modes (Lidzey 1995, Kena-Cohen 2008). Such one-dimensional cavities (Figure 16.15) generally consist of a planar, partial reflector pair sandwiching a sub-wavelength–thickness film of an arbitrary material. The cavity modes are traveling waves in the two planar dimensions (x and y in Figure 16.15), and form a discrete set in the perpendicular dimension (z). Illumination with a plane wave at an incidence angle θ with respect to the surface normal leads to excitation of a cavity mode with matching parallel wave vector ($k_\parallel$). For a fixed frequency ω, varying θ allows this condition to be met for all (discrete) cavity mode values. As such, dispersion measurements involve illumination of the cavity at varying θ and spectral analysis of the reflected and/or transmitted light. Cavity resonances are manifested as relatively sharp maxima (minima) in the transmitted (reflected) spectra, in analogy with a Fabry–Perot cavity in which only light of certain frequencies is transmitted. These resonances shift in wavelength as the incidence angle is varied. The dispersion, $\omega(k_\parallel)$, is obtained from the position of the resonant maxima or minima at each angle, and is clearly a function of the linear optical properties of the cavity material, expressed through a complex refractive index $n(\omega)$.

Figure 16.16a and b shows reflectivity spectra for varying incidence angles obtained from a 140 nm thick cavity containing a single anthracene crystal sandwiched between two SiN/SiO$_2$ DBR mirrors (Kena-Cohen 2008). Due to the fabrication process, the organic microcavity was formed over 500 µm wide, 2 cm long channels, bound by gold stripes. The channel width imposed an upper limit to the illumination spot size, requiring focusing optics to produce spots of diameter <300 µm. As these cavities contain an excitonic material, the dispersion curves show spectral signatures of strong exciton–cavity photon coupling (Weisbuch 1992). This is evidenced by anti-crossing in the dispersion curves (Figure 16.16c and d) traced out from the spectral minima and maxima, and physically represents the formation of normal modes of the two coupled oscillators. By adjusting the polarization of the illumination source, modes in which the electrical field is both parallel and perpendicular to the cavity plane, known respectively as transverse electric (TE) and transverse magnetic (TM) polarization modes (Chapter 1), may be accessed. Such polarization selectivity can be of importance in studying anisotropic materials.

Illumination of different spots across the sample revealed variations of tens of milli-electron volts in the measured polaritonic energies, a result of inhomogeneities in both the crystalline material and cavity parameters. This reinforces the necessity of probing sufficiently small areas to avoid inhomogeneous broadening of reflectivity or transmission spectra.

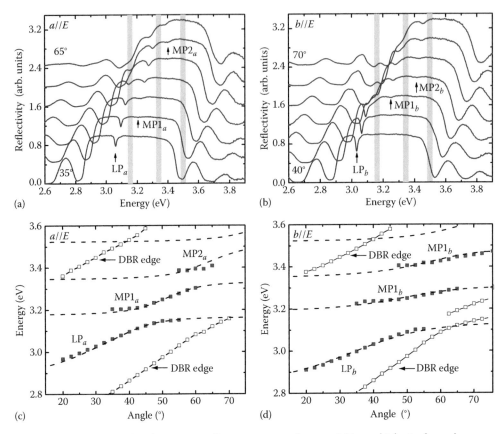

Figure 16.16 TM (p)-polarization reflection spectra from a 140 nm thick single anthracene crystal cavity with SiN/SiO$_2$ DBR mirrors, for various incidence angles, and electric field oriented parallel to the monoclinic anthracene crystal a (a) and b (b) axes. The lower polariton (LP$_{a,b}$) and middle polariton (MP1,2$_{a,b}$) branch reflectivity dips are indicated in the figures. Gray boxes indicate the position of bare anthracene film excitonic transitions. The spectral position of the polaritonic dips in (a) and (b) are plotted as a function of angle in (c) and (d), respectively (solid squares). Open squares: DBR stopband edges. Dashed lines: fitted curves using a four-body coupled harmonic oscillator Hamiltonian. (Reprinted from Kéna-Cohen, S., Davanço, M., and Forrest, S.R., Strong exciton-photon coupling in an organic single crystal microcavity, *Phys. Rev. Lett.*, 101, 116401. Copyright 2008, American Physical Society.)

16.5 TIPS WHEN BUILDING MICRO-OPTICAL SETUPS

In this final section, we outline a pair of useful procedures when aligning a micro-optical system.

16.5.1 Beam Alignment

This section describes the basic process of aligning an optical setup. As a prototypical example, we use the alignment of a collimated beam into a single-mode optical fiber, but the process involved is universal for tasks where both the position and slope/angle of an optical beam are important. The setup (Figure 16.17) consists of a detector, two mirrors (M1 and M2) and kinematic mounts, an optical fiber, an incoupling lens (usually an asphere, microscope objective, or high-power achromatic doublet), and a kinematic fiber coupling stage (three or five axes—usually three are sufficient). The first task is to set up the fiber stage by mounting the fiber tip at the manufacturer's specified focus of the incoupling optic (henceforth referred to

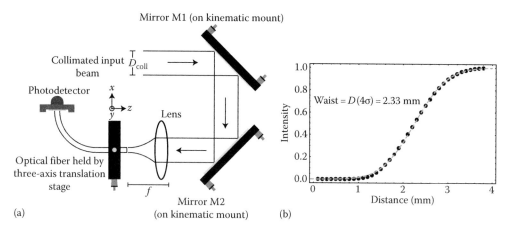

Figure 16.17 (a) Schematic of a beam alignment setup for coupling a collimated beam into a single-mode optical fiber. (b) Measurement of a collimated beam profile using the knife-edge technique.

as the lens). Recall that the optic should be chosen based on the collimated beam diameter, fiber mode field diameter, and wavelength of light, as discussed in Section 16.2. Then, the fiber position should be fixed in the xy plane so that it is at the center of the lens. For a five-axis stage, the angles should be set so that the axis of the fiber is aligned with the optical axis of the lens. Place M1 so that the collimated beam is centered on it and roughly align it so that the beam is directed onto the center of M2. Then align M2 so that the beam is centered on the lens, and check to see if the detector is showing at least a very weak signal—if a photodiode is used for detection, it may be necessary to use a high parallel resistance to amplify the measured voltage. If no signal is found, adjust M2 while monitoring the detected signal until power is detected.

The next part of the process is commonly referred to as beam walking. It is a procedure that iteratively aligns the beam to achieve the optimal position (i.e., on the core of the fiber) and slope in the x- and y-directions, and proceeds as follows: (1) Slightly adjust the x knob of M2, noting the adjustment direction. The signal should be reduced. (2) Adjust the x knob of M1 to optimize the signal and compensate for the signal loss due to the adjustment of M2. If the signal has increased compared to the value before M2's adjustment, then the beam is being walked in the correct direction and M2 should be slightly adjusted in the same manner as before, followed by adjustment of M1. (3) Iterate until the signal reaches a maximum. If the initial signal is not surpassed by adjustment of M1, then the beam should be walked in the opposite direction. Again, iterate until the signal reaches a maximum and plateaus. If slight adjustments are not resulting in increased signal strength, then use larger adjustments of M2. (4) Once this procedure has been completed for the x-direction, do the same for y. (5) After y is finished, redo the procedure for x. Go back and forth between x and y until no more gains in detected power can be made. At this point, the beam is aligned as well as possible for the settings of the lens.

To further optimize the incoupled power, one must adjust the distance between the lens and fiber. Adjust this distance to maximize the detected power, and then adjust M2 and look for a further increase. Continue adjusting the focal distance and M2, and iterate until no more gains are made. At this point, the beam should be optimally aligned. Minor improvement can be made by small adjustments of the x–y position of the lens (or fiber) along with adjustments of M2, but these gains are usually small and come at the cost of using the relatively insensitive actuators of the lens or fiber. While incoupling efficiency depends on the beam quality and optics, typical values should be near 50%. Efficiency far less than 50% indicates that a local maximum may have been reached, in which case it is best to start over and make larger initial adjustments of M1 and M2 to better span parameter space. If efficiency remains low, it likely means that a different focusing optic is required or that the collimation/shaping of the input beam needs adjustment.

16.5.2 Beam Profiling

The collimated beam size is important because it determines the appropriate lens focal length for coupling into a fiber or a spectrometer. A simple way to measure the transverse extent of a beam is the knife-edge method. The basic principle is to move a sharp metal edge across the beam while measuring the transmitted power. For a Gaussian beam centered at x_0, the detected power (neglecting diffraction) as a function of knife-edge position x_m is

$$P(x_m) = \frac{P_o}{2}\left[1 + Erf\left(\frac{x_m - x_o}{w/\sqrt{2}}\right)\right], \tag{16.11}$$

where

P_o is the power of the beam when the knife edge does not interfere

$Erf(x)$ is the error function, defined as $Erf(x) = \left(2/\sqrt{\pi}\right)\int_0^x e^{-x^2} dx$

w is the beam waist defined in Section 16.2.2.2

The setup consists of a razor blade mounted to a micrometer-driven translation stage, along with a broad area detector such as a power meter. Starting the measurement with the razor blade completely blocking the beam sets the background level, and the power is then noted as a function of blade position. For a Gaussian beam, the background-subtracted data should look similar to that in Figure 16.17b, which can then be fit to Equation 16.11 to determine the beam waist.

ACKNOWLEDGMENTS

We thank colleagues with whom we have built the characterization setups that informed the presentation in this chapter, including Oskar Painter, Paul Barclay, Stefan Strauf, and Stephane Kéna-Cohen. We also thank Antonio Badolato, our collaborator on near-infrared quantum dot studies, and Andrew Berglund for helpful comments on the chapter. Portions of the work described were performed at the NIST Center for Nanoscale Science and Technology, and partly supported by the NIST-CNST/UMD-NanoCenter Cooperative Agreement.

REFERENCES

Alivisatos, A.P. 1996. Semiconductor clusters, nanocrystals, and quantum dots. *Science* 271: 933–937.

Aspect, A., Roger, G., Reynaud, S., Dalibard, J., and Cohen-Tannoudji, C. 1980. Time correlations between the two sidebands of the resonance fluorescence triplet. *Physical Review Letters* 45: 617–620.

Badolato, A., Winger, M., Hennesy, K.J., Hu, E.L., and Imamoglu, A. 2008. Cavity QED effects with single quantum dots. *Comptes Rendus Physique* 9: 850–856.

Balistreri, M.L.M., Klunder, D.J.W., Blom, F.C. et al. 1999. Visualizing the whispering gallery modes in a cylindrical optical microcavity. *Optics Letters* 24: 1829–1831.

Basche, Th., Moerner, W.E., Orrit, M., and Talon, H. 1992. Photon antibunching in the fluorescence of a single dye molecule trapped in a solid. *Physical Review Letters* 69: 1516–1519.

Bayer, M. and Forchel, A. 2002. Temperature dependence of the exciton homogenous linewidth in $In_{0.60}Ga_{0.40}As/GaAs$ self-assembled quantum dots. *Physical Review B* 65: 041308.

Becker, W. 2005. *Advanced Time-Correlated Single Photon Counting Techniques*. Berlin, Germany: Springer.

Betzig, E. and Chichester, R.J. 1993. Single molecules observed by near-field scanning optical microscopy. *Science* 262: 1422–1425.

Coldren, L.A. and Corzine, S.W. 1995. *Diode Lasers and Photonic Integrated Circuits*. New York: John Wiley & Sons.

Delhaye, M. and Dhamelincourt, P. 1975. Raman microprobe and microscope with laser excitation. *Journal of Raman Spectroscopy* 3: 33–43.

Demtroder, W. 1998. *Laser Spectroscopy*. Berlin, Germany: Springer-Verlag.

Diels, J.-Claude., and Rudolph, W. 2006. *Ultrashort Laser Pulse Phenomena*. Amsterdam, the Netherlands: Elsevier (Academic Press).

Drexler, H., Leonard, D., Hansen, W., Kotthaus, J.P., and Petroff, P.M. 1994. Spectroscopy of quantum levels in charge-tunable InGaAs quantum dots. *Physical Review Letters* 73: 2252–2255.

Gerardot, B.D., Seidl, S., Dalgarno, P.A. et al. 2007. Contrast in transmission spectroscopy of a single quantum dot. *Applied Physics Letters* 90: 221106.

Grilli, E., Guzzi, M., Zamboni, R., and Pavesi, L. 1992. High-precision determination of the temperature dependence of the fundamental energy gap in gallium arsenide. *Physical Review B* 45: 1638–1644.

Gruber, A., Drabenstedt, A., Tietz, C., Fleury, L., Wrachtrup, J., and von Borczyskowski, C. 1997. Scanning confocal optical microscopy and magnetic resonance on single defect centers. *Science* 275: 2012–2014.

Gustafsson, A. 1998. Local probe techniques for luminescence studies of low-dimensional semiconductor structures. *Journal of Applied Physics* 84: 1715–1775.

Hecht, E. 1998. *Optics*. Reading, Boston, MA: Addison-Wesley.

Hell, S.W. 2009. Microscopy and its focal switch. *Nature Methods* 6:24–32.

Hobbs, P.C.D. 2000. *Building Electro-Optical Systems: Making It All Work*. New York: John Wiley & Sons.

Kasai, J. and Katayama, Y. 1995. Low-temperature micro-photoluminescence using confocal microscopy. *Review of Scientific Instruments* 66: 3738–3743.

Kéna-Cohen, S., Davanço, M., and Forrest, S.R. 2008. Strong exciton-photon coupling in an organic single crystal microcavity. *Physical Review Letters* 101: 116401.

Kimble, H.J., Dagenais, M., and Mandel, L. 1977. Photon antibunching in resonance fluorescence. *Physical Review Letters* 39: 691–695.

Krauss, G., Lohss, S., Hanke, T. et al. 2009. Synthesis of a single cycle of light with compact erbium-doped fibre technology. *Nature Photonics* 4:33–36.

Kurtsiefer, C., Mayer, S., Zarda, P., and Weinfurter, H. 2000. Stable solid-state of single photons. *Physical Review Letters* 85: 290–293.

Levi, A.F.J., Slusher, R.E., McCall, S.L., Tanbun-Ek, T., Coblentz, D.L., and Pearton, S.J. 1992. Room temperature operation of microdisk lasers with submilliamp threshold current. *Electronics Letters* 28: 1010–1012.

Lidzey, D.G., Bradley, D.D.C., Skolnick, M.S., Virgili, T., Walker, S., and Whittaker, D.M. 1995. Strong exciton–photon coupling in an organic semiconductor microcavity. *Nature* 395: 53–55.

Lòpez, J.F.G. and Vos, W.L. 2002. Angle-resolved reflectivity of single-domain photonic crystals: Effects of disorder. *Physical Review E* 66: 036616.

Mandel, L. and Wolf, E. 1995. *Optical Coherence and Quantum Optics*. Cambridge, U.K.: Cambridge University Press.

McCall, S.L., Levi, A.F.J., Slusher, R.E., Pearton, S.J., and Logan, R.A. 1992. Whispering-gallery mode lasers. *Applied Physics Letters* 60:289–291.

Michler, P., Imamoglu, A., Mason, M.D. et al. 2000. Quantum correlation among photons from a single quantum dot at room temperature. *Nature* 406: 968–970.

Michler, P. 2003. *Single Quantum Dots*. Berlin, Germany: Springer-Verlag.

Murray, C.B., Norris, D.J., and Bawendi, M.G. 1993. Synthesis and characterization of nearly monodisperse CdE (E = S, SE, TE) semiconductor nanocrystallites. *Journal of the American Chemical Society* 115: 8706–8715.

Moerner, W.E. and Orrit, M. 1999. Illuminating single molecules in condensed matter. *Science* 283: 1670–1676.

Moerner, W.E. and Fromm, D.P. 2003. Methods of single-molecule fluorescence spectroscopy and microscopy. *Review of Scientific Instruments* 74: 3597–3619.

Moreau, E., Robert, I., Manin, L., Thierry-Mieg, V., Gerard, J.-M., and Abram, I. 2001. Quantum cascade of photons in semiconductor quantum dots. *Physical Review Letters* 87: 183601.

Novotny, L. and Hecht, B. 2006. *Principles of Nano-Optics*. Cambridge, U.K.: Cambridge University Press.

Paschotta, R. 2008. *Encyclopedia of Laser Physics and Technology*. Berlin, Germany: Wiley-VCH.

Rakher, M.T. 2008. Quantum optics with quantum dots in microcavities. PhD thesis, University of California, Santa Barbara, CA.

Ribordy, G., Gautier, J.-D., Zbinden, H., and Gisin, N. 1998. Performance of InGaAs/InP avalance photodiodes as gated-mode photon counters. *Applied Optics* 37: 2272–2277.

Robbins, M.S. and Hadwen, B.J. 2003 The noise performance of electron multiplying charge coupled devices. *IEEE Transactions on Electron Devices* 50: 1227–1232.

Santori, C., Fattal, D., Vuckovic, J., Solomon, G.S., and Yamamoto, Y. 2002. Indistinguishable photons from a single-photon device. *Nature* 419: 594–597.

Santori, C., Fattal, D., Vuckovic, J., Solomon, G.S., Waks, E., and Yamamoto, Y. 2004. Submicrosecond correlations in photoluminescence from InAs quantum dots. *Physical Review B* 69: 205324.

Siegman, A. 1986. *Lasers*. Sausalito, CA: University Science Books.

Srinivasan, K., Barclay, P.E., Painter, O., Chen, J., Cho, A.Y., and Gmachl, C. 2003. Experimental demonstration of a high-Q photonic crystal microcavity. *Applied Physics Letters* 83: 1915–1917.

Srinivasan, K., Painter, O., Stintz, A., and Krishna, S. 2007. Single quantum dot spectroscopy using a fiber taper waveguide near-field optic. *Applied Physics Letters* 91: 091102.

Strauf, S., Hennessy, K., Rakher, M.T. et al. 2006. Self-tuned quantum dot gain in photonic crystal lasers. *Physical Review Letters* 96: 127404.

Vurgaftman, I., Meyer, J.R., and Ram-Mohan, L.R. 2001. Band parameters for III-V compound semiconductors and their alloys. *Journal of Applied Physics* 89: 5815–5875.

Weisbuch, C., Nishioka, M., Ishikawa, A., and Arakawa, Y. 1992. Observation of the coupled exciton-photon mode splitting in a semiconductor quantum microcavity. *Physical Review Letters* 69: 3314–3317.

Yariv, A. and Yeh, P. 2007. *Photonics: Optical Electronic in Modern Communications*. New York: Oxford University Press.

Yuan, Z.L., Kardynal, B.E., Stevenson, R.M. et al. 2002. Electrically driven single-photon source. *Science* 295: 102–105.

Chapter 17

NEAR-FIELD SCANNING OPTICAL MICROSCOPY

Ben Mangum, Eyal Shafran, Jessica Johnston,
and Jordan Gerton

CONTENTS

In this chapter, we will present an overview of near-field scanning optical microscopy (NSOM), a technique that surpasses the optical diffraction limit through the use of sub-wavelength apertures and other structures to confine optical fields to nanometer-scale dimensions. The technique is a type of scanning-probe microscopy, having roots in the development of scanning tunneling microscopy (STM) and atomic force microscopy (AFM). The chapter will focus on tip-enhanced (apertureless) NSOM due to its superior resolution and the unique challenges associated with its implementation. Furthermore, there is already an extensive body of literature dealing with aperture-type NSOM (Novotny and Hecht, 2006).

17.1 BACKGROUND AND INTRODUCTION

The possibility of performing optical microscopy with spatial resolution far below the optical diffraction limit is extremely attractive. For example, as the fundamental building blocks of biological structure, proteins have a characteristic size of ~10 nm. Understanding how these building blocks fit and work together requires a technique with spatial resolution on this scale that is also sensitive enough to identify individual molecules within a dense ensemble. While electron microscopy, x-ray crystallography, and nuclear magnetic resonance spectroscopy can yield structural information with exquisite detail, they are generally not suited for studies of complex molecular networks due to limitations in sample preparation or measurement procedures and their inability to identify individual molecules within a heterogeneous ensemble. Furthermore, both electron and x-ray imaging techniques can damage samples to the extent that these techniques can be altogether unsuitable for biological samples. Traditional optical microscopy/spectroscopy, on the other hand, is sensitive enough to identify single molecules but lacks the required resolution. Thus, optical techniques that can provide resolution at the nanometer scale are likely to reveal new mechanistic details in biological systems. Of course, a wide variety of nonbiological systems will also benefit from such techniques.

17.1.1 Historical Context of Near-Field Optics

The beginnings of near-field optical microscopy can be traced to a 1928 letter from an Irish scientist, E. H. Synge, to Albert Einstein wherein he outlined a scheme for surpassing the optical diffraction limit by using a small molecule, or scatterer, to probe the optical properties of a proximate sample (Novotny and Hecht, 2006). In his letter, Synge suggested that if the scatterer and a sample could be maintained at a separation far below the wavelength of light, the evanescent or "near-field" components of the scattered light would interact with the sample, rather than the propagating or "far-field" components. By scanning the sample relative to the scatterer and collecting the light emitted by or transmitted through the sample, an image could thus be built up that would not be subject to the limits imposed by diffraction of the far-field components. This idea was quite remarkable and in hindsight relates to the very nature of light–matter interactions; although the spatial extent of a propagating photon is defined by its wavelength as given by Heisenberg's uncertainty principle, the electromagnetic coupling between light and a material occurs on a much smaller length scale (Gerhardt et al., 2007). The ability of near-field optics to directly probe these short-range interactions is one of its most outstanding features.

Synge's original letter elicited a lukewarm response from Einstein (Novotny, 2007b), which prompted him to develop a revised scheme whereby the scattering particle was replaced by a sub-wavelength aperture in a thin metal plate or film. Very close to the aperture, confined optical fields reflect the shape of the aperture itself, rather than the far-field diffraction pattern. In this vision, which is very close to modern implementations of aperture-type near-field optical microscopes, the light transmitted through the aperture would excite optical processes (e.g., fluorescence, Rayleigh scattering, Raman scattering, etc.) in a sub-diffraction volume, and the resulting optical signal would be collected with a lens (e.g., microscope objective) and finally detected in the far field with a camera or other light-sensitive detector. Although a sub-wavelength aperture will transmit merely a fraction of the incident light, only this transmitted light will excite the sample, so background signals will be low. With Einstein's encouragement, Synge published this idea in 1928 in the classic journal, *The Philosophical Magazine* (Synge, 1928). In modern implementations of near-field microscopy, the aperture can be used either to illuminate the sample, as described earlier, or to collect the scattered near-field signal induced by a far-field source (e.g., focused laser beam) or both.

17.1.2 Development of Aperture-Type Near-Field Microscopy

Those familiar with near-field optics will recognize that both of Synge's ideas, namely, using a scatterer or an aperture to surpass the optical diffraction limit, have developed into distinct implementations of NSOM—apertureless and aperture-type NSOM, respectively. As Einstein predicted, aperture-type NSOM is, in fact, more easily realized, and thus, it was demonstrated first and was first to be developed into a commercial product (Figure 17.1). The first experimental demonstration was made with microwaves by Ash and Nichols in 1972. They used a 1.5 mm aperture and 10 cm waves to demonstrate a spatial resolution of $\sim\lambda/60$. The long delay between Synge's proposal and its first implementation was due to the extremely challenging technical difficulties associated with two key aspects of the technique: (1) maintaining a sub-wavelength vertical distance between the aperture and the sample surface and (2) scanning the sample or aperture laterally relative to the other with sub-wavelength precision. These tasks are obviously more difficult for optical wavelengths (400–800 nm), and not until the advent of scanning probe microscopy techniques in the early 1980s did it become possible to regulate the distance between a probe and a sample with sufficient precision. Finally in 1984, Pohl, Denk, and Lanz at IBM Rüschlikon achieved $\sim\lambda/20$ (25 nm) spatial resolution at $\lambda = 488$ nm with aperture-type NSOM (Pohl et al., 1984), while a group led by Aaron Lewis at Cornell independently achieved spatial resolution of $\sim\lambda/16$ (30 nm) at $\lambda = 488$ nm (Lewis et al., 1984).

In subsequent years, the capabilities of aperture-type NSOM were improved steadily by a number of groups and commercial instruments started appearing on the market in the mid 1990s. Although it is possible, in principle, to fabricate increasingly smaller aperture sizes, the resolution of aperture-type NSOM eventually becomes limited by two factors. First, the electromagnetic field confined within the metal-coated probe penetrates into the metal up

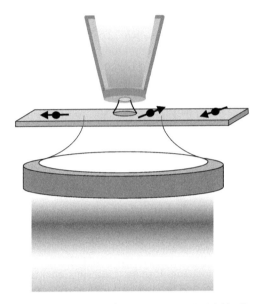

Figure 17.1 A modern implementation of aperture-type NSOM. The aperture is fabricated on the end of a metal-coated tapered optical fiber, which can be scanned over a sample surface with high precision.

to a characteristic length called the skin depth, $\delta = (\lambda/4\pi\sqrt{\varepsilon})$, where ε is the wavelength-dependent dielectric function of the metal (Chapter 1). Aluminum typically has the smallest skin depth in the visible regime, ~10 nm. Therefore, no matter how small its diameter, an aperture in an aluminum film has an effective size of at least twice the skin depth, or ~20 nm, thus setting the lower limit for spatial resolution. Secondly, the transmission of light through an aperture decreases with diminishing aperture diameter. For realistic experimental parameters, the transmission drops by almost a billion times for a 20 nm diameter aperture compared to one of 100 nm (Novotny and Hecht, 2006). Furthermore, light penetrating into the metal coating is partially absorbed within the skin-depth layer. The absorptive heating within the metal can become excessive to the point of destroying the tip if too much optical power is injected into the aperture (Stockle et al., 1999). Thus, for apertures smaller than 50–100 nm, the amount of light emerging from the probe is severely limited, making the optical signal too small to be practicably detected. Due to these limitations, the typical aperture size employed in NSOM is at least 50 nm and more characteristically 100 nm. In recent years, a number of efforts have attempted to increase the transmission through apertures by engineering their detailed shape (Thio et al., 1999). In particular, two-dimensional (2D) standing-wave resonances of surface plasmons, or localized surface plasmon polaritons, can be used to enhance optical transmission efficiency, thereby providing the potential for smaller effective aperture sizes and thus improving spatial resolution (Thio et al., 1999; Darmanyan and Zayats, 2003).

17.1.3 Development of Apertureless NSOM

To improve upon aperture-type NSOM resolution, apertureless techniques were developed whereby the aperture is replaced by a nanostructure, such as the tip of an AFM probe or a metal nanosphere, which is then illuminated with a diffraction-limited light source. Under appropriate conditions, the nanostructure redistributes the energy within the illumination profile and introduces steep gradients in the optical intensity, generally resulting in a region of elevated intensity at the distal end of the tip and an accompanying reduction in the intensity at

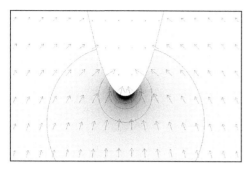 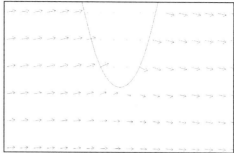

Figure 17.2 2D intensity plot showing a numerical simulation of a dielectric material in an otherwise uniform electric field. Here, a conical probe with a permittivity of 17.6 + 0.1i (Si at 532 nm) was used for the calculation. Darker colors indicate regions of higher electric field strength. The left panel shows a probe placed in a vertically oriented field. The right panel shows a probe placed in a horizontally oriented field. Note that there is a slight reduction in intensity near the apex compared to the far-field value. Contour lines show intensity intervals varying by a factor of 3. Arrows scale with the intensity and point in the direction of the field Δ.

other locations along the structure's surface. This region of enhanced optical intensity can then be used to modulate an optical signal, (e.g., fluorescence, Raman scattering, Rayleigh scattering, etc.), which is generally detected in the far-field using a standard lens-based microscope system. An image is obtained by correlating the detected signal with the relative position of the tip and sample while the tip or sample is rastered laterally. Since the region of enhanced intensity is generally confined to a very small volume surrounding the tip apex (Figures 17.2 through 17.4), an image with both far-field and near-field components is produced. The spatial resolution of the near-field component is generally limited only by the tip sharpness (Gerton et al., 2004) and can be as small as 5–10 nm.

In the most general sense, field enhancement results simply from a discontinuity in a component of the electric field amplitude on either side of an interface with different dielectric constants, as required by Maxwell's equations. In particular, the component perpendicular to the interface is larger in the region with lower dielectric constant since the electric

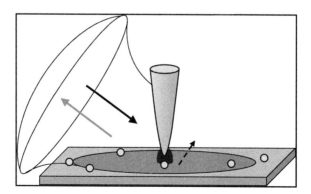

Figure 17.3 Schematic of the basic principle of apertureless NSOM. Solid arrows indicate direction of excitation light (dark arrow) and near-field signal (light arrow). The dashed arrow represents the polarization direction of excitation light. The small spheres generically represent nanoscale sample features, including individual particles or molecules. Note that it is possible to simultaneously illuminate several features with the far-field excitation source, while ideally only one feature is probed by the enhanced near field of the tip.

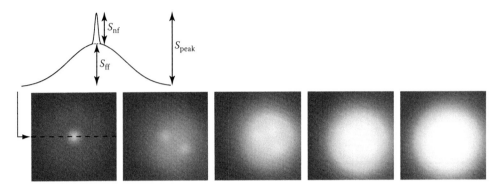

Figure 17.4 Simulated superposition of near-field and far-field signal components in apertureless NSOM. The signal profile shows the near-field signal (S_{nf}) superimposed onto the far-field background signal (S_{ff}). The image contrast, defined as the ratio (S_{nf}/S_{ff}), decreases from left to right as the number of molecules within the far-field illumination spot increases from one to five.

displacement field (Chapter 1) must be continuous across this interface (Jackson, 1999). This effect is amplified in regions of high curvature where the field line density is largest. Thus, by utilizing nanostructures properly aligned with an incident field, the enhancement can be substantial. This can be easily observed through the well-known example of a dielectric particle in an otherwise uniform electric field, as seen in Figure 17.2. This redistribution of electric field is known as the lightning-rod effect, where fields are enhanced most strongly at the regions of highest curvature. In metals, surface plasmons can further augment this field enhancement. Much recent work has been dedicated to optimizing the size and shape of such plasmonic nanoantennas for maximum field enhancement at particular wavelengths, as discussed in more detail in Section 17.2.

A generalized schematic of a modern-day implementation of apertureless NSOM showing a possible illumination and detection scheme is shown in Figure 17.3. Importantly, the incident field must have a strong component along the long axis of the probe (as in the left panel of Figure 17.2), which can be accomplished by illuminating from the side at shallow angles with linearly polarized light. In such a configuration, the probe is maintained in the center of the excitation beam while the sample can scan relative to the probe. Since both the tip and sample are illuminated by the excitation beam, both near-field and far-field signals are collected, resulting from the tip–sample interactions and direct far-field sample excitation, respectively.

One of the major technical hurdles in apertureless NSOM is the signal background caused by direct excitation by the illumination source. This background generally decreases image contrast and becomes more problematic for dense and/or complex samples (e.g., biological systems) for which an optical response is expected over the entire sample surface. Figure 17.4 illustrates this problem with a simulation where an increasing number of fluorophores are illuminated by the far-field background, while near-field signals come from one fluorophore at a time. Several strategies have been developed in recent years to overcome the background problem. Among the most notable is the use of nonlinear scattering processes (e.g., two-photon fluorescence) such that the optical signal adopts a higher order dependence on the optical intensity, thereby reducing the relative importance of the far-field background compared to the enhanced signal (Novotny, 2002). Another widely used strategy is to modulate the near-field signal, for example, by oscillating the nanostructured tip above the sample surface, followed by demodulation to suppress the background. Demodulation can be accomplished by interfering the scattered (near-field) light with a reference laser beam (in the case of elastic (Rayleigh) light scattering) or Fourier analysis and/or lock-in amplification of the detected optical signal. Modulation/demodulation strategies will be discussed in more detail in Sections 17.2 and 17.3.

17.1.4 Tip-on-Aperture Approach

Recently, Guckenberger and coworkers developed an approach that combines the benefits of aperture-type NSOM (low background) with apertureless NSOM (high resolution) (Frey et al., 2002, 2004). This is accomplished by fabricating a nanoscale metal tip at the periphery of an aperture-type probe. The tip is illuminated through the aperture, instead of a diffraction-limited far-field source, which greatly reduces the background. The light emerging from the aperture excites surface plasmons (Chapter 3) that travel from the base to the end of the tip, where they scatter and are reconverted to light. These tip-on-aperture probes were used to image individual fluorescent molecules dispersed on a glass surface with ~25 nm resolution. This approach has now been extended by van Hulst and coworkers (Taminiau et al., 2007). Using fluorescence from single molecules, they measured strong resonances in the strength of the field enhancement as a function of the length of the metal tip.

The tip-on-aperture approach, though promising, has a number of limitations. First, the probes are difficult to fabricate, requiring specialized techniques such as focused ion-beam milling and/or electron beam deposition. In addition, it seems the tips may be easily damaged even on nano-structured surfaces. Finally, the fabricated probes are not as sharp as commercial ones. As a consequence, the resolution of the tip-on-aperture approach (~25 nm) lags that of apertureless NSOM ($\leq$10 nm). As the tip-on-aperture approach is a very specialized technique being pursued only by a handful of experts worldwide, it will not be discussed further in this chapter. For more information, the reader is referred to the primary literature on the subject (Frey et al., 2002, 2004; Taminiau et al., 2007, 2008).

17.1.5 Topographical Artifacts in NSOM

Because NSOM is a scanning probe technique, it is subject to topography-induced artifacts. These artifacts are particularly common in aperture-type NSOM, where the optical signal arises from light emerging from the aperture, while the topography signal arises from the mechanical contact point between the probe and sample. The contact point may be a small protrusion or imperfection on the rim of the aperture, in which case the aperture would be lifted up by the height of the protrusion. For smooth samples, this obviously leads to a reduced optical signal, since the evanescent field emerging from the aperture weakens substantially with distance, but does not lead to imaging artifacts. For samples with substantial topographical features, however, imaging artifacts can result when the aperture is lifted off the surface as the aspersion scans over a topographical feature. For example, imagine that the optical signal is collected in transmission through the sample and that the sample features are reflective. As the aperture scans over a feature larger than its own diameter, the optical signal will decrease, yielding the lateral size of the feature itself. However, if the feature is substantially smaller than the aperture, the shape of the aperture will be reported. Additionally, as the tip scans over the feature, the aperture will get lifted further off the surface resulting in another reduction in the optical signal. This second feature is artifactual and does not reflect the optical properties of the sample, but rather its topography. For probes with multiple aspersions around the aperture rim the situation can become fairly complicated. These topography-coupled imaging artifacts can be reduced if the NSOM is operated such that the probe is scanned laterally at a constant height above the sample. In this case, the multiple-feature artifacts discussed earlier are suppressed, but the topography signal is not captured at all, resulting in a substantial loss of information. For more detailed information regarding topographical artifacts in aperture-type NSOM, the reader is referred to (Hecht et al., 1997; Novotny and Hecht, 2006).

These types of imaging artifacts are much less common in apertureless NSOM because the near-field optical signal primarily originates from the region on the probe that is closest to the sample, which is also generally the mechanical contact point for the topography signal. Still, optical images generated with apertureless NSOM can nevertheless contain features not directly related to the near-field coupling between the tip and sample and thus require substantial interpretation. Recently, the authors developed a technique that captures the full three-dimensional dependence of the optical signal on the tip–sample separation, which can, in principle, be used to eliminate those features not directly related to the near-field tip–sample coupling (Mangum et al., 2009), as discussed in Sections 17.3 and 17.4.

As previously mentioned, apertureless NSOM is currently a more active area of research compared to aperture-type techniques. For this reason, and because it is generally more difficult to implement, apertureless NSOM will be the focus of the remaining portion of this chapter.

17.2 THEORETICAL FOUNDATIONS

In apertureless NSOM, a diffraction-limited illumination source, usually a focused laser, is used to polarize the tip of a scanning probe, which then alters the local optical response of the sample. The focused laser also produces a direct optical response from the sample, leading to an image consisting of the superposition of near-field and far-field components, as illustrated in Figure 17.4. The separation between the near-field and far-field components becomes progressively more difficult as the complexity of the sample increases because the far-field background grows in proportion to the density of optically active sites, whereas the near-field signal stays roughly constant. There are two ways to deal with the background problem in apertureless NSOM: (1) to enhance the near-field component and (2) to suppress the far-field component. Both of these strategies will increase image contrast:

$$\text{Contrast} = \frac{S_{\text{nf}}}{S_{\text{ff}}}. \tag{17.1}$$

Note that the contrast can be negative if the near-field signal (S_{nf}) is negative, corresponding to a tip-induced reduction in the optical signal, such as fluorescence quenching (see below).

In general, the near-field component is the result of a competition between tip-induced enhancement of the optical field and various tip-induced suppression mechanisms. This section will discuss two field-enhancement mechanisms, namely, the lightning-rod effect and localized surface plasmon resonances. These two mechanisms are strongly dependent on the precise geometry of the near-field probe as well as on the tip material and the illumination wavelength and polarization. Tip-induced suppression mechanisms include fluorescence quenching and redirection of emission, as well as interference between excitation of a sample directly from the laser beam and indirectly via scattering from the tip. Of these, fluorescence quenching is the most important and thus will be the only one discussed in detail.

Modulating the near-field component of the optical signal, for example, by oscillating the tip above the sample surface, does not affect the far-field component. Thus, subsequent demodulation of the optical signal at the appropriate frequency suppresses the far-field background, which increases image contrast. The following section will discuss the theoretical increase in image contrast that can be achieved using a simple modulation/demodulation technique to suppress the background signal, without discussing a particular implementation of this method, which is left for Section 17.3.

17.2.1 Tip-Induced Enhancement of the Near-Field Signal

17.2.1.1 The Lightning-Rod Effect

When a dielectric material is placed in a uniform electric field, the applied field polarizes the material as the electrons and ions migrate toward their respective sides of the dielectric (Jackson, 1999). This charge separation creates an induced electric field, and the total field near the surface of the material is the superposition of both the applied field and the induced field. For a dielectric sphere, the induced field can be obtained analytically, as shown in Figure 17.5. Here, a uniform static field is applied along the vertical axis, but the calculation is also valid for an oscillating field with vertical polarization if the size of the sphere is much smaller than the wavelength. In this quasi-static approximation, retardation effects can be neglected and, at each point in time, the applied field can be considered uniform. At the (vertical) poles of the sphere, the total electric field is *enhanced* relative to the applied field, while the total field is reduced along the equator.

The maximum electric field at the poles of the sphere is given by

$$E_{peak} = E_0 \left(1 + 2 \frac{\varepsilon_r - 1}{\varepsilon_r + 2} \right), \tag{17.2}$$

where

E_0 is the applied field

ε_r is the permittivity of the sphere relative to that of the surrounding medium: $\varepsilon_r = \varepsilon_{dielctric}/\varepsilon_{media}$

In principle, a dielectric sphere can be used as an apertureless-NSOM probe if it can be scanned in close proximity to a sample. In this case, the enhanced field at the distal pole of the sphere can increase the optical response. As this response is generally proportional to the optical intensity or higher orders thereof depending on the particular scattering process (Novotny, 2002), the expression in Equation 17.2 must be raised to an appropriate power to find the expected enhancement in the scattering rate. For dielectric materials, the peak intensity can be enhanced by at most a factor of nine (when $\varepsilon_r \to \infty$) for this spherical geometry.

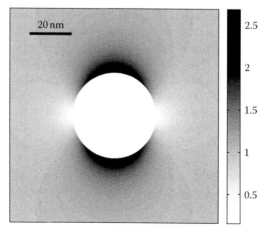

Figure 17.5 Electric field near a dielectric sphere. Darker colors indicate regions of higher electric field strength.

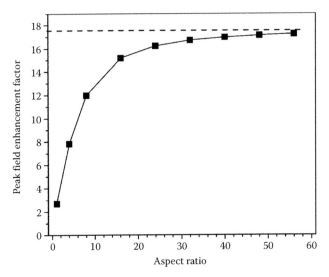

Figure 17.6 A 3D electrostatic finite element calculation of the maximum field enhancement for prolate spheroids of varying aspect ratios. Spheroids were given the permittivity of silicon (17.6 + 0.1i at 532 nm). Aspect ratio is calculated as the semimajor axis divided by the semiminor axis of the prolate spheroid (short axis was kept constant). The incident field is applied parallel to the long axis of the spheroid. The dashed line is the theoretical limit as the ratio approaches infinity, as found in Equation 17.3.

Although the spherical geometry can be solved analytically, it often does not accurately approximate the shape of many AFM tips. Furthermore, elongated geometries can yield significantly larger field enhancement. Figure 17.6 demonstrates this effect by plotting the maximum intensity enhancement from a 3D finite element calculation of prolate spheroids of increasing semi-axis ratios. In fact, Bohn et al. have shown that as the long axis of the spheroid approaches infinity, the intensity enhancement adopts the analytic form (Bohn et al., 2001):

$$E_{\text{peak}} = E_0 \cdot \varepsilon_r \Rightarrow I_{\text{peak}} = E_0^2 \cdot \varepsilon_r^2, \qquad (17.3)$$

which is reproduced well in the numeric calculations shown in Figure 17.6. This geometry-dependent electric-field enhancement, where the highest enhancement values occur in regions of highest curvature is known as the *lightning-rod effect*. It is important to remember that Equations 17.2 and 17.3 are strictly valid for static electric fields and thus can only be used for optical fields (resulting from either continuous wave or pulsed lasers) when retardation effects can be neglected, namely, when the size of the object is much smaller than the light wavelength. This, of course, prohibits rigorous application of Equation 17.3 for commercial AFM tips in NSOM, as the entire structure of a commercial tip, while very sharp at the apex, is an extended structure many times larger than optical wavelengths (~10 μm).

Nothing prohibits applying Equations 17.2 and 17.3 to metal nanoparticles as long as the size of the particle is no larger than the optical skin depth, $\delta = \lambda/(4\pi\sqrt{\varepsilon})$; otherwise, the bulk electrons are shielded and only the surface electrons experience the external driving field. In general, the dielectric function is complex valued and also frequency dependent, $\varepsilon_r(\omega) = \varepsilon_r'(\omega) + i \cdot \varepsilon_r''(\omega)$ equivalently, this is sometimes written as $\varepsilon(\omega) = \varepsilon_1(\omega) + i \cdot \varepsilon_2(\omega)$ (Chapter 1). Here, the imaginary part relates to light absorption, which ultimately results in energy loss through ohmic dissipation (Joule heating) (Anger et al., 2006). Metals generally have negative values for the real parts of their dielectric function throughout the visible spectrum (Chapter 3); this raises the possibility of exciting plasmon resonances (e.g., $\varepsilon_r = -2$ in Equation 17.2), which can massively increase the enhancement factor (see below).

More precise predictions of the field enhancement for realistic tip geometries, such as the conical or pyramidal tips that are commercially available, can be obtained by solving Maxwell's equations on a discrete grid using finite element analysis programs (Issa and Guckenberger, 2007; Taminiau et al., 2008). Alternatively, a discrete dipole approximation (DDA) (Kelly et al., 2003) or multiple multipoles method can also be used (Novotny et al., 1997). Such calculations have predicted intensity enhancement factors in excess of 1000 for metal tips and around 225 for dielectric tips (Novotny et al., 1997; Bohn et al., 2001). Such large enhancement factors, however, have never been observed; to the best of our knowledge, the largest *signal* enhancement factor reported for silicon tips is ~20 for fluorescence measurements of 4 nm diameter quantum dots (Gerton et al., 2004). While this represents a lower limit for the intensity enhancement factor, due to incomplete spatial overlap between the field enhancement volume and the sample, it falls well below predictions (Bohn et al., 2001). This discrepancy could be due to either rapid growth of oxide layers, which have a smaller permittivity, or irregular geometry.

17.2.1.2 Plasmon Resonances

Surface plasmons (Chapter 3) play a large role in near-field optics, as a material excited resonantly can generate enormous field enhancements beyond those expected from the lightning-rod effect alone. Optimal field enhancement requires the right combination of material, excitation wavelength, and geometry, which is also extremely important. For instance, if a sub-wavelength-sized metal sphere is illuminated with an excitation source near its plasmon resonance frequency, localized surface plasmons can greatly increase the enhancement factor. As the spherical particle is elongated along its polarization axis, the field can be enhanced even further (Sonnichsen et al., 2002). Moreover, given a certain geometry, the particle size also matters: if the particle is larger than the skin depth, the inner electrons will be shielded, resulting in reduced enhancement. For smaller particles with large surface area-to-volume ratios, electron collisions with the surface become a large source of plasmon damping, thus reducing plasmonic field enhancement (Link and El-Sayed, 1999; Mangum et al., 2009).

Bulk and surface plasmons can also play a major role in energy transfer, as discussed in more detail later in this chapter. In particular, a photoexcited particle (e.g., quantum dot or fluorophore) that would normally relax via radiative channels (fluorescence), can instead non-radiatively transfer its internal energy to a nearby plasmon-active structure. This results in a sharp reduction in the detected fluorescence rate (quenching), and also in the fluorescence lifetime (Yang et al., 2000; Trabesinger et al., 2002), as non-radiative decay channels become predominant. Recently, several groups have directly observed the competition between field enhancement and fluorescence quenching when using metal tips in NSOM (Anger et al., 2006; Kuhn et al., 2006; Bharadwaj and Novotny, 2007; Issa and Guckenberger, 2007; Mangum et al., 2009). The net signal depends on the details of this competition, which is a function of many parameters such as the tip geometry, material and size, and the light wavelength and polarization.

17.2.1.3 Optical Antennas

The combination of the lightning-rod effect and plasmon resonances naturally leads to the concept of designing nanostructures with strong, shape-specific resonances to drastically enhance the optical field. This is, in fact, a description of an optical antenna, which like their radio or microwave analogs, can be used to convert freely propagating electromagnetic waves into localized fields and vice versa. While any near-field probe can be considered an antenna inasmuch as it can locally focus light, we will more rigorously use the term "antenna" to describe a device that exhibits shape-specific resonances, which implies that they are made of metal.

There are a number of difficult challenges associated with scaling down antennas from the macro to the nanoscale as needed for optical field enhancement. For example, it is difficult to fabricate structures of this size using conventional lithography, so specialized techniques such as focused ion beam milling or electron beam lithography must be used. Importantly, complicated antenna geometries can be difficult to implement in NSOM because the antenna must be attached to the near-field probe in such a way that the largest field is at the apex of the tip. Furthermore, at optical frequencies, charge transport in nanoscale metallic structures can suffer from a number of damping mechanisms, in contrast to the ideal conductors envisioned in antenna theory for the microwave and radio wave regions of the electromagnetic spectrum. Thus, the design of efficient antennas at optical frequencies requires new theories, or at least rigorous adaptation of existing microwave theories, to account for this nonideal behavior.

At optical frequencies, the skin depth of a metal can be of the same order of magnitude as the antenna size. The penetration of electromagnetic waves into the antenna creates electron oscillations inside the metal, which tends to push the antenna resonance toward a higher frequency, and thus a shorter effective wavelength. For example, van Hulst and coworkers have shown that the resonant length of a linear monopole antenna is significantly shorter than the $\lambda/4$ predicted for an ideal conductor using classical antenna theory (Taminiau et al., 2007). Novotny has also modeled this effective wavelength shortening theoretically (Novotny, 2007a), which is very important for antenna design as it implies that optimized antenna sizes should be shorter than what traditional antenna theories project and are dependent on the shape of the tip and the properties of the metal.

A number of antenna designs have been used in NSOM to obtain large and confined field enhancement, thus obtaining optical resolution beyond the diffraction limit. For example, a $\lambda/4$ monopole antenna was accomplished with the tip-on-aperture approach developed by Guckenberger and coworkers (Frey et al., 2002, 2004; Taminiau et al., 2007). In this work, the antenna resonances were mapped by scanning an antenna over a single molecule while monitoring its rate of fluorescence emission and repeating the experiment for different length antennas. The observed fluorescence rate for similarly oriented molecules increased dramatically for the optimal antenna length. In a second experiment, the polarization of the florescence emission from single molecules was monitored while scanning an antenna in close proximity to the sample plane (Taminiau et al., 2008). When the antenna was directly over a molecule, its emission pattern changed to that of the coupled antenna-molecule system, illustrating that it is possible to redirect the dipole emission of a single quantum emitter to match that of a near-field antenna.

Another simple antenna geometry commonly employed is a single gold nanosphere attached to the end of a dielectric probe, such as a pulled glass fiber or an AFM tip (Anger et al., 2006; Kuhn et al., 2006). The spherical geometry yields plasmon resonance modes with strong field enhancement at the poles of the sphere along the polarization direction. These antennas have been used as described earlier to image single molecules and to study the competition between field enhancement and fluorescence quenching. As above, the emission rate of single molecules was recorded as the spherical nanoantenna was scanned in close proximity. As the antenna approached a molecule, the emission of the molecule initially increased due to field enhancement. At very short range (~10 nm), fluorescence quenching overwhelmed this enhancement, leading to a reduction in signal. Under similar illumination conditions using nonresonant, gold-coated AFM tips, only quenching was observed, demonstrating the importance of resonance effects (Mangum et al., 2009).

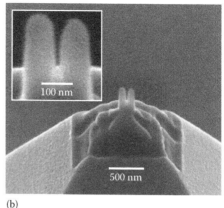

(a)   (b)

Figure 17.7 Electron micrograph of an aluminum bowtie antenna fabricated on an AFM probe (a) Top view (b) Side view. (Reprinted with permission from Farahani, J.N. et al., *Phys. Rev. Lett.*, 95, 4, 2005. Copyright 2005 American Physical Society.)

A more efficient antenna geometry is the "bowtie," where two triangular prisms face each other point-to-point with a small "feed" gap in between, as shown in Figure 17.7. The sharp corners of the triangles lead to strong lightning-rod field enhancement, which is augmented by a localized plasmon resonance at the appropriate wavelength. For small feed gaps, the two triangles strongly couple, leading to a shift in the combined plasmon resonance and further enhancement in the field within the gap. Farahani et al. (Farahani et al., 2005) have fabricated a bowtie antenna at the apex of an AFM tip from aluminum with a 40 nm feed gap. This antenna was scanned over individual semiconductor quantum dots while monitoring the fluorescence emission rate and lifetime. When the single emitter was directly below the feed gap, the antenna modified the radiative properties of the emitter, corresponding to a maximum count rate and a reduction of the excited-state lifetime. As a control, an aluminum-coated tip without the bowtie geometry also led to a decrease in the excited-state lifetime, but in this case no fluorescence enhancement was observed, as quenching was the dominant interaction.

17.2.2 Fluorescence Quenching

When an apertureless-NSOM probe is applied to a fluorescent sample, the field enhancement mechanisms discussed earlier can cause an increase in the detected fluorescence signal. Simultaneously, the tip can also open up additional channels for photoexcited fluorophores to relax back to the electronic ground state non-radiatively, thereby quenching the fluorescence. This is particularly important for metallic NSOM probes, which can respond to a wide range of wavelengths via dipole–dipole coupling similar to fluorescence (Förster) resonance energy transfer (FRET). In this process, it is thought that energy is transferred from fluorophore to tip via exchange of a virtual photon, which in turn excites plasmon waves in/on the metal. These plasmons can then dissipate the energy rapidly as heat within the tip. Thus, as a tip approaches a fluorescent sample, the local non-radiative relaxation rate (Γ_{NR}) increases, decreasing the apparent quantum yield, and results in a suppression of the detected emission (Novotny, 1996; Carminati et al., 2006). This fluorescence quenching may be accompanied by a change in the radiative rate, (Γ_R), where both Γ_{NR} and Γ_R depend on the orientation of the molecule transition dipole moment relative to the probe geometry (Carminati et al., 2006; Vukovic et al., 2008). This means that the total fluorescence lifetime, $\tau = (\Gamma_{NR} + \Gamma_R)^{-1}$, is altered near a metal surface. By using pulsed lasers and time-correlated single photon counting (TCSPC)

(Chapter 12), it is possible to measure τ directly, even for single molecules, by building up a histogram of fluorescence photon delay time following a laser pulse. This lifetime can then be plotted pixel by pixel to build up an image by scanning the NSOM probe across the surface, where the value of each pixel represents the fluorescence lifetime of the corresponding location. Since a metal tip will alter the lifetime, it can be used as a contrast mechanism in NSOM, as has been demonstrated previously (Trabesinger et al., 2002; Hu et al., 2003; Yoskovitz et al., 2008). This type of imaging can provide a great deal of information about near-field interactions between the tip and sample as it provides simultaneous topography, fluorescence, and lifetime data.

There are two general cases to consider with regard to quenching with metal tips: (1) quenching by tips with well-defined, closed geometries such as the spherical, bowtie, or monopole antennas described earlier; and (2) quenching by tips with open geometries such as the metal-coated pyramidal AFM probes available commercially. Closed geometries support localized plasmons with well-defined and relatively narrow resonance frequencies determined by the detailed geometry and material of the probe. In this case, the fluorescence quenching efficiency should depend very sensitively on the *emission* wavelength of the fluorophore, with maximum quenching at the plasmon resonance frequency. For such closed geometry tips, however, the field enhancement is also strongly wavelength dependent, so the competition between enhancement and quenching, and thus the net fluorescence signal, is delicately balanced and can be difficult to predict. Molecular fluorophores exhibit relatively small Stokes shifts; consequently, enhancement and quenching can both be quite strong near the plasmon resonance. Colloidal quantum dots, on the other hand, have a broad absorption profile extending into the UV. Thus, it is possible to have a large imbalance between field enhancement and quenching, as the excitation laser can be tuned far to the blue of a plasmon resonance, while the quantum dot emission can be at the resonance frequency. Varying the excitation wavelength toward the resonance frequency would then make it possible to study the competition between quenching and enhancement in detail. To our knowledge, such an experiment has not yet been reported.

For tips with open geometries, such as metal-coated pyramidal or conical probes, the situation can be quite different. These probes do not support localized plasmon resonances, but can still dissipate energy via damping of traveling plasmon waves. Thus, the balance between quenching and enhancement should be generally biased toward quenching in this case. The difference between the response of the fluorescence signal using tips with closed and open geometries has been demonstrated in a number of recent experiments. For example, Novotny and coworkers (Anger et al., 2006; Bharadwaj et al., 2007) have shown that for spherical gold tips, field enhancement is clearly evident leading to a strong enhancement in the fluorescence signal from single molecules, which is mitigated by quenching only when the tip is brought within ~5 nm of a molecule. Using gold-coated pyramidal tips, on the other hand, switches the balance strongly in favor of quenching, often leading to a complete lack of observable enhancement in the fluorescence signal at any distance scale, as demonstrated by a number of groups (Yang et al., 2000; Trabesinger et al., 2002; Kuhn et al., 2006; Mangum et al., 2009).

In any NSOM experiment, it is important to remember that the effect of the near-field probe on the sample can be nontrivial, especially for metallic probes. In particular, for fluorescent samples, the net detected signal is affected by field enhancement, quenching, and other possible mechanisms such as the redirection of fluorescence. Thus, determining the field enhancement factor of a particular probe is complicated in that it may be impossible to decouple any observed increase in fluorescence signals with any quenching that may also be occurring. In general, this would require a rather sophisticated model of the system, which is highly dependent on the probe geometry and material.

17.2.3 Suppressing the Far-Field Background Signal

The previous discussion has focused on tip-induced changes in the *near-field* signal, for both enhancement and quenching. As Equation 17.1 indicates, however, apertureless NSOM contrast can also be affected by suppressing the *far-field background* signal. There are two primary strategies to accomplish this: (1) to use nonlinear scattering processes (e.g., two-photon fluorescence) such that the signal adopts a higher order dependence on the optical intensity, thereby reducing the relative importance of the far-field background compared to an enhanced near-field signal, and (2) to modulate the near-field signal, for example, by oscillating the nanostructured tip above the sample surface, followed by demodulation to suppress the background.

17.2.3.1 Use of Nonlinear Scattering to Increase NSOM Contrast

If we assume a planar sample with a uniform coverage of molecules that are to be imaged, then the detected signal corresponding to the far-field illumination source will be

$$S_{ff} = \kappa A\, I_0^n, \tag{17.4}$$

where
  κ is a scaling factor that characterizes the total detection efficiency of the microscope
  A is the area of the laser focus spot at the sample plane
  I_0 is the (average) intensity of the laser across A
  n is the order of the scattering process used in the experiments (e.g., $n = 1$ for one-photon fluorescence, $n = 2$ for two-photon fluorescence, etc.)

Note that the most relevant measure of the background signal does not scale with the illumination area A but rather with the number of molecules within this area (see below). For the purpose of this discussion, however, these two quantities are proportional since we are assuming uniform coverage. The near-field signal originates from those molecules in a much smaller area near the tip, a, where the optical intensity is enhanced by a factor f:

$$S_{nf} = \kappa a\, (f I_0)^n. \tag{17.5}$$

Thus, the image contrast in this case is given by

$$C = \frac{S_{nf}}{S_{ff}} = \frac{a}{A}\, f^n. \tag{17.6}$$

Equivalently, the enhancement factor required for contrast greater than unity is $f > (A/a)^{1/n}$ (Novotny, 2002).

Clearly, the image contrast will improve for higher order processes. Ideally, the near-field illumination area is given by the tip sharpness, $a \sim (10\,\text{nm})^2$, and the far-field area is diffraction limited, which for a high numerical-aperture lens (NA ~ 1.4) is roughly $A \sim (250\,\text{nm})^2$. Inserting these numbers into Equation 17.6 yields the condition, $f > (625)^{1/n}$, to achieve contrast greater than unity. Note that while this prediction formally applies to the *intensity* enhancement factor, the analysis neglects tip-induced suppression of the signal (e.g., fluorescence quenching). To account for this, we can reinterpret f as the *signal* enhancement factor that results from the competition between tip-induced enhancement and suppression mechanisms.

Thus, for linear processes ($n = 1$) such as one-photon fluorescence, a signal enhancement factor greater than ~600 is needed to achieve sufficient contrast when imaging a uniformly distributed sample. This is well beyond any *observed* enhancement factors for either metal or dielectric (silicon) tips. For a second-order process the required enhancement is only ~25, a value that has been demonstrated for both metal and silicon tips using first-order processes (Gerton et al., 2004; Anger et al., 2006; Kuhn et al., 2006).

As indicated, the previous analysis is valid only for samples with a uniform distribution of particles/molecules where the implied spacing between them is smaller than the near-field resolution, which can be as small as ~10 nm. A more practical approach is to cast the criteria given earlier in terms of the number of molecules, N_A, within the far-field focus area, A. This is more applicable for samples that are not uniform, such as systems composed of linear arrays of fluorophores (e.g., polymers, J-aggregates, etc.) or ones that do not fill up the entire focal area. In addition, it is only of practical interest to consider samples where the minimum interparticle spacing is equal to the near-field resolution, since otherwise adjacent particles will not be resolved. This adjustment is easily made by the substitutions $a \rightarrow 1$ and $A \rightarrow N_A$ in Equations 17.4 through 17.6. Thus, the minimum enhancement factor now becomes

$$f > (N_A)^{1/n}. \tag{17.7}$$

17.2.3.2 Use of Signal Modulation to Increase NSOM Contrast

Although instructive, the previous analysis is incomplete in several respects. For example, it predicts that image contrast is independent of the illumination intensity, I_0, which is clearly unphysical in the limit $I_0 \rightarrow 0$. Most importantly, it neglects the possibility of improving contrast by employing appropriate data acquisition and analysis schemes. Below, we discuss a method commonly used in electronic signal processing, and more particularly in NSOM, to extract small signals from large backgrounds, namely, signal modulation (Knoll et al., 1999; Hillenbrand et al., 2000). Here, the discussion will be completely generic, but in Section 17.3, we will present a particularly powerful implementation of this method called single-photon near-field tomography (SP-NFT) (Mangum et al., 2009), which has been demonstrated for fluorescent samples but is also applicable in other situations.

Thus far, we have implicitly assumed that the tip is maintained above the sample at a distance (height) that optimizes the signal enhancement factor for the given conditions. In practice, however, the tip is often operated in tapping mode, where the probe is driven at its resonance frequency and undergoes rapid vertical oscillations. As the probe oscillates, the optical signal is modulated as the tip alternately approaches the sample and then retracts. The depth of the modulation depends on the strength of the near-field signal (i.e., enhancement and/or quenching) and also on the amplitude of oscillation (Xie et al., 2006; Mangum et al., 2008a). When the signal is averaged over timescales larger than the oscillation period, image contrast is actually reduced as the probe only spends a fraction of its time in the near-field interaction zone. However, if the optical signal is demodulated, the near-field portion can be efficiently extracted, thereby suppressing the background. Thus, when using tapping mode AFM imaging in apertureless NSOM, it is important to demodulate the acquired signal to achieve optimal contrast.

Various demodulation algorithms are widely used in many fields to recover small signals. Lock-in amplification is a particularly powerful technique that strongly suppresses background signals that fall outside a narrow frequency band and amplifies those that are in phase with a (sinusoidal) reference signal. When applied to continuous signals, the function of a lock-in amplifier is fairly easy to conceive: it multiplies the signal of interest by a sinusoidal reference and decomposes the resultant into in-phase and out-of-phase components. In this way, signals that are not at the same frequency as the reference are strongly suppressed. In many NSOM experiments, however, the detected signal is composed of a series of discrete electronic (e.g., TTL) pulses originating from a photodetector such as an avalanche photodiode (APD) or a photomultiplier tube (PMT). In this case, the operation of the lock-in amplifier may be more easily understood by considering each detected photon as a unit vector whose phase angle θ_i is given by the instantaneous phase of the tip oscillation signal (the reference) at the time of detection. In this phase-space picture, lock-in amplification is equivalent to vector addition of the detected photons transmitted through a narrow bandpass filter. Thus, the optical signal can be divided into near-field **NF** and far-field **FF** components, both of which are vector sums.

The far-field component **FF** results from an unbiased random walk in phase space, since the photon arrivals are completely uncorrelated with the tip oscillation phase. The length of this vector is thus proportional to the square root of the number of background photons and hence the square root of the unenhanced optical intensity (again, for a first-order process):

$$|FF| = \sqrt{\frac{\pi}{4} \beta \kappa I_0 N_A}, \tag{17.8}$$

where

N_A is the number of molecules/particles in the laser focus, as defined above

β characterizes the transmission of far-field photons through the lock-in bandpass filter

A typical value for β is 0.15, as determined experimentally by comparing the lock-in signal in the far-field illumination region with the sum signal in that region.

The near-field component comes from a biased (enhancement) or anti-biased (suppression) random walk about the particular oscillation phase corresponding to tip–sample contact, θ_p. For strong signal enhancement and an appropriately large tip oscillation amplitude, the near-field photons contributing to **NF** predominantly point in the direction θ_p, so their addition yields a vector whose magnitude is proportional to the number of near-field photons and hence proportional to the enhanced optical intensity (for a first-order process):

$$|NF| = \eta(\kappa f I_0), \tag{17.9}$$

where

κ and f are as defined previously

η takes into account the time the tip spends near the sample and the projection of the biased random walk onto the direction θ_p. Note that the expression within the parenthesis in Equation 17.9 is equivalent to Equation 17.5 for a first-order process with the substitution $a \rightarrow 1$ as justified earlier

The sum of these two components will yield a resultant vector whose magnitude, $|L| = |NF + FF|$, is the lock-in signal. To compute the image contrast expected with lock-in amplification, the average magnitude of the resultant vector must be found. This is done by adding the near-field and far-field components and averaging over all angles:

$$|L| = \sqrt{|NF|^2 + |FF|^2} = \sqrt{(\eta \kappa f I_0)^2 + \frac{\pi}{4} \beta \kappa I_0 N_A}. \tag{17.10}$$

When the tip is far from the sample, then $\mathbf{NF} = 0$ and the lock-in signal is given simply by the background, $|\mathbf{L}|_{bg} = |\mathbf{FF}|$. When the tip is directly on top of the sample, the peak lock-in signal $|\mathbf{L}|_{peak}$ is given by Equation 17.10. Thus, the lock-in contrast C_{LI} is given by

$$C_{LI} = \frac{|\mathbf{L}|_{peak} - |\mathbf{L}|_{bg}}{|\mathbf{L}|_{bg}} = \left[\frac{4\kappa I_0 (\eta f)^2}{\pi \beta N_A} + 1 \right]^{1/2} - 1. \tag{17.11}$$

The enhancement factor required for contrast greater than unity is then

$$f > \frac{1}{\eta} \sqrt{\frac{3\pi \beta N_A}{4\kappa I_0}}. \tag{17.12}$$

Comparing Equations 17.7 and 17.12 reveals some very important differences between the NSOM performance with and without signal modulation/demodulation. First, the required enhancement factor with demodulation scales with $\sqrt{N_A}$, rather than with N_A, the number of particles/molecules in the focal area. Second, the required enhancement factor is now inversely proportional to $\sqrt{I_0}$; thus, as the laser intensity increases, the required enhancement drops. Finally, the enhancement factor depends on the tip oscillation amplitude through η. As described in Mangum et al. (2008a), when the oscillation amplitude is too small, the signal modulation depth is compromised, while if the amplitude is too large, the tip spends only a small fraction of its time in the near-field interaction zone. Thus, there is an optimal oscillation amplitude that balances these two effects. Note that the only dependence of the contrast on oscillation frequency comes from the fact that the phase resolution for each photon is reduced for faster oscillation frequencies. Inserting typical values for κ, η, β, N_A, and I_0 into Equation 17.12 for experiments with fluorescent CdSe/ZnS quantum dots gives $f > 18$. Enhancement factors this high have already been demonstrated for silicon tips (no quenching) (Gerton et al.). Importantly, this model has been validated by comparing the predicted and observed image contrast for individual quantum dots as a function of the tip oscillation amplitude (Mangum et al., 2008a).

Figure 17.8 demonstrates the efficacy of the lock-in demodulation scheme. The figure shows an apertureless-NSOM image of CdSe–ZnS quantum dots on a glass surface without (a) and with (b) lock-in demodulation (Mangum et al., 2008a). The optical contrast is clearly improved significantly, revealing previously hidden details. These images were obtained using a silicon tip oscillating with an optimized amplitude of ~30 nm peak-to-peak and focused-TIR

(a)   (b)

Figure 17.8 Contrast improvement using lock-in demodulation. Tip-enhanced fluorescence image of CdSe–ZnS quantum dots on glass using a silicon tip in air. Panel (a) shows a simple photon sum image while panel (b) shows a lock-in demodulated image of the same data. (Reprinted from Mangum, B.D. et al., *Opt. Express*, 16, 6183, 2008a.)

(total internal reflection) illumination. The visible quantum dot density for this image is ~14 μm^2 and there is clearly sufficient contrast to increase the density further.

17.3 NUTS AND BOLTS OF APERTURELESS NSOM

This section will discuss a number of the most pertinent experimental issues associated with apertureless NSOM. The goal is to provide enough details so that newcomers to the field can obtain a clear idea of what it takes to successfully implement apertureless NSOM in the lab. Obviously, there is no sufficient space to give a comprehensive account of all the mundane details, so we will focus on the most important. In particular, we will discuss the strengths and weaknesses of various: (1) probe geometries and materials, (2) illumination and detection configurations, (3) methods for aligning the tip and laser, (4) feedback mechanisms for tip–sample distance control, and (5) data acquisition and analysis algorithms.

17.3.1 Probe Geometry and Material

In order to obtain the strongest near-field signal, optimization of the tip's enhancement factor is essential. There are generally three different types of tips employed in apertureless-NSOM experiments (Figure 17.9): commercial cantilever-based AFM tips made of silicon, silicon

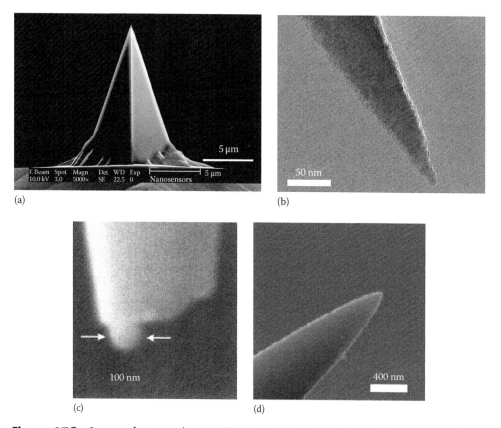

Figure 17.9 Survey of tips used in NSOM. Panels (a) and (b) show different views of commercially available cantilever-based Si tips. (Adapted from www.nanosensors.com/PPP-FM.htm.) Panel (c) shows a single gold nano-sphere attached to the end of a pulled glass fiber. (Reprinted from Kalkbrenner, T. et al., *J. Microsc. Oxf.*, 202, 72, 2001.) Panel (d) shows an example of an electrochemically etched gold wire. (Reprinted from Eligal, L. et al., *Rev. Sci. Instrum.*, 80, 3, 2009.)

nitride, or metal-coated silicon; spherical metal nanoparticles attached to the distal end of a pulled glass fiber or commercial cantilever-based AFM tip (Kalkbrenner et al., 2001); and electrochemically sharpened metal wires (gold, tungsten, silver, etc.) (Eligal et al., 2009). These various probes have different strengths and weaknesses, and the choice of one over another depends in large part on the particular optical process to be employed during the experiment.

17.3.1.1 Fluorescence NSOM

The key factor in choosing the most appropriate tip in fluorescence imaging is to optimize the competition between field enhancement and quenching, as described earlier. Although metal tips can generate extremely high field enhancement factors, particularly near a plasmon resonance frequency, they also quench fluorescence at very short range. This competes with field enhancement, thereby reducing the net fluorescence signal. Generally speaking, field enhancement is proportional to the real part of the *effective* polarizability (e.g., Equation 17.2, quantity in parentheses), while the imaginary part of the permittivity, ε'', is a predictor for the quenching efficiency (Anger et al., 2006). The precise dependence of quenching and enhancement on the complex permittivity is also very sensitive to geometry. Interestingly, it has recently been suggested that even in the absence of damping (i.e., $\varepsilon''(\omega) = 0$), a fluorophore can couple to a metal tip via intermediate-range (10–50 nm) excitation of surface plasmon traveling waves, which for an open-geometry tip (e.g., metal-coated cone or pyramid) results in a reduction of the fluorescence signal (Issa and Guckenberger, 2007).

For metals, both the effective polarizability and ε'' can become very large in magnitude near a plasmon resonance frequency (Chapter 3). Thus, it is difficult to predict what the net signal enhancement will be for an arbitrary geometry, and there are only a few rigorous calculations that have been compared with the experiment (Anger et al., 2006; Bharadwaj and Novotny, 2007; Vukovic et al., 2008) and only for a simple spherical geometry. Nonetheless, all fluorescence experiments thus far with metal tips have exhibited strong quenching, and thus reduced signal, at very short tip–sample separation distances. Tips composed of small metal spheres (attached to dielectric probes) can support strong localized plasmon resonances and can thus exhibit appreciable signal enhancement outside this quenching zone (see below). Metal-coated tips with extended geometries (e.g., metal-coated probes), on the other hand, do not support localized plasmon resonances and have exhibited no net enhancement of the fluorescence signal at any length scale (Yang et al., 2000; Trabesinger et al., 2002; Kuhn et al., 2006; Mangum et al., 2009).

Silicon has a complex permittivity of $\varepsilon_{Si} = 17.6 + 0.12i$ (at $\lambda = 532$ nm) (Adachi, 1988), indicating the potential for good fluorescence signal enhancement, with only minimal quenching. Furthermore, silicon tips can be made quite sharp, particularly compared to metal-coated tips, so the lightning-rod enhancement and resulting increase in spatial resolution should be quite good. Indeed, silicon tips have exhibited fluorescence signal enhancement factors as large as ~20 (Gerton et al., 2004) and spatial resolution as small as ~10 nm (Protasenko et al., 2002; Farahani et al., 2005; Ma et al., 2006; Xie et al., 2006; Mangum et al., 2008a; Mangum et al., 2009). As described earlier, this enhancement factor should provide adequate contrast to image even rather complex samples with high background signals. A literature search of dielectric constants at visible frequencies for commonly available materials indicates that silicon gives the largest enhancement factor, although reliable optical constants for some materials are difficult to obtain.

Evidently, silicon appears to be the best, or at least the most straightforward, probe material for fluorescence imaging, and there are a variety of probe geometries available, most of them cantilever based. Super-sharp silicon AFM probes have been used to obtain large signal enhancement in the past (Xie et al., 2006), but they also suffer from rapid wear, which leads to large variations in their performance. In our experience, standard silicon "tapping mode" probes yield very acceptable and repeatable results. One problem with silicon probes is the growth of oxide layers, which do not exhibit strong polarizability at optical frequencies as compared to Si. All silicon tips have some native oxide layer, but as tips age this layer thickens. Using fresh probes alleviates this problem somewhat, but it can be difficult to determine their exact date of manufacture. Storing tips in vacuum chambers is definitely recommended, and some manufacturers have begun shipping probes in hermetically sealed packages of fewer quantities to avoid unnecessary oxidation. Some studies have also indicated that contaminants from gel-pack off-gassing may also contribute to a reduction in enhancement factor (Protasenko et al., 2002). It is widely agreed that oxidized or contaminated tips may be "revived" to some extent via etching in hydrofluoric acid (HF) (Protasenko et al., 2002, 2004). Standard buffered oxide etch (BOE) procedures prescribe the rate at which silicon oxide layers can be eaten away (Wolf and Taubner, 1986). This procedure, while effective in removing oxide layers, can also dull the AFM tip, and it should thus be applied carefully and conservatively.

While more difficult to produce, metal nanospheres attached to dielectric tips (Kalkbrenner et al., 2001) can also yield excellent results (Anger et al., 2006; Kuhn et al., 2006; Bharadwaj et al., 2007). Because these probes can support localized plasmon resonances, they can yield very large field-enhancement factors that can overcome the reduced signal caused by fluorescence quenching beyond some critical tip–sample separation distance. This is clearly the case in Figure 17.10, where an 80 nm gold sphere is used to probe a vertically oriented single molecule; at very close range quenching overpowers the field enhancement. Thus, to optimize image contrast, these tips should be maintained at this critical height from the sample (~5 nm),

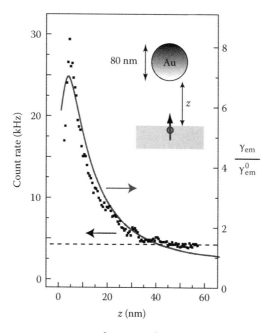

Figure 17.10 Fluorescence rate as a function of particle-surface distance z for a vertically oriented molecule. Dots represent experimentally observed count rates (left axis). The solid curve shows the theoretical normalized emission rate compared to free space, $\gamma_{em}/\gamma_{em}^0$ (right axis). The horizontal dashed line indicates the background level. (Reprinted with permission from Anger, P., Bharadwaj, P., and Novotny, L., *Phys. Rev. Lett.*, 96, 113002, 2006. Copyright 2006 by the American Physical Society.)

where the signal enhancement is maximized (Figure 17.10). To maintain a constant tip–sample gap, the AFM should be operated in shear force rather than tapping mode. When using tips composed of spherical nanoparticles, both the size of the particle and the metal to be employed should be chosen so that the localized plasmon resonance frequency is close to the absorption peak of the fluorophore: the resonance frequency is determined by both the size and permittivity of the particle. It has been shown that gold nanoparticles perform better at red wavelengths and silver nanoparticles at blue wavelengths (Bharadwaj and Novotny, 2007). Smaller particles yield higher resolution but also smaller enhancement (Link and El-Sayed, 1999; Anger et al., 2006): most reports utilize diameters in the 40–80 nm range (Anger et al., 2006; Kuhn et al., 2006; Novotny and Hecht, 2006; Bharadwaj and Novotny, 2007; Bharadwaj et al., 2007).

17.3.1.2 Scattering NSOM

It is more straightforward to choose an NSOM probe for nonfluorescent imaging modes, such as Rayleigh and Raman scattering. Here, quenching is generally not a problem, so the key requirement is maximizing field enhancement while maintaining the best spatial resolution possible. Although the metal nanosphere probes discussed earlier can yield reasonably large enhancement factors and nice analytical predictions, elongated metal probes can leverage the lightning-rod effect for even larger enhancement. Sharp protrusions on the surface of metal probes can also support localized plasmon resonances, occasionally leading to extremely high enhancement factors but poor repeatability. Many implementations of scattering NSOM utilize sharpened metal wires attached to tuning forks in combination with shear-force feedback to maintain a particular (small) tip–sample gap. Nevertheless, cantilever-based probes can more easily impart signal modulations via vertical oscillations of the probe, which increases image contrast as discussed in Section 17.2. On the other hand, most cantilever-based probes are metal-coated silicon and are therefore not as sharp as etched wires, thus compromising lightning-rod enhancement and spatial resolution. Furthermore, the open geometry of such metal-coated probes cannot fully leverage plasmon-based enhancement, because they do not support localized plasmon resonances. A recent report describes a method to impart vertical oscillations in the probe while maintaining shear-force feedback (Höppener et al., 2009), which can help recover signal modulations when using etched-wire probes.

17.3.2 Illumination and Detection Configurations

Standard microscope configurations generally illuminate a sample plane at normal incidence either from above or below. In either case, light can be collected by the same lens that focuses the light (reflection or episcopic mode) or by a lens on the side opposite the illumination (transmission or diascopic mode). In apertureless NSOM, however, since an AFM sits directly above the sample, the most common configurations include standard inverted episcopic illumination (in which the sample is illuminated through the substrate) as well as side illumination (Figure 17.3), which can be operated in either reflection or transmission mode. In the following, the term episcopic will refer to illumination and detection by the same lens at normal incidence in the inverted configuration, while side illumination schemes will be specified as such. The episcopic configuration is only suitable for transparent samples, while side illumination can be used for transparent or non-transparent samples.

To achieve field enhancement from either the lightning-rod effect or surface plasmon resonances, the polarization state of the excitation beam must match the orientation of the tip, as discussed in Section 17.2. This usually means that the polarization must have a strong vertical component (perpendicular to the sample plane) since the distal end of a spherical or elongated NSOM probe is adjacent to the sample. A notable exception is the bowtie antenna geometry, which lies in the sample plane, thus requiring horizontal polarization along the axis separating the prisms that form the feed gap. Generally, horizontal polarization can be achieved quite readily with any illumination/detection configuration, while vertical polarization may require more advanced preparation of the laser polarization state, particularly for the episcopic geometry.

17.3.2.1 Side Illumination and Detection

A typical side illumination and detection configuration is shown in Figure 17.3. In this scheme, a laser is focused into the gap between the tip and sample, which leads to a rather extended illumination spot at the sample plane. The vertical component of the polarization vector depends, of course, on the particular illumination angle and thus on the particular details of the setup. In general, a shallower illumination angle leads to stronger vertical polarization, but higher background as a larger area of the sample is illuminated. For non-transparent samples, this is essentially the only viable illumination and detection option. Note that the signal can be detected in the back-scattered geometry, as shown in Figure 17.3, or in a forward-scattered geometry or even from below if the sample is transparent.

Side illumination and detection is popular for Rayleigh scattering where the expected signal is relatively large, since the collection efficiency of the side lens is generally not as large as for the inverted episcopic configuration (see below). Also, when side illumination is coupled with forward-scattered signal detection, the un-scattered portion of the illumination can be used as a reference for demodulating the near-field signal (e.g., homodyne detection). In this configuration, side illumination is particularly susceptible to topographic artifacts caused by an elevation of the tip above the sample plane, due to a concomitant reduction in the homodyne signal, but can be reduced by operating in constant-height mode (Knoll et al., 1997).

17.3.2.2 Episcopic Illumination and Detection

A schematic diagram of several inverted episcopic configurations for apertureless NSOM is shown in Figure 17.11. This configuration is most appropriate for transparent samples where extremely high sensitivity is required since oil immersion objective lenses with numerical apertures (NA) approaching 1.5 can be used. Such objectives produce an extremely fine, diffraction-limited illumination area, which reduces the background signal, and also have very high collection efficiency. At first glance, however, it seems impossible to achieve vertical polarization in the episcopic configuration since it is a violation of Maxwell's equations for a propagating field to be polarized in the direction of propagation. However, when a large NA

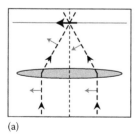 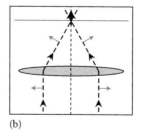 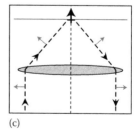

(a)   (b)   (c)

Figure 17.11 Raytracing schematic of several different epi-illumination configurations. Panels (a) represents linearly polarized light, yielding a purely horizontal resultant vector at the sample plane along the optic axis. Panel (b) demonstrates how radial polarization yields a purely vertical resultant vector at the focal plane along the optical axis. Panel (c) shows how allowing only a subsection of supercritical rays into the back aperture of the objective for linearly polarized light can also lead to a large axial component of light at the sample plane as the beam is totally internally reflected. This is achieved by applying a wedge-shaped mask to block all subcritical rays.

objective is used to focus a linearly polarized Gaussian (TEM_{00}) mode laser beam, those light rays originating from the periphery of the lens impinge on the sample with a large angle of incidence. The polarization vector for these peripheral rays can be tilted strongly toward the vertical direction. Within this subset of rays, those originating from diametrically opposed positions on the lens will have axial polarization components that cancel precisely at the center of the focal spot as seen in Figure 17.11a. However, away from the geometrical center the axial components do not cancel completely, which yields two lobes of significant axial intensity distributed symmetrically about the center of the focal spot. If the tip is positioned into one of these lobes, the axial field will induce enhancement as described earlier. This scheme for producing axial polarization is simply implemented but has the disadvantage of requiring the tip to be positioned off center within the focus.

Another way to achieve strong axial polarization is to use a higher order TEM_{01*} Gaussian mode, which exhibits radial polarization when collimated (Quabis et al., 2000; Dorn et al., 2003). When focused, the vertical polarization components from diametrically opposed peripheral rays precisely augment each other, thereby producing a strong axial-polarized field at the center of the laser spot, as shown in Figure 17.11b. Creating a radially polarized beam can be achieved using half-wave plates, twisted nematic crystals (Stalder and Schadt, 1996), or other more sophisticated techniques (Tidwell et al., 1993). For example, by carefully cutting and gluing half-wave plates into quadrants with fast-axis orientations as indicated in Figure 17.12, a beam with linear polarization can be converted into one with quasi-radial polarization. If this beam is then passed through a spatial filter, a pure radial mode can be extracted. In a similar but more continuous fashion, a commercial device using twisted nematic crystals can also produce a mode with nearly perfect radial polarization (Stalder and Schadt, 1996), which can be further refined via a spatial filter.

Interestingly, a TEM_{01*} Gaussian mode with radial polarization yields a smaller focal spot compared to a linearly polarized TEM_{00} beam (Quabis et al., 2000; Dorn et al., 2003). This decreases the background signal in proportion to the reduction in illumination area, but also makes it more difficult to align and maintain the tip at the center of the focal area. Using radial polarization, the tip must be stringently aligned with the optic axis to obtain optimal signal enhancement. In particular, using an oil immersion objective lens with NA = 1.4, the diffraction-limited focal spot has a diameter of ~250 nm with green ($\lambda \sim 532$ nm) illumination, and the tip needs to be aligned into the central ~100 nm to achieve strong and repeatable field enhancement. Mechanical and thermal drifts on the order of several tens of nanometer per minute can make it difficult to maintain tip–laser alignment with radial polarization. As described in the following, the task of maintaining good alignment of the tip and laser can be a substantial technical challenge in apertureless NSOM, particularly in the episcopic configuration.

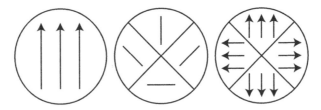

Figure 17.12 Cartoon diagramming the production of quasi-radial polarization. Linearly polarized light as shown in cross section on the left is incident on a custom half-wave plate, with the fast axis of each quadrant as shown in the middle panel. The result is a quasi-radial beam with polarization as shown on the right. True radial polarization may be obtained by spatially filtering this resultant beam.

Yet another scheme to achieve axial polarization within the episcopic configuration is to create a focal spot that undergoes TIR. This is accomplished by using a beam mask to block all subcritical rays being introduced into the back aperture of the objective. In this case, the light transmitted by the mask is internally reflected at the glass/air or glass/liquid interface, thus establishing an evanescent field at the sample interface. To avoid vector cancellation of the axial field at the center of the focal spot (Figure 17.11), a wedge-shaped mask is used, as shown in Figure 17.13. Thus, the evanescent field above the sample interface will be polarized primarily in the vertical direction when the appropriate direction of the initial laser polarization is chosen. In this case, the objective lens is asymmetrically under-filled, so the focal spot will be elongated along the symmetry axis of the wedge: an NA = 1.4 lens will typically yield a focal spot measuring ~1.5 × 0.5 (μm)^2 for green light.

This nearly diffraction-limited evanescent spot makes this configuration distinct from the more familiar wide-field TIR schemes, which typically use a prism, solid immersion lens, or more recently, a laser beam focused to the periphery of an NA objective, to create an evanescent field that extends across the entire field of view. The focused-TIR configuration is thus more appropriate for laser scanning and/or confocal applications. In apertureless NSOM, it provides three key benefits. First, the relatively small illumination spot reduces the far-field background signal compared to wide-field TIR. Second, the vertical polarization extends virtually over the entire illumination area, so the precise positioning of the tip within the focus

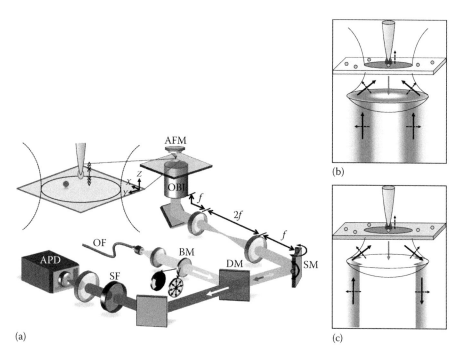

(a)

(b)

(c)

Figure 17.13 Experimental realization of a fluorescence apertureless-NSOM setup (a). An excitation beam exits an optical fiber (OF) and goes through a beam mask (BM) in either a wedge or radial configuration. The excitation beam is reflected by a dichroic mirror (DM) and off a scanning mirror (SM) before being focused through a microscope objective (OBJ). Signals are collected through the same path and directed onto an APD after passing through the appropriate spectral filters (SF). Panels (b) and (c) show ray diagrams for a radial and wedge beam mask, respectively. Solid arrows show the direction of beam propagation (dark arrows for excitation and lighter arrows for emission), while dashed arrows represent the polarization direction.

is not as critical as for radial polarization. Finally, it is easy to switch between axial and transverse polarization simply by rotating the initial laser polarization using a half-wave plate or a Pockels cell.

17.3.3 Aligning the Tip and Laser

Aligning the tip and laser is one of the dirty little secrets of apertureless NSOM that is not discussed in most publications. As indicated earlier, there are many technical challenges associated with positioning a sharp tip into a focal spot as small as ~250 nm diameter. This task requires some indicator of alignment, as well as the precision to perform it. Many apertureless-NSOM systems are essentially an AFM mounted atop some variation of an inverted microscope, most often including a high-precision piezo-actuated scanning stage capable of moving the sample relative to the tip. Alternatively or additionally, some systems have the ability to scan the tip relative to a stationary laser focus, while some can scan the laser. In general, the most versatile system, particularly with regard to tip–laser alignment, is one that has the ability to scan all three elements: the sample, the tip, and the laser.

The ability to move the laser and tip relative to each other must be accompanied with the ability to measure the degree of their alignment. Top view optics for imaging the AFM probe and sample from above can facilitate coarse alignment. This works best for AFM probes whose tip apexes are visible from directly above, where top view optics can yield sub-micron alignment with practice. Many AFM probes, however, are fabricated with the tip hidden below the cantilever, rather than protruding beyond it, limiting the alignment precision to several microns.

A more precise alignment method requires the presence of a measurable near-field optical signal embedded within the far-field background. For real-time alignment, the presence of a near-field signal is most easily detected with the use of a commercial lock-in amplifier. An offset between the far-field and near-field signals then indicates the degree of misalignment, and either the tip or laser can be adjusted to compensate. This method works best for relatively simple samples where an offset between the near-field and far-field background signals is clearly discernable. In addition, it is susceptible to null results when no near-field signal is observed for a reason distinct from misalignment. This procedure also requires pre-alignment of the tip and laser at least within the diameter of the focal area, which may or may not be possible using the top-view optics described earlier. A related alignment procedure requires co-registration of the optical and AFM images. This method can also work quite well, but only for samples with discernable topography.

For transparent samples, a high-precision method of alignment that is independent of the sample topography, complexity, or the optical accessibility of the tip from above is to use the episcopic optics below the tip and sample to co-image the tip apex and laser focus backscattered from the sample. If the tip itself produces a strong scattering signal that can be distinguished from scattering off the sample surface, then good alignment can be achieved using the excitation laser alone. This is most easily accomplished if the scattering signal from the tip is either spectrally distinct from the sample scattering (e.g., probe fluorescence, second harmonic generation, white-light generation, etc.) or is modulated by tip oscillations. In any case, the small depth of focus that accompanies high NA objectives can make it difficult to obtain sufficient scattering signal from an ultrasharp probe. To enhance sensitivity for the tip-scattered signal, it may be necessary to illuminate the tip with additional sources, possibly with spectral characteristics that are distinct from the excitation laser. Obviously, these spectral characteristics should not interfere with AFM operation nor influence the sample response. The additional sources may be applied from the side, from below along the excitation path, or possibly even from above. This method of co-imaging a scattering signal from the tip and the laser focus backscattered from the sample has the potential to yield tip–laser alignment with arbitrary precision but is also the most challenging to implement.

17.3.4 Tip–Sample Distance Control

In NSOM, the ability to scan a surface accurately with minimal damage to the sample is highly desired, while the main physical quantity of interest is the optical signal. We have already discussed the topographical artifacts that can result when conducting NSOM experiments in Section 17.1.5. In this section, we briefly compare two scanning techniques: vertical oscillations vs. horizontal oscillations (i.e., cantilever-based AFM—Figure 17.9a and b vs. tuning forks—Figure 17.9c). While there are particular advantages and disadvantages of either technique from purely a scanning standpoint, we focus primarily on the advantages and disadvantages only as they pertain to NSOM.

Shear-force AFM imaging utilizes a long probe attached to a tuning fork to scan the topography of a surface. A probe is attached to a tuning fork such that the tip oscillates laterally (parallel to the sample plane) at a well-defined lateral resonance frequency; as the tip approaches the sample, shear-force interactions between the probe and sample lead to changes in this resonance frequency. This shift in frequency can be used as the feedback parameter in order to ensure constant tip–sample separation. Shear-force probes are typically very rigid, which leads to very good sensitivity of the feedback loop, but also equates to a large settling time. This can lead to better tracking of the surface, but also requires slower scan rates.

If one desires to keep a constant tip–sample gap while measuring the optical signal, shear-force scanning is better than cantilever-based techniques because the lateral amplitudes are much smaller than the lowest possible vertical amplitudes of cantilever-based AFM. For aperture-type NSOM, it is indeed advantageous to keep the tip at a constant gap. This also minimizes any potential damage to the sample. On the other hand, for apertureless NSOM it is advantageous to vertically oscillate the tip (and thus modulate the near-field signal) in order to suppress the background. To implement this for tuning fork probes, a vertical oscillation scheme must be added—a more complex setup than that required by a cantilever-based system (Höppener and Novotny, 2008).

In cantilever-based AFM systems the probe vertically oscillates at or near its resonance frequency. Either the oscillation amplitude or the frequency is used as a feedback mechanism. The probes are not as stiff compared to tuning forks, leading to lower sensitivity of the feedback loop, but also faster settling times, allowing for faster scan rates. In this scheme, large vertical oscillations are easily achieved, which is ideal for modulation and subsequent demodulation of the near-field signal. Cantilever-based NSOM setups also allow for contact mode imaging, but this is not desirable, as demodulation schemes cannot be applied and this could potentially damage the tip and/or sample.

17.3.5 Optical Data Acquisition and Analysis

Optical data acquisition and analysis encompasses several key elements, which can be divided into three general classes: (1) the collection and detection optical train, including optical detectors, as shown in Figure 17.13 for the episcopic configuration, (2) the hardware used to process the optical signal after conversion into electronic form, and (3) the methods and algorithms used to analyze the data, including offline (software) programs. These three categories will be discussed in this section, once again focusing mainly on the episcopic configuration and more particularly fluorescence imaging. Furthermore, we will focus specifically on a particular data acquisition and analysis scheme called SP-NFT, which is quite powerful and versatile, and is additionally applicable to a range of different scattering mechanisms in NSOM.

Some key data-acquisition issues associated with operation of the AFM, such as the feedback mode and tip–laser alignment, have already been discussed earlier and will not be repeated here.

17.3.5.1 The Optical Train

Optical data acquisition begins with the microscope objective, which collects the optical signal and relays it to a detector. As described earlier, extremely high collection efficiencies can be achieved with transparent samples in the episcopic configuration by using oil-immersion objectives with NA values up to ~1.5. With side illumination, long working-distance air objectives can be employed with NA values up to ~0.8. For inelastic processes (e.g., fluorescence, Raman scattering), the collected signal is usually passed through a series of spectral filters, including a dichroic beamsplitter and a bandpass or longpass filter, to separate the scattered excitation light and ambient light from the emission signal.

As NSOM is inherently a scanning technique, the use of cameras for detection is not usually beneficial, except possibly to facilitate the spatial alignment of the tip and laser. Thus, PMTs and APDs are the most common detectors employed for NSOM. Both APDs and PMTs have very high sensitivity across the visible spectrum and into the near IR and also provide good temporal resolution. Furthermore, both can operate in photon-counting mode where each detection event leads to a short electronic burst, usually in the form of a TTL pulse. Dead times between pulses and secondary pulses (afterpulses) limit the maximum photon count rates for APDs to about 5×10^6 counts/s, while dark count rates limit the minimum detectable signal to about 50–100 counts/s. To make a 2D image, the photocounts are binned and tallied into spatiotemporal pixels as the tip and sample are scanned relative to each other in a raster pattern.

17.3.5.2 Hardware Processing of NSOM Signals

As photons are detected, the resulting electronic pulses can be processed with a range of hardware instrumentation. The simplest processing involves counting the number of pulses occurring in a particular time window during the raster scan and plotting this as a function of the scan position, thereby creating a 2D photon sum (or photon rate) image. This summing can be accomplished with a stand-alone photon counter, a commercial photon-counting card, or even a module that can be integrated directly into the AFM controller, thereby ensuring registration of the optical and AFM images.

As discussed extensively earlier, it is important in apertureless NSOM to discriminate between far-field and near-field signals. Oscillating the tip above the sample, such as in tapping-mode AFM, can be used to modulate the optical signal. It is then possible to demodulate the signal using a stand-alone photon counter to separate the photons into different bins corresponding to various ranges in tip height. This is accomplished by gating the photon counter in synchrony with the tip oscillation motion so that one gating window corresponds to the tip being closest to the sample in its oscillation trajectory and another to the tip being farthest from the sample (Yang et al., 2000). This method can be used to create three images corresponding to small tip–sample separations (near-field + far-field), large tip–sample separations (far-field), and their difference (near-field). In principle, the gating windows can be shifted to probe the optical response at various tip heights. This method, although elegant, is limited by the minimum width of the gate windows. Furthermore, it ignores the large majority of photons, namely, those that arrive outside the gate windows, which ultimately results in a loss of optical contrast.

Another way to demodulate the optical signal is to use a lock-in amplifier. Since typical oscillation frequencies of cantilever-based AFM probes are generally in the tens of kilohertz, many off the shelf lock-in amplifiers are well suited for such measurements. This was initially demonstrated for Rayleigh scattering (Inouye and Kawata, 1994; Zenhausern et al., 1994),

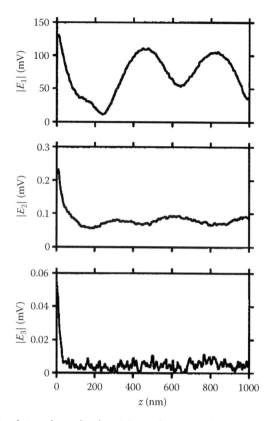

Figure 17.14 Optical signal amplitude $|E_n|$ vs. distance z between a PtIr-coated Si tip and Au sample, for different harmonic demodulation orders n. (Reprinted with permission from Hillenbrand, R. and Keilmann, F., *Appl. Phys. Lett.*, 80, 25, 2002.)

where the tip oscillation frequency and its harmonics were used as a reference signal. One key advantage of demodulating at higher harmonics is an increase in both contrast and resolution, due to a decrease in the effective field volume overlap with the sample as observed in the third-harmonic amplitude $|E_3|$ of Figure 17.14 (Hillenbrand and Keilmann, 2002). In Rayleigh scattering experiments, the photon flux is very high, and the use of a lock-in is thus ideal. As discussed in Section 17.2 and in Mangum et al. (2008a), a lock-in amplifier can also be used to effectively suppress the background signal in fluorescence experiments, even in the single-photon-counting regime (Xie et al., 2006; Mangum et al., 2008a; Höppener et al., 2009). In principle, lock-in demodulation can also be used with other scattering mechanisms, such as tip-enhanced Raman spectroscopy (TERS), IR scattering, and two-photon fluorescence.

17.3.5.3 Single-Photon Near-Field Tomography

Although a lock-in amplifier can be used in the single-photon-counting regime to efficiently suppress the far-field signal, the data processing removes all information about the individual signal photons. The range of possible signals that can be obtained from the lock-in amplifier is limited primarily to the total magnitude and phase at either the fundamental reference frequency or any of its higher harmonics averaged over the pixel acquisition time. Furthermore, it is extremely difficult to extract information about non-monotonic vertical variations in the near-field signal from the lock-in outputs. Recently, we developed a photon-counting technique that retains a complete record of each individual photon: its precise arrival time at the detector and, most importantly, the correlation between that arrival time and the phase of the tip oscillation trajectory (Gerton et al., 2004; Ma et al., 2006; Mangum et al., 2009). This technique enables a wide range of offline (software) analyses including phase-filtered lock-in algorithms (Ma et al., 2006; Mangum et al., 2008a,b, 2009) and height-resolved imaging (tomography), which can be used both to improve image contrast and to study the 3D form of tip–sample coupling. The technique relies upon the phase-correlation of single photons, and we therefore call it single-photon near-field tomography, or SP-NFT.

In SP-NFT, two data channels are simultaneously acquired: one corresponding to the photon arrival times and another representing the tip's oscillatory motion. In a current embodiment, a pair of data acquisition cards is used to sample (at 80 MHz) and record photon detection pulses from an APD and to generate timestamps for each tip oscillation and AFM line marker. The tip oscillation timestamps are obtained by using a homebuilt comparator circuit to generate an electronic pulse at the zero-crossing phase of the AC-coupled AFM deflection signal, which then triggers a time measurement from the data acquisition card. After acquiring these data channels, the tip oscillation phase corresponding to each photon arrival is computed and recorded to a computer disk. Each photon is also correlated with the lateral position of the probe using the AFM line markers to enable construction of a raster image.

When the pixel acquisition time is much longer than both the probe oscillation period and the inverse photon count rate, then a histogram of photon phase delays can be constructed for each lateral pixel. If the cantilever oscillation is harmonic, then the phase delays can be mapped to tip height after calibrating the tip oscillation amplitude, thereby producing a vertical approach curve. Alternatively, if the cantilever oscillations are anharmonic, then the AFM deflection signal can be recorded by an analog-to-digital card at a predefined sampling rate, enabling real-time measurement of the precise cantilever trajectory. The end result of this analysis is a 2D lateral ($x - y$) array of vertical approach curves, which represents a 3D map of the near-field coupling between the tip and sample. The photon sum for each three-dimensional pixel (voxel) can be normalized to its corresponding acquisition time. Thus, the value of a given voxel within the three-dimensional image space corresponds to the fluorescence count *rate* at that particular coordinate (x, y, z). Since the photon count rates are stored as a 3D array, the data can be sectioned arbitrarily.

Figure 17.15 shows a 3D rendering of the fluorescence from a 20 nm diameter dye-doped latex sphere as a function of the position of a silicon tip relative to the sphere. The 3D rendering was constructed from the intersection of particular $x - y$, $x - z$, and $y - z$ 2D sections. Field enhancement at the tip apex causes a strong increase in the fluorescence rate when the tip is immediately above the sphere, as indicated by the false color scale, which represents the fluorescence count rate. A 3D halo of mildly suppressed fluorescence signal surrounds the sphere; this is caused by interference at the sphere of the direct excitation light with that scattered from the tip. In any case, the SP-NFT technique makes it possible to study the coupling between tip and sample in three dimensions with nanoscale precision.

17.4 VARIATIONS AND APPLICATIONS OF APERTURELESS NSOM

There are many variations of apertureless NSOM including the use of fluorescence, Raman scattering, and elastic scattering, which are used to probe the optical response of a wide range of materials. In particular, different materials exhibit different optical properties depending on their electronic, phononic, and ro-vibrational structures. These various properties can be studied using different regions of the optical spectrum ranging from the visible through the IR, and even into the THz regime. Generally speaking, different regions of the spectrum probe a corresponding range of properties. For example, many electronic transitions have energies around 2 eV, well within the visible range, and are generally probed with fluorescence. Although IR fluorescence measurements are also possible (e.g., semiconducting carbon

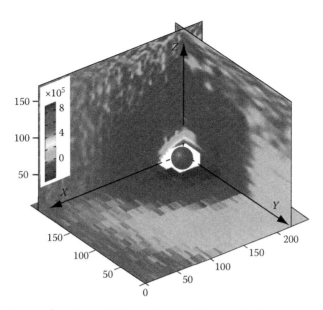

Figure 17.15 (See color insert.) 3D rendering of the near-field coupling between a Si tip and a fluorescent sphere. The white cutout region indicates the topographical volume traced by the AFM, while the false-color data represents the fluorescence count rate, as indicated by the inset (in counts/s). The spatial dimensions are in units of nanometers. (Reprinted with permission from Mangum, B.D. et al., *Nano Lett.*, 9, 3440, 2009.)

nanotubes have bandgaps in the IR), this spectral range is usually used to probe ro-vibrational transitions, such as the C–C stretch mode in organic materials. These ro-vibrational transitions can also be probed with visible light using Raman scattering, where the energy difference between the excitation laser and the scattered light reveals the IR transition energies. In recent work, THz radiation has been used to probe low-energy carrier distributions and dopants in semiconductors, as these long wavelengths couple to plasmons and phonons (Huber et al., 2008).

This section will explore some of the variations and applications of apertureless NSOM. It will first discuss some key experiments using fluorescence NSOM and then move on to some efforts that use elastic and inelastic (e.g., Raman) light scattering to probe the vibrational and phononic response of various materials. The experiments highlighted here will cover how NSOM techniques are utilized over a broad swath of the electromagnetic spectrum, from the THz through the visible.

17.4.1 Tip-Enhanced Fluorescence Microscopy

17.4.1.1 Measuring Spatial Resolution

Throughout the history of NSOM, a number of different metrics have been used to characterize spatial resolution. Most of these involve comparing the image width of a single particle with the known size of the object or measuring the width of a rising or falling edge as the tip passes over a sample. The difference between the image width and the known width of the object or edge then characterizes the spatial resolution. While this procedure is widely used, it is not particularly quantitative or rigorous. A proper measurement of spatial resolution involves the unambiguous detection of two closely spaced point sources, as per the Rayleigh criterion. The difficulty in making such a measurement with NSOM lies in the fact that the expected spatial resolution is on order of ~10nm. Single molecules are the most convenient objects small enough to serve as point sources. Of all the optical processes used in NSOM, only fluorescence has sufficient sensitivity to detect individual, point-like molecules, as discussed earlier. This arises from the extremely large matrix elements for electronic transitions compared with, for example, vibrational transitions within the molecule.

Making a self-consistent Rayleigh-like measurement of spatial resolution requires the two point sources to be separated by a known distance. Otherwise, image distortions could unknowingly bias the measurement. One way to provide such a length standard is to leverage the well-defined helical pitch of DNA to design an oligomer of known length and then to attach small (<1 nm) dye molecules to either end, forming fluorescent DNA dumbbells. Figure 17.16 shows a selection of fluorescence-NSOM images of individual DNA dumbbells (60 base

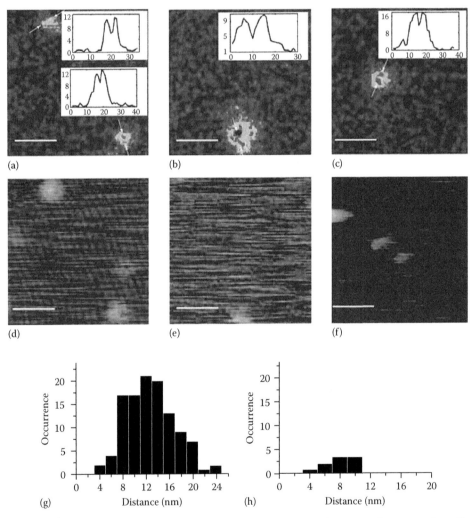

Figure 17.16 (a)–(c) Near-field images of DNA dumbbells. The insets show signal profiles through the image centers (indicated by arrows), where the horizontal axis is in pixels (1 pixel = 1.95 nm). (d)–(f) AFM images corresponding to images (a)–(c). Scale bars: 50 nm. (g) Histogram of distances between the resolved Cy3 molecules. (h) Histogram of distances between the two artifactual lobes of single Cy3 molecules. (Reprinted with permission from Ma, Z.Y., Gerton, J.M., Wade, L.A., and Quake, S.R., *Phys. Rev. Lett.*, 97, 2006. Copyright 2006 by the American Physical Society.)

pairs long) along with signal profiles through the two lobes corresponding to the positions of the two dye molecules (insets in (a) through (c)) (Ma et al., 2006). The middle row shows the corresponding AFM images: the DNA dumbbells are not large enough to be detected topographically. The bottom row shows a histogram of the measured center-to-center spacing between the image lobes of the dumbbells. The width of the histogram is caused by flexibility in the molecular linker used to attach the Cy3 dye molecules to the ends of the DNA and also by slight angles of the dumbbells relative to the horizontal sample plane. In addition, the individual dye molecules on either end of the dumbbell can exhibit double-lobed artifacts due to their particular orientation, which also introduces some variation in the measured dumbbell length. Regardless of possible length variations of the DNA, the spatial resolution of the NSOM is clearly ~10 nm for these conditions.

17.4.1.2 Measuring Biological Structure

Small fluorescent particles (fluorophores) are often used as labels to study biological structure. Common labels include fluorescent proteins, organic dyes, and semiconductor nanocrystals (colloidal quantum dots). These fluorophores can be attached to virtually any type of molecule or structure with great specificity using a variety of chemical or biochemical means. Once attached, these labels act as reporters to indicate the locations and/or transport of the target molecule or structure. Combining fluorescence labeling with sub-10 nm resolution is an important step in realizing detailed maps of protein structure on cellular surfaces. Another important step toward imaging cellular structure is the ability to measure in an aqueous environment. To the authors' knowledge, there has been only one demonstration of apertureless NSOM in an aqueous environment in combination with fluorescence labeling of a biological specimen. This experiment used NSOM probes composed of 80 nm diameter gold spheres attached to glass fibers (see above) to image fluorescently labeled calcium channels within an erythrocyte membrane, revealing an average channel spacing of ~90 nm, as shown in Figure 17.17 (Höppener and Novotny, 2008). One of the advantages of using one-photon fluorescence rather than higher order processes is that the large absorption cross section and relatively short excited-state lifetime leads to fast image acquisition. For example, the image shown in Figure 17.17b can be acquired in just a few minutes.

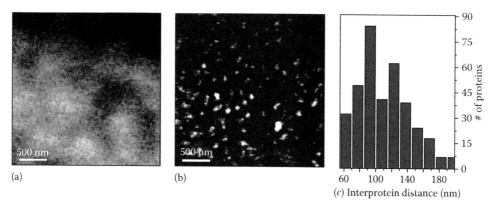

Figure 17.17 (a) Confocal fluorescence image of an erythrocyte plasma membrane immersed in H_2O. (b) Corresponding near-field fluorescence image showing individually resolved PMCA4 proteins. (c) Distribution of nearest-neighbor interprotein distances, revealing an average protein–protein separation of 90 nm. (Reprinted with permission from Höppener, C. and Novotny, L., *Nano Lett.*, 8, 642, 2008.)

17.4.2 Tip-Enhanced Raman Microscopy

Raman scattering is a two-photon process that involves the inelastic scattering of visible or near-IR light to probe vibrational states within a molecule or phononic modes within a material (Chapter 6). The process results in a spectrum of scattered wavelengths shifted in energy from the excitation laser reflecting the various ground-state ro-vibrational energy levels. The selection rules for Raman scattering are complimentary to those for IR spectroscopy, which probes the vibrational transitions directly (Chapter 6); materials that are Raman active are usually IR inactive and vice-versa. Both the IR and Raman spectra can provide a sensitive fingerprint for the chemical makeup of a material without the need for extrinsic labels.

Because the scattering process proceeds through a virtual level, it is quite weak and the data collection process is slow. The efficiency of Raman scattering is proportional to the square of the optical intensity, I^2, which results in extremely small cross sections on the order of 10^{-30}cm^2 (Hayazawa et al., 2002). TERS leverages strong field enhancement at the end of an NSOM probe to greatly increase the efficiency of this process. Since quenching is not an issue in Raman scattering, metal probes can be used for TERS; when measuring single wall carbon nanotubes, this can result in enhancement factors in the range of 10^3–10^4 (Hartschuh et al., 2005). Even with strong enhancement, TERS is still quite a slow process and image acquisition times can range from several minutes to tens of minutes.

In the near-field of the tip, the selection rules governing Raman transitions are relaxed, which results in TERS spectra that may contain additional features not found in a far-field Raman spectra (Hayazawa et al., 2003). Because the enhanced field is confined to a tight volume near the tip apex, TERS can probe variations in the vibrational response of a sample with very fine spatial resolution, limited primarily by the sharpness of the tip (Stöckle et al., 2000; Hayazawa et al., 2001, 2002; Demming et al., 2005; Anderson et al., 2007).

One nice demonstration of the capabilities of TERS is shown in Figure 17.18. Here, a single-walled carbon nanotube (SWCNT) on a glass surface was imaged using TERS. The Raman spectrum of a SWCNT reflects its particular chirality. In particular, semiconducting nanotubes can be distinguished from metallic ones via the spectral position of the radial breathing mode (RBM) peak (Anderson et al., 2007). The experiment was able to resolve with 40 nm resolution a discontinuity in the RBM frequency, thereby exposing a junction between two different chiralities.

17.4.3 Scattering NSOM

Like TERS, scattering NSOM (s-NSOM) does not require the sample to be labeled, but rather measures the tip–sample coupling directly. In s-NSOM, the tip interacts with its own image dipole within the material. Thus, when the tip is very close to a surface, the effective polarizability of the tip–sample complex takes on a value different from that of either the tip or the sample:

$$\alpha_{\text{eff}} = \alpha(1+\beta)\left(1 - \frac{\alpha\beta}{16\pi(\alpha+z)^3}\right)^{-1}, \tag{17.13}$$

where
    α is the polarizability of the tip dipole modeled as a polarizable sphere, $\alpha = 4\pi a^3 (\varepsilon_t - 1)/(\varepsilon_t + 2)$
    β is the dielectric response function of the sample, $\beta = (\varepsilon_s - 1)/(\varepsilon_s + 1)$
    z is the tip–sample separation distance
    ε_s is the permittivity of the sample
    ε_t is the permittivity of the tip
    a is the tip radius

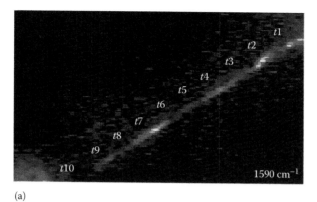

(a)

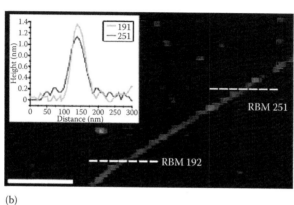

(b)

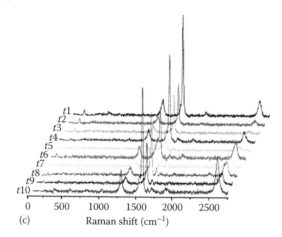

(c)

Figure 17.18 (See color insert.) TERS image of an SWCNT. Near-field Raman image (a) and corresponding topography image (b) of an isolated SWCNT, where the optical resolution was determined to be 40 nm. Also shown are a series of tip-enhanced Raman spectra (c) acquired along the length of the SWNT. From the recorded spectra, two resonant RBM modes are detected: one at 251 cm^{-1} corresponding to a semiconducting chirality and the second recorded from the lower section of the SWCNT at 192 cm^{-1}, corresponding to a metallic chirality. The inset of (b) displays two cross-sectional profiles acquired from both the upper and lower sections, respectively, revealing that the expected diameter change occurs as the SWCNT undergoes the transition from a semiconducting to metallic chirality. Scale bar denotes 200 nm and is valid for both (a) and (b). (Reprinted with permission from Anderson, N. et al., *Nano Lett.*, 7, 577, 2007.)

Since the scattering amplitude is dependent on the effective polarizability, the collected signal will reflect local variations in the permittivity of the sample (Knoll and Keilmann, 1999; Keilmann, 2004; Taubner et al., 2005; Huber et al., 2008). Additionally, as the scattering strength depends upon all layers located between the image dipole plane and the surface, subsurface imaging is also possible inasmuch as the evanescent near-field extends into the depth of a sample subsurface (Taubner et al., 2005).

Scattering NSOM can be performed in a variety of different spectral ranges. The earliest use of s-NSOM was with visible light (Rayleigh scattering) (Inouye and Kawata, 1994), but it has now been extended into the IR and THz regimes (Knoll and Keilmann, 1999; Keilmann, 2004; Taubner et al., 2005; Huber et al., 2008). This progression has been nicely summarized in a recent review article (Hayazawa et al., 2009). Since resolution is limited primarily by tip sharpness, for increasingly long wavelengths, near-field resolution outdoes the far-field by increasingly greater margins. For example, 10 nm resolution in the visible region corresponds to $\sim\lambda/63$ (Hillenbrand and Keilmann, 2002), but in the IR a resolution of 8 nm corresponds to $\sim\lambda/400$ (Raschke et al., 2005). More dramatically, moving into the THz regime can yield resolution on the order of 40 nm or $\sim\lambda/3000$ (Huber et al., 2008).

Quenching, photobleaching, and signal intermittency (blinking) are generally not a problem with s-NSOM in contrast to fluorescence imaging. Thus, the full measure of field enhancement can be leveraged to achieve very large signal enhancement factors. In addition, although s-NSOM suffers from a large background signal due to the oblique angle of illumination (see above), it can be alleviated to a degree by increasing the laser intensity. High-intensity illumination leads to a large scattering signal (despite being a small fraction of the total signal), which allows for more sophisticated detection techniques such as lock-in amplification at higher harmonics in tandem with interferometric schemes (Hillenbrand and Keilmann, 2000, 2002).

17.4.3.1 Visible (Rayleigh) Scattering

The visible range of the electromagnetic spectrum is remarkably diverse in terms of the range of optical response exhibited by various materials. For example, while metals generally have a strong optical response in the visible, particularly near plasmon resonance frequencies (see Section 17.2.1), most dielectric materials are transparent through the visible range. The response of metals in the visible is enhanced even further for nanoscale structures, since localized plasmon resonances can then lead to a rich spectrum of optical responses depending on the geometry of the constituent structures. Rayleigh scattering, by definition, arises from scattering from particles much smaller than the wavelength of light being employed and is a function of the particle polarizability. Thus, at visible wavelengths, variations in the polarizability can arise either from a boundary between different materials or from variations in the nanoscale geometry of a particular material. The strength of the scattering signal will reflect such differences in the polarizability as the tip is scanned over a boundary separating distinct regions. In addition, since the tip interacts with its image dipole within the material in s-NSOM, it can be used to image subsurface structures that do not register topographically.

To demonstrate subsurface optical contrast, Taubner and coworkers fabricated a sample consisting of gold nanoparticles ~25 nm in height on a silicon substrate and then were able to deposit polystyrene strips onto the sample in such a way that some of the gold nanoparticles were buried beneath a thin layer of polystyrene (Taubner et al., 2005). Figure 17.19 shows the simultaneously obtained topographic and s-NSOM signals, again obtained at $\lambda = 633$ nm after demodulation at the third harmonic of the tip oscillation frequency. The subsurface features are weaker, since they are further from the tip apex, and thus experience a smaller enhanced field. This also leads to image broadening for these features.

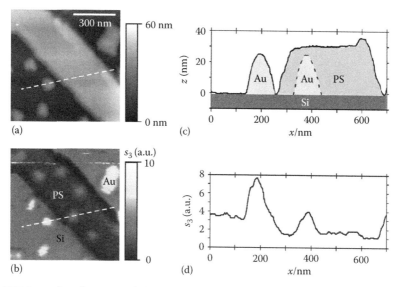

Figure 17.19 Subsurface near-field imaging with visible light (λ = 633 nm) of a sample consisting of Au islands on a Si surface, partly covered by a polystyrene (PS) strip. (a) Topography and (b) demodulated optical amplitude at the third harmonic of the tip oscillation frequency. (c) and (d) are line profiles extracted from images (a) and (b), illustrated with a sketch of the sample's structure. (Image adapted with permission from Taubner, T. et al., *Opt. Express*, 13, 8893, 2005.)

17.4.3.2 Infrared Scattering

One of the virtues of s-SNOM is the ability to explore different wavelength ranges; for example, operating s-SNOM at infrared frequencies can provide the ability to spectroscopically map vibrational resonances, thus offering chemical and structural nanoscale sensitivity (Taubner et al., 2005). Sometimes materials that show little or no contrast at visible wavelengths under s-NSOM yield excellent contrast at longer wavelengths. Materials exhibiting strong ro-vibrational modes should benefit from imaging at infrared frequencies—as activity of a ro-vibrational mode at a given energy level leads to a peak in the permittivity at the corresponding wavelength and thus a change in the s-SNOM signal (Equation 17.13). Figure 17.20 demonstrates this principle as a poly(styrene-b-2-vinylpyridine) (PS-b-P2VP) surface shows increased contrast when imaged with IR light (Raschke et al., 2005).

17.4.3.3 Terahertz (THz)

Moving to even lower energies, THz radiation is not only able to excite molecular vibrations and phonons, but also plasmons and electrons of nonmetallic conductors (Huber et al., 2008) (Chapter 11). This ability makes THz s-NSOM a valuable tool in the semiconductor industry as a device characterization technique, as it is sensitive to even the various levels of doping in Si. (Chen et al., 2003; Cho et al., 2005; Huber et al., 2008).

Of course working at THz frequencies has the challenges of creating a THz excitation source and the need for specialized detectors sensitive in this spectral region (Chapter 11). Nevertheless, even with these differences the technique is extremely similar to its s-NSOM counterparts at higher frequencies; the main principles of modulation/demodulation schemes and optimization of the probe remain the same, needing only to be optimized for this part of the spectrum. One key advantage of operating in this part of the spectrum is that THz s-NSOM has the capability to distinguish different materials by recognizing the changes in

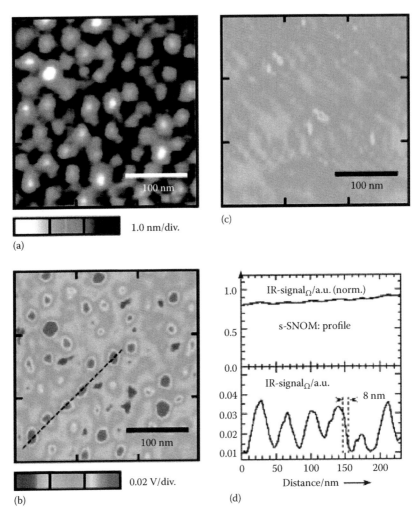

Figure 17.20 Simultaneously recorded (a) topographic and (b) infrared vibrational IR s-SNOM image at 3.39 μm (2950 cm⁻¹) of a PS-b-P2VP surface. The protruding regions corresponding to P2VP domains deliver a larger signal in accordance with their higher value for the complex dielectric constant. (c) Corresponding s-SNOM image of a similar sample region recorded at a wavelength of 632.8 nm with no distinct contrast. (d) IR s-SNOM spatial cross section along the dashed line in (b) (upper trace: raw data; lower trace: normalized and far-field-corrected data). The lower trace indicates a spatial resolution as high as 8 nm. (Reprinted with permission from Raschke, M.B. et al., *Chemphyschem*, 6, 2197, 2005.)

mobile carrier concentration (Huber et al., 2008). In a comparison of lower-energy s-NSOM techniques, THz illumination is unique in this ability to differentiate specific materials, surpassing other forms of illumination such as infrared and enabling further investigations into nonmetallic conductivity.

17.5 SUMMARY AND OUTLOOK

In summary, apertureless NSOM has proven to be extremely beneficial for use in imaging and understanding fundamental light–matter interactions on the nanometer scale. Paramount in these endeavors is the ability to achieve high spatial resolution, which has been demonstrated on a wide variety of samples in several distinct imaging modes—allowing for imaging or material characterization over a wide range of the electromagnetic spectrum. The various imaging modes available in NSOM also allow for both 3D mapping of light–matter interactions near a surface as well as imaging subsurface features. Apart from offering high spatial resolution, time-resolved measurements have added yet another dimension for investigation, as NSOM has been utilized to simultaneously achieve high spatial resolution and fluorescence lifetime data. Additionally, NSOM has been proven capable of providing spectroscopic information ranging from identification and localization of biomolecules and single proteins through the use of fluorescent tags to producing chemical fingerprints of individual nanotubes, with the ability of distinguishing between different types of polymers.

While the various iterations of NSOM have proven to be valuable scientific tools, near-field microscopy is still a young field with much yet to be explored. Many of the techniques highlighted in this chapter are recent advancements, thus leaving many opportunities to exploit their true power. For example, the 3D SP-NFT technique mentioned earlier is currently being used to map the energy-transfer landscape from single emitters to individual carbon nanotubes. Other NSOM contrast mechanisms are also being developed, such as rapidly rotating the polarization of the light incident on a near-field probe to allow for signal demodulation, which may serve as a great benefit for densely populated samples. Furthermore, additional data channels can be incorporated into NSOM by attaching biomolecules to probes, thus allowing for mapping of chemically specific binding interactions. Even though it is still in its experimental infancy, NSOM has emerged as a superior imaging technique and holds much promise for furthering our understanding of physics on the nanometer scale.

REFERENCES

Adachi, S. (1988) Model dielectric constants of Si and Ge. *Physical Review B*, 38, 12966.

Anderson, N., Hartschuh, A., and Novotny, L. (2007) Chirality changes in carbon nanotubes studied with near-field Raman spectroscopy. *Nano Letters*, 7, 577–582.

Anger, P., Bharadwaj, P., and Novotny, L. (2006) Enhancement and quenching of single-molecule fluorescence. *Physical Review Letters*, 96, 113002.

Barnes, W. L., Dereux, A., and Ebbesen, T. W. (2003) Surface plasmon subwavelength optics. *Nature*, 424, 824–830.

Bharadwaj, P., Anger, P., and Novotny, L. (2007) Nanoplasmonic enhancement of single-molecule fluorescence. *Nanotechnology*, 18, 044017.

Bharadwaj, P. and Novotny, L. (2007) Spectral dependence of single molecule fluorescence enhancement. *Optics Express*, 15, 14266–14274.

Bohn, J. L., Nesbitt, D. J., and Gallagher, A. (2001) Field enhancement in apertureless near-field scanning optical microscopy. *Journal of Optics Society American A*, 18, 2998–3006.

Bryant, G. W., de Abajo, F. J. G., and Aizpurua, J. (2008) Mapping the plasmon resonances of metallic nanoantennas. *Nano Letters*, 8, 631–636.

Carminati, R., Greffet, J. J., Henkel, C., and Vigoureux, J. M. (2006) Radiative and non-radiative decay of a single molecule close to a metallic nanoparticle. *Optics Communications*, 261, 368–375.

Chen, H. T., Kersting, R., and Cho, G. C. (2003) Terahertz imaging with nanometer resolution. *Applied Physics Letters*, 83, 3009–3011.

Cho, G. C., Chen, H. T., Kraatz, S., Karpowicz, N., and Kersting, R. (2005) Apertureless terahtertz near-field microscopy. *Semiconductor Science and Technology*, 20, S286–S292.

Darmanyan, S. A. and Zayats, A. V. (2003) Light tunneling via resonant surface plasmon polariton states and the enhanced transmission of periodically nanostructured metal films: An analytical study. *Physical Review B*, 67, 035424.

Demming, A. L., Festy, F., and Richards, D. (2005) Plasmon resonances on metal tips: Understanding tip-enhanced Raman scattering. *The Journal of Chemical Physics*, 122, 184716–184717.

Dorn, R., Quabis, S., and Leuchs, G. (2003) Sharper focus for a radially polarized light beam. *Physical Review Letters*, 91, 233901.

Eligal, L., Culfaz, F., Mccaughan, V., Cade, N. I., and Richards, D. (2009) Etching gold tips suitable for tip-enhanced near-field optical microscopy. *Review of Scientific Instruments*, 80, 033701.

Farahani, J. N., Pohl, D. W., Eisler, H. J., and Hecht, B. (2005) Single quantum dot coupled to a scanning optical antenna: A tunable superemitter. *Physical Review Letters*, 95, 017402.

Frey, H. G., Keilmann, F., Kriele, A., and Guckenberger, R. (2002) Enhancing the resolution of scanning near-field optical microscopy by a metal tip grown on an aperture probe. *Applied Physics Letters*, 81, 5030–5032.

Frey, H. G., Witt, S., Felderer, K., and Guckenberger, R. (2004) High-resolution imaging of single fluorescent molecules with the optical near-field of a metal tip. *Physical Review Letters*, 93, 200801.

Gerhardt, I., Wrigge, G., Bushev, P., Zumofen, G., Agio, M., Pfab, R., and Sandoghdar, V. (2007) Strong extinction of a laser beam by a single molecule. *Physical Review Letters*, 98, 033601.

Gerton, J. M., Wade, L. A., Lessard, G. A., Ma, Z., and Quake, S. R. (2004) Tip-enhanced fluorescence microscopy at 10 nanometer resolution. *Physical Review Letters*, 93, 180801.

Hartschuh, A., Qian, H. H., Meixner, A. J., Anderson, N., and Novotny, L. (2005) Nanoscale optical imaging of excitons in single-walled carbon nanotubes. *Nano Letters*, 5, 2310–2313.

Hayazawa, N., Inouye, Y., Sekkat, Z., and Kawata, S. (2001) Near-field Raman scattering enhanced by a metallized tip. *Chemical Physics Letters*, 335, 369–374.

Hayazawa, N., Inouye, Y., Sekkat, Z., and Kawata, S. (2002) Near-field Raman imaging of organic molecules by an apertureless metallic probe scanning optical microscope. *Journal of Chemical Physics*, 117, 1296–1301.

Hayazawa, N., Tarun, A., Taguchi, A., and Kawata, S. (2009) Development of tip-enhanced near-field optical spectroscopy and microscopy. *Japanese Journal of Applied Physics*, 48, 08JA02.

Hayazawa, N., Yano, T., Watanabe, H., Inouye, Y., and Kawata, S. (2003) Detection of an individual single-wall carbon nanotube by tip-enhanced near-field Raman spectroscopy. *Chemical Physics Letters*, 376, 174–180.

Hecht, B., Bielefeldt, H., Inouye, Y., Pohl, D. W., and Novotny, L. (1997) Facts and artifacts in near-field optical microscopy. *Journal of Applied Physics*, 81, 2492–2498.

Hillenbrand, R. and Keilmann, F. (2000) Complex optical constants on a subwavelength scale. *Physical Review Letters*, 85, 3029.

Hillenbrand, R. and Keilmann, F. (2002) Material-specific mapping of metal/semiconductor/dielectric nanosystems at 10 nm resolution by backscattering near-field optical microscopy. *Applied Physics Letters*, 80, 25–27.

Höppener, C., Beams, R., and Novotny, L. (2009) Background suppression in near-field optical imaging. *Nano Letters*, 9, 903–908.

Höppener, C. and Novotny, L. (2008) Antenna-based optical imaging of single Ca^{2+} transmembrane proteins in liquids. *Nano Letters*, 8, 642–646.

Hu, D., Micic, M., Klymyshyn, N., Suh, Y. D., and Lu, H. P. (2003) Correlated topographic and spectroscopic imaging beyond diffraction limit by atomic force microscopy metallic tip-enhanced near-field fluorescence lifetime microscopy. *Review of Scientific Instruments*, 74, 3347–3355.

Huber, A. J., Keilmann, F., Wittborn, J., Aizpurua, J., and Hillenbrand, R. (2008) Terahertz near-field nanoscopy of mobile carriers in single semiconductor nanodevices. *Nano Letters*, 8, 3766–3770.

Inouye, Y. and Kawata, S. (1994) Near-field scanning optical microscope with a metallic probe tip. *Optics Letters*, 19(3), 159–161.

Issa, N. A. and Guckenberger, R. (2007) Fluorescence near metal tips: The roles of energy transfer and surface plasmon polaritons. *Optics Express*, 15, 12131–12144.

Jackson, J. D. (1999) *Classical Electrodynamics* 3rd Edn., Hoboken, NJ: John Wiley & Sons.

Kalkbrenner, T., Ramstein, M., Mlynek, J., and Sandoghdar, V. (2001) A single gold particle as a probe for apertureless scanning near-field optical microscopy. *Journal of Microscopy-Oxford*, 202, 72–76.

Keilmann, F. (2004) Scattering-type near-field optical microscopy. *Journal of Electron Microscopy*, 53, 187–192.

Kelly, K. L., Coronado, E., Zhao, L. L., and Schatz, G. C. (2003) The optical properties of metal nanoparticles: The influence of size, shape, and dielectric environment. *Journal of Physical Chemistry B*, 107, 668–677.

Kittel, C. (2005) *Introduction to Solid State Physics*, Hoboken, NJ: John Wiley & Sons.

Knoll, B. and Keilmann, F. (1999) Mid-infrared scanning near-field optical microscope resolves 30 nm. *Journal of Microscopy-Oxford*, 194, 512–515.

Knoll, B., Keilmann, F., Kramer, A., and Guckenberger, R. (1997) Contrast of microwave near-field microscopy. *Applied Physics Letters*, 70, 2667–2669.

Kuhn, S., Hakanson, U., Rogobete, L., and Sandoghdar, V. (2006) Enhancement of single-molecule fluorescence using a gold nanoparticle as an optical nanoantenna. *Physical Review Letters*, 97, 017402.

Lewis, A., Isaacson, M., Harootunian, A., and Muray, A. (1984) Development of a 500 Å spatial resolution light microscope: I. light is efficiently transmitted through $\lambda/16$ diameter apertures. *Ultramicroscopy*, 13, 227–231.

Link, S. and El-Sayed, M. A. (1999) Size and temperature dependence of the plasmon absorption of colloidal gold nanoparticles. *Journal of Physical Chemistry B*, 103, 4212–4217.

Ma, Z. Y., Gerton, J. M., Wade, L. A., and Quake, S. R. (2006) Fluorescence near-field microscopy of DNA at sub-10 nm resolution. *Physical Review Letters*, 97, 260801.

Maier, S. A. and Atwater, H. A. (2005) Plasmonics: Localization and guiding of electromagnetic energy in metal/dielectric structures. *Journal of Applied Physics*, 98, 011101.

Mangum, B. D., Mu, C., and Gerton, J. M. (2008a) Resolving single fluorophores within dense ensembles: Contrast limits of tip-enhanced fluorescence microscopy. *Optics Express*, 16, 6183–6193.

Mangum, B. D., Mu, C., Ma, Z., and Gerton, J. M. (2008b) Single molecule contrast in tip-enhanced fluorescence microscopy. In Zayats, A. and Richards, D. (Eds.), *Nano-Optics and Near-Field Optical Microscopy*, Norwood, MA, Artech House.

Mangum, B. D., Shafran, E., Mu, C., and Gerton, J. M. (2009) Three-dimensional mapping of near-field interactions via single-photon tomography. *Nano Letters*, 9, 3440–3446.

Novotny, L. (1996) Single molecule fluorescence in inhomogeneous environments. *Applied Physics Letters*, 69, 3806–3808.

Novotny, L. (2002) Near-field optical characterization of nanocomposite materials. *Journal of the American Ceramic Society*, 85, 1057–1060.

Novotny, L. (2007a) Effective wavelength scaling for optical antennas. *Physical Review Letters*, 98, 266802.

Novotny, L. (2007b) The history of near-field optics. In Wolf, E. (Ed.) *Progress in Optics*. Amsterdam, the Netherlands: Elsevier.

Novotny, L., Bian, R. X., and Xie, X. S. (1997) Theory of nanometric optical tweezers. *Physical Review Letters*, 79, 645–648.

Novotny, L. and Hecht, B. (2006) *Principles of Nano-Optics*, Cambridge, U.K.: University Press.

Pohl, D. W., Denk, W., and Lanz, M. (1984) Optical stethoscope: Image recording with resolution lambda/20. *Applied Physics Letters*, 44, 651–653.

Protasenko, V. V., Gallagher, A., and Nesbit, D. J. (2004) Factors that influence confocal apertureless near-field scanning optical microscopy. *Optics Communications*, 233, 45–56.

Protasenko, V. V., Kuno, M., Gallagher, A., and Nesbitt, D. J. (2002) Fluorescence of single ZnS overcoated CdSe quantum dots studied by apertureless near-field scanning optical microscopy. *Optics Communications*, 210, 11–23.

Quabis, S., Dorn, R., Eberler, M., Glöckl, O., and Leuchs, G. (2000) Focusing light to a tighter spot. *Optics Communications*, 179, 1–7.

Raschke, M. B., Molina, L., Elsaesser, T., Kim, D. H., Knoll, W., and Hinrichs, K. (2005) Apertureless near-field vibrational imaging of block-copolymer nanostructures with ultra-high spatial resolution. *Chemphyschem*, 6, 2197–2203.

Sonnichsen, C., Franzl, T., Wilk, T., Von Plessen, G., Feldmann, J., Wilson, O., and Mulvaney, P. (2002) Drastic reduction of plasmon damping in gold nanorods. *Physical Review Letters*, 88, 077402.

Stalder, M. and Schadt, M. (1996) Linearly polarized light with axial symmetry generated by liquid-crystal polarization converters. *Optics Letters*, 21, 1948–1950.

Stöckle, R. M., Schaller, N., Deckert, V., Fokas, C., and Zenobi, R. (1999) Brighter near-field optical probes by means of improving the optical destruction threshold. *Journal of Microscopy-Oxford*, 194, 378–382.

Stöckle, R. M., Suh, Y. D., Deckert, V., and Zenobi, R. (2000) Nanoscale chemical analysis by tip-enhanced Raman spectroscopy. *Chemical Physics Letters*, 318, 131–136.

Synge, E. H. (1928) A suggested model for extending microscopic resolution into the ultra-microscopic region. *The Philosophical Magazine*, 6, 356–362.

Taminiau, T. H., Moerland, R. J., Segerink, F. B., Kuipers, L., and Van Hulst, N. F. (2007) $\lambda/4$ Resonance of an optical monopole antenna probed by single molecule fluorescence. *Nano Letters*, 7, 28–33.

Taminiau, T. H., Stefani, F. D., Segerink, F. B., and Van Hulst, N. F. (2008) Optical antennas direct single-molecule emission. *Nature Photonics*, 2, 234–237.

Taubner, T., Keilmann, F., and Hillenbrand, R. (2005) Nanoscale-resolved subsurface imaging by scattering-type near-field optical microscopy. *Optics Express*, 13, 8893–8899.

Thio, T., Ghaemi, H. F., Lezec, H. J., Wolff, P. A., and Ebbesen, T. W. (1999) Surface-plasmon-enhanced transmission through hole arrays in Cr films. *Journal of the Optical Society of America B-Optical Physics*, 16, 1743–1748.

Tidwell, S. C., Kim, G. H., and Kimura, W. D. (1993) Efficient radially polarized laser beam generation with a double interferometer. *Applied Optics*, 32, 5222–5229.

Trabesinger, W., Kramer, A., Kreiter, M., Hecht, B., and Wild, U. P. (2002) Single-molecule near-field optical energy transfer microscopy. *Applied Physics Letters*, 81, 2118–2120.

Vukovic, S., Corni, S., and Mennucci, B. (2008) Fluorescence enhancement of chromophores close to metal nanoparticles. Optimal setup revealed by the polarizable continuum model. *Journal of Physical Chemistry C*, 113, 121–133.

Wolf, S. and Taubner, R. N. (1986) *Silicon Processing for the Vlsi Era*, Sunset Beach, CA, Lattice Press.

Xie, C. A., Mu, C., Cox, J. R., and Gerton, J. M. (2006) Tip-enhanced fluorescence microscopy of high-density samples. *Applied Physics Letters*, 89, 143117.

Yang, T. J., Lessard, G. A., and Quake, S. R. (2000) An apertureless near-field microscope for fluorescence imaging. *Applied Physics Letters*, 76, 378–380.

Yoskovitz, E., Oron, D., Shweky, I., and Banin, U. (2008) Apertureless near-field distance-dependent lifetime imaging and spectroscopy of semiconductor nanocrystals. *Journal of Physical Chemistry C*, 112, 16306–16311.

Zenhausern, F., O'Boyle, M. P., and Wickramasinghe, H. K. (1994) Apertureless near-field optical microscope. *Applied Physics Letters*, 65, 1623–1625.

PART V

RECENT DEVELOPMENTS

RECENT DEVELOPMENTS IN SPATIALLY AND TEMPORALLY RESOLVED OPTICAL CHARACTERIZATION OF SOLID-STATE MATERIALS

Rohit P. Prasankumar and Antoinette J. Taylor

CONTENTS

The immense value of optical tools in characterizing materials has continually driven the development of innovative optical techniques and systems. These novel techniques can offer several advantages over existing optical characterization methods, such as extending the range of data that can be obtained on a given material, enabling one to obtain this data in a simpler manner, providing greater precision and accuracy in the measured data, or revealing information on a specific material functionality. The leading optics and applied physics journals thus report new optical techniques that address one or more of the issues mentioned earlier on a nearly weekly basis, one indicator of the sheer volume of activity in this area. This clearly makes it impossible to even attempt to describe the full range of experimental optical techniques in a single book. Therefore, in previous chapters, we have focused on mature techniques that are each uniquely capable of determining a given set of material properties, from the use of four-wave mixing to measure ultrashort dephasing times to the ability of near-field scanning optical microscopy (NSOM) to spatially resolve nanoscale features in a noncontact manner. However, several emerging methods can characterize previously inaccessible material properties, such as changes in crystal structure, electronic band structure, and carrier transport, at unprecedented time and length scales. This makes them worthy of inclusion in this book, albeit in an all-too-brief description.

In this chapter, we will thus give an overview of several optical techniques that are, on the whole, less developed than the methods that were comprehensively detailed in previous chapters. Spatially resolved techniques of this nature were overviewed in Sections 15.4 and 15.5 and therefore will not be discussed in detail here. However, we will describe advanced time-resolved techniques, such as ultrafast x-ray and electron diffraction (UF-XRD and UED) and ultrafast photoemission spectroscopy (UF-PES), which have garnered much attention over the past decade. Furthermore, a major focus in current experimental optics is to develop tools that can simultaneously provide high spatial and temporal resolution for comprehensive studies of biological, chemical, and physical systems. Therefore, we will also discuss some of the most exciting developments along these lines, including time-resolved near-field scanning optical microscopy (TR-NSOM) and ultrafast scanning tunneling microscopy (UF-STM). The descriptions given in this chapter will necessarily be brief, primarily focusing on the conceptual basis for each technique and the useful information that one might hope to obtain on a given sample. Our modest goal is to establish the ability of these novel techniques to reveal formerly unobtainable material properties, after which the extensive reference list can be consulted for more detailed information.

18.1 ADVANCED TIME-RESOLVED OPTICAL TECHNIQUES

Historically, the vast majority of techniques for temporally resolving carrier dynamics were first developed at optical frequencies, both due to the ready availability of ultrashort pulsed lasers at optical and near-infrared (IR) wavelengths and to the relative ease of working in this spectral range. However, the use of ultrashort pulses outside of the optical frequency range for temporally resolving carrier dynamics has rapidly gained traction in recent years. One reason for this is the existence of resonant transitions in materials at both lower ($<\sim 1\,eV$) and higher ($>\sim 6\,eV$) photon energies, such as the superconducting gap in high-T_c superconductors, intraband transitions in semiconductor nanostructures, and nuclear transitions in atomic and molecular systems. In addition, ultrashort pulses at different frequencies can directly probe material properties, such as structural and conductivity dynamics, that can only be indirectly accessed by optical pulses. Finally, methods for generating ultrashort pulses at other frequencies through nonlinear optical processes such as high-harmonic generation (HHG) and difference frequency generation have rapidly matured in recent years, resulting in the generation of single-cycle pulses at terahertz (THz) frequencies and the shortest pulses ever generated at extreme ultraviolet (UV) frequencies (<100 attoseconds [as] at extreme UV wavelengths) (Krausz and Ivanov, 2009). The generation and use of mid- to far-IR pulses is arguably more mature than ultrashort pulse generation at shorter wavelengths ($<\sim 200\,nm$) and is therefore covered in detail in Chapter 11.

In this section, we will provide an overview of techniques for probing materials with high photon energies ($h\omega \geq 6\,eV$ [$\sim 200\,nm$]), which reveal photoinduced changes in the crystal or electronic band structure on a femtosecond (fs) timescale, often with higher temporal and spatial resolution than at optical frequencies. We will begin by describing the measurement of structural dynamics in materials through UED and UF-XRD, two techniques that have shed much light on a variety of processes in condensed matter, such as nonthermal melting (Siders et al., 1999), photoinduced phase transitions (Baum et al., 2007;

Cavalleri et al., 2001b), and ultrafast carrier-driven nanoscale expansion (Yang et al., 2008). We will also describe the closely related techniques of ultrafast x-ray absorption spectroscopy (UF-XAS) and ultrafast electron crystallography/microscopy (UEC and UEM) in this section. The focus of the chapter will then switch to ultrafast angle-resolved photoemission spectroscopy (UF-ARPES), which is an exciting technique that enables the direct measurement of photoinduced changes in the electronic band structure on a femtosecond timescale.

Finally, it would be remiss of us not to briefly mention the tremendous progress that has taken place over the last 10 years in the generation and measurement of attosecond pulses at extreme UV wavelengths and their emerging use for tracking processes at the fundamental timescales of electron, lattice, and nuclear motion (Bucksbaum, 2007; Corkum and Krausz, 2007; Goulielmakis et al., 2007; Kapteyn et al., 2007; Krausz and Ivanov, 2009). This was made possible by several innovations in ultrafast technology, including carrier-envelope phase control (Chapter 7), HHG (Section 18.1.1.1), and attosecond pulse measurement techniques (Corkum and Krausz, 2007). This has enabled the generation of pulses as short as 78 attoseconds at extreme UV wavelengths (Goulielmakis et al., 2007; Krausz and Ivanov, 2009). To date, attosecond pulses have been used to measure ultrafast atomic-scale atomic and nuclear dynamics in atoms, molecules, and clusters. However, there have been relatively few applications of attosecond pulses in measuring dynamics in the condensed matter systems that are the focus of this book (Krausz and Ivanov, 2009), except when using UV probe pulses in UF-ARPES experiments. The rapid progress in this field makes it very likely that attosecond spectroscopy will find many applications in condensed matter physics, probing phenomena such as charge screening in solids, collective electronic motion in metallic nanosystems, and electron transfer across interfaces (Krausz and Ivanov, 2009), in years to come.

18.1.1 Ultrafast Structural Dynamics

The groundbreaking development of x-ray diffraction (XRD) and electron diffraction (ED) nearly a century ago revolutionized the measurement of the atomic-level structure of materials ranging from DNA to high-temperature superconductors. However, many natural phenomena, such as phase transitions, chemical reactions, and biological functions, are driven by structural changes that can occur on timescales as short as femtoseconds. A deep understanding of these phenomena therefore depends on the ability to track the material structure during the intermediate stages between stable states. This can be accomplished using UF-XRD and UED, the most well developed techniques for directly resolving dynamic changes in material structure after ultrashort pulsed photoexcitation. Both of these techniques are capable of resolving structural changes in a variety of material systems with high temporal and spatial resolution. Their respective advantages and limitations for examining structural dynamics in a given material can, in large part, be traced back to their parent techniques, ED and XRD. We will therefore briefly discuss ED and XRD here to provide a common basis for the descriptions of their time-resolved versions in the following sections.

ED and XRD are two of the most powerful methods for examining the static crystal structure of biological, chemical, and physical systems. Both methods rely on passing a beam of x-rays or electrons through a crystalline sample and observing the resulting diffraction pattern from which the crystal structure can be deduced. They are thus governed by many of the same equations. A general representation was developed for XRD by W. H. and W. L. Bragg, who formulated an expression to account for the intense peaks of reflected radiation observed for certain wavelengths and incident angles ("Bragg peaks") during the diffraction of incident x-rays by crystalline solids (Ashcroft and Mermin, 1976; Kittel, 1995). They regarded a crystal as being constructed of parallel planes of ions separated by a distance d. Constructive interference from these planes then occurs if the well-known Bragg condition is satisfied,

$$n\lambda = 2d\sin\theta \qquad (18.1)$$

where
θ is the incident angle
λ is the wavelength of the incident radiation
n is an integer

For a given wavelength and incidence angle, the spacing between different crystal planes can then be determined. It is clear from this condition that λ must be less than $2d$ to observe constructive interference, which means that λ must be in the hard x-ray range (<1 nm) for XRD experiments since d is typically several angstroms. Equation 18.1 is also valid for electrons (as well as neutrons), which typically have much higher energies (tens to hundreds of kiloelectronvolts) and correspondingly shorter wavelengths.

Generally, the diffraction intensity of a Bragg peak is proportional to the square of the structure factor (Ashcroft and Mermin, 1976; Bressler and Chergui, 2004; Chergui and Zewail, 2009; Kittel, 1995):

$$F_g = \sum_j f_j e^{-2\pi i g \cdot r_j} \qquad (18.2)$$

where
f_j is the atomic scattering factor
g is the scattering vector of the diffracted beam
r_j is the position of the jth atom in the unit cell

The structure factor describes the manner in which an incident beam is scattered from a crystal unit cell, with the specific scattering powers of the elements comprising the unit cell taken into account through f_j. This atomic scattering factor is different for electrons and x-rays since they interact with matter through different mechanisms.

As mentioned earlier, ED and XRD each have relative advantages and disadvantages for extracting the crystal structure of different materials. These are worth highlighting here since, to a large extent, they directly govern the relative properties of UF-XRD and UED. Perhaps the most significant difference is the ~5 orders of magnitude greater scattering cross section for electrons as compared to x-rays (Chergui and Zewail, 2009). This originates from the electron charge, which causes Coulomb forces between the incident electrons and the nuclei/valence electrons to dominate the interaction, as compared to x-rays that only couple to the core electrons that are spatially localized around the nucleus. The diffracted x-ray intensity thus depends on the electron density around each nucleus and is therefore sensitive to the underlying atomic structure (Cavalleri et al., 2001a; Rousse et al., 2001). In contrast, ED is largely a surface-sensitive technique since the large scattering cross section limits the penetration depth of the electrons into matter. The quasi-two-dimensional (2D) nature of ED typically results in diffraction "rings" from crystalline solids, as compared to the diffraction "spots" measured by XRD, which probes the three-dimensional (3D) bulk material (Ashcroft and Mermin, 1976). These properties strongly influence the types of materials that can be investigated by each technique; for example, samples for ED experiments must typically be <100 nm in thickness, which makes this an excellent technique for investigating the structural properties of nanostructures (down to single nanoparticles (NPs) [Yang et al., 2008]) but requires considerable sample preparation for the study of bulk samples (such as single crystals). In contrast, although the weak scattering cross section for x-rays limits their use in investigating certain samples, they have proven to be particularly useful in determining crystal and protein structures. The different strengths of these techniques have made ED the tool of choice for characterizing gases, surfaces, and thin crystals (Zewail, 2006), while XRD is more commonly used for characterizing solid-state materials and proteins/molecules that can be crystallized. More details about each of these techniques and their similarities and differences can be found in references (Ashcroft and Mermin, 1976; Bressler and Chergui, 2004; Chergui and Zewail, 2009; Kittel, 1995). These characteristics of ED and XRD are carried over to the time-resolved versions of these techniques, helping determine the applicability of each technique to a given materials system.

Finally, x-ray absorption spectroscopy (XAS) is another technique capable of measuring the structural properties of a variety of materials, through simply measuring the absorption of x-rays from a tunable source as a function of photon energy E (Bressler and Chergui, 2004; Chergui and Zewail, 2009). The resulting spectra typically reveal absorption that decreases with increasing photon energy, with discrete absorption edges due to the ionization of core-shell electrons at low energies (known as x-ray absorption near-edge structure [XANES]) and an oscillatory structure appearing at high energies, well above the photon energy above which all electrons are ionized (known as extended x-ray absorption fine structure [EXAFS]). These oscillations are due to interference effects between the incident and backscattered photoelectrons at nearest neighbor atoms, which allows one to deduce structural information from the normalized EXAFS spectrum. XANES gives both electronic and molecular structural information, although it is significantly more difficult to extract direct structural information from XANES than EXAFS (Bressler and Chergui, 2004). Importantly, XAS can be used to extract the structure of amorphous and liquid materials, which are typically difficult to examine with XRD (Chergui and Zewail, 2009).

18.1.1.1 Ultrafast X-Ray Diffraction and Absorption

The addition of time resolution to a standard x-ray diffraction experiment is not a new idea, as time-resolved XRD and XAS experiments with nanosecond time resolution were first demonstrated 20–25 years ago (Bressler and Chergui, 2004; Rousse et al., 2001), and time-resolved XRD experiments at synchrotron light sources can routinely achieve ~100 picosecond (ps) time resolution (Pfeifer et al., 2006). However, truly "ultrafast" XRD experiments have only been performed within the past ~10 years (Bressler and Chergui, 2004; Cavalleri et al., 2001a), primarily due to previous difficulties in generating sub-picosecond x-ray pulses. Several approaches were devised for overcoming these issues (described in the following), enabling the study of photoinduced structural changes such as nonthermal melting (Siders et al., 1999), acoustic phonon generation and propagation (Lindenberg et al., 2000), and interactions between electronic and lattice degrees of freedom in strongly correlated systems (Cavalleri et al., 2001b) with femtosecond time resolution.

The experimental setup for an UF-XRD experiment is conceptually similar to that of an optical pump–probe experiment (Chapter 9, e.g., Figure 9.1). Typically, a sub-100 fs pulse from an amplified Ti:sapphire laser is split into two parts, one to generate the x-ray probe pulse (the different methods for doing this are discussed in the following) and another to photoexcite the sample. The pump pulse then initiates a structural change that can be tracked by diffracting the x-ray probe pulse from the sample and measuring the diffraction intensity and/or angular position of the resulting Bragg peaks as a function of time delay (Bargheer et al., 2006). This is often accomplished by imaging the diffracted probe pulse onto an x-ray CCD camera; photoinduced structural changes in the sample at a given time delay can then be obtained by taking the ratio of the diffraction spectrum after photoexcitation to that taken prior to photo-excitation (Cavalleri et al., 2001b; Rose-Petruck et al., 1999).

Femtosecond x-ray sources generally fall into one of the two categories: tabletop laser-based sources or synchrotron-based sources. These sources have been extensively described in several review papers (Bargheer et al., 2006; Chergui and Zewail, 2009; Pfeifer et al., 2006; Rousse et al., 2001), and Bressler and Chergui (2004) provide a table comparing the relevant properties of the different x-ray sources discussed in the following as well as a useful discussion comparing plasma and synchrotron sources. To a large extent, the physics governing all ultrafast x-ray sources is the same: electrons are accelerated in an electric field to extremely high velocities, causing them to emit x-rays. However, there are significant disparities between different x-ray sources that are worth discussing. Laser-based plasma x-ray sources operate by focusing an intense femtosecond laser pulse (peak intensity $\sim 10^{17}$ W/cm^2) onto a metal target. This generates a plasma in which electrons are accelerated by the pulse electric field to near-relativistic velocities, emitting x-ray radiation with a spectrum containing intense line emissions characteristic of the material along with a weak continuum background (Bargheer et al., 2006; Cavalleri et al., 2001a; Chergui and Zewail, 2009). These are the only tabletop sources that can generate hard x-rays (energies >1 keV) with femtosecond pulse durations (limited by the driving laser pulse) at repetition rates from a few hertz to several kilohertz (Chergui and Zewail, 2009). However, the generated x-rays are spatially incoherent and emitted in all directions, making the maximum achievable flux on the sample heavily dependent on the focusing x-ray optics (Bargheer et al., 2006; Bressler and Chergui, 2004; Chergui and Zewail, 2009).

More recently, the technique of HHG has attracted much attention due to its ability to produce extremely short, temporally and spatially coherent sub-femtosecond pulses (Pfeifer et al., 2006). These ultrashort pulses are centered at "extreme UV" frequencies that are integer multiples of the fundamental frequency, up to ~ 500 eV (Chergui and Zewail, 2009). They are created by focusing amplified sub-10 fs laser pulses (each with energy typically ≥ 1 mJ) into a gas (e.g., neon, argon, or krypton), which ionizes electrons near the peak of the driving electric field. The electrons then accelerate in the field away from their parent ions until the field switches sign, causing the electrons to change direction and reaccelerate toward the ions. The kinetic energy gained through this electron motion, along with the ionization potential, is converted into photon energy when the electron recombines with its parent ion. This process occurs every half cycle of the laser field, resulting in an HHG spectrum with peaks at odd multiples of the driving laser frequency. These pulses have been very useful in studying nuclear and electronic dynamics in atoms and molecules (Krausz and Ivanov, 2009) but as yet cannot be used to track structural dynamics since photon energies >2 keV are required (from the Bragg condition (18.1) described earlier) (Bressler and Chergui, 2004; Chergui and Zewail, 2009).

Synchrotron-based x-ray pulses are generated when electron bunches accelerated to relativistic velocities in a storage ring emit x-rays upon interacting with a bend magnet or undulator. Although this method produces brighter x-ray pulses with wider tunability than any other technique, the electron bunches (and therefore the resulting x-ray pulses) are limited in duration to tens of picoseconds, necessitating the use of other approaches to reduce the pulse duration. One of these approaches is the "slicing" scheme developed at the Advanced Light Source (ALS) at Lawrence Berkeley National Laboratory (Schoenlein et al., 2000), which uses a femtosecond laser pulse to "slice" out a small part (~ 100 fs) of a co-propagating electron bunch that can then be used to produce an ultrashort x-ray pulse at a separate bend magnet or undulator. The disadvantage of this technique is that the vast majority of electrons in the original bunch are discarded in the slicing process, significantly reducing the brightness of the resulting x-ray pulse. Other methods for reducing the x-ray pulse duration from synchrotron sources are described in (Cavalleri et al., 2001a; Pfeifer et al., 2006; Schoenlein et al., 1996).

Finally, the use of free-electron lasers (FELs) based on linear electron accelerators to generate high-brightness x-ray pulses has very recently become possible (Bargheer et al., 2006; Bressler and Chergui, 2004; Chergui and Zewail, 2009). The first demonstration of this kind was the sub-picosecond pulse source (SPPS) at Stanford, which provided 10^6 photons (as compared to 10^4 photons from electron slicing) in a 100 fs pulse at a photon energy of 9 keV and a 10 Hz repetition rate (Chergui and Zewail, 2009). This has motivated the development of hard x-ray FELs, expected to provide up to 10^{12} x-ray photons/pulse, which are beginning operation now at Stanford and Hamburg (Chergui and Zewail, 2009). The Hamburg source has already enabled single-shot coherent soft x-ray diffraction imaging of laser-ablated nanostructures with 50 nm spatial resolution and 10 ps temporal resolution (Barty et al., 2008), which is the first demonstration of real-time imaging of photoinduced structural changes. This was previously impossible with existing sources due to the combination of their limited photon flux and the weak diffraction efficiency ($\sim 10^{-5}$) of x-ray photons when interacting with a sample. The high photon flux from x-ray FELs may also enable UF-XRD experiments on biological molecules that will allow researchers to track structural processes on a femtosecond timescale (Rousse et al., 2001).

The applicability of UF-XRD to the study of ultrafast structural dynamics in semiconductors was immediately evident once suitable sources were available. This resulted in the direct observation of effects such as acoustic phonon propagation and nonthermal melting using both laser-based plasma and synchrotron x-ray sources (Lindenberg et al., 2000; Siders et al., 1999). Here, we will discuss nonthermal melting as a unique example of the capabilities of UF-XRD with both classes of sources. This phenomenon occurs when an ultrashort pulse with photon energy greater than the band gap and fluence more than twice the melting threshold deposits energy into the system by photoexciting carriers. This causes an ultrafast solid-to-liquid phase transition (due to disordering introduced by the photoexcitation) that occurs before the energy in the electronic subsystem is transferred to the lattice through electron–phonon interactions (Chin et al., 1999; Siders et al., 1999). The first direct evidence of this phenomenon was reported in Chin et al. (1999). In this work, ultrashort x-ray pulses obtained from a synchrotron source were diffracted from an InSb crystal after femtosecond optical photoexcitation. The number of diffracted photons at sub-picosecond time delays decreased by ~6% due to the formation of a thin disordered layer, well before carrier–phonon thermalization could take place. A spectral shift of the diffraction peak due to lattice expansion was observed after ~2 ps, indicating that the energy in the electronic subsystem was transferred to the lattice on this timescale.

Similar experiments were performed on single crystal Ge films using a laser-based plasma source (Siders et al., 1999). Ultrafast optical excitation reduced the integrated XRD by 20% in the central part of the pump spot within 7 ps while the edge of the pump spot appeared equivalent to the un-pumped crystal (Figure 18.1). At 40 ps, the integrated diffraction from the central spot was further reduced to 50% of the un-pumped value, and the diffraction from the edge was reduced by 20%. The reduction in diffraction efficiency from the central spot at early times is significantly greater than that expected from that of a superheated solid (~10%), indicating that a ~35 nm layer of the film is disordered within a few picoseconds. The drop in diffraction efficiency on slower timescales occurs due to conventional thermal melting. Later UF-XRD experiments demonstrated that the generation of large amplitude coherent phonons drives the nonthermal melting process (Lindenberg et al., 2000). Overall, these experiments demonstrate the ability of UF-XRD to directly track nonthermal melting and have been followed by several other studies of this nature (Pfeifer et al., 2006 and references within) to gain more insight on this phenomenon.

Vanadium dioxide (VO_2) is a model strongly correlated electron system that undergoes an insulator-to-metal transition (IMT) upon heating beyond $T_M = 340$ K concurrently with a

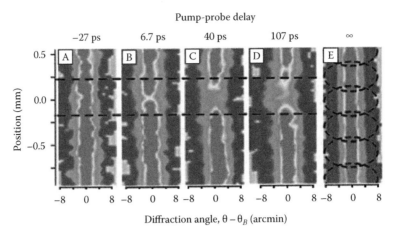

Figure 18.1 (See color insert.) Ultrafast XRD experiments on a Ge film at different time delays. The region of the sample photoexcited by the pump pulse is indicated by the dotted lines in (A)–(D). The image at infinite time delays (E) depicts six single-shot damage regions, clearly showing that the diffraction signal has almost completely recovered after several seconds. (From Siders, C. et al., Detection of nonthermal melting by ultrafast x-ray diffraction, *Science*, 286, 1340–1342, 1999. Reprinted with permission from AAAS.)

structural transition from a low-temperature (T) monoclinic to a high-T rutile phase. The nature of this phase transition has been a subject of much debate for many years, with some studies suggesting that it is structurally driven and others indicating that it is driven by electron localization. A wide range of time-resolved optical techniques have been used to address this question, including all-optical pump–probe spectroscopy (Chapter 9) (Becker et al., 1994; Cavalleri et al., 2004b; Kim et al., 2006), four-wave mixing (Lysenko et al., 2007a,b), optical-pump THz-probe spectroscopy (Hilton et al., 2007; Kübler et al., 2007), UF-XRD (Cavalleri et al., 2001b), UF-XRA (Cavalleri et al., 2004a, 2005), and UEC/UEM (Baum et al., 2007; Grinolds et al., 2006) (discussed in Section 18.1.1.2). A portion of this work is reviewed in Cavalleri et al. (2006). However, these studies, as well as many others using different optical and non-optical techniques, have failed to reach a definitive conclusion on the nature of the IMT in VO_2.

Here, we will discuss the UF-XRD experiments reported in Cavalleri et al. (2001b), which were the first direct observation of a solid–solid phase transition on an ultrafast timescale. An ultrafast laser-based plasma source was used to generate the x-ray probe pulses, which were then focused onto the bulk VO_2 sample, held at a temperature $T < T_M$. An ultrafast structural transition from the low-T monoclinic phase (corresponding to a diffraction angle of 13.9°, as observed in static XRD experiments) to the high-T rutile phase (diffraction angle 13.78°) was clearly observed through the development of a shoulder in the angle-dependent diffraction curves at the position of the equilibrium rutile phase within 300 fs (Figure 18.2a). Normalized diffraction curves, obtained by dividing the time-resolved diffraction curves (like those in Figure 18.2a) by the curve for the un-pumped sample, are shown in Figure 18.2b. These curves demonstrate that the high-T rutile phase is formed within ~1 ps in a ~50 nm thick layer near the surface; additional data (not shown) reveals that 250 nm of material is in the rutile phase after 10 ps. Although these experiments could not determine whether the phase transition in VO_2 is primarily structurally or electronically driven, they clearly demonstrated the utility of UF-XRD in characterizing correlated electron materials.

Ultrafast x-ray absorption spectroscopy with sub-picosecond time resolution is a much newer technique than UF-XRD; to the best of our knowledge, there have only been two studies of this nature on condensed matter systems to date (Cavalleri et al., 2005; Stamm et al.,

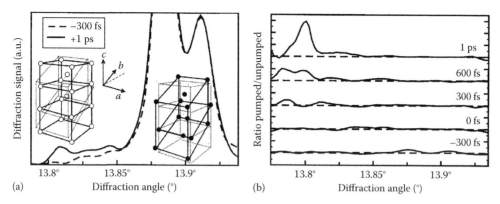

Figure 18.2 (a) Angle-dependent UF-XRD signals from VO_2 at $-300\,fs$ and $+1$ ps. The peaks at angles $>13.85°$ are associated with the low-T monoclinic phase, while the photoinduced peak near $13.8°$ is associated with the high-T rutile phase. (b) Normalized angle-dependent change in x-ray reflectivity at different time delays. (Reprinted with permission from Cavalleri, A., Tóth, C., Siders, C., Squier, J., Raksi, F., Forget, P., and Kieffer, J., *Phys. Rev. Lett.*, 87, 237401, 2001b. Copyright 2001 by the American Physical Society.)

2007). This is primarily due to the low tunability, low photon flux, and poor shot-to-shot stability of laser-based plasma sources (Chergui and Zewail, 2009), necessitating the use of synchrotron x-ray sources for time-resolved XAS experiments. Although there have been several demonstrations using these sources to probe condensed matter systems with sub-nanosecond time resolution (Bressler and Chergui, 2004; Chergui and Zewail, 2009), techniques for generating sub-picosecond x-ray pulses from these sources have only recently been developed as discussed earlier, and therefore their use for UF-XAS is still relatively unexplored. One might expect, however, that third generation x-ray FELs will be an excellent source for UF-XAS experiments once they commence regular operation, increasing the range of information that can be obtained about a variety of samples from ultrafast x-ray techniques.

18.1.1.2 Ultrafast Electron Diffraction, Crystallography, and Microscopy

The use of electrons as a probe for ultrafast structural dynamics has yielded a great deal of information about different biological, chemical, and physical systems (Zewail, 2006). Femtosecond electron-based techniques were pioneered by the group of A. H. Zewail at the California Institute of Technology in the early 1990s, when the concept was first proposed (Williamson and Zewail, 1991; Zewail, 1991), and since then investigated by several other groups, for example, Dwayne Miller's group at the University of Toronto (Harb et al., 2008; Sciaini et al., 2009; Siwick et al., 2003). Ultrafast electron diffraction, the first of these methods to be demonstrated, has primarily been applied to isolated molecular systems in the gas phase (Zewail, 2005). The next generation of UED systems was developed for ultrafast electron crystallography measurements on condensed matter, surfaces, and biological systems (Ruan et al., 2004; Zewail, 2005), after which ultrafast electron microscopy was developed to visualize dynamic structural changes with the spatial resolution of transmission electron microscopy (TEM) and the temporal resolution of ultrafast optical spectroscopy (Lobastov et al., 2005; Zewail, 2006). These techniques were initially applied to tracking dynamics from reagents to products in chemical reactions (Williamson et al., 1997) but since have been used to study ultrafast structural changes in a number of other systems, including several solid-state materials (discussed in more detail in the following). The advantages and disadvantages of using electrons vs. x-rays as structural probes, as discussed earlier, are largely derived from their "parent" techniques of ED and XRD. However, several additional concerns emerged when extending ED experiments to ultrashort timescales, including determining the temporal zero between the optical and electron pulses, minimizing space-charge effects that broaden the pulse duration and increasing the system sensitivity to small photoinduced changes (Srinivasan et al., 2003). These issues were successfully overcome, resulting in the capability to perform single-electron UED, UEC, and UEM experiments with sub-100 fs time resolution and sensitivity to photoinduced changes as small as 1% (Zewail, 2006). The combination of these powerful techniques has enabled the measurement of ultrafast structural dynamics in gases and condensed matter as well as substances (e.g., water, fatty acids) residing on crystalline surfaces.

In UED experiments, an ultrashort laser pulse is split into two parts, one to photoexcite the sample and the other to generate an electron pulse. The pump pulse then goes directly to the vacuum chamber to photoexcite the sample, while the probe pulse is directed onto a photocathode to generate electrons through the photoelectric effect. The electrons are then accelerated, collimated, and focused onto the sample, after which a CCD camera is used to detect the diffracted electrons. This basic concept also applies to UEC, although these complex experiments required a substantial redesign of earlier generation UED systems for improved sensitivity and time resolution; details are given in Ruan et al. (2004). Finally, UEM experiments essentially interface a TEM with femtosecond optical pulses to photoexcite the sample and generate the electron beam (with ~1 electron/pulse).

The experimental setup for all three techniques thus consists of four major components: (1) laser system, (2) vacuum chambers, (3) electron gun, and (4) CCD camera system. Here, we will give an overview of each component and their use in UED, UEC, and UEM; more detailed descriptions are given in Srinivasan et al. (2003) and Zewail (2006). The laser source in most contemporary implementations is an ultrafast Ti:sapphire laser system (Chapter 7) (although very recently the use of a 1036 nm laser system has been reported [Yurtsever and Zewail, 2009]). Amplified laser systems (repetition rates of 1 kHz, pulse energies of a few millijoule) are typically used for UED and UEC, but for UEM, the use of extremely weak femtosecond laser pulses from the oscillator alone was critical in achieving ultrashort single-electron pulses with minimal space-charge broadening (to be discussed in more detail in the following). The 800 nm Ti:sapphire pulses are frequency doubled (Lobastov et al., 2005) or tripled (Srinivasan et al., 2003) to generate the electron pulses; part of this beam is also split off to pump the sample.

The short penetration depth of electrons in matter and consequent surface sensitivity necessitates that ED experiments are performed in high vacuum. The first UED experiments used one chamber for the electron gun, the CCD detector, and the beam of sample molecules (Srinivasan et al., 2003). However, issues with arcing the electron gun (which can destroy the CCD) and instrument contamination from sample molecules motivated the use of several (up to four) separate vacuum chambers in subsequent UED, UEC, and UEM setups (Lobastov et al., 2005; Ruan et al., 2004; Srinivasan et al., 2003). The electron gun primarily consists of a negatively biased photocathode, which is back-illuminated with a femtosecond UV pulse to generate electrons through the photoelectric effect. The electrons are then focused with a magnetic lens assembly, and their direction is controlled with two pairs of aluminum deflection plates.

Ultrashort electron pulses can be significantly broadened by electron–electron repulsion (known as "space-charge broadening"), which often stretches a femtosecond electron pulse to several picoseconds (Dantus et al., 1994). Therefore, it is desirable to have as few electrons per pulse as possible to minimize these effects (~1000 electrons in a 1 ps pulse for UED experiments) (Zewail, 2006). This makes the scattering intensity significantly lower than in conventional ED experiments, which must be accounted for by improvements in detector sensitivity and data acquisition time. Higher time resolution was achieved by using a sensitive

CCD camera that is capable of detecting single electrons, leading to 1 ps time resolution in UED experiments. The long-range order of the crystalline samples measured in UEC experiments enabled the use of fewer electrons/pulse, resulting in 300 fs time resolution, and UEM single-electron imaging leads to sub-100 fs time resolution (Zewail, 2006).

These advances in the components comprising an ultrafast electron-based experiment have been critical in extending the ability to image ultrafast structural dynamics with extremely high spatial and temporal resolution. It was equally important to develop approaches for surmounting a number of experimental challenges, some of which were mentioned earlier (e.g., determining zero delay and detecting the small photoinduced changes induced by the pump pulse). The electron pulse duration is measured using streaking techniques in which two laser pulses with a known separation in time, ω_{sp}, are used to generate two electron pulses, which are then sent between a pair of metal deflector plates and subjected to a fast time-dependent electric field. This imparts a dispersive effect on the electron pulses, which separates them by a certain distance in pixels, D_{pix}, when impinging on a CCD detector. The pulse duration can then be calculated from ω_{sp} and D_{pix} (Srinivasan et al., 2003).

The temporal overlap of the laser and electron pulses is determined using an ion-induced lensing technique (Srinivasan et al., 2003; Zewail, 2006). Using the same UED setup, the laser pulse is used to ionize a gas through multiphoton processes. The resulting positively charged ions create a transient electric field that focuses the electron pulse as it passes through the plasma, only when it overlaps in time with the ionizing laser pulse. This can therefore be used as a measure of zero time delay between the laser and electron pulses in UED.

Another issue that affects experimental performance is velocity mismatch between the ultrashort laser and electron pulses, which affects the temporal resolution of the measurement, particularly for thick samples, since the velocity of the electron pulse is ~1/3 that of the laser pulse. This can be reduced by tuning the incidence angle between the electron and laser beams, along with using higher accelerating voltages to increase electron velocities (Srinivasan et al., 2003).

Finally, the development of the frame-reference method was invaluable in improving the sensitivity of UED to the small changes in the sample induced by the pump beam. In a UED experiment, the electrons scatter from all atoms in the sample, and, therefore, the recorded diffraction patterns consist of contributions from atoms both affected and unaffected by the pump pulse. To extract the photoinduced structural changes, a reference diffraction pattern (typically the signal at negative time delays, before the pump pulse) is acquired and subtracted from the diffraction pattern at each time delay to isolate the photoinduced structural changes, in much the same way as an optical chopper is used to isolate pump-induced changes in pump–probe experiments (Chapter 9). In addition, this also minimizes contributions from background scattering and systematic errors in the detection system.

The substantial effort that has been expended on developing ultrafast electron-based diagnostics of structural dynamics has now yielded an extremely powerful, versatile technique for interrogating a number of biological, chemical, and physical systems. Here, we will briefly discuss a few examples that align well with those previously discussed in this book, focusing on UEC and UEM applications since they have primarily been used to investigate condensed matter systems. We begin with a discussion of UEC and UEM measurements on VO_2 that were aimed at tracking transient structures during the photoinduced IMT (Baum et al., 2007; Grinolds et al., 2006), focusing primarily on the UEC experiments as they were performed on single crystals and therefore can be directly compared to the UF-XRD experiments described earlier in Section 18.1.1.1.

In these experiments, the temporal evolution of 16 Bragg spots that were observed in static ED experiments was tracked at room temperature after femtosecond photoexcitation. The resulting time-dependent traces fit into two general classes of behavior, with one

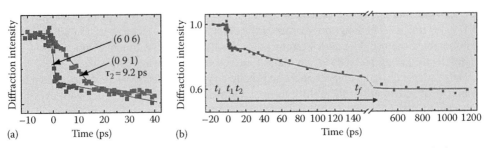

Figure 18.3 (a) Time-resolved intensity change of (606) (dark gray) and (091) (light gray) spots. (b) Time-resolved intensity change of the (606) spot on long timescales, revealing transitional structures during the phase transition. (Adapted from Baum, P., Yang, D., and Zewail, A., 4D visualization of transitional structures in phase transformations by electron diffraction, *Science*, 318, 788, 2007. Reprinted with permission from AAAS.)

set exhibiting an ultrafast (307 fs) decay and the other set showing a slower 9.2 ps decay (Figure 18.3a). The diffraction intensity continues to change over several hundred picoseconds due to sound-wave-induced shear motion (Figure 18.3b). From these time-dependent traces, it was deduced that the femtosecond component was due to lengthening of the V–V bond, the picosecond component due to local atomic displacements, and the sub-nanosecond component due to long-range shear motion. Importantly, the threshold for the photo-induced phase transition corresponds well with that for the thermally induced transition, suggesting that the structural pathway for both transitions is the same (Baum et al., 2007). In addition, UEM experiments were able to image the photoinduced phase transition in VO_2 films with high spatial and temporal resolution (Grinolds et al., 2006). Overall, these measurements expand on the earlier UF-XRD experiments by demonstrating the ability to track transient structures appearing during the transition from the monoclinic to the rutile phase. This should have general applicability to the study of strongly correlated systems, which has been demonstrated by the use of UEC and UEM to temporally resolve photo-induced melting of both metals (Siwick et al., 2003) and semiconductors (Yurtsever and Zewail, 2009), as well as to measure the role of structural dynamics in high-temperature superconductors after photoexcitation (Carbone et al., 2008; Gedik et al., 2007).

The applicability of ultrafast electron-based techniques to a completely different system is illustrated by the use of UEC to track ultrafast structural dynamics in ZnO nanowire (NW) ensembles (Yang et al., 2008). ZnO NWs were grown on sapphire substrates with a 100 nm GaN layer; the vertically aligned NWs were ~150 nm in diameter and ~2 μm in length and were spaced by 300 nm. Photoexcitation with 800 nm pulses creates carriers through three-photon absorption (3PA) or enhanced two-photon absorption (TPA) since the 1.55 eV pulse energy is less than half of the 3.37 eV band gap in ZnO and the 3.44 eV band gap in GaN. The intensity, width, and position of the Bragg spots resulting after diffraction of an electron pulse from the sample were tracked as a function of time delay (Figure 18.4). It is clear from this figure that the intensity and width recover within ~200 ps, but the c-axis lattice expansion (proportional to the displacement of the Bragg spots from their equilibrium positions) has a delayed rise time and recovers on a much longer timescale (~1 ns). It is worth noting that expansion of the NWs perpendicular to their growth direction is much smaller than expansion along the NW axis (not shown) (Yang et al., 2008). The magnitude of the expansion was 1%–2% at the highest pump fluences (~22.5 mJ/cm²), two orders of magnitude higher than the expected amount of thermal expansion. This anisotropic lattice expansion was attributed to the potential generated by the photoinduced carriers over the length of the NW. These experiments therefore demonstrated the potential of UEC for tracking structural dynamics

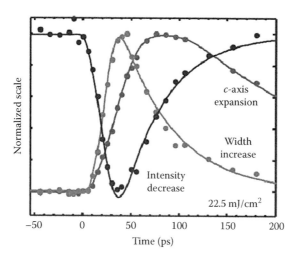

Figure 18.4 Time dependence of the diffraction intensity, width, and c-axis expansion of the (006) Bragg spot in ZnO NWs. (From Yang, D.-S., Lao, C., and Zewail, A.H., 4D electron diffraction reveals correlated unidirectional behavior in zinc oxide nanowires, *Science*, 321, 1660, 2008. Reprinted with permission from AAAS.)

in nanoscale systems, revealing phenomena that could not be directly observed using either time-integrated techniques such as ED or XRD or time-resolved techniques such as optical pump–probe spectroscopy (Chapter 9) or time-resolved photoluminescence (Chapter 12). The further development of UED, UEC, and UEM thus holds much promise for uncovering novel phenomena in a variety of condensed matter systems.

18.1.2 Ultrafast Photoelectron Spectroscopy

Much as UF-XRD and UED made it possible to directly measure structural dynamics in the time domain, ultrafast photoelectron spectroscopy (UF-PES) has made it possible to track electron dynamics in energy and momentum space. UF-PES is based on the powerful technique of angle-resolved photoemission spectroscopy, the most direct way of revealing the electronic band structure of a solid. ARPES makes use of the well-known photoelectric effect, which occurs when an incident photon with energy greater than the work function ejects an electron from a material. The electron momentum is also conserved in this process and can be measured by tracking the emission angle, giving access to the spectrum of occupied states below the Fermi level E_F (Figure 18.5). Time resolution can then be added to ARPES experiments by using one femtosecond pulse to photoexcite the sample and another pulse to eject electrons, allowing changes in the occupation of states within the band structure to be temporally resolved by changing the time delay between the two pulses. Furthermore, UF-PES can give information on unoccupied electronic states that is not available from conventional ARPES. This is the only technique capable of probing the dynamics of electrons as they relax through different points in momentum space in systems ranging from simple metals (Petek and Ogawa, 1997) to high-temperature superconductors (Perfetti et al., 2007), as described in more detail in the following.

The concept and experimental implementation of UF-PES are largely drawn from ARPES, making it worthwhile to describe ARPES in some detail. In a typical ARPES experiment, a quasi-monochromatic beam of light generated either by a synchrotron source or from a frequency harmonic of a femtosecond laser (discussed further in the following) is incident on a single crystal sample that is held under vacuum. Electrons are subsequently emitted in

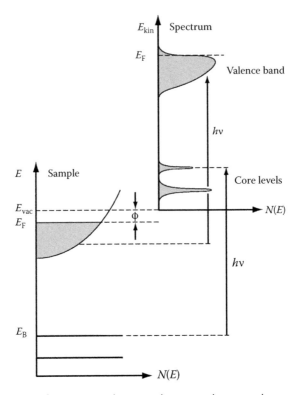

Figure 18.5 Principle of ARPES. A photon with energy $h\nu$ is incident on a solid and ejects electrons with a distribution given by the occupation of electronic levels in the solid below the Fermi level E_F. (Reprinted with permission from Damascelli, A., Hussain, Z., and Shen, Z., Angle-resolved photoemission studies of the cuprate superconductors, *Rev. Mod. Phys.*, 75, 473–541, 2003. Copyright 2003 by the American Physical Society.)

all directions from the sample through the photoelectric effect, after which they are collected by an electron spectrometer operating with high energy and momentum resolution. This is typically a hemispherical electron energy analyzer or a time-of-flight detector in which the electron time-of-flight from the sample to the detector is proportional to its kinetic energy (Bauer, 2005; Damascelli et al., 2003; Koralek and Dessau, 2006; Petek and Ogawa, 1997). The kinetic energy E_{kin} and angle θ of the emitted electrons can then be related to the electron binding energy E_B and crystal momentum p inside the solid by

$$E_{kin} = h\nu - \phi - |E_B| \tag{18.3}$$

$$p_\| = \hbar k_\| = \sqrt{2mE_{kin}} \cdot \sin\theta \tag{18.4}$$

where
  ν is the frequency of the incident photon
  ϕ is the work function of the material
  $k_\|$ is the parallel component of the electron wave vector (Damascelli et al., 2003)

Here, the momentum of the incident photon is neglected, which is a reasonable approximation for $h\nu < 100\,\text{eV}$. By analyzing the ARPES data with these equations, one can resolve the electronic band structure of the sample in momentum space.

Although ARPES is indispensable for gaining insight on the electronic band structure of a solid, it has some important limitations. Perhaps the most significant limitation is its surface sensitivity, governed by the electron mean free path in the solid. This is only a few monolayers at the typical photon energies used in synchrotron-based ARPES experiments (20–100 eV), a range where synchrotron sources have high flux while preserving relatively good momentum resolution (which is inversely proportional to the photon energy) (Damascelli et al., 2003; Koralek and Dessau, 2006). This necessitates the use of ultrahigh vacuum conditions (pressure $<5 \times 10^{-11}$ torr) and samples with extremely high-quality surfaces (often cleaved in situ) for performing ARPES experiments. Even after these precautions are taken, the electronic structure at the surface may be different from that in the bulk, which can complicate data interpretation.

This issue of surface sensitivity has been one of the primary factors driving the development of laser-based ARPES, which uses a frequency doubled, tripled, or quadrupled laser to operate at significantly lower photon energies ($h\nu \sim 6$–7 eV) (Carpene et al., 2009; Koralek and Dessau, 2006). These sources are a much better probe of the bulk electronic band structure since the penetration depth is approximately an order of magnitude greater at $h\nu \sim 6$–7 eV than at $h\nu \sim 20$–100 eV. These lower photon energies also substantially improve the momentum resolution as the momentum distribution of the emitted electrons is more widely dispersed in angle (Equation 18.4), effectively increasing the momentum resolution for a given angular resolution (Koralek and Dessau, 2006). Finally, the ability of laser-based ARPES systems to fit on a table top makes this technique more accessible to a wider range of researchers, particularly those that do not have access to a synchrotron source. However, electrons emitted after low photon energy excitation are susceptible to deflection by weak magnetic fields, necessitating careful shielding of the ARPES chamber. In addition, the low photon energies used in laser-based ARPES primarily probe energies near the Fermi level and cannot access the deeper electron levels that are important for understanding the chemical or magnetic state of the sample (Bauer, 2005). For systems in which this is important, the HHG process discussed earlier in Section 18.1.1.1 has recently been used to extend excitation wavelengths to higher photon energies in laser-based ARPES experiments; a detailed review is given in Bauer (2005).

The addition of time resolution to ARPES is a relatively straightforward extension of laser-based ARPES experiments (Bauer, 2005; Bovensiepen, 2007; Carpene et al., 2009; Petek and Ogawa, 1997). A portion of the beam from a femtosecond laser is split off before the sample (usually before nonlinear optical processes are used to generate higher frequencies) to photoexcite electrons, after which a time-delayed probe pulse is used to eject electrons as in a standard ARPES experiment. The photon energy of the pump pulse is always kept below the work function of the material ($h\nu_{pump} < \phi$), so that the pump excites a nonthermal carrier distribution but does not eject any electrons (except through undesirable multi-photon absorption). Time-resolved photoemission methods can then be split into two general categories depending on the probe photon energy. If $h\nu_{probe} > \phi$, then the probe pulse can eject photons from the sample irrespective of the presence of a pump pulse (as in standard time-integrated ARPES). Therefore, after pump photoexcitation the probe is sensitive to changes in both occupied and unoccupied states below E_F, revealing the electron distribution in energy and momentum space (Figure 18.6a); this is known as time-resolved one-photon photoemission (TR-1PPE). In contrast, if $h\nu_{probe} < \phi$, then both pump and probe photons are necessary to eject an electron from the sample, known as time-resolved two-photon photoemission (TR-2PPE) (Figure 18.6b). Here, the pump pulse excites electrons to intermediate states, after which the probe pulse ejects them into the vacuum. It is important to note that when $h\nu_{pump}$ and $h\nu_{probe}$ are different, then the choice of photon energies determines the intermediate states that are probed (Bovensiepen, 2007). More details are given in Bauer (2005), Bovensiepen (2007), Carpene et al. (2009), and Petek and Ogawa (1997).

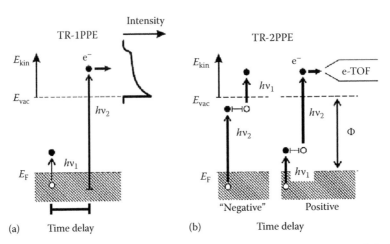

Figure 18.6 Schematic depicting the processes involved in TR-1PPE (a) and TR-2PPE (b). (Adapted with permission from Bovensiepen, U., *J. Phys. Condens. Matter*, 19, 083201, 2007.)

TR-2PPE has historically been the dominant technique in UF-PES, as TR-1PPE is more experimentally challenging due to its increased sensitivity to surface properties, space-charge effects, and competing electron emission from multi-photon absorption of the pump pulse (Carpene et al., 2009). However, in recent years both methods have been used to gain insight on materials, particularly since the use of HHG to generate probe photons gives access to time-resolved dynamics of deeper lying states (Bauer, 2005; Bovensiepen, 2007). Initial time-resolved photoemission studies focused on using TR-2PPE to study dynamics in metals and their surfaces (Haight, 1995; Petek and Ogawa, 1997). A particularly interesting class of experiments that attracted much interest when UF-PES was initially developed was the study of image potential state dynamics at metal surfaces (Haight, 1995; Petek and Ogawa, 1997; Schoenlein et al., 1988). These unique states are due to the Coulomb attraction between an electron outside the surface and its image charge in the solid, forming a one-dimensional (1D) hydrogen atom-like potential perpendicular to the surface at energies just below the vacuum potential E_{vac} (Figure 18.7a). The associated electron wave functions exist outside of the surface, with greater confinement for higher order wave functions (although electrons are nearly free in the parallel plane). The end result is that the image potential traps electrons in vacuum near the metal surface (essentially forming a 2D electron gas) as there are no available states near E_{vac} for them to occupy within the metal (Haight, 1995; Petek and Ogawa, 1997; Schoenlein et al., 1988).

TR-2PPE has been used to study image state dynamics in metals including Ag and Cu (Haight, 1995; Schoenlein et al., 1988). The first time-resolved photoemission experiments on image potential states in Ag were performed on the Ag(100) surface in which $n = 1$ and $n = 2$ image potential states have binding energies of 0.53 and 0.16 eV, respectively, relative to $E_{vac} = 4.43$ eV (Figure 18.7a) (Schoenlein et al., 1988). A mode-locked dye laser producing 55 fs, 620 nm (2 eV) pulses was frequency doubled to produce 4 eV pump photons, which excited electrons from E_F to the $n = 1$ state. A subsequent 2 eV probe photon ejected electrons from the $n = 1$ state into the vacuum. Figure 18.7b depicts photoelectron spectra as a function of time delay in which the main peak in kinetic energy at ~1.5 eV originates from the difference between $E_B = 0.53$ eV for the $n = 1$ state and $h\nu_{probe} = 2$ eV. The ability to measure extremely fast processes with TR-2PPE is illustrated through the time dependence of the 1.5 eV peak in Figure 18.7c, revealing that the lifetime of the $n = 1$ image state is between 15 and 35 fs. Finally, the dotted line in Figure 18.7c shows that the presence of oxygen on the surface significantly lowers the photoemission signal and reduces the image state lifetime (Schoenlein et al., 1988).

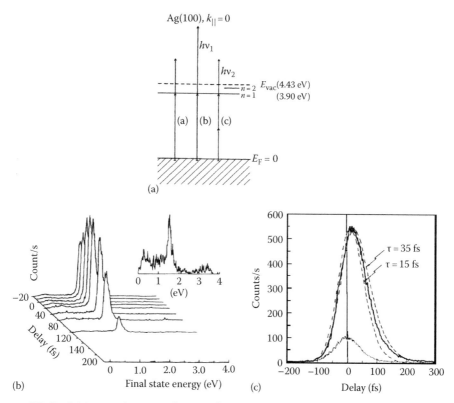

Figure 18.7 (a) Potential energy diagram for image potential states on the Ag(100) surface. Transitions (a) and (b) depict UV-VIS and UV-UV 2PPE, respectively, while (c) depicts three-photon photoemission. (b) Time-dependent photoelectron spectra measured on the Ag(100) surface. The inset shows the background spectrum measured at $t = 200\,fs$, before the pump pulse arrives. The 1.5 eV peak in the background spectrum is due to three-photon photoemission from the probe pulse. (c) Time-dependent signal measured at the $n = 1$ state on a clean Ag(100) surface (solid line) and on an oxygen-dosed surface (dotted line). The dashed lines are exponential functions with time constants of 15 and 35 fs convolved with the cross-correlation of the pump and probe pulses. (Adapted with permission from Schoenlein, R., Fujimoto, J., Eesley, G., and Capehart, T., *Phys. Rev. Lett.*, 61, 2596–2599, 1988. Copyright 1988 by the American Physical Society.)

As discussed earlier, one significant advantage of UF-PES over other ultrafast experiments is that it directly probes the occupation of electronic states in k-space and can examine their coupling to other degrees of freedom, important for superconductors and other strongly correlated systems. TR-1PPE was therefore used to study ultrafast electronic dynamics in single crystals of optimally doped $Bi_2Sr_2CaCu_2O_{8+}$ Δ (Bi-2212) ($T_c = 91$ K) to determine the electron–phonon coupling strength in this system (Perfetti et al., 2007), which is pertinent to the longstanding question of which quasiparticle is responsible for binding pairs together in high-temperature superconductors. Importantly, these experiments may indicate whether phonons are the pairing "glue" in these systems, as in conventional low-temperature superconductors, or whether another quasiparticle is responsible for pairing.

The 1.5 eV, 50 fs pump pulses and 6 eV, 80 fs probe pulses were generated from an amplified 30 kHz Ti:sapphire laser system and focused on the sample, after which the photoemitted electrons were detected normal to the surface in a time-of-flight spectrometer. It is worth noting that the well-known "kink" in the electronic dispersion that has been observed in standard ARPES experiments on high-T_c superconductors (Damascelli et al., 2003), believed due to interactions

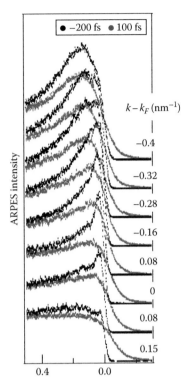

Figure 18.8 Time-resolved ARPES spectra acquired on Bi-2212 at different points in momentum space at t = −200 fs (before the pump pulse) (black) and t = 100 fs (gray). The sample temperature was 30 K. (Reprinted with permission from Perfetti, L., Loukakos, P., Lisowski, M., Bovensiepen, U., Eisaki, H., and Wolf, M., *Phys. Rev. Lett.*, 99, 197001, 2007. Copyright 2007 by the American Physical Society.)

between quasiparticles and bosonic modes, was not resolved here since the broad probe bandwidth and acceptance angle of the spectrometer limit the spectral resolution of the system.

Figure 18.8 depicts ARPES spectra acquired along the nodal direction in k-space on Bi-2212 at T = 30 K as a function of time, energy, and momentum (Perfetti et al., 2007). It is clear from this figure that the pump pulse introduces significant changes in the ARPES spectra by photoexciting a nonequilibrium electron distribution in Bi-2212. The electronic temperature T_e was extracted from this data, showing that the pump pulse increases T_e to 740 K within 50 fs, after which it drops to 300 K within ~100 fs and subsequently relaxes on a slower timescale of several picoseconds. Analysis of this data indicates that the hot electrons thermalize among themselves within the 50 fs pump pulse duration, after which they couple preferentially to a small subset of available "hot" phonon modes at early times (~110 fs). The hot electrons and phonons reach a common temperature of 300 K within 330 fs, after which they lose the remainder of their energy to the "cold" lattice on longer timescales (~2 ps), such that the sample surface reaches a local equilibrium at 60 K. At longer times, the remaining heat diffuses into the bulk material to return the sample to its original T = 30 K.

The existence of different timescales for quasiparticle relaxation is indicative of the existence of an energy bottleneck, which has previously been linked to the opening of the superconducting gap below T_c (Perfetti et al., 2007). However, in these experiments, a bottleneck is observed at T = 300 K, well above T_c, which suggests that this is not the operative mechanism in these experiments. An extended version of the two-temperature model (Chapters 8 and 9) was solved to gain more insight on electron–phonon coupling in this system. This model demonstrated that only 20% of the available phonon modes strongly couple to the electronic subsystem in Bi-2212. This is likely because only a few phonon branches are strongly coupled to the electrons, and in addition the coupling is extremely anisotropic. Therefore, the fast relaxation in Bi-2212 originates from coupling to the strongly coupled "hot" phonon modes, and the longer relaxation that leads to the energy bottleneck is due to the weakly coupled phonon modes. Overall, the weak electron–phonon coupling observed for the majority of phonon modes in these experiments suggest that phonons are not the sole pairing "glue" in high-T_c superconductors, although they may contribute to pairing in conjunction with other interactions. The ability of UF-PES to extract quantitative information on important material parameters (e.g., the electron–phonon coupling parameter) will be valuable in the study of other condensed matter systems; a powerful demonstration was recently given when TR-ARPES was used to unravel the role of the electron–phonon interaction in charge-density wave formation in $TbTe_3$ (Schmitt et al., 2008).

18.2 OPTICAL TECHNIQUES THAT COMBINE HIGH TEMPORAL AND SPATIAL RESOLUTION

The advanced time and spatially resolved optical techniques described throughout this book have each been extremely successful in extracting both static and dynamic properties of different condensed matter systems. However, the combination of temporal and spatial resolution in a single experiment would be extremely powerful, capable of shedding light on dynamics in individual nanostructures as well as tracking nanoscale processes through space and time. In addition, the majority of optical techniques discussed in Parts II and III of this book are only capable of spatially resolving features down to several hundred nanometers and will therefore average over any material properties that vary on smaller length scales (e.g., nanoscale inhomogeneities in strongly correlated materials or size variations in ensembles of NWs or quantum dots [QDs]). Finally, optical techniques are typically simpler, more versatile, and less prone to damaging samples than the electron and x-ray-based techniques described earlier in this chapter (which are also capable of high temporal and spatial resolution).

The development of techniques that combine temporal and spatial resolution in one experiment has therefore been an active area of research over the past decade, leading to a host of methods that are simultaneously capable of sub-picosecond temporal resolution and sub-micron spatial resolution (down to sub-100 fs, sub-100 nm resolution [Labardi et al., 2005]). These techniques can all be grouped under the general heading of "time-resolved microscopy," but there are many important variations within this category, many created by adding spatial resolution to the time-resolved techniques discussed in Part III and others created by adding time resolution to the spatially resolved techniques in Part IV. Although it is impossible to describe all of the different methods that exist for combining temporal and spatial resolution, we will attempt to survey some of the most prominent ones here.

18.2.1 Time-Resolved Optical Microscopy in the Far Field

The power and simplicity of optical microscopy (Chapter 15) made it a natural choice for the first studies that explored combining the high temporal resolution of ultrafast spectroscopy with the spatially resolved techniques discussed in Part IV. The resulting time-resolved microscopic techniques can generally be divided into two classes. Section 18.2.1.1 will discuss time-resolved wide-field microscopic techniques, which conceptually are akin to introducing time resolution to a standard microscope setup. Time-resolved scanning microscopic techniques, which have the common feature of tightly focusing the probe and/or the pump beams to a diffraction-limited spot on the sample, will be described in Section 18.2.1.2.

18.2.1.1 Time-Resolved Wide-Field Optical Microscopy

For many applications, the ideal experiment would be to add ultrafast time resolution to a standard bright-field optical microscope, as this would provide a great deal of information about a sample with relatively little experimental effort. In a pump–probe experiment (Chapter 9), the simplest implementation of this idea would consist of using the probe beam as the light source in the microscope and separately introducing the pump beam into the setup. By detecting the photoinduced changes in the probe beam with a CCD, one can temporally and spatially resolved dynamics in a sample with sub-picosecond temporal and sub-micron spatial resolution. This approach was first used in 1985 to image melting and evaporation dynamics of a silicon surface after intense femtosecond photoexcitation (Downer et al., 1985). In this work, a white light supercontinuum generated by focusing the 620 nm output of a 10 Hz dye laser into water was focused onto the Si surface, after which the reflected beam was collected by an objective and then imaged onto a photographic film. The 620 nm pump beam was separately focused onto the sample at normal incidence. In addition, the Si surface was translated after every laser shot to ensure that a fresh region of the sample was measured. Photoexcitation at a fluence of 0.5 J/cm^2 melted the Si surface, increasing the probe reflectivity within the pump spot in ~500 fs. On a timescale of tens of picoseconds, dynamics corresponding to the ejection of material from the Si surface were observed, after which the probe reflectivity recovered on a longer timescale of hundreds of picoseconds. These were the first time-resolved microscopy experiments on solid-state materials, to the best of our knowledge. Similar experiments were later performed to examine laser ablation (Sokolowski-Tinten et al., 1998; Zhang et al., 2007), photoinduced phase transitions in solids (Sokolowski-Tinten et al., 1998), and single perylene microcrystals (Tamai et al., 1993). Several variations of this technique have been developed, such as ultrafast imaging interferometry, developed to image surface deformations with 100 fs time resolution and ~1 μm spatial resolution (Temnov et al., 2004, 2006). Simultaneous frequency- and time-resolved single shot images of transient absorption in β-carotene were also obtained (Furukawa et al., 2004; Makishima et al., 2006). Finally, single-pulse magneto-optic Kerr effect (MOKE) microscopy was also developed recently to image magnetization dynamics with sub-picosecond temporal and 50 μm spatial resolution (Elazar et al., 2008).

The time-resolved microscopy experiments described earlier were quite successful in obtaining simultaneous sub-micrometer spatial resolution and sub-picosecond temporal resolution in a wide-field optical imaging experiment. However, these efforts were limited to measurements of relatively large (>1%) pump-induced reflectivity changes (Temnov et al., 2006), due to the inability to combine the standard lock-in amplification techniques that are used for detecting small pump-induced changes (Chapter 9) with CCD detection. This limits the applicability of these techniques to materials in which large changes in reflectivity or transmission, for example, occurring when the surface of a sample is melted, can be induced. In contrast, a significant fraction of pump–probe experiments aim to gently perturb the sample, resulting in normalized photoinduced changes ($\Delta T/T$, $\Delta R/R$) that are typically smaller than 10^{-3} (Hilton et al., 2006). Therefore, the development of a technique to rapidly acquire wide-field optical images with sub-picosecond temporal resolution and submicron spatial resolution, along with sensitivity to small ($<10^{-3}$) photoinduced changes, would significantly extend the capability of time-resolved microscopy to image dynamic processes ranging from vortex dynamics in superconductors to charge transport in neurons or semiconductor NWs. Recently, by utilizing a novel 2D smart pixel array detector in place of the CCD camera used in previous experiments, this has been achieved (Prasankumar et al., 2009).

In the first demonstration of this approach, known as ultrafast optical wide-field microscopy (UOWFM), 400 nm pump and 550 nm probe beams from an optical parametric amplifier

(OPA) (seeded by 50 fs pulses from a 100 kHz amplified Ti:sapphire system) were noncollinearly directed onto the sample in transmission, as in a conventional pump–probe setup, and a mechanical chopper modulated the pump beam. After the sample, the probe beam was imaged onto the 2D smart pixel detector array using a high-NA objective and a zoom lens with variable magnification. This 2D detector array, originally developed for optical coherence tomography, has previously been used to perform time- and wavelength-resolved spectroscopy (Bourquin et al., 2003) but has never been used for ultrafast optical imaging, to the best of our knowledge. The 144×90 pixel silicon detector array used here performs real time amplitude demodulation on each pixel and can achieve sensitivities comparable to those attainable through lock-in amplification; further details can be found in Bourquin et al. (2003).

A gold patterned amorphous Si sample was imaged onto the detector by blocking the pump beam and chopping the probe beam, enabling the setup to operate as a conventional optical microscope, and indeed the resulting images compared well to those measured by a standard optical microscope. The pump beam was then unblocked and chopped, after which images were acquired at different time delays, t, between the pump and probe pulses (Figure 18.9). As expected, there is no signal at negative time delays. By $t = 1$ ps an image is visible, indicating that photoexcited carriers in the amorphous Si film modify the probe transmission.

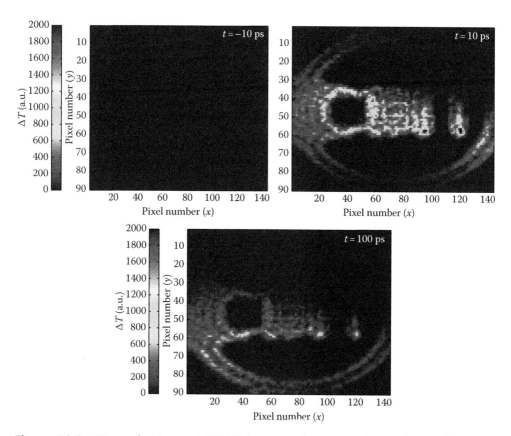

Figure 18.9 (See color insert.) UOWFM images of a patterned amorphous Si film at different time delays between the pump and probe pulses. (From Prasankumar, R.P., Ku, Z., Gin, A.V., Upadhya, P.C., Brueck, S.R.J., and Taylor, A.J., Ultrafast optical wide field microscopy, in *Conference on Lasers and Electro-Optics*, Baltimore, MD, 2009. With permission.)

The transmission change (proportional to the intensity on the pixels in the image) is maximum from $t = 1$–10 ps, after which it begins to decay, as observed in the $t = 100$ ps image.

The high sensitivity of this system was quantified by noting that with a pump power of 8 mW, a peak $\Delta T/T$ of ~6% was measured when the probe beam was focused onto a single pixel detector. Figure 18.9 reveals that this signal is spread over ~6000 pixels, indicating that each pixel detects a signal of $\Delta T/T$ ~10^{-5}; this is comparable to the typical performance of the system when using a standard single pixel InGaAs detector. In addition, each image is acquired in <1 s, demonstrating the rapid image acquisition capability of this technique. It is also worth noting that images can be acquired over a broad range of probe wavelengths (limited only by the silicon material used for the detector pixels), at temperatures down to 4 K, and in transmission or reflection on nearly any sample that can be studied with conventional bright-field optical microscopy. Finally, it should be straightforward to adapt this technique to work with other techniques such as phase contrast and dark-field microscopy (Chapter 15). UOWFM should therefore extend the ability of time-resolved microscopy to sensitively measure spatiotemporal dynamics in a variety of biological, chemical, and physical systems under low photoexcitation conditions (Prasankumar et al., 2009).

Along the same lines, a technique known as femtosecond Kerr-gated wide-field fluorescence microscopy has been developed to image the fluorescence (or equivalently, the photoluminescence [PL]; Chapter 5) from a sample with diffraction-limited spatial resolution and sub-100 fs time resolution (Gundlach and Piotrowiak, 2008). The key advance making this possible was to place a nonlinear Kerr gate medium (e.g., fused silica) between the sample and the CCD detector in a wide-field microscope. The Kerr gate operates by using a gate pulse to induce a transient intensity-dependent birefringence in a Kerr medium placed between two crossed polarizers. The luminescence emitted from the sample after femtosecond photoexcitation can only pass through the polarizers when it spatially and temporally coincides with the gate pulse within the Kerr medium. By imaging the PL onto the CCD camera as a function of time delay between the gate and excitation pulses, the spatial and temporal evolution of the luminescence can be mapped out with sub-100 fs temporal and diffraction-limited spatial resolution (Gundlach and Piotrowiak, 2008). It is also worth noting that a simple adjustment of the microscope enables the emission spectrum from a given sample position to be mapped as a function of time.

An important consideration when using this technique is that the maximum time delay that can be measured is limited by the gate on–gate off contrast ratio (which determines the noise that leaks through the crossed polarizers with no pump pulse), since at long time delays the amount of PL that passes through the crossed polarizers drops below the noise level. This technique was initially benchmarked by spatially and temporally resolving the emission from an ensemble of Au NPs, which revealed a 60 fs emission lifetime. Further measurements on a chemical solution demonstrated the ability to measure PL lifetimes up to 100 ps (Gundlach and Piotrowiak, 2008).

More recently, femtosecond Kerr-gated wide-field fluorescence microscopy has been applied to resolve the emission from single CdSSe nanobelts (Gundlach and Piotrowiak, 2008, 2009). Figure 18.10 depicts representative time-resolved PL images at different delay times after 350 nm, 30 fs photoexcitation, demonstrating that different nanobelts "turn on" their emission at different times. Further measurements on individual nanobelts within this ensemble revealed that the PL originated primarily from the tip of the wedge-shaped crystal, where the confinement enhanced amplified spontaneous emission. Overall, this powerful technique should provide much insight into the dynamics of luminescent nanostructures, with potential extension to other systems that exhibit ultrafast luminescence dynamics. One could also imagine combining femtosecond Kerr-gated fluorescence microscopy with UOWFM to simultaneously measure transient absorption and luminescence dynamics in the same sample (e.g., an individual semiconductor nanostructure).

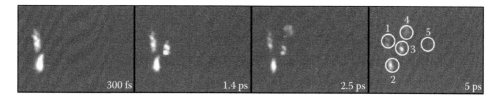

Figure 18.10 (See color insert.) Time-resolved PL images of CdSSe nanobelts as a function of time delay. (Reprinted with permission from Gundlach, L. and Piotrowiak, P., Ultrafast spatially resolved carrier dynamics in single CdSSe nanobelts, *J. Phys. Chem.* C., 113, 12162–12166, 2009. Copyright 2009 American Chemical Society.)

18.2.1.2 Time-Resolved Scanning Optical Microscopy in the Far Field

Time-resolved optical scanning microscopy, as with its wide-field analogue, is a conceptually simple technique, particularly when used in a pump–probe experiment (Rudolph et al., 2003). In this case, a standard pump–probe setup (e.g., Figure 9.1) is modified by introducing a microscope objective or another tight focusing, high numerical aperture (NA) optic to focus the probe (and sometimes the pump as well) onto the sample with a submicron spot size. The transmitted or reflected probe beam is then focused onto a standard single pixel detector. By two-dimensionally scanning the position of the sample in the plane normal to the incident beam at each time delay, one can track the evolution of carrier dynamics in space and time. Perhaps the most important issue to consider in these experiments is pulse broadening due to the large amount of group velocity dispersion (GVD) from the objective (which was not a significant issue in the majority of wide-field experiments discussed in Section 18.2.1.1 since the pump and probe were not focused onto the sample through an objective). This can significantly lengthen the pulse duration at the sample position; Jasapara and Rudolph (1999) and Larson and Yeh (2006) describe two different approaches for circumventing this problem.

A representative application of time-resolved optical scanning pump–probe microscopy is to examine the spatial dependence of carrier recombination in ion-implanted GaAs (Fujii et al., 2004) and Si (Othonos and Christofides, 2002). In these materials, the ion-introduced defects significantly reduce the carrier lifetime. Therefore, by measuring the lifetime as a function of position, one can image the defect density distribution in these samples. This technique has also been used to image exciton densities in organic crystals (Yago et al., 2008).

A relatively widespread variation of this technique is time-resolved magneto-optical Kerr effect (TR-MOKE) microscopy, which has become quite popular in recent years for spatially resolving magnetization dynamics in bulk materials (Ogasawara et al., 2009; Rizo et al., 2008), ensembles of magnetic nanostructures (Li et al., 2009), and individual nanostructures (Barman et al., 2006, 2007; Laraoui et al., 2007a,b). These different studies, although conducted with various laser sources (e.g., amplified Ti:sapphire lasers vs. Ti:sapphire oscillators) and samples (e.g., magnetic cylinders patterned to exhibit cavity-enhanced MOKE (Barman et al., 2006, 2007) vs. those retaining their intrinsic MOKE (Laraoui et al., 2007a,b; Li et al., 2009), share a similar experimental configuration. In particular, these TR-MOKE microscopy setups are essentially identical to the time-resolved scanning pump–probe setup described earlier, except that after the sample the MOKE signal (or Faraday rotation signal) is detected as described in Chapter 13 as a function of time delay and sample position. TR-MOKE microscopy has been demonstrated with a spatial resolution as high as ~200 nm (Li et al., 2009; Ogasawara et al., 2009) and a typical temporal resolution of 100–200 fs. In addition, TR-MOKE microscopy has also been adapted for operation in a high-field magnet at temperatures down to 4 K and field up to 7 T (Rizo et al., 2008). This technique has been used to temporally and spatially resolve phenomena such as ultrafast demagnetization and spin precession dynamics in individual ferromagnetic disks (Barman et al., 2006, 2007;

Laraoui et al., 2007a,b) and diffusion of spin-polarized electrons in GaAs/AlGaAs quantum wells (QWs) (Rizo et al., 2008). TR-MOKE microscopy therefore has great promise for shedding light on other space-and-time-varying processes in magnetic systems, particularly those relevant to spintronics (Rizo et al., 2008) and ultrafast magnetic switching devices (Li et al., 2009; Ogasawara et al., 2009).

Arguably the most prevalent time-resolved scanning microscopy technique is time-resolved micro-PL, used primarily to track dynamics in single nanostructures (e.g., isolated QDs or NWs). In most cases, the experimental setup is identical to that of a micro-PL setup, with the detector replaced by a time-correlated single-photon counting (TC-SPC) module (Chapters 12 and 16). This enables PL measurements with a time resolution of tens to hundreds of picoseconds and diffraction-limited (<1 μm at visible wavelengths) spatial resolution. An example of this technique was given in Chapter 16 (Figure 16.12d), where the PL from a single QD was temporally resolved after optical excitation. Time-resolved micro-PL has also been particularly useful in spatially, temporally, and spectrally resolving the emission from single semiconductor NWs (Glennon et al., 2009; Pemasiri et al., 2009; Titova et al., 2007), since the diffraction-limited excitation spot is comparable to or smaller than typical NW lengths (~1–30 μm). In these experiments, the excitation beam can be scanned along the length of the NW and time and spectrally resolved data can be taken at each position. This capability was critical in understanding exciton migration and localization in single CdSe NWs (Glennon et al., 2009).

The earlier discussion has touched on the use of time-resolved microscopy (primarily in the "scanning" configuration, not the "wide-field" configuration) to measure carrier dynamics in individual nanostructures. In fact, this is one of the most promising applications of time-resolved microscopy, as the increasing miniaturization of electronic and photonic devices has led to applications including the use of single NPs as nanoscale biolabels and transistors. These applications will rely critically on a deep understanding of the physics of isolated NPs and their interactions with the surrounding environment, which cannot be obtained from studies of NP ensembles due to the inhomogeneous broadening from their inherent size and shape distribution. This is relatively straightforward for time-resolved micro-PL experiments, as described earlier, because single photon counters (Chapters 12 and 16) can be used to detect the emitted signal and spectral filters can be used to remove the unabsorbed pump light. However, for pump–probe experiments and associated derivatives (e.g., time-resolved four-wave mixing, time-resolved MOKE, etc.), the challenge is to detect an extremely small photoinduced change in the signal upon a large background. In addition, even time-integrated absorption measurements are difficult because the NP is usually significantly smaller than the laser spot size, leaving the vast majority of the incident light unaffected by its presence. This light then becomes the undesirable background signal, which is difficult to remove since it has the same wavelength and polarization as the signal of interest (the photoinduced change).

Recently, several techniques have been developed to surmount these challenges and acquire both time-integrated and time-resolved absorption measurements on single NPs (in addition to the time-resolved magnetization dynamics on individual magnetic NPs described earlier) (Van Dijk et al., 2005b). Initial measurements of this kind focused on metal NPs due to their large absorption cross sections (Chapter 3), which made it easier to measure both static and transient absorption signals from single NPs. Several different methods were developed for measuring the static absorption from an individual metal NP, including dark-field microscopy (Itoh et al., 2001), differential interference contrast microscopy (DIC) (Matsuo and Sasaki, 2001; Stoller et al., 2006), total internal reflection microscopy (Pelton et al., 2006; Sönnichsen et al., 2000), confocal microscopy with interferometric detection (Lindfors et al., 2004), photothermal imaging (Berciaud et al., 2004; Boyer et al., 2002), and spatial modulation microscopy (in which the sample position is modulated at a certain frequency that is then sent to a lock-in amplifier for detection) (Arbouet et al., 2004; Muskens et al., 2006a,b). Using these techniques, the absorption of metal NPs as small as 2 nm in diameter can be measured (Muskens et al., 2006b).

Transient absorption measurements can also be performed on individual metal NPs by isolating a single NP using one of the techniques described earlier and then introducing a pump beam for NP photoexcitation. The first measurement of this kind, to the best of our knowledge, was reported in Itoh et al. (2001). A dark-field microscope was used to isolate scatted light from a single Au NP, revealing a peak near 580 nm due to a surface plasmon resonance (SPR) (Chapter 3). Degenerate pump–probe measurements at the SPR peak on a single Au NP revealed fast relaxation of the photoexcited electrons within 3 ps due to electron–phonon coupling. Similar dynamics were observed when spatial modulation microscopy was used to identify a single Ag NP and near-IR (850 nm) pump; visible (425 nm) probe experiments were subsequently performed. Typical photoinduced changes in transmission for a 30 nm Ag NP were $\Delta T/T \sim 10^{-4}–10^{-6}$ for pump powers of 180–480 μW. The measured data agreed well with the results from a two-temperature model, demonstrating that electron–phonon coupling dominates the early time dynamics (Muskens et al., 2006b). Finally, in another experiment, total internal reflection microscopy was used to measure the scattering spectrum from individual Au nanorods, revealing an SPR peak near 1.55 eV. Degenerate pump–probe experiments revealed that resonant SPR excitation creates an ultrafast nonlinearity with the same magnitude as that obtained from nonresonant excitation of the electron population, suggesting that the plasmons experience intensity-dependent damping (Pelton et al., 2006).

Another approach for measuring carrier dynamics in single NPs is to use a time-resolved common-path interferometer to measure electronic dynamics and acoustic oscillations in individual Au NPs (Van Dijk et al., 2005a,b, 2007). In these experiments, probe and reference pulses separated by a fixed time delay are used in a common-path interferometer set to operate near its dark fringe. The pump-induced change in the real or imaginary part of the NP dielectric constant changes the amplitude and/or phase of the probe electric field, which is then detected by the interferometer. This enables sensitive measurements of the pump-induced changes in the real and imaginary parts of the NP dielectric constant. When this technique was used to examine a single Au NP, a fast decay was observed in the time-resolved response, attributed to electronic heating (as in Itoh et al., 2001), and oscillations due to acoustic breathing modes were observed on longer timescales (Van Dijk et al., 2005a). In another study, a similar time-resolved interferometric technique was used to measure surface plasmon dephasing in single Ag NPs, revealing a dephasing time of 10 fs in 75 nm diameter NPs that agreed with previous measurements on NP arrays (Liau et al., 2001).

Far-field transient absorption measurements on single NPs have recently been extended to study dynamics in individual semiconductor QDs (Liau et al., 2001; Sotier et al., 2009; Wesseli et al., 2006) and NWs (Carey et al., 2009; Johnson et al., 2004; Song et al., 2008), which have a significantly smaller absorption cross section than metal NPs (and therefore smaller $\Delta T/T$ signals) (Carey et al., 2009). The first experiment of this kind, to the best of our knowledge, was performed on a single InAs/GaAs QD that was grown through molecular beam epitaxy (MBE) and isolated by depositing Al shadow masks on the sample with 200–450 nm diameter. By pumping in the continuum states and tuning the probe photon energy through both interband transitions and continuum states, photoinduced transmission changes of $\Delta T/T \sim 10^{-5}$ could be measured for pump powers as low as 2 fJ (Wesseli et al., 2006). Similar experiments were performed on single CdSe/ZnSe QDs (doped with 1 electron/dot and also isolated by a 100 nm metal aperture), which have a larger confinement potential than InAs/GaAs QDs (Sotier et al., 2009). In these measurements, the fluence of a pump pulse was tuned to resonantly excite only one electron–hole pair in the QD. This significantly decreased the fundamental exciton absorption, as measured by a tunable probe, after which optical gain

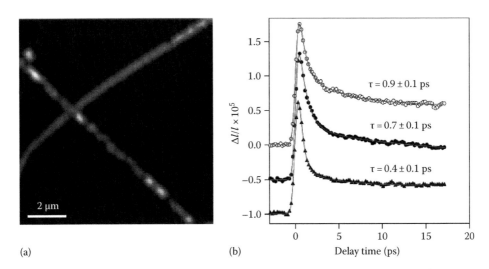

(a)   (b)   Delay time (ps)

Figure 18.11 (a) Transient absorption image of crossed CdTe NWs at $t = 0$. (b) Transient absorption signals from different points on a single CdTe NW, demonstrating the variation of fast time constant with position. (Reprinted with permission from Carey, C.R., Yu, Y., Kuno, M., and Hartland, G.V., Ultrafast transient absorption measurements of charge carrier dynamics in single II–VI nanowires, *J. Phys. Chem. C*, 113, 19077–19081, 2009. Copyright 2009 American Chemical Society.)

due to stimulated emission built up within ~15 ps and recovered on a much longer timescale of hundreds of picoseconds due to interband recombination. In addition, by appropriately tuning the pump and probe incident fluences, exactly one photon could be added or subtracted from the probe pulses, with significant implications for quantum information processing (Sotier et al., 2009).

Finally, spatially resolved transient absorption measurements on single CdTe and CdSe NWs (with diameters of ~38 nm and lengths of tens of micrometers) were recently reported (Carey et al., 2009). The pump and probe beams were tightly focused and scanned along individual NWs in these experiments, demonstrating the ability to obtain transient absorption images (Figure 18.11a). These experiments revealed differences in the dynamics as a function of position (Figure 18.11b), which were attributed to variations in the surface chemistry of the NWs. These experiments demonstrate the ability of time-resolved far-field optical scanning microscopy to extend beyond measurements on metal NPs and provide substantial insight into transient processes in single semiconductor nanostructures, which should find many applications in the near future.

Here, it is instructive to briefly compare the relative advantages of time-resolved wide-field and scanning optical microscopy. Time-resolved wide-field microscopy is capable of rapidly acquiring data since there is no spatial scanning of the sample or beam at each time delay; it should take essentially the same amount of time as a conventional pump–probe experiment. In addition, it can rapidly acquire time-resolved data over large sample areas, for example, ensembles of nanostructures (Figure 18.10) (Gundlach and Piotrowiak, 2009) or a bulk sample with micro-scale features (Figure 18.9) (Prasankumar et al., 2009). However, this also limits the minimum signal that can be measured since the probe (and often the pump) spot diameter is relatively large. The diffraction-limited spot sizes used in time-resolved scanning microscopy enable more of the incident light to interact with the sample, creating larger pump-induced absorption changes. These considerations dictate that mode-locked laser oscillators are typically used for time-resolved scanning microscopy measurements since the pump power is limited by the damage threshold of the sample, and, therefore, a laser amplifier offers no significant advantage over an oscillator when the beams are focused tightly. In fact, the low pulse repetition rates of an amplified system (and correspondingly reduced signal-to-noise ratio [SNR]) make a high-repetition-rate laser oscillator a better choice for nearly any time-resolved scanning microscopy experiment. In contrast, amplified laser systems must nearly always be used for wide-field experiments, since an oscillator typically does not have enough pulse energy to generate a measurable change in absorption on each pixel when both pump and probe are imaged onto a CCD or 2D array detector in a wide-field configuration. Therefore, the choice between time-resolved wide-field and scanning microscopy experiments will largely be dictated by the available laser systems and sample under consideration. It is important to remember that these are both far-field techniques and are thus limited to a spatial resolution of hundreds of nanometers. TR-NSOM is the only technique that has concurrently achieved true nanoscale spatial resolution in conjunction with ultrafast time resolution and will therefore be discussed in the next section.

18.2.2 Time-Resolved Near-Field Scanning Optical Microscopy

The many advantages of using all-optical techniques to characterize solid-state materials have led to the development of several methods for surpassing the well-known diffraction limit (see Chapter 15 for more detail), which restricts the maximum achievable spatial resolution with far-field optics to approximately half the incident wavelength of light. Near-field scanning optical microscopy (NSOM) (Chapter 17) has achieved the highest spatial resolution (<10 nm) for any all-optical technique to date. Therefore, it was natural to explore combining NSOM with time-resolved optical spectroscopy to achieve extremely high spatial and temporal resolution in a single experiment. This was done by several groups in the late 1990s, enabling them to track carrier transport through space and time as well as to isolate time-varying dynamics in individual nanostructures (Nechay et al., 1999; Siegner et al., 2001; Vasa et al., 2009).

As with several of the techniques discussed in this chapter, the concept behind TR-NSOM is actually quite simple as it is based on photoexciting a sample with ultrashort optical pulses and using a standard NSOM setup (to the best of our knowledge, only aperture-type NSOM has been used in TR-NSOM experiments to date) to probe the response at the nanoscale. However, there are several possible configurations in which the pump and probe beams can be arranged, depending on whether they are locally (through the NSOM tip) or globally (in the far field) incident on the sample (Nechay et al., 1999; Siegner et al., 2001). These include pumping through the NSOM tip and probing in the far field, pumping in the far field and probing through the tip, and both pumping and probing through the tip. In addition, the probe signal can be measured in reflection or transmission, and the NSOM can be operated in illumination or collection mode; most implementations use illumination mode and measure the transmitted probe due to the higher SNR in this configuration (Nechay et al., 1999). Finally, the pump and probe wavelengths can be varied to perform either degenerate (identical pump and probe wavelengths) or nondegenerate (different pump and probe wavelengths) pump–probe spectroscopy (Chapter 9).

A typical TR-NSOM setup, operating with the probe coupled into the NSOM fiber in illumination mode and a far-field pump, is shown in Figure 18.12 (Nechay et al., 1999; Siegner et al., 2001). In this setup, ~100 fs pulses from a Ti:sapphire laser oscillator at 800 nm and a repetition rate of 100 MHz are split into pump and probe beams. The pump beam is modulated at 1 MHz by an acousto-optic modulator (AOM) and then focused by a 40× objective to a ~10 μm spot on the sample. The probe is modulated at 1.05 MHz by a separate AOM and then sent into a prism compressor (Chapter 7) to pre-compensate the GVD from the NSOM fiber.

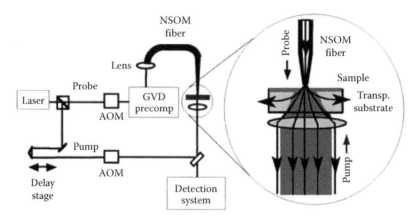

Figure 18.12   Schematic of a typical TR-NSOM setup, described in more detail in the text. (Reprinted with permission from Siegner, U. et al., *Meas. Sci. Technol.*, 12, 1847–1857, 2001.)

The probe is coupled into the NSOM fiber and transmitted through the sample, after which it is collected through the same objective used to focus the pump beam and detected by a low-noise avalanche photodiode and lock-in amplifier at the 50 kHz difference frequency. The spatial resolution of this particular setup was 150 nm, with a temporal resolution of 250 fs (Nechay et al., 1999; Siegner et al., 2001). The NSOM itself will not be described in detail here as a comprehensive description was given in Chapter 17.

The demanding nature of TR-NSOM experiments necessitates that several issues must be addressed in optimizing the experimental setup. The most important concern is SNR, due to the minimal power that can be transmitted through the NSOM fiber aperture (typically 10^{-3}–10^{-6} of the incident power) and the large background signals from the unaffected probe as well as the unabsorbed pump beam. The SNR in a TR-NSOM experiment can be increased by detecting the sum (Guenther et al., 1999) or difference (Nechay et al., 1999; Siegner et al., 2001) frequency signal after the pump and probe beams are chopped at different frequencies, appropriately designing the experimental geometry, and controlling the polarizations of both beams, much as in a standard pump–probe experiment (Chapter 9). Other issues include the influence of the GVD from the different components in the TR-NSOM setup, particularly the NSOM fiber, and potential nonlinear effects in the fiber (Nechay et al., 1999). The GVD can partially be compensated with a prism compressor (Figure 18.12), and lowering the input power to the fiber to a few milliwatts (which is also the limit set by the damage threshold of the fiber tip) can minimize nonlinear effects. However, this limits the maximum pump fluence when photoexciting the sample through the NSOM tip (Vollmer et al., 1999). It is also worth noting that most of the experimental issues with time-integrated NSOM discussed in Chapter 17, such as tip–sample interactions and the influence of sample topography on the measured image, remain with TR-NSOM. A more detailed discussion of these issues is given in Nechay et al. (1999) in which $\Delta T/T$ signals as small as 10^{-4} were detected after these concerns were identified and addressed.

The ability of TR-NSOM not only to time resolve carrier dynamics in a single nanostructure but also to resolve carrier transport in time and space made semiconductor quantum wires (QWR) a natural choice as one of the first systems to which TR-NSOM was applied (Achermann et al., 2000; Emiliani et al., 2000, 2001; Guenther et al., 1999; Siegner et al., 2001). In Achermann et al. (2000), degenerate TR-NSOM measurements using the setup in Figure 18.12 were performed on individual "V-groove" GaAs/AlGaAs QWRs (with diameters of 4.2 nm) grown by chemical vapor deposition on a GaAs substrate that was subsequently etched off after the samples were mounted on a glass disk. PL measurements estimated a QWR exciton transition energy of $h\nu \sim 1.59$–1.6 eV at room temperature.

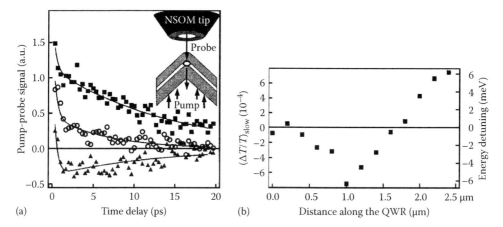

Figure 18.13 (a) TR-NSOM signal at hv = 1.596 eV (squares), 1.586 eV (circles), and 1.576 (triangles) at a fixed position on the QWR. (b) Slow component amplitude vs. position at a fixed hv = 1.578 eV and the associated energy detuning from the exciton resonance. (Reprinted with permission from Achermann, M., Nechay, B., Siegner, U., Hartmann, A., Oberli, D., Kapon, E., and Keller, U., Quantization energy mapping of single V-groove GaAs quantum wires by femtosecond near-field optics, *Appl. Phys. Lett.*, 76, 2695, 2000. Copyright 2000, American Institute of Physics.)

The pump pulse was used to photoexcite the QWRs from the far field, after which the probe pulse was sent through the NSOM tip to detect the pump-induced transmission changes. The femtosecond Ti:sapphire oscillator was tuned to different photon energies around the exciton resonance to examine time-resolved dynamics at a fixed position on a single QWR (Figure 18.13a). An initial fast decay of 300–400 fs is observed, likely due to exciton–phonon relaxation (Chapter 8). This is followed by a slower component (10–12 ps) that changes sign from negative to positive with increasing photon energy (the zero crossing is ~1.588 eV). The positive sign of this component at higher hv is due to bleaching, while the negative sign at lower hv is attributed to pump-induced broadening of the exciton resonance. It was then shown that the amplitude of the slow component $((\Delta T/T)_{slow})$ near the zero crossing is linearly proportional to the detuning from the exciton resonance energy E_x. This enabled E_x to be measured as a function of position along the QWR by extracting $(\Delta T/T)_{slow}$ at a fixed hv (Figure 18.13b). The variation in E_x with position is due to variation of the QWR thickness, causing quantization energy fluctuations along the wire; thickness variations of ~1 monolayer cause energy fluctuations of 12 meV. These experiments demonstrated the ability of TR-NSOM to resolve quantization energy fluctuations (and the associated thickness fluctuations) in QWRs (Achermann et al., 2000; Siegner et al., 2001).

In a different study, nondegenerate TR-NSOM measurements were performed at room temperature on GaAs QWRs (cross-sectional dimensions 6 × 13 nm) that were embedded in a GaAs QW, all grown on a GaAs substrate by MBE (Emiliani et al., 2000, 2001). The TR-NSOM setup was similar to that in Figure 18.12, except that the dynamics were examined in reflection, both with the pump incident on the sample from the far field (FFNF) and with the pump coupled into the NSOM fiber to locally photoexcite carriers in the QWRs (NFNF); the probe was coupled into the NSOM fiber in both cases, with resulting spatial and temporal resolutions of 200 nm and 200 fs, respectively. The results of these experiments are depicted in Figure 18.14, revealing that the $\Delta R/R$ signal decays very little in 50 ps after far field photoexcitation, while it decays by ~50% when pumped in the near field. The difference between the two signals was attributed to carrier transport out of the excitation volume in the near-field pumping case. A 1D diffusion model was used to determine that electrons and holes spatially separate on an ultrafast timescale, after which the decay of the $\Delta R/R$ signal in Figure 18.14b

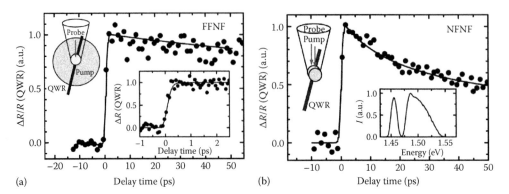

Figure 18.14 (a) Pump-induced reflectivity change in a single QWR in a far-field pump (hv = 1.52 eV) and near-field probe (hv = 1.46 eV) configuration (FFNF). (b) Pump-induced reflectivity change in a near-field pump (hv = 1.50 eV) and near-field probe (hv = 1.46 eV) configuration (NFNF). (Reprinted with permission from Emiliani, V., Guenther, T., Lienau, C., Notzel, R., and Ploog, K., Femtosecond near-field spectroscopy: Carrier relaxation and transport in single quantum wires, *J. Microsc.*, 202, 229–240, 2000. Copyright 2000 by the American Physical Society.)

is dominated by electron diffusion along the QWR axis. These experiments thus demonstrated the ability of TR-NSOM to separately track electron and hole transport in individual nanostructures with high spatial and temporal resolution, essentially providing a method for directly measuring the electron and hole mobilities in the time domain. Subsequently, this technique was extended to isolate dynamics from individual QDs with femtosecond time resolution (Guenther et al., 2002; Lienau, 2004; Unold et al., 2004). Overall, TR-NSOM is a challenging technique to implement due to the stringent SNR requirements, but one that can provide unique insight on spatiotemporal dynamics in condensed matter as well as other systems (such as plasmonics [Vasa et al., 2009]). Finally, it is worth noting that a newly developed method for adaptive control of the optical field on the nanoscale could provide another approach to perform time-resolved measurements in the near field (Aeschlimann et al., 2007; Brixner et al., 2005, 2006).

18.2.3 Ultrafast Scanning Tunneling Microscopy

The invention of the STM (Binnig et al., 1982) in 1981 revolutionized science through its ability to image surfaces with atomic spatial resolution, enabling researchers to visualize both sample topography and the local density of states (DOS) with <1 Å resolution (Binnig and Rohrer, 1987; Bonnell, 2000). This invention in turn stimulated the development of other scanning probe microscopies such as atomic force microscopy (AFM) (Bonnell, 2000) and NSOM (Chapter 17), giving researchers a suite of techniques to choose from when examining a sample at the nanoscale. More recently, STMs have even been used to manipulate individual atoms into specific positions on a substrate (Chen, 1993). The potential advantages of adding time resolution to the unparalleled spatial resolution of the STM were recognized fairly early on (Nunes Jr and Freeman, 1993), but over a decade elapsed before picosecond time resolution was achieved in conjunction with nanometer spatial resolution. This technique, known as UF-STM, is still in its infancy and has yet to be applied to a broad range of materials but offers great potential for concurrent sub-picosecond, sub-nanometer surface imaging that will give much insight on the dynamic properties of nanoscale excitations (Yarotski and Taylor, 2004).

A standard STM operates by placing a sharp metal tip in close proximity (<1 nm) to the surface to be imaged (with the space between the tip and sample known as the tunnel junction), with an applied voltage between the tip and sample that is less than the work functions of the tip and sample materials (Griffith and Kochanski, 1990). Electrons tunnel between the sample and the tip since the electron wave functions in the tip overlap those in the sample surface, generating a tunneling current that is proportional to the local DOS of the sample and the tip. Importantly, the current is exponentially dependent on the tip–sample separation, changing by an order of magnitude if the tip moves 1 Å from its original position, which illustrates the extremely high spatial resolution of STM (Yarotski and Taylor, 2004).

An STM is typically operated in one of two modes: constant height mode or constant current mode (Bonnell, 2000; Hansma and Tersoff, 1987). In constant current mode, the tip is scanned across the surface and the current is kept constant by using a feedback loop to adjust the tip height with a piezoelectric actuator at each point. Recording the tip height as a function of position in the plane gives an image of the sample topography at a constant surface charge density. In constant height mode, the voltage and tip height are held constant by allowing the current to change; the resulting image of the changes in current as a function of surface position can be related to the surface charge density (as can the height vs. position images acquired in constant current mode). Therefore, an STM can be used to map both sample topography and the local DOS (Bonnell, 2000; Griffith and Kochanski, 1990; Hansma and Tersoff, 1987). It is worth noting that constant height mode typically gives faster scan rates since the electronics can respond faster than the piezoelectric actuator that controls the sample–tip distance (Hansma and Tersoff, 1987).

Although there had been some previous investigations aimed at adding time resolution to STM, the field was revolutionized in 1993 when the three major approaches for performing UF-STM experiments were demonstrated over a period of only 11 days (Steeves and Freeman, 2002). These techniques all utilize the nonlinear response of the system to two laser pulses separated in time, essentially in a pump–probe configuration (Chapter 9) where the pump pulse initiates a process that affects the response of the sample to the probe pulses (Yarotski and Taylor, 2004). The nonlinearities can be either intrinsic to the tip–sample interaction, such as the exponential dependence of the tunneling current on the tip–sample distance (Freeman and Nunes, 1993), or external, such as the addition of an optoelectronic switch to the tip (Weiss et al., 1993). To date, although ultrafast optical spectroscopy and STM have separately achieved temporal and spatial resolutions of <10 fs and <1 nm, respectively, it has been difficult to simultaneously obtain high spatial (<1 nm) and high temporal (<1 ps) resolution in UF-STM, and each approach has its advantages and disadvantages, as described in the following.

Conceptually, the simplest approach to UF-STM is, arguably, time-resolved STM through tunneling distance modulation (Freeman and Nunes, 1993; Steeves and Freeman, 2002; Yarotski and Taylor, 2004). This technique is based on using a probe pulse to transiently modulate the distance between the tip and the sample, which lowers the tunneling impedance and thus gates the pump-induced change in the sample response at a given pump–probe delay. The nonlinear dependence of the tunneling current on tip–sample distance dictates that tip motions of only 1 Å give sufficient modulation. However, the piezoelectric actuators that are typically used to vary the tip–sample distance respond too slowly; an alternative was to use a magnetostrictive STM tip made of annealed nickel, threaded through a coil of copper wire. A short current pulse with a 300 ps rise time was then launched along the copper wire, creating a magnetic field that caused a magnetostrictive response in the tip. This in turn launched an acoustic pulse down the tip that modulated the tunnel junction. The time resolution of this technique was limited to nanoseconds due to the width of the acoustic pulse, which set the duration of the tip movement. Overall, this approach is attractive because it is independent of sample properties, with the same spatial resolution as in a conventional STM measurement and can temporally resolve any property that can be measured by time-integrated STM measurements. However, the inertia of the moving tip will always limit the time resolution, and, therefore, this approach has not been heavily explored since the initial demonstration (Freeman and Nunes, 1996; Steeves and Freeman, 2002; Yarotski and Taylor, 2004), as researchers have focused on developing UF-STM techniques with no moving parts.

Junction-mixing STM (JM-STM) has been the most widely used UF-STM technique due to its ability to achieve high spatial resolution (~1 nm) with good temporal resolution (~8 ps) (Freeman and Nunes, 1996; Nunes Jr and Freeman, 1993; Steeves and Freeman, 2002; Yarotski and Taylor, 2002, 2004). In this technique, the nonlinear dependence of the measured tunneling current on the voltage is utilized to mix the signal from the sample with an electrical probe pulse in the tunneling junction; conceptually, this is quite similar to an ultrafast cross-correlation measurement (Chapter 7) in which an unknown pulse is mixed with a known pulse through a nonlinear interaction, allowing one to determine the properties of the unknown pulse. Here, the unknown signal is any photoinduced change in the sample DOS, and the known signal is a voltage pulse generated on the surface with controlled properties (Yarotski and Taylor, 2004).

JM-STM was first tested on a sample that consisted of an Au transmission line in contact with two GaAs photoconductive switches (Nunes Jr and Freeman, 1993). Two identical optical pulses from a picosecond mode-locked dye laser with a controllable time delay between them were focused on the two switches, generating short voltage pulses that propagated along the transmission line to the region probed by the STM tip. The signal due to the overlap of the two pulses was then extracted by optically chopping both pulse trains at different frequencies and detecting the sum frequency signal as a function of time delay between the pulses. This was essentially an autocorrelation measurement that determined the time resolution of the system to be ~100 ps. More recently, ion-damaged GaAs, which has much shorter lifetimes, was used to reduce the duration of the voltage pulses to ~20 ps, and by using low-temperature-grown GaAs (LT-GaAs) and optimizing the transmission line design, a temporal resolution of 8 ps and concurrent spatial resolution of 1 nm was achieved (Yarotski and Taylor, 2002, 2004). The temporal resolution is primarily limited by the carrier lifetime in the photoconductive switch, the tip–sample capacitance, and the transmission line impedance (Steeves and Freeman, 2002; Yarotski and Taylor, 2004). These studies demonstrated the potential of JM-STM to be the technique of choice for UF-STM experiments since there are no fundamental limits on the temporal resolution, and it has already been shown to provide extremely high spatial resolution.

The final UF-STM technique, photoconductively gated STM (PG-STM), requires the integration of an ultrafast photoconductive switch with the STM tip (Steeves and Freeman, 2002; Weiss et al., 1993; Yarotski and Taylor, 2004). In this technique, an ultrashort pump pulse photoexcites the sample, which changes the tunneling current that passes through the tip when the probe pulse photoexcites the photoconductive switch. The dynamics of the excitation can therefore be temporally and spatially mapped by scanning the pump–probe delay at each point on the surface. Another viewpoint is to think of the pump pulse creating an excitation in the sample that dynamically evolves in time. This temporal evolution is sampled (or "gated") by the probe pulse as a function of the pump–probe delay.

PG-STM has several advantages over other UF-STM techniques, including straightforward extraction of the dynamic signal independent of the nature of the sample or the processes under investigation (Yarotski and Taylor, 2004). In addition, the time resolution of PG-STM is only limited by the laser pulse width and the lifetime of the photoconductive switch, which typically uses silicon-on-sapphire or LT-GaAs, both with sub-picosecond carrier lifetimes, as the active material. This has enabled a temporal resolution as high as 900 fs (Yarotski and Taylor, 2004), but capacitive coupling between the STM tip and the sample was believed to limit the best achievable spatial resolution to microns since the fields involved in this coupling extend much farther than the spatial extent of electrons tunneling into the tip

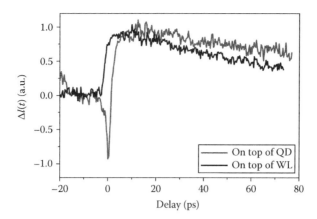

Figure 18.15 Time-resolved tunneling signals from an InAs QD (gray) and the WL (black). (Reprinted with permission from Yarotski, D. et al., *Appl. Phys. Lett.*, 81, 1143–1145, 2002.)

(Steeves and Freeman, 2002; Weiss et al., 1993; Yarotski and Taylor, 2004). However, the subsequent introduction of a novel LT-GaAs STM tip minimized these effects, enabling 20 nm spatial resolution with 2 ps time resolution (Donati et al., 2000; Yarotski and Taylor, 2004).

This unique PG-STM system was subsequently used to examine ultrafast carrier dynamics in InAs QDs (Figure 18.15) (Yarotski et al., 2002). The STM tip was positioned either directly above a QD or between dots, in which case it examined the wetting layer (WL), a 1.5 monolayer thick film of InAs that forms underneath the MBE-grown dots (Chapter 2). The output of a 120 fs, 800 nm, 82 MHz Ti:sapphire oscillator was split into pump and probe beams to excite the QD sample and the STM tip, respectively. The pump-induced tunneling signal, $\Delta I(t)$, was measured as a function of pump–probe delay on both single QDs and the WL. The signal from the WL is simply due to excitation of carriers and their subsequent recombination, as in a typical semiconductor. However, $\Delta I(t)$ measured for a QD reveals a large negative peak at early times, after which the signal turns positive and slowly decays. The $\Delta I(t)$ signal was shown to nearly exactly match the pump-induced change in reflectivity, $\Delta R/R$, taken with an 800 nm optical probe (not shown), suggesting that the same processes govern both signals (Yarotski et al., 2002).

In these QDs, electrons are captured from the WL into the QD within several hundred femtoseconds (corresponding to the rise time of the $\Delta R/R$ signal), causing the negative peaks in the $\Delta I(t)$ and $\Delta R/R$ signals. However, a potential barrier between the WL and QD valence band states delays hole capture for ~2 ps, when the signals become positive. Electron–hole recombination subsequently dominates carrier relaxation, occurring on an 890 ps timescale. These experiments thus demonstrated the ability of UF-STM to resolve carrier dynamics in individual InAs QDs, and furthermore, that the measured dynamics compared well with those measured by standard optical pump–probe spectroscopy on the QD ensemble. Although further optimization of this technique is needed, one might expect UF-STM to have a significant impact on future measurements of ultrafast dynamics in a wide variety of nanostructured materials.

18.3 CONCLUSION

This chapter has described a number of advanced optical techniques for characterizing solid-state materials, expanding on the topics of linear optical spectroscopy, time-resolved optical spectroscopy, and spatially resolved optical spectroscopy that have been discussed in detail in the previous sections of this book. Although these techniques are in relatively early stages of development, they have already been used to shed much light on the optical properties of a wide range of solid-state systems. In particular, the use of ultrashort x-ray and electron pulses to time-resolve structural dynamics in different materials has enabled researchers to directly observe processes such as melting and phase transitions on unprecedented timescales, while UF-PES has already been used to address important problems in condensed matter physics, such as the nature of the pairing quasiparticle in high-T_c superconductors. Similarly, the combination of temporally and spatially resolved optical spectroscopy, either measured in the near field or the far field, has allowed researchers to spatially resolve dynamics in a single nanostructure or simultaneously in many nanostructures.

Further development of these techniques, as well as others that were not discussed in this book and those that have yet to be envisioned, should give even more insight into the properties of solid-state materials, as well as other biological, chemical, and physical systems. Indeed, the numerous advantages of optical tools for characterizing materials, perhaps foremost the ability to measure both static as well as temporally and spatially varying material properties in a noncontact, nondestructive manner with sub-100 fs and sub-100 nm resolution, should continue to make optics the first tool utilized by experimenters for learning about material properties.

REFERENCES

Achermann, M., Nechay, B. A., Siegner, U., Hartmann, A., Oberli, D., Kapon, E., and Keller, U. (2000) Quantization energy mapping of single V-groove GaAs quantum wires by femtosecond near-field optics. *Appl. Phys. Lett.*, 76, 2695.

Aeschlimann, M., Bauer, M., Bayer, D., Brixner, T., García De Abajo, F. J., Pfeiffer, W., Rohmer, M., Spindler, C., and Steeb, F. (2007) Adaptive subwavelength control of nano-optical fields. *Nature*, 446, 301.

Arbouet, A., Christofilos, D., Del Fatti, N., Vallée, F., Huntzinger, J. R., Arnaud, L., Billaud, P., and Broyer, M. (2004) Direct measurement of the single-metal-cluster optical absorption. *Phys. Rev. Lett.*, 93, 127401.

Ashcroft, N. W. and Mermin, N. D. (1976) *Solid State Physics*, Fort Worth, TX: Harcourt Brace.

Bargheer, M., Zhavoronkov, N., Woerner, M., and Elsaesser, T. (2006) Recent progress in ultrafast x-ray diffraction. *ChemPhysChem*, 7, 783–792.

Barman, A., Wang, S., Maas, J., Hawkins, A. R., Kwon, S., Bokor, J., Liddle, A., and Schmidt, H. (2007) Size dependent damping in picosecond dynamics of single nanomagnets. *Appl. Phys. Lett.*, 90, 202504.

Barman, A., Wang, S., Maas, J. D., Hawkins, A. R., Kwon, S., Liddle, A., Bokor, J., and Schmidt, H. (2006) Magneto-optical observation of picosecond dynamics of single nanomagnets. *Nano Lett.*, 6, 2939–2944.

Barty, A., Boutet, S., Bogan, M. J., Hau-Riege, S., Marchesini, S., Sokolowski-Tinten, K., Stojanovic, N. et al. (2008) Ultrafast single-shot diffraction imaging of nanoscale dynamics. *Nat. Photonics*, 2, 415–419.

Bauer, M. (2005) Femtosecond ultraviolet photoelectron spectroscopy of ultra-fast surface processes. *J. Phys. D Appl. Phys.*, 38, R253–R267.

Baum, P., Yang, D. S., and Zewail, A. H. (2007) 4D visualization of transitional structures in phase transformations by electron diffraction. *Science*, 318, 788.

Becker, M. F., Buckman, A. B., Walser, R. M., Lepine, T., Georges, P., and Brun, A. (1994) Femtosecond laser excitation of the semiconductor-metal phase-transition in VO_2. *Appl. Phys. Lett.*, 65, 1507–1509.

Berciaud, S., Cognet, L., Blab, G. A., and Lounis, B. (2004) Photothermal heterodyne imaging of individual nonfluorescent nanoclusters and nanocrystals. *Phys. Rev. Lett.*, 93, 257402.

Binnig, G. and Rohrer, H. (1987) Scanning tunneling microscopy—From birth to adolescence. *Rev. Mod. Phys.*, 59, 615–625.

Binnig, G., Rohrer, H., Gerber, C., and Weibel, E. (1982) Tunneling through a controllable vacuum gap. *Appl. Phys. Lett.*, 40, 178.

Bonnell, D. (2000) *Scanning Probe Microscopy and Spectroscopy*, Hoboken, NJ: Wiley.

Bourquin, S., Prasankumar, R. P., Morgner, U., Kartner, F. X., Fujimoto, J. G., Lasser, T., and Salathe, R. P. (2003) High speed femtosecond pump-probe spectroscopy by use of a two-dimensional smart pixel detector array. *Opt. Lett.*, 28, 1588.

Bovensiepen, U. (2007) Coherent and incoherent excitations of the Gd (0001) surface on ultrafast timescales. *J. Phys.: Condens. Matter*, 19, 083201.

Boyer, D., Tamarat, P., Maali, A., Lounis, B., and Orrit, M. (2002) Photothermal imaging of nanometer-sized metal particles among scatterers. *Science*, 297, 1160–1163.

Bressler, C. and Chergui, M. (2004) Ultrafast x-ray absorption spectroscopy. *Chem. Rev.*, 104, 1781.

Brixner, T., García de Abajo, F., Schneider, J., and Pfeiffer, W. (2005) Nanoscopic ultrafast space-time-resolved spectroscopy. *Phys. Rev. Lett.*, 95, 093901.

Brixner, T., García de Abajo, F. J., Spindler, C., and Pfeiffer, W. (2006) Adaptive ultrafast nano-optics in a tight focus. *Appl. Phys. B*, 84, 89–95.

Bucksbaum, P. H. (2007) The future of attosecond spectroscopy. *Science*, 317, 766–769.

Carbone, F., Yang, D. S., Giannini, E., and Zewail, A. H. (2008) Direct role of structural dynamics in electron-lattice coupling of superconducting cuprates. *Proc. Natl. Acad. Sci.*, 105, 20161.

Carey, C. R., Yu, Y., Kuno, M., and Hartland, G. V. (2009) Ultrafast transient absorption measurements of charge carrier dynamics in single II-VI nanowires. *J. Phys. Chem. C*, 113, 19077–19081.

Carpene, E., Mancini, E., Dallera, C., Ghiringhelli, G., Manzoni, C., Cerullo, G., and De Silvestri, S. (2009) A versatile apparatus for time-resolved photoemission spectroscopy via femtosecond pump-probe experiments. *Rev. Sci. Instrum.*, 80, 055101.

Cavalleri, A., Chong, H., Fourmaux, S., Glover, T., Heimann, P., Kieffer, J., Mun, B., Padmore, H., and Schoenlein, R. W. (2004a) Picosecond soft x-ray absorption measurement of the photo-induced insulator-to-metal transition in VO_2. *Phys. Rev. B*, 69, 153106.

Cavalleri, A., Dekorsy, T., Chong, H., Kieffer, J., and Schoenlein, R. W. (2004b) Evidence for a structurally-driven insulator-to-metal transition in VO_2: A view from the ultrafast timescale. *Phys. Rev. B*, 70, 161102.

Cavalleri, A., Rini, M., Chong, H., Fourmaux, S., Glover, T., Heimann, P., Kieffer, J., and Schoenlein, R. W. (2005) Band-selective measurements of electron dynamics in VO_2 using femtosecond near-edge x-ray absorption. *Phys. Rev. Lett.*, 95, 067405.

Cavalleri, A., Rini, M., and Schoenlein, R. W. (2006) Ultra-broadband femtosecond measurements of the photo-induced phase transition in VO_2: From the mid-IR to the hard x-rays. *J. Phys. Soc. Jpn.*, 75, 011004.

Cavalleri, A., Siders, C. W., Sokolowski-Tinten, K., Toth, C., Blome, C., Squier, J. A., Von Der Linde, D., Barty, C. P. J., and Wilson, K. R. (2001a) Femtosecond x-ray diffraction. *Opt. Photonics News*, 12, 28–33.

Cavalleri, A., Tóth, C., Siders, C. W., Squier, J. A., Raksi, F., Forget, P., and Kieffer, J. C. (2001b) Femtosecond structural dynamics in VO_2 during an ultrafast solid-solid phase transition. *Phys. Rev. Lett.*, 87, 237401.

Chen, C. J. (1993) *Introduction to Scanning Tunneling Microscopy*, Oxford University Press, Oxford, England.

Chergui, M. and Zewail, A. H. (2009) Electron and x-ray methods of ultrafast structural dynamics: Advances and applications. *ChemPhysChem*, 10, 28–43.

Chin, A.H., Schoenlein, R. W., Glover, T. E., Balling, P., Leemans, W. P., and Shank, C. V. (1999) Ultrafast structural dynamics in InSb probed by time-resolved x-ray diffraction. *Phys. Rev. Lett.*, 83, 336–339.

Corkum, P. B. and Krausz, F. (2007) Attosecond science. *Nat. Phys.*, 3, 381–387.

Damascelli, A., Hussain, Z., and Shen, Z. X. (2003) Angle-resolved photoemission studies of the cuprate superconductors. *Rev. Mod. Phys.*, 75, 473–541.

Dantus, M., Kim, S. B., Williamson, J. C., and Zewail, A. H. (1994) Ultrafast electron diffraction 5: Experimental time resolution and applications. *J. Phys. Chem.*, 98, 2782–2796.

Donati, G. P., Rodriguez, G., and Taylor, A. J. (2000) Ultrafast, dynamical imaging of surfaces by use of a scanning tunneling microscope with a photoexcited, low-temperature-grown GaAs tip. *J Opt. Soc. Am. B*, 17, 1077–1083.

Downer, M. C., Fork, R. L., and Shank, C. V. (1985) Femtosecond imaging of melting and evaporation at a photoexcited silicon surface. *J. Opt. Soc. Am. B*, 2, 595–599.

Elazar, M., Sahaf, M., Szapiro, L., Cheskis, D., and Bar-Ad, S. (2008) Single-pulse magneto-optic microscopy: A new tool for studying optically induced magnetization reversals. *Opt. Lett.*, 33, 2734–2736.

Emiliani, V., Guenther, T., Lienau, C., Nötzel, R., and Ploog, K. H. (2000) Ultrafast near-field spectroscopy of quasi-one-dimensional transport in a single quantum wire. *Phys. Rev. B*, 61, 10583–10586.

Emiliani, V., Guenther, T., Lienau, C., Notzel, R., and Ploog, K. H. (2001) Femtosecond near-field spectroscopy: Carrier relaxation and transport in single quantum wires. *J. Microsc.*, 202, 229–240.

Freeman, M. R. and Nunes, G. (1993) Time-resolved scanning-tunneling-microscopy through tunnel distance modulation. *Appl. Phys. Lett.*, 63, 2633–2635.

Freeman, M. R. and Nunes, G. (1996) Ultrafast time resolution in scanning probe microscopy of repetitive phenomena. *Appl. Surf. Sci.*, 107, 238–246.

Fujii, Y., Horiuchi, K., Kannari, F., Hase, M., and Kitajima, M. (2004) Optical imaging of defect density distribution in ion-implanted GaAs using ultrafast carrier dynamics. *Jpn. J. Appl. Phys.*, 43, 184–185.

Furukawa, N., Mair, C. E., Kleiman, V. D., and Takeda, J. (2004) Femtosecond real-time pump–probe imaging spectroscopy. *Appl. Phys. Lett.*, 85, 4645.

Gedik, N., Yang, D.-S., Logvenov, G., Bozovic, I., and Zewail, A. H. (2007) Nonequilibrium phase transitions in cuprates observed by ultrafast electron crystallography. *Science*, 316, 425–429.

Glennon, J., Tang, R., Buhro, W., Loomis, R., Bussian, D., Htoon, H., and Klimov, V. I. (2009) Exciton localization and migration in individual CdSe quantum wires at low temperatures. *Phys. Rev. B*, 80, 081303.

Goulielmakis, E., Yakovlev, V. S., Cavalieri, A. L., Uiberacker, M., Pervak, V., Apolonski, A., Kienberger, R., Kleineberg, U., and Krausz, F. (2007) Attosecond control and measurement: Lightwave electronics. *Science*, 317, 769–775.

Griffith, J. E. and Kochanski, G. P. (1990) Scanning tunneling microscopy. *Annu. Rev. Mater. Sci.*, 20, 219–244.

Grinolds, M. S., Lobastov, V. A., Weissenrieder, J. and Zewail, A. H. (2006) Four-dimensional ultrafast electron microscopy of phase transitions. *Proc. Natl. Acad. Sci.*, 103, 18427.

Guenther, T., Emiliani, V., Intonti, F., Lienau, C., Elsaesser, T., Nötzel, R., and Ploog, K. H. (1999) Femtosecond near-field spectroscopy of a single GaAs quantum wire. *Appl. Phys. Lett.*, 75, 3500.

Guenther, T., Lienau, C., Elsaesser, T., Glanemann, M., Axt, V., Kuhn, T., Eshlaghi, S., and Wieck, A. (2002) Coherent nonlinear optical response of single quantum dots studied by ultrafast near-field spectroscopy. *Phys. Rev. Lett.*, 89, 057401.

Gundlach, L. and Piotrowiak, P. (2008) Femtosecond Kerr-gated wide-field fluorescence microscopy. *Opt. Lett.*, 33, 992–994.

Gundlach, L. and Piotrowiak, P. (2009) Ultrafast spatially resolved carrier dynamics in single CdSSe nanobelts. *J. Phys. Chem. C*, 113, 12162–12166.

Haight, R. (1995) Electron dynamics at surfaces. *Surface Science Reports*, 21(8), 275–325.

Hansma, P. K. and Tersoff, J. (1987) Scanning tunneling microscopy. *J. Appl. Phys.*, 61, R1.

Harb, M., Ernstorfer, R., Hebeisen, C. T., Sciaini, G., Peng, W., Dartigalongue, T., Eriksson, M. A., Lagally, M. G., Kruglik, S. G., and Miller, R. J. D. (2008) Electronically driven structure changes of Si captured by femtosecond electron diffraction. *Phys. Rev. Lett.*, 100, 155504.

Hilton, D. J., Prasankumar, R. P., Fourmaux, S., Cavalleri, A., Brassard, D., El Khakani, M. A., Kieffer, J. C., Taylor, A. J., and Averitt, R. D. (2007) Enhanced photosusceptibility near T_c for the light-induced insulator-to-metal phase transition in vanadium dioxide. *Phys. Rev. Lett.*, 99, 226401.

Hilton, D. J., Prasankumar, R. P., Trugman, S. A., Taylor, A. J. and Averitt, R. D. (2006) On photo-induced phenomena in complex materials: Probing quasiparticle dynamics using infrared and far-infrared pulses. *J. Phys. Soc. Jpn.*, 75, 011006.

Itoh, T., Asahi, T., and Masuhara, H. (2001) Femtosecond light scattering spectroscopy of single gold nanoparticles. *Appl. Phys. Lett.*, 79, 1667–1669.

Jasapara, J. and Rudolph, W. (1999) Characterization of sub-10-fs pulse focusing with high-numerical-aperture microscope objectives. *Opt. Lett.*, 24, 777–779.

Johnson, J. C., Knutsen, K. P., Yan, H. Q., Law, M., Zhang, Y. F., Yang, P. D., and Saykally, R. J. (2004) Ultrafast carrier dynamics in single ZnO nanowire and nanoribbon lasers. *Nano Lett.*, 4, 197–204.

Kapteyn, H., Cohen, O., Christov, I., and Murnane, M. (2007) Harnessing attosecond science in the quest for coherent x-rays. *Science*, 317, 775.

Kim, H.-T., Lee, Y. W., Kim, B.-J., Chae, B.-G., Yun, S. J., Kang, K.-Y., Han, K.-J., Yee, K.-J., and Lim, Y.-S. (2006) Monoclinic and correlated metal phase in VO_2 as evidence of the Mott transition: Coherent phonon analysis. *Phys. Rev. Lett.*, 97, 266401.

Kittel, C. (1995) *Introduction to Solid-State Physics*, Hoboken, NJ: Wiley.

Koralek, J. D. and Dessau, D. S. (2006) Lasers emerge as a tool for the direct study of electrons in solids. *Photon. Spectra*, 40, 72–80.

Krausz, F. and Ivanov, M. (2009) Attosecond physics. *Rev. Mod. Phys.*, 81, 163–234.

Kübler, C., Ehrke, H., Huber, R., Lopez, R., Halabica, A., Haglund, R., and Leitenstorfer, A. (2007) Coherent structural dynamics and electronic correlations during an ultrafast insulator-to-metal phase transition in VO_2. *Phys. Rev. Lett.*, 99, 116401.

Labardi, M., Zavelani-rossi, M., Polli, D., Cerullo, G., Allegrini, M., De silvestri, S., and Svelto, O. (2005) Characterization of femtosecond light pulses coupled to hollow-pyramid near-field probes: Localization in space and time. *Appl. Phys. Lett.*, 86, 031105.

Laraoui, A., Albrecht, M., and Bigot, J. Y. (2007a) Femtosecond magneto-optical Kerr microscopy. *Opt. Lett.*, 32, 936–938.

Laraoui, A., Venuat, J., Halte, V., Albrecht, M., Beaurepaire, E., and Bigot, J. Y. (2007b) Study of individual ferromagnetic disks with femtosecond optical pulses. *J. Appl. Phys.*, 101, 09C105.

Larson, A. M. and Yeh, A. T. (2006) Ex vivo characterization of sub-10-fs pulses. *Opt. Lett.*, 31, 1681–1683.

Li, J., Lee, M.-S., He, W., Redeker, B. R., Remhof, A., Amaladass, E., Hassel, C., and Eimüller, T. (2009) Magnetic imaging with femtosecond temporal resolution. *Rev. Sci. Instrum.*, 80, 073703.

Liau, Y. H., Unterreiner, A. N., Chang, Q., and Scherer, N. F. (2001) Ultrafast dephasing of single nanoparticles studied by two-pulse second-order interferometry. *J. Phys. Chem. B*, 105, 2135–2142.

Lienau, C. (2004) Ultrafast near-field spectroscopy of single semiconductor quantum dots. *Phil. Tran. Math. Phys. Eng. Sci.*, 362, 861–879.

Lindenberg, A. M., Kang, I., Johnson, S. L., Missalla, T., Heimann, P. A., Chang, Z., Larsson, J., Bucksbaum, P. H., Kapteyn, H. C., and Padmore, H. A. (2000) Time-resolved x-ray diffraction from coherent phonons during a laser-induced phase transition. *Phys. Rev. Lett.*, 84, 111–114.

Lindfors, K., Kalkbrenner, T., Stoller, P., and Sandoghdar, V. (2004) Detection and spectroscopy of gold nanoparticles using supercontinuum white light confocal microscopy. *Phys. Rev. Lett.*, 93, 37401.

Lobastov, V. A., Srinivasan, R., and Zewail, A. H. (2005) Four-dimensional ultrafast electron microscopy. *Proc. Natl. Acad. Sci. USA*, 102, 7069–7073.

Lysenko, S., Rúa, A., Vikhnin, V., Fernández, F., and Liu, H. (2007a) Insulator-to-metal phase transition and recovery processes in VO_2 thin films after femtosecond laser excitation. *Phys. Rev. B*, 76, 035104.

Lysenko, S., Vikhnin, V., Fernandez, F., Rua, A., and Liu, H. (2007b) Photoinduced insulator-to-metal phase transition in VO_2 crystalline films and model of dielectric susceptibility. *Phys. Rev. B*, 75, 075109.

Makishima, Y., Furukawa, N., Ishida, A., and Takeda, J. (2006) Femtosecond real-time pump-probe imaging spectroscopy implemented on a single shot basis. *Jpn. J. Appl. Phys. Part 1 (Regular Papers Short Notes And Review Papers)*, 45, 5986.

Matsuo, Y. and Sasaki, K. (2001) Time-resolved laser scattering spectroscopy of a single metallic nanoparticle. *Jpn. J. Appl. Phys. 1*, 40, 6143–6147.

Muskens, O. L., Christofilos, D., Del fatti, N., and Vallée, F. (2006a) Optical response of a single noble metal nanoparticle. *J. Opt. A-Pure Appl. Opt.*, 8, S264–S272.

Muskens, O. L., Del fatti, N., and Vallée, F. (2006b) Femtosecond response of a single metal nanoparticle. *Nano Lett.*, 6, 552–556.

Nechay, B. A., Siegner, U., Achermann, M., Bielefeldt, H., and Keller, U. (1999) Femtosecond pump-probe near-field optical microscopy. *Rev. Sci. Instrum.*, 70, 2758.

Nunes Jr, G. and Freeman, M. R. (1993) Picosecond resolution in scanning tunneling microscopy. *Science*, 262, 1029.

Ogasawara, T., Iwata, N., Murakami, Y., Okamoto, H., and Tokura, Y. (2009) Submicron-scale spatial feature of ultrafast photoinduced magnetization reversal in TbFeCo thin film. *Appl. Phys. Lett.*, 94, 162507.

Othonos, A. and Christofides, C. (2002) Spatial dependence of ultrafast carrier recombination centers of phosphorus-implanted and annealed silicon wafers. *Appl. Phys. Lett.*, 81, 856.

Pelton, M., Liu, M., Park, S., Scherer, N., and Guyot-Sionnest, P. (2006) Ultrafast resonant optical scattering from single gold nanorods: Large nonlinearities and plasmon saturation. *Phys. Rev. B*, 73, 155419.

Pemasiri, K., Montazeri, M., Gass, R., Smith, L. M., Jackson, H. E., Yarrison-Rice, J., Paiman, S. et al. (2009) Carrier dynamics and quantum confinement in type II ZB-WZ InP nanowire homostructures. *Nano Lett.*, 9, 648–654.

Perfetti, L., Loukakos, P., Lisowski, M., Bovensiepen, U., Eisaki, H., and Wolf, M. (2007) Ultrafast electron relaxation in superconducting $Bi_2Sr_2CaCu_2O_{8+\delta}$ by time-resolved photoelectron spectroscopy. *Phys. Rev. Lett.*, 99, 197001.

Petek, H. and Ogawa, S. (1997) Femtosecond time-resolved two-photon photoemission studies of electron dynamics in metals. *Prog. Surf. Sci.*, 56, 239–310.

Pfeifer, T., Spielmann, C., and Gerber, G. (2006) Femtosecond x-ray science. *Rep. Prog. Phys.*, 69, 443–505.

Prasankumar, R. P., Ku, Z., Gin, A. V., Upadhya, P. C., Brueck, S. R. J., and Taylor, A. J. (2009) Ultrafast optical wide field microscopy. *Conference on Lasers and Electro-Optics*. Baltimore, MD.

Rizo, P. J., Pugžlys, A., Liu, J., Reuter, D., Wieck, A. D., Van der Wal, C. H., and Van Loosdrecht, P. H. M. (2008) Compact cryogenic Kerr microscope for time-resolved studies of electron spin transport in microstructures. *Rev. Sci. Instrum.*, 79, 123904.

Rose-Petruck, C., Jimenez, R., Guo, T., Cavalleri, A., Siders, C. W., Raksi, F., Squier, J. A., Walker, B. C., Wilson, K. R., and Barty, C. P. J. (1999) Picosecond-milliangstrom lattice dynamics measured by ultrafast x-ray diffraction. *Nature*, 398, 310–312.

Rousse, A., Rischel, C., and Gauthier, J. C. (2001) Femtosecond x-ray crystallography. *Rev. Mod. Phys.*, 73, 17–31.

Ruan, C. Y., Vigliotti, F., Lobastov, V. A., Chen, S., and Zewail, A. H. (2004) Ultrafast electron crystallography: Transient structures of molecules, surfaces, and phase transitions. *Proc. Natl. Acad. Sci. USA*, 101, 1123.

Rudolph, W., Dorn, P., Liu, X., Vretenar, N., and Stock, R. (2003) Microscopy with femtosecond laser pulses: Applications in engineering, physics and biomedicine. *Appl. Surf. Sci.*, 208, 327–332.

Schmitt, F., Kirchmann, P. S., Bovensiepen, U., Moore, R. G., Rettig, L., Krenz, M., Chu, J. H., Ru, N., Perfetti, L., and Lu, D. H. (2008) Transient electronic structure and melting of a charge density wave in $TbTe_3$. *Science*, 321, 1649.

Schoenlein, R. W., Chattopadhyay, S., Chong, H. H. W., Glover, T. E., Heimann, P. A., Shank, C. V., Zholents, A. A., and Zolotorev, M. S. (2000) Generation of femtosecond pulses of synchrotron radiation. *Science*, 287, 2237.

Schoenlein, R. W., Fujimoto, J. G., Eesley, G. L., and Capehart, T. W. (1988) Femtosecond studies of image-potential dynamics in metals. *Phys. Rev. Lett.*, 61, 2596–2599.

Schoenlein, R. W., Leemans, W. P., Chin, A. H., Volfbeyn, P., Glover, T. E., Balling, P., Zolotorev, M., Kim, K. J., Chattopadhyay, S., and Shank, C. V. (1996) Femtosecond x-ray pulses at 0.4 angstrom generated by 90 degrees Thomson scattering: A tool for probing the structural dynamics of materials. *Science*, 274, 236–238.

Sciaini, G., Harb, M., Kruglik, S. G., Payer, T., Hebeisen, C. T., Heringdorf, F.-J. M. Z., Yamaguchi, M., Hoegen, M. H.-V., Ernstorfer, R., and Miller, R. J. D. (2009) Electronic acceleration of atomic motions and disordering in bismuth. *Nature*, 458, 56–60.

Siders, C. W., Cavalleri, A., Sokolowski-Tinten, K., Tóth, C., Guo, T., Kammler, M., Horn Von Hoegen, M., Wilson, K. R., Von Der Linde, D., and Barty, C. P. J. (1999) Detection of nonthermal melting by ultrafast x-ray diffraction. *Science*, 286, 1340–1342.

Siegner, U., Achermann, M., and Keller, U. (2001) Spatially resolved femtosecond spectroscopy beyond the diffraction limit. *Meas. Sci. Technol.*, 12, 1847–1857.

Siwick, B. J., Dwyer, J. R., Jordan, R. E., and Miller, R. J. (2003) An atomic-level view of melting using femtosecond electron diffraction. *Science*, 302, 1382.

Sokolowski-Tinten, K., Bialkowski, J., Cavalleri, A., Von der Linde, D., Oparin, A., Meyer-ter-Vehn, J., and Anisimov, S. I. (1998) Transient states of matter during short pulse laser ablation. *Phys. Rev. Lett.*, 81, 224–227.

Song, J. K., Willer, U., Szarko, J. M., Leone, S. R., Li, S., and Zhao, Y. (2008) Ultrafast upconversion probing of lasing dynamics in single ZnO nanowire lasers. *J. Phys. Chem. C*, 112, 1679–1684.

Sönnichsen, C., Geier, S., Hecker, N. E., Von Plessen, G., Feldmann, J., Ditlbacher, H., Lamprecht, B., Krenn, J. R., Aussenegg, F. R., and Chan, V. Z. H. (2000) Spectroscopy of single metallic nanoparticles using total internal reflection microscopy. *Appl. Phys. Lett.*, 77, 2949.

Sotier, F., Thomay, T., Hanke, T., Korger, J., Mahapatra, S., Frey, A., Brunner, K., Bratschitsch, R., and Leitenstorfer, A. (2009) Femtosecond few-fermion dynamics and deterministic single-photon gain in a quantum dot. *Nat. Phys.*, 5, 352–356.

Srinivasan, R., Lobastov, V. A., Ruan, C. Y., and Zewail, A. H. (2003) Ultrafast electron diffraction (UED)—A new development for the 4D determination of transient molecular structures. *Helv. Chim. Acta*, 86, 1763–1838.

Stamm, C., Kachel, T., Pontius, N., Mitzner, R., Quast, T., Holldack, K., Khan, S., Lupulescu, C., Aziz, E. F., and Wietstruk, M. (2007) Femtosecond modification of electron localization and transfer of angular momentum in nickel. *Nat. Mater.*, 6, 740–743.

Steeves, G. M. and Freeman, M. R. (2002) Ultrafast scanning tunneling microscopy. *Adv. Imag. Electron Phys.*, 125, 195–231.

Stoller, P., Jacobsen, V., and Sandoghdar, V. (2006) Measurement of the complex dielectric constant of a single gold nanoparticle. *Opt. Lett.*, 31, 2474–2476.

Tamai, N., Asahi, T., and Masuhara, H. (1993) Femtosecond transient absorption microspectrophotometer combined with optical trapping technique. *Rev. Sci. Instrum.*, 64, 2496.

Temnov, V. V., Sokolowski-Tinten, K., Zhou, P., and Von der Linde, D. (2004) Femtosecond time-resolved interferometric microscopy. *Appl. Phys. A: Mater. Sci. Process.*, 78, 483–489.

Temnov, V. V., Sokolowski-Tinten, K., Zhou, P., and Von der Linde, D. (2006) Ultrafast imaging interferometry at femtosecond-laser-excited surfaces. *J. Opt. Soc. Am. B*, 23, 1954–1964.

Titova, L. V., Hoang, T. B., Yarrison-Rice, J. M., Jackson, H. E., Kim, Y., Joyce, H. J., Gao, Q. et al. (2007) Dynamics of strongly degenerate electron-hole plasmas and excitons in single InP nanowires. *Nano Lett.*, 7, 3383–3387.

Unold, T., Mueller, K., Lienau, C., and Elsaesser, T. (2004) Space and time resolved coherent optical spectroscopy of single quantum dots. *Semicond. Sci. Technol.*, 19, 260–263.

Van Dijk, M. A., Lippitz, M., and Orrit, M. (2005a) Detection of acoustic oscillations of single gold nanospheres by time-resolved interferometry. *Phys. Rev. Lett.*, 95, 267406.

Van Dijk, M. A., Lippitz, M., and Orrit, M. (2005b) Far-field optical microscopy of single metal manoparticies. *Acc. Chem. Res.*, 38, 594–601.

Van Dijk, M. A., Lippitz, M., Stolwijk, D., and Orrit, M. (2007) A common-path interferometer for time-resolved and shot-noise-limited detection of single nanoparticles. *Opt. Express*, 15, 2273–2287.

Vasa, P., Ropers, C., Pomraenke, R., and Lienau, C. (2009) Ultra-fast nano-optics. *Laser Photon. Rev.*, 3, 483–507.

Vollmer, M., Giessen, H., Stolz, W., Ruhle, W. W., Ghislain, L., and Elings, V. (1999) Ultrafast nonlinear subwavelength solid immersion spectroscopy at T = 8 K. *Appl. Phys. Lett.*, 74, 1791–1793.

Weiss, S., Ogletree, D. F., Botkin, D., Salmeron, M., and Chemla, D. S. (1993) Ultrafast scanning probe microscopy. *Appl. Phys. Lett.*, 63, 2567–2569.

Wesseli, M., Ruppert, C., Trumm, S., Krenner, H. J., Finley, J. J., and Betz, M. (2006) Nonlinear optical response of a single self-assembled InGaAs quantum dot: A femtojoule pump-probe experiment. *Appl. Phys. Lett.*, 88, 203110.

Williamson, J. C., Cao, J. M., Ihee, H., Frey, H., and Zewail, A. H. (1997) Clocking transient chemical changes by ultrafast electron diffraction. *Nature*, 386, 159–162.

Williamson, J. C. and Zewail, A. H. (1991) Structural femtochemistry: Experimental methodology. *Proc. Natl. Acad. Sci.*, 88, 5021.

Yago, T., Tamaki, Y., Furube, A., and Katoh, R. (2008) Imaging of exciton absorption in perylene crystals by femtosecond-laser scanning microscopy. *Jpn. J. Appl. Phys.*, 47, 1400.

Yang, D.-S., Lao, C., and Zewail, A. H. (2008) 4D electron diffraction reveals correlated unidirectional behavior in zinc oxide nanowires. *Science*, 321, 1660.

Yarotski, D. A., Averitt, R. D., Negre, N., Crooker, S. A., Taylor, A. J., Donati, G. P., Stintz, A., Lester, L. F., and Malloy, K. J. (2002) Ultrafast carrier-relaxation dynamics in self-assembled InAs/GaAs quantum dots. *J. Opt. Soc. Am. B*, 19, 1480–1484.

Yarotski, D. A. and Taylor, A. J. (2002) Improved temporal resolution in junction-mixing ultrafast scanning tunneling microscopy. *Appl. Phys. Lett.*, 81, 1143–1145.

Yarotski, D. A. and Taylor, A. J. (2004) Ultrafast scanning tunneling microscopy: Principles and applications. *Topics Appl. Phys.*, 92, 57–98.

Yurtsever, A. and Zewail, A. H. (2009) 4D nanoscale diffraction observed by convergent-beam ultrafast electron microscopy. *Science*, 326, 708–712.

Zewail, A. H. (1991) Femtosecond transition-state dynamics. *Faraday Discuss. Chem. Soc.*, 91, 207–237.

Zewail, A. H. (2005) Diffraction, crystallography and microscopy beyond three dimensions: Structural dynamics in space and time. *Phil. Trans. Math. Phys. Eng. Sci.*, 363, 315–329.

Zewail, A. H. (2006) 4D ultrafast electron diffraction, crystallography, and microscopy. *Annu. Rev. Phys. Chem.*, 57, 65–103.

Zhang, N., Zhu, X., Yang, J., Wang, X., and Wang, M. (2007) Time-resolved shadowgraphs of material ejection in intense femtosecond laser ablation of aluminum. *Phys. Rev. Lett.*, 99, 167602.

INDEX

Printed and bound by CPI Group (UK) Ltd, Croydon, CR0 4YY

24/10/2024

01778288-0015